DISCARDED
JENKS LRC
GORDON COLLEGE

W9-BCK-917

Survey of Semiconductor Physics

Electrons and Other Particles in Bulk Semiconductors

Karl W. Böer

University of Delaware

JENKS L.R.C.
GORDON COLLEGE
255 GRAPEVINE RD.
WENHAM, MA 01984

VNR VAN NOSTRAND REINHOLD
New York

Copyright ©1990 by Van Nostrand Reinhold

Library of Congress Catalog Card Number 89-21473
ISBN 0-442-23793-6

All rights reserved. No part of this work covered by the copyright
hereon may be reproduced or used in any form or by any means
— graphic, electronic, or mechanical, including photocopying,
recording, taping, or information storage and retrieval systems —
without written permission of the publisher.

Printed in the United States of America

Van Nostrand Reinhold
115 Fifth Avenue
New York, New York 10003

Van Nostrand Reinhold International Company Limited
11 New Fetter Lane
London EC4P 4EE, England

Van Nostrand Reinhold
480 La Trobe Street
Melbourne, Victoria 3000, Australia

Nelson Canada
1120 Birchmount Road
Scarborough, Ontario M1K 5G4, Canada

16 15 14 13 12 11 10 9 8 7 6 5 4 3 2 1

QC
611
.B64
1990

Library of Congress Catologing-in-Publication Data

Böer, K. W. (Karl Wolfgang), 1926-
    Survey of semiconductor physics : electrons and other particles in
bulk semiconductors / Karl W. Böer.
        p.   cm.
    Includes bibliographical references.
    ISBN 0-442-23793-6
    1. Semiconductors.   I. Title.
QC611.B64 1990
537.6'22–dc20
                                                            89-21473
                                                            CIP

To my students,

who are endeavoring to further unravel the exciting puzzles of nature, and who are helping to create the foundation for the development of new and better semiconductor devices.

Preface

Any book that covers a large variety of subjects and is written by one author lacks by necessity the depth provided by an expert in his or her own field of specialization. This book is no exception.

It has been written with the encouragement of my students and colleagues, who felt that an extensive card file I had accumulated over the years of teaching solid state and semiconductor physics would be helpful to more than just a few of us. This file, updated from time to time, contained lecture notes and other entries that were useful in my research and permitted me to give to my students a broader spectrum of information than is available in typical textbooks.

When assembling this material into a book, I divided the topics into material dealing with the homogeneous semiconductor, the subject of this book, and the inhomogeneous semiconductor, the latter material left for a future volume. In order to keep the book to a manageable size, sections of tutorial character had to be interwoven with others written in shorter, reference style. The pointers at the right-hand page header will assist in discriminating the more difficult reference parts of the book from the more easy-to-read basic educational sections.

For reference purposes, I included more tables and figures than are necessary for a text, and have added several footnotes that will be helpful in reminding the reader of facts which are often difficult to locate.

The necessity of keeping the volume to a size that can easily be handled required a stringent selection of the material. Any such selection is subjective. I apologize to my colleagues who may not find their own subject of specialization represented in sufficient detail. I also apologize for my selection of references in the text. They cannot be comprehensive. The given references are meant as examples, and in a rapidly progressing field may not remain best choices. The reader should be aware of this fact, and after gaining entry into the field should search for more recent literature.

The book is organized around three material groups: the crystalline, amorphous, and superlattice-type semiconductors. Even though most of the basic effects are discussed for crystalline semiconductors, I have extended this discussion at times to the other materials.

When similar topics have to be covered in different parts of the book, a progressively advanced description is used, including more model details or a more sophisticated mathematical approach. The search for interconnecting sections is eased by cross-referencing, an extended Table of Contents, and a comprehensive word index. For a brief scan of topics, however, the short version of the Table of Contents and the List of Tables, both at the end of the book, may be consulted. A List of Symbols is included to assist in identifying the terms used. Multiple use of the same symbols was permitted (in order to avoid excessive subscripts) where confusion is unlikely. Conventional use of symbols was given preference whenever possible.

The sentences at the beginning of each chapter and the Summary and Emphasis at its end provide a quick overview.

Many figures have been relabeled or redrawn from the original, omitting nonessential features or less important curves of a family. Some figures were transposed into a coordinate system that permits easy comparison with other figures in the text. Tables included without source citations were compiled from data published in Landoldt-Börnstein (1955 and 1982). The author appreciates the permission received from publishers and authors to use their material for tables and figures. Copyright lines have been added when requested.

The book has many contributors, foremost are the active students who took up the challenge of extensively commenting upon the material. They helped to point out difficult passages, and assisted in several iterations of editing. The book, however, could not have been written without extensive assistance from many of my colleagues and the large number of specialists and referees employed by the publisher.

It is impossible to recall all the names of students and colleagues who gave invaluable advice and criticism, not only at the University of Delaware, but also during several extended visits to Stuttgart, Berlin, Hamburg, New Dehli, Osaka, and Sydney. Among others, I appreciate especially the constructive comments I received from P. Avakian, J. R. Beamish, I. Broser, S.-T. Chui, M. Crawford, S. G. Hudgens, C. S. Ih, R. V. Kasowski, J. J. Kramer, P. T. Landsberg, O. Madelung, D. G. Onn, S. T. Pantelides, K. Ploog, S. Y. Ren, J. M. Schultz, J. Singh, A. Suna, J. Tauc, R. Winston, D. M. Wood, and A. Zunger.

The dedicated work of S. Pruitt, D. Willette, M. Hobbs, L. Abrantes and J. Holowka of bringing the rough manuscript into a camera-ready form is gratefully acknowledged.

I also wish to acknowledge with gratitude the special assistance I received from the editors and staff of Van Nostrand Reinhold.

Finally, a special word of thanks is due to my family who supported me throughout this work and surrounded me with an environment of peace, love, and understanding.

An Encouragement to the Reader

In a rapidly developing field, a frequent updating of the text and the references is essential to avoid passages which are no longer acceptable as the most probable explanation of experimental observation, as new results often solve old puzzles. In a book as diverse as this, it is imperative to receive assistance from interested readers. These comments will alert us to such new developments, and will direct our attention to sections which need updating.

How to Use This Book

This book is intended to provide a general overview of the different fields in semiconductor physics and therefore does not offer the depth of description desirable for a specific class in any of the subfields. However, its manuscript, which from time to time was supplemented with material from other relevant textbooks, has served the author over many years as a useful outline for such classes. The manuscript was advantageously used in these classes since it provided easy access to a wide variety of reading material and enabled the author to broaden the horizons of his students beyond the specific subjects covered in typical textbooks. As such, it stimulated the interest of the students and permitted me to teach a complex subject with ample cross-referencing opportunities.

Based on this experience, five examples of selections of chapters to be used as skeleton outlines for specific classes are given on the following pages. If chapters are read out of sequence, guidance to prerequisite chapters is provided by numbers in square brackets at the end of each line in the example lecture outlines. The chapters given with titles in square brackets can be omitted without loss of cohesiveness. In addition, the pointer on top of the right-hand page will assist in selecting material by the degree of difficulty. For material that is easy to comprehend, the pointer will tend to the left. For more difficult material, or material that is given for reference, the pointer will tend to the right.

The homework problems are collected from a wide variety of topics. Some are of review (r) character to enhance retention. Other topics require mathematical solutions to connect partially presented material. When elementary in nature, such topics are identified by (e). There are topics (*) which require independent thinking and provide various degrees of challenge to the student. Some of the topics require additional literature (l), and are designed to familiarize the student with literature searches.

INTRODUCTION TO SEMICONDUCTING PHYSICS

CARRIERS IN SEMICONDUCTORS

SEMICONDUCTOR PHOTONICS

DEFECTS IN SEMICONDUCTORS

xiii

CRYSTALLINE, AMORPHOUS, AND
SUPERLATTICE SEMICONDUCTORS

Contents

List of Tables

PART I

BONDING AND STRUCTURE

Chapter 1

The Semiconductor

A semiconductor is determined by its chemical composition and atomic structure, which give it technically interesting, tunable electronic properties.

1.1 Introduction

A *semiconductor* is a solid with an electrical conductivity between that of a metal and an insulator. This conductivity is caused by *electronic particles*, such as electrons, holes, and polarons, referred to later in this book as *carriers* (of charge) which are set free by ionization. Such ionization can be produced thermally by light, other particles, or an electric field. It involves only a small fraction of the total number of atoms, with a density of free electronic particles typically on the order of $10^{14} \ldots 10^{18}$ cm^{-3}, compared to atomic densities of 10^{22} cm^{-3}.

The *changes in electronic properties* controlled by external means—such as light, applied voltage, magnetic field, temperature—or mechanical pressure make the semiconductor an interesting material for electronic devices, many of which have become familiar parts of our daily lives. Such devices include diodes, transistors, and integrated circuits in radios, TV sets, and computers; photosensors for streetlight control, as exposure meters in cameras, and for the identification of goods in stores; a large variety of sensors for measuring temperature, weight, and magnetic fields; and all kinds of electro-optical displays. It is almost impossible to think of a modern appliance that does not contain a semiconducting device. Our cars, homes, and offices have become filled with such devices, ranging from electronic clocks and telephones to controls, electronic typewriters, copying machines, solar cells, and new imaging films.

The first *semiconducting property* was reported by Faraday for silver sulfide in 1833; in 1851 Hittorf measured the semilogarithmic dependence of the conductivity on $1/T$ (temperature) in Ag_2S and Cu_2S. The first *application of a semiconductor as a device* was based on the observation of Braun (1874) that point contacts on some metal sulfides are rectifying; this became the well-known cat's whisker detector of the early twentieth century. Schuster noticed in 1874 that the contact of copper with copper oxide is rectifying; this observation became the basis for the copper oxide rectifier introduced in 1926 by Grondahl. The discoveries of the photoconductivity of selenium by Smith (1873) and the photovoltaic effect in the same material by Adams and Day (1876) led to the first photocells as discussed by Bergmann (1931, 1934). The term *semiconductor** was introduced much later by Königsberger and Weiss (1911). Only during the middle of this century did other materials, such as Ge, Si, CdS, GaAs, and several other similar compounds gain great interest. All initial studies involved crystalline materials; that is, their atoms are ordered in a three-dimensional periodic array.

At the same time (ca. 1950) the first *amorphous semiconductors* (α-Se) became important for electrophotography (Xerox^{TM}). A few years later with the discovery of electronic switching (Ovshinsky, 1968), another class of amorphous semiconductors, the chalcogenide glasses—and, more recently, with further development of solar cells, the amorphous Si alloys—gained substantial interest.

Some organic semiconductors, and, lately, new artificial compounds, the superlattices, are entering the vast inventory of semiconducting materials of technical interest. Today, we see the first

---

* The word itself was rediscovered at this time. It was actually used much earlier (Ebert, 1789) in approximately the correct context, and then again 62 years later by Bromme (1851). However, even after its more recent introduction, serious doubts were voiced as to whether even today's most prominent semiconductor, silicon, would not better be described as a metal (Wilson, 1931; Gudden, 1933; Meissner, 1935).

signs that new compounds can be *designed* to exhibit specific desired properties, such as a large carrier mobility, appropriate optical absorption, or higher temperature superconductivity.

Semiconductors have sparked the beginning of a new material epoch. Technology has evolved from the stone age through the bronze and iron ages into the age of semiconductors,* materials that are influencing culture and civilization to an unprecedented degree.

1.2 Chemical Aspects of Semiconductors

Only a small number of elements and simple compounds have semiconductive properties conducive for device development. Preferably, these materials are covalently bound, and are typically from group IV of the periodic system of elements or are compounds of group III with group V elements—see Mooser and Pearson, 1956, and Section 2.2. In these semiconductors, the atoms have a low number of nearest neighbors (*coordination number*). In tetrahedrally bound crystals the coordination number is 4. These semiconductors have *highly mobile carriers* which are desirable for many devices.

A number of important semiconducting properties of these devices are rather similar whether the material is in crystalline or amorphous form. This emphasizes the importance of the chemical aspect and justifies more attention to the subject, which will be provided in Chapter 2.

At closer inspection, however, almost every physical property of a semiconductor depends on the relative positioning of the atomic building blocks with respect to each other, i.e., on the structure of the semiconductor.

---

* Recently, this age was termed the silicon age (Queisser, 1985), in reference to the material now most widely used for semiconducting devices. Despite the great abundance of silicon in the earth's crust (27.5%, surpassed only by oxygen with 50.5%, and followed by aluminum with 7.3% and iron with 3.4%), and its dominance as the material of choice in the semiconductor industry with a total market of $26.6 billion in 1984, equivalent to approximately $3 per $1000 gross national product per capita as an average in the USA, Europe, and Japan (*source:* Motorola), other semiconductor materials (crystalline and amorphous) are now being identified which may show even greater potential for very large-scale application in the future.

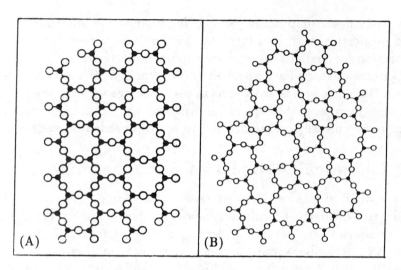

Figure 1.1: (A) Crystalline and (B) amorphous states of an A_2B_3 compound—e.g., As_2S_3 (after Zachariasen, 1932). ©American Chemical Society.

1.3 Structural Aspects of Semiconductors

As are all crystals, semiconductors can be thought of as composed of *unit cells*, which describes the smallest atomic building block within an ordered crystalline lattice with translational symmetry in all directions. In a crystalline semiconductor, these unit cells are packed in a close-fitting, three-dimensional array that results in *long-range periodicity*.

In an amorphous semiconductor, there is a similar packing of statistically slightly *deformed unit cells* with variations in interatomic distance and bond angle. By constructing such a semiconductor from such deformed unit cells, long-range periodicity is lost, and a variation in the coordination number occurs. The *short-range order* of an amorphous semiconductor, however, is similar to the crystalline state of the same material. A two-dimensional example of maintaining short-range order while losing long-range order is given in Fig. 1.1.

The consequences of the microscopic structure for the physical properties are well known for crystalline semiconductors and comprise major parts of this book, while the study of the behavior of amorphous semiconductors is still in its infancy. We have attempted to point out similarities and differences between the two states and to provide plausibility arguments where a rigorous treatment is still missing.

One of the reasons for a more manageable theoretical treatment of the crystalline state is its *periodicity*, which permits an analysis by replacing a many-body problem of astronomical extent (typically 10^{22} particles per cm^3) with a periodic repetition of one unit cell containing only a few relevant particles. Recent attempts to use a somewhat similar approach (e.g., a modified Bethe lattice approach—see Section 3.9.4) for the amorphous state have been relatively successful for theoretical prediction of some of the fundamental properties of amorphous semiconductors.

Further discussion will be postponed until the elements necessary for a more sophisticated analysis are properly introduced.

1.4 Electrical Aspects of Semiconductors

In several respects, the electrical aspects of semiconductors are unique and require a sophisticated knowledge of detail, which will be developed later in this book. Here, only a few general facts can be mentioned to emphasize some of the reasons which have sparked such enormous interest in this class of materials.

Semiconductors bridge a large gap of electrical conductivities between metals and insulators, however, more importantly, their electrical resistance* can change as a result of external forces, e.g., by applied voltage, magnetic field, light, mechanical stress, or a change in temperature. This is often accomplished by fabricating a *device* from such semiconductors when incorporating minute amounts of specific impurities (*dopants*) in a certain inhomogeneous pattern into the semiconductor. The electrical response of these devices makes them electronically *active*, as opposed to such passive elements as wires, insulators, or simple resistors made from thin metal layers.

Although the large field of semiconducting devices deals with the inhomogeneity aspect to a great extent, a thorough understanding of their operation requires a detailed analysis of the electronic and related properties of the homogeneous material. That is the topic of this book. The device aspects are dealt with in many reviews, most comprehensively by Sze (1981).

---

* The resistance is used here rather than the material resistivity because of the inhomogeneity of the electronic transport through most of the devices.

Summary and Emphasis

Semiconductors have become the basic elements of modern civilization, with an extremely large range of application, including high-speed communication and computation, and microminiaturization of integrated circuitry, providing the potential for artificial intelligence. An increasing variety of materials, distinguished by their chemical composition and atomic structure, is used to produce semiconducting devices and to perform specific device functions with increasing efficiency in smaller dimensions and at lower costs.

A substantially improved understanding of the physics of all relevant processes is essential for developing new materials designed and tailored for specific high-performance devices and to broaden the spectrum beyond currently available devices.

Exercise Problems

1.(r) List as many devices and appliances you can think of which use, or are based on, semiconductors or photoconductors and describe in general terms their performance. Include devices which absorb (detect) or emit light.

2.(r) List some of the materials used for semiconducting (or photoconducting) devices.

Chapter 2

Crystal Bonding

The bonding of atoms in semiconductors is accomplished by electrostatic forces and the tendency of atoms to fill their outer shells. Interatomic attraction is balanced by short-range repulsion due to strong resistance of atoms against interpenetration of core shells.

In this chapter, we will refresh our memory about the different types of bonding of condensed matter (solids), irrespective of whether they are crystalline or amorphous. This review is quite general and is not restricted to semiconductors.

The formation of solids is determined by the interatomic forces and the size of the atoms shaping the crystal lattice. The interatomic forces are composed of a far-reaching attractive and a short-range repulsive component, resulting in an *equilibrium distance* of vanishing forces at an interatomic distance r_e, at which the potential energy shows a minimum (Fig. 2.1). In binary compounds, this equilibrium distance r_e can be written as the sum of *atomic radii*

$$r_e = r_A + r_B, \tag{2.1}$$

where r_A and r_B are characteristic for the two atoms A and B (Fig. 2.2) and can be used when other binary compounds are formed with the same bonding type, containing A or B. For a more detailed discussion, see Section 2.9.

Attractive interatomic forces are predominantly electrostatic in character (e.g., in ionic, metallic, van der Waals, and hydrogen bonding) or are a consequence of sharing valence electrons to fill their outer shells, resulting in covalent bonding. Most materials show mixed bonding, i.e., at least two of these bond types contribute significantly to the interatomic interaction. In the better compound semiconductors, these mixed bondings are more covalent and less ionic. In other semiconductors, one of the other types may contribute, e.g., van der Waals bonding in organic crystals and metal-

7

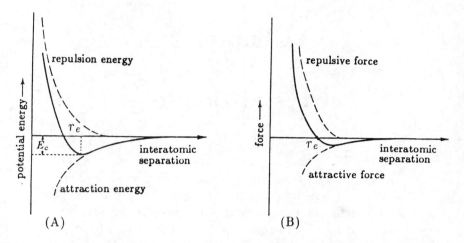

Figure 2.1: Interaction potential (A) and forces (B) between two atoms; r_e is the equilibrium distance; E_c is the bonding energy at $r = r_e$.

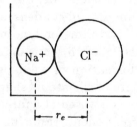

Figure 2.2: Na^+ anion and Cl^- cation shown as hard spheres in actual ratio of radii.

lic bonding in highly conductive semiconductors. These different bonding types will be discussed in the following sections.

The *repulsive interatomic forces*, called Born forces (see Born and Huang, 1954), are caused by a strong resistance of the electronic shells of atoms against interpenetration. The repulsive Born potential is usually modeled with a strong power law*

$$eV(r) = \frac{\beta}{r^m} \qquad \text{with} \qquad m \simeq 10 \ldots 12 \qquad (2.3)$$

---

* A better fit for the Born repulsion is obtained by the sum of a power and an exponential law:

$$V_{\text{Born}} = \frac{\beta}{r^m} + \gamma \exp\left(-\frac{r}{r_0}\right). \qquad (2.2)$$

r_0 is the *softness parameter*, listed for ions in Table 2.7 (pg. 25). For more sophisticated repulsion potentials, see Shanker and Kumar (1987).

with β the force constant [see Eq. (4.1)] and m an empirical exponent
—see Section 2.1. For ionic crystals the exponent m is somewhat
smaller ($6 < m < 10$).

2.1 Ionic Bonding

Ionic bonding is caused by Coulomb attraction between ions. Such
ions are formed by the tendency of atoms to complete their outer
shells. This is most easily accomplished by compounds between elements of group I and group VII of the periodic system of elements;
here, only one electron needs to be exchanged. For instance, in a
NaCl crystal, the Cl atom captures one electron to form a negative
Cl^- ion and the Na atom loses the single electron in its outer shell
to become a positive Na^+ ion. The bonding is then described by
isotropic (radial-symmetric) *nonsaturable* Coulomb forces attracting
as many Na^+ ions as space permits around each Cl^- ion, and vice
versa, while maintaining overall neutrality, i.e., an equal number of
positive and negative ions. This results in a closely packed NaCl lattice with a coordination number 6 (= number of nearest neighbors).

   The energy gain between two ions can be calculated from the
potential equation

$$eV = -\frac{e^2}{4\pi\varepsilon_0 r} + \frac{\beta}{r^m} \qquad \text{for} \quad r = r_e, \qquad (2.4)$$

containing Coulomb attraction (first term) and Born repulsion (second term). For an equilibrium distance $r_e = r_{Na^+} + r_{Cl^-} = 2.8$ Å,
which results* in a minimum of the potential energy of $eV_{min} \simeq$
-5 eV for a typical value of $m = 9$.

   In a crystal we must consider *all* neighbors. For example, in
an NaCl lattice, six nearest neighbors exert Coulomb attraction
in addition to 12 next-nearest neighbors of equal charge exerting
Coulomb repulsion, etc. This alternating interaction results in a
summation that can be expressed by a proportionality factor in the
Coulomb term of Eq. (2.4), the *Madelung constant* (Madelung, 1918).
For a NaCl crystal structure we have

$$A = \frac{6}{\sqrt{1}} - \frac{12}{\sqrt{2}} + \frac{8}{\sqrt{3}} - \frac{6}{\sqrt{4}} + \frac{25}{\sqrt{5}} - + \dots, \qquad (2.5)$$

---

   * β can be eliminated from the minimum condition $(dV/dr|_{r_e} = 0)$. One
obtains $\beta = e^2 r_e^{m-1}/(4\pi\varepsilon_0 m)$ and as *cohesive energy* $eV_{min} = -e^2(m - 1)/(4\pi\varepsilon_0 m r_e)$.

Table 2.1: Madelung constant for a number of crystal structures.

Crystal Structure	Madelung Constant
NaCl	1.7476
CsCl	1.7627
Zinc-blende	1.6381
Wurtzite	1.6410
CaF_2	5.0388
Cu_2O	4.1155
TiO_2 (Rutile)	4.8160

where each term presents the number of equidistant neighbors in the numerator and the corresponding distance (in lattice units) in the denominator. This series is only slowly converging. Ewald's method (the theta-function method) is powerful and facilitates the numerical evaluation of A. For NaCl, we obtain from (Madelung, 1918; Born and Landé, 1918):

$$eV = -A \frac{e^2}{4\pi\varepsilon_0 r_e} + \frac{\beta'}{r_e^m} \qquad (2.6)$$

with $A = 1.7476$, a lattice binding energy of $eV_{\min}^{(A)} = H^0(\text{NaCl}) = 7.948$ eV, compared to an experimental value of 7.934 eV. Here, β' and m are empirically obtained from the observed lattice constant and compressibility. The Madelung constant is listed for several AB-compounds in Table 2.1 (see Sherman, 1932).

The Born-Haber cyclic process is an empirical way of obtaining the lattice energy, i.e., the binding energy per mole. The process starts with the solid metal and gaseous halogen, and adds the heat of sublimation $W_{\text{subl}}(\text{Na})$ and the dissociation energy $(1/2)W_{\text{diss}}(\text{Cl}_2)$; it further adds the ionization energy $W_{\text{ion}}(\text{Na})$ and the electron affinity $W_{\text{elaff}}(\text{Cl})$ in order to obtain a diluted gas of Na^+ and Cl^- ions; all of these energies can be obtained experimentally. These ions can be brought together from infinity to form the NaCl crystal by gaining the unknown lattice energy $H^0(\text{NaCl})$. This entire sum of

Table 2.2: Lattice constants (a in Å) and ratio of lattice constants c/a for simple AB compounds (after Weißmantel and Hamann, 1979).[*]

NaCl		Structure		CsCl Structure		Zinc-blende		Wurtzite		c/a
AgF	4.93	NaBr	5.973	BaS	6.363	AlP	5.431	AgI	4.589	1.63
AgCl	5.558	NaI	6.433	CsCl	4.118	AlAs	5.631	AlN	3.110	1.60
AgBr	5.78	PbS	5.935	CsBr	4.296	AlSb	6.142	BeO	2.700	1.63
BaO	5.534	PbSe	6.152	CsI	4.571	BeS	4.86	CdS	4.139	1.62
BaS	6.363	PbTe	6.353	TlI	4.206	BeSe	5.08	CdSe	4.309	1.63
BaSe	6.633	RbF	5.651	TlCl	3.842	BeTe	5.551	GaN	3.186	1.62
BaTe	7.000	RbCl	6.553	TlBr	3.978	CSi	4.357	InN	3.540	1.61
CaO	4.807	RbBr	6.868	TlI	4.198	CdS	5.832	MgTe	4.529	1.62
CaS	5.69	RbI	7.341	NH$_4$Cl	3.874	CdSe	6.052	MnS	3.984	1.62
CaSe	5.992	SnAs	5.692	NH$_4$Br	4.055	CdTe	6.423	MnSe	4.128	1.63
CaTe	6.358	SnTe	6.298	NH$_4$I	4.379	CuF	4.264	TaN	3.056	—
CdO	4.698	SrO	5.156	TiNO$_3$	4.31	CuCl	5.417	ZnO	3.249	1.60
KF	5.351	SrS	5.582	CsCN	4.25	CuBr	5.691	ZnS	3.819	1.64
KCl	6.283	SrSe	6.022			GaP	5.447	NH$_4$F	4.399	1.60
KBr	6.599	SrTe	6.483			GaAs	5.646			
KI	7.066	TaC	4.454			GaSb	6.130			
LiF	4.025	TiC	4.329			HgSe	6.082			
LiCl	5.140	TiN	4.244			HgTe	6.373			
LiBr	5.501	TiO	4.244			InAs	6.048			
LiI	6.012	VC	4.158			InSb	6.474			
MgO	4.211	VN	4.137			MnS	5.611			
MgS	5.200	VO	4.108			MnSe	5.832			
MgSe	5.462	ZrC	4.696			ZnS	5.423			
NaF	4.629	ZrN	4.619			ZnSe	5.661			
NaCl	5.693					ZnTe	6.082			

[*]For explanation of the different crystal structures see Chapter 3.

processes must equal the heat of formation $W^0(\text{NaCl})$ which can be determined experimentally (Born, 1919, and Haber, 1919):

$$W^0_{\text{solid}}(\text{NaCl}) = \left\{ W_{\text{subl}}(\text{Na}) + W_{\text{ion}}(\text{Na}) \right.$$
$$\left. + \frac{1}{2} W_{\text{diss}}(\text{Cl}_2) + W_{\text{elaff}}(\text{Cl}) \right\} + H^0(\text{NaCl}). \tag{2.7}$$

In this equation a minor correction of an isothermal compression of NaCl from $p = 0$ to $p = 1$ (atm), heating it from $T = 0\,\text{K}$ to room temperature, and an adiabatic expansion of the ion gases to $p = 0$ has been neglected. The corresponding energies almost cancel. The error is $< 1\%$.

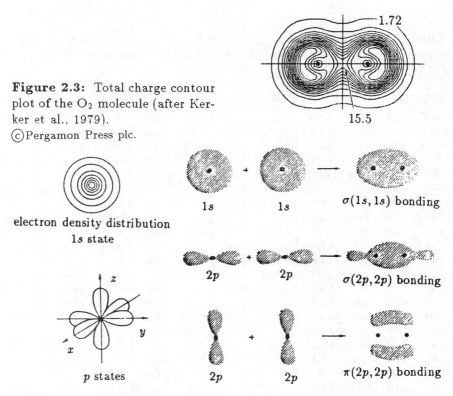

Figure 2.3: Total charge contour plot of the O_2 molecule (after Kerker et al., 1979). ©Pergamon Press plc.

electron density distribution
1s state

p states

$1s$ $1s$ $\sigma(1s, 1s)$ bonding

$2p$ $2p$ $\sigma(2p, 2p)$ bonding

$2p$ $2p$ $\pi(2p, 2p)$ bonding

Figure 2.4: Atomic and molecular electron density distribution for $\sigma(s)$, $\sigma(p)$, and $\pi(p)$ bonding—see Appendix A.4.8 for further explanation (after Weißmantel and Hamann, 1979).

A listing of lattice constants (for definition see Chapter 3) of a number of predominantly ionic AB-compounds is given in Table 2.2.

2.2 Covalent Bonding

Covalent bonding is caused by two electrons that are shared between two atoms: they form an *electron bridge* as shown in Fig. 2.3 for a diatomic oxygen molecule. This bridge formation can be understood quantum-mechanically by a nonspherical electron density distribution that extends between the bonded atoms. Examples of such density distributions are shown in Fig. 2.3 for an O_2 molecule and schematically in Fig. 2.4 for a molecule formation with electrons in a 1-s or 2-p state, e.g., for H_2 or F_2, respectively.

If an approaching atom of the same element has in its protruding part of the electron density distribution an unpaired electron with antiparallel spin, both eigenfunctions may overlap; the Pauli princi-

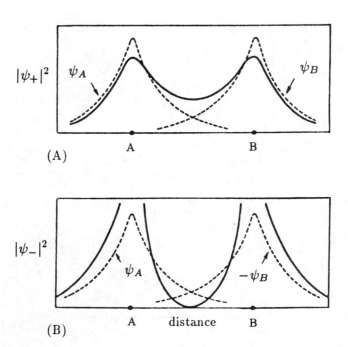

Figure 2.5: Wavefunctions of one-electron states [dashed curves—identical in (A) and (B)] and probability function to find one electron (solid curves) in (A), a bonding state, and (B), an antibonding state, showing finite and vanishing electron density at the center between atoms A and B for these two states, respectively [observe the plotting of $-\psi_B$ in (B)]. The picture of these two one-electron states shown here shall not be confused with the two-electron potential given in Fig. 2.6.

ple is not violated. Their combined wave function ($\psi_+ = \psi_A + \psi_B$) yields an increased electron density $|\psi_+|^2$ in the overlap region (see Fig. 2.5A); the result is an *attractive force* between these two atoms *in the direction of the overlapping eigenfunctions*. This is the state of lowest energy of the two atoms, the *bonding state*. There is also a state of higher energy, the *antibonding state*, with $\psi_- = \psi_A - \psi_B$ in which the spin of both electrons is parallel. Here the electrons are strongly repulsed because of the Pauli principle, and the electron clouds cannot penetrate each other; therefore, the electron density between both atoms vanishes (Fig. 2.5B). The resulting potential distribution as a function of the interatomic distance between two hydrogen atoms forming an H_2 molecule is given in Fig. 2.6. In this figure the ground state (bonding) S and the excited state (antibonding) A are shown. The figure also contains as center curve the

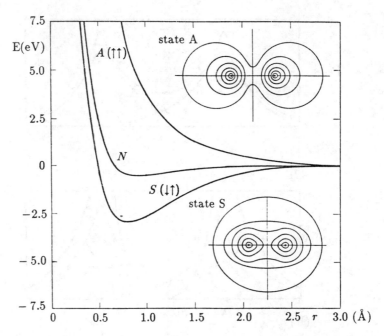

Figure 2.6: Potential energy for the two valence electrons of two co-valently bound hydrogen atoms approaching each other; (upper curve) antibonding state; (lower curve) bonding state; (middle curve) from free atom charge distribution bonding. Charge density distributions shown in the insert are for the two covalent states (after Kittel, 1966). ©John Wiley & Sons, Inc.

Table 2.3: Bond lengths relevant to organic molecules α-Si and related semiconductors (after Cotton and Wilkinson, 1972).

Bond	Bond length (Å)	Bond	Bond length (Å)
C$-$C	1.54	Si$-$Si	2.35
C$=$C	1.38	Si$-$H	1.48
C$=$C	1.42 (graphite)	Ge$-$Ge	2.45
C$\equiv$C	1.21	Ge$-$H	1.55
C$-$H	1.09 (sp^3)	C$-$Si	1.87

classical contribution of two H-atoms with a charge density corresponding to free atoms. Such bonding is small compared with the covalent bonding.

The bond length (center-to-center distance) between C-atoms in organic molecules decreases with increasing bonding valency as

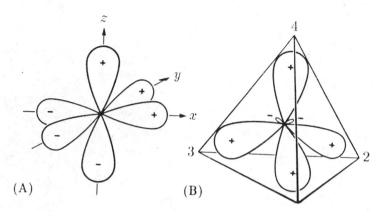

Figure 2.7: Linear combination (hybridization) of a $1s$ function (spherical) with $3p$ functions (A) results in four sp^3 functions (B) which extend towards the four tetrahedra axes 1–4, and result in strongly directional bonding with a bond angle of $109.47°$.

shown in Table 2.3. Other bond lengths typical for organic or similar molecules are also listed.

With additionally missing unpaired electrons in the outer shell, more than one atom of the same kind can be bound to each other. The number of bonded atoms is given by the following valency: monovalent atoms can form only diatomic molecules; divalent atoms, such as S or Se, can form chains; and trivalent atoms, such as As, can form two-dimensional (layered) lattices. Solids are formed from such elements by involving other bonding forces between the molecules, chains, or layers, e.g., van der Waals forces—see Section 2.5. Only tetravalent elements can form three-dimensional lattices which are covalently bound (e.g., Si).

2.2.1 Tetrahedrally Bound Elements

Silicon has four electrons in its outer shell. In the ground state of an isolated atom, two of the electrons occupy the s-state and two of them occupy p-states, with a $2s^2 2p^2$ configuration—see Appendix A.4.8. By investing a certain amount of promotion energy,* this $s^2 p^2$-configuration is changed into an sp^3-configuration, in which an unpaired electron sits in each one of the singly occupied orbitals

---

* The promotion energy is 4.3, 3.5, and 3.3 eV for C, Si, and α-Sn, respectively. However, when forming bonds by establishing electron bridges to neighboring atoms, a substantially larger energy is gained, therefore resulting in net binding forces. Diamond has the highest cohesive energy in

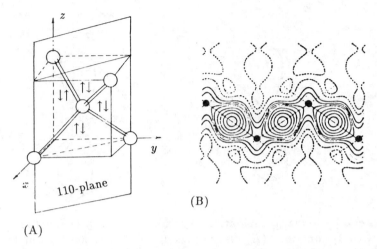

(A)   (B)

Figure 2.8: (A) Unit cell of diamond with pairs of electrons indicated between adjacent atoms; (B) electron density profile within the (110) plane (after Dawson, 1967).

with tetrahedral geometry (see Fig. 2.7). From the s-orbital and the three p-orbitals, four linear combinations can be formed, represented as $\sigma_i = 1/2(\varphi_s + \varphi_{px} \pm \varphi_{py} \pm \varphi_{pz})$, depending upon the choice of signs. This is referred to as *hybridization*, with σ_i as the *hybrid function* responsible for bonding. When we bring together a large number of Si atoms, they arrange themselves so that each of them has four neighbors in tetrahedral geometry as shown in Fig. 3.9. Each atom then forms four electron bridges to its neighbors, in which each one is occupied with two electrons of opposite spin, as shown for the center atom in Fig. 2.8A. Such bridges become evident in a density profile within the (110) plane shown for two adjacent unit cells in Fig. 2.8B.

In contrast to the ionic bond, the covalent bond is angular-dependent, since the protruding atomic eigenfunctions extend in well-defined directions. Covalent bonding is therefore a *directional* and *saturable bonding*; the corresponding force is known as a *chemical valence force*, and acts in exactly as many directions as the valency describes.

---

this series, *despite* the fact that its promotion energy is the largest because its sp^3-sp^3 C-C bonds are the strongest (see Harrison, 1980).

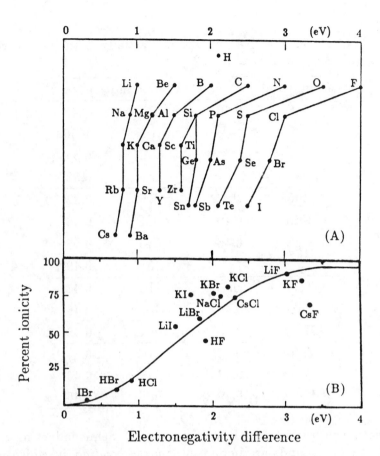

Figure 2.9: (A) Electronegativity of the elements with groups from the periodic table of elements identified by interconnecting lines; (B) ionicity of alkali halides and halide molecules as a function of the difference in electronegativity (after Pauling, 1960).

2.3 Mixed Bonding

Crystals that are bonded partially by ionic and partially by covalent forces are referred to as mixed-bond crystals. Most actual semiconductors have a fraction of covalent and ionic bonding components (see, e.g., Mooser and Pearson, 1956).

2.3.1 Tetrahedrally Bonded Binaries

By using the Grimm-Sommerfeld rule (see below) for isoelectronic rows of elements, Welker and Weiss (1954) predicted desirable semi-

Table 2.4: Static effective charges of partially covalent AB-compounds (after Coulson et al., 1962).

Compound	e^*/e	Compound	e^*/e
ZnO	0.60	BN	0.43
		AlN	0.56
		GaN	0.55
		InN	0.58
ZnS	0.47	BP	0.32
CdS	0.49	AlP	0.46
HgS	0.46	GaP	0.45
		InP	0.49
ZnSe	0.47	AlAs	0.47
CdSe	0.49	GaAs	0.46
HgSe	0.46	InAs	0.49
ZnTe	0.45	AlSb	0.44
CdTe	0.47	GaSb	0.43
HgTe	0.49	InSb	0.46

conducting properties for III-V compounds.* Semiconducting III-V and II-VI compounds are bound in a *mixed bonding*, in which electron bridges exist, i.e., the bonding is directed, but the electron pair forming the bridge sits closer to the anion. This *degree of ionicity* increases for these compounds with an increased difference in electronegativity (Fig. 2.9) from III-V to I-VII compounds and within one class of compounds, e.g., from RbI to LiF—see also Table 2.4.

The mixed bonding may be expressed as the sum of the wave functions describing covalent and ionic bonding

$$\psi = a\psi_{\text{cov}} + b\psi_{\text{ion}} \tag{2.8}$$

with the ratio b/a defining the *ionicity* of the bonding. This bonding can also be described as rapidly alternating between that of covalent and ionic. Over an average time period, a fraction of ionicity (b/a) results. The ionicity of the bonding can be described by a static

---

* Meaning compounds between one element of group III and one element of group V on the periodic system of elements.

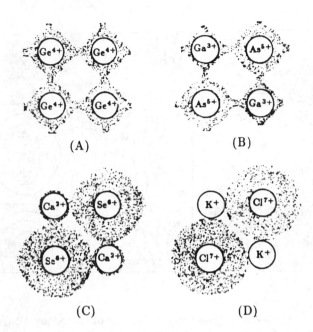

Figure 2.10: Schematic sketch of mixed bonding from nearly perfect covalent (A) in Ge to perfect ionic (D) in KCl. It shows diminishing bridge formation and increasing cloud formation of electrons around anions with increasing ionicity (after Ashcroft and Mermin, 1976).

effective ion charge e^*, as opposed to a dynamic effective ion charge (discussed in Section 11.2.1), which is less by a fraction on the order of b/a than in a purely ionic compound with the charge given by the valency. For instance, in CdS the divalent behavior of Cd and S could result in a doubly charged $Cd^{++}S^{--}$ lattice, while measurements of the electric dipole moment indicate an effective charge of 0.49 for CdS.

The static effective charge for other II-VI and III-V compounds is given in Table 2.4.

The effective charge concept can be confusing if one does not clearly identify the ionic state of the system. For instance, in the case of CdS, a purely ionic state is $Cd^{++}S^{--}$, as opposed to the covalent state of $Cd^{--}S^{++}$ (which is equivalent to the $Si^{\times}Si^{\times}$-configuration). In other words, the covalent state is that in which both Cd and S have four valence electrons and are connected to each other by a double bond. This must not be confused with the neutral $Cd^{\times}S^{\times}$

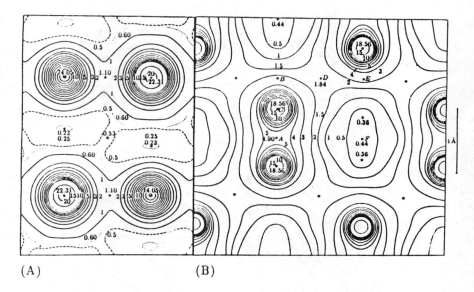

(A)   (B)

Figure 2.11: Electron density distribution obtained by Fourier analysis of x-ray diffraction pattern of (A) NaCl and (B) diamond (after Brill, et al., 1942).

configuration, which is a mixed-bonding state. The expression for the static effective charge (see Coulson et al., 1962) is

$$\frac{e^*}{e} = \frac{N(a/b)^2 - (8 - N)}{1 + (a/b)^2},\qquad(2.9)$$

with N as the valency. For $N = 2$, the effective charge vanishes when $a/b = \sqrt{3}$. For $N = 3$ in III-V compounds, e^* vanishes when $a/b = \sqrt{5/3}$, and for group IV semiconductors when $a = b$.

In crystals, low coordination numbers (typically 4) signify a considerable covalent contribution to the bonding.

The different degree of bridge formation in crystals with mixed bonding (Fig. 2.10) can be made visible by a Fourier analysis of x-ray diffraction from which the electron density distribution around each atom can be obtained. This is shown for a mostly ionic crystal in Fig 2.11A and for a mostly covalent crystal in Fig. 2.11B.

2.4 Metallic Bonding (Delocalized Bonding)

Metallic bonding can be understood as a collective interaction of a mobile electron fluid with metal ions. Metallic bonding occurs when the number of valence electrons is only a small fraction of the

Table 2.5: Ionic radii r_i and half the nearest-neighbor distances in metals r_m in Å (after Ashcroft and Mermin, 1976).

Metal	r_i	r_m	r_m/r_i	Transition Metal	r_i	r_m	r_m/r_i
Li	0.60	1.51	2.52	Cu	0.96	1.28	1.33
Na	0.95	1.83	1.93	Ag	1.26	1.45	1.15
K	1.33	2.26	1.70	Au	1.37	1.44	1.05
Rb	1.48	2.42	1.64				
Cs	1.69	2.62	1.55				

coordination number; then neither an ionic nor a covalent bond can be established. Metallic bonding of simple metals, e.g., alkali metals, can be modeled by assuming that each metal atom has given up its valence electron, forming a lattice of positively charged ions, submerged in a fluid of electrons. Between the repulsive electron-electron and ion-ion interactions and the attractive electron-ion interaction, a net attractive binding energy results, which is *nondirectional and not saturable*, and results in *close-packed structures* with high coordination numbers (8 or 12; Wigner and Seitz, 1933), but relatively wide spacing between the submerged metal ions (Table 2.5). Such metals have low binding energies (~ 1 eV/atom) and high compressibility. They are mechanically soft, since the nondirectional lattice forces exert little resistance against plastic deformation. This makes metals attractive for forming and machining.

In other metals, such as transition group elements, the bonding may be described as due to covalent bonds which rapidly hop from atom pair to atom pair. Again, free electrons are engaged in this resonance-type bonding. These metals have a higher binding energy of ~ 4 to 9 eV/atom and an interatomic distance that is closer to the one given by the sum of ionic radii (Table 2.5). They are substantially harder when located in the middle of the transition metal row, e.g., Mo and W (Ashcroft and Mermin, 1976).

In semiconductors with a very high density of free carriers, metallic binding forces may contribute a small fraction to the lattice bond, interfering with the predominant covalent bonding and usually weakening it, since these electrons are obtained by ionizing other bonds. Changes in the mechanical strength of the lattice can be observed in photoconductors (see Section 6.2.4) in which a high density of free

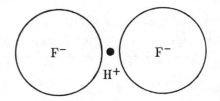

Figure 2.12: Hydrogen bonding between a positive hydrogen ion (proton) and two ions (coordination number 2).

carriers can be created by light (Gorid'ko et al., 1961). For more information, see Ziman (1969) and Harrison (1966).

2.5 Van der Waals Bonding

Noble gas atoms or molecules with saturated covalent bonds can be bound to each other by dipole-dipole interaction (Debye). The dipole is created between the nucleus (nuclei) of the atom (molecule) and the cloud of electrons moving around these nuclei, and forms a fluctuating dipole moment even for a spherically symmetrical atom. The interaction creates very weak, nonsaturable attractive forces. This results in low melting points and soft *molecule crystals*. The bonding energy can be approximated by

$$eV = -\frac{\alpha_a}{r^6} + \frac{\alpha_r}{r^{12}} \tag{2.10}$$

Van der Waals forces are the main binding forces of organic semiconductors (van der Waals, 1873).

2.6 Hydrogen Bonding

Hydrogen bonding (Fig 2.12) is a type of ionic bonding in which the hydrogen atom has lost its electron to another atom of high electronegativity. The remaining proton establishes a strong Coulomb attraction. This force is not saturable. However, because of the small size of the proton, hydrogen bonding is strongly localized, and spatially no more than two ions have space to be attracted to it. When part of a molecule, the hydrogen bond—although ionic in nature—fixes the direction of the attached atom because of space consideration. It should not, however, be confused with the covalent bonding of hydrogen that occurs at dangling bonds (see Section 22.4.1) in semiconductors, e.g., at the crystallite interfaces of polycrystalline Si or in amorphous Si:H.

2.7 Intermediate Valence Bonding

An interesting group of semiconductors are *transition metal compounds*. The transition metals have partially filled *inner* $3d$, $4d$, $5d$, or $4f$ shells and a filled outer shell that provides a shielding effect to the valence electrons. In these compounds the crystal field has a reduced effect. Some of these compounds show *intermediate valence bonding*. The resulting unusual properties range from resonant valence exchange transport in copper-oxide compounds (Anderson, 1986; Anderson et al., 1987) to giant magnetoresistance and very large magneto-optical effects in rare-earth semiconductors. For a review, see Holtzberg et al. (1980).

2.8 Other Bonding Considerations

Other, more subtle bonding considerations have gained a great deal of interest because of their attractive properties. These are related to magnetic and special dielectric properties, to superconductivity, as well as to other exotic effects.

For instance, dilute semimagnetic semiconductors such as the alloy $Cd_{1-\xi}Mn_\xi Te$ (Furdnya, 1982, 1986; Brandt and Moshchalkov, 1984; Wei and Zunger, 1986; Goede and Heimbrodt, 1988) show interesting magneto-optical properties. They change from paramagnetic ($\xi < 0.17$) to antiferromagnetic ($0.6 < \xi$) to the ferro- or antiferromagnetic behavior of MnTe; exhibit giant magneto-optical effects and bound magnetic polarons; and offer opportunities for opto-electric devices that are tunable by magnetic fields.

These materials favor specific structures and permit the existence of certain quasi-particles, such as small polarons or Frenkel excitons. These discussions require a substantial amount of understanding of the related physical effects, and are therefore postponed to a more appropriate section of this book (see also Phillips, 1973; Harrison, 1980; Ehrenreich, 1987).

2.9 Atomic and Ionic Radii

The equilibrium distances between atoms in a crystal define atomic radii when *assuming* hard-sphere atoms touching each other. In reality, however, these radii are soft with some variation of the electronic eigenfunctions and, for crystals with significant covalent fraction, with dependence on the angular atomic arrangement. However, for many crystals the hard-sphere radii are very useful for most lattice estimates.

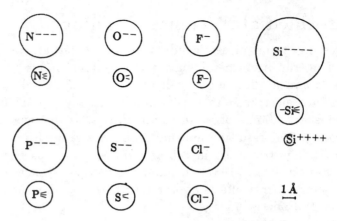

Figure 2.13: Scale drawing of rigid sphere atoms with different bonding character [ionic or covalent, identified by the appropriate number of minus signs (upper row) or valence lines (lower row), respectively].

When comparing the lattice constants of chemically similar crystals, such as NaCl, NaBr, KCl, and KBr, one can determine the radii of the involved ions (Na^+, K^+, Cl^-, and Br^-) if at least one radius is known independently. Goldschmidt (1927) used the radii of F^- and O^{--} for calibration. Consequent listings of other ionic radii are therefore referred to as *Goldschmidt radii*. These radii are independent of the compound in which the atoms are incorporated as long as they exhibit the same type of bonding. One distinguishes atomic, ionic, metallic, and van der Waals radii. Ionic radii vary with changing valency.

A list of the most important ion and atomic radii is given in Table 2.6. The drastic change in radii with changing bonding force (Mooser and Pearson, 1956) is best demonstrated by comparing a few typical examples for some typical elements incorporated in semiconductors (Fig. 2.13). For more recent estimates of tetrahedral covalent radii, see van Vechten and Phillips (1970).

The deviation from strict rigidity, i.e., the softness of the ionic spheres, is conventionally considered by using a softness parameter r_0 in the exponential repulsion formula [Eq. (2.2)]. This parameter is listed for a number of ions in Table 2.7.

This softness also results in a change of the standard ionic radii as a function of the number of surrounding atoms. A small correction Δ_m in the interionic distance is listed in Table 2.8. This needs to be considered when crystals with different coordination numbers m,

Table 2.6: Covalent (effective ionic charge $e^* = 0$) and standard ionic (identified by $\pm e^*$) radii in Å.

e^*		+1	0		+2	0		+3	0		+4	0	−4		0	−3		0	−2
	Li	0.68	1.34	Be	0.30	0.90	B	0.16	0.88	C		0.77	2.60	N	0.70	1.71	O	0.73	1.46
	Na	0.98	1.54	Mg	0.65	1.30	Al	0.45	1.26	Si	0.38	1.17	2.71	P	1.10	2.12	S	1.04	1.90
	K	1.33	1.96	Ca	0.94	1.74	Sc	0.68		Ti	0.60			As	1.18	2.22	Se	1.14	2.02
	Cu	0.96		Zn	0.74	1.31	Ga	0.62	1.26	Ge	0.53	1.22	2.72						
	Rb	1.48		Sr	1.10		Y	0.88		Zr	0.77			Sb	1.36	2.45	Te	1.32	2.22
	Ag	1.26		Cd	0.97	1.48	In	0.81	1.44	Sn		1.40	2.94						
	Cs	1.67		Ba	1.2		La	1.04		Ce	0.92			Bi	1.46		Po		2.30
	Au	1.37		Hg	1.10	1.48	Tl	0.95	1.47	Pb	0.84	1.46							

Table 2.7: Repulsive potential softness parameters (Eq. (2.2)) in Å (after Shanker and Kumar, 1987).

Ion	$r_0(th)$	$r_0(exp)$	Ion	$r_0(th)$	$r_0(exp)$
Li$^+$	0.069	0.042	F$^-$	0.179	0.215
Na$^+$	0.079	0.090	Cl$^-$	0.238	0.224
K$^+$	0.106	0.108	Br$^-$	0.258	0.254
Rb$^+$	0.115	0.089	I$^-$	0.289	0.315
Cs$^+$	0.130	0.100			

Table 2.8: Change of interatomic distance Δ_m (in Å) for compounds deviating from coordination number 6.

m	Δ_m	m	Δ_m	m	Δ_m	m	Δ_m
1	−0.50	4	−0.11	7	+0.04	10	+0.14
2	−0.31	5	−0.05	8	+0.08	11	+0.17
3	−0.19	6	0	9	+0.11	12	+0.19

Table 2.9: Mohs hardness.

Material	Chemistry	Lattice Type	Hardness
Talc	$Mg_3H_2SiO_{12 \cdot aq}$	Layer lattice	1
Gypsum	$CaSO_4.H_2O$	Layer lattice	2
Iceland spar	$CaCO_3$	Layer lattice	3
Fluorite	CaF_2	Ion lattice	4
Apatite	$Ca_5F(PO_4)_3$	Ion lattice	5
Orthoclase	$KAlSi_3O_8$	SiO_4 frame	6
Quartz	SiO_2	SiO_4 frame	7
Topaz	$Al_2F_2SiO_4$	Mixed ion-valency lattice	8
Corundum	Al_2O_3	Valency lattice	9
Diamond	C	Valency lattice	10

i.e., the number of surrounding atoms, are compared with each other (e.g., CsCl and NaCl).

With increasing atomic number, the atomic (or ionic) radius of homologous elements increases. The cohesive force therefore decreases with increasing atomic (ionic) radii. Thus, compounds formed by the same bonding forces, and crystallizing with similar crystal structure, show a decrease, for example, in hardness,* melting point, and band gap, but an increase in dielectric constant and carrier mobility—see the respective sections.

The ratio of ionic radii determines the preferred crystal structure of *ionic* compounds. This is caused by the fact that the energy gain of a crystal is increased with every additional atom that can be added per unit volume. When several possible atomic configurations are considered, the material crystallizes in a modification that maximizes the number of atoms in a given volume. This represents the state of lowest potential energy of the crystal, which is the most stable one. An elemental crystal with isotropic radial interatomic forces will

---

* This empirical quantity can be defined in several ways (e.g., as Mohs, Vickers, or Brinell *hardness*) and is a macroscopic mechanical representation of the cohesive strength of the lattice. In Table 2.9 the often used Mohs hardness is listed, which orders the listed minerals according to the ability of the higher-numbered one to scratch the lower-numbered minerals.

Table 2.10: Preferred lattice structure for AB-compounds with ionic binding forces (after Goldschmidt, 1927).

r_A/r_B	Preferred Stable Lattice
< 0.22	None
0.22 ... 0.41	Zinc-blende or Wurtzite
0.41 ... 0.72	NaCl lattice
> 0.72	CsCl lattice

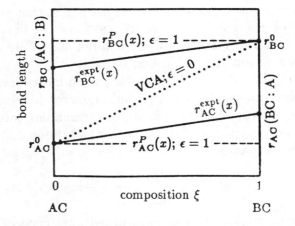

Figure 2.14: Variation of bond-length in an $A_{1-\xi}B_\xi C$ alloy for rigid atoms ($\varepsilon = 1$), virtual crystal approximation ($\varepsilon = 0$), and experimentally observed relaxation.

therefore crystallize in a close-packed structure. In a binary crystal, the ratio of atomic radii will influence the possible crystal structure. For isotropic nonsaturable interatomic forces, the resulting stable lattices are shown in Table 2.10 for different ratios of the ion radii— see following sections.'

When a substantial amount of covalent bonding forces are involved, the rules to select a stable crystal lattice for a given compound are more complex. Here atomic bond length and bond angles must be considered. Both can now be determined from basic principal density functions calculations—see Section 8.3.2. We can then define atomic radii from the turning point of the electron density distribution of each atom, and obtain an angular-dependent internal energy scale from these calculations (Zunger, 1979). Using axes constructed from these radii, one obtains well-separated domains in which only one crystal structure is observed for binary compounds (Zunger, 1979; Villars and Calvert, 1985).

2.9.1 Bond-Length Relaxation in Alloys

The lattice constant of alloys $A_{1-\xi}B_\xi C$ of binary compounds AC and BC interpolates according to the concentration

$$a(\xi) = (1 - \xi)a_{AC} + \xi a_{BC} \qquad (2.11)$$

when they crystallize with the same crystal structure (Vegard, 1921). However, the bond length between any of the three pairs of atoms is neither a constant, as suggested from the use of constant atomic radii (Pauling, 1960), nor a linear interpolation as shown by the dotted line in Fig. 2.14 for total relaxation of the bond of atom B in a different chemical environment AC (or of A in BC).

This nonrigidity of atoms is important when incorporating isovalent impurities into the lattice of a semiconductor (doping) and estimating the resulting deformation of the surrounding lattice. With the bond length r_{BC} within the AC lattice (see Table 2.11), one defines a relaxation parameter

$$\epsilon = \frac{r_{BC}(AC : B) - r_{AC}^o}{r_{BC}^o - r_{AC}^o}. \qquad (2.12)$$

The superscript o indicates the undisturbed pure crystal, the notation AC:B indicates B as doping element with a sufficiently small density incorporated in an AC compound, so that B-B interaction can be neglected.

This relaxation parameter can be estimated from the bond-stretching and bond-bending force constants α and β (see Table 2.12), according to Martins and Zunger (1984),

$$\epsilon = \frac{1}{1 + \dfrac{1}{6}\dfrac{\alpha_{AC}}{\alpha_{BC}}\left[1 + 10\dfrac{\beta_{AC}}{\alpha_{AC}}\right]}, \qquad (2.13)$$

yielding values of ϵ typically near 0.7—see Table 2.11; that is, isovalent impurity atoms behave more like rigid atoms ($\epsilon = 1$) than totally relaxed atoms ($\epsilon = 0$) in a virtual crystal approximation [Eq. (2.11)].

Summary and Emphasis

The interatomic forces are responsible for crystal bonding and are, to a large degree, electrostatic forces (Coulomb forces) between the

Table 2.11: Bond-length of isovalent impurity in given host lattice and bond-length relaxation parameter (after Martins and Zunger, 1984).

System	r_{BC}(AC:B)(Å)	ϵ	System	r_{BC}(AC:B)(Å)	ϵ
AlP:In	2.480	0.65	InP:Al	2.414	0.73
GaP:In	2.474	0.63	InP:Ga	2.409	0.73
AlAs:In	2.553	0.60	InAs:Al	2.495	0.74
GaAs:In	2.556	0.62	InAs:Ga	2.495	0.73
AlSb:In	2.746	0.61	InSb:Al	2.693	0.75
GaSb:In	2.739	0.60	InSb:Ga	2.683	0.74
AlP:As	2.422	0.65	AlAs:P	2.395	0.67
AlP:Sb	2.542	0.61	AlSb:P	2.444	0.73
AlAs:Sb	2.574	0.60	AlSb:As	2.510	0.71
GaP:As	2.414	0.62	GaAs:P	2.387	0.68
GaP:Sb	2.519	0.57	GaSb:P	2.436	0.73
GaAs:Sb	2.564	0.60	GaSb:As	2.505	0.70
InP:As	2.595	0.67	InAs:P	2.562	0.74
InP:Sb	2.700	0.60	InSb:P	2.597	0.79
InAs:Sb	2.739	0.64	InSb:As	2.667	0.75
ZnS:Se	2.420	0.70	ZnSe:S	2.367	0.78
ZnS:Te	2.539	0.67	ZnTe:S	2.407	0.78
ZnSe:Te	2.584	0.71	ZnTe:Se	2.502	0.74
β-HgS:Se	2.611	0.76	HgSe:S	2.553	0.80
β-HgS:Te	2.716	0.71	HgTe:S	2.579	0.82
HgSe:Te	2.748	0.74	HgTe:Se	2.665	0.80
ZnS:Hg	2.482	0.73	β-HgS:Zn	2.380	0.80
ZnSe:Hg	2.587	0.74	HgSe:Zn	2.494	0.78
ZnTe:Cd	2.755	0.70	CdTe:Zn	2.674	0.78
ZnTe:Hg	2.748	0.69	HgTe:Zn	2.673	0.78
γ-CuCl:Br	2.440	0.81	γ-CuBr:Cl	2.367	0.79
γ-CuCl:I	2.563	0.80	γ-CuI:Cl	2.407	0.76
γ-CuBr:I	2.585	0.79	γ-CuI:Br	2.500	0.76
C:Si	1.665	0.35	Si:C	2.009	0.74
Si:Ge	2.380	0.58	Ge:Si	2.419	0.63
Si:Sn	2.473	0.53	α-Sn:Si	2.645	0.70
Ge:Sn	2.549	0.55	α-Sn:Ge	2.688	0.67

Table 2.12: Bond-length (d), bond-stretching (α) and bond-bending (β) force constants, calculated from elastic constants (after Martin, 1970).

Crystal	$d(\text{Å})$	$\alpha(\text{N/m})$	$\beta(\text{N/m})$	Crystal	$d(\text{Å})$	$\alpha(\text{N/m})$	$\beta(\text{N/m})$
C	1.545	129.33	84.71	InP	2.541	43.04	6.24
Si	2.352	48.50	13.82	InAs	2.622	35.18	5.49
Ge	2.450	38.67	11.37	InSb	2.805	26.61	4.28
α-Sn	2.810	25.45	6.44	ZnS	2.342	44.92	4.81
SiC	1.888	88.	47.5	ZnSe	2.454	35.24	4.23
AlP	2.367	47.29	9.08	ZnTe	2.637	31.35	4.45
AlAs	2.451	43.05	9.86	CdTe	2.806	29.02	2.44
AlSb	2.656	35.35	6.79	β-HgS	2.534	41.33	2.56
GaP	2.360	47.32	10.46	HgSe	2.634	36.35	2.36
GaAs	2.448	41.19	8.94	HgTe	2.798	27.95	2.57
GaSb	2.640	33.16	7.23	γ-CuCl	2.341	22.9	1.01
				γ-CuBr	2.464	23.1	1.32
				γ-CuI	2.617	22.5	2.05

electrons and atomic nuclei. These are the basic elements for ionic and hydrogen bonding forces, but are also involved in metallic bonding and, as dipole-dipole interaction, in Van der Waals bonding. In addition, strong quantum-mechanical effects, determining specific orbitals, and Pauli exclusion are major contributing factors in covalent and metallic bonding, respectively. Comparatively minor additions contributing to crystal bonding stem from magnetic interaction due to electron spin. While the overlap of eigenfunctions of unpaired electrons with opposite spin provides the major contribution to the covalent attraction, the near-impermeability of all other electronic orbitals determines the rigidity of atoms in close proximity to each other and justifies the listing of atomic radii.

The type of bonding of a solid determines, to a large extent, its properties as a material, which is more or less desirable for a specific semiconducting device. Typically, the more covalently bonded semiconductors provide the more desired properties, e.g., low effective masses and high-carrier mobilities, as will be defined later. However, for specific devices requiring certain dielectric or magnetic properties, other bonding types are preferable. These will be discussed in the relevant sections of this book.

Exercise Problems

1.(e) The Madelung constant for an NaCl crystal is given in Eq. (2.5). Determine the Madelung constant for a CsCl crystal.

2.(e) Give 8 binaries and compare the lattice constant which you obtain from the appropriate Goldschmidt radii with the tabulated lattice constant.

(a) Determine the specific density of these compounds.

(b) Determine the side length of a cube containing 1 mol of sodium. (Be careful to take the proper atomic radius.)

3.(e) Estimate the Coulomb attractive force, the repulsive force, and the lattice energy for CsCl.

4.(1) Discuss bond-length relaxation resorting to original literature (Martins and Zunger, 1984).

5.(*) The bulk modulus B is given by Eq. (4.2). Show that in a NaCl crystal it is given by

$$B = \frac{e}{18r_0} \left. \frac{d^2 V}{dr^2} \right|_{r=r_0}.$$

With V(r) given by Eq. (2.4), show that

$$B(\text{NaCl}) = \frac{m-1}{18} \frac{\alpha e^2}{r_0^4} \quad \text{and} \quad m = 1 + \frac{18 B r_0^3}{V_{\text{ion}}(r_0)}.$$

6.(e) Calculate the equilibrium lattice constant and the lattice energy of solid Argon [use Eq. (2.10) with $\alpha_a = 1.6 \cdot 10^{-52}$ kcal cm^6 and $\alpha_r = 2.4 \cdot 10^{-97}$ kcal cm^{12}].

7.(r) Plot the atomic, ionic, and metallic radii as functions of the atomic number. Group together each column of the periodic system. Discuss the trend. Be as specific as you can.

Chapter 3

The Structure and Growth
of Semiconductors

The bonding forces and atomic sizes determine the ar-
rangement of the atoms in equilibrium in crystals and
thus determine the intrinsic properties of semiconduc-
tors. Nonequilibrium states can be frozen-in and deter-
mine the structure of amorphous semiconductors and
superlattices.

Many physical properties depend on the *periodicity* and *symmetry*
of the lattice that determines its *crystal structure.* A short summary
of the basic elements of the crystal structure is presented in this
chapter. For an extensive review, see DiBenedetto (1967), Newnham
(1975), Brown and Forsyth (1973), Barrett and Massalski (1980),
and Chernov (1984).

The easiest way to define the structure of a crystalline semiconductor (Fig. 3.1) is by its smallest three-dimensional building block,
the *unit cell.* From these unit cells, the ideal crystal is constructed by
three-dimensional repetition. The unit cell usually contains a small
number of atoms, from one for a *primitive unit cell* to a few atoms
for nonprimitive cells and compound crystals. In molecular crystals,
this number can be much larger and is usually a small multiple of
the number of molecules forming the crystal. This three-dimensional
periodic array of atoms is called the *crystal lattice.*

To define a unit cell, one introduces a three-dimensional *point*
lattice and adds to this imaginary lattice an atomic *basis* (or *motif*),
i.e., one, two, or more atoms in a specific arrangement for each point,
in order to arrive at the crystal lattice (Klug and Alexander, 1974,
Buerger, 1956). Figure 3.2 shows in two dimensions a crystal with a
basis containing two atoms.

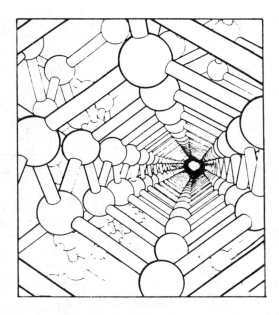

Figure 3.1: Diamond structure viewed along the [111] direction (after Pauling and Hayward, 1964).

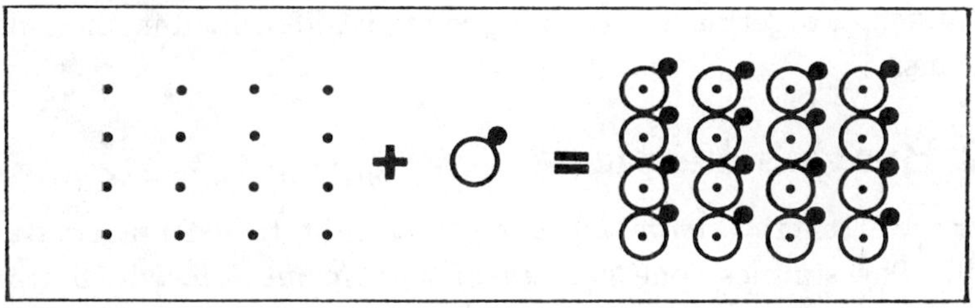

Figure 3.2: The crystal lattice: point lattice plus basis with two atoms.

3.1 Crystal Systems

A *coordinate system* is introduced so that its origin lies at the center of an arbitrary atom (or basis) and its axes point through the centers of preferably adjacent atoms (or basis) while best representing the symmetry of the lattice*—see Section 3.2.1. A *lattice vector* points from the origin along each axis to the center of the next *equivalent*

---

* However, there is not always a unique way to define this coordinate system—see Section 3.2.1. For mathematical reasons, an orthogonal system is preferred when possible.

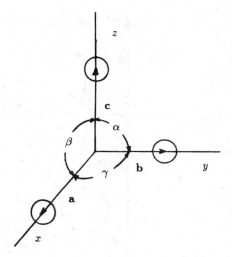

Figure 3.3: Coordinate system
and angles between lattice vectors
within a crystal.

atom,* or from the center of one basis to the center of the next. The
value of this vector is called the *lattice constant*.

All possible crystals can be ordered into seven *crystal systems*
according to the relative length of their lattice vectors and the angle
between these vectors—see Fig. 3.3. These crystal systems are listed
in Table 3.1, together with other properties identified in the following
sections.

3.2 Bravais Lattices

There are several *symmetry operations* that transfer a crystal into
itself. The simplest one is a *linear transformation*, which transfers
the lattice point t_0 into an equivalent lattice point t:

$$t = t_0 + n_1 a + n_2 b + n_3 c \qquad (3.1)$$

with a, b, and c as the lattice vectors in x-, y-, and z-directions,
respectively, and integers n_1, n_2, and n_3. This linear transformation
shifts the entire lattice by an integer number of lattice constants and
thereby reproduces the lattice. All lattices show linear transformation symmetry. A *unit cell* can now be defined as the smallest parallelepiped that forms the entire crystal when sequentially shifted by

---

* For example, from Na to Na in an NaCl crystal, and not from Na to
the next Cl ion. "Equivalent" refers to the neighborhood of this atom,
which must be identical to the atom at the origin.

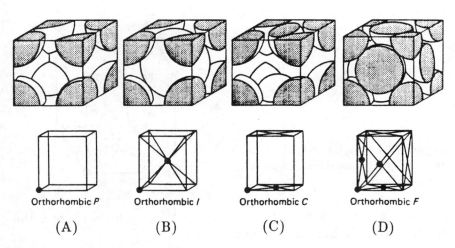

Orthorhombic *P*   Orthorhombic *I*   Orthorhombic *C*   Orthorhombic *F*

(A)   (B)   (C)   (D)

Figure 3.4: Unit cells of the orthorhombic Bravais lattice. (A) primitive *P*; (B) body-centered *I*; (C) base-centered *C*; and (D) face-centered *F*. (Upper row) fractional atoms shown within each unit cell; (Lower row) number of atoms per unit cell indicated.

a linear transformation [Eq. (3.1)]. There are only 14 different unit cells possible; they form 14 different lattices, called *Bravais lattices*.

In each of the crystal systems, there is one lattice with a unit cell that contains only one lattice atom,* the *primitive unit cell* (P in Table 3.1). In some crystal systems, there exist lattices with unit cells with more than one atom per cell. For example, in the orthorhombic system the extra atom(s) may sit in the center of the unit cell (*body-centered*, I), in the center of the base $[(\mathbf{a} + \mathbf{b})/2$, *base-centered*, C], or in the center of all faces† (*face-centered*, F), as shown in Fig. 3.4. All Bravais lattices are listed in the last column of Table 3.1.

3.2.1 The Primitive Cell

Occasionally one needs to describe the lattice as subdivided into primitive cells, while filling the entire space without voids. This can always be done; an example is presented in Fig. 3.5. The figure shows a face-centered cubic lattice with four lattice atoms in its unit cell. If the orthogonal system of crystal axes is replaced with one

---

* Since each corner is shared by 8 adjacent cells, only $\frac{1}{8}$ of each corner atom belongs to each cell. Therefore, with 8 corners one has $8 \times \frac{1}{8} = 1$ atom per primitive cell.

† Each surface is shared by two neighbor cells; for example, with 6 surfaces, there are $6 \times \frac{1}{2} = 3$ surface atoms per unit cell.

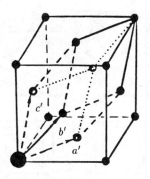

Figure 3.5: Face-centered cubic unit cell with inscribed trigonal primitive cell.

connecting the corner atom to the nearest face-centered atom, the crystal structure becomes trigonal with

$$\alpha' = \beta' = \gamma' = 60° \quad \text{and} \quad a' = b' = c' = \frac{a}{\sqrt{2}}.$$

The cubic system is usually preferred because of a simpler mathematical description, but the trigonal representation is totally equivalent to it. This example shows that for a given crystal the choice of a certain crystal system is not unique.

3.3 Crystal Classes (Point Groups)

The other symmetry operations, excluding any translation, are *rotation*, *reflection* (composed from rotation and inversion on a plane), and *inversion*, i.e., "reflection" at a point. Crystals that are distinguished by one or a combination of these can be divided into 32 different *crystal classes*. These symmetry operations are applied to the *basis* about a *point* of the Bravais lattice and therefore are also called *point groups*. The symmetry operations are usually identified by their *Schönflies* or *Hermann-Mauguin* symbol.

The Schönflies symbol identifies with capital letters, C, D, T, O the basic symmetry: *cyclic*, *dihedral*, *tetrahedral*, and *octahedral*. A subscript is used to identify the rotational symmetry, e.g., D_3 has three-fold symmetry. Another index, v, h, d is used for further distinction—see, e.g., Brown and Forsyth (1973).

The Hermann-Mauguin nomenclature indicates the type of symmetry directly from the symbol. It is a combination of numbers (n) and the letter m: n indicates rotational symmetry (for n = 2, 3, 4, or 6, an n-fold symmetry) and $\bar{n}$ denotes either an inversion ($\bar{1}$) or a roto-inversion (with a $\bar{3}$-, $\bar{4}$-, or $\bar{6}$-fold symmetry); m indicates a mirror plane parallel to, and $\frac{n}{m}$ perpendicular to, the rotational axis

Table 3.1: Crystal systems, point groups, and Bravais lattices.

Crystal Systems	Lattice Vector Relation	Lattice Angle Relation	Crystal Class Schönflies	Crystal Class Hermann-Mauguin	Bravais Lattices
Triclinic	$a \neq b \neq c$	$\alpha \neq \beta \neq \gamma$	C_1	1	P
			$\bar{C}_1$	$\bar{1}$	P
Monoclinic	$a \neq b \neq c$	$\alpha = \gamma = 90° \; \beta > 90°$	C_2	2	P,C
			$C_{1h}\,(C_2)$	m	P,C
			C_{2h}	$\frac{2}{m}$	P,C
Orthorhombic	$a \neq b \neq c$	$\alpha = \beta = \gamma = 90°$	C_{2v}	$2mm$	P,C,F,I
			$D_2\,(V)$	222	P,C,F,I
			$D_{2h}\,(V_h)$	$\frac{2}{m}\frac{2}{m}\frac{2}{m}$	P,C,F,I
Tetragonal	$a = b \neq c$	$\alpha = \beta = \gamma = 90°$	C_4	4	P,I
			S_4	$\bar{4}$	P,I
			C_{4h}	$\frac{4}{m}$	P,I
			C_{4v}	$4mm$	P,I
			$D_{2d}(V_2)$	$\bar{4}2m$	P,C,F,I
			D_4	422	P,I
			D_{4h}	$\frac{4}{m}\frac{2}{m}\frac{2}{m}$	P,I
Trigonal or Rhombohedral	$a = b = c$	$\alpha = \beta = \gamma \neq 90°$	C_3	3	C,R
			$S_6(C_{3i})$	$\bar{3}$	C,R
			C_{3v}	$3m$	H,C,R
			D_3	32	H,C
			D_{3d}	$\bar{3}\frac{2}{m}$	H,C,R
Hexagonal	$a = b \neq c$	$\alpha = \gamma = 90° \beta = 120°$	C_6	6	C
			C_{3h}	$\bar{6}$	C
			C_{6h}	$\frac{6}{m}$	C
			C_{6v}	$6mm$	C
			D_{3h}	$\bar{6}2m$	C,H
			D_6	622	C
			D_{6h}	$\frac{6}{m}\frac{2}{m}\frac{2}{m}$	C
Cubic or Isometric	$a = b = c$	$\alpha = \beta = \gamma = 90°$	T	23	P,F,I
			T_h	$\frac{2}{m}\bar{3}$	P,F,I
			T_d	$\bar{4}3m$	P,F,I
			0	432	P,F,I
			0_h	$\frac{4}{m}\bar{3}\frac{2}{m}$	P,F,I

with n-fold symmetry. Repetition of m or other symbols indicates the symmetry about the other orthogonal planes or axes.

All possible combinations of rotation, reflection, and inversion are listed in Table 3.1, with both symbols to identify each of the 32 point groups.

3.4 Space Groups

Combining the symmetry operations leading to the point groups with nonprimitive translation yields a total of 230 *space groups*. Alternatively, there are 1421 space groups when the ordering of spins is also considered (Birss, 1964). They include *screw axis* and *glide plane* operations; the former combines translation (shifting) with rotation, the latter combines translation with reflection.

The Schönflies symbol designates the different possibilities, arbitrarily arranged, of combining the symmetry operations by a superscript referring to the point group symbol (e.g., O_h^7 for Si).

In the Hermann-Mauguin symbol, the Bravais lattice identifier is added: A, B, C identify the specific base*, P, I, F; R (rhombohedric), and H (hexagonal). In addition, small letters, a, b, c, d, or n are appended to identify specific glide planes—namely, at $\frac{a}{2}, \frac{b}{2}, \frac{c}{2}, \frac{r+s}{4}$, and $\frac{r+s}{2}$, for a, b, c, d, and n, respectively, with r and s standing for any a, b, or c.

Typical element semiconductors have O_h symmetry, e.g., diamond O_h^7 (or Fd3m) for Ge and Si. Other binary semiconductors have C_{4v} (or 4mm), zinc-blende T_d^2 (or F43m), wurtzite C_{6v}^4 (P6$_3$mc), rock salt O_h^5 (or Fm3m), and other symmetry.

In summary, crystals are classified according to their lattice symmetry in four different ways, depending on the type of symmetry operation employed. This is shown in Table 3.2.

3.5 Crystallographic Notations

A *lattice point* is identified by the coefficients of the lattice vector pointing to it:

$$\mathbf{R}_n = n_1\mathbf{a} + n_2\mathbf{b} + n_3\mathbf{c}. \tag{3.2}$$

A lattice point is conventionally given by the three coefficients *without* brackets:

$$n_1 \ n_2 \ n_3.$$

A *lattice direction* is identified by a line pointing in this direction. When this line is shifted parallel so that it passes through the origin, the position of the nearest lattice point on this line, identified by the coefficients of Eq. (3.2) and enclosed in *square brackets*, defines this direction:

$$\left[n_1 \ n_2 \ n_3\right].$$

---

* A is the face spun between **b** and **c**, B between **a** and **c**, and C between **a** and **b**.

Table 3.2: Crystal Classification

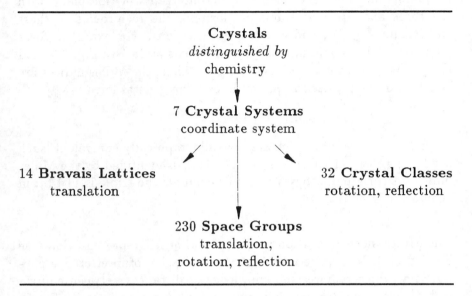

Crystals
distinguished by
chemistry

7 Crystal Systems
coordinate system

14 Bravais Lattices
translation

32 Crystal Classes
rotation, reflection

230 Space Groups
translation,
rotation, reflection

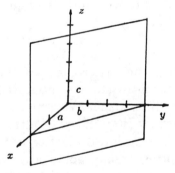

Figure 3.6: Example of a (2 1 0) plane.

Conveniently, one may reduce this notation by permitting simple fractions; for example, $[2\,2\,1]$ may also be written as $[1\,1\,\frac{1}{2}]$. Negative coefficients are identified by a bar: $[1\,1\,\bar{1}]$ is a vector pointing diagonally downward to the right.

Equivalent directions are directions which are crystallographically equivalent; for example, in a cube these are the directions $[1\,0\,0], [0\,1\,0], [0\,0\,1], [\bar{1}\,0\,0], [0\,\bar{1}\,0]$ and $[0\,0\,\bar{1}]$. All of these are meant when one writes $\langle 1\,0\,0 \rangle$; in general

$$\langle n_1\ n_2\ n_3 \rangle.$$

A *lattice plane* is described by *Miller indices*. These are obtained by taking the three coefficients of the intercepts of this plane with the three axes n_1, n_2, and n_3, forming the reciprocals of these coefficients $\frac{1}{n_1} \frac{1}{n_2} \frac{1}{n_3}$, and clearing the fractions. For example, for a plane parallel to **c** and intersecting the x-axis at **2a** (see Fig. 3.6) and the y-axis at **4b**, the fractions are $\frac{1}{2} \frac{1}{4} \frac{1}{\infty}$. Thus, the Miller indices are $(2\,1\,0)$ and are enclosed in parentheses. The general form is

$$(h\ k\ l).$$

A *family of planes* which are crystallographically equivalent [such as $(1\,1\,1), (\bar{1}\,1\,1), (1\,\bar{1}\,1), (1\,1\,\bar{1}), (\bar{1}\,\bar{1}\,1)$, etc.] is identified by the Miller indices in curly parentheses. For this example the triple is $\{1\,1\,1\}$; in general, it is

$$\{h\ k\ l\}.$$

The Miller indices notation is a *reciprocal lattice* representation (see Section 3.6). It is quite useful for the discussion of interference phenomena, which requires the knowledge of distances between equivalent planes. The distance between the $\{hkl\}$ planes is easily found; in a cubic system, it is simply

$$d_{hkl} = \frac{a}{\sqrt{h^2 + k^2 + l^2}}, \tag{3.3}$$

with a the lattice constant. In other crystal systems the expressions are slightly more complicated.* (See Warren, 1969; Zachariasen, 1967; von Laue, 1960; and James, 1954 for more details. A recent overview is given by Schultz, 1982).

The reciprocal lattice is a lattice in which each point relates to a corresponding point of the actual lattice by a reciprocity relation given below [Eqs. (3.6)–(3.10)].

---

* The general expression for the distance between two planes is given by

$$d^2_{hkl} = \frac{\begin{vmatrix} 1 & \cos\gamma & \cos\beta \\ \cos\gamma & 1 & \cos\alpha \\ \cos\beta & \cos\alpha & 1 \end{vmatrix}}{\dfrac{h}{a}\begin{vmatrix} h/a & \cos\gamma & \cos\beta \\ k/b & 1 & \cos\alpha \\ l/c & \cos\alpha & 1 \end{vmatrix} + \dfrac{k}{b}\begin{vmatrix} 1 & h/a & \cos\beta \\ \cos\gamma & k/b & \cos\alpha \\ \cos\beta & l/c & 1 \end{vmatrix} + \dfrac{l}{c}\begin{vmatrix} 1 & \cos\gamma & h/a \\ \cos\gamma & 1 & k/b \\ \cos\beta & \cos\alpha & l/c \end{vmatrix}}$$

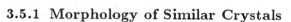

3.5.1 Morphology of Similar Crystals

When a specific chemical compound crystallizes in different crystal classes, it is called a *polymorph*. When crystals with the same structure are formed by compounds in which only one element is exchanged with a homologous element, they are referred to as *morphotrop*. When similar compounds crystallize in a similar crystal form, they are called *isomorph* when they also have other physical properties in common, such as similar cation to anion radii ratio and similar polarizability.

3.6 Reciprocal Lattice

As indicated above, the introduction of a reciprocal lattice is advantageous when one needs to identify the distance between equivalent lattice planes. This is of help for all kinds of interference phenomena, such as x-ray diffraction, the behavior of electrons when taken as waves, or lattice oscillations themselves. In a quantitative description, the relevant waves are described by wavefunctions (see Appendix A.4.5) of the type*

$$\varphi(\mathbf{k}, \mathbf{r}) = A \exp\left\{i\left(\mathbf{k} \cdot \mathbf{r} - \omega t\right)\right\} \tag{3.4}$$

where A is the amplitude factor, $\mathbf{r}$ is a vector in real space, and $\mathbf{k}$ is a vector in reciprocal space. Here, $\mathbf{k}$ is referred to as the *wave vector*, or *wavenumber*, if only one relevant dimension is discussed; the wave vector is normal to the wave front and has the magnitude

$$|\mathbf{k}| = \frac{2\pi}{\lambda} \tag{3.5}$$

with λ the wavelength. Since $\mathbf{k} \cdot \mathbf{r}$ is dimensionless, $\mathbf{k}$ has the dimension of reciprocal length. Multiplied by $\hbar$ ($= h/2\pi$, where h is the Planck's constant), $\hbar\mathbf{k}$ has the physical meaning of a momentum as will be shown in Section 7.2.3.

When $\mathbf{R}_n$ is a lattice vector [for ease of mathematical description we now change from $(\mathbf{a}, \mathbf{b}, \mathbf{c})$ to $(\mathbf{a}_1, \mathbf{a}_2, \mathbf{a}_3)$]:

$$\mathbf{R}_n = n_1 \mathbf{a}_1 + n_2 \mathbf{a}_2 + n_3 \mathbf{a}_3, \tag{3.6}$$

---

* This description is more convenient than an equivalent description, which in one direction reads $\varphi(x) = A \exp\{2\pi i(\frac{x}{\lambda} - \nu t)\}$.

one obtains the corresponding vector $\mathbf{K}_m$ in reciprocal space with the three fundamental vectors $\mathbf{b}_1$, $\mathbf{b}_2$, $\mathbf{b}_3$:

$$\mathbf{K}_m = m_1\mathbf{b}_1 + m_2\mathbf{b}_2 + m_3\mathbf{b}_3 \qquad (3.7)$$

where both sets of unit vectors are related by the orthogonal relation

$$\mathbf{a}_i \cdot \mathbf{b}_j = 2\pi\delta_{ij} \qquad \text{and} \qquad i, j = 1, 2, 3, \qquad (3.8)$$

where δ_{ij} is the *Kronecker delta symbol*

$$\delta_{ij} = \begin{cases} 1 & \text{for } i = j \\ 0 & \text{for } i \neq j. \end{cases} \qquad (3.9)$$

The orthogonal relation can also be expressed by

$$\mathbf{b}_1 = \frac{2\pi\mathbf{a}_2 \times \mathbf{a}_3}{\mathbf{a}_1 \times \mathbf{a}_2 \cdot \mathbf{a}_3}, \text{ etc. (cyclical)}, \qquad (3.10)$$

that is, every vector in the reciprocal lattice is normal to the corresponding plane of the crystal lattice and its length is equal to the reciprocal distance between two neighboring corresponding lattice planes (see Kittel, 1966). This definition is distinguished by a factor 2π from the definition of a reciprocal lattice found by crystallographers. This factor is included here to make the units of the reciprocal space identical to the wave vector units.

3.6.1 Wigner-Seitz Cells and Brillouin Zones

As knowledge about an entire crystal can be derived from the periodic repetition of its smallest unit, the *unit cell*, one can derive knowledge about the wave behavior from an equivalent cell in the reciprocal lattice. A convenient way to introduce this discussion is by examining the *Wigner-Seitz cell* rather than the unit cell itself.

A Wigner-Seitz cell is formed when a lattice point is connected with all equivalent neighbors, and planes are erected normal to and in the center of each of these interconnecting lines. An example is shown in Fig. 3.7, where for the face-centered unit cell ($\mathbf{a}_1$, $\mathbf{a}_2$, $\mathbf{a}_3$) the Wigner-Seitz cell is constructed; the plane orthogonal to and intersecting the lattice vector $\mathbf{a}_2$ is visible.

When such a Wigner-Seitz cell is constructed from the unit cell of the reciprocal lattice, the resulting cell is called the *first Brillouin zone*. It is the basic unit for describing lattice oscillations and electronic phenomena.

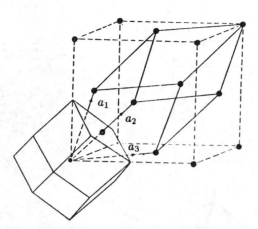

Figure 3.7: Face-centered cubic lattice (dashed lines) with primitive parallelepiped, and from it the derived Wigner-Seitz cell in real space, which is equivalent to the Brillouin zone in reciprocal space.

Most semiconductors crystallize with cubic or hexagonal lattices. The first Brillouin zones of these lattices are given in Fig. 3.8, and will be referred to frequently later in the book.

In these discussions, lattice symmetry is of great importance, and points about which certain symmetry operations can reproduce the lattice are often cited. These symmetry points can also be transformed into the reciprocal lattice and are identified here by specific letters. The most important *symmetry points* with their conventional notations are identified in the different Brillouin zones of Fig. 3.8. Γ is always the center of the zone ($k_x = k_y = k_x = 0$), and X is the intersection of the Brillouin zone surface with any of the main axes (k_x, k_y, or k_z) in any of the cubic lattices. The points Δ, Λ, and Σ lie halfway between Γ and X, Γ and L, and Γ and K, as shown in Fig. 3.8. The positions of the other symmetry points (H, K, L, etc.) can be obtained directly from Fig. 3.8.

The extent of the first Brillouin zone can easily be identified. For instance, in a primitive orthorhombic lattice with its unit cell extending to a, b, and c in the x-, y-, and z-directions respectively, the first Brillouin zone extends from $-\frac{\pi}{a}$ to $\frac{\pi}{a}$ in k_x-, from $-\frac{\pi}{b}$ to $\frac{\pi}{b}$ in k_y-, and from $-\frac{\pi}{c}$ to $\frac{\pi}{c}$ in k_z-direction. Since the wave equation is periodic in $\mathbf{r}$ and $\mathbf{k}$, all relevant information is contained within the first Brillouin zone.

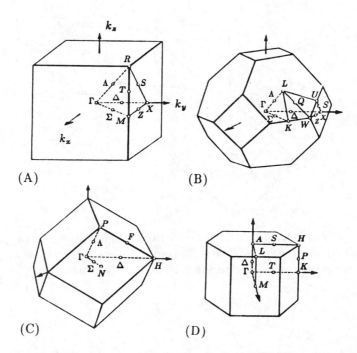

Figure 3.8: Brillouin zones for the three cubic lattices. (A) Primitive; (B) face-centered; (C) body-centered; and (D) for the primitive hexagonal lattice with important symmetry points and axes.

3.7 Relevance to Semiconductors

Lattice periodicity is one of the major factors in determining the band structure of semiconductors—see Chapter 7. The symmetry elements of the lattice are reflected in the corresponding symmetry elements of the bands, from which important qualitative information about the electronic structure of a semiconductor is obtained. Therefore the main features of the symmetry of some of the typical semiconductors are summarized below. A comprehensive review of element and compound structures is given by Wells (1984).

3.7.1 Elemental Semiconductors

Most of the important crystalline semiconductors are elements (Ge, Si) or binary compounds (III-V or II-VI) (Mooser and Pearson, 1956). They form crystals in which each atom is surrounded by

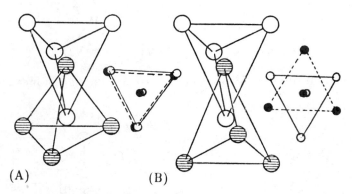

Figure 3.9: Two intertwined tetrahedra with (A) base triangles parallel (dihedral angle 0°), and (B) base triangles rotated by 60°.

four nearest neighbors,* i.e., they have a *coordination number* of 4. The connecting four atoms (*ligands*) surround each atom in the equidistant corners of a tetrahedron. The lattice is formed so that each of the surrounding atoms is again the center atom of an adjacent tetrahedron, as shown for two such tetrahedra in Fig.3.9. Of the two principal possibilities for arranging two tetrahedra, only one is realized in nature for elemental crystals: the *diamond lattice*, wherein the base triangles of the intertwined tetrahedra are rotated by 60°. Ge and Si are examples. In amorphous elemental semiconductors, however, both possibilities of arranging the tetrahedra are realized—see Section 3.9.3.

3.7.2 Binary Semiconducting Compounds

Binary III-V and II-VI compounds are formed by both tetrahedral arrangements which are dependent on relative atomic radii and preferred valence angles (see Section 2.2), although with alternating atoms as nearest neighbors. These compounds can be thought of as an element (IV) semiconductor after replacing alternating atoms with an atom of the adjacent rows of elements (III and V). Similarly, II-VI compounds can be created by using elements from the next-to-adjacent rows—see Fig. 3.10.

---

* There are other modifications possible. For example, seven for Si, of which four are stable at room temperature and ambient pressure (see *Landoldt-Börnstein*, 1982, 1987). Only Si I and α-Si are included in this book. Si III is face-centered cubic and a semimetal; Si IV is hexagonal diamond and is a medium gap semiconductor (see Besson et al., 1987).

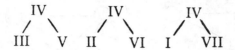

Figure 3.10: Binary compounds with semiconducting properties.

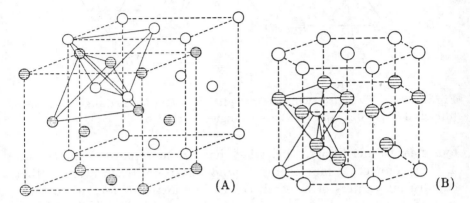

Figure 3.11: (A) Zinc-blende lattice (GaAs) constructed from two inter-penetrating face-centered cubic sublattices of Ga and As, with a displaced origin at $\frac{a}{4}, \frac{a}{4}, \frac{a}{4}$, with a the edge length of the elementary cube. (B) Wurtzite lattice (CdS) constructed from two intertwined hexagonal sub-lattices of Cd and S.

Aside from these classical AB-compounds, there are others that have interesting semiconducting (specifically thermoelectrical) properties. Examples include the II-V compounds (such as ZnSb, ZnAs, CdSb or CdAs), which have orthorhombic structures. For a review, see Arushanov (1986).

The diamond lattice for AB-compounds results in a *zinc-blende lattice* shown in Fig. 3.11A. Most III-V compounds, as for instance GaAs, are examples.

Unrotated interpenetrating tetrahedra, as shown in Fig. 3.9A, produce the *wurtzite lattice* (Fig. 3.11B) which can also be obtained for a number of AB-compounds. Examples include ZnS and CdS. ZnS can also crystallize in a zinc-blende modification. Under certain conditions, alternating layers of wurtzite and zinc-blende, each several atomic layers thick, are observed. This is called a *polytype*. Usually, the zinc-blende structure is more stable at lower temperatures and the wurtzite structure appears above a transition temperature (1053°C in CdS). With rapid cooling the wurtzite structure can be frozen-in.

Other structures of binary semiconductors include

- NaCl-type semiconductors, with PbTe as an example;
- cinnabar (deformed NaCl) structures, with HgS as an example;
- anti-fluorite silicide structures, with Mg_2Si as an example. These structures can be regarded as derived from the fcc lattice (Fig. 3.11A) with one of the two interstitial positions filled by the second metal atom, similar to the Novotny-Juza compounds (Section 3.7.3); and
- $A_3^I B^V$-structures, with Cs_3Si as an example.

For a review, see Parthé (1964), Sommer (1968), and Abrikosov et al. (1969).

3.7.3 Ternary, Quaternary Semiconducting Compounds

There are several classes of ternary and quaternary compounds with known attractive semiconducting properties. All have tetrahedral structures: each atom is surrounded by four neighbors. Some examples are discussed in the following sections. For a review, see Zunger (1986).

One can conceptually form a wide variety of ternary, quaternary, or higher compounds which have desirable semiconducting properties when one replaces within a tetrahedral lattice, subsequent to the original replacement shown in Fig. 3.10, certain atoms with those from adjacent rows, as given in Fig. 3.12. These examples represent a large number ($\sim$ 140) of such compounds and indicate the rules for this type of compound formation. For instance, a $II\text{-}III_2\text{-}VI_4$ compound can be formed by replacing 8 atoms of column IV first with 4 atoms each of columns II and VI, and consequently the 4 atoms of column II with one vacancy (0), one atom of column II, and two atoms of column III.

3.7.3A Ternary Chalcopyrites   Best researched are the *ternary chalcopyrites* $I\text{-}III\text{-}VI_2$; they are constructed from two zinc-blende lattices in which the metal atoms are replaced by an atom from each of the adjacent columns. In a simple example one may think of the two Zn atoms from ZnS as transmuted into Cu and Ga:

$$^{30}ZnS + {}^{30}ZnS \rightarrow {}^{29}Cu^{31}GaS_2,$$

with some deformation of the zinc-blende lattice since the Cu-S and Ga-S bonds have different strengths, and with a unit cell twice the size of that in the ZnS lattice (Fig. 3.13). For a review, see Miller et al. (1981).

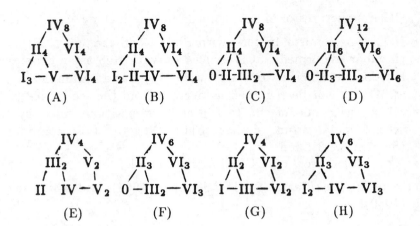

Figure 3.12: Construction of pseudobinaries (B), ternaries (A, C, D, E) pseudoternaries (G, H), and quaternaries (F) from element (IV) semiconductors (0 represents a vacancy, i.e., a missing atom at a lattice position).

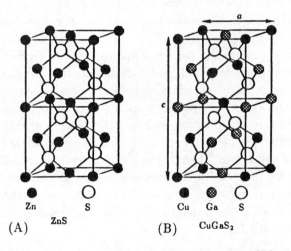

Figure 3.13: (A) ZnS (or GaP) double unit cell; (B) CuGaS$_2$ (or ZnGeP$_2$) unit cell.

3.7.3B Ternary Pnictides and ABC$_2$ Compounds

Other ternaries with good semiconducting properties are the *ternary pnictides* II-IV-V$_2$ (such as ZnSiP$_2$) which have the same chalcopyrite structure and, in a similar example, can be constructed from GaP by the transmutation

$$^{31}\text{GaP} + {}^{31}\text{GaP} \rightarrow {}^{30}\text{Zn}\,{}^{32}\text{SiP}_2.$$

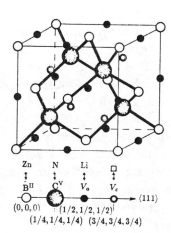

Figure 3.14: Unit cell of the Novotny-Juza compound.

Still another class with chalcopyrite structure is composed of the I-III-VI$_2$ compounds, of which $CuFeS_2$ is representative. (These structures are reviewed by Jaffe and Zunger, 1984.)

3.7.3C Novotny-Juza Compounds Interesting variations of this tetrahedral structure (see Parthé, 1972) are the *Novotny-Juza compounds*, which are partially filled tetrahedral interstitial I-II-V compounds (e.g., LiZnN). Here the Li atom is inserted into exactly one half of the available interstitial sites of the zinc-blende lattice (e.g., on V_a or on V_c as shown in Fig. 3.14). A substantial preference for the Li atom to occur at the site closer to the N atom (rather than the site next to the Zn—the lattice energy of this structure is lower by about 1 eV) makes this compound an *ordered* crystal with good electronic properties (Carlson et al., 1985, Kuriyama and Nakamura, 1987, Bacewicz and Ciszek, 1988). It should be noted, however, that the Zn atom is fourfold coordinated with N atoms, while the N atom is fourfold coordinated with Zn and fourfold coordinated with Li; therefore, it has eight nearest neighbors.

3.7.3D The Adamantine $A_nB_{4-n}C_4$ and Derived Vacancy Structures Examples of this class of $A_nB_nC_4$ structures with $n = 1$ or 3, such as A_3BC_4 or AB_3C_4, are the famatinites (e.g., Cu_3SbS_4 or $InGa_3As_4$) or lazarevicites (e.g., Cu_3AsS_4). With $n = 2$ this class reduces to ABC_2 (e.g., $CuGaAs_2$ or $GaAlAs_2$), and with $n = 4$ it reduces to the zinc-blende (ZnS) lattice. The layered sublattices can be ordered (e.g., in $CuGaAs_2$) or disordered (alloyed) as in $GaAlAs_2$, and are discussed in the following section. All of these compounds follow the octet $(8 - N)$ rule (see Section 3.9.1); they are fourfold coordinated (each cation is surrounded by four anions and vice versa).

The $8 - N$ rule determines how many shared electrons are needed to satisfy perfect covalent bonding (see Section 2.2) for any atom with N valency electrons, e.g., 1 for Cl with $N = 7$, 2 for S with $N = 6$, or 4 for Si with $N = 4$, requiring single, chain-like, or tetrahedral bondings, respectively.

Deviations from the $A_n B_{4-n} C_4$ composition may occur when including ordered vacancy compounds into this group, such as II-III$_2$-VI$_4$ compounds (e.g., $CuIn_2Se_4$) in which one of the II or III atoms is removed in an ordered fashion, resulting in defect famatinites or defect stannites.

An instructive generic overview of the different structures of tetragonal ternaries or pseudoternaries is given by Bernard and Zunger (1988) (Fig. 3.15). See also Shay and Wernick (1974); Miller et al. (1981), and the conference proceedings on ternary and multinary compounds.

3.7.3E Pseudoternary Compounds

Finally, one may consider *pseudoternary compounds* in which one of the components is replaced by an *alloy* of two homologous elements. For example, Ga replaced by a mixture of Al and Ga in GaAs yields $Al_\xi Ga_{1-\xi} As$; replacement of As by P and As yields $GaP_\xi As_{1-\xi}$. These pseudoternary compounds contain alloys of isovalent atoms in one of the sublattices.

When the two alloying elements are sufficiently different in size, preference for *ordering* exists for stoichiometric composition in the sublattice of this alloy. Substantial band gap bowing (see Section 9.2.1A) gives a helpful indication of predicting candidates for this ordering of stoichiometric compounds. Examples include $GaInP_2$, which shows strong bowing, where the Ga and In atoms are periodically ordered (Srivastava et al., 1985), Ga_3InP_4, or $GaIn_3P_4$ with similar chalcopyrite-type structures (see also Section 9.2.1A). Here again, the coordination number is four; each atom is surrounded by four nearest neighbors, although they are not necessarily of the same element.

A different class of such compounds is obtained when alloying with nonisovalent atoms, such as Si+GaAs.

The desire to obtain semiconductors with specific properties that are better suited for designing new and improved devices has focused major interest on synthesizing new semiconducting materials as discussed above, or using sophisticated growth methods to be discussed

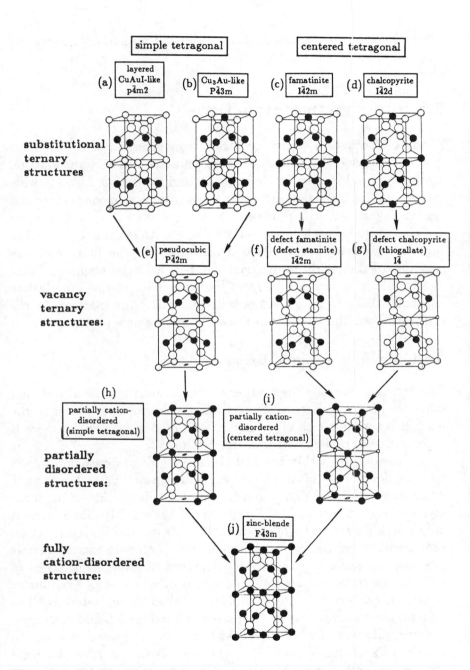

Figure 3.15: Structure of $A_n B_{4-n} C_4$ (adamantine) compounds (a)–(d) and their derived, ordered vacancy structures (e)–(g). Also included are cation-disordered structures including ordered vacancies (h) and (i) and the parent zinc-blende (with ordered or disordered sublattice). Vacancies are shown as open rectangles (after Bernard and Zunger, 1988).

below, aided by theoretical analyses to predict potentially interesting target materials (see Ehrenreich, 1987).

3.8 Superlattice Structures

Periodic alternation of one or a few monolayers of semiconductor A and B produces a composite semiconductor called a *superlattice*. Material A could stand for Ge or GaAs, and B for Si or AlAs. A wide variety of other materials including alloys of such semiconductors and organic layers can also be used.

The width of each layer could be a few Ångstroms in ultrathin superlattices to a few hundred Ångstroms. In the first case, one may regard the resulting material as a new artificial compound (Isu et al., 1987); in the second case, the properties of the superlattice approach those of layers of the bulk material. Superlattices in the range between these extremes show interesting new properties.

3.8.1 Superlattices and Brillouin Zones

The introduction of a new superlattice periodicity has a profound influence on the structure of the Brillouin zones. In addition to the periodicity within each of the layers with lattice constant a, there is superlattice periodicity with lattice constant l.

Consequently, within the first Brillouin zone of dimension π/a, a mini-Brillouin zone of dimension π/l will appear. Since l is usually much larger than a, e.g., $l = 10a$ for a periodic deposition of 10 monolayers of each material, the dimensions of the mini-Brillouin zone is only a small fraction (a/l) of the Brillouin zone and is located at its center with Γ coinciding. Such a mini-zone is of more than academic interest, since the superlattice is composed of alternating layers of different materials. Therefore, reflections of waves, e.g., excitons or electrons, can occur at the boundary between these materials. The related dispersion spectrum (discussed in Sections 5.2 and 9.3.1) will become substantially modified, with important boundaries at the surface of such mini-zones. It is this mini-Brillouin zone structure that makes such superlattices especially interesting; this will become clearer in later discussions throughout the book. A more detailed discussion of the mini-zones is inherently coupled with corresponding new properties, and is therefore postponed to the appropriate sections in this book.

3.8.2 Superlattice Deposition

With modern deposition techniques (e.g., vapor phase epitaxy), one is able to deposit onto a planar substrate (e.g., onto a cleaved single crystal of appropriate surface orientation) monolayer after monolayer of the same or a different material. [For a review of the deposition techniques using *molecular beam epitaxy*, see Ploog (1981), Joyce (1985), Gossard (1986), and Kelly and Kelly (1985), and for those using *metal-organic vapor phase epitaxy*, see Dapkus (1984) and Richter (1986).] More recently, interest has been stimulated in lateral superlattices.

3.8.3 Ultrathin Superlattices

Single or up to a few atomic layer sequential depositions can be accomplished (Gossard, 1986; Petroff et al., 1979) even between materials with substantial lattice mismatch, e.g., Si and Ge, GaAs and InAs (Fig. 3.16). The thickness of each layer must be thinner than the critical length beyond which dislocations (see Section 18.3) can be created. This critical length decreases with increasing lattice mismatch, and is on the order of 25 Å for a mismatch of 4% (see Section 26.2).

In ultrathin superlattices, the transition range between a true superlattice and an artificial new compound is reached. This opens an interesting field for synthesizing a large variety of compounds that may not otherwise grow by ordinary chemical reaction followed by conventional crystallization techniques.

Estimates as to whether or not such a spontaneous growth is possible have been carried out by estimating the enthalpy of formation of the ordered compound from the segregated phases. For instance, for a single layer $(GaAs)_1$-$(AlAs)_1$ ultrathin superlattice, the formation enthalpy from the components GaAs and AlAs is given by

$$\Delta Q = E_{GaAlAs_2} - (E_{GaAs} + E_{AlAs}). \qquad (3.11)$$

The formation enthalpy depends on the lattice mismatch. It is on the order of 10 meV for ultrathin superlattices with low mismatch (GaAs-AlAs) and about one order of magnitude larger for superlattices with large mismatch (such as GaAs-GaSb or GaP-InP), as shown in Table 3.3. The diatomic system Si-Ge, while having a large lattice mismatch, nevertheless shows a lower formation enthalpy, for reasons of lower constraint of the lattice.

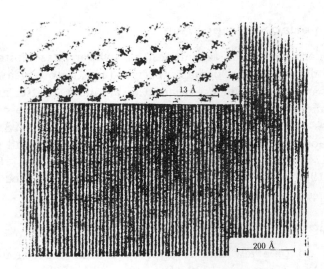

Figure 3.16: Transmission electron microscope image of an ultrathin superlattice of $(GaAs)_2$-$(AlAs)_2$ bilayers (after Petroff et al., 1979).

Table 3.3: Formation enthalpy for single-layer superlattices (after Wood et al., 1988; and Bernard et al., 1988).

Superlattice	Lattice Mismatch	Formation Enthalpy
$(GaAs)_1$-$(AlAs)_1$	0.1%	12.5 (meV/4 atoms)
(GaP_1)-$(InP)_1$	7.4%	91 (meV/4 atoms)
$(GaAs)_1$-$(GaSb)_1$	7.5%	115 (meV/4 atoms)

The formation enthalpy also decreases with increasing thickness of each of the layers (Wood et al., 1988). Therefore, the ultrathin superlattices of isovalent semiconductors are chemically unstable with respect to the segregated compounds. These always have a lower formation enthalpy. Alloy formation does not require nucleation necessary for crystal growth of the segregated phases. Therefore, alloy formation of GaAs-AlAs is the dominant degradation mechanism. Recrystallization is usually frozen-in at room temperature.

Superlattices with low lattice mismatch, however, are also unstable with respect to alloy formation, e.g., to $Ga_{1-x}Al_xAs$, which has a formation enthalpy between that of the superlattice and the segregated phases. In contrast, the alloy formation energy of semiconductors with large mismatch lies above that for ultrathin superlattices. They are therefore more stable (Wood and Zunger, 1988).

Several of these ultrathin superlattices can be grown under certain growth conditions *spontaneously* as an ordered compound, without artificially imposing layer-by-layer deposition. For instance, $(GaAs)_1(AlAs)_1$ grown near 840 K by Petroff et al. (1978) and Kuan et al. (1985), $(InAs)_1(GaAs)_1$ grown by Kuan et al. (1987), $(GaAs)_1(GaSb)_1$ grown by Jen et al. (1986), $(InP)_n(GaP)_n$ grown by Gomyo et al. (1987), and $(InAs)_1(GaAs)_3+(InAs)_3(GaAs)_1$ grown by Nakayama and Fujita (1985). All of these lattices grow as ordered compounds of the $A_nB_{4-n}C_4$-type (see Section 3.7.3D).

3.8.4 Intercalated Compounds

In crystals, such as graphite, which show a two-dimensional lattice structure, layers of other materials can be inserted between each single or multiple layer to form new compounds with unusual properties. This insertion of layers can be achieved easily by simply dipping graphite into molten metals, such as Li at 200–400°C. After immersion, the *intercalation* starts at the edges and proceeds into the bulk by rapid diffusion. Such interlayers can be, for example, halogens or alkali metals. Examples include KC_8 or LiC_6, compounds which are transparent (yellow) and show anisotropic conductivity and low-temperature superconductivity.

In the process of intercalation, the metal atom is ionized while the graphite layer becomes negatively charged. When immersed in an oxidizing liquid, the driving force to oxidize Li can be strong enough to reverse the reaction. This reversible process is attractive in the design of high-density rechargeable batteries when providing electrochemical driving forces.

Other layer-like lattices can also be intercalated easily. An example is TaS_2. Many of these compounds have extremely high diffusivity of the intercalating atoms. Some of them show a very large electrical anisotropy.

For a review, see Whittingham and Jacobson (1982).

3.8.5 Organic Superlattices

Well known are the *Langmuir-Blodgett films* (Langmuir, 1920, Blodgett, 1935), which are monomolecular films of highly anisotropic organic molecules, such as alkanoic acids and their salts which form long hydrophobic chains. One end of the chain terminates in a hydrophobic acid group. Densely packed monomolecular layers can be obtained while floating on a water surface; by proper manipulation, these layers can be picked up, layer by layer (Fig. 3.17), onto an ap-

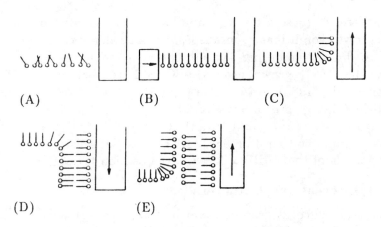

Figure 3.17: Langmuir-Blodgett technique to produce multilayer films of amphiphilic, i.e., either hydrophilic or hydrophobic molecules from a water surface in a head-to-head and tail-to-tail mode. (A) Monolayer on top of water surface; (B) monolayer compressed and ordered; (C) monolayer picked-up by glass slide moving upward; (D) second monolayer deposited by dipping of glass slide; and (E) third monolayer picked-up by glass slide moving upward.

propriate substrate, thereby producing a highly ordered superlattice structure; up to 10^3 such layers on top of each other have been produced. The ease in composing superlattices with a large variety of compositions makes these layers attractive for exploring a number of technical applications including electro-optical and microelectronic devices. For recent reviews, see Roberts, 1985 and Agarwal (1988).

3.9 Amorphous Structures

Although there is no macroscopic structure* discernible in amorphous semiconductors (*glasses* for brevity), there is a well-determined *microscopic order* in atomic dimensions, which for nearest and next-nearest neighbors is usually nearly identical to the order in the crystalline state of the same material. The *long-range order*, however, is absent (see Phillips, 1980).

In many respects, the glass can be regarded as a supercooled liquid. When cooling down from a melt, glass-forming materials undergo two transition temperatures: T_f, where it becomes possible to

---

* The surface of glasses, even at very high magnification, does not show any characteristic structure; after fracture, glasses show no preferred cleavage planes whatsoever.

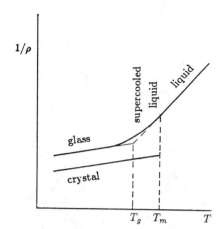

Figure 3.18: Schematics of the dependence of the reciprocal density as a function of the temperature.

pull filaments (honey-like consistency); and T_g, where form elasticity is established, i.e., the glass can be formed into any arbitrary shape —its viscosity has reached 10^{15} p, and its atomic rearrangement time is $\sim 10^5$ s. Only T_g is now used as the transition temperature and is identified in Fig. 3.18 (for a review see Jäckle, 1986). When plotting certain properties of a semiconductor—such as its specific density (Fig. 3.18), the electrical conductivity, and many others as a function of the temperature—a jump and break in slopes are observed at the melting temperature T_m when crystallization occurs. Such a jump is absent when cooling proceeds sufficiently fast and an amorphous structure is frozen-in.

Fast cooling (quenching) for typical glasses is already achieved with a rate < 1 deg/s, while many solids, including metals, become frozen-in liquids and remain amorphous at room temperature when this rate is $\sim 10^7$ deg/s, which can be achieved by *splat cooling* on fast rotating disks.

Near a transition temperature $T_g < T_m$, the slope gradually changes and, for $T < T_g$, the curves in Fig. 3.18 for a glass and a crystal of the same material run essentially parallel to each other.

Materials that have a large fraction of covalent bonding (see Section 2.2) show a tendency for glass formation. The liquid becomes significantly more viscous before crystallization takes place. The glass formation composition range is shown for some ternary compounds in Fig. 3.19. In this range, while still liquid, cross-linking of many atoms has already taken place, and the principal building blocks (see below) of the glass are established; however, they cannot adjust with sufficient rigor to produce long-range periodic-

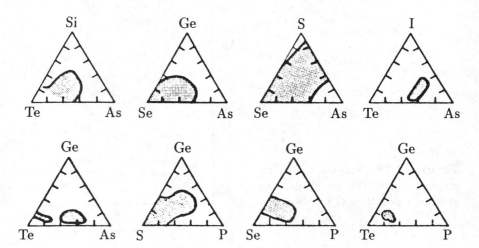

Figure 3.19: Approximate glass-forming (shaded) regions in a few ternary semiconductor alloy systems—a point within this triangle represents an alloy of the three components of a composition given by the normal to each side of the triangle (after Mott and Davis, 1979).

ity. Nevertheless, all bonds tend to be satisfied by attachment to an appropriate neighbor.

The resulting structure is composed of principal building blocks that join each other with slight deviation from the preferred inter-atomic angle and distance (*frustration*) and consequent relaxation of these relations within the building blocks. In contrast, during crystallization, such building blocks easily break up so that a larger crystallite can grow by sequentially adding atoms rather than entire building blocks. During glass formation, there is usually little tendency to form *dangling bonds*, i.e., bonds not extending between two atoms—see Section 3.9.5.

The structural analysis of an amorphous semiconductor can therefore be divided into two parts: the principal atomic building blocks, and the arrangement of these blocks to form the glass.

3.9.1 Principal Glass-Forming Building Blocks

Atomic semiconductors can be dealt with most easily since all of the neighbors are equivalent. Perfect crystalline order with tetra-hedral binding (fourfold coordination) requires the formation of six-member rings—see Fig. 1.1. Polk (1971) introduced *odd-numbered rings* (5 or 7), and thereby formed glasses with an otherwise tetra-hedrally coordinated arrangement of atoms around these building blocks. Such odd-numbered rings were recently confirmed in α-Si (Pantelides, 1987) and in α-C (Galli, 1988).

Comparing crystalline (c) and amorphous (α) structures of the same element (e.g., Ge), one sees that first- and second-neighbor distances (2.45 vs. 2.46 and 4.02 vs. 4.00 Å for c-Ge vs. α-Ge, respectively) are surprisingly similar, as is the *average* bond angle (109.5 vs. 108.5°). There is, however, a *spread* of $\pm 10°$ in the bond angle for the amorphous structure, resulting in an *average coordination number* of 3.7 rather than 4 for c-Ge (Etherington et al., 1982). The lower effective coordination number indicates a principal building block structure that is slightly less filled but without vacancies, which are ill-defined in amorphous structures (see Section 25.2). Hard sphere models, which would assist in defining sufficient space between the spheres as vacancies, must be used with caution since covalent structures can relax interatomic lattice spacing when relaxing bond angles (Waire et al., 1971).

Binary compounds are more difficult to arrange in such a fashion, since odd-member rings cannot be formed in an AB sequence without requiring at least one AA or BB sequence. Random network models, however, can also be made with larger *even-numbered rings*. Zachariasen (1932) suggested the first one for SiO_2-type glasses, which was shown for a two-dimensional representation in Fig. 1.1.

Many covalent polyatomic binary compounds containing chalcogenes, easily form semiconducting glasses such as As_2S_3, As_2Se_3, or Ge_xTe_y. The principal building blocks obey the $8 - N$ rule. For example, As with $N = 5$ is bonded to three Se atoms, while the Se with $N = 6$ in turn is surrounded by two As atoms in an As-Se-As configuration. Similarly, the Ge with $N = 4$ is surrounded by four Te atoms, while each Te atom with $N = 6$ has two Ge atoms as nearest neighbors in $GeTe_2$, similar to the SiO_2 configuration.

In some of these amorphous chalcogen compounds, however, the interatomic nearest neighbor distance is shorter and the coordination number is significantly lower than in the corresponding crystalline compounds (Bienenstock, 1985). A chalcogen-chalcogen pairing [e.g., by including Ge-Te-Te-Ge or an ethane-like $Ge_2(Te_{1/2})_6$ formation*] can distort the building blocks. The large variety of possible Ge_xTe_y building blocks, still fulfilling the $8 - N$ rule, created by replacing

---

* The "chemical formula" using $Te_{1/2}$ shows the symmetry of the Te binding, and indicates that on the other side of each of the Te atoms another Ge atom is bound.

Ge-Ge with Ge-Te or Te-Te bonds, is the reason that glasses of a continuous composition from pure Ge to pure Te can be formed (Boolchand, 1985).

3.9.2 Coordination Number and Constraints

In crystalline covalent semiconductors, the coordination number is given by the $8 - N$ rule, which is 4 for Si. For compounds one can define an *average coordination number*, drawing a shell in the atomic distribution function around an arbitrary atom and averaging (Fig. 3.20). These shells contain at nearest neighbor distance a maximum of $m = 4$ atoms for GaAs, $m = 3$ for GeTe (also for As), and $m = 2$ for a linear lattice such as Se. The average coordination number is $\overline{m} = 2.7$ for $GeTe_2$, and $\overline{m} = 2.4$ for As_2Se_3. In a crystal there are $m/2$ *constraints* per atom with respect to bond length, since two atoms share a bond. This can easily be fulfilled if $m/2 \leq 3$, since each atom can shift with respect to its neighbor in three dimensions. There are also $m(m-1)/2$ constraints with respect to the bond angle, since it is defined by three atoms. Therefore, bond length *and* bond angle are constrained only if

$$\frac{\overline{m}}{2} + \frac{\overline{m}(\overline{m} - 1)}{2} \leq 3, \quad \text{or} \quad \overline{m} \leq \sqrt{6} \simeq 2.4. \qquad (3.12)$$

Si- and GaAs-type semiconductors are *overconstrained*: a large internal strain prevents any significant deviation from its ordered, crystalline state. Not so Se or As_2Se_3. The former, with $\overline{m} = 2$, is *underconstrained*: it provides a large amount of freedom for deviation from uniformity in bond length and angle; therefore, it easily forms amorphous structures. The latter, As_2Se_3, needs only minor alloying to cause $\overline{m}$ to drop below 2.4, and therefore also forms a glass easily. Since Eq. (3.12) shows a quadratic dependency on $\overline{m}$, a glass-forming tendency is rather sensitive to a lowering $\overline{m}$ (Ovshinsky, 1976; Adler, 1985; Phillips, 1980).

3.9.3 Short-Range vs. Intermediate-Range Order

The above-described principal building blocks are also described as *intermediate-range order*. These blocks are composed of subunits, identified by the *short-range order* of a few atoms and characterized by bond lengths, bond angles (next-nearest neighbor distances), and site geometry. Intermediate-range order describes *third neighbor distances, dihedral angles, atomic ring structures,* and *local topology*. It distinguishes for tetra-, tri-, and divalent bonding

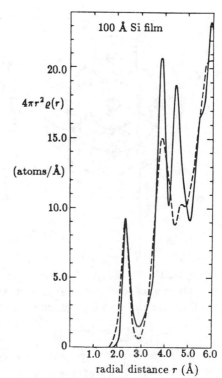

Figure 3.20: Distance distribution (radial distribution function) of atoms in amorphous (dashed) and crystalline (solid curve) Si layers of 100 Å thickness (after Moss and Graczyk, 1970).

truly three-dimensional (tetrahedral), two-dimensional (layer-like), and one-dimensional (chain-like) structures, respectively.

Intermediate-range order shows some interesting features that distinguish amorphous from crystalline states. For instance, monatomic group IV semiconductors crystallize only in the diamond lattice with a dihedral angle of 60°. Amorphous Ge, however, shows a dihedral angle of 0° (see Fig. 3.9A). The interesting feature of this structure is the disappearance of the third neighbor peak in diffraction analysis, which is observed at 4.7 Å for c-Ge. With a dihedral angle of 0° for α-Ge, this third neighbor distance is 4.02 Å; thus it is very close to, and nearly indistinguishable from, the second nearest neighbor at 4.0 Å (see Fig. 3.20).

3.9.3A EXAFS and NEXAFS Information about the structure that surrounds specific types of atoms can be obtained from the extended x-ray absorption fine structure (EXAFS). With synchrotron radiation a continuous spectrum of x-rays is available for investigating absorption or luminescence spectra which show characteristic edges when an electron of a specific atom is ex-

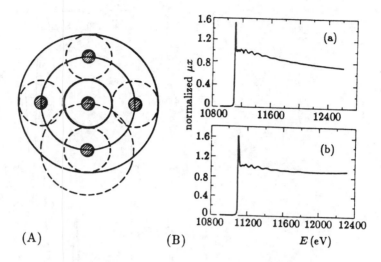

Figure 3.21: (A) EXAFS representation with electron wave emitted from center atom and scattered waves from surrounding atoms. (B) EXAFS for crystalline Ge (a) and for amorphous Ge (b) (after Stern, 1985).

cited from an inner shell into the continuum. Interference of such electrons with backscattered electrons from the surrounding atoms (Fig. 3.21A) results in a fine structure of the absorption beyond the edge (Fig. 3.21B). This results from interference between outgoing and reflected parts of the electron DeBroglie wave, as indicated by solid and dashed rings in the Fig 3.21B. This fine-structure, therefore, yields information about the distance to the surrounding atoms, as well as their number, and provides species identification of the neighbor atoms (Hayes and Boyce, 1985).

EXAFS do not require long-range periodicity, and therefore are useful in analyzing amorphous short-range structures.

When measuring x-ray fluorescence rather than absorption, the surrounding of specific impurities of low density can also be analyzed, since such fluorescence has a much lower probability of overlapping with other emission in the same spectral range.

In addition, near-edge x-ray fine structure (NEXAFS), within 30 eV of the edge, gives information from low-energy photoelectrons which undergo multiple scattering, and provides information on the average coordination number, mass of neighbor atoms, average distance, and their variations with temperature. For special cases, it also yields information on the angular distribution of the surrounding atoms. For a review, see Bienenstock (1985), Stern (1978, 1985).

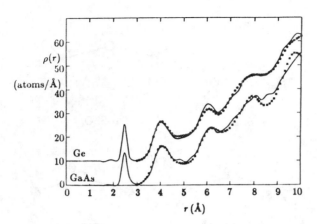

Figure 3.22: Radial density function of α-Ge (Temkin et al., 1973) and α-GaAs (Temkin, 1974) shown as solid curves and obtained from computation for a random network by Steinhardt et al., 1974 ($\circ$) and by Temkin, 1974 ($\times$).

3.9.4 Network Structures

The entire glass can be composed of a *network-like structure* from elements of intermediate-range order, or as a *matrix-like structure*, which is preferable for elemental semiconductors and will be explained in Section 3.9.5.

Network glasses are constructed from principal building blocks with *long-range disorder* added. Such disorder can be introduced in several ways by statistical variation of interatomic distances and bond angles.

The results of a calculation of the radial density distribution function of such random network structures are in satisfactory agreement with the experimental observation obtained from x-ray diffraction data and from EXAFS or NEXAFS, as shown for amorphous Ge and GaAs in Fig. 3.22.

Requirements for creating such a network were given by Bell and Dean (1972) and applied to α-SiO$_2$, α-GeO$_2$, and α-BeF$_2$. When starting from an Si(O$_{1/2}$)$_4$ unit, one proceeds with a covalent random network, connecting to it other Si(O$_{1/2}$)$_4$ units with twofold oxygen coordination, while requiring that:

- the bond angle of Si atoms must not deviate more than $\pm 10°$ from the ideal value of $109.47°$;
- all tetrahedra are corner-connected;

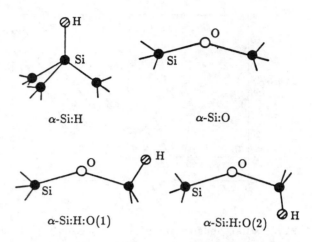

Figure 3.23: Atomic clusters occurring in an α-Si:O:H alloy.

- the bond angles of O atoms may spread by $\pm 25°$ from the ideal value of $150°$;
- there is equal probability for all dihedral angles;
- there is no correlation between bond angles at O atoms and dihedral angle; and
- there is complete space filling.

Modifications of these instructions yield slightly different networks. The relation to the dihedral angle (e.g., assuming some correlation) is an example of such modification.

A relatively simple infinite aperiodic network structure is called a *Bethe lattice* (Bethe, 1935, Peierls, 1935, see also Runnels, 1967). Another kind of network structure is the *fractal structure*, in which void spaces between more densely filled regions can be identified (see Mandelbrot, 1981).

3.9.5 Matrix Glasses; α-Si:H

Constructing an atomic amorphous semiconductor but relaxing the requirements for a fourfold coordination creates dangling bonds. These bonds could attract monovalent elements such as H or F. Alternatively, the tetravalent host atom could be replaced with an element of lower valency such as N or O.

When foreign atoms are introduced in a density that is large enough so that their interaction can no longer be neglected, we call this process an *alloy formation*; as such, we may include homologous elements (e.g., C in α-Si). In all such cases, we then satisfy the $8 - N$

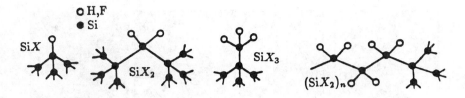

Figure 3.24: Local bonding variation in an α-Si:H alloy with higher densities of hydrogen.

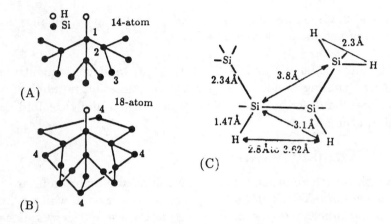

Figure 3.25: (A, B) Different atomic clusters to which an H-atom is attached. (C) Local order in α-Si:H with typical interatomic distances shown (after Menelle and Bellissent, 1986).

rule but achieve a greater degree of flexibility in constructing the amorphous *host matrix*, a reason why such alloys are easily formed.

In addition, we may include into such a host matrix more than one kind of atom, and thereby create more complex alloys, e.g., forming α-Si:O:H or α-Si:N:H by also incorporating oxygen or nitrogen into α-Si:H. Examples of such clusters are shown in Fig. 3.23.

For a theoretical analysis of the various possibilities of incorporating foreign atoms (e.g., in Si-H, SiH$_2$, or SiH$_3$ configurations as shown in Fig. 3.24), it is useful to consider larger *atomic clusters* from this network. Such clusters contain nearest, next-nearest, and higher-order neighbors from the alloyed atom, as shown in Fig. 3.25.

Experimental analyses using a variety of techniques (neutron scattering, small angle neutron scattering, EXAFS, etc.) indicate that the technically interesting α-Si:H contains hydrogen without substantially changing the dihedral angle. It is incorporated by decorating preexisting defects with hydrogen, thereby forming SiH or

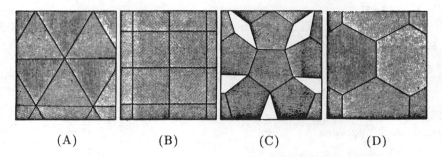

(A)   (B)   (C)   (D)

Figure 3.26: Space filling with one (A), (B), and (D) and two (C) pentagon and diamond-type tiles.

SiH_2 units (Menelle and Bellissent, 1986), and consequently eliminating dangling bonds. α-Si:H has a coordination number of ~ 3.7, and may be described as a mixture of fourfold and threefold coordinated atoms.

3.10 Quasicrystals

Materials with an order between crystalline and amorphous are called *quasicrystals* (Nelson, 1986). In contrast to a crystal, which is formed by a *periodic repetition of one unit cell*, these quasicrystals can be assembled by an *aperiodic repetition of two different unit cells*. One or both of these unit cells may have fivefold symmetry as first shown by Shechtman et al. (1984). This symmetry is forbidden for crystals since it cannot fill space without overlap or voids.

Quasicrystals have no long-range order* even though there is no variation of bond length or bond angle: all atoms fit into one of the two unit cells (Kramer and Neri, 1984; Levine and Steinhardt, 1984; Cahn et al., 1986; Henley, 1987; and Steinhardt, 1987). Similar quasicrystalline behavior can be modeled with specific (Fibonacci) sequences of superlattices (Todd et al., 1986) (see the following section).

In two dimensions, the arrangement of tiles into a completely filled pattern provides an instructive example. While only three types of regular polygons (namely triangles, squares and hexagons) can fill a given flat surface by themselves, the use of two types allows many other tile shapes, including one with fivefold symmetry, to fill

---

* This poses some interesting anomalies in solid-state properties which conventionally require periodicity, such as x-ray diffraction and electronic band structure (Sokoloff, 1985).

the same surface completely (Penrose tiling: Penrose, 1974; Nelson, 1986) (see Fig. 3.26).

3.11 The Growth of Semiconductors

The growth of a semiconductor is determined by four processes:
- transport of atomic particles to the surface
- diffusion of the particles across the surface
- attachment of each particle at a preferred surface site
- dissipation of heat developed during growth.

Each of these processes has a specific time constant which can be altered by the experimental conditions. Depending on these conditions, a wide variety of growth habits, morphologies, and degrees of order will ensue, which will be discussed below.

3.11.1 Nucleation

Any growth starts from a nucleus or crystal seed. Nucleation takes place when *supersaturation* is reached—that is a state in which the atomic particles are close enough to each other at a low enough temperature (below the melting point) so that an ordered arrangement of these atoms is energetically preferred. Whenever a few atoms find each other in an ordered state,* they are in a lower energy state and tend to grow when this nucleus is of an overcritical size. The surface tension of a statistically formed nucleus assists in growth by keeping atoms with the nucleus. Interplay between volume and surface energy defines a critical size above which regular growth starts.

3.11.2 Growth Habit

When the transport to the surface is slower than the diffusion across the surface, the growth from a nucleus is ordered and a microcrystallite is formed. When such growth occurs from one nucleus (seed) only, a *single crystal* will grow. Growth from statistically formed nuclei results in *polycrystals*.

When the transport to the surface surpasses the diffusion velocity at the surface, i.e., when the viscosity of the material near the surface is too large, the atomic particles do not have time to find their proper,

---

* Even after melting, many nuclei are not broken down and can act as growth centers when the temperature is lowered below the melting point T_m. Only at temperatures substantially above T_m are these small nuclei destroyed.

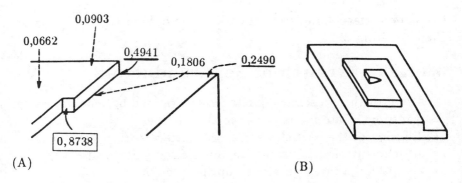

Figure 3.27: (A) Surface of a cubic alkali halide with atomic step and bonding energies at the various indicated sites. (B) Spiral growth around a screw dislocation in a cubic material.

ordered position before more arriving atoms block their movement and freeze in a highly disordered state, creating an *amorphous solid*. Such growth occurs more easily when the atomic building blocks are larger molecules, or are formed by covalent forces in which, aside from space filling, well-defined angle relations must also be fulfilled.

3.11.2A Single Crystal Growth   Crystal growth can easily be understood in a simple model of an ionic crystal (e.g., an alkali halide). The energy gained by attracting an ion pair on the surface of a crystallite depends on the position at the surface. It depends on the number of attracting neighbors and is lowest on a flat surface and highest at an inner edge (Fig. 3.27A). Consequently, near the melting point, where the balance between absorption and desorption of ions is only slightly shifted in favor of absorption, the *sticking* of ion pairs will be most persistent at inner corners with a bonding energy $= 0.8738e^2/r$, as shown in Fig. 3.27A; hence, growth will be fastest here. Thereby, all inner corners and inner edges will rapidly disappear, leaving only ideal, flat surfaces. Growth perpendicular to such surfaces has to await nucleation of a two-dimensional seed; this then becomes the growth rate limiting process.

A certain type of crystal defects, the screw dislocation (see Section 18.3.2), provides atomic steps which cannot be outgrown. Continued absorption of atoms at such steps results in *spiral growth* (Fig. 3.27B). Fast spiral growth around a screw dislocation results in a whisker growth (Levitt, 1970; Kittel, 1986). Such whiskers have a mechanical strength which is orders of magnitude larger than that of bulk crystals and approaches the theoretical maximum cohesion strength. Here, the central screw dislocation prohibits slips of lattice planes which otherwise would ease mechanical deformation.

3.11.2B Crystal Growth Techniques There are numerous techniques to grow single crystals, such as:

- vapor phase
- melt
- solution
- chemical reactions.

The *growth from the vapor phase* can proceed by simple evaporation *in vacuo*, in an inert atmosphere, or in a reacting atmosphere (the latter, e.g., to maintain stoichiometry of a dissociating compound, or to synthesize the desired compound).

Nucleation can be spontaneous (e.g., observed by sublimation of CdS, forming platelets), or by seeding. Vapor phase deposition onto an identical or similar lattice, called *epitaxy*, is well known. Vapor phase epitaxy may involve atomic or molecular beams *in vacuo* (*molecular beam epitaxy*), or may involve dissociation of an evaporated compound, e.g., an organometallic compound, with one of the components condensing.

Single crystal *growth from the melt* requires careful seeding, which can involve dipping of a cooled seed from the top surface and consequent slow pulling (*Czochralski method*). Here, heat removal through the grown crystal is the growth rate limiting process.

Another method uses the fact that the growth rate is often anisotropic, e.g., it is largest in *c*-direction of certain semiconductors. When random crystallization is started in a narrow tip at the bottom of a vessel, crystals aligned with their *c*-axis in the direction of the axis of the tip capillary will outgrow crystallites with other orientations and will act as an oriented seed for the melt in the wider, upper part of the vessel (*Bridgeman technique*).

Crystallization from a solution is often employed when using water or organic solvents, and is usually started by evaporating part of the solvent, or by lowering of the temperature, thereby achieving the supersaturation necessary for growth. However, many other solvents can be used, including liquid metals or other molten solids which show phase segregation before both components, the solvent and the dissolved material, solidify.

A wide variety of *chemical reactions* can assist the crystal growth process. These reactions, however, must proceed slowly enough to avoid incorporation of undesired reaction by-products, or to disturb the heat removal during crystallization. For this, we may use highly diluted reaction partners with consequent evaporation during a spray deposition of a fine mist onto a heated condensation plate. Another method involves gels containing the reaction partners and permitting only a slow interdiffusion to sufficiently slow down the chemical reaction, thereby permitting reasonable crystal growth.

In all, there is a wide variety of crystal growth methods. Often, a specific material requires its own individually designed growth technique to obtain single crystals of the desired quality, with few general rules applicable. Many reviews are available to describe this subject in detail, e.g., Buckley (1951), Laudise (1970), Goodman (1978), Holden and Morrison (1982), Pamplin (1980), and Brice (1986).

Summary and Emphasis

Crystal bonding and crystal structure are intimately related to each other.

The crystal structure is determined by the tendency to fill a given space with the maximum number of atoms under the constraint of bonding forces and atomic radii. This tendency is caused by the strong driving force of lowering the internal energy of a solid in thermal equilibrium, and the fact that bonding energy is delivered for each additional atom, which consequently is converted into heat and dissipated during the crystallization process. This causes the number of atoms per unit volume to be maximized, and therefore the structure to be ordered.

An amorphous structure results with no long-range order, when sufficient viscosity restricts atomic motion during fast enough cooling so that atomic building blocks cannot find their crystalline, ordered position before further motion becomes more restricted by their rigidified surrounding. In an amorphous structure the short-range order is much like that in a crystal, while long-range periodicity does not exist. This lack of periodicity has major consequences for the lack of interference effects. This modifies substantially the theoretical analysis of photon, phonon, and electronic band structure effects.

Superlattices, created by alternating deposition of thin layers of different semiconductors, substantially enrich the variety of semi-conducting materials, and show new and very attractive material properties with potential for new and improved devices.

The chemical composition and structure of a solid determines all of its intrinsic properties. These properties provide the basis from which any semiconducting device is formed. They may be compared to the architecture of a house, while its furniture equates to the specific treatments necessary to fabricate the device. The selection of an appropriate architecture is essential to blend with the chosen furniture for optimum appeal. Also, the selection of the most appropriate semiconductor is essential for obtaining devices of optimum performance. Recent developments indicate the potential to supplement perfection of a given material, as was done for silicon, with the tailoring of totally new materials to achieve superb performance. Modern crystal growth methods employing epitaxy provide a potential for better designed crystals and thus improved devices.

Exercise Problems

1.(e) Draw a coordinate system for a cubic crystal and identify the following: 100, 001, [111], $(23\bar{1})$.

2.(e) In a cubic lattice, [100] is perpendicular to (100). What direction is perpendicular to (111)? Did you discover a simple rule? Does this rule hold in general [e.g., for $(21\bar{1})$]? Or for lattices other than cubic? Consider a tetragonal lattice and give the direction normal to (111).

3.(e) Identify the plane with maximum density of atoms in a cubic primitive lattice: (201), (111), (112), or (110). Relate the density of atoms in (111) and (110) to each other.

4. CsCl looks much like Fig. 3.4B, with Cs in the center and Cl at each of the corners, except that it is cubic. What is its lattice structure? Why is it not body-centered cubic?

5.(e) Carbon crystallizes as diamond ($a = 7.91\text{Å}$) or as graphite (hexagonal $a = 2.461\text{Å}$ and $c = 6.701\text{Å}$). The density of diamond is 3.51 g/cm^3; and that of graphite is 2.25 g/cm^3. How many atoms does each of the cells contain? Draw the corresponding Bravais lattices.

6.(e) What is the angle between the [110] and [111] direction in a cubic lattice?

7.(e) Draw the planes $(01\bar{1}0)$ in a hexagonal lattice (there are three a-axes at an angle of 60°). How many members does this family of planes have? What is the simplest symbol of this family?

8.(e) Calculate the density of CdS.

9.(e) Derive the critical ratio for ionic radii listed in Table 2.10 for the three types of lattices.

10.(e) How many atoms are in the unit cell of Si?

    (a) What is the angle between the tetrahedral bonds of the atom?

    (b) What is the lattice constant of Si as obtained from the atomic radius and crystal structure?

11.(e) Give the coordinates of the interstitial positions in a body-centered cubic lattice.

12.(e) What is the distance between two adjacent planes (hkl) in an orthorhombic primitive crystal? [For a cubic primitive lattice it is given by Eq. (3.3).]

13.(r) What is the difference between a Wigner-Seitz cell and a Brillouin zone?

14.(e) Construct in two dimensions the first 4 Brillouin zones for a rectangular lattice with a and $b = 2a$.

15. Why is the unit cell of the ternary chalcopyrite twice as large as that of zinc-blende?

    (a) Give actual examples for various tetrahedrally bonded ternaries or quaternaries and discuss Fig. 3.12.

16.(e) Calculate the packing fraction (ratio of the volume of the Bravais parallelepiped to the volume of atomic spheres or fractions thereof contained therein) for the fcc, bcc, sc, and diamond lattices.

17. Show that the Wigner-Seitz cell for any two-dimensional Bravais lattice is either a hexagon or a rectangle.

18.(e) Show that the volume of a primitive Bravais lattice is $\mathcal{V} = \mathbf{a}_1 \cdot (\mathbf{a}_2 \times \mathbf{a}_3)$ and that of the reciprocal lattice is $(2\pi)^3/\mathcal{V}$.

PART II

PHONONS

Chapter 4

Elastic Properties

Spring-like interatomic forces allow macroscopic elastic deformations of the semiconductor and coupled microscopic oscillations of each atom. The macroscopic constants describing such deformation provide important clues for the microscopic forces.

Elastic properties of solids are determined by interatomic forces and the crystal structure of solids. These properties determine not only the macroscopic elastic behavior of a solid, but also provide the basic elements for all lattice oscillations.

The interatomic forces can be described in a first approximation as spring forces. This means that the force $\mathcal{F}$ is proportional to an atomic displacement $u = r - r_e$:

$$\mathcal{F} = \beta u, \tag{4.1}$$

with β the (spring) force constant. This holds for small displacements from the equilibrium position of each atom as long as the lattice potential can be approximated by a parabola (see Fig. 4.1 for values of r close to the equilibrium distance r_e). For $V = V_0 + \alpha(r - r_e)^2$ and $\mathcal{F} = dV/dr$, we obtain the linear relation [Eq. (4.1)] with $\beta = 2\alpha$

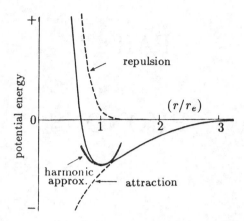

Figure 4.1: Interatomic potential and parabolic (harmonic) approximation. r_e is the equilibrium distance.

for the spring forces. For nonlinear effects, see Section 4.1.3 and Chapter 16.

In a three-dimensional lattice with interatomic forces restraining changes in the *bond length* and *bond angle*, a more complicated relation between the force and the *atomic displacement* in such a network must be considered. Since various force constants can be distinguished in such a three-dimensional lattice, the general relation between forces and *lattice deformation* will be analyzed first.

Microscopic parameters can be derived from macroscopic experiments by exposing a sample to specific *forces* (*stress*) and observing the *resulting deformation* (*strain*). The following section will deal briefly with stationary deformation of the semiconductor as a macroscopic continuum. In Section 4.2 we will then describe additional information which can be obtained from dynamic deformation.

4.1 Elastic Constants

The basis for deriving the elastic constants is a simple application of *Hooke's law* [Eq. (4.1)] to the different ways of applying mechanical forces to a semiconductor, which are summarized for isotropic systems and homogeneous strains in Fig. 4.2.

There are three macroscopic *elastic moduli* which can be expressed for a homogeneous medium by the following relations:

- the compression modulus $B(= 1/\kappa$ with κ the compressibility):

$$p = -B \frac{\Delta \mathcal{V}}{\mathcal{V}};$$ (4.2)

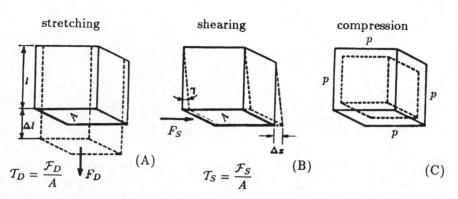

Figure 4.2: Three basic types of mechanical deformation of a cube. (A) stretching (in one dimension); (B) shearing; and (C) volume compression (or dilation).

- the linear elasticity modulus (*Young's modulus* or stretching modulus) E:

$$\mathcal{T}_D = E\frac{\Delta l}{l}; \qquad (4.3)$$

- and the shear modulus G:

$$\mathcal{T}_S = G\frac{\Delta x}{l} \qquad (4.4)$$

where p is the pressure, $\mathcal{V}$ is the volume and l is the length of the undeformed cube, $\mathcal{T}_D$ is the pull force, and $\mathcal{T}_S$ is the shear force.

In a typical semiconductor these moduli depend on the crystal orientation.

From this mechanical approach we can deduce important consequences for the different types of waves of interest in a semiconductor.

4.1.1 Stress-Strain Relations

The mechanical deformation depends on the type of stress applied. A parallelepiped (e.g., a crystallite) can be deformed by changing the lengths of its edges and the angle between them or, equivalently, by changing its unit cell accordingly. During such a deformation, an arbitrary point, described by a vector **r** from the origin, is displaced to **r'**, which in turn is given in component form as

$$x' = \left(1 + \frac{\partial u_1}{\partial x}\right) x + \left(1 + \frac{\partial u_1}{\partial y}\right) y + \left(1 + \frac{\partial u_1}{\partial z}\right) z$$

$$y' = \left(1 + \frac{\partial u_2}{\partial x}\right) x + \left(1 + \frac{\partial u_2}{\partial y}\right) y + \left(1 + \frac{\partial u_2}{\partial z}\right) z \qquad (4.5)$$

$$z' = \left(1 + \frac{\partial u_3}{\partial x}\right) x + \left(1 + \frac{\partial u_3}{\partial y}\right) y + \left(1 + \frac{\partial u_3}{\partial z}\right) z$$

where u_1, u_2, and u_3 are the components of the displacement vector, and assuming that $(\partial u_i/\partial x, \partial u_i/\partial y, \partial u_i/\partial z) \ll 1$. When expressed in tensor form, we have

$$\mathbf{r}' = \mathbf{r} + \underline{\mathbf{T}} \cdot \mathbf{r} \qquad (4.6)$$

with the components of $\underline{\mathbf{T}}$ given in Eq. (4.5).

A pure deformation without translation or rotation of the crystal is described by the symmetric part of this tensor

$$\underline{\mathbf{T}}_s = e_{ik} = \begin{pmatrix} 1 + \dfrac{\partial u_1}{\partial x} & \dfrac{1}{2}\left(\dfrac{\partial u_1}{\partial z} + \dfrac{\partial u_2}{\partial x}\right) & \dfrac{1}{2}\left(\dfrac{\partial u_1}{\partial z} + \dfrac{\partial u_3}{\partial x}\right) \\[2ex] \dfrac{1}{2}\left(\dfrac{\partial u_1}{\partial y} + \dfrac{\partial u_2}{\partial x}\right) & 1 + \dfrac{\partial u_2}{\partial y} & \dfrac{1}{2}\left(\dfrac{\partial u_2}{\partial z} + \dfrac{\partial u_3}{\partial y}\right) \\[2ex] \dfrac{1}{2}\left(\dfrac{\partial u_1}{\partial z} + \dfrac{\partial u_3}{\partial x}\right) & \dfrac{1}{2}\left(\dfrac{\partial u_2}{\partial z} + \dfrac{\partial u_3}{\partial y}\right) & 1 + \dfrac{\partial u_3}{\partial z} \end{pmatrix}$$

$$(4.7)$$

defining a set of strain coefficients e_{ik} with which the crystallite deformation is easily described. For instance, the new volume after application of hydrostatic pressure is given from $\mathcal{V}' = \mathbf{a}' \cdot \mathbf{b}' \times \mathbf{c}'$, when retaining only first-order terms, by

$$\mathcal{V}' = \mathcal{V}[1 - (e_{xx} + e_{yy} + e_{zz})]. \qquad (4.8)$$

(For more detail, see Joos, 1945.)

The strain coefficients e_{ik} relate to the six independent components of the forces ($\mathcal{T}_{xx}$, $\mathcal{T}_{yy}$, $\mathcal{T}_{zz}$ and $\mathcal{T}_{yz}$, $\mathcal{T}_{zx}$, $\mathcal{T}_{xy}$) in a tensor relationship

$$\underline{\mathbf{e}} = \underline{\mathbf{S}}\,\underline{\mathcal{T}}, \qquad (4.9)$$

with 36 *elastic compliance constants* s_{ik} (with $i, k = 1, \ldots 6$).

Often the inverse relationship is used

$$\underline{\mathcal{T}} = \underline{\mathbf{C}}\,\underline{\mathbf{e}} \quad \text{and} \quad \underline{\mathbf{C}} = \underline{\mathbf{S}}^{-1} \qquad (4.10)$$

with the *elastic stiffness constants* c_{ik} (with $i, k = 1, \ldots 6$). Equation (4.10) is also known as the generalized Hooke's law. Since $\mathcal{T}_{yz} = \mathcal{T}_{zy}$,

Table 4.1: Elastic stiffness constants for primitive lattices.

Crystal System	Independent Elastic Stiffness Constants							
Cubic	c_{11}	c_{12}					c_{44}		
Orthorhombic	c_{11}	c_{12}	c_{13}	c_{23}	c_{22}	c_{33}	c_{44}	c_{55}	c_{66}
Tetragonal	c_{11}	c_{12}	c_{13}			c_{33}	c_{44}		
Trigonal	c_{11}	c_{12}	c_{13}	c_{14}		c_{33}	c_{44}		
Hexagonal	c_{11}	c_{12}	c_{13}			c_{33}	c_{44}		

etc., and $e_{yz} = e_{zy}$, etc., both tensors can be reduced to $\underline{e}$ and $\underline{T}$ vectors with six components. This permits us to write a tensor relation between $\underline{e}$ and $\underline{T}$ where $\underline{C}$ and $\underline{S}$ are 6×6 tensors of second rank. The force components can be written as

$$T_{ik} = T_i' = \sum_{k=1}^{6} c_{ik} e_k \qquad (4.11)$$

with the convention $k = 1 \ldots 6$ equivalent to xx, yy, zz, yz, zx, and xy, respectively, for a single index (*Voigt notation*).

4.1.2 Symmetry Considerations

The number of elastic stiffness constants can be reduced from 36 because of two considerations:

1. the parabolicity of the lattice potential renders the c_{ik} symmetrical: $c_{ik} = c_{ki}$, leaving 21 independent constants;
2. lattice symmetry further reduces this number.

For instance, in a *cubic lattice* with cyclic interchange of x, y, and z, the same state is reproduced. This yields

$$c_{12} = c_{23} = c_{31}, \; c_{11} = c_{22} = c_{33} \quad \text{and} \quad c_{44} = c_{55} = c_{66}$$
$$c_{14} = c_{15} = c_{16} = c_{61} = c_{62} = c_{63} = c_{64} = c_{65} = 0$$

and an exceedingly simple matrix results:

$$\mathbf{T} = \begin{pmatrix} c_{11} & c_{12} & c_{12} & 0 & 0 & 0 \\ c_{12} & c_{11} & c_{12} & 0 & 0 & 0 \\ c_{12} & c_{12} & c_{11} & 0 & 0 & 0 \\ 0 & 0 & 0 & c_{44} & 0 & 0 \\ 0 & 0 & 0 & 0 & c_{44} & 0 \\ 0 & 0 & 0 & 0 & 0 & c_{44} \end{pmatrix} \begin{pmatrix} e_{xx} \\ e_{yy} \\ e_{zz} \\ e_{yz} \\ e_{zx} \\ e_{xy} \end{pmatrix} \qquad (4.12)$$

with only three independent elastic stiffness constants c_{11}, c_{12}, and c_{44}.

The independent elastic stiffness constants for some of the other crystal systems are given in Table 4.1.

The values of these elastic stiffness constants for the more important semiconductors are listed in Table 4.2. These stiffness constants depend on the lattice temperature (see Section 4.1.4); the room temperature values are given when not otherwise stated.

4.1.3 Third-Order Elastic Constants

In the previous discussion only the harmonic part of the interatomic potential was included; this limits the amplitude of deformation to the validity of Hooke's law. For nonlinear effects, which include thermal expansion and phonon scattering, the *third-order elastic constants* must be included (see Brugger, 1964). These can be obtained from the lattice potential with higher terms:

$$V = V_0 + \alpha_1(r - r_e)^2 + \alpha_2(r - r_e)^3 + \dots \qquad (4.13)$$

with the strain energy

$$E_S = E_{S0} + \frac{1}{2}\sum_{i,k} c_{ik}e_ie_k + \frac{1}{6}\sum_{i,k,l} c_{ikl}e_ie_ke_l \qquad (4.14)$$

defining the third-order elastic constants. These constants can be obtained from sound wave propagation (see Section 4.2.1) in a sufficiently prestressed semiconductor. In Table 4.3 the set of third-order elastic constants for a few semiconductors is given. (See McSkimmin and Andreatch, 1964, 1967; for a general review, see Hiki, 1981.)

4.1.4 Temperature Dependence

The expansion of the lattice with increased temperature leads to a reduction of interatomic forces and thus a reduction in the value of the elastic stiffness constants (Garber and Granato, 1975). Typical behavior is shown for Si in Fig. 4.3. Near 0 K this dependence disappears, shown as a horizontal tangent in Fig. 4.3—as required by the 3rd law of thermodynamics. The figure also shows that doping (Section 27.3) usually weakens the lattice and thereby reduces the elastic stiffness constants.

4.1.5 Information from Elastic Stiffness Constants

The elastic stiffness constants provide a wealth of information about the interatomic forces. Their magnitude reflects the strength of the

Table 4.2: Elastic stiffness constants for some important semiconductors* in 10^{11} dyn/cm^2.

Crystal	c_{11}	c_{12}	c_{44}	c_{13}	c_{33}	c_{55}	c_{66}
C (diamond)	10.764	1.252	5.774				
Si	16.577	6.393	7.962				
Ge	12.40	4.13	6.83				
AlAs	12.02	5.70	5.89				
AlSb	8.77	4.34	4.076				
GaP	14.050	6.203	7.033				
GaAs	11.90	5.38	5.95				
GaSb	8.834	4.023	4.322				
InP	10.11	5.61	4.56				
InAs	8.329	4.526	3.959				
InSb	6.669	3.645	3.020				
ZnO (F)	2.07	1.177		1.061	2.095	0.448	0.446
ZnO (D)	2.096	1.204		1.013	2.21	0.461	
hex-ZnS (F)	12.34	5.85	2.89	4.55	13.96		3.25
hex-ZnS (D)	13.03	6.90	2.74	5.28	14.34		
cub-ZnS	9.81	6.27	4.483				
ZnSe	9.009	5.34	3.96				
ZnTe	7.13	4.07	3.12				
CdS	8.31	5.04		4.62	9.48	1.533	
CdS (D)	8.38	5.11		4.50	9.653	1.577	
CdSe (F)	7.41	4.52	1.317	3.93	8.36		1.445
CdSe (D)	7.42	4.53		3.86	8.477		1.340
CdTe	5.33	3.65	2.04				
PbS	11.39	2.89	2.72				
PbSe	12.37	1.93	1.59				
PbTe	10.72	0.77	0.13				
HgSe	5.95	4.31	2.20				
HgTe	5.361	3.660	2.123				

*The indices D and F refer to the constant electrical displacement and constant electric field in materials (see Hanson et al., 1974).

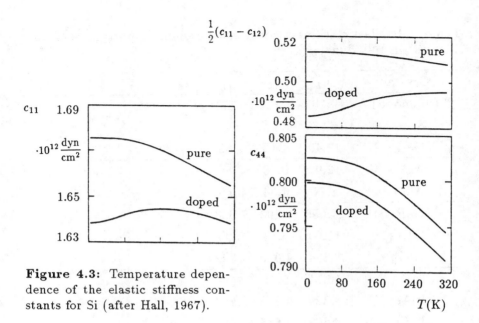

Figure 4.3: Temperature dependence of the elastic stiffness constants for Si (after Hall, 1967).

Table 4.3: Third-order elastic constants in 10^{11} dyn/cm^2 at room temperature (after Mitra and Massa, 1982).

Semiconductor	c_{111}	c_{112}	c_{123}	c_{144}	c_{166}	c_{456}
Si	-79.5	-44.5	-7.5	$+1.5$	-31	-8.6
Ge	-71.0	-38.9	-1.8	-2.3	-29.2	-5.3
GaAs	-67.5	-40.2	-0.4	-7.0	-32.0	-6.9
InSb	-31.4	-21.0	-4.8	$+0.9$	-11.8	$+0.02$

bond and, for a similar bonding type, shows the decreasing strength of the bond with increasing atomic number (atomic radius), as seen in Table 4.2 by comparing compounds in which only one of the homologous elements is varied, e.g., InP, InAs, and InSb.

Most direct information can be derived by expressing the compressibility in terms of c_{ik}. For instance, the compressibility in cubic crystals is given by

$$\kappa = \frac{3}{c_{11} + 2c_{12}}. \tag{4.15}$$

When binding forces are radially symmetric (ionic bond), one expects the *Cauchy relation* (Cauchy, 1828; see Zener, 1947; Leibfried, 1951; Musgrave, 1970) to hold

$$c_{44} = \frac{c_{11} - c_{12}}{2}. \qquad (4.16)$$

Deviation from this relation indicates an elastic anisotropy and is presented as anisotropy parameter $a = c_{11} - c_{12}/(2c_{44})$. Such anisotropy is given by *Keating's relation* (Keating, 1966) for monatomic cubic semiconductors:

$$c_{44} = \frac{c_{11} - c_{12}}{2} \cdot \frac{c_{11} + 3c_{12}}{c_{11} + c_{12}}. \qquad (4.17)$$

Other information can be obtained about the influence of lattice defects (Mason and Bateman, 1964) and about the influence of free electrons (Keyes, 1961) which reduce the strength of the lattice bonds. A decrease in all elastic stiffness constants is shown by Hall (1967).

4.2 Elastic Waves

Acoustic oscillations correspond to the macroscopic behavior of a solid responding to sound waves. At long wavelengths, this behavior can be described (Brown, 1967; Musgrave, 1970) by a continuum model in which the displacement u_i (in the harmonic approximation) follows the equation of motion:

$$\rho \frac{\partial^2 u_i}{\partial t^2} = \sum_k \frac{\partial T_{ik}}{\partial x_k} = \sum_j \sum_k \sum_l c_{ijkl} \frac{\partial^2 u_i}{\partial x_j \partial x_k}, \qquad (4.18)$$

where u_1, u_2, and u_3 are displacements in the x-, y-, and z-directions, respectively, ρ is mass density, and T_{ik} is the stress tensor (see Section 4.1.1). The c_{ik} are related to the components of the c_{ijkl} tensor. This relation is given by

$$c_{\alpha\beta} = c_{ijkl} \text{ for } 1 \leq \alpha, \beta \leq 3$$
$$c_{\alpha\beta} = \frac{1}{4} c_{ijkl} \text{ for } 4 \leq \alpha, \beta \leq 6$$
$$c_{\alpha\beta} = \frac{1}{2} c_{ijkl} \text{ for } 1 \leq \alpha \leq 3, 4 \leq \beta \leq 6 \text{ or } 1 \leq \beta \leq 3, 4 \leq \alpha \leq 6$$

and $\alpha = (ij), \beta = (kl)$ with i, j, k, l permutating for x, y, z, following the convention for $(\alpha$ or $\beta)=1 \ldots 6$, equivalent to xx, yy, zz, yz, zx, and xy, respectively. Equation (4.18) is the wave equation.

In a cubic crystal the relation [Eq. (4.18)] can be expressed as

$$\rho\frac{\partial^2 u_1}{\partial t^2} = c_{11}\frac{\partial^2 u_1}{\partial x^2} + c_{44}\left(\frac{\partial^2 u_1}{\partial y^2} + \frac{\partial^2 u_1}{\partial z^2}\right) + (c_{12} + c_{44})\left[\frac{\partial^2 u_2}{\partial x \partial y} + \frac{\partial^2 u_3}{\partial x \partial z}\right],$$
(4.19)

with similar equations for u_2 and u_3 obtained by cyclic exchange of (u_1, u_2, u_3) and (x, y, z). Their solutions can be written for the displacement vector in plane wave form:

$$\mathbf{u(r, t)} = \mathbf{A}\exp[i(\mathbf{q} \cdot \mathbf{r} - \omega t)] \qquad (4.20)$$

with $|\mathbf{q}| = 2\pi/\lambda$ and $\mathbf{q}$ the wave vector for lattice oscillations. Here $\mathbf{q}$ will be used consistently to set it apart from $\mathbf{k}$, the electron wave vector (see Section 7.2). In one dimension, q is referred to as the *wavenumber*. Introducing this *Ansatz* (German for "trial solution") into Eq. (4.19) yields the set of equations that connect ω and $\mathbf{q}$, called the *dispersion equations*—see Section 5.1.1.

4.2.1 Sound Waves in a Cubic Crystal

We now determine the crystallographic direction in which the sound wave propagates and whether it is induced as a compression (longitudinal) wave or as a shear (transverse) wave. Only in special directions (see below) are these waves purely longitudinal or transverse; otherwise, they have components of each. In either case, the resulting dispersion relations are easily obtained. For instance, when a longitudinal sound wave propagates in the $\langle 100 \rangle$ direction, one obtains from Eq. (4.19)

$$\rho\frac{\partial^2 u_1}{\partial t^2} = c_{11}\frac{\partial^2 u_1}{\partial x^2} \qquad (4.21)$$

which yields with the *Ansatz* $u_1 = A\exp[i(q_x x - \omega t)]$ the dispersion relation

$$\rho\omega^2 = c_{11}q_x^2. \qquad (4.22)$$

Rewriting Eq. (4.21) as

$$\frac{\partial^2 u_1}{\partial t^2} = \frac{1}{v_s^2}\frac{\partial^2 u_1}{\partial x^2} \qquad (4.23)$$

one obtains the sound velocity (also see Section 5.1.5B)

$$v_s = \frac{\omega}{q} = \sqrt{\frac{c_{11}}{\rho}}. \qquad (4.24)$$

For longitudinal and transverse acoustic waves, one has the simple relations

$$v_{s,\ell} = \sqrt{\frac{c_{11}}{\rho}} \qquad \text{and} \qquad v_{s,t} = \sqrt{\frac{c_{44}}{\rho}}, \qquad (4.25)$$

obtained from an equation similar to Eq. (4.21). In the $\langle 110 \rangle$ direction, the relationship is a bit more involved, since two transverse modes must be distinguished:

$$v_{s,\ell} = \sqrt{\frac{c_{11} + c_{12} + 2c_{44}}{2\rho}}; \quad v_{s,t_1} = \sqrt{\frac{c_{44}}{\rho}}; \quad v_{s,t_2} = \sqrt{\frac{c_{11} - c_{12}}{2\rho}}. \qquad (4.26)$$

Finally, in the $\langle 111 \rangle$ direction one obtains

$$v_{s,l} = \sqrt{\frac{c_{11} + 2c_{12} + 4c_{44}}{3\rho}} \quad \text{and} \quad v_{s,t} = \sqrt{\frac{c_{11} - c_{12} + c_{44}}{3\rho}}. \qquad (4.27)$$

Later, when electron scattering with longitudinal acoustic phonons is considered, c_l is used as an abbreviation for a longitudinal elastic constant, which equals c_{11} in the $\langle 100 \rangle$ direction and $(c_{11} + c_{12} + 2c_{44})/2$ in the $\langle 110 \rangle$ direction, etc.

These velocities are conventionally labeled v_1, v_5, and v_7 for longitudinal waves in the $\langle 100 \rangle$, $\langle 110 \rangle$, and $\langle 111 \rangle$ directions, respectively. The corresponding transverse waves are labeled v_2; v_3, v_4; and v_6, v_8 for the $\langle 100 \rangle$ (one wave); $\langle 110 \rangle$ (2 waves); and $\langle 111 \rangle$ (2 waves) directions, respectively. For more detail about wave identification, see *Landoldt-Börnstein* III, 17, 1982.

These velocities are given for a number of semiconductors in Table 4.4. Since they are related to the elastic stiffness constants, the sound velocities decrease with increasing temperature, as shown in Fig. 4.4. The change in crystal volume with temperature has the opposite influence on the sound velocity, but usually is a smaller effect.

With application of uniaxial stress, the lattice symmetry is lowered and new branches appear for the sound velocities, which split proportional to the applied stress.

4.2.1A Ultrasound Measurement of Elastic Constants

The sound velocities can be measured by using ultrasound transducers coupled to properly cut semiconductor platelets and measuring the time delay between an emitted pulse and its reflected echo. For a review, see Truell et al. (1969). This method can be used to determine experimentally the elastic stiffness constants from the sound

Table 4.4: Sound velocities (in 10^5 cm/s).

Material	v_1 long [001]/[001]	v_2 shear [001]/[110]	v_3 long [110]/[110]	v_4 shear [110]/[001]	v_5 shear [110]/[110]	v_6 long [111]/[111]	v_7 shear [111]/[110]	T (K)
Si	8.4332	5.8446	9.1333	5.8442	4.6740			298
GaP	5.847	4.131	5.238	3.345	2.476	6.648	3.466	300
GaAs	4.731	3.345	5.289	3.350	2.479	5.397	2.796	300
InSb	4.784	3.350	3.77	2.29	1.63	5.447	2.799	77
ZnSe	3.42	2.29	4.82	2.82	2.05	3.89	1.87	10
CdS	4.155	2.67				4.93	2.35	10
CdTe	4.25	1.76						4.2
	3.35	1.79						10

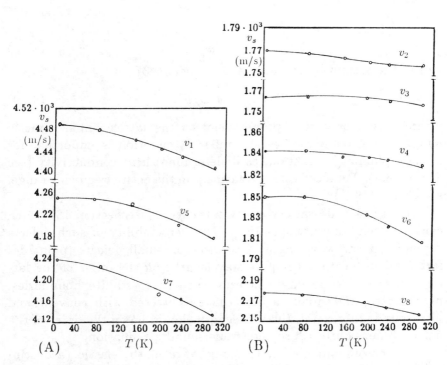

Figure 4.4: CdS sound velocities vs. temperature for: (A) different compression and (B) shear waves.

velocity for longitudinal and transverse acoustic waves in different crystallographic directions.

4.2.1B Sound Beam Mixing   At sufficient amplitudes, the non-linear part of the interatomic potential becomes important in sound waves; this permits two sound beams of the same or different frequency to mix, and results in the creation of a third beam with sum or difference frequency (Hiki and Mukai, 1973).

4.2.1C Sound Damping and Crystal Defects   Sound echos in good semiconducting samples can be observed after many reflections at the surfaces. Damping of the sound occurs because of inelastic scattering with phonons (see Chapter 5 for definition) and crystal defects. For interaction with phonons, considering the anharmonic part of the lattice potential, there are two extreme regimes: the Akhieser regime (Akhieser, 1939) with $\omega_s \tau \ll 1$, and the Landau-Rumer regime with $\omega_s \tau \gg 1$ (ω_s = sound frequency, τ = phonon relaxation time)—see Section 11.2. In the former, the sound attenuation is given by (Woodruff and Ehrenreich, 1961)

$$\gamma_A = \delta^2 \omega_s^2 T, \qquad (4.28)$$

and in the second by (Landau and Rumer, 1937)

$$\gamma_{L-R} = \delta^2 \omega_s T^4, \qquad (4.29)$$

where δ is a quantity* proportional to the lattice anharmonicity and T is the temperature. A similar attenuation is important for phonon-phonon interaction, as it determines heat conductivity (see Section 6.3). For a consistent treatment in the entire frequency range, see Guyer (1966).

There is a wide variety of crystal defects (see Section 18) which cause further sound attenuation. The detectability of such defects depends on the wavelength of the sound: small defects (point defects and dislocations) require wavelengths on the order of the lattice constant; while larger defects, such as crystallite boundaries, precipitates, and small cavities can be detected with conventional ultrasound. In general, the wavelength should be on the order of the defect dimension to cause detectable sound attenuation.

In a continuum model, the long wavelength, elastic (acoustic) waves are only one type of possible oscillations. This description will now be extended to more general lattice oscillations.

Summary and Emphasis

The elastic properties of a semiconductor can be obtained by assuming unit cells with atoms connected by springs which resist deformation relating to changes of interatomic distances and angles. Such response to external strain is conventionally described by a set of elastic stiffness constants c_{ik}. When the external strain exceeds the range in which the harmonic approximation of the interatomic potential is valid, higher order stiffness constants are used.

Elastic properties can be measured by stretching, shearing (bending), and volume compression, or kinetically by sound wave propagation. Different modes of sound waves propagate with different velocities from which all stiffness constants can be determined.

The description of a solid as a continuum responding to external strain yields a set of stiffness constants which, in turn, can be in-

---

* The constant δ can be expressed as $\Gamma^2 C_v/(3\rho v_s^3)$ with Γ the Grüneisen parameter [Eq. (6.26)], C_v the specific heat, ρ the density, and v_s the average sound velocity.

terpreted on the atomic scale, giving information about interatomic forces. These constants are useful to describe macroscopic and microscopic deformations of a semiconductor and its interatomic configuration. They are applied when interaction between local deformation (phonons—see the following chapter) and (e.g.) electrons are analyzed.

Exercise Problems

1.(e) How do spring constants and elastic stiffness constants relate to each other?

2.(*) A cubic crystal is pulled in the [001] direction. Its new axes are given by

$$
\left.
\begin{aligned}
a_1' &= (1 - \nu k + \beta k^2)a_1 \\
a_2' &= (1 - \nu k + \beta k^2)a_2 \\
a_3' &= (1 - \nu k + \beta k^2)a_3
\end{aligned}
\right\} \quad \text{with } k = \frac{T}{E},
$$

where ν is the Poisson's ratio, and β is the nonlinearity constant.

(a) Derive the Young's module E and the Poisson's ratio in terms of the c_{ik} (neglecting the β term).

(b) Derive β as a function of c_{ik} and third-order stiffness constants (see Hiki, 1981).

3.(*) Derive the velocities of transverse waves in cubic crystals in $\langle 110 \rangle$ and $\langle 111 \rangle$ directions.

4.(r) Derive the c_{ik} in an orthorhombic crystal (or a cubic crystal) from ultrasound experiments. Describe methodology and results. Draw a picture to show different wave forms.

5.(*) Discuss the influence of crystal symmetry on different acoustic waves.

6. Picture the tensor ellipsoid of the strain tensor and discuss its connection to stress and lattice deformation.

Chapter 5

Phonons

Atoms, coupled to each other in a crystal, allow a wide spectrum of oscillations characteristic for each semiconductor. Each mode of such collective oscillations is equivalent to a harmonic oscillator which can be quantized as a phonon. These phonons are responsible for all thermal properties and, when interacting with other quasiparticles, for the damping of their motion.

Phonons are one of the most important quasi-particles in solids. They have a decisive influence on most electronic and photonic properties of semiconductors. They are mostly related to losses and often crucially determine the performance of devices. For an understanding of their properties, we will first analyze the oscillatory behavior of lattices.

When an atom is coupled to another atom of the same mass in a diatomic molecule, it can oscillate in a vibrational mode with a well-defined eigenfrequency

$$\omega_0 = \sqrt{\frac{\beta}{M}} \tag{5.1}$$

with β as the spring (force) constant and M as its mass. For Coulomb interaction, the spring constant can be given easily, relating the force to the spatial derivative of the Coulomb potential and yielding, near the equilibrium distance a, for $\beta \simeq e^2/(4\pi\varepsilon_0 a)^2$.

When this same atom is imbedded in a monatomic crystal, a broad spectrum of oscillation is possible. In the following sections, we will analyze this vibration *spectrum*.

Here, such an oscillatory state may be envisaged by holding a lattice of steel balls interconnected by springs, as shown in a two-dimensional model in Fig. 5.1, and by wiggling it, setting the balls in a jiggling oscillatory motion. The motion of each individual ball, although oscillatory in nature, is a complicated one, with components

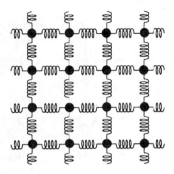

Figure 5.1: Two-dimensional representation of a lattice of steel balls held in place by interconnecting springs.

in all directions. It can be decomposed into a sum of many harmonic parts of a large number of frequencies. A systematic discussion in the following pages will show, however, that order can be brought into this complex picture.

5.1 Vibration Spectrum of Coupled Oscillators

The coupled lattice atoms in such a state of complex oscillation can be thought of as performing collective oscillations. Each mode into which it can be decomposed belongs to the entire crystal. Such modes can be regarded as quanta of elementary excitations, $\hbar\omega$, and are called *phonons*.* They are not localized; together they belong to the entire crystal, until a specific event (a *scattering event*) occurs where, temporarily, such localization takes place. In our steel ball model it could be demonstrated by hitting one of the balls with a small hammer, temporarily making it oscillate more orderly in its resonance frequency, until, by interaction with its neighbors, the more erratic, jiggling motion is restored.

In order to understand this concept better, an analysis of the classical equation of motion for a simple one-dimensional lattice is given first. In such a lattice, *longitudinal* and *transverse modes of oscillations* are possible. In pure form, these oscillations entail oscillations in the direction of, or perpendicular to, the atomic chain,

---

* An elastic wave can thus be regarded as a stream of phonons, in analogy to an electromagnetic wave which can be described as a stream of photons. Both quasi-particles are not conserved; phonons or photons can be created by simply increasing the temperature or the electromagnetic field.

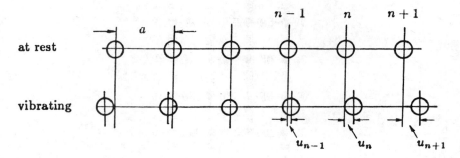

Figure 5.2: Linear chains of atoms with the upper chain at rest and the lower vibrating in a longitudinal mode.

respectively. In general, therefore, one can describe an arbitrary state of oscillations by a superposition of one longitudinal and two (perpendicular to each other) transverse branches of oscillations.

Pure longitudinal oscillations are described first.

5.1.1 Longitudinal Lattice Oscillations

In a *longitudinal* (compression) mode, the atoms in a monatomic chain oscillate, thereby changing the distance between each other, although, remaining entirely along the x-axis—see the lower row of Fig. 5.2. For small displacements, assuming a harmonic lattice potential,* the force T is proportional to the relative atomic displacement u (Hooke's law). For an arbitrary atom (index n), the forces acting on it are transmitted from its two neighbors ($n-1$ and $n+1$), and are given by (see Fig. 5.2)

$$T = \beta(u_{n-1} - u_n) - \beta(u_n - u_{n+1}) \tag{5.2}$$

with β the spring constant. Using Newton's law, the equation for motion of this atom is

$$M \frac{d^2 u_n}{dt^2} = T = \beta(u_{n+1} + u_{n-1} - 2u_n). \tag{5.3}$$

The solution of this equation must describe atomic oscillations, which are given by

$$u_n = A \exp i(q_n a - \omega t) \tag{5.4}$$

---

* In the harmonic approximation there is no exchange of energy between different modes of oscillation. Such exchange, necessary to return to equilibrium after any perturbation, *requires* anharmonicity.

with q as the wavenumber and na as the position (x) of the nth atom. Equation (5.4) is used as a trial solution and is introduced into Eq. (5.3) for u_n, u_{n+1}, and u_{n-1}. We then obtain the relationship between ω and q,

$$M\omega^2 = -\beta \left[\exp\left(iqa\right) + \exp\left(-iqa\right) - 2\right] = 2\beta\left[1 - \cos\left(qa\right)\right], \quad (5.5)$$

from which the *dispersion equation* is obtained:

$$\omega(q) = \pm 2\sqrt{\frac{\beta}{M}} \left|\sin\left(\frac{qa}{2}\right)\right|. \quad (5.6)$$

The $\pm$ sign refers to waves traveling to the left or to the right. Standing waves, as required for the boundary conditions [Eq. (5.7)] are obtained by superposition of the running waves with opposite propagation and equal amplitude.

The dispersion relation is pictured in Fig. 5.3.* The frequency increases linearly with q for $(qa/2) \ll 1$, as is expected for elastic waves in a homogeneous elastic material. However, $\omega(q)$ then levels off and reaches a saturation value $(\sin\frac{\pi}{2} = 1)$ when $q \to \frac{\pi}{a}$, i.e., when the wavelength of the oscillations $(\lambda = 2\pi/q)$ approaches twice the interatomic distance; the atomic character of the medium becomes apparent here—see below. The maximum frequencies $(\omega_{\max} = 2\sqrt{\beta/M})$ are typically on the order of 10^{13} Hz; the related maximum phonon energy $\hbar\omega_{\max}$ is on the order of 30 meV for typical semiconductors.

5.1.2 Transverse Lattice Oscillation

The same dispersion relation is obtained for transverse waves in a linear chain as long as the amplitude remains small and the force constants can be described by the same β. For these waves, however, the oscillations occur in a plane perpendicular to the x-axis. Since there are two of these planes orthogonal to each other, we distinguish two *transverse polarizations* of these oscillations. Figure 5.4 presents some examples of such *modes* of transverse vibrations in a very short chain.

As a boundary condition we can require that no energy is transferred from or to the outside; this is the adiabatic boundary condition

---

* To point out the similarity to the $E(k)$ diagram for electrons in a periodic potential (see Section 7.2), $\hbar\omega$ is plotted rather than ω as a function of q throughout the book. The limit of the wavenumber between $-\pi/a$ and $+\pi/a$ indicates the boundaries of the first Brillouin zone (Section 3.6.1). Extending the diagram beyond this interval provides no new information.

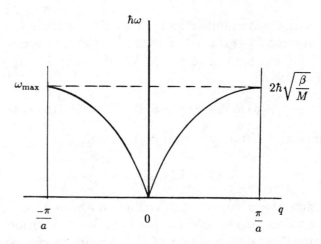

Figure 5.3: Dispersion relation (here for the energy $\hbar\omega$) as given by Eq. (5.6) for a linear chain of the same type of atoms with an interatomic distance a. The boundaries in q_x are those of the first Brillouin zone. The figure also contains a dash-dotted line, which represents the dispersion relation for a continuum.

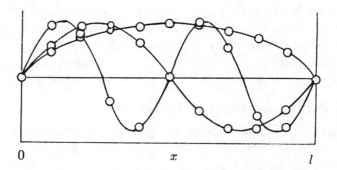

Figure 5.4: Three modes of transverse oscillation in a one-dimensional chain.

and requires nodes at the outer "surfaces" ($x = 0$ and $x = l$). With this condition, q can only attain discrete values q_n, with

$$q = q_n = \frac{2\pi n_q}{l} \quad n_q = 1, 2, \ldots \frac{l}{a} = N, \tag{5.7}$$

where N is the number of atoms in the chain. The upper limit of $n_q = l/a$ is given by the fact that the minimum wavelength is a when fulfilling the boundary condition of nodes at both ends of the chain. Smaller wavelength representations, although possible, describe no new type of atomic oscillation and are disregarded. Therefore, a

chain of N atoms has exactly N different modes for the given polarization of such oscillations.* Or, an arbitrary state of motion of such
chain can be decomposed by the superposition of $2N$ normal modes.
The factor of two stems from the two polarizations of transverse oscillations. Depending on the total energy (i.e., thermal energy) in
the chain, these different modes can be excited to a greater or lesser
degree; however, with oscillations of quantized energy steps

$$E_n = \hbar \omega_n \quad \text{and} \quad \omega_n = \omega(q_n); \tag{5.8}$$

each one of these steps represents a *phonon*. For a single oscillator
a larger excitation causes a larger amplitude at the same eigenfrequency, i.e., a larger energy and therefore a larger number of phonons,
each one of them having the same energy. The occupation of a specific mode is then given by the total energy within this mode divided
by $\hbar \omega_n$:

$$\frac{E_n^{(n)}}{E_n} = n + \frac{1}{2} \tag{5.9}$$

with n phonons residing in this mode. The term $1/2$ stems from the
zero-point energy, i.e., a remaining fraction of energy at $T = 0$ K.

In thermal equilibrium, the many different modes are occupied
with phonons according to the Bose-Einstein distribution function:
the higher the temperature, the more phonons that appear.

With one longitudinal and two transverse branches, we have a
total of $3N$ phonon modes in a linear lattice. This also holds true
for two- or three-dimensional *monatomic* lattices, where N is the
total number of lattice atoms (i.e., there is also a total of $3N$ phonon
modes).

Comparing the relative displacement of atoms in the three different modes shown in Fig. 5.4, we see that the displacement of adjacent
atoms with respect to each other becomes larger with decreased wavelength, i.e., with increased q. Therefore, more energy is contained
in modes with higher ω. This relation between phonon energy and

---

* Strictly speaking, there are only $N - 2$ different modes when requiring
that the surface atoms remain at rest. By bending the (long) linear chain
into a circle, one can obtain an equivalent condition by introducing cyclic
boundary conditions [requiring $u(x = 0) = u(x = l)$] to get around this
"$N - 2$" peculiarity (Born and von Kármán, 1912). For large N, however,
one always has $N - 2 \simeq N$. Such a cyclic boundary condition is also
necessary to permit running waves.

wavelength $\lambda = 2\pi/q$ is given by the dispersion relations [Eqs. (5.6) and (5.14)].

5.1.3 Transverse Oscillation in a Diatomic Lattice

In a lattice with a basis (see Section 3.1) (e.g., a diatomic linear chain with alternating masses), the equation of motion can be split into a set of two oscillatory equations

$$M_1 \frac{d^2 u_{2n}}{dt^2} = \beta(u_{2n+1} + u_{2n-1} - 2u_{2n}) \tag{5.10}$$

and

$$M_2 \frac{d^2 u_{2n+1}}{dt^2} = \beta(u_{2n+2} + u_{2n} - 2u_{2n+1}). \tag{5.11}$$

These require two wave equations for the displacements of light and heavy atoms with indices 1 and 2, respectively, and a' the nearest neighbor distance $= a/2$ with a the lattice constant:

$$u_{2n} = A \exp\left[i(2nqa' - \omega t)\right] \tag{5.12}$$

and

$$u_{2n+1} = B \exp\left[i\left(\{2n + 1\}qa' - \omega t\right)\right] \tag{5.13}$$

with amplitudes A and B. Introducing Eqs. (5.12) and (5.13) into Eqs. (5.10) and (5.11), and solving the ensuing secular equations yields the *dispersion equation* for a diatomic chain

$$\boxed{\omega_{\pm}^2 = \frac{\beta(M_1 + M_2)}{(M_1 M_2)}\left[1 \pm \sqrt{1 - \frac{4M_1 M_2 \sin^2(qa')}{(M_1 + M_2)^2}}\,\right].} \tag{5.14}$$

It has two solutions depending on the sign of the square root. The two corresponding $\hbar\omega(q)$ branches are shown in Fig. 5.5.

The upper branch is referred to as the *optical branch*, the lower as the *acoustic branch*. The amplitudes of the different sublattice oscillations near $q = 0$ for the same wavelength are the same for the acoustic branch

$$\frac{A}{B} \simeq \begin{cases} 1 & \text{for } q \to 0 \\ \infty & \text{for } q \to \dfrac{\pi}{a}. \end{cases} \tag{5.15}$$

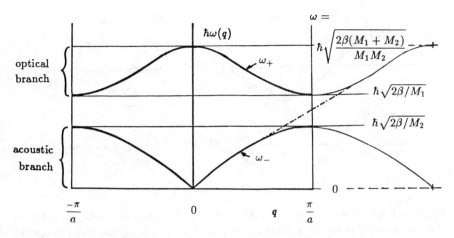

Figure 5.5: Dispersion relations for a diatomic chain of light (M_1) and heavy (M_2) atoms, indicating acoustic and optical branches with a forbidden frequency range in between. The frequencies at $q = 0$ and $q = \pm\pi/a$ are given at the right-hand ordinate. The dash-dotted curve part indicates the one branch when $M_1 = M_2$, and thus consequently increases the Brillouin force to $\pi/(a/2)$.

For $q = \pi/a$, the heavy atom is at rest, as shown in Fig. 5.6B, with amplitude $B = 0$. For the optical branch

$$\frac{A}{B} \simeq \begin{cases} -\dfrac{M_2}{M_1} & \text{for } q \to 0 \\[2mm] 0 & \text{for } q \to \dfrac{\pi}{a}. \end{cases} \tag{5.16}$$

Here, for $q = \pi/a$, the light atom is at rest with amplitude $A = 0$. In the latter case the lattices oscillate opposite to each other while the center of mass remains stationary; therefore, the lighter mass oscillates with larger amplitude, as indicated in Fig. 5.6.

If the lattice binding force is (partially) ionic, the optical branch of oscillation will show a large dipole moment interacting with electromagnetic radiation (photons); hence, the name *optical phonons*. The acoustic branch has a much smaller dipole moment since adjacent atoms oscillate with each other (Fig. 5.6A); therefore, optical stimulation is less effective for acoustic mode excitation. This mode is stimulated preferably by mechanical means such as sound waves.

At short wavelengths ($q \simeq \pi/a$), the heavy atoms are oscillating while the light atoms are at rest in the optical mode; the oscillation energy is large. At the same wavelength in the acoustic mode, only

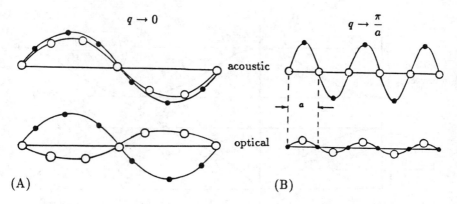

Figure 5.6: Acoustic and optical branches of transverse oscillations with the same wavelength in a diatomic lattice for: (A) long wavelength limits, and for (B) short wavelength limits ($\lambda = 2a$) with heavy and light atoms at rest for acoustic and optical modes, respectively.

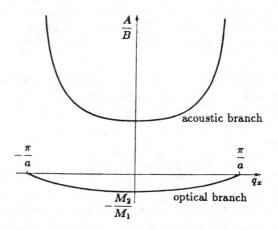

Figure 5.7: Amplitude ratio for light and heavy atoms as a function of the wavenumber (after Weißmantel and Hamann, 1979).

the light atoms are oscillating with larger amplitude but with lesser energy—see Fig. 5.7.

5.1.4 Phonon Spectra in a Three-Dimensional Lattice

In a three-dimensional lattice, the dispersion equation $\omega(\mathbf{q})$ is given by a hypersurface. A cut through such a surface in a specific crystallographic direction is conventionally presented, and is shown for the [100] direction in Fig. 5.8 for three typical semiconductors. One distinguishes optical and acoustic, and longitudinal and transverse

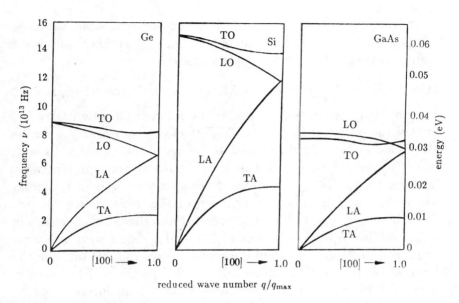

Figure 5.8: Phonon spectrum for Ge, Si, and GaAs in the [100] direction (after Landoldt-Börnstein, 1987).

branches. Monatomic lattices also show optical branches when they have more than one atom per unit cell (Si and Ge). Here we must distinguish between oscillations in which the atoms within the unit cell vibrate about their center of mass (optical), and oscillations in which the center of the unit cell vibrates (acoustic). Some materials undergo a phase transition between diatomic and monatomic unit cells (e.g., $hcp \rightarrow fcc$ transition), whereby the optical branch disappears during the transition. For example, Ca shows such a phase transition at 450°C.

In diatomic lattices there is a gap between optical and acoustic branches (GaAs), while in monatomic lattices there is none (Ge and Si). The phonon energies are higher in lattices with larger binding energies (compare Ge and Si in Fig. 5.8).

Optical phonon branches show a small $\omega(\mathbf{q})$ dependence near $\mathbf{q} = 0$, with a rather well-defined phonon energy $\hbar\omega_0 \simeq \hbar\omega(\mathbf{q} = 0)$. The transverse optical branch near $q = 0$ is important for optical absorption, while the longitudinal branch is more important for scattering with electrons—see Sections 32.2.1E and 32.2.1F for more detail.

Additionally, in anisotropic lattices one has to distinguish more complicated modes, such as involving bond stretching and bond bending between the different lattice atoms.

5.1.5 Phonon Dispersion in Microscopic Force Models

More sophisticated models consider the interacting potential $V(R)$ between the lattice atoms, from which the force constants $\beta_{\alpha i}$ can be determined by forming the spatial derivative of these potentials with respect to the deformation. The force constants are represented by a matrix, with $\alpha = 1, \ldots r$ atoms making up the basis of the point lattice and $i = 1, 2, 3$ for the direction of the three lattice vectors. When these force constants are introduced into the set of equations of motion similar to Eqs. (5.10) and (5.11), and a planar wave *Ansatz* is introduced, the resulting secular equations have as their solution the dispersion equation. The difference from the simple model treated in Section 5.1.3 is a more sophisticated potential for the interacting ions and the consideration of more than next-nearest neighbor interaction.

Depending on the interacting potential, we must distinguish the *rigid ion model*—which includes only ion-ion interaction, i.e., central forces except for nearest neighbor interaction, which may be viewed as noncentral (Rajagopal and Srinivasan, 1960)—and several more refined models, which include the influence of the electron shell (bond angle influence). For a review, see Bilz and Kress (1979).

Using the rigid ion model, Zdetsis (1977, 1979) has obtained excellent results for covalent crystals such as Ge, when incorporating up to the fifth nearest neighbor into the equation of motion.

Shell models, which include valence forces into the interaction potential, are used for covalent semiconductors. Turbino et al. (1972) have used such a potential to achieve good agreement with measured $\omega(q)$ curves for the group IV elements. Another shell model proposed by Jaswal (1975) successfully calculated the dispersion relation for III-V and II-VI compounds (Jaswal, 1977). Best agreement between the model computation and experiment has been obtained for III-V compounds by Weber (1977) with a *bond charge model*, and is shown for Si and Ge in Fig. 5.9. It should be pointed out, however, that all of these calculations use an empirical potential with several (usually 4–6) adjustable parameters.

More recently, with the advances of computation methods, *ab initio* calculations have become possible by tracing back lattice dynamics to electron-electron and electron-ion interaction, which together describe the interatomic forces. An example for the excellent agree-

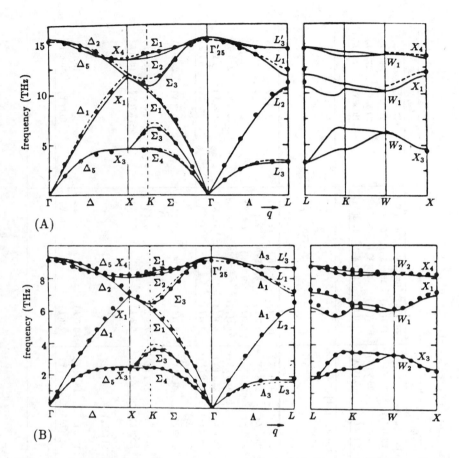

Figure 5.9: Phonon dispersion curves for: (A) Si, and (B) Ge. Solid curves: calculated by Weber (1977); filled circles: experimental values.

ment of deformable bond calculation with the experiment is given in Fig. 5.10 for GaAs (Kunc et al., 1975). (See also Section 8.3.2).

   The resulting branches of the dispersion equation are divided into two sets. In one set with three branches, $\hbar\omega$ goes to zero for vanishing **q**; these are the *acoustic branches*. The other set contains all others, the *optical branches*, as shown in the $\hbar\omega(\mathbf{q})$ diagram. There are three optical branches: one longitudinal and two transverse. However, if the number of atoms in the basis (p) is larger than two, there are $3(p-1)$ optical branches. For instance, in a crystal with 4 atoms per unit cell one has 3 acoustic and 9 optical branches. The latter ones are distinguished as A_1 and B (both nondegenerate), and E_1 and E_2 (both twofold degenerate). Only A_1 and E_1 are dipole active, i.e., they show up in absorption or reflection. A_1, E_1 and E_2 are Raman

Table 5.1: Phonon energies in meV given for the (characteristic point).

Material	$\hbar\omega_{TO}$			$\hbar\omega_{LO}$			$\hbar\omega_{TA}$			$\hbar\omega_{LA}$		
C (diam)	165 (Γ_{25})	150 ($L_{3'}$)	132 (X_4)	165 (Γ_{25})	155 ($L_{2'}$)		70 (L_3)	100 (X_3)		165 (Γ_{25})	147 (X_1)	125 (L_1)
Si	64.2 ($\Gamma_{25'}$)	60.7 ($L_{3'}$)	57.5 (X_4)	64.2 ($\Gamma_{25'}$)	52.1 ($L_{2'}$)		14.1 (L_3)	18.6 (X_3)			51 (X_1)	46.9 ($L_{2'}$)
Ge	37.3 ($\Gamma_{25'}$)	35.4 ($L_{3'}$)	33.8 (X_4)	37.3 ($\Gamma_{25'}$)		29.5 (X_1)	7.7 (L_3)	9.8 (X_3)			29.5 (X_1)	27.4 ($L_{2'}$)
α-Sn	24.8 ($\Gamma_{25'}$)	23.7 ($L_{3'}$)	22.8 (X_4)	24.8 ($\Gamma_{25'}$)	20.2 (L_1)	5.17 (X_3)	4.1 (L_3)	5.2 (X_3)			19.3 (X_1)	17.1 ($L_{2'}$)
BP	98.8 (Γ)			102 (Γ)								
AlN	83.0 (E_1)	81.5 (A_1),[Γ]		111 (E_1),[Γ]	110 (A_1),[Γ]							
AlP	54.3 (Γ)			61.9 (Γ)								
AlAs	44.9 (Γ)			50.1 (Γ)								
AlSb	39.4 (Γ)			42.0 (Γ)								
GaP	45.3 (Γ)	43.8 (L)	44.0 (X)	50.7 (Γ)	50.3 (L)	45.4 (X)	10.3 (L)	13.0 (K)	12.9 (X)		31 (X)	26.7 (L)
GaAs	33.0 (Γ)	22.3 (L)	31.4 (X)	35.2 (Γ)	29.5 (L)	29.8 (X)	7.7 (L)	8.8 (Σ_2)	9.8 (X)		28 (X)	25.8 (L)
GaSb	27.6 (Γ)	26.9 (L)	26.9 (X)	28.7 (Γ)	25.2 (L)	26.0 (X)	5.7 (L)	9.3 (W)	6.9 (X)			19.1 (L)
InP	38.0 (Γ)	39.3 (L_1)	40.1 (X_2)	42.7 (Γ)		41.4 (X_1)	6.8 (L_3)		8.5 (X_0)		24 (X_3)	20.7 (L)
InAs	26.9 (Γ)	26.7 (L)	26.7 (X)	29.5 (Γ)	25.1 (L)	25.1 (X)	5.4 (L)		6.6 (X)		19.8 (X)	17.2 (L)
InSb	22.9 (Γ_{15})	21.9 (L_3)	22.2 (X_5)	24.4 (Γ_{15})	19.9 (L_1)	19.6 (X_1)	4.1 (L_3)		4.6 (X_5)		17.2 (X_3)	15.7 (L_1)
ZnO	46.7 (A_1)	50.6 (E_1)		71.2 (A_1)	72.7 (E_1)							
ZnS(cubic)	34.3 (Γ)	35.9 (L)	39.2 (X)	43.2 (Γ)	41.8 (L)	40.9 (X)	8.7 (L)		11.1 (X)		26.2 (X)	24.2 (L)
ZnS(hex)												
ZnSe	25.4 (Γ)	25.6 (L)	25.6 (X)	31.0 (Γ)	27.8 (L)	27.7 (X)	7.2 (L)		8.7 (X)		23.6 (X)	22.5 (L)
ZnTe	21.9 (Γ)	21.5 (L)	21.5 (X)	25.6 (Γ)	22.3 (L)	22.8 (X)	5.2 (L)		6.7 (X)		17.7 (X)	16.9 (L)
CdO	32.5 (Γ)			65.5 (Γ)								
CdS	30.0 (Γ_{15})	5.3 (Γ_6)		37.9 (Γ_{15})	31.6 (Γ_6)							
CdSe	20.9 (Γ_{15})	4.2 (Γ_6)		25.7 (Γ_{15})								
CdTe	17.4 (Γ)	17.9 (L)	4.3 (X')	21.0 (Γ)	17.9 (L)	18.4 (X)	3.6 (L)					13.4 (L)
HgTe	14.6 (Γ)	15.9 (L)	16.6 (X)	14.9 (Γ)	18.1 (L)	16.9 (X)	2.3 (L)		2.0 (X)		10.0 (X)	10.6 (L)
CuBr	17.4 (Γ,γ)	15.0 (Γ,β)		21.0 (Γ,γ)								
CuI	16.5 (Γ)			18.6 (Γ)								

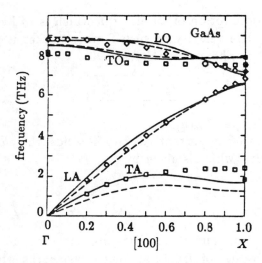

Figure 5.10: *Ab initio* calculation of the phonon spectrum of GaAs. Solid curve: full calculation; dashed curve: from quadrupled cell (after Kunc et al., 1975).

active—see Chapter 17; B is not optically active. A listing of the characteristic phonon energies is given in Table 5.1.

The conventional diagram within the first Brillouin zone shown in Fig. 5.9 is an intersection of the $\omega(\mathbf{q})$ hypersurface with planar surfaces connecting the listed symmetry points (see Section 3.6 and Fig. 3.8), and changing direction at each vertical line. A similar representation is conventionally used in the $E(\mathbf{k})$ diagram for the dispersion relation of electrons in the semiconductor—see Section 8.5.1.

5.1.5A Pressure Dependence of Phonon Spectrum With hydrostatic pressure the interatomic distance is reduced, resulting in an increase of ion-ion interaction and a reduction of electronic screening. This causes an increase of the optical phonon energy at the Γ-point of $E(q)$ with increasing pressure.

However, the pressure coefficient for TA modes at the Brillouin zone boundary (at X and L) is negative, which seems to be an intrinsic property of shear distortion in predominately covalent semiconductors that are involved in such modes (Martinez, 1980). The corresponding pressure coefficients are listed in Table 5.2.

5.1.5B Phonon Velocity A phonon $\hbar\omega$ has been introduced as related to lattice oscillation after localization with a certain frequency ω, i.e., a certain mode of a collective state of oscillations A phonon may also be seen as a quasi-particle when described as a wave packet, similar to the description of an electron which is to be discussed in Section 7.2.3. The velocity of such a wave packet is given by the *group velocity** or *phonon velocity*:

$$\mathbf{v}_g = \nabla_q \omega(\mathbf{q});$$ (5.20)

that is, by the slope of $\omega(q)$.

There is another important velocity in solids, the *phase velocity,†* or *sound velocity*, which is given by

$$v_s = \frac{\omega}{q}.$$ (5.22)

Both group and phase velocities are the same in the acoustic branch at low values of q, i.e., at long wavelengths where $\omega \propto q$— see Figs. 5.5, 5.6, and 5.9 near the Γ-point; however, they become substantially different where dispersion occurs.

With decreasing wavelength (increasing q), the phase velocity decreases but remains on the same order of magnitude as the sound

---

* The group velocity can be defined when at least two waves of slightly different frequencies interact and form a wave train with an envelope forming beats (Fig. 5.11). Since energy cannot flow past a node, one readily sees that the velocity with which energy is transmitted must equal the velocity with which the nodes move. Adding two waves with ω, q and $\omega + d\omega, q + dq$, one obtains for the superposition

$$u_1 + u_2 = (A_1 + A_2)\cos(\omega t - qx)\cos\left(\frac{t}{2}\,d\omega - \frac{x}{2}\,dq\right).$$ (5.17)

Thus the motion of the zero-phase point of the envelope

$$\frac{t}{2}\,d\omega - \frac{x}{2}\,dq = 0$$ (5.18)

yields for the velocity of the groups of waves

$$v_g = \frac{x}{t} = \frac{d\omega}{dq}$$ (5.19)

or, more generally, Eq. (5.20).

† The velocity in which the phase of a single wave moves; for a node, it is given by

$$\omega t - qx = 0,$$ (5.21)

resulting in the phase velocity given by Eq. (5.22); this is the velocity with which energy is transported in such a wave.

Table 5.2: Pressure coefficient of phonon modes at room temperatures in meV/kbar (after Martinez, 1980).

Material	$d\omega_{TO}/dp$	$d\omega_{LO}/dp$	$d\omega_{TA}(X)/dp$	$d\omega_{TA}(L)/dp$
C	0.445	0.09 ± 2.1		
Si	0.064	−0.027 ± 0.004	−0.019 ± 0.005	
Ge	0.057		−0.004 ± 0.003	
AlSb	0.083	0.085		−0.048
GaP	0.054	0.056	−0.011 ± 0.0005	−0.0093 ± 0.0008
GaAs	0.053	0.054		
GaSb	0.062	0.062		
ZnO	0.062		−0.012	
ZnS	0.077	0.053	−0.021 ± 0.002	−0.18
ZnSe	0.063	0.052	−0.019	−0.17
ZnTe	0.070	0.056	−0.015	−0.01 ± 0.002
CdS		0.056	−0.025	
CuCl(40K)	0.096	0.059		
CuBr(40K)	0.087	0.088		
CuI (40K)	0.089	0.073		

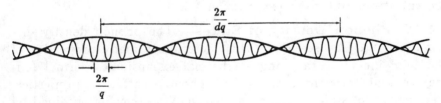

Figure 5.11: Wave train caused by superposition of two waves with slightly different frequencies, resulting in beats of the envelope function.

velocity near the Γ-point. At the surface of the Brillouin zone (e.g., at the X-point of a cubic semiconductor), the sound velocity is decreased slightly to $v_s(q = \frac{\pi}{a}) = \omega(q = \frac{\pi}{a})/(\pi/a)$. An estimate for the sound velocity is obtained by extrapolating the acoustic branch $\omega = v_s q$ to $q = \pi/a$ with $\hbar\omega(q = \pi/a) \simeq k\Theta$, yielding

$$v_s = \frac{k}{\hbar}\frac{a\Theta}{\pi} = 416.8\, \Theta(K)\, a(\text{Å}) \text{ (cm/s)}; \qquad (5.23)$$

with the Debye temperature Θ (see Section 6.1.2) on the order of 300 K, the average phonon velocity is on the order of 10^5 cm/s.

In contrast, the group velocity, given by the slope of $\omega(\mathbf{q})$, vanishes at $q = \pi/a$; at the surface of the Brillouin zone the waves have become standing waves, composed of two components moving in opposite directions with the same amplitude and velocity. This is the condition of Bragg reflection and will be discussed later in more detail for a similar problem dealing with electrons (Section 7.2.3).

In summary: in the diatomic, one-dimensional chain we have for the acoustic branch

$$v_g = v_s = a\sqrt{\frac{\beta}{2(M_1 + M_2)}} \quad \text{for } q \ll \frac{\pi}{a} \tag{5.24}$$

and

$$\left.\begin{array}{l} v_g = 0 \\[1.2em] v_s = a\sqrt{\dfrac{2\beta}{\pi^2 M_2}} \end{array}\right\} \quad \text{for } q \simeq \frac{\pi}{a}; \tag{5.25}$$

for the optical branch

$$\left.\begin{array}{l} v_g = 0 \\ v_s : \text{meaningless} \end{array}\right\} \quad \text{for } q \ll \frac{\pi}{a} \quad \text{and} \quad \text{for } q \simeq \frac{\pi}{a}. \tag{5.26}$$

In a three-dimensional lattice, similar results are obtained except for the anisotropic velocities (see Section 4.2.1).

5.1.5C Phonon Density of States   The phonon density distribution $N(\omega)d\omega$ [see Eq. (6.13)] determines many properties of the solid including mechanical (thermal expansion), thermal (energy content), electrical, and optical (phonon scattering) properties. The density of states $g(\omega)$, contained in $N(\omega)$, can be obtained by counting the number of *modes* occurring within a frequency interval $(\omega, \omega + d\omega)$.

It is instructive to derive the density of states analytically for the linear chain of atoms. Here the dispersion relation [Eq. (5.6)] has discrete normal modes for $q = q_n$ given in Eq. (5.7):

$$\omega(q_n) = \omega_0 \sin\left(\frac{q_n a}{2}\right) = \omega_0 \sin\left(\frac{\pi n_q a}{l}\right). \tag{5.27}$$

The density of states is given by the modes per frequency interval, which can be obtained by differentiation of Eq. (5.27), replacing $\cos(\pi n_q/l)$, and using $\sin^2 + \cos^2 = 1$:

$$\frac{d\omega}{dn_q} = \frac{\pi a}{l}\omega_0 \cos\left(\frac{\pi n_q a}{l}\right) = \frac{\pi a}{l}\omega_0 \sqrt{1 - \left(\frac{\omega}{\omega_0}\right)^2}. \tag{5.28}$$

Consequently, one has for the density of states of the linear chain the reciprocal of Eq. (5.28):

$$dn_q = g(\omega)\,d\omega = \frac{l}{\pi a \omega_0}\,\frac{d\omega}{\sqrt{1 - \left(\dfrac{\omega}{\omega_0}\right)^2}}. \tag{5.29}$$

In a three-dimensional lattice, and in more general terms, this density of states can be expressed by

$$g(\omega)d\omega = \mathcal{V} \int_{S_\omega} \frac{dS_\omega}{|\nabla_q \omega|}\,d\omega \tag{5.30}$$

where $\mathcal{V}$ is the volume, S_ω is the surface of constant energy,* dS_ω is the surface element of S_ω, and $\nabla_q \omega$ is the gradient of ω in **q**-space. In the nondispersive range, where $\omega \propto q$ for the acoustic branch [see Eq. (6.9)], we can show that

$$g(\omega_{\mathrm{ac}}) \propto \omega^2. \tag{5.31}$$

For higher values of q, and in optical branches, peaks occur in the density distribution when this gradient tends to zero (see also Section 6.1.2 and Fig. 6.2). These are the *critical points* and are identified by capital letters. The different branches are numbered in the $g(\omega)$ diagram starting from the acoustic branch. Most important are critical points in the center, or at the boundary of, the Brillouin zone (see Fig. 5.9): at Γ (for optical branches) and at X, K, and L (for identification see Fig. 3.8). The corresponding density of state function has spikes as shown in Fig. 5.12: many more modes per $d\omega$ interval occur, where the branch $\omega(q)$ is flat rather than in a range where ω varies steeply with q, as, for example, in the acoustic branch at the Γ-point.

Gilat and Raubenheimer (1966) obtained $g(\omega)$ by covering the first Brillouin zone with an evenly spaced mesh of points and determining the frequency gradients at each of these points. After sampling a large number of such frequencies, a smooth density distribution can be assembled (*root sampling technique*).

---

* This surface is similar to the Fermi surface in the Brillouin zone discussed in Section 9.1.4A. For phonons, however, the energy surface lies at much lower energies.

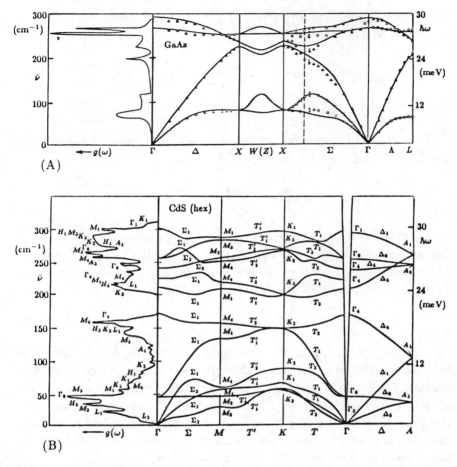

(A)

(B)

Figure 5.12: (A) Phonon dispersion curves of GaAs as measured from neutron diffraction and one phonon density of states as a function of frequency calculated from the rigid ion model (after Patel et al., 1984). (B) Same for CdS (after Nusimovici et al., 1970).

5.1.6 Local Phonon Modes

In the neighborhood of lattice defects (see Chapter 18), the binding forces and the mass for extrinsic defects are altered; thus the oscillatory behavior is locally modified. The eigenfrequency of such a defect (subscript d) is different from the most abundant lattice frequencies. In the simple isotropic case it is given by

$$\omega_{0,d} = 2\sqrt{\frac{\beta_d}{M_d}}. \tag{5.32}$$

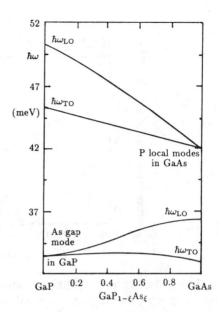

Figure 5.13: Different optical branches for $GaAs_\xi P_{1-\xi}$ mixed crystals as a function of the alloying parameter ξ. Local modes are identified for extreme values of ξ.

When the density of a specific defect is large enough and its eigenfrequency lies outside the allowed ranges of the intrinsic phonon spectrum,[*] a *local mode* can be experimentally distinguished (e.g., its IR absorption can be observed), which provides valuable information about the defect center. More detail is given in Section 11.4.1.

Often, the lattice oscillation surrounding a defect is described by radially expanding and contracting oscillations, i.e., by a *breathing mode*. For a review, see Mitra and Massa (1982). For local modes in highly disordered semiconductors see also Section 6.3.4.

5.1.6A Local Modes for Mixed Crystals When mixed crystals have a substantial difference in atom mass of the mixing elements, then separate branches for these elements appear, as in $GaAs_\xi P_{1-\xi}$ shown in Fig. 5.13. For $\xi \simeq 0$, the As mode is seen as a gap mode; for $\xi \simeq 1$, the P mode is seen as a local mode in GaAs.

5.1.6B Local Modes with Isotope Splitting High-resolution spectroscopy differentiates isotopes adjacent to an impurity. For

---

[*] For instance, if the mass of the defect atom is much smaller, the eigenfrequency can be higher than the highest lattice phonon frequency (*local mode*): for heavier atoms, its optical phonon may lie within the gap between the optical and acoustic branches (*gap mode*). In a *resonant mode*, the eigenfrequency of the foreign atom lies within the band of a phonon branch of the host lattice. This mode shows a greatly enhanced amplitude.

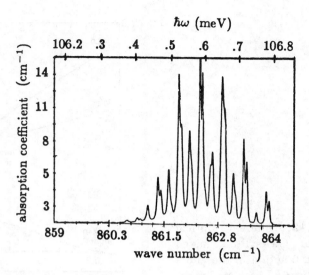

Figure 5.14: Infrared high-resolution local mode absorption of interstitial oxygen in Ge at 6 K; ν_3-band with 3.7 μeV (0.03 cm^{-1}) resolution (after Pajot and Clauws, 1986).

instance, an interstitial oxygen (mass 16) in a germanium lattice has two nearest neighbors (resulting in a defect of Ge_2O type) that determine its oscillatory modes. With five isotopes,* there are 33 combinations of $^MGe\,O^m\,Ge$ defects with $M + m$ different for each one. Each of these defects should cause an absorption made up of three lines spaced by 0.229 and 0.067 cm^{-1}. Not all of these features, however, have been observed. Figure 5.14 shows a typical optical absorption spectrum in which the $^{70}Ge_2O$, $^{70}GeO\,^{72}Ge$, $^{74}GeO\,^{76}Ge$ and $^{76}GeO_2$ components can be identified (Pajot and Clauws, 1986).

5.2 Phonons in Superlattices

A remarkable result of the periodic structure of a superlattice of alternating layers A and B with a superlattice constant l (= thickness of layers A + B) is its reduced dispersion relation with a boundary at $\pm\pi/l$ for the first superlattice Brillouin zone (*mini-zone*—see Section 9.3) rather than at $\pm\pi/a$ for the bulk lattice. Consequently, the phonon spectrum is *folded* at the mini-zone boundary rather than continued to the main Brillouin zone boundary (Fig. 5.15). Since the alternating materials in a superlattice have different atomic masses

---

* These isotopes and their relative abundance are ^{70}Ge (20.52%), ^{72}Ge (27.43%), ^{73}Ge (7.76%), ^{74}Ge (36.54%), and ^{76}Ge (7.76%).

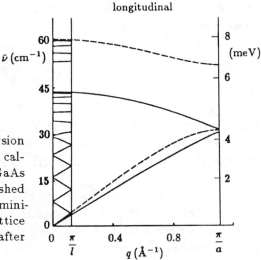

Figure 5.15: Phonon dispersion curves for longitudinal mode, calculated for a linear chain in GaAs (solid curve) and for AlAs (dashed curve), and shown folded for a mini-zone for a symmetrical superlattice with a periodicity of $l = 6a$ (after Colvard et al., 1985).

and force constants, there is, as in an AB-compound, a gap between each of the branches of the folded dispersion curve.

In comparison with the dispersion curve of the host lattice, several folded branches evolve from the acoustic branch; however, only the first folded branch is referred to as an acoustic branch of the superlattice. Several other branches are folded from the optical branch. The phonon dispersion spectrum is modified to a large degree from the bulk phonon behavior, as is expected considering the coupling of atom groups and the reflection at the boundaries between the superlattice layers.

This multibranch structure can be directly observed from phonon scattering experiments (see Section 17.1.4, Fig. 17.11, and Colvard, 1987). For a review, see Klein (1986).

5.3 Surface Phonons

Atoms at and near the surface are bound with lesser strength because of missing external neighbors. Consequently, there is a modification of the phonon spectrum near the surface that can be excited through surface waves (see below). These surface phonons have a lower frequency caused by a lower force constant than the optical phonons in the crystal bulk, and can be observed in the forbidden frequency range (Ibach, 1971).

In the acoustic branch, elastic surface waves propagate along the surface within a thin layer that is the thickness of the wavelength

(*Rayleigh waves*). The frequency of this surface wave can be derived from Maxwell's equation (see Madelung, 1981) resulting in

$$\omega_s = \omega_{TO} \sqrt{\frac{\varepsilon_{st} + 1}{\varepsilon_{opt} + 1}} \tag{5.33}$$

where ε_{st} is the static and ε_{opt} is the optical dielectric constant (see Chapter 14). This frequency lies between the longitudinal and transverse branches [compare with Eq. (11.38)] and can be observed in crystal plates of sufficient width (for silicon, see Ludwig, 1974; Krebes, 1977).

Such surface modes are of special interest in many devices in which phenomena close to the surface or to interfaces are important (e.g., in Si near a Si-SiO$_x$ boundary, or near heterojunctions). In small crystallites, powder, or a thin layer, the phonon spectrum is substantially modified and becomes increasingly determined by surface phonons as the crystallite diameter becomes smaller (see Ruppin and Englman, 1970; Otto, 1974).

5.4 Phonons in Amorphous Semiconductors

The vibrational spectrum of amorphous semiconductors is an important means for obtaining information about their structure. However, the description used previously in an $\hbar\omega(q)$ representation is no longer appropriate because of nonperiodicity: q is not a good quantum number for phonons in a glass. Long wavelength acoustic phonons, on the other hand, behave much like phonons in crystals.

Instead, *local mode phonons* are used for exploring the atomic structure. From the phonon spectrum, we obtain valuable information about the specific bonding character, topology, and local symmetry, thus describing the phonon spectrum as a *local* (vibrational) *density of states* (LDOS).

The different vibrational modes can be calculated using a molecular cluster model (see Section 48.2 and Lucovsky et al., 1983), wherein the frequencies of specific modes of atomic vibrations are calculated. For instance, in an As$_2$O$_3$ glass the arsenic atom sits at the top of a triangular pyramid and can vibrate in a mode stretching the pyramid along its symmetry axis or bending it orthogonally to the axis. The oxygen atom can vibrate in three independent modes: stretching, rocking, and bending its bonds (see Fig. 5.16). These vibrational modes are represented by a broad distribution of features in the local density of states and are given for As and O separately in

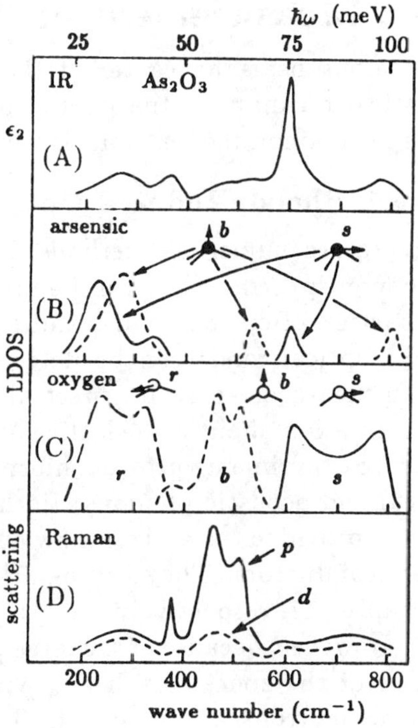

Figure 5.16: Calculated and measured vibrational spectra of α-As_2O_3. (A) Measured IR response; (B) calculated As modes; (C) calculated O modes with r, b, and s for rocking, bending, and stretching, respectively; (D) measured Raman spectrum, polarized and depolarized for p and d, respectively (after Lucovsky, 1985).

the middle two segments of Fig. 5.16, as calculated by Lucovsky et al. (1983). From comparison with the experiment (see Section 17.1.5), one observes the strongest contribution from stretching motions of both As and O atoms. In contrast, the Raman response (see Section 17.1.3) is dominated by bending modes of As and O atoms. In a more detailed analysis one can deduce that the specific glass structure is ring-like, with at least 25% of the atoms located in such rings (Lucovsky, 1985).

Local modes in tetrahedrally bound amorphous semiconductors can be described in a similar fashion by analyzing specific vibrational modes of an intermediate order molecular cluster (Section 3.9.4). We must distinguish alloy atoms, e.g., H in an α-Si host, or foreign atoms (impurities) with a lower density so that impurity-impurity interaction can be neglected in the latter material.

In many respects these local modes are similar to those in a crystalline host; however, the LDOS features in a glass are broader since the local environment is deformed to a larger extent and variety.

5.5 Measurement of Phonon Spectra

Phonons near the center of the Brillouin zone can be excited by optical techniques: the photon momentum is too small to shift the larger phonon momentum significantly from the center.

5.5.1 Phonon-Induced Optical Properties

There are numerous methods for obtaining information about the phonon spectra. They can be divided into methods which reveal the structure of optical modes and those that deal with acoustic modes. The optical mode can be detected by optical absorption or reflection measurements. Its spectral distribution yields the most direct information about these lattice oscillations. In addition, optical scattering experiments provide information about the most abundant optical and acoustic phonons. Both techniques are well developed and have provided extensive knowledge about the phonon spectrum in semiconductors. They will be discussed in detail in Section 11.2 and Chapter 17, respectively.

However, because of the large phonon momentum near the boundaries of the zone, only heavy particles can interact sufficiently. As such, neutrons are almost ideal probes for investigating the entire phonon spectrum, except for compounds in which thermal neutrons react strongly with the nucleus of one of the elements, e.g., ^{10}B or ^{113}Cd.

5.5.2 Inelastic Scattering with Neutrons

Slow neutrons (mass M_n) have an energy $(3/2)kT$ *and* a momentum (wavenumber)

$$q = \frac{M_n v}{\hbar} = \frac{\sqrt{3 M_n k T}}{\hbar} = 2.5 \cdot \sqrt{T} \cdot 10^7 \ (\text{cm}^{-1}), \qquad (5.34)$$

which, at low temperatures, are comparable to the total extent of the Brillouin zone. In scattering with phonons while conserving total energy and momentum, such neutrons are therefore able to probe the entire phonon spectrum. *Slow neutrons* can be obtained by filtering neutrons from a nuclear reactor through a beryllium plate at 4.2 K. *Monoenergetic neutrons* are obtained by diffraction from a single crystal. A collimated beam of these monochromatic neutrons is scattered by phonons within a semiconductor. From the angular distribution of the energy (i.e., the incident neutron energy plus phonon energy), we then obtain a complete $\omega(q)$ dispersion curve of all active branches of the phonon spectrum (see reviews by Dolling, 1974; Bührer and Iqbal, 1984). High-energy branches are inactive at temperatures much below the Debye temperature of the semiconductor.

Summary and Emphasis

The phonon spectrum can be divided into an intrinsic spectrum determined by the bulk of each individual semiconductor and additional features of the spectrum due to surfaces, impurities, and other defects. Such a spectrum can be determined by optical absorption or by scattering experiments, and can be calculated by a variety of approximations, most recently by *ab initio* methods from basic principles. Agreement between theory and experiment is good and yields important information about the detail of atomic oscillation. From the dispersion relation we can obtain directly the density of states of the different modes of oscillation, the knowledge of which is essential for identification of the energy density in the most important phonon branches.

Phonons are one of the most important types of quasi-particles in semiconductors. They are responsible for all thermal properties. They cause thermalization of a wide variety of events when interacting with other particles, such as electrons and holes. They supply the necessary damping (i.e., dissipation of energy into the thermal reservoir), thereby providing for optical absorption, electric energy dissipation, and solidification of liquids, to mention just a few. The phonon spectrum is part of the fingerprint of the specific semiconductor.

Exercise Problems

1.(e) Describe in an atomic picture the difference between longitudinal optical and acoustic oscillations as it is done in the text for the transverse oscillations (Fig. 5.6).

2.(e) In a linear atomic chain of 8 atoms with mass M, separated by 4 Å from each other:

(a) How many modes of oscillation exist?

(b) List the wavelength for each mode.

(c) What changes occur when each alternating atom has the mass $M/2$?

(d) If the energy gap starts at 10 meV, how far does it extend?

(e) How large is the force constant if the atomic chain is formed with alternating atoms of mass $M_1 = 10^{-22}\,g$ and $M_2 = 0.5\,M_1$, given the information in part (d)?

3.(1) How do neutrons interact with phonons? How is a typical neutron refraction experiment performed? Identify and check the appropriate literature.

4. What is the difference between group and phase velocities? When are they about equal to each other?

(a) Estimate the sound velocities for NaCl, CdS, and GaAs from listed values of c_{ik}. Compare these with the tabulated sound velocities. Comment.

(b) Estimate the sound velocity assuming Coulomb attraction and effective charges.

(c) Compare and discuss results from (a) and (b).

5.(e) Discuss the reasons why Bragg reflections occur at the boundaries of Brillouin zones.

6. Develop the dispersion relation for a linear chain of atoms with interatomic forces extending beyond nearest neighbors. Show that the dispersion equation can now be expressed as

$$\omega = \frac{2}{\sqrt{M}}\left\{\sqrt{\beta_1}\,\sin\left(\frac{qa}{2}\right)+\sqrt{\beta_2}\,\sin(qa)+\sqrt{\beta_3}\,\sin\left(\frac{3}{2}qa\right)+\ldots\right\}$$

where β_i is the spring constant connecting the ith neighbor.

7.(e) Plot group and phase velocities of phonons as a function of q for both optical and acoustic branches of a diatomic linear lattice.

8.(e) Derive the dispersion equation [Eq. (5.14)].

9.(e) The dispersion equation for a monatomic linear chain is given by $\omega_0|\sin(qa/2)|$ in Eq. (5.6).

(a) What is the meaning of ω_0?

(b) Show that the density of phonon states is given by

$$g(\omega) = \frac{2}{\pi a\sqrt{\omega_0^2 - \omega^2}}$$

(c) What is the meaning of the singularity at $\omega = \omega_0$?

(d) What is the name of this singularity?

10. Show that in a monatomic cubic crystal with lattice constant a, the density of phonon states has the form

$$g(\omega) = \frac{\omega^2}{2\pi^2}\left(\frac{M}{\beta_l a^2}\right)^{\frac{3}{2}} + \frac{\omega^2}{\pi^2}\left(\frac{M}{\beta_t a^2}\right)^{\frac{3}{2}}$$

where β_l and β_t are the spring constants for longitudinal and transverse deformation.

11.(*) Discuss Bragg reflections of phonons at the boundary of the Brillouin zone.

12.(1) The occupation of a single mode (n) is given by $E_n^{(n)}/E_n = n + \frac{1}{2}$ [Eq. (5.9)] with n phonons residing in this mode.

    (a) Show that $n = f_{BE}$ and discuss its implication for temperatures $kT > E_n^{(n)}$

    (b) The contribution $\frac{1}{2}$ in Eq. (5.9) stems from zero-point energy. Research this contribution and explain its implication.

    (c) What does zero-point energy mean?

13.   How many phonons $\hbar\omega_0$ (where ω_0 is the eigenfrequency of a vibrating oxygen molecule) are required to produce a vibrational amplitude of 10% of its equilibrium interatomic distance?

14.(*) Phonons in an amorphous semiconductor are very helpful for a better understanding of its structure.

    (a) Explain the different phonon modes in a specific amorphous semiconductor.

    (b) How do these modes compare to the phonon spectrum in a crystal?

    (c) How do they compare to localized mode of a crystal defect?

15.(r) Explain in your own words the different photomechanical effects.

Chapter 6

Phonon-Induced Thermal Properties

Phonons are responsible for all thermal properties of a solid, such as its heat content and transport. The anharmonic part of lattice oscillations causes thermal expansion. All these are integral contributions of the phonon spectrum; only at low temperatures, where part of the spectrum can be frozen-out, do they become partially spectrum-selective.

The three macroscopic thermal properties caused by phonons are:

- heat capacity,
- thermal expansion, and
- thermal conductivity (thermal diffusivity),

neglecting electronic contributions here.

Since these are easily observable effects, they were studied early, and rather general, often semi-empirical descriptions were given. These effects are of interest with regard to the interaction of electrons with phonons and the dissipation of energy, e.g., from laser excitation. The first set of processes is discussed in Chapters 32 and 33. The latter processes are kinetic in nature and are discussed in Chapters 48 and 49. In this chapter we will provide the essential basic information.

6.1 Heat Capacity

The heat capacity is a measure of the content of thermal energy, which is represented by all active lattice oscillations. The specific heat, described for constant volume,* is defined as

$$C_v = \left(\frac{\partial U}{\partial T}\right)_\mathcal{V} \tag{6.2}$$

where U is the total energy of all phonons in the solid.

In a monatomic semiconductor, each lattice atom is represented by a harmonic oscillator with an average energy of $kT/2$ per degree of freedom. Thus, with three degrees of freedom per atom, times two (for the kinetic and potential energies of the oscillator in a solid), we have $U = 3N_{\mathrm{Av}}kT$ for a solid with N_{Av} atoms per mole. From Eq. (6.2) we obtain for the specific heat the *Dulong-Petit law* (1819):

$$C_v = 3N_{\mathrm{Av}}k = 3R, \tag{6.3}$$

where R is the gas constant. This equation is approximately fulfilled at sufficiently high temperatures when all modes of oscillation are excited. The specific heat is given here in [Ws/(mol K)] and should be distinguished from the values often given in tables in units of [Ws/(cm^{3}K)] or [Ws/(gK)].

6.1.1 Einstein Model

The Einstein model assumes one mode of oscillation with one corresponding eigenfrequency only. At high enough temperatures, most of the oscillatory energy is indeed present in one kind of phonon, the transverse optical phonons, which have nearly the same frequency in the entire Brillouin zone. Therefore, there is some justification for identifying the oscillatory lattice energy with $U_0 = 3N\hbar\omega_0$ for a one-atomic semiconductor with three oscillators per atom (in the three lattice coordinates). In general, the Dulong-Petit law is given by $C_v = fR/2$, with f as degrees of freedom.

At lower temperatures these phonons *freeze out*, i.e., they can no longer be thermally excited. Phonons are *bosons*, i.e., particles with integer spin, and follow Bose-Einstein statistics. The occupation of

---

* Although the specific heat is measured more easily for constant pressure C_p, the difference between C_v and C_p is very small for solids and is given by

$$C_p - C_v = \frac{9\alpha^2 T\mathcal{V}_m}{\kappa} \tag{6.1}$$

where α is the thermal expansion coefficient, κ is the isothermal compressibility, and $\mathcal{V}_m$ is the molar volume.

the states is given by multiplication of U_0 with the Bose-Einstein distribution $f_{BE} = [\exp(\hbar\omega_0/\{kT\}) - 1]^{-1}$, yielding

$$U = U_0 f_{BE} = \frac{3N_{Av}\hbar\omega_0}{\exp\left(\dfrac{\hbar\omega_0}{kT}\right) - 1} \tag{6.4}$$

and therefore from Eqs. (6.2) and (6.3)

$$C_v = 3R\left(\frac{\hbar\omega_0}{kT}\right)^2 \frac{\exp\left(\dfrac{\hbar\omega_0}{kT}\right)}{\left\{\exp\left(\dfrac{\hbar\omega_0}{kT}\right) - 1\right\}^2}, \tag{6.5}$$

which represents the Einstein model of the specific heat (Einstein, 1907). Equation (6.5) shows the observed decrease of C_v with decreasing temperature as part of a quantum-mechanical phenomenon, the successive freeze-out of phonons, rather than a continuous decrease of oscillatory amplitudes.

6.1.2 Debye Model

Debye (1912) approached the problem from a different point of view, having recognized that various acoustic oscillatory modes are dominant at low temperature. With a distribution function $g(\omega)$ of these modes, the phonons are distributed over these modes according to the Bose-Einstein statistics. When neglecting optical modes, Debye obtained for the thermal energy content

$$U = \int_0^{\omega_D} \hbar\omega\, g(\omega) f_{BE}(\hbar\omega)\, d\omega = \int_0^{\omega_D} \frac{\hbar\omega\, g(\omega)\, d\omega}{\exp\left(\dfrac{\hbar\omega}{kT}\right) - 1}, \tag{6.6}$$

which requires an upper limit of integration ω_D, given by the fact that the total number of modes cannot exceed $3N$:

$$3N = \int_0^{\omega_D} g(\omega)\, d\omega. \tag{6.7}$$

Permitting all standing wave-type oscillations (or all running waves fulfilling cyclic boundary conditions), we obtain the distribution function by mode counting (see Section 5.1.5C), similar to that done for electron waves (see Section 27.2.1):

$$g(\omega)\, d\omega = \frac{V}{8\pi^3} 4\pi q^2\, dq. \tag{6.8}$$

With the group velocity $v_g = \partial\omega/\partial q$, and assuming a *linear dispersion relation* ($\omega = v_g q$) for a macroscopic continuous solid, we obtain

$$g(\omega)d\omega = \frac{\mathcal{V}}{2\pi^2}\frac{\omega^2}{v_g^3}d\omega = 9N\frac{\omega^2}{\omega_D^3}d\omega; \qquad (6.9)$$

the limiting *Debye frequency* ω_D can be replaced by the so-called *Debye temperature* Θ, according to

$$\hbar\omega_D = v_s\sqrt{\frac{6\pi^2\hbar^2 N}{\mathcal{V}}} = k\Theta. \qquad (6.10)$$

This yields the well-known Debye expression for the specific heat [from Eq. (6.2) using Eqs. (6.6), (6.9), and (6.10)]:

$$C_v = 9R\left(\frac{T}{\Theta}\right)^3\int_0^{\Theta/T}\frac{x^4\exp x}{(\exp x - 1)^2}dx \qquad (6.11)$$

which can be approximated by expansion of the exponent (see Joos, 1945 for detail) as

$$C_v = \begin{cases} 233.78R\left(\dfrac{T}{\Theta}\right)^3 & \text{for } T \ll \Theta \\[2ex] 3R\left[1 - 0.05\left(\dfrac{\Theta}{T}\right)^2\right] & \text{for } T \gtrsim 0.5\Theta. \end{cases} \qquad (6.12)$$

A better approximation includes, in addition to the cubic term at low T, a linear component if free electron contributions are significant and a quadratic term for layer-like lattices (Jandl et al., 1976).

A relatively good fit with the experiment can be obtained for some monatomic crystals by choosing the appropriate Θ—see Fig. 6.1. The Debye temperatures for a variety of solids are listed in Table 6.1. For other semiconductors (e.g., diatomic), the agreement is less satisfactory. We then evaluate the measured $C_v(T)$ with Eq. (6.11) but permit a *temperature-dependent* Debye temperature. The deviation from $\Theta = $ const. is a measure of the departure of the actual phonon spectrum from the assumed distribution—see Fig. 6.3.

6.1.3 Validity Range of the Approximations

In summary, we can describe the specific heat with Dulong-Petit's law reasonably well at temperatures well above the Debye temperature. With decreasing temperature, the freezing-out of the dominant

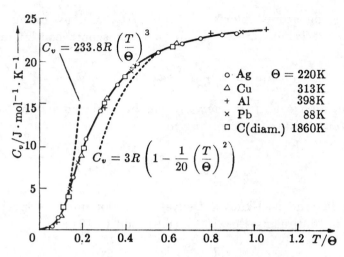

Figure 6.1: Specific heat of a few monatomic crystals as a function of the normalized temperature and the two branches of the approximation [Eq. (6.12)] (after Weißmantel and Hamann, 1979).

mode makes the major contribution. Here the Einstein approximation is a useful one, while at very low temperatures, contributions of the acoustic phonon branch dominate and the Debye approximation fits the experimental results remarkably well. However, the rapid change of the Debye temperature with the actual temperature, even at low temperatures, signals the need for a more detailed analysis which will be discussed in the following section.

6.1.4 Lattice Theory of the Specific Heat

There are several branches of the phonon spectrum. In the simplest case, following Eq. (6.9) these can be taken into account for the acoustic phonons by setting $1/v_g^3 = \frac{1}{3}(1/v_l^3 + 2/v_t^3)$, while still assuming a parabolic distribution. However, the actual phonon distribution function is much more complex.

Figure 6.2 shows the difference between the simple quadratic distribution function and one that is obtained from neutron scattering—see Section 5.5.2. Introducing this experimentally obtained distribution function into Eq. (6.6) and calculating with it, the specific heat yields a temperature-dependent Debye temperature. A significant deviation from $\Theta = const$ usually appears at low temperatures. The temperature dependence of the Debye temperature of Si is shown as an example in Fig. 6.3.

Table 6.1: Debye temperatures.

Material	Θ	Material	Θ	Material	Θ
C_{Dia}	2219	RbCl	165	PbTe	161
C_{Gra}	391	RbBr	131	CdS	215
Si	658	RbJ	103	CdSe	180
Ge	366	CsF	–	CdTe	162
Sn_{sc}	212	CsCl	168	TiC	923
Sn_{met}	202	CsBr	151	AlSb	263
Se	152	CsJ	127	GaAs	345
Te	143	NH_4Cl	266	GaSb	269
LiF	730	NH_4Br	223	InAs	248
LiCl	422	AgCl	146	InSb	200
LiBr	245	AgBr	150	SiO_2	470
LiJ	–	TlCl	125	TiO_2	760
NaF	491	TlBr	115	CaF_2	510
NaCl	321	MgO	946	$NaClO_3$	234
NaBr	225	ZnO	415	$NaBrO_3$	247
NaJ	164	ZnS	315	Cu_3Au	284
KF	336	ZnSe	273	Mg_3Cd	290
KCl	233	ZnTe	220	Bi_2Te_3	155
KBr	174	ZnCu	299	Fe_2O_3	660
KJ	132	PbS	228		
RbF	228	PbSe	–		

The density of state distribution function can be obtained by mode counting (Section 5.1.5C), from a variety of approximations (Section 5.1.5) and from basic principle computation.

The specific heat can then be calculated by integrating the temperature derivative of the product of density distribution and statistical distribution function

$$C_v = \frac{1}{V} \frac{\partial}{\partial T} \int_0^\infty \frac{\hbar \omega g(\omega)\, d\omega}{\exp\left(\dfrac{\hbar \omega}{kT}\right) - 1}. \tag{6.13}$$

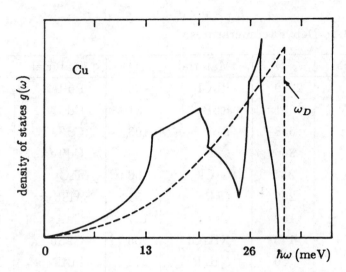

Figure 6.2: Density of states for phonons in copper as obtained from neutron scattering (solid curve) and as obtained from a Debye approximation (dashed curve), scaled to yield equal area under both curves; $\Theta = 344$ K (after Svenson et al., 1967).

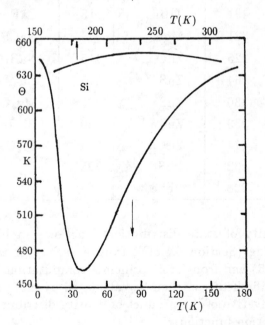

Figure 6.3: Temperature dependence of the Debye temperature of Si. Observe upper and lower scales for high and low temperatures (after Flubacher et al., 1959).

At low temperatures $(T \ll \Theta)$ only acoustic phonons with very long wavelengths $(\lambda \gg a)$ participate in the heat content of a solid.

As the Debye temperature is approached, optical phonons also become excited; initially, these are the optical phonons with lowest energy. They are near the boundary of the Brillouin zone and have a wavelength close to the lattice constant since $\hbar\omega_0(q = \pi/a) < \hbar\omega_0(q = 0)$, as shown in Figs. 5.8 and 5.9.

6.1.5 Specific Heat and Phase Changes

There are major changes in the measured specific heat at temperatures where phase changes occur (*configurational specific heat*). The most obvious change happens at the melting point between solid and liquid state, which will not be discussed further here. Others deal with first- or second-order phase changes (see below) at which new oscillatory modes are able to participate in the phonon spectrum. Examples are molecule crystals or compounds with radicals, such as SO_4, NO_3, ClO_4, etc., which can become free to rotate within the solid long before melting occurs. The solid reacts to such a change in degrees of freedom by a change in crystal structure, volume, and many other physical properties. Other phase changes below the melting temperatures are less drastic and involve only a change in lattice structure (symmetry).

All these changes cause a change in the phonon spectrum, and thus also a change in the specific heat. The specific heat develops jumps or spikes at these phase transformations of first or second order, respectively, which, in turn, can be used for the detection of these changes (see Fig. 6.4). Depending on the sign of the peak, we must distinguish exotherm or endotherm processes responsible for the phase transformation.

The order of the transition is defined by the degree of the lowest derivative of the Gibbs free energy $G = U - TS$ that shows a jump. Since $(\partial G/\partial T)_p = -S$ and $(\partial G/\partial p)_T = \mathcal{V}$, a first-order transition shows a discontinuity of entropy and volume at the transition temperature. Melting is one such transition. With $(\partial^2 G/\partial T^2)_p = -(\partial S/\partial T)_p = -C_p/T$ and $(\partial^2 G/\partial^2 p)_T = (\partial \mathcal{V}/\partial p)_T = -\kappa \mathcal{V}$, the specific heat or compressibility shows such discontinuity. Dielectric or magnetic ordering at the Curie or Néel temperature and the λ-point transformation of liquid helium are examples for higher-order phase transitions (Landau and Lifshitz, 1958).

With sensitive measuring methods (differential thermal analysis), small changes in lattice and defect structures can be detected, such

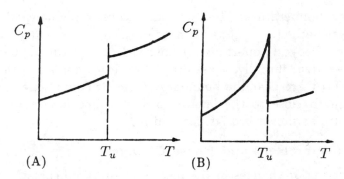

Figure 6.4: Specific heat as a function of the temperature for (A) first- and (B) second-order phase transitions.

as partial recrystallization, changes in dielectric and magnetic states involving their degree of order, and even in the density of point defects.

6.1.6 Specific Heat of Amorphous Semiconductors

At low temperatures many amorphous materials show an anomalous linear increase of the specific heat with temperature. This cannot be explained by a conventional phonon spectrum. However, it can be explained when atoms tunnel within the glass to different *metastable* positions when these different positions of atoms have nearly the same energy (*two-level tunneling system*). If the barrier between the different positions is low and thin enough, such transitions can be activated at rather low temperatures and result in a specific heat contribution of (Phillips, 1972)

$$C_v \propto T. \tag{6.14}$$

When plotting C_v/T^3 as a function of T, we observe a characteristic maximum near $T = 10\,\mathrm{K}$ which coincides with the plateau of the thermal conductivity in amorphous semiconductors (see Section 6.3.4) and may point to related causes (Yu and Freeman, 1986). Experimental data can be found in Pohl et al., 1974.

6.2 Thermal Expansion

The anharmonicity of the lattice potential and the resulting increased average amplitude of the lattice oscillations are responsible for the expansion of the semiconductor with increasing temperature. Crystal defect-related additional expansion is neglected here (see Sec-

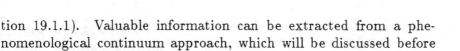

tion 19.1.1). Valuable information can be extracted from a phenomenological continuum approach, which will be discussed before the atomistic model is presented.

6.2.1 Phenomenological Description

Two thermal expansion coefficients are commonly used, a linear and a volume expansion coefficient:

$$\alpha = \frac{1}{l}\frac{\Delta l}{\Delta T} \quad \text{and} \quad \alpha_v = \frac{1}{\mathcal{V}}\frac{\Delta \mathcal{V}}{\Delta T}, \tag{6.15}$$

with $\alpha_v \simeq 3\alpha$ for isotropic materials. For anisotropic crystals the linear thermal expansion coefficient is a tensor, which is proportional to the strain tensor (see Section 4.1.1) and the temperature difference:

$$(e_{ik}) = (\alpha_{ik})\Delta T, \tag{6.16}$$

with the tabulated components in the main axis of the tensor ellipsoid (α_{xx}, α_{yy}, α_{zz}), usually referred to as (α_1, α_2, α_3). Because of the varied strengths of the lattice forces in different directions for anisotropic lattices, an anisotropy of α_i appears. Examples of strongly anisotropic (layer) semiconductors are BN, GaS, GaSe, InSe, TlSe, InBi (White et al., 1975), and some chalcopyrites such as $CuGaSe_2$ (Bodnar and Orlova, 1985). Semiconductors with a high degree of anisotropy, e.g., Se, show a minor contraction in the direction of strongest bonding while expanding in the direction of weakest bonding, with the volume expansion coefficient α_v usually positive* (see Section 6.2.1A). This fact provides an opportunity to design composite materials with essentially zero thermal expansion (Roy and Agrawal, 1985).

The relation between the strength of the lattice bonding and the thermal expansion yields a useful empirical rule, the *Grüneisen rule*, which connects the melting point T_m of isotropic materials with the thermal expansion and holds well for metals:

$$\left(\frac{\Delta l}{l}\right)_{T_m \,(\mathrm{K})} \simeq 0.07; \quad \text{hence} \quad \alpha \simeq \frac{0.07}{T_m}. \tag{6.17}$$

---

* Exceptions like H_2O, In, and δ'Pu are caused by rather unusual changes in the atomic structure of the solid. (Were it not for these changes in H_2O, lakes would freeze from the bottom up.) Another important anomaly of α_v is observed at very low temperatures (see Section 6.2.1A).

The maximum thermal expansion at the melting point is about 7% and is independent of the material. The melting point, in turn, is related to the Debye temperature and thus to the elastic stiffness constants, i.e., to the strength of bonding via another empirical relation

$$T_m = \gamma a^2 M \Theta^2, \tag{6.18}$$

where M is the atomic mass, γ is the Lindemann parameter (see Section 6.2.1A), and a is the interatomic distance (in a monatomic lattice). This *Lindemann relation* can be justified with a lattice dynamic model.

Table 6.2 lists the thermal expansion coefficients and other relevant parameters for several important semiconductors.

6.2.1A Lattice Dynamic Considerations   The average kinetic energy per atom as a linear harmonic oscillator is

$$E_{osc} = \frac{1}{2} M \left(\frac{d\bar{u}}{dt} \right)^2 = \frac{1}{2} M \omega^2 \bar{u}^2, \tag{6.19}$$

which is equivalent to $(1/2)kT$ for a linear oscillator, with $\bar{u} = fa$ and f as the fractional displacement from the equilibrium interatomic distance a. The oscillatory energy per atom in a solid is $3kT$; at high temperatures most of the oscillations are in the optical branch. Therefore, we obtain from Eq. (6.19) with $\hbar\omega = \hbar\omega_0 \simeq k\Theta$ for the melting point with $E_{osc} = 3kT_m$

$$T_m = \frac{k}{9\hbar^2} M \Theta^2 f_{max}^2 a^2, \tag{6.20}$$

defining f_{max} as the maximum displacement of atoms at the melting point. This equation is identical to the Lindemann relation [Eq. (6.18)] with $\gamma = k f_{max}^2/9\hbar^2$. Comparison with the experiment shows that for most solids $0.1 < f_{max} < 0.15$; that is, the maximum amplitude of the lattice oscillation is roughly 10% of the interatomic spacing at the melting temperature. This is equivalent to the Grüneisen rule since the rms of the linear oscillation amplitude is $f_{max}/\sqrt{2} \simeq 0.07$, and assuming that almost all of these oscillations are totally anharmonic, i.e., atoms are essentially rigid and any oscillation needs additional lattice space.

Any thermal expansion requires a nonharmonic contribution from the interatomic potential (see Fig. 4.1). Nonharmonic terms can be evaluated from higher-order elastic constants (Hiki, 1981). Consider-

Table 6.2: Thermal expansion coefficients α in $10^{-6}\,\mathrm{K}^{-1}$, lattice constant a in Å, density ρ in g/cm^3, and melting point T_m in K for some semiconductors.*

Crystal	α (300 K)	a (T,K)	a (600 K)	ρ (T,K)	T_m (K)
Si	2.59	5.43102 (296)		2.32900 (296)	1685
Ge	5.75		5.6603	5.3234 (298)	1210.4
AlAs	5.20	5.660 (291)	5.6790	3.760 (300)	2013
AlP		5.4635 (298)		2.40 (300)	2823
AlSb		6.1355 (292)		4.26 (300)	1338
GaP	4.65	5.4505 (300)	5.4742	4.138 (300)	1730
GaSb		6.09593 (298)		5.614 (300)	985
GaAs	6.63	5.65325 (300)	5.6800	5.3176 (298)	1513
InP	4.75	5.8687 (291)	5.8870	4.81 (300)	1335
InAs	4.52	6.0583 (298)	6.080	5.667 (300)	1215
InSb		6.47937 (298)		5.7747 (300)	800
Crystal	α(300K)	a(T,K)	c(T,K)	ρ(T,K)	T_m (K)
ZnO	$4.51_\perp 3.0_\parallel$	3.2495 (298)	5.2069 (298)	5.6753 (293)	2300
ZnS (cub)	6.8	5.4102 (298)			
ZnS (hex)		3.8226 (299)	6.2605 (299)	4.075	2103 (10.3 bar)
ZnSe (cub)	6.9	5.6676 (300)		5.266 (300)	1793
ZnTe (cub)	8.19 (283)	6.1037 (300)		5.636	1568
CdO (cub)		4.689 (300)		8.15	
CdS (hex)		4.1362 (298)	6.714 (298)		
CdS (cub)		5.818		4.82	1750
CdSe (hex)		4.2999 (297)	7.0109 (297)	5.81	1514
(cub)		6.052			
CdTe	4.8	6.486 (300)		5.87 (4)	1265

* In some compounds, decomposition before melting is avoided by an appropriate ambient at elevated pressure. Near the melting point, additional lattice expansion is observed because of intrinsic (Shottky or Frenkel) defect generation (see Section 27.1.1).

ing the third-order term in the lattice potential [see Eq. (4.13)]

$$E_{\mathrm{pot}} = E_0 + \beta_1 u^2 + \beta_2 u^3,\qquad (6.21)$$

we obtain for the average displacement of a lattice atom

$$\bar{u} = \frac{\displaystyle\int_{-\infty}^{\infty} u \exp\left(-\frac{E_{\mathrm{pot}}}{kT}\right)}{\displaystyle\int_{-\infty}^{\infty} \exp\left(-\frac{E_{\mathrm{pot}}}{kT}\right)}\,d\omega,\qquad (6.22)$$

which can be evaluated by introducing Eq. (6.21) into Eq. (6.22) and integrating, which yields

$$\bar{u} \simeq \frac{3}{4} \frac{\beta_2}{\beta_1^2} kT. \tag{6.23}$$

Using for the thermal expansion the average relative displacement

$$\alpha = \frac{d}{dT}\left(\frac{\bar{u}}{a}\right) = \frac{3}{4} \frac{\beta_2}{a\beta_1^2} k \tag{6.24}$$

(a is the interatomic distance for a cubic crystal), we obtain a first-order estimate of the thermal expansion coefficient in terms of the anharmonicity constant (β_2) of the lattice oscillation. In this approximation α is temperature-independent. Near the melting point, however, additional lattice expansion is observed because of intrinsic (Schottky or Frenkel) defect generation (see Section 19.1.1).

At lower temperatures, however, α decreases with decreasing T, much like the specific heat. This can be explained by the above theory by replacing the Dulong-Petit value for the lattice energy [used to obtain Eq. (6.20)] with the appropriate function $u(T)$ yielding $C_v(T)$ at lower temperatures (e.g., the Debye function). A direct possibility for introducing C_v is provided through the *Mie-Grüneisen theory* of the equation of state of solids (Grüneisen, 1926), which yields

$$\alpha = \frac{\Gamma}{3} \frac{C_v}{B\mathcal{V}}, \tag{6.25}$$

with Γ the *Grüneisen parameter*, which can be expressed as

$$\Gamma = -\frac{d\ln\omega}{d\ln\mathcal{V}} = -\frac{\mathcal{V}d\omega}{\omega d\mathcal{V}}. \tag{6.26}$$

For high enough temperatures, Γ is nearly constant (see Fig. 6.5). Equation (6.25) shows with $\alpha \propto C_v$ the main tendencies of the observed behavior of $\alpha(T)$, neglecting the slight temperature dependence of volume $\mathcal{V}$ and bulk inverse compressibility B [see Eq. (6.12)].

A more sophisticated lattice dynamic theory must consider the temperature dependences of Γ and C_v calculated for the different modes (i) of lattice oscillations; hence (Mitra and Massa, 1982):

$$\alpha(T) = \frac{\sum_i \Gamma_i C_v^{(i)}}{3B\mathcal{V}}. \tag{6.27}$$

This relation is the basis for further lattice dynamic analysis relating to Γ_i and $C_v^{(i)}$ (see Namjoshi et al., 1971).

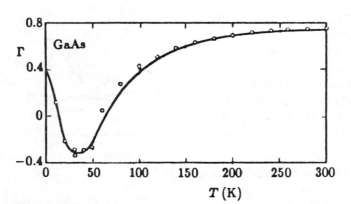

Figure 6.5: Grüneisen parameter as a function of the temperature for GaAs (after Soma et al., 1982). ©Pergamon Press plc.

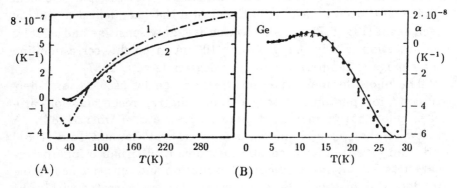

Figure 6.6: (A) Linear thermal expansion coefficients of Ge, GaAs, and ZnSe redrawn from Novika (1966) and Smith and White (1975). (B) Linear thermal expansion coefficient of Ge (expanded scale) for low temperatures.

6.2.2 Negative Thermal Expansion

At low temperature ($T < 0.2\Theta$), the thermal expansion coefficients of numerous semiconductors (IV, II-V, and II-VI compounds) become negative (observed first by Valentiner and Wallot, 1915); at still lower temperatures, a range of positive values of α often reappears (Daniels, 1962). This behavior, shown in Fig. 6.6, can be understood from the residual anisotropy of lattice oscillations that renders the Grüneisen parameters negative for low-frequency transverse acoustic modes in diamond and zinc-blende lattices (Barron, 1957—see also Section 6.2.1). This means a contraction of the crystal perpendicular to a longitudinal oscillation, which can yield a net volume contraction in an intermediate temperature range.

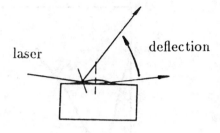

Figure 6.7: Optical detection of extremely small amounts of buckling due to inhomogeneous thermal expansion.

6.2.3 Photothermal Expansion

Phonons created as a result of optical excitation cause local heating, thereby expanding the lattice. Such expansion, in turn, can be used to detect the phonon generation. When such generation is inhomogeneous, e.g., when caused from a focused light spot, the local expansion causes a buckling of the crystal that can deflect a probing laser beam (Fig. 6.7). This deflection is highly sensitive and enables the detection of a buckling of only 10^{-4} Å in height, corresponding to a change in temperature of 10^{-5} degrees (Amer, 1987).

The photothermal expansion method can be used as loss spectroscopy for photoluminescence (nonradiative recombination produces phonons) or in photovoltaic devices, etc. (Amer, 1987). A review of photoacoustic methods is given by Patel and Tam (1981) and Tam (1986). Photoacoustic methods, a subgroup of photothermal effects, use modulated optical excitation and sensitive ac (acoustic) detection methods. Here acoustic waves are generated within the semiconductor.

6.2.4 Photomechanical Effect

Light of sufficient energy absorbed in a semiconductor produces free electrons or excitons. This reduces the bond strength of the lattice atoms from which ionization took place. Consequently, the lattice parameters related to the bond strength (i.e., to the elastic stiffness constants) change.

Except for very high excitation rates, only a very small fraction of the bonds are involved, and we observe only small changes in mechanical and thermal properties. We also observe a *photoplastic effect* when photogenerated carriers are trapped by dislocations, which thereby reduces their mobility (Upit and Manika, 1978). Some of these have been experimentally observed (Gorid'ko et al., 1961; Upit and Manika, 1978). This photomechanical effect needs to be distinguished from the *photoelastic effect* (see Feldman and Waxler, 1979) in which the index of refraction is changed as a result of a mechanical lattice deformation (see Section 20.3.2C).

6.3 Thermal Conductivity

Heat transport in a solid is caused by phonons and all other mobile quasi-particles which contain excess energy and interact with the lattice, such as electrons (see Sections 6.3.3A and 31.2), holes, polarons, excitons, and photons. In semiconductors with low carrier density, and in the absence of light and electric fields, the thermal energy transport is caused mainly by phonons. Such transport would propagate with sound velocity if phonons were not scattered. Such *ballistic phonon transport* is observed shortly after a heat pulse is applied at low temperatures before thermalization of phonons occurs (Knaak et al., 1986). However, the sample dimension is usually much larger than the mean free path of phonons. Consequently, the thermal energy flux is given by a diffusion-type of transport, carried by the random motion of phonons and directed by their gradient as the driving force:

$$\mathbf{Q} = -\kappa \nabla T \tag{6.28}$$

with κ as the *thermal conductivity*. The thermal current density can be derived from the phonon flux as a *particle current* of N phonons with an average velocity v_s. After integration over all angles of the phonon flux through an arbitrary surface normal to the flux, we obtain $\frac{1}{6} N v_s$. With an average thermal energy of $\frac{3}{2} kT$ per phonon, this results in a net thermal energy flux between two closely spaced parallel planes at T_1 and T_2 of

$$Q = \frac{1}{6} N_{\mathrm{Av}} v_s \cdot \frac{3}{2} k (T_2 - T_1), \tag{6.29}$$

with heat transfer taking place after an inelastic collision, i.e., after a distance of the mean free path for phonons λ is traversed. Replacing the temperature gradient with $(T_2 - T_1)/2\lambda$, and inserting the classical expression for the specific heat of such a phonon gas $C_v = \frac{3}{2} N k$, we obtain from Eqs. (6.28) and (6.29) the often-cited expression for the thermal conductivity (Debye, 1914):

$$\boxed{\kappa = \frac{1}{3} C_v v_s \lambda.} \tag{6.30}$$

Table 6.3: Room temperature thermal conductivity of alkali halides in cal/(deg cm s).

Compound		F	Cl	Br	I
	Atomic Weight	19.00	35.457	79.916	126.92
Na	22.997	**0.124**	0.090	0.012	0.010
K	39.096	0.057	**0.138**	0.023	0.030
Rb	85.48	?	0.007	**0.016**	0.014

The various modes (i) and branches (j) of the phonon spectrum contributing to κ can be accounted for by appropriate summation while retaining the relation of Eq. (6.30):

$$\kappa = \frac{1}{3} \sum_{i,j} C_{ij} v_{ij} \lambda_{ij}. \qquad (6.31)$$

The contribution of the different components of the *specific heat* introduces a temperature dependence of κ. The *phonon velocities* v_{ij} are group velocities ($\partial \omega / \partial q$) and are less temperature-dependent (Section 5.1.5B).

The critical parameter for the thermal conductivity is the *mean free path*, which requires a thorough study of the phonon scattering mechanisms (Section 6.3.2). For a review of experimental techniques, see Rowe and Bhandari (1986), and Tye (1969).

In anisotropic materials, the thermal conductivity can be anisotropic because of the anisotropy of the elastic constants and therefore of the phonon velocity. See de Goër et al. (1982).

6.3.1 Dependence of Thermal Conductivity on the Mass Ratio in Binaries

The heat conductivity depends on the phonon velocity and on scattering. The phonon velocity increases with increasing lattice bonding. Therefore, the thermal conductivity decreases with an increase of the atomic number when comparing elements within the same column of the periodic system.

When both atoms in a binary compound have substantially different masses, the thermal conductivity is decreased, as is the phonon velocity. Examples include the alkali halides, for which the room temperature thermal conductivities are listed in Table 6.3.

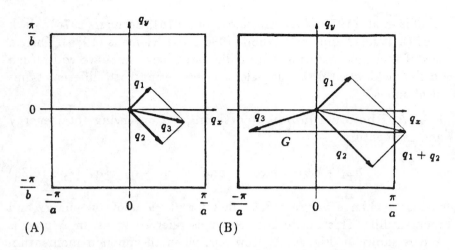

Figure 6.8: N- and U-processes [(A) and (B), respectively] for two phonons with momentum q_1 and q_2, producing a third with momentum q_3. The reciprocal lattice vector indicates the Bragg reflection at the zone boundary.

6.3.2 Phonon Scattering

Phonons can interact with each other because of the anharmonicity of the lattice potential, which permits phonon-phonon scattering. This is the most basic (*intrinsic*) scattering mechanism, limiting the thermal conductivity. Other scattering mechanisms may reduce the mean free path further according to

$$\frac{1}{\lambda_{ij}} = \sum_k \frac{1}{\lambda_{ijk}}, \tag{6.32}$$

where the index k identifies all scattering mechanisms. The more important ones are:

- phonon-phonon scattering,
- phonon scattering at point defects,
- phonon scattering at short- or long-range disorder,
- phonon scattering at line defects,
- phonon scattering at grain boundaries or surfaces, and
- phonon scattering with carriers.

Each one of these scattering categories must be distinguished with respect to the type of phonons involved. A large body of literature describes the different phenomena: e.g., Drabble and Goldsmid (1961), Steigmeier (1969), Touloukian et al. (1970), Childs et al. (1973),

Challis et al. (1975), Klemens and Chu (1976), Berman (1976), Slack (1979), Vandersande and Wood (1986), and Klemens (1986). Only a few of these mechanisms will be discussed here, since we will return to this field when discussing electron scattering with different types of phonons.

6.3.2A Phonon-Phonon Scattering   Conserving total energy and momentum,

$$\hbar\omega_1 + \hbar\omega_2 = \hbar\omega_3 \quad \text{and} \quad \mathbf{q}_1 + \mathbf{q}_2 = \mathbf{q}_3 \qquad (6.33)$$

yields no change *per se* in heat flow for phonons of low energy and momentum. This scattering process is referred to as an *N-process* and is shown in Fig. 6.8A. However, when the phonon momentum is large enough so that the summation $\mathbf{q}_1 + \mathbf{q}_2$ leads to a phonon with a momentum outside the first Brillouin zone, Bragg reflection at the zone boundary occurs and the resulting $\mathbf{q}_3$ has its direction essentially reversed. Mathematically, one accounts for the Bragg reflection by adding the reciprocal lattice vector $\mathbf{G}$ with the length of the Brillouin zone, and thereby obtains a resulting phonon within this Brillouin zone (see Fig. 6.8B). Peierls (1929, 1955) defined this as an *Umklapp process* (U-process). Only these processes contribute directly to a thermal resistance.

U-processes require a sufficient $\mathbf{q}$ of the initiating phonons, i.e., sufficient phonon population at higher $\hbar\omega$. This occurs at higher temperatures ($T \gtrsim \Theta/4$). More complex lattice structures result in more complicated Brillouin zones with more opportunities for Umklapp processes. Therefore, these crystals usually show a lower heat conductivity.

Leibfried and Schlömann (1963) give an estimate of the thermal conductivity, which increases hyperbolically with decreasing temperature:

$$\kappa \propto \lambda = \frac{12}{5}\sqrt[3]{4}\left(\frac{k}{h}\right)^3 \frac{Ma}{\left(\Gamma + \dfrac{1}{2}\right)^2}\frac{\Theta^3}{T} \simeq 5.76 \frac{Ma}{\left(\Gamma + \dfrac{1}{2}\right)^2}\frac{\Theta^3}{T}\ (\text{Å}),$$

$$(6.34)$$

where M is the atomic mass, a is the interatomic spacing, and Γ is the Grüneisen anharmonicity parameter [Eq. (6.26)]. It follows from Eq. (6.34) that the thermal conductivity at high temperatures increases rapidly with increasing Debye temperatures.

The increasing scattering at higher temperature is due to an increase in amplitude, and hence in anharmonicity of the oscillation, causing an increase in phonon-phonon scattering cross section.

At lower temperatures, U-processes freeze out. This results in a decrease of scattering and thus an increase of the thermal conductivity with decreasing temperature:

$$\kappa_U \propto \lambda = \frac{7}{4}\left(\frac{k}{h}\right)^3 \frac{Ma}{\left(\Gamma + \frac{1}{2}\right)^2} \frac{T^3}{\Theta} \exp\left(\frac{\Theta_U}{T}\right) \tag{6.35}$$

where $\Theta_U < \Theta$ (typically $\Theta_U \simeq 0.5\Theta$), in agreement with the experiment (Klemens, 1959). One should remember, however, that the Debye temperature relates to the energy of the phonons in the optical branch, while U-processes also occur in the lower-energy acoustic branch. (Klemens, 1958).

6.3.2B Scattering at Crystal Boundaries   If the mean free path exceeds the smallest distance between crystallite boundaries, then λ in Eq. (6.30) is replaced by the Casimir length L_C (Casimir, 1938)

$$\kappa = \frac{1}{3}C_v v_s L_C; \tag{6.36}$$

L_C is on the order of the shortest distance (e.g., platelet thickness) in the material (also see Fig. 6.11B) and depends on the roughness of the surface or the type of crystallite interface. Only rough surfaces cause back scattering and thereby increase the thermal resistance. Roughness relates to the wavelength of the scattered elastic wave (low-energy acoustic branch). At smaller dimensions of the roughness, some of the reflection will be specular: a surface appears smoother to longer wavelength waves. The acoustic phonon wavelength is typically on the order of 100 Å at 10 K. At such low temperatures, $\kappa(T)$ is then determined by $C_v(T)$ and therefore varies $\propto T^3$ (Fig. 6.11B). See also de Goër et al. (1965).

6.3.2C Phonon Scattering of Lattice Defects   At lower temperatures, phonon scattering at lattice defects starts to compete with the intrinsic phonon-phonon scattering and limits the thermal conductivity to a degree that depends strongly on the density and distribution of lattice imperfections.

Such scattering at lattice defects can be caused by the change in mass and elastic properties of the lattice surrounding the defect,

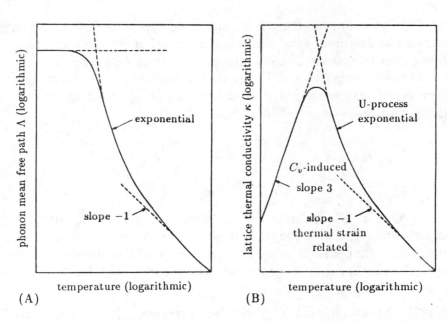

Figure 6.9: Typical temperature dependence of (A) the mean free path and (B) the thermal conductivity in semiconductors.

hence changing the wave dispersion relation. Phonon scattering at *point defects* can be compared to scattering of waves at obstacles. The relaxation time of phonons, scattered at foreign atoms of mass M_f, can be estimated as

$$\tau = \frac{4\pi v_s{}^3}{N_f a_f{}^3 \omega^4} \left(\frac{M_f - M_0}{M_0} \right)^2 \qquad (6.37)$$

where N_f is the density of foreign atoms, $a_f{}^3$ is their atomic volume, and ω is the frequency of the scattered phonons (Klemens, 1955). τ is approximately equal to the time between scattering events $\tau \simeq \lambda/v_s$. An increased scattering probability is obtained for phonons with a dominant wavelength of a similar dimension as the obstacle. Therefore, point defect scattering is more pronounced at high and intermediate temperatures, providing more short wavelength phonons at high q values, i.e., at higher energies. However, scattering at extended defects, such as dislocations (Sproul et al., 1959), stacking faults, colloids (Walton, 1967), voids (Vandersande, 1980), grain boundaries (Vandersande and Pohl, 1982), and surfaces, is also important at low temperatures.

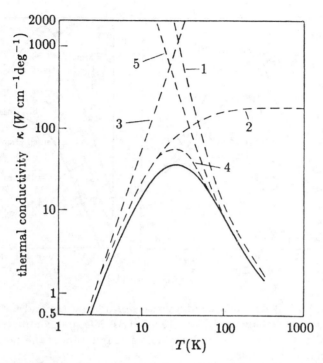

Figure 6.10: Typical temperature dependence of the thermal conductivity for Si (solid curve) compared with the different scattering contributions. (1) Umklapp scattering; (2) diffuse boundary and isotope scattering; (3) diffuse boundary scattering; (4) Umklapp, diffuse boundary and isotope scattering; and (5) Umklapp and isotope scattering (after Glassbrenner and Slack, 1964).

When, because of phonon scattering at these defects, the mean free path levels off at lower temperatures, the temperature dependence of the thermal conductivity is given only by that of the specific heat, i.e., by $C_v \propto (T/\Theta)^3$. At higher temperatures where the specific heat levels off, the thermal conductivity is determined by the increase in phonon-phonon scattering with temperature. The combined behavior is sketched in Fig. 6.9. The influence of additional scattering mechanisms on the thermal conductivity of Si is shown in Fig. 6.10.

Lattice disorder of an alloy type can substantially reduce the thermal conductivity. A similar reduction is observed in the common mixture of different isotopes in any lattice. An example for Ge is shown in Fig. 6.11A, and for SiGe alloys or polycrystals in Fig. 6.11B. Here, scattering at the surface of thin layers reduces the thermal

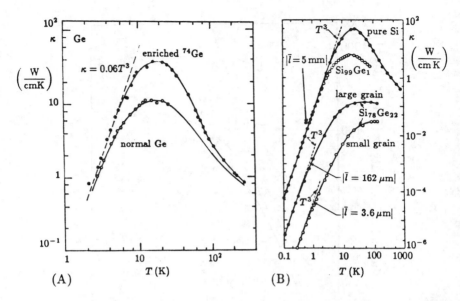

Figure 6.11: (A) Thermal conductivities of normal Ge and Ge enriched with ^{74}Ge (after Geballe and Hull, 1964). (B) Thermal conductivity of Si and SiGe alloys as single and polycrystals as a function of the temperature (after Kumar et al., 1985).

conductivity at lower temperatures. Phonon mean free paths up to 30 cm are observed at $T < 1$ K in pure Si (Vandersande and Wood, 1986).

Finally, one observes strong phonon absorption in the frequency range of local modes (see Section 5.1.6) of lattice defects (*resonant scattering*). This can lead to dips in the $\kappa(T)$ curve when, with increasing T, phonon branches above the local mode that could be populated are kept depleted by this scattering process. Strong interaction is also observed for paramagnetic impurities with resonant scattering involving *spin-lattice relaxation*, shown for the Mn doping in GaAs in Fig. 6.12B, curves 6 and 7 (Holland, 1964). See also Adilov et al. (1986) for doping with Ni.

6.3.3 Phonon Scattering in Semiconductors

Such scattering is discussed as scattering of phonons with electrons and other quasi-particles later (Sections 32.2.1, 32.3, and 33). This type of scattering reduces the thermal conductivity especially at low and medium temperatures.

The influence of very low density (*p*-type) doping (i.e., the introduction of desired impurities—see Chapter 18) and much higher *n*-

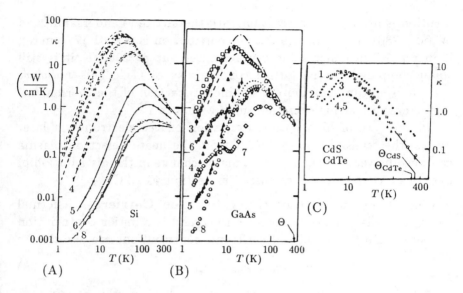

Figure 6.12: Influence of doping on the thermal conductivity (all in $W\,cm^{-1}\,K^{-1}$): (A) for p-type Si with doping densities of $1 \cdot 10^{13}$, $4.2 \cdot 10^{14}$, $4 \cdot 10^{15}$ and $4 \cdot 10^{16}$ cm^{-3} and for n-type Si with doping densities of $6.2 \cdot 10^{19}$, $1.7 \cdot 10^{20}$, $3 \cdot 10^{20}$ and $3.5 \cdot 10^{20}$ cm^{-3} for curves 1 – 8, respectively (reproduced from Steigmeier, 1969); (B) in GaAs with different doping at a density of $7 \cdot 10^{15}$, $7 \cdot 10^{15}$ (Te), $2.7 \cdot 10^{17}$ (Zn), $2.6 \cdot 10^{18}$ (Zn), $1.4 \cdot 10^{18}$ (Cd), $5 \cdot 10^{16}$ (Mn), $5 \cdot 10^{17}$ (Mn), $1.2 \cdot 10^{19}$ (Zn) cm^{-3} for curves 1–8, respectively; and (C) for CdS, curve 1; CdTe "pure" curves 2 and 3; and $1.3 \cdot 10^{20}$ (Mn) and $1.6 \cdot 10^{20}$ (Fe) cm^{-3} curves 4 and 5, (after Holland, 1964; Moore and Klein, 1969; and Jongler et al., 1980).

type doping on the thermal conductivity is shown for Si in Fig. 6.12A. The thermal conductivity of group IV and III-V semiconductors above 300 K is discussed by Logachev and Vasilev (1973).

With high doping densities, other thermal energy transport mechanisms may become important, such as the transport by ambipolar diffusion of electrons and holes or excitons (see Section 15.1), with consequent scattering or recombination at the cooler end of the sample. For a review of the theory, see Slack (1979); for an experimental review, see Parrott and Stuckes (1975) or Berman (1976).

IR photon emission from the hot end and reabsorption at the cold end, for example, by free carriers (Vandersande and Wood, 1986) also contributes to the thermal energy transport at high temperatures in partially transparent semiconductors (Waseda and Ohta, 1987). Transparency is needed for the transmission of photons; some absorption is needed for interaction with the lattice (Vandersande and Wood, 1986). In Ge this radiative contribution is $\simeq 0.01$ W/(cm K); it increases with increasing temperature, but at 2000 K it is still small compared to the lattice conductivity of Ge. This *radiative transmission* can contribute a significant fraction of the thermal energy transport at high temperatures with $\kappa \propto T^3$.

In the case of high optical excitation of free electrons via laser pulses, the transmission of energy from the heated electron plasma can become very large, causing a major change in the distribution of phonons. An example is discussed in Section 48.2.2C.

6.3.3A Thermal Conductivity by Free Carriers   Thermal energy can also be carried by electrons, and is a major contribution in metals. Both phonon and electronic contributions are additive:

$$\kappa = \kappa_{\mathrm{ph}} + \kappa_{\mathrm{el}}. \tag{6.38}$$

The electronic contribution can be estimated similarly to the phonon contribution, using the equivalent of Eq. (6.31):

$$\kappa_{\mathrm{el}} = \frac{1}{3} C_v^{(\mathrm{el})} v_{\mathrm{rms}}^2 \tau_n; \tag{6.39}$$

with $C_v^{(\mathrm{el})} = 2\,nk$ for a classical electron gas and $v_{\mathrm{rms}}^2 = 3kT/m_n$ we have

$$\kappa_{\mathrm{el}} = 2\,nk^2 (\frac{\tau_n}{m_n})T \tag{6.40}$$

where n, τ_n, and m_n are the electron density, their relaxation time, and effective mass, respectively. The electronic contribution is related to the electron conductivity [$\sigma = en\mu = e^2 n\tau_n/m_n$—see Eqs. (28.15) and (28.16)] by the Wiedemann-Franz law

$$\frac{\kappa}{\sigma} = 2 \left(\frac{k}{e} \right)^2 T \tag{6.41}$$

and gives a marked contribution only at higher carrier densities (Kittel, 1986). The factor 2 in Eq. (6.41) holds for nondegenerate semiconductors. At higher electron densities, and depending on the scattering mechanism, this factor varies between 2 and 4. The entire proportionality factor at the right side of Eq. (6.41) is called the Lorentz number $L = 2(k/e)^2$. For a strongly degenerate electron gas, $L = (\pi^2/3)(k/e)^2$; for more detail, see Smith (1978).

Bipolar thermal conductivity can have a major contribution in narrow-band gap semiconductors. Electrons and holes are moving

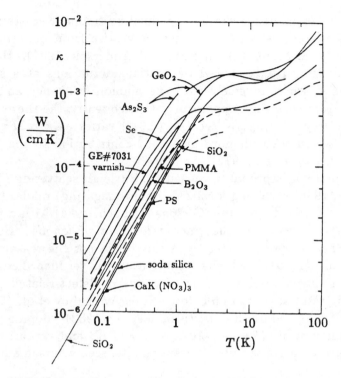

Figure 6.13: Low temperature thermal conductivities of various amorphous solids (after Stephens, 1973).

in the same direction without causing a net electric current. The thermal conductivity by bipolar diffusion can be larger by a factor of up to 10^2 than the electronic contribution alone (Vandersande and Wood, 1986).

6.3.4 Phonon Scattering in Amorphous Semiconductors

In strongly disordered solids the mean free path is typically on the order of a very few interatomic spacings (~ 10 Å) and is essentially independent of the temperature (Kittel, 1949). At low temperatures, the material behaves like an isotropic elastic medium for the long wavelength acoustic phonons.

Below a few degrees K, these phonons freeze out, and all amorphous solids behave remarkably uniformly. Their thermal conductivity has a plateau at $\simeq 10$ K, and at lower temperatures decreases proportionally to T^2. Above the plateau, the thermal conductivity increases linearly with temperature. This behavior is essentially independent of the material and any impurity (see Fig. 6.13).

This T^2-dependence can be explained by resonant scattering between *two-level states* (TLS) which may arise from quantum tunneling between atoms (Matsumoto and Anderson, 1981). Here the scattering length λ_{ph}, caused by spatial fluctuations at a scale of about 10 Å, varies inversely with the phonon frequency and hence also inversely with the temperature. Consequently, the thermal conductivity $\kappa = C_v v_s \lambda_{ph}/3$ [see Eq. (6.31)] varies as $\kappa \propto T^2$ in the range in which such phonons give a T^3-contribution to the specific heat (Anderson et al., 1972).

The plateau is probably caused by Rayleigh scattering (Yu and Freeman, 1986) involving localized (nonpropagating) modes (Akkerman and Maynard, 1985) or *fractons* (Orbach, 1984) with a phonon wavelength $\lambda \simeq l_\omega$ as the *fracton localization length*. However, neither the source of scattering in this range nor the reason for an increase in $\kappa(T) \propto T$ at higher temperatures is yet identified.

Recent literature gives a broad range of defect-related phonon scattering results; for example, for α-Si see Vakhabov et al. (1985)—doping with Se and Te; Radhakrishnan et al. (1982)—doping with O; Igamberdiev et al. (1983)—$\kappa(6 < T < 300\ K)$ and various doping. For a representative sample of recent results, see Anderson and Wolfe (1986).

Summary and Emphasis

Phonons provide the thermal energy content of a semiconductor. Only a negligible amount is added by the kinetic energy of free electrons and holes in normal semiconductors since their density is comparatively small. After knowing the phonon density distribution (a product of density of states and Bose-Einstein distribution), the thermal energy content is obtained by simple summation. The change of the thermal energy content with temperature, the specific heat, is directly accessible experimentally.

The transport of phonons within the semiconductor is a diffusion-type transport, driven by the gradient of the phonon density and described by a random walk of phonons between scattering events. The measured thermal conductivity in a temperature gradient provides experimental access to a variety of such microscopic scattering processes.

Thermal expansion is the result of the anharmonicity of interatomic potentials and, in contrast to many other nonharmonic phenomena (e.g., optical), presents itself already at low amplitudes of lattice oscillations (i.e., at relatively low temperatures).

Phonons populate at the acoustic branches almost exclusively at low temperatures, and only at temperatures approaching the Debye temperature do they extend with a significant fraction to include the optical branches. The thermal properties due to phonons, however, change smoothly with temperature because of the additive contribution of all phonons. A thorough knowledge of their distribution permits one to predict detail of the thermal properties, however, with limited sensitivity. Better tools to investigate phonon spectra are related to differential effects specific to certain phonon modes (e.g., optical effects).

Exercise Problems

1. Calculate the Debye temperature for NaCl in a one-dimensional model. Use ion radii from Table 2.6.
2. Discuss $C_v(T)$ and develop Eq. (6.12).
3.
   (a) Calculate the specific heat for a one-atomic two-dimensional lattice in the Debye approximation; show that it increases $\propto T^2$ at low temperatures.
   (b) Show that it increases $\propto T$ in a one-dimensional lattice.
4. Calculate the Debye temperature for the actual lattices of Si, Ge, GaAs, and NaCl, and compare these with the tabulated values. Comment.
5. Derive Eq. (6.1) from thermodynamic concepts.
6. Show that in a crystal with a density of states, given by Eq. (6.8), that the most probable phonon energy $\hbar\omega_{mp}$ is given by the condition $\exp[\hbar\omega_{mp}/(kT)]\{1 - \hbar\omega_{mp}/(2kT)\} = 1$. For $T < \Theta$ give an estimate for $\hbar\omega_{mp}$.
7.(*) Show that the energy content of a classical oscillation is kT with exactly half of its average energy as kinetic and half as potential energy (refer to the appropriate textbooks).
8.(*) In a binary compound the mass ratio plays an important role for the thermal conductivity.
   (a) Why?
   (b) Always?
   (c) List the different phonon scattering mechanisms and discuss their relative importance.

9.(*) What influence do paramagnetic impurities have on thermal conductivity? On its temperature dependence?

10.(*) Electrons and phonons in a perfect crystal are not localized. When scattering occurs, they temporarily localize; thereafter they delocalize again. Explain. For electrons a common analysis uses a wave packet and dispersing envelope function.

11. Show that the flux of particles (here phonons) through a planar surface normal to the flux is $\frac{1}{6} N \bar{v}$ with $\bar{v}$ their average velocity.

12.(*) The ratio of thermal conductivity (κ) to electrical conductivity (σ) due to free electrons is given by Eq. (6.41), the Wiedemann-Franz law.

    (a) Derive and explain its relation (advance to the relevant section in the book).

    (b) What is the heat content of a free electron gas (Boltzmann gas)?

    (c) Why does the temperature appear in the ratio of κ/σ?

13. Estimate the relative amplitude of lattice oscillations as a function of temperature (as a fraction of the Debye temperature).

14.(e) Follow explicitly the approximations yielding from Eq. (6.11) the temperature dependence of the specific heat given in Eq. (6.12).

PART III

ENERGY BANDS

Chapter 7

Elements of Band Structure

Characteristic for much of the electronic behavior in solids is the existence of energy bands, separated by band gaps.

Electronic transitions in energy and in space are the basic processes of interest in semiconductor physics. The first group is responsible for the large variety of excitation and de-excitation (i.e., recombination) processes; the second must be considered for carrier transport. Both are characterized by quantum-mechanical features: the spectrum of electronic energy states, called *eigenstates*, the distribution of electrons over these states, and the forces that cause changes in this distribution.

First, we will discuss the general principles that yield the spectrum of energy states typical for the solid semiconductor. It is possible to obtain the main features of this spectrum using two apparently different models. The first model starts from individual atoms and its immediate neighborhood, and expands with less and less attention to the atomistic structure the further one extends from the origin; this approach is here referred to as the proximity or chemical approach. The second model is at first view rather insensitive to the detailed properties of individual atoms, but considers the long-range periodicity of a crystalline lattice; we will call it the periodicity approach. Both yield similar qualitative results: energy spectra that show broad, permitted ranges of energy interspersed with forbidden ranges which, in space, extend as *bands* throughout the entire semiconductor.

The first approach is successfully used for amorphous semiconductors; and both approaches have been used for crystals. In this chapter, we will present both the approaches, will point to common features, and will expose some of the differences in the results. This discussion will start from a rather heuristic description and will introduce sequentially more sophisticated elements.

7.1 The Proximity (Chemical) Approach

In a very primitive model, the exchange of electrons between two atoms can be made plausible by considering the splitting of eigenstates of degenerate oscillators, i.e., oscillator states having the same eigenfrequency when they become coupled with each other. The addition of more atoms of the same kind at increased distances splits the energy levels into more levels which span a range of energies. If the levels are spaced closely enough, *Heisenberg's uncertainty principle* no longer permits distinction between the individual levels.* In this case, one obtains an *allowed energy range* in a large enough cluster of atoms, instead of a discrete energy level spectrum of a single atom or an aggregate of a few atoms—see Fig. 22.5. Since outer shell electrons can be exchanged more easily, the energy ranges created from valence electrons will be wider than the ranges created from the shielded inner electrons. The latter will more closely resemble the discrete eigenstates of isolated atoms. Since the same atoms behave alike, this allowed energy range extends throughout the crystal. In two dimensions (x, E) one therefore can draw allowed *energy bands* separated by *forbidden zones* (Fig. 7.1).

---

* Applying $\Delta E \Delta t \simeq \hbar$, and relating Δt to the time an electron resides at a sufficiently high energy level E_{ik} (later identified as belonging to an upper band), an uncertainty of ΔE results. The time Δt is related to scattering (see Section 28.1.4); the electron is removed from this level after $\lambda/v_{\mathrm{rms}} \simeq 10^{-12}$ s, yielding an uncertainty of $\simeq 1$ meV, which is on the same order as the splitting provided by only 10^4 atoms (assuming a band width of ~ 1 eV and an equidistant splitting of 1 level per added atom—that is, within a crystallite of < 100 Å diameter. With larger crystallites the splitting is even closer and results in a level continuum.

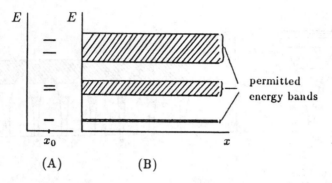

(A)   (B)

Figure 7.1: (A) Splitting of eigenstates when two atoms have approached each other. (B) Simple band model of a many-atom crystal.

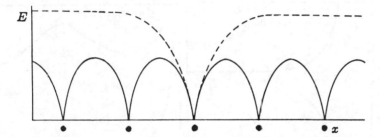

Figure 7.2: Potential energy of an electron in an atom (dashed curve) and an atomic cluster (solid curve).

In Fig. 7.1 the total electron energy is drawn disregarding the potential energy that an electron experiences when separated from an individual atom, which is shown in Fig. 7.2 for a single atom (dashed curve) and for a small one-dimensional cluster (solid curve). The band model emphasizes the *collective behavior*, i.e., the *sharing* of the electron among the atoms of the cluster. The potential distribution picture, on the other hand, emphasizes the *localization* of an electron within each potential funnel. Both pictures are valid: the band picture is more relevant for higher bands, while the potential picture is more relevant for lower (core) levels.

When the band picture is overlaid on the picture of the individual potentials of many adjacent atoms (Fig. 7.3), we recognize that the semiclassical approach, in which electrons may only move above potential barriers, is inappropriate, since bands which indicate electronic exchange exist well below the crest of the barriers. *Tunneling* through such barriers, i.e., a *quantum-mechanical* exchange, is the reason for the electron transfer—see Section 42.4.3.

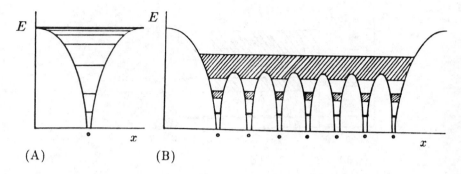

Figure 7.3: Potential energy and eigenstates of electrons in (A) an atom and (B) a small crystal.

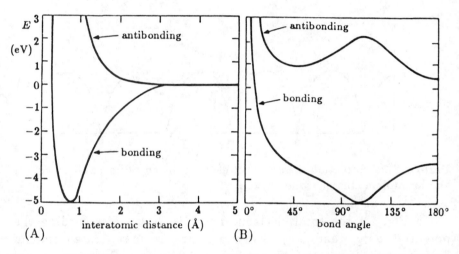

Figure 7.4: Electron energy as a function of: (A) the interatomic distance of a hydrogen molecule; and of (B) the bond angle in H_2O (after Adler, 1985).

This heuristic approach will now be expanded to make bands "plausible" in a more appropriate quantum-mechanical analysis. The analysis will be applied first to a cluster of atoms that form the building block of an amorphous semiconductor, and then to a periodic lattice.

7.1.1 Electronic Structure of Amorphous Semiconductors

The electron energy depends sensitively on the interatomic distance and bond angle, as shown for a simple H_2 and an H_2O molecule in Fig. 7.4. The bonding and antibonding curves reflect antiparallel and parallel spin of the electrons in the H molecule, respectively.

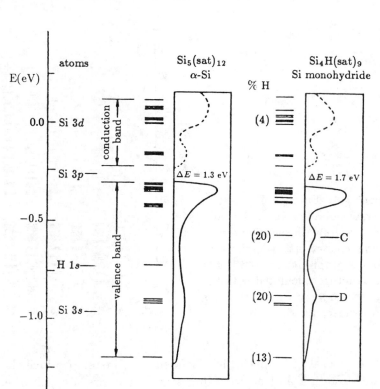

Figure 7.5: Electron energy level distribution for bonding (valence band) and antibonding (conduction band) states of a 17-atom cluster and optical density of states (see Section 13.1) for (A) α-Si, and (B) α-Si:H (after Johnson et al., 1980). The agreement with the experimental band gap of such a small cluster is spurious, however, and should not be overevaluated. The calculated band gap depends substantially on the boundary conditions.

The electronic structure of a solid can be obtained by starting from an arbitrary atom and including more and more neighbors in an appropriate configuration; the eigenfunctions of such a cluster are determined by solving its Schrödinger equation. This is referred to as a *tight-binding approach*. Solutions can be obtained numerically, using reasonable approximations (Reitz, 1955; Heine, 1980; Slater and Johnson, 1972—see also Chapter 8). The analysis can be described as that for a large molecule of, say, 20–50 atoms and delivers a spectrum of *energy eigenvalues* that, when the cluster is large enough, presents a valuable estimate of the upper bands.

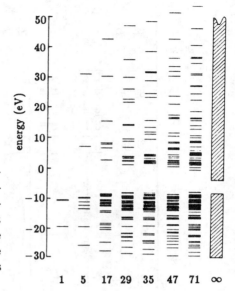

Figure 7.6: Energy of one-electron molecular orbitals for diamond clusters with successively increasing cluster size (numbers at bottom of graph), compared to the band structure calculated with the same parameters (after Watkins and Messmer, 1973).

The several Si atoms which form such a cluster produce the typical sp^3-bonding and antibonding states that are formed when the atoms are close enough—see Section 2.2 and Fig. 2.7. For a small cluster of Si atoms in an "amorphous configuration" (Section 3.9), we calculate an eigenvalue spectrum (Fig. 7.5A) and see that the proximity to other Si atoms significantly changes and splits the atomic levels (shown on the left). They are split by a large amount and are distributed unevenly in energy. A much larger number of atoms, however, is required to create a truly band-like level distribution.

This spectrum changes significantly when hydrogen is added to this cluster, which forms a bridging hydrogen structure (Ovshinsky and Adler, 1978). It removes states from the gap of α-Si and thereby increases the band gap from 1.3 eV for α-Si to ~ 1.7 eV for the technically more interesting, hydrogenated α-Si:H as shown by Eberhart et al. (1982).

When more atoms of the same kind are incorporated within such a cluster, more levels appear *within* the two bands, i.e., within the range of bonding and antibonding states. These bands have been labeled in Fig. 7.5 as the conduction and valence bands for the benefit of the experienced reader. For more detailed definitions, see Section 9.1. Such level distribution is shown for diamond C in a 35-atom molecule in Fig. 22.5, and for a successively larger cluster in Fig. 7.6.

The finite cluster approach always overestimates the band gap energy, since, by necessity, this approach omits states present which are far removed from the center that corresponds to the Γ-point of the Brillonin zone and, in direct band gap materials (Section 13.2) lie near the edges of the band gap.

From these examples, we can deduce that some information about the energy width of the upper energy bands and the band gap can be obtained from clusters containing only ∼ 50 atoms in an appropriate configuration. That is, the outer atoms must be kept artificially at positions they would attain when interacting with the surrounding atoms within a much larger amorphous network of atoms.

The level distribution within a band, however, is poorly represented by such a small cluster. The incorporation of many more atoms presents major computational problems for amorphous semiconductors; however, this problem becomes exceedingly simple in a periodic lattice of a crystalline semiconductor. For more reading, see Adler (1985).

7.2 The Periodicity Approach

The behavior of electrons in a semiconductor can be approximated by assuming that they are nearly *free electrons*, but interact with the *periodic potential* that simulates the lattice. In order to distinguish the influence of this periodic potential, one should first recall the behavior of a *free electron in vacuo*. This is determined by the solution of the *Schrödinger equation* (see Appendix A.4.2A):

$$\nabla^2 \psi + \frac{2m_0}{\hbar^2} E \psi = 0, \tag{7.1}$$

which can be described by an *electron wave*

$$\psi(\mathbf{r}) = A \exp(\pm i\, \mathbf{k} \cdot \mathbf{r}) \tag{7.2}$$

with A as an amplitude factor. The *wave vector* $\mathbf{k}$ relates to electron momentum and energy as

$$\mathbf{k} = \frac{m_0 \mathbf{v}}{\hbar} = \frac{\mathbf{p}}{\hbar}; \qquad E = \frac{m_0}{2} v^2 = \frac{p^2}{2m_0} = \frac{\hbar^2 k^2}{2m_0}, \tag{7.3}$$

or, more accurately, to the expectation value of the momentum given by

$$\langle \mathbf{p} \rangle = \int_{-\infty}^{\infty} \psi \frac{\hbar}{i} \nabla \psi^* \, d\mathbf{r} = \hbar \mathbf{k} \int_{-\infty}^{\infty} \psi \psi^* \, d\mathbf{r} = \hbar \mathbf{k}. \tag{7.4}$$

The wave vector **k** is the reduced wave vector—see the discussion later in this section and Fig. 7.12.

Hence, $E(\mathbf{p})$ or $E(\mathbf{k})$ is described by a three-dimensional paraboloid (by a parabola in one relevant coordinate) with one electronic parameter, the *electron rest mass.*

Equation (7.2) represents an electron wave with a wavelength, the *de Broglie wavelength,** of

$$\lambda_{DB} = \frac{2\pi}{k} = \frac{h}{|p|} = \frac{h}{m_0}\frac{1}{v} = 7.27 \frac{1}{v(\mathrm{cm/s})} \ (\mathrm{cm}) \qquad (7.5)$$

or, when introducing the electron energy from Eq. (7.3),

$$\lambda_{DB} = \frac{h}{\sqrt{2m_0 E}} = 12.26 \frac{1}{\sqrt{E(\mathrm{eV})}} \ (\mathrm{\AA}). \qquad (7.6)$$

An electron in the lattice, i.e., when it is exposed to a periodic potential, no longer behaves like a free particle: it experiences interference from the lattice potential when, with increasing electron energy, its de Broglie wavelength becomes comparable to the lattice constant. The ensuing *Bragg reflections* prohibit a further acceleration of the electron, described later in more detail. This simple discussion also indicates the existence of a finite energy range, the energy band in a semiconductor. Near the band edges the electron behaves to some extent like a free electron, i.e., like a classical particle. The quantum-mechanical nature becomes evident when it gains energy in an electric field or is forced to occupy higher states. At energies of 4 eV, the de Broglie wavelength is $\sim$ 6 Å, i.e., small enough to permit interference effects within the periodic potential of the lattice.

This plausibility argument can be substantiated by describing the electron with a wave equation, the Schrödinger equation (see Appendix A.4.2), and by introducing into the Schrödinger equation a periodic potential $V(\mathbf{r})$,

$$\nabla^2 \psi + \frac{2m_0}{\hbar^2}\left(E(\mathbf{k}) - V(\mathbf{r})\right)\psi = 0. \qquad (7.7)$$

---

* The de Broglie wavelength is on the same order of magnitude as the uncertainty distance obtained from Heisenberg's uncertainty principle $\Delta x \gtrsim \hbar/\Delta p_x$, which has the same form as λ_{DB}. This yields uncertainty distances of $\sim$ 10 Å for thermal (free) electrons at room temperature.

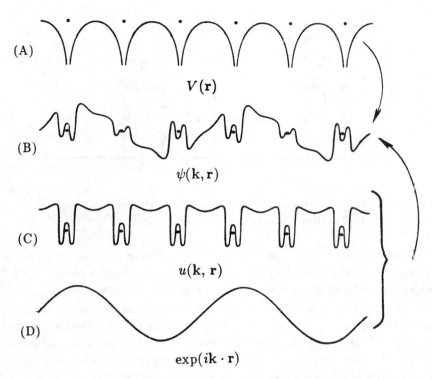

Figure 7.7: A schematic representation of electronic eigenstates in a crystal. (A) The potential plotted along a row of atoms; (B) A Bloch function; the state itself is complex but only the real part is shown. The Bloch function is composed of the product of (C) $u_k(\mathbf{r})$, which has the periodicity of the lattice; and (D) a plane electron wave, the real part of which is shown to construct the electron wave function (after Harrison, 1970).

The solutions of this Schrödinger equation are so-called *Bloch-functions* which can be expressed as a linear combination of waves

$$\psi_n(\mathbf{k}, \mathbf{r}) = u_n(\mathbf{k}, \mathbf{r}) \exp(i\,\mathbf{k} \cdot \mathbf{r}), \qquad (7.8)$$

with n as the band index specifying a certain band. The waves are plane waves with a space-dependent amplitude factor $u_n(\mathbf{k}, \mathbf{r})$, which shows lattice periodicity (Bloch's theorem, 1928; see Section 8).

A one-dimensional schematic representation is given in Fig. 7.7 to indicate the relationship between the lattice potential $V(r)$ and the Bloch function $\psi_n(\mathbf{k}, \mathbf{r})$, which contains $u_n(\mathbf{k}, \mathbf{r})$ and the plane wave function of the electron $\exp(i\mathbf{k} \cdot \mathbf{r})$ (Harrison, 1970), to construct the electron wave function.

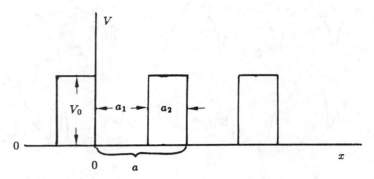

Figure 7.8: Kronig-Penney potential with V_0 the barrier height, a_1 and a_2 the well and barrier widths.

Inserting Eq. (7.8) into Eq. (7.7), we obtain the result that solutions exist only for certain ranges of the electron energy $E_n(\mathbf{k})$, which are interspersed with energy ranges in which real solutions do not exist. This confirms the previously obtained results: namely, that the energy spectrum in a solid consists of alternating allowed and forbidden energy ranges (energy bands). The periodic potential approach, however, gives additional information that can be demonstrated readily in a simple one-dimensional (x) model.

7.2.1 The Kronig-Penney Model

An enormously simplified periodic potential $V(x)$ is sufficient for introduction into Eq. (7.9) to show the typical behavior. This is the *Kronig-Penney potential* (Kronig and Penney, 1931),* which is shown in Fig. 7.8. Since the discussion of this behavior is rather transparent, it will be used here to introduce the reader to the basic features of the band model.

Introducing (7.8) into Eq. (7.7) for one relevant dimension, we see that $u(x)$ must satisfy

$$\frac{d^2 u}{dx^2} + 2ik\frac{du}{dx} - \left(k^2 - \frac{2m_0\left[E - V(x)\right]}{\hbar^2} \right) u = 0. \qquad (7.9)$$

---

* In one dimension, there are other periodic potentials for which the Schrödinger equation can be integrated explicitly. $V(x) = -V_0\mathrm{sech}^2(\gamma x)$ is one such potential, which yields solutions in terms of hypergeometric functions (see Mills and Montroll, 1970). The results are quite similar to the Kronig-Penney potential discussed later.

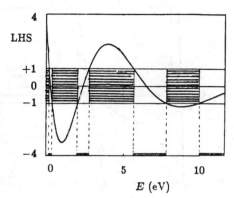

Figure 7.9: Left-hand side (LHS) of Eq. (7.12) as a function of E (contained in α and β), computed for $a_1 = 6$ Å, $a_2 = 1.2$ Å, and $V_0 = 10$ eV.

We can split Eq. (7.9) after the introduction of the Kronig-Penney potential into two differential equations: one for the bottom of the well, and one for the top of the barrier with a potential $V = V_0$. The solutions in each part can be expressed as the sum of two waves:

$$u_1(x) = A \exp\left[i\left(\alpha - k\right)x\right] + B \exp\left[-i\left(\alpha + k\right)x\right] \quad \text{for } 0 < x < a_1$$
$$u_2(x) = C \exp\left[i\left(\beta - k\right)x\right] + D \exp\left[-i\left(\beta + k\right)x\right] \quad \text{for } -a_2 < x < 0$$
$$(7.10)$$

where α and β are the k-values for a free electron *in vacuo*, for $V = 0$, and for a constant barrier potential V_0, respectively:

$$\alpha = \sqrt{\frac{2m_0 E}{\hbar^2}} \quad \text{and} \quad \beta = \sqrt{\frac{2m_0 (V_0 - E)}{\hbar}}. \qquad (7.11)$$

The integration constants can be determined by the continuity requirements of $u(x)$ and its first derivatives at $x = a_1$ and $x = a_2$, which yield*

$$-\frac{\alpha^2 - \beta^2}{2\alpha\beta} \sin(\alpha a_1) \sinh(\beta a_2) + \cos(\alpha a_1) \cosh(\beta a_2) = \cos(ka). $$
$$(7.12)$$

Eq. (7.12) provides the *dispersion relation* $E(k)$ (E is contained in α and β).

---

* For $E > V_0$, the square root in β becomes imaginary. Introducing $\gamma = i\sqrt{2m_0(E - V_0)/\hbar^2}$, and with $\sinh(i\gamma) = i \sin\gamma$ and $\cosh(i\gamma) = i \cos\gamma$, we obtain for higher electron energies a similar equation:

$$-\frac{\gamma^2 + \alpha^2}{2\alpha\gamma} \sin(\gamma a_2) \sin(\alpha a_1) + \cos(\gamma a_2) \cos(\alpha a_1) = \cos(ka).$$

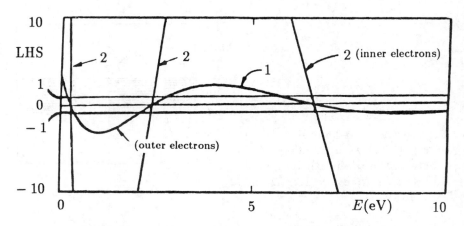

Figure 7.10: LHS as in Fig. 7.9, but for two different values of the parameter V_0 (10 eV for curve 1 and 40 eV for curve 2—other parameters as in Fig. 7.9), indicating the reduced width of the permitted bands for higher potential barriers, (i.e., for inner electrons that are more tightly bound).

The dispersion relation is the key to many discussions of electronic properties in solids. Since the wavenumber k is proportional to the electron momentum (Eq. (7.3)), the dispersion equation relates the electron energy to mass and velocity, both of which are essential for understanding the specific behavior of electrons in a semiconductor. This will be explained in detail in several of the following sections.

Equation (7.12) reveals that a sequence of allowed energy ranges is interspersed with forbidden energy ranges: energy gaps are formed when the left-hand side (LHS) of Eq. (7.12) exceeds ± 1, which are the limiting values of the equation's right-hand side. In Fig. 7.9 the hatched ranges show the energy gaps; no solution of the Schrödinger equation can be found here for real values of k. This picture describes a situation between a free electron *in vacuo*, where all energies are permitted, and an electron bound to an isolated atom, where the permitted energy ranges shrink to a set of discrete energy *levels*. The height and width of the potential barriers and wells (a_1, a_2, and V_0) determine whether an electron behaves more like an electron bound to a single atom (large a_2/a_1 and V_0) or more like a free electron *in vacuo* (small a_2 and V_0); see Fig. 7.10. In the latter example the permitted ranges extend over a wider energy range.

More information can be deduced from the $E(k)$ behavior within each of the permitted energy ranges shown in Fig. 7.11. At the bottom of the first permitted energy range, $E(k)$ is nearly parabolic.

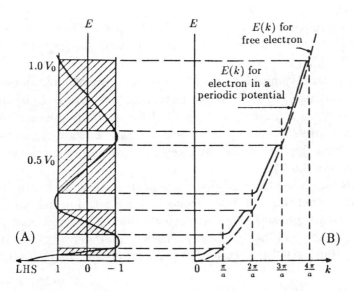

Figure 7.11: (A) As Fig 7.9, however, for a larger a_1/a_2 ratio. (B) $E(k)$ for a free electron (parabola) and for a Kronig-Penney potential in an extended wavenumber (k) representation.

Then E moves with increasing k through an inflection point, and, at the top of this range, becomes nearly parabolic again, but with a negative curvature.

Compared to the parabola of the free electron, the lower part of the $E(k)$ curve is raised. At the upper edge of the first allowed range, i.e., at $k = \pm\pi/a$, the curve coincides again with the free electron parabola. The next permitted band starts after a jump in E from E_1 to E_2 at $k = \pm\pi/a$, and has a similar $E(k)$ behavior as the first energy band, except that the curvatures are larger at the bottom and top of the band. The top is reached at $k = \pm2\pi/a$, where again a jump of E occurs, from E_3 to E_4, etc. (Fig. 7.11). This behavior continues for higher bands with broader allowed bands and narrower band gaps. Figure 7.11 also contains $E(k)$ for the free electron [Eq. (7.3)], which is parabolic in the entire $E(k)$ range.

This qualitative behavior is independent of the actual shape of the periodic potential as long as it has sufficient amplitude. Although periodicity of $V(x)$ is a necessary—but not sufficient—condition for energy bands with interspersed forbidden gaps, it so happens that in solids, for inner shell electrons, the potential barriers are sufficiently high to cause rather narrow, lower bands. Electrons at sufficiently high energies occupy wider bands and behave more like free electrons:

they can move readily through the lattice. They will, however, be subject to interference with the periodic lattice potential (see Section 7.2.3).

When analyzing the effect of a three-dimensional periodic potential and using a real lattice potential, the actual $E(\mathbf{k})$ behavior becomes more complex; however, it still maintains the basic features of *energy bands* interspersed with *band gaps*. The gaps, however, disappear from higher bands because of an overlap of permitted bands (see Section 8.5.3).

This fundamental behavior is the basis for the electronic behavior of semiconductors, and is described in more detail in many textbooks of solid state physics, e.g., Anderson (1963), Ashcroft and Mermin (1976), Bube (1976), Callaway (1976), Fletcher (1971), Harrison (1980), Haug (1972), Kittel (1985), Kittel (1986), and Ziman (1972).

7.2.2 Periodicity of E(k); Reduced k-Vector

A general feature of the solutions of the Schrödinger equation is the periodicity of $E(\mathbf{k})$, given in Fig. 7.12B. This figure shows the periodicity in $\mathbf{k}$ with a period length of $k_x a = 2\pi$. This means that a shift of the solution $E(k_x)$ by $2\pi/a$ in k_x represents the same behavior. This is indicated in Fig. 7.12 which contains as Fig. 7.12A a copy of Fig. 7.11, and shows explicitly the periodicity in Fig. 7.12B. Any full segment of the periodic representation is a *reduced k-vector representation*. It is shown within the first *Brillouin zone* in Fig. 7.12C; i.e., within $-\pi/a < k_x < \pi/a$. It will be explained for a three-dimensional lattice in Section 8.5.1.

The reduced $E(k)$ shows an alternating sign of the curvature at the edge of each band at $k = 0$. It is positive for the first band, negative for the second, etc. This interesting peculiarity occurs in real crystals in a somewhat similar fashion, although is more complex because of a multiplicity of bands, as will be discussed in Section 9.1.5.

It is instructive to look at an enlarged detail of Fig. 7.12 as shown in Fig. 7.13D. This figure can be constructed from two parabolas of free electrons, shifted by $\frac{2\pi}{a}$ —a situation that can be thought of by inserting a lattice into the vacuum, although with vanishing lattice potential (an *empty lattice*—see Section 8.5.2). The electron in each reference system is described by its corresponding parabola (subfigure B). When interacting through a periodic perturbation potential of amplitude U_0, the crossing of both E(k) parabolas is

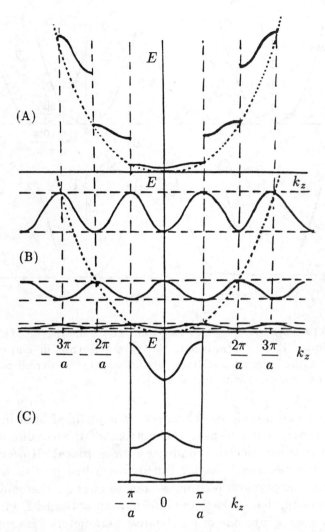

Figure 7.12: Comparison between: (A) extended wavenumber k; (B) periodic, and (C) reduced wavenumber representations of $E(k)$.

eliminated and a splitting occurs with a gap of the order of $|2U_0|$, as shown in subfigures C and D.

7.2.3 Newtonian Description of a Quasi-Free Electron

In many discussions about electron behavior in solids, a classical particle picture is used rather than the quantum-mechanical one of a wave packet. It is often more intuitive. Electrons behave like little balls, "sliding down" a potential hill, and "scattering" upon

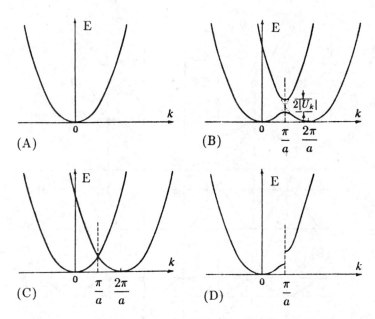

Figure 7.13: E(k) for (A) free electron. (B) Splitting of E(k) when a small periodic field is introduced. (C) Free electron in empty lattice. (D) E(k) of the original parabola, disturbed by the periodic potential perturbation (compare with Fig. 7.12A).

collision with an atomic lattice defect. It is justified by using Bohr's *correspondence principle* near the bottom of the conduction band (the band in which electron conduction takes place). However, since not all of the electron behavior can be described by this model, as explained in the previous section, we can account for the modification by incorporating (see below) the information obtained from its dual nature as a wave into one of its classical parameters—its mass.

An electron, regarded as a classical (Newtonian) particle, has a momentum

$$p = m_0 v \quad \text{and a kinetic energy} \quad E = \frac{m_0}{2} v^2 = \frac{p^2}{2m_0}. \quad (7.13)$$

Relativistic effects are excluded here (i.e., $v \ll c$ is assumed): the electron mass is its rest mass m_0. The velocity of such a particle changes with time in response to an acting force $\mathcal{F}$ (*Newton's second law*):

$$\frac{dp}{dt} = m_0 \frac{dv}{dt} = \mathcal{F}. \quad (7.14)$$

On the other hand, an electron *in vacuo*, regarded as a wave, has

a momentum $\quad p = \hbar k \quad$ and an energy $\quad E = \dfrac{p^2}{2m_0} = \dfrac{\hbar^2 k^2}{2m_0}. \quad$ (7.15)

When exposed to a force, such as that supplied by an electric field F, and with $\mathcal{F} = -eF$, its momentum increases accordingly:

$$\frac{dp}{dt} = \hbar \frac{dk}{dt} = \mathcal{F}. \qquad (7.16)$$

When the electron is described as a *wave packet*, its velocity is the *group velocity.**.

$$v_g = \frac{d\omega}{dk} = \frac{1}{\hbar} \frac{dE}{dk} \qquad (7.17)$$

Applying Newton's law to such an electron wave packet [in a relation similar to Eq. (7.14)],

$$m_0 \frac{dv_g}{dt} = \mathcal{F}, \qquad (7.18)$$

and with

$$\frac{dv_g}{dt} = \frac{1}{\hbar} \frac{d}{dt} \left(\frac{dE}{dk} \right) = \frac{1}{\hbar} \frac{d^2 E}{dk^2} \frac{dk}{dt} = \frac{1}{\hbar^2} \frac{d^2 E}{dk^2} \mathcal{F}, \qquad (7.19)$$

we see by comparison with Eq. (7.18) that the factor preceding $\mathcal{F}$ has the dimension of an inverse mass. This factor is proportional to the curvature of $E(k)$.

---

* In an infinite crystal the electron (when not interacting with a localized defect) is not localized and is described by a simple wavefunction (i.e., having one wavelength and the same amplitude throughout the crystal). The probability of finding it is the same throughout the crystal ($\propto \psi^2$). When localized, the electron is represented by a superposition of several wavefunctions of slightly different wavelengths. The superposition of these wavefunctions is referred to as a *wave packet*. A moving electron is represented by a moving wave packet $\psi = \dfrac{1}{2\delta k} \displaystyle\int_{k-\delta k}^{k+\delta k} u(x, k) \exp i(kx - \omega t) dk$ which quickly spreads out over time. It has its maximum at a position $\bar{x} = \dfrac{1}{\hbar} \dfrac{\partial E}{\partial k} t$, yielding for the group velocity, i.e., the velocity of the maximum of this wave packet $v_g = \dfrac{\partial \bar{x}}{\partial t} = \dfrac{1}{\hbar} \dfrac{\partial E}{\partial k}$. With $E = \hbar\omega$, we obtain $v_g = \dfrac{d\omega}{dk}$.

7.2.3A The Effective Mass   If we want to retain the Newtonian behavior, we have to replace the electron mass m_0 in Eq. (7.18) with the *effective electron mass* when comparing Eqs. (7.18) and (7.19):*

$$m_n = \frac{\hbar^2}{\dfrac{d^2 E}{dk^2}}. \qquad (7.20)$$

This effective mass contains the peculiarities of the interaction of the electron with the lattice (the subscript n distinguishes this mass from the rest mass). However, a possibly important part caused by the adiabatic approximation (see Section 8.1), is missing. The influence of this part is discussed in Section 27.1.2 and can be described by a different effective mass—the polaron mass.

From Fig. 7.11, we see that the effective electron mass at the lower edge of the third band, here assumed to harbor free electrons (see Section 9.1.1), is smaller than the rest mass of a free electron, since the *curvature* of $E(k)$ is larger here. At higher energies within the band this curvature decreases, changes sign, and, at the upper edge of the band, becomes negative (as shown in Fig. 7.14). Consequently, the effective electron mass increases, becomes infinite near the center of an allowed band, and changes sign there. Coming from negative infinity, the effective electron mass returns to a finite but negative value which, at the top of the band, is on the same order of magnitude as at the bottom of the band (Figs. 7.12 and 7.14).

This behavior is repeated in the next band, except that the sign sequence is exchanged. Here the effective mass is negative at $k = 0$; however, the effective electron mass is always positive at the bottom of any band and negative at the top. For lower bands, i.e., narrower bands, the value of the effective mass becomes larger at the band edge.

When electrons accelerate substantially above the lower edge of the band in sufficiently high fields, the de Broglie wavelength of the electron becomes smaller and comparable to the interatomic

---

* For the electron behavior, only *expectation values* can be given. In order to maintain Newton's second law, we continue to use $\hbar k$ [Eq. (7.15)], which is no longer an electron momentum. It is well-defined within the crystal and is referred to as *crystal momentum*. We then separate the electron properties from those of the crystal by using $d^2 E/dk^2$ to define its *effective mass*.

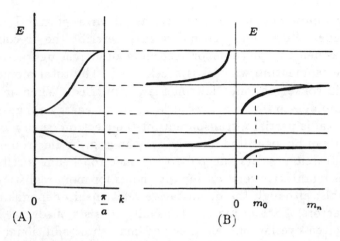

Figure 7.14: (A) Typical $E(k)$ for two simple bands, and (B) derived effective *electron* masses within these permitted energy bands. [Actually, one determines $m_n(k)$; this graph is turned 90° to show its relation to the band model shown at the left.]

lattice spacing. Here, *interference effects* of the electron wave with the periodic lattice potential become important: Bragg reflection becomes more prevalent, while more and more frequency components of the wave packet are reflected. Therefore, further acceleration will become more difficult to achieve; in the Newtonian model, the effective mass of the electron increases until, near the center of the band, further acceleration stops. When the energy of the electron is raised above the center of the band, the electron will *decelerate* in the direction of the electric field until it reaches the top of the band, where it will come to a standstill. The electron wave has then reached a perfect diffraction condition.* It can be described as a *standing*

---

* In theory, the electron will continue to accelerate in the opposite direction to the field and lose energy, thereby descending in the band, and the above-described process will proceed in the reverse direction until the electron has reached the lower band edge, where the entire process repeats itself. This oscillating behavior is called the *Bloch oscillation*. Long before the oscillation can be completed, however, scattering (see Chapter 32) interrupts the process. Whether in rare cases (e.g., in narrow mini-bands of superlattices or ultrapure semiconductors at low temperatures) such Bloch oscillations are observable, and whether they are theoretically justifiable in more advanced models (Krieger and Iafrate, 1986) is still controversial. In three-dimensional lattices, other bands overlap and transitions into these bands complicate the picture.

wave, composed of incoming and refracted waves of exactly the same amplitude. With some caution we may describe the "recoil" of the lattice as being responsible for absorbing an increasing fraction of the electron momentum when it is accelerated. The total momentum is thus still conserved, and Newton's law is fulfilled. When an electron wave impinges on a thin crystal layer in an energy range in which the crystal is partially transparent for the electron, such momentum transfer can be measured directly by changing the electron energy so that diffraction occurs and part of the electron beam is reflected.

This qualitative relationship also holds for more realistic periodic potentials, although the quantitative relationship depends on many other factors. Each of the bands usually consists of several branches which often overlap one another and may show additional extrema (saddle points) in the first Brillouin zone, making the dependence of the effective mass on the electron energy more complicated. Near the band edge (for electrons), only one—perhaps degenerate—$E(k)$ branch is present in typical semiconductors, so that the above description holds rather well. This branch can be split, for example, by crystal anisotropy or electric or magnetic fields. This will be discussed in Section 13.3.

In summarizing the much more involved behavior of an electron in such a realistic band, we may wonder if we gained a more intuitive picture using the particle model. If we recognize, however, that the electron will mostly reside close to the bottom of the band, usually within a few kT, the model is then quite helpful for an analysis of a number of basic processes. The electron will behave here like a particle with a constant effective mass; the value of this effective mass depends on the actual lattice potential, i.e., on the chemical and crystallographic nature of the material because these determine the shape of $E(k)$—see Section 27.1.2.

In Section 30.3.2 we present a more detailed description of the effective mass for the application of this concept to carrier transport in typical semiconductors.

7.3 Periodicity vs. Proximity Approach

The proximity of a sufficient number of atoms and the periodic lattice structure of a crystal, both lead to allowed energy bands interspaced by band gaps. We may use one or the other picture to obtain further information about the band structure.

The *periodic lattice structure approach* is more suited for obtaining the specific $E(\mathbf{k})$ structure of the inner part of the band (near $k = 0$) which cannot be obtained from the proximity approach. It reflects the symmetry of the lattice and permits one to obtain the results in the most economical fashion. Its results, however, are restricted to periodic lattices, i.e., to crystals. This refers specifically to interference phenomena involving diffraction from further-than-nearest and next-nearest neighbor distances. These distances, however, can still be discerned in the x-ray diffraction of amorphous semiconductors, and therefore may also be expected to influence electron behavior further away from $k = 0$.

The *proximity approach* can be used to obtain some information about the inside of the bands for first orientation. However, the inadvertent inclusion of artificial states at the surface of the cluster and the requirement for an extremely large cluster size to provide band states close to the band edges have been the handicaps of this approach.

A *supershell approach* is sometimes used to avoid some of the shortcomings of the periodic lattice and proximity approaches. This approach takes a cluster of sufficient size and repeats it *periodically* until the entire crystal volume is filled. In this fashion the mathematical methods developed for studying periodic lattices can be used, while certain elements of an amorphous structure are included in the cluster. The error due to the forced adjustment of each cluster can be minimized by increasing the size of the cluster.

Many of these results are important for understanding the behavior of metals (e.g., overlapping bands), but will not be discussed here. Other results relate to semiconductors, including semiconductor-metal transitions (see Section 34.1). Some heuristic examples of near-band edge properties are given below.

7.3.1 Band-Edge Fuzzing (Deviation from Periodicity)

The ideal periodicity of a crystal lattice can be modified for a number of reasons, among them lattice oscillations or displaced lattice atoms. An amorphous semiconductor, for example, may be described by having frozen-in large fluctuations of the interatomic distances and bond angles. In some respects such structures are "almost crystalline," but with slightly changing lattice parameters. A lattice with a different lattice constant causes a different $E(\mathbf{k})$ with a different width of allowed bands and gaps. Therefore, we expect variations of the band

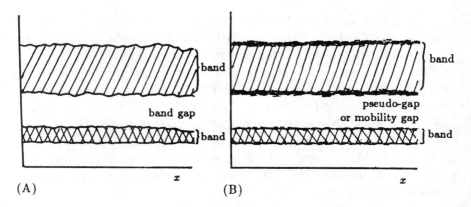

Figure 7.15: (A) Perturbed and (B) fuzzed-out band edges in a crystal with phonons and in an amorphous semiconductor.

edges in time and space.* Rather than being perfectly straight, the band edge becomes perturbed. Over a time average, the band edge appears to be fuzzed-out (Fig. 7.15). For further detail see Chapter 25.

7.3.2 Discrete Defect Level in the Band Gap

When the deviation from the ideal lattice structure is sufficiently large, the eigenstate of a *disordered atom* may lie within the band gap. A plausibility argument may be obtained from the proximity model.

Assume an extra atom is incorporated in an *interstitial site* of the lattice (later discussed in Section 19.1.1). This extra atom is much closer to its neighbors; the exchange frequency is substantially larger than that for the nearest neighbors which yield the largest exchange frequency in an ideal lattice (equivalent to the band edges). Thus the eigenstates of this interstitial atom, here an *intrinsic point defect*,

---

* This concept must be used with caution, since **k** is a good quantum number only when electrons can move without scattering over at least several lattice distances. That is certainly not the case in most amorphous semiconductors near the "band edge" (see Section 40.4). However, at higher energies further inside the band, there is some evidence that the mean free path (Section 28.1.5) is much larger than the interatomic distance even in amorphous semiconductors. In bringing the two approaches together, the argument presented here lacks rigor and has plausibility only in terms of correspondence.

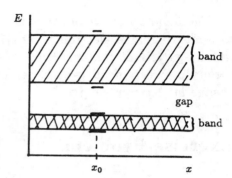

Figure 7.16: Simple intrinsic (i.e., chemically identical to the atoms of the crystal) atomic interstitial in an idealized lattice.

and *its* nearest neighbors lie outside of the allowed bands of the ideal lattice, i.e., within the band gaps (Fig. 7.16).

Energy states within the gap are *localized* at the position of this lattice defect (x_0 in Fig. 7.16), and play an important role in localizing (*trapping*) electrons in real crystals (see Section 27.3) and in amorphous semiconductors. It also becomes reasonable to expect an energy distribution of such localized (*trap*) levels in the gap near the band edge, when taking into consideration that in crystalline and amorphous semiconductors a wide variety of lattice imperfections and lattice parameter variations are observed. In Chapter 24 we will return to this level distribution near the band edge; but first, a more sophisticated analysis of the three-dimensional periodic lattice will be presented.

Summary and Emphasis

The typical band structure with alternating bands and band gaps is characteristic for all solids in contrast to isolated atoms which show a discrete level spectrum. The detail of the band structure, however, depends on the chemistry of the material and its atomic structure (symmetry). Deviation from a periodic structure predominantly influences the energy range near the band edges while it has little influence near the center of the bands.

Electrons near the lower edge of a band in a periodic lattice behave akin to electrons *in vacuo*. The influence of the lattice can be expressed as an effective mass which, for typical semiconductors, is smaller than the electron rest mass at the band edge and increases with increasing distance from the band edge.

In disordered or amorphous semiconductors the band edge is fuzzy and the electronic states become localized when extended sufficiently beyond the band edge.

The band structure of solids is the singlemost significant feature to understand the electronic behavior of semiconductors. The detailed analysis of this structure has provided insight into a large variety of effects dealing with electronic transport and any induced change in electron energy.

Exercise Problems

1.   The acceleration of carriers is intimately related to the effective mass.

   (a) Introduce the effective mass into the de Broglie wavelength and compare it with the lattice constant of better semiconductors.

   (b) What is the result with respect to diffraction conditions if, following an external force, the electron climbs higher in a nonparabolic band?

   (c) Describe the motion of an electron in the conduction band with a single, idealized $E(k)$ as given in the upper band of Fig. 7.14 under the influence of an external field and without scattering.

   (d) Draw a vector diagram showing the acceleration as a function of the energy within this conduction band.

   (e) What do you conclude about the ability of an electric field to move electrons substantially above the band edge (as in the case of high fields, capable of impact ionization). Electrons must reach at least a height within the conduction band well in excess of the band gap. Develop your suggestions carefully.

2.(e) Show that the group velocity is twice the phase velocity for a free electron gas. Discuss its meaning.

3.(*) The effective mass is a function of the energy and changes substantially toward the center of the band.

   (a) Monovalent metals have a partially filled valence band. One can deduce that *all* electrons within this band contribute to the conductivity. Assuming again a simple $E(k)$ dispersion within such a band (draw it), what do you conclude about the effective mass distribution and the effective mass at the Fermi surface? (Careful!)

   (b) What could this mean to the response of carriers at various energies within the band to an external force?

4.   Verify Eq. (7.12) as a condition that must be fulfilled if the solutions of the Schrödinger equation for the one-dimensional *Kronig-Penney potential*,

$$u_1(x) = A \exp[i(\alpha - k)x] + B \exp[-i(\alpha + k)x] \text{ for } 0 < x < a_1$$

$$u_2(x) = C \exp[i(\beta - k)x] + D \exp[-i(\beta + k)x] \text{ for } a_1 < x < a_2$$

and their slopes are continuous at $x = 0$ and $x = a_1$. Remember, $u(x)$ is a periodic function with periodicity $a_1 + a_2$.

(a) The amplitude of the Kronig-Penney potential enters decisively into the solution of the Schrödinger equation. Discuss the solution [Eq. (7.12)] in the limit $V_o \to 0$.

5.(e) Calculate the ground state for an electron confined in a one-dimensional well of width d. Give d when the ground state is at kT for room temperature.

6.(r) The effective mass can be described in terms of the curvature of $E(k)$ or as a function of the momentum. Explain the differences and reasons to use one or the other model.

7.   A useful formula relates the effective mass to the width of the nth band gap (E_{gn}):

$$m^* = \frac{m_o}{1 + \dfrac{4E_n}{E_{gn}}}, \qquad (7.21)$$

with $E_n = [\hbar^2/(2m_o)](n\pi/a)^2$. Derive this relation from $E(k)$, which can be approximated near band edges by (McKelvey, 1966):

$$E(k) = \frac{\hbar^2}{4m_o} \left\{ k^2 + \left(k - \frac{2\pi n}{a}\right)^2 \pm \sqrt{\left[k^2 - \left(k - \frac{2\pi n}{a}\right)^2\right]^2 + \left(\frac{2mE_{gn}}{\hbar^2}\right)^2} \right\}$$

where E_{gn} is the band gap between the nth and the $(n-1)^{st}$ band. Hint: expand the square root near the edge of the band.

8.(*) Could you transfer knowledge about the band structure obtained from crystals to glasses, and where does the similarity break down? Can the effective mass picture be used in glasses? Explain your reasoning carefully.

Chapter 8

Quantum Mechanics of Electrons in Crystals

The band structure of semiconductors reveals most of their intrinsic properties. Its key is the dispersion relation $E_n(\mathbf{k})$ for the various bands.

In order to understand electronic transitions and the electron transport in a semiconductor, one must analyze its three-dimensional band structure, which is determined by solving the Schrödinger equation with a three-dimensional lattice potential for the actual crystal. To obtain such a potential and to solve the Schrödinger equation amounts to solving a *many-body problem*, because it involves all the lattice atoms and all of their electrons. It requires several sets of approximations for the problem to become manageable. The limited space for this chapter permits only a brief enumeration of the different topics rather than an in-depth discussion. Further relevant reading includes Herman and Skillman (1963), Fletcher (1971), Ziman (1971), Phillips (1973), Bassani et al. (1975), Ashcroft and Mermin (1976), and Harrison (1980). In this chapter, following the conventional description, cgs units are used.

In the previous sections a crude periodic potential was presented for a one-dimensional, one-electron model, which yielded some basic results for the band model. In this section we will take the opposite approach. Starting from the most general model, we will simplify it step by step until it can be solved mathematically. This treatment will provide some insight into the present understanding of the band structures.

8.1 The Schrödinger Equation

It is relatively easy to write down in a general form the Schrödinger which describes the many-body problem of n electrons in a lattice of

N atoms, including atomic motion within the crystal, but excluding spin-orbit interaction and relativistic effects:

$$-\sum_{i=1}^{n} \frac{\hbar^2}{2m_0} \nabla_i^2 \Phi - \sum_{k=1}^{N} \frac{\hbar^2}{2M_k} \nabla_k^2 \Phi + \sideset{}{'}\sum_{i,j=1}^{n} \frac{e^2}{2r_{ij}} \Phi$$

$$+ V_{\text{ion-ion}}(\mathbf{R_1}, \mathbf{R_2}, \ldots)\Phi + V_{\text{el-ion}}(\mathbf{r_1}, \mathbf{r_2}, \ldots, \mathbf{R_1}, \mathbf{R_2}, \ldots)\Phi = E\Phi. \tag{8.1}$$

The prime at the summation sign indicates that summation over the same indices ($i = j$) is to be excluded. Here R_i are the ion coordinates and r_i the electron coordinates; Φ is the wavefunction for the system of atoms and electrons. The five terms on the left-hand side describe: the kinetic energy of the electrons; the kinetic energy of the atoms; the potential energy due to the electron-electron interaction; the potential energy due to the interatomic interaction; and, finally, the potential energy due to the electron-atom interaction. A major problem is the extremely large number of terms in the sums of Eq. (8.1). A first step usually made to reduce the complexity is to separate the chemically inert core and valence electrons and to neglect a deformation of the ion core.*

For a reasonable first approximation, one takes into consideration the large ratio of atomic to electron masses. This lets the lattice oscillations appear to stand still for the much faster, more easily accelerated electrons and permits the use of an *adiabatic approximation* (Born and Oppenheimer, 1927; Wigner and Pelzer, 1932). With the *Ansatz*

$$\Phi(\mathbf{R_1}, \mathbf{R_2}, \ldots, \mathbf{r_1}, \mathbf{r_2} \ldots) = \varphi(\mathbf{R_1}, \mathbf{R_2}, \ldots)\psi(\mathbf{R_1}, \mathbf{R_2}, \ldots, \mathbf{r_1}, \mathbf{r_2}, \ldots), \tag{8.2}$$

a separation of the atomic and electronic eigenfunctions can be achieved. The resulting *Schrödinger equation for electrons* in a lattice with atoms at rest is

---

* The latter assumption is questionable when explaining certain dielectric properties with strong polarization (Chapter 16). If a core–valence electron separation is *not* made, all potentials in Eq. (8.1) are simple Coulomb potentials. Otherwise, the effective interaction potentials must be obtained.

$$-\sum_{i=1}^{n}\frac{\hbar^2}{2m_0}\nabla_i^2\psi + \sum_{i=1}^{n}{}'\frac{e^2}{2r_{ij}}\psi$$

$$+ V_{\text{el-ion}}(\mathbf{R_1}, \mathbf{R_2}, \ldots, \mathbf{r_1}, \mathbf{r_2}, \ldots)\psi = E'(\mathbf{R_1}, \mathbf{R_2}, \ldots)\psi. \tag{8.3}$$

and the *Schrödinger equation* for the oscillating atoms is

$$-\sum_{k=1}^{N}\frac{\hbar^2}{2M}\nabla_k^2\varphi$$

$$+ \left[V_{\text{ion-ion}}(\mathbf{R_1}, \mathbf{R_2}, \ldots) + E'(\mathbf{R_1}, \mathbf{R_2}, \ldots) \right]\varphi = E\varphi. \tag{8.4}$$

Here the total electron energy (E') is added to the potential energy of the lattice. Two terms responsible for electron-lattice interaction

$$-\sum \frac{\hbar^2}{2M_k}\varphi\nabla_k^2\psi - \sum \frac{\hbar^2}{M_k}\nabla_k\varphi\nabla_k\psi$$

are neglected in Eq. (8.4) and can be used in a perturbation approach to induce transitions between stationary solutions of Eqs. (8.3) and (8.4) (Kubo, 1952). A further simplification to the Schrödinger equation for electrons [Eq. (8.3)] can be introduced when the ion motion *induced* by the *electron configuration* is neglected, yielding:

$$-\sum_{i=1}^{n}\frac{\hbar^2}{2m_0}\nabla_i^2\psi + \sum_{i=1}^{n}{}'\frac{e^2}{2r_{ij}}\psi + V_{\text{el-ion}}(\mathbf{r_1}, \mathbf{r_2}, \ldots)\psi = E\psi, \tag{8.5}$$

with $V_{\text{el-ion}}$ as the *periodic potential* of the lattice atoms. The coordinates of electrons are the variables, while ions are assumed to be at their average positions. Lattice oscillations therefore do not enter this analysis.

The most drastic step is the reduction of the electronic Schrödinger equation to a *one-electron approximation* by separating (see Appendix A.4.9A)

$$\psi(r_1, r_2, \ldots, r_n) = \psi(r_1)\psi(r_2)\ldots\psi(r_n). \tag{8.6}$$

The assumption that each electron experiences essentially the same potential of an averaged distribution of all other electrons yields a single set of terms in the Schrödinger equation:

$$-\frac{\hbar^2}{2m_0}\nabla^2\psi + \left[V_{\text{el-ion}}(\mathbf{r}) + V_{\text{el-el}}(\mathbf{r}) \right]\psi = E\psi. \tag{8.7}$$

The entire problem has now been reduced to the problem of a single electron moving independently of all other electrons in a static potential composed of a perfect periodic potential of the lattice and an average potential describing its interaction with all the other electrons.

The main goal in the analysis of the realistic behavior of electrons in a crystal lattice is to obtain the potential $V_{\text{el-ion}} + V_{\text{el-el}}$. With the potential known, the problem becomes one of numerically integrating Eq. (8.7). Several methods for accomplishing this are known and will be summarized in Section 8.3.

The electron behavior can then be described in one of two fashions, depending on the kind of problem to be discussed.

(1) After assuming a simple periodic function for the potential, the wavefunction for the electron in higher bands (conduction bands—see Section 9.1) is best expressed as the *Bloch function*

$$\psi_n(\mathbf{k}, \mathbf{r}) = u_n(\mathbf{k}, \mathbf{r}) \exp(i\,\mathbf{k} \cdot \mathbf{r}) \qquad (8.8)$$

where n is the band index and $u_n(\mathbf{k}, \mathbf{r})$ is periodic with the lattice periodicity (*Bloch theorem**). These Bloch functions are plane waves that are modulated with lattice periodicity.

(2) For tighter bound states (lower bands), one often uses to better advantage the *Wannier functions*, which are defined as wave packets of the Bloch functions

$$\varphi_n(\mathbf{r} - \mathbf{r}_0) = \frac{1}{\sqrt{\mathcal{V}}} \sum_{\mathbf{k}} u_n(\mathbf{k}, \mathbf{r}) \exp(i\,\mathbf{k} \cdot \mathbf{r}) \exp(-i\,\mathbf{k} \cdot \mathbf{r}_0), \qquad (8.9)$$

and are localized near the lattice site $\mathbf{r}_0$; $\mathcal{V}$ is the crystal volume.

This description is important, as it assists in finding $V(\mathbf{r})$ in terms of simpler quantities.

8.2 The Crystal Potential

The initial effort of the band theory is devoted to obtaining the *crystal potential* $V(\mathbf{r})$. The crystal potential has two contributions, the electronic part and the ionic part [see Eq. (8.7)].

---

* The Bloch theorem states that nondegenerate solutions of the Schrödinger equation in a periodic lattice are also solutions after translation by a lattice vector, with the amplitude function having lattice periodicity $u_{k,\beta}(\mathbf{r}) = u_{k,\beta}(\mathbf{r} + \mathbf{r}_0)$; $\mathbf{r}_0$ is any translation vector which reproduces the Bravais lattice.

8.2.1 Electronic Contribution

The *electronic contribution* in Eq. (8.7) can be approximated by the solution of the corresponding Hartree equation, describing the **Coulomb interaction** only:

$$\left[-\frac{\hbar^2}{2m_0} \nabla^2 - \sum_i \frac{Ze^2}{|\mathbf{r} - \mathbf{R}_i|} - V_{\text{el-el}}(\mathbf{r}) \right] \psi_n(\mathbf{r}) = E_n \psi_n(\mathbf{r}) \quad (8.10)$$

with

$$V_{\text{el-el}}(\mathbf{r}) = \sum_\beta{}' e^2 \int \frac{|\psi_\beta(\mathbf{r}_j)|^2}{|\mathbf{r} - \mathbf{r}_j|} \, d\mathbf{r}_j; \quad (8.11)$$

the prime at the sum indicates summation over all β except $\beta = n$ in Eq. (8.11).

The Hamiltonian of the Hartree equation contains its own eigenfunction. Its solution therefore involves first guessing an approximate solution $\psi(\mathbf{r})$ that describes the probability of finding an electron at the position $\mathbf{r}$, and a consequent iteration by introducing Eq. (8.11) into the Schrödinger equation [Eq. (8.7)]. This equation is then solved, using the resulting eigenfunction ψ as the input function in Eq. (8.11) and continuing until convergence is achieved (**self-consistent field method**).

However, the *symmetrical* many-electron wavefunction [Eq. (8.6)] is incompatible with the Pauli principle. A many-electron wavefunction obeying the Pauli principle can be constructed by using an *antisymmetric* normalized product of all of the one-electron eigenfunctions $\psi_i(\mathbf{r}_j)$ in a *Slater determinant* (see Appendix A.4.9)

$$\psi(\mathbf{r}_1, \mathbf{r}_2, \ldots) = \frac{1}{\sqrt{n!}} \begin{vmatrix} \psi_1(\mathbf{r}_1) & \psi_2(\mathbf{r}_1) & \cdots & \psi_n(\mathbf{r}_1) \\ \psi_1(\mathbf{r}_2) & \psi_2(\mathbf{r}_2) & \cdots & \psi_n(\mathbf{r}_2) \\ \vdots & \vdots & \vdots & \vdots \\ \psi_1(\mathbf{r}_n) & \psi_2(\mathbf{r}_n) & \cdots & \psi_n(\mathbf{r}_n) \end{vmatrix}. \quad (8.12)$$

There are $n!$ possibilities for distributing n indistinguishable electrons over n states. The determinant vanishes if two electrons occupy the same state, since two rows of the determinant are then identical.

Fock introduced an additional electronic exchange interaction, considering the *Pauli exclusion principle* (Fock, 1930; Corson, 1951), using Eq. (8.11):

$$\left[-\frac{\hbar^2}{2m_0} \nabla^2 - \sum \frac{Z_i e^2}{|\mathbf{r} - \mathbf{R}_i|} - V_{\text{el-el}}(\mathbf{r}) - V_{\text{ex}} \right] \psi_n(\mathbf{r}) = E_n \psi_n(\mathbf{r})$$

$$(8.13)$$

with an extra term, the *exchange interaction*

$$V_{\text{ex}}\psi_n(\mathbf{r}) = -e^2 \sum_{\beta} \psi_{\beta}(\mathbf{r}) \int \frac{\psi_{\beta}^{*}(\mathbf{r}_j)\psi_n(\mathbf{r}_j)}{|\mathbf{r} - \mathbf{r}_j|}\, d\mathbf{r}_j. \qquad (8.14)$$

This is a Coulomb term and arises from the correlated motion of electrons with consideration of the antisymmetry of the wavefunction.

This poses some difficulties, as it renders the potential a nonlocal operator (it is an integral operator). Many attempts deal with a more suitable way of using the Hartree-Fock concept to arrive at an appropriate potential (Löwdin, 1956; Pratt, 1957; Slater, 1953; Wood and Pratt, 1957). Other methods take care of electron-electron interaction more adequately (Brueckner, 1955; Pines, 1956; Bohm et al., 1957; Hubbard, 1957). A self-consistent calculation, carried out to the desired degree, however, poses computational difficulties.

The Hartree or Hartree-Fock band structure calculation overestimates substantially the band gaps.

More recently, attention has been given to a **density functional method**, where the electronic contribution, separated into a Coulomb and an exchange correlation (xc) term, depends on the ground state electron density distribution $n(\mathbf{r})$ (McWeeny 1957; Chirgwin, 1957). This can be done using a formalism of Hohenberg and Kohn (1964); see also Kohn and Sham (1965):

$$V\left[\varrho(\mathbf{r})\right] = V_{\text{ext}}(\mathbf{r}) + V_{\text{Coul}}\left[\varrho(\mathbf{r})\right] + V_{\text{xc}}\left[\varrho(\mathbf{r})\right], \qquad (8.15)$$

where V_{ext} is the external (to the electron) potential imposed by the ions, and V_{xc} represents the correction potential including exchange and *correlation*.

All of the more recent first-principles calculations are based on this density functional method—see Section 8.3.2. In a Bloch function description, for example, the ground state charge density

$$\varrho(\mathbf{r}) = \sum_{n} \sum_{\mathbf{k}} |\psi_n(\mathbf{k}, \mathbf{r})|^2 \qquad (8.16)$$

is a sum over occupied bands n and wave vectors $\mathbf{k}$ in the entire Brillouin zone of the crystal. Since $\psi_n(\mathbf{k}, \mathbf{r})$ depends on $V[n(\mathbf{r})]$, and $n(\mathbf{r})$ depends on $\psi_n(\mathbf{k}, \mathbf{r})$, these calculations must be carried out in a self-consistent manner.

The density profile can be plotted in a two-dimensional representation as equidensity contour lines which are similar to those obtained from a Fourier analysis of x-ray diffraction data. It permits distinction of electrons from different bands. Examples for O_2, NaCl, and diamond are given in Figs. 2.3 and 2.11.

8.2.2 Pseudopotentials

The exact ionic potential shows very large amplitudes near the center of each ion and thereby creates substantial and unnecessary computational problems. These problems can be avoided by using pseudopotentials which avoid these potential spikes as indicated below. This part of the potential is of importance for inner core electrons but usually not for valence or conduction band electrons. Introduced by Prokofjew (1929) and Fermi (1934), and applied to atoms by Hellmann (1935), the pseudopotential became a major means to provide the most important input to the Hamiltonian *relevant* for valence electrons. In the late 1950s and 1960s (Phillips and Kleinman, 1959; review of Harrison, 1966), it was shown that valence electrons are effectively excluded from the ion core of an atom by an almost exact balance between two strong forces: the Coulomb attraction to the core, and the quantum-mechanical repulsion from the core electrons (*exclusion*). This is also known as the Phillips cancellation theorem (Phillips and Kleinman, 1959). The resulting net force can be described by a rather weak, attractive **pseudopotential** (Ziman, 1964; Harrison, 1966), which has lattice periodicity

$$V(\mathbf{r}) = \sum_{\mathbf{k}} \sum_{\alpha} V_\alpha(\mathbf{k}) S_\alpha(\mathbf{k}) \exp(i\mathbf{k} \cdot \mathbf{r}), \qquad (8.17)$$

with α as the type of atom in the unit cell, V_α as the *atomic* pseudopotential of atom α, and S_α as a structure factor. A typical plot of an *atomic* pseudopotential is shown in Fig. 8.1. The repulsive branch starts at $\sim 1/2$ of the bond length. The amplitude of the pseudopotential is much reduced compared to the atomic (Coulomb) potential. The use of the smaller amplitude periodic lattice potential permits the application of a perturbation formalism. For reviews, see Heine and Weaire (1970), Bassani and Giuliano (1972), and Cohen (1984).

A more sophisticated approach to the crystal potential is interwoven with methods for solving the wave equation; therefore, we will return to this subject in the following section.

8.3 Solving the Crystal Wave Equation

After a suitable crystal potential $V(\mathbf{r})$ is determined, the Schrödinger equation containing this potential [Eq. (8.7)] must be solved under

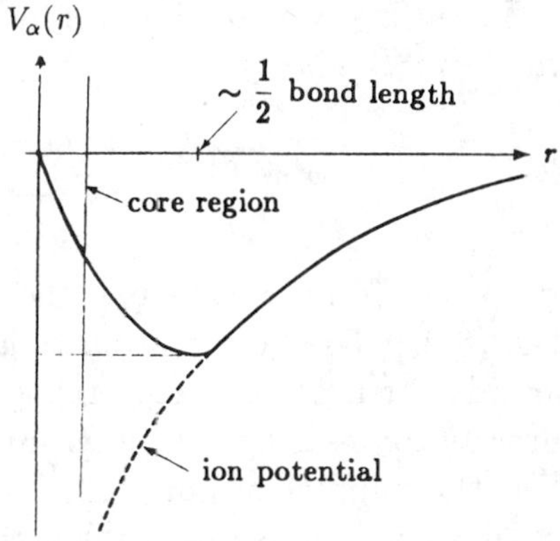

Figure 8.1: Typical behavior of the pseudopotential compared to the Coulomb-attractive ion potential, which dominates further away from the core.

the appropriate boundary condition* to obtain the dispersion relation $E_n(\mathbf{k})$. A wide variety of methods for efficiently computing such solutions have been developed, all of which are based on variational principles. Usually, this wavefunction is expanded in terms of *trial functions*, followed by a *variation of the expansion coefficients*.

This technique proposed by Ritz (see Morse and Feshbach, 1953) replaces the problem of solving the wave equation with the simpler one of solving equivalent *secular equations*. With modern computers, most of these methods can be used for efficient computation and lead to similar results.

8.3.1 Expansion Methods

Two different classes of expansion methods can be distinguished:

- those that expand the electron eigenfunctions as Bloch functions with expansion coefficients to satisfy the Schrödinger equation, and
- those that expand the electron eigenfunctions as solution functions of the Schrödinger equation, with expansion coefficients so that the solutions satisfy the boundary conditions.

8.3.1A The Tight-Binding Method   The electronic eigenfunction in a crystal lattice can be approximately derived from the atomic eigenfunctions (Heitler and London, 1927; Bloch, 1928; Slater, 1951, 1953) of each lattice atom in the unit cell and consequent periodic repetition. This is the basis for the *tight-binding approximation* in which the crystal wave function is composed of a linear combination

---

* Namely, $\mathbf{k}$ must be real and $\psi(\mathbf{r})$ periodic with lattice periodicity $\psi(\mathbf{r}) = \psi(\mathbf{r} + \hat{n}\mathbf{a}_i)$ (*Born–van Karman boundary condition*).

of the eigenstates of the free atoms. Properly set up, these are the *Bloch tight-binding sums*:

$$\psi_n(\mathbf{k}, \mathbf{r}) = \frac{1}{\sqrt{N}} \sum_{\mathbf{r}_j} \varphi_n(\mathbf{r} - \mathbf{r}_j) \exp(i\mathbf{k} \cdot \mathbf{r}_j), \qquad (8.18)$$

composed from $\varphi_n(\mathbf{r} - \mathbf{r}_j)$, the Wannier functions of the free atom [Eq. (8.9)]. Unlike the atomic eigenfunctions, the Wannier functions already form a complete orthogonal set, and therefore can be used directly to describe the electronic eigenfunctions in the crystal. Here N is the number of unit cells in the crystal. The tight-binding sums are used for analyzing low energy states (deep lying bands), where the eigenfunctions are rather localized near the nucleus of each atom and are barely disturbed by surrounding atoms. The method can be simplified when the actual atomic eigenfunctions are replaced by a similar but simpler complete set of orthonormalized functions (Kane, 1976).

Overlap integrals describe the interaction between the atoms that causes broadening of the bands. For higher bands, one must consider the fact that the atomic states extend over much larger than inter-atomic distances and the overlap integral becomes very sensitive to the tail of the atomic potential. This fundamental difficulty limits the applicability of the tight-binding method to deep states and core bands.

8.3.1B   The Orthogonalized Plane Wave Method   A simple set of basis functions are plane waves, as suggested by Sommerfeld and Bethe (1933). This *plane wave method* uses wave functions

$$\psi(\mathbf{k}, \mathbf{r}) = \frac{1}{\sqrt{N\mathcal{V}}} \sum_j a_j(\mathbf{k}) \exp\left[i(\mathbf{k} + \mathbf{k}_j) \cdot \mathbf{r}\right]; \qquad (8.19)$$

where a_j is the expansion parameters and $\mathcal{V}$ is the volume of the unit cell; however, Eq. (8.19) is only slowly converging.

An improvement suggested by Herring (1940) uses information obtained for the core states from the tight-binding approximation. Higher energy (valence and conduction) states are orthogonal to these core states (*orthogonalized plane wave method*), and can be used with more rapid convergence (for a review, see Bassani, 1966).

8.3.1C   Pseudopotential Methods   With a pseudopotential (see Section 8.2.2) rather than the real lattice potential, we lose information about the bands of core electrons which are of little interest to

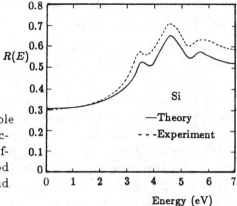

Figure 8.2: Agreement achievable between the optical reflection spectrum of Si calculated with the self-consistent pseudopotential method and experiment (after Philipp and Ehrenreich, 1963).

semiconductor behavior, but gain the simplicity of describing the solutions with pseudo-wavefunctions, which are slowly varying and can be approximated more easily by superposition of fewer terms. The resulting eigenvalues for the upper (valence and conduction) bands, however, are the same as obtained with the exact potential extending into the core region.

The ease of arriving at quantitative results, which can be compared with the experiment, permits the repeated readjustment of the pseudopotential in another trial, until agreement with the experiment is satisfactory (*empirical pseudopotential method*, as discussed by Cohen and Heine, 1970).

It should be noted that the *empirical* pseudopotential method determines a crystal potential by fitting the *band structure* to the experiment. However, it cannot be made self-consistent, since the screening given by the second and third terms of Eq. (8.15) are not included.

The agreement that can be achieved between theory, when fitted empirically, and experiment [e.g., using the reflectivity spectrum (see Section 13.1.1) as a check] is rather good (Fig. 8.2). However, when the so-adjusted potential is used to calculate the electron density distribution, there is only fair agreement with the distribution obtained from x-ray diffraction. For examples of recent application of empirical pseudopotential methods, see Chelikowsky and Phillips (1978), and Ihm and Cohen (1980).

First-principles pseudopotentials suitable for computing valence electron total energies and the full band structure are now available and show substantially improved agreement with the experiment (Bachelet et al., 1982; Ihm et al., 1979). See Section 8.3.2.

8.3.1D Semi-Empirical Nearest Neighbor Tight-Binding Theory   A theoretical shortcut to obtain the chemical trend of band structures of tetrahedrally bound semiconductors was introduced by Harrison (1973) and extended by Vogl et al. (1983). Here, one uses a limited set of orbitals, usually one *s*- and three *p*-localized pseudo-orbitals plus one excited state *s**, the latter for also obtaining indirect gap features. These are adjusted to fit optical band gaps in constructing a pseudoHamiltonian, somewhat similar to the conventional empirical pseudopotential method. The resulting band structure is easily obtained and is in reasonable agreement with the one obtained for such semiconductors from pseudopotentials. When the matrix elements of this model are fixed by the atomic energies of the lattice constituents and by a set of universal constants, certain chemical trends of the electronic structure of zinc blende and diamond semiconductors can be predicted.

8.3.1E Cellular Method   Another method is the *cellular method* (Wigner and Seitz, 1933, 1934), in which the lattice is divided into Wigner-Seitz cells (see Section 3.6.1) and the Schrödinger equation is solved within each cell

$$\psi(\mathbf{k}, \mathbf{r}) = \sum_{l=0}^{\infty} \sum_{m=-l}^{l} c_{lm}(\mathbf{k}) Y_{lm}(\theta, \varphi) R_l(E, r) \tag{8.20}$$

with $Y_{lm}(\theta, \varphi)$ as spherical harmonics and $R_l(E, r)$ as the solution of the radial wave equation given by

$$\frac{1}{r^2} \frac{d}{dr} \left(r^2 \frac{dR_l}{dr} \right) + \left[\frac{2m_0}{\hbar^2} (E - V(r)) - \frac{l(l+1)}{r^2} \right] R_l = 0. \tag{8.21}$$

Solutions can easily be obtained when using appropriate boundary conditions* (ψ continuous at r_B and $d\psi/dr|_{r_B} = d\psi/dr|_{r_B + R} = 0$) with nearly spherical symmetry (dominant contribution from the spherical core), and the Wigner-Seitz cell (boundary at r_B) containing only one atom in its center (Bell, 1953). More recently, more accurate calculations of the diamond were carried out, using this method (Leite et al., 1975).

---

* The difficulty of the cellular method resides in determining *appropriate* boundary conditions from only a few known points on the surface of the Wigner-Seitz cell (Shockley, 1938).

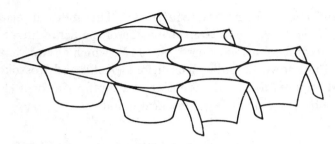

Figure 8.3: Muffin-tin potential, showing the typical form of the approximated potential surrounding each atom.

8.3.1F Augmented Plane Wave Method The difficulties with insufficiently known boundary conditions in the cellular method can be reduced by an *augmented plane wave method*, in which the potential is prescribed by a muffin-tin type (Fig. 8.3). For a primitive lattice and spherical symmetry up to one half of the interatomic spacing (r_0), we have

$$\psi(\mathbf{k}, \mathbf{r}) = \sum_{l=0}^{\infty} \sum_{m=-1}^{l} c_{lm}(\mathbf{k}) Y_{lm}(\theta, \varphi) R_l(E, r) \eta(r - r_0)$$
$$+ \sum_{j} b_j(\mathbf{k}) \exp\left[i\left(\mathbf{k} + \mathbf{k}_j\right) \cdot \mathbf{r}\right] \left\{1 - \eta(r - r_0)\right\} \qquad (8.22)$$

$$\text{with} \quad \eta(r - r_0) = \begin{cases} 0 & \text{for } r > r_0 \\ 1 & \text{for } r < r_0. \end{cases} \qquad (8.23)$$

Equation (8.22) has a tight-binding and a plane wave contribution. This method was suggested by Slater (1937) and has become very useful for analyzing electronic states for closely packed materials (metals). However, it yields energy gaps of semiconductors which are too small.

The muffin-tin aspect of the potential (Fig. 8.3) can be modified by assuming an adjustable flat potential to fit the experimental results (Loucks, 1967; Dimmock, 1971).

The potential can be generalized to the linearized augmented plane wave method (Anderson, 1975; Krakauer et al., 1981; Wei and Krakauer, 1985). There are no shape approximations of the potential, and one treats all electrons, valence and core, in an equal fashion. This is the present state-of-the-art method for band calculations (see Section 8.3.2).

8.3.1G Green's Function Method   This method, also referred to as *KKR method*, uses a transformation of the Schrödinger equation into an integral equation (Korringa, 1947), which circumvents the difficulty of unknown boundary conditions (Kohn and Rostoker, 1954), but also requires a crystal potential in a muffin-tin form (Fig. 8.3).

One obtains $\psi(\mathbf{k}, \mathbf{r})$ in an integral equation

$$\psi(\mathbf{k}, \mathbf{r}) = \int_{\mathcal{V}} G\left[\mathbf{r} - \mathbf{r}', E(\mathbf{k})\right] V(\mathbf{r}')\psi(\mathbf{k}, \mathbf{r}') \, d\mathbf{r}', \qquad (8.24)$$

where $V(\mathbf{r}')$ is the muffin-tin potential and $G_{\mathbf{k}}(\mathbf{r} - \mathbf{r}', E(\mathbf{k}))$ is the Green's function. It is determined as solution of

$$\left(\frac{\hbar^2}{2m}\nabla^2 + E\right) G(\mathbf{r} - \mathbf{r}', E(\mathbf{k})) = \delta(\mathbf{r} - \mathbf{r}')$$

$$\text{with} \qquad G(\mathbf{r} + \mathbf{r}_0, E(\mathbf{k})) = G(\mathbf{r}, E(\mathbf{k}))\exp(i\mathbf{k} \cdot \mathbf{r}_0) \qquad (8.25)$$

as the boundary condition. The Green's function can be expressed as the expansion

$$G(\mathbf{r} - \mathbf{r}', E(\mathbf{k})) = -\frac{1}{\mathcal{V}}\frac{\sum_n \exp\left[i\left(\mathbf{k} + \mathbf{k}_n\right) \cdot \left(\mathbf{r} - \mathbf{r}'\right)\right]}{\left[\hbar^2/\left(2m\right)\right]\left(\mathbf{k} + \mathbf{k}_n\right)^2 - E(\mathbf{k})} \qquad (8.26)$$

(see Friedman, 1956). The Green's function method is closely related to the augmented plane wave method and can be used to obtain the same results for muffin-tin type of potentials.

A substantial reduction of computation is possible when the analysis is restricted to the immediate neighborhood of certain symmetry points in the crystal, with the expansion [Eq. (8.26)].

Employing the Green's function method, and using a dynamically screened Coulomb interaction, yields satisfactory results for the ground state energy (band gap) for a number of semiconductors and insulators (Louie, 1986). For a more general introduction of Green's function, see the Appendix (Section A.4.3).

8.3.1H The k·p Method   This method was introduced by Bardeen (1938) and was used to explore the band structure in the vicinity of $\mathbf{k} = 0$ or other critical points $\mathbf{k}_0$ of the Brillouin zone (Seitz, 1940; see Section 8.5.1). It is based on a perturbation approach with the standard $\hbar\mathbf{k} \cdot \mathbf{p}/m$ term (with $\mathbf{p} = -ih\nabla$) in the Hamiltonian, applied to the functions $u_n(\mathbf{k}, \mathbf{r})$ rather than to the full Bloch function in the neighborhood of the critical point. Introducing this function into the one-electron Schrödinger equation one obtains

$$\left[-\frac{\hbar^2}{2m_0}\nabla^2 + V(\mathbf{r}) + \frac{\hbar}{m_0}\mathbf{k}\cdot\mathbf{p} + \frac{\hbar^2 k^2}{2m_0} \right] u_n(\mathbf{k},\mathbf{r}) = E_n(\mathbf{k})u_n(\mathbf{k},\mathbf{r}).$$

$$(8.27)$$

Here the $u_n(\mathbf{k},\mathbf{r})$ are a complete, orthogonal set of eigenfunctions in real crystal space. This set is called the $\mathbf{k}\cdot\mathbf{p}$ *representation*. The first two terms in the Hamiltonian of Eq. (8.27) define the unperturbed Hamiltonian $\mathbf{H}_0$, and the last two terms can be developed as small perturbations in the vicinity of the given $\mathbf{k}_0$. Thus Eq. (8.27) may be written as

$$\left[\mathbf{H}_0 + \frac{\hbar}{m_0}(\mathbf{k} - \mathbf{k}_0)\cdot\mathbf{p} + \frac{\hbar^2}{2m_0}(\mathbf{k} - \mathbf{k}_0)^2 \right] u_n(\mathbf{k},\mathbf{r}) = E_n(\mathbf{k})u_n(\mathbf{k},\mathbf{r}).$$

$$(8.28)$$

Written in matrix form, the number of $\mathbf{k}\cdot\mathbf{p}$ matrix elements is greatly reduced because of symmetry considerations. These matrix elements

$$\mathbf{M}_{jn}(\mathbf{k}_0) = \int \psi_n^*(\mathbf{k}_0,\mathbf{r})\mathbf{p}\psi_n(\mathbf{k}_0,\mathbf{r})\,d\mathbf{r} \qquad (8.29)$$

can be evaluated easily [ψ_n is the Bloch function—see Eq. (8.9)]. They contain only a small number of parameters which are taken mostly from the experiment.

The eigenvalues near a characteristic point $\mathbf{k}_0$ can be expressed as

$$E_n(\mathbf{k}) = E_n(\mathbf{k}_0) + \frac{\hbar^2(\mathbf{k} - \mathbf{k}_0)^2}{2m_0} + \frac{\hbar^2}{m_0^2}\sum_{j,n}{}'\frac{(\mathbf{k} - \mathbf{k}_0)\cdot\mathbf{M}_{jn}(\mathbf{k}_0)}{E_n(\mathbf{k}_0) - E_j(\mathbf{k}_0)},$$

$$(8.30)$$

where the prime at the sum indicates summation over all n and j with $n \neq j$.

Equation (8.30) can be simplified when interactions between only two bands are of interest [e.g., the valence and conduction bands (see Section 9.1)] and the energy difference between these is small compared to the difference with all other bands. Then we can write

$$E_n(\mathbf{k}) = E_n(\mathbf{k}_0) + \frac{\hbar^2}{2}\sum_{i=1}^{3}\frac{(k_i - k_0)^2}{m_i} \quad \text{with } i = x,y,z \qquad (8.31)$$

and can express the effective mass (see Section 7.2.3) as

$$\frac{1}{m^*} = \frac{1}{m_0} \pm \frac{2|\mathbf{M}_{nj}(\mathbf{k}_0)|^2}{m_0^2\,(E_j(\mathbf{k}_0) - E_n(\mathbf{k}_0))}, \qquad (8.32)$$

with $+$ or $-$ for the upper or lower band, respectively.

One can use Eq. (8.30) for the *deviation* of $E(\mathbf{k})$ *from parabolicity* near a critical point $\mathbf{k}_0$

$$E(\mathbf{k} - \mathbf{k}_0) = -\frac{E_g}{2} + \frac{\hbar^2(\mathbf{k} - \mathbf{k}_0)^2}{2m_0} \pm \frac{1}{2}\sqrt{E_g^2 + \frac{4\hbar^2(\mathbf{k} - \mathbf{k}_0)^2|\mathbf{M}_{nj}(\mathbf{k}_0)|^2}{m_0^2}}$$

$$(8.33)$$

which yields with Eq. (8.32):

$$E(\mathbf{k} - \mathbf{k}_0) = -\frac{E_g}{2} + \frac{\hbar^2(\mathbf{k} - \mathbf{k}_0)^2}{2m_0} \pm \frac{E_g}{2}\sqrt{1 + \frac{2\hbar^2(\mathbf{k} - \mathbf{k}_0)^2}{E_g}\left(\frac{1}{m^*} - \frac{1}{m_0}\right)}$$

$$(8.34)$$

with m^* as the effective mass (assumed to be the same in all directions) in the conduction $(+)$ or valence $(-)$ band (see Section 9.1).

This illustrates the usefulness of the $\mathbf{k} \cdot \mathbf{p}$ method in a simple example. The use of this method for band structure analysis was initiated by Kane (1956) and has been carried on by many others.

Examples of $E(\mathbf{k})$ determined by the $\mathbf{k} \cdot \mathbf{p}$ method are given by Cardona and Pollak (1966); see also the review by Kane (1982).

8.3.2 Ab Initio Calculations

With the substantial increase in computer power, nonempirical *ab initio* calculations have replaced most of the classical approximations.* In these *first-principle* local density functional *calculations*, which are capable of deriving the complete band structure and total

---

* The following abbreviations are commonly used

EPM: Empirical pseudofunctional method
KKR: Korringa Kohn Rostocker (Green's function) method
LAPW: Linear augmented plane wave method
LCAO: Linear combination of atomic orbitals
LMTO: Linear muffin-tin orbitals
MB: Mixed basis
OPW: Orthogonal plane wave method
PSF: Pseudofunctional method
PW: Plane wave

energy of any equilibrium atomic configuration of a semiconductor, one distinguishes Hamiltonians:

(1) all electrons, core included, with potentials of the form $-Z/r$ (Z = all electrons of a given atom). This includes as *old methods*

  (a) Green's function method (KKR)
  (b) Orthogonal plane wave method (OPW) , and
  as *new methods*

  (A) Linear augmented plane wave method (LAPW) (Wei and Krakauer, 1985),
  (B) Linear muffin-tin orbital method (LMTO) (Anderson, 1975; Skriver, 1983),
  (C) Pseudofunctional method (PSF, PW) (Kasowski et al., 1986),
  (D) mixed basis (PW and atomic orbitals) (Bendt and Zunger, 1982).

These methods can be used for any element.

As an alternative one has Hamiltonians for:

(2) Valence electrons only, with pseudopotentials for the atom. The Coulomb part of the potential is of the form $-Z_{\text{val}}/r$. This includes as *old methods* which fit the pseudopotential to the experiment

  (a) Empirical pseudofunctional method (EPM)(see e.g., Cohen and Heine, 1970).
  As a *new method* which computes the pseudopotential from the microscopic theory using either a pure plane wave basis

  (A) Plane wave method (PW) (Zunger and Cohen, 1978b)
  (B) or a mixed basis (MB) (Ho and Louie, 1979).

These methods are applicable for main row elements only.

Presently, mostly used are the LAPW, LMTO, and PSF methods for all-electron eigenfunctions.

The Schrödinger equation is solved for

$$\left[-\frac{\hbar^2}{2m_0} \nabla^2 + V_{\text{cry}}(\mathbf{r}) \right] \psi_n(\mathbf{k}, \mathbf{r}) = E_n(\mathbf{k}) \psi_n(\mathbf{k}, \mathbf{r}) \qquad (8.35)$$

with $\psi_n(\mathbf{k}, \mathbf{r})$ as the Bloch function for the one-electron eigenstates. The total crystal potential can be described within the density functional framework as

$$V_{\text{cry}}(\mathbf{r}) = V_{\text{ext}}(\mathbf{r}) + V_{\text{scr}}[\varrho(\mathbf{r})]. \qquad (8.36)$$

The electron density is given as

$$\varrho(\mathbf{r}) = \sum_n \sum_{\mathbf{k}} f_{nk} |\psi_n(\mathbf{k}, \mathbf{r})|^2 \qquad (8.37)$$

where f_{nk} is the occupation function of states of wave vector $\mathbf{k}$ within the band with index n.

The external potential can be given either in the all-electron framework "ae" by

$$V_{\text{ext}}^{(\text{ae})}(\mathbf{r}) = -\sum \sum \frac{Z_\alpha}{|\mathbf{r} - \mathbf{r}_0 - \mathbf{R}_\alpha|} \qquad (8.38)$$

where Z_α is the total number of electrons of the atom, $\mathbf{r}_0$ is a Bravais lattice vector, and $\mathbf{R}_\alpha$ is a basis site vector. In the pseudopotential framework "ps" the external potential is given by

$$V_{\text{ext}}^{(\text{ps})} = \sum_{\mathbf{R}_\alpha} \sum_{\mathbf{r}_0} \sum_l \omega_{\alpha,l}(\mathbf{r} - \mathbf{r}_0 - \mathbf{R}_\alpha) \hat{P}_{l,\alpha} \qquad (8.39)$$

with $\hat{P}$ as the projection operator for the angular momentum l about the basis site $\mathbf{R}_\alpha$. The pseudopotential $\omega_{\alpha,l}$ can be decomposed as

$$\omega_{\alpha,l} = -\frac{Z_\alpha^v}{r} + V_{\alpha,l}(\mathbf{r}) \qquad (8.40)$$

where Z_α^v is the number of valence electrons and $V_{\alpha,l}(\mathbf{r})$ is a short-range repulsive core potential that depends on basis site ($\mathbf{R}_\alpha$) and angular momentum (l) to distinguish the element's chemistry within one row of elements (Zunger and Cohen, 1978b).

For practical purposes, in order to achieve better convergence of the self-consistent iteration using such pseudopotentials with development in a plane wave basis, one needs atomic pseudopotentials whose Fourier components fall rapidly with the wave vector. This is achieved by smearing out the charge density, replacing Z_α^v/r with $(Z_\alpha^v/r)\text{erf}(\sqrt{\beta}\,r)$, yielding

$$\omega_{\alpha,l}(r) = -\frac{Z_\alpha^v}{r}\text{erf}(\sqrt{\beta}\,r) + \left[v_{\alpha,l}(r) - Z_\alpha^v \text{erfc}(\sqrt{\beta}\,r) \right] ; \qquad (8.41)$$

where erf and erfc are the error function and complementary error function, and β is the smearing parameter, typically of the order of $2\,(\text{a.u.})^{-2}$ (Baur et al., 1983).

8.4 Relativistic Effects

Electrons described as Bloch electrons are free (except for scattering) to move within the periodic potential of the crystal; their velocity in a semiconductor is on the order of the thermal velocity ($\sim 10^7$ cm/s), and relativistic effects can be neglected. However, this is no longer true for electrons that move in the strong local field near the nuclei, with velocities approaching the velocity of light. Here relativistic terms in the wave equation must be carried. One replaces the Schrödinger equation with the *Dirac relativistic equation* (Rose, 1961), which can be written for the upper two-component spinor as

$$
\left[\left\{ -\frac{\hbar^2}{2m_0}\nabla^2 + V(\mathbf{r}) + \frac{1}{8m_o^3 c^2}\nabla^4 - \frac{\hbar^2}{4m_0^2 c^2}\nabla V(\mathbf{r})\cdot\nabla \right\} \underline{1} - \right.
$$
$$
\left. \frac{i\hbar^2}{4m_0^2 c^2}\underline{\sigma}\cdot(\nabla V(\mathbf{r})\times\nabla) \right]\psi = E\psi, \tag{8.42}
$$

with $\underline{1}$ as the unity matrix, and $\underline{\sigma}$ as the Pauli operator matrices.* There are three additional terms. The first and second additional terms represent the relativistic correction to kinetic energy and potential (the *Darwin correction*), and the third term represents the spin-orbit coupling: this term describes the interaction of the electron spin with the magnetic moment of the electron in its orbit. It reduces the symmetry, causes removal of some degeneracies of valence band states, and thereby determines *spin-orbit splitting* of valence bands.

The spin-orbit splitting becomes more pronounced for heavy elements with larger nuclear charges. At the Γ-point, the valence band of Si with $z = 14$ splits by 0.04 eV, while for Ge with $z = 32$ the valence band splits by 0.29 eV (see Table 9.8 for a listing of such splittings). The atomic spin-orbit splitting is the same as that in a crystal since the interaction occurs deep in the atomic core, where the surrounding atoms of the crystal have little influence.

---

* The components of $\underline{1}$ and $\underline{\sigma}$ are

$$
\underline{1} = \begin{pmatrix} 1 & 0 \\ 0 & 1 \end{pmatrix} \text{ and } \sigma_x = \begin{pmatrix} 0 & 1 \\ 1 & 0 \end{pmatrix}, \ \sigma_y = \begin{pmatrix} 0 & -i \\ i & 0 \end{pmatrix}, \ \sigma_z = \begin{pmatrix} 1 & 0 \\ 0 & -1 \end{pmatrix}.
$$

Relativistic corrections to the band structure calculation have been made by Soren (1965), Loucks (1965), Onodera and Okazaki (1966), and Pay-June Lin-Chung and Teitler (1972).

8.5 Band Structure of Three-Dimensional Lattices

The dispersion relation $E_n(\mathbf{k})$ for three-dimensional lattices yields additional bands, which may or may not overlap. In any one crystallographic direction, however, each band shows single-valued curves. To get a better perception of the topography of this $E_n(\mathbf{k})$ structure, which is a three-dimensional hypersurface in the four-dimensional $(E, \mathbf{k})$-space, we start our discussion from a two-dimensional display of $E(\mathbf{k})$ in the first Brillouin zone.

8.5.1 Brillouin Zones

As shown in the one-dimensional example in Section 7.2, the translational symmetry of the lattice permits a *reduced representation* of the dispersion relation $E(k)$ (see Fig. 7.12) with a periodicity $(2\pi/a$ in the k_x-direction) related to the *reciprocal lattice periodicity*. This periodicity is maintained in three dimensions. For example, one has within a primitive orthorhombic unit cell:

$$-\frac{\pi}{a} < k_x < \frac{\pi}{a}, \quad -\frac{\pi}{b} < k_y < \frac{\pi}{b}, \quad \text{and} \quad -\frac{\pi}{c} < k_z < \frac{\pi}{c}.$$

For a complete discussion, the momentum vector $\mathbf{k}$ can be restricted to this *reduced* cell, the first Brillouin zone, which contains all relevant information (see Section 3.6).

$E(\mathbf{k})$ can be plotted easily in a one-dimensional lattice (see Fig. 7.12). The $E(\mathbf{k})$ behavior of the first two bands of a two-dimensional structure is shown in Fig. 8.4; it presents one curved surface for each of the bands. The center ($k_x = k_y = 0$) is denoted by Γ. Cuts of this surface with a plane parallel to E show a different $E(\mathbf{k})$ behavior, depending on the orientation of the plane in the k_x- and k_y-directions. However, because of symmetry of the assumed square lattice, a cut normal to the k_x-direction must result in the same $E(\mathbf{k})$ as a cut at the same k-value normal to the k_y-direction. Another cut at $45°$ between k_x and k_y shows a different $E(\mathbf{k})$ from the aforementioned, but the same $E(\mathbf{k})$ for all equivalent $45°$ cuts. All essential elements of the $E(\mathbf{k})$ surfaces are contained in such cuts.

The intersection of the major crystallographic symmetry axes with the surface of the first Brillouin zone are identified in Fig. 8.4A

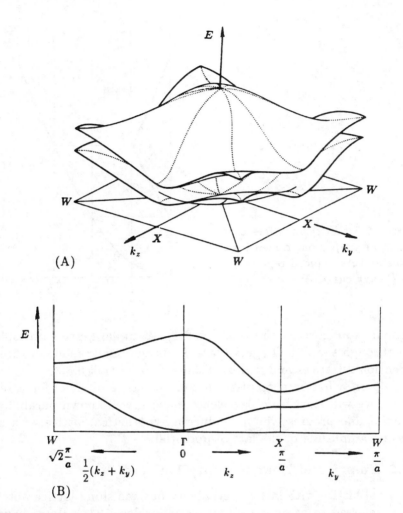

Figure 8.4: (A) $E(\mathbf{k})$ surfaces for two bands in a hypothetical two-dimensional square lattice, with symmetry points X and W indicated at the boundary of the first Brillouin zone (bottom plane). (B) $E(\mathbf{k})$ in a conventional one-dimensional (k) representation. The diagram is broken at the vertical lines representing symmetry points W, Γ (center of zone), and X (after Harrison, 1970).

by the appropriate letters* (X for the first and W for the second in the example given above). $E(\mathbf{k})$ in a two-dimensional representation shown in Fig. 8.4B is plotted along a $\mathbf{k}$-vector connecting these points

---

* See Section 3.6: Γ for the center: $k = (0, 0, 0)$; in a cubic lattice for the point L: $k = \frac{2\pi}{a}(\frac{1}{2}, \frac{1}{2}, \frac{1}{2})$; for the point X : $k = \frac{2\pi}{a}(1, 0, 0)$.

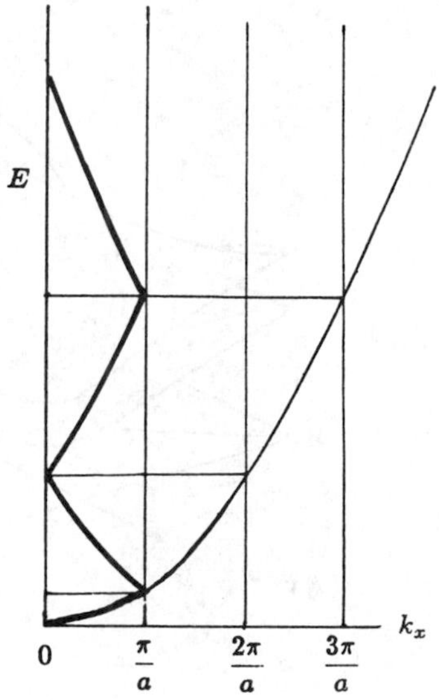

Figure 8.5: $E(\mathbf{k})$ for a free electron in an empty one-dimensional lattice and its reduced representation (heavy curves) between 0 and π/a.

(compare subfigures A and B of Fig. 8.4). A similar representation of cuts through the $E(\mathbf{k})$ hypersurface is conventionally made in three-dimensional lattices and will be discussed in Section 8.5.3.

To gain a better perception about the shape of $E(\mathbf{k})$ for a real lattice, we will start from free electrons with well-known paraboloid behavior and progressively introduce a hypothetical lattice with increasing amplitude of the lattice potential.

8.5.2 Empty and Nearly Empty Lattices

A crystal lattice with lattice periodicity but vanishing lattice potential is referred to as an *empty lattice* (Shockley, 1938). Introducing a very small lattice potential, with just enough of an amplitude to influence slightly the $E(\mathbf{k})$ behavior obtained for the empty lattice, defines this as a *nearly empty lattice*. The development of $E(\mathbf{k})$ with an increasing amplitude of the lattice potential is instructive in understanding the origin of the different bands in an actual crystal.

Starting with a free electron *in vacuo*, one obtains the well-known parabolic $E(\mathbf{k})$ behavior (see Section 7.2), redrawn in Fig. 8.5. Adding to the model an empty, one-dimensional lattice with lattice constant a does not change $E(\mathbf{k})$. However, one can now insert multiples of $\frac{\pi}{a}$ on the k_x-axis and fold the diagram to a reduced $E(\mathbf{k})$ representation, as indicated by the heavy curves in Fig 8.5. The $E_n(\mathbf{k})$ diagram given in Fig. 8.6A is obtained by replacing the one-dimensional lattice with the empty face-centered cubic lattice of the same geometry as the Ge lattice (Herman, 1958). The index n indicates the different branches of $E(\mathbf{k})$. The comparison of Fig. 8.6A with Fig. 8.5 shows that several $E(\mathbf{k})$ curves (bands) overlap in the empty Ge lattice while they do not in the empty primitive cubic lattice; this overlap is caused by the additional lattice point in the Ge lattice. The discussion of an empty lattice provides a useful method of determining the sequence of the different bands since such a sequence does not depend on the lattice potential.

The periodic crystal potential, switched-on at a very small amplitude, yields $E_n(\mathbf{k})$ of the *nearly empty lattice*, shown in Fig. 8.6B. A deformation and splitting in the $E_n(\mathbf{k})$ dispersion relation becomes visible. With the knowledge that the unit cell of the diamond lattice contains 2 atoms, and thus 8 valence electrons, one concludes that the lowest four $E_n(\mathbf{k})$ curves, occupied by two electrons each due to spin degeneracy, belong to one set of bands. This set of bands will later be identified as the valence band, disregarding degeneracy, and the curves above the fourth belong to the next higher set of bands, later defined as the conduction band (Section 9.1). The band gap (shaded region in Fig. 8.6) is expected above the fourth and below the fifth $E_n(\mathbf{k})$ curves. The actual band gap is the smallest distance between the highest point of the fourth and the lowest point of the fifth curves (heavy dash-dotted curve) in Fig. 8.6B. These points are not necessarily on top of each other (Fig. 8.10).

By comparing Figs. 8.6A with B, one obtains guidance for the developing band structure. The splitting of the bands and the avoidance of a cross-over can be clearly seen in the left diagram between Γ and X. The actual band structure is shown in Fig. 8.7 for the full amplitude of the periodic lattice potential of Ge.

8.5.3 The Band Structure of Typical Semiconductors

A computation of the band structure of Ge by Herman (1958) is shown in Fig. 8.7. This figure also shows the common notation of the bands introduced by Bouckaert et al. (1936)—see Bassani and Pastori Parravicini, 1975, for a review. Further modifications of $E_n(\mathbf{k})$, due to the introduction of spin-orbit interaction, can be seen by comparing Figs. 8.8A and B.

8.5.3A Symmetry of E(k) $E_n(\mathbf{k})$ has the same point group symmetry as the crystal to which the Brillouin zone belongs (see

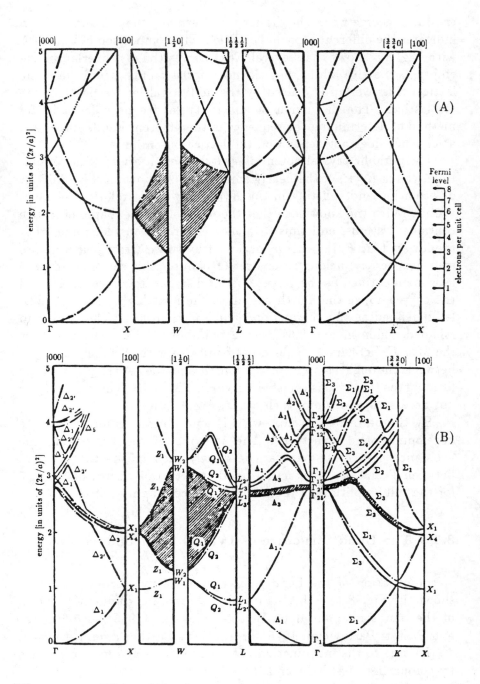

Figure 8.6: $E(\mathbf{k})$ of (A) the empty Ge lattice, and (B) the nearly empty Ge lattice, disregarding spin-orbit splitting (schematic) (after Herman, 1958).

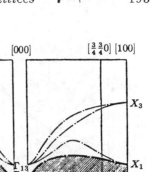

$[000]$ $[100]$ $[1\tfrac{1}{2}0]$ $[\tfrac{1}{2}\tfrac{1}{2}\tfrac{1}{2}]$ $[000]$ $[\tfrac{3}{4}\tfrac{3}{4}0]$ $[100]$

Figure 8.7: Band structure diagram for germanium; the spin-orbit splitting is omitted for transparency. The hatched region represents the region between the uppermost valence band $E(\mathbf{k})$ and the lowest conduction band $E(\mathbf{k})$, and has no other meaning *per se*. The smallest distance (at Γ) represents the *direct band gap* (see Section 13.2). The distance between the absolute minimum of $E_c(k)$ at L and its absolute maximum of $E_v(k)$ at Γ represents the *indirect band gap*—see Section 13.2 (after Herman, 1958).

Bassani et al., 1975). In general, one has in a nonreduced Brillouin zone representation

$$E_n(\mathbf{k}) = E_n(\mathbf{k} + \mathbf{K}) \qquad (8.43)$$

$$E_n(\mathbf{k}) = E_n(-\mathbf{k}) \qquad (8.44)$$

$$E_n(\mathbf{k}) = E_n(\alpha\mathbf{k}) \qquad (8.45)$$

with $\mathbf{K}$ as a lattice vector in the reciprocal lattice and α identifying any point group operation, e.g., a rotation. The relation (8.44) is also known as *Kramer's theorem*.

Consequently, for nondegenerate $E_n(\mathbf{k})$ at the center (Γ) or at the surface of the Brillouin zone, $E_n(\mathbf{k})$ must have an extremum, as can be seen from fulfilling Eqs. (8.43) and (8.44) simultaneously.

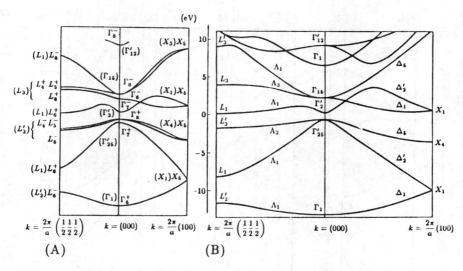

Figure 8.8: Energy bands for Ge. (A) with, and (B) without spin-orbit splitting (after Cardona and Pollak, 1966).

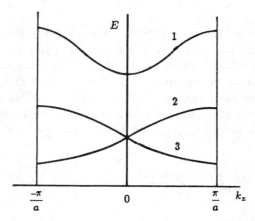

Figure 8.9: Symmetry-related information about the shape of $E_n(\mathbf{k})$ at $k = 0$ and at the surface of the Brillouin zone.

However, noninteracting branches $E_n(\mathbf{k})$ may cross,* thus permitting a finite slope at the symmetry point, as shown for the Γ-point for curves 2 and 3 in Fig. 8.9.

---

* The crossing $E_n(\mathbf{k})$ must belong to different symmetry states that cannot interact with each other. States of the same symmetry interact and cannot cross (noncrossing rule). Examples of crossings can be seen for Si at X_1 and for noncrossings at Δ for GaAs.

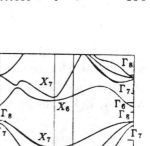

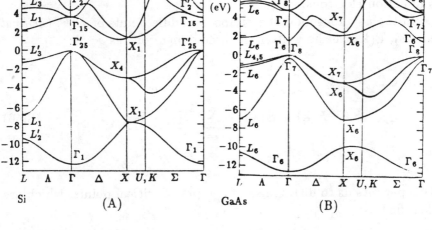

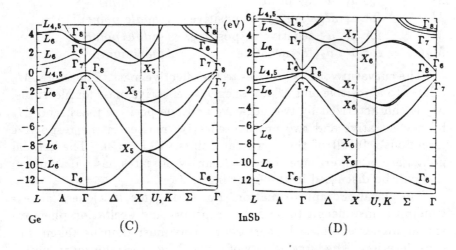

Figure 8.10: Energy bands of Si, Ge, GaAs, and InSb in subfigures (A)–(D), respectively, as calculated with empirical nonlocal pseudopotentials. Energy scale in eV (after Chelikowsky and Cohen, 1976).

Extrema of $E_n(\mathbf{k})$, however, are not limited to the center or surfaces of the Brillouin zone, as Fig. 8.10 shows. There is a maximum of $E(\mathbf{k})$ along Σ and along Δ (see Figs. 3.8 and 8.10) for one of the Γ_{15} and Γ_2' branches. In Si, there is, e.g., a secondary minimum at the bottom of the conduction band at $\sim 0.8\frac{\pi}{a}k_x$. Other minima are at the equivalent points—see Fig. 9.10. Such minima are referred to as *satellite minima*, and play an important role in the semiconductive properties of the material: they determine the properties of the conduction electrons.

In the neighborhood of extrema or saddle points of $E(k)$ (*critical points*), $E(\mathbf{k})$ can be written in three dimensions as

$$E(\mathbf{k}) = E(\mathbf{k_0}) + \sum_{i=1}^{3} a_i \left(k_i - k_{0i} \right)^2. \qquad (8.46)$$

This permits us to distinguish four types of critical points, which are classified as:

M_0	for a_1, a_2, a_3 positive	(*minimum*)
M_1	for a_1 negative a_2, a_3 positive	(*saddle point*)
M_2	for a_1, a_2 negative a_3 positive	(*saddle point*)
M_3	for a_1, a_2, a_3 negative	(*maximum*).

The curvature of $E(k)$ relates to the effective mass. Therefore, M_0 relates to a positive and M_3 to a negative effective mass for electrons. M_2 is characterized by two negative and one positive mass, and M_1 by one negative and two positive effective masses in the respective directions. In all of these critical points, $\nabla_k E = 0$. The critical points are of importance: here the density of states has a maximum; this will be discussed in the following section.

The parabolic approximation [Eq. (8.46)] for $E(\mathbf{k})$ near an extremum is insufficient to describe transport and excitation phenomena at higher energies; higher-order terms must then be taken into consideration. The large variety of bands (Fig. 8.10), however, necessitates restricting this discussion to the most important ones. These will be identified in Section 9.1, while the discussion on the shape of these bands will be postponed to Section 9.1.4.

Any further discussion of the different sets of energy bands shown in Fig. 8.10 requires an understanding of the symmetry properties of the electronic states, which is not the topic of this book. A systematic introduction can be found, for example, in Bassani and Pastori Parravicini (1975).

8.5.3B Density of States   Bands originate from the splitting of atomic eigenstates. Therefore, each band contains N states, where N is the number of atoms in the crystal times a degeneracy factor ν_D

(for a simple band $\nu_D = 2$). The density of states $g(E)$ per energy interval dE is then defined by*

$$\nu_D N = \int_{E_{\min}}^{E_{\max}} g(E)dE. \qquad (8.47)$$

For computational purposes, it is more convenient to express g as a function of $\mathbf{k}$ rather than of E with

$$g(E)dE = \begin{cases} g(k)\dfrac{dk}{dE}dE & \text{in one dimension} \\[2mm] g(\mathbf{k})\dfrac{1}{|\nabla_k E(\mathbf{k})|}dE & \text{in three dimensions.} \end{cases} \qquad (8.48)$$

Within the first Brillouin zone, one can easily follow the filling of the band with electrons. Since electrons are fermions with a spin of $1/2$, they can fill each state only to a maximum of two electrons, one with spin up, the other with spin down. The states fill *consecutively* near $T = 0$; the highest states filled at $T = 0\,\mathrm{K}$ identifies the Fermi energy, which is also referred to as the *Fermi level* (in ideal semiconductors the Fermi level lies near the mid-gap—see Section 27.2.2A).

A single band of type 1 in Fig. 8.9 fills from the center (Γ) to complete first a small sphere (Fig. 9.7), which, with continued filling, will become deformed. The degree of deformation depends on the strength of the lattice forces (see Section 9.1.4A, Fig. 9.8). With S_E the total area of the enclosing surface to which this filling proceeded, we can then express the volume of the k-space between E and $E + dE$ by the integral over a closed surface at E:

$$\oint_E \frac{dS_E}{|\nabla_k E(\mathbf{k})|} dE. \qquad (8.49)$$

Since the volume in the entire first Brillouin zone, given here for a cubic crystal, is $(2\pi/a)^3$ and a crystal of volume $\mathcal{V}$ has $\nu_D \mathcal{V}/a^3$ electron states per band, there are $8\pi^3/(\nu_D \mathcal{V})$ electron states in this band; therefore

$$g_n(E)dE = \frac{\nu_D}{(2\pi)^3} \oint \frac{dS_E}{|\nabla_k E_n(\mathbf{k})|} dE. \qquad (8.50)$$

---

* Compare with a similar calculation of the density of states for phonons given in Section 5.1.5C.

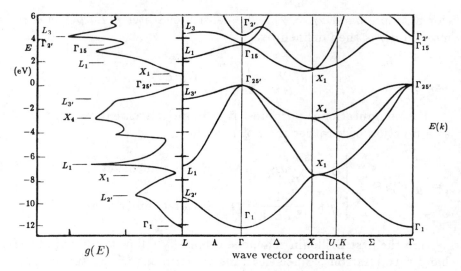

Figure 8.11: Band structure $E(\mathbf{k})$ and density of state distribution $g(E)$ in valence and conduction bands of Si, with corresponding symmetry points identified. The density of state distribution is turned by $90°$ from the conventional representation to relate directly to the $E(\mathbf{k})$ representation (after Chelikowsky and Cohen, 1976; Chelikowsky et al., 1973).

From Eq. (8.50) one sees that the increment of the density of states is steepest where the slope of $E(k)$ is the smallest. This is the case at or near the critical points described in Section 8.5.3A.

The band index n is added for distinction, since with more than one band and with overlapping bands the density of states is additive:

$$g(E)dE = \sum_n g_n(E)dE. \tag{8.51}$$

The total density of states contains a substantial amount of structure stemming from the critical points of the different bands, as shown in Fig. 8.11. This figure shows clearly the interrelationship between $E(k)$ for each band (right part of the figure) with the density distribution (left part). The listed symmetry points will assist in finding the related branches.

For most semiconducting properties, only the density of states near the edges of the band gap is important. This density can be easily estimated in a parabolic $E(\mathbf{k})$, approximation, and will be discussed in more detail in Section 27.2.1.

Summary and Emphasis

The analysis of the atomic and electronic behavior of semiconductors rests on the solutions of the appropriate Schrödinger equation. This equation contains the interaction potential of all involved particles. This interaction potential can be approximated in a number of different ways and can now be computed from basic principles.

The solution of the Schrödinger equation for a one-electron approximation is usually based on variational principles by expanding the wave function in terms of a set of orthogonal trial functions, followed by variation of the expansion coefficients. Selection of the most appropriate trial functions and sufficiently accurate boundary conditions reduces the computational effort and yields results in reasonable agreement with the experiment.

Methods to analyze multi-electron and nonadiabatic problems are developed, and provide guidance for the understanding of higher-order particle interaction which will be discussed later in the book.

The band structure and related density of states is computed for most of the common semiconductors, and becomes available now for a number of new materials with promising semiconducting properties.

The rapid development of computational techniques has changed the quantum mechanical analysis of solid-state processes from providing the qualitative foundation to a quantitative description of a large number of processes involving particle interactions, e.g., such as between lattice atoms (ions) and electrons. This has led to a quantitative description of the band structure and the prediction of stable new compounds before they were synthesized. It permits prediction of their electronic and other properties, and will become an essential tool in designing new materials for solid-state devices.

Exercise Problems

1.(*) Expand on the explanation given in the text referring to two terms dealing with electron-lattice interaction and neglected in the adiabatic approximation. For what type of processes do these terms become important?

2.(*) There are two terms neglected in the adiabatic approximation:

$$\sum \frac{\hbar^2}{2M_k} \nabla_k^2 \psi \quad \text{and} \quad V_{\text{ion}}(\mathbf{R_1}, \mathbf{R_2}, \ldots)\psi.$$

Expand on the explanation given in the text as to why these terms are justifiably neglected.

3.   List and discuss the contributions to the potential term in the one-electron Schrödinger equation.

4.(*)   Explain the principal differences between conventional solution methods of the Schrödinger equation and the Green's function method.

5.   What is the difference between the recent *ab initio* methods and previous approximations?

6.(r)   List spin-orbit splitting for 10 typical semiconductors. Show and explain the trend with atomic numbers in element semiconductors and binary compounds.

7.   Discuss the evolution of the band structure from the empty lattice dispersion shown in Figs. 8.6 and 8.7.

8.(e)   Discuss the effective mass of electrons near the four types of critical points $M_0, \ldots M_3$.

9.(e)   Discuss the relation of the density of state distribution to the band structure shown in Fig. 8.11. Be specific in identifying the critical points.

Chapter 9

Bands and Band Gaps in Solids

Valence and conduction bands and the band gap in between these bands are of key interest for all semiconducting properties. The change of these bands with changing external parameters or chemical composition provides illuminating insight into the electronic behavior of solids.

We will now return to a more simple description of the most important bands in semiconductors, the valence and conduction bands and their dependence on various material and external parameters.

9.1 Valence and Conduction Bands

A set of bands (subbands), created by the splitting and hybridization (see below) of the ground state of valence electrons, taken together are referred to as the *valence band*. The number of states contained in this band is given by the multiplicity of its atomic state (typically 4 for sp^3-hybridization), multiplied by the number of atoms creating this band, e.g., the number of atoms in an entire ideal crystal. In real crystals this number is reduced since the electron scattering limits the coherence length of the electron wave (i.e., the length in which quantum-mechanical interaction can take place). With a mean free path λ (see Chapter 28) the number of atoms responsible for the band level splitting is of the order of $(\lambda/a)^3$ for a primitive cubic lattice with lattice constant a. Each of these levels is broadened by collision broadening [Eq. (39.2)]; therefore, even for λ approaching a, the result is bands rather than discrete levels.

The formation of such bands is often shown as it evolves from the spectrum of isolated atoms when they are brought together to form the crystal lattice. Their levels split with decreasing interatomic distance, as shown in Fig. 9.1 and discussed in Section 7.1. There are several possibilities for the energy and electron distribution over

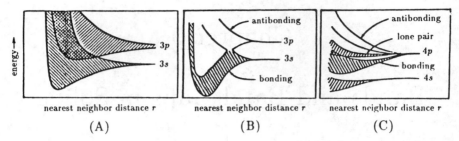

Figure 9.1: Electronic energy distribution for a crystal as a function of the interatomic distance. (A) Overlap of s- and p-bands in main group metals. (B) sp^3-hybridization in covalent tetravalent elements for $r \lesssim r_x$; r_x is the crossover distance. (C) Lone pair semiconductors such as Se.

these levels, depending on the crystal bonding and structure. Three relatively simple examples are given in Fig. 9.1.

Subfigure A shows the splitting and *overlap* of the s- and p-bands of a main group metal. The atomic states remain nearly unchanged when the atoms approach each other (here s- and p-bands do not mix) to form a metal.

Subfigure B shows the splitting for a covalent (tetravalent) crystal, e.g., for a *thatogen*, such as C (diamond) or Si. Near the crossover point, a *hybridization* into sp^3-states occurs, providing four sp^3-states for the lower band and four sp^3-states for the upper band. A *band gap* appears between these two hybrid bands—see Sections 9.2 and 13.1.1.

Subfigure C shows the lone pair configuration typical for higher valency atoms, e.g., a *pnictogen*, such as P or As, or a *chalcogen*, such as Se or Te,* where some electrons in the p-state do not participate in the bonding states. This is the case for $N > 4$, since antibonding states always increase more in energy than bonding states decrease when atoms approach each other (see Fig. 9.1C). Therefore, a *pair* of electrons when nonbonded (*lone pair*) has lower energy than when split into one bonding state electron and one anti-bonding state electron; only a total of four electrons can occupy all of the bonding orbitals.

Bands obtained from such splitting, which are completely or partially filled with valence electrons, are called *valence bands*.

---

* AB-compounds containing these elements are referred to as pnictides or chalcogenides.

9.1.1 Insulators and Semiconductors

In a covalent monatomic crystal the four valence electrons fill each of the levels in the **valence band.** In such a totally filled valence band, electronic conduction is impossible, since electrons can only move by an exchange: for every electron moving in one direction, exactly one electron must move in the opposite direction; there is no free momentum space (*Pauli principle*) for a net electron transport.* Diamond is an example of a simple crystal with tetravalent atoms and a totally filled valence band. It is therefore an *insulator.*

The band above the valence band has the same number of states, but contains no electrons with vanishing excitation. It is an *empty band.*

In materials in which this band is relatively close to the valence band (i.e., in materials with a narrow **band gap** between these bands, such as germanium with a band gap of $E_g = 0.64$ eV), thermal excitation at room temperature will bring a number of electrons into this upper band and partially fill it, although to a very small fraction of the total level density. These electrons can easily gain energy from an external electric field as there are enough free levels available adjacent to each of these electrons; a *net electron transport* (see next section) in the direction of the electric field can take place. Electronic conduction occurs; therefore, this band is referred to as the **conduction band.** Germanium acts as an insulator for vanishing excitation and as a conductor with sufficient thermal excitation: it is called a **semiconductor,** and is distinguished from an insulator by a somewhat narrower band gap.

This description of electronic transport in a partially filled band needs a more precise discussion, which will be given in Chapter 27. However, in order to understand many of the following sections, a heuristic description is supplied below.

9.1.1A Electrons and Holes   When an electron is lifted from the valence band into the conduction band, an empty state is simultaneously created in the valence band. Just as the electron can move in the conduction band, so can the empty state in the valence band. Since it is surrounded by electrons, which can move into the empty state, the empty state moves in the opposite direction of the electron.

---

* In a quantum-mechanical picture any $E(\mathbf{k})$ state represents a certain mass and velocity. In a filled band, all of them add up to zero.

The valence band behavior somewhat resembles the filling of theater seats on the parquet. When all these seats are filled, people still can move, but only by exchanging seats. When the balcony (the conduction band) is opened, some people from the front of the parquet may move up to occupy seats in the balcony, leaving their empty seats below. Consequently, people in the parquet may move toward the stage, giving the impression that empty seats move in the opposite direction. However, only people are moving toward the stage, following the force of attraction. Chairs remain fixed to the floor; their *state* of being empty or filled is moving.

The situation is somewhat similar in a semiconductor: only electrons are moving, not atoms or ions. Ionic conductivity is neglected here. Thus an empty state deserves its own name: a **hole**. It behaves much like an electron with an effective mass similar in value to the electron mass and not that of an ion.

As shown earlier (Section 7.2.3), the electron within the conduction band is free to move when considering its mass as an effective mass. To distinguish it from an electron *in vacuo*, it is also referred to as a *Bloch electron*. The holes are also free to move. Both contribute to the electrical conductivity and are therefore called *carriers* (of this current).

Electrons tend to occupy the lowest energy states; that is, they fill a band from the bottom up and occupy states near the bottom of the conduction band designated as E_c. Holes will consequently, like soap bubbles in water, bubble toward its upper surface: they will collect at the upper edge of the valence band designated as E_v.

When an electron is accelerated in the direction of a mechanical force, for instance, by accelerating the semiconductor in the Tolman experiment, the hole is accelerated in the opposite direction. Therefore, its effective mass m_p has the sign opposite to the mass of an electron:

$$m_p = -\frac{\hbar^2}{\left.\frac{\partial^2 E}{\partial k^2}\right|_{E_v}} \quad \text{and} \quad m_n = +\frac{\hbar^2}{\left.\frac{\partial^2 E}{\partial k^2}\right|_{E_c}}; \qquad (9.1)$$

this is illustrated in Fig. 9.2. At the top of the valence band, $E(k)$ has a *negative curvature* that results in a *negative effective mass for electrons*. However, since the effective mass of the hole has the opposite sign as the electron [Eq. (9.1)], its effective mass is *positive*

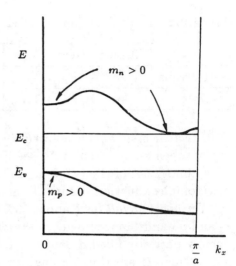

Figure 9.2: $E(k)$ for a simple valence and conduction band; signs of the effective masses of holes and electrons are indicated.

here, as is the effective mass of an electron at the bottom of the conduction band:

$$m_p(E_v) > 0, \quad m_n(E_c) > 0. \tag{9.2}$$

Finally, the charge of an electron is negative* by convention; therefore, the charge of a hole (i.e., of an atom with a missing electron) is positive. This permits the distinction between a particle flux under the influence of an electric field† and the electric current caused by electrons and holes. The particle flux proceeds in opposite directions: electrons move in the direction of the electric field, and holes move against it. However, both electron and hole currents have the same sign and are therefore additive. Table 9.1 summarizes the typical properties of electrons and holes in a semiconductor.

9.1.2 Metals

In *monovalent metals*, the valence band is only partially filled; 50% of the *s*-states are occupied. Therefore, electronic conduction takes place even with vanishing excitation.

---

* A positive charge was arbitrarily related to the charge of a glass rod, rubbed with silk (by Benjamin Franklin); this charge was not caused by an added electron as it became known later, but by a missing electron on the glass rod. This electron was removed by the silk.

† In this and all chapters dealing with electrical conductivity, the electric field is identified as F and the energy as E.

Table 9.1: Typical properties of electrons and holes.

	Charge	Near	m^*	$\partial^2 E/\partial k^2$	Mobility	$j_{n,p}(F)$
Electrons	$-e$	E_c	$+m_n$	positive	$-\mu_n$	positive
Holes	$+e$	E_v	$+m_p$	negative	$+\mu_p$	positive

Divalent metals show allowed eigenvalues in the range of the overlap between s- and p-bands*—see Fig. 9.1A. In this range, therefore, only about 25% of the two s- and the six p-states are filled, and thus electronic conduction occurs at vanishing excitation.

Trivalent metals have about 63% of their states in the overlapping s- and p-bands unoccupied. In *transition metals*, inner shell electrons occupy partially filled d-bands.

All metals are thus distinguished from semiconductors by their substantial ($\geq 50\%$) fraction of free states in the highest partially occupied band with vanishing excitation. This band is therefore both a valence band, containing the valence electrons, and a conduction band where all *metallic conduction* takes place.

This simple model for distinguishing nonmetals from metals by the complete or incomplete filling of bands is due to Wilson (1931). The model is still valid for most cases, except where magnetic properties or strong binding properties (polarons—see Section 27.1.2) interfere, as, for example, in NiO.

9.1.3 Semimetals and Narrow Gap Semiconductors

Metals that show a very small overlap of conduction and valence band exhibit weak electronic conduction. The density of states near the Fermi level (see Section 27.2.2A) is very small and only the relatively few electrons there control the conduction. Only these electrons can be scattered. For reviews, see Ziman (1969), or Cracknell and Wong (1973). These materials are termed *semimetals*; examples are graphite and bismuth.

On the other hand, semiconductors with very small band gaps show *relatively* high conductivity to other semiconductors. By some

---

* The difference between metals, where the overlap range is allowed, and semiconductors, where the overlap range is forbidden (band gap), depends on *Wigner's rules* (Wigner, 1959), which state that eigenstates belonging to different symmetry groups of the Hamiltonian cannot mix (metals). In semiconductors they do mix, yielding sp^3-hybridization for Si.

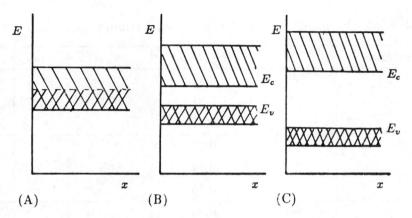

Figure 9.3: Empty (hatched) and filled (cross-hatched) states in the upper bands of (A) a metal; (B) a semiconductor; and (C) an insulator. The energies at the top of the valence band and the bottom of the conduction band are identified as E_v and E_c, respectively.

alloying, application of pressure, or even at elevated temperature, the band gap can vanish, turning such semiconductors into metals. Examples include gray tin with a gap of 0–0.08 eV (suggested by magneto-optical experiments; Pidgeon, 1969), and lead chalcogenides and their alloys with other II-VI compounds, or with SnTe and mercury chalcogenides. For a review, see Tsidilkovski et al. (1985).

The lead chalcogenides are characterized by a small band gap with a maximum of the valence band and a minimum of the conduction band at the L-point rather than the Γ-point for most other direct gap semiconductors (Section 13.2) with E_g and m_n as shown in Table 9.2. Alloyed with CdTe ($E_g = 1.5$ eV), the band gap varies with composition, but can dip below zero (to -0.1 eV) because of bowing (see Section 9.2.1A) for an alloy of PbTe containing a few percent Cd (Schmit and Stelzer, 1973).

A typical change of the $E(\mathbf{k})$ behavior near the Γ-point is shown in Fig. 9.4 for $\mathrm{Hg}_{1-\xi}\mathrm{Cd}_\xi\mathrm{Te}$ as a function of the hydrostatic pressure or the composition (Kane, 1979). The band gap can be expressed by an empirical relation (Hansen et al., 1982)

$$E_g(\xi, T) = -0.302 + 1.93\xi + 5.35 \cdot 10^{-4}(1-2\xi)T - 0.81\xi^2 + 0.832\xi^3 \text{ (eV)}$$

$$(9.3)$$

for $0 < \xi < 1$ and $4.2 < T < 300$ K. The gap also changes linearly with pressure, with a pressure coefficient of 10^{-2} eV/kbar.

Table 9.2: Band parameters for lead chalcogenides.

Material	T (K)	E_g (eV)	$m_{n\parallel}$	$m_{n\perp}$
PbS	290	0.41		
	77	0.31	0.11	0.11
	4	0.29		
PbSe	290	0.27		
	77	0.17	0.05	0.08
	4	0.15		
PbTe	290	0.32		
	77	0.22	0.025	0.22
	4	0.19		

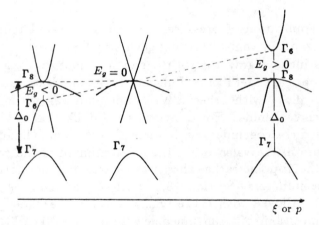

Figure 9.4: Conduction and valence bands of $Hg_{1-\xi}Cd_\xi Te$ as a function of pressure or composition (schematic).

Another interesting property is observed when alloying PbTe and SnTe. The L_6^+ and L_6^- bands are inverted; the L_6^- band is the conduction band in PbTe, while it is the valence band in SnTe, and vice versa for the L_6^+ band. In an alloy of $Pb_{1-\xi}Sn_\xi Te$ with $\xi = 0.62$, both bands touch at 300 K. For $\xi > 0.62$, the alloy becomes a metal (see Fig. 9.5). Such materials are also referred to as *gapless semiconductors* when the top of the valence band touches the bottom of the conduction band.

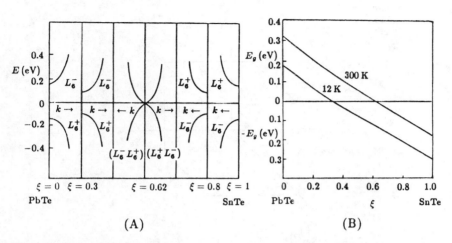

$\xi = 0$ $\xi = 0.3$ $\xi = 0.62$ $\xi = 0.8$ $\xi = 1$
PbTe   SnTe

PbTe   SnTe

(A)   (B)

Figure 9.5: (A) Conduction and valence bands in a schematic $E(k)$ representation of $Pb_{1-\xi}Sn_{\xi}Te$ for different alloy ratios ξ. (B) Band gap as a function of alloy ratio for $T = 12$ and 300 K (after Dimmock et al., 1966).

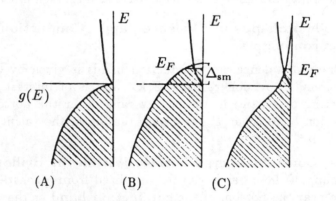

(A)   (B)   (C)

Figure 9.6: Density of states of valence and conduction bands with electron filling indicated for: (A) gapless semiconductor; (B) semimetal; and (C) gapless semiconductor with overlapping tail states (Section 24.1).

These examples demonstrate the possibility of designing very narrow band gap materials or semimetals by alloying two narrow band gap semiconductors with each other (see Fig. 9.6). The density of states in semimetals at the Fermi level is very small compared to that of a metal. In gapless semiconductors, it is zero (see Fig. 9.4 at $\xi = 0.16$ for $T = 0$ K).

A special case of a gapless semiconductor occurs when tailing band states overlap with the conduction band (Fig. 9.6C). Tailing

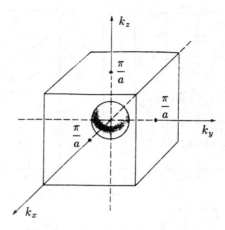

Figure 9.7: Equi-energy surface for small electron energies within the first Brillouin zone for a primitive cubic lattice.

states are band states which extend into the band gap due to the disturbed neighborhood of lattice defects as discussed in Chapter 25.

9.1.4 The Shape of Valence and Conduction Bands in Semiconductors

The shape of valence and conduction bands is given by the three-dimensional dispersion relation $E(\mathbf{k})$, and is of interest for semiconducting properties in the energy range near the bottom of the conduction band (near E_c) and near the top of the valence band(s) (near E_v).

9.1.4A Constant Energy Surface Within the Brillouin Zone
The shape of these bands can be visualized from a constant energy surface near the bottom of the **conduction band** or the top of the valence band. This surface can be identified by sequentially filling the band* as described in Section 8.5.3B. Electrons that populate states within $E(\mathbf{k})$ up to a certain energy E_1 [i.e., to the corresponding $k_1(E_1)$] are contained in a small sphere in the center of the zone (Fig. 9.7) that grows parabolically in radius with increasing degree of filling—that is, with an increasing value of k (see Section 27.2.1). $k(E)$ is single-valued and monotonic within each band.

---

* Each band contains a large number of energy levels (see Section 8.5.3B). Increasing numbers of electrons first fill the levels at the lowest energy (for $T = 0$) and then successively higher and higher energies. This process is referred to as band filling. See also Sections 9.1 and 9.2.1C.1.

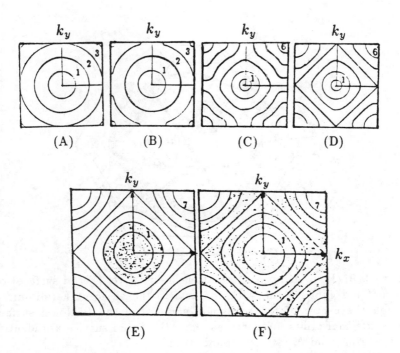

Figure 9.8: Two-dimensional cut through a family of equi-energy surfaces for a primitive cubic crystal with increasing binding potential from (A)–(D). Filling with electrons is indicated by shading in (E) and (F). Spherical surfaces indicate quasi-free behavior of electrons. Deviations from the sphere indicate substantial lattice influence.

When starting from the center, $E(\mathbf{k}_1)$ becomes progressively more deformed with higher energy (higher k). A two-dimensional representation (Fig. 9.8) demonstrates more clearly some typical shapes of *equal energy surfaces* in a simple cubic lattice. Subfigures A–D are distinguished by increasing binding forces, showing increasingly more deviation from circles.

A similar development is observed near the top of the **valence band**: when it becomes filled with holes, the upper part of the Brillouin zone becomes depleted down to a certain energy E_1' with the effect of rounding down the sharp edges of the Brillouin zone.

Depending on the density of electrons within the band, a large fraction of the Brillouin zone is filled. Figure 9.9 illustrates such a Brillouin zone with a partially filled valence band of a metal. This figure also shows the connection to another neighboring zone. The periodicity of $E(\mathbf{k})$ is indicated at the upper left of Fig. 9.9. The zones are of the same order, first Brillouin zones. Higher Brillouin

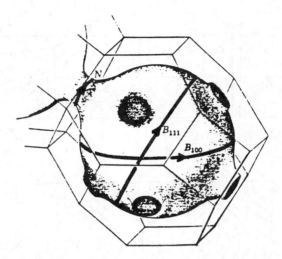

Figure 9.9: Equi-energy surface at the Fermi level (Fermi surface) of a face-centered cubic metal, with contacts to the adjacent Fermi surfaces through the center of the hexagonal surfaces of the Brillouin surfaces. Three extremal orbits for electrons along the Fermi surface are identified by B_{111}, B_{100}, and N (after Pippard, 1965).

zones are not relevant to the discussion of semiconducting properties. In contrast to the one-dimensional $E(k)$ representation of Fig. 7.12, where several curves in the first Brillouin zones are shown, in the k-space representation only one such equivalent surface can be drawn without confusion. It is the valence band in Figs. 9.8E and F, 9.9, and 9.10C, and the conduction band in Figs. 9.7 and 9.10A and B.

In semiconductors, the equi-energy surfaces close to the top of the valence band (nearly full Brillouin zone) and at the bottom of the conduction band (nearly empty Brillouin zone) are of interest for the electron transport. They are shown in Fig. 9.10 for Ge and Si.

Figure 9.10A shows for Si six small ellipsoids in the $\langle 100 \rangle$ direction which are centered at $\sim \pm 0.8 k_x$, $\pm 0.8 k_y$, and $\pm 0.8 k_z$. Figure 9.10B indicates for Ge in the $\langle 111 \rangle$ direction eight ellipsoids which are centered at the L-points (see Fig. 3.8). There are four such L minima per Brillouin zone. Other semiconductors having their lowest minima outside the Γ-point are GaP and AlSb. Most other semiconductors have their lowest conduction band minimum at (000).

Figure 9.10C shows the warped equi-energy surface, near the upper edge of the valence band for heavy holes (see Section 9.1.5B) in Si and Ge. This equi-energy surface fills almost the entire Brillouin zone. Its shape is shown without the surrounding Brillouin zone

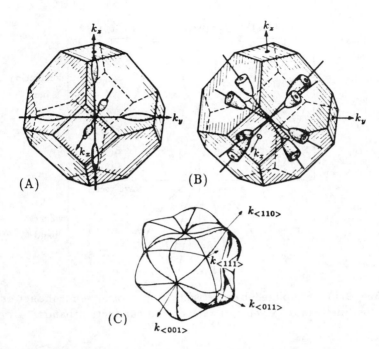

Figure 9.10: Brillouin zone of (A) Si and (B) Ge with equi-energy surfaces near the conduction band edge and (C) equi-energy surface of the valence band for heavy holes in Si or Ge near the upper band edge (after McKelvey, 1966; Seeger, 1973).

in nearly the same scale as in Figs. 9.10A and B. The filling of the valence band with holes is greatly exaggerated; otherwise, the Brillouin zone with a flat outer surface and only slightly rounded edges would show.

The *shape of these equi-energy surfaces* is intimately related to the three-dimensional shape of the bands $(E(\mathbf{k}))$—see Section 8.5.3. The anisotropy is evident from Fig. 9.10.

The influence of this anisotropic $E_n(\mathbf{k})$ behavior, and therefore of the effective mass in Si and Ge on carrier transport, is discussed in Section 33.4.1A. We will first analyze the anisotropy of the effective mass in more detail.

9.1.5 The Effective Mass in Real Bands

In Section 7.2.2 the effective mass is defined as proportional to the inverse of the second derivative of $E(k)$, i.e., of the curvature of $E(k)$.

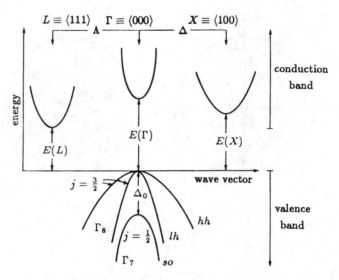

Figure 9.11: Simplified band structure of typical semiconductors near the most important extrema of valence and conduction bands.

The anisotropy of $E(k)$ in an actual crystal makes the introduction of a tensor relation for the effective mass necessary:

$$m_n = \underline{\mathbf{m}}^*_{ik} = \left(\!\!\left(\frac{\hbar^2}{\dfrac{\partial^2 E}{\partial k_i \partial k_k}} \right)\!\!\right). \qquad (9.4)$$

This tensor ellipsoid, when transformed to main axes, contains usually only very few nonvanishing components. We will identify these below for some of the more important semiconductors.

Figure 9.11 shows the important minima of typical conduction bands and maxima of the valence bands. In most semiconductors, the relevant conduction band minimum lies at the Γ-point (direct gap), or at or near the L- or X-point (indirect gap as discussed in Section 13.2). The energies of these minima are listed in Table 9.3. The relevant maxima of the valence band are at the Γ-point. Their energies are listed in Table 9.5.

We will first discuss the shape of the conduction and valence bands near the band edge, and thereby provide simple expressions for the respective effective masses.

9.1.5A The Conduction Bands   The relationship of $E(\mathbf{k})$ in the vicinity of the conduction band minima near the band edge can be described as

conduction band at, Γ :
$$E_c(\mathbf{k}) = \frac{\hbar^2}{2m}k^2 \qquad (9.5)$$

satellite conduction band:
$$E_c(\mathbf{k}) = \frac{\hbar^2}{2}\left(\frac{(k_x - k_{0x})^2}{m_\parallel} + \frac{(k_y - k_{0y})^2 + (k_z - k_{0z})^2}{m_\perp}\right), \qquad (9.6)$$

relating $E(\mathbf{k})$ directly to the corresponding effective masses.

Satellite minima of $E(\mathbf{k})$ at $\mathbf{k}_0 \neq 0$ show ellipticity, such as those of the indirect gap semiconductors Ge, Si, and GaP; so do the higher satellite minima of others III-IV compounds, such as, e.g., GaAs. For symmetry reasons, these ellipsoids are ellipsoids of revolution about the main axes; hence, we distinguish only two effective masses, parallel and orthogonal to these axes, $m_\parallel$ and $m_\perp$, respectively [Eq. (9.6)]. The main axes of these ellipsoids of revolution lie in $\langle 100 \rangle$ direction for Si and in $\langle 111 \rangle$ direction for Ge.

Another example is the camel's back conduction band of some III-V compound semiconductors (see Fig. 9.12); it can be described by

$$E(\mathbf{k}) = \frac{\hbar^2 k_\parallel^2}{2m_\parallel} + \frac{\hbar^2 k_\perp^2}{2m_\perp} - \sqrt{\left(\frac{\Delta^c}{2}\right)^2 + \Delta_0^c \frac{\hbar^2 k_\parallel^2}{2m_\parallel}} \qquad (9.7)$$

where $k_\parallel$, $m_\parallel$ and $k_\perp$, $m_\perp$ are the components parallel and perpendicular to [100], and Δ_0^c describes the nonparabolicity; ΔE and Δ^c are identified in Fig. 9.12.

The energies of the conduction band minima and the corresponding effective masses are listed in Table 9.3. Band gaps of some higher compound semiconductors are listed in Table 9.4.

9.1.5B The Valence Band   The valence band develops from the twofold degenerate s-states and the sixfold degenerate p-states of an atom forming sp^3-hybrids. Without spin-orbit interaction it breaks into two bands at $\mathbf{k} = 0$ (Γ_1 and $\Gamma_{25'}$, as shown in Fig. 8.7). With spin-orbit interaction, the upper edge $\Gamma_{25'}$ breaks into two sets of bands: one set with a total momentum of $3/2$ is fourfold degenerate at $\mathbf{k} = 0$; this degeneracy can be removed in anisotropic crystals with crystal field coupling, for example, in hexagonal CdS.

The other set is twofold degenerate with $j = 1/2$ and shifted by the spin-orbit splitting energy Δ_0 (Figs. 8.8 and 9.11); it is referred

Table 9.3: Electron effective masses in units of m_0, conduction bands in eV (at 0 K) with characteristic point and band numbers identified, and camel's back parameters in eV (Δ^c and Δ_0^c).

Crystal	1. Indir. Gap	$m_{n\parallel}$	$m_{n\perp}$	2. Direct Gap	m_n	3. Gap	$m_{n\parallel}$	$m_{n\perp}$	Δ^c	Δ_0^c or ΔE
Si	$(X_1)1.16$	0.191	0.916	$(\Gamma_2)4.19$		$(K_3)1.70$				
Ge	$(L_6)0.76$	1.57	0.081	$(\Gamma_7)0.898$	0.038					
GaP	$(\Delta_1)2.35$	0.91	0.25	$(\Gamma_1)2.895$	0.093	$(L_1)2.64$			$(\Delta)0.355$	$(\Delta)0.433(\Delta_0^c)$
GaAs	$(L_6)1.82$	1.9	0.075	$(\Gamma_6)1.519$	0.067	$(X_6)2.03$	1.8	0.257	$(X)0.304$	$(X)0.009(\Delta E)$
GaSb	$(L_6)1.22$	0.95	0.11	$(\Gamma_6)0.86$	0.041	$(X_6)1.72$	1.2	0.250	$(X)0.178$	$(X)0.025(\Delta E)$
InP	$(L_6)2.19$			$(\Gamma_6)1.50$	0.077	$(X_6)2.44$				
InAs	$(L_6)1.53$			$(\Gamma_6)0.37$	0.024	$(X_6)2.28$				
InSb	$(L_6)1.03$	0.09		$(\Gamma_6)0.25$	0.0136	$(X_6)1.71$				
ZnS_{cub}				$(\Gamma_1)3.78$	0.34					
ZnS_{hex}				$(\Gamma_1)3.91$	0.28					
ZnSe	$(L_6)3.96$			$(\Gamma_6)2.82$	0.16	$(X_6)4.57$				
ZnTe				$(\Gamma_6)2.394$	0.122					
CdS				$(\Gamma_6)2.583$	0.21					
CdSe				$(\Gamma_7)1.829$	0.112					
CeTe	$(L_6)2.82$			$(\Gamma_6)1.475$	0.096	$(X_6)3.48$				

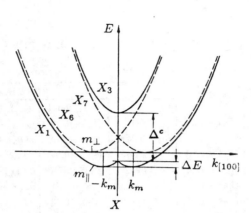

Figure 9.12: Camel's back conduction band in GaP along the Δ-axes near the Brillouin zone boundary at X. Dashed and solid curves for diamond and zinc-blende structures, respectively.

Table 9.4: Band gaps of some higher compound semiconductors.

Crystal	Indirect	E_g	Direct	E_g
Mg_2Si	$\Gamma_{15} \to X_1$	0.6	$\Gamma_{15} \to \Gamma_1$	2.17
Mg_2Ge	$\Gamma_{15} \to X_3$	0.74	$\Gamma_{15} \to \Gamma_1$	1.64
Mg_2Sn	$\Gamma_{15} \to X_3$	0.36	$\Gamma_{15} \to \Gamma_1$	1.2
β-LiSb*	$\Gamma_{15} \to X_1$	0.74	$\Gamma_{15} \to \Gamma_1$	2.2
K_3Sb*	$\Gamma_{15} \to X_1$	1.39	$\Gamma_{15} \to \Gamma_1$	0.56
Cs_3Sb*	$\Gamma_{15} \to X_1$	1.75	$\Gamma_{15} \to \Gamma_1$	1.02

*Calculated by Wei and Zunger (1986).

to as the *spin-orbit split-off band* "so" with an effective mass m_{so}. For $\mathbf{k} \neq 0$ the degeneracy of the upper band is removed. Here, the band splits into two bands, each of which is twofold degenerate. The one with lower curvature is the *heavy hole band* "hh" with an effective mass m_h; the other with larger curvature is the *light hole band* "lh" with an effective mass m_l. The valence band energies at a number of symmetry points are given in Table 9.5 for some typical semiconductors. Near the Γ-point ($\mathbf{k} = 0$) the Hamiltonian describing the three sets of upper valence bands can be written as (D'yakonov and Perel', 1971)

$$\mathbf{H} = -(A + 2B)k^2 + 3B(\mathbf{k} \cdot \hat{\mathbf{L}})^2 - \frac{1}{3}\Delta_0(\underline{\sigma} \cdot \hat{\mathbf{L}}) + \frac{1}{3}\Delta_0 \qquad (9.8)$$

Table 9.5: Upper (except Γ_6) valence bands in eV in typical semiconductors.

Semiconductor	Γ-related		Maxima	L-related	X-related
Si	$(\Gamma_1) - 12.5$	$\Gamma_{25} = (\Gamma_8)0$	$(\Gamma_7) - 0.044$	$(L_3) - 2.82$	$(X_4) - 6.27$
Ge	$(\Gamma_6) - 12.7$	$(\Gamma_8)0$	$(\Gamma_7) - 0.29$	$(L_{6,5}) - 1.43$	$(X_5) - 3.29$
GaP	$(\Gamma_1) - 13.0$	$\Gamma_{15} = (\Gamma_8)0$	$(\Gamma_7) - 0.08$	$(L_3) - 1.1$	$(X_5) - 2.7$
GaAs	$(\Gamma_7) - 12.6$	$(\Gamma_8)0$	$(\Gamma_7) - 0.35$	$(L_{4,5}) - 1.2$	$(X_7) - 2.87$
GaSb	$(\Gamma_6) - 12.0$	$(\Gamma_8)0$	$(\Gamma_7) - 0.756$	$(L_{4,5}) - 1.1$	$(X_7) - 2.37$
InP	$(\Gamma_6) - 11.4$	$(\Gamma_8)0$	$(\Gamma_7) - 0.21$	$(L_{4,5}) - 0.94$	$(X_7) - 2.06$
InAs	$(\Gamma_6) - 12.7$	$(\Gamma_8)0$	$(\Gamma_7) - 0.43$	$(L_{4,5}) - 0.9$	$(X_7) - 2.37$
InSb	$(\Gamma_6) - 11.6$	$(\Gamma_8)0$	$(\Gamma_7) - 0.82$	$(L_{4,5}) - 0.95$	$(X_7) - 2.24$
ZnS	$(\Gamma_6) - 0.8$	$\Gamma_{15} = (\Gamma_8)0$	$(\Gamma_7) - 0.067$	$(L_3) - 1.4$	$(X_5) - 2.5$
ZnSe	$(\Gamma_6) - 12.5$	$\Gamma_{15} = (\Gamma_8)0$	$(\Gamma_7) - 0.45$	$(L_{4,5}) - 0.76$	$(X_7) - 1.96$
ZnTe	$(\Gamma_3) - 4.5$	$\Gamma_{15} = (\Gamma_8)0$	$(\Gamma_7) - 0.97$	$(L_3) - 1.1$	$(X_5) - 1.3$
CdS	$(A_{5,6}) - 0.5$	$(\Gamma_5)0$	$(\Gamma_6) - 0.8$	$(L_{2,4} - 1.4$	$(M_4) - 0.7$
CdSe		$(\Gamma_9)0$	$(\Gamma_7) - 0.416$		
CdTe	$(\Gamma_6) - 11.1$	$(\Gamma_8)0$	$(\Gamma_7) - 0.89$	$(L_{4,5}) - 0.65$	$(X_7) - 1.60$

with

$$A = -\frac{\hbar^2}{4}\left(\frac{1}{m_l} + \frac{1}{m_h}\right), \quad B = \frac{\hbar^2}{4}\left(\frac{1}{m_l} - \frac{1}{m_h}\right), \quad (9.9)$$

where $\hat{\mathbf{L}}$ is the angular momentum matrix, $\underline{\sigma}$ is the Pauli operator spin matrix, and Δ_0 is the spin-orbit splitting.

Only the light and heavy hole bands need to be considered when Δ_0 is larger than certain energies, e.g., kT or the ionization energy of shallow acceptors (see Section 21.1).

For these two bands, the Hamiltonian reduces considerably. Luttinger (1956) gives it in a form that does not require spherical symmetry

$$\mathbf{H} = \frac{\hbar^2}{2m_0}\left[\left(\gamma_1 + \frac{5}{2}\gamma_2\right)\nabla^2 - 2\gamma_3(\nabla \cdot \mathbf{J})^2 \right.$$
$$\left. + 2(\gamma_3 - \gamma_2)(\nabla_x^2 J_x^2 + \nabla_y^2 J_y^2 + \nabla_z^2 J_z^2)\right], \quad (9.10)$$

where $\mathbf{J}$ is a pseudovector representing the spin momentum operator (Bir and Pikus, 1972) and γ_i are the Luttinger parameters.

For $\gamma_3 = \gamma_2$, the eigenvalues of this Hamiltonian yield two parabolic bands

$$E_{lh} = \frac{\gamma_1 + 2\gamma_2}{2m_0}\hbar^2 k^2 \quad \text{and} \quad E_{hh} = \frac{\gamma_1 - 2\gamma_2}{2m_0}\hbar^2 k^2 \quad (9.11)$$

Table 9.6: Hole effective masses, valence band splitting,* and Luttinger valence band parameters.

Crystal	γ_1	γ_2	γ_3	$\Delta_0\,(T(K))$	Δ_1	$m_h(100)$	$m_h(110)$	$m_h(111)$	m_l	m_{so}
Si	4.285	0.339	1.446	0.045 (10)		0.537			0.153	0.234
Ge	13.38	4.28	5.69	0.297 (10)		0.284	0.352	0.376	0.044	0.095
GaP	4.05	0.49	1.25	0.08 (100)		0.419		0.997	0.16	0.465
GaAs	6.95	2.25	2.86	0.34 (10)		0.51			0.082	0.154
GaSb	13.3	4.4	5.7	0.76		0.28			0.05	
InP	5.15	0.94	1.62			0.56		0.60	0.12	0.121
InAs	20.4	8.3	9.1	0.38 (1.5)		0.35		0.43	0.026	
InSb	3.25	−0.20	0.90	0.85 (100)		0.34	0.42	0.45	0.016	
ZnS$_{cub}$				0.067 (80)		1.76			0.169	
ZnS$_{hex}$				0.086 (77)	0.055	$1.4_{\parallel}$	$0.49_{\perp}$			
ZnSe	4.3	1.14	1.84	0.403 (80)						
ZnTe	3.9	0.83	1.30	0.97 (80)						
CdS				0.062 (77)	0.027	0.64			0.64	
CdSe				0.416 (77)	0.039	≥ 1	0.45		0.9	
CdTe	5.3	1.7	2.0	0.811 (80)		0.72	0.81	0.84	0.13	

*The two top valence bands are split at $k = 0$ by Δ_1 due to crystal field coupling; $\Delta_1 \neq 0$ in anisotropic crystals, such as hexagonal structures. Δ_0 is the spin-orbit splitting. Δ_0 and Δ_1 are given in eV.

for light and heavy holes with effective masses

$$m_l = \frac{m_0}{\gamma_1 + 2\gamma_2} \quad \text{and} \quad m_h = \frac{m_0}{\gamma_1 - 2\gamma_2}. \tag{9.12}$$

With γ_3 substantially different from γ_2, the two sets of valence bands become warped and are determined from the extensive dispersion equation:

$$E_{l,m} = \frac{\hbar^2}{2m_0}\left[\gamma_1 k^2 \pm \sqrt{4\gamma_2^2 k^2 + 12(\gamma_3^2 - \gamma_2^2)(k_x^2 k_y^2 + k_y^2 k_z^2 + k_z^2 k_x^2)}\right] \tag{9.13}$$

with the + and − signs for light and heavy hole bands, respectively. The degree of warping can be judged from the tabulated values of γ_i (Table 9.6). For example, it is small for Ge and GaAs ($\gamma_2 \simeq \gamma_3$) and much larger for Si and InP (also see Fig. 9.14).

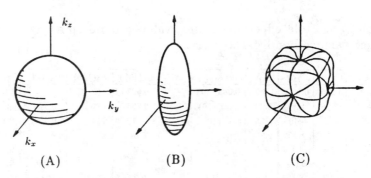

Figure 9.13: Typical shapes of surfaces of constant energy: (A) spherical; (B) ellipsoidal; and (C) warped.

Often an equivalent description of these two bands is given.*

$$E_\pm(k) = Ak^2 \pm \sqrt{(Bk^2)^2 + C(k_x^2 k_y^2 + k_y^2 k_z^2 + k_z^2 k_x^2)}, \qquad (9.15)$$

with the $+$ and $-$ signs for the light and heavy hole bands, respectively. The band parameters can be interpreted as A giving the average curvature, B giving the splitting between heavy and light hole bands, and C describing the warping. These parameters are related to the *Luttinger band parameters* by

$$\frac{\gamma_1}{m_0} = -\frac{2}{\hbar^2}A, \quad \frac{\gamma_2}{m_0} = -\frac{1}{\hbar^2}B, \quad \frac{\gamma_3}{m_0} = -\frac{1}{\sqrt{3}\hbar^2}\sqrt{C^2 + 3B^2}. \ (9.16)$$

The different shapes of the surfaces of constant energy for the different conduction and valence bands are summarized in Fig. 9.13.

An impression about the degree of warping for Si can be obtained from Fig. 9.14, which shows cuts through the $E(k)$ surface in different crystallographic directions (Pantelides, 1978).

9.1.5C Probing Bands with Cyclotron Resonance When applying a strong enough magnetic field, the band shape can be probed by forcing the electrons into circles perpendicular to the direction of

---

* There are also Dresselhaus parameters L, M, and N (Dresselhaus et al., 1955) to describe the valence band. They are related to the Luttinger parameters by

$$\gamma_1 = -2m_0(L + 2M)/3; \quad \gamma_2 = -2m_0(L - M)/6; \quad \gamma_3 = -2m_0 N/6.$$
$$(9.14)$$

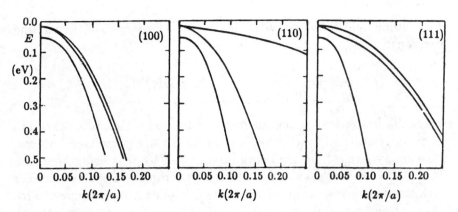

Figure 9.14: $E(k)$ diagram for the valence bands of Si near the Γ-point in three symmetry directions (after Pantelides, 1978).

the magnetic induction (for vanishing electric field), following the Lorentz force:

$$\vec{\mathcal{F}} = \hbar \frac{d\mathbf{k}}{dt} = e\,(\mathbf{F} + \mathbf{v} \times \mathbf{B}).\qquad (9.17)$$

These circling electrons are accelerated by the magnetic induction with a centripetal acceleration

$$\mathbf{a} = \frac{\hbar}{m^*}\frac{\partial \mathbf{k}}{\partial t} = \frac{e}{m^*}\,(\mathbf{v} \times \mathbf{B}) = \frac{e}{\hbar m^*}\,(\nabla_k E(\mathbf{k}) \times \mathbf{B}),\qquad (9.18)$$

this acceleration is a measure of the effective mass, m^*, hence of the band shape at the specific electron energy (see below). The accelerated electrons can interact with a high frequency, small amplitude probing electromagnetic field, and show a resonance absorption—the *cyclotron resonance* (in the classical limit)* when the ac-frequency of

---

* Here cyclotron resonance is discussed within the same band, and quantum effects are neglected. This can be justified when, neglecting scattering, each electron describes full circles which have to be integers of its De Broglie wavelength [Eq.(7.6)]. This integer represents the quantum number n_q of the circle; and for the magnetic induction discussed here it is a large number. Resonance means absorption (or emission) of one quantum $\hbar\omega_c$, hence changing n_q by $\Delta n_q = \pm 1$, which is the selection rule for cyclotron transitions. Since $\Delta n_q \ll n_q$, a change in circle diameter is negligible; hence the classical approach is justified. At higher fields the circles become smaller; and when approaching atomic size, the quantum levels (Landau levels–see Section 31.3.2) become wider-spaced and a quantum mechanical approach is required. For reviews, see Lax (1963), Mavroides (1972), and McCombe and Wagner (1975).

the probing field coincides with the *cyclotron frequency* of the circling electrons:

$$\omega_c = \frac{eB_0}{m^*} = 17.84 \frac{m_0}{m^*} B_0 \ (\mathrm{GHz/kG}),$$ (9.19)

where B_0 is the stationary magnetic induction. This resonance absorption is quite distinct when the electrons are permitted to complete many circles* before being interrupted by scattering (see Section 32). Since the path of an electron to complete a cycle is smaller with the decreasing radius of the circle and scattering is reduced with the decreasing density of defects and phonons (see Section 32), cyclotron resonance measurements are usually performed at high magnetic fields in materials of high purity and at low temperature. Excessively high magnetic fields, however, cause inconveniently large high-resonance frequencies (for B_0 of 10 kG, one obtains $\nu = \omega_c/2\pi \simeq 100$ GHz).

Although electrons along all cross sections of the Fermi surface perpendicular to the direction of the magnetic field cause resonance absorption, absorption *maxima* are observed at the extrema of the cross section (belly or neck) because of the higher electron density here. In addition, one observes so-called dog-bone cross sections, in which electrons circle between four adjacent Fermi surfaces near the neck of each surface, as well as many other cross-section shapes depending on the direction of the magnetic field and the shape of the Fermi-surfaces.

In semiconductors, the cyclotron resonance can be used to probe the shape of the bands near the edge of the conduction or valence band. Since the cyclotron resonance frequency depends only on the effective mass, its measurement yields the most direct information about its behavior (see Section 9.1.5D—and reviews by Smith, 1967; and Pidgeon, 1980).

---

* The circle diameter $(\pi v_n/\omega_c)$ is typically of the order of 10^{-3} cm for a magnetic induction of 10 kG; here v_n is the thermal velocity of an electron. In metals, however, one also has to consider the skin penetration of the probing electromagnetic field. The skin depth of a metal is usually a very small fraction of the circle diameter, so that the probing ac-field can interact only at the very top part of each electron cycle close to the surface. This enhances information about near-surface behavior in metals, while in semiconductors, probing extends throughout the bulk.

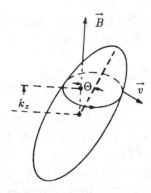

Figure 9.15: Magnetic orbit in k-space near the bottom of the conduction band in a satellite valley of Si, with arbitrary orientation of the magnetic induction.

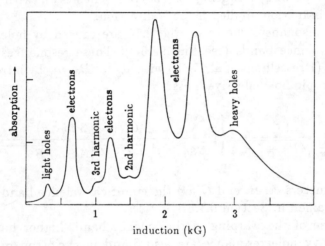

Figure 9.16: Cyclotron resonance absorption in Si near 4 K at 23 GHz with magnetic induction parallel to the [111] axis (after Dexter et al., 1956). The higher harmonics identified belong to the heavy hole resonance.

9.1.5D Measurement of Effective Masses with Cyclotron Resonance The effective mass is obtained directly from the cyclotron frequency [Eq. (9.19)]. By changing the relative alignment of the magnetic induction and the crystal axis, one can probe the anisotropy of the effective mass, such as that caused by the elliptical satellite bands—see Fig. 9.13B.

The resonance frequency in such elliptical bands is

$$\omega_c = e\sqrt{\frac{B_x^2}{m_y m_z} + \frac{B_y^2}{m_z m_x} + \frac{B_z^2}{m_x m_y}}, \qquad (9.20)$$

with the components of a matrix tensor (m_x, m_y, m_z) in diagonal form—see Section 27.2.2B. With $m_\parallel = m_x$ and $m_\perp = m_y = m_z$, the longitudinal and transverse effective masses, one obtains from Eq. (9.20) for the resonance frequency

$$\omega_c = eB\sqrt{\frac{\cos^2\theta}{m_\perp^2} + \frac{\sin^2\theta}{m_\parallel m_\perp}}, \qquad (9.21)$$

where θ is the angle between $\mathbf{B}_0$ and the principal axis of the $E(\mathbf{k})$ ellipsoid (Fig. 9.15).

Figure 9.16 shows the measured resonance spectrum in Si at a constant probing frequency of 23 GHz (cm-wave), with varying magnetic induction applied in [111] direction.

Other resonances shown in Fig. 9.16 are caused by holes in the different valence bands (see Fig. 9.10C). These resonances can be expressed (Dresselhaus et al, 1955) with $\omega_c = eB_0/m_p$ by introducing an anisotropic hole effective mass:

$$\frac{m_p}{m_0} = \frac{1}{A \pm \sqrt{B^2 + C^2/4}}\left\{1 \pm \frac{C^2\left(1 - 3\cos^2\theta\right)^2}{64\sqrt{B^2 + C^2/4}\left[A \pm \sqrt{B^2 + C^2/4}\right]}\right\}.$$
$$(9.22)$$

The parameters A, B, and C are the empirical valence band parameters explained in Section 9.1.5B

Because of the warping of the valence band, higher harmonics of the heavy hole resonance are also found in cyclotron resonance (Fig. 9.16).

9.1.5E The Conduction Band at Higher Energies   The parabolic approximation is no longer sufficient when a significant fraction of the electrons is at higher energies in the band—for example, at elevated temperatures, high electric fields, after optical excitation, or when pushed up by high doping—see Sections 33.2, 45.5.4, and 24.3.1. With increasing E, the band curvature decreases; hence, usually m^* increases, introducing an *energy-dependent effective mass*. When expressed by $m_n(T)$, the changes are relatively small between $0\,\mathrm{K}$ and $300\,\mathrm{K}$ (1–5%).

This dependency is shown in Fig. 9.17 as a function of the temperature for Si and as a function of the energy above the band edge for Ge.

When a more accurate description of the band shape is needed beyond $E_c(k = 0)$, we can use an expression obtained from the $\mathbf{k} \cdot \mathbf{p}$ theory.

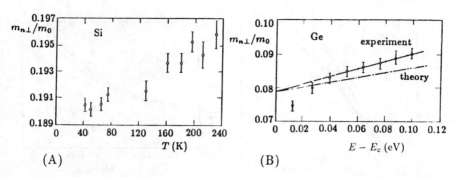

Figure 9.17: (A) Effective transverse electron mass as a function of the temperature for Si (after Oussel et al., 1976) and (B) as a function of the energy for Ge (after Aggarwal et al., 1969).

For the shape of the conduction band of GaAs, for example, we have:

$$E(k \text{ near } \Gamma_6) = \frac{\hbar^2 k^2}{2m_n} + (\alpha + \beta s)k^4 \pm \gamma \sqrt{s - 9t}\, k^3 \qquad (9.23)$$

with

$$s = \frac{k_x^2 k_y^2 + k_y^2 k_z^2 + k_z^2 k_x^2}{k^4} \quad \text{and} \quad t = \frac{k_x^2 k_y^2 k_z^2}{k^6}. \qquad (9.24)$$

The second term in Eq. (9.23) describes the deviation from parabolicity, and the third term describes a slight band warping of the conduction band. In GaAs there is also a slight spin splitting of the conduction band (Rössler, 1984). See Fig. 9.18.

9.1.6 The Momentum Effective Mass

The effective mass of a carrier is conventionally described in relation to Newton's second law, yielding the well-known relation involving the curvature of $E(k)$—see Section 7.2.3. Another way to introduce this mass in a semiconductor with a spherical band of arbitrary shape is through the relationship between the carrier velocity and the pseudo-momentum:

$$v_i = \sum_j \frac{\hbar}{m_{ij}} k_j \quad \text{with} \quad v_i = \frac{1}{\hbar} \frac{\partial E}{\partial k_i}. \qquad (9.25)$$

In a spherical band, therefore, the *momentum effective mass* is a scalar:

$$\boxed{\frac{1}{m^*} = \frac{1}{\hbar^2} \frac{1}{k} \frac{\partial E}{\partial k}.} \qquad (9.26)$$

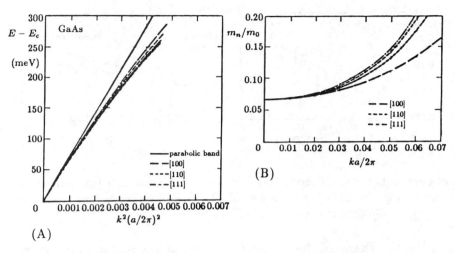

Figure 9.18: (A) $E(\mathbf{k})$ in the lowest conduction band of GaAs as a function of the square of the wave vector. (B) Effective electron mass as a function of the wave vector with same identification. The two curves for the [110] direction show the spin splitting of the conduction band (after Rössler, 1984). ©Pergamon Press plc.

In a parabolic $E(k)$, this definition is identical to the conventional one, relating the effective mass to the second derivative of $E(k)$. The common description is satisfactory as long as the discussion is restricted to the energy range near the band edge, where parabolicity is a reasonable approximation. At higher energies the momentum effective mass is more appropriate (Zawadzki, 1982), and it will therefore be used in the following section.

9.1.7 The Effective Mass at Higher Energies

A large fraction of the electrons can reach substantially higher energies within a band when the Fermi level is shifted into the conduction band by higher doping. Since the curvature decreases with increasing distance from the lower band edge, the effective mass increases. In semiconductors with a very low effective mass (here one has a low density of states near the band edge) one can reach this condition at moderate doping levels, for example, with a donor density in excess of $10^{17}\,\mathrm{cm}^{-3}$ for InSb (see Section 24.3.1) as can be seen by comparing upper and lower abscissae of Fig. 9.19.

Kane (1957) estimates the shape of the conduction band as a function of the wave vector for a three-band model (Γ_6, Γ_7, and Γ_8) near $k = 0$ for InSb. In GaAs the corrections are somewhat smaller

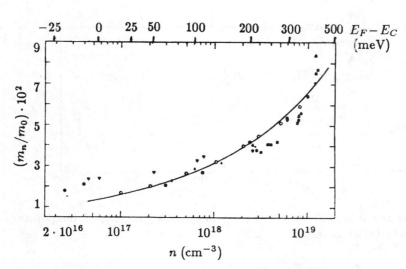

Figure 9.19: Effective masses of electrons in InSb at 300 K as calculated from the three-band Kane model (solid curve) and measured by various authors as a function of the position of the Fermi level below and inside the conduction band (after Zawadzki, 1974).

(Vrehen, 1968) because of a larger band gap. Thus, one obtains for the nonparabolic conduction band

$$E(k) = \frac{\hbar^2 k^2}{2m_n^0} - \left(1 - \frac{m_n^0}{m_0}\right)\left(\frac{\hbar^2 k^2}{2m_n^0}\right)^2 \left\{ \frac{3E_g^2 + 4\Delta_0 E_g + 2\Delta_0^2}{E_g(E_g + \Delta_0)(3E_g + 2\Delta_0)} \right\};$$

(9.27)

where the energy is normalized to $E = 0$ at E_c; Δ_0 is the spin-orbit splitting; and m_n^0 is the effective mass at the bottom of the conduction band, which can be expressed (Kane, 1957) as

$$\frac{1}{m_n^0} = \frac{1}{m_0} + \frac{4P^2}{3\hbar^2 E_g}\frac{\Delta_0 + \frac{3}{2}E_g}{\Delta_0 + E_g}.$$

(9.28)

Here P is the matrix element connecting the conduction band with the three valence bands.

The effective mass slightly above the bottom of the conduction band can be approximated as

$$m_n(E) = m_n^0\left(1 + \frac{2E}{E_g}\right) \simeq m_n^0\sqrt{1 - \left(\frac{v_g}{v}\right)^2},$$

(9.29)

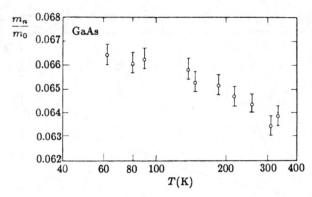

Figure 9.20: Electron effective mass as a function of temperature for GaAs (after Blakemore, 1982).

with $v = \sqrt{E_g/(2m_n^0)}$ and v_g as the group velocity [Eq. (7.17)] (Landsberg, 1987). Equation (9.29) is in fair agreement with the experiment for InSb (Fig. 9.19).

With increasing temperature, one must also consider the lattice expansion and consequent relative shift of the different bands if they are near enough to E_c. This influences the effective mass as a function of temperature and, in the given example (GaAs), causes a reduction in m_n with increasing temperature rather than an increase with increasing band filling—see Fig. 9.20.

9.2 The Band Gap

In *semiconductors*, the valence band is separated from the conduction band by a relatively narrow band gap (for further distinction, see Section 13.1.1); in insulators, the gap is much wider (Fig. 9.3). The distinction between semiconductors and insulators is arbitrary at a band gap of ~ 2 eV. A wide variety of materials provides a continuous transition of behavior from that of *insulators* and *wide, narrow, and zero band gap semiconductors* to metals.

The band gap shows a distinctive trend that is seen for various AB-compounds in Fig. 9.21 (also see Section 9.1, Tables 9.2, 9.3, and 9.7): it decreases with decreasing ionicity of the lattice binding forces and decreases steeply in the same class of compounds for homologous components with increasing atomic number, e.g., increasing ionic radius (interatomic spacing) or decreasing binding energy—see Phillips (1970). We must be careful, however, when using this trend for predicting the band gap of unknown compounds, since substan-

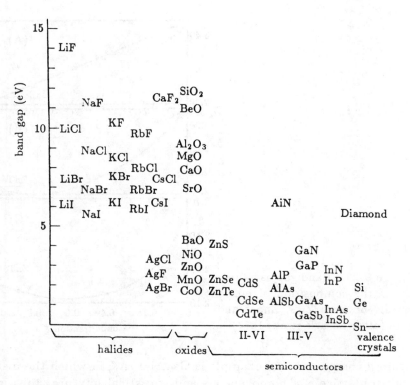

Figure 9.21: Band gaps for various AB-compounds and element semi-conductors (after Hayes and Stoneham, 1984) ©John Wiley & Sons, Inc.

tial deviations occur between materials with the gap at the Γ-, L-, or X-point in $E(k)$.

9.2.1 Band Gap Variation

A continuous variation of the width of the band gap can be achieved by *alloy* formation, or by application of hydrostatic pressure, which thereby changes the lattice composition or the lattice constant and thus the band gap (see Section 9.2.1B). Alloys* can be formed between similar metals or by mixing similar, homologous elements

---

* The conventional term *alloy* of metals also encompasses crystallite mixtures of nonintersoluble metals, such as lead and tin (solder). Here, however, only materials within their solubility ranges are discussed. The *Hume-Rothery rule* identifies these metals as having similar binding character, similar valency, and similar atomic radii (Hume-Rothery, 1936). Corresponding guidelines apply to the intersolubility of cations or anions in compounds.

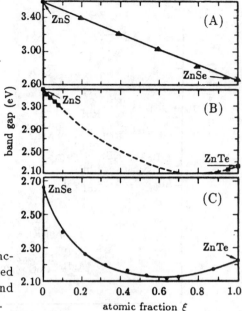

Figure 9.22: Band gap as a function of the composition of mixed crystals between ZnS, ZnSe, and ZnTe (after Larach et al., 1957).

within a compound. An example is $Ga_\xi Al_{1-\xi}As$, in which the metal atom sublattice is a homogeneous and statistical mixture (substitutional alloy) of Ga and Al, with a resulting band gap between that of GaAs and AlAs. Another example is $CdS_\xi Se_{1-\xi}$, in which the nonmetal sublattice is alloyed. There is an infinite number of such pseudo-compounds, which provides the possibility of creating any desired band gap. For quaternary alloys such as $Ga_\xi Al_{1-\xi}P_\eta As_{1-\eta}$, see Pearsall (1982).

9.2.1A Band Gap Bowing Alloy formation between components AC and BC produces a linear interpolation of the band gap

$$\overline{E}_g(\xi) = E_g(AC) + \xi\,[E_g(BC) - E_g(AC)] \qquad (9.30)$$

only if the alloying atoms (A and B) have nearly identical binding forces to atom C, and have nearly the same atomic radii. The bar over E_g indicates the averaging over various local configurations with different numbers of nearest neighbors of a certain atom. An example is $ZnS_\xi Se_{1-\xi}$ (Fig. 9.22A). If the radii are substantially different, a strong *bowing* of the band gap is observed, as shown for $ZnS_\xi Te_{1-\xi}$ and $ZnSe_\xi Te_{1-\xi}$ in Figs. 9.22B and C (Bernard and Zunger, 1987). For a review, see Jaros (1985).

Table 9.7: Bowing parameter b in eV (computed from Richardson, 1973.)

	GaAs	AlSb	InP	GaSb	InAs	InSb	ZnS	ZnSe	ZnTe	CdTe
GaP	0.38	2.31	1.36	3.51	2.67	5.68	0.87	1.41	3.51	2.96
GaAs		0.44	0.22	1.44	0.84	2.86	1.20	0.76	1.55	0.76
AlSb			0.01	0.005	0.05	0.92	3.64	1.88	1.17	1.55
InP				0.49	0.22	1.55	2.12	0.92	0.84	0.46
GaSb					0.008	0.24	4.08	1.85	0.44	−0.24
InAs						0.52	3.13	1.25	0.35	0.05
InSb							6.20	3.21	0.90	0.44
ZnS								0.60	3.02	2.07
ZnSe									0.90	0.02
ZnTe										−0.57

The bowing is described with an empirical bowing parameter b

$$\overline{E}_g(\xi) = E_g(\xi) + b\xi(1 - \xi); \qquad (9.31)$$

this bowing parameter is listed in Table 9.7 for a selection of II-VI and III-V compounds. Bowing is caused by a change of the lattice energy of the alloy, which is due to chemical and structural differences when a lattice atom is replaced by an alloying atom. The change in band gap is given by

$$\Delta E_g = \Delta E_g^{\text{chem}} + \Delta E_g^{\text{struct}}. \qquad (9.32)$$

The varying chemical nature of the alloying atom may be expressed by its different *electronegativity* "en" and *hybridization* "pd" when forming the lattice bond:

$$\Delta E_g^{\text{chem}} = \Delta E_g^{en} + \Delta E_g^{pd}. \qquad (9.33)$$

The changes in structure are induced by different *bond lengths* (u) and *tetragonal (bond angle) distortion* (η) (see below):

$$\Delta E_g^{\text{struct}} = \Delta E_g^{u} + \Delta E_g^{\eta}. \qquad (9.34)$$

The shifting of different conduction band minima by alloying can be substantially dissimilar from each other and consequently may change their relative position. An example is shown in Fig. 9.23 for the Γ-, X-, and L-points of $Ga_{1-\xi}Al_\xi As$.

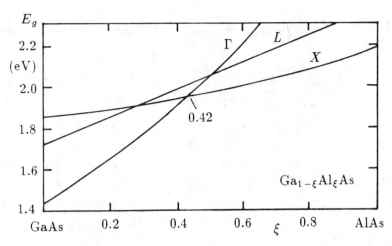

Figure 9.23: Observed changes of Γ, X, and L conduction band minima of $Ga_{1-\xi}Al_\xi As$ as a function of composition at room temperature (from data of Saxena, 1981).

The structural changes are more easily discussed when starting from a 50% alloy. When ordered, this alloy can be described as a chalcopyrite. For example, an ordered $Zn_{0.5}Cd_{0.5}S$ can be described as $ZnCdS_2$, which in structure is similar to $CuGaS_2$. The bond alternation is given by an anion displacement parameter $u = (\overline{AC}^2 - \overline{BC}^2)/a^2 + 1/4$ with (A, B, C) standing for the components (Zn, Cd, S in the given example), and $\overline{AC}$ as the average distance between atoms A and C, etc. The tetragonal distortion is described by the ratio $\eta = c/2a$. The more the measured u in the actual alloy deviates from the ideal ratio 1/4 and η from 1 for an ideal chalcopyrite, the larger is the contribution from $\Delta E_g^{\text{struct}}$ to the bowing.

In ternary compounds, e.g., chalcopyrites, little bowing is observed when the anion sublattice is alloyed (e.g., $CuInS_\xi Se_{1-\xi}$, $CuInSe_\xi Te_{1-\xi}$, and $CuInS_\xi Te_{1-\xi}$), while there is substantial bowing when the cation sublattice is alloyed (e.g., $CuGa_\xi In_{1-\xi}Se_2$ or $Cu_\xi Ag_{1-\xi}GaSe_2$). For reviews, see Martins and Zunger (1986) and also Jaffe and Zunger (1986).

Strong bowing indicates a large change in lattice energy, which acts as a driving force for ordering, that is, for compound formation when a stoichiometric atomic ratio is reached rather than the formation of a statistical alloy. Examples are SiGe, $GaInP_2$, or Ga_2AsP (Jen et al., 1986; Srivastava et al.; 1986, Ourmazd and Bean 1985).

9.2.1B Band Gap Dependence on Temperature and Pressure

The band gap (and other band features—see Section 13.1.2) change with temperature and pressure:

$$\Delta E_g = \left(\frac{\partial E_g}{\partial T}\right)_p \Delta T + \left(\frac{\partial E_g}{\partial p}\right)_T \Delta p. \tag{9.35}$$

Temperature-induced changes are due to:

- changes in lattice constants and
- changes in electron-phonon interaction.

The first term of Eq. (9.35), however, is closely related to the second term, which gives the changes of the **band gap under pressure**. The latter can be divided into several contributions, solely related to a change in the lattice constant. These changes influence the optical behavior due to changes in:

- matrix element, which depends on $1/a$ (lattice constant in a cubic crystal);
- density of states inducing changes in the effective mass;
- energy of electronic levels;
- plasma frequency containing a changed density of dipoles and a changed effective charge; and
- phonon frequency as the lattice stiffens with increasing pressure (anharmonicity of oscillations).

The changes in **electron-phonon interaction** dominating the first term of Eq. (9.35) are more involved. They have attracted substantial interest and can be divided into three different approaches:

(1) the approach suggested by Fan (1951): involving an electron self-energy term that arises from spontaneous emission and reabsorption of a phonon; this approach was expanded by Cohen (1962) to include intervalley scattering;

(2) the approach suggested by Antončik (1955): introducing a temperature-dependent structure factor *(Debye-Waller factor),**

---

* This Debye-Waller factor (W) is related to the probability of phonon emission during electron or x-ray diffraction and is given in the Debye approximation by

$$p = \exp(-2W) = \exp\left[-\frac{6E_R}{k\Theta}\left\{\frac{1}{4} + \frac{T}{\Theta}\int_0^{\Theta/T}\frac{x\,dx}{\exp(x)-1}\right\}\right], \tag{9.36}$$

where E_R is the recoil energy $= Mv^2/2$.

that is experimentally accessible from the temperature dependence of the Bragg reflections;

(3) the approach suggested by Brooks (1955) and refined by Heine and van Vechten (1976): relating to a change in lattice vibrations from ω to ω' when an electron is excited from the valence into the conduction band:

$$E_g(T) \simeq E_g(0) - kT \sum_i \ln\left(\frac{\omega_i'}{\omega_i}\right). \qquad (9.37)$$

See Lautenschlager et al. (1985) for recent work.

Finally, the **influence of uniaxial stress** lowers the crystal symmetry, and thus removes the degeneracy. Uniaxial stress experiments can be designed to yield information on level and effective mass symmetries, and on deformation potentials. These changes will be discussed in Section 21.3.2.

Since the influences of temperature and pressure are many-fold and involve all bands, it is not possible to describe features which apply universally to all semiconductors. A few examples will be presented here. More comprehensive literature includes a large number of odd cases that behave substantially differently from the given examples, e.g., referenced by Martinez (1980).

The band gap usually decreases with increasing temperature—see Fig. 9.24. Exceptions are the lead chalcogenides, which show an increase of the gap with temperature; this band gap is determined at the L-point—see Fig. 9.5.

In a wide temperature range, the band gap changes linearly with temperature:

$$E_g = E_{g0} + \beta_E T, \qquad (9.38)$$

where β_E is typically in the -10^{-4} eV/deg range—see Table 9.8. For lower temperature, however, $|\beta_E|$ decreases and vanishes for $T \to 0$ according to the third law of thermodynamics.

The temperature dependence of the gap over a larger temperature range can be approximated by (Varshni, 1987)

$$E_g = E_{g0} - \frac{\alpha T^2}{\beta + T} \qquad (9.39)$$

with α and β as empirical parameters listed by Varshni (1967).

The change of the band gap due to pressure is expressed as

$$\Delta E_g = \left(\frac{\partial E_g}{\partial p}\right)_T \Delta p; \qquad (9.40)$$

Table 9.8: Direct (d) and indirect (i) band gaps for various semiconductors and their temperature and pressure coefficients.

Crystal	$E_g(0\ \text{K})$	$E_g(300\ \text{K})$	$\dfrac{dE_g}{dT}\left(\dfrac{\text{meV}}{\text{K}}\right)$	$\dfrac{dE_g}{dp}\left(\dfrac{\text{meV}}{\text{kbar}}\right)$
Si	$1.1695(i)$	$1.110(i)$	-0.28	-1.41
Ge	$0.744(i)$	$0.664(i)$	-0.37	5.1
α-Sn	0		-0.5 (2nd)	
GaP	$2.350(i)$	$2.272(i)$	-0.37	10.5
GaAs	$1.519(d)$	$1.411(d)$	-0.39	11.3
GaSb	$0.812(d)$	$0.70(d)$	-0.37	14.5
InP	$1.4236(d)$	$1.34(d)$	-0.29	9.1
InAs	$0.418(d)$	$0.356(d)$	-0.34	10.0
InSb	$0.2352(d)$	$0.180(d)$	-0.28	15.7
ZnS_{cub}	$3.78(d)$	$3.68(d)$	-0.47	-5.8
ZnS_{hex}	$3.91(d)$	$3.8(d)$	-0.30	
ZnSe	$2.820(d)$	$2.713(d)$	-0.45	0.7
ZnTe	$2.391(d)$	$2.26(d)$	-0.52	8.3
CdS	$2.585(d)$	$2.485(d)$	-0.41	4.5
CdSe	$1.841(d)$	$1.751(d)$	0.36	5
CdTe	$1.606(d)$	$1.43(d)$	-0.54	8

[*]For an explanation of the difference between direct and indirect band gaps, see Section 13.2.

however, it cannot be linearized in a wider pressure range. It often involves several bands which are influenced by different pressure coefficients that may even have different signs. As an example, the pressure dependence of the gap in GaAs is given in Fig. 9.25A. The gap first increases, shows a maximum at $\sim 6 \cdot 10^4$ atm, then decreases again with increasing pressure. The lower valley in the conduction band in GaAs (Γ-point) increases while the $\langle 100 \rangle$ valley decreases with pressure. At pressures above $8 \cdot 10^4$ atm, GaAs becomes an indirect band gap material.

Pressure coefficients near 1 atm are typically on the order of 10 meV/kbar. They are listed in Table 9.8 for some semiconductors. They are usually positive for $\Gamma\langle 000 \rangle$ and $L\langle 111 \rangle$ valleys, and are

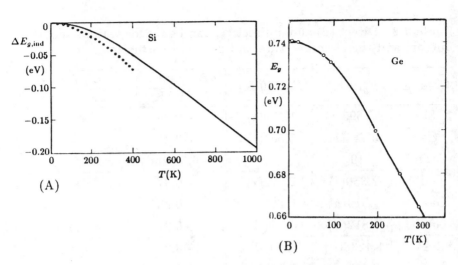

Figure 9.24: Decrease of the band gap with increasing temperature: (A) in Si (after Lautenschlager et al., 1985), solid line = theory, circles = experimental; and (B) in Ge (after MacFarlane et al., 1957).

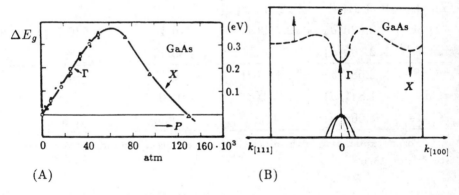

Figure 9.25: (A) Pressure dependence of the band gap of GaAs, and (B) indicating the trend (arrows) of $E(k)$ with increasing pressure (after Edwards et al, 1959). ©Pergamon Press plc.

negative for valleys along $\langle 100\rangle$—see Fig. 9.25B. A more extensive list of the temperature and pressure dependence is given in Table 9.9.

9.2.1C Band Gap Narrowing with Heavy Doping   The band gap in heavily doped semiconductors is substantially reduced. There are many reasons for the reduction of the band gap, summarized in an extensive review by Abram et al., (1978); see also Mahan (1980) and van Overstraeten (1982). The more important phenomena which

Table 9.9: Temperature and pressure dependence of the band gap.

	$\dfrac{dE_g^{(ind)}}{dT}$ (10^{-4}eV/K)	$\dfrac{dE_g^{(dir)}}{dT}$ (10^{-4}eV/K)	$\dfrac{dE_g^{(ind)}}{dp}$ (10^{-6}eV/bar)	$\dfrac{dE_g^{(dir)}}{dp}$ (10^{-6}eV/bar)
Si	$E_g^{(ind)} = 1.17 + 1.059\cdot10^{-5}T - 6.05\cdot10^{-1}T^2$ (eV)			-1.41
Ge			-5.1	$+15.3$
AlAs	-3.6	-5.1		
AlSb	-5.3	-3.5	2.8	10.2
GaP	$E_g^{(ind)} = 2.338 - 6.2\cdot10^{-4}\dfrac{T^2}{T-460}$	$E_g^{(dir)} = 2.895 - 0.1081\left[\coth\left(\dfrac{164}{T}\right) - 1\right]$	$E_g^{(ind)} = 2.14 - 0.24p - 0.048p^2$ (p in kbar)	$E_g^{(dir)} = 2.76 + 0.97p - 0.35p^2$ (p in kbar)
GaAs		$E_g^{(dir)} = 1.519 - 5.408\cdot10^{-4}\dfrac{T^2}{T+204}$		$E_g^{(dir)} = 1.45 + 0.0126p - 3.77\cdot10^{-5}p^2$ (p in kbar)
GaSb	$E_g^{(ind)}(L) = 2.035 - 0.019\dfrac{T^2}{94+T}$	-3.5	-8.8	14
InP		-3		8.0
InAs				-11.4
InSb		$E_g^{(dir)} = 0.236 - 0.06\dfrac{T^2}{500+T}$		15

influence the band gap at high doping densities $(N_d > 10^{17}\text{ cm}^{-3})$ are given below:

(a) The exchange energy of electrons due to their fermion nature. It tends to keep the electrons with parallel spin away from each other, but is attractive to electrons with away opposite spin, resulting in a net attractive term. As a consequence, one obtains a lowering of the conduction band edge δE_{cex} relative to the Fermi level in equilibrium and at $T = 0$:

$$\delta E_{\text{cex}} = -\frac{e^2}{\pi \varepsilon \varepsilon_0} \Lambda_\delta k_F \simeq B_{\delta e} \left(\frac{N_d(\text{cm}^{-3})}{10^{18}} \right)^{\frac{1}{3}} \quad (\text{meV}); \qquad (9.41)$$

the wave number at the Fermi-surface is given by

$$k_F = \left(\frac{3\pi^2 N_d}{\nu_D} \right)^{\frac{1}{3}}, \qquad (9.42)$$

assuming total ionization of uncompensated donors of density N_d and ν_D the degeneracy factor of the band; Λ_δ and $B_{\delta e}$ are numerical factors listed in Table 9.11 (Mahan, 1980).

(b) The attractive interaction between free electrons and charged donors, causing another reduction in the conduction band edge:

$$\delta E_{\text{ced}} = -\frac{e^2}{8 \varepsilon \varepsilon_0 \lambda_{TF}} \simeq C_{\delta e} \left(\frac{N_d(\text{cm}^{-3})}{10^{18}} \right)^{\frac{1}{6}} \quad (\text{meV}), \qquad (9.43)$$

with the Thomas Fermi screening length

$$\lambda_{TF} = \pi^{\frac{2}{3}} \sqrt{ \frac{\varepsilon_{\text{st}} \varepsilon_0 \hbar^2}{m_n e^2 (3 N_d)^{\frac{1}{3}}} } \qquad (9.44)$$

which is responsible for screening at sufficient doping density—see Section 9.2.1C.1. Also see Landsberg et al., 1985.

(c) The exchange energy for holes, causing a relative increase of the valence band edge by

$$\delta E_{\text{vhx}} = \frac{m_p e^4 \sqrt{\omega_p}}{\pi (\varepsilon \varepsilon_0)^2 \hbar} J \simeq D_{\delta p} \left(\frac{N_d(\text{cm}^{-3})}{10^{18}} \right)^{\frac{1}{4}} \quad (\text{meV}), \qquad (9.45)$$

with ω_p as the *plasma frequency* (Section 12.1.1) and J as an integral (Mahan, 1980) of nearly constant value ($J \simeq 0.8$).

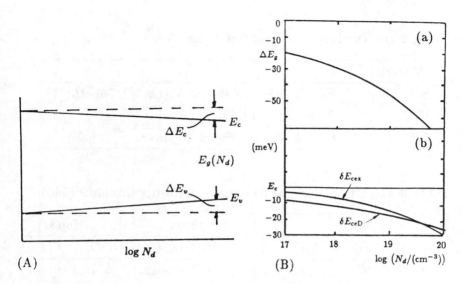

Figure 9.26: (A) Change of the band gap with increasing donor density (schematically). (a) Relative lowering of the conduction band edge. (b) Band gap narrowing as a function of the density of shallow donors in Si (after Mahan, 1980).

(d) The interaction between holes and donors is repulsive, and thus causes a relative lowering of the valence band edge by:

$$\delta E_{\text{vhd}} = -0.48 \frac{e^2}{\varepsilon \varepsilon_0} N_d^{\frac{1}{3}} \simeq E_{\delta p} \left(\frac{N_d(\text{cm}^{-3})}{10^{18}} \right)^{\frac{1}{3}} \quad (\text{meV}). \quad (9.46)$$

The total change in the band gap is obtained as the sum of all of these contributions:

$$\Delta E_g = \delta E_{\text{cex}} + \delta E_{\text{ced}} + \delta E_{\text{vhx}} + \delta E_{\text{vhd}} = \Delta E_c + \Delta E_v. \quad (9.47)$$

The influence of the first two contributions on the conduction band is shown in Fig. 9.26A for Si. The total reduction of the band gap as a function of the doping density is shown for Si in Fig. 9.26B.

Pantelides et al. (1985) estimated that multivalley interaction and density fluctuation cause a further reduction in the electrically obtained band gap, which is on the order of 50 meV at $N_d \simeq 10^{20} \text{cm}^{-3}$ in GaAs. This agrees with the optically determined band gap reduction via absorption, luminescence, or photoluminescence spectroscopy (Wagner, 1984). See also Berggren and Sernelius (1984).

Simple empirical relations to approximate the change in band gaps with doping are given for three semiconductors in Table 9.10.

Table 9.10: Doping-dependent band gap.

Material	
Si	$E_g = 1.206 - 0.0404 \ln(N_a(\mathrm{cm}^{-3})/7.5 \cdot 10^{16})$
GaAs	$E_g = 1.45 - 1.6 \cdot 10^{-8} \sqrt[3]{n(\mathrm{cm}^{-3})}$
InP	$E_g = 1.344 - 2.25 \cdot 10^{-8} \sqrt[3]{n(\mathrm{cm}^{-3})}$

Table 9.11: Parameters for band gap narrowing (Mahan, 1980).

Crystal	ν_D	Λ_δ	Electrons			Holes	
			$A_{\delta F}$	$B_{\delta e}$	$C_{\delta e}$	$D_{\delta p}$	$E_{\delta p}$
Si	6	0.95	3.3	−6.5	−12.1	−13.1	6.1
Ge	4	0.84	6.6	−4.9	−6.0	−8.2	4.5

Shift of the Fermi Level into the Band at High Doping Densities. When the doping density N_d exceeds the effective level density N_c at the lower edge of the conduction band, the Fermi level moves from the band gap into the conduction band. This results in an "effective widening" of an optical gap, since, for example, optical excitation from the valence band can only proceed to empty states above the Fermi level. The shift is given for parabolic bands, using for the density of states in these bands Eq. (27.32) and $n \simeq N_d$:

$$\delta E_{cF} = \frac{\hbar^2}{2m_{\mathrm{dsn}}}\left(\frac{3\pi^2 N_d}{\nu_D}\right)^{\frac{2}{3}} \simeq A_{\delta F}\left(\frac{N_d(\mathrm{cm}^{-3})}{10^{18}}\right)^{\frac{2}{3}} \qquad (\mathrm{meV}), \quad (9.48)$$

where ν_D is the degeneracy factor of the conduction band and m_{dsn} is the density of state mass for electrons (see Section 27.2.2B). This equation is valid when all donors are ionized and uncompensated. A similar expression can be obtained for high doping with acceptors, replacing m_{dsn} with the density of state mass for holes m_{dsp}, and ν_D with the degeneracy factor for the valence bands. However, the larger effective mass of holes in the heavy hole band makes this effect less favorable. Table 9.11 lists $A_{\delta F}$ and ν_D for Si and Ge. Such a shift is substantial even at moderate electron densities for semiconductors with low effective mass (e.g., InP). See Fig. 24.5 for the optical manifestation of this shift (Burstein-Moss effect).

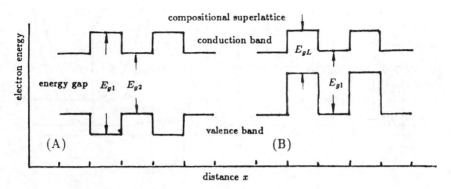

Figure 9.27: Periodic potential for (A) a superlattice of type I and (B) a superlattice of type II.

9.3 Bands in Superlattices

Permitted bands separated by band gaps occur as a consequence of the quantum-mechanical properties of electrons in a periodic potential. This remains true in superstructures. The additional periodicity in the lattice potential can be provided by alternating layers of materials with different band gaps, i.e., in *superlattices*—see Esaki and Tsu (1970). Some of the effects observed in a superlattice are already present in a single layer that provides a two-dimensional *quantum well confinement* of the electronic eigenfunctions—also see Section 42.4.3E and Fig. 42.19. For a comprehensive review on single- and multilayer structures, see Ando et al. (1982).

9.3.1 Mini-Bands

When a semiconductor with a larger band gap is interspaced with another one of a smaller gap, the former acts as a barrier for electrons in the conduction band of the latter. To obtain the actual barrier height, it is important to know the valence band edge offset (Kroemer, 1983), which is influenced by the substrate material and the deposition sequence. The periodic alternation of such layers produces a potential (Fig. 9.27) of the same form as the previously discussed Kronig-Penney potential* (Section 7.2). The resulting eigenvalue

---

* The alternating potential shown in Fig. 9.27A is of type I, i.e., a minimum of $E_c(x)$ coincides with a maximum of $E_v(x)$. Both minima and maxima coincide in a type-II superlattice (Fig. 9.27B). An example for type II is the $Ga_\xi In_{1-\xi}As$ and $GaAs_\eta Sb_{1-\eta}$ superlattice. For values of ξ and η below 0.25, the valence band of the former extends above the

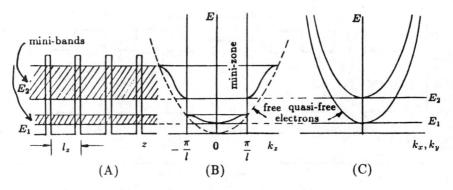

Figure 9.28: (A) Mini-bands and (B) mini-zones in the conduction band in the k_z-direction for a superlattice. Carriers are confined in the z-direction in the lower mini-bands. There is no confinement in the x- and y-directions. In the k_x- and k_y-directions, the ordinary band picture applies, but with the band minimum lifted to the respective mini-band minimum, shown in subfigure (C).

spectrum is similar to the spectrum of free electrons exposed to the periodic potential of a crystal, except that now the periodic potential is imposed on *Bloch electrons* with an effective mass m_n, and the potential has a lower amplitude and a larger period length than the periodic potential in a bulk lattice. Consequently, within the conduction band, one observes a subband structure of *mini-bands* located within the valleys of this band; the higher mini-bands extend beyond the height of the potential barriers. The lower mini-bands are separated by *mini-band gaps* (Fig. 9.28) in the direction of the superlattice periodicity z. Within the plane of the superlattice layers (x, y), however, the electron eigenfunction experiences only the regular lattice periodicity. Therefore, the dispersion relations $E(k_x)$ and $E(k_y)$ are much like those for the unperturbed lattice except for the mixing with the states in the z-direction; this results in lifting the lowest energy (at $k = 0$) of the $E(k)$ parabola above E_c of the bulk well material (Fig. 9.28B and C). The second mini-band results in a second, shifted parabola, etc.

The mini-band structure is a direct band gap structure, independent of whether the host (well) material has a direct or indirect band gap (see Section 13.2). It permits optical transitions from the

conduction band of the latter, resulting in quasimetallic behavior. For a review of type-II superlattices, see Voos and Esaki (1981).

valence band, which has a similar mini-band structure, to the lowest conduction mini-band state at $k = 0$ (Fig. 9.31B).

Variations of the period length, barrier width, and barrier height change the width of the allowed mini-bands and the interfacing mini-gaps. Using the Kronig-Penney model, these are determined by Eq. (7.12); however, m_0, contained in α and β [see Eq. (7.11)], is replaced by m_n and the lattice parameters a_1 and a_2 are replaced by the superlattice well and barrier widths l_1 and l_2. This yields an implicit equation for the band edges [i.e., for $\cos(ka) = 1$ in Eq. (7.12)]:

$$\frac{\alpha^2 - \beta^2}{2\alpha\beta} = ctg(\alpha l_1)ctgh(\beta l_2). \qquad (9.49)$$

The resulting mini-band structure and $E(k)$ dispersion relation shown in Fig. 9.28B is very similar to the band structure shown in Figs. 7.3 and 7.11 for the periodic crystal potential, except that the edge of the Brillouin zone in the k_z-direction (mini-zone) lies at $\pm\pi/l$ with $l = l_1 + l_2$. Typically, l is on the order of $5\ldots50$ lattice constants; thus the mini-zone is only a small fraction $(1/5\ldots1/50)$ of the Brillouin zone of the host lattice. In contrast, the Brillouin zone in the k_x- and k_y-directions extends to the full width $\pm\pi/a$ and $\pm\pi/b$. There are no mini-gaps in the x- and y-directions.

Figure 9.29 shows the computed widths of mini-bands and intermittent gaps as a function of the period length $(2l)$ for a symmetrical well/barrier structure with a barrier height of 0.4 eV. For $l_1 = l_2 = 40$ Å, the lowest band is rather narrow and lies at 100 mV above the well bottom. The second band extends from 320 to 380 mV. Higher bands $(E_3, E_4 \ldots)$ overlap above the top of the barrier.

Increasing the thickness or the height of the barrier layer reduces the tunneling through the barriers. The electronic eigenfunctions within each separated well can be estimated easily [see Section 22.1.1, Eq. (22.5)], yielding a level rather than mini-band spectrum with

$$E_n = \frac{\hbar^2\pi^2}{2m_n l_1^2}n_q \quad \text{with} \quad n_q = 1, 2, \ldots \qquad (9.50)$$

The quantum number gives the number of electron half-wavelengths in the confined state (Fig. 9.30). The dependence of the energy of these states on the quantum well thickness $(E_n \propto 1/l_1^2)$ can easily be verified from optical absorption (Section 13.4.1) and gives a beautiful confirmation of the quantum-mechanical model. The width

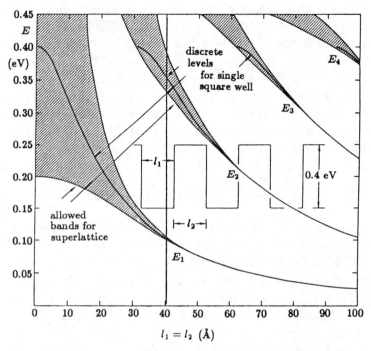

Figure 9.29: Computed mini-bands for a symmetrical superlattice (after Esaki, 1985).

of these levels is lifetime broadened and can be estimated from the uncertainty relation

$$\Delta E_n \Delta t \simeq \frac{\Delta E_n l_1}{v_{\mathrm{rms}}} \mathcal{T}_e \simeq \hbar, \qquad (9.51)$$

with the tunneling probability $\mathcal{T}_e$ through such barriers [Section 42.4.3, Eq. (42.61)] given by

$$\mathcal{T}_e \simeq 16 \exp\left\{ -\sqrt{\frac{8m_n}{\hbar}} (\Delta E_c - E_n) l_2 \right\}, \qquad (9.52)$$

where ΔE_c is the barrier height and E_n is the energy of the level from which tunneling takes place.

A more sophisticated approach in dealing with superlattices of various dimensions resorts to a quantum-mechanical description of the periodic superlattice as given, for example, by Schulman and McGill (1981); also see review by Bastard and Brum (1986) and Section 9.3.

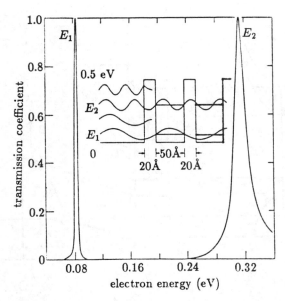

Figure 9.30: Electron eigenstates in a quantum well and relative transmission coefficients for a double barrier of 0.5 eV height and 20 Å width. Well width $l_1 = 50$ Å(after Esaki, 1986). ©IEEE.

The $E(k)$ behavior in the mini-bands of the *valence band* is a bit more complicated, since we must distinguish between light and heavy holes which result in two sets of mini-bands. Considering excited states in the valence band, we observe a crossing of states between heavy and light hole bands, as shown in Fig. 9.31. This results in a mixing between these states (Collins et al., 1987). As shown on an enlarged scale in Fig. 9.31C, this indicates that the dispersion curves *cannot* cross, and that the interaction (mixing) that takes place near the points of *intended* crossing results in a splitting, making the top of the upper band light-hole-like, and its bottom part heavy-hole-like, and vice versa for the lower band—see also Fig. 11.5. Such mixing can be observed directly by applying an electric field or uniaxial stress that changes the energy of such states (Section 44.8, Fig. 44.28).

Superlattices that show such a beautiful illustration of quantum mechanical behavior have been fabricated from a number of semiconductor pairs that have little lattice mismatch. The best researched is the GaAs/Ga$_x$Al$_{1-x}$As couple. The height of the barrier can be changed by varying the concentration of Al in Ga$_x$Al$_{1-x}$As; typically, it is a few tenths of an eV (Cho, 1971; Woodall, 1972; Chang et al., 1973).

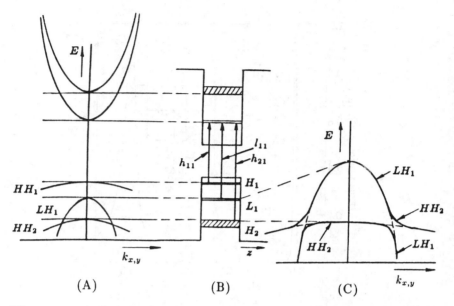

Figure 9.31: (A) Dispersion relation for electron and hole mini-bands; (B) optical excitation transition into the ground state of the conduction band; (C) Light and heavy hole dispersion relation at enlarged scale.

We will return in later chapters to these superlattices to show that mini-bands are observed at the predicted energy ranges, and we will discuss other important properties of these structures (see Section 13.4.1 and Chapter 39).

9.3.2 Bands in Ultrathin Superlattices

As the width of the layers in superlattices become thinner and thinner, the superlattice structure finally disappears and is replaced by the electronic structure of a single compound. A distinction between a true superlattice and a bulk semiconductor can be made when all band gaps between mini-bands disappear and the density of states increases monotonically from the band edge into the band.

For instance, stacking single layers of GaAs and AlAs in the [100] direction results in a $(GaAs)_1$-$(AlAs)_1$ structure identical to bulk $GaAlAs_2$. Consequently, the band structure must be the same. Extending this discussion to the symmetrical $(GaAs)_n$-$(AlAs)_n$ superlattices, Batra et al. (1987) have shown that for $n \geq 3$ the band alignment of the valence band becomes staggered with hole confinement in GaAs. However, electrons are confined in AlAs for $3 < n < 10$; only for $n > 10$ are electrons also confined in GaAs, as is expected for

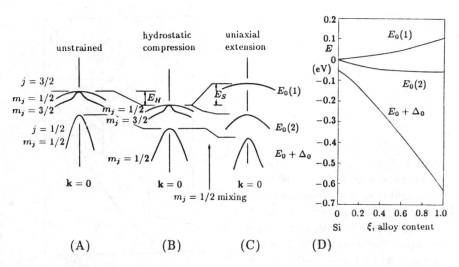

Figure 9.32: Valence bands of a $Ge_\xi Si_{1-\xi}$ alloy. (A) unstrained; (B) with hydrostatic compression; and (C) with uniaxial extension as in a strained superlattice which removes the degeneration of the $J = 3/2$ $m_j = 1/2$ and $m_j = 3/2$ bands at $k = 0$. (D) Magnitude of the splitting and shift of the three valence bands (alloy deposited on Si) as function of the alloy composition at $T = 300$ K (after Pearsall et al., 1986).

thicker superlattices with near-bulk gap properties for each layer— see also Kamimura and Nakayama, 1987. For $Si_n Ge_m$ ultrathin superlattices, see Pearsall et al. (1987) and Froyen et al. (1987).

An interesting strained-layer superlattice is composed of Si and $Si_\xi Ge_{1-\xi}$ layers (Abstreiter et al. 1985; People, 1986). A thin strained layer of a $Si_\xi Ge_{1-\xi}$ alloy on top of Si shows a substantial splitting of the valence bands, as indicated schematically in Fig. 9.32 and a reduction of the band gap, which could not be obtained otherwise. For estimates of the $Ge_n Si_m$ superlattices, see Froyen et al. (1987). It is shown that in such a strained layer superlattice the direct band gap (at Γ) is lowered and can approach the indirect gap, making it in essence a new direct band gap material.

9.3.3 Density of States in Mini-Bands

When proceeding from a level structure of an isolated well to mini-bands in a superlattice with sufficiently permeable barriers, we can follow the broadening in the density of states. It has a staircase character for the isolated well. Each level can be occupied by the number of electrons given by its degeneracy multiplied by the number of atoms in the wells. When significant tunneling becomes possible,

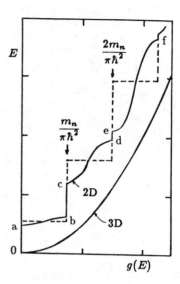

Figure 9.33: Density of states for electrons in a band (3D), electrons in a quantum well (dashed staircase), and electrons in mini-bands of a superlattice, with mini-bands extending between a and b, c and d, and e and f (after Esaki, 1985).

each level splits into bands, and the staircase behavior (dashed steps) becomes somewhat softened (Fig. 9.33).

The effective density of states near the bottom of the first mini-band (for thin enough barriers, so that the mini-band width is less than several kT) is given by

$$N_{2c} = \frac{m^* kT}{\pi \hbar^2},\qquad(9.53)$$

measured in cm^{-2}. The subscript $2c$ refers to the two-dimensionality of the structure and to a mini-conduction band. Equation (9.53) can be derived in a similar fashion as N_c in three dimensions (see Section 27.2.2A).

9.3.4 Two-, One-, Zero-Dimensional Quantum Well Structures

The previously described superlattices can be regarded as quantum-well layers, i.e., as an array of two-dimensional quantum wells. Alternating the deposited material in a second dimension yields *quantum wires*; alternating deposition in all three dimensions yields *quantum boxes* (Fig 9.34A). The density of state distribution of free electrons in the mini-conduction bands becomes more confined with decreasing dimension of the quantum-well structures, as shown in Fig. 9.34C, when the barriers are high and wide enough.

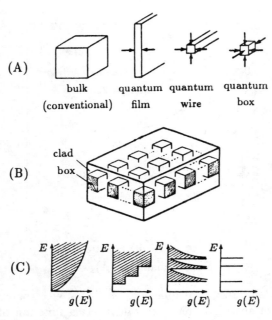

Figure 9.34: (A) Bulk and quantum well structures; (B) example of a quantum box array; (C) distribution of the density of states of quasi-free electrons as a function of the energy for bulk and quantum well structures corresponding to (A) (after Asada et al., 1986). ©IEEE.

9.3.5 Electronic States in Microcrystallites

The electronic structure of very small isolated crystallites, e.g., as a suspension in a liquid dielectric, may in some respects* be compared with a zero-dimensional quantum well (see Section 9.3.4) with infinite barriers. The band gap increases from that of the bulk when the crystallite size decreases below a few hundred Å.

Electrons and holes are confined to the crystallites; therefore, the binding energy of this electron-hole pair (exciton) is increased compared to the band gap of the bulk material (see Section 15.3).

For a crystallite of radius R, the *confinement energy* becomes comparable to the gap energy for $R < 100$ Å. The energy for the lowest $1s$ state can be approximated as

$$E_{1s} = E_g + \frac{\hbar^2}{8R^2}\left(\frac{1}{m_0} + \frac{1}{m_i}\right) - \frac{1.8e^2}{4\pi\varepsilon\varepsilon_0 R}, \qquad (9.54)$$

---

* However, with an important condition missing—the periodic barrier of similar thickness and height.

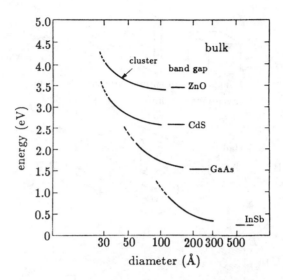

Figure 9.35: Computed dependence of the lowest electronic excited state ($1s\,(\Gamma_8 \to \Gamma_9)$) of spherical microcrystallites as a function of their radii, compared to their bulk material band gap (after Brus, 1986). ©IEEE.

where m_i is the isotropic hole mass. The first term is the band gap (Section 9.2); the second term represents the ground state of a quantum well (the crystallite)—see Section 22.1.1, Eq. (22.5); the third term represents a minor Coulomb correction due to the fact that screening in such a small crystallite is rather small. Figure 9.35 shows the computed energy of the lowest excited state as a function of the radius of the microcrystallite (after Brus, 1986). Experimental data showing such trends in PbS are given by Wang et al. (1987).

9.4 Bands in Amorphous Semiconductors

The band structure in amorphous semiconductors cannot be determined in the same fashion as for the crystalline state (described in Sections 8–8.5) since long-range periodicity is missing and **k** is no longer a good quantum number. Therefore, an $E(\mathbf{k})$ diagram cannot be drawn for amorphous materials. Consequently, the effective mass picture, which depends on an analysis of $E(\mathbf{k})$, cannot be used in its classical form.

Nevertheless, there is strong experimental evidence from optical absorption and reflection spectroscopy and from photoemission, that similarities exist between the band structure in amorphous and crystalline states of the same material.

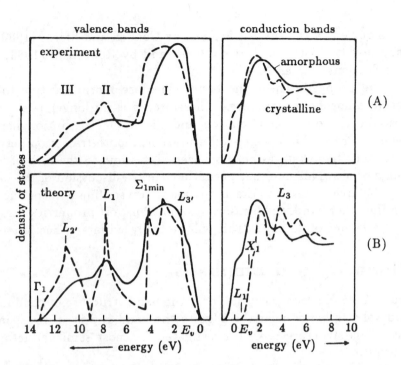

Figure 9.36: Electron density of state distributions of amorphous and crystalline Ge. (A) Experimental data derived from photoemission; (B) theory (after Eastman et al., 1974).

This can be understood by recognizing that the major features of the electronic properties of a solid are determined by short-range order, as proposed by Ioffe and Regel (1960) and shown more convincingly by Weaire and Thorpe (1971) for amorphous Si and Ge, using a tight-binding model.*

Densities of state distributions for amorphous and crystalline Ge are shown in Fig. 9.36—as obtained experimentally (A), as compared with the theoretical distribution (B). The distribution of the amorphous state is much smoother. The sharp van Hove singularities due to long-range order are absent, while several of the major features remain in both states.

The more refined tight-binding calculation of Bullett and Kelly (1975) shows a substantially improved agreement with the experiment.

---

* A tight-binding model, however, provides information only about the gross properties of a semiconductor (crystalline or amorphous).

A review of the subject is given by Connell and Street (1980), in the book by Mott and Davis (1979), and by Robertson (1983) (see also Sections 13.5 and 25).

The properties near the band edges are more sensitive to the actual amorphous structure. Therefore, it is no longer possible to describe the dispersion relation and the level distribution here in general terms. There seems to be a rather smooth transition between extended (band) states and localized states due to the lack of long-range order. This does not permit a cohesive discussion of the band-edge behavior similar to the discussion for crystalline semiconductors. On the other hand, the measurement of transport properties suggests the existence of an edge. This will be analyzed later—Section 40.4.

Summary and Emphasis

For semiconducting purposes, only the band structure of conduction and valence bands near the edge to the band gap is of direct interest. This structure is well investigated for most semiconductors of interest.

It has nearly spheric equi-energy surfaces for the conduction band at Γ and ellipsoids of rotation at X and L. The lowest minimum of $E(\mathbf{k})$ of the conduction bands and the highest maximum of $E(\mathbf{k})$ of the valence bands determine the band gap. In the valence band two subbands need to be considered; the light and heavy hole bands, degenerated at Γ in isotropic semiconductors and split by the crystal field in anisotropic ones. In addition, the split-off spin-orbit band may become important at higher hole energy, especially for low atomic number elements where such splitting is relatively small.

The $E(\mathbf{k})$ behavior of the bands near the band edges is described by the effective mass of electrons or holes, and is directly accessible to cyclotron resonance measurements.

The band structure and the band gap are influenced by external parameters. The band gap usually shrinks with increasing temperature and expands with increasing pressure. It can be changed by alloying, where, for alloying with similar elements, the gap interpolates linearly between the value of the pure compounds. With dissimilar element alloying, a substantial bowing is observed. The band gap is also changed (decreased) by heavy doping.

In superlattices, anisotropic carrier confinement in mini-bands occurs. The effective gap and band width of mini-bands can be varied easily by changing layer width and barrier height of the superlattice.

With reduced layer width, the electronic behavior of such superlattices approaches that of simple chemical compounds, assuming bulk properties.

The bands in amorphous semiconductors near the band edge are ill-defined, as k is no longer a good quantum number since long-range periodicity is missing. Nevertheless, the density of state distribution shows significant similarities to that of the same material in the crystallite state. Near the band edge, a smooth transition between extended and localized states occurs.

Most of the action in semiconducting devices revolves around the edges of the valence and conduction bands. Selection of appropriate semiconductors for well-performing devices requires extensive knowledge of $E(\mathbf{k})$ and the influence of internal (doping) and external (pressure, temperature, fields) parameters on $E(\mathbf{k})$. It is expected that with advances of our knowledge, the design of specifically tailored semiconductors with optimized properties will become practical, replacing some of the presently used common semiconductors.

Exercise Problems

1.(r) If one could artificially increase the lattice spacing of a group-critical distance r_c at which sp-hybridization starts, what electrical properties would you expect from the resulting hypothetical crystal (see Fig. 9.1B)? How could one measure that the crystal is in such a state? Are there known crystals which show such behavior?

2.(r) For the band behavior:
   (a) In a semiconductor with simple bands (extrema at $k = 0$) and sp-hybridization as shown in Fig. 9.1B, what changes in band gap would you expect with increasing hydrostatic pressure and with increasing temperature as they relate to changes in the lattice constant?
   (b) What changes occur considering side valleys?
   (c) How do valence bands react?

3.(e) For the Burstein-Moss shift of the band edge:
   (a) Explain, by resorting to the meaning of the effective mass, why the Burstein-Moss effect is easily observable in InSb (see also Section 24.3.1).
   (b) Why would such an observation be difficult in an n-type semiconductor with $m_n = m_0$? Calculate the necessary donor density to observe a shift of 25 meV. What other

effect would interfere, and to what extent? (Give a quantitative estimate.)

4.(r) Describe quantitatively mini-bands in GaAs/Ga$_{.5}$Al$_{.5}$As:
   (a) as a function of well width for constant barrier width;
   (b) as a function of barrier width for constant well width;
   (c) discuss $E(\mathbf{k})$ ∥ and ⊥ to the layers;
   (d) discuss the density of state function and its relevance.

5.(*) Describe the differences between gapless semiconductors, semimetals, and gapless semiconductors with overlapping tail states (see Section 24.1) in respect to conduction electrons and holes, effective mass, temperature, and pressure dependence.

6.   Compare and discuss the degree of warping of valence bands for Ge, Si, GaAs, and InP using the Luttinger parameters listed in Table 9.6.

7.   Discuss systematic variations of the band gap of binaries shown in Fig. 9.21.

8.(e) How does the degeneracy of a band enter the density of states? At the Γ-point? At the minima of satellite valleys? What is the degeneracy at E_c for Ge? For Si?

9.(l) How are different bands labeled? What types of critical points do you know? How are they identified and labeled?

10.(e) Show, by solving the Schrödinger equation for a single layer quantum well, that normal to the well there are eigenstates with $E_n = E_c + \alpha n_q^2$, and within the well layer there are quasi-free electron states with $E = E_c + E_n + \hbar^2(k_x^2 + k_y^2)/(2m_n)$.

PART IV

PHOTONS

Chapter 10

Basics of Optical Spectroscopy

The interaction of light with matter is described by Maxwell's equations from which the basic optical material parameters can be obtained.

The interaction of electromagnetic radiation (photons) with semiconductors provides major insight into the electronic and phononic structure of these solids. Such interactions can be described as resonant and nonresonant.

Resonant absorption concerning ions is observed in the infrared part of the spectrum; resonant absorption dealing with bound electrons is observed at shorter wavelengths and is usually separated from the ionic component by a wavelength range with low optical absorption, except for semiconductors with almost zero band gap. Both types of absorption describe *intrinsic* properties of the semiconductor. In addition, a large variety of *extrinsic* ionic and electronic resonance transitions of semiconductor defects can be distinguished. These optically induced transitions are extensively used for the identification of such defects.

Nonresonant interactions can be observed between photons and free electrons, except at high densities of free electrons when they act jointly as plasmons.

The interaction between photons and the semiconductor can be *elastic*, without absorption, or *inelastic*, with absorption. The latter requires damping of the excited state, by either absorption or emission of phonons or by collisions with electrons or other quasiparticles.

The resonant transitions will be described for ions in Section 11.2 and for electrons in Section 12.2.1. The description in this chapter is based on solutions of Maxwell's equations, which deal with the resonant transitions in a *phenomenological* way.

Comparison with the experiment requires the transformation of experimentally accessible quantities, such as reflectance and optical transmissivity, into quantities obtained as a result of a theoretical analysis, such as the set of optical constants. This relationship is summarized in Section 10.2.

10.1 Phenomenological Theory

In this section, the interaction of electromagnetic radiation with a semiconductor is described in a classical model. We begin with *Maxwell's equations* which result in the dispersion equation, yielding the optical parameters of the semiconductor (its complex dielectric constant) as a function of the frequency of the electromagnetic radiation (the energy of the photons). For more information, see Moss (1961a), or Palik (1985).

10.1.1 Reflection, Transmission, and Absorption

Light impinging on a semiconductor is subject to a number of optical interactions before it is absorbed. First, a fraction of the light is reflected at the outer surface; another fraction is scattered by crystal imperfections, phonons, and other quasi-particles; then a fraction of the light within the semiconductor is absorbed by various elementary excitation processes. The unabsorbed fraction is transmitted and exits through the semiconductor's surfaces after partial reflection.

Initially, let us regard the semiconductor as a **continuum**, represented by four parameters which can be measured macroscopically: μ, ε, ϱ, and σ, the magnetic permeability, the dielectric constant, the space charge density, and the electric conductivity, respectively. With these parameters, the relationship between absorption, reflection, and transmission can be obtained from *Maxwell's equations* with the proper boundary conditions:

$$\nabla \times \mathbf{E} = -\mu\mu_0 \frac{\partial \mathbf{H}}{\partial t} \tag{10.1}$$

$$\nabla \times \mathbf{H} = \varepsilon\varepsilon_0 \frac{\partial \mathbf{E}}{\partial t} + \sigma \mathbf{E} \tag{10.2}$$

$$\nabla \cdot \mathbf{E} = \frac{\varrho}{4\pi\varepsilon\varepsilon_0} \tag{10.3}$$

$$\nabla \cdot \mathbf{H} = 0. \tag{10.4}$$

Here $\mathbf{E}$ is chosen as the electric field vector rather than $\mathbf{F}$ as in other chapters of this book; $\mathbf{H}$ is the magnetic field vector.

10.1.1A Nonabsorbing Dielectrics   For homogeneous, non-magnetic, nonconductive dielectrics, one obtains with $\varrho = \sigma = 0$, $\mu = 1$, $\mu_0 = 1/(\varepsilon_0 c^2)$ from Eqs. (10.1) and (10.2):

$$\nabla \times \nabla \times \mathbf{E} = \nabla \left(\nabla \cdot E \right) - \nabla^2 E = -\nabla \times \left(\frac{1}{\varepsilon_0 c^2} \frac{\partial \mathbf{H}}{\partial t} \right)$$
$$= -\frac{1}{\varepsilon_0 c^2} \frac{\partial}{\partial t} \left(\nabla \times \mathbf{H} \right), \tag{10.5}$$

which yields the undamped wave equation for the electric vector

$$\nabla^2 \mathbf{E} = \frac{\varepsilon}{c^2} \frac{\partial^2 \mathbf{E}}{\partial t^2} \tag{10.6}$$

and a similar one for the magnetic vector. Assuming a plane wave entering the dielectric in the x-direction with linear polarization in the y-direction, one has with $\mathbf{E} = (0, E_y, 0)$ from Eq. (10.6)

$$\frac{\partial^2 E_y}{\partial x^2} = \frac{\varepsilon}{c^2} \frac{\partial^2 E_y}{\partial t^2}, \tag{10.7}$$

which can be solved with the trial solution

$$E_y = f(x)\exp(-i\omega t), \tag{10.8}$$

where $f(x)$ is the amplitude function. Substitution of Eq. (10.8) into Eq. (10.7) yields

$$\frac{d^2 f}{dx^2} + \frac{\varepsilon\omega^2}{c^2} f = 0 \tag{10.9}$$

with the solution

$$f(x) = A\exp\left[\pm i\left(\frac{\omega x}{v}\right)\right], \tag{10.10}$$

where

$$v = \frac{c}{\sqrt{\varepsilon}} = \frac{c}{n_r}. \tag{10.11}$$

Here $n_r = \sqrt{\varepsilon}$ is the index of refraction and c is the light velocity *in vacuo*. Thus, with the $+$ sign in Eq. (10.10), Eq. (10.8) describes a plane wave traveling in $+$ x-direction with a phase velocity v and amplitude A:

$$E_y = A \exp\left[i\omega\left(\frac{x}{v} - t\right)\right] = A \exp\left[i\omega\left(\frac{n_r}{c}x - t\right)\right]. \tag{10.12}$$

The **energy flow** in this wave is given by the *Poynting vector* (a vector in the direction of the wave propagation)

$$\mathbf{S} = \mathbf{E} \times \mathbf{H}. \tag{10.13}$$

Here, we assume $\mathbf{E} = (0, E_y, 0)$ and $\mathbf{H} = (0, 0, H_z)$. The **energy density** is given by

$$W = \frac{1}{2}(\varepsilon\varepsilon_0 E^2 + \mu\mu_0 H^2). \tag{10.14}$$

With an equal amount of energy in the electrical and magnetic component, one obtains for the total energy density twice the energy represented by the electrical vector:

$$W = \varepsilon\varepsilon_0 E_y^2. \tag{10.15}$$

10.1.1B Semiconductors with Optical Absorption   The introduction of a finite conductivity $\sigma = \sigma(\omega)$ produces a damping contribution [second term of Eq. (10.16)] on the electromagnetic wave in a semiconductor, resulting in a finite optical absorption.* From Eqs. (10.2) and (10.5), we obtain the *damped wave equation* for the electric vector

$$\nabla^2\mathbf{E} = \frac{\varepsilon}{c^2}\frac{\partial^2\mathbf{E}}{\partial t^2} + \frac{\sigma}{\varepsilon_0 c^2}\frac{\partial\mathbf{E}}{\partial t}. \tag{10.16}$$

Using the same trial solution as given in the previous section for $f(x)$, we obtain

$$\frac{d^2 f}{dx^2} + \frac{\omega^2}{c^2}\left(\varepsilon - i\left(\frac{\sigma}{\varepsilon_0\omega}\right)\right)f = 0 \tag{10.17}$$

---

* One can understand this by equating damping with transfer of energy into heat, and optical absorption with extraction of this energy from the radiation field. Such absorption occurs even outside a specific electronic or ionic resonance absorption—see Chapter 11.

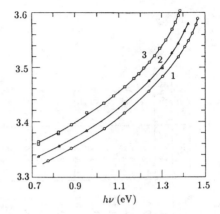

Figure 10.1: Refractive index of GaAs as a function of the photon energies with the temperature as family parameter: $T = 103, 187$, and 300 K for curves 1...3, respectively (after Marple, 1964).

which has a solution that can be written exactly as Eq. (10.10). This yields a plane wave traveling in the x-direction

$$E_y = A \exp\left[i\omega(\frac{\tilde{n}}{c} x - t)\right], \qquad (10.18)$$

except that the index of refraction used to describe the ratio c/v [Eq. (10.11)] is now complex and is identified as $\tilde{n}$. This *complex index of refraction*

$$\tilde{n} = \frac{c}{v} = \sqrt{\tilde{\varepsilon}} = n_r + i\kappa \qquad (10.19)$$

is related to the complex dielectric constant $\tilde{\varepsilon}$ in a similar fashion as given in Eq. (10.11), and contains as its real part the previously defined index of refraction and as its imaginary part the *extinction coefficient*. The *complex dielectric constant* $\tilde{\varepsilon}$ is given by

$$\tilde{\varepsilon} = \varepsilon' + i\varepsilon'' \quad \text{with} \quad \varepsilon' = \varepsilon \quad \text{and} \quad \varepsilon'' = \frac{\sigma}{\varepsilon_0\omega}, \qquad (10.20)$$

with its imaginary part related to the conductivity. From Eqs. (10.19) and (10.20), we obtain the important relations for an optically absorbing (damping) homogeneous continuum:

$$
\boxed{
\begin{aligned}
\varepsilon' &= n_r^2 - \kappa^2 \\
\varepsilon'' &= \frac{\sigma}{\varepsilon_0\omega} = 2n_r\kappa
\end{aligned}
}
\quad \text{or} \quad
\boxed{
\begin{aligned}
n_r^2 &= \frac{1}{2}\left[\varepsilon' + \sqrt{\varepsilon'^2 + \varepsilon''^2}\right] \\
\kappa^2 &= \frac{1}{2}\left[-\varepsilon' + \sqrt{\varepsilon'^2 + \varepsilon''^2}\right].
\end{aligned}
}
\qquad (10.21)
$$

The optical parameters for a number of typical semiconductors are listed in Tables 10.1 and 10.2. The temperature-dependence of n_r is shown for the example of GaAs in Fig. 10.1.

Table 10.1: Optical constants for element and III-V compound semiconductors for $E > E_g$.

Crystals:	Si	Ge	GaP	GaAs	GaSb	InP	InAs	InSb	CdTe	α-Si*	hν
n_r	hν										hν
	3.673 (1.5eV)	4.653	3.178	3.666	4.388	3.456	3.714	4.418	2.98	3.9	1.5eV
	3.906 (2.0eV)	5.588	3.334	3.878	5.239	3.549	3.995	4.194	2.99	4.2	2.0eV
	4.320 (2.5eV)	4.340	3.605	4.333	4.312	3.818	4.364	3.570	3.14	4.5	2.5eV
	5.222 (3.0eV)	4.082	4.081	4.509	3.832	4.395	3.197	3.366	3.37	4.4	3.0eV
	5.610 (3.5eV)	4.020	5.050	3.531	3.785	3.193	3.008	3.511	2.89	3.8	3.5eV
	5.010 (4.0eV)	3.905	3.790	3.601	3.450	3.141	3.313	2.632	2.39	2.8	4.0eV
	2.452 (4.5eV)	1.953	3.978	3.913	1.586	3.697	3.194	1.443	2.43	2.1	4.5eV
	1.570 (5.0eV)	1.394	3.661	2.273	1.369	2.131	1.524	1.307	2.48	1.7	5.0eV
	1.340 (5.5eV)	1.380	1.543	1.383	1.212	1.426	1.282	1.057	–	1.3	5.5eV
	1.010 (6.0eV)	1.023	1.309	1.264	0.935	1.336	1.434	0.861	–	1.1	6.0eV
κ	0.005 (1.5eV)	0.298	0.000	0.080	0.344	0.203	0.432	0.643	0.32	0.08	1.5eV
	0.022 (2.0eV)	0.933	0.000	0.211	1.378	0.317	0.634	1.773	0.35	0.46	2.0eV
	0.073 (2.5eV)	2.384	0.006	0.441	2.285	0.511	1.786	2.221	0.53	1.1	2.5eV
	0.269 (3.0eV)	2.145	0.224	1.948	2.109	1.247	2.034	1.994	0.86	2.0	3.0eV
	3.014 (3.5eV)	2.667	0.819	2.013	2.545	1.948	1.754	2.517	1.52	2.8	3.5eV
	3.586 (4.0eV)	3.336	2.171	1.920	3.643	1.730	1.799	3.694	1.71	3.0	4.0eV
	5.082 (4.5eV)	4.297	2.180	2.919	3.392	2.186	3.445	2.894	1.67	2.9	4.5eV
	3.565 (5.0eV)	3.197	3.631	4.084	2.751	3.495	2.871	2.441	2.06	2.8	5.0eV
	3.302 (5.5eV)	2.842	3.556	2.936	2.645	2.562	2.344	2.333	–	2.5	5.5eV
	2.909 (6.0eV)	2.774	2.690	2.472	2.416	2.113	2.112	2.139	–	2.3	6.0eV

*Sensitive to preparation and hydrogen content. Values of CdTe and α-Si from Palik, 1985.

Table 10.1: (Cont'd.) Optical constants for element and III-V compound semiconductors for $E > E_g$.

Crystals:	Si	Ge	GaP	GaAs	GaSb	InP	InAs	InSb	ZnSe	CdTe	$h\nu$
ε' $h\nu$											
= 1.5eV	13.488	21.560	10.102	13.435	19.135	11.904	13.605	19.105	6.4	11.5	= 1.5eV
2.0eV	15.254	30.361	11.114	14.991	25.545	12.493	15.558	14.448	6.0	12.2	2.0eV
2.5eV	18.661	13.153	12.996	18.579	13.367	14.313	15.856	7.811	6.8	13.2	2.5eV
3.0eV	27.197	12.065	16.601	16.536	9.479	17.759	6.083	7.354	7.7	15.0	3.0eV
3.5eV	22.394	9.052	24.833	8.413	7.852	6.400	5.973	5.995	7.8	14.5	3.5eV
4.0eV	12.240	4.123	9.652	9.279	− 1.374	6.874	7.744	− 6.722	8.5	17.0	4.0eV
4.5eV	−19.815	−14.655	11.073	6.797	− 8.989	8.891	− 1.663	− 6.297	9.5	10.2	4.5eV
5.0eV	−10.242	− 8.277	0.218	−11.515	− 5.693	− 7.678	− 5.923	− 4.250	8.0	7.5	5.0eV
5.5eV	− 9.106	− 6.179	−10.266	6.705	− 5.527	4.528	− 3.851	− 4.325	5.6	4.6	5.5eV
6.0eV	− 7.443	− 6.648	− 5.521	− 4.511	− 4.962	− 2.681	− 2.403	− 3.835	6.5	–	6.0eV
ε''											
1.5eV	0.038	2.772	0.000	0.589	3.023	1.400	3.209	5.683	0	0.2	1.5eV
2.0eV	0.172	10.427	0.000	1.637	14.442	2.252	5.062	14.875	0	0.8	2.0eV
2.5eV	0.630	20.695	0.046	3.821	19.705	3.904	15.592	15.856	0.5	2.1	2.5eV
3.0eV	2.807	17.514	1.832	17.571	15.738	10.962	13.003	13.421	1.5	2.4	3.0eV
3.5eV	33.818	21.442	8.268	14.216	19.267	12.443	10.550	17.673	1.9	7.6	3.5eV
4.0eV	35.939	26.056	16.454	13.832	25.138	10.871	11.919	19.443	2.2	11.0	4.0eV
4.5eV	24.919	16.782	17.343	22.845	10.763	16.161	22.006	8.351	3.8	13.8	4.5eV
5.0eV	11.195	8.911	26.580	18.563	7.529	14.896	8.752	6.378	8.5	14.8	5.0eV
5.5eV	8.846	7.842	10.974	8.123	6.410	7.308	6.008	4.931	6.2	6.5	5.5eV
6.0eV	5.877	5.672	7.041	6.250	4.520	5.644	6.005	3.681	7.8	–	6.0eV

Table 10.2: Refractive index as a function of the photon wavelength for some semiconductors at 300 K for $E < E_g$ (from *American Institute of Physics Handbook*, 1963—for more extensive data, see Palik, 1985).

	$\lambda(\mu\text{m})$	n_r	$\lambda(\mu\text{m})$	n_r	$\lambda(\mu\text{m})$	n_r
Si	1.3570	3.4975	2.4373	3.4408	5.50	3.4213
	1.3673	3.4962	2.7144	3.4358	6.00	3.4202
	1.3951	3.4929	3.00	3.4320	6.50	3.4195
	1.5295	3.4795	3.3033	3.4297	7.00	3.4189
	1.6606	3.4696	3.4188	3.4286	7.50	3.4186
	1.7092	3.4664	3.50	3.4284	8.00	3.4184
	1.8131	3.4608	4.00	3.4255	8.50	3.4182
	1.9701	3.4537	4.258	3.4242	10.00	3.4179
	2.1526	3.4476	4.50	3.4236	10.50	3.4178
	2.3254	3.4430	5.00	3.4223	11.04	3.4176
Ge	2.0581	4.1016	2.998	4.0452	8.66	4.0043
	2.1526	4.0919	3.3033	4.0369	9.72	4.0034
	2.3126	4.0786	3.4188	4.0334	11.04	4.0026
	2.4374	4.0708	4.258	4.0216	12.20	4.0023
	2.577	4.0609	4.866	4.0170	13.02	4.0021
	2.7144	4.0552	6.238	4.0094		
GaAs	0.78	3.34	13.0	2.97	17.0	2.59
	8.0	3.34	13.7	2.895	19.0	2.41
	10.0	3.135	14.5	2.82	21.9	2.12
	11.0	3.045	15.0	2.73		
InSb	7.87	4.00	12.98	3.91	17.85	3.85
	8.00	3.99	13.90	3.90	18.85	3.84
	9.01	3.96	15.13	3.88	19.98	3.82
	10.06	3.95	15.79	3.87	21.15	3.81
	11.01	3.93	16.96	3.86	22.20	3.80
	12.06	3.92				

Using Eq. (10.19), the wave equation can be rewritten as

$$E_y = A \exp i\omega\left(\frac{n_r}{c}x - t\right)\exp\left(-\frac{\omega\kappa}{c}x\right) \qquad (10.22)$$

and shows the damping factor in the second exponential. Using a more conventional expression $\exp(-\alpha_o x)$ for the damping of the energy flux, with α_o as the *optical absorption coefficient* for the *energy density*, we obtain by comparison with the second exponential in Eq. (10.22)

$$\alpha_o = \frac{2\omega\kappa}{c} = \frac{4\pi}{\lambda}\kappa; \qquad (10.23)$$

here the energy flow is given by the product of the electric and magnetic vectors, thus producing a factor of 2 in the exponent. However, H_z is phase-shifted by δ with $\tan\delta = \kappa/n_r$. From Eqs. (10.21) and (10.22), we also obtain

$$\alpha_o = \frac{\sigma}{\varepsilon_0 n_r c}, \qquad (10.24)$$

which shows the direct connection of α_o with the electrical conductivity. This conductivity is to be taken at the optical frequency and needs further explanation.

10.1.1C The Complex Electrical Conductivity   The right-hand side of Eq. (10.2) of Maxwell's equations is considered as total current

$$\mathbf{j} = \sigma\mathbf{E} + \varepsilon\varepsilon_0\frac{\partial\mathbf{E}}{\partial t} = (\sigma + i\omega\tilde{\varepsilon}\varepsilon_0)\mathbf{E}. \qquad (10.25)$$

The second term is obtained after Fourier transformation from the time into the frequency domain. When introducing for $i\omega\tilde{\varepsilon}\varepsilon_0$ a *complex conductivity*, we obtain

$$\tilde{\sigma} = \sigma' + i\sigma'' = \varepsilon_0\omega(\varepsilon'' - i\varepsilon') \qquad (10.26)$$

or

$$\sigma' = \varepsilon''\varepsilon_0\omega = \sigma_d$$
$$\sigma'' = -\varepsilon'\varepsilon_0\omega = -(n_r^2 - \kappa^2)\varepsilon_0\omega. \qquad (10.27)$$

Observe that the real part of $\tilde{\sigma}$ is proportional to the imaginary part of $\tilde{\varepsilon}$ and vice versa. The conductivity σ_d is proportional to the *displacement current*, caused by bound electrons surrounding each atom core, which are slightly displaced and oscillate out of phase with the acting electric field.

Table 10.3: Dielectric constant tensor.

Crystal System	Characteristic Symmetry	Number of Independent Coefficients	Form of tensor Showing Independent Coefficients
Cubic	4 3-fold axes	1	$\begin{pmatrix} \varepsilon & 0 & 0 \\ 0 & \varepsilon & 0 \\ 0 & 0 & \varepsilon \end{pmatrix}$
Tetragonal Hexagonal Trigonal	1 4-fold axis 1 6-fold axis 1 3-fold axis	2	$\begin{pmatrix} \varepsilon_1 & 0 & 0 \\ 0 & \varepsilon_1 & 0 \\ 0 & 0 & \varepsilon_3 \end{pmatrix}$
Orthorhombic	3 mutually perpendicular 2-fold axes; no axes of higher order	3	$\begin{pmatrix} \varepsilon_1 & 0 & 0 \\ 0 & \varepsilon_2 & 0 \\ 0 & 0 & \varepsilon_3 \end{pmatrix}$
Monoclinic	1 2-fold axis	4	$\begin{pmatrix} \varepsilon_{11} & 0 & \varepsilon_{13} \\ 0 & \varepsilon_{22} & 0 \\ \varepsilon_{13} & 0 & \varepsilon_{33} \end{pmatrix}$
Triclinic	A center of symmetry or no symmetry	6	$\begin{pmatrix} \varepsilon_{11} & \varepsilon_{12} & \varepsilon_{13} \\ \varepsilon_{12} & \varepsilon_{22} & \varepsilon_{23} \\ \varepsilon_{13} & \varepsilon_{23} & \varepsilon_{33} \end{pmatrix}$

The current can now be rewritten. When expressed by the dielectric function, Eq. (10.25) reads

$$\begin{aligned} \mathbf{j}(\omega) &= \left\{ \sigma_0 + \omega\varepsilon_0\varepsilon''(\omega) + i\omega\varepsilon_0\varepsilon'(\omega) \right\} \mathbf{E}(\omega) \\ &= \left\{ \sigma_0 + \sigma_d(\omega) + i\omega\varepsilon_0(n_r^2 - \kappa^2) \right\} \mathbf{E}(\omega). \end{aligned} \tag{10.28}$$

10.1.1D Dielectric Polarization For later discussions of microscopic models, it is advantageous to introduce the dielectric polarization. The electric displacement $\mathbf{D}$, field strength $\mathbf{E}$, and polarization $\mathbf{P}$ are related by

$$\mathbf{D} = \varepsilon_0 \mathbf{E} + \mathbf{P} \tag{10.29}$$

and

$$\mathbf{P} = \varepsilon_0 \tilde{\chi}\mathbf{E} = \varepsilon_0(\tilde{\varepsilon} - 1)\mathbf{E} \tag{10.30}$$

with $\tilde{\varepsilon}$ as the complex dielectric constant. Here, only a linear relationship between $\mathbf{P}$ and $\mathbf{E}$ is given. Higher terms are of importance at higher fields and give rise to nonlinear optical effects, which are discussed in Chapter 16. Equation (10.30) defines the susceptibility $\tilde{\chi}$ as the proportionality constant between the polarization **P** and the electric field **E**. From Eqs. (10.29) and (10.30) we have

$$\mathbf{D} = \varepsilon_0 \tilde{\varepsilon} \mathbf{E} = \varepsilon_0 (1 + \tilde{\chi}) \mathbf{E}. \qquad (10.31)$$

In an anisotropic material, ε and χ have tensor form with components

$$P_i = \varepsilon_0 \chi_{ij} E_j \quad \text{and} \quad D_i = \varepsilon_0 \varepsilon_{ij} E_j. \qquad (10.32)$$

This matrix relationship must be used when applying Maxwell's equations to an anisotropic medium. For instance, from *Poisson's equation* $\nabla \cdot \mathbf{E} = \varrho/(\varepsilon \varepsilon_0)$, we have for the *ij*-component

$$\varepsilon_0 \varepsilon_{ij} \frac{\partial E_j}{\partial x_i} = \varrho. \qquad (10.33)$$

Depending on the crystal symmetry, the dielectric constant tensor can be reduced, and contains at most six independent coefficients—see Table 10.3. This dielectric constant will be the subject of further discussion later in this book—see Chapter 14 and Section 48.2.1.

10.2 Measurement of Optical Parameters

The parameters obtained from an analysis of the optical phenomena are the complex dielectric constant (or index of refraction) and the complex conductivity, as well as the amplitude and polarization of electromagnetic waves after interacting with the semiconductor.

On the other hand, the parameters measured directly are the intensities of reflected and transmitted radiation and its change of the state of polarization. It is the purpose of this section to summarize briefly the most relevant interrelation of these two sets of parameters, thereby permitting a quantitative comparison between theory and experiment. For more detail, see Stern (1968), or Palik (1980).

10.2.1 Reflectance and Transmissivity in Dielectrics

The index of refraction, as well as the extinction and absorption coefficients can be related to the amplitude and polarization of reflected and transmitted optical waves which can be measured directly. The relationship between these waves can be obtained from the wave equation and the boundary conditions at the interface between two media (see Moss, 1961a).

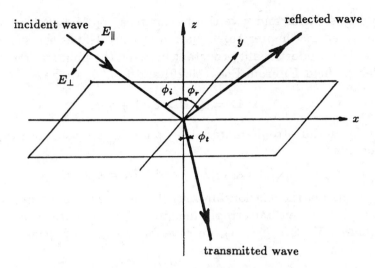

incident wave $E_\parallel$
reflected wave
$E_\perp$
z
y
ϕ_i ϕ_r
x
ϕ_t
transmitted wave

Figure 10.2: Coordinate system with light beams and interface.

The electric vector of the incident wave (denoted by subscript i) can be described by its two components normal and parallel to the plane of incidence at $z = 0$ (Fig. 10.2):

$$(E_y, E_z)_i = (E_\perp, E_\parallel)_i \exp\left[i\omega\left\{\frac{n_{r1}}{c}\left(x\sin\phi_i + z\cos\phi_i\right) - t\right\}\right],$$
(10.34)

with the corresponding components of the reflected (index r) and transmitted (index t) waves

$$(E_y, E_z)_r = (E_\perp, E_\parallel)_r \exp\left[i\omega\left\{\frac{n_{r1}}{c}\left(x\sin\phi_r + z\cos\phi_r\right) - t\right\}\right]$$
(10.35)

$$(E_y, E_z)_t = (E_\perp, E_\parallel)_t \exp\left[i\omega\left\{\frac{n_{r2}}{c}\left(x\sin\phi_t + z\cos\phi_t\right) - t\right\}\right].$$
(10.36)

The coordinate system is chosen so that the interface is normal to the z-axis and cuts z at $z = 0$. The x- and y-axes are chosen so that the normal-to-the-incident wavefront lies in the xz-plane; the angles are identified in Fig. 10.2.

From the condition that at the plane of incidence ($z = 0$) the tangential component of the electrical field of all three waves must be the same, we require that the corresponding exponents must be equal: $n_{r1} \sin \phi_i = n_{r1} \sin \phi_r = n_{r2} \sin \phi_t$. From here, *Snell's law* can be deduced:

$$\boxed{\begin{aligned} \phi_i &= \phi_r \\ n_{r1} \sin \phi_i &= n_{r2} \sin \phi_t. \end{aligned}} \tag{10.37}$$

For $n_{r1} > n_{r2}$ this yields an angle of total reflection (i.e., for $\phi_t = 90°$), with $\phi_i = \phi_c$ the critical angle,* given by

$$\boxed{\sin \phi_c = \frac{n_{r2}}{n_{r1}}.} \tag{10.39}$$

Total reflection is sometimes used in photosensing devices for light trapping by properly shaping the surface in order to increase the optical path within the device for more optical absorption, and thereby increasing the photosensitivity.

The amplitudes of the tangential components of the reflected and transmitted waves must be continuous when passing through $z = 0$; hence

$$(E_{\|i} - E_{\|r}) \cos \phi_i = E_{\|t} \cos \phi_t \tag{10.40}$$

$$E_{\perp i} + E_{\perp r} = E_{\perp t} \tag{10.41}$$

$$n_{r1}(E_{\perp i} - E_{\perp r}) \cos \phi_i = n_{r2} E_{\perp t} \cos \phi_t \tag{10.42}$$

$$n_{r1}(E_{\|i} + E_{\|r}) = n_{r2} E_{\|t}. \tag{10.43}$$

The first set of two equations is obtained from the electric vector components; the second set is obtained from the magnetic vector components.

---

* A *Brewster angle* is defined as the angle under which no component $E_\|$ is reflected. Here, $\sin \phi_i = \cos \phi_t$, hence

$$\tan \phi_B = \frac{n_{1r}}{n_{2r}}. \tag{10.38}$$

After solving this set of four equations for the four components of the electric vector, we obtain *Fresnel's equations*:

$$
\begin{aligned}
E_{\|r} &= E_{\|i}\frac{n_{r2}\cos\phi_i - n_{r1}\cos\phi_t}{n_{r2}\cos\phi_i + n_{r1}\cos\phi_t} \\[2mm]
E_{\perp r} &= E_{\perp i}\frac{n_{r1}\cos\phi_i - n_{r2}\cos\phi_t}{n_{r1}\cos\phi_i + n_{r2}\cos\phi_t} \\[2mm]
E_{\|t} &= E_{\|i}\frac{2n_1\cos\phi_i}{n_{r1}\cos\phi_t + n_{r2}\cos\phi_i} \\[2mm]
E_{\perp t} &= E_{\perp i}\frac{2n_{r1}\cos\phi_i}{n_{r2}\cos\phi_t + n_{r1}\cos\phi_i}
\end{aligned}
\tag{10.44}
$$

which are generally valid and are the basis for all following discussions.

All measurable quantities are related to the *energy* flux, i.e., for the incident and reflected waves to the Poyntings vector. For the incident and reflected waves we have

$$
W_a = \varepsilon_0 n_{r1}^2 E_a^2 \quad \text{for} \quad a = i, r.
\tag{10.45}
$$

For the transmitted wave we have

$$
W_t = \varepsilon_0 n_{r2}^2 E_t^2.
\tag{10.46}
$$

The reflectance* and transmittance are defined by the ratios of the energy flux to the incident energy flux normal to the interface:

$$
R = \left(\frac{E_r}{E_i}\right)^2 \quad \text{and} \quad T = \left(\frac{E_t}{E_i}\right)^2 \frac{\cos\phi_t}{\cos\phi_i},
\tag{10.47}
$$

which can be computed from the normal and parallel components given by the Fresnel equations. The resulting formulae become rather lengthy and confusing. Some simplified cases better demonstrate the typical behavior. These simplifications include

(1) air as the first medium with $n_{r1} = 1$ and $\sigma_1 = 0$;
(2) a nonabsorbing second medium with $\sigma_2 = 0$; and
(3) that the incident wave is normal to the interface.

---

* Reflectance (etc.) is used rather than reflectivity since it is not normalized to the unit area; this is similar to the use of the word resistance (not normalized) vs. resistivity, distinguishing between the suffixes -*ance* and -*ivity*.

With assumptions (1) and (2), we obtain from Eqs. (10.47) and (10.44) for the two components of the reflected beam

$$R_\perp = \frac{\sin^2(\phi_t - \phi_i)}{\sin^2(\phi_t + \phi_i)} \quad \text{and} \quad R_\| = \frac{\tan^2(\phi_t - \phi_i)}{\tan^2(\phi_t + \phi_i)} \tag{10.48}$$

and the transmitted beam

$$T_\perp = \frac{\sin 2\phi_i \sin 2\phi_t}{\sin^2(\phi_i + \phi_t)} \quad \text{and} \quad T_\| = \frac{\sin 2\phi_i \sin 2\phi_t}{\sin^2(\phi_i + \phi_t)\cos^2(\phi_i + \phi_t)}. \tag{10.49}$$

When assumptions (2) and (3) hold, we obtain the well-known relations for the reflected and transmitted beams:

$$\boxed{\begin{aligned} R = R_\perp = R_\| &= \left(\frac{n_{r1} - n_{r2}}{n_{r1} + n_{r2}}\right)^2 \\ T = T_\perp = T_\| &= \frac{4 n_{r1} n_{r2}}{\left(n_{r1} + n_{r2}\right)^2}. \end{aligned}} \tag{10.50}$$

10.2.2 Reflectance and Transmittance in Semiconductors

In semiconductors, condition (2) of the previous section no longer holds; there is absorption. However, in spectral ranges in which n_r is much larger than κ, we obtain still rather simple approximations, which for the reflected wave are:

$$R_\perp \simeq \frac{(n_r - \cos \phi_i)^2 + \kappa^2}{(n_r + \cos \phi_i)^2 + \kappa^2} \quad \text{and} \quad R_\| \simeq R_\perp \frac{(n_r - \sin \phi_i \tan \phi_i)^2 + \kappa^2}{(n_r + \sin \phi_i \tan \phi_i)^2 + \kappa^2}. \tag{10.51}$$

These reflectance components are shown for a set of n_r and κ in Fig. 10.3. There is a substantial amplitude difference for low angles of incidence between the parallel and perpendicular polarized components. The ratio of these components shows a maximum at the *Brewster angle* given by

$$\tan^2 \phi_B \simeq n_r^2 + \kappa^2. \tag{10.52}$$

Reflection at this angle can be used to obtain nearly linearly polarized light from the normal component (the amplitude of the parallel component is negligible)—see Fig. 10.3.

For normal incidence, parallel or normal components can no longer be distinguished in reflectance:

$$\boxed{R_0 = R_\perp = R_\| = \frac{(n_r - 1)^2 + \kappa^2}{(n_r + 1)^2 + \kappa^2}.} \tag{10.53}$$

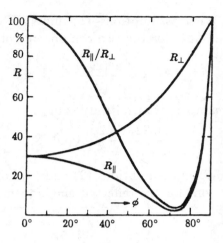

Figure 10.3: Polarized components of the reflectance parallel and perpendicular to the semiconductor surface as a function of the incident angle, computed for $n_r = 3$ and $\kappa = 1$ (Moss, 1961a).

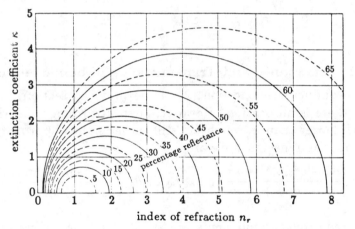

Figure 10.4: Relation between the normal reflectance and the optical constants for a single surface reflectance [Eq. (10.53)].

Equation (10.53) is the equation of a circle centered at $n = (1 + R_0)(1 - R_0)$ with a radius of $2\sqrt{R_0}/(1 - R_0)$, as shown in Fig. 10.4.

For a **semiconductor plate** with two planar surfaces separated by a distance d, we must consider reflection from both surfaces, as shown in Fig. 10.5. The result of the *series* of sequential reflections, indicated by the subscript Σ, is

$$R_\Sigma = R_0 \left[1 + \frac{T_0^2 (1 - R_0)^2}{1 - R_0^2 T_0^2} \right] \quad \text{and} \quad T_\Sigma = T_0 \left[\frac{(1 - R_0)^2}{1 - R_0^2 T_0^2} \right],$$

$$(10.54)$$

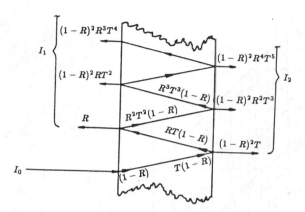

Figure 10.5: Multireflection in a semitransparent slab.

with T_0 as the transmittance through the slab at its first pass under normal incidence without reflection. For a perfectly transmitting slab, we consequently obtain

$$R_\Sigma = \frac{2R_0}{1 - R_0} \simeq 2R_0. \tag{10.55}$$

In a more precise analysis, a phase shift δ between both reflected waves must be included, which, for normal incidence, is given by

$$\delta_o = \frac{2\pi n_r d}{\lambda}. \tag{10.56}$$

The total reflectivity of this plate at normal incidence is given by

$$R_\Sigma = \frac{\sinh^2(\alpha_o d/2) + \sin^2 \delta_o}{\sinh^2(\alpha_o d/2 + \gamma_o) + \sin^2(\delta_o + \psi_o)}, \tag{10.57}$$

where the auxiliary functions γ_o and ψ_o are given by $\gamma_o = \ln \sqrt{1/R_0}$, with R_0 from Eq. (10.53), and $\psi_o = \tan^{-1}\left(2\kappa/[n_r^2 + \kappa^2 - 1]\right)$; α_o is the optical absorption coefficient [Eq. (10.23)]. The transmissivity through this plate is given by

$$T_\Sigma = \frac{\sinh^2 \gamma_o + \sin^2 \psi_o}{\sinh^2(\alpha_o d/2 + \gamma_o) + \sin^2(\gamma_o + \psi_o)}. \tag{10.58}$$

For vanishing absorption ($\kappa = 0$, $\psi_o = 0$), we confirm from Eqs. (10.57) and (10.58) that the sum of transmissivity and reflectivity is 1.

The interference pattern from the superposition of front and back reflections makes an evaluation of κ (or α_o) in thin planar layers

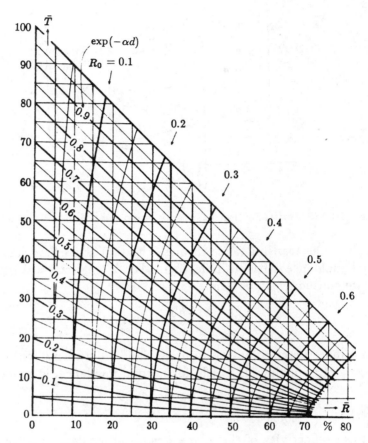

Figure 10.6: Abac chart for obtaining α_o and R_0 from measured values of $\overline{T}$ and $\overline{R}$.

difficult. However, one can deduce the index of refraction from the ratio of maximum to minimum amplitudes of the transmitted light for adjacent extrema

$$\frac{T_{0,\max}}{T_{0,\min}} = \frac{(n_r^2 + 1)^2}{4n_r^2} \tag{10.59}$$

as long as the absorption term can be neglected—see Moss, 1961, for inclusion of α_o.

In order to determine the absorption coefficient, compromises must be made to average out the interference pattern by making either the surfaces slightly nonplanar (rough), or the light slightly polychromatic. This yields an "average" reflectivity

$$\overline{R} = R_0 \left(1 + \overline{T} \exp\left(-\alpha_o d\right)\right) \qquad (10.60)$$

and an "average" transmissivity

$$\overline{T} = \frac{(1 - R_0)^2 \exp(-\alpha_o d)}{1 - R_0^2 \exp(-2\alpha_o d)} \simeq (1 - R_0)^2 \exp(-\alpha_o d), \qquad (10.61)$$

with R_0 given by Eq. (10.53). From the measured values of $\overline{T}$ and $\overline{R}$, the more relevant values of α_o and R_0, and hence of κ and n_r, can be obtained. This is done most easily with an Abac chart, as shown in Fig. 10.6.

The reflectance and absorption coefficient of some typical semiconductors is given in Table 10.4.

10.2.3 Modulation Spectroscopy

For analytical purposes, modulating one of the parameters of a semiconductor while investigating the optical response is a very powerful technique. The light itself (frequency or intensity), temperature, pressure, mechanical stress (uniaxial), and electric or magnetic field have all been employed as parameters to be modulated. As optical response, the reflectance is most often used by measuring its relative change $\Delta R/R_0$ or its higher derivatives (see Section 42.4.3H) as a function of the wavelength. Absorption within the band is too strong to be observed, except for extremely thin platelets.

The modulation is detected with a phase-sensitive lock-in technique, which is extremely sensitive and permits detection of rather small signals ($< 10^{-5}$)—see Cardona (1969), Seraphin (1973), and Enderlein (1977).

With this technique it is relatively easy to detect changes in spectral ranges of high optical absorption. For instance, we can measure the energy difference between the critical points within the valence and conduction bands. This method is analyzed below.

From the reflectance equation [Eq. (10.50)]

$$R_0 = \left(\frac{n_r - n_{ra}}{n_r + n_{ra}}\right)^2, \qquad (10.62)$$

we obtain the modulated signal of the reflected light beam, which is proportional to the changes in the complex dielectric constant; n_{ra}

Table 10.4: Absorption constant and reflectance of element and III-V compound semiconductors for $E > E_g$.

Crystals		Si	Ge	GaP	GaAs	GaSb	InP	InAs	InSb	$h\nu$
	$h\nu$									
	= 1.5 eV	0.327	0.419	0.272	0.327	0.398	0.305	0.337	0.406	= 1.5 eV
	2.0 eV	0.351	0.495	0.290	0.349	0.487	0.317	0.370	0.443	2.0 eV
	2.5 eV	0.390	0.492	0.320	0.395	0.484	0.349	0.454	0.447	2.5 eV
	3.0 eV	0.461	0.463	0.369	0.472	0.444	0.427	0.412	0.416	3.0 eV
R	3.5 eV	0.575	0.502	0.458	0.425	0.485	0.403	0.371	0.474	3.5 eV
	4.0 eV	0.591	0.556	0.452	0.421	0.583	0.376	0.393	0.608	4.0 eV
	4.5 eV	0.740	0.713	0.461	0.521	0.651	0.449	0.566	0.598	4.5 eV
	5.0 eV	0.675	0.650	0.580	0.668	0.585	0.613	0.583	0.537	5.0 eV
	5.5 eV	0.673	0.598	0.677	0.613	0.592	0.542	0.521	0.563	5.5 eV
	6.0 eV	0.677	0.653	0.583	0.550	0.610	0.461	0.448	0.572	6.0 eV
	1.5 eV	0.78	45.30	0.00	12.21	52.37	30.79	65.69	97.79	= 1.5 eV
	2.0 eV	4.47	189.12	0.00	42.79	279.43	64.32	128.43	359.46	2.0 eV
	2.5 eV	18.48	604.15	1.63	111.74	579.07	129.56	452.64	562.77	2.5 eV
α_o	3.0 eV	81.73	652.25	68.26	592.48	641.20	379.23	618.46	606.27	3.0 eV
(in 10^3	3.5 eV	1069.19	946.01	290.40	714.20	902.86	691.21	622.13	892.82	3.5 eV
cm^{-1})	4.0 eV	1454.11	1352.55	880.10	778.65	1477.21	701.54	729.23	1497.79	4.0 eV
	4.5 eV	2317.99	1960.14	994.27	1331.28	1547.17	996.95	1571.19	1320.24	4.5 eV
	5.0 eV	1806.67	1620.15	1839.99	2069.81	1394.02	1771.52	1455.26	1237.01	5.0 eV
	5.5 eV	1840.59	1584.57	1982.53	1636.68	1474.51	1428.14	1306.62	1300.55	5.5 eV
	6.0 eV	1769.27	1686.84	1635.71	1503.20	1469.28	1285.10	1284.15	1300.85	6.0 eV

is the index of refraction for an external medium in contact with the semiconductor. Using the *Maxwell relation*

$$n_r^2 = \varepsilon_{opt} = \varepsilon \quad \text{and} \quad n_{ra}^2 = \varepsilon_{a,opt} = \varepsilon_a, \qquad (10.63)$$

such changes can be expressed as (Re = real part)

$$\frac{\Delta R}{R_0} = Re\left[\frac{2n_{ra}}{n_r(\varepsilon - \varepsilon_a)}\Delta\varepsilon\right]. \qquad (10.64)$$

Equation (10.64) can also be written as

$$\frac{\Delta R}{R_0} = Re\left[(\alpha_S - i\beta_S)\Delta\tilde{\varepsilon}\right] = \alpha_S\Delta\varepsilon' - \beta_S\Delta\varepsilon'', \qquad (10.65)$$

where α_S and β_S are the *Seraphin coefficients* (Seraphin and Bottka, 1965) and $\Delta\tilde{\varepsilon} = \Delta\varepsilon' + i\Delta\varepsilon''$ is the change in the complex dielectric constant due to the modulation.

A more useful equation for comparing the experimental results with a theoretical analysis is the inverse relation

$$\Delta\varepsilon = \frac{\dfrac{\Delta R}{R_0} + 2i\Delta\theta}{\alpha_S - i\beta_S} \qquad (10.66)$$

with

$$\Delta\theta = -\frac{\omega}{\pi}P\int_0^\infty \left[\frac{\Delta R}{R_0}(\omega')\right]\frac{d\omega'}{\omega'^2 - \omega^2}, \qquad (10.67)$$

where P indicates the principal part of the integral. The integral can be evaluated by a Kramers-Kronig analysis—see Section 11.1.4 and Aspnes (1980).

10.2.3A Relation to Band-to-Band Transitions   The analysis of the modulation reflection spectrum is involved. It permits identification of the energy and type of critical points. As an example, we will give below a short outline of how the energy difference from the valence band to an extremum of an upper conduction band can be obtained from typical reflection signals.

The complex dielectric constant for band-to-band excitation in a one-electron approximation is given by (see Section 13.1.1 and, e.g., Petroff, 1980; Aspnes, 1980):

$$\varepsilon(E,\gamma) = 1 + \frac{4\pi e^2 \hbar^2}{m_0^2 E^2}A_0^2\sum_{k,c,v}|\mathbf{e}\cdot\mathbf{M}_{cv}(\mathbf{k})|^2\left[\frac{1}{E_c(\mathbf{k}) - E_v(\mathbf{k}) - h\nu - i\gamma}\right.$$
$$\left. + \frac{1}{E_c(\mathbf{k}) - E_v(\mathbf{k}) + h\nu + i\gamma}\right], \qquad (10.68)$$

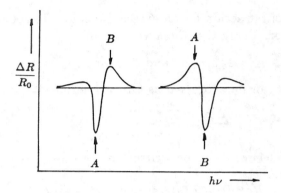

Figure 10.7: Typical line shape of a reflection signal caused by a simple band-to-band transition.

with $\mathbf{M}_{cv}$ as the momentum matrix element, $\mathbf{e}$ is the electric vector of the impinging electromagnetic wave, A_0 as its amplitude, and γ as an empirical damping parameter. Equation (10.68) can be simplified for a direct transition into one parabolic band to

$$\varepsilon(E, \gamma_d) = \frac{e^2 \hbar^2 A_0^2}{\pi^2 m_0^2 E^2} |\mathbf{e} \cdot \mathbf{M}_{cv}|^2 \int \frac{d^3 k}{E_c(k) - E_v(k) - h\nu - i\gamma} \quad (10.69)$$

with

$$E_c(\mathbf{k}) - E_v(\mathbf{k}) = E_g + \frac{\hbar^2}{2} \left(\frac{k_x^2}{m_{xx}} + \frac{k_y^2}{m_{yy}} + \frac{k_z^2}{m_{zz}} \right). \quad (10.70)$$

From Eq. (10.69) we deduce that an isolated critical point (of the type M_0 or M_3—Section 13.1) can be described by

$$\varepsilon(E, \gamma) = A\gamma^{-n} \exp(i\theta)(E - E_g + i\gamma)^n \quad (10.71)$$

where A is an amplitude factor, θ is a phase projection factor, and n is an exponent which is $-1/2$, 0 (logarithmic), or $1/2$ for one-, two-, or three-dimensional critical points—see Section 13.1. The gap E_g, i.e., the energy difference between critical points in any of the valence and conductions bands related to the initiated transition, and the damping γ can be determined from the line shape, which typically shows three extrema near the critical point (Fig. 10.7). With subscripts A and B identifying the two largest extrema, E_g and γ, can be estimated according to (Aspnes, 1980)

$$E_g = E_A + (E_B - E_A)f(\xi) \quad (10.72)$$

and

$$\gamma = (E_B - E_A)g(\xi) \simeq E_B - E_A, \qquad (10.73)$$

with $0.8 < g(\xi) < 1.2$ and $f(\xi)$ monotonically increasing from 0 to 1 with increasing $\xi = (\Delta R/R_0)_A/(\Delta R/R_0)_B$. Hence, E_g always lies closest to the largest peak in the reflection spectrum: the closer to the peak, the larger the ratio of their magnitude.

Summary and Emphasis

In this chapter we have shown that the interaction of electromagnetic radiation with a semiconductor can be described by two parameters; the complex dielectric constant and conductivity. They can be derived from Maxwell's equations, applied to the semiconductor as a continuum, without considering its atomic structure.

These parameters are closely associated with the index of refraction, the extinction coefficient (or to the optical absorption coefficient), and the displacement current. These, in turn, can be related to directly measured quantities, such as the transmitted and reflected light as a function of the wavelength, impinging angle, and polarization.

The so obtained optical constants connect to the microscopic theory of optically induced transitions, and involve a wide variety of quasi-particles, such as electrons, holes, excitons, phonons, polaritons, etc.—particles which we will discuss in detail in later chapters. We will analyze in the next two chapters the interrelationship between the phenomenological description given here and the microscopic theory of elementary excitation processes.

Optical absorption and reflection spectroscopy provides direct experimental access to the key parameters of the semiconductor describing the interaction with electromagnetic radiation. The parameters can be obtained as a result of a microscopic theory of all radiation-induced transitions, and thereby provide the basis for a quantitative comparison between such theory and the experiment.

Modern methods, including modulation spectroscopy, have increased the experimental sensitivity by many orders of magnitude and have opened a wide range of opportunities for an improved understanding of photon-induced transitions in solids. The optical analysis has become one of the most useful analytical techniques for semiconductor material improvement.

Exercise Problems

1.(e) Antireflecting coating greatly enhances the transmission of light into a semiconductor (important for photovoltaic devices). The reflectivity for normal incidence of a semiconductor with index of refraction n_{r2}, covered with a transparent layer of thickness d_1 and index of refraction n_{r1}, is given by

$$R_0 = \frac{r_1^2 + r_2^2 + 2r_1 r_2 \cos\theta}{1 + r_1^2 r_2^2 + 2r_1 r_2 \cos\theta} \qquad (10.74)$$

$$\text{with } r_1 = \frac{n_{r0} - n_{r1}}{n_{r0} + n_{r1}}; \quad r_2 = \frac{n_{r1} - n_{r2}}{n_{r1} + n_{r2}}$$

and θ given by

$$\theta = \frac{2\pi n_{r1} d_1}{\lambda}. \qquad (10.75)$$

(a) Determine the condition for d_1, when the reflectivity becomes a minimum.

(b) Calculate the spectral distribution $1.35 < \lambda < 11.0$ μm of the reflectivity when Si is coated with a thin layer of sapphire optimized in thickness for $\lambda = 1.35$ μm. Use values from Tables 10.2 and Appendix A.5.

(c) Give the relation for the index of refraction n_{r1} of the antireflecting layer for vanishing reflection at optimum thickness.

(d) From Table 10.5 select the best antireflecting coating to minimize reflection for Si at its band edge.

2.(e) Derive the wave equation in a semiconductor of finite conductivity, [Eq. (10.18)] and

(a) Discuss the optical refraction and absorption $\alpha_o(\lambda)$ for Si, in the wavelength range $1.35 < \lambda < 11.0$ μm, using the values for n_r as given in Table 10.1.

(b) Discuss the ω dependence of $\tilde{\sigma}$ in the same wavelength range.

(c) At what electron density would the dc-conductivity be equal to the displacement conductivity at the band edge of Si?

3.(r) Calculate and list the reflectance under normal incidence of Si, Ge, GaAs, and GaSb (using n_r and κ). Discuss the trend. Check it for four other semiconductors.

4.(e) Derive the reflectance and transmission of a planar plate [Eqs. (10.54), using Fig. 10.5]. Discuss the error made by not including the phase shift [Eq. (10.56)].

Table 10.5: Refractive indices of antireflecting coatings (after Green, 1982).

Material	n_r	Material	n_r
MgF_2	1.3...1.4	Si_3N_4	1.9
SiO_2	1.4...1.5	TiO_2	2.3
Al_2O_3	1.8...1.9	Ta_2O_5	2.1....2.3
SiO	1.8...1.9	ZnS	2.3...2.4

5. Discuss means to obtain linearly polarized light from elliptically polarized (natural) light. Give quantitative conditions.

6. The absorption coefficient changes over many orders of magnitudes near the absorption edge. Describe as accurately as possible how you would measure $\alpha_o(\lambda)$ of an unknown semiconductor in this range. Describe the experimental setup, the specimen, and its preparation, and what you would expect to measure.

7. What measurements are best suited to obtain the optical constants ε' and ε''. Describe quantitatively how to investigate an Si platelet in the range $0.01 < h\nu < 10$ eV. What type of light source and analyzing instruments would you use?

8.(r) Describe Brewster angle and related angles of similar effects and their usefulness in measuring certain optical parameters.

9.(r) Why are most band structure experiments performed by analyzing the spectral distribution of the reflected rather than the absorbed light?

10.(e) With a light flux of 100 mW/cm²:
(a) How large is the photon flux at 5000 Å?
(b) How large is the photon flux at 5000 Å within a band width of 100 Å in the solar spectrum (use blackbody radiation at 6000 K)?

Chapter 11

Photon-Phonon Interaction

The interaction of photons with lattice oscillations provides very valuable information about the optical phonon branches.

The continuum model presented in the previous sections will now be refined by introducing the atomic microstructure of the semiconductor. In this chapter we will deal with the interaction of photons with lattice oscillation. A description of the relation between electric fields and the lattice polarization provides the foundation for the understanding of this interaction.

11.1 Electric Fields and Lattice Polarization

The dielectric polarization enhances a local electric field $\mathbf{E}_{\mathrm{loc}}$ in the presence of an external field $\mathbf{E}_e$, neglecting space charge effects and expressed as applied voltage divided by distance between electrodes (V/d):

$$\mathbf{E}_{\mathrm{loc}} = \mathbf{E}_e + \mathbf{E}_{\mathrm{surf}}, \qquad (11.1)$$

where $\mathbf{E}_{\mathrm{surf}}$ is the field at the center of a microscopic cavity produced by the induced charges at its surface. This field can be evaluated in an isotropic crystal with a spherical cavity yielding

$$\mathbf{E}_{\mathrm{surf}} = \frac{1}{3\varepsilon_0}\mathbf{P}. \qquad (11.2)$$

For anisotropic crystals the derivation of the local field is more complex (Mueller, 1935; Jackson, 1962). From Eqs. (10.28), (11.1), and (11.2) we obtain the relation between the local and applied fields:

$$\mathbf{E}_{\mathrm{loc}} = \frac{\varepsilon + 2}{3}\mathbf{E}_e. \qquad (11.3)$$

When using the fact that the total polarization per unit volume is composed of N_j atoms with individual polarizability α_j, we obtain

$$\mathbf{P} = \varepsilon_0 \mathbf{E}_{\text{loc}} \sum_j N_j \alpha_j = \varepsilon_0 \frac{\varepsilon + 2}{3} \mathbf{E}_e \sum_j N_j \alpha_j. \qquad (11.4)$$

With Eq. (10.28), we obtain the *Clausius-Mossotti relation*

$$\boxed{\frac{\varepsilon - 1}{\varepsilon + 2} = \frac{1}{3} \sum_j N_j \alpha_j = \frac{1}{3\mathcal{V}} \sum_j \alpha_j} \qquad (11.5)$$

which relates the dielectric constant with the atomic polarization.

The polarizability in ionic crystals is additive. For instance, in a binary compound AB, we have

$$\alpha_{AB} = \alpha_A + \alpha_B. \qquad (11.6)$$

When one atomic polarizability is known (e.g., for Li obtained by Pauling, 1927), one can derive all other ionic polarizabilities from Eqs.(11.5) and (11.6). For a recent review see Shanker et al. (1986).

11.1.1 Ionic and Electronic Polarizability

In (partially) ionic crystals, the atomic polarizability α_j can be divided into an ionic part α_i, determined by the relative shift of oppositely charged ions, and an electronic part α_e, determined by the relative shift of the electrons. The static dielectric constant ε_{st} contains both parts and can be obtained from Eq. (11.5):

$$\frac{\varepsilon_{\text{st}} - 1}{\varepsilon_{\text{st}} + 2} = \frac{1}{3}(N_i \alpha_i + N_e \alpha_e). \qquad (11.7)$$

At high frequencies, where only the electrons can follow a changing external field, we therefore have for the optical dielectric constant

$$\frac{\varepsilon_{\text{opt}} - 1}{\varepsilon_{\text{opt}} + 2} = \frac{1}{3} N_e \alpha_e; \qquad (11.8)$$

consequently, we relate the *ionic polarizability* to the difference of Eqs. (11.7) and (11.8)

$$\alpha_i = \frac{3}{N_i} \left(\frac{\varepsilon_{\text{st}} - 1}{\varepsilon_{\text{st}} + 2} - \frac{\varepsilon_{\text{opt}} - 1}{\varepsilon_{\text{opt}} + 2} \right) \qquad (11.9)$$

a quantity which is on the order of $1/N_i \simeq 10^{-24}$ cm^3.

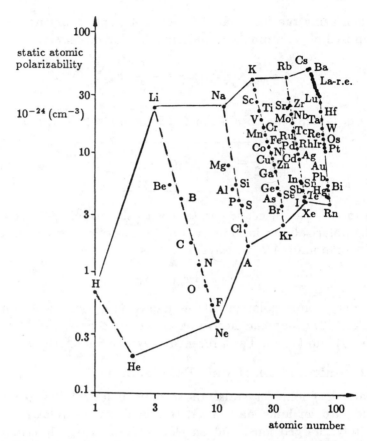

Figure 11.1: Atomic polarization as a function of the atomic number (after Jonscher, 1983).

Table 11.1 gives a list of some ionic polarizabilities. It shows an increase of these polarizabilities with increasing number of electrons within each row of elements (see also Fig. 11.1). For the same electronic shell (e.g., Na^+ and F^-, or K^+, Ca^{++}, and Cl^-) it also increases with increasing nuclear charge.

11.1.2 Piezoelectricity and Electrostriction

Crystals lacking inversion symmetry become electrically polarized when they are elastically strained (*piezoelectricity*), or they change their lattice constant when exposed to an external electric field (*electrostriction*). This effect is used in electromechanical transducers. Examples for such materials are SiO_2 (α-quartz), ADP ($NH_4H_2PO_4$), and KDP (KD_2PO_4). A much smaller quadratic

Table 11.1: Crystalline state polarizability of ions in 10^{-24} cm^3 (after Jaswal and Sharma, 1973; Boswarva, 1970). ©Pergamon Press plc.

Ion	α	Ion	α
Li$^+$	0.029	F$^-$	0.876
Na$^+$	0.285	Cl$^-$	3.005
K$^+$	1.149	Br$^-$	4.168
Rb$^+$	1.707	I$^-$	6.294
Cs$^+$	2.789		
Mg^{++}	0.094	O^{--}	1.657
Ca^{++}	1.157	S^{--}	4.497
Sr^{++}	1.795	Se^{--}	5.686
Ba^{++}	3.188	Te^{--}	9.375

(in **E**) effect is observed in other ionic crystals due to the non-harmonicity of the lattice forces.

If the center of positive and negative charges in a piezoelectric crystal does not coincide under zero external field, the crystal may show a spontaneous electric polarization and is called *ferro-electric*.

The spontaneous polarization may be compensated by charges from the atmosphere absorbed at the surface. In some crystals such compensation may be relieved after heating, when spontaneous polarization again becomes observable. These crystals, of which Turmalin is an example, are called *pyroelectric*. In other materials, e.g., organic waxes, such a net dipole moment may be frozen-in during solidification. These are called *electrete* (Gutman, 1948).

Such state of ordered molecular dipoles is stabilized by its lower total energy. The order is destroyed at temperatures exceeding the critical temperature T_c (*Curie temperature*). Examples of ferro-electric crystals with Curie temperatures (K) and saturation polarization (μ Coul/cm^2)— separated by a comma in parentheses— are KDP (213,9), barium titanate (393,26), and potassium niobate (712,30). For more information see Kittel (1966) and the literature cited there.

11.1.3 Time- and Frequency-Dependent Dielectric Response

The previous discussion of the dielectric polarization indicates a time-delayed response of the semiconductor to changes in the electric field.

Such delay is caused by the time it takes for the different dielectric displacements to adjust; inertia requires that any response cannot be instantaneous. Phenomenologically, we can express the total response as

$$\mathbf{D}(t) = \varepsilon_0 \mathbf{E}(t) + \mathbf{P}(t) = \varepsilon_0 \left[1 + \chi(t)\right] \mathbf{E}(t) \qquad (11.10)$$

with $\mathbf{D}(t)$ the dielectric displacement. The relation between the changing electric field $\mathbf{E}(t)$ causing a change in polarization $\mathbf{P}(t)$ can be related by a *dielectric response function* $f(t)$. Such kinetic responses are discussed in Chapter 47. With respect to optical phenomena, it is more appropriate to transform Eq. (11.10) from the time into the frequency domain by means of a Fourier transformation:

$$P(\omega) = \int_{-\infty}^{\infty} \frac{P(t) \exp(-i\omega t)}{\sqrt{2\pi}} \, dt \quad \text{or} \quad E(\omega) = \int_{-\infty}^{\infty} \frac{E(t) \exp(-i\omega t)}{\sqrt{2\pi}} \, dt.$$

$$(11.11)$$

The frequency-dependent polarization is determined by the frequency-dependent susceptibility which can be expressed in terms of the response function

$$\tilde{\chi}(\omega) = \chi'(\omega) - i\chi''(\omega) = \int_{0}^{\infty} f(t) \exp(-i\omega t) \, dt \qquad (11.12)$$

with

$$\chi'(\omega) = \int_{0}^{\infty} f(t) \cos(\omega t) \, dt \quad \text{and} \quad \chi'(\omega) = \chi'(-\omega) \text{ (even)}$$

$$\chi''(\omega) = \int_{0}^{\infty} f(t) \sin(\omega t) \, dt \quad \text{and} \quad \chi''(\omega) = -\chi''(-\omega) \text{ (odd)}.$$

$$(11.13)$$

or similar relations between $\tilde{\varepsilon}(\omega)$ and f(t). It is the purpose of a microscopic theory to provide such a response function as will be shown for a simple example in Section 11.2.

11.1.4 The Kramers-Kronig Relations

As shown above, $\chi'(\omega)$ and $\chi''(\omega)$, or ε' and ε'' are related to each other. With the help of the Hilbert transformation,

$$P \int_{-\infty}^{\infty} \frac{\sin(\omega t)}{\omega - \omega_0} \, d\omega = \pi \cos(\omega_0 t) \qquad (11.14)$$

such relation can be further developed, as shown below. P is the *Cauchy principal value* of the integral: in order to avoid the singularity at $\omega = \omega_0$, $P\int_{-\infty}^{\infty} \ldots d\omega$ is given by $\lim_{\delta \to 0} \int_{-\infty}^{\omega_a - \delta} \ldots d\omega + \int_{\omega_a + \delta}^{\infty} \ldots d\omega$.

We can eliminate the unknown dielectric response function in Eq. (11.13) by inserting Eq. (11.14) into the first equation of Eqs. (11.13)

$$
\begin{aligned}
\chi'(\omega_a) &= \int_0^{\infty} f(t) \frac{1}{\pi} P \int_{-\infty}^{\infty} \frac{\sin(\omega t)}{\omega - \omega_a} \, d\omega \, dt \\
&= \frac{1}{\pi} P \int_{-\infty}^{\infty} \frac{d\omega}{\omega - \omega_a} \int_{-\infty}^{\infty} f(t) \sin(\omega t) \, dt;
\end{aligned}
\tag{11.15}
$$

the extension of the lower integration limit over ω from 0 to $-\infty$ is permitted for causality reasons: its contribution must be equal to zero. When inserting $\chi''(\omega_a)$ for the second integral in Eq. (11.15) we obtain

$$
\chi'(\omega_a) = \frac{1}{\pi} P \int_{-\infty}^{\infty} \frac{\chi''(\omega)}{\omega - \omega_a} \, d\omega
\tag{11.16}
$$

and in a similar fashion

$$
\chi''(\omega_a) = \frac{1}{\pi} P \int_{-\infty}^{\infty} \frac{\chi'(\omega)}{\omega - \omega_a} \, d\omega.
\tag{11.17}
$$

These equations can easily be transformed* into the more familiar forms of the Kramers-Kronig integrals

$$
\chi'(\omega_a) = \frac{2}{\pi} P \int_0^{\infty} \frac{\omega \chi''(\omega)}{\omega^2 - \omega_a^2} \, d\omega \quad \text{and} \quad \chi''(\omega_a) = \frac{2\omega_a}{\pi} P \int_0^{\infty} \frac{\chi'(\omega)}{\omega^2 - \omega_a^2} \, d\omega.
\tag{11.18}
$$

When using $\bar{\varepsilon}$, we obtain the corresponding Kramers-Kronig relations (Kramers and Kronig, 1929; see, e.g., Toll, 1956; Moss, 1961):

$$
\boxed{\varepsilon'(\omega_a) = 1 + \frac{2}{\pi} P \int_0^{\infty} \frac{\omega \varepsilon''(\omega)}{\omega^2 - \omega_a^2} \, d\omega}
\tag{11.19}
$$

and

$$
\boxed{\varepsilon''(\omega_a) = -\frac{2\omega_a}{\pi} P \int_0^{\infty} \frac{\varepsilon'(\omega)}{\omega^2 - \omega_a^2} \, d\omega,}
\tag{11.20}
$$

---

* This follows from

$$
P \int_{-\infty}^{\infty} \frac{f(x)}{x - a} \, dx = P \int_0^{\infty} \frac{x \left[f(x) - f(-x) \right] + a \left[f(x) + f(-x) \right]}{x^2 - a^2} \, dx.
$$

where ω_a is the arbitrary frequency at which ε' and ε'' are evaluated. Equations (11.19) and (11.20) show not only that ε' and ε'', and thus the extinction coefficient and the dielectric constant depend on each other, but that their values at a given frequency ω_a depend on the behavior of ε' and ε'' in the *entire frequency range*. This interdependence is amplified* when the contributing transition at ω is a resonance transition and lies close to the evaluated transition ω_a.

The dispersion relation given here relates the dispersion process to absorption processes in a single integral formula (Toll, 1956). For instance, it permits us to determine the dispersion at any frequency if we know the absorption in the entire frequency range. However, frequency ranges far away from the range to be evaluated, i.e., far away from ω_a, have little influence.

It also makes immediately clear that there is no dispersion ($\varepsilon' = 1$) if there is no absorption ($\varepsilon''(\omega) = 0$). The shape of $\varepsilon'(\omega)$ depends in a sensitive way on the shape of $\varepsilon''(\omega)$: for instance, an edge in the absorption corresponds to a maximum in the dispersion, while an isolated maximum in absorption corresponds in a decline of ε' with ω.

The interrelation of $\varepsilon'(\omega)$ and $\varepsilon''(\omega)$ in the entire frequency range leads to a number of interesting sum rules.

11.1.5 Sum Rules

Sum rules are helpful in checking the consistency of the approximation used, e.g., if all important transitions are included in the Kramers-Kronig relation. The more important ones are (Landau and Lifshitz, 1958):

$$\int_0^\infty \{n_r(\omega) - 1\}\, d\omega = 0, \qquad (11.21)$$

i.e., the index of refraction, averaged over all frequencies must be equal to one.

When absorption from free electrons is involved (see Section 12.2.1),

$$\int_0^\infty \omega\varepsilon''(\omega)\, d\omega = \frac{\pi}{2}\omega_p^2 \qquad (11.22)$$

---

* Such interdependence can be visualized by considering a row of coupled pendula and forcing one of them to oscillate according to a given driving force. All other pendula will influence the motion, the more so, the closer the forced oscillation is to the resonance frequency of the others.

and

$$\int_0^\infty \omega k(\omega)\, d\omega = \frac{\pi}{4}\omega_p^2, \tag{11.23}$$

where ω_p is the plasma frequency given by Eq. (12.4). The sum rule [Eq. (11.22)] is equivalent to the f-sum rule for atoms (Kronig, 1926). The sum rules indicate that strong emission in one part of the spectrum must be compensated by additional absorption in the same or other parts of the spectrum (Stern, 1963).

A review over more recently discovered sum rules which are helpful for absorption and dispersive processes is given by Smith (1985).

11.2 Semiconductors with Ionic Oscillations

Let us now follow microscopically the interaction of the electromagnetic radiation with atomic oscillations. The interaction can be described classically, and provides a rather simple illustration of connecting the microscopic lattice parameters with optical constants.

In semiconductors with ionic lattice forces, certain lattice oscillations couple strongly with electromagnetic radiation. These are the optical oscillations near $q = 0$ (see Section 5.1.3). They are enhanced by an external electromagnetic field of comparable frequency. In the configuration described previously, i.e, light traveling in x-direction with the electric vector in y-direction, these are *transverse optical lattice oscillations in the center* of the Brillouin zone ($\lambda \gg a$).

These lattice oscillations were discussed in Section 5.1.3 by analyzing the equation of motion. Now an *external force* $e_r E$, caused by an electric field E, is introduced which interacts with the ionic charges of the lattice. Since most semiconductors are partially ionic and partially covalent, the effective charge e_r instead of the electron charge e is used here—see Section 11.2.1. Counteracting the external force are the elastic restoring forces β, and the damping friction term γ which prevents unchecked energy extraction from the external field at resonance. As in classical mechanics the friction is proportional to the velocity. The friction constant is related to absorption (inelastic scattering events) and is equal to $1/\tau$, where τ is the lifetime of the given state, or, as will be discussed later, the appropriate relaxation time (see Section 30.3.3). The equation of motion now reads

$$M_r \left(\frac{d^2 u}{dt^2} - \gamma \frac{du}{dt} + \frac{\beta}{M_r} u \right) = e_r E = e_r E_0 \exp\left[i\left(\mathbf{q} \cdot \mathbf{r} - \omega t \right) \right], \tag{11.24}$$

with a planar electromagnetic wave of frequency ω and wave vector $\mathbf{q}$.

Equation (11.24) has a plane wave solution

$$u(t) = \tilde{A} E_0 \exp\left[i\left(\mathbf{q} \cdot \mathbf{r} - \omega t\right)\right] \qquad (11.25)$$

with the complex amplitude factor

$$\tilde{A}(\omega) = A' + iA'' = \frac{e_r}{M_r} \frac{1}{\omega_{\text{TO}}^2 - \omega^2 + i\gamma\omega}. \qquad (11.26)$$

The derivation of Eq. (11.26) is straightforward and can be found in many textbooks. The relevant eigenfrequency ($\omega_{\text{TO}} \simeq \sqrt{\beta/M_r}$) is the transverse optical frequency at the center of the Brillouin zone (see Section 5.1.3), and describes maximum interaction with the planar electrical wave.

This solution [Eqs. (11.25) and (11.26)] is equivalent to the solution given by Eq. (10.17) with $E_y(t)$ now represented by the ionic displacement $u(t)$.

The macroscopic interaction with light can be described by the dielectric polarization. A crystal containing N ionic oscillators per cm^3 shows a polarization of

$$P = N e_r u - \varepsilon_0(\varepsilon_{\text{opt}} - 1)E, \qquad (11.27)$$

with u as given by Eq. (11.25) and the electronic polarization separated in the second term. When using the complex dielectric constant, P can also be described by

$$P = \varepsilon_0(\tilde{\varepsilon} - 1)E. \qquad (11.28)$$

After eliminating P from Eqs. (11.27) and (11.28), we obtain for the complex dielectric constant

$$\tilde{\varepsilon} = \varepsilon' - i\varepsilon'' = \varepsilon_{\text{opt}} + \frac{\omega_p^2}{\omega_{\text{TO}}^2 - \omega^2 + i\gamma\omega}, \qquad (11.29)$$

where ω_p is introduced as a characteristic frequency, the *ionic plasma frequency**

$$\omega_p = \sqrt{\frac{N e_r^2}{M_r \varepsilon_0}} = 2.94 \cdot 10^{13} \sqrt{\frac{N}{10^{22}} \frac{20}{M_r^*} \frac{e_r}{e}} \quad (\text{s}^{-1}), \qquad (11.30)$$

with M_r^* the effective atomic weight [Eq. (11.40)].

---

* ω_p has the same form as the plasma frequency for electrons [Eq. (12.4)], except N is the density and M_r the mass of phonons.

We can now separate the real and imaginary parts of Eq. (11.29), and obtain

$$\varepsilon'(\omega) = n_r^2 - \kappa^2 = \varepsilon_{opt} + \omega_p^2 \frac{\omega_{TO}^2 - \omega^2}{\left(\omega_{TO}^2 - \omega^2\right)^2 + \gamma^2 \omega^2}, \qquad (11.31)$$

which is proportional to the index of refraction (for $\kappa^2 \ll n_r^2$), and

$$\varepsilon''(\omega) = 2 n_r \kappa = \omega_p^2 \frac{\omega \gamma}{\left(\omega_{TO}^2 - \omega^2\right)^2 + \gamma^2 \omega^2}, \qquad (11.32)$$

which is proportional to the optical absorption (for $n_r \simeq$ const).

Both ε' and ε'' are given as functions of the frequency of the impinging optical wave in Fig. 11.2. It shows an absorption peak at the transverse optical frequency ω_{TO}, with a half-width equal to the damping factor γ (other useful relations are appended):

$$\Delta\omega_{1/2} = \gamma \quad \text{or} \quad \frac{\Delta\omega}{\omega} \simeq \tan\delta = \frac{\varepsilon''}{\varepsilon'} \simeq \frac{\gamma}{\omega_{TO}} \qquad (11.33)$$

The real part of the dielectric constant shows a transition from the optical dielectric constant at high frequencies ($\omega \gg \omega_{TO}$) to the static dielectric constant for low frequencies ($\omega \ll \omega_{TO}$). The amplitudes at the extrema and their interrelationship are given in Fig. 11.2.

The relations derived here are rather general and show a strong absorption at the eigenfrequency of dipole oscillators. The half-width of the absorption peak is proportional to the damping. With $1/\gamma$ given by the lifetime of the state, the half-width of the line is also referred to as its *natural line width*. Often the *Lorentzian line shape function* $g(\omega)$ is used to describe the spectral distribution of the optical absorption near ω_{TO} (see Section 20.3.1A):

$$g(\omega) = \frac{\gamma}{2\pi} \frac{1}{(\Delta\omega)^2 + \left(\dfrac{\gamma}{2}\right)^2}; \qquad (11.34)$$

with $g(\omega)$, Eqs. (11.31) and (11.32) can be approximated for $\omega \simeq \omega_{TO}$ as

$$\varepsilon' = \varepsilon_{opt} + \frac{\pi \omega_p^2}{\omega_{TO}} \frac{\Delta\omega}{\gamma} g(\omega) \quad \text{and} \quad \varepsilon'' = \frac{\pi \omega_p^2}{2\omega_{TO}} g(\omega). \qquad (11.35)$$

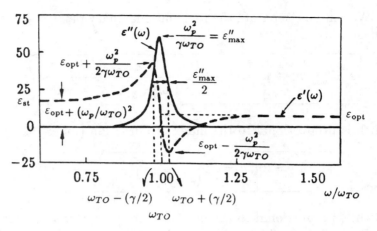

Figure 11.2: Real (dashed curve) and imaginary (solid curve) parts of the dielectric constant calculated for a hypothetical crystal from Eqs. (11.31) and (11.32) with $\varepsilon_{\text{opt}} = 12$, $\omega_p/\omega_{\text{TO}} = 1.7$, and $\gamma/\omega_{\text{TO}} = 0.05$.

A few additional relations are instructive. For low frequencies, we obtain from Eq. (11.31) the relation

$$\varepsilon'(0) = \varepsilon_{\text{st}} = \varepsilon_{\text{opt}} + \frac{\omega_p^2}{\omega_{\text{TO}}^2}. \tag{11.36}$$

Therefore, Eq. (11.29) can also be written as

$$\tilde{\varepsilon} = \varepsilon_{\text{opt}} + \frac{\varepsilon_{\text{st}} - \varepsilon_{\text{opt}}}{1 - \dfrac{\omega^2}{\omega_{\text{TO}}^2} - i\dfrac{\gamma\omega}{\omega_{\text{TO}}^2}}. \tag{11.37}$$

The corresponding relation for a *longitudinal wave* (which is not influenced by a transverse electromagnetic wave) can be obtained by setting $\tilde{\varepsilon} = 0$ in Eq. (11.29)—see Fig. 11.3. Neglecting damping ($\gamma = 0$), we obtain from Eq. (11.37) for the optical longitudinal wave at $q = 0$ ($\omega = \omega_{\text{LO}}$) the useful *Lyddane-Sachs-Teller relationship* (Lyddane et al., 1959)

$$\frac{\omega_{\text{LO}}}{\omega_{\text{TO}}} = \sqrt{\frac{\varepsilon_{\text{st}}}{\varepsilon_{\text{opt}}}}. \tag{11.38}$$

For resonances including damping, the Lyddane-Sachs-Teller equation requires some modification—see Chang et al. (1968). Both

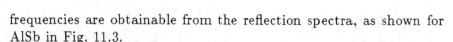

frequencies are obtainable from the reflection spectra, as shown for AlSb in Fig. 11.3.

In covalent monatomic crystals, with vanishing dipole moment, $\varepsilon_{st} = \varepsilon_{opt}$; hence $\omega_{TO} = \omega_{LO}$, which for these materials is usually denoted as ω_0.

11.2.1 Effective Charges

The ionicity of a lattice bond with partially ionic and partially covalent binding forces can be identified by a **dynamic effective charge**. Such effective charge was used in the preceding section. Since kinetic effects are involved, this charge is slightly different from the static effective charge introduced in Section 2.3. Three types of **dynamic effective charges** are used. The **Callen effective charge** (Callen, 1949) is given as

$$e_c = \omega_{TO} \sqrt{\varepsilon_0 \frac{\varepsilon_{st} - \varepsilon_{opt}}{\varepsilon_{opt}^2} M_r a^3} \qquad (11.39)$$

where a is the interionic distance and M_r is the **reduced ion mass**[*]:

$$\frac{1}{M_r} = \frac{1}{M_+} + \frac{1}{M_-}. \qquad (11.41)$$

The **Szigetti effective charge** is related to e_c by

$$e_s = e_c \frac{3\varepsilon_{opt}}{\varepsilon_{st} + 2}. \qquad (11.42)$$

The **macroscopic transverse effective charge** is related to the other effective charges by

$$e_t = \frac{\varepsilon_{opt} + 2}{3} e_s = \frac{\varepsilon_{opt} + 2}{\varepsilon_{st} + 2} \varepsilon_{opt} e_c. \qquad (11.43)$$

All of these dynamic effective charges are smaller than e for typical semiconductors. Crystals with partially covalent and partially ionic binding forces can be described as crystals composed of ions, but with a fractional (i.e., an effective) charge. For example, the parameters for GaAs have the following values: the lattice constant

---

[*] Sometimes the *effective atomic weight* is used:

$$M_r^* = \frac{M_r}{M_H}, \qquad (11.40)$$

with M_H the mass of the hydrogen atom.

is 5.65 Å; the reduced mass is $3.02 \cdot 10^{-30}$ Ws^{3}cm^{-2}; $\varepsilon_{\text{opt}} = 10.8$; $\varepsilon_{\text{st}} = 12.5$; $\omega_0 = k\Theta/\hbar = 1.308 \cdot 10^{11}\Theta = 5.45 \cdot 10^{13}$ s^{-1}; and, using Eq. (11.36), $\omega_{\text{TO}} = 5.863 \cdot 10^{13}$s^{-1}; hence, $e_c/e = 0.17$. The relative Szigetti effective charge of GaAs is $e_s/e = 0.38$.

Occasionally, one uses the **effective field** to describe the electron-lattice interaction:

$$F_{\text{eff}} = \frac{em_n k\Theta}{4\pi\hbar^2} \frac{1}{\varepsilon_0\varepsilon^*};\tag{11.44}$$

this field can be interpreted as the Debye potential $(k\Theta/e)$ acting across $4\pi a_{\text{qH}}$ with a_{qH} the quasi-hydrogen radius [Eq. (15.3)], using ε^*, however, rather than ε; typically, F_{eff} is on the order of 10 kV/cm.

11.2.2 IR Absorption, Reflection, Reststrahlen

The optical eigenfrequencies of typical semiconductor lattices are on the order of 10^{13} s^{-1}; that is, they are in the 20...60 μm IR range. As discussed in the previous section, only the TO branch of a diatomic lattice with nonvanishing dipole moment is *optically active*; it absorbs light.

Since the absorption coefficient is rather large, it is difficult to measure this coefficient in absorption: it requires a very thin layer. The change in the reflectance is more easily accessible. For a sufficiently thick sample, so that neglecting the reflection at the back surface is permissible, the coefficient is given by Eq. (10.53). The reflectance shows a steep rise at ω_{TO}, followed by a sharp drop-off with a minimum at $\omega = \omega_{\text{LO}}$ (Fig. 11.3). The reflectance spectrum, although similar to $\varepsilon'(\omega)$, shows distinct differences in its quantitative behavior, as seen by comparing Eq. (11.31) with Eq. (11.59) and the dashed curve in Fig. 11.2 with Fig. 11.3.

The observed reflectivity spectrum for typical diatomic semiconductors follows much of the simple theory given above, as shown in Fig. 11.3B.

The range of high optical reflection can be used to obtain nearly monochromatic IR light when employing multiple reflections (Fig. 11.4). In each reflection, one loses a fraction of the nonreflected (i.e., the transmitted) light leaving very little of it after several of these steps. This method is called the *Reststrahlen* method (from "remaining beam" in German). The *Reststrahlen* frequency is identified by ω_{TO} (see Fig. 11.3), since for reasonable damping factors

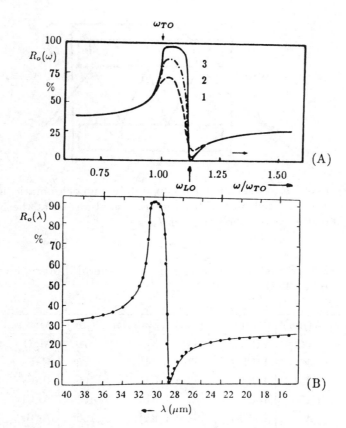

Figure 11.3: (A) Reflectance of a semi-infinite plate in the range of an absorption maximum for a hypothetical crystal with $\varepsilon_{opt} = 12$, $\varepsilon_{st} = 15$, $\omega_p/\omega_{TO} = 1.7$, and $\gamma/\omega_{TO} = 0.05$, 0.02, and 0.004 for curves (1)–(3), respectively (after Hass, 1964). (B) Measured IR reflection spectrum for AlSb (after Turner and Reese, 1962).

the reflection maximum lies only slightly above ω_{TO}, which can be approximated by

$$\omega_{TO} \simeq \sqrt{\frac{\beta}{M_r}}. \tag{11.45}$$

Here β can be estimated for an ionic crystal from the Coulomb potential

$$\beta \simeq \frac{Z^2 e^2}{4\pi\varepsilon_0 a_0^2}, \tag{11.46}$$

with Z as the ionic charge and a_0 as the interionic equilibrium distance. M_r is given by Eq. (11.41).

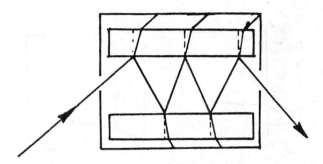

Figure 11.4: *Reststrahlen* set-up to obtain monochromatic light near ω_{TO}. The unreflected part of the spectrum is absorbed at the black wall of the box after penetration through each slab.

Table 11.2: *Reststrahl* wavelengths for several binary compounds (after Bube, 1988).

Material	$\lambda\,(\mu m)$	Material	$\lambda\,(\mu m)$	Material	$\lambda\,(\mu m)$	Material	$\lambda\,(\mu m)$
SiC	12.5	LiF	33	KCl	70	MgO	25
GaAs	38	LiCl	53	KBr	88	CaF_2	39
ZnS	35	LiBr	63	KI	100	BaF_2	53
ZnSe	47	NaF	42	RbF	65	AgCl	100
CdS	37	NaCl	61	RbI	135	AgBr	126
CdTe	70	NaBr	76	CsCl	100	TlCl	158
LiII	17	KF	53	CsI	158	TlBr	233

Table 11.2 lists the *Reststrahl wavelength* for a number of crystals. As can be see from Eqs. (11.45) and (11.46), the *Reststrahl* frequency increases, i.e., its wavelength decreases with increasing ionic charge, decreasing interionic distance, and decreasing reduced ion mass.

11.3 The Dispersion Relation Near q=0

In the previous section, the interaction of an electromagnetic wave with TO phonons at $q = 0$ was discussed. We will now analyze this interaction more precisely. In actuality, photons have a finite momentum h/λ which is very small compared to the extent of the Brillouin zone $\hbar\pi/a$. For IR light, the ratio $a/2\lambda$ is on the order of 10^{-5}.

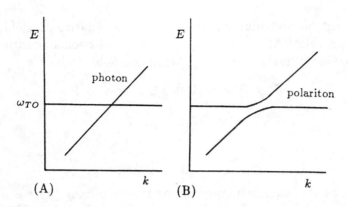

Figure 11.5: Dispersion relation $E(k)$ of simple oscillatory phenomena; (A) without and (B) with interaction.

When enlarging the $E(q)$ diagram, shown for phonons in Fig. 5.8, for the TO branch near $q = 0$, we obtain a horizontal line. For the photon we have another straight line according to

$$\boxed{p_{\text{phot}} = \frac{h}{\lambda} = \frac{h\nu}{c} = \hbar k.}$$ (11.47)

Both lines cross as shown in Fig. 11.5A. In Fig. 11.5, k is used instead of q to emphasize the interaction with light.

Since there is interaction between the photon and the phonon, we observe a characteristic split in the dispersion spectrum according to the *von Neumann noncrossing principle*. This split occurs near the $E(k)$ value where the dispersion curves of the two quasi-particles would cross if they did not interact with each other. This is shown schematically in Fig. 11.5B, and will become more transparent from a quantitative analysis.

11.3.1 The Phonon-Polariton

Since photons and phonons are bosons, the interaction between the two leads to a state that can no longer be distinguished as a photon or a phonon. To emphasize this, a new name is used to describe this state: the *polariton* or, more precisely, the *phonon-polariton*, when mixing between a *phonon* and a photon state occurs, to set it apart from the *exciton-polariton* (Section 15.1) or *plasmon-polariton* (Section 17.2.1).

The set of governing equations for such polaritons, yielding the dispersion relation, can be obtained from Maxwell's equations and the equation of motion in the polarization field

$$\nabla \times \mathbf{H} = \varepsilon\varepsilon_0 \dot{\mathbf{E}} + \sigma \mathbf{E} = \varepsilon\varepsilon_0 \dot{\mathbf{E}} + \dot{\mathbf{P}} \qquad (11.48)$$

$$\nabla \times \dot{\mathbf{E}} = -\mu\mu_0 \dot{\mathbf{H}} \qquad (11.49)$$

$$\ddot{\mathbf{P}} + \omega_0^2 \mathbf{P} = \chi\varepsilon_0 \mathbf{E}, \qquad (11.50)$$

with ω_0 as the eigenfrequency of the free oscillator (here a phonon, later in our discussion it could be an exciton or a plasmon). These equations define the electric, magnetic, and polarization fields. Assuming linear polarized planar waves, propagating in the x-direction with $\mathbf{E} = (0, E_y, 0)$, $\mathbf{H} = (0, 0, H_z)$, and $\mathbf{P} = (0, P_y, 0)$, we obtain after insertion of

$$E_y = E_{y0} \exp\left[i\left(kx - \omega t\right)\right] \qquad (11.51)$$

$$H_z = H_{z0} \exp\left[i\left(kx - \omega t\right)\right] \qquad (11.52)$$

$$P_y = P_{y0} \exp\left[i\left(kx - \omega t\right)\right] \qquad (11.53)$$

into Eqs. (11.48)–(11.50), the following governing equations for $\omega(k)$:

$$\omega\varepsilon\varepsilon_0 E_y + \omega P_y - k H_z = 0 \qquad (11.54)$$

$$k E_y - \omega\mu\mu_0 H_z = 0 \qquad (11.55)$$

$$\chi\varepsilon_0 E_y + (\omega^2 - \omega_0^2) P_y = 0. \qquad (11.56)$$

These yield the dispersion equation for polaritons after eliminating E_y, P_y, and H_z:

$$\omega^4 - \left(\omega_0^2 + \frac{\chi}{\varepsilon} + \frac{k^2}{\varepsilon\varepsilon_0\mu\mu_0}\right)\omega^2 + \frac{k^2\omega_0^2}{\varepsilon\varepsilon_0\mu\mu_0} = 0. \qquad (11.57)$$

Entering the resonance frequencies for phonons and using the Lyddane-Sachs-Teller relation [Eq. (11.38)], we obtain for semiconductors with $\mu = 1$ the *phonon-polariton equation* (after using $\varepsilon_0\mu_0 c^2 = 1$):

$$\boxed{\omega^4 - \omega^2\left(\omega_{\text{LO}}^2 + \frac{c^2 k^2}{\varepsilon_{\text{opt}}}\right) + \frac{c^2 k^2}{\varepsilon_{\text{opt}}}\omega_{\text{TO}}^2 = 0.} \qquad (11.58)$$

The resulting dispersion curves are shown in Fig. 11.6.

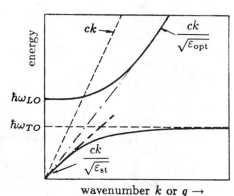

Figure 11.6: Dispersion curve for the phonon-polariton with appropriate slopes indicated. Dashed line ck is for photons *in vacuo*.

wavenumber k or $q \rightarrow$

The slope of the longitudinal branch approaches $c/\sqrt{\varepsilon_{\text{opt}}}$, i.e., it has a slightly larger slope than the slope of the transverse branch near $k = 0$, which is $c/\sqrt{\varepsilon_{\text{st}}}$.

The energy of the phonon-polariton is typically on the order of 50 meV. The first direct observation of the lower branch of this dispersion spectrum was published by Henry and Hopfield (1965) and is shown for GaP in Fig. 11.7. The measurement involves Raman scattering and will be discussed in more detail in Section 17.1.3F. The upper branch was measured for GaP by Fornari and Pagannone (1978). For more on the polariton see Section 17.1.3.

11.4 One- and Multiphonon Absorption

The *one-phonon spectrum* provides information only very near to the Γ point, i.e., the center of the Brillouin zone—see Section 11.2. A *multiphonon* interaction, however, yields a wealth of information about the phonon spectrum in the entire Brillouin zone, since the second phonon can provide the necessary momentum shift.

The second phonon interacts because of the anharmonicity of the lattice potential, which results in a *weak* optical absorption. Such an interaction takes place when all three (or more) particles fulfill energy and momentum conservation:

$$h\nu = \sum_i s_i \hbar \omega_i \quad \text{and} \quad \sum_i s_i q_i = 0, \qquad (11.59)$$

where $s_i = \pm 1$, and positive or negative signs in the sum represent simultaneous absorption or emission of phonons. An example for a two-phonon absorption with momentum conservation is shown in Fig. 11.8.

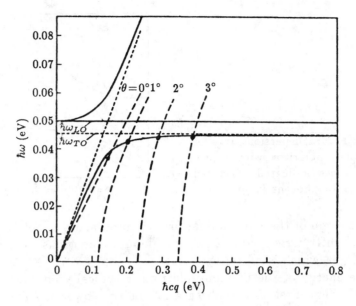

Figure 11.7: Dispersion curves of the LO and TO phonons of GaP and the 6328 Å photon (dashed) at long wavelengths (small q values), and of the polariton (solid curves). Raman scattering angles are indicated (after Henry and Hopfield, 1965). Observe that the edge of the Brillouin zone is at $(\pi/a = 5.76 \cdot 10^7 \text{ cm}^{-1})$ $\hbar cq = 1137$ eV, i.e., far to the right in the scale given for the abscissa.

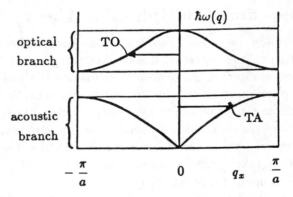

Figure 11.8: Schematics of a two-phonon absorption from a photon (of $k = 0$) with conservation of momentum.

Stronger absorption peaks are obtained at critical points—that is, where the slopes of the two (or more) $\omega(\mathbf{q})$ branches of the involved phonons are nearly horizontal; in other words, when the density of such phonons is relatively high—see Section 5.1.5C.

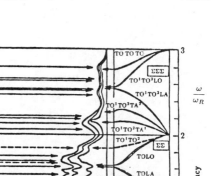

Figure 11.9: Multiphonon absorption in Si correlated to the combined density of states along the Δ, Λ, and Σ axes, showing maxima near the critical points L, X, W, and Σ; measured at three different temperatures; $\omega_R = 15.5 \cdot 10^{12}\,\mathrm{s}^{-1}$ used for normalizing the ordinate, is the Raman frequency (after Johnson, 1959; Bilz et al., 1963).

The multiphonon processes are observable in materials in which the much stronger one-phonon absorption does not occur because of selection rules, e.g., in Si and Ge, as the latter would hide the weaker multiphonon absorptions. In Fig. 11.9 a typical multiphonon absorption spectrum is given for Si. The correlation with the phonon dispersion spectrum is indicated by horizontal arrows. In the lower and upper part of the figure are two-phonon and three-phonon processes, respectively. The corresponding positions are indicated in the dispersion curves at the left and right sides of the absorption spectrum.

11.4.1 Local Mode Absorption

Another important example of ion-induced optical absorption is caused by local mode ionic resonances—see Section 5.1.6. This absorption provides some of the most direct information about certain defect centers. For instance, an Si atom can be incorporated in a GaAs lattice on a Ga or As site as a donor or acceptor, respectively.

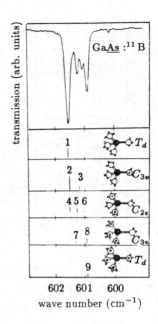

Figure 11.10: IR absorption spectrum of an $^{11}B_{As}$ substitutional in GaAs with ^{69}Ga and ^{71}Ga identified as open or dotted circles. Point group symmetry and corresponding line split identified in lower panels. Relative line heights represent probability of occurrence (after Talwar et al., 1986).

The lattice resonances, observed as local vibrational modes, are distinctly different in both cases (47.6 or 49.5 meV, respectively), since the bonding forces are different for an Si atom surrounded by Ga atoms or by As atoms, and provide means for positive identification. In addition, the fact can be used that two Ga isotopes [^{69}Ga (60.4%) and ^{71}Ga (39.6%)] are statistically distributed, surrounding the Si_{As} substitutional. This results in characteristic anisotropies, splitting the resonance into three lines which can be resolved (Thies et al., 1983). Other changes can be detected when using ^{29}Si or ^{30}Si instead of ^{28}Si (Talwar et al., 1986).

In GaAs:$^{11}B_{As}$, four lines of the local vibrational modes can be resolved (Fig. 11.10) by high-resolution Fourier-transformed infrared absorption spectroscopy, and are related to the isotope arrangement surrounding the B substitutional (Talwar et al., 1986).

The resonance structure can be calculated by using molecular modes, the Bethe lattice approximation, or the Green's function method. Comparison of the results with the experiment lays the groundwork for a better understanding of the actual incorporation of such defect centers into the lattice, and also permits observation of changes in the binding forces when the charge is altered in the defect center. Furthermore, it indicates softening or stiffening of bonds when a given impurity occupies different lattice sites and assumes different charge character (Talwar et al., 1986).

Summary and Emphasis

We have shown that photons and TO phonons interact strongly near $q = 0$, resulting in a large absorption, which is characteristic for the atomic mass, bonding force and effective ionic charge of the specific semiconductor. This absorption is located in the $10 - 100$ μ range, and is referred to as *Reststrahl* absorption.

In semiconductors with vanishing ionic charges, photons cannot excite single phonons. However, a much weaker optical absorption is observed by simultaneous interaction with two or more phonons, thereby offering for such materials the opportunity to obtain additional (to neutron spectroscopy) information about the entire phonon-dispersion spectrum.

Still more important is the local mode spectroscopy which offers detailed information about the lattice surrounding of defects which have modes in the gap between acoustic and optical modes of the host lattice. High-resolution spectroscopy permits distinction between different isotope neighbors and provides means for the identification of such defects.

Phonon spectroscopy offers access to direct information about the vibrational spectrum of a semiconductor and many of its defects. Such information is helpful in judging bonding forces and effective ionic charges of crystal lattices and the local structure of certain crystal defects. Such knowledge, aside from its basic interest, is of great value for device optimization.

Exercise Problems

1.(e) Derive the Lyddane-Sachs-Teller relationship [Eq. (11.38)].

2.(*) Discuss the sum rules in relation to optical absorption and emission. Check the literature (e.g., Smith, 1985) for more sum rules and discuss their physical meanings.

3.(e) The field in a spherical microscopic cavity is usually given as

$$\mathbf{E}_{\text{surf}} = \frac{4\pi}{3}\mathbf{P}. \qquad (11.60)$$

The different factor given in Eq. (11.2) is due to a different system of units used. Which are the two systems and how are they related? Elaborate (see Appendix A.1).

4.(l) Equation (11.2) needs to be modified when the shape of the cavity is nonspherical.
(a) Why are other shapes of interest?

(b) What are the corresponding modifications?

5.(*) When discussing ionic polarization and the related dispersion relation, we did not explicitly mention electronic polarization. What is its influence in the range of *Reststrahl* absorption?

6.(*) What reasons are there to discuss the dielectric response in the time domain, and what reasons entice preferred discussion in the frequency domain?

7.(e) Develop the amplitude function Eq. (11.26) with the trial solution Eq. (11.25) from Eq. (11.24).

8. Reconcile Eq. (11.27) with the division between ionic and electronic parts of the polarization given in Section 11.1.1.

9.(e) Show that the half-width of a Lorentzian absorption line is equal to the damping factor in Eq. (11.24).

10.(e) The Lyddane-Sachs-Teller relationship is developed without damping. What influence would damping have?

11.(e) Derive the spring constant [Eq. (11.46)] from the Coulomb potential and give these for five compounds of Table 11.2. Compare these with the tabulated elastic stiffness constants after appropriate transformation.

12. Estimate the line spacing for the four lines of local modes in GaAs:$^{11}B_{As}$ with the two isotopes ^{69}Ga and ^{71}Ga (see Fig. 11.10) for the defect geometry.

Chapter 12

Photon–Free-Electron Interaction

The interaction of photons with free electrons or holes in the respective bands reveals the basic differences of the inter and intraband excitation, the differences between single and multiple particle processes, and the differences between resonance and nonresonance absorption.

The optical excitation of electrons or holes within their respective bands may involve a large variety of excitation mechanisms and multiparticle influences during the excitation process, such as:

- direct (vertical) transition from one into another branch of the same band;
- indirect transition with phonon assistance;
- transitions followed by other inelastic scattering processes;
- transitions involving single electrons;
- transitions involving collective effects of electrons (plasmons);
- resonance and nonresonance transitions; and
- transitions between subbands created by a magnetic field.

These processes will be discussed in the following sections.

12.1 Free-Electron Resonance Absorption

The absorption from free electrons in the conduction band or holes in the valence band can result in resonant or nonresonant absorption. The former is discussed first because of the similarities in the mathematical treatment with the ionic oscillations discussed earlier.

The logical extension of the field-induced lattice oscillation is the polarization of the *electronic shell* surrounding each lattice atom in an external electromagnetic field. In a classical model, the shift of the electron cloud with respect to the nucleus is proportional to the electric field and given by the electronic polarizability α_e. The resulting dispersion formula is similar to Eq. (11.29), except for a different eigenfrequency and damping factor relating to the electronic shell. However, the computation of these parameters requires a quantum mechanical analysis, considering electronic transitions between different bands. This will be discussed in Section 13.1.1. On the other hand, collective electronic resonances can be well described in a classical model and will be discussed first. When the electron density is large enough, such electron-collective effects cause resonance absorption, which for semiconductors, lie in the IR range beyond the band edge.

12.1.1 Electron Plasma Absorption

In an external field, the electrons act jointly when they are shifted as an entity with respect to the ionized donors. The field E_x exerts a force $-eE_x$ and causes a shift by Δx from their on-the-average neutral position, creating a net charge at the two outer surfaces, which can be calculated by integrating the Poisson equation

$$\delta E_x = \frac{\varrho}{\varepsilon \varepsilon_0} \Delta x = \frac{ne}{\varepsilon \varepsilon_0} \Delta x. \tag{12.1}$$

The induced field δE_x counteracts the driving field E_x; that is, $e\delta E_x$ acts as restoring force, causing oscillations of the electron-collective, following the equation of motion (Pines, 1963):

$$nm_n \frac{d^2 \Delta x}{dt^2} - nm_n \gamma \frac{d\Delta x}{dt} = -ne\delta E_x = -\frac{n^2 e^2}{\varepsilon_{\text{opt}} \varepsilon_0} \Delta x. \tag{12.2}$$

We now have introduced the optical dielectric constant, since we expect oscillations between the band edge and the *Reststrahl* frequency. With a driving external field, this equation can be modified as a damped harmonic oscillator equation [similar to Eq. (11.24)]

$$m_n \left(\frac{d^2 \Delta x}{dt^2} - \gamma \frac{d\Delta x}{dt} + \omega_p^2 \Delta x \right) = eE_x \exp(i\omega t). \tag{12.3}$$

The damping of the electron gas can be related to the energy relaxation time* ($\gamma = 1/\tau_e$); ω_p is the *plasma frequency* [obtained by comparing Eqs. (12.2) and (12.3)]:

$$\omega_p = \sqrt{\frac{ne^2}{\varepsilon_{\text{opt}} \varepsilon_0 m_n}}. \tag{12.4}$$

Quantized, $\hbar\omega_p$ is referred to as a *plasmon*:

$$\hbar\omega_p = 38.9 \sqrt{\frac{n}{10^{16}} \frac{10}{\varepsilon_{\text{opt}}} \frac{m_0}{m_n}} \quad (\text{meV}). \tag{12.5}$$

Equation (12.3) has an oscillatory solution with an amplitude similar to Eq. (11.26):

$$\Delta x = \frac{eE_x}{m} \frac{1}{\omega_p^2 - \omega^2 - \dfrac{i\omega}{\tau_e}}. \tag{12.6}$$

This amplitude can be related to the polarizability by

$$P_x = \frac{ne}{\varepsilon_{\text{opt}}} \Delta x. \tag{12.7}$$

From Eqs. (12.6), (12.7), and (11.28) with (12.4), we obtain for the complex dielectric constant

$$\tilde{\varepsilon} = \varepsilon' + i\varepsilon'' = \varepsilon_{\text{opt}} + \frac{\omega_p^2}{\omega_p^2 - \omega^2 - \dfrac{i\omega}{\tau_e}}, \tag{12.8}$$

which can be separated into real and imaginary parts as

$$\varepsilon' = \varepsilon_{\text{opt}} + \omega_p^2 \left(\frac{\left(\omega_p^2 - \omega^2\right) \tau_e^2}{\left(\omega_p^2 - \omega^2\right)^2 \tau_e^2 + \omega^2} \right) \tag{12.9}$$

and

$$\varepsilon'' = \omega_p^2 \frac{\omega\tau_e}{\left(\omega_p^2 - \omega^2\right)^2 \tau_e^2 + \omega^2}. \tag{12.10}$$

---

* The energy relaxation time is used here: the additional energy is dissipated by plasmons. The increment in electron energy is on the average larger than the LO phonon energy. For further justification see Section 33.2.2.

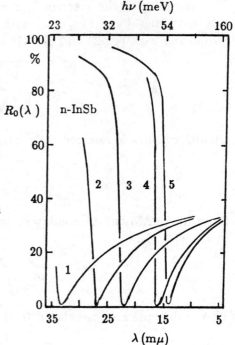

Figure 12.1: Reflection spectrum of *n*-type InSb which shows the typical dispersion behavior given in Fig. 11.2. Family parameters are the different electron densities: $n = 3.5 \cdot 10^{17}$, $6.5 \cdot 10^{17}$, $1.2 \cdot 10^{18}$, $2.8 \cdot 10^{18}$, and $4 \cdot 10^{18}$ for curves 1–5, respectively; $T = 295$ K (after Spitzer and Fan, 1957).

Equations (12.9) and (12.10) have a form similar to Eqs. (11.31) and (11.32), except that the resonance frequency is given by the plasma frequency ω_p, and the damping factor can be identified as the reciprocal electron relaxation time τ_e.

In the frequency range where $n_r^2 \gg \kappa^2$ (see Fig. 10.6), we can experimentally obtain ω_p and ε_{opt} by measuring the reflectance [Eq. (10.53)]

$$R_0 \simeq \frac{(n_r - 1)^2}{(n_r + 1)^2} \quad \text{or} \quad n_r = \frac{1 + \sqrt{R_0}}{1 - \sqrt{R_0}}. \tag{12.11}$$

The index of refraction as a function of ω is given by

$$\boxed{n_r^2 \simeq \varepsilon_{\text{opt}}\left(1 - \frac{\omega_p^2}{\omega^2}\right).} \tag{12.12}$$

Table 12.1: Valence electron plasmon energies.

Material	$\hbar\omega_p$ (eV)	Material	$\hbar\omega_p$ (eV)	Material	$\hbar\omega_p$ (eV)
Si	16.4 ... 16.9	GaP	16.6	InP	14.8
Ge	16 ... 16.4	GaAs	15.8	InSb	12.8

The frequencies $\omega_{\min}$ and $\omega_{\max}$ at which the reflection minimum and maximum occur (Fig. 12.1) are related to ω_p and ε_{opt} as [using Eqs. (12.12) and (12.11)]

$$\omega_{\min} = \omega_p \sqrt{\frac{\varepsilon_{\text{opt}}}{\varepsilon_{\text{opt}} - 1}} \quad \text{and} \quad \omega_{\max} \simeq \omega_p, \tag{12.13}$$

for $R_0 \ll 1$ and $R_0 \simeq 1$ with $n_r \simeq 1$ and $n_r \simeq 0$, respectively. With known ε_{opt}, we have a useful method for determining ω_p, and with it, the effective carrier mass in ω_p [Eq. (12.4)], provided the carrier density is known by independent means. Typical reflection spectra observed for InSb are shown in Fig. 12.1, with reflection minima shifting to higher energies with increasing carrier density according to Eqs. (12.4) and (12.13).

12.1.2 Valence Electron Plasma Absorption

The plasmon absorption discussed before is caused by free electrons in the conduction band. In addition, we observe valence electron plasmons when all the valence electrons oscillate with respect to the cores. The mathematical theory is quite similar to the one given in the previous section, except that in the expression for the plasmon frequency the density of all valence electrons (typically 10^{23} cm^{-3}) is entered. Each of these electrons is elastically bound to the core ions. This results in a plasmon frequency of $\sim 10^{16}$ s^{-1}, and a plasmon energy of ~ 10 eV.

The valence plasmon absorption can be measured as distinct losses when electrons penetrate through a thin layer of the semiconductor. We must distinguish between longitudinal and transverse plasmons. The former interact strongly with electrons, the latter with electromagnetic radiation. Typical plasmon energies are given in Table 12.1.

12.1.3 Charge Density Waves

In materials with a high density of free electrons (metals or degenerate semiconductors) the conduction electron charge density, which is usually constant in a homogeneous material, can undergo a wave instability. Then the charge density becomes sinusoidally modulated in space (Overhauser, 1978) with an extra periodicity not related to the lattice periodicity:

$$\varrho(\mathbf{r}) = \varrho_0(\mathbf{r})\left[1 + A\cos\left(\hat{\mathbf{q}}\cdot\mathbf{r} + \phi\right)\right] \qquad (12.14)$$

where A is the amplitude (typically ~ 0.1) and $\hat{\mathbf{q}}$ is the wave vector of the charge density wave

$$\hat{q} \simeq \frac{2p_F}{\hbar} \qquad (12.15)$$

with p_F as the momentum at the (here assumed to be spherical) Fermi surface. In Eq. (12.14) ϕ is the phase. The charge density wave is caused by the interaction between the electrons, which can be described by the exchange energy (Pauli principle) and the correlation energy (electron-electron scattering). Wave formation tends to reduce these contributions. The Coulomb interaction opposes the above-mentioned effects and suppresses wave formation. Additional interaction with the lattice cancels part of the suppression; hence in lattices with small elastic moduli, which could interact more readily, such charge density waves are more likely to occur. Here, the lattice ions are also slightly displaced with an average amplitude of $\sim 0.01\ldots 0.1$ Å and with the periodicity of the charge density wave (Overhauser, 1978).

These waves are observed, e.g., in TaS_2 and $TaSe_2$, as two satellites to the Bragg reflection, caused by the slight lattice displacement (Wilson et. al., 1975).

A phase modulation, when quantized, can yield low energy collective excitation spectra $\hbar\omega$ from

$$\phi = \phi(\mathbf{r}, t) \propto \sin(\hat{\mathbf{q}}\cdot\mathbf{r} - \omega t). \qquad (12.16)$$

These are called *phasons* and are observed in $LaGe_2$. At low temperatures they could have a measurable effect on electron scattering (Huberman and Overhauser, 1982).

12.2 Nonresonant Free-Electron Absorption

At lower densities of free electrons, the plasma frequency is shifted into and below the lattice resonances, and the amplitude of the absorption, which is $\propto \omega_p^2$, is much reduced. Far away from the resonance transition, we observe the *nonresonant* part of the *absorption* tailing from the resonance peak through the extrinsic optical absorption range and extending toward the band edge.

12.2.1 Free-Carrier Absorption

The optical absorption and reflection induced by free carriers outside of the resonance absorption provide valuable information about carrier relaxation and, for holes, about the band structure near $\mathbf{k} = 0$.

12.2.1A Dispersion Relation for Free Carriers The relationship between free carriers and an external field is the same as developed in Section 12.1.1 (Lax, 1963), except that here the restoring forces are negligible and electrons respond only to external electromagnetic forces and damping:

$$m_n \left(\frac{d^2 \Delta x}{dt^2} + \gamma \frac{d\Delta x}{dt} \right) = -eE_x \exp\left(-i\omega t \right). \tag{12.17}$$

This provides an independent opportunity to measure by optical absorption the electron scattering, and will be discussed in more detail in the following section. Equation (12.17) has the solution

$$\frac{d\Delta x}{dt} = v_x = -\frac{eE_x}{m_n} \frac{\tau_e}{1 - i\omega\tau_e}. \tag{12.18}$$

This approach is similar to the discussion in Section 11.2, and is used for educational purposes in order to arrive directly at the complex conductivity. One can use this velocity to define an electronic current[*]

$$\tilde{j}_x = -en\tilde{v}_x = \tilde{\sigma}E_x, \tag{12.19}$$

with a complex conductivity

$$\begin{aligned}
\tilde{\sigma} = \sigma' + i\sigma'' &= \frac{e^2 n\tau_e}{m_n(1 - i\omega\tau_e)} \\
&= \frac{e^2 n}{m_n} \left[\frac{\tau_e}{1 + \omega^2\tau_e^2} + i\frac{m\omega\tau_e^2}{1 + \omega^2\tau_e^2} \right].
\end{aligned} \tag{12.20}$$

Introducing Eq. (12.20) into (10.26), we obtain with (12.4):

$$\varepsilon' = n_r^2 - \kappa^2 = \varepsilon_{\text{opt}} \left(1 - \frac{\omega_p^2}{\omega^2} \frac{\omega^2\tau_e^2}{1 + \omega^2\tau_e^2} \right) \tag{12.21}$$

---

[*] This is an ac-current at the frequency ω which, for higher ω (i.e., for $\omega \gtrsim 1/\tau_e$), is substantially different from the dc-current, as indicated in Section 10.1.1C.

and

$$\varepsilon'' = 2n_r\kappa = \varepsilon_{\text{opt}} \frac{\omega_p^2}{\omega^2} \frac{\omega\tau_e}{1 + \omega^2\tau_e^2}, \qquad (12.22)$$

which is identical to the resonance case, but for $\omega \gg \omega_p$.

The optical absorption coefficient due to free electrons is

$$\boxed{\alpha_o = \frac{2\omega\kappa}{c} = \frac{\omega_p^2}{n_r c} \frac{\tau_e}{1 + \omega^2\tau_e^2},} \qquad (12.23)$$

which for high enough frequencies ($\omega\tau_e \gg 1$) becomes the classical $\alpha_o \propto \lambda^2$-relation (using $\omega = 2\pi c/\lambda$):

$$
\begin{aligned}
\alpha_o &= \alpha_{o0}\frac{\tau_0}{\tau_e} = \frac{\omega_p^2\lambda^2}{4\pi^2 n_r c^3 \tau_e} \\
&= 3\cdot 10^{-4}\frac{n}{10^{16}[\text{cm}^{-3}]}\frac{m_0}{m_n}\frac{10}{\varepsilon_{\text{opt}}}\frac{10^{-13}}{\tau_e}\frac{\lambda^2[\mu m]^2}{n_r}
\end{aligned} \qquad (12.24)
$$

with τ_0 as a normalization factor ($\tau_0 = 1s$). Such behavior is often observed for the free-electron absorption in the extrinsic range beyond the band edge (Fig. 12.2).

In the following section we will show that this λ^2-relation, however, is modified by the specific scattering mechanism of the carriers when this scattering depends on the energy of the carrier within the band; here $\tau_e = \tau_e(\lambda)$, and α_o becomes a more complicated function of λ.

The free-electron absorption is rather small near the absorption edge for semiconductors with normal electron densities, and becomes observable in the IR only at higher levels of doping ($n > 10^{17}$ cm^{-3}— see Fig. 12.2A) when the probing frequency is closer to the plasma frequency—see Section 12.1.1.

12.2.1B Free-Electron Absorption as a Function of the Scattering Mechanisms   In the previous section, we discussed the absorption by a simple one-photon process with free electrons acting as a collective. Examining the $E(\mathbf{k})$ behavior of a typical conduction band, we recognize that individual optical, i.e., essentially vertical, direct transitions of electrons in the conduction band are rather rare events. They are restricted to transitions at specific values of k, which find an allowed $E(k)$ for the given photon energy [$h\nu_d$ (1) in Fig. 12.3] and a sufficiently large occupation probability of the ground state from which such a transition (1) starts. Only from the minimum of the conduction band to a band above it are such

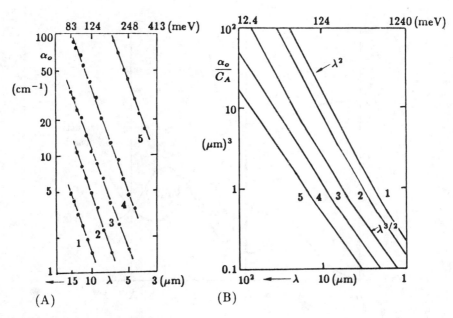

Figure 12.2: (A) Optical absorption of free electrons with predominant acoustic phonon scattering beyond the band edge of InAs at room temperature with the electron density as family parameter: $0.28, 1.4, 2.5, 7.8$, and $39 \cdot 10^{17}$ cm^{-3} for curves 1–5, respectively (after Dixon, 1960). (B) Optical absorption of free electrons as in (A), at $T = 600, 300, 77, 20$, and 4.2 K for curves 1–5, respectively, with C_A a proportionality factor (after Seeger, 1973).

transitions plentiful [at the Γ-point with $h\nu_d$ (3) in Fig. 12.3]. For most other photon energies, only indirect transitions are possible— see Section 13.2. They require the supply or emission of a phonon in addition to the absorption of the photon, as shown in Fig. 12.3 as transition $h\nu_i$ (2).

The transition probability for such indirect transitions can be calculated from a second-order perturbation theory including the Hamiltonian for a photon and the Hamiltonian for a phonon transition. This is necessitated by the fact that three particles are involved in the transition: a photon, an electron, and a phonon or a lattice defect. Such types of transitions are discussed in Chapter 16 which deals with nonlinear optics—see also Fan et al. (1956) and Schiff (1968). The interaction is much like that of the scattering of an electron with a phonon: the electron is first elevated by the photon to a virtual state within the conduction band.

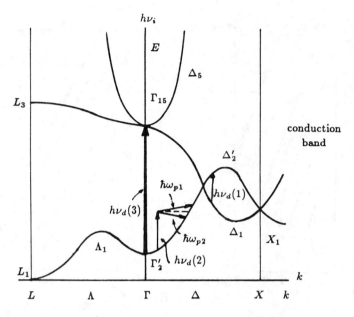

Figure 12.3: Direct ($\Gamma'_2 \to \Gamma_{15}$) and indirect transitions for free electrons within the conduction band ($h\nu_d$ and $h\nu_i$ respectively). The indirect transition requires, in addition to the absorption of a photon $h\nu_d$, the absorption ($\hbar\omega_{p1}$) or emission ($\hbar\omega_{p2}$) of a phonon of the proper energy and momentum.

The mathematical formulation of the theory for the scattering with different types of phonons is similar to the analysis of the electron scattering for carrier transport (Section 32.2.1), or for higher energies within the band (Chapter 33—see also Seeger, 1973). In a first approximation, the optical absorption coefficient can be expressed by the classical free-electron absorption α_o [Eq. (12.24)], and by replacing the electron relaxation time with the one relating to the predominant scattering mechanism, as analyzed in Section 32.2.1. This permits the replacement of the yet unknown τ_e in Eq. (12.24) with an appropriate expression—see Seeger, 1973.

For instance, we obtain for predominant *acoustic deformation potential scattering* [see Eq. (32.17)]

$$\alpha_{o,ac,E} = \alpha_{o,ac} \frac{\hbar\omega}{4kT} \sqrt{\frac{T}{T_e}} \sinh\left(\frac{h\nu}{2kT_e}\right) K_2\left(\frac{h\nu}{2kT_e}\right), \qquad (12.25)$$

where T_e is the electron temperature* (see Section 33.2.1) and $\hbar\omega$ is the energy of the absorbed or emitted acoustic phonon; K_2 is a modified Bessel function (Watson, 1944).

When only small deviations from thermal equilibrium are considered ($T_e \simeq T$) and scattering occurs with low-energy phonons ($h\nu \ll kT$), than the factor following $\alpha_{o,ac}$ approaches 1 and $\alpha_{o,ac,E} \simeq \alpha_{o,ac}$ with

$$\alpha_{o,ac} = \alpha_{o0} \frac{2^{7/2}(m_n kT)^{3/2}\Xi_c^2}{3\pi^{3/2}\hbar^4 c_l} \qquad (12.26)$$

and α_{o0} given by Eq. (12.24). Here Ξ_c is the deformation potential for the conduction band, and c_l is the appropriate elastic stiffness constant. Also, $\alpha_{o,ac}$ shows the classical λ^2 behavior (contained in α_{o0}) as indicated in Fig. 12.2A.

At higher energies (for $h\nu \gg kT$), however, one obtains

$$\alpha_{o,ac,E} \simeq \frac{\alpha_{o,ac}}{2}\sqrt{\frac{\pi}{2}\frac{\hbar\omega_{LO}}{kT}} \propto \lambda^{3/2}, \qquad (12.27)$$

which gives a somewhat lower slope (when the emission of many phonons becomes possible), which is indicated in Fig. 12.2B.

There are substantial differences for the scattering of electrons which are excited into higher band states. From here, scattering is greatly enhanced by rapid generation of LO phonons—see Section 33.3.2. This scattering can be described as the scattering of a *heated electron gas* with T_e much larger than T.

The absorption coefficients for other phonon scattering mechanisms show a wavelength dependence similar to the acoustic one for long-wavelength excitation. However, at low temperatures, it becomes nonmonotonic as soon as the electron energy exceeds that of the rapidly generated optical phonons (Fig. 12.4A). At low temperature LO phonons are frozen-out; a low scattering and therefore a low absorption results. When the electrons gain energy above $\hbar\omega_{LO}$ with decreasing λ of the exciting light, the relaxation time decreases dramatically, and therefore $\alpha_{o,opt,E}$ increases. At high temperatures this resonance is hidden since there are enough electrons thermally excited near $\hbar\omega_{LO}$.

---

* An elevated (above the lattice temperature) electron temperature is used here to indicate the occupancy of states higher in the conduction band by electrons. This concept will be discussed in more detail later—see Section 33.2.1.

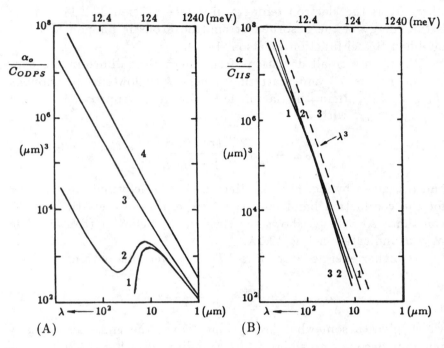

Figure 12.4: (A) Optical absorption of free electron computed for predominant optical deformation potential scattering with $\Theta = 720$ K and C_{ODPS} a proportionality factor of dimension cm^{-2} at 36, 77, 300, and 850 K for curves 1–4, respectively. (B) Optical absorption of free electrons computed for ionized impurity scattering with C_{IIS} a proportionality factor of dimension cm^{-2} at 40, 77, and 300 K for curves 1–3, respectively (after Seeger, 1973).

The maximum is due to resonance scattering when the electron energy within the band equals that of the LO phonon. This non-monotonic behavior is observed for *optical deformation potential scattering* as well as for *polar optical scattering*, for which the wavelength dependence increases from λ^2 to $\lambda^{2.5}$ for higher energies.

For lower temperatures ($kT < \hbar\omega_{LO}/2$) near thermal equilibrium ($T_e = T$) one obtains for the ratio of optical and acoustic deformation potential scatterings:

$$\frac{\alpha_{o,\mathrm{opt}}}{\alpha_{o,\mathrm{ac}}} = \frac{4}{\sqrt{\pi}} \left(\frac{D_o v_l}{\Xi_c \omega_0} \right)^2 \sqrt{\frac{T}{\Theta}} . \qquad (12.28)$$

where D_o and Ξ_c are the optical and acoustic deformation potential constants and v_l is the longitudinal sound velocity. Here Θ is the Debye temperature.

Ionized impurity scattering (Wolfe, 1954) causes a stronger dependence on the wavelength, which for higher energies approaches λ^3:

$$\alpha_{o,I} = \alpha_{o0} \frac{N_I Z^2 e^4}{3\sqrt{2}\pi^{3/2}(\varepsilon\varepsilon_0)^2 (mkT_e)^{1/2}\hbar\omega} \sinh\left(\frac{\hbar\omega_{LO}}{2kT_e}\right) K_0\left(\frac{\hbar\omega_{LO}}{2kT_e}\right) \propto \lambda^3;$$

$$(12.29)$$

where K_0 is a modified Bessel function (Watson, 1944). The factor following α_{o0} is closely related to the Conwell-Weisskopf relaxation time—see Section 32.2.4. The computed absorption behavior for scattered ionized impurities is shown in Fig. 12.4B.

With different scattering mechanisms, the optical absorption coefficients for each one of them is added [the inverse of the relaxation times are added—see Eq. (12.24)] in order to obtain the total absorption:

$$\alpha_o = \sum_i \alpha_{o,i}. \qquad (12.30)$$

When studying a family of absorption curves at different temperatures, we can obtain some information about the predominant scattering mechanism, especially about scattering at higher electron energies for semiconductors with sufficient electron densities. We will return to this subject in Section 33.7.1.

12.2.1C Free-Hole Absorption The absorptions of free electrons and holes follow similar principles. However, although most electron transitions are indirect, requiring the emission or absorption of phonons (Fan, 1967), the most prevalent long-wavelength transitions for holes are direct, as shown in Fig. 12.5A. Therefore, under otherwise similar conditions, the absorption coefficient for holes is much larger than that for electrons.

Closely spaced valence bands (Fig. 12.5) provide possibilities for a wide spectrum of direct transitions for very long wavelengths. In addition, the change from heavy to light holes, detected by simultaneous conductivity measurements, assists in identifying the specific feature in the absorption spectrum.

Steps in the optical absorption spectrum are observed when the energy of the holes is sufficient to permit optical phonon relaxation—see Fig. 12.5B. Such a step was calculated for a single threshold

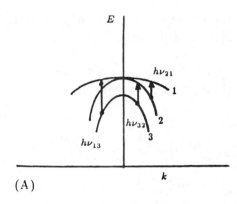

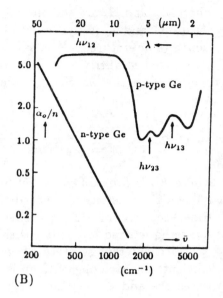

Figure 12.5: (A) Valence band transitions indicated. (B) Optical absorption cross section as a function of wavelength (wavenumbers) for n- and p-type Ge (after Kaiser et al., 1953).

in the electron energy for $\hbar\omega_{LO}$ emission—see Section 12.2.1B and Fig. 12.4A.

12.3 Carrier Dispersion; Electric/Magnetic Fields

The dispersion equation with an additional magnetic field is obtained from Maxwell's equations, exactly as in Section 10.1.1B from Eq. (10.16), except that the magnetic field causes a change in wave propagation or polarization and therefore requires a vector relation for the amplitude function $\mathbf{E}(x)$, resulting in

$$\nabla^2 \mathbf{E}(x) + \frac{\omega^2}{c^2}\left(\underline{\varepsilon} - i\left(\frac{\underline{\sigma}}{\varepsilon_0 \omega}\right)\right)\mathbf{E}(x) = 0, \qquad (12.31)$$

where $\underline{\varepsilon}$ and $\underline{\sigma}$ are the *dielectric* and the *conductivity matrices*, respectively. The dielectric matrix is not influenced by the magnetic field. For isotropic crystals, we can simply use the dielectric constant, requiring, however, matrix notation to permit matrix calculus. For a magnetic induction in the z-direction, the $\underline{\sigma}$ matrix simplifies to

$$\underline{\sigma} = \begin{pmatrix} \sigma_{xx} & \sigma_{xy} & 0 \\ \sigma_{yx} & \sigma_{yy} & 0 \\ 0 & 0 & \sigma_{zz} \end{pmatrix}. \qquad (12.32)$$

The components of this matrix are obtained from the equation of motion, which is identical to Eq. (12.17), except for the addition of the Lorentz force $e(\mathbf{v}_D \times \mathbf{B})$:

$$m^* \left(\frac{d\mathbf{v}_D}{dt} + \frac{\mathbf{v}_D}{\tau_m} \right) = -e(\mathbf{E} + \mathbf{v}_D \times \mathbf{B}). \qquad (12.33)$$

With $\mathbf{E} = \mathbf{E}(x) \exp(i\omega t)$ and $en\mathbf{v}_D = \underline{\sigma}\mathbf{E}$, we obtain

$$\sigma_{xx} = \sigma_{yy} = \sigma_0 \frac{1 - i\omega\tau_m}{(1 - i\omega\tau_m)^2 + (\omega_c\tau_m)^2} \qquad (12.34)$$

$$\sigma_{xy} = -\sigma_{yx} = \sigma_0 \frac{\omega_c\tau_m}{(1 + i\omega\tau_m)^2 + (\omega_c\tau_m)^2} \qquad (12.35)$$

$$\sigma_{zz} = \sigma_0 \frac{1}{1 + i\omega\tau_m}, \qquad (12.36)$$

with $\omega_c = eB/m^*$ as the cyclotron frequency and $\sigma_0 = en\mu_n$ as the dc-conductivity. Given the magnetic induction $\mathbf{B} = (0, 0, B_z)$, we can distinguish from the general dispersion equation*

$$k^2 \mathbf{E} = \left(\frac{\omega}{c} \right)^2 \underline{\varepsilon}\mathbf{E} + i\frac{\omega}{\varepsilon_0 c^2}\underline{\sigma}\mathbf{E} \qquad (12.37)$$

a number of cases depending on the relative orientation of the propagation ($\mathbf{k}$) and polarization ($\mathbf{E}$) of the light interacting with the semiconductor. These cases are identified in Table 12.2.

For the **longitudinal case** one has

$$k_z^2 E_x = \left(\frac{\omega}{c} \right)^2 \varepsilon_{\text{opt}} E_x + i\frac{\omega}{\varepsilon_0 c^2}(\sigma_{xx} E_x + \sigma_{xy} E_y), \qquad (12.38)$$

$$k_z^2 E_y = \left(\frac{\omega}{c} \right)^2 \varepsilon_{\text{opt}} E_y + i\frac{\omega}{\varepsilon_0 c^2}(-\sigma_{xy} E_x + \sigma_{xx} E_y). \qquad (12.39)$$

For circular polarization we have

$$(n + i\kappa)^2_\pm = \varepsilon_{\text{opt}} + i\frac{\sigma_\mp}{\varepsilon_0 \omega} \quad \text{with} \quad \sigma_\mp = \sigma_{xx} \mp i\sigma_{xy} \qquad (12.40)$$

---

* Obtained from Maxwell's equations with

$$\frac{\varepsilon}{c^2}\ddot{\mathbf{E}} + \frac{\sigma}{\varepsilon_0 c^2}\dot{\mathbf{E}} = -\nabla \times (\nabla \times \mathbf{E})$$

after insertion of

$$\mathbf{E} = eE_0 \exp\left[i(\mathbf{k} \cdot \mathbf{r} - \omega t)\right].$$

Table 12.2: Relative orientation of **B**, **k**, and **E**.

Magnetic Induction		$\mathbf{B} = (0,0,B_z)\uparrow$	
Wave Propagation		longitudinal	transverse
		Faraday configuration	Voigt configuration
		$\mathbf{k} = (0,0,k_z)\uparrow$	$\mathbf{k} = (0,k_y,0)\rightarrow$
Electric Vector	linear	$\mathbf{E} = (0,E_y,0)\rightarrow$ or	$\mathbf{E} = (0,0,E_z)\uparrow$ or
Polarization	linear	$\mathbf{E} = (E_x,0,0)\nearrow$	$\mathbf{E} = (E_x,0,0)\nearrow$
	circular	$E = E_x \pm iE_y$	

and, after separation of real and imaginary parts,

$$\varepsilon'_\pm = n_r^2 - \kappa^2 = \varepsilon_{\text{opt}}\left(1 - \frac{1}{\omega}\frac{\omega \pm \omega_c}{[(\omega \pm \omega_c)\tau_m]^2 + 1}\right) \tag{12.41}$$

$$\varepsilon''_\pm = 2n_r\kappa = \varepsilon_{\text{opt}}\frac{\omega_p}{\omega}\frac{\omega_p\tau_m}{[(\omega \pm \omega_c)\tau_m]^2 + 1}, \tag{12.42}$$

with the $+$ or $-$ sign for right- or left-polarized light, respectively, and ω_p the plasma frequency [Eq. (12.4)].

For the **transverse case** there are three dispersion equations:

$$k_y^2 E_x = \left(\frac{\omega}{c}\right)^2 \varepsilon_{\text{opt}}E_x + i\frac{\omega}{\varepsilon_0 c^2}(\sigma_{xx}E_x + \sigma_{xy}E_y) \tag{12.43}$$

$$0 = \left(\frac{\omega}{c}\right)^2 \varepsilon_{\text{opt}}E_y + i\frac{\omega}{\varepsilon_0 c^2}(-\sigma_{xy}E_x + \sigma_{xx}E_y) \tag{12.44}$$

$$k_y^2 E_z = \left(\frac{\omega}{c}\right)^2 \varepsilon_{\text{opt}}E_z + i\frac{\omega}{\varepsilon_0 c^2}\sigma_{zz}E_z, \tag{12.45}$$

which yield for *parallel polarization* ($\mathbf{E}\|\mathbf{B}$):

$$\tilde{\varepsilon}_\| = (n_r + i\kappa)^2_\| = \varepsilon_{\text{opt}} + i\frac{\sigma_{zz}}{\varepsilon_0\omega}, \tag{12.46}$$

with

$$\varepsilon' = \varepsilon_{\text{opt}}\left(1 - \frac{(\omega_p\tau_m)^2}{(\omega\tau_m)^2 + 1}\right); \quad \text{and} \quad \varepsilon'' = \varepsilon_{\text{opt}}\frac{\omega_p}{\omega}\frac{\omega_p\tau_m}{(\omega\tau_m)^2 + 1}, \tag{12.47}$$

and for the *perpendicular polarization* ($\mathbf{E}\perp\mathbf{B}$):

$$\tilde{\varepsilon}_\perp = (n_r + i\kappa)^2_\perp = \varepsilon_{\text{opt}} + i\frac{1}{\varepsilon_0\omega}\left(\sigma_{xx} + i\frac{\sigma_{xy}^2}{\varepsilon_{\text{opt}}\varepsilon_0\omega + i\sigma_{xx}}\right) \tag{12.48}$$

with

$$\varepsilon' = \varepsilon_{\mathrm{opt}} \left(1 - \frac{(\omega_p \tau_m)^2 \beta}{(\omega \tau_m)^2 \beta^2 + \alpha^2} \right), \quad \varepsilon'' = \varepsilon_{\mathrm{opt}} \frac{\omega_p}{\omega} \frac{\omega_p \tau_m \alpha}{(\omega \tau_m)^2 \beta^2 + \alpha^2};$$

$$(12.49)$$

and with the auxiliary functions

$$\alpha = 1 + \frac{(\omega \tau_m)^2 \omega_c^2}{\omega^2 + \left[(\omega^2 - \omega_p^2) \tau_m \right]^2} \text{ and } \beta = 1 - \frac{(\omega^2 - \omega_p^2) \tau_m^2 \omega_c^2}{\omega^2 + \left[(\omega^2 - \omega_p^2) \tau_m \right]^2}.$$

$$(12.50)$$

There are two characteristic frequencies entering the dispersion equation: the plasma (ω_p) and the cyclotron (ω_c) frequencies, which determine possible resonances. The damping is determined by the appropriate relaxation time and gives the width of the resonance peak. These equations describe all possible interactions of electromagnetic radiation with a semiconductor while exposed to a dc-magnetic induction. Examples for such interactions are cyclotron resonance, magnetoplasma reflection, and the Faraday and Voigt effects. These will be discussed in the following sections. For more detail, see Madelung (1978) and Roth (1982).

12.3.1 Cyclotron-Resonance Absorption

Cyclotron-resonance absorption occurs when the frequency of the impinging electromagnetic radiation (usually in the microwave range) equals the cyclotron frequency (eB/m^*). It is measured with $\mathbf{k} \| \mathbf{B}$ and linear polarized $\mathbf{E} = (E_x, 0, 0)$ radiation, which renders Eq. (12.38)

$$k_z^2 = \left(\frac{\omega}{c} \right)^2 \varepsilon_{\mathrm{opt}} + i \frac{\omega}{\varepsilon_0 c^2} \sigma_{xx} \quad \text{with} \quad \varepsilon_{\mathrm{opt}} = \frac{1}{\varepsilon_0 \omega} \sigma_{xx}^{(R)} \qquad (12.51)$$

and yields for the absorption [from Eq. (12.34)]

$$2 n_r \kappa = \frac{\sigma}{\varepsilon_0 \omega} \frac{1 + (\omega_c^2 + \omega^2) \tau_m^2}{\left[1 + (\omega_c^2 + \omega^2) \tau_m^2 \right]^2 + 4 \omega^2 \tau_m^2}. \qquad (12.52)$$

Cyclotron resonance, effectively determines ω_c, which is a direct measure of the effective mass. By rotating the semiconductor with respect to $\mathbf{B}$, one obtains m^* as a function of the crystallographic orientation. This was discussed in more general terms in Section 9.1.5C and will be expanded for the electrical transport in Section 31.2.3.

In addition to the conventional method of observing cyclotron resonance by microwave absorption, one can detect the resonances optically by exciting carriers near the band edges. This will be discussed in Section 17.3.

In **superlattices**, the confinement of cycling electrons within each well can be easily detected by the optical method described before (Cavenett and Pakulis, 1985): the resonances become angle-dependent and are pronounced only within the plane of the super-lattice, where electrons can follow the Lorentz force.

12.3.2 Faraday Effect

The dispersion relation for circular polarized light [Eqs. (12.40) and (12.46)] with $\mathbf{k}\|\mathbf{B}$ shows that the propagation velocity $\omega\sqrt{\tilde{\varepsilon}_{\pm}}/c$ is different for right- or left-hand polarization. Therefore, a linear polarized light beam, composed of an equal fraction of left- and right-polarized components, experiences a turning of its polarization plane with progressive traveling through the semiconductor with a *Faraday angle*, defined by

$$\theta_F = \frac{1}{2}(\kappa_+ - \kappa_-)d = \frac{\omega}{2c}\left\{\sqrt{\varepsilon - \frac{i\sigma_+}{\varepsilon_0\omega}} - \sqrt{\varepsilon - \frac{i\sigma_-}{\varepsilon_0\omega}}\right\}d. \quad (12.53)$$

This angle can be evaluated, using Eqs. (12.40), (12.34), and (12.35); for $1/\tau_m \ll (\omega_c, \omega_p)$ it yields

$$\theta_F = \frac{360°}{2\pi}\frac{ne^3B}{m_n^2\sqrt{\varepsilon_{\text{opt}}}\varepsilon_0 2c(\omega^2 - \omega_c^2)}d \quad \text{for } \omega \gg \omega_c$$

$$= \frac{1.64°}{\sqrt{\varepsilon_{\text{opt}}}}\frac{(n/10^{18}\text{ cm}^{-3})(\lambda/10\,\mu\text{m})^2(B/\text{kG})(d/\text{cm})}{(m_n/m_0)^2(1 - \omega_c^2/\omega^2)}. \quad (12.54)$$

The quantity $\theta_F/(Bd)$ is known as the *Verdet coefficient* and is proportional to $n\lambda^2$. The proportionality of the Verdet coefficient with λ^2 is shown in Fig. 12.6 for n-InSb with the electron density and temperature as the family parameter. It shows an increase of the Verdet coefficient with carrier density in agreement with Eq. (12.54). It also shows an increase with decreasing temperature due to a temperature-dependent effective mass (see below). With known carrier density, the Verdet coefficient yields the value of the effective mass. The changes in m_n, as a function of T and n, are due to changes in the electron distribution within the band, indicating that the effective mass increases for electrons higher in the conduction band—see Section 9.1.7.

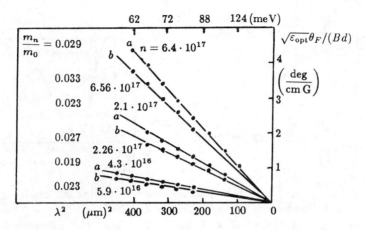

Figure 12.6: Faraday rotation in n-InSb with electron density and temperature as the family parameter. The corresponding average effective mass ratios for the six curves are shown at the left side of the figure; curves a and b are for $T = 77$ and 290 K, respectively (after Pidgeon, 1962).

Faraday rotation for electrons is opposite in sign to the rotation for holes. In a multiband semiconductor, both carriers must be considered. For instance, with heavy and light holes, the quantity n/m_n^2 in Eq. (12.54) must be replaced by $n_l/m_l^2 + n_h/m_h^2$ for holes.

12.3.3 Magnetoplasma Reflection

A relatively simple means to determine one or both of the characteristic frequencies ω_c and ω_p—and thereby the effective mass, carrier type, and carrier density—is to measure the spectral distribution of the reflectivity with $\mathbf{k}\|\mathbf{B}$ or $\mathbf{k}\perp\mathbf{B}$, i.e., for Faraday or Voigt configuration, respectively.

12.3.3A Faraday Configuration Arranging the propagation of the interacting light parallel to the magnetic induction and using circular polarized light causes the index of refraction to vanish [i.e., for $n_r^2 \gg \kappa^2$, the reflectivity shows a maximum—see Eq. (12.41)] when

$$\omega = \omega_p \pm \frac{1}{2}\omega_c + \frac{\sqrt{\varepsilon}}{8}\frac{\omega_c^2}{\omega_p} \qquad (12.55)$$

for left $(-)$ or right $(+)$ polarized light. Consequently, a shift in the plasma edge by $\pm\omega_c/2$ is seen (when $\omega_p \gg \omega_c$) when a magnetic induction is applied. This is shown in Fig. 12.7A for InSb and can be used to determine ω_c and thereby the effective mass.

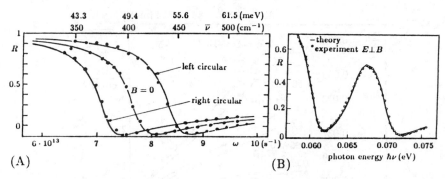

Figure 12.7: (A) Shift of the plasma absorption edge in Faraday configuration in n-InSb at room temperature for $n = 10^{18}$ cm^{-3}, $\tau_e = 2.8 \cdot 10^{-13}$ s, and $m_n/m_0 = 0.035$ at $B = 0$ and ± 25.4 kG. The curves are calculated (after Palik et al., 1962). (B) Transverse magnetoplasma reflection (Voigt configuration) in InSb at room temperature for $n = 1.8 \cdot 10^{18}$ cm^{-3}, $\tau_e = 3.6 \cdot 10^{-13}$ s, and $B = 35.2$ kG (after Wright and Lax, 1961).

12.3.3B Voigt Configuration   With plane-polarized light at normal incidence to the crystal surface and the magnetic vector parallel to the surface, we obtain for $\mathbf{E} \perp \mathbf{B}$ as a condition for minimum reflection $(n_r \to 0)$ from Eq. (12.49), neglecting the damping term:

$$\omega^4 \left(\frac{\varepsilon_{\mathrm{opt}} - 1}{\varepsilon_{\mathrm{opt}}} \right) - \omega^2 \left\{ \left(\frac{\varepsilon_{\mathrm{opt}} - 1}{\varepsilon_{\mathrm{opt}}} \right) (\omega_p^2 + \omega_c^2) + \omega_p^2 \right\} + \omega^4 = 0, \tag{12.56}$$

which has two minima at

$$\omega_1^2 \simeq \omega_p^2 \left(\frac{\varepsilon_{\mathrm{opt}}}{\varepsilon_{\mathrm{opt}} - 1} \right) + \varepsilon_{\mathrm{opt}} \omega_c^2 \quad \text{and} \quad \omega_2^2 \simeq \omega_p^2 - \omega_c^2 (\varepsilon_{\mathrm{opt}} - 1). \tag{12.57}$$

They are shown in Fig. 12.7B for n-InSb, from which the cyclotron frequency, and thereby the effective mass, can be obtained. Once the cyclotron frequency is known, the plasmon frequency can also be obtained, and thereby the carrier density.

Summary and Emphasis

In this chapter we have shown that the dispersion due to free electrons in the conduction band (or holes in the valence band) can be used to determine optically the density, effective mass, and relaxation time of free carriers. Most important is the measurement of the effective mass and its anisotropy, which can be determined by simple changes of the relative alignment of crystal, electrical, and magnetic fields. The other parameters can often be obtained more easily by electrical measurements; however, the optical determination may be helpful to remove ambiguities, e.g., for ambipolar semiconductors or other unusual cases.

In addition, optical measurements can be performed when electrical contacts cannot be applied, or electrical measurements are otherwise impeded.

Free-carrier absorption and reflection with or without magnetic fields provide important information about carrier parameters—most significantly about the effective carrier mass, its anisotropy, and its dependence on other variables, such as carrier density and temperature. The knowledge of the effective carrier mass is essential for the evaluation and analysis of a large variety of semiconducting and electro-optical device properties.

Exercise Problems

1.(r) Discuss electron plasma absorption for electrons in conduction and in valence bands. Separate it from free-electron or free-hole absorption. Separate that from the absorption related to the complex conductivity near the *Reststrahl* absorption.

2.(r) What do we learn about electron relaxation from quasi-free electron absorption?

3.(r) What influence on optical absorption of quasi-free electrons has a magnetic field?

4.(*) How are plasmons of valence electrons and plasmons of holes in the valence band distinguished from each other? Describe the difference for a metal and a p-type semiconductor.

5.(*) Why is the optical dielectric constant used in the expression for plasmons [Eq. (12.4)]. Is this always correct? What are the determining factors for making the decision which ε to choose?

6.(e) In Eqs. (12.6) and (12.8) – (12.10), the energy relaxation time is used. Instead, one also could use the momentum relaxation time. What are the criteria to use one or the other? In what wavelength range would the use of τ_m be more appropriate?

7.(e) Where in the dispersion spectrum is $n_r^2 \gg \kappa^2$. Give $n_r(\omega)$ and $\kappa(\omega)$ for the example shown in Fig. 11.2. Is this condition fulfilled for the plasmon reflection spectrum shown in Fig. 12.1?

8.(r)   Describe in your own words the relationship between resonance and nonresonant absorption of free carriers. Give a quantitative relation for making this distinction.

9.(e)   Relate the derivation of the dispersion relations, Eqs. (12.21) and (12.22), shown in Section 12.2.1A to the derivation of Eqs.

(11.31) and (11.32), shown in Section 11.2.

10.   Prove that the half-width of a line with Lorentzian shape is given by the inverse relaxation time (i.e., by the damping factor γ). Relate this to Eqs. (12.21) and (12.22).

11.(r) Discuss the optical nonresonant absorption of free electrons as a function of the phonon energy for interaction with acoustic and optical phonons. Where in the Brillouin zone do such transitions occur? Draw a to-scale diagram for a typical semiconductor $(m_n = 0.1m_0)$ near $k = 0$.

12.(e) Analyze the following magneto-optical effects:

(a) Explain the difference between Faraday rotation and magneto-plasma reflection in Faraday and in Voigt configuration.

(b) How does one determine separately ω_p and ω_c from such measurements?

Chapter 13

Band-to-Band Transitions

*Optically induced band-to-band transitions are res-
onance transitions and provide the most direct information about the band structure of semiconductors.*

Such optical band-to-band transitions cannot be described in
a classical model discussed in the previous chapters; it requires a
careful quantum mechanical analysis—see Bassani (1966, 1975). One
distinguishes direct band-to-band transitions, which are essentially
vertical transitions in E($\mathbf{k}$), and indirect transitions which involve
phonons, permitting major changes of $\mathbf{k}$ during the transitions. We
will first discuss the direct transitions.

13.1 Photons in Band-to-Band Transitions

Light of sufficiently short wavelength with

$$\mathbf{E} = A_0 \mathbf{e} \exp\{i(\mathbf{k}_0 \cdot \mathbf{r} - \omega t)\} \tag{13.1}$$

initiates electronic transitions from one to another band. To avoid
confusion with the energy of an optical transition, E, we have used
A_0 here as the amplitude. The electric polarization vector is $\mathbf{e}$, and
$\mathbf{k}_0$ is the wave vector of the light, traveling in $\mathbf{r}$-direction.

The number of optically induced transitions at the same $\mathbf{k}$ (i.e.,
neglecting the wave vector of the photon) between band μ and band
ν is proportional to the square of the momentum matrix elements
given by

$$\mathbf{e} \cdot \mathbf{M}_{\mu\nu}(\mathbf{k}) = \mathbf{e} \cdot \int_{\mathcal{V}} \psi_\mu^*(\mathbf{k}, \mathbf{r})(-i\hbar\nabla)\psi_\nu(\mathbf{k}, \mathbf{r})\, d\mathbf{r}, \tag{13.2}$$

with the integral extending over the crystal volume $\mathcal{V}$. The proportionality factor for a specific transition is

$$P_{\mu\nu} = \frac{1}{h}\left(\frac{eA_0}{m_0 c}\right)^2 \delta(E_\mu(\mathbf{k}) - E_\nu(\mathbf{k}) - h\nu) \tag{13.3}$$

325

where δ is the Dirac delta function. The factor $1/h$ indicates the quantum nature of the transition, $e/(m_0 c)$ stems from the interaction Hamiltonian between light and electrons, and A_0 from the amplitude of the light (value of the Poynting vector: A_0^2). The delta function switches on this contribution when a transition occurs from one state to another, i.e., when $E_\mu(\mathbf{k}) - E_\nu(\mathbf{k}) = h\nu$. No broadening of any of these transitions is assumed. Close proximity to adjacent transitions and Kramers-Kronig interaction makes such broadening consideration unnecessary, except for very pronounced features.

After integration over all states within the first Brillouin zone and all bands between which the given photon $h\nu$ can initiate transitions, we obtain for the number of such transitions per unit volume and time:

$$W(\nu) = \sum_{\mu,\nu} \int_{Bz} \frac{2}{(2\pi)^3} P_{\mu\nu} |\mathbf{e} \cdot \mathbf{M}_{\mu\nu}(\mathbf{k})|^2 \, d\mathbf{k}$$

$$= \frac{2}{(2\pi)^3 h} \left(\frac{eA_0}{m_0 c} \right)^2 \sum_{\mu,\nu} \int_{Bz} |\mathbf{e} \cdot \mathbf{M}_{\mu\nu}(\mathbf{k})|^2 \delta(E_\mu(\mathbf{k}) - E_\nu(\mathbf{k}) - h\nu) \, d\mathbf{k},$$

$$(13.4)$$

integrated over the entire Brillouin zone (Bz). The factor $a^3/(2\pi)^3$ normalizes the $\mathbf{k}$-vector density within the Brillouin zone; $a^3 = \mathcal{V}$ cancels from the integration of Eq. (13.2); the factor 2 stems from the spin degeneracy. Equation (13.4) follows directly from a perturbation theory, and is often referred to as *Fermi's golden rule*. The matrix elements vary little within the Brillouin zone; therefore, we can pull these out in front of the integral. This leaves only the delta function inside the integral. The integral identifies the sum over all possible transitions which can be initiated by photons with a certain energy $h\nu$, and it is commonly referred to as the *joint density of states* between these two bands. It is given by

$$J_{\mu\nu}(\omega) = \frac{2}{(2\pi)^3} \int_{Bz} \delta(E_\mu(\mathbf{k}) - E_\nu(\mathbf{k}) - h\nu) \, d\mathbf{k}$$

$$= \frac{2}{(2\pi)^3} \int \frac{dS}{\left| \nabla_k \left(E_\mu(\mathbf{k}) - E_\nu(\mathbf{k}) \right) \right|_{E_\mu - E_\nu = h\nu}}, \qquad (13.5)$$

where dS is a surface element of $\mathbf{k}$-space, with the surface depicted by

$$E_\mu(\mathbf{k}) - E_\nu(\mathbf{k}) = h\nu; \qquad (13.6)$$

here μ stands for any one of the conduction bands, and ν for any one of the valence bands.*

For each of the transitions, $\mathbf{k}$ is constant; it is a *direct transition*. When the slopes of $E(\mathbf{k})$ of both bands are different, Eq. (13.4) is fulfilled only for an infinitesimal surface area. In contrast, a larger area about which such transitions can take place appears near points at which both bands have the same slope, i.e., near *critical points*— see van Hove (1953) and Phillips (1956). Here

$$\nabla_k E_\mu(\mathbf{k}) = \nabla_k E_\nu(\mathbf{k}) = 0 \quad \text{or} \quad \nabla_k E_\mu(\mathbf{k}) - \nabla_k E_\nu(\mathbf{k}) = 0, \quad (13.8)$$

and the expression under the integral of Eq. (13.5) has a singularity. By integration this results in a kink in the joint density of states. The singularity related to Eq. (13.5) is referred to as a *van Hove singularity*.

The types of critical points are referred to as $M_0 \ldots M_3$, $P_0 \ldots P_2$, and Q_0 or Q_1 in three-, two-, and one-dimensional $E(\mathbf{k})$ representations. Two- and one-dimensional representations are applicable to superlattices or quantum wire configurations—see Section 9.3.5.

They describe the relative curvature of the lower and upper bands with respect to each other; in a first-order expansion we can express this by

$$E_\mu(\mathbf{k}) - E_\nu(\mathbf{k}) = E_0 + \frac{\hbar^2}{2} \left(s_1 \frac{k_x^2}{m_{rx}} + s_2 \frac{k_y^2}{m_{ry}} + s_3 \frac{k_z^2}{m_{rz}} \right), \quad (13.9)$$

which may be interpreted as a flattened-out lower band (E_0) and the accordingly changed curvature of the upper band describing the energy difference between both. Here s_1, s_2, and s_3 represent the signs ($= \pm 1$) and m_{rx}, m_{ry}, m_{rz} have the dimension of a mass but are here only a proportionality factor measuring the *relative* band curvature.

---

* The rewriting of Eq. (13.5) can be understood from the behavior of the *Dirac delta function*

$$\int g(x)\delta[f(x)]\, dx = g(x_0) \left| \frac{df(x)}{dx} \right|_{x=x_0}^{-1} \quad (13.7)$$

with $f(x_0) = 0$.

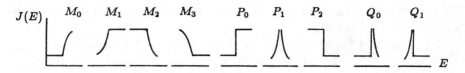

$J(E)$ M_0 M_1 M_2 M_3 P_0 P_1 P_2 Q_0 Q_1 E

Figure 13.1: Joint density of state representation near critical points (schematic) in three (M_i), two (P_i), and one (Q_i) dimensions of $E(\mathbf{k})$ with subscripts identifying the number of negative s_i's in Eq. (13.9).

The combination of signs identifies the type of critical point,[†] with a schematic representation shown in Fig. 13.1.

Matrix elements are mainly influenced by *selection rules*, which are similar to those in atomic spectroscopy. The symmetry properties of ψ_μ and ψ_ν, e.g., whether even or odd under reflection or inversion, and whether they are the same or different from each other, determine whether the matrix element has a finite value or vanishes; judgment can be rendered on group theoretical arguments—see Bassani (1975). Comparing the results of such an analysis with experimental observation of strong or weak optical absorption provides quite convincing arguments by assigning certain features of the absorption spectrum to the appropriate critical points for *calibration* of the $E(\mathbf{k})$ behavior. This augments the previously described computation and eliminates its weakest point: namely, the fact that quantitative results cannot be obtained when crystal potentials are not known with sufficient accuracy. Recent *ab initio* calculation, however, can provide the necessary input for satisfactory agreement without adjustable parameters.

13.1.1 Band-to-Band Optical Absorption Spectrum

The optical absorption coefficient α_o (see Section 10.1.1B) is proportional to the number of optical transitions per volume and time elements. The coefficient can be calculated from simple optical principles (see Bassani et al., 1975): it is given by the absorbed energy per unit time and volume, $h\nu W(\nu)$, divided by the energy flux $2\pi A_0^2 \nu^2 \varepsilon_0 n_r / c$. The energy flux is equal to optical energy density given by the square of the wave amplitude per wavelength interval

---

[†] $+++$, $++-$, $--+$, $---$ for $M_0 \ldots M_3$ as maximum, saddle point, saddle point, and minimum, respectively, in three dimensions; $++$, $+-$, or $--$ for P_0, P_1, or P_2 for maximum, saddle point, or minimum in two dimensions, respectively; and $+$ or $-$ for Q_0 or Q_1 for maximum or minimum in one dimension, respectively.

$2\pi A_0^2 \nu^2 n_r^2 / c^2$, divided by the light velocity within the semiconductor c/n_r. With Eq. (13.4) we now obtain:

$$\alpha_o(\nu) = \frac{2\pi e^2}{\varepsilon_0 n_r c m_0^2 \nu} \sum_{\mu,\nu} \int_{Bz} \frac{2}{(2\pi)^3} |\mathbf{e} \cdot \mathbf{M}_{\mu\nu}(\mathbf{k})|^2 \delta(E_\mu(\mathbf{k}) - E_\nu(\mathbf{k}) - h\nu) \, d\mathbf{k}.$$

(13.10)

For a further analysis of the integral we need more knowledge about $E(\mathbf{k})$. This will be done near the fundamental absorption edge in Section 13.1.4, where some simplifying assumptions for the valence and conduction bands are introduced. For a short review see, Madelung (1981).

With the availability of synchrotron radiation sources.* light sources of sufficient power and stability have become available to perform absorption and reflection spectroscopy for higher bands in the vacuum UV and soft x-ray range (Petroff, 1980). Since the optical absorption here is very strong, requiring extremely thin crystals for absorption spectroscopy which are difficult to obtain with sufficient crystal perfection, it is preferable to obtain such spectra in reflection.

A more sensitive set of methods relates to modulation spectroscopy (see Section 10.2.3) where a crystal variable, e.g., the electric field, temperature, or light intensity, is modulated and the changing reflection signal is picked up by look-in technology.

To compare the optical properties with the band structure, one must now express the transitions in terms of the complex dielectric constant $\tilde{\varepsilon} = \varepsilon' + i\varepsilon''$. The optical absorption constant α_o is then obtained from $\varepsilon'' = \alpha_o n_r c / (2\pi\nu)$ [see Eqs. (10.21) and (10.23)], with

$$\varepsilon'' = \frac{e^2}{\varepsilon_0 m_0^2 \nu^2} \sum_{\mu,\nu} \int_{Bz} \frac{2}{(2\pi)^3} |\mathbf{e} \cdot \mathbf{M}_{\mu\nu}(\mathbf{k})|^2 \delta(E_\mu(\mathbf{k}) - E_\nu(\mathbf{k}) - h\nu) \, d\mathbf{k}.$$

(13.11)

This representation has the advantage of not depending on the index of refraction, which causes a distortion of the calculated $\varepsilon''(\nu)$ distribution. With the first Kramers-Kronig relation [Eq. (11.19)] and

---

* Electrons at very high speed (100–500 MeV) emit a polychromatic radiation in a very wide frequency range (from visible to x-ray) when accelerated in a magnetic field (synchrotron or large storage-ring).

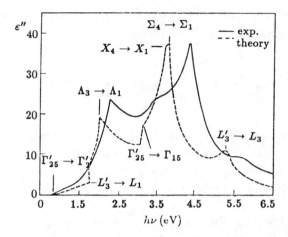

Figure 13.2: Optical absorption spectrum of Ge (after Brust et al., 1962). The reference to vertical transitions from various valence to conduction bands at critical points corresponds to the notation shown for Si in Fig. 8.8. The dashed curve is experimental; the solid curve is computed by the empirical pseudofunctional method and fit to the experiment.

$\omega_a = (E_\mu - E_\nu)/\hbar$, we obtain for ε' from Eq. (13.11) after conversion of the δ-function (see Bassani et al., 1975)

$$\varepsilon' = 1 + \frac{8\pi e^2}{m_0^2} \sum_{\mu,\nu} \int_{Bz} \frac{2}{(2\pi)^3} \frac{|\mathbf{e} \cdot \mathbf{M}_{\mu\nu}(\mathbf{k})|^2}{\dfrac{[E_\mu(\mathbf{k}) - E_\nu(\mathbf{k})]}{\hbar}} \frac{dk}{\dfrac{[E_\mu(\mathbf{k}) - E_\nu(\mathbf{k})]^2}{\hbar^2} - \omega^2}.$$

$$(13.12)$$

As an example, the $\varepsilon''(\nu)$ spectrum is given in Fig. 13.2 for Ge. It shows several shoulders and spikes that can be connected to critical points in the corresponding theoretical curve, such as the L, Γ, and X transitions [compare with $E(\mathbf{k})$ of Ge shown in Fig. 8.10]. The theoretical spectrum was obtained by numerically sampling $E_\mu(\mathbf{k}) - E_\nu(\mathbf{k})$ in the entire Brillouin zone at a large number of $\mathbf{k}$ values. By ordering and adding the transitions that occur at the same value of $h\nu$, possibly involving different band combinations, and repeating for other values of $h\nu$, we obtain the joint density of states. These yield the given $\varepsilon''(\nu)$ as shown in Fig. 13.2 (Phillips, 1966). The biggest peaks relate to ranges in $E(\mathbf{k})$ where both bands involved in the transition run parallel to each other for an extended k-range.

The spectral distribution between 1.5 and 6 eV of ε' and ε'' for Si is shown in Fig. 13.3. These distributions relate [see Eqs. (10.21)

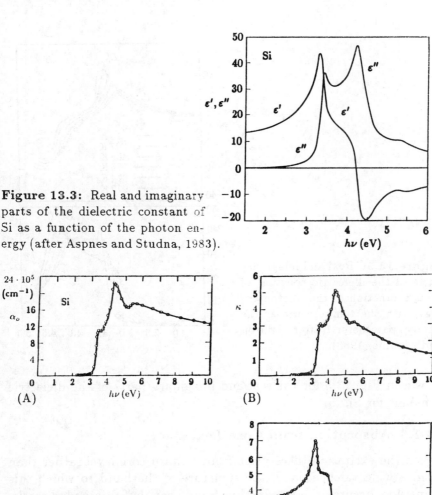

Figure 13.3: Real and imaginary parts of the dielectric constant of Si as a function of the photon energy (after Aspnes and Studna, 1983).

Figure 13.4: Optical constants of Si as a function of the photon energy: (A) absorption coefficient; (B) extinction coefficient; and (C) refractive index (after Carlson, 1957; Philipp and Taft, 1960).

and (10.23)] to the experimentally obtained optical constants shown in Fig. 13.4A–C.

13.1.2 Temperature Influence

The influence of temperature on the principal band gap was discussed in Section 9.2.1B. An example for changes within higher bands is shown in Fig. 13.5 for the real and imaginary parts of the dielectric constant for Si.

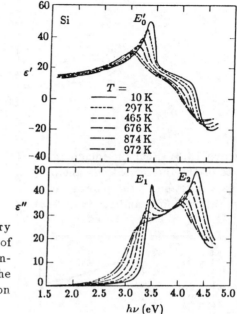

Figure 13.5: Real and imaginary parts of the dielectric constant of Si as a function of the photon energy with the temperature as the family parameter (after Jellison and Modine, 1983).

shown in Fig. 13.5 for the real and imaginary parts of the dielectric constant for Si.

13.1.3 Absorption from Core Levels

When the excitation takes place from a sharp core level rather than from a wide valence band, the structure of the band to which this transition occurs, the conduction band, can be obtained directly. When comparing excitation spectra from core levels with those from the valence band, however, a difference occurs because of the different relaxation of the excited state which is stronger when the hole is localized at the core than when it is more widely spread out in the valence band (Zunger, 1983).

The joint density of states involving excitation from a core level is identical with the density of states in the conduction band. A typical example for a core-to-conduction band spectrum of the reflectivity is given in Fig. 13.6. The experimental curves usually show less detail than those obtained from band structure calculations (Martinez et al., 1975b). This may be caused by electron-hole interaction and local field effects which are insufficiently accounted for in the single electron approximation used for band calculation (Hanke and Sham, 1974). Also, in older calculations an error in the threshold energies

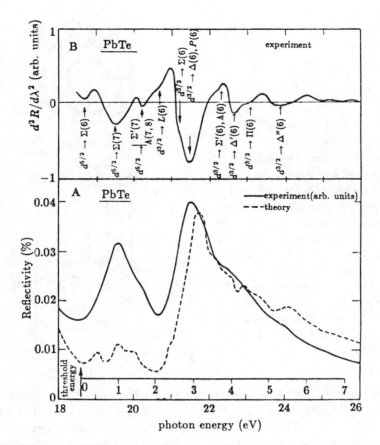

Figure 13.6: Reflectivity of PbTe in the conduction band range with excitation from the core levels $d^{3/2}$ and $d^{5/2}$ (distance from the conduction band edge ~ 18.6 eV—see shifted scale.) The upper part of the figure shows the second derivative of the reflectivity and exposes more structure of the reflection spectrum. The transitions to the corresponding critical points are identified (after Martinez et al., 1975).

of typically ~ 0.5 eV is encountered, which is due to insufficient knowledge of the lattice potential at that time.

13.1.4 The Fundamental Absorption Edge

Attention will now be focused on the energy range near the threshold for valence-to-conduction band transitions, the *fundamental absorption edge*. Usually, several excitation processes are possible between different subbands for a given optical excitation energy. For a photon energy slightly exceeding the band gap, transitions between the different valence bands into different $E(k)$ values of the conduction

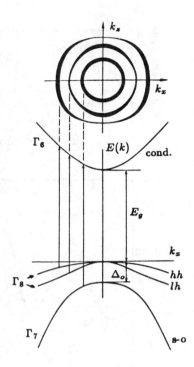

Figure 13.7: Optical vertical excitation transitions for monochromatic photons near the band edge of GaAs near $k = 0$ from the heavy hole (hh), light hole (lh), and spin-orbit split-off (so) valence bands into the conduction band. The upper diagram indicates the resulting electron distribution in k-space. The flattened outer rings indicate the strongly warped hh and the lesser warped lh valence bands (see Section 9.1.5B); the split-off valence band is not warped (after Lyon, 1986).

band are possible near $k = 0$, as shown in Fig. 13.7. The distribution of these electrons is represented by nearly spherical shells (upper part of Fig. 13.7). Deviations from spherical shells occur because of the warping of the valence bands. Thermalization will average the excited electron distribution to approach a Boltzmann distribution (see Section 27.2.2).

The transition between the top of one valence band to the bottom of the lowest conduction band when both lie at $k = 0$ is discussed first. This transition is responsible for the *direct optical absorption edge*. The joint density of states is estimated from a parabolic approximation of both bands:

$$h\nu = E_c(\mathbf{k}) - E_v(\mathbf{k}) = E_g + \frac{\hbar^2 k^2}{2m_n} + \frac{\hbar^2 k^2}{2m_p} = E_g + \frac{\hbar^2 k^2}{2m_r}. \quad (13.13)$$

With E_g as the band gap energy and m_r as a reduced carrier mass

$$\frac{1}{m_r} = \frac{1}{m_n} + \frac{1}{m_p}, \quad (13.14)$$

we obtain for the joint density of states*

$$J_{cv} = \frac{1}{2\pi^2} \left(\frac{2m_r}{\hbar^2}\right)^{3/2} \sqrt{h\nu - E_g}. \tag{13.15}$$

After introducing J_{cv} into Eq. (13.10), we obtain for the optical absorption coefficient near the band edge (Moss et al., 1973):

$$\alpha_{o,cv} = \frac{4\pi^2 2^{3/2} e^2 m_r^{3/2}}{m_0^2 \varepsilon_0 n_r h^2 c} \sqrt{h\nu - E_g} |\mathbf{e} \cdot \mathbf{M}_{cv}|^2. \tag{13.16}$$

The momentum matrix element can be approximated for transition from the three valence bands in zinc-blende-type semiconductors across the band gap (near E_g) from the Kane estimate (Kane, 1957) when using

$$|\mathbf{e} \cdot \mathbf{M}_{cv}|^2 = \left(\frac{m_0}{\hbar}\right)^2 P^2 \tag{13.17}$$

where P is the interband (momentum matrix) parameter, which can be obtained from the $\mathbf{k} \cdot \mathbf{p}$ perturbation theory (Section 9.1.7)

$$P^2 = \left(\frac{1}{m_n^0} - \frac{1}{m_0}\right) \frac{3}{2} \hbar^2 \frac{E_g + \Delta_0}{3E_g + 2\Delta_0} E_g \tag{13.18}$$

[compare with Eq. (9.28)]. Here Δ_0 is the spin-orbit splitting energy. With known m_n, E_g, and Δ_0, we obtain a numerical value for P, and therefore for the matrix element†

$$|\mathbf{e} \cdot \mathbf{M}_{cv}|^2 = \frac{3}{2} \frac{m_0}{m_n} (m_0 - m_n) \frac{E_g + \Delta_0}{3E_g + 2\Delta_0} E_g$$

$$= 2.186 \cdot 10^{-55} \frac{m_0}{m_n} \left(\frac{m_0 - m_n}{m_0}\right) \frac{E_g + \Delta_0}{3E_g + 2\Delta_0} E_g \left(\frac{W^2 s^4}{cm^2}\right). \tag{13.20}$$

---

* Recognizing that Eq. (13.5) yields $J_{cv} = \frac{2}{(2\pi)^3} \frac{d}{d(h\nu)} \left(\frac{4\pi}{3} k^3\right)$; using then Eq. (13.13), we obtain Eq. (13.15).

† The momentum matrix element with dimensions $W^2 s^4 cm^{-2}$ should not be confused with the often used oscillator strength

$$f_{cv} = \frac{2}{3m_0 h\nu} |\mathbf{e} \cdot \mathbf{M}_{cv}|^2, \tag{13.19}$$

which is dimensionless and on the order of one, while the matrix element is not. The factor 1/3 in Eq. (13.19) is due to averaging, with $|M_x|^2 = |M_y^2| = |M_z|^2 = 1/3|\mathbf{M}|^2$.

When it is introduced into Eq. (13.16), we obtain for $h\nu \simeq E_g$:

$$\alpha_{o,\mathrm{cv}} = \frac{12\pi^2 e^2 \sqrt{2m_0}}{\varepsilon_0 n_r h^2 c} \left(\frac{m_n m_p}{m_0(m_n + m_p)} \right)^{3/2} \frac{m_0 - m_n}{m_n} f(E_g)$$

$$= 5.92 \cdot 10^7 \left(\frac{m_r}{m_0} \right)^{3/2} \frac{m_0 - m_n}{m_n} \frac{3}{n_r} f(E_g) \qquad (13.21)$$

with $\quad f(E_g) = \dfrac{E_g + \Delta_0}{3E_g + 2\Delta_0} \sqrt{h\nu - E_g} \qquad (13.22)$

which increases proportional to the square root of the energy difference from the band edge:

$$\boxed{\alpha_{o,\mathrm{cv}} = \alpha_{o,\mathrm{dir}} \propto \sqrt{h\nu - E_g}.} \qquad (13.23)$$

13.2 Direct and Indirect Transitions

The optical transitions discussed in the previous section are *direct transitions*; they involve a direct optical excitation from the valence to the conduction band, using only photons. This is represented by a vertical transition in the reduced $E(\mathbf{k})$ diagram, since the momentum of a photon, which is added to or subtracted from the electron momentum when a photon is absorbed or emitted, is negligibly small: the ratio of photon to electron momentum can be estimated from $p_{\mathrm{ph}}/p_{\mathrm{el}} = (h\nu/c)/(\hbar k)$ which, at the surface of the Brillouin zone, is $\simeq (h/\lambda)/(\hbar\pi/a) = 2a/\lambda \simeq 10^{-3}$ for visible light.

Nonvertical transitions are possible when, in addition to the photon, *phonons* are absorbed or emitted during the transition. These transitions are *indirect transitions*. These phonons can provide a large change in the electron momentum $\mathbf{k}$ (the phonon momentum is identified by $\mathbf{q}$):

$$\mathbf{k}' = \mathbf{k} \pm \mathbf{q}. \qquad (13.24)$$

Because of the necessity of finding a suitable phonon for the optical transition to obey energy and momentum conservation for the transition, the probability of an indirect transition is less than that of a direct transition by several orders of magnitude. An estimate of the matrix elements for phonon-assisted electron transitions follows the corresponding selection rules including the symmetry of the phonon (Lax and Hopfield, 1961).

Indirect transitions identify the band gap for materials where the lowest energy minimum of the conduction band is not at the

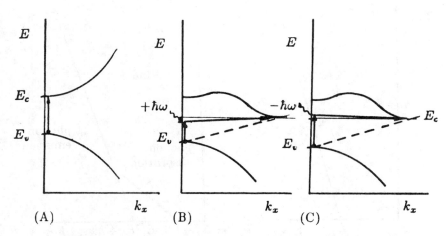

Figure 13.8: (A) Direct and (B) indirect transitions with absorption; and (C) with emission of a phonon.

same **k** as the highest maximum of the valence band. For instance, the highest maximum of the valence band is at **k** = 0, but the lowest minimum of the conduction band is at **k** ≠ 0 (Fig. 13.8). These materials are *indirect band gap semiconductors*; Ge, Si, GaP, and AgCl are examples. Most other semiconductors are *direct gap semiconductors*.

When, together with the photon, a phonon is absorbed, we obtain for the absorption constant

$$\alpha_o^{(\text{abs})} = \frac{2\pi e^2 f_{\text{BE}}|M|^2}{m_0^2 \nu c n_r} \int\int_{Bz} \frac{2 d\mathbf{k}_1 \, d\mathbf{k}_2}{(2\pi)^3(2\pi)^3} \delta(E_c(\mathbf{k}_2) - E_v(\mathbf{k}_1) - h\nu + \hbar\omega)$$

(13.25)

where f_{BE} is the Bose-Einstein distribution function for the phonon [Eq. (27.26)], and M is the matrix element for the simultaneous absorption of a phonon and a photon—see Bassani and Pastori Parravicini (1975). Using a parabolic band approximation and integrating over the delta function, we obtain

$$\alpha_o^{(\text{abs})} = \frac{2\pi e^2 f_{\text{BE}}|M|^2}{m_0^2 \nu c n_r} \frac{1}{8(2\pi)^3} \left(\frac{2m_p}{\hbar^2}\right)^{3/2} \left(\frac{2m_n}{\hbar^2}\right)^{3/2} (h\nu - E_g + \hbar\omega)^2.$$

(13.26)

A similar value for the absorption constant is obtained when a phonon is emitted, except $\hbar\omega$ is replaced by $-\hbar\omega$ and f_{BE} is replaced by $1 - f_{\text{BE}}$. The total optical absorption results as the sum

$$\alpha_{o,\text{ind}} = \alpha_o^{(\text{abs})} + \alpha_o^{(\text{em})}.$$

(13.27)

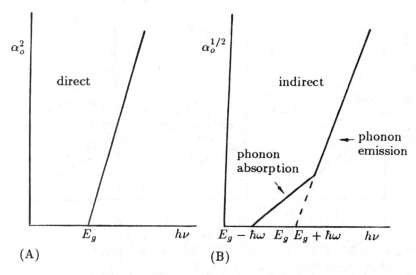

Figure 13.9: Simple theoretical behavior of the absorption coefficient for (A) direct and (B) indirect transitions.

It has a quadratic dependence on the photon energy (for more detail, see Moss et al., 1973):

$$\alpha_{o,\text{ind}} \propto f_{\text{BE}}(h\nu + \hbar\omega - E_g)^2 + (1 - f_{\text{BE}})(h\nu - \hbar\omega - E_g)^2,$$

(13.28)

where $\hbar\omega$ is the energy of a phonon of proper momentum and energy. The two branches of indirect transitions [Eq. (13.28) and Fig. 13.9] show a different temperature dependence. The phonon absorption branch vanishes at low temperatures, when the phonons are frozen-out.

Weak indirect transitions, resulting in a factor $\sim 10^3$ lower absorption, can be observed only in a wavelength range where they do not compete with direct transitions. Indirect transitions are followed at higher energies by direct transitions with a secondary band edge. Figure 13.10 shows an example for Ge where the indirect band edge is preceding the direct edge.

13.2.1 Allowed and Forbidden Transitions

The matrix element [Eq. (13.2)] contains two terms; the second one was neglected in the previous discussions. The one described before relates to an *allowed transition*, and has a value close to 1 for its oscillator strength; here the selection rules are fulfilled. The other term relates to a *forbidden transition*; it represents a transition in

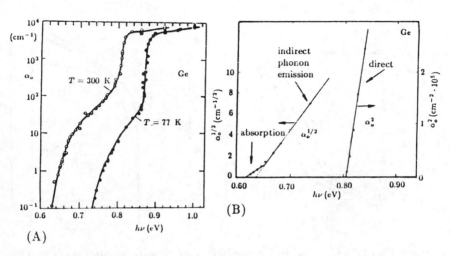

Figure 13.10: (A) Absorption coefficient near the band edge for Ge (after Dash and Newman, 1955). (B) Analysis of the lower $(10^{-1}\ldots10^1\ \text{cm}^{-1})$ and upper $(10^1\ldots5\cdot10^2\ \text{cm}^{-1})$ part of the absorption range at 300 K, giving evidence for indirect and direct absorption edges following each other at 0.66 and 0.81 eV, respectively (after Bube, 1974).

which the Bloch function of the electron wave is orthogonal to the electric polarization of the light. The forbidden transition becomes finite but small when the electron momentum changes slightly during a transition from the ground to the excited state, which is caused by the finite, small momentum of the photon. It becomes important for the agreement between theory and experiment for the absorption spectrum further away from the band edge (Johnson, 1966). This is indicated in Fig. 13.11B for the example of InSb.

The matrix element for such a forbidden transition may be calculated from the $\mathbf{k}\cdot\mathbf{p}$ perturbation theory with

$$\mathbf{M}_{cv}(\mathbf{k}) = (\mathbf{k} - \mathbf{k}_0)|\nabla_{\mathbf{k}}M_{cv}(\mathbf{k})|_{\mathbf{k}=\mathbf{k}_0} \tag{13.29}$$

These transitions become more allowable for larger values of $\mathbf{k}$, and therefore provide a larger contribution to the optical absorption (see Fig. 13.11B).

Forbidden transitions are always additional components for both direct and indirect transitions. The absorption coefficient for these transitions varies as a function of photon energy as (Bardeen et al., 1956):

$$\alpha_{o,\text{dir,forb}} = A_{\text{df}}(h\nu - E_g)^{3/2} \tag{13.30}$$

$$\alpha_{o,\text{ind,forb}} = A_{\text{if}}(h\nu \pm \hbar\omega - E_g)^3. \tag{13.31}$$

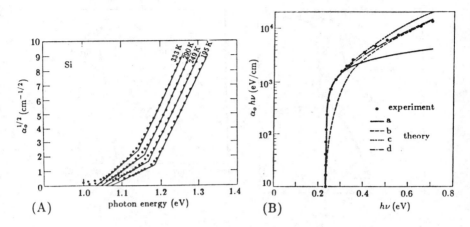

Figure 13.11: (A) Absorption coefficient of Si with temperature as the family parameter indicating decreasing contribution of the phonon absorption branch with decreasing T (after MacFarlane and Roberts, 1955). (B) Absorption coefficient near the band edge of InSb; (a) direct allowed transitions; (b) direct forbidden transitions; (c) as curve (a), however, corrected for nonparabolic bands; (d) as curve (c), however, corrected for the k-dependence of matrix elements (after Johnson, 1966).

The proportionality coefficient for forbidden direct transitions is given for parabolic bands as

$$A_{\mathrm{df}} = \frac{8\pi^3 e^2 (2m_r)^{3/2}}{3m_0^2 n_r \varepsilon_0 ch^5 \nu} |\nabla_{\mathbf{k}} \mathbf{M}_{\mathrm{cv}}|^2_{\mathbf{k}=\mathbf{k}_0} \qquad (13.32)$$

(Moss et al., 1973).

13.2.2 Multiphonon Assistance of the Optical Excitation

In Section 13.2 we discussed indirect transitions in which the phonon provides a significant momentum exchange and thereby permits access for an excited valence electron from the Γ point to a lower satellite valley.

Direct transitions can also be assisted by TO phonons or by an LO and TO two-phonon process according to selection rules, thereby permitting optical absorption with a photon energy below the absorption edge.

Usually, an exciton spectrum lies below the absorption edge— see Section 15.1. Phonons will interact with these excitons and can provide a more structured tail of the optical absorption, as observed

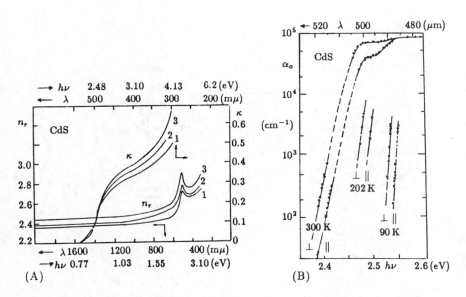

Figure 13.12: (A) Refractive index and extinction coefficient for CdS measured at 300, 415, and 450 K for curves 1–3, respectively (after Khawaja and Tomlin, 1975). (B) Absorption edge of CdS measured in polarized light at 300, 202, and 90 K (after Dutton, 1958).

in GaP by Gershenzon et al. (1962). This will be discussed in Section 15.1—see Fig. 15.12.

13.2.3 Transitions from Different Valence Bands

Transitions from the two upper valence bands, which are split at $\mathbf{k} = 0$ *in anisotropic lattices*, can be separated by using polarized light. For instance, the band edge transitions from the two valence bands, commonly identified as V_1 and V_2, which are separated by crystal field splitting in CdS, can be identified. When polarized light with its electric vector perpendicular to the c-axis is used, transitions from the V_1 band are allowed, while only transitions from V_2 are allowed when the electric vector is parallel to the c-axis (Fig. 13.12). Transitions from the spin-orbit split-off band overlap, and are more difficult to separate because they cannot be turned off individually.

13.2.3A Anisotropy of the Optical Absorption
In anisotropic semiconductors, certain band degeneracies in valence and conduction bands are removed and the optical absorption depends on the relative orientation of the crystal with respect to the light beam and its polarization. Only such transitions which have a component

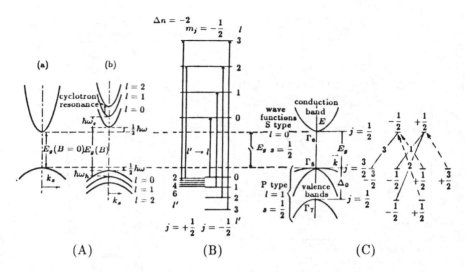

Figure 13.13: (A) Schematics of single bands for $k \simeq 0$ transitions (a) without and (b) with a magnetic induction. (B) Splitting of these bands into Landau levels with possible major transitions indicated for the different ladders. In addition, the spin splitting multiplicity is indicated for the first step in the ladder. (C) Valence band splitting without magnetic field and allowed transitions for σ^+ (shown as solid lines) and σ^- (shown as dashed lines) transitions in zinc-blende crystals (absorption with intensity ratio indicated at center of arrows) (after Weisbuch and Hermann, 1977).

of the dipole orientation in the direction of the electrical vector of the light can be excited. Consequently, the absorption spectrum becomes angle dependent: it is anisotropic. An example for such anisotropy related to valence band splitting was given in the previous section.

13.3 Band-to-Band Magneto-Absorption

In a magnetic field, band degeneracies are lifted. Band states are split into different *Landau levels*. Such splitting is proportional to the magnetic induction. It is described in more detail in Section 31.3.1.

Direct band-to-band transitions at $k \simeq 0$ in a magnetic field are then given by transitions between the different Landau levels of the valence and conduction bands.

The splitting of the bands is shown in Fig. 13.13A for a simple two-band model. The allowed transitions near $k = 0$ are shown in Fig. 13.13A and B, and are given by

$$\Delta E = E_g + \left(l + \frac{1}{2}\right)\frac{\hbar e B_z}{m_r} + m_j g \mu_B B_z + \frac{\hbar^2 k^2}{2 m_r} \qquad (13.33)$$

where l is the orbital, m_j is the azimuthal and s is the spin quantum number; μ_B is the Bohr magneton, and m_r is the reduced effective mass $(m_n^{-1} + m_p^{-1})^{-1}$. Here g is the Landé g-factor (see also Section 31.3.2), which is derived from the g-tensor (Yafet, 1963); it is dependent on the relative direction of the magnetic induction to the principal axis of the conduction band ellipsoid:

$$g = \sqrt{g_\parallel^2 \cos^2 \theta + g_\perp^2 \sin^2 \theta} \quad \text{with} \quad g_\parallel = g_{zz} \quad \text{and} \quad g_\perp = g_{xx} = g_{yy}.$$

The transitions are subject to *selection rules* which depend on the polarization of the radiation. These selection rules are

$$\text{for } \mathbf{E} \| \mathbf{B} : \quad \Delta k_z = 0, \ \Delta l = 0, \ \Delta m_j = 0$$
$$\text{for } \mathbf{E} \perp \mathbf{B} : \quad \Delta k_z = 0, \ \Delta l = 0, \ \Delta m_j = \pm 1,$$

(13.34)

for right and left circular polarized light with, $\Delta m_j = +1$ and -1 indicated as σ^+ and σ^-, respectively. The intensity ratio of the different transitions is indicative of the original degree of spin orientation. For instance, for equidistributed spin of conduction electrons, the intensity ratio of absorption for right polarized light $\left(-\frac{3}{2} \rightarrow -\frac{1}{2} \text{ to } -\frac{1}{2} \rightarrow +\frac{1}{2}\right)$ is 3:1.

There are four possibilities at $k = 0$ for the angular momentum quantum number $(j = m_l + s)$ of the valence band: $j = -3/2, -1/2, 1/2,$ and $3/2$. Those of the conduction band are $-1/2$ and $+1/2$. For $s = -1/2$, we have transitions from $j = +1/2$ and $-3/2$. For $s = 1/2$, we have transitions from $j = +3/2$ and $-1/2$. The permitted transitions are shown in Fig. 13.13B.

The general behavior at room temperature can be seen in Fig. 13.14A with Landau levels identified. They shift linearly with the magnetic induction (Fig. 13.14B) and permit a measurement of the reduced effective mass m_r^{-1} from the slopes, and the direct band gap from an extrapolation to $B \rightarrow 0$.

At low temperatures, the transitions from the light and heavy hole bands and from the spin split-off band are resolved. An example is shown in Fig. 13.15 (Zawadzki and Lax, 1966). An analysis of such spectra also provides information about band anisotropies (Sari, 1972). For a review, see Mavroides (1972).

13.4 Superlattice Band-to-Band Absorption

One of the most direct confirmations of the simple quantum-mechanical model of quasi-free conduction electrons in a periodic well struc-

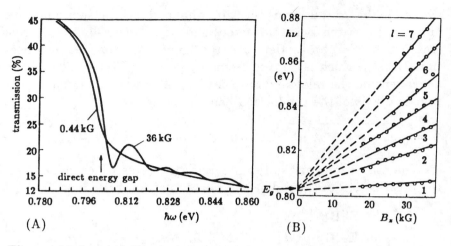

Figure 13.14: (A) Transmission spectrum of Ge at 300 K, 36 kG, and 0.4 kG. (B) Principal minima of these magneto-oscillations as a function of the magnetic induction (after Zwerdling et al., 1957).

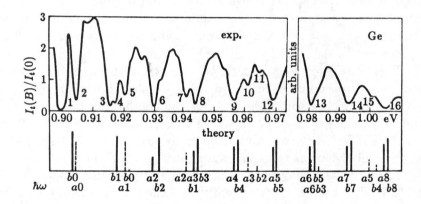

Figure 13.15: Observed and calculated Landau transition spectra for the direct gap in Ge at 4.2 K and 38.9 kG with $\mathbf{B}\|\mathbf{E}$ (after Roth et al., 1959).

ture is obtained from the measurement of the optical absorption spectrum in superlattices.

Superlattices are characterized by mini-bands that develop from the valence and conduction bands of the well material, whose positions and widths are determined by the height and width of the interfacing barrier layers. Optically induced transitions give an instructive picture of these mini-band structures, and reflect selection rules.

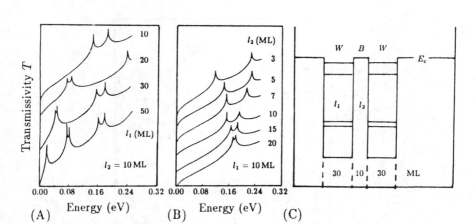

Figure 13.16: Optical transmission spectrum of two GaAs quantum wells coupled by one $Al_{0.35}Ga_{0.65}As$ barrier layer with (A) variation of the well thickness l_1 and (B) variation of the barrier thickness l_2 expressed in numbers of monolayers (ML) and showing the split into two levels for each quantum state. (C) Schematics of the well and barrier geometry for 30/10/30 ML structure—third curve in A (after Torabi et al., 1987).

A development of such mini-bands is shown in Fig. 13.16. Two quantum wells are formed by well (W) layers of GaAs, and are separated by one barrier (B) layer of $Al_{0.35}Ga_{0.65}As$. In a sequence of experiments the well or barrier width was varied.

Each quantum well contains a set of energy levels that can be calculated from a single square well potential with infinite barrier height, yielding

$$E_n = \frac{\hbar^2 \pi^2}{2m_n l_1^2} n_q^2, \quad \text{with } n_q = 1, 2, \ldots \quad (13.35)$$

which shows a shift toward higher energies and wider spacing of the lines with decreased *well width* l_1 [see Eq. (22.5)]. In addition, the lines split and the splitting increases with reduced *barrier width*, as shown in Fig. 13.16B.

With decreasing barrier width, these lines also become broader (Schulman and McGill, 1981); here, the interaction between the wells becomes more probable (see Section 22.1.1).

In both subfigures the width of one layer is kept constant (here at 10 monolayers $\simeq 28\,\text{Å}$) while changing the width of the other layer. The observation agrees with the calculation of a simple one-dimensional square well/barrier potential.

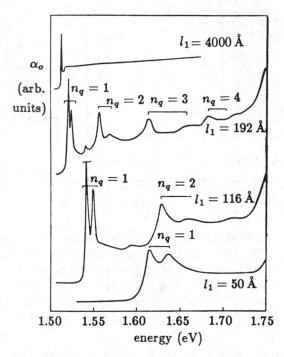

Figure 13.17: Optical absorption spectrum of GaAs/GaAlAs superlattices with different lattice constants. The doublets are due to transitions from light and heavy hole mini-bands. The $l_1 = 4000$ Å sample shows bulk properties with the exciton peak at 1.513 eV—see Fig. 15.5 (after Dingle et al., 1974).

When more wells in a periodic superlattice structure can interact (see Section 22.1.1), the lines spread to *mini-bands*: due to the interaction of a sufficient number of layers, the discrete features shown in Fig. 13.16 are broadened into continuous mini-bands, as indicated in a sequence of curves in Fig. 13.17 and discussed in the following section. The mean free path of electrons, however, gives their coherence length across which superlattice periodicity is recognized.

13.4.1 Absorption in Compositional Superlattices

When the well width is very large, the optical absorption spectrum for superlattices is identical to that of the bulk (well) material (Fig. 13.17, uppermost curve).

With decreasing well width, the onset of optical absorption shifts to higher energies, and relatively sharp lines appear within the band

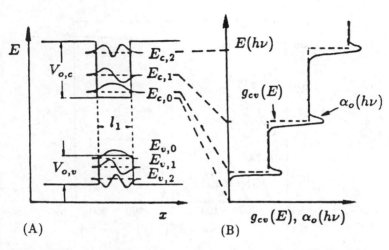

Figure 13.18: (A): Envelope functions of three eigenstates in a single quantum well. (B): Corresponding joint density of states from valence to conduction "band" transitions (dashed curve) and absorption coefficient including a simple excitonic feature preceding each step (solid curve).

of the well material, as shown in the second and third curves. The lines correspond to the eigenstates of conduction electrons in this well, which for a simple square well are given in Eq. (13.35), indicating a $1/l_1^2$ dependence (Dingle, 1975; Miller et al., 1980; Pinczuk and Worlock, 1982). The doublets shown in this figure are due to transitions from the light and heavy hole mini-bands which were omitted in Fig. 13.18 for clarity.

The lines become broader and mini-bands develop as the wells and the barrier become thinner (lowest curve).

The envelope functions of the different eigenstates of valence and conduction electrons within the well are shown in Fig. 13.18A for a deep well with wide barrier. For the optical transitions, the figure indicates maximum overlap between eigenfunctions of the same subband index, while transitions between subbands of different indices have rather small transition probabilities. This is the basis for the corresponding selection rules.

The joint density of state function (Fig. 13.18B) is step-like for superlattices with wide barriers. These steps become smooth when mini-bands develop and broaden with decreasing barrier width—see Fig. 9.33.

Due to excitonic absorption (see Section 15.3), peaks occur slightly below each subband threshold, which causes a modified, in-

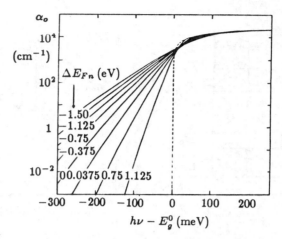

Figure 13.19: Calculated changes in the absorption coefficient of GaAs n-i-p-i doping superlattice, with changing injection rate of carriers indicated by a change in the split of quasi-Fermi levels as family parameter. For $(N_d l_n, N_a l_p) > 5 \cdot 10^{12}$ cm^{-2} and $l_n = l_p = 750$ Å. Dashed curve shows the absorption coefficient without excitonic absorption (after Döhler, 1987).

creased absorption at each lower edge shown by the solid curve of Fig. 13.18B. This is in general agreement with the experiment—see also Section 15.3

The confinement of the electron eigenfunctions in narrow mini-bands yields high oscillating strength for optical transitions. The optical absorption into the mini-bands is a direct one, even though the well material may have an indirect band gap.

13.4.1A Zero-Dimensional Superlattice

A strong confinement of the eigenfunction in quantum wires or quantum boxes causes a further increased optical absorption probability between allowed states (see Fig. 9.34C).

A treatment by Brodsky (1980), of α-Si:H as producing barriers between submicroscopic islands of Si, incorporates typical superlattice features, namely *zero-dimensional* Si *boxes*. Such a model can explain the observed increase in optical absorption and a band edge shift toward higher energy of α-Si:H compared to crystalline Si: each quantum box absorbs light according to its mini-bands. The natural variation of the box size eliminates any structure in the absorption spectrum and makes it appear as an absorption edge.

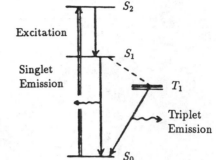

Figure 13.20: Singlet/triplet exciton schematics with ground state S_0, excited singlet states S_1 and S_2, and triplet state T_1.

13.4.2 Absorption in Doping Superlattices

Doping superlattices show several differences from compositional superlattices:

(1) only one absorption peak is observed, since the wells and barriers are usually wider than in compositional superlattices and the wells are rather shallow;

(2) a Franz-Keldysh shift (see Section 42.4.3H) of the absorption edge is observed, resulting in a broadening of the absorption onset, since the field at the barrier edges extends over a substantial material width; and

(3) the absorption onset can be shifted with changing carrier density, as shown in Fig. 13.19. The depth of the well can be modulated by carrier injection—see Section 45.6.4. For more detail, see Ruden (1987) and Döhler (1987).

The mini-bands are influenced to a minor extent by hydrostatic pressure, and substantially by uniaxial stress, as well as electric or magnetic fields.

13.5 Optical Band Gap of Amorphous Semiconductors

The optical band gap of amorphous semiconductors is less well-defined than in a crystalline material (Section 9.4), due to substantial tailing of defect states into the band gap (Mott and Davis, 1979; Sayakanit and Glyde, 1987). Usually, an *effective band gap* is taken from the optical absorption spectrum plotted in an $(\alpha_o h\nu)^{1/2}$ vs. $(h\nu - E_g)$ presentation and extrapolated to $\alpha_o = 0$, or—for practical purposes in thin films, presented as E_{40}—the energy is used at which α_o reaches a value of 10^4 cm^{-1}. Plotting the absorption coefficient vs. band gap energies in a semi-logarithmic graph, we obtain a straight

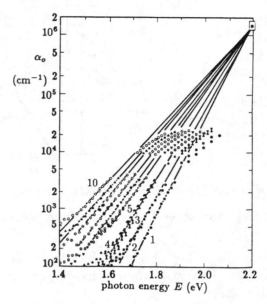

Figure 13.21: Optical absorption of α-Si:H as a function of the photon energy with measurement temperature T_M = 1.7, 151, and 293 K or treatment temperature T_H = 523, 723, 748, 823, 853, 873, and 898 K for curves 1–10, respectively, indicating an additive effect of temperature and structural disorder (after Cody et al., 1981).

line and deduce from its slope the *Urbach parameter* E_0 (see also Chapters 24 and 25)

$$\alpha_o = \alpha_{o0} \exp\left(\frac{h\nu - E_g}{E_0}\right), \qquad (13.36)$$

which gives the steepness of the level distribution near the band edge.

A particularly striking example of band tailing of α-Si:H is given in Fig. 13.21 which demonstrates the relation between an increased severity of disorder and the slope of the band tailing (Cody et al., 1981). Such increased severity is caused by a heat treatment of α-Si:H at temperatures above 700 K, when hydrogen is released: thereby dangling bonds are created which act as major defects—see Section 25.2). The tailing, described by Eq. (13.36), increases from E_0 = 50 to 100 meV with increasing treatment temperature.

Another example for hydrogenated (or fluorinated) amorphous Si and $Si_{1-\xi}Ge_\xi$ alloys is shown in Fig. 13.22. The α-Si:H shows a relatively steep absorption edge, indicating direct absorption; the slope decreases with alloying.

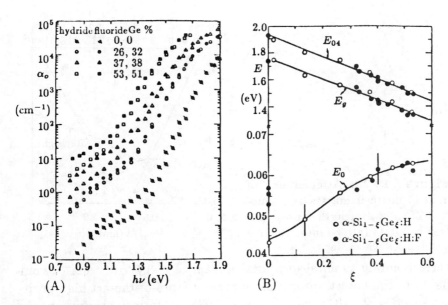

Figure 13.22: (A) Optical absorption spectra of α-$Si_{1-\xi}Ge_\xi$:H and α-$Si_{1-\xi}Ge_\xi$:F. (B) Variation of the band gap energies with composition (after Mackenzie et al., 1985).

The hydrogenated $Si_{1-\xi}Ge_\xi$ alloys show little bowing in spite of the large lattice mismatch between Si and Ge. This is characteristic for one-atomic semiconductors which show less restraint against lattice deformation than two-atomic or higher-atomic lattices—see also Section 9.4 and Chapter 25.

13.5.1 Extrinsic Absorption in Glasses

The absorption of amorphous materials below the band edge decreases rapidly, following the Urbach tail, and reaches values that can be much lower than those for high purity crystals. It is difficult to make crystals with very low defect densities, while in amorphous structures most defects, i.e., wrong bonds or impurities, are incorporated into the host material and yield a much lower optical absorption far from the absorption edge. This makes glasses good candidates for *optical fibers in communication systems*.

Fused silica (SiO_2) is presently used at a length of up to 30 km between repeater stations with an absorption of $\sim$ 0.16 dB/km*

---

* 1 dB (*decibel*) is equal to the logarithm (base ten) of the ratio of two *power* levels having the value 0.1; that is: $\log_{10}(p_1/p_0) = 0.1$, or

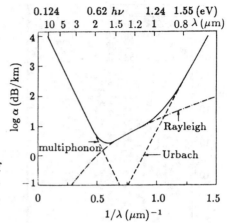

Figure 13.23: Attenuation of light in intrinsic glasses as a function of the wavelength for a hypothetical glass (after Lines, 1986). ©Annual Reviews, Inc.

corresponding to an absorption constant of $\simeq 3.7 \cdot 10^{-7}$ cm^{-1}, compared to the lowest absorption of present state-of-the-art high purity GaAs of $\sim 2 \cdot 10^{-3}$ cm^{-1} (Lines, 1986). Optical absorption and pulse spreading because of finite dispersion $(\partial n_r/\partial\lambda)$ are the limiting factors for the length of an optical transmission line (Lines, 1986). Absorption losses are usually caused by larger inclusions, imperfections, and impurities, and require purification in the 10^{13} cm^{-3}, i.e., in the parts per billion (ppb) range. The remaining losses in SiO$_2$ fibers are due to water contamination, which is difficult to remove.

When fibers are prepared well enough so that only intrinsic phenomena limit the absorption, then three effects have to be considered:

(1) the Urbach tail, responsible for electronic transitions, and extending in energy from the band gap down;

(2) multiphonon absorption, extending from the *Reststrahl* frequency up; and

(3) Rayleigh scattering (see Section 17.4), due to density fluctuation in the midfrequency range, as shown in Fig. 13.23.

The lowest intrinsic absorption values predicted for SiO$_2$ are 0.1 dB/km at $\lambda = 1.55$ μm with $\sim 90\%$ of the attenuation stemming from Rayleigh scattering, $\sim 10\%$ from multiphonon absorption, and negligible contribution from the Urbach tail.

$p_1/p_0 = 1.259$. For the example given above of 0.16 dB/km, one measures a reduction of the light intensity by 3.8% per km. The intensity is reduced to 33% after 30 km.

Summary and Emphasis

In this chapter we have sketched the theory of optical transitions between the valence and conduction bands in rather general terms. For transitions near the band edge this theory can be simplified with an effective mass approximation, assuming parabolic band shapes and arriving at quantitative expressions for the absorption constant as function of the photon energy.

Dependent on the conduction band behavior one has strong direct or weak indirect transitions at the band edge. In addition, a contribution of forbidden transitions modifies the absorption further away from the band edge.

Dependent on the slope of the optical absorption in the intrinsic range near the band edge, direct or indirect transitions can be identified unambiguously, and, with the help of an external magnetic field, the effective mass can be determined from the period of magneto-oscillations of the near-edge absorption due to Landau splitting of the band states.

Deviations from the ideal, periodic crystal lattice provide tailing states extending beyond the band edge, usually as an Urbach tail which decreases exponentially with distance from the band edge.

The behavior of the optical absorption near the band edge provides valuable information about the predominant transitions (direct or indirect) in the intrinsic range, and about the degree of disorder in the extrinsic range. An important application of the knowledge of intrinsic transitions relates to the design of efficient solar cells, and of extrinsic transitions to the development of better optical fibers, as examples.

Exercise Problems

1. Give the number of optical transitions per unit volume and time:
    (a) as function of the joint density of states;
    (b) in terms of the expansion of Eq. (13.9);
    (c) and determine m_r in Eq. (13.9) for $m_n = 0.1$ and $m_p = 0.5m_0$.

2. Explain the mathematical transformation shown in Eq. (13.5) from the hint given in the footnote and with Eq. (13.7).

3.(e) Explain the behavior for critical points shown in Fig. 13.1 from the integral given in Eq. (13.5).

4.(e) Copy Figs. 8.8 and 13.2 with the same scale and show explicitly the corresponding transitions, causing the characteristic features in the optical absorption spectrum. Identify the characteristic points with corresponding symbols.

5.(e) Give ε'' in notation corresponding to Eq. (13.12).

6. Plot the temperature dependence of the band gap (Fig. 9.24) and of other characteristic features of the absorption spectrum shown in Fig. 13.5. Discuss your observation.

7.(e) Derive the joint density of states for simple parabolic bands given in Eq. (13.15) from the general expression Eq. (13.5).

8.(*) Derive the explicit expression for the proportionality factor of indirect transitions [Eq. (13.28)]. Where in this expression is the indication that indirect transitions are less probable than comparative direct transition?

9.(*) Transitions from different valence bands in anisotropic semiconductors can be easily distinguished (Fig. 13.12B). How would you recognize transitions from the split-off spin-orbit valence band? Explain the differences.

10.(r) Magneto-electric effects of conduction electrons were explained in Section 12.3. In Section 13.3, magneto-optical effects are discussed. Explain the basic, identical features and point out differences.

11.(r) Optical spectra reveal a great deal about widely different topics in crystal physics:

    (a) What are the characteristic differences in absorption spectra of direct and indirect band gap materials?

    (b) What do you observe with changing temperatures?

12.(1) How large is the spin-orbit splitting of CdS? Which bands does it affect? What is the reason for such splitting? Is there a spin-orbit splitting of the conduction band?

13.(r) Why does the temperature coefficient of the band gap vanish for $T \rightarrow 0\,\mathrm{K}$? Explain for a microscopic model and by using thermodynamics.

14.(e) Explain the development of mini-bands from electronic states in single quantum wells. Show the progressive development of the density of state function.

Chapter 14

The Dielectric Function

The dielectric constant—or better, the dielectric function determines—the screening of the electric field by the surrounding material.

The dielectric constant can be distinguished as a transverse dielectric constant, related to interaction with electromagnetic radiation, and a longitudinal one, related to charge screening. The dielectric screening can be divided into electronic and ionic parts. Both parts have been described in preceding chapters. We will review them here from a common point of view. The transverse dielectric constant is influenced by resonance transitions. It becomes a function of the optical excitation frequency, as shown for ε' in Fig. 14.2. It is no longer a constant, as indicated in a limited frequency interval by the use of ε_{st} or ε_{opt}. Therefore, it is better called the *dielectric function.*

14.1 Longitudinal and Transverse Dielectric Constants

Most of the discussion in the literature deals with the longitudinal dielectric constant, which is responsible for free and bound charge screening and is important for the field distribution seen by carriers near impurities, for plasmon interaction, etc. Much less attention has been given to the transverse dielectric constant which couples with transverse phonons, even though it is responsible for the screening seen by an interacting electromagnetic field, i.e., for optical properties. One of the reasons is the fact that most of this interaction takes place near the center of the Brillouin zone, with negligible change of the momentum vector, and it can be shown that near $\mathbf{q} = 0$ the longitudinal and transverse dielectric constants are the same (Sharma and Auluck, 1981):

$$\varepsilon_L(\mathbf{q} = 0, \omega) = \varepsilon_T(\mathbf{q} = 0, \omega). \tag{14.1}$$

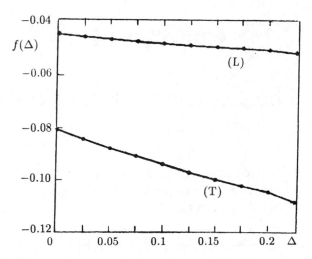

Figure 14.1: Correction function $f(\Delta)$ [Eq. (14.2)] for the longitudinal (L) and transverse (T) dielectric constant near $\omega = 0$ (after Sharma and Auluck, 1981).

This justifies the use of the common $\varepsilon = \varepsilon_L$ for the analysis of most optical data.

When the wave vector deviates from $\mathbf{q} = 0$, however, both dielectric constants become different. For the static dielectric constant, Sharma and Auluck (1981) have evaluated ε_L and ε_T, and obtained for the real part

$$\varepsilon'_{L,T} = \varepsilon'(\mathbf{q} = 0, \omega = 0) - 24 \left(\frac{E_p E_F}{E_g^2 k_F} \right)^2 q^2 \, |f_{T,L}(\Delta)| \qquad (14.2)$$

where E_p is the plasmon energy $(= \sqrt{4\pi\hbar^2 e^2 n/m})$, E_F is the Fermi energy $(= \hbar^2 k_F^2/(2m))$, k_F is the wave number at the Fermi-surface, assumed to be spherical. The band gap is E_g, $\Delta = E_g/(4E_F)$, and $f_{T,L}(\Delta)$ is the correction function for transverse or longitudinal interaction, shown in Fig 14.1. Both ε_L and ε_T decrease proportional to q^2, however, with ε_T becoming smaller than ε_L.

14.2 Dielectric Screening, Function of Frequency

At frequencies above the band gap $(h\nu > E_g)$ electronic band-to-band transitions take energy from the electric (electromagnetic) field and are therefore responsible for the screening, which, dependent

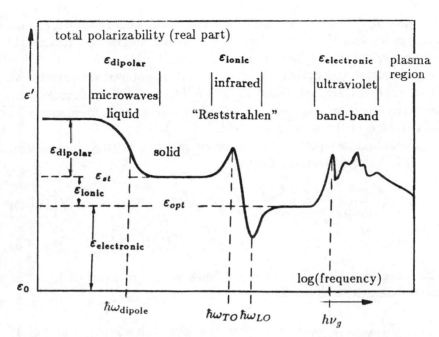

Figure 14.2: Frequency dependence of the real part of the dielectric constant, with ionic and electronic resonances indicated. The structure within the band-to-band transition is indicative of the multilevel transition.

on the excitation spectrum, is a rather complicated function of the frequency.

For $h\nu < E_g$, only few electrons are free; their influence on the dielectric constant is usually negligible. Impurity atoms with electronic resonances in this range are too few to markedly influence $\tilde{\varepsilon}$. Hence, for $\hbar\omega_{TO} < h\nu < E_g$, we have a nearly constant dielectric constant: $\varepsilon' = \varepsilon_{opt}$ and $\varepsilon'' = 0$.

At lower frequencies, ionic motion can follow the ac-electric field with resonances in the *Reststrahl* range. For $h\nu < \omega_{TO}$, the real part of the dielectric constant ε' is given by ε_{st}. The major change in ε occurs near ω_{TO} [Eq. (11.36)], which was discussed in Section 11.2 (Fig. 11.2).

When melting, the semiconductor's molecular dipoles become free to rotate at rather low frequencies, causing another step in the dispersion relation shown in Fig. 14.2. A preceding step is observed in compounds containing radicals such as NO_3, SO_4, etc., which can become free to rotate before melting occurs.

In summary, the tabulated values of the dielectric constants are restricted to the nonresonance frequency ranges of the spectrum:

the optical dielectric constant between band edge and *Reststrahl* frequencies and the static dielectric constant below the *Reststrahl* frequency, but above melting-related transitions.

The static and optical dielectric constants of various crystals at room temperature are listed in Table 14.1. Near resonances, the dielectric function changes in a complex fashion with frequency and polarization of the interacting field.

The imaginary part of the dielectric function is given by [see also Eq. (13.11), and Bassani, 1966]

$$\varepsilon''(\mathbf{q}, \omega) = \frac{4\pi^2 e^2}{\varepsilon_0 m_0^2 \omega^2} \sum_{v,c,k'} \left| \mathbf{e} \cdot \mathbf{M}_{kk'}^{cv} \right|^2 \delta\left(E_c(\mathbf{k}+\mathbf{k}') - E_v(\mathbf{k}) - \hbar\omega\right) f\left[E_v(\mathbf{k})\right] \cdot$$
$$\cdot \left\{ 1 - f\left[E_c(\mathbf{k}+\mathbf{k}')\right] \right\}. \qquad (14.3)$$

The integral in Eq. (13.11) is left here as a sum over all $\mathbf{k}'$. The matrix element is given by

$$\mathbf{M}_{kk'}^{cv} = \int_V \phi_c^*(\mathbf{k}+\mathbf{k}', \mathbf{r}) \exp(i\mathbf{q} \cdot \mathbf{r}) \mathbf{p} \phi_v(\mathbf{k}, \mathbf{r}) d^3\mathbf{r} \qquad (14.4)$$

with $\mathbf{e}$ as the polarization vector of the electric field, $\mathbf{p}$ as the electron momentum vector, $f[E(\mathbf{k})]$ as the distribution function in $\mathbf{k}$-space, and ϕ as the Bloch function for valence (v) or conduction (c) band.

The real part of the dielectric function can be obtained from Eq. (14.3) by using Kramers-Kronig relations (see Section 11.1.4):

$$\varepsilon'(\mathbf{q}, \omega) = 1 + \frac{2}{\pi} P \int_0^\infty \frac{\omega' \varepsilon''(\mathbf{q}, \omega') d\omega'}{\omega^2 - \omega'^2} \qquad (14.5)$$

The evaluation of Eqs. (14.3) and (14.5) is involved. For a simplified approach, see Penn (1962). For a general overview, see Stern (1963).

There are, however, occasions where it is more appropriate to consider dielectric screening as a function of the wave vector.

14.2.1 Dielectric Screening as Function of Wave Vector

In the neighborhood of a lattice defect, one uses the dielectric constant to describe the influence of the screening of the surrounding lattice on the electric potential extending from such a defect. When the polarization of this lattice, farther away from the center by at least a few lattice constants is concerned, the use of the static dielectric constant ε_{st} is justified.

Table 14.1: Static and optical dielectric constants of various crystals.[*]

	ε_{st}	ε_{opt}		ε_{st}	ε_{opt}		ε_{st}	ε_{opt}
C	5.7	5.7	ZnO	8.6	4.0	PbS	19	18.5
Si	11.9	11.9	ZnS	8.6	5.2	PbSe	280	25.2
Ge	16.0	16.0	ZnSe	8.33	5.90	PbTe	450	36.9
AlSb	11.22	9.88	ZnTe	9.86	7.28	CdF	7.78	2.40
GaP	11.0	9.1	TlCl	37.6	5.1	SiC	10	6.65
GaAs	13.1	11.1	TlBr	35.1	5.4	Cu_2O	7.1	6.2
GaSb	15.69	14.44	CuCl	6.3	3.7	LiF	9.3	1.9
InP	12.6	9.6	CuBr	7.0	4.4	KCl	4.49	2.20
InAs	14.61	11.8	CuI	7.1	5.5	KBr	4.52	2.39
InSb	17.88	15.68	AgCl	9.5	3.97	KI	4.68	2.68
CdS	8.42	5.27	AgBr	10.6	4.7	RbCl	4.58	2.20
CdSe	9.3	6.1				RbI	4.55	2.61
CdTe	10.3	6.9				CsI	6.32	3.09

[*]For many applications such a simple representation of ε is much too coarse. For a listing of $n(\omega)$ and $k(\omega)$ in a very large frequency range from vacuum UV to IR, see Palik (1985).

When the interaction in closer vicinity of the center is considered, so that less screening from the lattice occurs, the dielectric constant decreases. For an estimate of such a *dielectric function*, one considers the eigenstates between which such interaction takes place, yielding as the screening function an expression similar to Eq. (13.12) (Cohen, 1963):

$$\varepsilon'(\mathbf{k}, \mathbf{k}') = 1 + \frac{4\pi e^2}{k^2} \sum_{v,c,k'} \frac{|\mathbf{e} \cdot \mathbf{M}_{kk'}^{cv}|^2}{E_c(\mathbf{k} + \mathbf{k}') - E_v(\mathbf{k})}, \qquad (14.6)$$

where $M_{kk'}^{cv}$ is the matrix element for transitions from $\mathbf{k}$ in the valence band to $\mathbf{k}'$ in the conduction band. This expression shows more clearly the decrease of $\varepsilon' \propto k^{-2}$ for larger values of $\mathbf{k}$—see Fig. 14.3.

The screening function was computed for Si first by Nara (1965) and for other semiconductors by Walter and Cohen (1970) and Brust (1972). An approximation was given by Penn (1962):

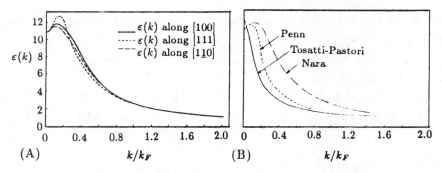

Figure 14.3: Dielectric constant as a function of the wavenumber (A) according to numerical computations by Nara (1965) and (B) in comparison to analytical approximations by Penn (1962) and Tosatti and Pastori Parravicini (1971). ©Pergamon Press plc.

$$\varepsilon'(k) = 1 + \frac{\left(\dfrac{\hbar\omega_p}{E_g^*}\right)^2 \left(1 - \dfrac{E_g^*}{4E_F}\right)}{\left[1 + \dfrac{E_F}{E_g^*}\left(\dfrac{k}{k_F}\right)^2 \sqrt{1 - \dfrac{E_g^*}{4E_F}}\right]^2} \qquad (14.7)$$

where ω_p is the plasma frequency, E_g^* is the energy difference between bonding and antibonding states in the valence and conduction bands (Table 14.2), E_F is the Fermi energy for valence electrons, and $k_F = k(E_F)$; for screening by valence electrons in an isotropic crystal, k_F is given by

$$k_F^3 = 3\pi^2 n_V$$

where n_V is the density of valence electrons. See also Tosatti and Pastori Parravicini (1971).

The behavior of the dielectric constant for Si as a function of the wave number was computed by several groups and is shown in Fig. 14.3. Aside from a small maximum near $k = 0$, it shows a rapid decrease toward $\varepsilon = 1$ for higher k-values, indicating that the capability of the lattice to screen decreases $\propto k^{-2}$, that is, with higher excitation into the band. Since $k \propto 1/r$, it also indicates, that as expected, the screening decreases rapidly as one approaches the immediate proximity of the defect center.

Table 14.2: Electronic crystal parameters: Fermi energy of valence electrons $E_{F,v}$, plasmon energy of valence electrons, fraction of ionicity,* dielectric average band gap E_g^*, and Thomas-Fermi screening length of valence electrons (after van Vechten, 1980).

Crystal	$E_{F,v}$ (eV)	$\hbar\omega_p$ (eV)	f_1 (%)	E_g^* (eV)	λ_{TF} Å	Crystal	$E_{F,v}$ (eV)	$\hbar\omega_p$ (eV)	f_1 (%)	E_g^* (eV)	λ_{TF} Å
C	28.9	31.2	0	13.5	0.39	BeO	25.3	28.3	62.0	18.8	0.40
Si	12.5	16.6	0	4.77	0.48	BeS	15.6	19.7	28.5	7.47	0.45
Ge	11.5	15.6	0	4.31	0.49	BeSe	14.3	18.4	26.1	6.57	0.46
α-Sn	8.72	12.7	0	3.06	0.52	BeTe	12.0	16.1	16.9	4.98	0.49
SiC	19.4	23.2	17.7	9.12	0.43	MgTe	9.04	13.0	55.6	4.80	0.52
BN	28.1	30.6	25.7	15.2	0.39	ZnO	17.6	21.5	65.5	12.5	0.44
BP	17.8	21.7	5.8	7.66	0.44	ZnS	12.6	16.7	62.3	7.85	0.48
BAs	16.1	20.1	2.6	6.47	0.45	ZnSe	11.4	15.6	62.3	6.98	0.49
AlN	19.3	23.0	44.6	11.0	0.45	ZnTe	9.91	14.0	59.9	5.66	0.51
AlP	23.4	16.5	38.8	6.03	0.47	CdS	10.8	14.9	67.9	7.01	0.50
AlAs	11.4	15.5	43.2	5.81	0.47	CdSe	9.94	14.0	68.4	6.42	0.51
AlSb	9.77	13.8	43.3	4.68	0.51	CdTe	8.75	12.7	67.5	5.40	0.53
GaN	18.2	22.1	45.2	10.3	0.44	HgS	10.8	14.9	79.0	8.20	0.50
GaP	12.4	16.5	32.8	5.76	0.48	HgSe	9.91	14.0	68.0	6.06	0.51
GaAs	11.5	15.6	31.0	5.20	0.48	HgTe	8.75	12.6	65.2	4.95	0.52
GaSb	9.82	13.9	26.0	4.12	0.51	CuF	20.3	23.9	76.6	18.1	0.42
InN	14.9	18.9	49.6	8.36	0.46	CuCl	12.6	16.7	74.6	9.60	0.48
InP	10.7	14.8	42.1	5.16	0.50	CuBr	11.1	15.2	73.5	7.35	0.49
InAs	10.1	14.2	35.9	4.58	0.51	CuI	10.1	14.1	69.2	6.61	0.51
InSb	8.76	12.7	32.7	3.76	0.52	AgI	8.77	12.8	77.3	6.48	0.52

*With f defined by $f_1 = (C/E_g^*)^2$ with $C = 1.5e^2(Z_b/r_b - Z_a/r_a)\exp(-r_{ab}/\lambda_{TF})$, Z_i and r_i the valency and covalent radii, and E_g^* the distance of bonding and antibonding band states in valence and conduction bands.

14.3 Empirical Screening Parameters

An empirical formula for the dielectric function was given by Nara and Morita (1966)

$$\frac{1}{\varepsilon(k)} = \frac{Ak^2}{k^2 + \alpha^2} + \frac{Bk^2}{k^2 + \beta^2} + \frac{1}{\varepsilon_{st}}\frac{\gamma^2}{k^2 + \gamma^2} \qquad (14.8)$$

with adjustable parameters which, for Si, have the values $A = 1.175$, $B = -0.175$, $\alpha = 0.7572$, $\beta = 0.3123$, and $\gamma = 2.044$.

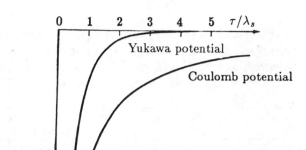

Figure 14.4: Coulomb and screened Coulomb potentials.

Often, instead of the dielectric function one uses an *effective dielectric constant* given by

$$\frac{1}{\varepsilon^*} = \frac{1}{\varepsilon_{\text{opt}}} - \frac{1}{\varepsilon_{\text{st}}} \qquad (14.9)$$

when the screening close to a defect center is concerned. This effective dielectric constant expresses the difference between the electronic polarization already accounted for in a Bloch electron picture and the lattice polarization including ionic displacements—see Section 27.1.2A.

14.3.1 Screened Coulomb Potential

When the Coulomb potential, screened by the dielectric function extends too far from a center to give reasonable agreement with the experiment because of other interfering charges, one introduces a stronger screened Coulomb potential. Such screening can be caused by additional charged carriers that are interfering within the reach of the Coulomb potential. The long-range part of the Coulomb potential can be reduced by an empirical exponential factor

$$V(r) = \left(\frac{e}{4\pi\varepsilon_{\text{st}}\varepsilon_0 r} \right) \exp\left(-\frac{r}{\lambda_s} \right). \qquad (14.10)$$

This potential is called the *Yukawa potential*, with λ_s the screening radius, and is shown in Fig. 14.4 compared with the unscreened Coulomb potential.

When the screening is caused by a relatively small density of free electrons surrounding the center, e.g., a charged defect, λ_s is defined as a *Debye-Hückel type** length*:

$$\lambda_D = \sqrt{\frac{\varepsilon_{st}\varepsilon_0 kT}{e^2 n}} = 381\sqrt{\frac{\varepsilon_{st}}{10}\frac{T}{100}\frac{10^{16}}{n}}\ (\text{Å}). \qquad (14.11)$$

Here, the Boltzmann statistic is applied for the screening electrons. In contrast, λ_s is defined as a *Thomas-Fermi length* when a higher carrier density requires the application of the Fermi-Dirac statistics:

$$\lambda_{TF} = \pi^{2/3}\sqrt{\frac{\varepsilon_{st}\varepsilon_0 \hbar^2}{3^{1/3}m_n e^2 n^{1/3}}} = 11.6\sqrt{\frac{\varepsilon_{st}}{10}\frac{m_0}{m_n}\left(\frac{10^{18}}{n}\right)^{1/3}}(\text{Å}).$$

$$(14.12)$$

Finally, a combination of dielectric and electronic screening was proposed by Haken (1963) who introduced as an effective dielectric constant

$$\frac{1}{\varepsilon_{\text{eff}}} = \frac{1}{\varepsilon_{st}} + \frac{1}{2}\left\{\frac{1}{\varepsilon_{\text{opt}}} - \frac{1}{\varepsilon_{st}}\right\}\left[\exp\left(\frac{r_e}{r_{\text{pe}}}\right) + \exp\left(\frac{r_h}{r_{\text{ph}}}\right)\right] \qquad (14.13)$$

with

$$r_{\text{pe}} = \sqrt{\frac{\hbar}{2m_n\omega_{\text{LO}}}} \quad \text{and} \quad r_{\text{ph}} = \sqrt{\frac{\hbar}{2m_p\omega_{\text{LO}}}} \qquad (14.14)$$

the polaron radii for electrons and holes [see Section 27.1.2A and Eq. (27.5)]; r_e and r_h are the relative distances of the electron and hole from their respective center of gravity within an exciton—see Section 15.1.

By introducing empirical parameters, an improved agreement between theory and experiment is obtained and the results can be given in an analytical form.

Summary and Emphasis

The dielectric constant is one of the most important material parameters describing the ability of a semiconductor to screen an electric field. Since the local electric field determines most of the semiconducting properties, the dielectric constant is contained in almost ev-

---

* In semiconductors, mostly free electrons rather than ions (as in the Debye-Hückel theory) provide the screening. This characteristic length in semiconductor physics is more commonly called the *Debye length*.

ery quantitative description of electrical and optical phenomena in semiconductors.

We have summarized our knowledge about ε in this chapter and emphasized its character as a *function* [of the frequency and of the momentum vector (or space)]. One distinguishes the longitudinal and transverse dielectric function. The former is mostly used. This is justified near $\mathbf{q} = 0$ where both coincide. We presented some overview of the general behavior of the dielectric function. It is influenced strongly by resonance transitions, in the neighborhood of which it changes rapidly with the frequency and can attain rather high values there. It also decreases with increasing $\mathbf{k}$ vector (the transverse dielectric function decreases steeper than the longitudinal one) and approaches 1 at the boundary of the Brillouin zone.

Sufficient knowledge of the appropriate value of ε is essential for the quantitative description of any electro-optical phenomenon in semiconductors. In many cases the use of ε_{st} or ε_{opt} is insufficient and needs to be replaced by the knowledge of ε at the appropriate frequency or $\mathbf{k}$ vector. In the absence of such knowledge, often interpolative or empirical formulae are used.

Exercise Problems

1.(r)   Compare the frequency dependence of the dielectric function of monatomic and diatomic semiconductors. Discuss.

2.   Relate the k-scale shown in Fig. 14.3 to the conventional scale of the Brillouin zone.

3.(*)   The calculation of the Fermi-energy implies a certain approximation. What approximation? When applying this model to the valence band of a semiconductor, $E_{F,v}$ in Table 14.2 does not give the width of the valence band. Why?

4.(r)   Compare the dielectric constants for similar semiconductors and analyze the trend. Discuss your findings and indicate possible reasons.

5.(l)   What is the significance of the parameters α, β, and γ in Eq. (14.8). Why would one want to include such parameters to describe $\varepsilon(\mathbf{k})$? Check with the original literature.

Chapter 15

Excitons and Associates

Optical band-to-band absorption can produce an electron and a hole in close proximity which attract each other and can form a hydrogen-like bond state, the exciton. Its spectrum of ground and excited states, as well as that of molecules or higher associates of excitons, gives valuable information about the electronic structure of the semiconductor.

The transitions discussed in the previous chapters neglect the Coulomb interaction of the electron hole pairs created during the absorption of a photon with band gap energy. With photons of $h\nu \simeq E_g$, both electron and hole do not have enough kinetic energy at low temperatures to separate. They form a bound state. This state can be modeled by an electron and a hole, circling each other much like the electron and proton in a hydrogen atom, except that they have almost the same mass*; hence, in a semi-classical model, their center of rotation lies closer to the middle on their interconnecting axis (Fig. 15.1). This bound state is called an *exciton*.

These excitons have a significant effect on the optical absorption close to the absorption edge. This will be the main topic of this chapter.

---

* This causes the breakdown of the adiabatic approximation. The error in this approximation is on the order of the fourth root of the mass ratio. For hydrogen this is $(m_n/M_H)^{1/4} \simeq 10\%$ and is usually acceptable. For excitons, however, the error is on the order of 1 and is no longer acceptable. This is relevant for the estimation of exciton molecule formation and is discussed in Section 15.1.2E.

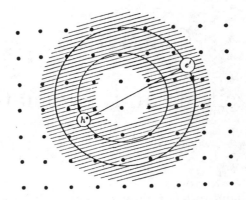

Figure 15.1: Exciton with center of mass slightly shifted toward the heavier hole.

15.1 Exciton Absorption

The Hamiltonian to describe the exciton can be approximated as

$$H = H_0 + U \quad \text{with} \quad U = -\frac{e^2}{\varepsilon \varepsilon_0 |\mathbf{r}_e - \mathbf{r}_h|}, \quad (15.1)$$

with the additional term U describing the Coulomb interaction between the electron and hole. The Schrödinger equation $H\psi = E\psi$ has as eigenvalues a series of quasi-hydrogen states

$$E_{\text{exc},nq} = R_H \frac{m_r}{m_0} \frac{1}{\varepsilon_{\text{st}}^2} \frac{1}{n_q^2} \quad \text{with} \quad R_H = \frac{m_0 e^4}{2(4\pi\varepsilon_0\hbar)^2} = 13.6 \text{ eV} \quad (15.2)$$

with R_H as the Rydberg energy, m_r as the reduced exciton mass given by $m_r^{-1} = m_n^{-1} + m_p^{-1}$ for simple parabolic bands, and n_q as the principal quantum number. The exciton radius is a quasi-hydrogen radius

$$a_{\text{exc},nq} = a_H \frac{\varepsilon_{\text{st}}}{m_r/m_0} n_q^2 \quad \text{with} \quad a_H = \frac{4\pi\varepsilon_0\hbar^2}{m_0 e^2} = 0.529 \text{ Å} \quad (15.3)$$

where a_H is the Bohr radius of the hydrogen atom. For a review, see Bassani et al. (1975), Haken (1976), and Singh (1984).

Depending on the reduced exciton mass and dielectric constant, one distinguishes between *Wannier-Mott excitons*, which extend over many lattice constants and are free to move through the lattice, and *Frenkel excitons*, which have a radius comparable to the interatomic distance. The latter exciton becomes localized and resembles an atomic excited state—for more detail, see Singh (1984). For the large Wannier-Mott excitons, the screening of the Coulomb potential is appropriately described by the static dielectric constant which is used in Eq. (15.2).

When the lattice interaction is stronger, the electron-hole interaction can be described by an effective dielectric constant $\varepsilon^* = (\varepsilon_{opt}^{-1} - \varepsilon_{st}^{-1})^{-1}$, which provides less shielding. A further reduction of the correlation energy was introduced by Haken [Haken, 1963—see Eq. (14.13)]

$$U = -\frac{e^2}{\varepsilon_0|\mathbf{r}_e - \mathbf{r}_h|}\left\{\frac{1}{\varepsilon_{st}} + \frac{1}{2}\left(\frac{1}{\varepsilon_{opt}} - \frac{1}{\varepsilon_{st}}\right)\left[\exp\left(\frac{r_e}{r_{pe}}\right) + \exp\left(\frac{r_h}{r_{ph}}\right)\right]\right\}.$$
(15.4)

The radius of the exciton consequently shrinks, and the use of the effective mass becomes questionable. Here a tight-binding approximation becomes more appropriate to estimate the eigenstates of the exciton, which is now better described as a Frenkel exciton.

For both types of excitons, one obtains eigenstates below the band gap by an amount given by the binding energy of the exciton. In estimating the binding energy, the band structure of valence and conduction bands must be considered, entering into the effective mass and dielectric function. Such structures relate to light and heavy hole bands, energy and position in $\mathbf{k}$ of the involved minima of the conduction band, and other features determining band anisotropies. This will be explained in more detail in the following sections. We will first discuss some of the general features of Frenkel and Wannier-Mott excitons.

15.1.1 Frenkel Excitons

Frenkel excitons (Frenkel, 1931—also see Landau, 1933) are observed in ionic crystals with relatively small dielectric constants, large effective masses, and large coupling constants, and in molecular crystals (see below). These excitons show relatively large binding energies, usually in excess of 0.5 eV, and are also referred to as tight-binding excitons.

Figure 15.2 shows the absorption spectrum of KCl with two relatively narrow Frenkel exciton absorption lines. They relate to the two valence bands at the Γ-point that are shifted by spin-orbit splitting. The doublet can be interpreted as excitation of the Cl^- ion representing the valence bands in KCl. This absorption produces tightly bound excitons, but does not produce free electrons or holes; that is, it does not produce photoconductivity as does a higher

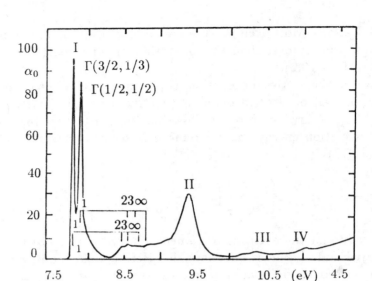

Figure 15.2: Absorption spectrum of KCl at 10 K with two narrow exciton peaks identified as transitions at the Γ-point. The hydrogen-like series are due to the Coulomb tail of the potential (after Tomiki, 1969.)

energy absorption. The excited state of the Cl^- is considered the Frenkel exciton—see Kittel (1966). It may move from one Cl^- to the next Cl^--ion by quantum-mechanical exchange. The longer-range Coulomb potential of the exciton permits additional excited states which have a hydrogen-like character, although with higher binding energy (~ 1 eV) than in typical semiconductors because of a large effective mass and relatively small dielectric constant. An extension of a tight-binding potential with a Coulomb tail is observed in a large variety of lattice defects (see Section 22.1.2), and provides characteristics mixed between a deep level and a shallow level series. It creates a mixture of properties with Frenkel and Wannier-Mott contributions—see Tomiki (1969).

The strength of the exciton absorption substantially exceeds that of the band edge which coincides with the series limit ($n = \infty$). The features to the right of this limit in Fig. 15.2, labeled with roman numerals, result from excitation into the higher conduction bands.

Other Frenkel excitons are observed in molecular crystals, such as in anthracene, naphthalene, benzene, etc., where the binding forces within the molecule (covalent) are large compared to the binding forces between the molecules (van der Waals). Here localized excited states within the molecules are favored.

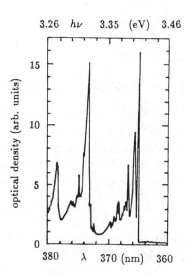

Figure 15.3: Optical absorption spectrum of singlet-triplet excitons in a tetrachlorobenzene crystal plate-let at 4.2 K, measured with unpolarized light (after Nikitine et al., 1961). ©Pergamon Press plc.

15.1.1A Singlet and Triplet Exciton States

If more than one electron is involved in the excited state, we can distinguish singlet and triplet excited states while the ground state is always a singlet state (see Fig. 13.20). In recombination, the singlet-singlet transition is allowed (it is a luminescent transition), while the triplet-singlet transition is spin-forbidden. Consequently, the triplet state has a long lifetime, depending on possible triplet/singlet mixing.

Such singlet and triplet excitons are common in organic semiconductors and have been discussed extensively—see Pope and Swenberg (1982) and Clarke (1982). Their importance in inorganic semiconductors in the neighborhood of crystal defects has been recently recognized (Cavenett, 1984), and also in layered semiconductors (GaS and GaSe—Cavenett, 1980). The most common configuration for singlet/triplet states in inorganic semiconductors is the neutral associate of a donor and acceptor (see Section 21.2.3).

An example of a molecular, singlet and triplet Frenkel exciton is shown in Fig. 15.3 for tetrachlorobenzene (see Nikitine et al., 1961). When the unit cell contains more than one identical atom or molecule, than an additional splitting of the excited eigenstates occurs. This is referred to as *Davidov splitting*, and is observed in organic molecules.

For the exciton transport, we have to distinguish between triplet and singlet excitons—for a short review, see Knox (1984). For the latter, a dipole-dipole interaction via radiation, i.e., luminescence and reabsorption, contributes to the exciton transport. Such a mechanism is negligible for triplet excitons, which have a longer lifetime and therefore much lower luminescence. For exciton diffusion, see Section 36.1.

15.1.2 Wannier-Mott Excitons

Wannier-Mott excitons are found in most of the typical semiconductors and extend over many lattice constants—see Wannier (1937) and Mott (1938). Their eigenstate can be represented as a product of Bloch states of the electron and the hole u_c and u_v, with ϕ an appropriate envelope function—see Section 9.1.1A:

$$\psi = \phi(\mathbf{r}_e, \mathbf{r}_h) u_c(\mathbf{r}_e) u_v(\mathbf{r}_h). \tag{15.5}$$

In simple parabolic bands the function $\phi(\mathbf{r}_e \, \mathbf{r}_h)$ can be expressed in the form

$$\phi_{nq}(\mathbf{K}, \mathbf{R}) = \frac{1}{\sqrt{N\mathcal{V}}} A_{nq}(R) \exp\{i\mathbf{K} \cdot \mathbf{R}\} \tag{15.6}$$

where $\mathbf{K}$ is the wave vector of the exciton ($= \mathbf{k}_e + \mathbf{k}_h$), $\mathbf{R}$ is the center of mass (Fig. 15.1) coordinate, $\mathcal{V}$ is the volume, and $A_{nq}(\mathbf{R})$ is the normalized quasi-hydrogen envelope function (Fig. 21.1):

$$A_{nq}(R) = \frac{1}{\sqrt{\pi a_{\text{exc},nq}^3}} \exp\left(-\frac{R}{a_{\text{exc},nq}}\right) \quad \text{with} \quad R = |\mathbf{r}_e - \mathbf{r}_h|. \tag{15.7}$$

The eigenvalues of this wave function are given in Eq. (15.2). For a short review, see Elliott (1982) or Knox (1984).

The minimum optical energy needed to excite an electron and to create an exciton is slightly smaller than the band gap energy:

$$\boxed{E_{g,\text{exc}} = E_g - \frac{m_r}{m_0 \varepsilon_{\text{st}}^2} R_H \frac{1}{n_q^2} + \frac{\hbar^2 K^2}{2(m_n + m_p)} \quad \text{with } n_q = 1, 2, \dots;}$$

$$\tag{15.8}$$

the second term is the quasi-hydrogen binding energy [Eq. (15.2)], and the third term is the kinetic energy due to the center of mass motion of the exciton. For a listing of exciton energies, see Table 15.1.

The Wannier-Mott exciton is mobile and able to diffuse—see Section 36.1. Since it has no net charge, it is not influenced in its motion by an electric field* and does not contribute directly to the electric current.

---

* It is, however, influenced by a field gradient.

Table 15.1: $1S$-exciton energies [in eV; in parenthesis: $T\,(\mathrm{K})$].

Material	Exciton Ground State	Binding Energy (meV)	Biexciton Ground State	Binding Energy (meV)
Si	1.1545 (1.6)	14.3 (1.8)	2.3077 (1.6)	1.4 (1.6)
Ge	0.7405 (2.1)	4.2		
GaAs	1.5150 (1.8)	4.9		
InP	1.4186 (1.6)	5.1		
	A Exciton		*B* Exciton	
ZnO	3.3758 (4)	61 (1.6)	3.3810 (4)	59
ZnO$_\perp$	3.3776 (4)	61 (1.6)	3.3912 (4)	59
ZnS$_{\mathrm{hex}}$	3.8715 (1.8)		3.8998 (1.8)	
ZnSe	$A_\parallel$ 2.8034 (4.2)	19.9 (1.6)	$A_\perp$ 2.8020 (2)	
ZnTe	$A_\parallel$ 2.3812 (2)	132 (1.6)	$A_\perp$ 2.3808 (2)	
CdS$_\perp$	2.5528 (4.2)	27 (1.8)	2.5682 (4.2)	27 (1.8)
CdS$_\parallel$	2.5541 (4.2)		2.5688 (1.8)	
CdSe	$\parallel$ 1.8262 (1.6)	15 (1.8)	$\perp$ 1.8390 (80)	16 (1.8)
	$\perp$ 1.8134 (80)			
CdTe	$A_\parallel$ 1.59638 (10)	11 (1.6)	$A_\perp$ 1.59573 (10)	11 (1.6)

The ionization energy of these excitons in typical semiconductors is on the order of 10 meV (Thomas and Timofeev, 1980); hence, kT at room temperature is sufficient to dissociate most of them.

The principal quantum number n_q defines s-states which contribute to electric dipole transitions in direct gap semiconductors with allowed transitions, while p-states ($l = 1$) contribute to dipole-forbidden transitions (see Section 15.1.1A). With introduction of symmetry breaking effects, such as external fields, external stresses or those in the neighborhood of crystal defects (Gilason et al., 1982), and moving slightly away from $\mathbf{k} = 0$, the other quantum numbers, l and m, need to be considered. This results in a more complex line spectrum (see Section 15.1.2A.1). Exciton transitions with $n_q \to \infty$ merge into the edge of the band continuum (Fig. 15.4).

At low temperatures, excitons have a major influence on the optical absorption spectrum.

This can be seen from the matrix elements M_{cv} for transitions from near the top of the valence band to the vicinity of the bottom of the conduction band. When considering exciton formation, the band-to-band transition matrix elements [e.g; Eq. (13.20)] are modified by multiplication with the eigenfunction of the exciton $\phi_{n\mathbf{k}}$—see Eq. (15.5) and the following section.

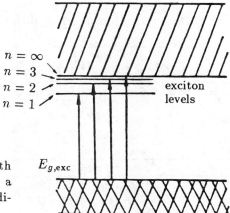

$n = \infty$
$n = 3$
$n = 2$
$n = 1$

exciton levels

Figure 15.4: Band model with exciton levels that result in a hydrogen-like line spectrum for direct gap semiconductors.

$E_{g,\text{exc}}$

15.1.2A Direct Gap Excitons For a single photon absorption at the Γ-point in a direct band gap material, the matrix element is given by

$$M_{\text{cv}}^{(\text{exc})} = M_{\text{cv}}\phi_{nq}(r = 0) \tag{15.9}$$

with M_{cv} given by Eq. (13.20) and

$$\phi_{nq} = \sum_{\mathbf{K}} A_{nq}(\mathbf{K})\exp(i\,\mathbf{K}\cdot\mathbf{R}), \tag{15.10}$$

where $A(\mathbf{K})$ is the Fourier transform of the envelope function $A(\mathbf{R})$ which is given in Eq. (15.7) for isotropic semiconductors. $A_{nq}(\mathbf{K})$ is given by

$$A_{nq}(\mathbf{K}) = \frac{2\pi a_{\text{exc},nq}^2}{\sqrt{\pi a_{\text{exc},nq}^3}}\frac{1}{\left[1 + (K a_{\text{exc},nq})^2\right]^2}. \tag{15.11}$$

The strength of the absorption is proportional to the square of the matrix element, which yields

$$\alpha_{o,\text{vc},\text{exc}} = \alpha_{o,\text{vc}}|A_{nq}(K = 0)|^2 \tag{15.12}$$

where $\alpha_{o,vc}$ is the valence-to-conduction band optical absorption coefficient neglecting excitons. For exciton states* below the band gap it follows

$$|\phi_{nq}(K = 0)|^2_{h\nu < E_g} = \begin{cases} \dfrac{1}{\pi a_{qH}^3 n_q^3} & \text{allowed} \\[2ex] \dfrac{n_q^2 - 1}{\pi a_{qH}^5 n_q^5} & \text{forbidden} \end{cases} \qquad (15.13)$$

for the indicated type of transitions (Section 13.1.4) and with $a_{qH} = a_H \varepsilon_{st} m_0 / m_r$. For higher excited states, the line intensity decreases proportional to n_q^{-3} or $n_q^{-5}(n_q^2 - 1)$ for allowed or forbidden transitions. The optical absorption per center is spread over a large volume element of radius $a_{qH} n_q^2$, therefore the corresponding matrix element is reduced accordingly. The line spacing given by

$$E_{g,\text{exc}}^{(\text{dir})} = E_g - E_{\text{exc},nq}; \qquad (15.14)$$

E_g is also the line limit (Fig. 15.4). Hence, we expect one strong line for the ground state in absorption, followed by much weaker lines for the excited states which converge at the absorption edge—see Fig. 15.5 A and C.

Electric dipole-forbidden (ls) transitions† are observed in only a few semiconductors. Most extensively investigated is Cu_2O with d-like valence bands, which has two series of hydrogen-like levels— the yellow (superscript y) and the green (superscript g) series from the Γ_7 and Γ_8 valence bands, respectively. The observed level spectrum is given in Fig. 15.6, with

$$E_{nq}^{(y)} = 2.1661 - \frac{0.0971}{n_q^2} \text{ eV} \quad \text{and} \quad E_{nq}^{(g)} = 2.2975 - \frac{0.1565}{n_q^2} \text{ eV}$$

$$(15.15)$$

for the p-levels of these two series. According to the second case of Eq. (15.13), the series starts with $n_q = 2$ (see Fig. 15.5A for Cu_2O).

---

* Only for s-states $\phi_{nq}(K = 0) \neq 0$.

† Strictly, such transitions cannot occur at $\mathbf{k} = 0$, however, a slight shift because of the finite momentum of the photon permits the optical transition to occur because of a weak electric quadrupole coupling (Elliot, 1961). Such transitions can also be observed under a high electric field using modulation spectroscopy (Washington et al., 1977). Dipole-forbidden transitions are easily detected with Raman scattering, which follows different selection rules—see Section 17.1.3.

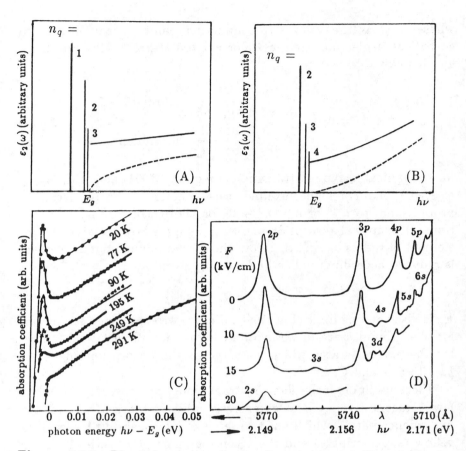

Figure 15.5: Direct band gap dipole allowed (A and C) and dipole-forbidden (B and D) transitions including exciton excitation (i.e., effects of Coulomb interaction—solid curves) and excluding this (dashed curves in A and B). Examples are (C) for direct transition in Ge (after McLean, 1963) and (D) for Cu_2O at 4 K (after Grosman, 1963). Subfigure (C) shows the decrease of absorption at higher temperatures where excitons can no longer exist. Subfigure (D) shows that with sufficient perturbation by an electric field, the electric dipole selection rule is broken and the *s*-transitions are also observed.

In addition, there are two dipole-allowed excitons in the blue and violet range of the spectrum from the two valence bands into the higher conduction band Γ_{12} (Compaan, 1975). Another material showing forbidden exciton spectra is SnO_2.

In contrast to the strongly absorbing Frenkel exciton with a highly localized wave function in the ground state, the intensity of the Wannier-Mott exciton lines are reduced by $(a/a_{qH})^3$: the larger the

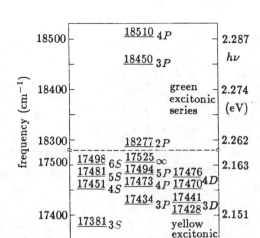

Figure 15.6: Yellow and green dipole-forbidden exciton series in Cu_2O (after Yu, 1979).

quasi-hydrogen radius a_{qH} is compared to that of the corresponding atomic eigenfunction a, the weaker is the corresponding absorption line.

Applying the Kramers-Kronig relation (mixing of exciton and band states—see Section 11.1.4), one obtains a substantially increased absorption for $h\nu > E_g$ near the band edge. Following Eq. (15.12) after summation over all $\mathbf{K}$ values, and recognizing that the Coulomb interaction affects the absorption, we have (see Madelung, 1978):

$$|\Sigma_K A_{nq}(K)|^2_{h\nu > E_g} = \gamma_e \frac{\exp \gamma_e}{\sinh \gamma_e} \quad \text{with } \gamma_e = \pi \sqrt{\frac{E_{exc,nq}}{(h\nu - E_g)}}. \quad (15.16)$$

Consequently, it follows that, with exciton contribution, a semiconductor with a direct band gap between spherical parabolic bands has an increased absorption given by (for detail see Bassani et al., 1975)

$$\alpha_{o,cv,exc}^{(\mathrm{dir})} = \alpha_{o,cv} \begin{cases} \gamma_e \dfrac{\exp \gamma_e}{\sinh \gamma_e} & \text{allowed} \\[3mm] \gamma_e \dfrac{\left[1 + (\gamma_e/\pi)^2\right] \exp \gamma_e}{\sinh \gamma_e} & \text{forbidden}. \end{cases} \quad (15.17)$$

The Coulomb interaction of the electron and hole influences the relative motion, and the optical absorption in the entire band edge range is thereby enhanced, as shown in Fig. 15.5. For an advanced discussion, see Beĭnikhes and Kogan (1985).

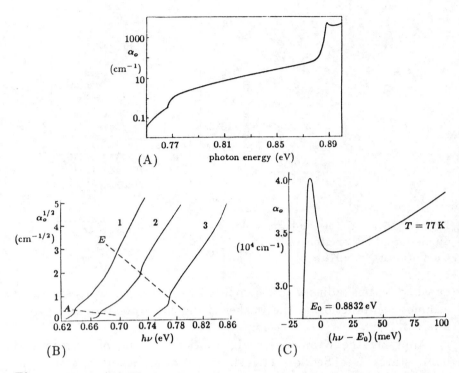

Figure 15.7: (A) Weak indirect and strong direct excitonic transitions in Ge. (B) Detail of the indirect exciton transition preceding the band edge ($\Gamma_{8v} \rightarrow L_{6c}$) of Ge at 291, 195, and 4.2 K for curves 1–3, respectively (after MacFarlane et al., 1957). Dashed lines indicate threshold for absorption (A) or emission (E) of a phonon. (C) Direct exciton at $k = 0$ and band-to-band transition from Γ_{8v} into Γ_{7c} of Ge at 77 K.

In semiconductors with high ε_r and low m_r, however, only the first exciton peak is usually observed: in GaAs the relative distance $E_g - E_{\text{exc}}(\mathbf{K} = 0)$ is only 3.4 meV. Higher absorption lines, which are too closely spaced, are reduced in amplitude and merge with the absorption edge in most direct gap III-V compounds.

A line spectrum including higher excited states can be observed more easily when it lies adjacent to the reduced absorption of forbidden transitions. It is also easier to observe in materials with a somewhat higher effective mass and lower dielectric constant to obtain a wide enough spacing of these lines. Well-resolved line spectra of higher excited states can be observed when they do not compete with other transitions or are broadened beyond recognition—see Sections 20.3.2 and 44.3.3.

For indirect band gap semiconductors, transitions at $\mathbf{K} = 0$ proceed into a higher band above the band gap, and a direct gap exciton line associated with these transitions can be observed. Compared with a direct gap semiconductor, however, this line shows resonance line broadening (see also Section 21.1.7). As an example, the exciton line near the direct transition at the Γ-point of Ge at 0.883 eV for 77 K is shown in Fig. 15.7A and C.

Complexity of Exciton Spectra.   An exciton has several degrees of freedom influencing its spectrum:

(a) its translational motion, expressed by the translational wave vector $\mathbf{K}$;

(b) its relative motion of electron and hole, expressed by its hydrogenic quantum numbers n_q, l, and m; and

(c) its spin degree of freedom, expressed by its total spin quantum number; and

(d) nonspherical bands.

Except for case (a), which was included in our previous description, there is often a high degree of degeneracy arising from the energy bands. However, there are many possibilities for degeneracy-breaking events, which will be discussed below.

The simple hydrogen-like model described above must be modi-
real semiconductors because of several contributions (Flohrer
79):

isotropy of effective masses and dielectric constants;
ting of degeneracy of conduction or valence bands;
ectron-hole exchange interaction;
xciton-phonon interaction;
ction of local mechanical stress or electrical fields; and
nteraction with magnetic fields.

anisotropy of the effective masses produces excitons,
ed in the direction in which the mass is smallest. A compres-
n in the direction of the largest effective mass reduces a_{qH} in this
irection by a factor of less than 2 and increases E_{qH} up to a factor
of 4 (Shinada and Sugano, 1966). The *reduced exciton mass*, entering
the expression for the exciton energy in Eq. (15.8) is given by

$$\frac{1}{m_r} = \frac{2}{3}\frac{1}{m_r^{\perp}} + \frac{1}{3}\frac{\varepsilon_{\perp}}{\varepsilon_{\parallel}}\frac{1}{m_r^{\parallel}} \qquad (15.18)$$

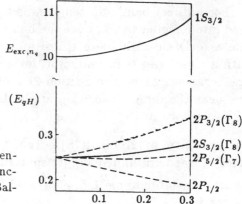

Figure 15.8: Exciton binding energy for *s*- and *p*-states as a function of the reduced mass (after Baldereschi and Lipari, 1973).

The meaning of $\perp$ and $\parallel$ depends on the crystal structure. For instance, in *Wurtzite-type* semiconductors $\parallel$ means parallel to the *c*-direction. For calculation of the reduced effective mass, we distinguish six effective masses: $m_n^\perp$, $m_n^\parallel$, $m_{pA}^\perp$, $m_{pA}^\parallel$, $m_{pB}^\perp$, and $m_{pB}^\parallel$ with the *A* and *B* valence bands split by the crystal field with Γ_9- and Γ_7-symmetry, respectively, neglecting the spin-orbit split-off band, the *C* band.

The **lifting of band degeneracies** is observed when band splitting occurs; e.g,. in an anisotropic crystal or, following the applica tion of mechanical stress, an electric or a magnetic field. The line spectrum can be distinguished with respect to transiti different valence bands, which result in different exciton li with different spacing because of a different reduced mass— tion 15.1.2A.1. Taking into consideration splitting and w the valence bands, one obtains a splitting of the *p*-like exciton (Baldereschi and Lipari, 1973). This is shown in Fig. 15.8 as a fun tion of the reduced effective mass. The reduced mass m_r can in turn be expressed as a function of the Luttinger valence band parameters

$$m_r = \frac{6\gamma_3 + 4\gamma_2}{5} \frac{m_n m_0}{m_0 + m_n \gamma_1}. \tag{15.19}$$

For a review, see Rössler (1979), and Hönerlage et al. (1985).

The **electron-hole exchange interaction** causes a splitting of the fourfold degenerate $A(n_q = 1)$ exciton into two Γ_6 and Γ_5 exciton states, each twofold degenerate. The fourfold $B(n_q = 1)$ exciton splits into the Γ_5 exciton, which is twofold degenerate, and into the nondegenerate Γ_1 and Γ_2 exciton states. The $A(\Gamma_6)$ and $B(\Gamma_2)$ states are not affected by the exchange interaction.

Table 15.2: Parameters* for the A and B excitons in CdS. All energies are in meV (after Flohrer et al., 1979).

$\varepsilon_{\perp,t} = 8.42$	$m_{pA}^{\perp} + m_{n}^{\perp} = (0.8 \pm 0.1)m_0$	$m_{rA}^{\perp} = (0.16 \pm 0.03)m_0$	$\Delta_{\perp}^{t} \simeq 0.2$
$\varepsilon_{\parallel,t} = 8.92$	$m_{pA}^{\parallel} + m_{n}^{\parallel} = (2.7 \pm 0.2)m_0$	$m_{rA}^{\parallel} = (0.185 \pm 0.03)m_0$	$\Delta_{\parallel}^{t} \simeq 0.2$
$\varepsilon' = 7.3$			$\Delta_{\perp}^{l} \simeq 1.9$
$m_{rA} = 0.155m_0$	$m_{pA} = 0.8m_0$	$\gamma_A = 0.2$	$3j_0 = 1$
$m_{rB} = 0.173m_0$	$m_{pB} = 1.6m_0$	$\gamma_B = -0.33$	$9j_1 = 2.1$
$\Delta E_{aA} = -0.09$	$\Delta E_{wA} = -0.23$	$\Delta E_{aB} = -0.27$	$\Delta E_{wB} = -0.40$
$E_{qH,A} = 28.0$	$E(\Gamma_5^l - \Gamma_5^t)_A = 1.9$	$E(\Gamma_5^t - \Gamma_6)_A = 0.3$	
$E_{qH,B} = 31.3$	$E(\Gamma_5^l - \Gamma_5^t)_B = 1.2$	$E(\Gamma_5^t - \Gamma_1)_B \le 0.1$	

*The parameters Δ are introduced by Flohrer et al., 1979; the parameters j_0 and j_1 are introduced by Cho, 1979 for the derivation of the excitonic eigenstates. They are listed here for completeness. ε' is the background dielectric constant.

As an example, the model parameters for hexagonal CdS are given in Table 15.2.

Exciton-phonon interaction depends strongly on the coupling, which is weak for predominantly covalent semiconductors, intermediate for molecular crystals, and strong for ionic crystals (alkali halides). Excitons can interact with phonons in a number of different ways. One of these interactions is the reason for exciton scattering—discussed in Section 36.1. One distinguishes exciton interaction (for reviews, see Vogl, 1976; Yu, 1979) by:

- nonpolar optical phonons via the deformation potential (a short-range interaction—Loudon, 1963);
- longitudinal optical phonons via the induced longitudinal electrical field (Fröhlich interaction—Fröhlich, 1954);
- acoustic phonons via the deformation potential for a large wave vector, since for $q \simeq 0$ the exciton experiences a nearly uniform strain resulting in a near-dc-field, resulting in no interaction with the electrically neutral exciton (Kittel, 1963); and
- piezoelectric acoustic phonons via the longitudinal component similar to the Fröhlich interaction (Mahan and Hopfield, 1964).

Another interaction involves three-particle scattering among phonons, phonons, and excitons, such as the relevant Brillouin and Raman scattering—see Yu (1979), Reynolds and Collins (1981) and Chapter 16.

In lattices with a large coupling constant, the exciton-phonon interaction can become large enough to cause **self-trapping** of an

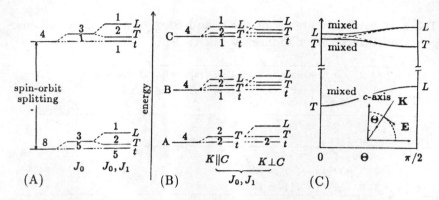

Figure 15.9: Exciton levels split from the spin triplet states (t) into dipole-allowed spin singlets and to the longitudinal (L) and transverse (T) dipole allowed states due to exchange effects. (A) Cubic, (B) uniaxial crystals, and (C) mixed longitudinal and transverse modes of the A and B excitons (after Cho, 1979).

exciton (Kabler, 1964). This can be observed in predominantly ionic crystals with a large band gap, and occurs because of a large energy gain due to distortion of the lattice by the exciton. It results in a very large increase of the effective mass of the *phonon-dressed exciton*. This is distinguished from a polaron by the short-range interaction of the exciton dipole, compared to the far reaching Coulomb interaction of the electron/polaron (see Section 27.1.2). The exciton is a Frenkel-exciton.

The self-trapped exciton may be a significant contributor to photochemical reactions—see Section 46.1.1. It is best studied in alkali halides—see Section 22.4.1. For a review, see Rashba (1976) and Toyozawa (1980).

In anisotropic crystals with an external perturbation, we must consider the relative direction of the exciting optical polarization $\mathbf{e}$, the exciton wave vector $\mathbf{K}$, and the crystallographic axis $\mathbf{c}$. One distinguishes σ and π modes when $\mathbf{e}$ is $\perp$ or $\parallel$ to the plane of incident light, respectively (*transverse and longitudinal excitons*). When the incident angle $\theta \neq 90°$, mixed modes appear (Fig. 15.9C). The resulting exciton lines in cubic and hexagonal systems for $\mathbf{k} \parallel \mathbf{c}$ and $\mathbf{k} \perp \mathbf{c}$ are given in Fig. 15.9A and B.

Many of the band degeneracies are lifted when the crystal is exposed to internal or external perturbation. Internally, this can be done by alloying (Kato et al., 1970), and, externally, by external

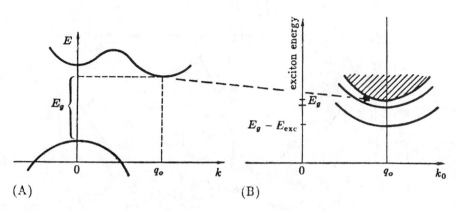

Figure 15.10: Dispersion relation for indirect band gap excitons. (A) Typical band dispersion. (B) Satellite minimum with ground and first excited state parabolas of indirect excitons.

fields such as mechanical stress, electrical or magnetic fields (reviewed by Cho, 1979). Examples will be given in Section 20.3.2.

15.1.2B Indirect Gap Excitons Indirect gap excitons are associated with optical transitions to a satellite minimum at $K \neq 0$. The energy of an indirect exciton is

$$E_{g,exc}^{(indir)} = E_g^{(indir)} - E_{exc,nq} + \frac{\hbar^2}{2m_r}(K - q_0)^2, \qquad (15.20)$$

with m_r as the reduced mass between the valence bands and the satellite minima of the conduction band. In order to compensate for the electron momentum k_0 at the satellite minimum, the transition requires an absorption or emission of a phonon of appropriate energy and momentum. The phonon momentum q_0 is equal to k_0—see Fig. 15.10.

During the indirect excitation process, an electron and a hole are produced with a large difference in wave vectors. Such excitation can proceed to higher energies within the exciton dispersion using a slightly higher photon energy. The excess in the center of mass momentum is balanced by only a small change in phonon momentum. Therefore, we observe an onset for each branch of appropriate phonon processes for indirect band gap transitions, following selection rules, rather than a line spectrum for direct band gap material. Here, relaxing phonon processes are not observed since they have a much smaller probability.

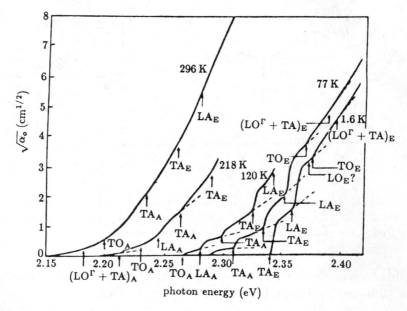

Figure 15.11: Indirect exciton transition: in GaP more threshold energies can be identified for indirect, phonon-assisted excitons. The additional absorption edges connected to different types of phonons or phonon pairs are identified accordingly. Curves measured at 296, 218, 120, 77, and 1.6 K, respectively (after Dean and Thomas, 1966).

One obtains for the absorption coefficient caused by these indirect transitions

$$
\alpha_{o,cv,exc}^{(indir)} \propto \alpha_o^* \Bigg\{ f_{BE} \sqrt{h\nu - E_g^{(indir)} + E_{exc,nq} + \hbar\omega_q} + \\
+ (f_{BE} + 1) \sqrt{h\nu - E_g^{(indir)} + E_{exc,nq} - \hbar\omega_q} \Bigg\},
$$

(15.21)

where α_o^* is a proportionality factor. This factor includes the absorption enhancement relating to the square of the exciton envelope function, and f_{BE} is the Bose-Einstein distribution function. The two terms describe transitions with absorption and emission of a phonon, respectively. The allowed spectrum consequently has two edges for the ground state of the exciton, plus or minus the appropriate phonon energy. It is shown in Fig. 15.7B for the indirect gap of Ge. Excited exciton states disappear because of the strong $(1/n_q^3)$ dependence of the oscillator strength. The branch caused by phonon

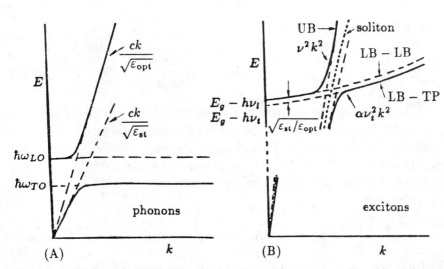

Figure 15.12: Comparison between (A) a phonon- and (B) an exciton-polariton (schematic and not to scale) with upper (UB) and lower branch (LB) longitudinal (LP) and transverse polaritons (TP). In addition the soliton branch (see Nelson, 1981) is indicated by the dotted curve.

absorption also disappears at lower temperatures as less phonons are available.

In Fig. 15.11, a more structured series of square-root shaped steps is shown for GaP. Here different types of phonons are involved. The corresponding energies of the absorbed or emitted phonons are 12.8, 31.3, and 46.5 meV for TA, LA, and LO phonons in GaP, respectively.

15.1.2C Exciton-Polaritons In some of the previous sections, the interaction of light with electrons was described by near band-to-band transitions close to $K = 0$, creating direct band gap excitons. This interaction requires a more precise analysis of the dispersion relation. The exciton and photon dispersion curves are similar to those for phonons and photons (Fig. 15.12). Both show the typical splitting due to the von Neumann noncrossing principle. In this range, the distinction between a photon and the exciton can no longer be made. The interacting particle is a polariton or, more specifically, an *exciton-polariton*.

Entering the resonance frequency for *excitons* into the polariton dispersion equation [Eq. (11.58)], and including spatial dispersion ($\hbar^2 K^2 \omega_t^2 / m_r^2$), we obtain the *exciton-polariton equation*

$$\nu^4 - \nu^2 \left(\nu_l^2 + \frac{c^2 k^2}{\varepsilon^*} \right) + \frac{c^2 K^2}{\varepsilon^*} \nu_t^2 = 0 \qquad (15.22)$$

with a kinetic energy term

$$h\nu_{(l,t)} = h\nu_{\text{exc}(l,t)} + \frac{\hbar K^2}{2(m_n + m_p)}. \tag{15.23}$$

The ability of the exciton to move through the lattice represents propagating modes of excitation within the semiconductor. They are identified by the term $\propto K^2$ and have a group velocity ($\propto \partial E/\partial K$) on the order of 10^7 cm/s. Consequently, the dielectric constant is wave-vector dependent and can be written as

$$\varepsilon^* = \varepsilon_{\text{opt}} \left(1 + \frac{\nu_l^2 - \nu_t^2}{\nu_t^2 - \nu^2 + \beta K^2 - i\gamma\nu} \right) \quad \text{with } \beta = \frac{h\nu_t}{(m_n + m_p)}. \tag{15.24}$$

There is now an additional spatial dispersion term βK^2 in the denominator in contrast to an otherwise similar equation for the phonon-polaritons, which does not have spatial dispersion.

The polariton dispersion equation has several branches. These branches depend on crystal anisotropy and the relative orientation of the polarization of light, exciton **K**-vector, and crystallographic axes. As many as four lower and two upper branches are predicted and observed by single and two-photon excitation processes. An example is shown for CdS in Fig. 15.13, with upper and lower branches for longitudinal and transverse excitons pointed out. For a review, see Hörnerlage et al. (1985).

In addition, a *soliton* branch (Fig. 15.12—Nelson, 1981) can appear when nonlinear contributions in the lattice equation of motion [Eq. (5.3)] are considered.

Solitons.   A soliton is a wave that preserves its shape and velocity upon collision with other soliton waves; it is a localized traveling wave and can be obtained as a special solution to partial differential equations of a special kind (Scott et al., 1973). Such differential equations describe the oscillations of lattice points when considering nonlinear force constants (*Toda lattice equations*—Toda, 1970). Soliton solutions remain localized while traveling, in contrast to common solutions of nonlinear differential equations which employ a perturbation analysis and result in *wave packets* which smear out when traveling as a result of lattice dispersion [$v(\omega)$]. Solitons become technically important when transmission over long distances is essential, e.g., light pulses in fiber optics. High pulse rate (1 Tbit/s) fixes ω at a high value, and transmissions over long distances (30 km)

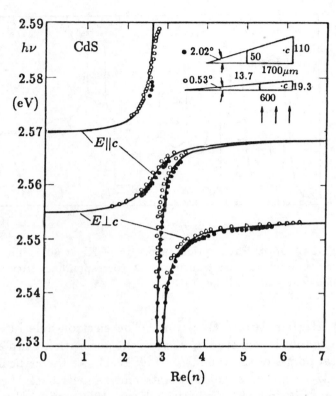

Figure 15.13: Energy of A and B exciton polaritons as a function of the real part of the index of refraction for $\mathbf{E}\|$ and $\perp$ to $\mathbf{c}$ in CdS compared to theoretical curves. Dimensions shown inside the wedges in the insert are in μm (after Broser and Rosenzweig, 1980). ©Pergamon Press plc.

are predicted for solitons (Hasegawa and Kodama, 1981). See also Section 13.5.1.

Surface Polaritons. *Surface polariton modes* progress in a thin layer near the semiconductor surface, and can be used to reveal properties of the crystal near the surface. Grazing incident light or reflection measurement is used for their detection (Hopfield and Thomas, 1963).

The region near the surface cannot be penetrated by *bulk excitons* to a thickness of either the space charge region at the surface or the bulk exciton diameter, whichever is larger (Altarelli et al., 1979). These excitons are reviewed by Fischer and Lagois (1979) and are one of the topics of *Carrier Transport in Inhomogeneous Semiconductors* (in preparation).

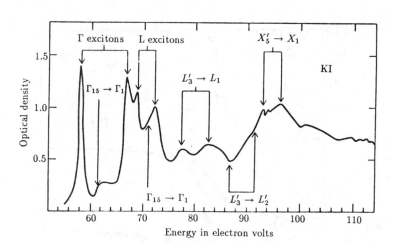

Figure 15.14: Optical absorption spectrum of KI at 80 K indicating Γ and L excitons and specific transitions at corresponding critical points (after Greenaway and Harbeke, 1968).

15.1.2D Higher Band Excitons The electron-hole interaction yielding excitons is well studied close to the absorption band edge (at the critical points of the type M_0—Fig. 13.1). At the critical points M_1 and M_2, which are saddle points and camel's back, *hyperbolic excitons* are possible (Phillips, 1964; Kane, 1969), and influence the absorption or reflection spectrum (Toyozawa et al., 1967). They can cause distinct structures in the indirect absorption range (Glinskii et al., 1979). An example of such higher band excitons is given in Fig. 15.14 for KI with corresponding critical points identified.

15.1.2E Trions and Biexcitons An exciton can be weakly bound to an electron or another exciton to form a *trion* or *biexciton*, respectively (Lampert, 1958; Moskalenko, 1958).

The trion is composed of either two electrons and a hole (this one is similar to an H^- ion) or two holes and an electron. Therefore, it is negatively or positively charged. Observed first by Thomas and Rice (1977), their binding energy in Ge is on the order of 0.2 meV. This trion has an effective mass of about 20% more than the sum of the free electron and hole masses. Its radius is about 50% larger than that of an exciton, and its ionization energy is about 10% of the ionization energy of excitons.

Biexcitons are similar to a hydrogen molecule (when $m_p \gg m_n$) or a positronium molecule (when $m_p \simeq m_n$). The binding energy of a biexciton is higher than that of a trion and is typically of the order of $10 \ldots 20\%$ of an exciton. It decreases with increasing ratio of m_n/m_p (Akimoto and Hanamura, 1973). This is similar to the decrease in relative binding energy from a hydrogen molecule $E_{H_2}/R_H = 4.7/13.6 = 0.35$ to the relative binding energy of a positronium molecule $E_{x_2}/E_x = 0.13/6.8 = 0.02$. Such biexcitons have been observed by Hanamura and Haug (1977) and Thewalt (1978) in Si. A high density of excitons favor the formation of biexcitons. However, the low binding energy requires low temperatures which, in turn, favors further condensation into an electron-hole liquid (Section 34.1). This condensation can be suppressed by applying a uniaxial stress in Ge or Si (Gourley and Wolfe, 1978); larger biexciton signals are then observed (Kulakovskii and Timofeev, 1977). Further discussion of *excitonic molecules* is provided by Kulakovskii et al. (1985).

Biexcitons can be observed readily in CuCl and CuBr or in ZnS and CdS. See Haken and Nikitine (1975) and Ueta and Nishina (1976).

The line shape of trions and biexcitons (see Section 44.3.3C), observed by luminescence or scattering experiments, is typically asymmetric because of the recoil energy when they recombine: the line has a larger low-energy tail. During this process there is always a remaining partner that can take up part of the energy as kinetic energy. For a review, see Hanamura (1976) and Hönerlage et al. (1985).

At higher densities, excitons or excitonic molecules can no longer exist, but form an electron-hole plasma, as described in Section 34.1.

15.1.2F Bound Excitons An exciton becomes trapped at a lattice defect when the interaction to the defect becomes larger than its thermal energy. Its eigenfunction can be substantially perturbed by the core potential of the defect. For an analysis, a knowledge of the specific defect is essential. This discussion will therefore be postponed to Sections 21.2 and 21.2.3.

15.2 Excitons in Magnetic Field

When the magnetic field is small enough so that the ground-state energy of the exciton E_{qH} is large compared to the cyclotron resonance energy $\hbar\omega_c$, the field can be taken as a small perturbation and analyzed similar to the Zeeman effect. The Schrödinger equation for

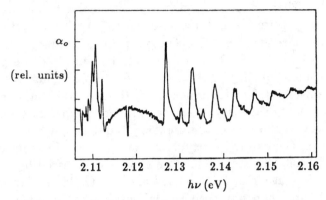

Figure 15.15: Optical absorption spectrum of GaSe near the band edge at 49.5 kG magnetic induction ($\mathbf{k}\|\mathbf{B}\perp\mathbf{E}$) with free and bound exciton lines below 2.115 eV and Landau levels above 2.12 eV, including bond state satellites below each Landau level (after Brebner et al., 1973).

the envelope function of the exciton in a magnetic field in z direction can then be approximated as

$$\left[\frac{\hbar^2}{2m_r}\nabla^2 - \frac{e^2}{\varepsilon\varepsilon_0 r} + \frac{e^2}{8m_r c^2}B_z^2\left(x^2 + y^2\right)\right]\phi(\mathbf{r}) = E'\phi(\mathbf{r}), \quad (15.25)$$

with

$$E' = E - E_g - \frac{\hbar^2 K^2}{2(m_n + m_p)} - \frac{em_0\hbar}{2cm_r}B, \quad (15.26)$$

which yields exciton states below each Landau level.

A typical optical spectrum is shown for GaSe in Fig. 15.15 with excitonic lines at low energies and corresponding satellites below each Landau level at higher energies—see Bassani and Pastori Parravicini (1975), Cavenett (1981), and Stradling (1984).

15.2.1 Two-Dimensional Excitons

In anisotropic semiconductors the dielectric constant and effective mass are anisotropic and result in ellipsoidal excitonic eigenfunctions. In extreme cases, the excitons may have disk shape; that is, they are essentially two-dimensional with an increase up to a factor of 4 in binding energy (Shinada and Sugano, 1966). An example is given in the following section.

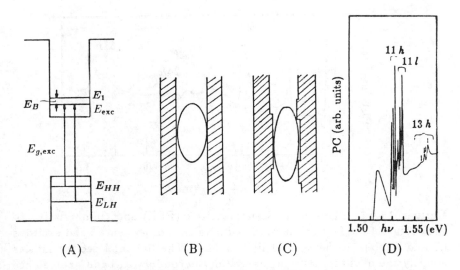

Figure 15.16: (A) Exciton state for light and heavy holes with exciton binding energy E_B indicated. (B) Exciton state confined between two barrier layers in a superlattice. (C) Exciton confined between rough barriers. (D) Photoconductivity (PC) spectrum for $GaAs/Ga_{0.25}Al_{0.75}As$ ($l = 146$ Å) at 2 K. Excitons identified with nmL or nmH relating to the nth and m^{th} electron and hole (light = l, heavy = h) band, respectively. Each maximum within a series is caused by a well-size fluctuation of one monolayer (after Yu et al., 1987). ©Pergamon Press plc.

15.3 Excitons in Superlattices

Exciton absorption in superlattices causes the near band edge features shown in Fig. 13.17. These Wannier-Mott excitons are observed in bulk material only at low temperatures, but remain visible to much higher temperatures in superlattices. The substantially increased lifetime of excitons at higher temperatures is due to the increased exciton binding energy, which is caused by the two-dimensional confinement of excitons in each well. The confinement results in elliptical orbits with a highly compressed coordinate in the direction normal to the superlattice plane. Here the orbiting electron and hole approach each other closely, which causes the increase in their binding energy. The Initial work was done by Dingle et al. (1974). For a review, see Ploog and Döhler (1983) and Miller and Kleinman (1985).

Excitons in superlattices have a higher binding energy when the well thickness is smaller than the diameter of an exciton in bulk material. For the extreme case of a perfect two-dimensional confinement, the eigenvalues of the two-dimensional Schrödinger equation

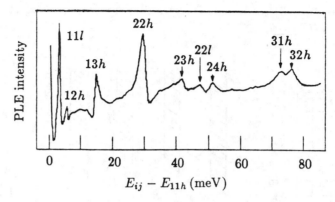

Figure 15.17: Photoluminescence exciton (PLE) spectrum relative to the energy of the first conduction and h-valence or l-valence band excitons (h and l relate to heavy and light holes). The first and second indices identify the quantum level in the well of the conduction band and valence band, respectively. Well width = 220 Å, T = 5 K, single GaAs well AlGaAs barrier (after Koteles et al., 1987).

have been calculated, resulting in an increase of the binding energy of up to a factor of 4 (Shinada and Sugano, 1966).

In the mini-band picture (Fig. 15.16), the exciton is shown as a line below the lower edge of the first mini-band ($E_1 - E_{\text{exc}}$). Its bonding energy is indicated as E_B. The exciton energy, however, lies *above* the gap energy of the well material. We distinguish light (l) and heavy (h) hole excitons, and excitons relating to the first or higher electron mini-bands. An example is given in Fig. 15.17 for a single GaAs/AlGaAs quantum well: excitons are shown combining up to the third "conduction band level" with up to the fourth "valence band level."

The line width of the exciton absorption is given (in addition to the broadening by alloy fluctuation—see Section 20.3.2D) by its relaxation time (see Section 49.2.1) and, when severely confined, is also given by the quality of the well interfaces. Roughness in these interfaces causes additional broadening (Bajaj, 1987). At very low temperatures, well-size fluctuations are resolved as different spikes separated by $\lesssim 1$ meV, as shown by Yu et al. (1987)—Fig. 15.16C.

15.3.1 Superlattice Excitons in Perturbing Fields

Electric and uniaxial stress fields cause characteristic changes in the exciton spectrum of superlattices, and are best detected in single

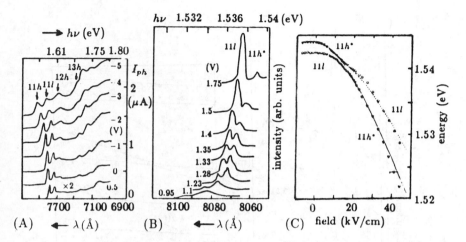

Figure 15.18: Shift of exciton peaks with external electric field in GaAs/Al$_{0.3}$Ga$_{0.7}$As single quantum wells. (A) Well width 80 Å, $T = 10$ K, applied voltage across 2100 Å. (B, C) Anticrossing of $2p$ and $2s$ (not resolved) state of $11H$ with $1s$ state of $11L$ of well width 160 Å at 4.3 K. Applied voltage as family parameter (after Collins et al., 1987).

quantum wells where the line spectrum is rather sharp and higher quantum transitions can be followed without ambiguity.

The electric field perpendicular to the layers causes a *Stark shift* toward longer wavelength. New peaks become visible, caused by transitions which were forbidden without perturbation, e.g., such from the mth level of the valence band to the nth level of the conduction band with $m \neq n$. Such transitions become permitted because of a field-induced deformation of electron and hole eigenfunctions which now overlap. It is shown in Fig. 15.18A for $12\,h$ and $13\,h$ excitons which are not observed at zero bias.

The changes in peak position with the electric field illustrate the *anticrossing of two levels*, demonstrating the von Neumann noncrossing rule when the states interact with each other—Fig. 15.18B and C.

Similarly, shifts and anticrossing of levels are observed when uniaxial stress is applied, as shown in Fig. 15.19 for a 220 Å wide quantum well. For earlier works, see Miller et al. (1985).

Summary and Emphasis

In this chapter we have discussed excitons which are quasi-particles resembling hydrogen atoms, except that the positively charged part-

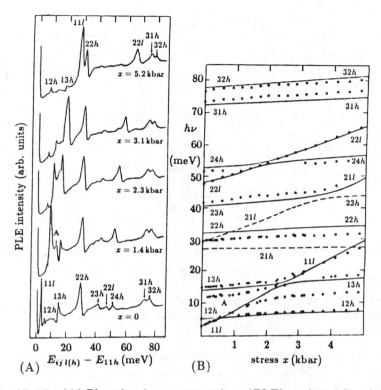

Figure 15.19: (A) Photoluminescence exciton (PLE) peaks with uniaxial stress parallel to [100] (family parameter) measured at $T = 5$ K in a single GaAs/Al$_{0.35}$Ga$_{0.65}$As quantum well of 220 Å width. Energy plotted relative to the 11 h peak. (B) Exciton peak position as a function of the stress. Dashed curves are theoretical, and could not be experimentally observed (after Koteles et al., 1987).

ner is not a heavy proton but a hole of about the same mass as an electron. Excitons show a hydrogen-like excitation spectrum, however, because of the perturbation from the lattice, other than s-states are also observed in optical transitions. In a free hydrogen atom all optical transitions are equivalent to transitions between s-states.

One distinguishes large (Wannier-Mott) and small (Frenkel) excitons. Large excitons, typically with a diameter > 3 lattice constants, have a small ionization energy, typically on the order of 20 meV, while small excitons have ionization energies on the order of 1 eV. The former is found in typical, mostly covalent semiconductors, the latter in ionic or molecule (organic) crystals. This distinction as to size and ionization or bonding energy is typical and can also be found in other quasi-particles, e.g., in polarons—see Section 27.1.2.

Excitons combine an electron and a hole within their mutual Coulomb field; consequently, many kinds of excitons are observed depending on the type of hole within the light, heavy, or split-off spin-orbit band, and on the type of electron: from a band at Γ—a direct exciton, or from a satellite valley—an indirect exciton. One also observes higher energy excitons when they are created by direct-vertical [in $E(k)$] transitions connecting bands at higher energies than E_g.

All of these excitons have a profound effect on the optical absorption spectrum by permitting absorption slightly below the corresponding edge, and by substantially increasing the absorption near the edge, but inside the intrinsic, band-to-band range.

Excitons are quasi-particles of major influence on the optical spectrum near the absorption edge. They permit significant insight into the bonding type of the semiconductor, the dielectric function, and effective mass, all of which influence the spectrum. In superlattices, they can be used as an optical probe of the perfection of the interlayer boundaries on an atomic scale, and can also be used to analyze in detail the influence of a wide variety of field perturbations. As such, they offer an important potential for analytical purposes. Other, even more significant applications will be discussed in later chapters.

Exercise Problems

1. What are the binding forces for molecular excitons and for bound excitons?
   (a) Explain the differences in the potential for both excitons.
   (b) What differences and what similarities exist between Frenkel and Wannier-Mott excitons?

2.(*) How does the lattice symmetry enter into the description of Wannier-Mott excitons?
   (a) Be explicit for the ground state in Si.
   (b) Compare this to the ground state of shallow donors and to shallow acceptors.

3.(e) Calculate and list the ionization energy for 10 typical semiconductors of your choice.
   (a) Observe and discuss the trend.
   (b) Does the effective mass and the static dielectric constant show a similar trend?
   (c) What are the reasons for these trends?

4.(e) Show the Fourier transformation from $A_{nq}(R)$ to $A_{nq}(K)$.

5.(r) Why are the transitions from p- to s-states dipole-forbidden?

6. Explain in your own words the reasons for the difference between the line spectrum for direct excitons and for edge-like absorption of indirect excitons?

7.(*) What implications do the different wave sectors of holes and electrons at a satellite minimum have for the corresponding exciton?

8.(e) Summarize the effects which an anisotropic effective mass has on excitons. Compute examples for four semiconductors known to have large anisotropies in their effective mass.

9.(*) In what ways can you imagine phonons affecting eigenstates of excitons?

10.(e) Why is the Bose-Einstein distribution function included in the expression for optical absorption with indirect excitons [Eq. (15.21)]?

11. What are longitudinal and transverse excitons (polaritons)?
    (a) Why is there a translational term in the dispersion equation?
    (b) What is the shape of the $E(K)$ curve for larger values of **K** in the transverse branch?
    (c) What are hyperbolic excitons?

12. Why do biexcitons have a higher ionization energy than trions?

13.(*) Explain the optical absorption spectrum of GaSe excitons in a small magnetic field as shown in Fig. 15.15.
    (a) Compute the value of the reduced effective mass?
    (b) What can you conclude about $E(K)$?
    (c) What causes the asymmetry of lines in Fig. 15.15?

14.(e) Are excitons in superlattices direct or indirect excitons?
    (a) How does the well thickness enter into the exciton spectrum?

15.(e) Describe the difference between excitons in a quantum well and a superlattice. Why can they be observed at room temperature?

Chapter 16

Nonlinear Optical Effects

Nonlinear optical effects are high amplitude effects, employing the nonparabolicity of lattice potentials. The resulting nonharmonicity of oscillations permits the mixing of different signals with corresponding changes in frequency and amplitude.

Nonlinear optical effects are based on the fact that, at sufficiently high amplitudes, all solid-state oscillations become anharmonic (Rabin and Tang, 1975). This amplitude range is readily accessible with high intensity laser light* (Bloembergen, 1982) or with excitation near resonances where high amplitude oscillations can easily be stimulated. Anharmonic oscillations require higher order terms in the dielectric constant or dielectric susceptibility; i.e., the displacement $\mathbf{D}$ and the polarization $\mathbf{P}$ are no longer linear functions of the electric field $\mathbf{E}$.

These nonlinearities permit active interaction between two or more photons with a large variety of technically interesting phenomena of light mixing, rectification, and amplification. It opens a host of opportunities in the field of nonlinear optics to develop new devices and to design new experiments for analytical purposes. We will discuss here the basic elements of nonlinear optical effects. For

---

* Material destruction is avoided by monochromatic irradiation in a wavelength range of little absorption and by the use of short pulses. Material destruction can occur by simple lattice heating, by dielectric breakdown (10^9 W/cm^2 is equivalent to 10^6 V/cm oscillation amplitude and achieved by focusing a 10^3 W laser on a 10 μm diameter spot), by stimulated Brillouin scattering with intense multiphonon absorption, or by self-focusing. Typical destruction thresholds are 100 MW/cm^2 for a 100 ns pulse in a low-absorbing range, and 1 kW/cm^2 for a cw laser beam. (In comparison, sunlight on the earth's surface transmits 100 mW/cm^2). For more on damage, see Kildal and Iseler (1976).

more information, see reviews by Zernike and Midwinter (1973) and Chemla and Jerphagnon (1980).

16.1 Electronic Effects

Nonlinear effects are mostly investigated in the frequency range between atomic and electronic resonance absorption, i.e, between *Reststrahl* absorption and the band edge. Here the electronic contribution can relate to valence electrons or to free electrons if their density is high enough. We will discuss the influence of valence electrons first.

16.1.1 Nonresonant Effects of Valence Electrons

The anharmonicity is caused by deviations from the parabolic potential, which was used for small oscillation amplitudes to yield a polarization proportional to the field

$$\mathbf{P} = \varepsilon_o \chi \mathbf{E} = \varepsilon_0 (\varepsilon - 1) \mathbf{E}. \tag{16.1}$$

At higher amplitudes, the corresponding polarization has higher terms in $\mathbf{E}$ in the polarization equation, which is known as *hyperpolarization*. It is expressed as an expansion in $\mathbf{E}$, containing the higher harmonics:

$$
\begin{aligned}
\mathbf{P}(\omega_i) = {} & \chi^{(1)}(\omega_i) \cdot \mathbf{E}(\omega_i) \\
& + \sum_{j,l} \chi^{(2)}(-\omega_i; \omega_j, \omega_l) : \mathbf{E}(\omega_j)\mathbf{E}(\omega_l) \\
& + \sum_{j,l,m} \chi^{(3)}(-\omega_i; \omega_j, \omega_l, \omega_m) \vdots \mathbf{E}(\omega_j)\mathbf{E}(\omega_l)\mathbf{E}(\omega_m) + \dots ,
\end{aligned}
\tag{16.2}
$$

where the susceptibility $\chi^{(i)}$ is an $i+1$ rank tensor containing waves of possibly the same or different frequencies $\omega_i, \omega_j, \omega_l$, etc.*

---

* Sometimes an inverted relation between E and P is used; for instance, for the second-order term, one obtains

$$E_i(\omega_3) = \sum_{i,j,k} \delta_{i,j,k}(-\omega_3; \omega_1, \omega_2) P_j(\omega_1) P_k(\omega_2) \tag{16.3}$$

with

$$\chi^{(2)}_{ijk}(-\omega_3; \omega_1, \omega_2) = \sum_{l,m,n} \chi^{(1)}_{il}(\omega_3) \chi^{(1)}_{jm}(\omega_1) \chi^{(1)}_{kn}(\omega_2) \delta_{lmn}(-\omega_3; \omega_1, \omega_2).$$

$$\tag{16.4}$$

Table 16.1: Nonlinear susceptibilities (in 10^{-8} cm/V) for a number of AB compounds (after Moss et al., 1973). ©John Wiley & Sons, Inc.

Material	d_{14}	Material	d_{14}	Material	d_{33}	d_{31}	d_{15}
GaP	1.0	ZnS_{cub}	0.3	ZnS_{hex}	0.37	0.19	0.21
GaAs	3.7	ZnSe	0.8	CdS	0.44	0.26	0.29
GaSb	6.3	ZnTe	0.9	CdSe	0.55	0.29	0.31
InAs	4.2	CdTe	1.7				

Often the third-rank susceptibility tensor $\chi_{ijk}^{(2)}$ is replaced by the tensor d_{im} with the convention

$$i = 1\ldots 3 \text{ for } x, y, x \text{ and } m = 1 \ldots 6 \text{ for } xx, yy, zz, yx, zx, xy \quad (16.5)$$

respectively. The values of the nonlinear susceptibility for some of the semiconductors are given in Table 16.1. For recent evaluation of $\chi^{(n)}$, see Choy and Byer (1976) and Kurtz et al. (1978).

Such nonlinearities represent the ability of interaction of electromagnetic waves of different frequencies with each other. The influence of two electromagnetic waves (ω_1 and ω_2), acting simultaneously on a crystal with anharmonic oscillations, can be expressed as:

$$m_0\left(\frac{d^2 u}{dt^2} - \gamma\frac{du}{dt} + \omega_0^2 u + q_2 u^2\right) = eE_x^{(1)}\exp(i\omega_1 t) + eE_x^{(2)}\exp(i\omega_2 t), \quad (16.6)$$

assuming that all waves are polarized in x direction, with ω_0 as the appropriate resonance frequency, and q_2 as the nonharmonicity factor [compare with Eq. (12.3)]. The oscillation $u(\omega)$ now contains components of $\pm\omega_1 \pm \omega_2$, the sum and differences of the original frequencies:

$$u(\omega) = A(\pm\omega_1, \pm\omega_2)\exp\{i(\pm\omega_1 \pm \omega_2)t\}. \quad (16.7)$$

The advantage of this representation is that the δ values are found to be nearly independent of ω, while the χ values are not. The value of δ is observed to be $\simeq 5 \cdot 10^{-16}$ $(Vcm^3)^{-1}$, within a factor of 2 for most semiconductors while χ values vary over four orders of magnitude (Miller, 1964; Wynne 1971). The insensitivity of δ to a specific semiconductor probably relates to the fact that integral properties, depending on overall bonding, determine δ rather than properties depending on the detail of the energy structure (Friedel and Lannoo, 1973).

When using a specific combination of ω_1 and ω_2, e.g., $\omega_1 + \omega_2$ with $\omega_2 = \omega_1$, we obtain in the conventional manner (see Section 10.1.1A) for the amplitude of the second harmonic

$$A(2\omega_1) = -q_2 \frac{e^2 \left(E_x^{(1)}\right)^2}{4m_0^2} \frac{1}{\omega_0^2 - (2\omega_1)^2 + 2i\gamma\omega_1} \frac{1}{[\omega_0^2 - \omega_1^2 + i\gamma\omega_1]^2}.$$
(16.8)

From the polarization relation $\{P(2\omega_1) = eNu(2\omega_1)\}$, one obtains the appropriate equation for the susceptibility. In general, one has

$$\chi_{xxx}^{(2)}(-\omega_3; \omega_1, \omega_2) = q_2 \frac{e^3 N}{\varepsilon_0 m_0} \cdot \frac{1}{D(\omega_3)} \cdot \frac{1}{D(\omega_1)} \cdot \frac{1}{D(\omega_2)} \qquad (16.9)$$

with $D(\omega_i) = \omega_0^2 - \omega_i^2 + i\gamma\omega_i$. For higher-order susceptibilities, $\chi^{(n)}$, the $D(\omega_i)$ term is repeated $n + 1$ times. In the extrinsic range between band edge and *Reststrahl* energy (see below), one has with $(\omega, \omega_1, \omega_2,) \ll \omega_0$ a simple relationship for the polarization: $\chi_{xxx}^{(2)} = q_2 e^3 N / (m_0 \varepsilon_0 \omega_0^6)$. This relation can be transformed by using Eq. (11.31) for $(\kappa, \gamma) \to 0$ into

$$\chi_{xxx}^{(2)} = \frac{(n_r^2 - 1)e}{m_0 \omega_0^4} q_2. \qquad (16.10)$$

In order to replace the anharmonicity parameter q_2, we can use the displacement u which is obtained from

$$m_0 \omega_0^2 u = -m_0 q_2 u^2.$$

Here, restoring force and nonlinear force are equal to each other—see Eq. (16.6). This yields $q_2 = \omega_0^2 / u$ and, consequently,

$$\chi_{xxx}^{(2)} = \frac{(n_r^2 - 1)e}{m_0 \omega_0^2 u} = 1.76 \cdot 10^{-9} (n_r^2 - 1) \left(\frac{10^{16}}{\omega_0}\right)^2 \left(\frac{1\text{Å}}{u}\right) \text{ (cm/V)}. \qquad (16.11)$$

Typical displacements for observing unharmonicities are on the order of the covalent radii (1Å). It should be noted, however, that restoring forces $\propto u^2$ imply that crystals with inversion symmetry cannot produce second- (even) order harmonics.

16.1.2 Nonlinear Polarization of Free Electrons

A free-carrier gas has a statistical center of symmetry, thereby excluding second-order effects in polarization. Free carriers contribute, however, to third-order effects in the susceptibility. These can be distinguished from other effects by their dependency on the carrier density.

Nonlinear polarization of free carriers is caused by nonparabolicity of conduction or valence bands and by the field dependence of some of the carrier scattering mechanisms (Rustagi, 1970; Wang and Ressler, 1969, 1970).

The third-order susceptibility due to free carriers (fc) can be given as

$$\chi_{fc}^{(3)} = \chi_{np}^{(3)} + \chi_{sc}^{(3)} + \chi_{if}^{(3)} \tag{16.12}$$

with the subscripts "np", "sc", and "if" standing for nonparabolicity of the respective band, carrier scattering, and interference between the two processes. In a simple (Kane) model, the nonparabolic contribution can be estimated (Wolf and Pearson, 1966) as

$$\chi_{np}^{(3)}(-\omega_3; \omega_1, \omega_2) = n \frac{e^4}{\varepsilon_0 m_n^2 E_g \omega_1 \omega_2 \omega_3^2} \frac{1 + \dfrac{8}{5}\dfrac{E_F}{E_g}}{\left(1 + 4\dfrac{E_F}{E_g}\right)^{5/2}} \tag{16.13}$$

with E_F as the Fermi-energy. For $E_F \ll E_g$ and semiconductors of $m_n = 0.1 m_0$ and $E_g \simeq 1$ eV, we have $\chi_{np}^{(3)} \simeq 10^{-33} n$ (cm^2/V^2), which, for moderate carrier densities, can approach the values of $\chi^{(3)}$ due to valence electrons.

16.2 The Different Mixing Effects

In semiconductors with no inversion symmetry, the second- (or higher-) order susceptibility is large enough for relatively efficient (> 10%) interaction of two or more photons (phonons) with each other, often without invoking resonance transitions in the solid. These interactions can be used to generate light of a different wavelength; others can be used for light amplifiers or detectors of light beams. Such interactions consist of:

(1) harmonic generation by two identical photons $\propto \chi^{(2)}$ with $\omega_i = \omega_j$;

(2) third harmonic generation by three identical photons $\propto \chi^{(3)}$ with $\omega_j = \omega_i = \omega_m$;

(3) sum and difference frequency mixing $\propto \chi^{(2)}$ with $\omega_i = \omega_j \pm \omega_l$;

(4) parametric amplification, closely related to mixing* $\propto \chi^{(3)}$;

(5) optical rectification $\propto \chi^{(2)}$ with $\omega_i = -\omega_j$; and

(6) self-focusing of light $\propto \chi^{(3)}$ with $\omega_j = -\omega_l = \omega_m$.

Some of these multiphoton interactions can be used as analytical tools and are important for gaining new insight about the oscillatory behavior of solids, which is not otherwise obtainable. For example, multiphonon absorption can yield excited states which cannot be excited otherwise, because of selection rules. A few examples are given below. For a review, see Byer and Herbst (1977) or Warner (1975).

Most of these experiments are based on the mixing of several photons or of photons and phonons in order to probe levels with different symmetries. As an example, we can compare a typical linear absorption experiment in which the parity of the excited state must be opposite to that of the ground state (since the dipole moment is odd and causes the final state to change its parity), with a nonlinear two-photon interaction in which the parity of the excited state remains the same since the dipole matrix element appears twice. Consequently, one can observe the same parity transitions compared to the ground state with nonlinear mixing experiments.

There is a very wide variety of experiments in which nonlinear optical effects are used whenever two or more quasi-particles, such as photons, phonons, polaritons, or electrons are involved. Several of these will be discussed in other section of this book, e.g., multiphonon absorption (Section 11.4), photon scattering (Chapter 17), and the Burstein-Moss shift (Section 24.3.1). In each of these cases, and sometimes implicit, a nonlinearity of the dielectric polarization is involved, i.e, ε and χ are a function of the acting field.

An important topic is the nonlinear *solid-state spectroscopy*, which is an extension of the well-known one-photon spectroscopy. It deals with all types of interactions accessible for single photons, and probes the higher amplitude behavior; it also makes bulk regions of crystals accessible for excitation in ranges of very high optical absorption (see the following section) or permits stimulations of tran-

---

* Amplification is achieved when a small signal at ω_s is mixed with a strong laser pump beam at ω_p and results in the creation of an additional beam at ω_i, the *idler frequency*, according to $\omega_p = \omega_s + \omega_i$. In this process, energy from ω_p is pumped into ω_s and ω_i; consequently, ω_s is amplified.

sitions which are forbidden for simple photon or phonon processes, as discussed before.

Because of the limited conversion efficiency for multiphonon processes, nonlinear spectroscopy requires experimental skill; and for sufficiently accurate results, it is restricted to such semiconductors with proper symmetry which can be grown in large enough crystals of sufficient optical perfection.

An interesting range for nonlinear spectroscopy is that in which the sum, or difference of energy of two photons, equals a specific resonance transition. Here, high amplitudes can be reached with rather modest input signals (see below).

16.2.1 Up-Conversion and Difference Mixing

Since the polarization is low when far from a resonance transition, and higher-order susceptibilities rapidly decrease with increasing order, we need a high intensity laser for most experiments to create signals above noise. Near resonances, however, the efficiency for higher-order interaction dramatically increases. The absorption of single photons in this range is usually too high to reach regions beyond the near-surface layer. These difficulties can be avoided by *up-conversion*, i.e., by mixing two low energy photons. These photons enter the crystal bulk without appreciable absorption, while the effect of an up-converted photon with an energy near or at the electron resonance with large absorption can now be studied within the bulk.

Such up-conversion is commonly used to reach exciton-polaritons (see Section 15.1.2C), or characteristic points in band-to-band transitions.

Difference mixing can bring the resulting photon into the phonon absorption range, i.e., the *Reststrahl* range, where the bulk photon-phonon or phonon-polariton interactions can be studied (see Kildal and Mikkelsen, 1973 and Section 17.1.3F).

The transfer of energy between the two beams of light for mixing depends on the coherence length of the interaction. The energy transfer follows the conservation-of-photon relation given by the Manley-Rowe equations (Manley and Rowe, 1959):

$$\frac{1}{\nu_1}\frac{dI_1}{dz} = \frac{1}{\nu_2}\frac{dI_2}{dz} = -\frac{1}{\nu_3}\frac{dI_3}{dz}. \qquad (16.14)$$

In *sum frequency generation* ($\nu_3 = \nu_1 + \nu_2$), both lower energy laser beams lose power, the sum of which is gained by the higher frequency beam. In *difference frequency generation* ($\nu_2 = \nu_3 - \nu_1$), the higher frequency beam (ν_3) loses power while both lower frequency beams (ν_1 and ν_2) gain the equivalent amount. This effect can be used to amplify the amplitude of a low intensity beam. The frequency ν_2 can be generated with high efficiency (approaching 50%) and can be amplified when a cavity is provided and tuned to ν_2. This constitutes a parametric oscillator.

16.2.2 Walk-Off

In order for the different beams to remain in phase throughout the crystal, the index of refraction for the involved frequencies must remain the same. Even away from resonances, however, this is usually not the case. As a result, the beams with frequency ν_1 and ν_3 will *walk off* in slightly different directions, limiting the *coherence length* (the length along which two light beams can interact when they remain parallel to each other and in phase) to

$$l_{\text{coh}} = \frac{\lambda}{4\left(n_r(\nu_2) - n_r(\nu_3)\right)} \tag{16.15}$$

with λ as the fundamental wavelength for second harmonic generation [$\nu_1 = \nu_2$ and $\nu_3 = 2\nu_2$ in Eq. (16.14)]. By carefully adjusting the incident angle θ with respect to the optical axis, however, one can match for two beams the refractive index for the fundamental beam of one, to the extraordinary beam of the other, if birefringence is strong enough to compensate dispersion. This is called *birefringence phase matching* and has been analyzed by Midwinter and Warner (1965) for uniaxial crystals and by Hobden (1967) for biaxial crystals.

Other methods of phase matching include matching in optically active media, Faraday rotation, and using anomalous dispersion; see the literature listed in the review by Chemla and Jerphagnon (1980). Phase matching is important for efficient nonlinear optical devices.

16.2.3 Mixing and dc-Fields

Nonparabolic parts of the polarization can be tested by an external dc-field that causes a sufficient relative shift of the bound valence electrons relative to the ion cores. This could be regarded as a prestressing of the lattice atoms in an external electric field and testing these prestressed atoms with the electromagnetic irradiation.

As a result, one observes an increase in the power of second harmonic generation with applied external field:

$$P(2\omega) \propto E_{dc}^2. \tag{16.16}$$

The inverse of this process—namely, the frequency difference generation of two identical photons—

$$\omega_1 - \omega_1 = 0, \tag{16.17}$$

is called *optical rectification* and generates a dc-field or, in an external circuit, a dc-current.

16.2.4 Conversion Efficiencies

Depending on the material, rank of susceptibility, the frequencies' proximity to a resonance, and the acting field amplitude, the conversion efficiencies for nonlinear processes can vary from less than 1% to values in excess of 80%. The efficiency of a phase-matched harmonic generation was evaluated by Armstrong et al. (1962) and is given by

$$\eta(\nu \to 2\nu) = \frac{I(2\nu)}{I(\nu)} = \tanh^2(\varepsilon A_0 l_{coh}) \tag{16.18}$$

where A_0 is the amplitude of the incident laser, and l_{coh} is the coherence length. Total power transfer into the up-converted beam is theoretically possible; experimentally, more than 50% efficiency has been achieved.

For transitions close to resonances, giant oscillator strengths are sometimes observed, e.g., when creating excitonic molecules—see Section 15.1.2E. In CuCl up to fourth-order excitation and cascade recombination was observed for such multiphoton processes (Maruani et al., 1978.).

16.3 Electro-Optical Effects

Electro-optical effects refer to a change in the dielectric function with an applied electric field. Such changes can be derived from the anharmonic oscillator model. When such a change is linear with the electric field, it is referred to as the *Pockels effect*. When it is quadratic, it is called the *Kerr effect*.

16.3.1 The Pockels Effect

When in Eq. (16.6) a constant electric field E_0 is added on top of the electromagnetic radiation, we have similar conditions as indicated earlier for nonharmonic lattice oscillations—see Section 4.1.3. Even for small amplitude oscillations, i.e., low light intensities, the range of anharmonic oscillations can be reached by sufficient prestressing— here by a sufficient dc-field. This lowers the symmetry of centro-symmetric crystals, so that these also show second harmonic generation, e.g., seen in Si.

Such prestressing causes a shift in the resonance frequency

$$\Delta(\omega_0^2) = \omega_0^2 - (\omega_0')^2 = 2q_2 \frac{eE_0}{m\omega_0^2} \tag{16.19}$$

and results in a change in the index of refraction

$$\Delta n_r = \frac{(n_r^2 - 1)\Delta(\omega_0^2)}{2n_r(\omega_0^2 - \omega^2)}. \tag{16.20}$$

This change in n_r is anisotropic because of the acting field, and renders the optical properties of isotropic semiconductors anisotropic. Such anisotropy is expressed by the index of refraction ellipsoid

$$\frac{x^2}{n_1^2} + \frac{y^2}{n_2^2} + \frac{z^2}{n_3^2} + \frac{2yz}{n_4^2} + \frac{2zx}{n_5^2} + \frac{2xy}{n_6^2} = 1 \tag{16.21}$$

with indices as identified in relation (16.5). Conventionally, the anisotropic index of refraction relates to the field as

$$\frac{1}{n_m^2} = \frac{1}{n_r^2} + \sum_i r_{mi} E_i \tag{16.22}$$

with the electro-optical coefficient r_{mi}. We can show easily in x-direction that r_{xx} and χ_{xxx} are related as $n_r^4 r_{xx} = -2\chi_{xxx}$ [using Eqs. (16.10), (16.19) and (16.20)] and, in a general form (Franken and Ward, 1963)

$$n_m^4 r_{mi} = -2d_{im}. \tag{16.23}$$

Such anisotropy results in different light velocities dependent on the relative orientation of the dc-field and light beam polarization (with propagation perpendicular to the dc-field). It causes a rotation of the plane of polarization of the light while traveling through the medium.

When rotation by $\pi/2$ is achieved, a light beam of linear polarization is fully switched on when exiting through a polarizing filter with orientation normal to the polarization of the incident beam. The field necessary to achieve this switching is

$$E_{\pi/2} = \frac{\lambda}{n_r^3 r_{41} L}. \tag{16.24}$$

The field to accomplish this switching is on the order of 10^4 V/cm for $\lambda = 1\text{m}\mu$, $n_r \simeq 3$, $L = 3$ cm, and $r_{41} \simeq 10^{-10}$ cm/V.

16.3.2 The Kerr Effect

The Kerr effect depends on introducing a third-order term in displacement $m_0 q_3 u^3$ into Eq. (16.6). Here, the index of refraction changes proportional to quadratic terms in the dc-field. Such an effect is the only electro-optic nonlinear effect in centro-symmetric semiconductors. The corresponding change in the dielectric constant is given by

$$\Delta n_r = -q_3 \frac{3e^2(n_r^2 - 1)}{2n_r m_0^2 \omega_0^4 (\omega_0^2 - \omega^2)} E_0^2,$$

obtained similarly as in Eq. (16.20). Centro-symmetric semiconductors in which the Kerr effect can be observed are the element semiconductors such as Si and Ge. The Kerr effect is well known in certain organic liquids such as nitrobenzene and CS_2. Its field-induced birefringence is often used for modulation or fast switching of a light beam.

Summary and Emphasis

Nonlinear optical effects are the result of electromagnetically induced oscillations which couple with each other due to the anharmonic terms in the potential. These can be reached with higher amplitudes beyond the harmonic range or by prestressing with an applied dc-field. Nonlinear effects can be analyzed in terms of a field-dependent dielectric function, i.e., a second- and higher-order polarization.

Such nonlinear effects are required for the mixing of two or more excitation processes (photons, phonons, polaritons), resulting in new states with the sum or differences of incident energies. They can involve transitions far from resonances (virtual states) or near resonances, the latter with much increased conversion efficiencies.

These nonlinear effects can be used for analytical purposes—that is, to excite the bulk of a semiconductor in the intrinsic range of high optical absorption via up-conversion or down-conversion of two photons in the extrinsic (low absorption) range, resulting in photons in a resonance (high absorption) range, or by a host of scattering phenomena to be discussed in the following chapter.

Nonlinear spectroscopy has a significant impact on semiconductor physics by permitting transitions which are forbidden in conventional spectroscopy; by permitting a wide range of multiphoton spectroscopy; by initiating coherent excitation; and by laser action in previously inaccessible ranges.

Exercise Problems

1.(r) List the general principles involved in nonlinear optics.
2.(r) Discuss the different effects in multiphoton mixing; compare these to the effects known to you in radio-wave mixing.
3. Describe in your own words parametric amplification and the relevant conditions which must be fulfilled.
4.(e) In scattering of particles, energy and momentum conservation need to be fulfilled. In nonlinear optics equivalent rules must be obeyed. Which are these?
5. Second harmonic generation was one of the first demonstrations of nonlinear interaction of light waves in solids. Describe the effect, its observation, and the necessary conditions to be fulfilled.
6.(l) Search the literature for recent publications on the Kerr-effect *in solids*. Review the paper and describe the findings in your own words.

Chapter 17

Photon Scattering

Photon scattering is a powerful method to study elementary excitation processes.

Scattering of photons with other quasi-particles is another *nonlinear* effect, and provides substantial insight into various properties of the semiconductor. The optical absorption is based on an interaction with the dipole moment described by the matrix element; in contrast, scattering is based on interaction with the polarizability tensor. There is a multitude of possibilities for scattering of photons on phonons. These, as well as scattering processes with other quasi-particles, will be discussed in the following sections.

17.1 Elastic and Inelastic Scattering

During a scattering event, energy and momentum must be conserved. If the scattering particle absorbs only a small amount of energy compared to the photon energy, while a substantial amount of momentum is transferred, we refer to an *elastic scattering* event. When a significant amount of energy and momentum are transferred, we speak of an *inelastic scattering* event. When this energy transfer occurs to or from the phonon, we speak of Stokes or anti-Stokes scattering, respectively, with

$$E_1 = E_0 \pm \hbar\omega \qquad (17.1)$$

and

$$\mathbf{k}_1 = \mathbf{k}_0 \pm \mathbf{q}; \qquad (17.2)$$

where $\mathbf{q}$ indicates a phonon, and $\mathbf{k}$ indicates a photon. The angle between $\mathbf{k}_1$ and $\mathbf{k}_0$ is determined by the experimental setup: the direction of incoming light and the offset position of the scattered light detector. The selection of $\mathbf{q}$ is then automatic, fulfilling Eq. (17.3). The subscripts 0 and 1 denote the incident and scattered beams, respectively. The vector diagram in Fig. 17.1 indicates the influence of the scattering angle θ on the selection of the phonon with appropri-

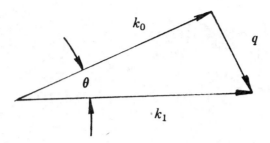

Figure 17.1: Vector diagram of momentum conservation during the scattering process.

ate momentum **q** (wavelength) to maintain momentum conservation in the scattering process:

$$q = \sqrt{k_1^2 + k_0^2 - 2k_1 k_0 \cos \theta}. \tag{17.3}$$

This indicates an important peculiarity of the scattering process. Since the energies or momenta of the scattering particles can be vastly different, substantially different paths in the Brillouin zone are traversed. With

$$|\mathbf{k}| = \frac{2\pi}{\lambda} \qquad \text{and} \qquad 0 < q < \frac{\pi}{a} \simeq 10^8 \text{ cm}^{-1}, \tag{17.4}$$

we see that photons of energy comparable to phonons (IR), or of visible light, have a k-vector of $10^3 \ldots 10^5$ cm^{-1}; this is very small compared to the extent of the Brillouin zone. These photons therefore can probe only the range very close to the center of the Brillouin zone near $q = 0$, i.e., the long wavelength part of the phonon spectrum. *X-ray photons* can provide significantly higher momenta up to the zone boundary for $\lambda = 2a$. However, the energy transmitted by the phonon is only a very small fraction of the x-ray energy and poses difficulties in detecting the small relative Stokes or anti-Stokes shift of such interaction. On the other hand, substantially larger shifts in the phonon momentum can be obtained by scattering with neutrons, as discussed in Section 5.5.2.

Elastic scattering of light on phonons, i.e., scattering on density fluctuation, resulting in fluctuation of n_r and κ, is called *Rayleigh scattering*. As it represents a very small effect in solids, Rayleigh scattering will only be mentioned briefly in Section 17.4. *Inelastic scattering* caused by *acoustic phonons* is identified as *Brillouin scattering*; inelastic scattering caused by *optical phonons* is called *Raman scattering*. In most of these scattering events an intense light beam

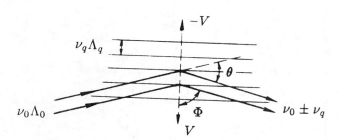

Figure 17.2: Brillouin scattering by a moving acoustic wave, with Bragg reflection at an angle θ.

impinges on a solid, while only a small amount of light is scattered by phonons.

17.1.1 Brillouin Scattering

Brillouin scattering (Brillouin, 1922) can be understood on the basis of classical arguments. The coherent propagation of transverse acoustic phonons as sound waves causes local density waves, and thus waves of small changes in the index of refraction, to which a photon field is strongly coupled when the Bragg conditions are fulfilled (Fig. 17.2).

The light beam is scattered at the Bragg angle θ and contains two components: one photon with the TA phonon frequency added (anti-Stokes), and another with this phonon frequency subtracted (Stokes) as a result of the Doppler shift:

$$h\nu_0 \pm h\Delta\nu = h\nu_0 \pm \hbar\omega_{TA} = h\nu_0 \pm 2h\nu_0 \frac{v_s}{c} n_r \sin\frac{\theta}{2}. \qquad (17.5)$$

Here v_s is the sound velocity and c is the light velocity. Such a Brillouin shift is shown in Fig. 17.3 and is a direct measure of the phonon frequency and the sound velocity. Thus, it yields information on elastic constants, their anisotropy, and various other properties related to the interaction of acoustic phonons with other low energy excitation phenomena (Pine, 1972). The width of the two Brillouin components yields information on a variety of relevant damping processes, such as carrier-induced damping or structural relaxation (Balkanski and Lallemand, 1973).

Brillouin scattering is also used to investigate *acoustoelectric effects* (Hutson and White, 1962) and moving *acoustoelectric domains*. The latter are caused by carriers in piezoelectric materials when their drift velocity starts to surpass the sound velocity, causing packets of

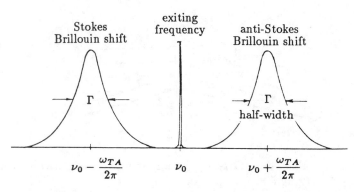

Figure 17.3: The Brillouin scattering spectrum with Stokes and anti-Stokes scattering components.

acoustic waves to move through the semiconductor (Conwell, 1967; Mayer and Jørgensen, 1970).

17.1.1A Resonant Brillouin Scattering When the wavelength of monochromatic light approaches an electronic transition, *resonant Brillouin scattering* can be observed (Weisbuch and Ulbrich, 1978). Scattering with a phonon of sufficient energy to permit such an electronic transition is then enhanced. Fine tuning to achieve this transition can be achieved by changing the laser frequency or the band gap by slight temperature variation (Pine, 1972—see also polariton scattering below). For a review, see Yu (1979).

Intense acoustical waves can also be produced by acoustoelectric domain generation (Conwell, 1967). Strongly enhanced scattering, expressed by a large scattering cross section, is observed when the band edge or other critical points are approached—see, e.g., Hamaguchi, et al. (1978).

17.1.2 Resonant Scattering with Exciton-Polaritons

If the impinging light has an energy below, but close to, a free exciton line in direct band gap semiconductors (see Section 15.1), the phonon can supply the missing energy, and *resonant scattering with exciton polaritons* occurs.

When the energy supplied by a photon lies below the $1s$ exciton, one observes one backward scattered line at $h\nu_{1s}$ (Fig. 17.4) with a transition $k_2 \to k_2'$. When the frequency of the photon lies above $h\nu_{1s}$, four Stokes-shifted lines are expected, for which energy and momentum conservation is fulfilled. Phonons near the center of the Brillouin zone interact strongly with optical radiation, creating an

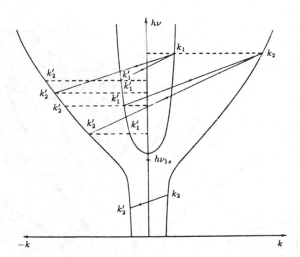

Figure 17.4: Dispersion curves of free exciton-polaritons, indicating the Stokes processes of Brillouin scattering between different branches.

exciton-polariton state. This is a mixed state that is created between the photon field and the electronic polarization field, as opposed to a mixed state with the ionic polarization field for the phonon-polariton. Scattering occurs between the exciton-polariton and a longitudinal acoustic phonon; therefore, the scattering is a Brillouin scattering process.

In contrast to the phonon-polariton scattering previously discussed, the lower branch is not flat but parabolic in its upper part, since the exciton is mobile and can acquire kinetic energy which causes an $E \propto k^2$ behavior. In addition, there are several lower branches according to the different excited states of the exciton—see Fig. 17.5; for a review, see Yu (1979).

A measured spectrum,* showing several of these branches from the ground and excited states of the exciton, is shown in Fig. 17.5. Some of these relate to interaction with LA phonons, others with TA phonons, as indicated in the figure. We will return to the spectrum of excitons-polaritons when discussing hyper-Raman scattering—see Section 17.1.3F.

---

* The analysis of the measured reflection spectrum as a function of the wavelength and incident angle is rather involved (Broser et al., 1978). A relatively simple method for measuring the central part of the exciton-polariton spectrum in transmission through a prismatic crystal was used by Broser et al. (1978). See Section 15.1.2C.

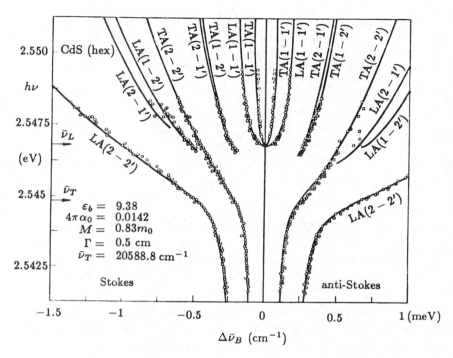

Figure 17.5: Brillouin shift of the scattered laser light in CdS as a function of the laser energy. Theoretical transitions refer to the indexing shown in Fig. 17.4. Parameters: dielectric constant $\varepsilon_b = 9.38$, oscillator strength $4\pi\alpha_o = 0.0142$, excitonic mass $m_r = 0.83m_0$, phenomenological damping constant $\gamma = 0.5$ cm, transverse exciton energy 2.5448 eV, and longitudinal exciton energy 2.5466 eV. The numerals 1 and 2 refer to inner and outer branch polaritons (after Wicksted et al., 1984).

17.1.2A Stimulated Brillouin Scattering   Stimulated Brillouin scattering can be achieved at high optical intensities, where longitudinal acoustic waves can be amplified (Chiao et al., 1964). The energy for the parametric amplification (see Chapter 16 and Section 44.7) is supplied from the Stokes component of the Brillouin scattering.

17.1.3 Raman Scattering

Raman scattering occurs with emission or absorption of optical phonons—for a review, see Balkanski (1980). While Brillouin scattering is very sensitive to the scattering angle (the acoustic phonon energy changes linearly with $\mathbf{q}$ near $\mathbf{q} = 0$), the classical Raman scattering is not. At a well-defined optical phonon energy, added to or subtracted from the impinging photon, scattering occurs independent of the scattering angle since all scattering takes place near the Γ-point ($\mathbf{q} = 0$) of the Brillouin zone.

The selection rules for light *scattering* are different from the selection rules for optical *absorption*. Therefore, a different set of optical phonons is *Raman active* at the center of the Brillouin zone—see Mitra (1969). Specifically, for crystals with inversion symmetry, even parity excitations are Raman active, while odd parity transitions are observed in IR absorption. The set of Raman-active oscillations can be identified by group theoretical rules according to the crystal symmetry. Consequently, the intensity distribution and polarization of different Raman lines, observed in anisotropic semiconductors, depend on the angle and polarization of the impinging light *relative to the crystal orientation*—see Birman (1974) and Poulet and Mathieu (1970)

Raman scattering is a relatively rare process. Its probability can be estimated from the polarizability tensor (Born and Huang, 1954). The scattering efficiency is estimated as

$$\eta_s = \frac{3h^4 \nu_s^4 L\, d\Omega}{\rho c^4 \omega_{\mathrm{TO}}} |\alpha_R|^2 \times \begin{cases} f_{\mathrm{BE}}(\omega_{\mathrm{TO}}) + 1 & \text{for Stokes} \\ f_{\mathrm{BE}}(\omega_{\mathrm{TO}}) & \text{for anti-Stokes} \end{cases} \quad (17.6)$$

where ρ is the density, L is the sample length for radiation of the scattered light of frequency ν_s, emitted into the solid angle $d\Omega$; α_R is the first derivative polarizability $\partial \chi^{(1)}/\partial \Delta x$, or *Raman polarizability*, which relates to the polarization by $P = \chi^{(1)} E + N\alpha_R E u$ (Wynne, 1974); and $f_{\mathrm{BE}}(\omega_{\mathrm{TO}})$ is the phonon population given by the Bose-Einstein distribution function. This efficiency is strongly dependent on the frequency of the exciting light and, through $f_{\mathrm{BE}}(\omega_{\mathrm{TO}})$, on the temperature. Typically, it is on the order of 10^{-6} to 10^{-7}; thus, one needs a strong monochromatic light source to generate an observable (faint) scattering signal.

A typical Raman spectrum is shown for AlSb in Fig. 17.6A. By choosing the polarization with respect to the lattice orientation, discrimination between TO and LO phonons is possible. The ratio of scattering efficiencies for parallel and normal polarization of the scattered beam with respect to the polarization of the exciting beam assists in analyzing more complex Raman spectra (Loudon, 1964). Another example, given in Fig. 17.6B, shows the typical development of the scattering spectra when Ga is added to AlSb to make it a ternary alloy. These spectra can be explained by a system of

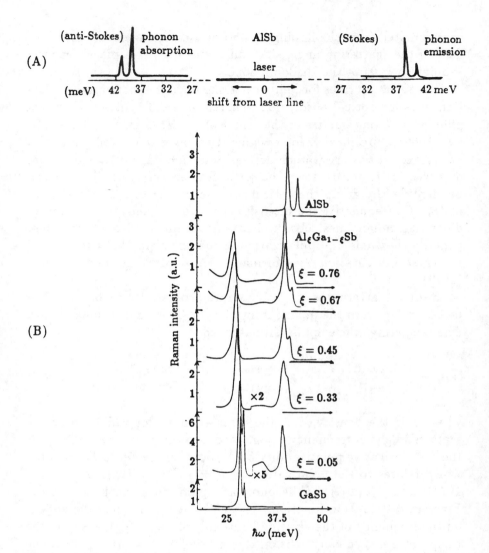

Figure 17.6: (A) Raman spectrum of AlSb (after Mooradian and Wright, 1966). (B) Dependence of the Stokes branch on the composition of $Al_\xi Ga_{1-\xi} Sb$ (after Charfi et al., 1977). ©Pergamon Press plc.

two coupled modes of lattice oscillation (Jahne, 1977), and provide valuable information on crystal structure and composition.

17.1.3A Polar and Nonpolar Raman Scattering   We must distinguish scattering on lattice vibrations which are associated with a dipole moment (*polar modes*) and such on *nonpolar modes*. The latter are vibrations in covalent crystals, and those in crystals with

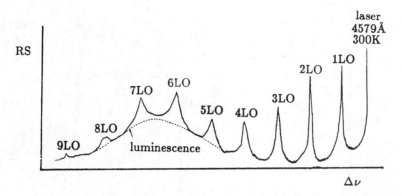

Figure 17.7: Raman scattering (RS) spectrum of CdS at 300 K with laser excitation at 4579 Å (after Leite et al., 1969).

an ionic bonding fraction that do not possess a dipole moment, i.e., long wavelength acoustic modes—$q \ll \pi/a$.

For a discussion of scattering on nonpolar phonons, see Born and Huang (1954) or Cochran (1973). The discussion of polar mode scattering is more involved (Poulet, 1955). Both treatments are substantially different from each other (Hayes and Loudon, 1978).

17.1.3B First- and Higher-Order Raman Scattering First- and second-order Raman scattering are distinguished by the emission or absorption of one or two phonons. In second-order Raman scattering, both phonons may be emitted or absorbed, or one may be emitted and another one absorbed, giving a Stokes and an anti-Stokes component. In addition, we distinguish sequential and simultaneous scattering events involving two phonons. Second-order Raman scattering provides access to the entire Brillouin zone, since the second phonon can deliver the necessary momentum. In nonpolar modes, it has been used to obtain information on the deformation potential (Carles et al., 1977), which is important for electron-phonon scattering. A review of second- and higher-order Raman scattering is given by Spitzer and Fan (1957). A spectrum for an unusually large number of resolved multiphonon scattering events in CdS is shown in Fig. 17.7.

17.1.3C Raman Scattering from Local Modes Raman scattering from local modes of optical phonons yields the frequency of these modes related to certain crystal defects. Thereby, it has given a great deal of information about these defects, including the symmetry of their nearest lattice environment, the mass and bonding

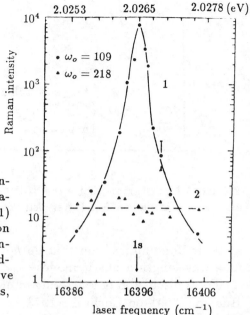

Figure 17.8: Resonance enhancement of the first-order Γ_{12}^- Raman line (at 13.5 meV, curve 1) in Cu_2O, inducing the $1s$ exciton excitation near 2.0365 eV. The enhancement is absent at the second-order feature (at 26.9 meV, in curve 2) (after Compaan and Cummins, 1973).

force of these impurities—sensitive even for isotope distinction—and their tendency to form defect associates (Barker and Sievers, 1975; Hayes and Loudon, 1978).

Raman scattering can also be modified or enhanced by surface interaction, e.g., by reduced bonding surface plasmons—see Burstein, et al. (1979).

17.1.3D Resonant Raman Scattering Resonant Raman scattering provides important information about the symmetry of certain transitions. It is observed with substantial enhancement of the scattering cross section for band-to-band transitions, free and bound excitons, and for polaritons when the energy of the initiating light is slightly below the energy for the transition which then is enabled by the additional phonon. In Fig. 17.8 such a resonant enhancement is shown for a first-order odd-parity Γ_{12}^- phonon to the $1s$ exciton (see Section 15.1.2) at the Γ point in Cu_2O. The exciton transition itself has even-parity and is therefore electric dipole-forbidden. An enhancement is absent in the even-parity two-phonon process.

Other dipole-forbidden (i.e., not observable in optical absorption) levels of the s- and d- yellow series in Cu_2O (Fig. 15.6) can be excited by resonant scattering. In contrast, dipole-allowed optical

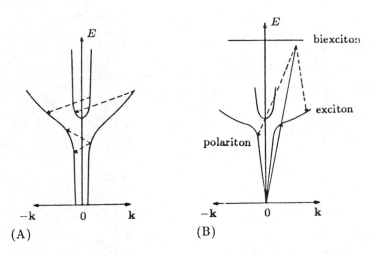

Figure 17.9: (A) Resonance Brillouin scattering of exciton-polaritons compared to (B) hyper-Raman scattering via virtual biexcitons.

transitions interact only with even-parity phonons for a resonant scattering (Washington et al., 1977).

Another type of resonant scattering can be observed as an edge-type excitation of the dipole-forbidden $1s$ yellow exciton in Cu_2O with a two-phonon process. In a first excitation, an intermediate dipole-allowed exciton is produced which then decays to a $1s$ exciton with emission of another phonon (Yu et al., 1973).

17.1.3E Stimulated Raman Scattering   Stimulated Raman scattering is observed at higher incident intensities. Here, a two-photon up-conversion can take place.

When two laser beams are used with $\omega_2 - \omega_1 = \omega_{TO}$, maximum amplification occurs. The intensity of the scattered beam can reach the threshold for lasing.

17.1.3F Hyper-Raman Scattering   The hyper-Raman scattering yields information about the exciton-polariton dispersion relation, similar to the resonant Brillouin scattering discussed in Section 17.1.2. However, more intense optical excitation is required to induce a two-photon excitation of a *virtual biexciton**—Fig. 17.9B; see also Section 15.1.2E. Hyper-Raman scattering, when using two photons of frequencies ν_1 and ν_2, each having an energy slightly below the

---

* A state close to an actual biexciton state which immediately decays into other states.

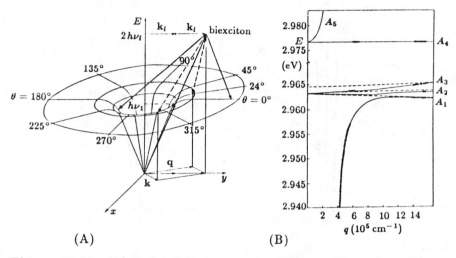

(A)   (B)

Figure 17.10: (A) Hyper-Raman scattering illustrated by a three-dimensional schematic of one lower polariton branch. Exciting photon: $h\nu_l(k_l)$; virtual biexciton: $2h\nu_l(2k_l)$; backscattered polariton: solid arrow at 180° paired to solid arrow at 0° for leftover polariton; forward scattered pair: dashed arrows as alternatives. (B) Dispersion curves of 5 branches of polaritons in CuBr for $e\|[001]$ and $K\|[110]$ (after Hönerlage et al., 1985).

band gap energy, and propagating with wave vectors k_1 and k_2 inside the semiconductor, creates a new intermediate state with

$$\mathbf{k}_1 + \mathbf{k}_2 = \mathbf{K} \quad \text{and} \quad h\nu_1 + h\nu_2 = E(\mathbf{K}). \qquad (17.7)$$

If the energy $h\nu_1 + h\nu_2$ is close to the resonant state (the biexciton), one observes a strongly enhanced transition—see Fröhlich (1981). The virtual biexciton decays into two quasi-particles, one of which is observed, while the other remains in the crystal to conserve energy and momentum. There are three possibilities:

$h\nu_1 + h\nu_2 = E_{LP}(\mathbf{K}_L) + E_{LP}(\mathbf{K}_L)$ (2 lower branch polaritons),

$h\nu_1 + h\nu_2 = E_{LP}(\mathbf{K}_L) + E_{UP}(\mathbf{K}_U)$ (1 lower, 1 upper polariton),

$h\nu_1 + h\nu_2 = E_{LP}(\mathbf{K}_L) + E_{Le}(\mathbf{k})$ (1 lower polariton, 1 upper exciton).

For all of these alternatives the condition

$$\mathbf{k}_1 + \mathbf{k}_2 = \mathbf{K}_i + \mathbf{K}_j \qquad (17.8)$$

is fulfilled.

Figure 17.10A depicts the creation of a biexciton from two photons (solid long arrow up) $h\nu_1 + h\nu_1 = 2h\nu_1$ with $\mathbf{k}_1 + \mathbf{k}_1 = 2\mathbf{k}_1$; both photons are provided by the same laser. This biexciton decays into two lower polaritons (pair of arrows down) which are offset by an angle to fulfill momentum conservation and land at different points ($h\nu$) on the polariton surface. The point of landing depends on the energy of the initiating photon pair and the angle of observation of the emitted (scattered) photon (solid arrows for backscattering, dashed arrows for forward scattering). This makes hyper-Raman scattering a three-photon process. It is determined by the third-order term in the susceptibility.

The entire polariton spectrum is obtained by changing the energy of the exciting light (the resulting virtual biexciton), the angle and energy of the emitted photon, and calculating energy and momentum of the leftover polariton. Several polariton branches have consequently been observed (Fig. 17.10B).

Hyper-Raman scattering follows selection rules other than those for normal Raman scattering or IR absorption. It thereby yields additional information about the lattice vibrational spectrum, e.g., about dipole modes in centro-symmetrical lattices which are forbidden in normal Raman scattering. It also permits excitation deep inside a crystal. For recent reviews, see Denisov et al. (1987) and Hönerlage et al. (1985).

17.1.4 Raman Scattering in Superlattices

Direct evidence for the folding of phonon branches in the minizone (see Section 5.2) of superlattices can be obtained from Raman scattering. Here new doublets are seen, as presented in Fig. 17.11. These doublets appear for the second and higher folded branches of the phonon dispersion curves, as shown in the insert of this figure.

The dispersion relation within the folded lower branches for LA modes normal to the superlattice interfaces can be approximated by (Rytov, 1956)

$$\omega_{m\pm} = v_s^\perp \left| \frac{2\pi\hat{m}}{l} \pm q \right|, \tag{17.9}$$

where $\hat{m}$ is the order of the branch, $v_s^\perp$ is the sound velocity normal to the layer, and l is the layer thickness. There is no folding, however, of the branches parallel to the superlattice layers, i.e., in y-directions. Thus, phonons propagating in this direction do not show the additional Raman doublets, as shown in the lower curve in Fig. 17.11.

One observes Brillouin scattering from the first branch of the folded spectrum. Raman scattering is noted from the higher branches which stem from folding the LA branch of the bulk lattice, as well as

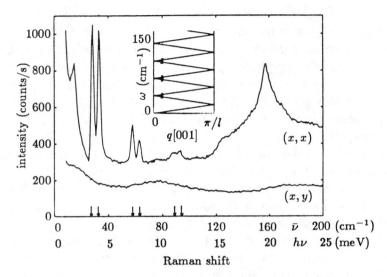

Figure 17.11: Raman spectrum of superlattice of 42 Å GaAs and 8 Å $Al_{0.3}Ga_{0.7}As$ with doublets by folded phonons as shown in insert. Arrows indicate theoretical peak positions obtained for a superlattice as identified by x-ray diffraction. The lower curve, measured in polarization parallel to the superlattice (y-direction) does not show the folded phonon doublets. Insert: Mini-zone with experimental values $\times$ (after Colvard et al., 1985).

from the LO branches (Jusserand et al., 1983). The dispersion relation for LO phonons is essentially flat, giving one shifted frequency per branch.

The frequency shift of the Raman signal depends on the period of the superlattice, as can be seen from Fig. 5.15 by changing the width of the mini-zone. This breaks up the bulk LO branch at different points (Jusserand et al. 1984). For a review on Raman scattering in superlattices, see Abstreiter (1986) and Abstreiter et al. (1986).

17.1.5 Raman Scattering in Glasses

Raman scattering in glasses does not follow the selection rules for anisotropic crystals.* Hence, a wider spectrum of Raman transitions is observed, although some of the gross features are similar to the ones obtained in the same material in crystalline form.

---

* In glasses, one cannot plot Brillouin zones; there is a breakdown of q-conservation, i.e., all momenta can contribute during scattering, causing substantial broadening.

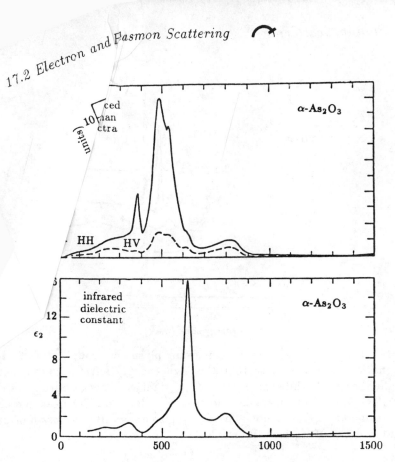

Figure 17.12: Reduced Raman and optical IR spectra for amorphous As_2O_3 (after Galeener et al., 1979).

A typical Raman spectrum is shown in Fig. 17.12 for amorphous As_2O_3 and compared to the IR optical dispersion distribution. It is obvious that both spectra expose different features which can be associated with the various oscillatory modes of the different As-O bonds: rocking, stretching, and bending of an As_4O_6 cluster molecule. Some of these are optically active, while others are Raman active—see Section 5.4, Fig. 5.16, for identification of the oscillatory modes. For more detail, see the review of Galeener et al. (1983).

17.2 Electron and Plasmon Scattering

When the density of free carriers exceeds 10^{15} cm^{-3}, one considers the collective effect of these carriers as *plasmons* (see Section 12.1.1), i.e., their ability to oscillate as an entity against the resting lattice. Here the amplitude becomes large enough to compete markedly with other IR absorptions.

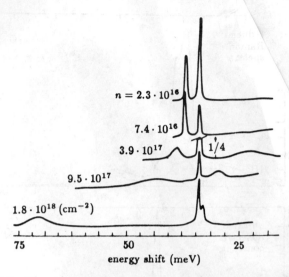

$$n = 2.3 \cdot 10^{16}$$

$$7.4 \cdot 10^{16}$$

$$3.9 \cdot 10^{17} \quad 1/4$$

$$9.5 \cdot 10^{17}$$

$$1.8 \cdot 10^{18} \ (\text{cm}^{-2})$$

75   50   25

energy shift (meV)

Figure 17.13: Raman scattering from plasmons coupled with LO phonons in GaAs. Observe that the high energy (left) LO peak shifts to higher energies, while the lower energy (right) LO peak approaches the center TO peak with increasing electron density. The TO peak remains at the same energy; it is not affected by plasmons (after Mooradian and Wright, 1968). ©Pergamon Press plc.

Longitudinal plasmon modes can scatter with other quasi-particles such as LO phonons and electrons (Harper et al., 1973). The frequency of these plasmons increases with the square root of the electron density (see Eq. (12.4): $\omega_p = \left[ne^2/(\varepsilon_{\text{opt}}\varepsilon_0 m_n)\right]^{1/2}$), and is on the order of the optical phonon frequencies at a carrier density between 10^{15} and 10^{17} cm^{-3}. Resonance effects occur which are similar to those of the photon-TO phonon interaction leading to polaritons. However, the completely longitudinal plasma oscillation cannot interact with transverse electromagnetic radiation. After application of a magnetic field, a transverse component also appears, and such interaction becomes possible (Wherrett and Firth, 1972). Furthermore, interaction of plasmons with LO phonons can change the plasmon mode, permitting interaction with photons and polaritons. These events will be discussed in the following section.

17.2.1 Plasmon-Polariton Scattering

Plasmons can be observed by Raman scattering of light (Fig.17.13) and show the expected line shift with changing carrier density. Their

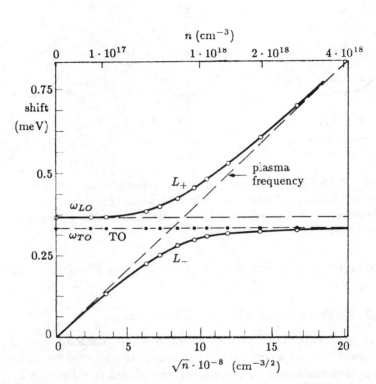

Figure 17.14: Energy of the LO phonon interacting with plasmons in numerous GaAs samples of different electron density (donor density), showing the split into two branches in the resonance range. The TO phonon is not affected (after Mooradian, 1968.)

interaction with LO phonons leads to the typical split of the dispersion curves for polaritons, as shown in Fig. 17.14.

This set of dispersion curves can be obtained from the polariton equation [Eq. (11.58)], considering plasmons instead. The LO resonance splits into two branches $\hbar\omega_+$ and $\hbar\omega_-$, given by

$$\omega_\pm^2 = \frac{1}{2}\left(\omega_{\text{LO}}^2 + \omega_p^2\right) \pm \frac{1}{2}\sqrt{(\omega_{\text{LO}}^2 + \omega_p^2)^2 - 4\omega_p^2\omega_{\text{TO}}^2}, \qquad (17.10)$$

and shown in Fig. 17.14. The coupled mode energies can be obtained from the Raman spectra shown in Fig. 17.13. The dots show the unchanged TO branch, and the circles show the two measured branches of the LO phonon as a function of the electron density, which causes the change of the plasmon frequency. The multiparticle interaction involves a phonon, a photon, and a plasmon (Patel

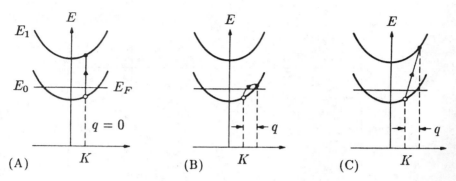

Figure 17.15: Electronic excitations in superlattices: (A) intersubband photon excitation; (B) intrasubband phonon excitation; (C) intersubband indirect (photon plus phonon) excitation (after Pinczuk and Abstreiter, 1989).

and Slusher, 1968). For more information, see Platzman and Wolff (1973).

17.2.2 Raman Scattering by Electrons

Free-single-electron scattering, as opposed to plasmon scattering in semiconductors, is a small effect. It was first observed by Mooradian (1968). It is enhanced by band structure effects (see Section 17.2.2A) and yields the electron distribution within the conduction band: the scattering yields the Doppler shift caused by the electrons with a velocity v. Hence, from the observed shape of the scattered light spectrum, one obtains the velocity distribution.

Scattering of single electrons may also involve resonant transitions from deep (e.g., rare earth) or shallow centers. In the first case, the electronic states are confined within the unit cell; in the second case, these states are spread out over a much larger volume element, with properties largely determined by the host. From these scattering experiments one obtains information about the excitation energies and the center symmetry. IR absorption and Raman scattering are complementary experiments. Scattering signatures are more abundant.

17.2.2A Free-Electron Scattering in Superlattices In superlattices, a large resonant enhancement is observed for direct transitions into excited mini-band states. There is a large joint density of states for intersubband transitions in a wide range of k because of the parallel shift of $E(k)$ as shown in Fig. 17.15A. Other indirect scattering transitions are also observed as shown in subfigures B and C (Pinczuk and Abstreiter, 1989).

17.3 Cyclotron and Spin-Flip Resonance Scattering

With a magnetic field applied, electrons are forced into orbits with a characteristic frequency, the cyclotron frequency ω_c. The band states split into Landau levels with spacing $\hbar\omega_c$. The ensuing mixed collective mode frequency in Voigt configuration (see Section 12.3.3B) is now composed of three frequencies:

$$\omega_\pm^2 = \frac{1}{2}(\omega_p^2 + \omega_c^2 + \omega_{LO}^2) \pm \frac{1}{2}\sqrt{(\omega_p^2 + \omega_c^2 + \omega_{LO}^2)^2 - 4(\omega_c^2\omega_{LO}^2 + \omega_p^2\omega_{LO}^2)}.$$

(17.11)

Consequently, the resonances become magnetic field dependent. These resonances are given by $\omega_c^2 + \omega_p^2 = \omega_{LO}^2$ (Palik and Furdyna, 1970). In Faraday configuration, strong absorption occurs at $\omega = \omega_c$ or $\omega = \omega_{TO}$, the former is independent of lattice coupling—see Section 12.3.1. In addition to the resonant Raman scattering at ω_c, two-photon resonances occur when $\omega_1 - \omega_2 = 2\omega_c$ (Patel and Slusher, 1968).

Also, resonant scattering occurs when the difference of photon frequencies is equal to a spin-flip frequency. Such *spin-flip Raman scattering* can be observed using small-signal gain techniques (parametric amplification) described in Section 16.2—Brueck and Mooradian (1973). At sufficient pump power one can achieve spin-flip laser output (Nguyen and Burkhardt, 1976). For a review, see Hayes and Loudon, 1978.

17.3.1 Spin-Flip Raman Scattering

The spin-flip scattering is caused by two initial light beams which create spin precessions at the difference frequency. From this precession, a third photon scatters coherently. This spin-flip Raman scattering should, however, be distinguished from magnetic dipole radiation (Brown and Wolff, 1972), or from nonlinearities induced by the magnetic field (Nguyen and Bridges, 1972).

Another type of spin-flip resonance scattering can be initiated when the spin-flip resonance ($\hbar\omega_0 = g/\mu_B\mathbf{B}$) is directly initiated by microwaves and a laser is scattered consequently (Romestain et al., 1974).

The Zeeman spin-flip mechanism has the largest effective scattering cross section, enhanced from an electron by $[m_0/m_n]^2$. Laser action can further be enhanced by up to 10^3 by initiating resonant transitions close to the band edge, yielding a total enhancement up to

10^6 (Smith et al., 1977). For a review of spin-flip Raman scattering see Pidgeon (1980).

17.3.2 Raman Line Shape Analysis

Optical Raman scattering produces lines which are displaced from the initiating light by the characteristic Raman energy, and show a line shape specific for the scattering event. Therefore, numerous attempts have been made to obtain more information about the event from the measured line shape. These attempts relate to the lifetime of the scattering quasi-particle—resulting in a symmetrical line broadening, to its kinetic energy—resulting in a one-sided tail, to existing built-in or external fields, and other, yet less understood, reasons—see Romestain and Weisbuch (1980). We will discuss the basic elements which cause line broadening in Chapter 20.

17.4 Rayleigh Scattering

Rayleigh scattering is a well-known scattering effect in media where large density fluctuations occur, such as in gasses. It is an elastic scattering phenomenon: that is, it proceeds without changes in frequency of the scattered photon. The scattering amplitude increases with decreasing wavelength of the scattered light.*

In solids the Rayleigh component can usually be neglected, except near critical points where density fluctuations can become rather large, e.g., when electron-hole condensation starts to occur. Frozen-in density fluctuations in glasses, although very small, provide transparency limitations for fiber optics because of such Rayleigh scattering.

17.5 Photon Drag

A collective effect of photon scattering with mobile particles can be observed as a *photon drag*, i.e., a radiation pressure. This pressure of a light beam acting on a free electron gas causes a slight shift of

---

* Rayleigh scattering is responsible for the blue light of the sky by scattering the short wavelength component of the sunlight on density fluctuations of the earth's atmosphere. It produces an absorption coefficient

$$\alpha_o = \frac{8\pi}{3} \frac{(n_r - 1)^2}{N} \cdot \frac{1}{\lambda^4},$$ (17.12)

where N is the density of air molecules and λ is the wavelength of the light.

the electron gas away from the incoming light. This causes a minor polarization, which can be observed as a potential difference between the front and back electrodes. This potential difference is called the *photon drag voltage*.

In a somewhat global form, the field created by a light flux ϕ—which is absorbed in a crystal platelet of thickness d, yielding an average generation rate of ϕ/d (W/cm^3)—can be estimated as

$$\frac{\phi}{cd} = enF \qquad (17.13)$$

where c is the light velocity. When integrated over a platelet-thickness, it results in a photon drag voltage of

$$V_{\text{phdr}} = -\frac{\phi}{encd}, \qquad (17.14)$$

a value that increases with decreasing carrier density. With lowering n, however, the hole density increases according to $p = n_i^2/n$ [see Eq. (27.72)], creating a hole photon drag with reversed sign of the photon drag voltage, as observed in Si by Gibson et al. (1970).

Assuming the holes remain trapped, the photon drag voltage can also be estimated from the classical equation of motion for a photon-generated electron gas (Bloembergen, 1965):

$$V_{\text{phdr}} = \frac{1}{4\pi^2}\frac{\mu_n}{\varepsilon_0 c^2}\frac{\phi_o}{\alpha_o}\frac{1}{\nu^2\tau_m^2}[1 - \exp(-\alpha_o d)]. \qquad (17.15)$$

It is on the order of ~ 1 mV for a laser beam of incident flux $\phi_o = 10^5$ W/cm^2 and for a semiconductor with $\mu_n \simeq 10^3$ cm^2/Vs, $\alpha_o = 1$ cm^{-1}, platelet thickness $d = 10$ cm, and $\nu^2\tau_m^2 = 10^3$ at a laser frequency $\nu \simeq 10^{14}$ s^{-1}.

17.6 Scattering at Larger Crystal Defects

Scattering on larger lattice defects such as grain boundaries may have a significant influence on optical properties due to the backscattering of light in polycrystalline semiconductors. This prevents some of the light from penetrating deep into the material, and thereby reduces photoelectronic effects.

Other scattering relates to solid-state phase segregation, wherein the scattering particles are of the size of the wavelength or larger. Here information about the size and polarizability can be extracted from the scattering, e.g., colloidal scattering and absorption.

The size of microcrystallites, e.g., in polycrystalline Si, can also be obtained from Raman scattering by assuming that the spatial extent of microcrystallites translates into an averaging of phonons over the corresponding range of $\mathbf{q}$-vectors.

In general, scattering can be seen as a loss mechanism that diverts part of the active photon spectrum away from potential carrier generation. In rare cases, it can also be observed as an active mechanism that changes the photon energy sufficiently to permit a transition resulting in carrier generation.

Summary and Emphasis

We have shown that photon scattering can be divided into elastic and inelastic scattering events, the latter providing important insight into the properties of the scattering quasi-particles.

Scattering with acoustic (Brillouin) and optical (Raman) phonons, by electrons, plasmons, and exciton-polaritons has been discussed. Each of the scattering events requires the interaction of the impinging electromagnetic wave with the scattering quasi-particle. Such interaction requires the nonparabolicity of the interaction potentials, i.e., anharmonicity.

The scattering is a small effect, demanding a high intensity of the primary light source—usually a laser. When its frequency or its second harmonic is close to a resonance frequency, the energy of the scattering particle can be used to shift the frequency of the scattered photon into the resonance, thereby increasing the scattering probability by orders of magnitude and presenting the opportunity for long wavelength, tunable lasing.

Properties which are accessible from the scattering experiment are the anisotropic spectrum of sound velocities from the Brillouin scattering, effective mass, and energy, [i.e., $E(\mathbf{k})$ or $E(\mathbf{q})$ for electrons, photons, or phonons] and from the line shape, the lifetimes of the involved quasi-particles. With a different set of selection rules, spectra become accessible which cannot be observed by single photon absorption experiments.

Scattering experiments are very powerful in disclosing the symmetry and the eigenfrequencies of some of the most important quasi-particles. Such experiments have been included in the recent set of analytical tools for semiconductors. They have also opened a new field of devices, e.g., producing strong and tunable long wavelength laser radiation. It is expected that, with further development of the field, a wider variety of analytic machines will be developed. These machines may also provide spatial resolution to identify semiconductor inhomogeneities as well as their composition.

Exercise Problems

1. Discuss selection rules in light of parity consideration and their influence on multiple photon transitions.

2.(e) Draw to scale a diagram that shows the momentum relation of a phonon scattering with a neutron in Si. Assume reasonable energy and momentum values. Explain your reasoning.

3.(r) Summarize the differences between Brillouin, Raman, and Rayleigh scattering. What are their characteristics? What does one learn from these scatterings?

4. Explain possible angle dependency of Raman scattering.
    (a) Why is Raman scattering independent of the angle between the incoming and the scattered light beam?
    (b) There is an angle dependence between incoming beam and crystal axis in anisotropic crystals. Explain.

5.(*) Describe quantitatively the conditions for resonant Brillouin scattering.

6.(e) Light with a photon energy of 2.4 eV passes through a crystal with an index of refraction of 1.5. Plot the angle and energy relation of phonons generated by first-order Brillouin scattering for a sound velocity of $5 \cdot 10^5$ cm/s.

7. Discuss Brillouin and hyper-Raman scattering of polaritons:
    (a) Explain the differences between the Brillouin resonance scattering of exciton polaritons and the hyper-Raman scattering of the same quasi-particles.
    (b) There is an angle dependence in hyper-Raman scattering. Does this fact disagree with the statement that Raman scattering is not angle-dependent?

8.(*) How does higher order Raman scattering compare with multiphonon optical absorption? What are the selection rules in both cases?

9. Draw to scale the $E(q)$ dispersion relation for GaAs and the mini-zones given in Fig. 17.11.
    (a) Identify the respective branches.
    (b) Where would you expect the lowest and the higher lines in the upper Raman spectrum?

(c) Can you obtain the well-layer thickness from the spectrum? Compare it with the values given in the figure caption.

10.(e) What changes do you expect in the plasmon polariton spectrum of Fig. 17.13 when a magnetic field is applied?

    (a) Give a magnetic field which produces a line within the figure, but is well separated from the plasmon lines.

    (b) How does this line shift with magnetic induction?

    (c) What are the appropriate selection rules?

11.(e) Discuss photon drag in Si.

    (a) Draw V_{phdr} as functions of the donor density for $\phi/d = 10^{20}$ cm^{-3}s^{-1}.

    (b) How does the intrinsic density influence the photon drag voltages?

    (c) Assuming 100 mW homogeneous excitiation at 1 eV in a 1 mm thick slab of Si with quantum efficiency 1, calculate the conversion efficiency of a solar cell based on photon drag only and assume 100% collection efficiency of the generated carriers.

12.(e) Connect Eqs. (17.14) and (17.15) for the photon drag voltage with each other. Identify the different assumptions used.

PART V

DEFECTS

Chapter 18

Crystal Defects and Interfaces

Semiconducting properties of most interest are predominantly caused by crystal defects.

A large variety of crystal defects determine the electronic properties of semiconductors. Some of them provide donors or acceptors. Others are responsible for carrier scattering or recombination; still others trap carriers and influence the space charge that determines the carrier transport. In this introductory chapter, typical examples of these defects will be discussed. We will also consider briefly here the crystal boundaries, which are a departure from the perfect infinite solid and hence fall into a special class of crystal defects. We distinguish:

- point defects, such as single atoms or vacancies in an otherwise near-ideal lattice;
- line defects (dislocations);
- surface defects relating to external or internal crystal surfaces; and
- volume defects, relating to usually small, three-dimensional inclusions (precipitates) or defect associates.

These defects will be enumerated in the following sections.

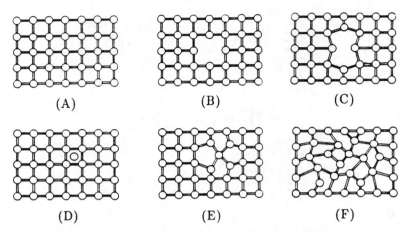

Figure 18.1: (A) Two-dimensional representation of cubic lattice and some of its defects; (B) ideal vacancy; (C) relaxed vacancy; (D) ideal interstitial; (E) interstitialcy, i.e., interstitial atom forming bridges between adjacent host atoms; and (F) extended interstitial with substantial lattice relaxation beyond nearest neighbors.

18.1 Point Defects

Point defects are the main class of defects that act as donors or acceptors, or, when their energy levels are farther separated from the bands, as traps or recombination centers. They are also important as scattering centers, especially when charged with respect to the host lattice. We distinguish:

A. Intrinsic (native) lattice defects, such as:
  a. vacancies (a missing lattice atom, Fig. 18.1B),
  b. interstitials (an additional host atom within the lattice, Fig. 18.1D), and
  c. antisite defects in compound semiconductors, e.g., in an AB compound where an A atom occupies a B site.
B. Impurities located at a
  a. (substitutional) lattice site, or at an
  b. interstitial site.
C. The lattice surrounding of such a defect will show some degree of relaxation, such as:
  a. neighbor position relaxation (Fig. 18.1E) and other special cases, including
  b. split interstitials (the *interstitialcy*) in which a host atom is replaced by two symmetrically displaced atoms.
D. Small defect associates are also counted as point defects such as:

a. defect pairs—
  α. divacancies,
  β. impurity associated with an intrinsic defect,
  γ. two impurities associated with each other;
b. higher defect associates, containing
  α. three defects or
  β. more than three defects in close proximity.
A large variety of defects can be distinguished according to their
E. geometrical arrangement with respect to each other
  a. forming an anisotropic center or
  b. forming distant pairs, distinguished by their distance within the lattice while still interacting.
Finally, all defects containing an impurity are distinguished by their
F. chemical identity, which relates to the
  a. defect center and to the
  b. host lattice.
It is, however, often more appropriate to identify the chemical identity with respect to the host lattice and to distinguish

G. host-related defects, identified as
  a. isovalent impurities (from the same column of the periodic system—also called isoelectronic);
  b. isocoric impurities* (from the same row of the periodic table, i.e., having the same core);
  c. substitutionals with $\Delta z = \pm 1$ (z is the chemical valency);
  d. substitutionals with $\Delta z = \pm 2$ or more;
  e. amphoteric defects,† in which the sign of Δz can change;
  f. transition metals, identified as a separate group because of their outer shell screening, giving more individuality to the defect and mostly yielding deeper levels.

---

* An isocoric P in a Si lattice can be thought of as "created" by adding to a lattice atom a proton, i.e., a point charge, and an extra electron (the donor electron, Section 18.2), thereby creating the most ideal hydrogen-like defect. Any other hydrogen-like donor, e.g., As or Sb in Si, is of different size, causing more lattice deformation and a substantially different core potential (see Section 21.1).

† That is, a defect that can act as a donor or acceptor (see Section 18.2) depending on the chemical potential of the lattice (influenced, e.g., by optical excitation or other doping).

The distinctions given above will become the major guidelines for the decision of which

H. approximations are best suited for a theoretical analysis to describe the defect with sufficient accuracy, such as

    a. an effective mass treatment, suited for shallow defects, or

    b. other approximations involving the core potential (deep centers), and

    c. substantial lattice relaxation.

This list is by no means complete, but gives an impression of the great abundance of defects present in real crystals. Each of these defects contributes to the wealth of electronic eigenstates, most of them as levels in the band gap. Therefore, it is not surprising that only a small fraction of these defects is presently unambiguously identified, a fact that influences the selection of examples discussed with confidence below (Lannoo and Bourgoin, 1981). For experimental methods of defect identification see Bourgoin and Lannoo (1983).

The identification of such levels in the gap is most directly accomplished by optical experiments with or without external fields, using optical absorption or luminescence as a measure of the transitions between two of the levels, or involving (photo) conductivity to provide information about transitions between the levels and one of the bands. More recently, biased junctions have been as a means to shift quasi-Fermi levels and thereby defect center occupation for positive identification. A discussion of these effects is given in Chapters 43, 45, and 48 and Section 46.3. In the following section, we present a short review on the most important electronically active point defects for semiconductors, and on the charge character of such defects with respect to the crystal lattice.

18.2 Donors and Acceptors

Foreign atoms which substitute for an intrinsic lattice atom are most easily described when they are from an adjacent column of the periodic system of elements with respect to the replaced atom, e.g., a boron or phosphorus atom replacing a silicon atom in a silicon crystal.

The P atom is pentavalent, and thus has one electron more in its outer shell than the host lattice atoms (Fig. 18.2A). When the P atom replaces one of the atoms of an Si lattice, this extra electron is only loosely bound and can easily be emitted by thermal ionization into the conduction band. Therefore, the P acts as a *donor* in the

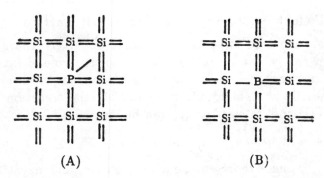

Figure 18.2: Substitutional phosphorus P as donor and boron B as acceptor in subfigures (A) and (B), shown within a covalently bound Si lattice with extra or missing electron (line) to complete bonding at the P or B atom, respectively.

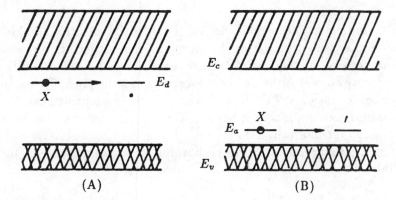

Figure 18.3: Band model of a typical semiconductor with a donor (A) and an acceptor (B), indicating the relative charge character with and without charge carrier (prime = negative, dot = positive, and cross = neutral relative to the surrounding lattice).

Si host crystal; it produces a level close to the conduction band, as shown in Fig. 18.3A. The energies E_d and E_a are assigned to the un-ionized donor and acceptor.

In a similar fashion, the incorporation of a B atom on an Si site causes the deficiency of an electron (Fig. 18.2B). This missing electron can be regarded as a hole bound to a B atom. The hole can be replaced by an electron from the valence band. In the hole picture, the B atom becomes ionized (Fig. 18.3B); it produces a free hole and acts as an *acceptor*, with a level close to the valence band—the energy of these defect states is discussed in Section 21.1.

18.2.1 Defect Notation Within the Host Lattice

When the donor is ionized, it becomes positively charged. The change in charge character plays an important role in later discussions on carrier capture and scattering. Such charge relations can be followed easily in a chemical representation: the "reaction" between a donor (D) and an electron (e) can be written as

$$D^\times \leftrightharpoons D^\bullet + e', \tag{18.1}$$

where $^\times$, $^\bullet$, and $'$ represent neutral, positive, and negative charge characters with respect to the lattice. This notation is referred to as *Kröger-Vink notation*,* and we will use it consistently in this book—see Hayes and Stoneham (1984). If we want to be more descriptive with respect to the chemical nature of the defect, we write

$$P_{Si}^\times \leftrightharpoons P_{Si}^\bullet + e', \tag{18.2}$$

which indicates the atomic defect (P) with its position within the lattice (i.e., substituting an Si atom) as the subscript. In general, point defects are identified by the symbol of the defect, with the lattice site on which the defect is located as a subscript. For example, a chlorine vacancy (V) in a NaCl lattice is identified as V_{Cl}, a potassium ion replacing a sodium ion in the same lattice as K_{Na}, and a copper interstitial as Cu_i.

In a similar fashion, the recharging of an acceptor can be described by

$$A^\times \leftrightharpoons A' + h^\bullet \tag{18.3}$$

or

$$B_{Si}^\times \leftrightharpoons B_{Si}' + h^\bullet, \tag{18.4}$$

where $h^\bullet$ represents a hole in the valence band.

18.2.1A Substitutionals in AB-Compounds

Substitutionals in an AB-host lattice act similarly to substitutionals in element semiconductors, except one distinguishes whether an anion or a cation is being replaced. For instance, replacement of a divalent

---

* The notation of charges with respect to the neutral lattice was introduced by Kröger, Vink, and Schottky (see Schottky and Stöckmann, 1954). This notation should not be confused with the charge identification used in an ionic lattice, e.g., Na^+Cl^-. Inclusion of a Cd^{++} instead of a Na^+ ion makes the cadmium ion singly positively charged with respect to the neutral lattice; hence it is identified here as $Cd_{Na}^\bullet$ when referenced specifically as a lattice defect.

Cd ion in a CdS crystal by a trivalent In ion, denoted by $In_{Cd}^{\times}$, yields:

$$In_{Cd}^{\times} \leftrightharpoons In_{Cd}^{\bullet} + e'. \qquad (18.5)$$

This results in a shallow donor, as does the replacement of a sulfur ion from group VI with a halogen ion from group VII. On the other hand, the replacement of Cd with an alkali-metal ion, or of S with a group V element like P or As, produces an acceptor. However, the incorporation of one type of defect (in CdS the incorporation of donors) is often easier than for the oppositely charged defect (acceptors in CdS), rendering the material preferably *n*- or *p*-type (CdS is *n*-type). Intrinsic compensation (see Section 19.2.6) is one reason for this preference. An interesting case occurs for some trivalent compounds. For example, in GaAs, the replacement of Ga with the tetravalent Si results in

$$Si_{Ga}^{\times} \leftrightharpoons Si_{Ga}^{\bullet} + e' \qquad (18.6)$$

acting as shallow donor, while the same Si, replacing a trivalent As ion,

$$Si_{As}^{\times} \leftrightharpoons Si_{As}' + h^{\bullet} \qquad (18.7)$$

results in a shallow acceptor. Depending on the growth condition, one or the other is preferred, and the material becomes either *n*- or *p*-type—see Section 27.3.1.

18.2.1B Vacancies and Interstitials   Vacancies and interstitials are denoted in a similar fashion. Metal-ion interstitials usually act as donors. For example, in CdS,

$$Cd_i^{\times} \leftrightharpoons Cd_i^{\bullet} + e'. \qquad (18.8)$$

Metal-ion vacancies act as acceptors:

$$V_{Cd}^{\times} \leftrightharpoons V_{Cd}' + h^{\bullet}. \qquad (18.9)$$

Nonmetal-ion vacancies usually act as donors:

$$V_S^{\times} \leftrightharpoons V_S^{\bullet} + e'. \qquad (18.10)$$

A first and rather simplified judgment as to whether the defect more readily becomes positively or negatively charged is easily rendered in ionic compounds. Semiconductors with mixed bonding character can also be judged by following the $8 - N$ rule—see Section 3.9.1. The rule maintains that elements tend to complete their outer shell by sharing electrons with neighbor atoms; surplus electrons are donated to the lattice; and missing electrons are attracted from elsewhere in the lattice. Interstitial metal ions tend to donate their valence electron(s), consequently reducing their radius, and cause less lattice deformation. Other, more complex aspects of point defects will be discussed in the following chapters. We will now give a short overview of line defects.

18.3 Line Defects

In the previous sections, *point defects* were discussed, i.e., defects that are located at a given atom in the lattice and involve their immediate surroundings. In this section we introduce one-dimensional *line defects* called *dislocations*, which often extend throughout the entire lattice. There are two main types of dislocations:

- edge dislocations and
- screw dislocations.

Other ways to distinguish dislocations will be discussed below—see Read (1953), Weertman and Weertman (1960), or Cottrell (1964); for theoretical analyses, see Nabarro (1967), and Hirth and Lothe (1968).

Dislocations have a major influence on certain types of crystal growth and on electron transport, and are therefore discussed here— see Friedel (1964), and Hull (1969). Other important influences include mechanical deformation (yield strength) and the formation of grain boundaries—see Section 18.4.

18.3.1 Edge Dislocations

Edge dislocations (Taylor, 1934; Orowan, 1934) are formed by an extra lattice plane inserted into a part of a crystal as shown in Fig. 18.4A. The dislocation line extends along the terminating edge of the inserted plane. Such dislocations are produced when a crystal is exposed to nonuniform mechanical stress large enough to result in plastic deformation along a glide plane. In Fig 18.4, a {001} plane is indicated. Another preferred glide plane in cubic crystals is the {110} plane.

18.3.2 Screw Dislocations

A *screw dislocation* (Burgers, 1939—Fig. 18.4B) shows a step at the outer surface, while only a slight lattice deformation exists surrounding the dislocation within the crystal. The screw dislocation is important for crystal growth (Frank, 1949) since adherence of atoms is

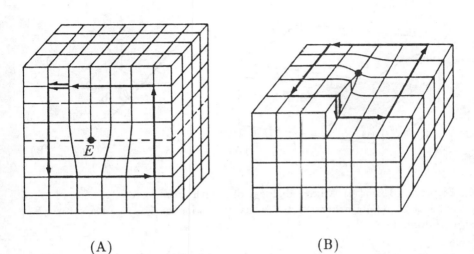

(A)   (B)

Figure 18.4: (A) Additional plane partially inserted into a simple cubic crystal, resulting in an edge dislocation E. The glide plane is indicated by a dashed line. (B) (Left-handed) screw dislocation in a simple cubic crystal producing a step at the crystal surfaces. The Burgers vector is identified by a double lined arrow.

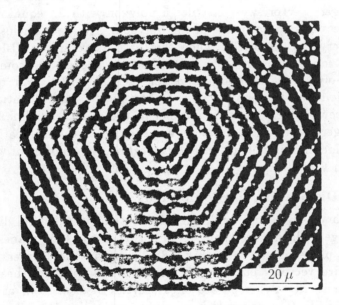

Figure 18.5: Spiral growth shown on the (0001) surface of an SiC crystal; screw dislocation in the center of the picture (after Sunagawa, 1975).

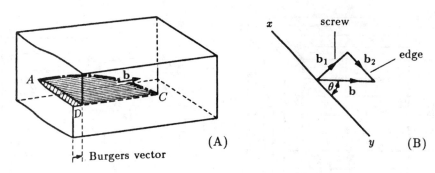

Figure 18.6: (A) Mixed dislocation, with pure screw at A and pure edge at C. (B) Burgers vector **b** involves an edge (**b$_2$**) and a screw (**b$_1$**) component.

substantially enhanced at an inside surface edge—see Section 3.11.2A and Fig. 3.27. A screw dislocation continuously maintains such an edge during growth by forming a growth spiral (Fig. 18.5).

18.3.3 The Burgers Vector

Dislocations are identified by their *Burgers vector*. In order to define the Burgers vector, we construct a polygon with an equal number of lattice steps on each side of the polygon. The path is closed in a perfect crystal. However, when this path surrounds a dislocation it is no longer closed. The lattice vector needed for completion of the polygon is the Burgers vector—see Fig. 18.4. For an edge dislocation, the Burgers vector is orthogonal to the dislocation line; for a screw dislocation, it is parallel to the dislocation line. When the Burgers vector is at an angle $0 < \theta < 90°$ to the dislocation line such dislocation is called a *mixed dislocation*, an example of which is shown in Fig. 18.6.

18.3.4 Dislocations in Compounds

In a binary (AB) compound, edge dislocations in the [110] direction are energetically favored, but require *two* extra (110) planes: an a and a b plane as shown in Fig. 18.7A. Again, the Burgers vector is a lattice vector. There is the possibility of splitting these adjacent planes into two separate ones (*partial dislocations*), as shown in Fig. 18.7B. The separation of planes maintains the correct AB-sequence above and below the dislocation line, except for a jump of composition between both partial dislocations. Such partial dislocations have a lower energy than the unit edge dislocation and therefore are favored. In more complex crystals, more than two extra planes

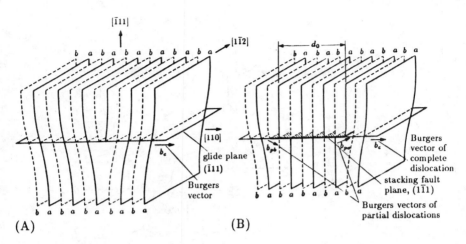

Figure 18.7: (A): Unit edge dislocation in an AB-compound in the [110] direction of a face-centered cubic crystal. (B): Extended dislocation of the same compound of two Shockley partial dislocations separated by a stacking fault (plane b continues as plane a etc. above the partial dislocation plane).

are required to restore periodic order. Therefore, the Burgers vector becomes larger the more complex the crystal structure is.

18.3.4A Partial Dislocations in Semiconductors

The preferred slip plane in tetragonally bound semiconductors is the $\{111\}$ plane which could lie either between the closely spaced planes, called *glide set*, or between the wider spaced planes, called *shuffle set* (Ba or bB in Fig. 18.8). These are 60° dislocations (Shockley, 1953), and can best be visualized by cutting out a lattice slab and rejoining the displaced atoms along the dashed lines 1-5-6-4 for the glide set, or along 1-2-3-4 for the shuffle set. With such an operation, an extra lattice plane is inserted below 5-6.

Along such dislocations a row of dangling bonds would appear, as shown in a perspective view in Fig. 18.9. Dangling bonds are expected to effectively trap electrons with major influence on electrical properties (Labush and Schröter, 1980). However, reconstruction of the dislocation core eliminates most of these dangling bonds with substantial reduction of the electrical influence of such dislocations (Hirsch, 1985).

The 60° dislocation consists of two partials (a 30° and a 90° partial), each of which is capable of reconstruction. Figure 18.10B shows the reconstruction of the 30° glide partial; whereas Fig. 18.11A

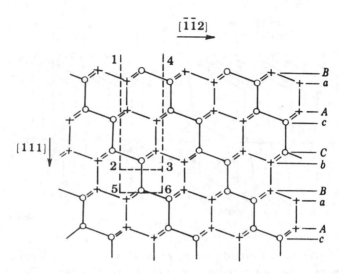

Figure 18.8: Projection of an Si-lattice normal to $(1\bar{1}0)$. Circles represent atoms in the paper plane, crosses lie in the next plane below. The (111) plane is normal to the paper (also normal to the direction [111]) and would appear as a horizontal trace. The dashed auxiliary lines indicate a cut-out of a lattice slab for creating a 60° dislocation just below it (after Johnson, 1981).

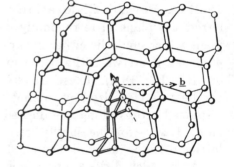

Figure 18.9: 60° dislocation on a shuffle plane of an Si crystal with a row of dangling bonds (after Schröter, 1979).

shows dangling bonds along a 90° glide partial, and Fig. 18.11B its reconstructed core.

The 90° glide partial, however, shows an interesting alternative of bonding (upper and lower middle part of Fig. 18.11C) which has the same energy. In the transition region, a dangling bond is created. Such a defect is also referred to as an *antiphase defect*, and can move as a *soliton* along the reconstructed glide partial (Heggie and Jones, 1983). For more detail, see the literature cited in Hirsch (1985).

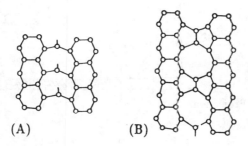

Figure 18.10: (A) 30° glide partial dislocation in Si shown in an (111) plane; (B) reconstructed (after Jones, 1981).

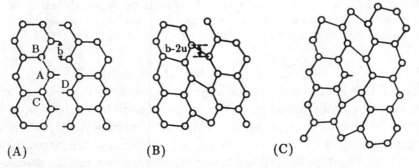

(A)   (B)   (C)

Figure 18.11: (A) 90° glide partial dislocation in Si in an (111) plane; (B) reconstructed; (C) antiphase defect (soliton) (after Jones, 1981).

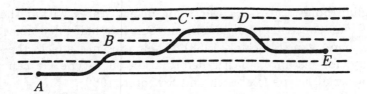

Figure 18.12: Movement of a dislocation in a glide plane by nucleation of a pair of kinks—at C and D (after Hirth and Lothe, 1968).

18.3.4B Dislocation Kinks In the process of creation or motion of dislocations, kinks may be formed by part of the dislocation shifting within a glide plane by one lattice plane from its original position (Fig. 18.12). At these kinks, dangling bonds exist where reconstruction cannot be completed (Hirsch, 1985).

18.3.4C Dislocation Climb and Jog If the last line of atoms along an edge dislocation is removed, e.g., by diffusion to the crystal surface, then the dislocation has climbed by one atomic spacing.

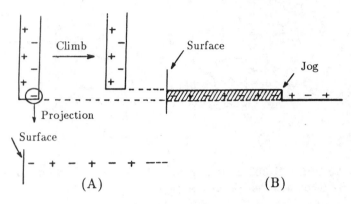

Figure 18.13: (A) Removal of the last atom in an edge dislocation causes a climb by one atomic space. (B) Partial removal causes a jog; this view is perpendicular to subfigure A (after Hirth and Lothe, 1968).

However, the diffusion of only a few atoms from the edge is more likely. This results in a *climb* of only a fraction of the dislocation, with a *jog* to the undisturbed part (Fig. 18.13). Climb by nucleation of jog pairs is equivalent to dislocation motion in perpendicular direction by nucleation of kink pairs. Widening of the distance between jog pairs usually requires diffusion of the interspacing line of atoms to the surface.

18.3.5 Dislocations and Electronic Defect Levels

Dangling bond states are electronic defect states within the band gap. Nonreconstructed dislocations result in a half-filled defect band in the band gap (Shockley, 1953). Their presence affects the carrier density; for instance, plastic deformation on n-type Ge can render the material p-type (Labusch and Schröter, 1980).

Core reconstruction reduces the density of dangling bonds dramatically so that the remaining ones act as isolated deep level point defects. In addition, the lattice deformation near the reconstructed core causes defect levels in the band gap.

Electron paramagnetic resonance and its angular dependence yield information about the density and orientation of defects with unpaired spins, i.e., dangling bonds. Experiments with plastically deformed Si at $650\,°C$ indicate that only $0.2\ldots2\%$ of the available sites are not reconstructed, and have unpaired spins at dangling bond centers (Weber and Alexander, 1983). Kinks, curved dislocations, or other special features, such as jogs or nodes, are the suspected sites of the remaining dangling bonds (Osip'yan, 1983).

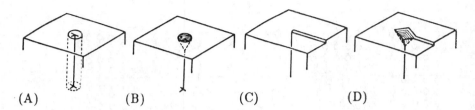

(A)   (B)   (C)   (D)

Figure 18.14: Formation of etch pits at the surface surrounding an edge (A,B) and a screw dislocation (C,D).

Deep level transient spectroscopy (DLTS—see Section 47.4.1B) indicates that filling of deep levels at dislocations, which lie in closer proximity to each other, produces a Coulomb barrier. This impedes consequent trapping of adjacent centers and can be recognized in the DLTS signal (Koeder et al, 1982). Acceptor levels at $E_c - E_a \simeq$ 0.35 and 0.54 eV, and a donor level at $E_d - E_v \simeq$ 0.4 eV, have been detected in plastically deformed Si (Kimerling and Patel, 1973; Weber and Alexander, 1983).

Luminescence and photoconductivity measurements yield additional information about deep levels associated with (mostly reconstructed) dislocations, kinks, jogs, etc. For a review, see Weber and Alexander (1983), Mergel and Labusch (1982), and Suezawa and Sumino (1983).

However, the unique identification of specific dislocation-related defects is difficult: with deformation at elevated temperatures, other defects and defect associates are formed, the signatures of which cannot easily be separated.

18.3.5A Dislocation Counting Dislocations can be made visible by several methods (Hull, 1975). One of the easiest is via surface etching, i.e., removing parts of the lattice surrounding the dislocation, which are under stress and therefore dissolve more easily. The surface etching method is shown schematically in Fig. 18.14. Since most of the dislocations extend throughout the entire crystal, we obtain their density by counting the *etch pits* at the surface. The best Si crystals contain less than 1 etch pit/cm^2. Typical commercial semiconductors have a dislocation density of $10^2 \ldots 10^4$ etch pits/cm^2.

Other methods which involve the direct observation of dislocations include: transmission electron microscopy of thin layers;

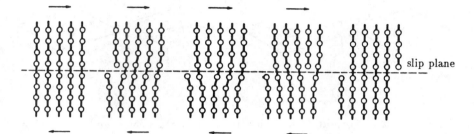

Figure 18.15: Glide motion of an edge dislocation along a slip plane.

EBIC,* relating to their electronic properties; reflection or transmission imaging, relating to mechanical stress of the surrounding; x-ray diffraction; tunneling; and field ion microscopy. Indirect observation is possible by using decoration, which uses precipitates along dislocation lines to increase contrast and also to permit light microscopic observation (Hull, 1969).

18.3.6 Motion and Creation of Dislocations

All dislocations can move conservatively if the motion is parallel to the Burgers vector, or, more generally, when the dislocation moves within its slip plane. Such motion is called *glide*, shown in a two-dimensional representation in Fig. 18.15, and is the essential part in a plastic deformation of a crystal.

The motion of an edge dislocation normal to both its Burgers vector and its dislocation line is called *climb*. It is nonconservative, meaning it causes a change of the height of the inserted plane, i.e., atoms must move to or from the dislocation core (dislocation line).

A gliding dislocation that encounters an obstacle—such as another crossing dislocation, a vacancy, a foreign atom, or other crystal defects—can become pinned at such an obstacle, i.e., the obstacle hinders dislocation movement. For the example given in Fig 18.16, the dislocation must "climb over" the obstacle in order to continue moving. Therefore, glide *and* climb are important in plastic deformation of crystals—see Cottrell (1958), or Hirth and Lothe (1968).

18.3.6A Dislocation Velocity   When a shear stress is exerted parallel to the Burgers vector of an edge dislocation, it is unlikely that the entire dislocation moves step by step as shown in Fig 18.15.

---

* Electron-beam-induced conductivity, used in a scanning electron microscope (Heydenreich et al., 1981).

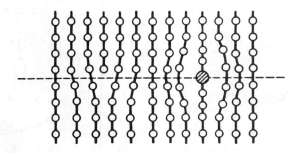

Direction of "attempted" dislocation motion

Figure 18.16: Strain field surrounding a foreign atoms hinders motion of a dislocation along a glide plane.

When drawing the edge dislocation in a plane normal to that shown in Fig 18.15, we can picture a more probable sequence of events, in which part of the dislocation moves one lattice spacing, and connects to the remaining part with kinks (Fig. 18.12). This type of movement is most probable when some pinning of the original dislocation occurs along its length (at A and E in Fig. 18.12).

The dislocation velocity is then determined by the nucleation rate and motion of double kinks, as well as by the interaction of such kinks with localized lattice. Such interaction is known to depend on the electrical charge of these defects, and is influenced by the position of the Fermi- (or quasi-Fermi) level (Patel and Chaudhuri, 1966). For instance, the velocity of dislocations under otherwise identical conditions is $\sim$ 50 times larger in n-type than in p-type Ge. Other deep level centers, e.g., oxygen, cause pinning of the dislocation.

This dislocation motion is more complicated when partial dislocations need to be considered. Generation of double kinks in the two partials may be correlated (Wessel and Alexander, 1977).

Some partials (e.g., with a Burgers vector inclined to a stacking fault) cannot glide; these are called *sessile* (Frank, 1949). Other dislocations which can move easily are called *glissile*.

18.3.6B The Frank-Read Source   The application of sufficient shear stress produces additional dislocations from a dislocation pinned at the two ends (*Frank-Read source*—Frank and Read, 1950), and forced to bow by mechanical deformation perpendicular to the direction of the dislocation (Fig. 18.17). In this way *dislocation loops*

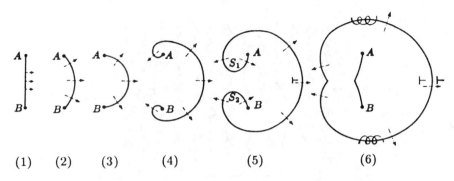

Figure 18.17: (1) Frank-Read source with progressively growing dislocation loop (2)–(5) and separation (6) to repeat this process (1)–(6).

can be formed sequentially, and the process repeats itself periodically as long at the stress persists (see Read, 1953).

18.3.6C Electroplastic Effects   Edge dislocations in covalent semiconductors are partially charged, since in addition to some remaining dangling bonds, the stress field surrounding the dislocation creates traps for carriers, and the termination of an extra layer in an ionic crystal carries an inherent charge (Petrenko and Whitworth, 1980). The movement of such a dislocation is therefore influenced by other charges.

The charge density per unit length of the dislocation can be changed when carriers are injected or optically generated. In addition, other defect centers can also be recharged. As a consequence, the ability of dislocations to climb is changed. Usually, this results in an increase of the mechanical strength, i.e., in its resistance to plastic deformation; photoquenching (see Section 45.8.1) causes a decrease in mechanical strength—see Osip'yan et al. (1986). This effect is referred to as the *electroplastic effect* and also, depending on how the change in carrier density is induced, as the *photoplastic effect* or the *cathodoplastic effect*. These effects are rather pronounced in II-VI compounds, and were first observed by Osip'yan and Savchenko (1968) in CdS.

The spectral distribution of the photoplastic effect is nearly identical to that of the photoconductivity, indicating a direct relation to the electron density. Further related to it is an increased charging of defects, which tends to pin the dislocation movement. A change in their charge distribution is also a possible mechanism for pinning of dislocations, thereby decreasing the plasticity.

18.3.6D Dislocation Currents, Exoemission, and Tribolumi-nescence   Inversely, the mechanical deformation of semiconductors can set carriers free when moving dislocations pass through charged defect centers. This results in an induced current, the *dislocation current*, which is proportional to the rate of plastic deformation.

Electrons set free near the surface can penetrate through it and are observed as emitted electrons, so-called *exoelectrons.* Such *exoemission* has long been known to occur when solids are deformed *in vacuo*, e.g., by scratching, bending, or breaking. Exoemission is reviewed by Mints et al. (1976).

Parallel to the emission of exoelectrons, *triboluminescence* is observed. Triboluminescence is a deformation-induced luminescence in which the electrons, set free by moving dislocations, recombine via luminescent centers (see Walton, 1977).

All of these effects need *plastic* deformation, as a hydrostatic compression does not produce luminescence (Scarmozzino, 1971). Time-resolved emission spectroscopy shows short (typically $\sim 0.1\,\mu s$) pulses during deformation, coinciding with dislocation current pulses and with birefringence kinetics which indicate near-surface discharges with dislocation movement. In addition, a weaker and continuous luminescence of typical luminescence centers is observed as a result of the recombination of generated free carriers (Bredikhin and Shmurak, 1979).

18.3.6E Disclinations   While dislocations are related to symmetries of translation, *disclinations* are related to symmetries of rotation. Both types of defects are sources of internal strain.

Disclinations can be used to describe frustrated systems resulting in curved lattice regions which are common in amorphous or organic semiconductors. For more information, see the review of Kléman (1985) and the original literature cited therein.

18.4 Surface Defects, Planar Faults

There are several types of two-dimensional defects:

- stacking faults,
- low angle grain boundaries,
- twin boundaries,
- crystallite boundaries,
- heteroboundaries,
- strained layer systems,
- surfaces (ideal and with absorption or chemisorption layers), and

- metal/semiconductor boundaries.

These defects have a significant influence on the electrical properties of real semiconductors. Most of this influence is caused by defect levels at the boundaries (surfaces), which are usually charged and attract compensating charges in the adjacent crystal volume (*space charges*).

In spite of their importance, however, these defects will only be mentioned briefly here since they deal with profound inhomogeneities which are not the topic of this book. They will be included in *Carrier Transport in Inhomogeneous Semiconductors* (in preparation). For more information on this subject, see the *Handbook of Semiconductors* (North Holland).

18.4.1 Stacking Faults

Stacking faults are the least disordered types of interface defects. They occur by alternating between wurtzite and zinc-blende structures: to form the lattice from atomic layers stacked on top of each other in the c-direction, we start from a hexagonal ("close packed") layer (A), and deposit the next identical layer (B) offset and turned by 60° to touch each second interspace at $\frac{1}{4} \frac{1}{4} \frac{1}{4}$. For the deposition of a similarly offset third layer, we have the choice of turning it back by 60° to the original position (A), resulting in a wurtzite lattice, or turning it forward by another 60° to offset it from the bottom layer by 120° (C), resulting in a zinc-blende lattice—see Section 3.7. The layer sequence ABABAB therefore characterizes wurtzite; the layer sequence ABCABC characterizes zinc-blende. A sequence ABABABCAB consequently identifies a *stacking fault*. A thin slice of a stacking fault can be seen as being imbedded between two partial dislocations, i.e., the boundary of a stacking fault is a partial dislocation—see Nabarro (1967).

ZnS and SiC are examples for materials that show such stacking faults. There can be a long-range periodicity in these stacking faults, up to a few hundred Å in SiC, which produces so-called *polytype structures*.

18.4.2 Grain Boundaries

One distinguishes several types of grain boundaries, the more important ones will be discussed in the following sections.

18.4.2A Low-Angle Grain Boundaries   Low-angle grain boundaries show a very small tilt between two regular crystallites and can

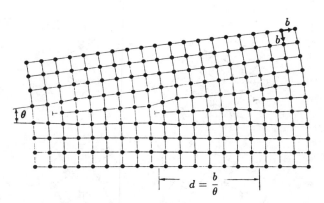

Figure 18.18: Low-angle grain boundary with an array of edge disloca-tions separated by seven lattice planes.

be regarded as an array of dislocations (Burgers, 1940; Bragg, 1940), as shown in Fig. 18.18. The angle between the two grains (at most a few degrees) is given by

$$\theta \simeq b/d \qquad (18.11)$$

where b is the length of the Burgers vector and d is the dislocation spacing.

18.4.2B Large-Angle Grain Boundaries   *Twin boundaries* are special angle boundaries between two identical crystallites. The least disturbed twin boundary is that of a stacking fault, as discussed in Section 18.4.1, by proceeding from an AB sequence to a BC sequence.

Other twin boundaries are under an angle at which each second (or third) atom falls onto a lattice site as shown in Fig. 18.19. It can therefore be regarded as an array of vacancy lines similar to an array of edge dislocations in low-angle grain boundaries.

Other crystallite boundaries may occur under a wide variety of angles and may incorporate at the interface a variety of disorders, including vacancies, dislocations, and liquid-like structures with a high degree of local stress (Fig. 18.20).

Grain boundaries can have a major influence on semiconducting properties by trapping carriers and creating compensating space charge layers. Some of the effects related to carrier mobility will be discussed in Section 37.3.

18.4.2C Heteroboundaries   The heteroboundary between two crystals of different composition is of great technical importance for heterojunction devices. It is characterized by a mismatch in sev-eral parameters, most importantly the lattice constant, the *electron*

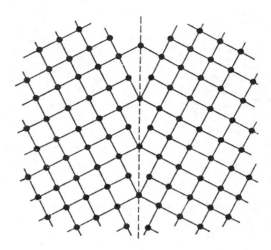

Figure 18.19: Twin grain boundary in the (112) surface of a cubic crystal.

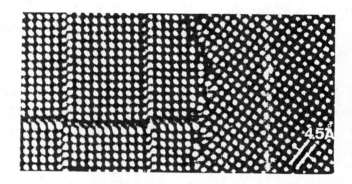

Figure 18.20: Crystal boundary of Au_4Mn taken with ultrahigh resolution microscopy which makes the atomic array visible (after Jouffrey, 1984).

*affinity** (see Bauer et al., 1983), and the energy gap. The mismatch of expansion coefficients is of interest when a device is produced at temperatures substantially different from room temperature, or is exposed to temperature cycling during operation. It causes curling

---

* The electron affinity of a semiconductor is defined as the energy difference from the lower edge of the conduction band to the vacuum level, i.e., the energy gained when an electron is brought from infinity into the bulk of a crystal, resting at E_c. It should be distinguished from the electron affinity of an atom, which is equal to the energy gained when an electron is brought from infinity to attach to an atom and forms an anion.

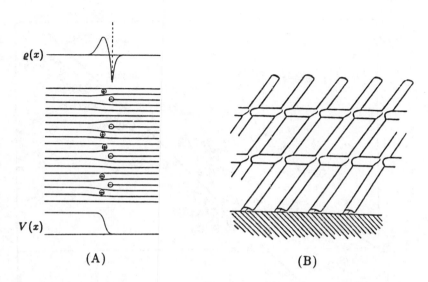

Figure 18.21: (A) Mismatch dislocations with compensating space charges, resulting in a space charge double layer and potential step. (B) Spatial distribution of the space charge double layer surrounding the dislocation network in the form of half cylinders (after Böer, 1983).

of thin layer laminates or may generate major interface defects; in extreme cases, it results in delamination.

The lattice mismatch is accommodated through a network of dislocations (Fig. 18.21B). The strain produced by the mismatch is only partially relieved by the mismatch dislocation network. The remaining strain decreases linearly with the distance from the heterojunction, and depends on the heterogeometry. Usually, it results in an opposite strain at the outer surfaces. This can best be seen for a lattice mismatch in one direction within the interface plane. It results in a bending of the crystal, with a neutral plane near its center. It has compression and expansion at the hetero-interface and the surface, respectively.

The mismatch dislocation network is usually charged and compensated by a space charge region, resulting in a potential step. This step is typically on the order of $\pm 0.1 \ldots 0.5$ eV (Böer, 1983) as shown in Fig 18.21A.

The compensating space charge is located close to each dislocation line, surrounding it in the form of a half-cylinder. The closeness is caused by the high density (small Debye length). The half, rather than full, cylinder is caused by the different probabilities of incorporating compensating defects in the different semiconductors. In

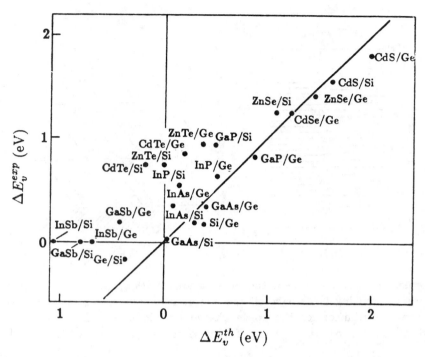

Figure 18.22: Experimental values for the jump of the valence band of various heterojunctions obtained from photoemission (ΔE_v^{exp}), compared with the calculation from the Harrison model (ΔE_v^{th}). Best agreement is obtained for junctions with high degree of lattice match (after Bauer et al., 1983).

addition, there is a potential step caused by the difference in electron affinities and modified by the actual dipole moment at the crystal interface (Kroemer, 1975; Harrison, 1977). The dipole moment varies with changing crystallographic directions of this interface.

The difference in electron affinities is considered to be the major contribution in causing the step in the conduction band at the interface (Anderson, 1962). Such linearization, however, can give misleading results. When considering the crystallographic direction of the interface and mismatch dislocation network, additional changes in step height by ± 0.5 eV can be encountered.

The difference between electron affinities can be obtained experimentally from photoemission measurements, where the threshold energy directly yields the difference from the top of the valence band to the vacuum level

$$\Delta E_v = E_{g1} - E_{g2} - \Delta E_c. \qquad (18.12)$$

Table 18.1: Photoelectrically determined workfunction of metals in eV (preferred values after Michaelson, 1972).

Metal	Workfunction	Metal	Workfunction	Metal	Workfunction
Ag (100)	4.75	Ga	4.5	Sb	4.08
(111)	4.81	Hg	4.53	Sr	2.35
Al	4.08	In	3.8	Ta	4.19
Au	4.82	Ir	5.3	Th	3.35
Be	3.92	Mg	3.68	Ti	4.45
Bi	4.24	Mn	3.76	Tl	3.36
C	4.81	Mo	4.3	U	3.63
Ca	2.71	Ni	5.01	V	3.77
Cd	4.07	Os	4.55	W (001)	4.52
Cm	4.4	Pb	4.14	(110)	4.58
Co	4.2	Pd	4.79	(111)	4.39
Cr	4.37	Pt	5.32	(112)	4.65
Cs	1.81	Rb	2.16	(211)	4.50
Fe α	4.70	Rh	4.57	(310)	4.35
β	4.62	Sn β	4.50	Zn (0001)	4.26
γ	4.68	γ	4.38	Zr	4.21

In Figure 18.22 the jumps of the valence bands of various heterojunctions are given, and are compared with the prediction of the Harrison model, using linear combination of atomic orbitals. There is reasonable agreement for couples with low lattice mismatch, but only fair agreement is observed and a deviation of up to 1 eV for semiconductor couples with larger lattice mismatches. For a review of recent results, see Katnani and Margaritondo (1983).

18.4.2D Metal/Semiconductor Interfaces Metal/semiconductor interfaces are special types of heterojunctions. The electrical properties of all large-area *metal contacts* are determined by this interface; in point contacts, high-field tunneling effects (see Section 42.4.3) dominate the electrical behavior. The most important parameter of a metal/semiconductor interface is its *relative*

456 Crystal Defects and Interfaces

Table 18.2: Intrinsic workfunction of semiconductors in eV (after Zunger, 1986; and Smith, 1949). ©John Wiley & Sons, Inc.

Material	Workfunction	Material	Workfunction
Si	5.2	ZnS	7.5
Ge	4.5	ZnSe	6.82
GaP (110)	5.9 ... 6.01	ZnTe	5.76
GaAs (110)	5.49 ... 5.56	CdS	7.26
InP (110)	5.69 ... 5.85	CdSe	6.62
		CdTe	5.78

workfunction ψ_{MS}, often referred to as the *barrier height* or *contact potential*,* which determines the electron density at this interface:

$$n(x_{MS}) = N_c \exp\left(-\frac{e\psi_{MS}}{kT}\right),\qquad(18.13)$$

where N_c is the effective level density at the band edge of the semi-conductor [Eq. (27.32)]. Like the difference in electron affinities† in a heterojunction, there are many factors which determine the barrier height ψ_{MS}, such as the chemical nature of metal and semiconductor and the actual structure of the interface. In first approximation, ψ_{MS} is the difference between the workfunctions of the metal (see Table 18.1) and the semiconductor (see Table 18.2).

$$\psi_{MS} \simeq \psi_M - \psi_S,\qquad(18.14)$$

which determines the electron exit from these materials into the vacuum. It is, however, modified by the actual double layer at the crystallographic interface. Chemical reaction and crystallographic

---

* The contact potential can be measured when both materials are facing each other in vacuum at a temperature sufficient for some electron emission. The material with lower workfunction will emit more electrons, consequently charging the higher workfunction material until the current between both materials vanishes. The relative potential difference, so established between the two materials, is the *contact potential*.

† The workfunction is the energy from the Fermi level to the vacuum level; the electron affinity is the energy from the conduction band edge to the vacuum level. The former is used in connection with metals, the latter preferably in connection with semiconductors.

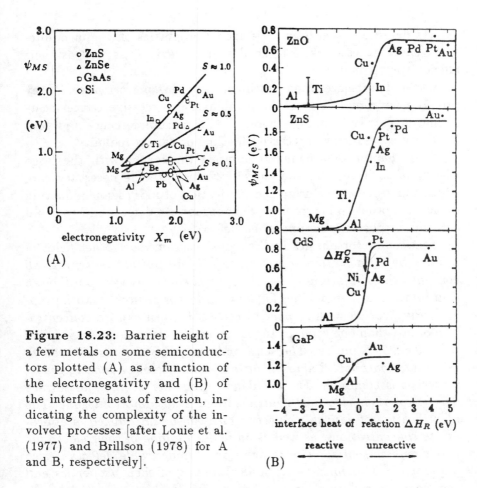

Figure 18.23: Barrier height of a few metals on some semiconductors plotted (A) as a function of the electronegativity and (B) of the interface heat of reaction, indicating the complexity of the involved processes [after Louie et al. (1977) and Brillson (1978) for A and B, respectively].

ordering of the interface region, as well as the pinning of the Fermi level by interface defects (Spicer et al., 1980), have a significant effect on ψ_{MS}. The barrier height of a few metals on some semiconductors is shown in Fig. 18.23.

Space charge effects near the metal/semiconductor contact influence the carrier transport in this region to a large degree. For further studies, see Spenke (1958), Henisch (1957), Rhoderick (1978), and Sharma (1984).

Summary and Emphasis

In this chapter we enumerated the most important crystal defects, such as point, line, and surface defects. Some of these defects are beneficial, such as donors or acceptors; other beneficial defects will be identified later, being responsible, e.g., for luminescence. These defects determine the desired electronic properties of a semiconductor. They are essential for the design of devices, as well as their efficient performance.

Other defects are often detrimental to this performance. They promote nonradiative carrier recombination, excessive carrier scattering, or carrier trapping, thereby reducing luminescence and conductivity by reducing the carrier lifetime or carrier mobility.

This chapter is designed to impress the reader with the large variety of such defects, and thereby to indicate the necessary care which must be taken during the fabrication of most semiconducting devices, in order to incorporate the appropriate defects and to avoid as much as possible the undesired defects.

Semiconductor physics is distinguished from most other fields of physics by its attention to defects which are present in very small concentrations. Materials, described by chemists as very pure, often contain too many impurities for semiconductor devices to work properly. Purification with respect to certain impurities to a concentration of less than one atom in 10^8 is sometimes desired, corresponding to 99.999,999% purity. The semiconductor, therefore, is referred to as 8-niner material. Requirement for 6-niner, semiconducting grade, materials is common. The avoidance altogether of line and surface defects, or their electrical neutralization, is an important goal. The identification of desirable defects which can be introduced in appropriate concentrations, as well as an understanding of their electronic behavior, is essential. The recognition of undesirable defects is also imperative. This includes an understanding of their properties and the means to reduce their density—or, for residuals, to reduce their detrimental effect.

Exercise Problems

1.   Identify the following defects in the notation given in Section 18.2.1:
   (a) Si host with: Li, H, or F interstitials;
         Zn, Mg, B, Sb, or O substitutionals.
   (b) CdS host with: Li, Be, or Cu interstitials;
         Cu, Zn, or In substitutionals for Cd;
         As, Se, or Cl substitutionals for S.
   (c) Which of these would act as donors, which as acceptors?

2.(e) Describe and apply the $8 - N$ rule to a number of different substitutional donors and acceptors in III-V compounds.

3.   How would the atomic size influence the incorporation of impurities as substitutionals or interstitials?

4.(e)   Semiconducting grade material usually contains 1 ppm (part per million) intrinsic defects; 1 impurity at this level, 3 other impurities at 100 ppb (parts per billion), and 8 foreign elements at 10 ppb.

   (a)   What is the average distance between anyone of these defects?

   (b)   On the average, how many lattice atoms lie between them in an Si crystal?

5.(l)   How many dangling bonds exist in an ideal edge dislocation spanning perpendicularly across a 1 cm^3 cube of Si? Most of them are electrically inactive by reconstruction of the dislocation. Check the recent literature and explain.

6.(*)   Discuss the bonding energies listed in Fig. 3.27 at the top of an NaCl crystal with an atomic step. Redraw this figure and show ion positions. Give quantitative estimates considering only nearest and next-nearest neighbors.

7.(e)   Discuss the grain boundary shown in Fig. 18.19 for an AB-crystal. Give the twin angles for cubic crystals. What problem would occur for ionic crystals? Are there angles at which this problem does not exist? Give two of these angles for hexagonal (CdS-type) crystals.

8.(e)   Draw in a three-dimensional representation an edge dislocation with a kink and a jog. Describe the importance of both for dislocation motion within a real crystal.

Chapter 19

Creation and Motion of Point Defects

Point defects can be created or moved within a semi-conductor at elevated temperatures. Their density can be controlled by an appropriately designed heat-treatment program.

Point defects, such as vacancies and interstitials, can be incorporated into a given crystal lattice by thermodynamic creation, while dopants are usually incorporated by diffusion from a source.* Both types of incorporation will be discussed in this chapter. We will first follow an exceedingly simple model to provide some orientation for the inexperienced reader. We then will provide more in-depth evaluations for actual semiconductors and their intrinsic defect-related centers.

19.1 Thermodynamics of Lattice Defects

In the previous section, different kinds of point defects were enumerated. Now we turn to the thermal creation of such defects, and to their interaction with other lattice defects or quasi-particles. Such interaction creates new types of defects which are derived from the previously discussed isolated defects.

---

* In contrast, *purification* can be accomplished by diffusion of impurities into a sink. With the solubility of impurities being a function of the temperature, a temperature gradient can be used as a driving force for purification. A more effective means is the use of the boundary between the liquid and solid phase, using the fact that the solubilities in these two phases are substantially different (a measure of which is the *segregation coefficient*). Zone refining is a well-established technique to achieve such purification—see de Kock (1982).

460

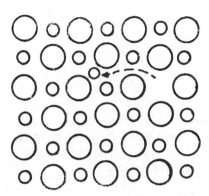

Figure 19.1: Frenkel disorder: cation vacancies and equal amount of cation interstitials.

19.1.1 Intrinsic Lattice Defects

With increased temperature, lattice vibrations (discussed in Chapter 3) become more vigorous, which makes the creation of *intrinsic lattice defects*, i.e., vacancies and interstitials, more probable. In addition, because of the anharmonicity of lattice vibrations, the lattice expands and thereby facilitates such defect creation.

19.1.1A Frenkel Disorder One type of thermodynamic defect is created by an atom moving from a lattice site into an interstitial position and leaving behind a vacancy. In thermodynamic equilibrium, the density of these defects is determined by the equality of thermal generation of these defect pairs and recombination, i.e., an interstitial finds a vacancy and "recombines." From size consideration, cation interstitials and vacancies are more probable. These are referred to as *Frenkel pairs*, and the corresponding lattice disorder as *Frenkel disorder* (Frenkel, 1926)—see Fig. 19.1.

19.1.1B Schottky Disorder Another type of disorder is generated at the surface of a crystal, where an atom from the bulk moves to the surface, creating a vacancy which in turn diffuses deeper into the crystal bulk.

In an ionic AB-compound, both types of (A and B) vacancies must be created in equal amounts to avoid preferential charging of the surface, or the bulk after vacancy diffusion. In fact, such charging will occur initially with every vacancy formed, but will make it more difficult to form another vacancy of the same type adjacent to it. It will also make it easier to place a vacancy of the opposite type adjacent to it, which in essence balances their density. The resulting disorder, i.e., an equal amount of anion and cation

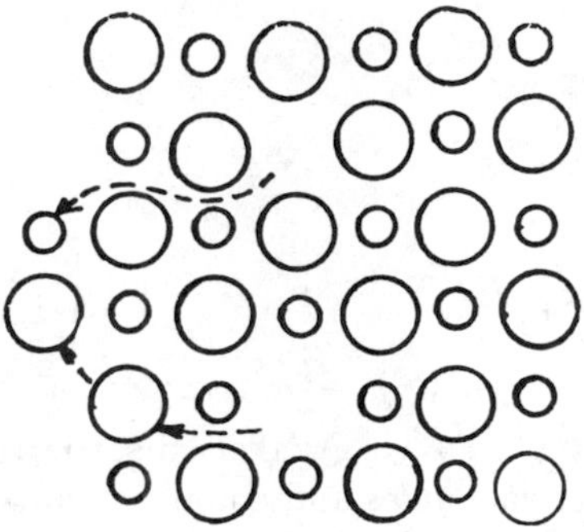

Figure 19.2: Schottky disorder: cation and anion vacancies in equal densities distributed throughout the crystal bulk.

vacancies distributed throughout a crystal, is referred to as *Schottky disorder* (Schottky, 1935)—see Fig. 19.2.

19.1.1C Intrinsic Defect Densities   The density of Frenkel- or Schottky-type *intrinsic defects* can be obtained from thermodynamic considerations: the *Helmholtz free energy*

$$F = U - TS \qquad (19.1)$$

due to these defects must be minimized. As a simple example we will discuss Schottky defects in an elemental crystal, i.e., single vacancies. Assuming that the concentration of these vacancies is low enough so that they are created independently of each other, the energy of n of these vacancies is given by

$$U = nE_S, \qquad (19.2)$$

where E_S, the *Schottky energy*, is the energy required to take a single atom from the crystal volume and put it at the crystal surface—for more detail, see Section 19.1.1G. The configurational entropy*

$$S^{\text{config}} = k \ln W \qquad (19.3)$$

is described by the total number of possibilities of selecting n indistinguishable atoms and moving them to the surface from a crystal containing N atoms:

$$W = \frac{N(N-1)\ldots(N-n+1)}{n!} = \frac{N!}{(N-n)!n!}. \qquad (19.4)$$

---

* For simplicity we have neglected here the vibrational part of the entropy ($S = S^{\text{config}} + S^{\text{vib}}$). These contributions are considered later in this chapter (Section 19.1.1E). The vibrational part results in an increase of the intrinsic defect density.

Using the *Stirling approximation*

$$\ln n! \simeq n \ln n - n, \qquad (19.5)$$

we obtain

$$S = k \ln W = k\left[N \ln N - N - \{(N-n)\ln(N-n) - (N-n)\}\right.$$
$$\left. - (n \ln n - n)\right]. \qquad (19.6)$$

When minimizing F at a given temperature, we obtain from Eqs. (19.1)–(19.6)

$$\left(\frac{\partial F}{\partial n}\right)_T = E_S - kT \ln \frac{N-n}{n} = 0; \qquad (19.7)$$

and for $n \ll N$

$$n = N \exp\left(-\frac{E_S}{kT}\right). \qquad (19.8)$$

Since, for a crystal of unit volume, n is equal to the *density* of vacancies N_V, and N is the *density* of lattice atoms N_L, we have for the density of Schottky defects in a monatomic crystal in thermal equilibrium

$$\boxed{N_V = N_L \exp\left(-\frac{E_S}{kT}\right).} \qquad (19.9)$$

In a diatomic lattice, for neutrality reasons, an equal amount of anion and cation vacancies must be formed.* Since the probability of forming ion pairs W_p is the square of the probability of forming single vacancies ($W_p = W^2$), it follows that

$$\left(\frac{\partial F}{\partial n}\right)_T = E_S - kT \ln \left(\frac{N-n}{n}\right)^2 = 0. \qquad (19.10)$$

Hence, for $n \ll N$ and for $n = N_S$, the density of Schottky defects in a diatomic lattice is given by

$$\boxed{N_S = N_L \exp\left(-\frac{E_S}{2kT}\right).} \qquad (19.11)$$

---

* This simple model of a pair-wise defect formation maintaining stoichiometry will be modified later (Section 19.2.3) to permit slight changes in the stoichiometry, thereby making the crystal n- or p-type. Inversely, the creation of such intrinsic defects can be enhanced or suppressed depending on the position of the Fermi-level, i.e., depending on doping.

In a similar fashion, the density of *Frenkel pairs* is determined, resulting in

$$N_F = \sqrt{N_L N_i} \exp\left(-\frac{E_{Fr}}{2kT}\right), \qquad (19.12)$$

where N_i is the density of interstitial sites, and E_{Fr} is the Frenkel energy, i.e., the energy to take one cation and put it at a (distant) interstitial site.

A more sophisticated approach involving the grand canonical ensemble that permits an analysis of more complicated cases can be found in Landsberg and Canagaratna (1984).

19.1.1D Vacancies and Interstitials with Lattice Relaxation

In actual semiconductors the primitive picture given in the previous section needs to be modified to take into consideration

- the different actual positions within the lattice which can be occupied by an interstitial, and
- the relaxation of the lattice surrounding a vacancy or an interstitial.

For example, within the Si lattice there are three different sites to place a self-interstitial (H, T, and B), shown for a $(01\bar{1})$ plane in Fig. 19.3, as well as two possibilities for the split configuration S and S'. They have different formation enthalpies, which also depend on the occupancy, i.e., the charge character of the center (Car et al., 1984).

The relaxation of the surrounding lattice may be minor or can be very substantial, as indicated schematically in Fig. 18.1C and F, respectively.

19.1.1E The Entropy Contribution to Realistic Defects

There are two contributions for an increased entropy of realistic intrinsic defects:

- the extended relaxation of the lattice surrounding such a defect; and
- the change in oscillatory frequencies of the surrounding lattice.

The extended relaxation permits numerous similar configurations with nearly identical energy. This causes a substantial increase in the configurational entropy, and thereby yields a much larger pre-exponential factor determining the density of such defects in equilibrium—see Lannoo and Bourgoin (1981).

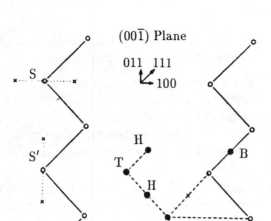

Figure 19.3: Open circles indicate lattice positions of Si atoms in a $(01\bar{1})$ plane. Solid circles indicate interstitials, T with tetragonal symmetry, H with hexagonal symmetry, and B, a bond center site. S and S' are interstitialcy positions (crosses) (after Car et al., 1984).

The vibrational part of the entropy is given in its most simple form as

$$\Delta S^{\text{vib}} = k \ln \left[\sum_i \frac{\nu_{io}}{\nu_i'} \right] \tag{19.13}$$

where ν_{io} and ν_i' are the original and changed frequencies of the undisturbed lattice and the lattice surrounding the defect. There are two major modes considered near the defect: the *breathing mode*, with all surrounding atoms moving in and out in phase; and a *vector-like motion*, including the defect (interstitial). When considering a somewhat reduced binding of the lattice surrounding the defect, we expect $\nu_{io} > \nu_i'$ and therefore a positive entropy contribution. This contribution also increases the pre-exponential factor and therefore causes a somewhat larger density of intrinsic defects (Talwar et al., 1980). For a recent review of *ab initio* calculation of native defects, see Jansen and Sankey (1989).

19.1.1F Antisite Defects Antisite defects are intrinsic lattice atoms placed on a wrong lattice site, e.g., a B atom on an A site B_A in an AB-lattice. Such antisite defects may be regarded as substitutional impurities except that they are supplied by a large reservoir of sublattice B atoms. Their energy of formation has two contributions: one from the reduction of the band gap due to disorder; and one from an electronic contribution causing a shift in the Fermi energy, the latter depending on doping.

19.1.1G The Formation Energy of Vacancies   The formation energy* of vacancies E_S is obtained at sufficiently high temperature from the exponential change of the electrical conductivity with temperature when these vacancies are electrically active as the dominant acceptors—see Section 27.3.3. The formation energies are typically in the 2...6 eV range.

A rough estimate about the formation energy E_S of vacancies can be made from the microscopic equivalent of a macroscopic cavity

$$E_S = AH_S, \qquad (19.14)$$

where A is the surface area of the cavity of the vacancy volume, and H_S is the macroscopic surface energy (Brooks, 1963; Friedel, 1967). This simple model provides fair agreement with the experiment and can be improved by considering the anisotropy of the cavity, which is not spherical, and of the surface energy (van Vechten, 1980).

The same cavity model gives some indication of a slight disparity in the density of V_A and V_B in an AB-compound, since the formation energy increases with increasing cavity radius. As net result, we observe a strong preference of n- or p-type semiconductivity when $r_A > r_B$ or $r_A < r_B$, respectively, in lattices with a substantial ionic bonding component; here A is the cation, B is the anion.

In Table 19.1 additional information is given for some of the II-VI compounds. These show substantial preference of nonstoichiometry because of the asymmetry of the formation energy of cation and anion vacancies.

For a more thorough evaluation of the formation energy, we must consider

- dangling bonds of the surrounding lattice atoms,
- reconstruction of the surrounding lattice, and
- lattice relaxation beyond nearest neighbors.

These effects are lattice specific, and require an individual analysis for each type of vacancy—see literature cited in Pantelides (1986).

---

* Often, the enthalpy H_S is cited rather than the energy; however, with $\Delta H_S = \Delta E_S + p\Delta \mathcal{V}$ and negligible volume changes in the solid, both are almost identical at room temperature. Near the melting point, a $\Delta \mathcal{V}/\mathcal{V} \simeq (1/3)(\Delta l/l) \simeq 2\%$ change in volume [Eq. (6.17)] may be considered.

Table 19.1: Melting point (K), minimum vapor pressure (atm), equilibrium constant, and surplus element of some II-VI compounds (after Lorent, 1967).

II-VI Compounds	T_m	$P_{min}(T_m)$	Log K_p	Surplus
ZnS	2100	3.7	0.85	Zn
ZnSe	1790	0.53	-1.65	Zn
ZnTe	1570	0.64	-1.4	Te
CdS	1750	3.8	0.9	Cd
CdSe	1512	0.41	-2.0	Cd
CdTe	1365	0.23	-2.75	Te
HgTe	943			

For example, the Si-vacancy is discussed by Watkins (1986). In contrast to vacancies in other materials, it cannot be frozen-in by rapid quenching at measurable densities because of its high mobility. It is efficiently produced by radiative electron damage at low temperatures. In order to avoid rapid recombination with the co-produced and highly mobile interstitial, trapping of the interstitial at group III atoms in p-type Si can be employed (Watkins, 1974). For the electron structure of the Si-vacancy, see Section 22.4.2. For an estimation of the energy of vacancy formation, see Car et al. (1985) and Table 19.2.

19.1.1H Formation Energy of Divacancies Vacancies in a covalent crystal are attracted to each other. This can be interpreted as being caused by the reduction of the surface area, which thereby reduces their surface energy. This process is similar to that which causes air bubbles in water to coalesce. Estimated accordingly, the binding energy of divacancies in Si is 1.03 eV, compared to a measured value of 1.2 eV (Ammerlaan and Watkins, 1972). In a microscopic model, one realizes that in a divacancy only six bonds are broken, rather than eight in two separate vacancies.

19.1.2 Frozen-In Intrinsic Defect Densities

The analysis presented in Section 19.1.1C assumes thermodynamic equilibrium, but does not ask how long it takes to achieve equilibrium. At sufficiently high temperatures, equilibrium is usually

Table 19.2: Activation energies for vacancies and interstitials of Si, P and Al or combined with vacancies in intrinsic, n- and p-type silicon in eV (after Car et al., 1985).

Species	E_{int}	Species	E_{int}	E_n	Species	E_{int}	E_p
$V_{Si}^{\bullet\bullet}$	5.1	$(P_{Si}V_{Si})^{\bullet}$	3.2	3.8	$(Al_{Si}V_{Si})^{\bullet}$	3.9	3.2
$V_{Si}^{\bullet}$	4.7	$(P_{Si}V_{Si})^{\times}$	3.0	3.0	$(Al_{Si}V_{Si})^{\times}$	3.5	3.5
$V_{Si}^{\times}$	4.2	$(P_{Si}V_{Si})'$	3.2	2.6	$(Al_{Si}V_{Si})'$	3.5	4.1
V_{Si}'	4.5	$(P_{Si}Si_i)^{\times}$	5.7	5.7	$(Al_{Si}Si_i)^{\times}$	4.0	4.0
$Si_i^{\bullet\bullet}$	5.6	$P_i^{\times}$	5.1	5.1	$Al_i^{\bullet\bullet}$	4.3	3.0
$Si_i^{\bullet}$	5.5	P_i'	5.2	4.6	$Al_i^{\bullet}$	3.9	3.1
$Si_i^{\times}$	5.7	P_i''	5.6	4.5	$Al_i^{\times}$	3.9	3.9
SD_{exp}	4.1–5.1	P_{exp}	3.7	3.1	Al_{exp}	3.5	?

reached faster than the temperature is changed. While cooling a crystal, the density of defects* will start to lag behind the value corresponding to the temperature. Finally, at low enough temperatures, it will not decrease appreciably within the time of the experiment. This "frozen-in" defect density $N_{f\text{-}i}$ identifies—via Eqs. (19.9), (19.11), or (19.12)—a *freezing-in temperature*. For example, for Frenkel defects:

$$N_{f\text{-}i} = \sqrt{N_i N_L} \exp\left(-\frac{E_{Fr}}{2kT_{f\text{-}i}} \right). \qquad (19.15)$$

Freezing-in occurs because of the reduced mobility of lattice defects with decreasing temperature, which finally makes it impossible for them in the given time frame to find a defect partner for recombination, e.g., for an interstitial to find a vacancy.

For example, the interatomic potential distribution of a Frenkel defect in a simple cubic monatomic lattice is shown in Fig. 19.4. In the figure, $x = x_V$ is the position of the vacancy; $x_{i1}, x_{i2}, x_{i3} \ldots$ are the positions of interstitial sites; and ΔE_i is the activation energy

---

* In materials in which two types of defects need to be considered, two different freezing-in temperatures appear and, because of conservation and neutrality considerations, a more complex behavior is expected (Hagemark, 1976).

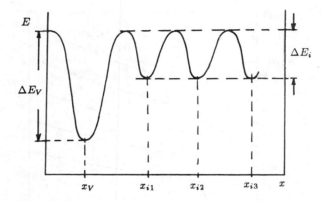

Figure 19.4: Atomic potential for an atom on a vacancy or interstitial position (schematic).

for interstitial motion. The time required for an atom to move from one to the adjacent interstitial site can be estimated from

$$\tau_i = \frac{2\pi}{\omega_i} \exp\left(\frac{\Delta E_i}{kT}\right),$$

(19.16)

where ω_i is an effective oscillation frequency of the interstitial atom.

In a crystal with a density of N_{Fr} Frenkel defects, the average time for a recombination event to occur is given by

$$\tau_r = \frac{1}{4\pi D_i r_i N_{Fr}},$$

(19.17)

where r_i is the reaction distance, which here is equal to the distance between adjacent interstitial sites a_i. When using Eq. (19.70) for the diffusion constant with $\alpha = 1/6$, we obtain

$$\tau_r = \frac{3}{a_i^3 \omega_i \sqrt{N_i N_L}} \exp\left(\frac{\Delta E_i + E_{Fr}/2}{kT}\right).$$

(19.18)

With ω_i on the order of 10^{13} s^{-1}, $\Delta E_i = 1$ eV and $E_{Fr} = 2$ eV, we obtain a recombination time on the order of one year at 500 K. After increasing the temperature to 700 K, the time constant is reduced to about 10 s, and the defect density comfortably follows a sufficiently slowly decreasing temperature.

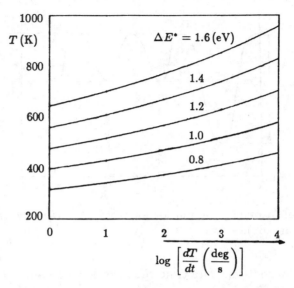

Figure 19.5: Freezing-in temperature as a function of the cooling rate with ΔE^* as the family parameter.

The freezing-in temperature is a function of the rate of cooling. Freezing-in is reached when this cooling rate equals the rapidly decreasing rate of recombination defined by $\partial T / \partial \tau_r$:

$$\frac{dT}{dt} = \frac{\partial T}{\partial \tau_r} = -\frac{kT}{\Delta E_i + E_{Fr}/2} \cdot \frac{T}{\tau_r}. \qquad (19.19)$$

From Eqs. (19.18) and (19.19), we obtain a well-defined freezing-in temperature contained in τ_{rec}, which increases with increasing cooling rate

$$T_{\text{f-i}} = \frac{\Delta E_i + E_{Fr}/2}{k} \cdot \frac{1}{\ln\left[\dfrac{kT_{\text{f-i}}}{\Delta E_i + E_{Fr}/2} \cdot \dfrac{a_i^3 \omega_i \sqrt{N_i N_L} T_{\text{f-i}}}{3\left(-\dfrac{dT}{dt}\right)}\right]}. \qquad (19.20)$$

With the values of the above-given example and a cooling rate of 1 deg/s, the value of the logarithmic term is $\simeq 25$. Thus, we obtain as a rough estimate, with $\Delta E^* = \Delta E_i + E_{Fr}/2$:

$$\boxed{T_{\text{f-i}} \simeq 400 \Delta E^*(\text{eV}) \cdot \left(1 + \frac{1}{30} \ln \frac{dT}{dt}\right) (\text{K}).} \qquad (19.21)$$

Since T_{f-i} is logarithmically dependent on the cooling rate, the freezing-in temperature changes little with a normal variation of cooling rates except for rapid quenching—see Fig. 19.5.

19.2 Defect Chemistry

The creation of different types of defects is often interrelated—see Section 19.1.1. For instance, for Schottky disorder of an AB-compound, vacancies of both A and B ions are produced in equal densities. The quasi-neutrality relation prevents the creation of one charged defect in substantial excess over the others.

The creation of lattice defects can be described similarly to the creation of a chemical compound in the presence of other reaction partners. The governing relation is the *mass action law*, which, for a reaction

$$A + B \rightleftharpoons AB, \tag{19.22}$$

can be written as

$$[A][B] = [AB]K_{AB} \tag{19.23}$$

where the square brackets indicate concentrations and K_{AB} is the mass action law constant

$$K_{AB} = K_{AB,0} \exp\left(-\frac{\Delta G}{kT}\right), \tag{19.24}$$

with ΔG as the change in Gibbs free energy*:

$$\Delta G = \Delta U + p\Delta V - T\Delta S = \Delta H - T\Delta S, \tag{19.25}$$

where ΔU is the change in *internal energy* due to changes in the interaction potential of the vacancy with the surrounding lattice, and ΔV is the often negligible change in volume. Both are usually combined as the change in *enthalpy* $\Delta H = \Delta U + p\Delta V$. ΔS is the change in *entropy*:

$$\Delta S = \Delta S_{basic}^{config} + \Delta S_{extend}^{config} + \Delta S^{vibr}. \tag{19.26}$$

The basic configurational entropy part is described in Section 19.1.1C; the parts dealing with extended lattice relaxation and with lattice vibration are described in Section 19.1.1E.

---

* In Section 19.1.1C, the Helmholtz free energy was used, which is related to the Gibbs free energy by $G = F + pV$. Since some of the reactions involve an interaction with a gas atmosphere, the more general notation is used here.

In solids, defect chemistry description is advantageous because it illustrates the interconnection of the different defects. We will now take into consideration that these defects "react" with each other and thereby change their charge and position in the lattice, i.e., their "defect chemical" composition.

For example, for the Schottky disorder, the defect chemistry notation can be written as

$$Ge_{Ge} \leftrightarrows V_{Ge} + Ge_{\text{surface}}. \tag{19.27}$$

Neglecting the difference between Ge_{Ge} and Ge_{surface}, they cancel in Eq. (19.27), yielding

$$0 \leftrightarrows V_{Ge}. \tag{19.28}$$

The corresponding mass action law reads

$$N_S = [V_{Ge}] = K_v = K_V \exp\left(-\frac{\Delta G}{kT}\right). \tag{19.29}$$

Disregarding a change in volume, and with $\Delta H = E_S$ as the Schottky energy, Eq. (19.29) can be rewritten as

$$N_S = K_0 K_V \exp\left(-\frac{\Delta H}{kT}\right). \tag{19.30}$$

This is equivalent to Eq. (19.9), derived earlier, with $K_V = N_L$ and $K_0 = \exp(\Delta S^{\text{vibr}}/k)$, a factor accounting for the vibrational entropy part in ΔG.

19.2.1 Reaction Partners in Solids

Solid-state reactions take place between crystal defects, carriers, and external partners, such as a gas atmosphere. Types of reactions to be considered in solid-state defect chemistry include:

- intrinsic defect formation;
- changing of stoichiometry by interaction with a gas atmosphere of one of the components;
- doping with foreign defects;
- defect associate formation;
- formation of ionized defects; and
- creation of free carriers.

A few examples of some typical reactions are given below, along with an analysis that shows the interdependence of the different reactions. For more detail, see Kröger (1964).

19.2.2 Defect Chemistry Involving Neutral and Ionized Defects

For example, a Frenkel disorder generates metal-ion interstitials which act as donors and metal-ion vacancies which act as acceptors. The reaction equation for the Frenkel disorder is

$$M_M^\times + V_i^\times \rightleftharpoons M_i^\times + V_M^\times \; ; (E_{Fr}). \qquad (19.31)$$

An "$\times$" is added to the chemical symbol of each defect, representing its neutral state with respect to the lattice; the energy necessary to achieve this transition is appended in parentheses.

These defects can be ionized according to

$$M_i^\times \rightleftharpoons M_i^\bullet + e' \; ; (\tilde{E}_d) \qquad (19.32)$$

and

$$V_M^\times \rightleftharpoons V_M' + h^\bullet \; ; (\tilde{E}_a) \qquad (19.33)$$

with $\tilde{E}_d = E_c - E_d$ and $\tilde{E}_a = E_a - E_v$. In addition, we create intrinsic carriers (see Section 27.3.2E):

$$0 \rightleftharpoons e' + h^\bullet \; ; (E_g). \qquad (19.34)$$

These reactions can be described by the following set of mass action law equations:

$$\frac{[M_i^\times][V_M^\times]}{[M_M^\times][V_i^\times]} = K_{M,i} \propto \exp\left(-\frac{E_{Fr}}{kT}\right), \qquad (19.35)$$

$$\frac{[M_i^\bullet]n}{[M_i^\times]} = K_D \propto \exp\left(-\frac{\tilde{E}_d}{kT}\right), \qquad (19.36)$$

$$\frac{[V_M']p}{[V_M^\times]} = K_A \propto \exp\left(-\frac{\tilde{E}_a}{kT}\right), \qquad (19.37)$$

and

$$\frac{np}{N_c N_v} = K_i \propto \exp\left(-\frac{E_g}{kT}\right), \qquad (19.38)$$

with the connecting quasi-neutrality condition:

$$n + [V_M'] = p + [M_i^\bullet]. \qquad (19.39)$$

The value of the defect chemical approach becomes apparent when we recognize that a crystal defect can participate in various reactions—e.g., V_M in Eqs. (19.31) and (19.33)—and can form different species, $V_M^\times$ and V_M'. Without ionization, the density of all vacancies is given by Eq. (19.12) or (19.35); with ionization, only the density of neutral vacancies is given by Eq. (19.35). Ionized vacancies drop out of Eq. (19.35) and are governed by Eq. (19.37). Therefore, the total amount of vacancies increases.

In general, several of these reactions influence each other and cause a shift in the density of the reaction partners. In order to obtain the densities of all partners, the system (19.35)–(19.39) must be solved simultaneously. In Section 19.2.7 we will introduce an instructive approximation for accomplishing this.

19.2.3 Changing of Stoichiometry

In this example, we will show how the heat treatment of an AB-compound in an atmosphere of one of the components can change the stoichiometry of this compound. For example, with Schottky defects we have*

$$0 \rightleftharpoons V_A^\times + V_B^\times \; ; (E_S). \tag{19.40}$$

The treatment of the AB-compound in a gas of A makes AB grow, and consequently causes an increase in the relative density of B vacancies

$$A_{gas} \rightleftharpoons A_A + V_B^\times \; ; (E_{Ag}), \tag{19.41}$$

which yields for equilibrium

$$[V_A^\times][V_B^\times] = K_S \propto \exp\left(-\frac{E_S}{kT}\right) \tag{19.42}$$

and

$$\frac{[A_A^\times][V_B^\times]}{p_A} = K_{Ag} \propto \exp\left(-\frac{E_{Ag}}{kT}\right). \tag{19.43}$$

Therefore, we obtain for the change in stoichiometry δ, with AB$\rightarrow$ A$_{1+\delta}$B, the expression

$$\delta = [A_A^\times] - [B_B^\times] = [V_B^\times] - [V_A^\times], \tag{19.44}$$

which can be evaluated using Eqs. (19.42) and (19.43).

Another possibility is the reduction of A vacancies according to $A_A^\times \rightleftharpoons A_{gas} + [V_A^\times]$, and consequently an increase in A interstitials or of antisites A_B. In actual compound crystals the densities of both vacancies are not the same, since their formation energy differs. This difference must be taken into account for any real crystal—see Section 19.1.1G.

---

* Or, for an $A_n B_m$ compound, we have $0 \rightleftharpoons n V_A^\times + m V_B^\times$ with $[V_A^\times]^n [V_B^\times]^m = K_S$.

19.2.4 Doping in Equilibrium with a Foreign Gas

In a similar fashion, the treatment of a semiconductor in a vapor of a foreign atom C with a partial pressure p_C causes a change of the intrinsic defect density. The interaction takes place via the neutrality condition. For instance, if the foreign atom is a metal atom replacing a lattice atom

$$C_{gas} \leftrightarrows C_A{}^\times \quad \text{with} \quad \frac{[C_A{}^\times]}{p_A} = K_{Cg}, \quad (19.45)$$

and acting as a donor

$$C_A{}^\times \leftrightarrows C_A{}^\bullet + e' \quad \text{with} \quad \frac{n[C_A{}^\bullet]}{[C_A{}^\times]} = K_D, \quad (19.46)$$

it will tend to reduce the density of intrinsic defects with the same charge, and to increase the density of intrinsic defects with the opposite charge, since they are interrelated by

$$n + [V_B{}^\times] = p + [V_A{}^\bullet] + [C_A{}^\bullet]. \quad (19.47)$$

In more general terms, a foreign atom incorporated as a donor tends to reduce the intrinsic donor densities and to increase the density of intrinsic acceptors. Vice versa, intrinsic and extrinsic acceptors cause an increase in the equilibrium density, i.e., the solubility of the foreign donors.

This important interrelationship is called *compensation*. It is used to increase doping densities otherwise limited by low solubility. This is technically important, e.g., for tunnel diodes.

19.2.5 Formation of Defect Associates

When crystal defects are able to form defect associates, these associates constitute a new species and influence the balance of the unassociated defects. For instance, in an AB-compound with Schottky disorder, the metal vacancies may form *dimers*, i.e, two vacancies as nearest neighbors:

$$2V_A^\times \leftrightarrows V_{A2}^\times \; ; (E_2), \quad (19.48)$$

with E_2 as the binding energy of these dimers.*. With the formation of $V_{A2}^\times$, a new species is created involving metal vacancies. The total

---

* With proper charging, these dimers could be regarded as equivalent to a nonmetal molecule B_2 sitting on a lattice site of a cluster of four vacancies. Such molecules are often covalently bound and therefore have a substantial binding energy; hence, they have a high probability of occurring.

amount of A vacancies is thus increased. Therefore, the balance between V_A and V_B in the Schottky disorder is shifted.

19.2.6 Compensation with Fixed Density of Donors

In the following example, an impurity is present at a fixed density within a monatomic semiconductor together with Schottky defects acting as acceptors. The extrinsic substitutional dopant acts as a donor with a density $[F_M]_{total}$. The following reactions take place, following the corresponding mass action laws:

$$[F_M]_{total} = \text{const} \qquad (19.49)$$

$$0 \rightleftharpoons [V_M^\times] ; \ (E_S) \qquad [V_M^\times] = K_0 \exp\left(-\frac{E_S}{kT}\right) \qquad (19.50)$$

$$V_M^\times \rightleftharpoons V_M' + h^\bullet; \ (\tilde{E}_a) \qquad p[V_M'] = N_v[V_M^\times] \exp\left(-\frac{\tilde{E}_a}{kT}\right) \qquad (19.51)$$

$$F_M^\times \rightleftharpoons F_M^\bullet + e'; \ (\tilde{E}_d) \qquad n[F_M^\bullet] = N_c[F_M^\times] \exp\left(-\frac{\tilde{E}_d}{kT}\right) \qquad (19.52)$$

$$0 \rightleftharpoons np; \ (E_g) \qquad np = N_c N_v \exp\left(-\frac{E_g}{kT}\right). \qquad (19.53)$$

Two conditional equations, the *conservation of mass equation*

$$[F_M]_{total} = [F_M^\times] + [F_M^\bullet] \qquad (19.54)$$

and the *quasi-neutrality equation*

$$n + [V_M'] = p + [F_M^\bullet], \qquad (19.55)$$

force an interaction between the different reaction partners, i.e, force compensation. We will analyze this interaction in Section 19.2.7A.

19.2.7 Brouwer Approximation

Before discussing the solution to the problem given in the previous section, we will indicate here how to analyze the typical system of equations encountered in defect chemistry.

The governing equations for defect chemical reactions are coupled with conservation of mass or charge equations. They contain the sum of densities, e.g., the neutrality equation given in Section 19.2.2:

$$n + [V_M'] = p + [M_i^\bullet]. \qquad (19.56)$$

Since each of these densities varies exponentially with temperature, mostly with different slopes, they are usually of substantially differ-

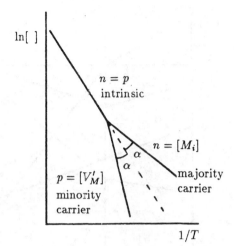

Figure 19.6: Brouwer diagram for a simple example of a semiconductor with one dominant intrinsic donor and free carriers.

ent magnitudes. Therefore, we can neglect one of the terms on each side, and thereby can distinguish four cases:

$$[V'_M] = [M_i^{\bullet}], \tag{19.57}$$

$$n = [M_i^{\bullet}], \tag{19.58}$$

$$[V'_M] = p, \tag{19.59}$$

$$\text{or} \quad n = p. \tag{19.60}$$

Depending upon the relative magnitude of the different activation energies, we have different temperature ranges in which the validity of these neutrality approximations changes over from one to the other case. For example, at low temperatures, Eq. (19.58) holds; with increasing temperature, Eq. (19.60) becomes valid. In each of these temperature ranges, the governing equation can be given explicitly. For instance, for $n = [M_i^{\bullet}]$, we obtain from Eq. (19.36)*

$$n = [M_i^{\bullet}] = \sqrt{[M_i^{\times}]K_D} = \sqrt{[M_i^{\times}]} \exp\left(-\frac{\tilde{E}_d}{2kT}\right), \tag{19.61}$$

and have $p = [V'_M]$ with Eq. (19.37), if $E_a < E_d$. At higher temperature, we obtain from $n = p$ with Eq. (19.38):

$$n = p = \sqrt{N_c N_v K_i} = \sqrt{N_c N_v} \exp\left(-\frac{E_g}{2kT}\right). \tag{19.62}$$

---

* In order to avoid exact compensation, we must also consider some extrinsic donors to make the donors predominant (see Section 27.3.2).

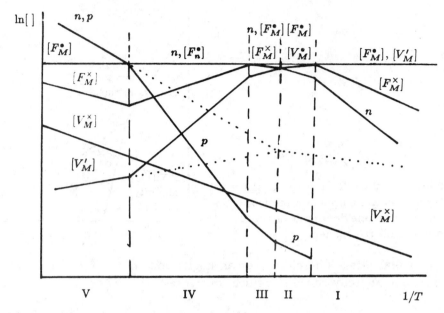

Figure 19.7: Brouwer diagram for a monatomic semiconductor with fixed density of foreign donors and intrinsic vacancies acting as acceptors (see text). Dotted curve occurs if no foreign doping is present ($[F_M]_{\text{total}} = 0$) (modified from Kröger, 1964).

In each of these temperature ranges, the densities can be represented as straight lines in a $\ln[\ ]$ vs. $1/T$ plot, which shows the dominant interaction. In the intermediate temperature range, a smooth transition occurs. Such a transition is neglected and replaced by sharp breaks in the Brouwer approximation (Brouwer, 1954). An illustration for such a relation is shown in Fig. 19.6 for the example given above.

In the Brouwer diagram, majority (n) and minority (p) carrier densities show a split with the same angle (α) in the semilogarithmic plot (Fig. 19.6). The obtained results are similar to those obtained in Section 27.3.2D; however, the carrier depletion range is neglected here.

19.2.7A Brouwer Diagram of Compensation with Fixed Donor Density   When carriers and atomic reaction partners are involved, the resulting Brouwer diagram becomes more instructive. For instance, in a monatomic semiconductor with Schottky defects and a fixed density of donors (given in Section 19.2.6), several of these defects interact. We see from Eqs. (19.49)–(19.55) that five out of eight possible temperature ranges occur:

$$
\begin{array}{lll}
\text{I} & [F_M]_{\text{total}} = [F_M^{\bullet}]; & [V_M'] = [F_M^{\bullet}] \\
\text{II} & = [F_M^{\times}]; & [V_M'] = [F_M^{\bullet}] \\
\text{III} & = [F_M^{\times}]; & n = [F_M^{\bullet}] \\
\text{IV} & = [F_M^{\bullet}]; & n = [F_M^{\bullet}] \\
\text{V} & = [F_M^{\bullet}]; & n = p.
\end{array}
\qquad (19.63)
$$

The corresponding Brouwer diagram is given in Fig. 19.7. It shows that the density of the neutral intrinsic vacancy $[V_M^{\times}]$ is not influenced by any of the other reactions. It extends as a straight line through all temperature ranges. All other reactions show interdependencies.

In the temperature ranges I, IV, and V, essentially all foreign atoms are ionized ($[F_M]_{\text{total}} = [F_M^{\bullet}]$). At low temperatures (range I), $[F_M^{\bullet}]$ is equal to the density of ionized acceptors $[V_M']$. At higher temperatures (range III), $[F_M^{\bullet}]$ is equal to the free-electron density; this is the depletion range. In range V, we observe $n > [F_M^{\bullet}]$; this is the intrinsic range with $n = p$. There are two intermediate ranges where the density of ionized donors is depressed (II and III); the density of charged acceptors starts to decline (II) because of competition with electrons in the quasi-neutrality condition.

The density of charged intrinsic acceptors dramatically increases at low temperatures (I) because of the incorporation of foreign donors, permitting neutrality via $[F_M^{\bullet}] = [V_M']$. When the electron density increases (II and III), the density of ionized acceptors decreases until n takes over (IV) in the neutrality condition. With n larger than the donor density, i.e., when $n = p$, $[V_M{}']$ approaches the density it would have in the undoped case. The recombination of free holes also causes a reduction in $[F_M{}']$. Therefore, the neutral donor density increases again.

This example shows the rather complex interdependency of the different densities, and presents a good illustration of an instructive analysis. The examples presented in these sections, which dealt with defect chemistry, have been simplified somewhat in order to indicate the principles involved, rather than to illustrate actual material behavior. For further reading, see original literature in Kröger (1964).

19.3 Diffusion of Lattice Defects

Although this book deals with homogeneous semiconductors, and diffusion requires a spatial gradient, the basic concepts of the diffusion of lattice defects shall be mentioned here briefly since they are prerequisites for actual doping—where foreign atoms are supplied by diffusion from the surface, or when intrinsic Schottky defects are created by outdiffusion of lattice atoms to the surface.

Diffusion is determined by *Fick's first law*, which relates the diffusion current j_i to the diffusion tensor D_{ik} and the density gradient

$$j_i = -\sum_{k=1}^{3} D_{ik} \frac{\partial N}{\partial x_k}, \tag{19.64}$$

where N is the density of the diffusing lattice defect. Diffusion must also follow the continuity equation (*Fick's second law*)

$$\frac{\partial N}{\partial t} = -\sum_{i} \frac{\partial}{\partial x_i} j_i = \sum_{ik} \frac{\partial}{\partial x_i} D_{ik} \frac{\partial N}{\partial x_k}. \tag{19.65}$$

For small defect densities, we can disregard defect interaction, i.e, D_{ik} is independent of N, and we have

$$\frac{\partial N(\mathbf{x}, t)}{\partial t} = \sum_{ik} D_{ik} \frac{\partial^2 N(\mathbf{x}, t)}{\partial x_i \partial x_k}, \tag{19.66}$$

with D as a symmetrical second rank tensor following the point group symmetry of the crystal. For a cubic (isotropic) crystal, we have

$$\frac{\partial N(\mathbf{x}, t)}{\partial t} = D \nabla^2 N(\mathbf{x}, t). \tag{19.67}$$

The diffusion equation can be integrated most easily by employing a Laplace transformation. For a simple case of initial conditions, with the dopant deposited at the surface ($x = 0$) and no depletion of dopants for a *one-dimensional semi-infinite sample*:

$$\begin{aligned} N &= N_0 \text{ at } x = 0 \text{ for } t \geq 0 \\ N &= 0 \quad \text{ at } x > 0 \text{ for } t = 0 \end{aligned} \tag{19.68}$$

yielding as the solution of Eq. (19.67)

$$N(x, t) = N_0 \left[1 - \mathrm{erf} \left\{ \frac{x}{\sqrt{4Dt}} \right\} \right], \tag{19.69}$$

the well-known error function distribution of dopants as the diffusion experiment proceeds.

In a very simple example of foreign atoms diffusing on interstitial sites, we relate the diffusivity to the spacing between interstitial sites a and the jump frequency ν_j by

$$D = \alpha a^2 \nu_j, \tag{19.70}$$

where α is a geometry-related factor. For a cubic lattice and interstitial diffusion, one has $\alpha = 1/6$. The jump frequency depends, via a Boltzmann factor, on the temperature

$$\nu_j = \nu_0 \exp\left(-\frac{\Delta E_i}{kT}\right), \tag{19.71}$$

where ΔE_i is the height of the saddle point between adjacent lattice atoms over which the diffusion proceeds. We express the diffusivity as

$$\boxed{D = D_0 \exp\left(-\frac{\Delta E_i}{kT}\right)} \quad \text{with} \quad \boxed{D_0 = \alpha a^2 \nu_0} \tag{19.72}$$

and ν_0 as an effective jump frequency.

The path of each of the diffusing atoms is that of a random walk, ignoring correlation effects, with $\overline{R^2}$ the mean square total displacement given by

$$\overline{R^2} = \hat{n} a^2 = \alpha D t, \tag{19.73}$$

where $\hat{n}$ is the number of interstitial jumps. The average distance traveled by the diffusing atom is simply $\sqrt{\overline{R^2}}$.

The diffusion coefficients for intrinsic and a number of extrinsic defects in Si and GaAs are given in Table 19.3.

19.3.1 Types of Diffusion

In the previous section, a simple diffusion of a foreign or lattice atom from interstitial to adjacent interstitial position was assumed, following a potential as shown in Fig. 19.4. In actual crystals, such a path is only one of several which contribute to the diffusivity—see Flynn (1972). Additional possibilities exist, and one distinguishes the following types of lattice diffusion:

(a) a simple exchange mechanism,
(b) a ring-type exchange mechanism,
(c) a vacancy-induced diffusion (vacancy mechanism),
(d) a lattice relaxation mechanism,

Table 19.3: Diffusion coefficients (in cm^2/s) for solid semiconductors near their melting points (after Shaw, 1973)—also see the Appendix, Table A.4.

Intrinsic Diffusion

Material	Anion Diffusion Coefficient	Cation Diffusion Coefficient	
Silicon	9×10^{-12}		
Germanium	6×10^{-12}		Relatively slow
GaAs	2×10^{-12}	$\sim 10^{-12}$	
InSb	1.3×10^{-14}	2.3×10^{-14}	
CdTe	1×10^{-9}	1×10^{-7}	
ZnSe	$\sim 10^{-8}$	$\sim 10^{-7}$	Relatively fast
AgI	Slow	$\sim 10^{-5}$	

Impurity Diffusion Coefficients in Si and GaAs

Diffusion Coefficent		Species in Si	Species in GaAs
Very fast	$> 10^{-5}$	Cu, Li	Cu, Li
	$10^{-6} - 10^{-5}$	Na, K, Fe, He, Au	Ag, O
		Co, Ni, Zn, Cr, Mn	
	$10^{-7} - 10^{-6}$	Ni, S	
	$10^{-8} - 10^{-7}$	Ag	
	$10^{-9} - 10^{-8}$	O, C	
Slow	$10^{-11} - 10^{-9}$	Group-III Acceptors and Group-V Donors	

(e) the previously described simple interstitial diffusion, or the interstitial diffusion with

(f) collinear displacement or

(g) noncollinear displacement (interstitialcy mechanism),

(h) a dumbbell interstitial mechanism,

(i) a crowdion mechanism, and

(j) motion with alternating recharging (athermal motion).

Nine simple examples of these mechanisms are shown schematically in Fig. 19.8 for a two-dimensional square lattice—see also Gösele (1986). For a comprehensive review of some earlier concepts, see Flynn (1972).

In actual semiconductors the diffusion processes are more involved since in all of them lattice relaxation, and in several of them

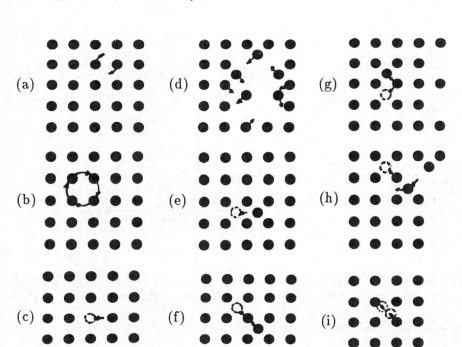

Figure 19.8: Different types of lattice diffusion for (a)–(i) given in the text.

recharging, is involved. Such recharging makes diffusion possible at much lower temperatures (see below).

In addition, there are usually several interstitial positions possible, offering different paths for interstitial diffusion—see Fig. 19.3.

19.3.2 Diffusion in Covalent Semiconductors

Self-diffusion *at high temperatures* in Si follows Arrhenius' law, with an exponent in the 5 eV range. However, the exponential carries a very large pre-exponential factor indicating a larger entropy contribution, probably due to substantial lattice relaxation, as well as a significant change in vibrational modes, due to force constant reduction. For an estimate of formation and migration energies through the different interstitial paths (Fig. 19.9), see Car et al. (1984, 1985).

Hydrogen as H• is a fast diffuser in *p*-type Si. It preferentially moves on an interstitial B-B path, meandering between the relaxing (by 0.4 Å) Si-atoms. In *n*-type Si, the neutral (or H′) hydrogen may be more stable with diffusion along a T-H path, as indicated in Fig. 19.9—see Van de Walle et al. (1989).

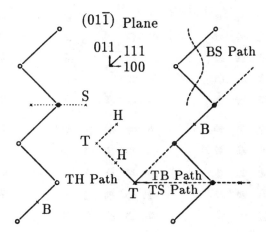

Figure 19.9: Various interstitial migration paths in Si. B: bond-centered, H: hexagonal, S: split interstitial, and T: tetrahedral position. In addition to the shown paths in (110), a path ST in [100] and TBTH in [111] directions are possible (after Car et al., 1984).

Impurities diffuse while associating with a vacancy, as interstitials, or combining with a lattice atom, *together* sharing a lattice site. A *concerted exchange mechanism*, suggested by Pandey (1986), appears less favorable for impurity diffusion. Depending on the charge character of these defects, the activation energy changes substantially when the semiconductor is *n*- or *p*-type. For instance, in Si the dopants P and Al associate with vacancies and diffuse as associates.

A recent analysis of Nichols et al. (1989), using interstitial in-jection from the surface of Si by oxidation (Tan et al., 1983), indicates that the diffusion of B, P, and As in Si is interstitial-mediated, while the diffusion of Sb seems to be vacancy-mediated. All of these have an activation energy of approximately 2.5 eV (Nichols et al., 1989).

The energy contour for interstitial boron diffusion with identification of the different interstitial sites corresponding to Fig. 19.9 is shown in Fig. 19.10.

Deep centers, especially some of the transition metal ions are known to be fast diffusers (Weber, 1983). They easily form associates with intrinsic defects mediating their diffusions; e.g., gold associates with interstitials, vacancies, or other dopants—see Stolwijk et al. (1983) and Lang et al. (1980).

19.3.2A Athermal Diffusion   When intrinsic defects in Si are produced by electron radiation at low temperatures (~ 20 K), they

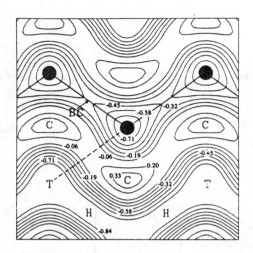

Figure 19.10: Total energy profile for a neutral B_i in Si with site labels as in Figs. 19.3 and 19.9. Energy contour lines in eV. Filled circles: normal Si position (after Nichols et al., 1989).

are observed to migrate with very low activation energies of ~ 0.2 eV (Watkins, 1986). This is in contrast to high-temperature (thermal) diffusion data, which indicates generation and motion of intrinsic defects in Si with an activation energy on the order of 3...5 eV (Frank, 1981). Such vacancies or interstitials are known to act as deep level defects. With sufficient free electrons and holes available from the preceding electron irradiation, they can be alternately recharged as they move to different sites, in turn stimulating the next jump—*Bourgoin and Corbett mechanism* (Bourgoin and Corbett, 1972).

Other alternate diffusion processes which proceed at very low temperatures may also involve dopants, e.g., aluminum (Troxell et al., 1979) or boron (Troxell and Watkins, 1980), Zn-O pairs in GaP or Fe-B pairs in Si—see Pantelides (1986) for more information.

19.3.3 Ionic Conductivity

If the diffusing atom is charged (ion), and an external field is applied, an electric current results—the ionic current. For ionic diffusion, an ionic mobility is defined by the Nernst-Einstein relation

$$\mu_i = \frac{e}{kT} D_i \qquad \text{and} \qquad \sigma_i = \frac{e^2 N_i}{kT} D_i, \qquad (19.74)$$

where σ_i is the ionic conductivity and N_i is the density of, for instance, singly charged interstitials. See Crank (1953), Shewmon

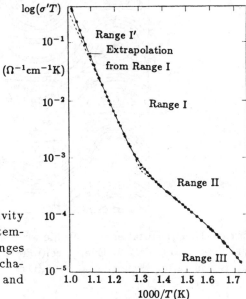

Figure 19.11: Ionic conductivity of NaCl as a function of the temperature with four distinct ranges of a different conduction mechanism identified (after Kirk and Pratt, 1967).

(1963), and Glyde (1967). The ionic drift current density is consequently given by

$$j_i = \sigma_i F = e N_i \mu_i F. \tag{19.75}$$

In typical semiconductors, the contribution of an ionic conductivity is negligibly small compared to the electronic conductivity, except for some unusual cases given in Section 19.3.3A. Figure 19.11 gives an example for NaCl with $E_g = 9$ eV and a negligible electron conductivity of $\sigma_n \simeq 10^{-4}\,\Omega^{-1}\,\text{cm}^{-1}$ at 1000 K. At temperatures up to 625 K (range III), diffusing divalent impurity-cation vacancy complexes seem to determine the current. In range II, these complexes are dissociated; in range I, intrinsic cation vacancies dominate; and closer to the melting point (range I') additional Schottky defects influence the current (Hayes and Stoneham, 1984).

19.3.3A Superionic Conductivity In a few semiconductors, often above a phase transition into a high-temperature modification, the ionic conductivity increases dramatically to a value near $1\,\Omega^{-1}\,\text{cm}^{-1}$. A typical example is AgI, shown in Fig. 19.12. Here the modification changes from hexagonal (β-AgI) to body-centered cubic (α-AgI) at $T_c = 147°$C, with a concurrent increase of the ionic conductivity by almost four orders of magnitude. At the high conduc-

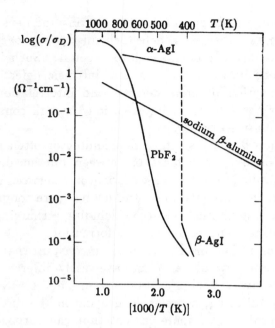

Figure 19.12: Conductivity as a function of the inverse temperature for three types of fast ion conductors (after Hayes and Stoneham, 1984). ©John Wiley & Sons, Inc.

tivity phase, the Ag ions are highly mobile in between an essentially rigid iodide sublattice (Gebhard et al., 1980).

Semiconductors with extremely large ionic conductivity are of high technical interest for fuel cell electrodes or electrolytes and as an ion monitor. Such materials include β-alumina, NiS, FeS$_2$, ZrO$_2$/Y$_2$O$_3$, Li$_3$N, PbF$_2$, and others. For a short review, see Hayes and Stoneham (1984); or for a more extensive review, see Gurevich and Ivanov-Shits (1988).

Summary and Emphasis

Intrinsic defects, and associates of these defects, are formed at elevated temperatures in thermodynamic equilibrium. The density of vacancies and interstitials increases following an Arrhenius law with an activation energy for typical semiconductors in the 2...5 eV range, and with a very large pre-exponential factor indicating substantial entropy contribution.

With decreasing temperature, one reaches a freezing-in temperature below which the annealing of these defects can no longer follow any reasonable cooling rate, leaving a residual density of these intrinsic defects which are frozen-in.

The thermodynamic approach of estimating the density of point defects can be extended to a defect chemistry approach in which *various interactions* can be taken into account. Such interactions include reactions between extrinsic and intrinsic defects, associate formation, ionization of these defects, and interrelation with free carriers. All of these force an interrelation through conservation of particles and quasi-neutrality.

Incorporation of defects into a semiconductor often necessitates diffusion from or to outer surfaces, or between different defects. Such diffusion is described by a random walk of atoms between neighboring sites. However, this diffusion may also involve more complicated site exchanges of atoms during each step, including recharging of defects during alternating steps, and associate formation.

The diffusion is measured by a diffusion constant that has matrix form in anisotropic crystals, and can have vastly different magnitude dependent on activation energies, temperature, and crystal structure. When ions are diffusing, a preferred drifting in the direction of an electric field results in an ionic current that can surpass electronic currents at elevated temperatures in wide gap materials or in some semiconductors, known as superionic conductors, with conductivities in the $1 \ \Omega^{-1} \ cm^{-1}$ range.

Incorporation of intrinsic and extrinsic point defects is a sensitive function of temperature and other thermodynamic parameters. To a large extent, they can be controlled by proper design of environment and treatment. This often requires consideration of some rather complex reactions, in which a multitude of processes can play an important role. Near room temperature, the freezing-in of various high-temperature equilibria plays a significant role for many important device parameters. In fact, almost all devices contain inhomogeneous doping distributions which must remain frozen-in during the life of the device.

Exercise Problems

1.(e) Calculate the density of Schottky defects of Si and GaAs, and draw it as a function of the temperature.

2.(e) Develop an expression for the densities of cation and anion vacancies, and of cation interstitials in an AB crystal with Schottky *and* Frenkel disorder when E_s and E_{Fr} have nearly the same value.

3.(*) Calculate the Schottky energy for producing a cation and an anion vacancy in an NaCl crystal, assuming perfect ionic bonding and deposition of the NaCl pair at a flat surface. Ignore any lattice relaxation. Search for the necessary data in the tables of this book.

4.(1) Check the recent literature for the activation energy of self-diffusion of intrinsic defects in GaAs. Give the frozen-in density of these defects in GaAs for a cooling rate of 10 deg/s. What are the freezing-in temperatures?

5. What is the driving force to produce intrinsic defects (Schottky or Frenkel) at elevated temperatures? Explain.

6.(1) How does the minimum vapor pressure point relate to non-stoichiometry in AB compounds? Draw a diagram.

7.(e) Derive Eq. (19.20) from the conditions given in Section 19.1.2.

8.(*) Give the complete set of defect chemical equations for the creation of Schottky defects in GaAs considering different formation energies for anion and cation vacancies. Assume shallow donors and acceptors for these vacancies with appropriate effective masses. Estimate the degree of nonstoichiometry at 600 K at a frozen-in interaction with the environment.

9. Give the average distance traveled for a hypothetical frozen-in atom with $\Delta E_i = 1$ eV, $a = 2$ Å, $\nu_0 = 10^{13}$ s^{-1} in a cubic crystal at 300 K after $1, 100$, and $10^6 s$.

10.(*) Give the average distance traveled for a P-atom in Si at 400 K (search for all relevant parameters in the appropriate tables in this book).

11.(e) If the concentration of a donor is given by a delta function in the center plane at x_0 of a long and narrow cylinder:

$$N(x, t = 0) = \frac{C_0}{A} \delta(x - x_0),$$

show that the density distribution of donors after diffusion for a time t has the form

$$N(x,t) = \frac{C_0}{A\sqrt{4\pi Dt}} \exp\left\{ -\frac{(x - x_0)^2}{4Dt} \right\}$$

where C_0 is a constant and A is the cross section of the cylinder.

12.(r) Nonstoichiometry is of great importance to compound semiconductors.

(a) CdS is n-type because of a preferred deviation from stoichiometry. In order to explain n-type behavior, it must be rich in which of the two elements?

(b) Such deviation can be explained by preferred interstitial or vacancy formation. Which of the two elements are involved? Why do these defects act as donors?

(c) If only vacancy formation would be involved, what arguments could be made to indicate preferred n-type behavior? Calculate the vacancy energies, using rationalized covalent radii.

(d) What is your prediction for the conductivity type of CdSe and CdTe?

13.(e) Develop the formula for the density of Frenkel defects in thermodynamic equilibrium.

14.(e) Draw the Brouwer diagram for the example given in Section 19.2.4 (doping in equilibrium with a foreign gas). Assume values for the different activation energies. Discuss the consequences deduced from the diagram.

Chapter 20

Optical Absorption at Lattice Defects

The optical absorption spectrum of lattice defects provides the most direct information about the electronic properties of these defects.

There is a large variety of lattice defects, the electronic properties of which will be discussed in Chapters 21 and 22. In preparation for this discussion, we insert here a chapter on the optical absorption spectrum which presents the most direct information on the electronic defect structure. Only a few examples are given to represent two typical classes of point defects. Attention will be given to the general optical absorption behavior, rather than to the detail. Any experimental technique identifying the defect will be discussed later, when analyzing the electronic properties of the specific defects.

Optical transitions at a lattice defect are determined by the ground and excited energy states of the defect center, their oscillator strength, and the influence of the surrounding lattice. This results in an *absorption spectrum* with lines (bound-to-bound transitions) at a certain energy, with a certain *absorption constant* (the strength of absorption) and a certain *line shape*. Each of these will be discussed in the following sections. Bound-to-free transitions resulting in an absorption edge will be discussed later in Sections 20.3.4 and 45.1.1.

20.1 Energy Position of Absorption Line

The most obvious difference between various types of defects is the energy of a specific absorption line. It allows distinction between the different defects and will be discussed in Chapters 21 and 22. However, since an understanding of the optical absorption of lattice defects requires a knowledge of the basic elements of its electronic structure in relation to the band structure of the surrounding lattice, these elements will be given here in a simplified form for the two major classes of point defects: the shallow and deep level defects.

20.1.1 Shallow Level Defects

Point defects (impurities) that result in shallow levels are described by a hydrogen-like model; that is, the defect is replaced by a proton, the extra proton of the donor and one electron, the extra electron that circles the proton at a distance of several lattice constants. Such a quasi-hydrogen donor can easily be discussed in a semiempirical form. For a more detailed analysis, see Section 21.1.

The eigenstates of this donor can be described as that of a hydrogen atom, although, with a modified field, reduced by ε_{st}, and with an electron of modified mass m_n instead of m_0. When using the Rydberg energy, which is the ionization energy of the hydrogen atom $R_H = E(n_q = 1) - E(n_q = \infty) = R_H \left(\frac{1}{1} - \frac{1}{\infty} \right) = -13.6$ eV, the eigenstates of the donor can be expressed by

$$E_{qH}^{(n)} = R_H \frac{m_n}{m_0} \left(\frac{1}{\varepsilon_{st}} \right)^2 \left(\frac{1}{n_q} \right)^2 = -13.6 \frac{m_n}{m_0} \left(\frac{1}{\varepsilon_{st}} \right)^2 \left(\frac{1}{n_q} \right)^2 \text{ (eV)}.$$
(20.1)

The corresponding quasi-Bohr radius is

$$a_{qH}^{(n)} = a_0 \varepsilon_{st} \frac{m_0}{m_n} n^2 = 0.5292 \, \varepsilon_{st} \frac{m_0}{m_n} n^2 \quad (\text{Å});$$
(20.2)

where n_q is the principal quantum number.

The use of the quasi-hydrogen model obtained from the effective mass approximation is justified when the quasi-Bohr radius is large compared to the lattice constant: the bound electron "circles" the center well within the lattice of the host, and the Coulomb forces at this distance are screened by ε_{st}. Here, the extent of the solution in k-space is small compared to π/a. Then only k-vectors near the band minimum contribute, and the use of a well-defined effective mass is justified.*

For typical semiconductors with $m_n \simeq 0.1 m_0$ and $\varepsilon_{st} \simeq 10$, we obtain a quasi-Bohr radius of $\simeq 50$ Å, which is much larger than the lattice constant, and an ionization energy of $\simeq 10$ meV, i.e., a very

---

* The quasi-hydrogen energy can be expressed as a function of the quasi-Bohr radius (for $n_q = 1$):

$$E_{qH} = \frac{\hbar^2}{2m_n} \frac{1}{a_{qH}^2} \quad \text{or} \quad E_{qH} = \frac{\hbar^2 k_r^2}{2m_n}$$
(20.3)

for $k_r = 1/a_{qH}$. This is a familiar expression identifying the circling electron in a parabolic and isotropic conduction band.

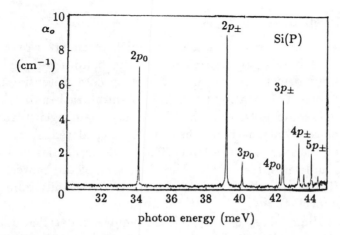

Figure 20.1: Hydrogen-like absorption spectrum from the $1s$ (A_1) ground state into the excited states of P-donors in Si at 4 K (after Jagannath and Ramdas, 1981). (See Fig. 21.6A for the corresponding level spectrum).

small fraction of the band gap. This center is indeed a shallow donor, which is intimately connected to the conduction band.

20.1.1A Optical Absorption at a Shallow Level Defect
There are three possibilities for an optical absorption at such a defect; the first two of them are:

(1) the transition of its electron into an excited state, or
(2) after its electron is removed, the excitation of an electron from the valence band into its ground or excited state.

The first possibility requires very little energy and lies in the far IR part of the spectrum as shown for an acceptor (P) in silicon in Fig. 20.1.

The second possibility lies close to the band edge:

$$h\nu_l = E_{\mathrm{qH}} \quad \text{or} \quad h\nu_h = E_g - E_{\mathrm{qH}} \qquad (20.4)$$

with the subscripts l or h for low or high energy transitions to the same center.

In addition there are

(3) transitions from a localized state of the defect center into the continuum states of the band. These transitions (see also Section 21.1.6) are edge-shaped and distinctly different from the line-type of absorption listed before. The experimental observations are discussed in Sections 20.3.4 and 45.1.1.

20.1.2 Deep Level Defects

Deep levels (subscript dd) are levels which often are closer to the center of the band gap than to one of the band edges. They are created from defect centers in which their core potential plays a dominant role. This is a tight-binding potential, rather than the far-reaching Coulomb potential. In calculating their eigenstates, both conduction and valence bands have to be considered. Examples of such centers are most of the transition metal impurities and certain vacancies and self-interstitials. The chemistry of the center, as well as its lattice surrounding, and its phonons substantially influence its electronic level energy.

The binding energy of deep centers requires a detailed quantum-mechanical analysis, which is discussed in Chapter 22. The results cannot be generalized in a simple form, as possible for shallow centers.

Since the binding energy of deep centers is usually much larger than typical phonon energies, the phonons can play a role in the related transitions. This can occur via two mechanisms:

(1) adding to, or subtracting from, the excitation energy (phonon-assisted transition, phonon replica—see Sections 44.5.2A and 45.5.2):

$$h\nu = E_{dd} \pm \hat{n}\hbar\omega \qquad (20.5)$$

for $\hat{n}$ phonons involved, a process with decreasing probability for each additional phonon emitted $(-)$ or absorbed $(+)$, or

(2) the center relaxes (by emitting phonons) when an electron is emitted into, or captured from, a band state, thereby changing the charge character of the defect state. Here, the maximum of the emission or absorption line is given by an equation similar to Eq. (20.5)

$$h\nu = E_{dd} \pm S\hbar\omega_b; \qquad (20.6)$$

except that S is the Huang-Rhys factor which gives the number of phonons involved in the center relaxation, and ω_b is a characteristic (optical) phonon frequency, typically of a breathing mode—see Section 20.1.2A.1. For alkali halides, S is on the order of 20.

In contrast to phonon-assisted transitions where phonons need to be supplied (a temperature-dependent process), here phonons are created while the center relaxes. Therefore, the shift given in Eq. (20.6) is independent of the temperature—see Fig. 20.11.

20.1.2A Defect Center Relaxation The process of defect center relaxation needs further explanation as it can substantially change
the energy of the defect level within the gap. Furthermore, energies
between the band edges and the center no longer add up to the band
gap, as they do for shallow level defects indicated in Eq. (20.4).

These processes can be qualitatively understood by recognizing
that a change in the charge of a defect causes a change in the bonding
to the neighbor atoms, and consequently a change in the relative
position of these atoms. This has an influence on the electronic
eigenstates of the defect. Such changes take much more time than,
e.g., a capture of an electron in the defect. Consequently, the defect
relaxes after capture and, with the change in atomic configuration,
the electronic defect level also changes, as the electronic eigenstate
relaxes.

A similar process takes place when an electron within the defect
is excited from the ground state into an excited state with a different
wavefunction. Consequently, the excited state requires a somewhat
larger space, causing a change in bonding to the neighboring atoms,
and resulting in a change of the atomic configuration. Another
component of this excitation process can be seen as a change from
a core configuration toward a more hydrogen-like one. This causes
polarization changes, which further result in shifts of the positions of
the surrounding atoms. Finally, with defect ionization, the electron
now belongs to the conduction band, and the defect level relaxes to
the unoccupied state.

While the recombination or optical excitation occurs instanta-
neously, shifting of the neighboring lattice atoms takes time and is
described as a relaxation of the lattice into a new equilibrium position. Since it is the relative atomic distance to the neighbors that
is changing, the configuration coordinate diagram that pictures this
relation is most helpful.

The Configuration Coordinate Diagram. We will briefly
sketch here the basic elements of the configuration coordinate diagram analysis in preparation for an understanding of the influence
of lattice relaxation on deep level defects.

The configuration coordinate representation is used to depict
the coupling between the elastic lattice energy and the electron-
lattice interaction energy. In a simple harmonic approximation of the

lattice energy, the Hamiltonian is given by

$$H_0 = \frac{1}{2} M \left(\frac{du}{dt} \right)^2 + \frac{1}{2} \beta u^2 = \frac{1}{2M} (P^2 + M^2 \omega^2 Q^2), \qquad (20.7)$$

where u is the lattice distortion and β is the restoring force constant—see Chapter 5. This is conventionally expressed in normal coordinates P (momentum $P = M du/dt$) and Q (distortion $Q = u = R - R_0$). Assuming a linear perturbation (Huang and Rhys, 1950; Lax, 1952; Pekar, 1953; O'Rourke, 1953) of the oscillation mode caused by an electron transition and resulting in a shift in the equilibrium configuration,* we have for the total Hamiltonian

$$H = H_0 - TQ, \qquad (20.8)$$

which can be transformed by a simple coordinate shift $\hat{Q} = Q - T/\beta$ (with T the electronic interaction force) into

$$H = \frac{1}{2} M \dot{\hat{Q}}^2 + \frac{1}{2} \beta \hat{Q}^2 - \frac{1}{2} \frac{T^2}{\beta}. \qquad (20.9)$$

Consequently, the energy as a function of the distortion coordinate is again given by parabolas; the elastic energy by the usual parabola with its minimum energy $E = 0$ at $Q = 0$. The total energy is given by the displaced $(\hat{E}, \hat{Q})$ parabola, with its minimum at $(E - T^2/2\beta, Q + T/\beta)$, and with

$$\hat{E} = E_{\text{relax}} = -\frac{T^2}{2\beta} \qquad (20.10)$$

the relaxation energy after capture of an electron, as shown in Fig. 20.2 by the solid parabola.

The electronic interaction force T is proportional to a coupling constant S (the Huang-Rhys factor—see below), and the spring constant β which is proportional to the square of the breathing mode

---

* The basis for this analysis is the adiabatic approximation which is described in Section 23.3. Here, the Hamiltonian is split into a part that deals with electrons and atoms at fixed lattice positions, multiplied by a function of displaced lattice atoms. The Hamiltonian can then be written as a sum of the electronic part and the interacting ionic part. This ionic part results in shifted, harmonic oscillations.

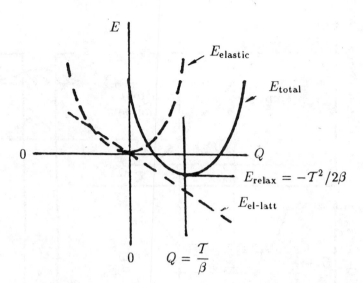

Figure 20.2: Configuration coordinate representation with elastic and electron-lattice interaction energies (dashed curve) and resulting total energy (solid curve).

frequency (close to the LO phonon frequency) ω_b^2 [see Eq. (5.6)]. Consequently, we obtain

$$\hat{Q} \propto \frac{S}{\omega_b^2} \tag{20.11}$$

In actuality, however, we must take into consideration higher order terms in the lattice distortion and electron-lattice interaction. They cause a deformation of the displaced parabola in addition to its shift.

The Huang-Rhys Factor. The electron-lattice coupling in the neighborhood of a tightly bound defect is determined by the interaction between electrons and appropriate phonons, and is given by the *Huang-Rhys factor S*. It is related to the coupling constant α_c of a free electron in an ideal lattice—see below and Section 27.1.2B. The Huang-Rhys factor can be interpreted as the number of phonons emitted while a defect center relaxes after it has captured an electron or after it is optically excited and the electron is removed (see Fig. 20.3—Huang and Rhys, 1950):

$$S = \frac{E_r - E_c}{\hbar\omega_b} = \frac{\frac{1}{2}M_r\omega_b^2(Q_2 - Q_1)^2}{\hbar\omega_b}, \tag{20.12}$$

where M_r is the reduced mass [Eq. (11.41)], E_r is the relaxation energy, $\hbar\omega_b$ is the energy of the emitted phonons, and Q_2 and Q_1 are

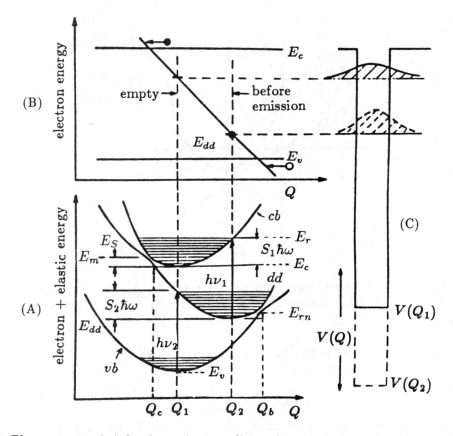

Figure 20.3: (A) Configuration coordinate diagram (schematic) of a deep level center with $E_{dd}(Q)$ minimum at Q_2 between conduction band $E_c(Q)$ and valence band $E_v(Q)$ minima at Q_1. Optical excitation of the filled deep level center takes place from its minimum at $E_{dd}(Q_2)$ to the conduction band at $E_c(Q_2)$ with $h\nu_1$, followed by (fast) relaxation with S_1 phonons to $E_c(Q_1)$, the bottom of the conduction band (see Section 20.3.3A). When the deep center is empty, the excitation occurs from $E_v(Q_1)$ to $E_{dd}(Q_1)$ with $h\nu_2$, and consequent relaxation of S_2 phonons to $E_{dd}(Q_2)$. (B) The corresponding band diagram. (C) The shape of a simplified core potential with well depths $V(Q_1)$ and $V(Q_2)$ and schematics of the corresponding envelope function (modified from Henry, 1980).

the configuration coordinates for the minimum energy in the excited and ground states, respectively—see below.

The Huang-Rhys factor can also be expressed by

$$S_i = \frac{(Q_2 - Q_1)^2}{2 \dfrac{\hbar}{M_r \omega_b}};$$

(20.13)

that is, as the ratio of the square of the total displacement divided by twice the mean square of the amplitude of the zero-point oscillation $\hbar/(M_r \omega_b)$.

Defect Level Relaxation. In the configuration coordinate diagram (Fig. 20.3A), the electronic plus elastic energy of the defect center and its nearest neighbors is plotted as a function of the distance between the defect and one representative neighbor (Q). The energy of the defect center, with an electron occupying it, is given by a curve (dd) with minimum at Q_2. When this electron is in the conduction band (cb) and the hole is in the valence band (vb), the minima are at Q_1. Optical (electronic) transitions occur from the relevant minimum without changing Q; consequent relaxation occurs with emission of S_i phonons, thus changing Q until the new minimum is reached. S_i is the corresponding Huang-Rhys factor. A typical example is explained in the caption of Fig. 20.3. More information on this subject is given in Section 20.3.3A, when the line shape is discussed.

The sum of the optical excitation transitions from the center to the conduction band, and from the valence band to the same center $h\nu_1 + h\nu_2$, can consequently exceed the band gap energy by a substantial amount:

$$h\nu_1 + h\nu_2 = E_g + S_1 \hbar \omega_b + S_2 \hbar \omega_b.$$

(20.14)

The two Huang-Rhys factors are a measure of the coupling of an electron *trapped* at the center to the lattice. This needs to be distinguished from α_c, which is the coupling factor of a *free electron* to the lattice. Typically, α_c is on the order of 0.1 for predominantly covalent compounds and is much smaller than S, which can be on the order of 10 for the same compound, depending on the type of center, i.e., electron localization, giving enough time for the ensuing polarization. The stronger the coupling, the more phonons that can be emitted in the process of relaxation.*

Recombination Center Relaxation. The process of capturing a conduction band electron into such a deep center shall be described here briefly, since it can be understood qualitatively in the configuration coordinate diagram used above to describe the excitation process. This recombination is the inverse to a *thermal excitation*, described first.

When the defect atom is thermally excited to oscillate (curve *dd*) with sufficient amplitude to cross Q_c in Figs. 20.3A and B, its electronic eigenstate can effectively mix with the excited electron states at E_m within the band. The trapped electron can then pass from the defect state to band states (Fig. 20.4A). Inversely, when it is in the conduction band it can pass from band states into the defect state. There, the electron can sequentially emit phonons until a significant fraction of band gap energy is dissipated, and it has settled into the relaxed state of this defect at Q_2—Henry (1980).

One of the important *band-to-band recombination* processes takes place via sequential capture of an electron and a hole by the same lattice defect, with relaxation for a substantial fraction of the band gap. The process needs a lattice defect that results in a very deep level with strong lattice coupling. Therefore, such a defect center is also called a *recombination center*.

An electron trapped at the *recombination center* has the ability to recombine with a hole when thermally excited in the center to E_{rn}, with a transition at Q_b, and consequent relaxation by phonon emission to E_v at Q_1.

This *elastic relaxation* must be distinguished from an *inelastic relaxation* in which the defect center moves into a *metastable* position. Such an inelastic relation is related to a photochemical reaction (Chapter 46). The inelastic relaxation may also be field-enhanced, and can be observed by a dielectric loss analysis using *Cole-Cole plot* responses (Hayes and Stoneham, 1984).

---

* This coupling is related to the depth of the electron levels. Centers with strong bonding (strong coupling) are more effective in "pushing the surrounding lattice atoms apart" when the electron is excited to a higher energy state. Eigenfunctions of shallow levels have the tendency to "slide over" the surrounding atoms when excited, by permitting the electron to circle within the surrounding lattice, thereby exerting comparatively little force on the surrounding atoms.

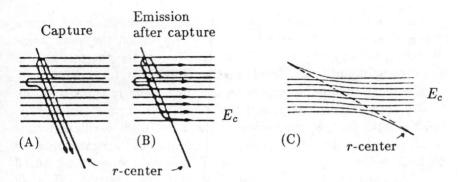

Capture

Emission
after capture

(A)

(B)

E_c

(C)

r-center

E_c

r-center

Figure 20.4: (A) Capture of an electron from band states (spacing grossly accentuated) to the recombination center. (B) Inverse process of reemission of the electron into the conduction band (after Henry, 1980). (C) Noncrossing of recombination center and band states (after Kayanuma and Fukuchi, 1984).

In the adiabatic limit the capture cross section of the recombination center can be evaluated from a semiclassical model (Kayanuma and Fukuchi, 1984), as

$$s_r = s_\infty \exp\left(-\frac{E_m}{kT}\right) \quad \text{with} \quad s_\infty = \frac{\omega_b}{2\pi} \frac{1}{N_c v_{\text{rms}}}. \quad (20.15)$$

Here s_∞ is the cross section of the center in a simple gas kinetic approach [see Section 32, Eq. (28.18)], E_m is the thermal activation energy (Fig. 20.4B), ω_b is the lattice distortion eigenfrequency (in a breathing mode at the recombination center), N_c is the density of states near the band edge, and v_{rms} is the thermal rms velocity of the electrons. This yields (Sumi, 1983):

$$s_r = 6.5 \cdot 10^{-14} \left(\frac{\hbar\omega_b}{40 \text{ meV}}\right) \left(\frac{300 \text{ K}}{T}\right)^2 \left(\frac{m_n}{m_0}\right) \exp\left(-\frac{E_m}{kT}\right) \quad (\text{cm}^2).$$
$$(20.16)$$

When is the Adiabatic Approximation Justified? The transition of the electron from a band state to an excited state of the recombination center is pictured in Fig. 20.4A in the adiabatic approximation. On its way to capture, the electron has a finite probability to be reemitted into band states (Fig. 20.4B). The net capture probability was evaluated by Henry and Lang (1977) to be 60% of the initial capture transition. In the nonadiabatic limit, reemission of trapped carriers can be neglected (Sumi, 1983).

The use of the adiabatic approximation, however, is not appropriate, when during the electronic transition within the band, some relaxation already has taken place. This lowers the thermal excitation threshold and avoids level crossing (Fig. 20.4C).

The decision whether or not to use the adiabatic approximation can be answered by comparing the natural electron lifetime ($\tau_e \simeq \hbar/\Delta E_B$, with ΔE_B as the half-band width) with the oscillatory time $\tau_L \simeq 1/\omega_b$—i.e., the time in which the potential of the center changes markedly. With $\tau_e \ll \tau_L$, the adiabatic approximation can be used since the electron during its lifetime encounters a certain deformed state of the center rather than a changing one.

The *adiabaticity parameter*

$$\gamma_a = \frac{E_S^2}{\sqrt{SkT}\,\hbar\omega_b} \tag{20.17}$$

is a more sophisticated measure of this relation, and indicates that at sufficiently high temperatures with $\gamma_a \ll 1$, the adiabatic approximation is acceptable (Sumi, 1983).

For the discussion of transitions with major center relaxation, we distinguish three closely related approximations: the *adiabatic approximation*; the *Condon approximation* (Kubo, 1952), in which the electron-lattice interaction is linearized; and the *static approach* (Markham, 1955), in which the electronic wavefunctions are independent of the defect coordinates, while the lattice wavefunction depends on the electronic state.

A listing of earlier literature and a critical discussion of the different approximations is given by Peuker et al. (1982). For further development, see Sumi (1983) and Kayanuma and Fukuchi (1984).

There are other processes which can change the energy of the center, such as Jahn-Teller distortion and the influence of mechanical, electric, and magnetic fields. All of these will be discussed in the appropriate sections of Chapter 22.

20.1.3 Shallow and Deep Centers

In summary, of these two types of defects (shallow and deep ones), the shallow electronic eigenstates are significantly influenced by the lattice environment of the defect. This environment is expressed in terms of the effective mass and dielectric constant. The deep eigenstates are influenced by the chemistry of the defect, the lattice, expressed by its valence and conduction band, and the lattice coupling, i.e., by phonon interaction.

The resulting optical excitation spectrum would be hopelessly complicated were it not for that often—for various reasons—only a few lines appear in the spectrum. Two such reasons are:

(1) the lines can have vastly different strengths, causing only the strongest lines to be recognized, while most of the weaker lines disappear in a broad and often unstructured baseline absorption; and

(2) many of the lines from excited states of a large variety of defects lie almost on top of each other. These lines are caused by quasi-hydrogen states, which only depend on the host lattice via ε_{st} and m_n and not on the chemistry, i.e., the core potential of the defect (Grimmeiss, 1986).

20.2 Strength of Optical Absorption Lines

A typical absorption spectrum of lattice defects shows a variety of lines with different shapes and amplitudes. The amplitude or *strength of the absorption* is measured by the absorption constant α_o, which is given by the optical cross section s_o for the specific transition and the density of the centers:

$$\alpha_o = s_o N_d \quad \text{or} \quad \alpha_o = s_o J_{fi}. \tag{20.18}$$

Here N_d is the density of the defect states, and J_{fi} is the joint density of states for an excitation involving bands, e.g., for a transition from a defect level into the band. The cross section for an absorption of a photon by the defect is in turn defined by the ratio of $h\nu$ times the optical transition rate [see Eq. (13.4)]

$$r_{fi} = \frac{1}{\pi h} \left(\frac{e A_0}{m_0} \right)^2 \left| \mathbf{e} \cdot \mathbf{M}_{fi,\text{qH}}^{(nq)} \right|^2 \delta(E_f - E_i - h\nu_{fi}) \tag{20.19}$$

to the energy flux W of the incoming optical radiation with vector potential $\mathbf{A} = A_0 \mathbf{e} \exp\{i(\mathbf{k} \cdot \mathbf{r} - 2\pi\nu t)\}$ (see Section 13.1.1)

$$W = 2\pi\nu^2 A_0^2 n_r \varepsilon_0 c, \tag{20.20}$$

which results in an optical cross section of

$$s_o = h\nu \frac{r_{fi}}{W} = \frac{e^2}{2\pi^2 m_0^2 c n_r \varepsilon_0 \nu} \left| \mathbf{e} \cdot \mathbf{M}_{fi,\text{qH}}^{(nq)} \right|^2 \delta(E_f - E_i - h\nu_{fi}). \tag{20.21}$$

This optical cross section is not very useful for practical evaluation since it assumes an ideal transition represented by a δ-function. Therefore, we need to be more specific as to an actual transition, which involves damping that causes a finite line width and strength; both of them are intimately coupled. We replace the δ-function by a more realistic line-shape function, which will be discussed in the next section:

$$g(\nu - \nu_o) = \frac{\gamma}{2\pi} \frac{1}{(h\nu - |E_f - E_i|)^2 + \dfrac{\gamma}{2}} \qquad (20.22)$$

where γ is the damping constant ($= 1/\tau$ with τ the lifetime in the excited state), yielding for the optical cross-section

$$s_o = \frac{e^2}{\pi^3 m_0^2 c n_r \varepsilon_0 \nu} \left| \mathbf{e} \cdot \mathbf{M}_{fi}^{(nq)} \right|^2 \frac{\gamma}{(h\nu - |E_f - E_i|)^2 + \dfrac{\gamma}{2}}. \qquad (20.23)$$

Another specific assumption needs to be made to evaluate the momentum matrix element—see Dexter (1958). For instance, for a transition between bound states of a hydrogen-like defect,* we have

$$\mathbf{M}_{fi} = \sum_{\mathbf{k,k'}} \int_{\mathcal{V}} \Psi_{nq}^*(\mathbf{k'}) \left[\mathbf{e} \cdot (-i\hbar\nabla) \right] \Psi_{nq}(\mathbf{k}) \, d\mathbf{r}; \qquad (20.25)$$

the impurity function can be expressed as an expansion in Bloch electron eigenfunctions

$$\Psi(\mathbf{k}) = \sum_{\mathbf{k}} c_c(\mathbf{k}) \psi_c(\mathbf{k}, \mathbf{r}) = F(\mathbf{r}) \psi(\mathbf{k_0}, \mathbf{r}); \qquad (20.26)$$

with $F(\mathbf{r})$ as the envelope function (see Appendix A.4.3A). This permits a substantial simplification: since the Bloch function is always the same, the matrix element changes only with changes in the envelope function. Therefore, we can discuss the strength of the absorption lines from an analysis of the envelope function only, and from corresponding selection rules. Equation (20.25) can then

---

* Here, the field to influence the transition of the spread-out eigenfunction is modified by the dielectric constant. Hence, Eqs. (20.19)–(20.21) need to be multiplied by the ratio of the *modified field*

$$\frac{F_{\text{eff}}}{F_0} = 1 + \frac{n_r^2 - 1}{3} + \ldots \qquad (20.24)$$

with terms beyond the Lorentz local field ratio (disregarded here) accounting for higher multipole interaction and other exchange effects. The effective field effect can be ignored for deep levels of centers which are localized between nearest neighbor atoms.

be simplified by separating the matrix element of the unperturbed crystal:

$$\mathbf{M}_{fi} = \int_{\mathcal{V}} F_f(\mathbf{r})\psi^*(\mathbf{k}_0, \mathbf{r}) p_A F_i(\mathbf{r})\psi(\mathbf{k}_0, \mathbf{r})\, d\mathbf{r}$$

$$= \int_{\mathcal{V}} |\psi(\mathbf{k}_0, \mathbf{r})|^2 F_f(\mathbf{r}) p_A F_i(\mathbf{r})\, d\mathbf{r} \tag{20.27}$$

with $p_A = \mathbf{A} \cdot (-i\hbar\nabla)/A_0$. From

$$F(\mathbf{r}) = \sum_k c_c(\mathbf{k}) u(\mathbf{k} = 0, \mathbf{r}) \exp(i\mathbf{k}\cdot\mathbf{r}) \tag{20.28}$$

and with the coefficients

$$c_c(k) = \frac{8\sqrt{\pi}}{\sqrt{\mathcal{V}} a_{\mathrm{qH}}^{5/2} \left\{ k^2 + \left(\dfrac{1}{a_{\mathrm{qH}}}\right)^2 \right\}^2}, \tag{20.29}$$

where a_{qH} is the quasi-hydrogen radius. We obtain the matrix element of the optical transition at the defect center by multiplying the matrix element for ideal lattice transition with the coefficient $c_c(k)$ (Eagles, 1960; Zeiger, 1964):

$$|\mathbf{M}_{fi,\mathrm{qH}}^{(nq)}|^2 = |\mathbf{M}_{fi}|^2 \{c_c(k)\}^2 . \tag{20.30}$$

Hence, we observe an enhanced absorption that extends up to k-values on the order of $1/a_{\mathrm{qH}}$ [see Eq. (20.29)]. The optical absorption caused by such a transition within a hydrogen-like donor has been discussed by Callaway (1963) and Zeiger (1964).

The transition is controlled by selection rules, i.e., by symmetry considerations involving the defect and its surrounding lattice—see Chapter 18.

20.2.1 The Oscillator Strength

Another way of looking at an optical absorption is to start from the free center (an isolated defect atom), which interacts with electromagnetic radiation through its electric ($\mathbf{R} = e\sum_i \mathbf{r}_i$) or magnetic ($\mathbf{J} = \sum_i \mathbf{l}_i + 2\mathbf{s}_i$) dipole moment for i electrons in the center.

The matrix elements of these dipole operators relate to the corresponding momentum matrix elements as:

$$|\mathbf{R}_{fi}| = \int \psi_f \mathbf{R}\psi_i \, d\mathcal{V} = 2\pi m(\nu_f - \nu_i)\int \psi_f \mathbf{P}\psi_i \, d\mathcal{V}. \tag{20.31}$$

For centers with inversion symmetry, the matrix elements are $\neq 0$ only for transitions between states of uneven parity, e.g., from s- to p-states. Even-parity transitions are electric dipole-forbidden, but magnetic dipole-allowed. These *selection rules* are determined by the symmetry of the center, which can be easily given for an isolated center. The *oscillator strength* of such an isolated center is given by

$$f_{fi} = \frac{4\pi m(\nu_f - \nu_i)}{3h} |\mathbf{R}_{fi}|^2 = \frac{1}{3\pi mh(\nu_f - \nu_i)} |\mathbf{P}_{fi}|^2, \qquad (20.32)$$

and the optical cross section can be expressed as a function of the oscillator strength:

$$s_o = \frac{\pi e^2}{mc} f_{fi} \frac{\gamma}{(h\nu - |E_f - E_i|)^2 + \dfrac{\gamma}{2}} \quad \text{with} \quad \gamma = \frac{1}{\tau}. \qquad (20.33)$$

When embedding this isolated center into a lattice, the velocity of light is reduced to c/n_r and the effective field is increased by the lattice polarization (Lorentz-Lorentz) to $F_{\text{eff}} = F_0(n_r^2 + 2)/3$. When the density of these centers (N) is small enough to neglect their interaction, the total absorption α_o relates to N times f_{fi} (Smakula, 1930) as

$$N f_{fi} = \alpha_o n_r \gamma \left(\frac{F_o}{F_{\text{eff}}}\right)^2 C^* \qquad (20.34)$$

where C^* is a constant, and $n_r(F_o/F_{\text{eff}})^2$ is typically on the order of 1. This gives a very simple formula to estimate the density of defect centers from the product of absorption constant α_o^{max} at the line maximum and line width ΔE, related to γ—see Section 20.3.1A:

$$f_{fi}N \simeq 1.3 \cdot 10^{17} \alpha_o^{\text{max}} \, (\text{cm}^{-1}) \Delta E \, (\text{eV}) \qquad (20.35)$$

For known defect center densities, the oscillator strength can be determined. For deep centers, f_{fi} is on the order of 1; for F-centers in alkali-halides, $f_{fi} \sim 0.5$.

20.3 Line Shape of Electronic Defect States

The optical absorption of defect states by typical resonance absorption has a Lorentzian line shape, with a line width depending on damping. Other composed lattice resonances have Gaussian shape and are discussed in Sections 20.3.2C and 20.3.2D. Still other line shapes are related to the specific excitation and deexcitation mechanisms involving phonons, and are discussed in Section 20.3.3.

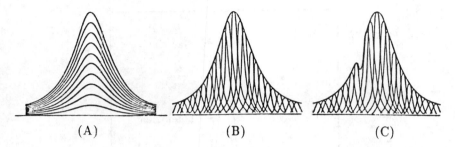

$$(A) \qquad\qquad (B) \qquad\qquad (C)$$

Figure 20.5: Schematic representation of line widths: (A) homogeneously broadened line from more and more identical centers; (B) inhomogeneously broadened line from superposition of different homogeneous components; and (C) removal of one of the components by photochemical changes (hole burning) or bleaching (after Stoneham, 1969).

20.3.1 Homogeneous Lines

All lines created by the superposition of identical contributions from any one of the involved centers are called *homogeneous lines*. Such a superposition is shown in Fig. 20.5A. Typical examples are listed in the following two sections.

20.3.1A Lifetime Broadening   The line shape can be caused by a finite lifetime of the electron in the excited state. It can be described by the damping term in the resonance equation responsible for the absorption, which is of *Lorentzian type*—see Section 11.2 and Eq. (11.34). It is given by the line-shape function*

$$g(\nu - \nu_0) = \frac{\gamma}{2\pi} \frac{1}{(\nu - \nu_0)^2 + \left(\dfrac{\gamma}{2}\right)^2}. \qquad (20.36)$$

It yields a line width at half of its maximum strength:

$$\Delta_{1/2} = \gamma. \qquad (20.37)$$

There is another typical line shape, the *Gaussian line shape*, which is observed for deep center relaxation—see Section 20.3.3. However, it is often related to inhomogeneous broadening (see Sec-

---

* The line shape function enters into the absorption cross section [Eq. (20.21)]; that is, $(1/h)g(\nu - \nu_{fi})$ replaces $\delta(E_f - E_i - h\nu_{fi})$ after introducing the damping term—see also Eq. (20.22).

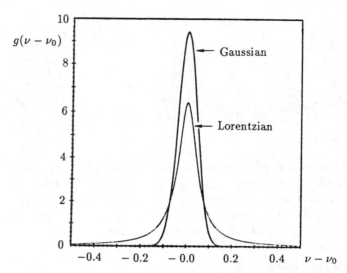

Figure 20.6: Line shape function according to Eqs. (20.42) and (20.44) with $\nu_0 = 10^{12}$ s^{-1} and $\gamma = 1$ s^{-1}.

tion 20.3.2). The line-shape function for a Gaussian line is given by

$$g(\nu - \nu_0) = \frac{1}{\sqrt{2\pi}\gamma} \exp\left[-\left(\frac{\nu - \nu_0}{2\gamma}\right)^2\right] \qquad (20.38)$$

with a line width at half of its maximum strength

$$\Delta_{1/2} = 2\sqrt{2\ln 2}\,\gamma = 2.355\gamma. \qquad (20.39)$$

Both line shapes are distinguished by their tails (the Gaussian line shows a steeper decline) as shown in Fig. 20.6.

The damping factor in Eqs. (20.36) and (20.38) can be related to the lifetime of the excited state, and in simple cases is given by $\gamma \simeq 1/\tau$.

In defect centers, the intrinsic line width is determined by the natural lifetime in the excited state. This lifetime in turn can be determined by

- recombination, or
- nonradiative bound-carrier-phonon interaction.

Both contributions depend on the actual type of defect center. For all practical reasons, the first contribution is usually small compared to the phonon interaction.

In shallow centers and at low temperatures (< 20 K) the emission of acoustic phonons is the limiting process. For instance, a transition from $2p_0$ to $1s$ of a quasi-hydrogen defect yields for such process a line width of

$$\Delta\nu \simeq 10^{-3} \frac{\Xi^2}{\rho a_{qH}^3 v_s^3} \nu_{fi}, \qquad (20.40)$$

where ρ is the density of the semiconductor, Ξ is the deformation potential, a_{qH} is the quasi-hydrogen radius, and v_s is the sound velocity. For a donor in Ge this line width $h\Delta\nu$ is estimated to be on the order of 10 μeV. In stress-free ultrapure Ge with (H,O) donors, it was observed (Haller et al., 1986) to be approximately 8.5 μeV.

In contrast to the homogeneously broadened line, which is rarely seen,* we usually observe inhomogeneously broadened lines when slightly shifted narrow lines are superimposed. This is shown in Fig. 20.5B, and will be discussed in Section 20.3.2.

20.3.1B Influence of the Magnetic Field The optical absorption of defect centers depends on a number of external influences. Most of these cause inhomogeneous broadening and will be discussed later. The magnetic field, however, can have an influence on the homogeneous line width. A high magnetic inductance compresses the defect eigenfunctions (see Section 21.3.4), causing a reduction in the scattering cross section and thereby an increase in the lifetime of hydrogen-like defects. This causes a *reduction* in the homogeneous line width.

Such a line narrowing is shown in Fig. 20.7 for InSb, which has a low effective mass, resulting in extremely large quasi-hydrogen radii. Consequently, this results in a large magnetic compression at only a moderate magnetic inductance. For more detail, see Stradling (1984).

20.3.2 Inhomogeneous Broadening

Most line broadening is due to the superposition of emissions from different centers of the same impurity atom, which differ from each other because of slightly variant lattice environments. Such environments may differ by built-in strain or electric fields, or by proximity to other defects. *Inhomogeneous line broadening* (Stoneham, 1969) is shown in Fig. 20.5B, and can be detected by bleaching one sub-

---

* Only in ultrapure semiconductors with vanishing internal stress and vanishing electric fields, the lines of isolated impurities have their natural width—see Jagannath et al. (1981) and Haller et al. (1986).

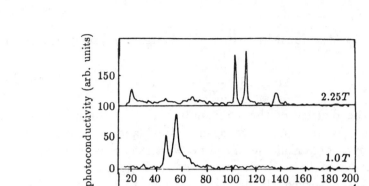

Figure 20.7: (1) Cyclotron resonance and (2) $1s \to 2p^+$ transition of a hydrogen-like impurity in InSb at 4.2 K, showing a shift and substantial narrowing of the lines with increased magnetic inductance from 10 to 22.5 kG (after Stradling, 1984).

group of these centers with a narrow laser line (see subfigure C and Section 20.3.2A), which can be detected with a consequent probing scan. This bleaching of a narrow line is often referred to as *hole burning*.

20.3.2A Hole Burning   The width of the hole, burned with a narrow line-width laser, depends on the transition probability of the defect center in a statistically arranged surrounding. With a short exposure, followed immediately by probing (echo), only such centers with highest excitation probability are bleached. This results in a narrow line width of the bleached "hole." When given a longer bleaching time, statistical fluctuations permit a wider distribution to be bleached, making the line width of the "hole" wider.

20.3.2B Electric Field Broadening   An electric field causes a *Stark effect* shift and splitting of the electronic eigenstates of the defect—see Section 21.3.3A. For hydrogen-like defects with large orbitals, this effect is rather large. It can be estimated by including the external field into the Hamiltonian of the effective mass equation (Appendix A.4.3A):

$$\left(-\frac{\hbar^2}{2m^*} \nabla^2 - \frac{e^2}{4\pi\varepsilon_{st}\varepsilon_0 r} - eV_{ext}(r) \right) \psi_e = E\psi_e. \qquad (20.41)$$

The eigenvalues for Eq. (20.41) in a uniform electric field
$(\rightarrow V_{\text{ext}} = [\mathbf{F}/e] \cdot \mathbf{r})$ are

$$E_{\text{qH}}(F) = E_{\text{qH}}^{(0)} \left\{ -\frac{1}{n_q^2} + \frac{3}{2} \frac{F a_{\text{qH}}^{(0)}}{E_{\text{qH}}^{(0)}} n_q (n_1 - n_2) - \left(\frac{F a_{\text{qH}}^{(0)}}{2 E_{\text{qH}}^{(0)}} \right)^2 \frac{n_q^4}{8} \times \right.$$

$$\left. \times \left[17 n_q^2 - 3 (n_1 - n_2) - 9 m^2 + 19 \right] + \dots \right\},$$

$$(20.42)$$

with the first term representing the undisturbed quasi-hydrogen solution, and the second and third terms representing the linear and
quadratic Stark effects, respectively. The superscript (0) identifies
the ground state for quasi-hydrogen solutions, n_1 and n_2 are two
integers (≥ 0) fulfilling $n_q = n_1 + n_2 + |m| + 1$, and m is the or-
bital quantum number—here the component in $\mathbf{F}$-direction. Such
an external (to the lattice defect) field can also be produced by
other charged centers in sufficiently close proximity. Then, $V_{\text{ext}}(r)$
is the potential distribution modified through such charged centers
$(= \sum \{ e_i / (4 \pi \varepsilon_{\text{st}} \varepsilon_0 | \mathbf{R}_i - \mathbf{r} |) \})$.

The statistical distribution of defects throughout the lattice causes a variation of this effective field from defect to defect. Consequently, it creates an accumulated broadening of the line, instead of
a well-defined shift and splitting of the hydrogen-like spectrum.

The probability of finding an electric field F^* is given (Larsen,
1976) by

$$P(F^*) = \frac{2}{\pi} F^* \int_0^\infty r \exp \left(\frac{4}{15} (2 \pi r)^{3/2} \right) \sin(F^* r) \, dr. \qquad (20.43)$$

Here F^* is expressed by the once-integrated Poisson equation $(\varrho \bar{r} / (\varepsilon_{\text{st}} \varepsilon_0)$.
After introducing the average distance between defects $(\bar{r} = N_i^{-1/3})$,
we obtain

$$F^* = \frac{e N_i^{2/3}}{\varepsilon_{\text{st}} \varepsilon_0} \qquad (20.44)$$

where N_i is the total density of charged centers. This probability
function [Eq. (20.43)] is given in Fig. 20.8A, and shows a maximum
at $5 e N_i^{2/3} / (\varepsilon_0 \varepsilon_{\text{st}}) \simeq 420$ V/cm for $N_i = 10^{13}$ cm^{-3} and $\varepsilon_{\text{st}} = 10$.

The strong broadening of higher excited states is shown in
Fig. 20.8B for the $1s \rightarrow 3p$ transition, peaking near 5.3 meV. Their
larger eigenfunctions are more prone to overlap into adjacent regions
of higher fields from neighboring Coulomb centers when compared to

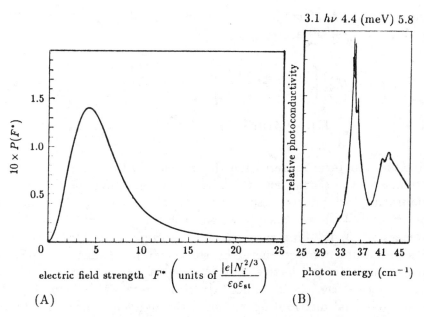

Figure 20.8: (A) Probability distribution for finding an average field F^* produced by a random distribution of point charges of density N_i at a given lattice point (Holtzmark distribution). (B) Photoconductivity distribution of high purity GaAs at 1.57 K (instrument resolution 8.6 μeV), showing substantial broadening of the ~ 5.3 meV peak due to electric fields from neighboring Coulomb centers (after Larsen, 1976).

the narrower $1s \rightarrow 2p$ near 4.3 meV. For further detail, see Larsen (1976).

20.3.2C Mechanical Stress Broadening   Mechanical stress due to dislocations or other statistically distributed defects surrounding a specific type of lattice defect contributes to the line broadening of this defect in a manner similar to electric field broadening. Such stresses, however, have a stronger influence on the structure of the conduction band. Hence, they will preferably broaden the donor states created by mixing with such band states—see Section 21.3.2. However, other states may be influenced also, as, e.g., nondonor-like states can be perturbed by a large *deformation potential*.

20.3.2D Line Broadening with Compositional Disorder
The line widths of hydrogenic defects (excitons and impurities) are substantially broader in semiconductor alloys than in their unalloyed components. As an example, an exciton line width is analyzed here.

Figure 20.9: Line width broad-
ening of the A exciton in CdS$_{1-\xi}$Se$_\xi$
due to compositional disorder.
The curve is calculated from Eq.
(20.49) with $\Delta E_g = 0.728\,\text{eV}$, $b = 0.31\,\text{eV}$, $a_{\text{qH}} = 25\,\text{Å}$, $c = 1.1$ (after
Goede et al., 1978).

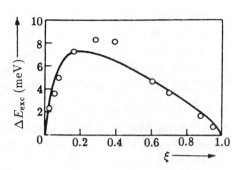

The compositional disorder in random alloys causes a local variation of the exciton energy. Depending on the actual composition of
the alloy within the volume of the exciton,

$$\mathcal{V}_{\text{exc}} = \frac{4\pi}{3} \langle r_{\text{exc}}^3 \rangle = 10\pi a_{\text{qH}}^3, \qquad (20.45)$$

one observes a variation of ε_{st} and m_n, which in turn causes a change
in a_{qH} and E_{qH} (Goede et al., 1978).

The probability of finding n atoms A of the alloy $A_{\bar\xi}B_{1-\bar\xi}C$ in
the volume of an exciton is given by

$$P(n, N_{\text{exc}}) = \binom{N_{\text{exc}}}{n} \bar\xi^n \left(1 - \bar\xi\right)^{(N_{\text{exc}}-n)} \qquad (20.46)$$

with N_{exc} as the number of cation sites in the exciton volume $\mathcal{V}_{\text{exc}}$,
and $\xi = n/N_{\text{exc}}$. For sufficiently large N_{exc}, one can approximate
Eq. (20.46) with a Gaussian distribution:

$$P(\xi) = \frac{1}{\sqrt{\pi}\, N_{\text{exc}} \sigma} \exp\left[-\frac{(\xi - \bar\xi)^2}{\sigma^2} \right], \qquad (20.47)$$

which yields a width at half of its maximum value:

$$\Delta_{1/2} = 2\sqrt{\ln 2}\, \sigma = 2\sqrt{2\ln 2} \sqrt{\frac{\bar\xi(1 - \bar\xi)}{N_{\text{exc}}}}. \qquad (20.48)$$

This results in a half-width of the exciton line

$$\Delta_{1/2}^{\text{exc}} = 2c\sqrt{\ln 2}\, \sigma \frac{\partial E_g}{\partial \xi}$$

$$= 2c\sqrt{2\ln 2} \left\{ \Delta E_g - b\left(2\bar\xi - 1\right) \right\} \sqrt{\frac{\bar\xi\left(1 - \bar\xi\right)}{10\pi\sqrt{2}}} \left(\frac{a}{a_{\text{qH}}}\right)^3,$$

$$(20.49)$$

where $\sqrt{2}\,a^3$ is the volume of the primitive cell, c is a fitting parameter on the order of 1 (here $\simeq 1.1$), and $\Delta E_g = E_g(\xi = 0) - E_g(\xi = 1)$.

The experimentally observed line broadening of exciton luminescence agrees reasonably well with the given estimates, as shown in Fig. 20.9 (see also Singh and Bajaj, 1986).

Line Broadening with Heavy Doping.   A very similar line broadening due to the statistical fluctuation of interdefect distances is observed with heavy doping. In a simplified approach, the line width can be estimated as

$$\Delta = \Delta^{(0)} \exp\left(-\frac{\bar{r}}{a_{\mathrm{qH}}}\right), \qquad (20.50)$$

with a_{qH} as the quasi-Bohr radius of the hydrogen-like defect, and $\bar{r} = N_d^{-1/3}$ as the mean distance between the defects. $\Delta^{(0)}$ is the normalized line width (Mott, 1974).

20.3.3 Deep Level Defect Line Broadening

Deep level defect centers can show a much larger line broadening when the coupling with lattice phonons is strong. The spectrum of some of these deep level defects shows, even at low temperatures, a broad Gaussian line shape of typically several $\hbar\omega_{\mathrm{LO}}$ half-widths. This signature is an indication of a strong coupling of the defect center with the surrounding lattice.

Strong lattice coupling can yield line widths of several hundred meV. In contrast, narrow line widths of the zero phonon absorption of $\lesssim 1$ meV, with well-separated phonon replica are observed for other centers with only minor lattice coupling.

The magnitude of the lattice coupling is a function of

(1) the host lattice: the coupling is increased with higher ionicity of the lattice binding forces and lower elastic force constant;
(2) the defect center: point defects show a stronger coupling than defect associates; and
(3) the electronic eigenstates of the center: deep defect levels show a stronger coupling than shallow states.

In a simplified picture, we may observe a deep defect center with strong coupling that tends to push the surrounding lattice atoms apart when an electronic excitation occurs, rather than to extend a hydrogen-like orbit over more of the surrounding lattice. The changes in hydrogen-like orbits have little effect on the position and polarization of the neighboring atoms, while the deformation of the lattice surrounding of an expanding, tightly bound center has a larger effect. In strong interaction with the surrounding lattice, many phonons are created by the relaxation that follows such an electronic excitation, and have a major effect on its excitation energy and line width. The basic principles involved are discussed in the following section.

20.3.3A Franck-Condon Principle, Lattice Relaxation The processes involved during excitation of a defect center with strong lattice coupling are described in a configuration coordinate diagram as introduced in Section 20.1.2A.1. However, the configuration co-ordinate diagram is applied here to the ground and excited states of a deep center (Fig. 20.10). It pictures the electronic *and* elastic lattice energies as a function of the relative distance between the defect center and its nearest surrounding lattice atoms. The center's ground state is identified by A_0, within a parabola $E(\mathbf{Q})$ near its minimum. Horizontal lines indicate specific *phonon-excited* states. With *electronic excitation*, the parabola is displaced. It is shifted in energy by the electronic excitation energy and in interatomic distance to a larger equilibrium distance Q_{B0}; the excited defect atom has a increased effective radius—Fig. 20.10B.

In Figure 20.10A the vibrational eigenfunction $\phi(Q)$ is indicated from the adiabatic approximation [Chapter 8, Eqs. (8.1) and (8.2)]. It can be shown (see Section 20.3.3C—Iadonisi, 1984) that in addition to the normal selection rules for the electronic part, the optical transition probability depends on the vibrational part of the eigenfunctions [see Eq. (20.59)] and is largest when the overlap integral

$$\int \phi_f^*(Q)\phi_i(Q)\, dQ \tag{20.51}$$

is a maximum. This is the case for an approximately vertical transition (*Franck-Condon Principle*—Fig. 20.10A). For the original discussion of the Frank-Condon principle, see Condon and Morse (1929).

With changes from this optimum transition energy, the transition probability drops off gradually, resulting in a rather broad line width (Fig. 20.10B).

After optical excitation, the vibrational state is also excited and will relax to its minimum energy—B_0 in Fig. 20.10B. The difference between the initial excited state and the relaxed state represents the *Franck-Condon shift* (Lax, 1952) between optical and thermal

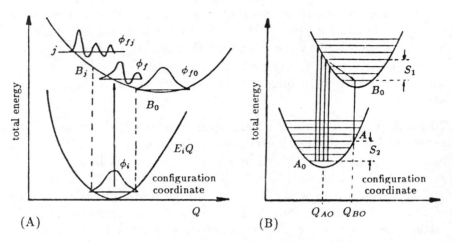

Figure 20.10: Configuration coordinate diagram showing the ground and excited states of a defect with strong lattice coupling and typical excitation and recombination transitions. (A) Ground and excited vibrational states indicated. (B) Franck-Condon principle showing larger optical energy than needed for thermal excitation from A_0 to B_0, and indicating absorption line shape ($A_0 \rightarrow B$) with multiple transitions possible. Horizontal lines indicate vibrational states of the system.

excitation. The latter, as an indirect transition, can proceed directly from minimum to minimum, and thus needs less energy than the optical excitation. In a similar fashion, light emission (luminescence) proceeds vertically from the relaxed excited state to the ground electronic state (A in Fig. 20.10B), and then relaxes to its minimum total energy state A_0.

The shift between optical excitation ($A_0 \rightarrow B$) and emission ($B_0 \rightarrow A$), *the Franck-Condon transitions*, is called the *Stokes shift*.

20.3.3B Strong, Medium, and Weak Coupling The strength of the coupling can be measured by the number of phonons emitted during a relaxation process. This is represented by S_1 after optical excitation, and S_2 after a photon emission (Fig. 20.10B), with S_i as the Huang-Rhys factor—see below and Section 20.1.2A.2.

The total energy of the defect is given by

$$E_e = E_{e0} + E_Q = E_{e0} + \frac{1}{2}\beta(Q - Q_0)^2, \qquad (20.52)$$

where E_{e0} is the electronic ground-state energy, E_Q is the elastic energy, and β is the force constant. The relaxation energy in state B, following a transition from state A_0, is expressed by (see Fig. 20.10B)

$$E_Q = \frac{1}{2}\beta(Q_{A0} - Q_{B0})^2 = \frac{1}{2}M_r\omega_b^2(Q_{A0} - Q_{B0})^2, \qquad (20.53)$$

where $\omega_b^2 = \beta/M_r$ and M_r is the reduced mass of the center. The Huang-Rhys factor is given by Eq. (20.12).

When $S > 6$, we speak of *strong coupling*; when $S \ll 1$, we have *weak coupling*. The line-shape is substantially different in each case.

Strong Coupling.   For strong coupling, the line shape function is given by a *Poisson distribution* of the phonon spikes at low temperatures

$$g(\nu - \nu_0) = \sum_\Delta \left\{ \frac{S_1^\Delta}{\Delta!} \exp\left(-S_1\right) \right\} \delta\left(\nu - \nu_0 - \frac{\omega_b}{2\pi}\Delta\right), \qquad (20.54)$$

with Δ as the number of phonons emitted at the specific transition, and the delta function presenting the zero-phonon line (for $\Delta = 0$) as well as the phonon replica. This results in a spectrum of spikes separated by $\hbar\omega_b$. In actuality, each of these phonon lines is broadened, and all replica melt into one broad feature for high values of the Huang-Rhys factor.

At higher temperatures ($kT > \hbar\omega_b$), the line-shape function converts to a *Gaussian distribution*

$$g(\nu - \nu_0) = \frac{1}{\sqrt{4\pi S_1 \hbar\omega_b kT}} \exp\left[-\frac{(h\nu - h\nu_0 - S_1\hbar\omega_b)^2}{4S_1\hbar\omega_b kT} \right] \qquad (20.55)$$

where $h\nu_0$ is the optical electronic transition $(A_0 \rightarrow B)$ in Fig. 20.10B and ω_b is the frequency of the characteristic phonon mode. Individual phonon lines are not resolved but contribute to a broad line of width

$$\Delta_{1/2} = \begin{cases} 2\sqrt{2\ln 2}\sqrt{S_1}\hbar\omega_b & \text{for } kT < \hbar\omega_b \\ 2\sqrt{2\ln 2}\sqrt{S_1\hbar\omega_b kT} & \text{for } kT > \hbar\omega_b. \end{cases} \qquad (20.56)$$

It is the strong coupling that permits interaction with a wide phonon spectrum. It broadens each individual line, so that even at very low temperatures only one broad feature appears in absorption and emission. This is shown in absorption and, shifted by approximately twice the relaxation energy, in emission for the F-centers of KBr in Fig. 20.11.

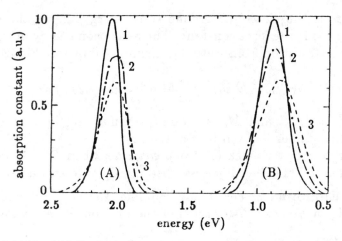

Figure 20.11: (A) Optical absorption and (B) luminescent emission spectra of F-centers in KBr for $T = 20$, 78, and 150 (185) K for curves 1–3, respectively (after Gebhardt and Kuhnert, 1964).

Weak Coupling.   For weak coupling the parabola for the excited electron state (B in Fig. 20.10) is not shifted significantly from the ground-state parabola. This causes the most probable transition to be the minimum-to-minimum zero-phonon line with a narrow line width.

In addition, Stokes-shifted lines with phonon emission and anti-Stokes lines with photon absorption occur as *phonon replica*. The line-shape function here is represented by

$$g(\nu - \nu_0) = (1 - S)\delta(\nu - \nu_0)$$
$$+ \frac{1}{2}(S + S_i)\delta\left(\nu - \nu_0 - \frac{\omega_b}{2\pi}\right)$$
$$+ \frac{1}{2}(S - S_i)\delta\left(\nu - \nu_0 + \frac{\omega_b}{2\pi}\right)$$
$$+ \text{higher replica}$$

$$(20.57)$$

where each δ-function is lifetime-broadened, as discussed in Section 20.3.1. S_i is the Huang-Rhys factor given by Eq. (20.13), and

$$S = S_i \coth\left(\frac{\hbar\omega_b}{kT}\right). \qquad (20.58)$$

For weak coupling, these replica are obtained from Eq. (20.57) as a series of spikes.

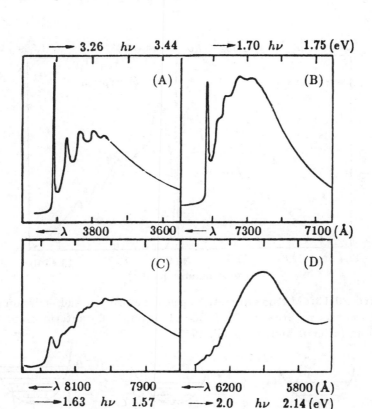

Figure 20.12: Optical absorption spectra of R_2-centers at 4 K for LiF, KCl, KBr, and NaCl in subfigures (A)–(D), respectively, showing decreasing structure with increasing lattice coupling (after Fitchen et al., 1963).

Intermediate Coupling.   With increasing intermediate coupling, the relative strength of the zero-phonon line gradually decreases, and the phonon replica broaden, which renders them less distinct. Finally, all features combine into a band, which then becomes the broad absorption "line" of the defect for strong coupling. This development is shown in Fig. 20.12 for the R_2-center—which is an associate of 3 F-centers (see Section 22.4.1)—in different alkali halides. These defect associates behave as one somewhat larger defect. This results in an intermediate behavior between a deep center and a center with a more extended electron eigenfunction.

With increasing coupling, the features in absorption and emission become mirror-symmetric (Fig. 20.13): the zero-phonon line in absorption finally disappears, both maxima become more and more

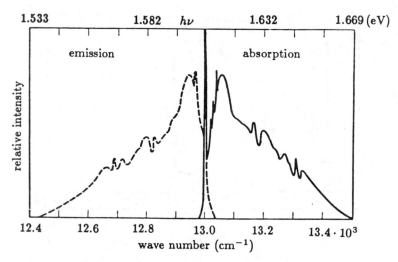

Figure 20.13: Mirror-symmetric optical absorption and emission specta for *R*-centers in SrF$_2$ at 20 K as an example for intermediate lattice coupling (after Beaumont et al., 1972).

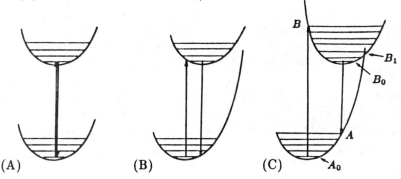

Figure 20.14: Ground and excited states in a configuration coordinate diagram for coupling increasing from subfigures (A) to (C).

separated (Fig. 20.11), and the distinction between different phonon replica becomes completely washed out (Fig. 20.12).

The degree of coupling depends on the elastic and electric properties of the host. A simple empirical relationship was observed between relaxation energy and the electronegativity (χ) of the host anions (Baranowski, 1979). This dependence is shown for the zinc chalcogenides in Fig. 20.15.

20.3.3C Quantitative Estimates from the Adiabatic Approximation The first quantitative evaluation of the optical absorption of deep centers with strong coupling was given by Huang

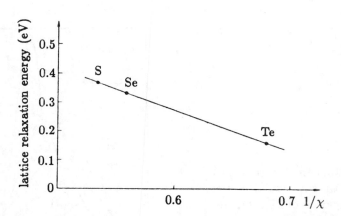

Figure 20.15: Lattice relaxation energy (Franck-Condon shift) as a function of the reciprocal electronegativity of S, Se, and Te in the corresponding zinc chalcogenides (after Baranowski, 1979).

and Rhys (1950) for the F-center absorption in alkali halides. Following the adiabatic approximation with $\phi(\mathbf{R}, \mathbf{r}) = \varphi(\mathbf{R})\psi(\mathbf{R}, \mathbf{r})$ [see Eq. (8.2)], we separate electron and phonon eigenfunctions. This permits us to express the matrix elements for the optical transition as the product of the matrix elements for the excitation from the electronic ground (μ') to the excited state (μ'') and the overlap integral between the vibrational wavefunctions:

$$\mathbf{M}_{\mu' n'}^{\mu'' n''} \simeq \int \psi_{\mu'}^*(-i\hbar\nabla)\psi_{\mu''}\, d\mathbf{r} \int \varphi_{\mu' n'}^*(\mathbf{R})\varphi_{\mu'' n''}(\mathbf{R})\, d\mathbf{R} \quad (20.59)$$

where $\psi_{\mu'}$ and $\psi_{\mu''}$ are the electronic eigenfunctions, and $\varphi_{\mu' n'}$ and $\varphi_{\mu'' n''}$ are the vibrational wavefunctions with the electron in its initial and final states, respectively. The absorption constant within the optical frequency range $\nu, \nu + \Delta\nu$ is given by

$$\alpha_o(\nu) = \frac{8\pi^2 N_{dd} e^2}{3hn_r cm^2 \nu\varepsilon_0} |\mathbf{M}_{\mu'\mu''}|^2 \mathcal{F}(\nu) \quad (20.60)$$

with $\mathbf{M}_{\mu'\mu''}$ as the matrix element for the electronic transition [first integral in Eq. (20.59)], N_{dd} as the density of deep centers, and $\mathcal{F}(\nu)$ given as the *sum* of the overlap integrals

$$\mathcal{F}(\nu) = \frac{1}{\Delta\nu} \sum_{n''}^{\nu,\nu+\Delta\nu} \left| \int \varphi_{\mu' n'}^*(X)\varphi_{\mu'' n''}(X)\, dX \right|^2. \quad (20.61)$$

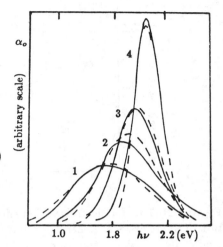

Figure 20.16: Theoretical (dashed) and experimental (solid) optical absorption curves of F-centers in KBr at 870, 470, 300, and 28 K for curves 1–4, respectively (after Huang and Rhys, 1950).

The first factor in Eq. (20.60) determines the amplitude of the optical absorption; the second factor $\mathcal{F}(\nu)$ determines the line shape which can be approximated by the line-shape function given in Eq. (20.55).

There is reasonable agreement between the Huang-Rhys theory and experimental observations of F-centers in KBr (see Fig. 20.16) with three adjustable parameters. These include a large Huang-Rhys factor ($S \simeq 25$), which is characteristic of such F-centers. For a review of the measurements of the related oscillator strengths, see Huber and Sandeman (1986).

20.3.3D Line Shape of Resonant States   Resonant states are those that overlap with a band, thereby substantially reducing the lifetime of an electron in the resonant state. Excitations into *resonant states* (Fig. 20.17) can be distinguished by the shape of the absorption peak, which can be approximated by (included in the absorption cross section):

$$s_{\mathrm{res}}(h\nu) = \frac{a + b(h\nu - E_0)}{(h\nu - E_0)^2 + \gamma_d^2}, \qquad (20.62)$$

where $1/\gamma_d$ is the lifetime of the electron in the resonant state, and a and b are the empirical parameters (Toyozawa et al., 1967; Velicki and Sak, 1966). Resonant peaks are observed which relate to shallow donors, acceptors, or to excitons. For example, Onton (1971) for such resonant states in GaP. Centers possessing resonant states act also as giant scattering centers for free electrons with an energy near $E = E_0$ (Fig. 20.17).

Figure 20.17: Conduction bands with minima at $\mathbf{k} = 0$ and $\mathbf{k} \neq 0$. The state E_0 lies in the band gap and is a bound state with negligible line width. The state E_1 originates from the unperturbed state E_1^0 and is a resonant state with greatly expanded line width, as the lifetime of an electron in this state is substantially reduced.

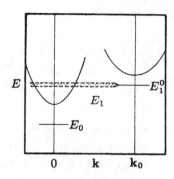

20.3.4 Photoionization Edge Shape

Excitation from a level directly into a band that results in free carriers is referred to as *photoionization*; the optical absorption spectrum is edge-like—see Section 45.1.1. The spectral distribution of the absorption is usually expressed in terms of the photoionization cross section $s_n(h\nu)$ for electrons or $s_p(h\nu)$ for holes [see Eq. (20.21)].

20.3.4A Photoionization from Deep Centers Photoionization, i.e., the transition from a trapped to a free-electron state, can be calculated from the deep level eigenfunction, the Bloch function of free carriers, and the appropriate perturbation operator. The ground state of the deep level defect can be approximated by

$$\psi_i(r) = \sqrt{\frac{\alpha}{2\pi}} \frac{\exp(-\alpha r)}{r} \quad \text{with } \alpha^2 = \frac{2m_T E_i}{\hbar^2} \tag{20.63}$$

as its initial state, where E_i is the ionization energy and m_T is a pseudomass of the electron within the deep center, used as an adjustable parameter of the theory. When ionized, the electron in the conduction band is described by a plane wave

$$\psi_f(r) = \sum_k u_n(\mathbf{k}) \exp(i\mathbf{k} \cdot \mathbf{r}) \tag{20.64}$$

as its final state. With these, first ignoring any lattice relaxation (for this, see Section 20.3.4C), we can obtain (Lucovsky, 1965; Grimmeiss and Ledebo, 1975) for the ionization cross section into a parabolic and isotropic band

$$s_n(h\nu) = \frac{8\pi\nu_D}{3} \frac{e^2\hbar}{\varepsilon_0 n_r c} \sqrt{E_i} \frac{\sqrt{m_T m_n}}{m_H^2} \frac{(h\nu - E_i)^{3/2}}{h\nu \left[h\nu + E_i \left(\dfrac{m_0}{m_n} - 1 \right) \right]^2}.$$

(20.65)

Here ν_D is the degeneracy of the occupied impurity state, and m_H is an effective carrier mass defined by the perturbation operator. With $m_T \simeq m_H \simeq m_0$, Grimmeiss and Ledebo (1975) obtained agreement with photoionization of electrons from O-doped GaAs into the conduction band.

20.3.4B Photoionization from Shallow Centers

When considering photoionization from shallow (hydrogen-like) donors, we start with a somewhat better-known eigenfunction of this defect—see Section 20.2. We then evaluate a transition matrix element with final states that are the eigenstates of the conduction band.

We also have to consider the occupancy of the involved states: it requires an occupied ground state, described by the Fermi distribution, and a free excited state—see Section 24.3.1. Let us look at transitions from filled acceptor levels into the conduction band as an example. We can express for free electrons $k(E)$ from

$$\frac{\hbar^2 k^2}{2m_n} = h\nu - (E_g - E_a).$$

(20.66)

This yields for the matrix element with Eqs. (13.20), (20.30), and (20.29)

$$|\mathbf{M}_{fi,\mathrm{qH}}|^2 = \frac{3}{2\mathcal{V}} \frac{m_0^2}{m_r} \frac{E_g(E_g + \Delta_0)}{3E_g + 2\Delta_0} \frac{32\pi a_{\mathrm{qH}}^3}{\left[1 + \dfrac{2m_n a_{\mathrm{qH}} \{h\nu - (E_g - E_a)\}}{\hbar^2} \right]}.$$

(20.67)

The joint density of states is given by the product of the density of acceptors and the density of band states near the conduction band edge:

$$J_{a,c} = N_a \frac{\mathcal{V}^2 (2m_n)^{3/2} \sqrt{h\nu - (E_g - E_a)}}{\hbar^3}.$$

(20.68)

The optical absorption coefficient [Eqs. (20.18) and (20.21)] can now be obtained from

$$\alpha_{o,qH,c} = \frac{e^2}{6\varepsilon_0 n_r c m_0^2 \nu \mathcal{V}} |\mathbf{M}_{fi,qH}|^2 J_{a,c};\qquad (20.69)$$

yielding a lengthy equation when inserting Eqs. (20.67) and (20.68) into Eq. (20.69). However, for semiconductors with a small effective mass and a large quasi-hydrogen radius, the second term of the sum in square brackets of Eq. (20.67) can be neglected, and the absorption constant can be approximated as

$$\alpha_{o,qH,c} = \frac{64\pi^2 \nu_D \sqrt{2m_n}\, e^2 a_{qH}^3 N_a}{3\varepsilon_0 n_r c \hbar^2} \frac{E_g + \Delta_0}{3E_g + 2\Delta_0} \sqrt{h\nu - (E_g - E_a)}$$

$$\simeq 1.1 \cdot 10^3 \nu_D \left(\frac{3}{n_r}\right)\left(\frac{a_{qH}}{10\,\text{Å}}\right)^3 \left(\frac{N_a}{10^{16}\,\text{cm}^{-3}}\right)\sqrt{h\nu - (E_g - E_a)},$$

$$(20.70)$$

where ν_D is the degeneracy of the conduction band. The absorption coefficient increases steeply with increasing quasi-hydrogen radius, and linearly with the density of acceptors.

The optical absorption has an *edge-like*, square root shape whenever bands are involved, rather than appearing as lines when the absorption occurs from a ground to an excited state within the center (see Fig. 20.1). When the transition proceeds into a band, photoconductivity is usually observed. Therefore, most of the discussion relating to such optical excitation can be found in Chapter 45.

20.3.4C Phonon Broadening of the Ionization Edge   For many centers, the phonon coupling plays a major role in broadening this edge of the optical absorption cross section $s_n(h\nu)$. Here s_n can be expressed approximately (see Noras, 1980, or Kopylov and Pikhtin, 1975) as

$$s_{n,\text{ph}} = \frac{C}{\sqrt{\pi}} \int_0^\infty \left\{ \frac{(\sqrt{2}\,\Gamma x)^b}{h\nu \left(\sqrt{2}\,\Gamma x + E_i \dfrac{m_T}{m_n}\right)^a} \exp\left[-\left(x - \frac{h\nu - E_i}{\sqrt{2}\,\Gamma}\right)\right] \right\} dx,$$

$$(20.71)$$

with Γ as the phonon broadening factor

$$\Gamma = \sqrt{S(\hbar\omega_b)^2 \coth\left(\frac{\hbar\omega_b}{2kT}\right)},\qquad (20.72)$$

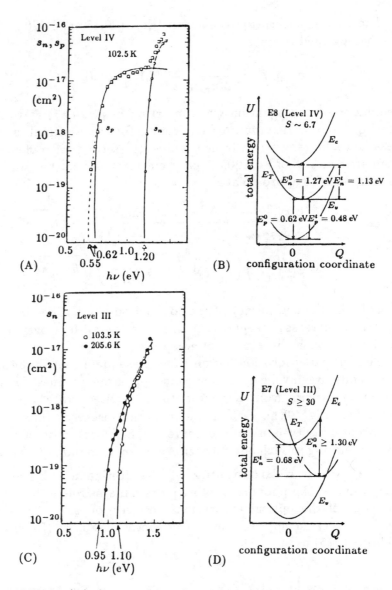

Figure 20.18: (A) Cross section for photoionization of deep impurity centers in CdTe at 102 K. Solid curve theoretical according to Eq. (20.71) with $C = 6.4 \cdot 10^{-17}$ and $5.4 \cdot 10^{-15}$ cm^{2}eV$^{3/2}$ for holes and electrons respectively, $a = 2$, $b = 3/2$, $m_n = 0.11 m_0$, $m_p = 0.35 m_0$, $m_T = 0.35 m_0$, $E_n^0 = 1.27$ eV, $E_p^0 = 0.62$ eV, and $S = 6.7$ for $\hbar\omega_c = 21$ meV. (B) Corresponding configuration coordinate diagram; (C) and (D) the same as for (A) and (B), but for a different deep center in CdTe (possibly a doubly ionized Cd interstitial) with $E_n^0 = 0.66$ eV, $E_p^0 = 1.23$ eV, and $S \simeq 30$ (after Takebe et al., 1982).

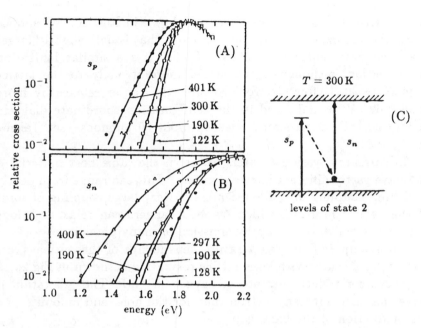

Figure 20.19: Optical cross section for the second electron state of oxygen in GaP for (A) electrons and (B) holes with the temperature as family parameter. (C) Band model of GaP with the O-level shown (after Henry and Lang, 1977).

where S is the Huang-Rhys factor; the exponents chosen appropriately for the corresponding model are

$$a = \begin{cases} 2 & \text{for } \delta\text{-function (short-range) potential} \\ 4 & \text{for Coulomb (long-range) potential.} \end{cases} \tag{20.73}$$

$$b = \begin{cases} 3/2 & \text{for forbidden transition} \\ 1/2 & \text{for allowed transition.} \end{cases} \tag{20.74}$$

This broadened edge-type appearance is shown in Fig. 20.18 for photoionization of deep centers in CdTe.

Several of these centers are identified, each of them with different coupling to the surrounding lattice. For instance, the center with a capture cross section of $\sim 10^{-17}$ cm^2 for holes and $\sim 10^{-16}$ cm^2 for electrons, as shown in Fig. 20.18A and B, indicates a coupling* to the

---

* This is obtained from the lattice relaxation of 0.14 eV ($= 1.27 - 1.13$ or $= 0.62 - 0.48$ eV) with breathing mode phonons of $\hbar\omega_b \simeq 21$ meV after capture of a hole or an electron, as indicated in Fig. 20.18A and B.

lattice with $S \simeq 6.7$. Another center, shown in Fig. 20.18C and D, has a photoionization energy for electrons that is only slightly larger than the first center (~ 1.3 eV). It also has a similar ionization cross section, although it is coupled more strongly to the lattice with $S \simeq 30$. After electron capture, the center relaxes by more than 0.62 eV, as indicated in the configuration coordinate diagram of Fig. 20.18D, and thereby releases about 30 phonons—see Takebe et al., 1981.

In summary, various centers—even in the same host material—can have vastly different lattice coupling. This can range from nearly vanishing coupling of shallow centers to very large coupling of some of the deep centers, in which trapped carriers can relax to bridge almost the entire band gap by emission of many phonons.

An example of the temperature dependence of the optical cross section for a center with large lattice relaxation is shown in Fig 20.19 for the second electron state of oxygen in GaP. The relaxation is larger for the captured electron than for the hole, and amounts to a major fraction of the band gap.

Summary and Emphasis

Optical absorption at lattice defects is distinguished by bond-to-bond absorption, resulting in a line spectrum, and by bond-to-free electron absorption, causing an edge-like spectrum. The line spectrum in turn is characterized by the energy, amplitude, and shape (width) of each line. The edge spectrum is characterized by its onset energy and its slope.

The line or edge energy provides important information about the electronic structure of the defect center. The line amplitude permits judgment on the density of these centers or joint density of states, and their symmetry relation within the lattice, matrix elements, and selection rules. The line shape relates information on the lifetime of the excited states and center-to-lattice coupling for homogeneously broadened lines, as well as on the center perturbation due to defect-sensitive stress and electric fields of the surrounding lattice for inhomogeneously broadened lines.

Two major classes of defects result in substantially different types of absorption spectra: shallow-level, hydrogen-like defects and deep-level, tight bonding defects. The shallow-level defects have electronic eigenfunctions which are heavily influenced by mixing with one band, the conduction band for donors, and the valence band for acceptors.

The deep-level centers have eigenfunctions of excited states which mix with *both* valence *and* conduction bands. The shallow levels also show little relaxation after a change in excitation or in charging of the center, i.e, the lattice atoms readjust only to a minor degree. Deep-level defects, in contrast, show a much larger lattice relaxation after a change in excitation or in charging of the centers. However, the degree of relaxation—the number of phonons emitted during relaxation—differs greatly with the type of defect and the specific electronic state. One may picture the difference between such eigenfunctions as those which "press" surrounding atoms apart and those which "extend over" surrounding atoms. The latter have much less influence on their equilibrium position.

The most direct signature of a lattice defect is its optical spectrum which permits identification of several classes of these defects. A detailed analysis of energy, amplitude, and shape of the spectrum and its change with external parameters is the key to the understanding of its electronic behavior. This in turn is essential for the discussion of electrical and optical properties of semiconductors. Most of the analytical information is obtained at very low temperatures, where the influence of phonons is minimized and the line spectra become more distinct.

A systematic correlation study with doping and other means to change the defect structure will help to unravel the mystery of many of the lines which are still unidentified. In the future, this will provide the means for a spectral analysis that may become even more powerful than the one commonly used to identify atoms in the gas phase.

Exercise Problems

1.(r) Describe the characteristic elements of a configuration coordinate representation. For what type of defect will it be used advantageously? Why?

2.(r) What are contributing effects for a homogeneous and an inhomogeneously broadened absorption line? What determines the width of the homogeneous line? How can one distinguish the two types experimentally?

3.(e) What is the difference between a Gaussian and a Lorentzian line shape? Draw a typical example for both at the same ν_0 and γ. Give two examples for both types of absorption.

4.(*) How does the Poisson distribution of phonon replica influence the line shape at higher temperatures and for higher lattice coupling ($S > 6$)? Give quantitative arguments.

5.(r) Discuss line shape and phonon replica for deep centers with different values of the Huang-Rhys factor.

6.(r) Discuss the influence of the temperature on the line shape of deep centers for different Huang-Rhys factors.

7.(*) Discuss line spectrum, amplitude distribution, and line shape of a quasi-hydrogen donor.
   (a) Give quantitative arguments.
   (b) Discuss the dependence of this spectrum and line shape on the density of donors. Which lines will be most influenced? Be quantitative.

8.(e) Derive the amplitude factor of the optical cross section given in Eq. (20.21). Discuss each contribution.

9. Discuss the extent of the wavefunction of hydrogen-like shallow defects in k-space for the ground state and excited states using Eqs. (20.26) and (20.29).

10.(e) Give the relationship between the dimensions of the oscillator strength and the momentum matrix element.

11.(*) How are the Franck-Condon principle and deep-level relaxation related? Explain the limitation of the adiabatic approximation.

12.(e) Search the recent literature for the identification of a deep center in a III-V compound and discuss the work.

Chapter 21

Shallow Level Centers

Shallow level defects are intimately connected with the adjacent band and have a hydrogen-like defect level spectrum.

In the previous chapter we have presented a general overview of the different defect levels and their optical absorption spectra. We will now discuss in more detail the defect level spectrum of shallow centers. These centers have eigenfunctions that extend beyond their neighbor atoms and mix only with the nearest band states. We will indicate that the ground state is influenced by the chemistry, how it is to be considered, to what degree the quasi-hydrogen approximation can be used, and what refinements are necessary to obtain a better agreement with the experiment.

In the following chapter we will then give a more detailed description of the deep level centers. In contrast, these have highly localized eigenfunctions of their ground states, mix with conduction *and* valence bands, and require a more thorough knowledge of the core potential for the calculation of their eigenvalues.

21.1 Hydrogen-Like Defects

Shallow level defects can be described as hydrogen-like defects—see Section 20.1.1. As an example, we will discuss here the electronic states of a substitutional donor, such as a phosphorus atom on a lattice site in a silicon host crystal. The P atom becomes positively charged after it has given its electron to the host. This electron, now near the bottom of the conduction band, is a quasi-free Bloch electron with an energy

$$E(k) = E_c + \frac{\hbar^2 k^2}{2m_n}. \tag{21.1}$$

Near the $P_{Si}^{\bullet}$ center, the electron can become localized. Its new eigenstate can be calculated by solving the Schrödinger equation

$$H\psi = E\psi \quad \text{with the Hamiltonian} \quad H = H_0 - \frac{e^2}{4\pi\varepsilon_{st}\varepsilon_0 r} \quad (21.2)$$

where H_0 is the unperturbed Hamiltonian of the host lattice, to which the attractive Coulomb potential of the defect is added. This potential is modified by the screening action of the host, which is expressed by the static dielectric constant.

This is in contrast to a Bloch electron in an ideal lattice, which interacts only with the electronic part of the lattice and therefore involves the optical dielectric constant ε_{opt}. When trapped, the electron becomes localized near the defect and causes a shift of the surrounding ions according to its averaged Coulomb potential. Therefore, the static dielectric constant is used here.

The eigenfunctions to H_0 are Bloch functions. They form a complete orthonormalized set—see Appendix A.4.2C. The solutions to H can be constructed near the defect from a wave packet of Bloch functions

$$\psi = \sum_{n,k} c_n(\mathbf{k})\psi_n(\mathbf{k},\mathbf{r}) \simeq \sum_{k} c_c(\mathbf{k})\psi_c(\mathbf{k},\mathbf{r}). \quad (21.3)$$

The summation over several bands with index n is dropped, since as shallow levels—here for a donor—their eigenfunctions are constructed primarily from eigenfunctions of the nearest band only: $\psi_c(\mathbf{k},\mathbf{r})$ are the Bloch functions of *conduction band electrons*, appropriate for the description of the *donor*

$$\psi_c(\mathbf{k},\mathbf{r}) = u_c(\mathbf{k},\mathbf{r})\exp(i\mathbf{k}\cdot\mathbf{r}). \quad (21.4)$$

As will be verified below, the eigenfunctions of such shallow level defects extend over several lattice constants, thus restricting $\mathbf{k}$ to values close to the center of the Brillouin zone. Since $u(\mathbf{k})$ changes only slowly with $\mathbf{k}$, we can pull $u_c(\mathbf{k} = \mathbf{k}_0,\mathbf{r})$ as constant from the sum for $\mathbf{k} \simeq 0$, or near any of the minima of $E(\mathbf{k} \simeq \mathbf{k}_0)$, and introduce with*

$$\psi(\mathbf{k},\mathbf{r}) = \left\{ \sum_{k} c_c(\mathbf{k})\exp(i\,\mathbf{k}\cdot\mathbf{r}) \right\} u_c(\mathbf{k}_0,\mathbf{r}) = F(\mathbf{r})\psi_i(\mathbf{k}_0,\mathbf{r}) \quad (21.5)$$

---

* This can easily be seen at the Γ-point for $k_0 = 0$: Here we have

$$\Psi(\mathbf{k} = 0,\mathbf{r}) = u(0,\mathbf{r})\exp(i\,0\cdot r) = u(0,\mathbf{r}).$$

an *envelope function*

$$F(\mathbf{r}) = \sum_{\mathbf{k}} c_c(\mathbf{k}) \exp(i\,\mathbf{k} \cdot \mathbf{r}), \qquad (21.6)$$

where $\psi_i(\mathbf{k}_0, \mathbf{r})$ is the Bloch function in the ith minimum of $E(\mathbf{k})$. The envelope function satisfies an appropriately modified Schrödinger equation for the quasi-hydrogen model

$$\left(-\frac{\hbar^2}{2m_n} \nabla^2 - \frac{e^2}{4\pi\varepsilon_{st}\varepsilon_0 r} \right) F(\mathbf{r}) = (E - E_c)F(\mathbf{r}) \qquad (21.7)$$

with the energy normalized to the edge of the conduction band and with an effective mass m_n for Bloch electrons near this band edge. This Schrödinger equation is identical to that for a hydrogen atom, but for an electron of effective mass m_n in a medium of dielectric constant ε_{st}. Therefore, the solution can be transcribed directly from that of a hydrogen atom, and yields for the envelope eigenfunction of the $1s$ ground state:

$$F(r) = \frac{1}{\sqrt{\pi\, a_{qH}^3}} \exp\left(-\frac{r}{a_{qH}} \right) \qquad (21.8)$$

with
$$a_{qH} = \frac{4\pi\varepsilon_{st}\varepsilon_0\hbar^2}{m_n e^2} = \frac{\varepsilon_{st}\, m_0}{m_n} a_0, \qquad (21.9)$$

with an effective Bohr radius a_{qH}, the *quasi-hydrogen radius*, and $a_0 = 0.529$ Å, the Bohr radius of the hydrogen atom. This envelope function is shown as the dashed curve in Fig. 21.1. The total wavefunction (solid curve) shows the modulation with the rapidly oscillating Bloch function with a period length of the lattice constant a.

In semiconductors with several equivalent minima (Si, Ge), the wavefunction becomes a sum of contributions from each of the minima:

$$\sum_j \alpha_j F_{jc}(\mathbf{r}) u_{jc}(\mathbf{k}_{j0}, \mathbf{r}).$$

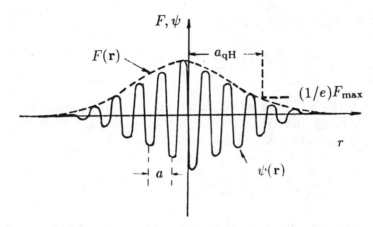

Figure 21.1: Relationship between the envelope function $F(r)$ and the wave function $\psi_n(r)$ of a Bloch wave packet for an electron localized near a hydrogen-like impurity. Here, a denotes the lattice constant.

The resulting eigenstates of the envelope function are bound states below the lowest free states in the conduction band E_c, and are given by a quasi-hydrogen energy spectrum*

$$E_{\text{qH}}^{(n)} = E_c - \frac{m_n e^4}{32\pi^2(\varepsilon_{\text{st}}\varepsilon_0)^2\hbar^2}\left(\frac{1}{n_q}\right)^2. \tag{21.10}$$

The dispersion behavior in k-space can be obtained from the Fourier transform of the envelope function:

$$F(\mathbf{r}) = \int F(\mathbf{k})\exp(i\,\mathbf{k}\cdot\mathbf{r})\,dr \tag{21.11}$$

---

* n_q is the *principal quantum number*, describing the entire energy spectrum for a simple hydrogen atom. All other states are degenerate. Therefore, in a pure Coulomb potential, this quantum number is the only one that determines the energy of a hydrogen level. When deviations from this spherical potential appear in a crystal, the $D = \Sigma l(l+1) = n_q^2$ degeneracy of each of these levels is removed, and the energy of the s-, p-, d-,... states are shifted according to $R_H/(n_q+l)^2$. The importance of these transitions is discussed in Section 21.1.3. To further lift the remaining degeneracies of the magnetic quantum number, a magnetic field must act (see Section 21.3.4).

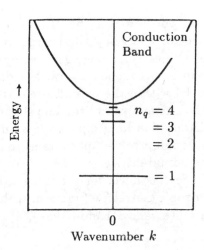

Figure 21.2: Ground and excited states of the donor level in the $E(k)$ diagram, indicating the extent of these levels in k-space. The ground state has the smallest radius; hence, its extension in k-space is the largest.

which yields for the ground state [compare with Eq. (21.8)]

$$F(k) = \frac{8\sqrt{\pi}}{a_{qH}^{5/2}} \frac{1}{\left[k^2 + \left(\frac{1}{a_{qH}} \right)^2 \right]^2} \tag{21.12}$$

and indicates that the wave packet extends in k-space approximately to $k \simeq 1/a_{qH}$. That is, $c(k)$ [see Eq. (21.6)] is $\sim$ constant for k up to $1/a_{qH}$ and decreases rapidly ($\propto 1/k^4$) for $k > 1/a_{qH}$. For higher excited states, the extent in k shrinks proportionally to $1/n_q^2$ as a_{qH} increases $\propto n_q^2$—see Eq. (20.2) and Fig. 21.2.

21.1.1 Charge Density Distribution, ESR, and ENDOR

The envelope function yields the charge distribution of such a shallow defect

$$\varrho(\mathbf{r}) = \frac{1}{\mathcal{V}} \int F^*(\mathbf{r})F(\mathbf{r})d^3\mathbf{r}. \tag{21.13}$$

A direct means to check such charge distribution is by analyzing the electron nuclear double resonance (ENDOR) signal with the assistance of an electron spin resonance (ESR) line width analysis (see Fehrer, 1959.)

The ENDOR technique is based on changes of the ESR signal caused by the spin-flip of appropriate nuclei (e.g., Si^{29}) within the reach of the electron cloud of the defect and induced by an external electromagnetic field. This *hyperfine interaction* and related super-hyperfine interactions are well-known tools for analyzing the actual lattice environment of a localized deep center when the lines are resolved—see Sections 21.3.4C and 21.3.4D. The latter can also be used for analyzing the charge distribution of shallow centers where such lines can no longer be separated. Here, they are inhomogeneously broadened by various distances and densities of the active nuclei within the extended electron cloud of the shallow defect center. The analysis of the line shape of the observed resonances, and of certain bleaching dips while saturating specific spin-flip resonances, although involved, permits us to obtain the envelope function within the actual lattice.

For instance, for a hydrogen-like donor in Si, it confirms an anisotropic* envelope function, which for the donor electron within the satellite valley (Kohn, 1957), is given by:

$$F^{(x)}(\mathbf{r}_l) = F(\mathbf{r}_l)_{\text{isotr}} \sqrt{\frac{a^{*3}}{a^2 b}} \; \frac{\exp\left[-\left(\frac{x_l^2}{(nb)^2} + \frac{y_l^2 + z_l^2}{(na)^2} \right) \right]}{\exp\left(-\frac{r_l}{na^*} \right)} \qquad (21.14)$$

with $a = 25$ Å, $b = 14.2$ Å, $a^* = 21$ Å, $n = \sqrt{0.029/E_i}$, and $E_i = E_c - E_d$ as the ground state ionization energy. The x-axis is aligned with the direction of $\mathbf{k}$. Similar expressions are given for $F^{(y)}$ and $F^{(z)}$—see Section 21.1.3 and Eq. (21.17).

21.1.2 The Chemical Identity

In the simple hydrogen-like approximation, the chemical identity of the donor is totally lost. The identity of the host is provided by ε_{st} and m_n.

In contrast to this theory, the ground state is observed to depend significantly on the chemical identity of the donor, such as the different elements of group V impurities (P, As, Sb, or Bi) or monovalent metal interstitials such as Li—see Section 21.1.9. The approximation used before is too coarse to show such dependency. In Section 21.1.4, a number of refinements will be discussed, which will address the complexity of the ground state.

It is remarkable, however, that *excited states* (*p*-states) of these shallow donors lose their chemical identity and are rather well explained by the simple theory given above, provided the correct ef-

---

* The anisotropy is observed because of the satellite band ellipsoid resulting in an anisotropy of the effective masses with m_l and m_t in the main axis.

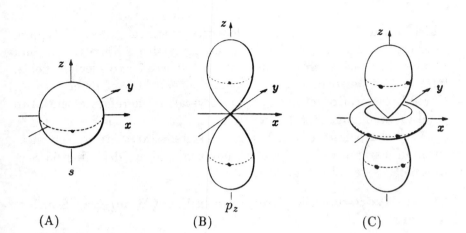

Figure 21.3: Shape of the eigenfunction of atomic orbitals s-, p-, and d-type for subfigures (A), (B), and (C), respectively.

fective mass and dielectric constant (see Section 14.2.1) are used, as will be explained in Section 21.1.3. There are several reasons why an improved agreement is obtained for higher states:

(a) higher s-states extend to much larger diameters* ($\propto n_q^2$); and

(b) p-states show a node of the wavefunction near the core (see Fig. 21.3), making the wavefunction less sensitive to the actual potential near the core region.

One needs extremely pure crystals, however, to avoid significant overlap of higher-state eigenfunctions with wave functions of other impurities which would cause a perturbation of these excited states. The eigenfunctions of excited states of such shallow impurities often extend beyond 1000 Å or more, interacting with each other when the distance between them is less than 1000 Å, equivalent to a density of $(1/1000 \text{ Å})^3 \simeq 10^{15}$ cm^{-3}. This requires ultrapure crystals and controlled doping in the < 100 ppb range. In addition, native defects,

---

* For instance, when $a_{\mathrm{qH}} \simeq 50$ Å for the $1s$ state, it is 200 Å for the $2s$ and 450 Å for the $3s$ states, making the hydrogenic effective mass approximation a much improved approximation. In addition, in semiconductors where ε/m^* is already very large, e.g., in GaAs with $\varepsilon_{\mathrm{st}} m_0/m_n = 192.5$, resulting in $a_{\mathrm{qH}} = 101.9$ Å $\simeq 18a$, this approximation is quite good for the $1s$ state. In GaAs, it results in $E_{\mathrm{qH}} = 5.83$ meV, while the experimental values vary from 5.81 to 6.1 mV for GaAs:Si and GaAs:Ge. For more comparisons between theory and experiment, see Bassani et al. (1974).

which are usually frozen-in at densities much in excess of 1 ppm, interfere by influencing the surrounding lattice. Finally, line and surface defects, which produce internal stresses and electric fields, interfere by perturbing the electronic eigenvalues of defect states. Careful preparation of near perfect crystals is therefore essential to yield unambiguous results.

Very sharp lines of many higher excited states are indeed measured in ultrapure silicon and germanium, and are described in Section 45.5.2—see also Fig. 20.1.

21.1.3 Hydrogen-Like Donors in Indirect Band-Gap Semiconductors

Most free electrons in indirect semiconductors are in one of the side valleys rather that at $k = 0$. The energy of such electrons is given by

$$E(\mathbf{k}) = E_c^{(i)} + \frac{\hbar^2}{2} \left\{ \frac{k_x^2 + k_y^2}{m_l} + \frac{k_z^2}{m_t} \right\} \tag{21.15}$$

where m_l and m_t are the longitudinal and transverse effective masses due to the ellipsoidal shape of these valleys. The effective mass Schrödinger equation for the envelope function in one of these valleys, here in the z-direction, is

$$\frac{-\hbar^2}{2} \left\{ \frac{1}{m_t} \left(\frac{\partial^2}{\partial x^2} + \frac{\partial^2}{\partial y^2} \right) + \frac{1}{m_l} \frac{\partial^2}{\partial z^2} \right\} F^{(z)}(\mathbf{r}) - \frac{e^2}{\varepsilon_{st}\varepsilon_0 r} F^{(z)}(\mathbf{r})$$

$$= (E - E_c) F^{(z)}(\mathbf{r}), \tag{21.16}$$

In addition, in noncubic semiconductors the dielectric constant is anisotropic (*anisotropic shielding*). The corresponding Schrödinger equation for the envelope function is given by

$$-\frac{\hbar^2}{2} \left\{ \frac{1}{m_t} \left(\frac{\partial^2}{\partial x^2} + \frac{\partial^2}{\partial y^2} \right) + \frac{1}{m_l} \frac{\partial^2}{\partial z^2} \right\} F_j(\mathbf{r})$$

$$- \frac{e^2}{\sqrt{\varepsilon_\perp \varepsilon_\|} \sqrt{x^2 + y^2 + \frac{\varepsilon_\perp}{\varepsilon_\|} z^2}} F_j(\mathbf{r}) = (E - E_c) F_j(\mathbf{r}), \tag{21.17}$$

with $\|$ and $\perp$ denoting directions parallel or perpendicular to the z-direction, and aligned with the crystallographic c-direction. It shows substantial anisotropy of the envelope function in its 1s state as an ellipsoid rather than a sphere—see Eq. (21.14). The nomenclature

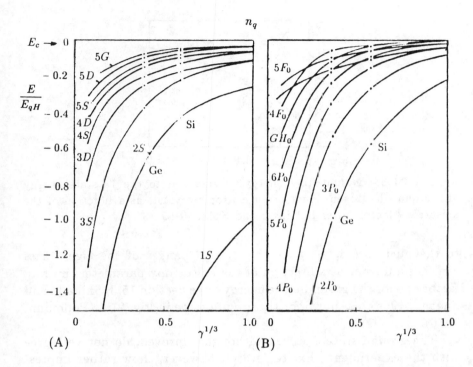

Figure 21.4: Energy levels of donor states calculated with hydrogenic effective mass approximation as a function of the anisotropy parameter [Eq. (21.18)] for (A) s-like and (B) p-like states. The limits, $\gamma = 1.0$ and $\gamma = 0$, indicate isotropic and two-dimensional semiconductors, respectively. The dots show the levels for Ge and Si (after Faulkner, 1969.)

$1s$, $2s$, etc., is usually retained for these states, to which the eigenfunctions would converge in the limit of vanishing anisotropy. Quasi-Bohr radius and effective Rydberg energy are defined by using m_t and $\sqrt{\varepsilon_\perp \varepsilon_\parallel}$ rather than m_n and $\varepsilon_{\mathrm{st}}$.

The degree of changes of various excited states as a function of the degree of sidevalley anisotropy* γ can be seen from Fig. 21.4. The values of γ for Ge and Si are marked by dots in the figure. Higher anisotropy compresses the eigenfunction in one direction and thereby increases the binding energy as the electron is forced closer

---

* Often the anisotropy factors γ, γ_a, or α are used with

$$\gamma = \frac{m_t}{m_l} \quad \text{or} \quad \gamma_a = \frac{m_\perp \varepsilon_\perp}{m_\parallel \varepsilon_\parallel} = 1 - \alpha. \qquad (21.18)$$

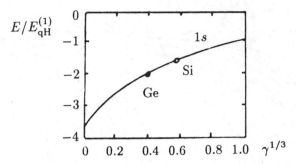

Figure 21.5: Ground-state energy (normalized to the 1s state energy of the quasi-hydrogen donor in an isotropic crystal) as a function of the anisotropy factor (after Luttinger and Kohn, 1955).

to the nucleus ($m_t > m_n; \varepsilon_\perp > \varepsilon_{st}$). Changes of the eigenstates can be significant as a function of the anisotropy parameter (up to a factor of 4 for the ground-state energy—see Section 15.3; Shinada and Sugano, 1966), as shown for the ground state in Fig. 21.5 (Pollmann, 1976).

The ground-state energies, although improved, do not yet agree with the experiment. Excited states, however, show rather impressive agreement, as indicated in Fig. 21.6. In this figure, all experimental values are shifted so that the $2p_0$ level agrees with the theory. This figure also includes an alkali interstitial and a group VI element. The $S^{\bullet}$-defect yields an energy four times higher because of the double charge of the center. This example shows that the agreement goes beyond the original quasi-hydrogen list of shallow donors in group IV semiconductors. Recent measurements summarized by Grimmeiss (1987) extend this list even further when crystals are carefully prepared to avoid built-in (electric and stress) fields.

For donors, the *chemical shift* of the ground state is largest in Si, namely 111 meV between B and In. The shift is much less severe in semiconductors with smaller effective mass and larger dielectric constant. In Ge this shift is smaller by a factor of ~ 100. Several other approximations, discussed in the following section, are used to obtain improved donor state energies (Luttinger and Kohn, 1955; Faulkner, 1969; also, see review by Baldereschi and Lipari, 1973; and Bassani et al., 1974).

21.1.4 Hydrogen-Like Ground State, Chemical Shift

The disagreement of the ground-state calculation in the hydrogenic effective mass theory with the experiment is substantial.

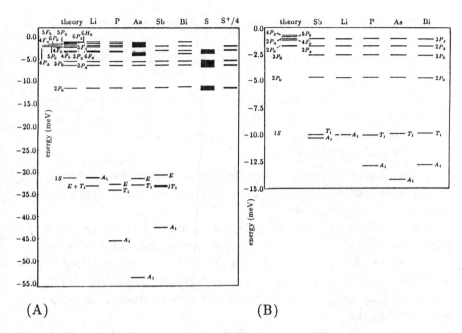

Figure 21.6: (A) Energy levels of shallow donors in silicon. The theoretical level distribution is obtained with $\varepsilon = 11.4$. The experimental values are shifted so that the $2p_0$ levels line up with the theoretical level. The experimental uncertainty is indicated as the width of the levels. (B) Same as (A) for Ge (after Faulkner, 1969).

Immediately apparent from Fig. 21.6 is the fact that the donor $1s$-state is split into three (or two) lines for Si (or Ge). This is caused by coupling between the states relating to the six (or four) equivalent sidevalleys of the conduction band in Si (or Ge), which was first analyzed by Morita and Nara (1966). The splitting varies for different donors and is by far the largest contribution to the observed deviation of the ground state from the simple effective mass hydrogen-like model. We will discuss this contribution in Section 21.1.4A.

Other contributions to the chemical shift deal with the short-range actual potential of the impurity. Such central cell corrections to the effective mass theory included various attempts to:

- use modified dielectric screening for estimating the short-range potential;
- consider the strain field from the misfit of the impurity into a substitutional site (Morita and Nara, 1966);
- use short range model potentials with adjustable parameters;

- introduce local pseudopotentials; and, with recent *ab initio* calculations
- consider the influence of lattice relaxation surrounding the impurity.

There are many other, more subtle models, which are reviewed by Stoneham (1975). We will deal with the more important aspects of the central cell correction to the hydrogenic potential in the following section (see also Pantelides, 1978).

21.1.4A Band Mixing   In element semiconductors, the mixing of hydrogen-like donor states with different subbands results in the splitting of the ground state.

Mixing through *intervalley interaction* in Si (Ge) (reviewed by Pantelides, 1978) removes the degeneracy of the sixfold degenerate ground state (E_0) in the neighborhood of the defect, splitting it into three (two) levels of symmetry Γ_1, Γ_{12}, and Γ_{15} (Γ_1 and Γ_{15}). The corresponding energies for Si:P (experimental values from Aggarwal and Ramdas, 1965) are:

$$E_c - E_{\Gamma_1} = E_0 - \lambda - 4\mu \ (A_1 \text{ singlet}) = 45.3 \text{ meV}$$

$$E_c - E_{\Gamma_{12}} = E_0 - \lambda + 2\mu \ (T_2 \text{ triplet}) = 33.7 \text{ meV} \qquad (21.19)$$

$$E_c - E_{\Gamma_{15}} = E_0 + \lambda \qquad (E \text{ doublet}) = 32.3 \text{ meV}$$

The A_1 singlet is an *s*-like state, and the T_2 triplet is a *p*-like state, with A_1 and T_2 being irreducible representations of the tetrahedral group T_d. For Ge, we have (experimental values from Reuszer and Fisher, 1964):

$$E_c - E_{\Gamma_1} = E_0 - 3\lambda' \ (A_1 \text{ singlet}) = 14.2 \text{ meV}$$
$$E_c - E_{\Gamma_{15}} = E_0 + \lambda' \quad (T_2 \text{ triplet}) = 10.0 \text{ meV} \qquad (21.20)$$

where λ and μ are the matrix elements for transitions from $(k_0, 0, 0)$ to $(0, k_0, 0)$ and $(-k_0, 0, 0)$, respectively, in Si, and where $\pm\lambda'$ is the matrix element between the impurity and its mirrored position at $(\pm\frac{1}{2}, \frac{1}{2}, \frac{1}{2})$ in Ge. These matrix elements are computed by Baldereschi (1970); see also Bassani et al. (1974).

Mixing of hydrogen-like states with several bands is considered when the bands have their edges at similar energies:

- different equivalent subbands (as discussed above),
- different nonequivalent subbands (e.g., in GaAs or GaSb), and
- different degenerate bands at $k = 0$ (valence bands for most semiconductors).

Table 21.1: Binding energy of donors in Si (meV) (after Pantelides, 1978).

Semiconductor	$1s$	State					
		$2p^0$	$2s$	$2p^\pm$	$3p^0$	$3s$	$3p^\pm$
Si (theor.)	31.27	11.51	8.82	6.40	5.48	4.75	3.12
Si (P)	45.5; 33.9; 32.6	11.45		6.39	5.46		3.12
Si (As)	53.7; 32.6; 31.2	11.49		6.37	5.51		3.12

Mixing with *nonequivalent valleys* is important when the valleys are closely spaced in energy to E_c (Vul' et al., 1970). A detailed analysis was performed by Altarelli and Iadonisi (1971). The coupling increases when, e.g., with hydrostatic pressure, the energy separation between the Γ- and X- minima in GaAs decreases (Costato et al., 1971). Castner (1970) presents indications that higher valleys and saddle points contribute to the split ground-state energies in Si.

The progressive improvement of the theoretical description of hydrogen-like defects is indicated in Fig. 21.7. The corresponding energies are listed in Table 21.1.

There are optical transitions possible between these split ground states. Three absorption lines due to transitions from the $1s(A_1)$ to the $1s(T_2)$ state are observed, while the transitions to $1s(E)$ are forbidden; such transitions can be observed, however, under uniaxial stress (Bergman et al., 1986).

In addition, one needs to use the proper dielectric constant $\varepsilon(k)$ at the position k of the minima of the sidevalleys (Baldereschi, 1970).

21.1.4B Short-Range Potential Corrections The energy of the impurity ground state also depends on the net impurity potential. In its most simple form, it is given as the difference between the Coulomb potential of the impurity and the lattice atom that is replaced and the central cell potential:

$$V(\mathbf{r}) = \left(Z_{\text{host}} - Z_{\text{imp}}\right) \frac{e^2}{4\pi\varepsilon\varepsilon_0 r} + V_{cc}(\mathbf{r}). \qquad (21.21)$$

The central cell potential V_{cc} extends over the range of the unit cell. It contains the short-range components due to the chemical individuality of the defect and the nonlinearity of the polarization in the direct neighborhood of the impurity.

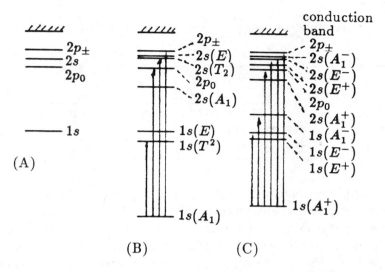

Figure 21.7: Lower energy states of a donor in Si: (A) in one-valley treatment; (B) in multivalley treatment and T_d symmetry; and (C) in multivalley treatment with appropriate D_{3d} symmetry (after Grimmeiss, 1986).

In first approximation, one connects the central cell potential asymptotically with the Coulomb potential, however, using a k-dependent ε near the impurity.

For the central cell potential, various approximations are made (Csavinski, 1965; Morita and Nara, 1966). When fulfilling proper orthogonality requirements on the core orbitals of the impurity, such impurity potentials can be used for isocoric impurities only in a generalized effective mass approximation to obtain better fitting ground-state energies of shallow-defect centers (Sah and Pantelides, 1972, 1974). For nonisocoric impurities, see Section 21.1.4C.

A central cell correction of the potential near the core is reviewed by Stoneham (1975). It is included in an effective impurity potential

$$
V_{\text{pseudo}}(\mathbf{r}) =
\begin{cases}
\dfrac{e^2 Z(r)}{4\pi \varepsilon_{\text{st}} \varepsilon_0 r} & \text{for } r > r_0 \\[2ex]
\displaystyle\int \dfrac{d\mathbf{k}}{(2\pi)^3} \dfrac{V'(\mathbf{k})}{4\pi \varepsilon(\mathbf{k}) \varepsilon_0} \exp(i\,\mathbf{k}\cdot\mathbf{r}) & \text{for } r < r_0,
\end{cases}
\tag{21.22}
$$

where $V'(\mathbf{k})$ is the Fourier transform of the unscreened impurity potential, and $\varepsilon(\mathbf{k})$ is the wave vector dependent dielectric function—see Fig. 14.3. $Z(r) = Z_0$ is used here for $r > r_0$, with Z_0 as the point charge of the impurity (1 for As, 2 for Se, etc.), and r_0 as an adjustable arbitrary boundary between the inner and outers region of the approximation. This is a simple example for an often-used treatment. Namely, it is the splitting of the region of interest into an inner region which is treated in more detail, and an outer region, with a Coulomb potential asymptotically approaching the unperturbed lattice potential. Boundary conditions between these two regions are of importance. For better approximations, the inner region should be made as large as possible, and should contain more than just the neighbor atom of the defect center (the Keating potential extends to the fifth neighbor—see Stoneham, 1986).

21.1.4C Local Pseudopotentials   Instead of computing the true impurity potential, which could yield the set of quasi-hydrogen levels of interest, and also deeper core levels of limited interest, one usually proceeds to estimate a *local pseudopotential** by which only outer shell electrons are influenced.

The *pseudopotential formalism* as applied to impurities can best be explained (Phillips and Kleinman, 1959; Austin et al., 1962) by comparing the Schrödinger equation containing the true impurity potential $V(\mathbf{r})$

$$\left[-\frac{\hbar^2}{2m_0}\nabla^2 + V(\mathbf{r})\right]\psi_n = E_n\psi_n \qquad (21.23)$$

with the Schrödinger equation containing the pseudopotential

$$\left[\frac{\hbar}{2m_0}\nabla^2 + V_{\text{pseudo}}(\mathbf{r})\right]\phi_n = E'_n\phi_n. \qquad (21.24)$$

Bassani and Celli (1961) have shown that the eigenvalues E_n and E'_n for Eqs. (21.23) and (21.24) are nearly identical for wavefunctions $\psi_n = \psi_c$ and $\phi_n = \phi_c$ for the conduction band, i.e., if only the behavior of outer electrons is of interest.

Therefore, it is sufficient to obtain the shallow energy states of an impurity when composed of conduction band eigenfunctions by using the pseudopotential, rather than the true impurity potential. A short review on how to obtain the proper pseudopotential is given by Pantelides (1978).

---

* Such a local pseudopotential is used near an impurity as opposed to the nonlocal pseudopotential used for band-structure analysis (see Section 8.2.2).

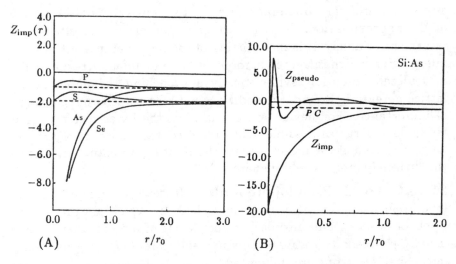

Figure 21.8: (A) Effective impurity charge $Z(r)$ modifying the Coulomb potential $V(r) = Z(r)e^2/(4\pi\varepsilon_{st}\varepsilon_0 r)$ to present the "true potential" for substitutional impurities in Si, as contrasted with $Z(r) = Z = 1$ or $= 2$ for a single or double point charge (PC), and shown as horizontal dashed lines; (B) Z_{imp} as in subfigure A for As in an extended scale. Z_{pseudo} is the corresponding pseudopotential with cancellation of the strong core part that is only relevant for core levels (after Pantelides, 1975).

An example of the potential distribution for several mono- and divalent hydrogen-like impurities in Si, as computed by Pantelides (1975), is given in Fig. 21.8A. It can be seen that isocoric impurities (P and S) have nearly point-charge Coulomb behavior, since the contribution from the core potential is similar to the replaced host atom and therefore cancels. Other elements with a different atomic core, however, show a major variation of the atomic potential in the core region indicated for As and Se. For these impurities, $V_{pseudo}(r)$ remains closer to the behavior of a point charge, as shown for As_{Si} in Fig. 21.8B.

Pseudopotentials were used to calculate the eigenvalues for a number of shallow level defects in Si (Pantelides and Sah, 1974) and compared to ENDOR data (Schechter, 1975).

21.1.4D Model Potentials   Simple *model potentials* have been used with moderate success (Abarenkov and Heine, 1965; Appapillai and Heine, 1972). These potentials contain the essential features of the pseudopotential: namely, a rather smooth core potential that gives the chemical identity of the impurity, and a Coulomb tail

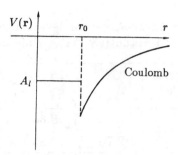

Figure 21.9: Abarenkov-Heine model potential of a shallow impurity.

that provides good agreement with higher states of quasi-hydrogen impurities. An example is the Abarenkov-Heine potential

$$V(\mathbf{r}) = \begin{cases} \sum_l A_l P_l & \text{for } r < r_0 \\ -\dfrac{e^2}{4\pi\varepsilon_{\mathrm{st}}\varepsilon_0 r} & \text{for } r > r_0, \end{cases} \tag{21.25}$$

where P_l is the angular projection operator and A_l are the empirical energy constants. This model potential is shown in Fig. 21.9, and yields reasonable agreement with the experiment if the summation in Eq. (21.25) includes higher terms in l (Baldereschi and Lipari, 1976).

21.1.4E Lattice Relaxation   The lattice surrounding a substitutional impurity relaxes. This was recently computed using norm-conserving pseudopotentials to describe the interaction between core and valence electrons, and the local density approximation for exchange and correlation interaction with an *ab initio* calculation with no adjustable parameters. For instance, incorporating a substitutional B atom relaxes the four surrounding Si-atoms inward by 0.21 Å (9%) and moves the B atom slightly (0.1 Å) off center toward the plane with three Si atoms, for a slight tendency of a threefold coordination. The *total energy* gain by this relaxation is 0.9 eV (Denteneer et al., 1989). A small fraction of it contributes to the chemical shift of the electronic energy of the hydrogef-like ground state.

21.1.5 Hydrogen-Like Acceptors

In principle, the hydrogenic effective mass theory for a shallow acceptor is much like that for a donor, except that three bands must be considered: the light and heavy hole valence bands and the split-off spin-orbit band. With larger split-off energies Δ compared to kT, the contribution from the split-off band to the ground-state energy can be neglected. In addition, band warping causes an anisotropy of the effective masses, which needs to be considered. For example, the

Table 21.2: Acceptor levels in (meV) computed from point charge screened potentials (after Baldereschi and Lipari, 1973).

Material	$1S_{3/2}$ (Γ_8)	$2S_{3/2}$ (Γ_8)	$P_{1/2}$ (Γ_6)	$P_{3/2}$ (Γ_8)	$P_{5/2}$ (Γ_8)	$P_{5/2}$ (Γ_7)
Si	31.56	8.65	4.18	12.13	8.51	5.86
Ge	9.73	2.89	0.61	4.30	2.71	2.04
AlSb	42.45	12.40	3.35	18.46	12.00	8.22
GaP	47.40	13.69	4.21	19.17	13.04	9.42
GaAs	25.67	7.63	1.60	11.38	7.20	5.33
GaSb	12.55	3.77	0.650	5.74	3.59	2.61
InP	35.20	10.53	1.97	15.89	9.98	7.32
InAs	16.31	5.00	0.420	7.91	4.76	3.63
ZnS	175.6	51.98	11.65	77.62	49.55	35.37
ZnSe	110.2	32.98	6.07	50.04	31.47	22.68
ZnTe	77.84	23.07	5.09	34.72	22.32	15.36
CdTe	87.26	26.42	3.70	41.43	25.85	17.68

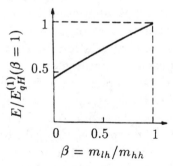

Figure 21.10: Energy of the acceptor ground state as a function of the effective mass ratio between light and heavy holes (after Gel'mont and D'yakonov, 1971).

characteristic decay length of the eigenfunction in the (111) direction in Ge:Ga is 92 Å, and in the (100) direction is 87 Å.

The *ground-state energy* for acceptors depends on the ratio between light and heavy hole masses ($\beta = m_{pl}/m_{ph}$), as computed by Gel'mont and D'yakonov (1971) and shown in Fig. 21.10. It decreases nearly linearly with decreasing β.

Holes in each of these bands contribute to the *excited state level spectrum* of the acceptors, and thereby produce a greater wealth of levels. This is shown in Fig. 21.11, with levels identified as

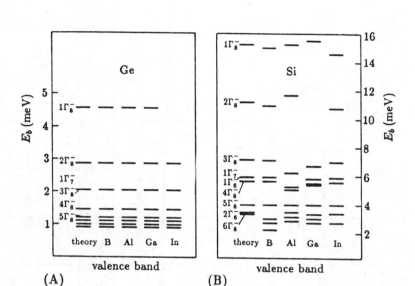

Figure 21.11: (A) Ground and (B) excited states of acceptors in Ge and Si (after Lipari and Baldereschi, 1978; experimental data from Haller and Hansen, 1974). ©Pergamon Press plc.

corresponding to the light and heavy hole bands of symmetry Γ_8 and spin-orbit split band of symmetry Γ_7. The quantum number preceding the band notation in Fig. 21.11 is the sum of quantum numbers, indicating successively higher excited states.

In addition, the valence bands are anisotropic, requiring the use of the appropriate Luttinger parameters, also called the inverse effective mass constants—see Luttinger and Kohn (1955) and Luttinger (1956). For a review, see Bassani et al. (1974) and Pantelides (1978).

Baldereschi and Lipari (1973) have computed a number of acceptor levels (Table 21.2) in various semiconductors with point charge screened potential. These computations provide the correct trend, but the so-obtained levels still show some differences to the experimental values given in Table 21.3.

21.1.6 Bound and Resonant States

The bound electron can be described as a wave packet of Bloch states with vanishing group velocity, which is *localized* at the lattice defect. These Bloch functions must be centered around critical points where $\nabla_k E(\mathbf{k})$ vanishes—see Section 8.5.3A and Callaway (1976).

At higher energies and multiple bands, one obtains a permitted excited state of the defect center within the gap with similar fea-

Table 21.3: Observed shallow acceptor levels in meV (after Bassani, et al. 1974).

Semiconductors	Acceptors	Transitions								Ionization energy
		1st	2nd	3rd	4th	5th	6th	7th	8th	
Ge	B	6.24	7.57	7.94	8.69	9.06	9.32	9.65	9.81	10.47
	Al	6.59		8.27	9.02		9.67	10.02	10.15	10.80
	Ga	6.74	8.02	8.44	9.19		9.84	10.17	10.31	10.97
	In	7.39	8.42	9.08	9.86	10.20	10.48	10.81	10.96	11.61
	Tl	8.87	9.83	10.57	11.32	11.65	11.92	12.26	12.43	13.10
Si	B	30.38	34.53	38.35	39.64 39.91	41.52	42.50	42.79	43.27	44.5
	Al	54.88	58.49		64.08 64.96 65.16	66.28	66.75	(67.1)	67.39	68.5
	Ga	58.23		67.12	67.95 68.25 68.43	69.85	70.49	(70.8)	71.11	72
	In	141.99	145.79	149.74	150.80 151.08	(152.8)	153.27		153.97	155
AlSb	Zn	24.66	26.21	27.17 27.53 27.87	29.41	30.9	32.2			33
InSb	Cd	5.7	6.94	7.31	7.93					
	Ag	25.5	26.5	27.3	27.9					

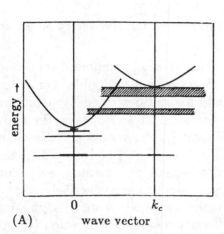
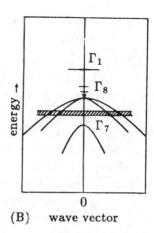

Figure 21.12: Schematic of (A) a donor or (B) an acceptor with excited localized states within the gap and resonant states within the band (after Bassani and Pastori Parravicini, 1975).

tures as previously described. Alternately, one finds states which are related to higher bands (e.g., the X-band in GaP or the spin-orbit split-off valence band in Si) and observes an overlap of these states with lower bands (Fig. 21.12). Here, the eigenfunctions form a *resonant state* within such a band composed of running Bloch waves (Bassani et al., 1969, 1974). The resonant states have a width that depends on the exchange integral between localized and nonlocalized states. The broadening of the resonant state occurs because the electron has a much reduced lifetime in the quasi-hydrogen state relating to the upper band before it tunnels to the lower band and relaxes to its $E(k)$ minimum. This is easier for higher energies; here the lifetime is smaller, and therefore the broadening is larger—see Fig. 21.12A. Similar features are obtained for localized and resonant states of an acceptor relating to the spin-orbit split-off band, and tunneling into the light and heavy hole bands (Fig. 21.12B). Resonant states with substantial broadening have been observed experimentally—see Onton et al. (1967), Bassani et al. (1969), Onton (1971), and Onton et al. (1972).

21.1.7 Shallow Defects in Compound Semiconductors

The shallow donor or acceptor in a compound semiconductor is more complex than in an elemental semiconductor for two reasons:

(1) the interaction of the electron or hole with the alternatingly charged ions of the lattice, and

(2) the differentiation between incorporating the defect on an anion or cation site.

Although the degree of ionicity of good compound semiconductors is small, the above-stated effects are not negligible. For example, isocoric acceptors in GaP should be well described by point-charge quasi-hydrogenic models—see Section 21.1.4. However, the two isocoric acceptors, GaP:Zn$_{Ga}$ and GaP:Si$_P$, have substantially different ground-state energies—64 and 204 meV, respectively. This difference cannot be explained by site-dependent screening, even though neighboring anions are expected to screen more effectively since they are surrounded by more electrons. It needs a more sophisticated analysis, beyond that of the effective mass approximation (Bernholc and Pantelides, 1977).

The degree of ionicity in compound semiconductors also determines the coupling of electrons with the lattice, i.e., with phonons; it is described by Fröhlich's coupling constant α_c [Eq. (28.6)]. Such interaction can be included by considering, instead of a Bloch electron, a *polaron* (see Section 28.1.2) to interact with the defect center, with an effective mass $m_{pol} = m_n/(1 - \alpha_c/6)$—see Eq. (28.10). The eigenstates of a hydrogen-like defect in compound semiconductors can be estimated as

$$E_{qH}^{(c)} = \left(1 + \frac{\alpha_c}{6} + \frac{\alpha_c}{24}\frac{E_{qH}}{\hbar\omega_{LO}}\right) E_{qH} \qquad (21.26)$$

(see Sak, 1971) where E_{qH} is the quasi-hydrogen energy [Eq (20.1)]. However, this approximation is not sufficient to explain the observed variations of the ground-state energies indicated above.

The modified hydrogenic effective mass approximation describes reasonably well the level spectrum of *excited states* of shallow donors and acceptors in compound semiconductors with a sufficiently large ε/m_p ratio, i.e., for many III-V and II-VI compounds—see Grimmeiss (1987).

21.1.8 Higher Charged Coulomb-Attractive Centers

With higher charges, a Coulomb-attractive center has its eigenfunctions closer to the core and requires more attention to core correction. Therefore, its ground state is more akin to deep level centers—see Chapter 22.

Higher charged centers can be created by substitutional impurities. These are further away from the group of the replaced element, and act similarly to a hydrogen-like donor or acceptor, except that

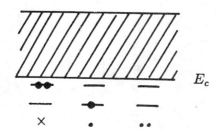

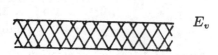

Figure 21.13: Double donor (e.g., S in Si) with corresponding charge character relative to the neutral lattice.

more electrons or holes are donated. For instance, if in a Si lattice one of its atoms is replaced by a sulfur, selenium, or tellurium atom, two electrons can be donated. Since the first electron is bound to a doubly charged center, it is bound at a level four times as deep (charge $Z = 2$):

$$E_i = \frac{Z^2 e^4}{2(4\pi\varepsilon_0\hbar)^2} \frac{m_n}{\varepsilon_{st}^2} = R_H \frac{Z^2}{\varepsilon_{st}^2} \frac{m_n}{m_0}. \tag{21.27}$$

The second electron behaves like an electron attached to an ordinary hydrogen-like donor. Hence the substitutional sulfur can be represented by a double donor with two levels (Fig. 21.13). Occupancy, however, determines which of the two levels is active: when filled with two electrons, only the shallow level is active; ionized once, the other electron becomes more strongly bound and the deeper level is active. From the quasi-hydrogen model, we estimate ionization energies of ~ 50 meV and ~ 120 meV for the second donor level in Ge and Si, respectively, compared to 10 or 32 meV for the first level. The actual energies for S in Si are 302 and 587 meV for the first and second ionization levels (Grimmeiss et al., 1980). The larger energies indicate tight binding, which makes the hydrogen-like approximation less accurate and requires central cell potential consideration. We will therefore return to these centers in Chapter 22.

A similar behavior is expected and observed for *two-level acceptors* from substitutional group II elements, e.g., Zn or Cd in Ge or Si with a second acceptor level at ~ 25 or ~ 210 meV, respectively, compared to 10 or 53 meV for the first level.

Elements from further-removed groups are more difficult to implant as substitutionals. Anions are often too large for the host lat-

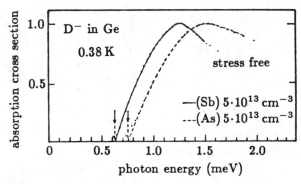

Figure 21.14: Photoionization cross section for D^- states in ultrapure Ge with As or Sb doping at 0.38 K. The low energy threshold indicates the binding energy of the D^- center (after Narita, 1985). ©Pergamon Press plc.

tice; cations tend to become more easily incorporated as interstitials (see below).

In an AB-compound the situation is more complex, as illustrated by incorporating a group IV element such as Sn (which is two groups removed from Cd) as a substitutional into CdS. Replacing the Cd ion, it acts as a two-level donor; replacing an S ion, it acts as a two-level acceptor—see also Sections 44.5.2 and 44.6.1C. This ambiguity makes it difficult to predict the behavior of such type of dopants without additional information.

21.1.8A Over-Charged Donors or Acceptors One observes doubly charged normal donors (or acceptors) when the neutral donor (or acceptor) can trap an additional electron (or hole). These are known as D' or $A^\bullet$ centers, and can be compared to an H^- ion (Faulkner, 1969).

The binding energy of the additional carrier is very small (0.54 and 1.7 meV for hydrogen-like impurities in Ge and Si, respectively—see Lampert, 1958), and results in a very large radius of the quasi-hydrogen eigenfunction, requiring high purity to avoid complications due to overlap.

The ionization energy can be measured from the photoconduction threshold (see Fig. 21.14), and shows a chemical shift for As'_{Ge} and Sb'_{Ge}. Anisotropy and multivalley effects are responsible for the deviations from the simple H^--ion model estimates.

Another type of center develops when a doubly charged donor (or acceptor), as described in the previous section, traps another electron (or hole); it becomes over-charged and binds three carriers. Because of the Pauli principle, this center does not have an isolated atom (such as He$^-$) as an analogue, which exists only in the metastable $(1s)(2s)(2p)$ state. On the other hand, the over-charged $Z = 2$ center can exist in a $(1s)^3$ binding state (McMurray, 1985). The ground-state energy of such a center requires central cell consideration as well as radial and angular correlation between the trapped carriers.

21.1.9 Metal-Ion Interstitials

Cation interstitials, either extrinsic or intrinsic, usually behave as donors. An example was given in Section 21.1.4 (Fig. 21.6) with Li as an interstitial in Si—see Reiss et al. (1956) and Haller et al. (1981). The metal atom on an interstitial position prefers to donate its valence electron(s). When it has no counterpart to form a charge-compensating bond, it can be described as a hydrogen-like donor. Since ionized cations are usually much smaller than anions, they are more easily incorporated on interstitial sites. Therefore, interstitial donors are more readily observed than interstitial acceptors.

The ground-state level of the interstitial depends on its site of incorporation—see, e.g., Fig 19.3 and Jansen and Sankey (1986); for interstitials in II-VI compounds, see Watkins (1977). Depending on the valency of the incorporated metal ion, they can act as single- or multilevel donors.

In a simple hydrogen-like model, higher excited states of these impurities cannot be distinguished from the classical substitutional donor. The chemical shift of the ground state, however, is substantial. In a pseudopotential approximation (Pantelides, 1975), a complication arises since there is no cancellation, such as for the potential of a replaced host atom in a substitutional impurity.

Self-interstitials in elemental semiconductors are rather deep centers, and will be discussed in Section 22.4.3.

21.2 Excitons Bound to Impurity Centers

When a neutral impurity is incorporated, it becomes charged when it traps a carrier. This now-charged defect acts as a Coulomb-attractive center, which can in turn trap a carrier of the opposite sign and thereby form a quasi-hydrogen state (Thomas et al., 1966). This state can be described as an exciton (see Section 15.1) bound to a neutral defect center (Hopfield et al., 1966).

The binding energy of such a center is the difference between the free-exciton energy and the bonding energy of the exciton to this center, which is typically on the order of 10 meV (Faulkner, 1968). There are numerous neutral centers to which such excitons can be bound, such as deep centers (e.g., isoelectronic defects) or shallow centers, such as neutral hydrogen-like donors or acceptors. We will discuss both types briefly in the following sections.

21.2.1 Excitons Bound to Shallow Donors or Acceptors

Here, we can distinguish excitons bound to neutral or to ionized donors or acceptors. Binding to neutral centers at first view shows some similarities to the binding at deep level centers, although the bonding mechanism is different. While the first carrier binds to a deep center entirely through short-range interaction, consequently attracting an oppositely charged carrier into a quasi-hydrogen orbit, here both carriers (the exciton) are attracted simultaneously into an H_2-like quasi-hydrogen state. In some respects, this state may be compared to a trion, i.e., two electrons and a hole attracted to an ionized donor, or two holes and an electron attracted to an ionized acceptor—see Section 15.1.2D. For a review, see Dean (1984).

21.2.1A Excitons Bound to Ionized Donors   It is relatively easy to calculate the bonding of an exciton to an ionized donor. The Hamiltonian (e.g., for an I_3 center in CdS) can be written as

$$E_{\text{exc}}^{(b)} = \frac{\hbar}{2m_p}\nabla_p^2 - \frac{\hbar}{2m_n}\nabla_n^2 + \frac{e^2}{4\pi\varepsilon\varepsilon_0 r_p} - \frac{e^2}{4\pi\varepsilon\varepsilon_0 r_n} - \frac{e^2}{4\pi\varepsilon\varepsilon_0 r_{\text{np}}}, \quad (21.28)$$

where r_p and r_n represent the distances of the trapped hole and electron from the donor, and r_{np} is their relative distance. The energy of the bound exciton is sensitive to the effective mass ratio m_n/m_p. Above a critical value of ~ 0.45, no bound exciton states are observed: the kinetic energy of the hole then becomes too large to be bound to the neutral donor. Such $D^\bullet, X$ states therefore exist in CdS but are absent in Si and GaP, where the mass ration is larger than 0.45 (Rotenberg and Stein, 1969; see also Dean and Herbert, 1977). For an ionized acceptor, the mass ratio condition is inverted and therefore is not fulfilled for most semiconductors. Consequently, excitons are usually not bound to ionized acceptors.

21.2.1B Excitons Bound to Neutral Donors, Acceptors
The bound exciton states on neutral acceptors $(A^\times, X)$ or on neutral donors $(D^\times, X)$ are not limited by the effective mass ratio. Their

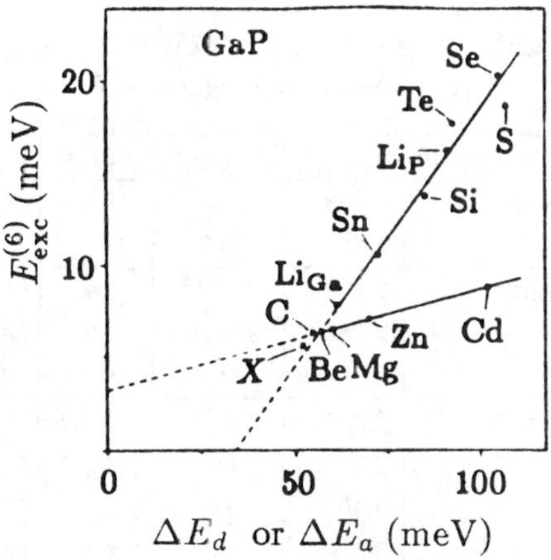

Figure 21.15: Binding energy of excitons to neutral donors or acceptors in GaP as a function of the ionization energy of single donors or acceptors (after Dean, 1984).

binding energy is proportional to the ionization energy of donors or acceptors, as shown in Fig. 21.15—see Dean, 1973. This empirical relation is referred to as *Hayne's rule* (see e.g., Halsted and Aven, 1965) and is given by (Halsted, 1967):

$$E_{\text{exc}}^{(b)} = a + b\Delta E_a \qquad \text{or} \qquad E_{\text{exc}}^{(b)} = c + d\Delta E_d. \tag{21.29}$$

In GaP the value of a is positive, while c is negative with $b \simeq 0.1$ and $d \simeq 0.2$. The intersects are related to the increase of the relative charge in the central cell when the exciton is bound; consequently, a and b are close to zero in Si (Dean, 1971; Haynes, 1960).

When calculated in an H_2-molecule approximation, the binding energy of $(D^\times, X)$ or $(A^\times, X)$ are about 30% of the ionization energy of the corresponding donors or acceptors. These energies are listed in Table 21.4 for a number of II-IV compounds.

In addition to the binding energy of such exciton, lines corresponding to the excited states of the donors or acceptor, and suggested by Thomas and Hopfield (1962b), can be observed (Thewalt et al., 1985).

21.2.2 Excitons Bound to Isoelectric Centers

These centers have a central core potential term but no Coulomb field for binding of the exciton—see Section 22.4.7. There is a large variety of such centers, for instance GaP:N$_P$.

The exciton bound to such centers is referred to as an *isoelectronically bound exciton* (Hopfield et al., 1966): an attractive core potential of the center *tightly binds* an electron which, in turn, can bind a hole into a quasi-hydrogen orbital. This electron-hole pair

Table 21.4: Bound exciton lines in II-IV compounds (after Taguchi and Ray, 1983). ©Pergamon Press plc.

Compound	E_L(ev)	E_B meV	$T(K)$	Assignment	Energy Level	Symmetry
ZnS	3.758	34.4	8	$(A^\times, X)$	$E_v + 1.22$eV	
	3.724	68.3	8			
ZnSe	2.799	2-3	4.2	$(D^\times, X); V_{Se}^\times$	$E_c - 0.02$eV	
	2.780	19-22	4.2	$(A^\times, X); V_{Zn}^\times$	$E_v + 0.28$eV	
				V_{Zn}''	$E_v + 1.1$eV	C_{3v}
ZnTe	2.377	4	1.7-4.2	$(D^\times, X)$		
	2.375	6	1.7-4.2	$(A^\times, X); V_{Zn}^\times$	$E_v + 0.06$eV	T_d
	2.362	19	1.7-4.2	$(D^\bullet, X); V_{Te}^\bullet$	$E_c - 0.03$eV	
	2.360	21	1.7-4.2	$(D^\times, X)$		
				V_{Zn}''	$E_v + 0.7$eV	
CdS	2.5471	5	4.2	$(D^\times, X)$		
	2.536	16	4.2	$(A^\times, X); V_{Cd}^\times$	$E_v + 0.8$eV	
				V_{Cd}	$E_v + 0.8$eV	
CdSe	1.822	4	4.2	$(D^\times, X); Cd_i^\times$		
	1.817	9	4.2	$(A^\times, X); Se_i^\times$		
CdTe	1.594	1	1.7-4.2	$(D^\times, X); V_{Te}^\times$	$E_c - 0.018$eV	
	1.590	4	1.7-4.2	$(A^\times, X); V_{Cd}^\times$	$E_v + 0.06$eV	C_{3v} or C_{2v}
	1.587	8		$(D^\bullet, X)$		
				V_{Cd}''	$E_v + 0.6$eV	

is equivalent to an exciton bound to the center, while it acts as an acceptor because of its asymmetry in carrier binding. Similar defect centers can be expected when first attracting a hole into a deep neutral center, and then an electron into the resulting Coulomb attractive defect, which would act as a donor-like defect center with the bound exciton.

21.2.3 Excitons Bound to Near and Distant Pairs

An example of a defect pair that can bind an exciton is the Be-Be center in Si. Here two Be atoms replace one Si atom and produce an axial field that causes a characteristic splitting of the ground state of the center. Such splitting is helpful in identifying excited states of the same center (see below).

The basic line structure of excited states of this bound exciton therefore behaves like that of an acceptor (e.g., B in Si), but is shifted by the binding energy of the exciton (3 meV) as shown in Fig. 21.16A and inset (d). The excited states are related to valence band odd-parity transitions $1\Gamma_8^-$, $2\Gamma_8^-$, and $1\Gamma_6 + 1\Gamma_7^-$ (Lipari and

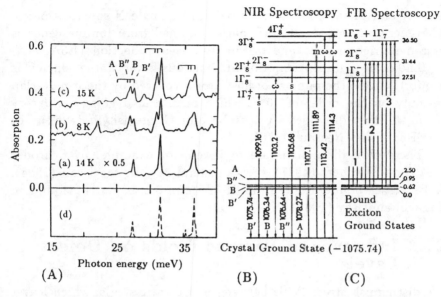

Figure 21.16: (A) Induced optical absorption at FIR of the Be–Be complex in Si. Brackets indicate the predicted absorption line quadruplets from the bound exciton ground state [labeled in (B)] to the three excited states shown in (C). Lower graph (d) shows the excited states of an Si:B acceptor shifted by the binding energy of the exciton (3 meV) to indicate coincidence in position and line intensity with the upper graphs. (B) Luminescence spectrum to the crystal ground state for annihilation of the bound exciton (NIR). This spectrum also contains odd-parity transitions. (C) Excited states of the bound exciton from its ground state, which is split due to the axial field of the Be–Be center (after Thewalt et al., 1985). ©Pergamon Press plc.

Baldereschi, 1978). The fine structure of the ground state (B, B$'$ and B$''$) becomes visible in all of these transitions in the far infrared (FIR) at slightly elevated temperatures, when these states become populated, as shown in Fig. 21.16A (a) and (b) for the excited states of the bound exciton.

In addition, these transitions can also be identified by the recombination radiation of this exciton, as indicated in the near infrared (NIR) line scheme in Fig. 21.16B (Thewalt et al., 1985).

Another isoelectronic system extensively studied is AgBr:I$_{Br}$ (Czaja and Baldereschi, 1979).

The eigenstates of the exciton, bound to certain pairs, depends on the intrapair distance. For instance, excitons bound to N–N pairs in GaP (Thomas and Hopfield, 1966; Cohen and Sturge, 1977) are observed for pairs with binding energies between 160 meV for nearest and 21 meV for farthest pairs, i.e., single N substitutionals. The energy spectrum of such pairs obtained from luminescence is extremely rich in lines, and will be discussed in Section 44.5.2.

The *optical absorption* of bound excitons is very large, since it is proportional to the number of host atoms covered by the *n-p* overlap (usually on the order of 10^3), which results in a *giant oscillatory strength*—first discussed by Rashba and Gurgenishvili (1962); see also Henry and Nassau (1970).

Theoretical discussions are reviewed, for example, by Schröder (1973) and Herbert (1977). For a general review, see Dean (1984). A listing of bound excitons in II-VI compounds can be found in Taguchi and Ray (1983).

21.3 Influence of Various Fields on Defect Levels

We distinguish stress fields between hydrostatic stress, which leaves the symmetry of the lattice unchanged, and uniaxial stress, which induces changes of the lattice symmetry. Electric fields can be external, usually homogeneous fields, and built-in fields which are space-charge-induced and show gradients. External magnetic fields again are homogeneous, while such fields exerted from spin- or orbit-coupling contain inhomogeneous components on the defect-atomic scale. We will discuss the influence of these fields on shallow level centers in the following sections.

Although the influence of electric and magnetic fields is much larger in shallow levels because of the large volume occupied by their eigenfunctions, there are also substantial influences on some deep level centers. These, however, are more complex in nature. Furthermore, some deep centers can also show rather extended wavefunctions, e.g., the A_1 center in Si which extends 70% up to the seventh neighbor shell.

21.3.1 Influence of Hydrostatic Pressure

Hydrostatic pressure does not influence the impurity potential appreciably, nor does it change the position of deeper bands. There is a much smaller shift of the valence band than for the conduction band for most semiconductors.

With increasing pressure a decreasing lattice constant causes characteristic changes, which are most pronounced for the conduction bands and are observable by a change in the band gaps. In

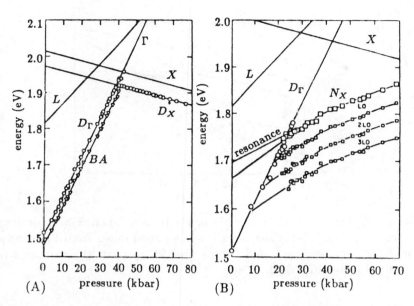

Figure 21.17: Luminescence of GaAs at 5 K. (A) Shallow defect levels in GaAs are connected to the lowest band (Γ below, and X above 41 kbar); D_Γ and D_X are excitons bound to donors, and BA is the band-to-acceptor luminescence emission. (B) Deep N-levels in GaAs are not connected with the lowest conduction band. Shown also are phonon replicas ($LO \ldots 3LO$) and resonance states of the N-levels below 22 kbar within the Γ-continuum (after Wolford, 1986).

cubic semiconductors, the energy separation from valence to conduction band increases at Γ and L (see Fig. 21.17 for GaAs) but decreases at X. These shifts occur with a similar pressure coefficient of approximately 10 μeV/bar in IV and III-V semiconductors (Paul, 1966).

Shallow donor states shift together with the conduction bands and can be identified as such by their connection to respective bands with an identical shift. Such a shift can be observed by the recombination luminescence of excitons, which are bound to such donors and shown in Fig. 21.17A. Changes of the connection from the Γ-band to the X-band are observed near the crossover pressure ($\sim$ 41 kbar for GaAs). These bound exciton lines are given here to exemplify shallow defects. Free excitons and conduction band-to-acceptor luminescence behave similarly.

Near the crossover pressure, the Γ-related shallow levels can become resonant with the X_1 minima. This can be observed by a large

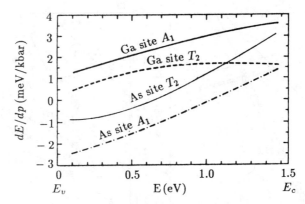

Figure 21.18: Shift of deep levels in GaAs as a function of hydrostatic pressure. A_1 and T_2 type levels are s-like pulled down from the conduction band (antibonding), and p-type pushed up from the valence band (bonding) (after Ren et al., 1982).

increase in carrier scattering to *resonance scattering* when the X_1 minimum is shifted to coincide with the donor level. At this pressure a minimum in the carrier mobility is observed (see Section 32.3.1A; Kosicki and Paul, 1966).

Other effects occur when, for example, in Ge the minimum goes from L_1 to Δ_1 at $p \simeq 50$ kbar. The degeneracy is increased from four- to tenfold, and is decreased to sixfold when the Δ_1 minimum is lowered below L_1. The consequence is a splitting from two sublevels (E_{Γ_1} and $E_{\Gamma_{15}}$) into five, and then back to three sublevels—E_Γ, $E_{\Gamma_{12}}$, and $E_{\Gamma_{15}}$ (Shimuzu, 1965).

Deeper donor levels show a lesser influence from hydrostatic pressure since they are more dependent on the core potential, which is little changed. Deep levels also are connected to several bands which tend to reduce and partially cancel their individual contributions. This can be seen in nitrogen-doped GaAs at pressures in excess of 20 kbar, as shown in Fig. 21.17B (Wolford et al., 1979). Below ~ 20 kbar the N-levels become resonant levels within the Γ-band and can no longer be observed by photoluminescence. A theoretical estimate of the shift of deep levels in GaAs was given by Ren et al. (1982) and is shown in Fig. 21.18.

Acceptors are insignificantly influenced by hydrostatic pressure, as it has a lesser influence on the valence band structure near $k = 0$.

Quantum well states and **superlattice states** show the connection of direct transitions with the Γ-band in GaAs $Al_\xi Ga_{1-\xi}As$ heterostructures (Fig. 21.19). Above the crossover pressure, indirect

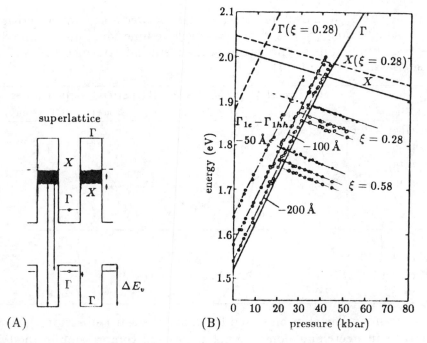

Figure 21.19: (A) Schematics of the GaAs/Al$_\xi$Ga$_{1-\xi}$As superlattice structure with transitions from Γ- and X-bands indicated. (B) Photoluminescence follows the pressure dependence of the direct band gap for the Γ-transition, although shifted toward higher energies due to confinement. Above the crossover pressure $\Gamma \to X$, indirect transitions are shown with phonon replica for the superlattice structure. These transitions occur from X-electrons inside the barrier to Γ-holes within the well (see subfigure A) (after Wolford, 1986.)

transitions within the barrier are observed, including phonon replica, which now shift parallel to the X-band with pressure and composition (Fig. 21.19B). For more information, see Wolford (1986).

21.3.2 Influence of Uniaxial Stress

Uniaxial stress lowers the symmetry of the semiconductor, depending on the relative direction of the stress with respect to the crystal axes. Consequently, the band structure becomes more complicated. The band degeneracies are removed, producing a splitting and a shift of the bands, as indicated for the six, without stress, equivalent valleys in Si and for the four valleys in Ge in Table 21.5. The shift is proportional to the stress, with the deformation potentials Ξ_d and Ξ_s for dilatation (subscript d) in the direction normal to the main axes

Table 21.5: Removal of degeneracies between equivalent critical points and donor states (s and p_0 envelope functions) under uniaxial stress, yielding new symmetries of the resulting deformed valleys (after Bassani et al., 1974).

c-band Minima at	Cubic Crystal (T_d Group)	[100] Stress (D_{2d} Group)	[111] Stress (C_{3v} Group)	[110] Stress (C_{2v} Group)
(100)		$X_1 + X_3$		$\Sigma_1 + \Sigma_2$
($\bar{1}$00)				$+ \Sigma_3 + \Sigma_4$
(010)	$\Gamma_1 + \Gamma_{12}$		$2\Lambda_1 + 2\Lambda_3$	
(0$\bar{1}$0)	$+\Gamma_{15}$ (Si)			
(001)		$X_1 + X_2$		
(00$\bar{1}$)		$+X_5$		$2\Sigma_1$
$(\frac{1}{2}, \frac{1}{2}, \frac{1}{2})$			Λ_1	$\Sigma_1 + \Sigma_3$
$(\frac{\bar{1}}{2}, \frac{1}{2}, \frac{1}{2})$				
$(\frac{1}{2}, \frac{\bar{1}}{2}, \frac{1}{2})$	$\Gamma_1 + \Gamma_{15}$	$X_1 + X_3$	$\Lambda_1 + \Lambda_3$	
$(\frac{1}{2}, \frac{1}{2}, \frac{\bar{1}}{2})$	(Ge)	$+X_5$		$\Sigma_1 + \Sigma_4$

of the sidevalley ellipsoids, and for uniaxial shear (subscript s). This results in stretching along the main axes and compression in the two normal directions—see Fritzsche (1962), Pollak (1965), and Cardona (1969). The deformation potentials are proportionality factors for the energy shift which, in cubic materials, is given by:

$$\Delta E^{(j)} = \sum_{\alpha, \beta} (\Xi_d \delta_{\alpha\beta} + \Xi_s \tilde{k}_\alpha \tilde{k}_\beta) u_{\alpha\beta} \qquad (21.30)$$

with the strain components $u_{\alpha\beta}$ (Herring and Vogt, 1956). Here $\delta_{\alpha\beta}$ is the Kronecker δ-symbol, and $\tilde{k}_\alpha, \tilde{k}_\beta$ are the components of the unit k-vector on the α- and β-axes. Uniaxial stress in the [111] direction for Si and in the [100] direction for Ge, however, results in no splitting, since the symmetry of the valleys remains unchanged under such a stress—see Table 21.5.

Shallow impurities relating to these bands split accordingly. In Si, for instance, the $1s$ donor state, which is split into A_1, T_2, and E (Γ_1, Γ_{12}, and Γ_{15}) because of intervalley interaction, splits further when uniaxial stress is applied in [100] direction. The Γ_{12}- and Γ_{15}-levels split and the Γ_1-level shifts. The splitting is proportional to the stress, and the shifting, as a second-order effect, is proportional to the square of the stress. This is shown in Fig. 21.20A; see also Wilson and Feher (1961).

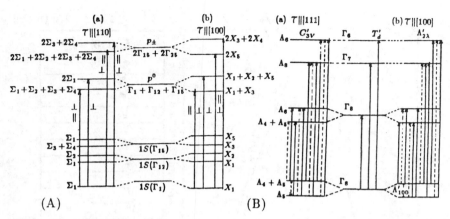

Figure 21.20: Schematic representation of shifts and splitting of (A) donor and (B) acceptor states in Si under uniaxial stress $\mathcal{T}$. Dashed line transitions for $\mathbf{E}\|\mathcal{T}$, solid lines for $\mathbf{E}\perp\mathcal{T}$ (after Bassani et al., 1974).

The influence of uniaxial stress on **acceptor states** is determined by the degeneracy of the valence bands at $\mathbf{k} = 0$. The states relating to Γ_6 and Γ_7 are shifted, and the Γ_8-state is shifted and split. The higher excited states are influenced accordingly (Rodriguez et al., 1972). The transitions are polarization-dependent, and are shown in Fig. 21.20B.

Such changes of the level spectrum can be observed in optical absorption or luminescence, and are helpful in identifying the levels with respect to their symmetry—see Section 15.3. For a review, see Rodriguez et al. (1972).

An example of the shifting and splitting of levels of the Se donor in Si (see also Section 21.1.8), is shown in Fig. 21.21 for the transmission spectrum. The allowed $^1T_2$ level splits with increasing stress and avoids crossing with the ^1E and $^3T_2$ levels, which are forbidden at low stress and become allowed with increasing linear stress in the [110] direction.

21.3.3 Influence of an Electric Field

An electric field acts in a manner similar to uniaxial stress by a shifting and splitting of the defect levels due to a lowering of the crystal symmetry. Although the changes are similar, they are smaller than the ones described in Section 21.3.2. The ground state typically shifts by 100 μeV for fields of 10 kV/cm; the shift of the first excited state for the same field is on the order of 1 meV. The effect is similar to the atomic *Stark effect*.

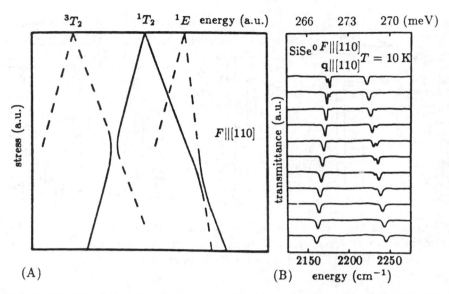

Figure 21.21: (A) Schematic energy diagram of the Se donor in Si (solid curves: observed). (B) Transmission spectra for increasing (from 1 to 11) uniaxial [110] stress as family parameter with lower and upper branch $^1T_2$ transitions allowed at low stress, changing to $^3T_2$ and ^1E transitions, respectively, at higher stress (after Bergman et al., 1986).

21.3.3A The Stark Effect   A splitting of degenerate eigenstates proportional to the applied homogeneous field was observed by Stark (1914) in hydrogen (*linear Stark effect*). This effect is due to the superposition of a perturbation term $e\mathbf{F}\cdot\mathbf{r}$ in addition to the Coulomb potential $-e^2/r$. A comprehensive description of the classical effect can by found by Herzberg (1937) or Sommerfeld (1950). Because of the larger quasi-hydrogen radius of shallow defect levels, the corresponding splitting within a semiconductor takes place at much lower electric fields. For an example, a shift of 0.1 meV requires $\sim 10^4$ V/cm for the ground state and only $\sim 10^3$ V/cm for the $2s$ excited state of shallow donors. An interpretation of the Stark effect for excited states, however, is complicated when the extended wavefunction starts to overlap with band states—see Franz-Keldysh effect, Section 42.4.3H. A Stark effect splitting is shown in Fig. 21.22 (Blossey, 1970). The normalizing field is the so-called *ionization field* of such defects, given by

$$F_i = \frac{E_{\mathrm{qH}}}{e a_{\mathrm{qH}}} \qquad (21.31)$$

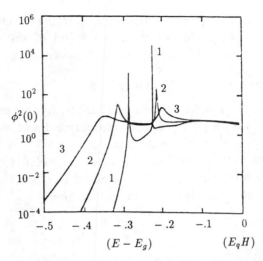

Figure 21.22: Stark-effect splitting of $n = 2$ hydrogen-like level at normalized fields $F = 0.01, 0.016$, and $0.025 F_i$ for curves $1 \ldots 3$, respectively (after Blossey, 1970).

Table 21.6: Ionization fields given in (kV/m) by Eq. (21.31) for quasi-hydrogen defects in some compound semiconductors (after Blossey, 1970).

AlSb	12	InP	7.8	ZnS	200	CdS	140
GaAs	5.7	InAs	0.70	ZnSe	75	CdSe	60
GaSb	1.0	InSb	0.08	ZnTe	47	CdTe	31

which is listed for some compound semiconductors in Table 21.6. It is of similar magnitude to the effective field describing interactions with phonons and given by Eq. (11.44).

In the field of other defects or electrons, the Stark effect is more complicated. These defects will already cause some random splitting because of the statistical distribution of their relative distance from each other, except for very low doping densities (Guichar et al., 1972). Therefore, the external field has only a minor influence, and requires a larger field strength, resulting in a dependence $\propto F^2$, the *quadratic Stark effect*. In addition to the described splitting and shift, the perturbation through the field permits optical transitions which, without a field, are forbidden by selection rules. For more information, see Blossey (1970).

For acceptors, one observes that the splitting of the Γ_8-related *acceptor* states is much larger than the shifts of the Γ_6- and Γ_7-related levels (Kohn, 1957; White, 1967; Blossey, 1970).

21.3.4 Influence of a Magnetic Field

The influence of the magnetic field on shallow, hydrogen-like defects can be divided into an effect on the conduction band states and the effect on the eigenstates of the defect.

As discussed earlier (Section 9.1.5C), the magnetic field forces electrons into orbits, circling with cyclotron frequency [Eq. (9.19)]. When the magnetic field is small enough so that

$$\hbar\omega_c \ll E_{qH}^{(n)} \tag{21.32}$$

where $E_{qH}^{(n)}$ is the quasi-hydrogen bonding energy of the nth level, the influence of this magnetic field can be regarded as a small perturbation. This case is discussed first.

When the field is larger, the band states split into distinct Landau levels. Here, additional effects need to be considered that will be discussed later.

For the small field approximation, we obtain a Zeeman splitting of the degenerate levels of the quasi-hydrogen defect and a diamagnetic shift, similar to the observation in isolated atoms. Within a semiconductor, the magnetic field removes *all* degeneracies, including the Kramer's degeneracy due to time reversal, and consequently causes the most extensive splitting of defect levels (Condon and Shortley, 1959; Haug, 1972).

21.3.4A The Zeeman Effect

The splitting of lines of the spectrum of atoms in a longitudinal or transverse magnetic field was first observed by Zeeman (1897) and is termed the *Zeeman effect*. The Zeeman effect is traditionally related to single atoms—see Herzberg (1937). Within a semiconductor, the influence of the surrounding lattice must be taken into consideration. In hydrogen-like defects, the electron orbits are very large. Consequently, a much smaller magnetic induction causes a major splitting.

Defect levels in semiconductors show Zeeman splitting, which relates to the defect quantum number and the appropriate Zeeman levels of the impurity (Bassani et al., 1974). At higher magnetic induction, which can easily be reached for excited states, the wavefunction becomes severely compressed normal to **B**, which causes an increase in the binding energy of the electron at the defect (Baldereschi and

Bassani, 1970). In addition, the band states split into Landau levels, causing further complication—see below and Section 31.3.

The influence of the magnetic field can be included in the Schrödinger equation by replacing the momentum operator $-i\hbar\nabla$ with the operator $-i\hbar\nabla - e\mathbf{A}$, with the vector potential

$$\mathbf{A} = \frac{1}{2}(\mathbf{B} \times \mathbf{r}) \qquad (21.33)$$

and $\mathbf{B}$ the magnetic induction. This yields in the effective mass approximation [see Eq. (21.33)]

$$\left(-\frac{\hbar^2}{2m_n}\nabla^2 - \frac{e\hbar}{m_n}(\nabla \cdot \mathbf{A}) + \frac{e^2}{2m_n}(\mathbf{A} \cdot \mathbf{A}) - \frac{e^2}{4\pi\varepsilon_{st}\varepsilon_0 r}\right)F(\mathbf{r}) = (E - E_c)F(\mathbf{r}).$$
$$(21.34)$$

The second term gives *normal Zeeman splitting*, which is linear in B for all $l \neq 0$ levels:

$$\Delta E = \frac{e}{2m_n}\mathbf{B} \cdot \mathbf{L} = \pm\hbar\omega_c, \qquad (21.35)$$

where $\mathbf{L}$ is the angular momentum operator. The splitting occurs into $2l + 1$ equidistant lines.

In addition, a diamagnetic term, stemming from the third term in Eq. (21.34), is due to the interaction of the electron spin with the magnetic induction (*anomalous Zeeman effect*). This produces a quadratic correction

$$\Delta E_d = \frac{e^2}{8m_n}(\mathbf{B} \times \mathbf{r})^2 \simeq \left(\frac{(\hbar\omega_c)^2}{8E_{qH}}\right)_s \qquad (21.36)$$

which is given here explicitly for the s-states (van Vleck, 1932). The quadratic term is the only magnetic field dependence for s- states which show no normal Zeeman splitting ($l = 0$). It produces a compression of the wavefunction, and thereby results in an increase in the binding energy. However, in some magnetic semiconductors (e.g., in $Hg_{1-\xi}Mn_\xi Te$ with $\xi > 0.1$) it causes a reduction in the ionization energy of shallow acceptors, by making the bands anisotropic and the hole effective mass for heavy holes close to that of light holes. This brings the acceptor closer to the band and causes ionization, i.e., a boil-off of holes (rather than the usual trapping, i.e., a freeze-out of carriers) with magnetic field. At the appropriate temperature (e.g., 1.4 K in $Hg_{1-\xi}Mn_\xi Te$) it can result in a giant negative magnetoresistance: $\Delta\varrho/\varrho_0 = 10^7$ at 70 kG—see Furdnya (1985).

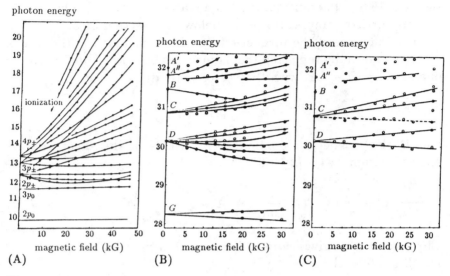

Figure 21.23: Absorption lines as a function of the magnetic induction in Ge: (A) for a P donor; (B) for a divalent Zn acceptor in Faraday geometry; and (C) the same in Voigt geometry ($E \perp H \| [100]$) with conventional notation of G, D, C, B, and A for the excited states of the lower excitation lines of this acceptor (after Horii and Nisida, 1970; Moore, 1971).

At high magnetic field ($\hbar\omega_c \gg E_{qH}$), the fourth term in Eq. (21.34), can be disregarded, and we obtain directly the Landau levels in the conduction band [Eq. (31.53)] as solution of Eq. (21.34).

There is a large body of experimental results showing the influence of the magnetic induction on defect levels. For example, in Ge, a magnetic field in the [111] direction will split the p^0 donor level in Voigt geometry (Section 12.3.3B) into two levels; only one is allowed in Faraday geometry—see Section 12.3.3A. The $p^{\pm}$ levels split into four levels, all of which are allowed in Faraday geometry. This splitting has helped to identify these and higher excited states (Horii and Nisida, 1970), as is shown for P-donors in Ge in Fig. 21.23A.

The splitting of acceptor impurity states is more complicated because of the band degeneracy and spin-orbit coupling; with a magnetic field, all degeneracies are consequently removed. For instance, a $\Gamma_8 \rightarrow \Gamma_8$ transition shows a splitting of the D^{-}term which can be well separated from the others into eight lines—six of which are observed in Faraday configuration and two in Voigt configuration. As many as eight Faraday and four Voigt lines are possible. This is shown in Fig. 21.23B and C for the Zn acceptor in Ge—see also

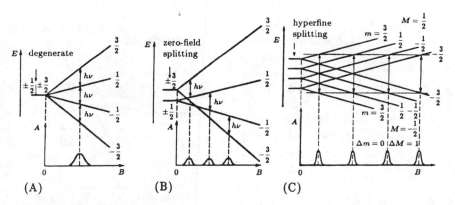

Figure 21.24: Typical types of spin-flip resonances for (A) a degenerate line; (B) a line with spin-orbit splitting; and (C) a line including interaction with unpaired nuclear spin.

Solpangkat et al. (1972) and Carter et al. (1976). For reviews, see Kaplan (1970) and Hasegawa (1969).

21.3.4B Magnetic Resonances at Lattice Defects   Electron spin-flip resonances can be measured in point defects with unpaired spin.

Spin-flip resonance absorption is related to the Zeeman splitting discussed in the previous section. Spin-flipping resonance of defect levels must be distinguished from the spin resonance of conduction electrons between different Landau levels (Gornick and Tsui, 1976). In a typical spin doublet splitting, the upper branch relates to the electronic state with the spin parallel to the magnetic induction; for the lower state the spin is antiparallel. In thermal equilibrium, more lower energy states are filled; a transition from the lower to the upper state can be initiated by supplying the necessary energy difference between these two states. This energy usually lies in the microwave range, where the corresponding resonance absorption can be observed (Poole, 1967; Abragam and Bleaney, 1976). The absorption permits a rather sensitive determination of the magnitude of the Zeeman splitting. Examples of typical spin-flip absorptions are shown in Fig. 21.24. For a review of such resonances in III-V compounds, see Varshni (1967) and in II-VI compounds, see Schneider (1967).

Spin-flip resonances can be detected *optically* when the microwave-induced transition occurs from the lower energy spin-up to the higher energy spin-down states. This also results in corresponding changes of the polarization of absorption or emission (luminescence) in the presence of the magnetic field. For *normal Zeeman transi-*

Figure 21.25: Magnetic resonance transition for donor-acceptor pair in CdS, causing an increase of the σ_- transition from the $S = 1/2$ donor to the $J = 3/2$ acceptor state (mixing with the uppermost valence band of CdS).

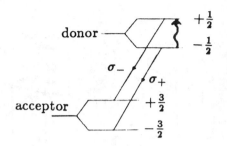

Figure 21.26: Optically determined magnetic resonance emission spectrum of the O^- center in GaP with its level scheme due to splitting by hyperfine interaction with the Ga nucleus shown in the lower part (after Gal et al., 1979).

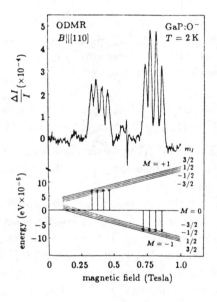

tions in longitudinal observation (Faraday geometry), one sees two components which are left and right circular-polarized. In transverse observation (Voigt geometry), one sees three components with linear polarization, of which one is parallel and two are perpendicular to the field polarized. These are known as π and σ components, for parallel and perpendicular (the latter for *"senkrecht,"* German) polarization. Such changes occur in single defects or in defect associates (Cavenett, 1981) as indicated in Fig. 21.25 for a donor-acceptor pair.

21.3.4C Hyperfine Splitting When the electrons in a defect center interact with the magnetic momentum of the nucleus, one observes *hyperfine splitting* of these levels, which provides information on the chemical identity of the center. As an example, the optically detected hyperfine magnetoresonance signal of oxygen-doped GaP is shown in Fig. 21.26.

21.3.4D Super-Hyperfine Splitting When the electrons of a defect center interact with the nuclei of the surrounding lattice atoms, one observes a *super-hyperfine*, or *ligand hyperfine splitting*. This type of splitting provides valuable information on the structure of the immediate neighborhood of the defect center (Spaeth, 1986), and is used to analyze the eigenfunction, i.e., the electron distribution of defect centers (ENDOR)—see Section 21.1.1.

Summary and Emphasis

The *major features* of shallow level defects can easily be described by a quasi-hydrogen spectrum, modified only by the dielectric constant and the effective mass of the host semiconductor. This relatively simple relationship holds surprisingly well for higher excited states of a large variety of such defects, while the ground state shows substantial deviations according to the chemical individuality of the defect center. Such individuality can be explained by considering the core potential and the deformation of the lattice after incorporating the defect.

Band anisotropies and the interaction between valleys cause the lifting of some of the degeneracies of the quasi-hydrogen spectrum. Local stress and electric fields cause additional splitting. Because of far-reaching interaction between these defects through such fields, the line spectrum is usually substantially, inhomogeneously broadened. Narrow lines can only be observed in ultrapure, strain-free materials.

The dependence of levels on hydrostatic pressure can be used to identify those which are connected to one band only. Shallow level defects are distinguished therefore from so-called deep level centers which originate from tightly bound centers, and connect with both bands; only accidentally may they lie close to one of them.

The influence of uniaxial stress and electric or magnetic fields can be used for further identification, and through selection rules relate to crystal-induced field-distorted symmetries.

Shallow defect centers play a dominant role as donors and acceptors in nearly all semiconducting devices. Their defect-level spectrum provides the key for their unambiguous identification with respect to their chemical identity and their specific incorporation within the host lattice. The sensitivity of shallow defect levels to internal fields permits their use to checking the degree of perfection of ultrapure and nearly stress-free semiconductors.

Exercise Problems

1.(e) Coulomb-attractive centers can interact over long distances when they are in an excited state.
  (a) Estimate the radii of higher excited states of quasi-hydrogen donors in Si, and at what densities overlap with other such donors would occur.
  (b) A more effective screening of the Coulomb potential is considered by introducing the Yukawa potential [Eq. (14.10)]. In which way would this influence the results of the previous problem? Give a quantitative description detailing your reasoning. Use the Debye length as screening parameter; assume all donors to be ionized.

2.(e) Give an explicit formula connecting a_{qH}, ε_{st}, and m^* to distinguish validity ranges of hydrogen-like and tight-binding approximations. Use a sphere containing six lattice atoms as an arbitrary boundary between both approximations in a cubic diatomic crystal with a lattice constant of 2 Å.

3.(*) Is the elementary approach suggested in Problem No. 2 justified? What changes in ε_{st} and m^* would be appropriate? Expand the previous estimates with a more sophisticated model.

4.(e) Discuss the influence of the anisotropy of the electron effective mass in Si and Ge for the quasi-hydrogen levels in your own words. Describe the 1-s electron state in a primitive hydrogen-like picture.

5.(e) Perform the Fourier transformation to obtain from the envelope function in real space $F(r)$ [Eq. (21.8)] the envelope function in momentum space $F(k)$ [Eq. (21.11)].
  (a) Discuss the difference of these two representations and their physical meaning.
  (b) Compute the extent of $F(k)$ in k-space for 1s-, 2s-, and 3s-states in GaAs.
  (c) Compute the extent of $F(\mathbf{r})$ and $F(\mathbf{k})$ for the 1s-state in Si, including the anisotropy of m_n and ε_{st}.

6.(l) Consult the original literature and describe in your own words the ENDOR analysis. How does it relate to obtaining information about the envelope function? Be specific.

7.(*) Describe the differences in the use of actual or pseudopotentials for an isocoric and a nonisocoric donor.

8. Describe resonant acceptor states. Are there resonant acceptor states if there would be only a light and a heavy hole band and no split-off spin-orbit band?

9.(e) Analyze Table 21.2 and identify the different trends you can find for acceptors. Add to the list the corresponding dielectric constants and effective masses. What do you observe in relation to these trends?

10.(l). Why are bound exciton or band-to-acceptor spectra used in Fig. 21.17 rather than spectra connected directly with a donor? Check the recent literature for similar information and report the findings.

11.(r) Calculate the ionization energy for the 2-s state of quasi-hydrogen donors in four typical semiconductors.

12.(*) Describe the differences between the Zeeman effect of isolated atoms and of donors in a semiconductor.

Chapter 22

Deep Level Centers

Deep level centers are connected to conduction and valence bands and often provide a preferred path for carrier recombination or act as deep traps.

In general, deep level centers require a tight-binding analysis in which, at least for the ground state, the wavefunction remains localized close to the core of the defect. They cannot be described by a hydrogenic effective mass approximation. Deep levels, however, do not necessarily require a large binding energy.[*] They are connected to the conduction *and* valence bands, i.e., these deep trap levels do not follow one specific band when perturbed by alloying or the application of hydrostatic pressure. Their central core potential dominates their behavior at the ground state, or they have unsaturated inner shells in transition metal impurities, permitting electronic transitions here. Specifically, such defects may or may not be charged relative to the lattice; they may be isoelectronic or isovalent.

The levels of these centers are described by a short-range potential. The pseudopotential method is an advantageous tool for determining $V(\mathbf{r})$. In addition, the deformed lattice environment must be considered—see Section 22.4.1, and for Jahn-Teller distortion, see Section 22.3. The deep states extend throughout the entire Brillouin zone. States from both bands and all near-band gap valleys are necessary to construct the ground-state electron eigenfunction of this center. The resulting deep defect levels communicate with both bands and act as deep traps for electrons or holes (Section 43.2.1A) or as recombination centers (Section 43.2.1C). A review of such deep centers can be found by Queisser (1971), Stoneham (1975), Lannoo and Bourgoin (1981), and Pantelides (1978, 1986).

---

[*] Deep levels also appear in narrow band gap materials (see Lischka, 1986).

In contrast to shallow level centers, which are easily identified chemically in their specific lattice environment and are rather well understood in their electronic level structure, it is much more difficult to identify a specific deep level lattice defect and to describe theoretically its electronic behavior. Except for a large variety of centers in ionic crystals, which were identified earlier (the well-known class of F-centers in alkali halides), most deep centers in semiconductors are still described by a combination of letters and numbers (Hayes and Stoneham, 1984), given to them by the authors who started their analysis; these centers are identified by their spectral signature.

Only recently, and by a concerted effort of various experimental methods, was it possible to identify some of them unambiguously. These methods include optical absorption, luminescence emission and excitation spectroscopy, electron-spin resonances, optical detection of magnetic resonances (ODMR), ENDOR, EXAFS, and deep level transient spectroscopy (DLTS). An important contribution was the improved growth techniques of ultrapure and stress-free crystals, which eliminate disturbing influences of the defect environment. In spite of new results, other deep level centers seem to escape an identification that is agreeable to all of us. The main problem with deep centers is their tendency to form associates or to incorporate into their structure major lattice deformation of decisive consequences.

In such a complex situation, we will introduce stepwise certain elements characteristic of deep centers.

22.1 Mathematical Models of Deep Level Centers

First, we will show that the eigenvalue spectrum of tight-binding defects is substantially different from hydrogen-like centers. For its instructional value, a very simple, one-dimensional example of a deep potential well connected to *one band only* is presented first.

22.1.1 Square Well Potential

A means of introducing the chemical individuality of a center is provided by assuming a rectangular one-dimensional well of depth $-V_0$ and width $2a$. A very similar approach is used to evaluate a level spectrum in two-dimensional quantum wells, in superlattices, and in Schottky barriers (Sections 9.3 and 42.4.3E). The steady-

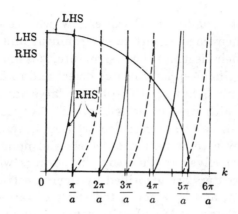

Figure 22.1: Left-hand (LHS) and right-hand sides (RHS) of Eqs. (22.2) (first equation: solid curve; second equation: dashed curve) with solutions indicated by dots (intersections of LHS and RHS).

state electron behavior is described by a solution of the Schrödinger equation

$$\frac{d^2\psi}{dx^2} - k^2\psi = 0 \quad \text{with} \quad k^2 = \frac{2m\left[E - V(x)\right]}{\hbar^2}. \tag{22.1}$$

Here V is used again as potential energy (eV), following the conventional use. With arguments similar to those in Section 7.2, we can show that solutions of Eq. (22.1) exist for k-values that are solutions of the transcendental equations:

$$\sqrt{\mu^2 + k^2} = k\tan(ka) \quad \text{or} \quad \sqrt{\mu^2 + k^2} = -k\cot(ka) \tag{22.2}$$

$$\text{with} \quad \mu^2 = \frac{2m|V_0|}{\hbar^2}. \tag{22.3}$$

These solutions can be obtained graphically from the intersection of the left- and right-hand sides of Eqs. (22.2), shown in Fig. 22.1. With discrete values of k obtained as solutions of Eq. (22.2), the permitted values of E inside the well are discrete and real. These are given by

$$E_n = \frac{\hbar^2}{2m_0}k_n^2 + V_0, \tag{22.4}$$

with $k_n \simeq n\pi/a$ for lower values of n—see Fig. 22.1. The electron restmass is used here since the electron remains close to the center

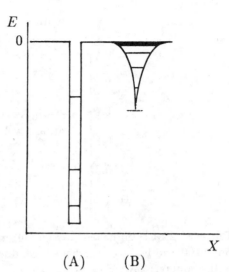

Figure 22.2: Electron eigenstates (A) in rectangular well and (B)
in a Coulomb-attractive well of a
quasi-hydrogen defect.

(A)   (B)

and does not move through the lattice. The eigenstates of such wells
are

$$E_n \simeq \frac{\hbar^2 \pi^2}{2m_0 a^2} n_q^2 + V_0 \quad \text{with} \quad n_q = 1, 2, \ldots \qquad (22.5)$$

and *increase quadratically* with n_q. The individuality of each center
is given by a different a and V_0.

In contrast, the eigenstates of a simple *hydrogen-like defect* are
given by

$$E_n = \frac{m_n e^4}{2\hbar^2 (4\pi \varepsilon_{st} \varepsilon_0)^2} \frac{1}{n_q^2} \quad \text{with} \quad n_q = 1, 2, \ldots, \qquad (22.6)$$

and decreases $\propto 1/n$, with eigenstates *converging toward the continuum* of free states at the edge of the conduction band—see Fig. 22.2B.

The use of a better central cell potential $V(r)$ of the defect
in the Schrödinger equation would yield more realistic results for
the deeper defect level spectrum. However, the inclusion of *both*
valence and conduction bands into the model of deep centers is more
important. In fact, it is this connection to both bands that permits
the distinction between deep and shallow levels, as will be discussed
in the following section. This will replace the linear relation between
well depth (V_0) and the depth of the ground state with a much
compressed relationship, as indicated in the example shown for the
square well in Figs. 22.3 and 22.16; see Sections 22.1.2 and 22.4.6.

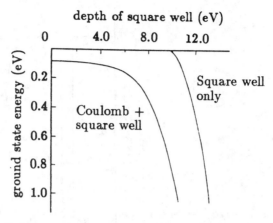

Figure 22.3: Influence of an increasingly attractive square well potential, added to a screened Coulomb potential $\varepsilon = 10$, $m_n = 0.1$, well diameter 5 Å (after Vogl, 1981).

22.1.2 Coulomb Tail and Deep Center Potential

In addition to the central cell potential, we have to consider the long-range Coulomb potential of charged deep centers. The Coulomb tail determines higher excited states of these centers and renders them hydrogen-like, similar to shallow centers. Consequently, one or several deep levels are observed, followed by a series of hydrogen-like shallow levels close to the respective bands (see Grimmeiss, 1987). The model potential combination of a square well and Coulomb potential describes the ground state of centers with the Coulomb potential predominating until, with increasing well depth, the short-range part of the potential becomes very large (> 10 eV in Fig. 22.3). The ground state of the center then shows typical deep level behavior, here calculated properly with interaction of valence *and* conduction bands.

The *atomic electronegativity** can be used as an indicator for the depth of the square well representing the core potential. Figure 22.4 shows the experimentally observed chemical trend: namely, a flat branch for Coulomb-dominated centers, and a steeply decreasing branch for core-dominated centers, which have a deeper well po-

---

* The atomic electronegativity is defined as the difference between the *s*-energies of host and impurity atoms for donors, and the respective *p*-energies for acceptors.

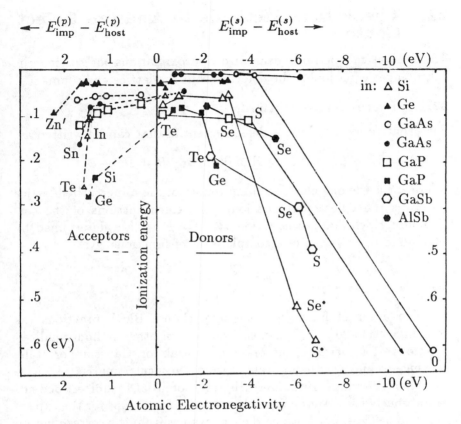

Figure 22.4: Ionization energy of donors and acceptors in a selection of host lattices, plotted versus the difference of *s*- or *p*-energies, respectively, of impurity and host atoms. Deeper levels are identified. For identification of shallow levels, see Vogl (1981).

tential. The reason for this empirical relationship will become clear in the discussion of Section 22.4.6.

The most important results obtained for deep level defects can be summarized as follows:

- deep centers are connected to both valence and conduction bands;
- the energy of the deep level varies at least an order of magnitude less than the impurity potential; and
- although the short-range potential is dominating, the eigenfunctions of some of the deep level impurity centers, such as substitutional chalcogens in Si or III-V compounds, extend well beyond nearest neighbors and do not change much with the chemistry of the impurity if incorporated at the same site (Ren et al., 1982).

22.2 Theoretical Methods to Analyze Defect Centers

The theoretical methods deal with approximations for solving the Schrödinger equation of the defect within the lattice environment.

22.2.1 Perturbative Methods

Perturbative methods use a defect potential that can be written as

$$V = V^0 + U \quad \text{with the Hamiltonian} \quad H = H^0 + U \qquad (22.7)$$

where H^0 is the one-electron Hamiltonian of the unperturbed lattice and U is the defect perturbation. The eigenfunctions of the corresponding Schrödinger equation $H\psi_\nu = E_\nu\psi_\nu$ are determined by expanding ψ_ν in terms of a complete set of functions ϕ_λ:

$$\psi_\nu = \sum_\lambda F_\lambda \phi_\lambda. \qquad (22.8)$$

As such a set of functions, one may choose Bloch functions (as done in Section 20.1.1), Wannier functions, or other orthonormalized functions—for instance, simple exponentials or Gaussian orbitals. The eigenvalues are then obtained from the secular matrix.

When the range of the perturbation potential is shorter, one advantageously uses Wannier or other *localized* functions for the expansion of ψ_ν. Thus, one obtains the eigenvalues from the corresponding Koster-Slater (1954) determinant (see also Bassani et al., 1969; Jaros and Brand, 1976).

22.2.2 Cluster Calculation

The eigenstates of a deep center can be estimated by considering only the atoms in its neighborhood, i.e., in an atomic cluster (Messmer and Watkins, 1973).

Cluster calculations (Section 7.1.1) are carried out by calculating the eigenfunctions of such a group of atoms, treating it as a large molecule. When initially calculating it with atoms from the ideal crystal, and then inserting the impurity into its center, one obtains information on its energy level structure. Although these cluster calculations are easily implemented, they converge slowly with cluster size and the results are very sensitive to conditions at the cluster boundary. Moreover, the defect-level energy is not very accurate, and corresponding changes of band states are difficult to obtain. An example for a diamond crystal is given in Fig. 22.5. A level splits into

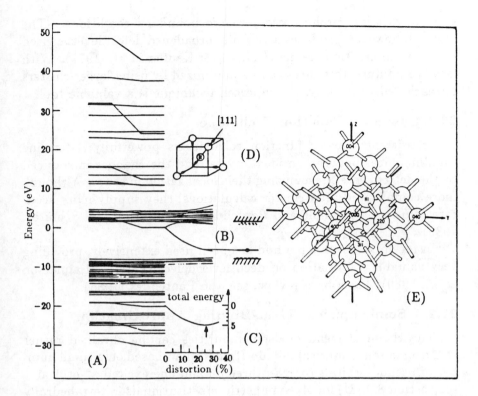

Figure 22.5: (A) Electron energy level distribution for the cluster of 35 carbon atoms (left half of A) depicted in (E) and including a nitrogen atom in its center (right half of A); (B) the deformed spectrum when the carbon atoms surrounding the center nitrogen atom relax; (C) Jahn-Teller shift as a function of the lattice relaxation (minimum total energy at 26% distortion); (D) unit cell; and (E) 35 atom cluster (after Watkins and Messmer, 1970).

the band gap when a nitrogen atom is incorporated into this cluster. In addition, a large shift to lower energies is seen in Fig. 22.5, when lattice relaxation in the neighborhood of the N atom is permitted (Jahn-Teller distortion—see Sturge, 1967). The replacement of a carbon atom by a nitrogen atom causes a substantial trigonal distortion of the four neighbor atoms.

22.2.3 Supercell Technique

Related to a cluster calculation is the supercell technique, in which the defect center is placed periodically in an otherwise perfect crystal. This technique replaces the questionable boundary condition for a cluster with less problematic *periodic* boundary conditions. The method, however, produces artificially broadened defect levels caused by defect interaction (Louie et al., 1976; Kauffer et al., 1977). With large computers, this effect can be minimized by using larger clusters for each cell. Currently, the supercell technique is a valuable tool.

22.2.4 Green's Function Technique

Self-consistent Green's function calculations powerfully determine the differences between an ideal crystal and the changes introduced by the defect center, recognizing their *localization in space*. Although more complicated than cluster calculations, they supply more accurate solutions (Bernholc and Pantelides, 1978; Baraff and Schlüter, 1980).

In recent years, this method has been used extensively, providing very valuable information on deep level defect centers (Hjalmarson et al., 1980). For a brief review, see also Pantelides (1986).

22.2.5 Semiempirical Tight-Binding Approximation

The chemical trend of deep impurities can be obtained rather well from a semiempirical pseudo-Hamiltonian based on a small number of pseudo-orbitals (one s-, three p-, and one excited s^*-orbital—see Section 8.3.1D) for A_1-symmetric substitutionals in tetrahedrally bonded semiconductors. The approximation is based on the band orbital model of Harrison (1973), and adopted by Vogl et al. (1983), to reproduce valence and conduction band structures of semiconductors. Hjalmarson et al. (1980) employed the same model to obtain information of the chemical trend on deep levels of substitutional impurities—see Section 22.4.6.

22.2.6 The Jahn-Teller Effect

When defects with a high degree of symmetry are incorporated in a lattice which distorts this symmetry, degenerate states of the defect are split. In addition, the symmetry of the perfect lattice is reduced in the neighborhood of the defect, causing further splitting; and at least one of the states will be lower than the degenerate state of the undisturbed defect (*Jahn-Teller theorem*, Jahn and Teller, 1937). For further information, see Bersuker (1984), and in paramagnetic crystals, see Bates (1978).

One distinguishes *static* and *dynamic Jahn-Teller effects*, the former is described above, the latter is caused by anisotropies induced from the different modes of the oscillating lattice.

In addition, a linear and quadratic contributions can be differentiated, depending on the degree of distortion from the surrounding lattice (Sturge, 1967).

22.3 Crystal Field Theory

Some qualitative information about the electronic behavior of deep centers can be obtained by starting from the electron eigenvalue spectrum of the isolated impurity atom in *vacuo*. Then one can determine to what extent this spectrum is influenced after the atom is exposed within the crystal to the field of the surrounding atoms, the *crystal field*. This crystal field is used then as a perturbation.

Such description is relatively simple when the symmetry of the surrounding lattice environment is known. When introduced as a substitutional impurity without lattice relaxation, the symmetry of the lattice environment is that of the undisturbed crystal.* This causes splitting of degenerated energy levels of the free atom: the eigenfunctions of any free atom *in vacuo* must be invariant against rotation and reflection, resulting in a large degeneracy of the eigenvalues. However, this is no longer true within a crystal, where the point group of the lattice determines the remaining degeneracies with lesser symmetry.

The *crystal field theory* deals only with a symmetry-related influence of the surrounding atoms but neglects the effects of the neighboring valence electrons. Therefore, it specifically addresses electrons in deeper shells that are partially filled and are shielded from the influence of other valence electrons. Such impurities are transition metal atoms.

For an illustrating example, the level splitting is discussed for an atom with two d-electrons (for instance, Ti, Zr, or Th) substituting for an atom of a host with O_h (cubic) symmetry, such as Si or Ge. Each of these electrons has 10 states available with $l = 2$, $m = -2$, -1, 0, 1, and 2, and $s = \pm 1/2$, resulting in 45 different states for the two electrons, distinguished by their quantum numbers L and S:

one	^1S-state	with	$L = 0$	$S = 0$
nine	^3P-states	with	$L = 1$	$S = 1$
five	^1D-states	with	$L = 2$	$S = 0$

---

* However, in actuality, deformations of the surrounding lattice result, with consequent lowering of the symmetry.

| twenty-one | ^3F-states | with | $L = 3$ | $S = 1$ |
| nine | ^1G-states | with | $L = 4$ | $S = 0$ |

The splitting of these levels becomes transparent after sequentially introducing electron-electron interaction and crystal field. This is shown in Fig. 22.6. Electron-electron interaction results in a splitting of the d^2-level into five levels, distinguished by L is shown in the second column of Fig. 22.6. The addition of the crystal field results in a further splitting of the ^1D (into 2), ^3F (into 3), and the ^1G (into 4 levels), as shown in the third column of Fig. 22.6. When the interaction between different levels is finally taken into consideration, many levels are shifted substantially (without further splitting), presenting a reordered level arrangement as given in the fourth column of Fig. 22.6.

It is instructive to reverse the sequence of a hypothetical interaction by first neglecting the electron-electron interaction*: one obtains the d^2-level splitting by crystal field interaction first. This results in a split into one e_g level and one t_{2g} level for each d-electron, i.e., for the two d-electrons into three levels with two electrons in e_g or t_{2g}, or one electron in each e_g and t_{2g} states, as shown in the sixth column of Fig. 22.6. The index g indicates even parity. Adding the electron-electron interaction further splits $(e_g)^2$ into three, $(e_g)^1(t_{2g})^1$ into four, and $(t_{2g})^2$ into three levels, as shown in the fifth column of Fig. 22.6. Finally, the level interaction causes level shifting, and yields the same results as for the previously discussed center column.

Additional splitting is caused by spin-orbit coupling. The corresponding optical spectra of some of these transition metal impurities in wide band gap material are therefore quite complicated—see Hellwege (1970), McClure (1961), and Schläfer and Gliemann (1967). In semiconductors of smaller band gap, only a few of these levels fall within the gap and are easily identifiable—see Ludwig and Woodbury (1962), Milnes (1983), and Bates and Stevens (1986).

By including the short-range defect potential and the host crystal symmetry in the neighborhood of the defect, and using Green's function techniques, Pantelides (1978) arrives at improved estimates for the deep defect energy levels.

---

* The use of lower- and upper-case letters to describe the states gives a good example to distinguish between one- and multi-electron states. The latter includes electron-electron interaction.

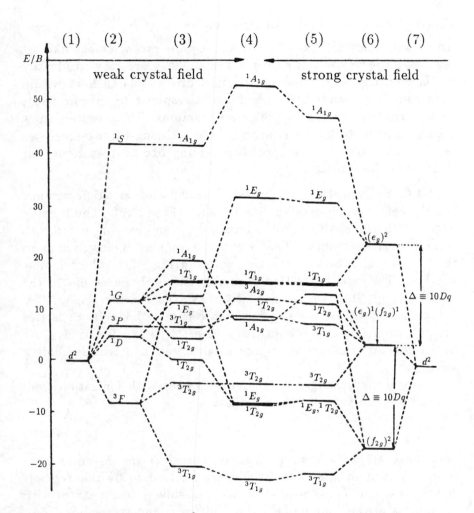

Figure 22.6: Splitting of d^2-levels in cubic lattices: (1) and (7) unperturbed atoms; (2) electron-electron interaction only; (3) crystal field interaction added; (5) electron-electron interaction and crystal field; and (6) crystal field only (after Schläfer and Gliemann, 1967).

22.4 Examples of Deep Centers

A great variety of deep centers exist in semiconductors; however, only a very few of them can be unambiguously identified. Such identification was done very early for a variety of deep defect centers in alkali halides. For the purpose of illustration, these will be given first as examples. We will then discuss some of the more important deep level defects in typical semiconductors.

22.4.1 Vacancies in Alkali Halides

Anion vacancies act like donors, and *cation vacancies* act like acceptors, however, both are deep centers. They are easy to identify by their optical absorption spectrum of broad isolated lines within wide band gaps, and their unambiguous response to specific treatments, which stimulate unique defect reactions. These centers show a large amount of lattice relaxation (large Huang-Rhys factor—see Section 21.1.4B) when recharged, providing excellent examples for electron-lattice interaction.

22.4.1A F-Centers   The classical example of an anion vacancy is an F-center in an alkali-halide crystal* (Fig. 22.7). The missing negative charge of the anion is replaced by an electron in order to restore local neutrality. That electron is not as tightly bound to the vacancy as it was to the Cl^- anion in Fig. 22.7, which is now missing. The level associated with this defect therefore lies in the band gap; in NaCl with a gap of 7.5 eV, the F-center lies 2.7 eV below the conduction band. It becomes deeper with increasing lattice binding strength. This is indicated in Fig. 22.7B as a function of the lattice constant that decreases monotonically with increasing binding strength.

The vacancy changes its charge character with ionization from neutral to positive, relative to the lattice

$$V_{Cl}^{x} \leftrightarrows V_{Cl}^{\bullet} + e'$$ (22.9)

and thus acts as a donor. *Excited states of the F-center* have been observed by Lüty (1960) and are referred to by the symbols K, L_1, L_2, and L_3. These levels are probably resonant states with X minima of the conduction band (Chiarotti and Grassano, 1966). They are located with large energy spacings of ~ 0.6 eV for each consecutive L_i level in KCl, which are typical for deep centers.

Replacing the missing ion with an electron does not completely restore the ideal lattice periodicity. The resulting lattice perturbation produces another level which lies closer to the respective band:

---

* F-centers ("Farb" centers: German for color centers) were the first lattice defects correctly identified and described by their electronic structure by Pohl and coworkers (see the review by Pohl, 1938). Later associates of two, three, or four F-centers were observed and referred to as M-, R-, and N-centers (N_1 in planar and N_2 in tetrahedral arrangement)—see Schulman and Compton (1962).

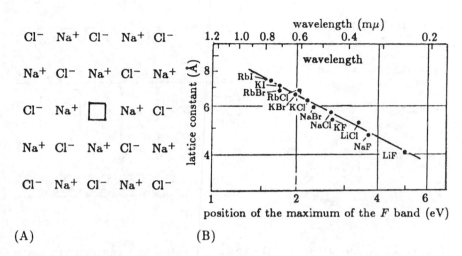

Figure 22.7: (A) F-center in a sodium chloride crystal. (B) Position of the maximum of the F-center absorption as a function of the lattice constant (after Mollwo, 1931).

a second carrier can be trapped in this level. The center, derived from an F-center, is called, **F′-center**, and returns to an F-center when ionized:

$$F' = F^{\times} + e' . \qquad (22.10)$$

In the example of KBr, the ionization energy of the F′-center is 1.4 eV, which is substantially less then the ionization energy of the F-center—2.05 eV.

In addition, the surrounding lattice near the vacancy is also perturbed; its eigenstates split off into levels in the band gap. When the vacancy is empty, the resulting perturbation is larger than when this vacancy has an electron trapped in it. These levels are shown schematically in Fig. 22.8, and are called α- or β-**bands**. Such levels are observed in alkali halide crystals. They result in a decrease of the band gap in the immediate neighborhood of the defect for KBr from 6.55 to 6.44 and 6.15 eV, respectively, at 90 K. These can be interpreted as excitons trapped at the empty vacancy or at the F-center.

22.4.1B Other Centers in Alkali Halides

A cation vacancy in an alkali halide is called a **V-center**, and acts as an acceptor. It is deeper than the F-center in the same crystal and less sharp, probably because of stronger lattice relaxation or of distant pair formation. Nearest-neighbor associates of such centers are termed V_2- and V_3-centers (Seitz, 1954).

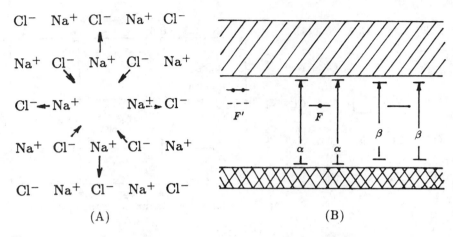

$$\text{Cl}^- \quad \text{Na}^+ \quad \text{Cl}^- \quad \text{Na}^+ \quad \text{Cl}^-$$

$$\text{Na}^+ \quad \text{Cl}^- \quad \text{Na}^+ \quad \text{Cl}^- \quad \text{Na}^+$$

$$\text{Cl}^- \leftarrow \text{Na}^+ \qquad \text{Na}^\pm \rightarrow \text{Cl}^-$$

$$\text{Na}^+ \quad \text{Cl}^- \quad \text{Na}^+ \quad \text{Cl}^- \quad \text{Na}^+$$

$$\text{Cl}^- \quad \text{Na}^+ \quad \text{Cl}^- \quad \text{Na}^+ \quad \text{Cl}^-$$

(A)   (B)

Figure 22.8: Disturbance of the anion and cation lattice around an anion vacancy. The displacement is indicated by arrows in subfigure A. The resulting split-off levels near valence and conduction band are shown as β in subfigure B. A smaller perturbation is observed if an electron is trapped, i.e., near an F-center, resulting in a split-off α in subfigure B. These perturbations represent the corresponding energies of the trapped excitons (Section 21.2).

There are more intrinsic centers which can be formed. These include the $\mathbf{V_k}$**-center**, the **I-center**, and the **H-center**, all of which are related to halogen defects (see e.g., Castner et al., 1958) and are, respectively, a self-trapped hole, a halogen ion interstitial, and a halogen molecule on a single halogen lattice site (Itoh, 1982). A few of these *color centers* are shown in Fig. 22.9. Each of these defects has a characteristic signature in optical absorption, luminescence, or in spin resonance. The wide band gap and tight-binding of defects in alkali halides permit separate identification, and give convincing evidence of a wealth of intrinsic defects, which are much more difficult to identify in common semiconductors. For a review, see Schulman and Compton (1962), Fowler (1968), Lüty (1973), Farge (1973), Williams (1978), Itoh (1982), and Hayes and Stoneham (1985).

22.4.2 Vacancies in Covalent Crystals

One of the most important centers in covalent crystals is the vacancy; it permits an estimate of the resulting changes when a substitutional impurity replaces the vacancy. An understanding of its configuration and electronic structure is a prerequisite for the analysis of substitutionals. The change in the charge distribution when introducing a

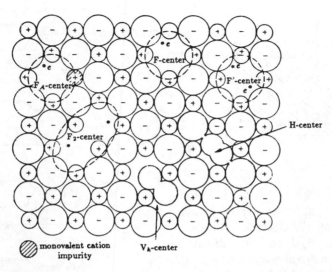

Figure 22.9: Some of the color centers in alkali halides (after Hayes and Stoneham, 1984). ©John Wiley & Sons, Inc.

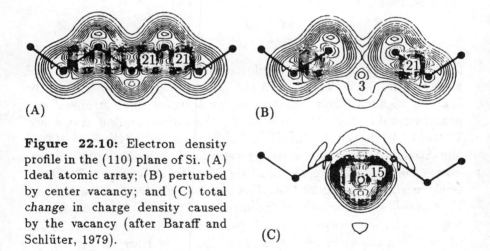

(A)

(B)

Figure 22.10: Electron density profile in the (110) plane of Si. (A) Ideal atomic array; (B) perturbed by center vacancy; and (C) total *change* in charge density caused by the vacancy (after Baraff and Schlüter, 1979).

(C)

vacancy into the Si lattice is shown in Fig. 22.10. We will return to this topic in Section 22.4.6.

Vacancies in covalent crystals result in deep levels which can have several occupation states. Typically, one distinguishes five charge states of the vacancy $V^{\cdot\cdot}$, $V^{\cdot}$, $V^{\times}$, V', and V''; two of them, $V^{\cdot}$ and V', are observed in Si in spin resonance, and two by diffusion experiments (Watkins, 1968, 1976). The breaking of covalent bonds results in *dangling bonds*, which group with the vacancy to

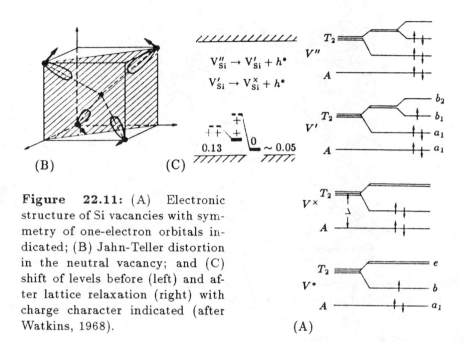

Figure 22.11: (A) Electronic structure of Si vacancies with symmetry of one-electron orbitals indicated; (B) Jahn-Teller distortion in the neutral vacancy; and (C) shift of levels before (left) and after lattice relaxation (right) with charge character indicated (after Watkins, 1968).

form molecular orbitals—a singlet state with a_1 and a triplet state with t_2 symmetry. These yield the ground states of the differently charged vacancies by population of the states with electrons of appropriate spin (Fig. 22.11). For Si, the corresponding states are $V^{\bullet\bullet}(a_1^2)$, $V^{\bullet}(a_1^2 t_2)$, $V^{\times}(a_1^2 t_2^2)$, $V'(a_1^2 t_2^3)$, and $V''(a_1^2 t_2^4)$—see Hayes and Stoneham (1984). A strong tetragonal Jahn-Teller distortion lowers the symmetry to D_{2d}, and significantly shifts and splits these levels as shown in Fig 22.11A. The unpaired electron in the $V_{Si}^{\bullet}$ and V_{Si}' states is spread equally over four or two of the surrounding atoms, respectively. The entire electronic behavior of the vacancy can be explained in a one-electron model (Pantelides and Harrison, 1976).

The states $V^{\bullet}$ and V' are observed by ESR resonances, $V^{\times}$ and V'' are indirectly observed after photo-excitation, and the $V^{\bullet\bullet}$ level is seen to shift below the $V^{\bullet}$ level, which is typical for a negative-U behavior—see Section 22.4.10A. This behavior was predicted by Baraff et al. (1980) and later experimentally confirmed by Watkins (1984). See also Stoneham (1975), Lipari et al. (1979), and Jaros et al. (1979).

The $V_{Si}^{\bullet}$ state is metastable and disproportionates to $V_{Si}^{\times}$ and the stable $V_{Si}^{\bullet\bullet}$ (Fig. 22.11), which lies $\simeq 0.13\,\mathrm{eV}$ above E_v. This center releases *two* holes when excited, since the $V_{Si}^\bullet$ center is shallower (0.05 eV) and dissociates immediately after creation.

In diamond, a number of very sharp lines appear which are related to the vacancy and identified as $GR1\dots GR8$ (Collins, 1981). At present, their origin is not completely understood.

A convenient method* for producing such vacancies is by bombardment with relatively fast electrons (Loferski and Rappoport, 1958) at cryogenic temperatures, and trapping the cogenerated, highly mobile interstitial (Section 19.1.1) at other lattice defects, e.g., group III atoms (Watkins, 1968). The threshold energy for producing vacancies with electrons in Ge and Si is 14.5 eV and 12.9 eV, respectively.

Vacancies in *binary semiconductors* require multi-electron approximations. While the states of the silicon vacancy are sp^3 hybrids, the state of a Ga vacancy in GaP are p-like.

22.4.3 Self-Interstitials

Elemental semiconductors such as Si and Ge have a rather loosely packed lattice with coordination number 4 and sufficient space to accommodate interstitials of the host lattice, i.e., *self-interstitials*.

There are several possibilities for incorporating an additional Si atom in an Si lattice, as shown in Fig. 19.3. Three of them are indicated as T, H, and B and have tetragonal, hexagonal, and bond-centered geometry. Two, indicated as S and S′, have a split configuration: the lattice atom has moved to one of the split positions while the interstitial atom moves to the other. The two equally displaced atoms are sometimes also referred to as an *interstitialcy*. After incorporation, the surrounding atoms relax to shifted positions.

The formation energy of the interstitials depends on their specific site as well on their charge state. This is shown in Fig. 22.12, with the formation energy plotted as a function of the position of the Fermi level (Car et al., 1984). Such dependency indicates that the recharging of an interstitial makes it more stable at a different position. Alternating recharging, e.g., by recombination with excess carriers followed by thermal ionization at the changed position, can therefore stimulate diffusion—see Section 19.3.2A.

The $Si_i^\bullet$ center is metastable; when recharging and relaxing into the $Si_i^{\bullet\bullet}$ center, it shows a negative U-character. Si interstitials are

---

* Quenching from high temperature is not fast enough to freeze-in measurable densities of vacancies (Watkins, 1986).

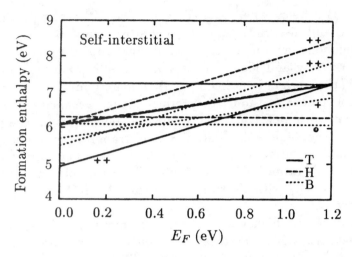

Figure 22.12: Si self-interstitial formation enthalpy as a function of the position of the Fermi level for the different interstitial position and charging states (after Car et al., 1984). More recently, a slightly lower (~ 1 eV) formation enthalpy was calculated by the same authors.

deep level centers with levels 0.6 and 0.8 eV below the conduction band.

22.4.4 Antisite Defects

Antisite defects are identified in some of the III-V compounds, e.g., As_{Ga}. This center is believed to be responsible for compensation of electrically active defects, consequently causing a reduction in semiconductivity. The defect by itself, or as an associate with As_i, is probably the so-called EL2 center in GaAs—see also Section 22.4.9.

In compound (AB) semiconductors, one distinguishes A_B and B_A as possible antisite defects. In higher compounds, such antisite formation is often more probable and presents a larger variety of defects: six in ABC-compounds, although only a few of them are energetically preferred.

22.4.5 Hydrogen in Silicon

Hydrogen in Si requires special attention. It is known to passivate many deep centers, and is therefore often used with great benefit for device fabrication (Pearton et al., 1987). It can be introduced into the Si lattice in a variety of ways (Seager et al., 1987). It diffuses easily and is known to attach itself to dangling bonds (Pearton et al., 1987). It strongly reduces the conductivity of p-type and weakly reduces it in n-type Si (Sah et al., 1983; Bergman et al., 1988). However, H also induces defect levels of its own in the band gap.

Substantially improved insight into the structure of defects has been gained after more reliable energy profiles, including lattice relaxation, were computed. An example of such an energy surface for an H-atom in a (111) plane of a boron-doped Si crystal through the three bond-minima positions is shown on the cover of this book (Denteneer et al., 1989).

Recently, it has been shown that hydrogen atoms are preferably introduced in p-type Si as H$^{\bullet}$ at a B-site (Fig. 19.9) after lattice relaxation by 0.4 Å in the bond direction (Van de Walle et al., 1989). For a review of recent work, see Patterson (1988).

In n-type Si, hydrogen is preferably incorporated as H$^{\times}$ or H$'$. The H$^{\times}$ seems to find several shallow minima near C- or T-sites with little barriers in between, easing interstitial diffusion. The H$'$ is more stable at a T-site, which is a high electron density location. The H$^{\times}$, H$'$ center is a deep donor and seems to act as a negative-U center—see Section 22.4.10A.

The incorporation, diffusion, and recharging of hydrogen interstitials resemble to some extent the corresponding properties of self-interstitials. They also depend sensitively on the position of the Fermi level—see also Section 19.3.2.

22.4.6 Substitutional Defects Replacing a Vacancy

Substitutionals can be considered as impurities replacing a vacancy. Inserting an Si atom into a vacant site of an Si host can be considered as having its s and p orbitals interacting with the A_1 and T_2 states of the vacancy, producing bonding and antibonding states which merge with valence and conduction bands, respectively. Inserting an impurity with substantially lower energy orbitals will produce only a small shift in the levels, causing an impurity-like hyper-deep bonding level in or below the valence band, and a vacancy-like antibonding level slightly lower than the vacancy states usually within the band gap.* For impurities with energy levels higher than Si, the corresponding levels lie within the conduction band, and slightly above the vacancy states for the bonding-antibonding pair. The vacancy states are limiting cases for donor- or acceptor-like states

---

* The A_1 state of the acceptor is always strongly bound and lies within or below the valence band, while the T_2 states may emerge from the conduction band into the gap.

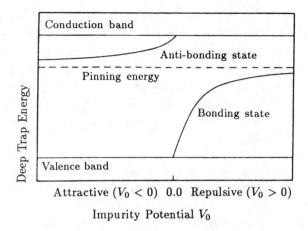

Figure 22.13: Depth of deep donor and acceptor states as function of their impurity binding potential V_0. The asymptotic "pinning energy" is that of the electron state of the vacancy (after Vogl, 1981).

to which both approach asymptotically with increasing depth of the binding potential of the electron or hole in the impurity.

This indicates how the influence of two bands compresses the spectrum of ground states of deep level from $\partial E/\partial V \simeq V_0$ in the simple one-band model (Section 22.1.1) to the actual $|\partial E/\partial V|$ that is rapidly decreasing with increasing $|V_0|$, where V_0 is the depth of the potential well. This is shown schematically in Fig. 22.13.

This behavior is deduced from Green's function calculation using semiempirical tight-binding Hamiltonians (Hjalmarson et al., 1980). It has been used to calculate the chemical trend of numerous deep level centers as a function of the impurity potential. These levels are shown in Fig. 22.14. For the impurities to the right of the intersect of the curves in Fig. 22.14 with the conduction band, the impurity potential is not large enough to offer a bond state. Here, the effective mass approximation for hydrogen-like states yields shallow levels connected to one band only (the conduction band in this example). For a short review, see Dow (1985).

When using an impurity potential comprised of a deep well and a Coulomb tail (e.g., like the Abarenkov-Heine potential) with

$$V(r) = \lambda V_0(r) + \frac{e^2}{4\pi\varepsilon\varepsilon_0 r} \quad \text{with} \quad V_0(r) = \begin{cases} V_0 & \text{for } r \leq r_0 \\ 0 & \text{for } r \geq r_0 \end{cases} \quad (22.11)$$

and a deep enough well to create a bound ground state in the gap, one can follow the transition from a deep level center with essentially a

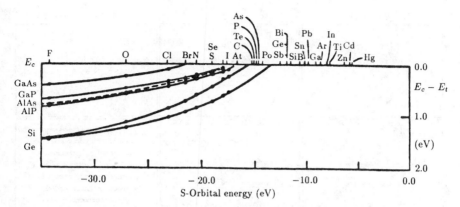

Figure 22.14: Calculated ionization energies $E_c - E_t$ for deep level substitutional impurities on anion sites in semiconductors listed at left margin (after Hjalmarson et al., 1980). Impurities with a lower value of the binding potential than shown by the intersection of the curves with E_c lie as resonant "deep" states *within* the conduction band.

vacancy-like charge density distribution to a hydrogen-like spread out distribution with decreasing impurity potential (Pantelides, 1986), as shown in Fig. 22.15.

22.4.6A Impurity and Site Symmetry   The chemical identity of an impurity (i.e., its size, bonding type, and valency) is responsible for changes of the local symmetry after the incorporation of an impurity. Some of these changes are related to the symmetry-breaking lattice relaxation, known as the Jahn-Teller effect and discussed in Section 22.2.6. Others deal with the strength and angle relation of the bonding forces of the impurity in relation to the tetrahedrally arranged available dangling bonds of the impurity.

The arrangement of atoms nearest to an impurity has an important influence on the density of state (DOS) distribution of the resulting levels. This influence is strong enough to make the DOS of substitutional S-donor look rather similar to an interstitial Si atom, both in a tetrahedral environment (Vigneron et al., 1982).

22.4.7 Isoelectronic Defects

Isoelectronic defects are formed by substitutionals from the same column of elements as the host atom. Different homologous elements have a different energy spectrum, since the long-range potential—the Coulomb term—cancels: the replacing atom has the same valency. The remaining central cell potential reflects the chemical identity of the center. Replacing a middle-row host atom (e.g., P in GaP) with

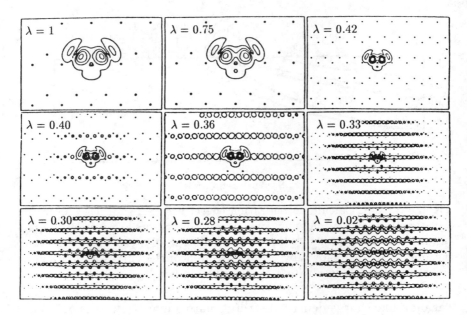

Figure 22.15: Charge-density distribution of a substitutional impurity in Si, computed from a potential similar to Eq. (22.11) for decreasing strength of the short-range potential from $\lambda = 1$ to $\lambda = 0.02$ in the nine subfigures. $\lambda = 1$ corresponds to the T_2 state of an Si vacancy (after Vigneron et al., 1982).

a highest-row atom (here N), one obtains a defect that acts as an electron trap with binding energy of 10 meV—see Section 44.5.2. Replacing it with a lower-row atom (here Bi), the defect acts as a hole trap with a binding energy of 38 meV (Dean et al., 1969). Replacement with mid-range atoms usually does not produce electronic defect centers, but instead produces *alloys* with the respective sublattice: As forms a $GaAs_{1-\xi}P_\xi$ mixed crystal, as does Sb, which forms $GaSb_{1-\xi}P_\xi$. There are no corresponding levels in the band gap.

When traps are created, some of them can be rather deep; for instance, ZnTe:O, with $E_c - E_t = 0.4$ eV, and CdS:Te, with $E_t - E_v = 0.19$ eV (Cuthbert and Thomas, 1967). An extensive review of the experimental observations of isoelectronic traps is given by Dean (1973). Theoretical models are quite sensitive to central cell approximations (Faulkner, 1968; Baldereschi and Hopfield, 1972; Jaros and Brand, 1979), and are not yet able to predict the binding energy beyond an order of magnitude.

In contrast to the hydrogenic effective mass treatment of shallow donors or acceptors, where the central cell corrections are often minor compared to the far-reaching Coulomb contributions, such short-range potentials are dominant. Furthermore, minor deviations (by 1%) in the estimated potential can result in major changes (by a factor of 2) of the electronic eigenstates of the defect.

Extensive theoretical and experimental work has been done with GaP. In Section 22.4.8A, GaP:O_P is discussed as one example. For reviews, see Stoneham (1975) and Pantelides (1978).

22.4.8 Chalcogens in Si

Chalcogens (O, S, Se, and Te), incorporated in Si, act as deeper donors. They easily form associates, especially oxygen. These will be discussed in Section 23.1.1A. Single substitutional donors are observed for neutral and singly charged S, Se, or Te: they act as double donors. The charge density distribution of these donors has been calculated by Ren et al. (1982).

The ground state of neutral and single ionized donors is rather deep (Table 22.1), and shows a substantial chemical shift. In addition, the ground 1s-state is split into A, E, and T_2 states, which are non-degenerate, doubly, and threefold degenerate, respectively. The p-states are shallow and follow rather well the effective mass (hydrogen-like) approximation, as can be seen by comparison with the last column of Table 22.1.

The small difference of the ionization energy of 1s (A_1) states between S and Se is similar to the minor variance of the ionization energy of the free atoms, compared to a more substantial difference to Te.

The larger hydrostatic pressure coefficient of the ground state compared to the hydrogen-like centers, indicates the connection to conduction bands (in addition to valence bands). This shift is a result of a substantial shift in respect to the X-valley. For a recent review, see Grimmeiss and Janzén (1986).

Oxygen-related centers are part of the family of so-called *thermal donors*,* some of which may be isolated oxygen centers, incorporated as interstitials, or combined with Si as an interstitialcy or a substitutional. For original literature on the different types of incorporation, see Wagner et al. (1984).

---

* Due to the thermal nature of incorporation, i.e, during a heat treatment between 350 and 555 °C in Czochralski-grown Si.

Table 22.1: Binding energy (in meV) and pressure coefficient (in meV/Pa) of ground and excited states of neutral and singly ionized chalcogens (after Wagner et al., 1984).

	$S^\times$	$Se^\times$	$Te^\times$	$S^\bullet/4$	$Se^\bullet/4$	$Te^\bullet/4$	E_{qH}
$1s(A_1)$	318.2	306.5	198.7	153.3	148.3	102.8	31.27
$1s(T_2)$	34.6	34.5	39.2	46.5	41.5	44.3	
$1s(E)$	31.6	31.2	31.6				
$2p_0$	11.4	11.5	11.5	11.4	11.5	11.8	11.75
$2s$	9.37	9.3	9.7				8.83
$2p_\pm$	6.4	6.4	6.3	6.43	6.4	6.4	6.40
$3p_0$	5.46	5.47	5.5				5.48
$3p_\pm$	3.12	3.12	3.12				3.12
$3d_0$		3.8	4.0				3.75
$4p_\pm$	2.2	2.2	2.1				2.19
$5p_\pm$		1.5					1.44
							−0.05
$\partial E(1s)/\partial p$ $(\times 10^{-8})$	−1.7	−1.8	−0.9	−2.05	−2.1	−1.2	(typical)

22.4.8A Oxygen in GaP   The oxygen in GaP is one of the more extensively studied defect centers, since it is of practical interest for light-emitting diodes. Incorporated as a substitutional of phosphorus, it yields two deep centers stemming from the same defect:

$$O_P^\bullet + e' \rightarrow O_P^\times \quad \text{and} \quad O_P^\times + e' \rightarrow O_P'. \quad (22.12)$$

The first center with $(E_c - E_d)^\bullet \simeq 0.8$ eV has a rather small lattice coupling with a Huang-Rhys factor of $S \simeq 3$: it relaxes to $(E_c - E_d)^\times = 0.96$ eV. The second center with $(E_c - E_d)^\times \simeq 0.6$ eV relaxes to $(E_c - E_d)' \simeq 2.03$ eV; that is, it has a very large Huang-Rhys factor of $S \simeq 30$. The corresponding band diagram is shown in Fig. 22.16. For a review, see Dean et al. (1983).

The strong lattice relaxation (see Section 20.1.2, Fig. 20.3, and Section 20.3.3B) that occurs when a second electron is captured produces a lower level than that for the captured first electron. This is indicative of a *negative U-center*, which will be discussed in Section 22.4.10. More recent observations indicate a much more complex behavior of the different oxygen-related centers, a discussion

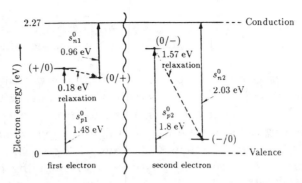

Figure 22.16: Band diagram of GaP with one- and two-electron O_P-states at 300 K, obtained from photocapacity measurements (after Dean et al., 1983).

of which is beyond the scope of this book. For a review, see Dean (1986).

With the incorporation of an acceptor in addition to O_P (e.g., Zn_{Ga}), the spectrum becomes very complicated. It shows a large series of additional lines which are due to donor-acceptor pairs—see Section 44.5.1A.1. For a review, see Jaros and Dean (1983).

22.4.9 Other Deep Defect Centers ($EL2, DX$)

There is a large variety of other impurity centers which cause deep levels in the band gap. Many of these are yet unidentified with respect to their chemical origin, despite substantial technical interest in some of these centers to produce lowly conductive semiconductors for field-effect transistors.

An example of such centers is the so-called *EL2 defect*, which plays a key role in creating semi-insulating GaAs, and presents a level near the center of the band gap. Its origin is probably related to an anion antisite defect—see Section 22.4.4.

Another defect in III-V compounds is often referred to as the DX center: D, since it acts like a donor and, X, because it does not behave effective-mass-like and is probably associated with an unknown defect X. These centers show major lattice relaxation and have deep center characteristics. They are involved in carrier trapping and recombination traffic, and are responsible for persistent photoconductivity and large Stokes shifts (Lang and Logan, 1977).

The DX centers are most pronounced in heavily doped n-type III-V compounds and alloys (e.g., AlGaAs or GaAsP—Lang, 1986), and are related to the chemical nature and concentration of dopants, specifically involving S, Se, Te, Si, and Sn. The DX centers probably relate to interstitial configuration (for a review, see Bhattacharya, 1988).

A very large group of deep centers are those involving transition metals. Many of them act as recombination centers. Others, such as Cu in II-VI compounds, are efficient activators for luminescence or for sensitizing the semiconductor for high-gain photoconductivity. This will be discussed Sections 44.6.1 and 45.2.2B. For a review of Fe-, Cr-, and Cu-related centers, see the corresponding chapters in Pantelides (1986).

Many of these deep level centers are probably related to more complex defect structures involving an impurity, as well as adjacent lattice defects such as vacancies, interstitials, and antisites.

22.4.10 Negative-U Centers

In Sections 22.4.2, 22.4.3, and 22.4.8A, examples are given for a negative-U center. U is the *Hubbard correlation energy* that was introduced by Hubbard (1963) as an energy penalty when two electrons with opposite spin occupy the same site. For a free atom, U is the difference between the ionization energy and the electron affinity; typically, it is on the order of 10 eV. Embedded in a crystal lattice, U is greatly reduced by lattice shielding and interaction to generally 0.1...0.5 eV. For most crystal defects, U is positive; this means that a defect, which has several charge states, has the higher charged state closer to the related band. This ordering of levels is easily understood once it is recognized that a second electron is less bound to a defect than the first because the two electrons repulse each other.

Consequently, a negative-U center indicates that with one electron already trapped, the second one is even more attracted. This can only happen when the first electron has polarized the defect configuration sufficiently so that the second one is trapped into a substantially different defect environment: for this, a substantial lattice interaction (relaxation) is required. The electron-lattice interaction can be expressed as

$$V(u) = -\lambda u(n_\uparrow + n_\downarrow) + \frac{\beta u^2}{2}, \qquad (22.13)$$

with u as the atomic displacement. For equilibrium ($\partial^2 V / \partial u^2 = 0$), we obtain for the energy of single and double occupancies $-\lambda^2/2\beta$ and $-2\lambda^2/\beta$, respectively, with $n_\uparrow$ and $n_\downarrow = 0$ or 1 as spin occupancy numbers, λ as the electronic lattice coupling, similar to α_c, however not dimensionless, and β, the elastic restoring term. After adding the always-positive normal Hubbard correlation energy, we have

$$U = U_0 - \lambda^2/\beta, \qquad (22.14)$$

which is a defect property and can become negative for large electron-lattice coupling and small lattice restoring forces. In effect, the lattice near the defect site now harbors a bipolaron (see Section 28.1.2), as suggested by Anderson (1975). This can be described as forming an *extrinsic Cooper pair*, similar to the Cooper pair formation in a metal (Section 35.1.1). Here, however, the *defect center* assists such pair-formation, consequently yielding a substantially increased binding energy (U). For calculation of λ, c, and U, see Baraff et al. (1980).

As a result, the negative-U center is not stable in its singly occupied state, since energy (U) can be gained by trapping another carrier. Consequently, the singly occupied shallow state is not observed after the filling process (e.g., via optical excitation) is switched off and sufficient time has passed for relaxation. With continued excitation, however, both the shallow and deep states can be observed (Watkins, 1984). In addition to the GaP:$O_P^\times$ center, the $V_{Si}^\bullet$ and the $Si_i^\bullet$ are known to have negative U character and many more deep centers are probably of a similar type (Watkins and Troxell, 1980).

22.4.10A Negative-U in Chalcogenide Glasses   Some optical and electron-spin resonance behavior of chalcogenide amorphous semiconductors can be explained by assuming negative-U centers (Street and Mott, 1975). Such centers were identified by Kastner et al. (1976) as valence alternation pairs. For instance, dangling bonds in broken chains can be bound to an adjacent unbroken chain by a valence alternation, shown here as a charge disproportionation

$$2Se_{(1)}^\times \rightarrow Se'_{(1)} + Se_{(3)}^\bullet; \qquad (22.15)$$

the two dangling Se atoms in such a chain disproportionate to form a negatively charged dangling bond in onefold coordination, shown as subscript, and one positively charged Se atom attached to the neighboring chain with threefold coordination (see Fig. 22.17), in an exothermic reaction: the charged pair has a *lower* energy than the neutral defects. The Coulomb energy between the two electrons in the negatively charged defect is more than compensated by the lattice energy. Therefore, this center is a negative-U center.

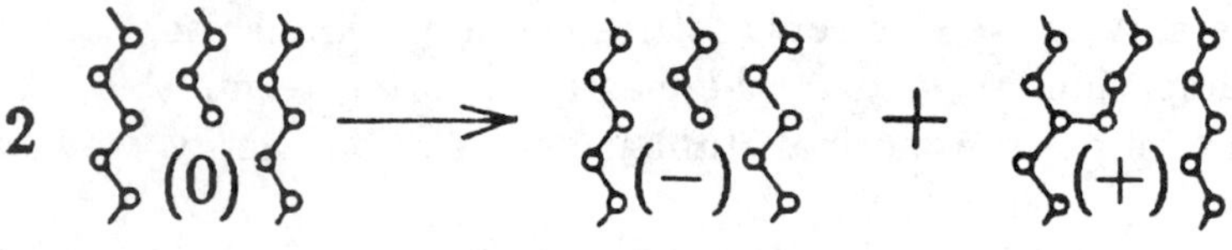

Figure 22.17: Negative-U model by disproportionation at a dangling bond with charge character indicated (after Street and Mott, 1975).

The existence of negative-U centers in amorphous semiconductors can explain the pinning of the Fermi level by such defects without showing a high density of uncompensated spins, which would otherwise be expected when compensation occurs between ordinary donors and acceptors (Fritzsche, 1976).

22.4.11 Shallow/Deep Center Instabilities

There are centers which show metastability as a shallow or as a deep center. Examples are metal-acceptor pairs in covalent crystals or bond excitons at a donor (see Section 21.2) in AlGaAs. These centers can be explained as having a negative-U character. When occupied with one electron, they behave as a typical shallow center, but turn into a deep center when a second electron is trapped.

22.4.11A Metastable Lattice Relaxation When a carrier is captured by a deep center, it relaxes often to a substantially lower energy. When the center has a bond, higher energy state, the degree of lattice relaxation changes and could drive the state into a metastability. The return to the ground state then requires an activation energy. The $EL2$ defect in GaAs has such a state (Chantre et al., 1981). For a discussion of these metastabilities, see Hamilton et al. (1988).

22.4.12 Transition Metal Impurities

Transition metal impurities include $3d$, $4d$, and $5d$ transition metals, $4f$ rare earth, and $5f$ actinides. All of these elements have unsaturated inner shells. Their outer-shell electrons shield part of the interaction with the host lattice. The center is described by a *tight-binding approximation*, i.e., like an individual atom, but perturbed by the surrounding lattice field—see Section 22.3.

The properties of the center are determined by its chemistry and its character relating to the host lattice, such as

- its *site character*, as *interstitial* or *substitutional* within a host lattice of given symmetry;

Table 22.2: Occupancies of d orbitals for octahedral coordination (see Fig. 22.19).

Number of d Electrons	High-Spin		Low-Spin	
	t_2	e	t_2	e
1	↑		↑	
2	↑↑		↑↑	
3	↑↑↑		↑↑↑	
4	↑↑↑	↑	↑↓↑↑	
5	↑↑↑	↑↑	↑↓↑↓↑	
6	↑↓↑↑	↑↑	↑↓↑↓↑↓	
7	↑↓↑↓↑	↑↑	↑↓↑↓↑↓	↑
8	↑↓↑↓↑↓	↑↑	↑↓↑↓↑↓	↑↑
9	↑↓↑↓↑↓	↑↓↑	↑↓↑↓↑↓	↑↓↑

- its *charge character* in relation to the *oxidation state*, conventionally identified by its remaining valence electrons, e.g., Cr^{3+} when replacing Ga^{3+} in a GaAs lattice, and its *charge state, relative to the host lattice*, e.g., $(Cr^{3+})^{\times}_{Ga}$ neutral, $(Cr^{2+})'_{Ga}$, or $(Cr^{4+})^{\bullet}_{Ga}$ as negatively or positively charged when incorporated in a GaAs lattice on a Ga site;
- its *spin character*, as *high-spin*, when the unsaturated d-shell electrons have preferably parallel spin (Hund's rule), resulting in a high electron-electron and weak crystal-field interaction, and as *low-spin*, when electrons have more antiparallel spin,* resulting in weak electron-electron and stronger crystal-field interactions (see Table 22.2); and

---

* With parallel spins there are less alternatives to populate states of equal energy (less degeneracy). With antiparallel spin orientation, there is more degeneracy, giving the opportunity to the Jahn-Teller splitting to produce an even lower level. For transition metal impurities, Hund's rule, and for Si-vacancies, the Jahn-Teller effect produces the lower ground state (Zunger, 1983).

- its *transition character* with respect to an optical excitation, for an *intracenter transition* with charge conservation at each center, or an *ionization* with charge transfer to the conduction or valence band or to another center.

The solubility of $3d$ impurities in Si is extremely low, typically 10^{14} cm^{-3}. This makes positive identification difficult since the solubility limit lies below the threshold of conventional analytic chemistry. The $3d$ elements are preferably incorporated in interstitial sites of Si, while $5d$ elements are mostly substitutionals.

The solubility of Cu and Ni in Si, however, is much larger, up to $5 \cdot 10^{17}$ cm^{-3}. Transition metal atoms are gettered in Si by incorporated oxygen, probably by stress-enhanced diffusion toward such centers. Cu decoration of dislocations which attracted oxygen is an example (Goorsley, 1989).

In III-V and II-VI compounds, the solubility of the $3d$-elements is somewhat higher, typically 10^{17} cm^{-3}. An exception is Mn, which forms continuous solid solutions with II-VI compounds: it produces a dilute magnetic semiconductor.

22.4.12A Site Character Incorporated as a substitutional, the transition metal impurity entails interaction with the vacancy states. This means that certain states hybridize strongly with the corresponding states of the vacancy, e.g., the $d(T_2)$ state of Cr$_{Si}$ with the T_2 state of the vacancy (see Zunger and Lindefelt, 1983).

Interstitials of the Zn to Ti group have little interaction of their d orbital with the crystal state.

22.4.12B Charge Character It is characteristic of these impurities that many of them have several stable charge characters (Fig. 22.18), yielding levels in the band gap with a relatively small energy difference between them, typically 0.5 eV rather than 2–3 eV. Each of these levels split due to the crystal field and the Jahn-Teller effect (dynamic and cooperative—see Bates and Stevens, 1986). The strain, induced by the defect, causes a change in symmetry surrounding each defect center. The tetrahedral symmetry of GaAs becomes orthorhombic with incorporation of Cr^{3+}, and tetragonal with Cr^{2+} (Bates and Stevens, 1986). In Si, interstitial $3d$ impurities cause an increase in distance of the four nearest neighbors, and a decrease in distance of the six next-nearest neighbors. This makes the lattice surrounding the $3d$ impurity nearly tenfold coordinate (Lindefelt and Zunger, 1984).

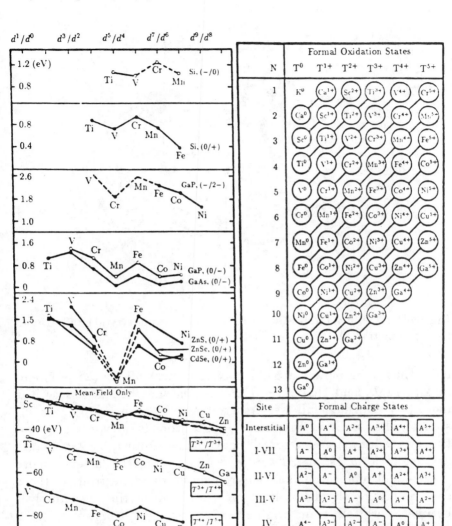

(A) (B)

Figure 22.18: (A) Ionization energy for $3d$ transition elements, free ions and as impurities in various compounds. (B) Formal oxidation states with total number N of electrons in $3d$ and $4s$ shells and free ion charges (upper diagram). Corresponding charges relative to the semiconductor lattice for cation substitutional incorporation (lower diagram) (after Zunger, 1986b).

Significant concerns include the identification of the ground state and the proper sequencing of higher excited states. A number of empirical rules are used in the classical discussions (Kaufmann and Schneider, 1983). Important factors are the *Hund rules*, which re-

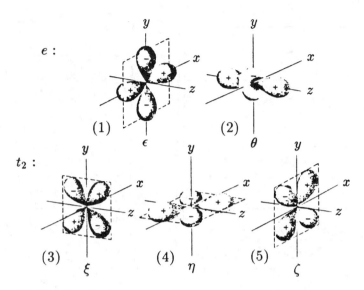

e :

t_2 :

(1)

(2)

(3)

(4)

(5)

Figure 22.19: d-orbitals of wavefunctions in e symmetry (1) and (2) and in t_2 symmetry (3)–(5).

quire that the ground state has maximum multiplicity $(2S + 1)$ and maximum L, and that $J = L \pm S$ when the shell is less $(+)$ or more $(-)$ than half full (see Ashcroft and Mermin, 1976). For instance, when a Cr atom replaces a Ga atom in n-type GaAs and traps an additional electron, it may change from a $3d^5 4s^1$ configuration to a $3d^5 4s^2$ or $3d^3 4s^2 4p^2$ configuration; the latter is more Ga-like and more probable. The remaining three electrons in the d shell determine the defect level spectrum and distribute themselves between e orbitals, transforming to $(2z^2 - x^2 - y^2)/\sqrt{6}$, $(x^2 - y^2)/\sqrt{2}$ (fourfold degenerate) and t_2 orbitals, transforming to yz, zx, xy (sixfold degenerate); lower-case letters indicate one-electron states—see Fig. 22.19.

The ground states of the differently charged Cr substitutional centers in GaAs are

$$(\mathrm{Cr}^{4+})^{\bullet} \quad - \quad (3d^2) \quad E - E_v = 0.45 \text{ eV}, \quad s_n = 9 \cdot 10^{-17} \text{ cm}^2$$
$$(\mathrm{Cr}^{3+})^{\times} \quad - \quad (3d^3) \quad E - E_v = 0.74 \text{ eV}, \quad s_n = 10^{-17} \text{ cm}^2$$
$$(\mathrm{Cr}^{2+})' \quad - \quad (3d^4) \quad E - E_c = 0.12 \text{ eV}, \quad \text{inside cond. band.}$$

Many other centers in GaAs as well as other semiconductors are identified and reviewed by Clerjaud (1985) and Zunger (1986b). A list of the different oxidation states and charges relative to the

Table 22.3: Observed excitation energies (eV) in cation substitutionals 3d-transition metal doped semiconductors (after Zunger, 1986).

Ground State Impurity / Host	$^3A_2$ Ti^{2+}	$^4T_1$ V^{2+}	$^5T_2$ Cr^{2+}	$^6A_1$ Mn^{2+}	^5E Fe^{2+}	$^4A_2$ Co^{2+}	$^3T_1$ Ni^{2+}
ZnS		$0.53(^4T_2)$	$0.64(^5E)$	$2.34(^4T_1)$	$0.44(^5T_1)$	$0.46(^4T_2)$	$0.54(^3T_2)$
	$1.21\ (^3T_1)$	$1.14(^4A_2)$	$1.36(^3T_2)$	$2.53(^4T_2)$	$2.07(^3A_2)$	$0.77(^4T_1)$	$1.13(^3A_2)$
		$1.39(^4T_1)$	$1.75(^3T_1)$	$2.67(^4E)$	$2.14(^3A_1)$	$1.76(^4T_1)$	$1.52(^3T_1)$
ZnSe	$0.74\ (^3T_1)$	$0.50(^4T_2)$	$0.68(^5E)$	$2.31(^4T_1)$	$0.34(^5T_2)$	$0.43(^4T_2)$	$0.50(^3T_2)$
	$1.22\ (^3T_1)$	$1.08(^4A_2)$	$1.61(^3T_2)$	$2.47(^4T_2)$	$1.26(^3T_1)$	$0.78(^4T_1)$	$1.10(^3A_2)$
		$1.24(^4T_1)$	$1.85(^3T_1)$	$2.67(^4E)$		$1.67(^4T_1)$	$1.46(^3T_1)$
ZnTe	—	—	$0.68(^5E)$	$2.3\ (^4T_1)$	$0.31(^5T_2)$	$0.72(^4T_1)$	—
				$2.4\ (^4T_2)$		$1.44(^4T_1)$	
				$2.6\ (^4E)$			
CdS	$0.40\ (^3T_2)$	0.61	$0.66(^5E)$	—	$0.32(^5T_2)$	$0.68(^4T_1)$	$0.51(^3T_2)$
	$0.71\ (^3T_1)$	1.08				$1.73(^4T_1)$	$1.01(^3A_2)$
	$1.22\ (^3T_1)$	$1.80(^4T_1)$					$1.58(^3T_1)$
CdSe	$0.38\ (^3T_2)$	$0.43(^4T_2)$	$0.62(^5E)$	—	$0.37(^5T_2)$	$0.37(^4T_2)$	$0.52(^3T_2)$
	$0.62\ (^3T_1)$	$0.86(^4A_2)$			$0.29(ZPL^*)$	$0.35(ZPL)$	$(0.45, ZPL)$
	$1.18\ (^3T_1)$	$1.31(^4T_1)$				$0.68(^4T_1)$	$0.99(^3A_2)$
						$1.61(^4T_1)$	$1.42(^3T_1)$
							$(1.35, ZPL)$
CdTe	$0.35\ (^3T_2)$	$0.82(^4A_2)$	$0.63(^5E)$	$2.2\ (^4T_1\ )$	$0.28(^5T_2)$	$0.37(^4T_2)$	—
	$0.62\ (^3T_1)$	$1.11(^4T_1)$		$2.5\ (^4T_2\ )$		$0.72(^4T_1)$	
	$1.14\ (^3T_1)$					$1.44(^4T_1)$	
GaP	$0.60\ (^3T_2)$		$0.87(^5E)$	$1.34(^4T_1)$	$0.41(^5T_2)$	$0.56(^4T_2)$	$0.82(^3T_2)$
	—	1.07		$1.53(ZPL)$		0.87-1.24	$1.24(^3T_1)$
						$(^4T_1)$	
						$1.50(^4T_1)$	$1.43(^3A_2)$
GaAs	$0.565(^3T_1\ )$		0.9	—	$0.37(^5T_2)$	$0.5\ (ZPL)$	$1.15(^3T_1)$
						$(^4T_2)$	
	$0.66\ (^3T_1, ^3T_2)$	1.03	$0.84(ZPL, ^5E)$			0.87-1.05	
						$(^4T_1)$	
	$1.04\ (^3T_1)$	0.69				$1.40(ZPL)$	
						$(^4T_1)$	
InP	—	—	$0.76(ZPL, ^5E)$	—	$0.35(^5T_2)$	$0.47(ZPL)$	—
						$(^4T_2)$	
						$0.92(^4T_1)$	—
						$0.79(ZPL)$	
						0.78-0.92	

* ZPL = Zero photon line

© John Wiley & Sons, Inc.

semiconductor is shown in Fig. 22.18B. The ionization energy of $3d$-transition metal ions is given in Fig. 22.18A.

22.4.12C Transition Character The transitions at the transition metal dopant can be described as intracenter transitions; for instance

$$(e)^2 t_2 + h\nu \rightarrow e(t_2)^2 \tag{22.16}$$

Table 22.4: Ionization energies (eV) of substitutional 3d-transition metal dopants in compound semiconductors and interstitial 3d dopants in Si. Where levels are not yet observed, but predicted in valence or conduction bands, or in the gap, this is indicated (VB, CB, or gap) (after Zunger, 1986). ©John Wiley & Sons, Inc.

Material	Ti	V	Cr	Mn	Fe	Co	Ni	Transition Type
ZnS	(gap)	2.11	1.0, 1.74	(VB)	2.01	(gap)	0.75	(×/•)
	—		2.78, 2.48	—	(gap)		2.46	(ı/×)
ZnSe	1.75	(gap)	0.46	(VB)	1.1, 1.25	~0.3	0.15	(×/•)
	—		2.26, 2.1	—			1.85	(ı/×)
ZnTe	(gap)	(gap)		(VB)		0.86	(VB)	(×/•)
	—	—	1.4	—	(gap)	—	~1	(ı/×)
CdS	~2.21	≳1.8	1.47	(VB)	(gap)	(gap)	≥0.27	(×/•)
CdSe	1.69	~1.5	0.64	(VB)	0.64	0.22	0.32	(×/•)
	—	—		—	—	—	1.81	(ı/×)
CdTe	~0.55	0.74		(VB)	0.13	—	(VB)	(×/•)
	—	—	1.34	—	—	—	0.92	(ı/•)
GaP	—	—	0.5	—	—	—	—	(×/•)
	(gap)	1.55	1.12	0.4	0.86	0.41	0.5	(ı/×)
	—	—	1.85	—	2.09	2.02	1.55	(ı/ıı)
GaAs	—	—	0.32-0.45 0.68	—	—	—	—	(×/•)
	1.07, 1.29	1.38	0.74	0.11	0.49	0.16	0.22, 0.35	(ı/×)
	—	—	1.57	—	(CB)	1.67	1.13	(ı/ıı)
InP	—	(CB)	0.94-1.03	0.22	0.8	0.24-0.32	—	(ı/×)
	—	0.2	—	—	—	—	—	(×/•)
Si	0.89	0.72	0.95 0.38	0.75	0.385	—	—	(×/•)
	1.09	1.01	—	1.06	—	—	—	(ı/×)
	0.25	0.30	—	0.25	—	—	—	(•/••)

or, when carriers from one of the bands are involved, as ionization, such as, e.g.,

$$(e)^3 + h\nu \rightarrow (e)^4 + h^\bullet. \tag{22.17}$$

The corresponding energies for intracenter transitions are listed in Table 22.3 and for ionization in Table 22.4—see also Pantelides and Grimmeiss (1980).

22.4.12D The Energy of Levels   The crystal-field splitting can be estimated in tetrahedral symmetry as

$$\Delta = |E(e) - E(t_2)| = -\frac{4}{15} \frac{Ze^2}{R} \frac{\langle r^4 \rangle}{R^4}, \tag{22.18}$$

Table 22.5: Effective crystal-field splitting of $3d$ impurities in binary semiconductors (after Zunger, 1986b).

Impurity	Host	Δ_{eff}(eV)	Impurity	Host	Δ_{eff}(eV)
Ni	ZnS	0.520	Fe	ZnS	0.430
	ZnSe	0.510		ZnSe	0.41
	GaAs	0.91		InP	0.43
	GaP	0.97		GaAs	0.44
Co	ZnS	0.453		GaP	0.45
	ZnSe	0.459	Mn	ZnS	0.402
	InP	0.575		ZnSe	0.400
	GaAs	0.590		GaP	0.52
	GaP	0.608	Cr	ZnS	0.540
				ZnSe	0.540
				InP	0.64
				GaAs	0.65
				GaP	0.67

where R is the distance to the nearest neighbor (ligand) and $\langle r^4 \rangle$ is the expectation value of r^4 for the $3d$ wavefunction (Hayes and Stoneham, 1984). Here Δ increases with increasing covalency, higher charge of the transition metal, and higher transition series.

The effective crystal-field splitting is given in Table 22.5 for some of the $3d$ impurities in a few binary compounds.

The multi-electron levels in the *irreducible representation* of the tetrahedral T_d point group are a_1, a_2, e_2, t_1, and t_2—see Bassani et al. (1974) or Madelung (1981). In Fig. 22.20, an example for the transition from the lower ${}^5T_2$ $(t^2 e^2)$ to the excited ${}^5E(t^3 e^1)$ state is given for GaAs:$(Cr^{2+})'$ (Clerjaud, 1985). A list of possible splittings of the free transition metal ion states after incorporation in a semiconductor of T_d symmetry is given in Table 22.6. In addition, one has to consider for splitting due to spin-orbit, lattice-phonon, and the Jahn-Teller effect interactions.

The relationship of the level spectrum to the host lattice and the chemical identity of the transition metal have been analyzed using a self-consistent quasiband crystal-field method, and employing a density-functional Green's function approach, introduced by

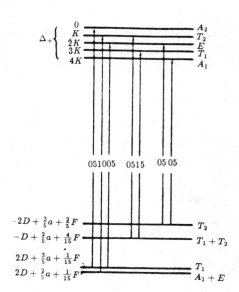

Figure 22.20: Level scheme for GaAs:$(Cr^{2+})'$ with weight factors for competing transitions (not to scale); see Fig. 22.23 (after Clerjaud, 1985).

Table 22.6: Splitting of free transition metal ion terms into many electron terms after incorporation into semiconductors of T_d symmetry at lattice (solid underlining) or interstitial (dashed underlining) sites.

Free-ion many-electron terms	T_d crystalline many-electron terms	Ground state	
		S	J
S Terms			
$Fe^{3+}, Mn^{2+}, Cr^+; d^5, {}^6S$	${}^6A_1(t^3e^2)$	$\frac{5}{2}$	$\frac{5}{2}$
D Terms			
$Sc^{2+}; d^1, {}^2D$	${}^2T_2(t^1e^0) + {}^2E(e^1t^0)$	$\frac{1}{2}$	
$Mn^{3+}, Cr^{2+}; d^4, {}^5D$	${}^5T_2(t^2e^2) + {}^5E(t^3e^1)$	2	$1, 2, \underline{3}$
$Co^{3+}, Fe^{2+}; d^6, {}^5D$	${}^5T_2(t^4e^2) + {}^5E(t^3e^3)$	2	$0, \frac{1}{2}, 1$
$Cu^{2+}, Ni^+; d^9, {}^2D$	${}^2T_2(t^5e^4) + {}^2E(t^6e^3)$	$\frac{1}{2}$	$\frac{1}{2}, \frac{3}{2}$
F Terms			
$V^{3+}, Ti^{2+}; d^2, {}^3F$	${}^3T_1(t^2e^0) + {}^3T_2(t^1e^1) + {}^3A_2(e^2t^0)$	1	
$Cr^{3+}, V^{2+}; d^3, {}^4F$	${}^4T_1(t^1e^2) + {}^4T_2(t^2e^1) + {}^4A_2(t^3e^0)$	$\frac{3}{2}$	$\frac{1}{2}, \frac{3}{2}, \underline{\frac{5}{2}}$
$Ni^{3+}, Co^{2+}, Fe^+; d^7, {}^4F$	${}^4T_1(t^5e^2) + {}^4T_2(t^4e^3) + {}^4A_2(t^3e^4)$	$\frac{3}{2}$	
$Cu^{3+}, Ni^{2+}, d^8; {}^3F$	${}^3T_1(t^4e^4) + {}^3T_2(t^5e^3) + {}^3A_2(t^6e^2)$	1	$\underline{0}, 1, 2$

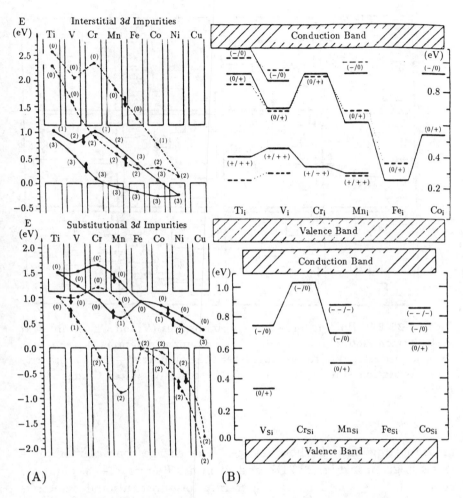

Figure 22.21: (A) Ground-state energies of neutral $3d$ impurities in a Si host with level occupancy in parentheses. (B) Calculated (solid curves) and observed (dashed) $3d$-impurity levels in Si at interstitial and lattice (substitutional) sites (after Beeler et al., 1985).

Lindefelt and Zunger (1982). This method was first applied to $3d$-transition metal impurities in an Si host lattice (Zunger and Lindefelt, 1983), and yields results in agreement with the experiment when spin polarization is included (Beeler et al., 1985; for comparative remarks, see Zunger, 1986b). Typical examples are shown in Fig. 22.21, identifying t_2 and e states, which are contained within the bands for some of the transition metal impurities. Their energy sequence is inverted when comparing interstitial and substitutional

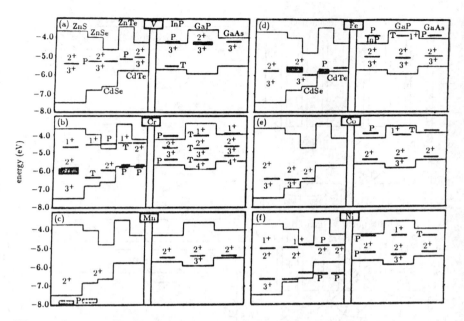

Figure 22.22: Binding energies of $3d$ impurities (V and Cr) in different host semiconductors [only identified in the upper diagrams (a) and (d)]. Energies are related to the vacuum level. The numbers $1+, 2+, \ldots$ indicate stable oxidation states. P: predicted; T: tentative (after Caldas et al., 1985).

incorporation into the host lattice. The charge character of these levels is given in Fig. 22.21B, and identifies donors ($' \rightarrow \times$) and acceptors ($\cdot \rightarrow \times$). The level spectrum for substitutional defects is less structured, and no deep levels are obtained for Ti and Fe. For $4d$ impurities, see Beeler et al. (1986); for other transition metals in III-V compounds, see Clerjaud (1985).

The investigations of Zunger (1985) have shown some universality of binding energies within the gap. When normalized to the vacuum level using the intrinsic workfunction of the host (see Table 18.2), all binding energies of the same $3d$ impurity in different host crystals are approximately the same—see Fig. 22.22.

Transition metal atoms in covalent crystals show only minor net charges, i.e., a minor fraction of ionic bonding, and tend to approach a noble metal configuration. In Ni this is best achieved. Its extremely high diffusivity in Si ($D = 10^{-4}$ cm^{2}s^{-1}) may be an indicator of this fact as it renders the Ni atom small and inert.

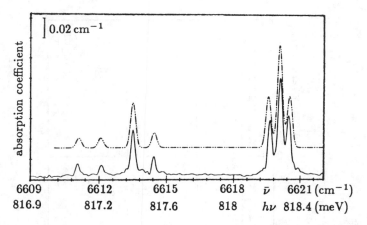

Figure 22.23: Optical absorption spectrum of Cr^{2+} in GaAs at 6 K (after Williams et al., 1982). The broken curve is a theoretical estimation using spin Hamiltonian parameters from EPR measurements (after Clerjaud, 1985).

The ability of transition metal impurities to trap carriers is used for compensation, i.e., to produce less conductive semiconductors.

Much of the earlier work to identify the level spectrum was done by electron-paramagnetic resonance (EPR)—Ludwig and Woodbury (1962). For a recent review, see Weber (1983).

22.4.12E Optical Absorption, Transition Metal Dopants

In Fig. 22.23, a typical optical absorption spectrum of Cr^{2+} in GaAs is shown at 6 K, which can be understood as $^5T_2 \rightarrow {}^5E$ transitions. The corresponding zero-phonon transitions are identified in Fig. 22.20.

In semi-insulating GaAs, ionizing transitions from the Cr^{3+} center are observed in addition to those listed above:

$$(Cr^{3+})^{\times}_{Ga} + e'_{VB} + h\nu \rightarrow (Cr^{2+})'_{Ga}. \qquad (22.19)$$

The shape of the Cr^{2+} absorption band is explained by phonon coupling and lattice strains in a tetragonal site; the Jahn-Teller energy of the $^5T_2$ state is 75 meV, and of the ^5E state is 6.2 meV (see Deveaud et al., 1984). For a review of other absorption spectra of transition metals, see Clerjaud (1985).

Most of the deep center optical transitions have been studied for transition metals and rare earths in II-VI compounds, which show well-localized states due to the strong interaction of $3d$ electrons. Some of these are described by crystal field theory (Griffith, 1964), as

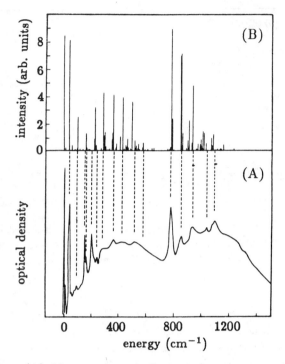

Figure 22.24: (A) Measured optical absorption of ZnSe:Co at 1.5 K. (B) Prediction of the Jahn-Teller model for transitions within the $3d^7$ configuration of Co^{2+} (after Uba and Baranowski, 1978).

a result of the splitting of d states by the field from nearest neighbors (Hennel, 1978). The Jahn-Teller shift (see Uba and Baranowski, 1978) has been used to correlate some of the rather complex features of the optical absorption of ZnSe:Co (Fig. 22.24).

The optical absorption spectrum of transition metal impurities in wide-gap crystals is rich in absorption lines due to several charge states that are available within the band gap for a variety of these elements, and various crystal-field splittings.

The absorption of some transition metal impurities is responsible for the color of the host crystal. For example, the red color of ruby is due to Cr^{3+} in Al_2O_3, and the blue color of sapphire is due to charge transfer from Fe^{2+} to Ti^{4+} in Al_2O_3.

Transition Metal Impurities in Alloys. The transition energies of $3d$ impurities tracks with the vacuum level of the host semiconductor. In an alloy $A_\xi B_{1-\xi} C$, the transition energy is given by

$$E(\xi) = E(0) + \overline{\alpha}\xi \quad \text{with} \quad \overline{\alpha} = \phi(\xi = 1) - \phi(\xi = 0), \quad (22.20)$$

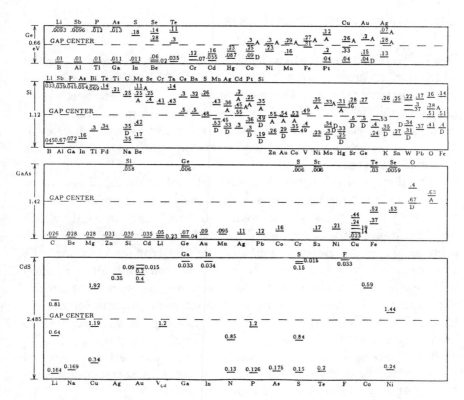

Figure 22.25: Measured ionization energies of various impurities in Ge Si, GaAs, and CdS from top to bottom at 300 K. The band gaps shown are 0.66, 1.12, 1.42, and 2.485 eV, respectively, for these semiconductors. Centers below or above the mid-gap are acceptors or donors, respectively, except when otherwise indicated by a D or an A.

using a linear variation of the slope $\overline{\alpha}$ with the intrinsic semiconductor workfunction ϕ on composition ξ.

22.4.13 The Level Spectrum of Various Point Defects

A summary of the different energies of some point defects is shown in Fig. 22.25 for the four semiconductors: Ge, Si, GaAs, and CdS (Sze, 1981; Landoldt-Börnstein, 1982). This figure has to be used with some caution. The reference to certain elements often refers to the *involvement* of the element, but possibly in conjunction with other defects as an associate. A review of defect levels in Si is given by Chen and Milnes (1980).

Summary and Emphasis

Deep level defect centers have tightly bound electrons in small orbits which, for the ground state, often do not extend beyond the distance to the next neighbors of the semiconductors. The electronic eigenfunctions mix with both conduction and valence bands. Many of the deep defect levels relax substantially after defect recharging or excitation, causing a significant change in the equilibrium position of the surrounding atoms.

The degree of relaxation is measured by the Huang-Rhys factor which gives the number of breathing mode phonons released during the relaxation process. With weak coupling, there are only a few phonons interacting. These are discernible as phonon-replicas of the original zero-phonon transition line. With stronger coupling, the spectrum becomes broadened to a degree that individual phonon lines are no longer identifiable. Furthermore, the spectral features, the largely broadened lines, are substantially shifted between optical absorption and emission; this shift is typically on the order of 20 breathing phonons (~ 0.5 eV).

When charged, such deep level centers may have higher excited states which have quasi-hydrogen character with corresponding orbits extending well into the surrounding lattice.

The deep centers act as deep traps for either electrons or holes. They may also act as recombination centers when their relaxation is significant, and thereby form a bridge between conduction and valence band for recombination traffic.

As traps and recombination centers, these deep level defects are usually detrimental for semi- and photoconducting devices, by reducing their response and rendering them slow—as will be discussed in the appropriate chapters. However, specific types of deep traps can be beneficial, by storing some carriers for a long time. This is important for sensitizing photoconductors or by accumulating the effect of x-ray or nuclear reaction in solid-state dosimeters. Other deep centers are essential for many applications in luminescence, or to render semiconductor layers low-conductive for field-effect transistors.

A wide variety of such deep centers, mostly transition metal impurities, exist. Within this group, these centers have vastly different defect level behaviors. Most important, aside from the level energy and the center capture cross section, is the degree of relaxation which distinguishes the utility of the center or the need of avoiding it in efficient devices.

Exercise Problems

1.(e) Derive the solutions of the Schrödinger equation for a one-dimensional square well of depth V_0 and width $2a$.
   (a) Compute the eigenvalues for $V_0 = -10$ eV and $a = 2$ Å.
   (b) Compute the value of the lower eigenstates as functions of V_0 for $-20 < V_0 < -5$ eV and for $a = 2$ Å.
   (c) At what value of V_0 is the first excited state a bound state?

2.(r) Describe in your own words the Jahn-Teller effect, as well as its influence on typical ground and excited states of a substitutional impurity defect in a cubic lattice.

3.(*) Describe the influence of the crystal field on d^2 levels of transition elements in isotropic semiconductors.

4.(r) List and describe the intrinsic point defects in alkali halides.

5.(r) Describe the differences between a positive- and a negative-U center.
   (a) Give examples for both.
   (b) What semiconductors would be favorable candidates to harbor negative-U centers?

6.(r) Describe the differences between vacancies in ionic and in element-covalent crystals. Analyze the multicharge character of vacancies.

7.(r) Summarize the electronic behavior of isovalent substitutional defects.

8.(*) What information can be used to identify the ground state among the different possibilities of arranging electrons in a partially occupied inner shell of a transition element?

9.(r). Why has Ni special properties as a dopant? What are these properties?

10.(l) Check the recent literature concerning $EL2$ centers. Currently, what is the best model to describe this center?

Chapter 23

Defect Associates

Many atomic crystal defects interact with each other, and thereby form defect associates with their own set of defect properties.

There is a very large group of point defects, which easily form associates, with their nearest neighbor position energetically preferred. There can be associates of two or more defects, as well as larger clusters causing phase segregation.

In addition, there are defects in a heavily disturbed crystal environment which may be classified as associates of impurities with intrinsic defects, or as clusters of intrinsic defects.

In this chapter, we will give a short overview with a few typical examples of how these centers are created, and what their main properties are.

23.1 Defect Center Pairs

Defects may arrange themselves as nearest neighbors or, statistically, in different distant pair positions. These defect pairs can act as one center if the distance between them provides sufficient overlap of their electronic eigenfunctions. Different intrapair distances, however, produce different level spectra. Consequently, this causes a great abundance of defect levels which can be observed in the optical spectra of such defects.

23.1.1 Donor-Acceptor Pairs

The formation of *donor-acceptor pairs* is intensively studied. Since ionized donors and acceptors are Coulomb-attractive to each other, they occupy preferably nearest lattice sites at low temperatures.

In addition to electrostatic attraction discussed below, space-filling aspects also create attractive forces. For instance, incorporation of an As donor in Si causes a slight shrinkage of the lattice

$$\frac{\Delta a}{a N_{\text{As}}} \simeq -0.4 \cdot 10^{-24} \text{ cm}^3 \tag{23.1}$$

while inclusion of a Ga acceptor dilates the lattice

$$\frac{\Delta a}{a N_{\text{Ga}}} \simeq 0.8 \cdot 10^{-24} \text{ cm}^3 \tag{23.2}$$

resulting in a strain-induced attraction of Ga and As within the Si lattice. Normally, a distribution of these pairs is found with various distances between them.

Such donor-acceptor pairs may be composed of cation vacancies or impurities that act as acceptors, as well as substitutionals, cation interstitials, anion vacancies, or other impurities that act as donors. See Williams (1968), Dean (1973), and Taguchi and Ray (1983) for II-VI compounds.

Donor-acceptor pairs in GaP are well-investigated. One distinguishes substitutionals of the same sublattice, such as a C_P-O_P (carbon or oxygen on a phosphorus site) pair acting as acceptor and donor, respectively, and are referred to as type-I pairs. An example of a type-II pair is Zn_{Ga}-O_P in GaP, in which the two sublattices are involved.

The energy of donor-acceptor pairs is influenced by their relative distance r_{da} and can be expressed as

$$E_{\text{da}} = E_g - (\tilde{E}_d + \tilde{E}_a) - \frac{e^2}{4\pi \varepsilon^* \varepsilon_0 r_{\text{da}}} - E^*(r_{\text{da}}), \tag{23.3}$$

where $\tilde{E}_d$ and $\tilde{E}_a$ are the distance of donor or acceptor levels from the respective band edges. The pair energy is substantially influenced by the Coulomb attraction between donor and acceptor, and by a correction energy term $E^*(r_{\text{da}})$, caused by overlap of donor and acceptor eigenfunctions. The latter term is much reduced at larger distances.

At low temperatures ($T < 10$ K), the absorption lines are sharp enough to distinguish a large number of pairs with different pair distance. This distance can be identified in crystallographic notation by the position of both atoms—the first assumed at 000, the second at $n_1 n_2 n_3$, with $\hat{m}$ the *shell number*:

$$2\hat{m} = n_1^2 + n_2^2 + n_3^2. \tag{23.4}$$

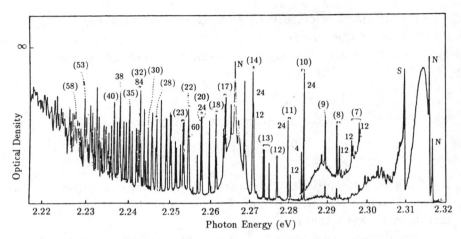

Figure 23.1: Donor-acceptor pair spectrum with $\Delta E_d + \Delta E_a = 157.5$ meV of type-II pairs Mg$_{Ga}$-S$_P$ in GaP at 1.6 K. Pairs up to $\hat{m} = 62$ are identified (in parentheses), as are some lines of excitons bound to N$_P$; see Section 21.2 (after Dean et al., 1969).

The separation distance in a cubic lattice with lattice constant a is then given for type-I pairs or for type-II pairs

$$(r_{da})_I = \sqrt{\frac{3\hat{m}}{2}}\, a, \quad \text{or} \quad (r_{da})_{II} = \sqrt{\left(\frac{3\hat{m}}{2}\right) - \left(\frac{5}{16}\right)}\, a. \quad (23.5)$$

The absorption lines of the donor-acceptor pair are identified sequentially according to $\hat{m}$. A typical spectrum showing the wealth of distinguished lines is given in Fig. 23.1. The energy as a function of the relative pair distance is shown in Fig. 23.2. We see that for higher pairs ($\hat{m} > 15$), the fit neglecting the perturbation term E^* is excellent. For less distant pairs, however, this term is necessary to provide a better agreement between theory and experiment.

In addition to donor-acceptor pairs, we distinguish *isoelectronic center pairs*, e.g., N-N pairs in GaP. We will return in Section 44.5.2 to these important pairs when we discuss their influence on the luminescence spectrum.

Other pairs which have been the subject of extensive research are the chalcogen pairs in Si.

23.1.1A Chalcogen Pairs in Si

Molecular pairs of S, Se, or Te act as single, deep centers with slightly lower depth than the ground states of single chalcogen substitutionals (Table 23.1, compare with

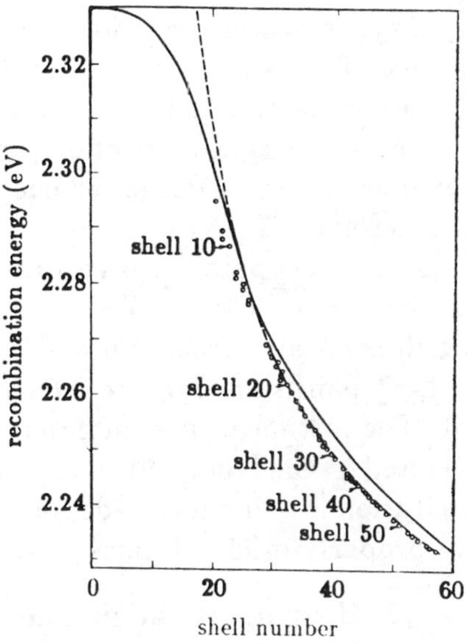

Figure 23.2: Energy of donor-acceptor pairs $C_{Ga} - S_P$ in GaP as a function of the shell number and fit with Eq. (23.3); dashed curve without and solid curve with $E^*(r_{da})$ consideration (after Hayes and Stoneham, 1984). ©John Wiley & Sons, Inc.

Table 23.1: Binding energies (in meV) of chalcogen pairs in Si (after Wagner et al., 1984).

	$S_2^{\times}$	$Se_2^{\times}$	$Te_2^{\times}$	$S_2^{\bullet}/4$	$Se_2^{\bullet}/4$
$1s(A_{1g})$	187.5	206.4	158.0	92.5	97.3
(E_u)	26.4	31.14	33.0	37.3	30.9
(A_{1u})	31.2	25.8	25.6	23.9	23.2
(E_g)	34.4	33.2			
$2p_0$	11.4	11.6	11.5	11.7	12.3
$2p\pm$	6.3	6.5	6.3	6.4	6.68
$3p_0$	5.54	5.54	5.44	5.45	5.83
$3p\pm$	3.12	3.12	3.12		3.12
$3d_0$	3.92	3.89			
$4p\pm$	2.2	2.2	2.1		
$5p\pm$		1.51	2.1		

Table 22.1). Again, the higher excited states lie very close to the corresponding hydrogen-like states.

These pairs are incorporated in $\langle 111 \rangle$ axial symmetry and show four $1s$ ground states: A_{1g}, E_g, E_u and A_{1u}—see Janzén et al. (1985). There are also mixed pairs observed, e.g., Si:S;Se or Si:Se;Te.

Oxygen centers are more difficult to identify because of the large variety of oxygen associates in *thermal donors*—see Section 22.4.8. There are interstitial pairs of oxygen and an oxygen interstitialcy replacing one Si atom proposed as possible diatomic oxygen centers. For more information on chalcogen pairs and higher associates, see Wagner et al. (1984).

Still other pairs involve vacancies. For instance, a *divacancy* of Cd in CdS with two holes attached is equivalent to an S_2 molecule, which is covalently bound and probably rather stable.

Any pair is *anisotropic*, even when embedded in an isotropic crystal. For instance, in a cubic crystal a nearest-neighbor pair can be oriented in the [100], [010], or [001] directions. Consequently, optical excitation is polarization-dependent. This can be observed by bleaching properly oriented pairs of such defects—see Section 20.3.2A.

23.1.2 Hydrogen Pairing in Silicon

Hydrogen is known to diffuse at moderate temperatures ($\leq 400\ ^{\circ}C$) as an interstitial into Si. It dramatically reduces p-type conductivity, while it has almost no effect on n-type conductivity (Pankove et al., 1985). The spatial distribution of H tracks that of acceptors (Johnson, 1985); an electrically neutral hydrogen-acceptor pair is probably formed via compensation and consequent Coulomb attraction (Pantelides, 1986):

$$H_i^{\times} + h^{\bullet} \rightarrow H_i^{\bullet} \quad \text{and} \quad H_i^{\bullet} + A' \rightarrow (AH)^{\times}. \qquad (23.6)$$

Only recent computation of the total energy profiles surrounding some impurities (e.g., B or Be) has helped to identify the character and site arrangement of such defect associates with hydrogen. The total energy depends critically on the specific interstitial position of H and the relaxation of the surrounding lattice.

The disassociation energy of the B-H associate with H at the BM site, according to $(BH)^{\times} \rightarrow B' + H^{\times}$ with $H^{\times}$ at a BC site, was estimated at 0.59 eV. This value is probably too low to explain the observed reactivation of B-acceptors in hydrogenated Si at $\sim 150\ ^{\circ}C$ (Johnson, 1985; Denteneer et al., 1989). The BM site lies at about 43% off B, between B and Si in the (110) plane. The BC site lies halfway between Si atoms in an undisturbed lattice.

Other pairs in Si involve a variety of elements. Those contained in the atmosphere or supplied during crystal growth are of special interest (Taguchi and Ray, 1983). Examples are pairs containing hydrogen or oxygen as a partner (Haller, 1986). Because of their

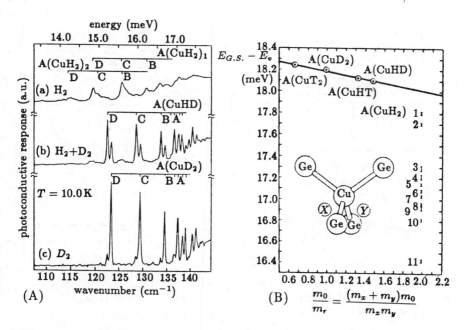

Figure 23.3: (A) Photothermal ionization spectra of copper-dihydrogen (or isotopes) acceptors in ultrapure Ge at 10 K. (B) Copper-dihydrogen complex (x and y stands for H, D, or T) and binding energies ($E_{G.s}$ is the ground-state energy of the acceptor) as a function of the reciprocal reduced mass (after Haller et al., 1986).

abundance, and the great variety of such pairs, little is yet known about them, despite highly sensitive methods of detection, e.g., photothermal ionization spectroscopy—see Section 45.5.2.

23.2 Triple and Higher Defect Centers

The variety of defects involving more than two atoms is even greater, and only the first steps of identification have been made, e.g., Cu,H,H, Cu,H,D, or C,H,T isotope combinations discussed by Haller (1986) and shown in Fig. 23.3.

Other impurities tend to order themselves and surround intrinsic lattice defects, such as four Li atoms surrounding a lattice vacancy in Si and creating a more complex defect center. Such a center is observed after irradiation of Li-doped Si with 2 MeV electrons (DeLeo et al., 1984). The resulting defect level spectrum of such centers can be rather complicated, and can be measured in emission (luminescence) as well as in absorption at sufficient density of these centers (Lightowlers and Davis, 1985). More information about the

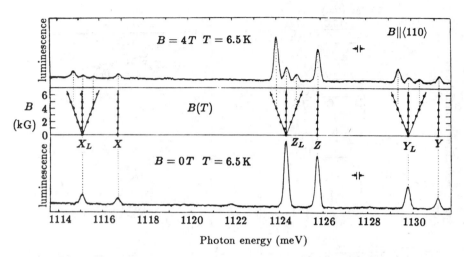

Figure 23.4: Bound-exciton spectrum related to the $(4\text{Li-V}_{\text{Si}})$ complex, which shows three triplet-singlet line pairs, revealed after application of a magnetic field (after Lightowlers and Davis, 1985). ©Pergamon Press plc.

center can be drawn from bound excitons and their behavior in a magnetic field (Fig. 23.4). For further discussion of these bound excitons, see Section 21.2.2.

Large defect associates, e.g., $(\text{Cu}_{\text{Cd}}\text{V}_{\text{S}})_n$ in CdS can finally lead to phase segregation of a small precipitate, such as, in this example, copper in CdS. A similarly well-known phase segregation is that of silver in silver halides. It can be described as a large association of F-centers, and is the basis for the photographic process—see Section 46.2.

Clustering often occurs when we want to obtain high doping densities. This develops near the solubility limit. Occasionally, the segregation occurs in the melt, and the associates remain separate because of the charging. This *colloidal* distribution can remain during solidification, and in glasses is known to be responsible for specific spectral absorption (coloration), which is different from bulk absorption, and sensitively depends on particle size. Very high densities of defects can be created by radiation damage.

23.3 Defect Clusters Due to Radiation Damage

The defect associates created by radiation damage are different from the previously discussed associates. They are more of a cluster of

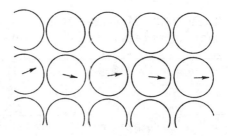

Figure 23.5: Energy transfer in a focused collision along a row of atoms with successively more aligned secondary impacts.

highly disordered atoms, and can best be described by following their specific generation mode, as discussed below.

Radiation with energetic particles, such as electrons, nuclei, or ions results in the creation of lattice defects if the energy and momentum of the impacting particle are sufficiently large. The maximum transferred energy E_{max} is given by

$$E_{max} = E_i \frac{4 M_i M}{(M_i + M)^2} \qquad (23.7)$$

where E_i and M_i are the energy and mass of the impacting particle, and M is the mass of the lattice atom. The energy transfer depends on the incident angle

$$E(\theta) = E_{max} \sin^2 \left(\frac{\theta}{2} \right). \qquad (23.8)$$

Here θ is the scattering angle in the center-of-mass reference frame.

One distinguishes three cases:

- For a head-on collision, $\theta = \pi$, and the impacting particle is reflected back.
- When the impacting angle is almost aligned with a low index crystallographic direction, i.e., of closely packed atoms, the target atom moves in a direction even closer to the low index direction. The process continues with more and more alignment of forward motion. This is referred to as a *focusing collision* (Fig. 23.5). It proceeds until a crystal defect or the opposite surface is reached, where reflection, radiation damage, or ejection occurs (Wedell, 1980; Überall and Sáenz, 1985).
- When the angle with a low index direction is larger, the target ion moves nearly perpendicular to the path of the incident particle.

The actual collision (scattering) event is more angle-dependent when far-reaching Coulomb forces provide the interaction, e.g., for protons or α-particles. This event does not depend on θ for so-called

Table 23.2: Measured atomic displacement energies (in eV) in semiconductors (after Hayes and Stoneham, 1984).*

IV		III-V		II-VI (Fourfold Coordination)	
C	25 Graphite	GaAs	9/9.4	ZnO	30–60/60–120
	35–80 Diamond	InP	6.7/8.7	ZnS	7–9/15–20
Si	13	InAs	6.7/8.3	ZnSe	7–10/6–8
Ge	13–16	InSb	5.7/6.6	CdS	2–7/8–25
Sn	12			CdSe	6–8/8–12
				CdTe	5.6–9/5–8

*The value(s) before the slash refers to the first atom, and the value(s) after the slash to the second atom of a compound. A range is indicated for diamond, Ge, and II-VI compounds. ©John Wiley & Sons, Inc.

hard-sphere collisions, i.e., for neutrons or fast ions. Here, θ is the angle between impacting and scattered particle trajectories. The first type of collision is described as *Rutherford scattering*, with a collision cross section of

$$\sigma(\theta) = \frac{R^2}{\left[2 \sin \left(\frac{\theta}{2} \right) \right]^4} \tag{23.9}$$

where R is the distance of closest approach of both particles.

The result of a particle impact can be divided into metastable atomic displacements (i.e., radiation damage) and electronic ionization.

The minimum energy necessary for displacing a lattice atom after an impact is typically on the order of 5 . . . 50 eV, and is shown in Tables 23.2 and 23.3 for some of the semiconductors.

From energy and momentum conservation, we estimate that the minimum energy for *impacting electrons* is on the order of 100 keV to cause radiation damage. Atoms displaced in such a manner have insufficient energy to cause secondary damage; however, displacement occurs preferably in the neighborhood of lattice defects. For electron-induced radiation damage, the relevant characteristic energies are listed in Table 23.4.

Table 23.3: Displacement energy thresholds for A or B Frenkel pair formation by radiation damage calculated (experimental) (after van Vechten, 1980).

Crystal AB	$E_d(A)$ (eV)	$E_d(B)$ (eV)	Crystal AB	$E_d(A)$ (eV)	$E_d(B)$ (eV)
SiC	27.6	26.1	BeO	28	64 (76)
			MgO		53 (60)
AlAs	13.7	19.5	ZnO	18.5	41.4 (57)
AlSb	11.1	16.2	ZnS	12.1	27.5
GaN	24.3	32.5	ZnSe	10.8 (10)	24.8
GaP	13.7	19.2	ZnTe	9.2 (7.4)	21.5
GaAs	12.4 (15-17)	17.6	CdS	11.0 (8)	24.3
GaSb	10.1	14.7	CdSe	10.1	22.4
InP	12.2	16.5	CdTe	8.7	19.5
InAs	10.8	14.9			
InSb	9.1	12.9			

Bombardment with *protons or neutrons* transfers substantially more energy, and results into numerous secondary damage events, which are wider-spaced for protons, and in a more compact region for neutrons. A similar, compact region of high disorder is observed after bombardment with *high energy ions*—see Fig. 23.6.

Ion bombardment from a gas discharge creates disorder in a region close to the surface. Ion implantation leaves even larger defect clusters. Figure 23.7 provides a graphic illustration of possible lattice defects after an impact with a fast ion. Careful posttreatment annealing is required to restore sufficient order to reobtain attractive semiconductive properties (i.e, high mobility and low recombination rates), since these defect clusters usually act as major scattering or recombination centers. For a short summary, see Hayes and Stoneham (1984).

Atomic displacement at lower energy can be achieved with photons after ionization, which, in conjunction with multiphonon events, cause photochemical reactions, and is discussed in Chapter 46.

Table 23.4: Electron damage in II-VI compounds with bonding energy (E_{bond}), electron threshold energy (T_d), atomic displacement energy (E_d), recoil energy for 100 keV electrons (E_{recoil}), and activation energy fo vacancy diffusion (E_{diff}) (after Taguchi and Ray, 1982). ©Pergamon Press plc.

Compound	E_{bond}(eV)	T_d(keV)	E_d(eV)	E_{recoil}(keV)	Vacancy	E_{diff}(eV)
ZnS	5.4	240	9.9	7.9	V'_{Zn} (330K)	1.04
		185	15.0		V_S (673K)	
ZnSe	4.9	195	7.6	5.4	V'_{Zn} (373K)	1.26
		240	8.2		V_{Se} (150K)	
ZnTe	4.7	110	4.2	3.1	V'_{Zn} (340K)	0.78
		300	6.7		V'_{Te} (200K)	0.73
CdS	5	290	7.3	5.4	V'_{Cd} (300K)	
		115	8.7		V'_S	
CdSe	4.9	320	8.1	4.1	V'_{Cd}	
		250	8.6		V'_{Se}	
CdTe	4.3	235	5.64	3.3	V'_{Cd} (353K)	0.8
		340	7.79		V'_{Te} (120K)	0.2

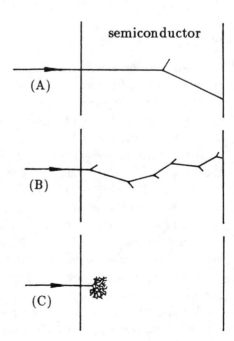

Figure 23.6: Typical damage tracks after bombarding a semiconductor with: (A) electrons; (B) protons or light ions; and (C) fast neutrons or heavy ions.

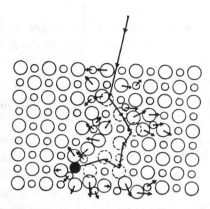

Figure 23.7: Schematics of a variety of different disorder processes, initiated by the impact of a fast ion (after Weißmantel and Hamann, 1979).

Summary and Emphasis

Defect center associates are plentiful in semiconductors and determine a large fraction of the mostly deeper defect level spectrum within the band gap. Only a few of these associates have been identified unambiguously at the present time.

Lattice defects can form associates, when their electronic eigenfunctions overlap and cause a significant perturbation of their eigenstates. Deep level centers, which have tightly bound electrons, form associates of a substantially modified excitation spectrum between the nearest and next-nearest pairs. Shallow level defects, however, form a very large variety of distant pairs because of the far-reaching Coulomb potential. This causes a rich level spectrum in which up to 100 lines can be identified, each one of which is characterized by a well-defined interdefects distance.

Pairs of defects are anisotropic. The specific symmetry of these and higher defect associates aids in a positive identification of such centers.

Still larger, often less-defined associates are formed by a variety of radiative damage experiments. Different impacting particles have their own characteristic signature of the resulting damage.

Defect associates play an important part in the defect spectrum of semiconductors. These associates influence to a large degree their electronic and optical behavior by acting as traps and recombination centers or as scattering centers for carrier transport. Unambiguous identification, creation, and disassociation of specific types of such defects will gain more importance as we better understand their role in enhancing or limiting the performance of devices.

Exercise Problems

1.(r) List and describe the properties of various intrinsic defect associates in alkalide halides.

2.(e) Calculate and draw the type-II donor-acceptor pair energy as a function of the shell number in a cubic lattice, with lattice constant $a = 2\text{Å}$, $\varepsilon_{st} = 10$, and $E^*(r_{da}) = 0$.

    (a) What changes occur for a type-II donor-acceptor pair?

    (b) What changes occur for an element semiconductor?

3.(e) Give energy and momentum conservation law for electron damage of Si. Estimate the threshold electron energy and compare with the tabulated value.

4.(e) Discuss the scattering cross section as a function of the angle between impacting and scattered particle trajectories. Derive Eq. (23.9). Draw the angle dependence.

5. Discuss the Haynes rule and try to give a plausibility argument for the different slopes.

6.(1) Discuss the influence of hydrogen doping on the electrical properties of Si.

    (a) Why is there a difference of hydrogen in n-type and p-type Si?

    (b) What kind of associate formations are known in Si, involving hydrogen?

    (c) How is hydrogen introduced into Si crystals?

Chapter 24

Defect States and Band States

Disordered lattices show a high density of states, extending from the band edge of ideal crystals and decreasing exponentially into the band gap.

Most point defects in an otherwise ideal lattice can be regarded as missing lattice atoms, which are replaced by lattice defects. For every defect that creates one or more levels in the band gap, the same number of levels that would have been created by the missing lattice atom within the bands are missing. Moreover, as indicated in Section 21.1.8, the lattice atoms surrounding the lattice defect relax into shifted positions, and also create levels which could be shifted into the band gap. Finally, thermal vibration of the lattice atoms gives rise to a perturbation of the band edge. All of these states contribute to perturbations near the band edge, and will be discussed in this chapter.

24.1 Band Tailing

For a crystal with a large concentration of point defects, the density of states inside each band is reduced. An equal number of levels extend from the band edge into the band gap as a consequence of the theorem of the conservation of eigenstates—the *golden rule*; see Bassani and Pastori Parravicini (1975).

Such a *tail of band states* into the band gap is often referred to as a *Lifshitz tail* (Lifshitz, 1964). It is well pronounced in heavily doped and amorphous semiconductors, and can be experimentally observed from the spectral distribution of the optical absorption (Shklovskii and Efros, 1984). For disorder due to dopants, having a correlation length on the order of interatomic spacing, the absorption coefficient shows an exponential decline and can be expressed as

$$\alpha_o = \alpha_{o0} \exp\left(\frac{h\nu - E_g}{E_0}\right). \qquad (24.1)$$

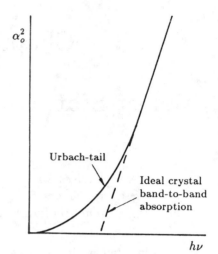

Figure 24.1: Optical absorption spectrum of a typical direct band-gap semiconductor with the absorption constant α_o proportional to the extended density of states in the Urbach tail.

Here, α_{o0} and E_0 are empirical parameters depending on the semiconductor and its defect structure as it relates to preparation, doping, and treatment of the semiconductor. This dependence of the optical absorption is widely observed, and is referred to as the *Urbach tail*—see Fig. 24.1 (Urbach, 1943). A detailed analysis of the optical behavior is involved (Casey and Stern, 1976). Some aspects of it were already discussed in Section 24.3. For a review, see Sa-yakanit and Glyde (1987).

There are several approaches for estimating the level distribution caused by a statistical distribution of point defects in an otherwise ideal lattice. The general behavior can be obtained from a semi-classical model evaluated by Kane (1963)—see also Bonch-Bruevich (1962) and Keldysh and Proshko (1964). Using an independent electron model, considering the lattice by replacing m_0 with m_n, and ignoring any change in the electron kinetic energy due to the impurities, the density of states is locally perturbed by the potential of lattice defects $V(\mathbf{r})$. When this potential lowers the conduction band, as a Coulomb-*attractive* center does, there are additional levels accessible to quasi-free electrons at energies below the unperturbed band edge. Near a *repulsive* center, however, there are fewer levels. That is, we assume that the density of states is given by the usual expression for a Fermi gas [Eq. (27.8)], however, with an energy scale shifted by the potential energy $eV(\mathbf{r})$ at any point $\mathbf{r}$, yielding

$$g(E, \mathbf{r}) = \frac{1}{2\pi^2} \left(\frac{2m_n}{\hbar^2} \right)^{3/2} \sqrt{E - eV(\mathbf{r})}. \qquad (24.2)$$

Integrating over the entire volume, we obtain the density of states $g(E)$. This integration can be replaced by an integration over an actual potential distribution

$$g(E)\, dE = \frac{1}{2\pi^2} \left(\frac{2m_n}{\hbar^2} \right)^{3/2} \int_{-\infty}^{E} \sqrt{E - eV}\, f(V)\, dV, \qquad (24.3)$$

with $f(V)$ as the distribution function of the potential. $V(\mathbf{r})$ is assumed to be variable with a Gaussian distribution around a mean potential V_0 (see Chapter 25), i.e.,

$$f(V) = \frac{1}{\sqrt{\pi \overline{V^2}}} \exp \left\{ -\frac{(V - \overline{V})^2}{\overline{V^2}} \right\}, \qquad (24.4)$$

where $e^2 \overline{V^2}$ is the mean square potential energy:

$$e^2 \overline{V^2} = \frac{e^4}{(4\pi \varepsilon \varepsilon_0 r_0)^2} \frac{4\pi}{3} r_0^3 N, \qquad (24.5)$$

N is the density of charged centers, and r_0 is the Debye screening length—see Section 14.3.1. Consequently, we obtain from Eqs. (24.3) and (24.4)

$$g(E)\, dE = \frac{1}{2\pi^2} \left(\frac{2m_n}{\hbar^2} \right)^{3/2} \int_{-\infty}^{E} \frac{\sqrt{E - eV}}{\sqrt{\pi \overline{V^2}}} \exp \left\{ -\frac{(V - \overline{V})^2}{\overline{V^2}} \right\} dV.$$
$$(24.6)$$

At high energy within a band $(E \gg eV)$, Eq. (24.6) yields the unperturbed density of states, e.g., for the conduction band:

$$g(E)\, dE = \frac{1}{2\pi^2} \left(\frac{2m_n}{\hbar^2} \right)^{3/2} \sqrt{E - E_c}\, dE. \qquad (24.7)$$

That is, high within a band, there are no changes in the density of states compared with the ideal lattice. For $E < E_c$, however, the density of states is modified to

$$g(E)\, dE = \frac{1}{4\pi^2} \left(\frac{m_n}{\hbar^2} \right)^{3/2} \sqrt{\overline{E^2}}\, (E_c - E)^{-3/2} \exp \left(-\frac{(E - E_c)^2}{\overline{E^2}} \right) dE,$$
$$(24.8)$$

with $\overline{E^2} = e^2 \overline{V^2}$. Equation (24.8) indicates that $g(E)$ decreases exponentially $\left(\propto \exp(-E^2) \right)$ below the edge of the conduction band—see Fig. 24.2. For a review of more rigorous approaches, see Shklovskii and Efros (1984).

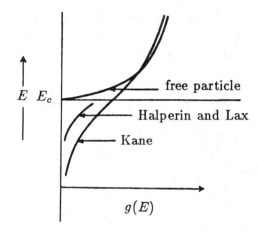

Figure 24.2: Density of states for quasi-free electron near the edge of the conduction band (E_c) in an ideal lattice, and Kane tail of states extending into the band gap in a lattice with random point defects (after Kane, 1963). The figure also contains the results of the Halperin and Lax approximation (1966, 1967).

Halperin and Lax (1966, 1967) developed a quantum-mechanical theory for deeper tail states, based on statistical fluctuation of the potential of lattice defects. They apply a variational method that maximizes the density of states with respect to the wavefunction containing statistically fluctuating potentials of Coulomb centers, which are assumed to be of the Yukawa screened Coulomb type. The Halperin and Lax theory gives the level distribution in the tail as

$$g(E)\, dE = c(\nu)\exp\left\{-\frac{a_{qH}^2}{16\pi N \lambda_{scr}^5}b(\nu)\right\} dE, \qquad (24.9)$$

with $\nu = 2m_n|E|\lambda_{scr}^2/\hbar^2$, a_{qH} as the quasi-Bohr radius, and λ_{scr} as the screening length.* The value of $b(\nu)$ is approximately 10 for $\nu = 1$, increases $\propto \nu^2$ for $\nu \gg 1$, and decreases $\propto \nu^{1/2}$ for $\nu \ll 1$; $c(\nu)$ is a proportionality factor that varies only slowly with ν compared to the exponential. The Halperin and Lax distribution is also shown in Fig. 24.2, and decreases more rapidly than the Kane approximation.

Other more rigorous estimates of the density of state distribution (Sa-yakanit and Glyde, 1980, 1982) near the band edge also

---

* For a self-consistent determination of the screening, which depends on the carrier density, which in turn depends on the level density, which again is influenced by the screening length—see Hwang and Brews (1971).

use randomly distributed Coulomb-attractive centers and employ a Feynman's path-integral method (Feynman, 1965). They obtain analytic results for $g(E)$, which can be written as

$$g(E) \propto \exp\left(-\frac{E_L^2}{2\overline{E}^2}c(\nu)\nu^{n(\nu)}\right). \tag{24.10}$$

Here, $\nu = (E_c - E)/E_L$; $E_L = \hbar^2/(2m_nL^2)$ is the energy to localize an electron within the correlation distance L; and $c(\nu)$ is a slowly varying function of the order of $1/10$. The exponent n depends on this correlation distance (Sa-yakanit, 1979) as

$$n(\nu) = \frac{32\nu}{(\sqrt{1+16\nu}-1)(\sqrt{1+16\nu}+7)} = \begin{cases} 2 & \text{for } \nu \to \infty \text{ or } L \to \infty \\ 0.5 & \text{for } \nu \to 0 \text{ or } L \to 0. \end{cases} \tag{24.11}$$

With a correlation distance of the defects typically between 1 and 10 Å, n is usually bracketed between 1 and 1.2, yielding with $c(1) \simeq 1/(18\sqrt{3})$

$$g(E) \propto \exp\left[-\frac{E_L^2 c(1)}{2\overline{E}^2}\left(\frac{E_c - E}{E_L}\right)^n\right] \simeq \exp\left(-\frac{E - E_1}{E_0}\right) \tag{24.12}$$

and

$$E_0 \simeq \frac{2\overline{E}^2}{E_L c(1)} \simeq 0.06\frac{\overline{E}^2}{E_L}. \tag{24.13}$$

For semiconductors with $m_n \simeq m_0$, we have $\overline{E}^2/E_L \simeq 1\,\text{eV}$; hence, $E_0 \simeq 0.06\,\text{eV}$. E_0 is identical with the characteristic energy for the Urbach tail—see Chapter 25.

These results confirm the tailing nature for deeper tail states, calculated numerically by Halperin and Lax. They approach the Kane approximation in the classical limit, and result in an improved agreement with the experiment (Sritrakool et al., 1985, 1986).

For heavily doped semiconductors with lower effective mass ($m_n < 0.1\,m_0$), the correlation length is on the order of the screening length, i.e., typically 20–100 Å. The characteristic energy of such semiconductors is typically one to two orders of magnitude smaller than given above. Here the extent of the Urbach tail is much reduced in agreement with the experiment (Sritrakool et al., 1986).

24.2 Disorder Effects in Semiconductor Alloys

Semiconductor alloys contain homologous elements at a nonstoichiometric ratio—see Section 9.2.1. Alloys can be formed in element

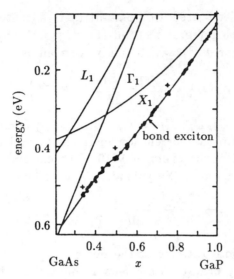

Figure 24.3: Band minima positions and ground state energies for N-donor (+) (after Jaros and Brand, 1979) and bound exciton (zero-phonon) line (after Wolford et al., 1976).

semiconductors such as $Si_\xi Ge_{1-\xi}$, in binary compounds such as $Al_\xi Ga_{1-\xi}As$, or in higher compounds such as $CuInSe_{2(1-\xi)}S_{2\xi}$, etc. Except for stoichiometric ratios [e.g., $Al_\xi Ga_{1-\xi}As$ for $\xi = 1/4$, $1/2$, and $3/4$ (Kuan et al., 1985), or $GaInP_2$—see Section 3.7.3], the atoms in the alloyed sublattice are statistically arranged. This causes random fluctuation of the composition in the microscopic volume elements of the crystal, and results in a broadening of resonances that depend on composition. Such resonances can be due to oscillating lattice atoms (phonon spectrum) or to specific electronic eigenfunctions (band edge, exciton spectrum, and defect excitation spectra). They are caused by the different masses, radii, and binding forces of the substituting atoms, which influence the lattice force constants, microscopic symmetry, band gap, dielectric constant, and other derived material parameters. Although the relative variation of some of these parameters is small, the result of such a fluctuation can result in sizable effects.

The shift of shallow levels from one band, e.g., the Γ point in GaAs, to a different one, the X point in the indirect gap of GaP, is shown in Fig. 24.3 for various degrees of alloying in $GaP_\xi As_{1-\xi}$.

The shift of the optical absorption edge caused by the changing composition was discussed in Section 9.2.1A. A fluctuation in ξ

results in band-gap fluctuations, which result in extended or localized states close to the band edge $E_g(\bar{\xi})$.

The exciton state also varies from position to position, depending on the actual (local) values of ξ:

$$E_{exc}(\mathbf{r}) = E_g(\mathbf{r}) - E_b(\mathbf{r}), \qquad (24.14)$$

where E_b is the binding energy. Thereby, it results in a line shift, shown in Fig. 24.3, and a line broadening. The influence on the line width was discussed in Section 20.3.2D. This shift is distinctively different from the shift of a donor with pressure, where the energy difference to the lowest band does not change–see Fig. 21.17. For a review, see Jaros (1985).

24.3 Impurity Influence, Fundamental Absorption

In *heavily doped* semiconductors, the absorption edge is shifted and deformed due to several reasons:

- tailing of band states into the band gap;
- shrinking of the band gap because of many-body effects (Section 9.2.1C); and
- partial filling of the conduction band states with electrons (Burstein-Moss effect—Section 24.3.1).

Such changes can be detected via excitation spectroscopy (Wagner, 1985), with an optical transition probability from filled states in or near the valence band to empty states in the conduction band. The band-to-band transitions have been discussed in Section 13.1, and the level-to-band transitions in Section 20.3.4. For reasons relating to highly disordered lattices, we will deviate here from this treatment as explained below.

Since the translational symmetry is broken by the random potential of the impurities, $\mathbf{k}$ is no longer a good quantum number. Consequently, it is more appropriate to use the energy as a label, yielding for the absorption coefficient (Abram et al., 1978)

$$\alpha_o(\nu) = \frac{\pi e^2 \hbar^2}{m_0^2 \nu c n_r} \int f_n(E) f_p(E+h\nu) P_i(E,\ E+h\nu) N_v(E) N_c(E+h\nu) dE,$$

$$(24.15)$$

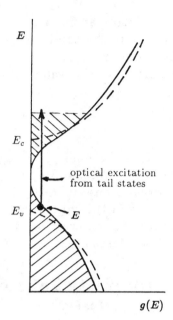

Figure 24.4: Optical transition from tailing states of the valence band to free states above the Fermi level within the conduction band.

where the probability P_i can be expressed by the sum of the matrix elements in a small energy interval around E and around $E + h\nu$:

$$P_i(E, E + h\nu) = \frac{1}{3} \sum \left| \int \psi_E \nabla_j \psi_{E+h\nu} dp \right|^2 . \qquad (24.16)$$

Examples of some quantitative estimates of such probabilities are given by Lasher and Stern (1964), Casey and Stern (1976), and Berggren and Sernelius (1981).

$N_v(E)$ and $N_c(E + h\nu)$ are the density of state functions near the respective band edges. These band edges are deformed from the ideal band distribution by the tailing of states, due to band perturbation from the random impurity potential as described in Section 24.1.

The occupancy factor, given by the Fermi distributions $f_n(E)$ and $f_p(E + h\nu)$, accounts for the probability of finding occupied states near the valence band edge (E) and empty states near the conduction band edge $(E + h\nu)$. The tailing of valence band states and partial filling of conduction band states with heavy doping result in an asymmetric excitation, e.g., for n-type material from the tail of the valence band states to states above the Fermi level. These upper states may be shifted to lie well within the conduction band—Fig. 24.4.

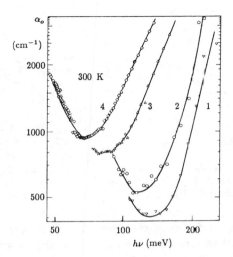

Figure 24.5: Band edge shift of HgTe due to the Burstein-Moss effect, caused by Al doping at densities of 0.75, 1.45, 3.12, and $3.8 \cdot 10^{18}$ cm^{-3} for curves 1 – 4, respectively (after Verie, 1967).

24.3.1 The Burstein-Moss Effect

In semiconductors with a low effective mass, the density of states near the lower edge of the conduction band [Eq. (27.18)] can be so low that even with moderate donor doping (Section 27.2), the lower states in the conduction band become filled. Hence, the Fermi level (Sections 27.1.1B and 27.2.2A) can be significantly shifted above this band edge (Section 9.2.1C).* Since an optical excitation can proceed only into free states, this filling results in a shift of the absorption edge toward higher energies, causing a *band-gap widening* (see Moss, 1961). The shift, referred to as *Burstein-Moss shift*, can be estimated from the position of the Fermi level—see Section 27.1.1B:

$$E_{g\text{B-M}} \simeq E_{\text{go}} + (E_F - E_c). \tag{24.17}$$

With this effect, a fine-tuning of the absorption edge can be achieved, which is used to produce optical filters of a rather precisely determined long wavelength cut-off. An example is given in Fig. 24.5 for HgTe.

---

* For instance, in InSb with $m_n = 0.0116$, the effective density of states is $N_c \simeq 3 \cdot 10^{16}$ cm^{-3}; hence doping with a shallow donor density in excess of 10^{17} cm^{-3} will cause a significant filling of conduction band states.

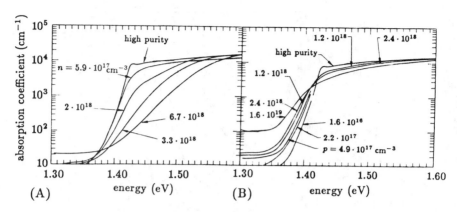

Figure 24.6: Absorption coefficient for (A) *n*-type and (B) *p*-type GaAs measured at room temperature with doping densities as family parameter (after Casey et al., 1975).

The *band edge narrowing*, explained in Section 9.2.1C, and the tailing of band states, discussed in Section 24.1, complicate this picture, as they have the opposite effect of the Burstein-Moss shift. In GaAs, both effects can be seen. The change of the optical absorption with higher doping densities, measured by Casey et al. (1975), shows the tailing (Fig. 24.6). In *n*-type GaAs, the Burstein-Moss shift predominates because of the lower effective mass for electrons, causing a significant shift of the Fermi level into the conduction band. For similar doping densities in *p*-type GaAs, however, the heavier hole mass permits only a negligible Burstein-Moss shift. Therefore, the shrinking of the gap and tailing state transitions provide the major cause for the changes in the absorption spectrum: the absorption edge shifts in the opposite direction, and the effective band gap narrows—see Fig. 24.6.

Summary and Emphasis

A high density of lattice defects causes states from within the bands to shift into the band gap, thereby forming an exponential tail of states, extending from the band edges. These tailing states are produced in addition to the defect states, discussed in previous chapters, and are created from the immediate lattice environment which is disturbed by the defect.

Such tailing states can be observed as *Urbach tails* in optical absorption extending from the band edge into the extrinsic range.

They are a measure of the degree of lattice disorder, resulting from intrinsic defects, or from a high level of doping.

A specific type of disorder relates to alloy formation within a sublattice of a semiconducting compound. This causes, in addition to the shift of the band (absorption) edge, a flattening of the edge, due to the local statistical arrangement of the atoms of the alloying components.

High donor doping influences the position of the Fermi level, and may shift it into the conduction band, thereby causing a shift of the optical absorption edge to higher energies, opposite to the Urbach tail shift. This Burstein-Moss shift can become very substantial in semiconductors with a low effective mass, where dopant densities in the 10^{18} cm^{-3} range can result in substantial (0.1 eV) shifts in the absorption edge.

Urbach tailing of the optical absorption and the Burstein-Moss shift of the band edge are significant factors which, respectively, influence the absorption of optical fibers and of certain optical filters. Lifshitz tailing of band states has a major influence on carrier transport at higher doping levels. A better understanding of these effects will offer the potential for the improved design of devices which are influenced by such states.

Exercise Problems

1.(e) Calculate the Urbach parameter E_0 [Eq. (24.1)] for the experimental absorption curves shown in Figs. 24.6B and 13.21.

2.(*) Discuss the consequences of the E_0 values determined in the above assignment with respect to potential fluctuations $\overline{E^2}$ — see Eq. (24.12).

3.(r) Review the different influences on the shift of the band gap due to alloying, as given in Sections 9.2.1, 24.2, and 20.3.2D.

4.(e) Calculate the position of the Fermi level within the conduction band for $N_d = 10^{18}$ cm^{-3}, using Eqs. (27.18) and (27.23), in four semiconductors in which you suspect the largest Burstein-Moss shift.

5.(r) Explain the factor in the absorption coefficient preceding the integral [Eq. (24.15)]. Identify the origin of each component.

6.(r) Explain the physical reasons why, for a given doping, the Burstein-Moss effect is larger for a semiconductor with a small electron effective mass. Explain in terms of $E(k)$.

Chapter 25

Defects in Amorphous Semiconductors

Defects in amorphous semiconductors can be defined with respect to specific classes of such semiconductors. Most of these defects are rather gradual displacements from an ideal surrounding.

In the preceding chapter, we have shown that heavily disordered crystalline semiconductors have band tails extending into the band gaps. In some respects, *amorphous semiconductors* may be described as heavily disordered crystals. Therefore, it is reasonable to explain some of their properties also with the concept of tailing states into the band gap (Mott, 1969).

Band tails, induced by disorder, are often referred to as *Lifshitz tails* (Lifshitz, 1964). When the degree of disorder is very large, the tailing states from valence and conduction bands may overlap. It was originally believed that the transfer of electrons from donor- into acceptor-like states would pin the Fermi level near the center of the band gap (Fig. 25.1). However, the small density of unpaired spins and other observed transport properties (see Section 40.4) do not support this explanation. A negative correlation energy of defects, caused by negative U-centers, is probably the reason for such pinning (Kastner et al., 1976). We will explain this behavior in Section 25.2.

Such behavior is suggested for the *chalcogenide glasses*, which are alloys containing elements of group VI and others, such as of groups IV and V, as the main glass-forming components; for instance, α-$Te_{40}As_{35}Si_{15}Ge_7P_3$, a well-known material because of its switching* capability (Cohen et al., 1969). In these materials, it is nearly impossible to move the Fermi level from its near-midgap position by doping. Some chemical modification, i.e, changing its conductivity

---

* Bipolar devices made from such materials are capable of switching at high speed from a low to a high conducting state at a critical bias (Ovshinsky, 1968).

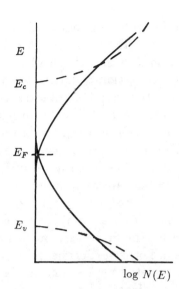

Figure 25.1: Strong tailing of states into the band gap and pinning of the Fermi level in the gap center in a semiconducting glass.

to become extrinsic while maintaining the band gap, however, was achieved in a few chalcogenide glasses by adding modifying elements. These should not be confused with dopants, since the necessary concentration of the modifier is relatively large to become effective. As such modifiers, the transition metals Ni, Fe, and Co can be used, and also in some instances, W, B, or C (Ovshinsky, 1977, 1980). For a review, see Adler (1985).

However, *tetrahedrally bound amorphous semiconductors*, such as α-Si:H, behave substantially different: these materials, when properly prepared, react easily to doping with a shift in their Fermi level much like a crystalline semiconductor. These semiconductors can be made n-type or p-type by doping with donors or acceptors, respectively. Nevertheless, α-Si:H also has a strong tailing of defect states into the band gap, although these tails do not overlap significantly near the center of the band gap. For an example of such band tailing in α-Si:H, see Fig. 13.21.

Before exploring the specific nature of defects in amorphous semiconductors, it is prudent to emphasize the distinction of these two classes of semiconductors.

25.1 Classes of Amorphous Semiconductors

There are at least two different classes of covalent amorphous semiconductors, distinguished by their electronic reaction to changes in the defect structure. The first class of *amorphous chalcogenides* includes multicomponent alloys, mentioned in the previous section, as well as the less complex compounds, such as α-As_2Se_3 and monatomic α-Se. Typically, these chalcogenides have a low average coordination number, below 2.4. For comparison, the average coordination numbers of some crystals with relatively low values are given here, e.g., 4 for Si, Ge, GaAs, etc., 3 for As or GeTe, 2.7 for $GeTe_2$, 2.4 for As_2Se_3, and 2 for Se (Adler, 1985).

The other class is composed of *tetrahedrally bound amorphous semiconductors*, such as α-Si:H, α-GaAs, and α-$CdGeAs_2$, with an average coordination number > 2.4. A third class, somewhat in between these two, contains α-P and α-As.

Aside from a pinning of the Fermi level due to overlapping tail states for the first class, and sensitivity to doping for the second class, there are other experimental distinctions for these two major groups. These relate to the strength of *electron-spin resonance* (ESR) signal, which indicates the existence of unpaired spin electrons (charged defects); the steepness of the *optical absorption edge*, which indicates the degree of band tailing; and other properties relating to carrier transport and discussed, as discussed in Section 40.4.

25.2 Defect Types in Amorphous Semiconductors

Defects are easily identified in *crystal lattices*, where vacancies, interstitials, and even small deviations from the periodic structure can be identified. This is much more difficult in *amorphous semiconductors*, where deviations from the *average bond length a* (see Table 2.3), *bond angle* θ, and *coordination number* $\hat{m}$ are the principal defect features. In general, many defects in glasses are of a gradual rather than distinct nature, and may be classified into

- local strain-related defects, i.e., variations of a, θ, and $\hat{m}$,
- deviation from an optimal bonding configuration,
- incorporation of small concentrations of impurities,
- dangling or floating bonds,
- microcrystallite boundaries, observed in some glasses, and
- variably sized small voids.

Most of these defects cause changes in the electron energy spectrum by deforming the band edge and extending states into the band gap. Depending on the type of amorphous semiconductor, the ensuing defect spectrum may extend nearly exponentially from the band edge into the band gap, the Urbach tail, or may produce well-defined—

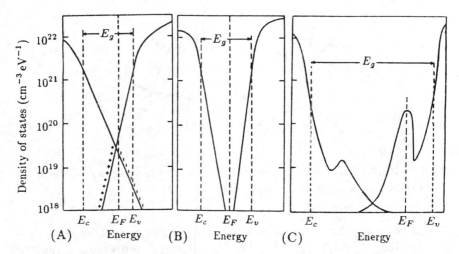

Figure 25.2: Exponential distribution of states into the band gap: (A) in chalcogenide glasses; (B) in tetravalent glasses; and (C) in tetravalent glasses with distinct peaks.

although broad—peaks of the distribution function within the gap (Fig. 25.2).

25.2.1 Strain-Related Defects

The *local strain-related defects* may be seen as similar to acoustic phonon-induced deformations (see Section 32.2.1C) of the band edge. Here, however, they are caused by a stationary, frozen-in strain, often also with a larger amplitude than for thermal phonons. Each of these stretched bonds or deformed bond angles can produce a level in the band tail when the deformation from the ideal values is sufficiently large.

In tetrahedrally bound semiconductors, deviations of the bonding angle and coordination number can also result in different types of bonding between neighboring atoms. For instance, in amorphous Si, an sp^3-, an sp^2-hybrid, or a p^3-configuration produces a neutral, positively, or negatively charged dangling bond, while an s^2p^2-hybrid produces a twofold coordinated Si atom. These bonds may be formed to relieve some of the stress. The corresponding bond angles are 109.5° (sp^3), 120° (sp^2), and 95° (p^3 and s^2p^2) (Adler, 1985).

Often, these defect centers are identified as $A_{\hat{m}}^e$ with A identifying the chemical specie: T for tathogen, P for pnictogen, and C for chalcogen, i.e., elements of the IV, V, or VI groups, respectively; $\hat{m}$

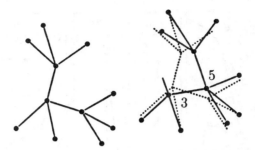

Figure 25.3: Pair creation (and recombination) of dangling (3) and floating (5) bonds (after Pantelides, 1988).

is the coordination number and e is the charge character relative to the lattice ($\bullet\bullet$, $\bullet$, $\times$, $'$, or $''$). For instance, $T_4^\times$ represents a neutral, fourfold coordinated Si atom, while $T_3^\times$ describes a neutral dangling bond, both for the sp^3 ground state. As possible defects in α-Si, all of $T_2^{\bullet\bullet}$, $T_2^{\bullet}$, $T_2^\times$, T_2', T_2'', as well as $T_3^{\bullet}$, $T_3^\times$ and T_3', are being considered as centers with lower-than-normal coordination, and as possible alternatives for local stress-relief.

25.2.2 Under- and Over-Coordinated Defects

There is evidence that two types of "intrinsic defects" are prevalent in α-Si:H—the threefold coordinated Si-atom, which is equivalent to a vacancy in crystalline Si, and the fivefold *over-coordinated* Si-atom, equivalent to a self-interstitial. These two types can be created in pairs, as indicated in Fig. 25.3, somewhat similar to a Frenkel pair creation. This reaction may be initiated by light and provides a possible mechanism for the Staebler-Wronski effect—see Section 46.3.2 (Pantelides, 1987).

25.2.3 Dangling and Floating Bonds

Dangling bonds do not seem to play a major role in most semiconducting glasses of technical interest. The elimination of these dangling bonds can be obtained, for example, by H or F in α-Si:H or α-Si:F. Such removal results in the major differences between the amorphous Si and the amorphous Si:H or Si:F alloys. The description as an alloy is used here since a large atomic fraction ($> 10\%$) of H or F is incorporated. As a consequence of the dangling bond removal, α-Si:H or α-Si:F can be doped and turn n- or p-type, similar to crystalline Si, as shown in Fig. 25.4A.

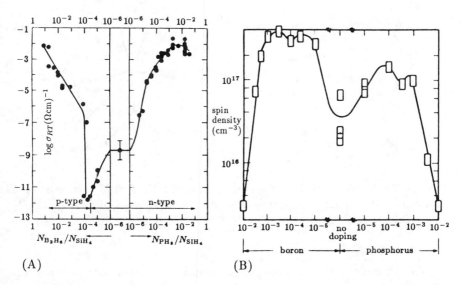

Figure 25.4: (A) Room temperature electrical conductivity of α-Si:H as a function of the phosphine or diborane concentration during deposition (after LeComber and Spear, 1976). (B) Spin density optically induced in α-Si:H as a function of doping gas pressure (after Knights et al., 1977). ©Pergamon Press plc.

The dominant intrinsic defect in α-Si:H, the D-*center*, characterized by paramagnetic resonance ($g = 2.0055$), was initially assigned to the dangling bond, i.e., a threefold coordinated Si-atom. Recently, however, evidence is growing that the center may be related to a floating bond, i.e., the fivefold coordinated Si-atom (Pantelides, 1988). The close relation of these two centers is shown in Fig. 25.5. In contrast to the dangling bond, the floating bond is highly mobile and is of interest to interstitial-mediated diffusion—see Section 19.3.2.

25.2.4 Deviation from Optimal Bonding Configuration

The *deviations from an optimal bonding configuration* occur predominately in chalcogenide glasses, and may be understood by comparing the relative bonding strengths of various bonds. For example, in amorphous $Ge_\xi Te_{1-\xi}$, a configuration wherein the stronger Ge-Ge bond appears most frequently while the number of the weaker Te-Te bond is minimized in the entire material, may be termed an *ideal amorphous $Ge_\xi Te_{1-\xi}$ structure*. Any deviation from it may be identified as a defect of the structure, and has a lower overall binding energy.

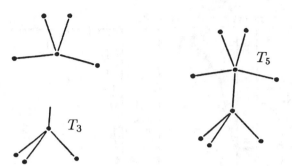

Figure 25.5: Relationship of threefold (T_3) and fivefold (T_5) coordinated Si-atoms with possible interconversion of these dangling and floating bonds (after Pantelides, 1988).

Another deviation from optimal bonding relates to a *valence alternation*, in which the valency of nearby atoms is changed. Charged dangling bonds may also be formed—see also Section 25.2.1. However, these bonds exactly compensate each other and form "pairs," so that no electron-spin resonance signal results—see Section 25.2.7. As charged, but compensated defects, they effectively pin the Fermi level.

In chalcogenide glasses, *defect states with negative correlation energy*, i.e, negative U-centers, can occur (Anderson, 1975; Ovshinsky, 1976b). These states can bind two electrons, the second with a larger binding energy than the first—see Section 22.4.10A. This can be explained when a strong electron-lattice interaction exists, and the energy released by the lattice deformation near the polarized defect is larger than the Coulomb repulsion of the second from the first electron. As a result, we expect in a system with N defect states and n electrons that $n/2$ states are doubly occupied in the lowest energy state, causing a pinning of the Fermi level. This also explains why these materials are diamagnetic—see Section 31.3.2.

25.2.5 Doping in Semiconducting Glasses

When the density of *foreign atoms* is small enough, and the atoms are not incorporated as part of the glass-forming matrix in tetragonal glasses, the ensuing defect may result in a distinct level similar to that in a crystalline semiconductor. Depending on the actual surrounding, however, the resulting energy level of a deep level defect is different. The same defect may have even a donor- or an acceptor-like character, or, with external excitation, act as a recombination center in a different microscopic environment of the host.

In these tetrahedrally bound amorphous semiconductors, e.g., in α-Si:H, the chemical nature of shallow defects determines their electronic defect behavior with less ambiguity. These act as *dopants*, with a similar effect as in crystals, though with a more complicated configuration. Incorporation of a P atom is likely to occur in an sp^3 configuration rather than a p^3 bonding. It is observed to shift the Fermi level to within 0.1 eV of the conduction band edge (or better, the electron mobility edge—see Section 40.4.2) and thereby act as an effective donor. The incorporation of a B-atom can be accomplished in an sp^3- or sp^2-configuration, with bond angles of 109.5° or 120° and a coordination number of 4 or 3, respectively, or as a complex with bridging H atoms (Adler, 1985). It acts as an acceptor. Other local bonding configurations can occur with the incorporation of N (pnictogen), O, or S (chalcogen), which result in different sets of donor levels within the gap.

25.2.6 Microcrystalline Boundaries and Voids

Microcrystallite boundaries seem to be out of order for true glasses. However, there is some evidence that in a few semiconducting glasses precrystallization takes place. This is possibly the result of imperfect growth techniques, and caused by the formation of small crystallite nuclei. In this case, the internal strain of the nuclei is relaxed by the creation of a boundary (microsurfaces) that resembles crystal boundaries with localized strain.

Small voids of variable size can be identified in the center of large-number rings (see Fig. 1.1) in certain glasses. These defects show some similarities to vacancies in a crystal lattice, although they vary in size and bond reconfiguration.

25.2.6A Recrystallization of Amorphous Si   Amorphous Si layers on crystalline Si recrystallize with an activation energy of ~ 2.5 eV (Lietoile et al., 1982). Such recrystallization may occur through diffusion of dangling bonds (Mosley and Paesler, 1984), mediated by floating bonds (Pantelides, 1989), converting all rings into six-member rings.

25.2.7 Spin Density of Defects

Electron-spin resonance yields additional information about a defect— see Section 21.3.4B. When this defect possesses an electron unpaired

with another one of opposite spin (short *unpaired spin*), the resulting
magnetic momentum can be picked up by a spin-flip electromagnetic
resonance experiment. The frequency and line shape of the resonance
is influenced by the surrounding of the defect and yields more detailed
information in crystalline solids. This hyperfine structure is washed
out in amorphous semiconductors. However, from the density of unpaired spins, we still obtain valuable information. This density is
small ($\sim 10^{16}$ cm^{-3}) in α-Si:H, and verifies a low density of dangling
bonds, $T_3^\times(sp^3)$, which represent unpaired spins (see Adler, 1985).
Coexistent with $T_3^\times$ centers are pairs of $T_3^\bullet$ and T_3' centers, which
have compensated spins. Their total energy depends on the relative
distance (Kastner et al., 1976), and has its minimum value when
they are nearest neighbors (*intimate charge transfer defect*—Adler
and Joffa, 1977).

The spin density can be used as a measure of uncompensated
donors or acceptors, and is shown in Fig. 25.4B. However, the observed decrease of unpaired spin density for high doping densities
causes some problems in explanation.

In contrast, nonhydrogenated α-Si shows spin densities which are
substantially larger and increase with damaging ion bombardment
up to 10^{19} cm^{-3}. With subsequent annealing, the spin density is
reduced as expected (Stuke, 1976).

25.3 Defect Spectrum in Amorphous Semiconductors

There is a continuous transition from band states that are *not localized* to *localized states* of major defects within the band gap. Depending on the degree of disturbance of the lattice potential, the
eigenstate of an arbitrary host atom may lie in the band, near the
band edges, or further separated within the gap.

An important question relates to the distribution of these levels
in a typical amorphous semiconductor. More specifically, we are
interested when such levels can be considered part of the band, and
when they become defect levels in the band gap. This distinction is
relatively easy in a crystalline semiconductor, in which band states
can be occupied by electrons that are described by nonlocalized
Bloch wavefunctions, as opposed to gap states described by localized
electron eigenfunctions. In amorphous semiconductors, this is not
possible since **k** is no longer a good quantum number.

One way to approach this question (Anderson, 1958) is to change from a strictly periodic potential, representing a crystal and yielding bands separated by gaps in the classical sense, to a perturbed potential, which can be made less and less periodic, and analyze the resulting eigenfunctions. This approach is illuminating, and will be discussed in the following section.

25.3.1 The Anderson Model

Anderson starts from a *three-dimensional* Kronig-Penney potential, which yields simple bands interspaced with band gaps, much like the one-dimensional case discussed in Section 7.1. The band width ΔE_B can be expressed as

$$\Delta E_B = 2\hat{m}I, \tag{25.1}$$

where $\hat{m}$ is the coordination number and I is the transfer integral

$$I = \int \psi^*(\mathbf{r} - \mathbf{R}_n) H \psi(\mathbf{r} - \mathbf{R}_{n+1}) d\mathbf{r}, \tag{25.2}$$

which can be approximated in the form

$$I = I_0 \exp\left(-\frac{r}{r_0}\right); \tag{25.3}$$

it decreases with increasing well depth $V = (1/e)I$. Here r is the distance from the well center and r_0 is the fall-off radius of the transfer integral. For a hydrogen-like potential well, I_0 is given by

$$I_0 = \left[\frac{3}{2}\left(1 + \frac{r}{r_0}\right) + \frac{1}{6}\left(\frac{r}{r_0}\right)^2\right] \frac{\sqrt{2me^2 E_0}}{\hbar} \tag{25.4}$$

with E_0 as the ground state in a single well and

$$r_0^2 = \frac{\hbar^2}{2mE_0}. \tag{25.5}$$

With periodic wells, the ground state E_0 broadens to a band with band width ΔE_B—see Eq. (25.1). It results in a rather narrow band when the wells are sufficiently deep and spaced widely enough.

This model has a wide range of application. For instance, we can use the results to explain that donors, when spaced close enough to each other, produce a narrow band rather than a sharp ground state level. The application of this model to amorphous semiconductors will become evident in Section 25.3.1B.

Let us first explore what happens when the model is applied to a periodic lattice potential, however, with changes from the strict periodicity, introduced by changing the interatomic distance or the potential. Anderson superimposes a random potential V with a spread of $\pm \Delta V/2$ onto the average well depth V_0 (Fig. 25.6B), and consequently obtains a broader level distribution with tails beyond the original band edges. When ΔV is very small compared to the well depth, only small deviations from the periodic Bloch-type solutions occur. These result in some scattering of essentially free Bloch electrons within this band, with a mean free path given by (Mott and Massey, 1965):

$$\lambda = \frac{\hbar}{\pi} \left(\frac{2}{\Delta V} \right)^2 \frac{v_e}{a^3 g(E)} \quad \text{with} \quad v_e = \frac{\hbar}{m} k. \quad (25.6)$$

When $\lambda \gg a$, all levels are extended band levels.* The fluctuating potential results only in a slight perturbation of the band edges.

25.3.1A Anderson Localization From Eq. (25.6) we see that the mean free path decreases with increasing spread of the fluctuating potential. When the mean free path is reduced to the distance between the wells, all states within the band become localized. An estimate for the relative mean free path can be obtained from Eq. (25.6), using the classical formula for $g(E)$:

$$g(E)\, dE = \frac{1}{2\pi^2} \left(\frac{2m}{\hbar^2} \right)^{3/2} \sqrt{E}\, dE, \quad (25.7)$$

yielding

$$\frac{\lambda}{a} = 32\pi \left(\frac{I}{\Delta V} \right)^2 = \frac{8\pi}{\hat{m}^2} \left(\frac{\Delta E_B}{\Delta V} \right)^2. \quad (25.8)$$

Localization occurs when $\lambda/a = 1$, i.e., when ΔV increases to $0.7 \Delta E_B$ for $\hat{m} = 6$. The electron is no longer free to move within the band but is localized within the radius of any one atom.

Anderson determines, as a criterion for localization, a more stringent decrease of the wavefunction, with increasing distance from a center, so that the remaining overlap is insufficient for diffusion from

---

* It should be noted that the definition of a band state is related to the coherence length of an electron wave, which is essentially the same as the mean free path λ (see Section 25.3.1A for more details).

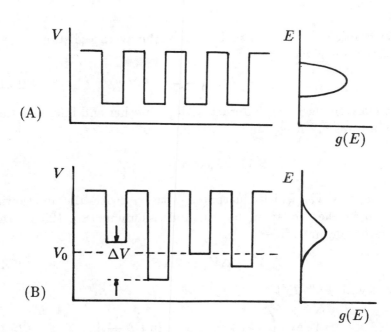

Figure 25.6: Anderson model: (A) periodic potential and resulting level distribution; and (B) Anderson potential with random potential V added to potential well depth and resulting level distribution.

neighbor to neighbor of an electron in such a center. This yields for an *Anderson localization*

$$\Delta V_A \gtrsim \frac{\sqrt{8\pi}}{\kappa_A \hat{m}} \Delta E_B, \tag{25.9}$$

with a numerical factor $\kappa_A \simeq 6$. Others have obtained values of κ_A between 1.3 and 5. A review of these estimations is given by Thouless (1974).

25.3.1B Anderson-Mott Localization Mott has applied the Anderson idea to randomly distributed defects as they may exist in heavily doped crystals, or for defect states in amorphous semiconductors, which results in an *impurity band* of width ΔE_i—see Sections 40.2 and 40.4.7. Instead of a random distribution of the potential well depth, a random distribution of centers *in space* is now assumed. If the density of the randomly placed centers is sufficiently large, we can think of these as forming an amorphous semiconductor with lateral disorder.

With a density of N_i of such centers, the average distance of any two of these is given by

$$r_i = \sqrt[3]{\frac{2}{N_i}}.$$

(25.10)

When close enough, they interact with each other and influence their eigenstates as given by the transfer integral

$$I_i \propto V_i = V_0 \exp\left(-\frac{r_i}{r_0}\right).$$

(25.11)

Equating this V_i, which fluctuates with r_i, with the fluctuating potential in the Anderson model, and assuming [Eq. (25.9)] that localization occurs when

$$V_i \simeq \Delta V_A \simeq 2\Delta E_i,$$

(25.12)

we have with Eq. (25.1):

$$V_0 \exp\left(-\frac{r_i}{r_0}\right) \simeq 4\hat{m} V_0 \exp\left(-\frac{r_A}{r_0}\right).$$

(25.13)

For an average coordination number $\hat{m} \simeq 5$, we obtain from Eq. (25.13)

$$\frac{r_A - r_i}{r_0} \simeq \ln(4\hat{m}) \simeq 3.$$

(25.14)

A somewhat lower coordination number for covalent crystals or amorphous semiconductors has only minor influence on the numerical value of condition (25.17). With an Anderson-Mott density, below which localization occurs,

$$(N_i)_{\text{loc}} = \left(\frac{4\pi}{3} r_A^3\right)^{-1},$$

(25.15)

we now obtain from Eq. (25.14) with Eqs. (25.10) and (25.15):

$$(N_i)_{\text{loc}} = N_{AM} \simeq 8 \cdot 10^{-3} r_0^{-3}.$$

(25.16)

Assuming hydrogen-like centers, this yields with $r_0 = a_{qH}$ the condition for Anderson-Mott localization—see Mott and Davis (1979):

$$\boxed{N_{AM}^{1/3} a_{\text{qH}} \simeq 0.2.}$$

(25.17)

This is essentially the same condition that was used for an insulator-metal transition when the density N described donors in a crystalline semiconductor—see Section 40.2.2B.

25.3.1C Band Tails and Localization We will now apply this concept to an amorphous semiconductor. Most of the states well within the conduction or valence band are similar to the states within a crystal. This can be justified by the measured $g(E)$ distribution— see Section 9.4. We may consider the tailing states extending into the band gap as disorder, resulting in a continuous distribution of states. Deeper states occur less frequently because the centers which produce such states are less frequent. When deep enough, each type of defect center will produce a localized level. Only if these centers are close enough will the corresponding levels broaden to bands, and electron transport can take place within such levels. Overlapping levels and narrow bands will all melt into the tailing states. These states are no longer localized.

In summary, when the disordered atoms are close to each other, the resulting states are band-like and not localized. When the spacing between disordered centers exceeds $\simeq 5$ times the quasi-hydrogen radius, localization occurs, and the resulting levels can be described as isolated levels in the band gap.

In an amorphous semiconductor, there seems to be a smooth transition between ideal and disordered states; hence, it is difficult to define a band edge in the classical sense. However, with the help of the localization criterion, a pseudo-band-edge can be defined as the energy of the defect level at which localization occurs. This will be explained in more detail in Section 40.4, when carrier transport is discussed—see also Götze (1981).

Summary and Emphasis

Two main classes of semiconducting glasses can be distinguished: the amorphous chalcogenides, which are often alloys of several elements; and the tetrahedrally bound amorphous semiconductors, of which α-Si:H is a typical representative.

Intrinsic defects in both classes are more gradual than in crystals, where vacancies and interstitials can be identified unambiguously. The intrinsic defects in semiconducting glasses may be compared with frozen-in local stresses, causing more or less local deformation.

Extrinsic defects (impurities) in chalcogenide glasses are often absorbed, and become part of the glass matrix without producing well-defined defect levels in the band gap. Even at larger densities they are not able to shift the Fermi-level from its pinned position close to the center of the band gap.

In tetrahedrally bound glasses, however, many impurities act as donors or acceptors, similarly as in crystalline semiconductors, and can easily render these materials *n*- or *p*-type.

The most prevalent feature in the defect level distribution of all glasses is a pronounced tail of states, extending into the band gap which conceals the actual band edge. Instead, another edge can be defined between extended and localized states, which, for typical semiconducting glasses with $a_{\text{qH}} = 30$ Å, lies at a level density of $\simeq 10^{18}$ cm^{-3}—see Eq.(25.17).

Although gradual in nature, the defects in most semiconducting glasses extend as tails well into the gap and have a dominating influence on electrical and optical properties. Saturating dangling bonds, with H or F in α-Si:H or α-Si:F, clean out most of the band gap in tetrahedrally bound amorphous semiconductors and permit well-defined doping, resulting in devices of high technical interest. In wide-gap oxide glasses, the exponential tail continues on to very low values in the center of the gap, with a level density there that lies well below that of crystalline, wide gap-materials. This permits optical transmission over very long distances, with possible limits near 30 km in optical fibers used for communication.

Exercise Problems

1.(r)   List the different types of intrinsic defects in semiconducting glasses in order of decreasing severity of lattice distortion or deviation from ideality.

2.(*)   Explain the connection between transfer integral and band width within a defect level band.

3.   What type of electron velocity is used in Eq. (25.6)? Explain why.

4.(*).   What is the reasoning for a mean free path of electrons in a narrow defect level band to yield Eq. (25.6)?

5.(*)   Derive Eq. (25.9) for the conditions given in the text.
    (a)   This is a very coarse approximation. Why?
    (b)   What can be done to improve the approximation?

6.(e)   Verify Eq. (25.17).

7.(*)   The application of the Anderson-Mott localization, as presented in Sections 25.3.1A and 25.3.1B, to an amorphous semiconductor is risky. List the critical elements and discuss the applicability.

Chapter 26

Defects in Superlattices

Doping of superlattices has a significant effect on their electrical properties, which are substantially different from that of homogeneous semiconductors. It provides attractive alternatives for device application.

There are significant differences between defects in superlattices and similar defects in bulk semiconductors. They relate to two different classes of defects: those which comprise an intimate part of superlattices; and those, such as point defects, which influence electronic and optical properties, similar to dopants.

The first class of defects relates to the superlattice interfaces in compositional superlattices or to periodic changes of doping, resulting in so-called *doping superlattices*: when the doping is done in alternating layers of an otherwise homogeneous material, one also obtains superlattice properties. These doping superlattices are discussed in Section 26.3, although the doping is no longer considered a defect within these superlattices.

Compositional superlattices have been discussed previously in this book. Lattice mismatch between layers of different composition significantly influences the properties of such superlattices, and is discussed in Section 26.2.

Finally, individual layers of the superlattice can be doped. Because of confinement, such dopants play a special role. We will discuss the influence of such doping first.

26.1 Defects in Compositional Superlattices

Donors or acceptors are preferably introduced in the large band-gap material. The free carriers produced here consequently move into the potential wells created by the smaller gap material. This results in a unique separation of doped material with few carriers and undoped material with a substantial density of carriers. This structure has

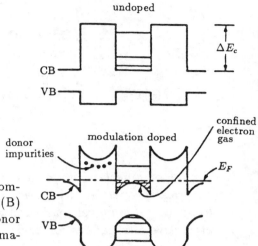

Figure 26.1: (A) Undoped compositional superlattices and (B) the same superlattice with donor doping in the large band-gap material.

obvious advantages for the carrier transport within the wells, i.e., in a high-purity undoped material, and permits very high carrier mobilities (Dingle et al., 1978), as will be discussed in Section 39.2.

As a consequence of the redistribution of carriers, a space charge is created in each layer: the depleted wide gap layer, if n-type, is charged positively; the lower gap layer, enriched with electrons, is charged negatively. Consequently, a bending of the bands occurs as shown in Fig. 26.1, and results in a somewhat more complicated periodic potential $V(x)$—see Döhler (1986).

26.1.1 Isolated Lattice Defects in Superlattices

When the extension of the wavefunction of a hydrogen-like impurity becomes comparable to the well thickness of the superlattice, an influence due to the difference in effective mass, dielectric constant and, most importantly, when the defect is introduced into the well material, a wavefuntion confinement, due to the potential barriers, occurs. This is reviewed by Bastard and Brum (1986), and Abstreiter et al. (1986).

The anisotropic confinement between the barriers reduces the quasi-Bohr radius of the impurity normal to the superlattice plane, and thereby increases the electronic binding energy of the defect. This change is relatively large when the impurity sits close to the barrier or when the well thickness is less than the Bohr radius and the barriers are sufficiently high. There are indications that a shallow-

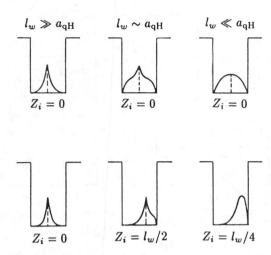

Figure 26.2: Schematics of the eigenfunctions of a hydrogen-like ground state in a quantum well of different width l_w relative to a_{qH} (upper row) and of different proximity to the well wall in a wider well (lower row). Z_i is the position offset of the defect from the center plane ($Z_i = 0$) (after Bastard and Brum, 1986). ©IEEE.

deep level transition occurs when the distance of the defect to the well barriers becomes small enough (Ren et al., 1988).

However, when $l_w \ll a_{qH}$, the defect eigenfunction extends beyond the well, the barriers become permeable—if thin enough, and the binding energy decreases again.

In Fig. 26.2, changes of the ground state of the wavefunction in such a superlattice are shown when the well becomes thinner (upper row) or the defect is located closer to the right barrier wall (lower row). Substantial deformation from the hydrogen-like behavior, shown undisturbed in the left column, becomes visible.

In addition, resonance effects occur with impurity states derived from higher subbands and overlapping with the continuum of the barrier material (Priester et al., 1984).

Acceptors in wells of superlattices were investigated by Gammon et al. (1986). The properties are a bit more complex because of the split of heavy and light hole subbands, which causes a splitting of the corresponding hydrogen-like levels of the acceptor.

26.1.1A δ-Function Doping When doping is confined to a single atomic layer within a superlattice, doping subbands are formed within a V-shaped potential well surrounding this monolayer (Zren-

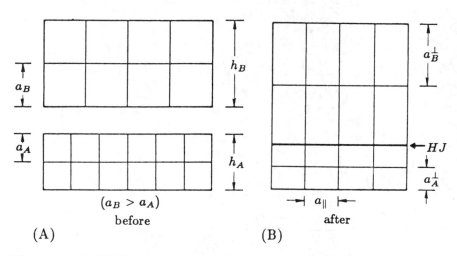

Figure 26.3: (A) Two thin layers of unstrained materials with substantial lattice mismatch (upper part) and (B) after connection with each other (HJ), showing a substantial tetragonal lattice deformation without stress relief into a dislocation network.

ner et al., 1984) and carrier saturation effects occur (Zrenner and Koch, 1986; Ploog, 1987).

26.2 Strained-Layer Superlattices

Another type of lattice imperfection plays an important role: the large built-in strains due to some lattice mismatch in compositional superlattices. Such strains result in the formation of dislocations for an extended lattice. Consequently, such types of superlattices have similarities with a sequence of heterojunctions, containing arrays of mismatch dislocations.

In sufficiently thin alternating layers, however, this strain cannot be relieved. The strain can no longer be considered a lattice defect but becomes an intimate part of the superlattice. There is a critical distance necessary for the creation of mismatch dislocation. This critical thickness d_c is a function of the lattice mismatch Δa, and is evaluated for a number of semiconductor pairs (van der Merve, 1978; Fiori et al., 1984; Fritz et al., 1985; Osbourn, 1986).

A first estimation of the critical thickness is given by

$$d_c \simeq \frac{a}{\Delta a} \frac{b}{4\pi(1+\nu)} \left\{ \ln\left(\frac{d_c}{b}\right) + 1 \right\} \qquad (26.1)$$

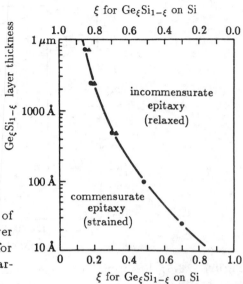

Figure 26.4: Critical thickness of a layer of $Ge_\xi Si_{1-\xi}$ on Si (lower scale) or on Ge (upper scale) for creation of dislocations (after Pearsall et al., 1986).

where b is the length of the Burgers vector and ν is the Poisson ratio ($\simeq 0.3$). Typically, with $b \simeq 4$ Å and $\Delta a / a \simeq$ 2–5%, one estimates a critical thickness on the order of 100 Å, below which mismatch dislocations are not formed.

Large strains from lattice pairs with substantial mismatch cause major tetragonal deformations (see Fig. 26.3), which cause changes in lattice constant and band structure. This provides additional flexibility in designing a wide variety of superlattices with different band interconnections. Of special interest is the strain-induced splitting of the valence band; it causes a change in hole masses (Osbourn, 1986), which is otherwise difficult to achieve. Various devices made from strained-layer superlattices are reviewed by Osbourn et al. (1986); see also Sections 3.8.3 and 9.3.2.

Figure 26.4 gives an example of the $Ge_\xi Si_{1-\xi}$ superlattice, with a dividing curve between the strained superlattice regime (lower left) and the relaxed regime (upper right) in which mismatch dislocations can be formed. Such strained-layer superlattices grow at an appropriate temperature: at too low temperatures, amorphous layers are grown; at too high temperatures, interdiffusion takes place.

26.2.1 Superlattice-Induced Phase Changes

Under large hydrostatic pressure of 100–300 kbar, the open tetrahedral structure with a coordination number 4 of many semiconductors

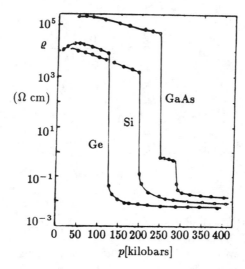

Figure 26.5: Resistivity of some semiconductors as a function of the hydrostatic pressure with phase transitions; recent results put transitions at 30–50% lower pressures (after Minomura and Drickamer, 1962). ©Pergamon Press plc.

can be changed in a first-order phase transition to a body-centered tetragonal structure with coordination number 6 and a 15–20% volume reduction (Hanneman et al., 1964). As a result, the band gap is dramatically reduced, and some transitions result in a metallic phase as shown in Fig. 26.5 (see also Froyen and Cohen, 1983). A few of these high-pressure modifications can be retained metastably at atmospheric pressure (Kasper and Richards, 1964).

Embedding such a semiconductor between layers of a material with a different lattice constant creates a strained-layer superlattice, which enhances such stabilization. After a transformation is induced by high pressure, the reverse transformation is suppressed, resulting in a hysteresis. In some cases, for example, the GaAs/AlAs superlattice remains stable down to room pressure in the tetragonal structure, after inducing it at 155 kbar (Weinstein, 1987), provided the individual layer is thin enough: 19 Å in the given example.

26.3 Doping Superlattices

A periodic modulation of the bands, which is characteristic for superlattices, can also be achieved by a periodic alternation of n- and p-type doping (first proposed by Döhler, 1972). Such a *doping superlattice* is shown in Fig. 26.6. For better separation, i layers (undoped,

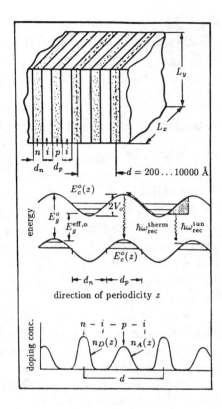

Figure 26.6: Doping superlattice with alternating n- and p-type layers of the same host material, interfaced with thin undoped (i) layers (after Döhler, 1986). ©Pergamon Press plc.

or compensated *intrinsic layers*) can be inserted between each of the n- and p-type layers.

This periodic structure shows a similar type of minibands as the compositional superlattice, however, with minibands that are wider for shorter superlattice constants d and lower barriers. Because doping superlattices depend on the Debye length for their change in space charge, their lattice constant is usually larger than that of compositional superlattices, and typically is on the order of 300 to 3000 Å.

A lowering of the barriers can be achieved by increasing the carrier densities in the n- and p-conducting layers. Such an increase, in turn, can be produced by light, resulting in optical carrier generation, or by an electric field resulting in carrier injection. The positive and negative space charges in the n- and p-type layers, respectively, are thereby reduced.

As a consequence, the electrical and optical properties of the doping superlattice become *tunable* by a changing carrier density. Such tunability includes a change in the effective band gap, the carrier lifetime, the luminescence spectrum, and other optical parameters (Döhler, 1986); for more detail, see Sections 44.8, 45.6.2, and 45.6.4A. Doping superlattices have been produced in GaAs, Si, InP, PbTe, and other semiconductors (Ruden, 1987).

Summary and Emphasis

The doping of compositional superlattices modifies the periodic potential by including space charge effects, causing some modification in the miniband spectrum. Such modifications can be varied by a change of the free carrier density within each well, which in turn can be influenced by light or carrier injection. Consequently, the electrical and optical properties become tunable. Such tunability is enhanced in doping superlattices, which are produced by the alternating doping of thin n- and p-type layers in an otherwise homogeneous material.

Compositional superlattices with lattice mismatch contain at the interfaces a mismatch dislocation network. However, when the layers are thin enough, such a network cannot develop and a dislocation-free strained-layer superlattice is formed. The built-in strain may cause phase transitions, or stabilize modifications, which may not be stable in bulk semiconductors.

The level structure of isolated lattice defects is changed when these defects are located close to the interface of the layers. This influence is significant for shallow level defects, which, when incorporated within well layers, experience substantial compression of their eigenfunctions, and thereby an increase in the bonding energy of electrons.

Doping can be restricted to a single atomic layer, thereby permitting the creation of a doping superstructure.

Tailored doping in superlattices permits more precision in the design of devices, and opens a new dimension by creating materials with tunable properties.

Exercise Problems

1.(r) Draw the different alternatives of periodic p- and n-type doping in compositional type-I and type-II superlattices.

2.(*) Discuss the significance of δ-function doping for possible device utilization.

3.(*) Estimate the influence of barrier layers on the electronic eigenfunctions of hydrogen-like donors in GaAs/ GaAlAs superlattices of 50 Å well and barrier width.

4. In doping superlattices, often an *i*-layer is inserted between each *n*- or *p*-type layer. Give the reason for such an insertion.

5.(r) Discuss the influence of carrier injection on the properties of a doping superlattice.

6.(r) Review the influence of confinement in superlattices for excitons and other lattice defects.

7. The critical thickness required for a layer in a cubic lattice to avoid dislocation formation was given in Eq. (26.1). How would this relation change in a lattice with orthorhombic structure?

8.(*) Review the properties of strained-layer superlattices and discuss the relevance of band-edge offset and of symmetry influence.

9.(l) Search the literature for two-dimensional hydrogen and summarize its relation to the topics discussed in this chapter.

10.(r) Review the probabilities of superlattice degradation by interdiffusion and recrystallization.

   (a) Compare strained-layer superlattices with those having minimal lattice mismatch.

   (b) Compare compositional and doping superlattices with respect to degradation.

Chapter 27

Equilibrium Statistics
of Semiconductors

Most electrons and holes in a semiconductor relate to defects in the host material: they are generated from, scattered by, or recombine through them. Almost all technically important semiconductor properties are defect-controlled. Their influence on carrier densities in equilibrium can be obtained from statistics.

Electrons and holes are the most important particles in semiconductors. Only in rare cases are they generated from the semiconductor itself, i.e., from the *host material*. By far, most of these carriers originate from *lattice defects*. In addition, the transport properties of these carriers, discussed in the following chapter, are determined almost exclusively by lattice defects, including phonons. In comparison to other fields of physics where the major effects are related to basic material properties, this is a rather unusual fact.

Nevertheless, the influences of the host material on the electronic properties are important and will be discussed first. The dominating influence of lattice defects on the density of electrons and holes will then be evaluated. In this chapter we will take an equilibrium statistical approach for a quantitative analysis of the carrier distribution as a function of temperature and doping.

27.1 The Intrinsic Semiconductor

An *intrinsic semiconductor* is a pure semiconductor with no dopants. There is negligible influence from lattice defects on its carrier density. The electronic properties of such a semiconductor are determined by the *mutual* generation or recombination of electrons and holes.

668

27.1.1 Electron and Hole Densities in Equilibrium

Normally, semiconductors are exposed only to thermal excitation. If they are kept long enough at a constant temperature, *thermodynamic* (short "thermal") *equilibrium* is established. For the following discussion, it is assumed that such equilibrium is always established, no matter how long it takes to reach equilibrium—see Section 27.2.3A. In this section we will not deal with kinetic effects.

When equilibrium is reached, the same temperature T, determining the distribution function, characterizes all subsystems, e.g., electrons, holes, and phonons (i.e., lattice atoms). The density distribution of electrons or holes in the conduction or valence band can be obtained from simple statistical arguments as the product of the level density $g(E)$ and the statistical distribution function $f(E)$:

$$n(E)dE = g(E)f(E)dE. \tag{27.1}$$

The density distribution for holes, i.e., of missing electrons, is:

$$p(E)dE = g(E)\left[1 - f(E)\right]dE. \tag{27.2}$$

Since the probability of finding an electron at energy E is proportional to $f(E) \leq 1$, with $f(E) = 1$ as the certainty, the probability of finding an electron *missing* is simply $1 - f(E)$.

27.1.1A Level Distribution Near the Band Edge   In Section 8.5.3 the level distribution (density of states) within a band was derived in general terms. For the purpose of carrier transport, one is interested in an explicit expression of this distribution. Such an expression can be given easily near the edge of the conduction or valence band.

In an isotropic parabolic band, the dispersion equation of Bloch electrons is given by

$$E(k) = E_c + \frac{\hbar^2}{2m_n}k^2 + \dots . \tag{27.3}$$

These electrons are confined within the semiconductor. This can be described as confinement within a "box," for definiteness a cube of dimensions l, and requires standing wave boundary conditions with nodes at the box surface.* This prevents the escape of electrons from

---

* Or cyclic boundary conditions; here energy and particle number are conserved by demanding, that with the passage of a particle out of a sur-

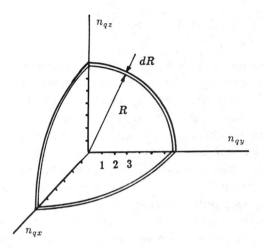

Figure 27.1: Spheres of constant energy in quantum number space.

the box since no energy can be transmitted beyond a node. For the three components of the wave vector, one therefore has

$$k_i = \frac{\pi}{l_i}\hat{n}_i \quad \text{with} \quad \hat{n}_i = 0, 1, 2, \dots \qquad (27.4)$$

with $i = (x, y, z)$. The wave vector and therefore the energy are represented by a set of discrete values given by the triplet of integers $\hat{n}_x$, $\hat{n}_y$, $\hat{n}_z$. The requirement of finding a triple of integers is sometimes referred to as second quantization. In addition, m_s is used to identify the spin quantum state. The energy increases monotonically with these integers.

At low temperatures, the energy states are filled sequentially; each state can be occupied by a maximum of two electrons with opposite spin (*Pauli principle*).

In an $\hat{n}$ space, as shown in Fig. 27.1, one accounts for the sequential filling of these states up to a radius $R = R(\hat{n}_x, \hat{n}_y, \hat{n}_z)$, which is determined by the number of available electrons. The number of states G within an energy range of E to $E + dE$ is obtained from the volume of the spherical shell of the octant, permitting only positive values of $\hat{n}_i$

$$G_n^{dR} = \frac{1}{8}\frac{4\pi}{3}(R_2^3 - R_1^3) = \frac{\pi}{6}d^3R = \frac{\pi}{2}R^2\,dR. \qquad (27.5)$$

---

face, an identical one enters from the opposite surface (Born-von Karman boundary condition). The two conditions are mathematically equivalent.

Since $R^2 = \hat{n}_i^2$, and using Eq. (27.4), we can replace R with k and obtain the number of states in the momentum interval dk:

$$G_k^{dk} = \frac{\pi}{2} \cdot \frac{l^3}{\pi^3} k^2 \, dk \qquad (27.6)$$

within $l^3 = \mathcal{V}$, the volume of the semiconductor. Finally, replacing k with the energy from Eq. (27.3), we obtain

$$G_E^{dE} = \frac{\mathcal{V}}{2\pi^2} k^2 \, dk = \frac{\mathcal{V}}{2\pi^2} \left(\frac{2m_n}{\hbar^2} \right)^{3/2} \frac{1}{2} \sqrt{E - E_c} \, dE \qquad (27.7)$$

as the number of states between E and $E + dE$. Dividing by the crystal volume and permitting double occupancy, i.e., permitting for spin up and down for each state, we obtain the density of states for electrons:

$$g_n(E) \, dE = \frac{1}{2\pi^2} \left(\frac{2m_n}{\hbar^2} \right)^{3/2} \sqrt{E - E_c} \, dE. \qquad (27.8)$$

This density is zero at the lower edge of the conduction band and increases proportionally to the square root of the energy near the band edge (Fig. 27.2).* A similar square-root dependence results for the density of states near the upper edge of the valence band, where a quasi-free hole picture can be applied. Replacing m_n with m_p in Eq. (27.8), and shifting the energy axis by $E_g = E_c - E_v$, yields for holes:

$$g_p(E) dE = \frac{1}{2\pi^2} \left(\frac{2m_p}{\hbar^2} \right)^{3/2} \sqrt{E_v - E} \, dE. \qquad (27.9)$$

The result of this simple model is carried through most of the common discussions of carrier transport in solids. It is valid near both edges of the band, and is an acceptable approximation as long as the bands are parabolic in $E(k)$. For modification at higher energies, see Sections 9.1.5 and 9.1.7.

---

* Here and in several of the following figures, the energy axis is plotted vertically in order to facilitate comparison with the band model even though $g(E)dE$ is the dependent and E the independent variable.

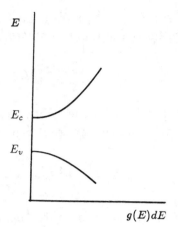

Figure 27.2: Square-root dependence of the density of states near the band edge, where the quasi-free electron or hole model can be applied.

27.1.1B Statistical Distribution Functions Electrons are fermions with spin $\pm \frac{1}{2}$ and follow the *Fermi-Dirac distribution* function (for derivation, see any classic text, e.g., McKelvey, 1966)

$$f_{\mathrm{FD}}(E) = \frac{1}{\exp\left(\dfrac{E - E_F}{kT}\right) + 1}, \qquad (27.10)$$

which is shown for a family of curves in Fig. 27.3 with the temperature as family parameter. This distribution function has a box-like behavior, occasionally referred to as the "Fermi ice-block," for $T = 0$ with complete filling ($f_{\mathrm{FD}} = 1$) of all levels for $E < E_F$, and complete depletion ($f_{\mathrm{FD}} = 0$) for $E > E_F$. Here E_F is the *Fermi energy*. For $T > 0$, the degree of filling decreases exponentially with increasing E, and reaches 50% at the Fermi energy: the corners of the box are rounded off—"the ice-block melts." E_F is defined by $f_{\mathrm{FD}}(E_F) = 0.5$.

For the evaluation of the Fermi-Dirac distribution at energy (several kT) above the Fermi energy, the distribution can be approximated by a *shifted Boltzmann distribution*

$$f_B(E) = \exp\left(-\frac{E - E_F}{kT}\right). \qquad (27.11)$$

The degree of deviation from the Fermi-Dirac distribution is indicated in Fig. 27.3B.

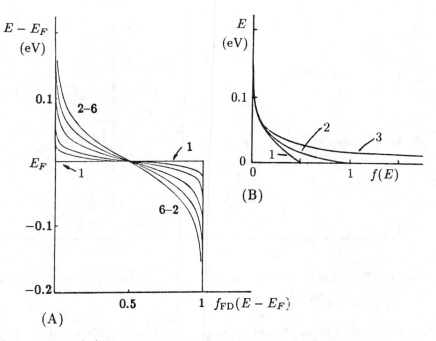

Figure 27.3: (A) Fermi-Dirac distribution function with the temperature as family parameter: 0, 50, 100, 200, 300, and 400 K for curves 1–6, respectively. (B) Comparison of Fermi-Dirac, Boltzmann, and Bose-Einstein distributions for curves 1–3, respectively, at $T = 300$ K. f_{FD} is drawn at a scale shifted by E_F.

At this point it is educational to compare these distributions with the *Bose-Einstein distribution*

$$f_{BE}(E) = \frac{1}{\exp\left(\dfrac{E}{kT}\right) - 1},$$
(27.12)

which describes the distribution of bosons, which have integer spin, such as phonons, photons, or excitons. The distribution has a singularity at $E = 0$; that is, it increases dramatically ($f_{BE} \to \infty$) as $E \to 0$, while for the same argument $f_B \to 1$ and $f_{FD} \to 0.5$ (for $E - E_F \to 0$). This is shown in Fig. 27.3B. The tail of all three distributions for $E/kT > 3$ is practically identical.

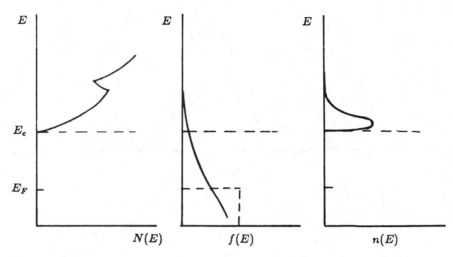

Figure 27.4: Level, Fermi-Dirac, and electron density distributions within the lower part of the conduction band.

27.1.1C Electron Distribution   The electron distribution $n(E)$ within a band is obtained as the product of level density and Fermi distribution:

$$n(E)\, dE = g_n(E) f_{\mathrm{FD}}(E)\, dE, \qquad (27.13)$$

and is shown schematically in Fig. 27.4 for the lower part of the conduction band. The electron distribution is zero within the band gap since the level distribution vanishes in the gap for an ideal semiconductor.

Because of the steep decrease of the Fermi-Dirac distribution function, most of the conduction electrons are located very close to the bottom of the conduction band: 95% of the electrons are within $3\,kT$ from E_c, except for high electric fields when substantial electron heating occurs—see Section 33.2.1. Since a typical band width is on the order of 10 eV, i.e., $\sim 400\,kT$ at room temperature, we can assign to the vast majority of the conduction electrons in a semiconductor the same effective mass.*   Therefore, it is justified to replace the

---

* This is justified since m_n does not change much near the bottom of the conduction band (see Fig. 7.14); moreover, m_n *increases* with increasing E, which usually renders higher energy electrons less important for a number of low-field transport properties.

electron distribution function $n(E)$ with a simple electron density in the conduction band

$$n = \int_{E_c}^{\infty} g_n(E) f_{\text{FD}}(E) dE. \tag{27.14}$$

Because of the steep decrease of $f(E)$ with E, the integration to the upper edge of the conduction band is replaced with ∞, with negligible error in the result. When evaluating Eq. (27.14), we can rewrite this equation by defining an effective level density at the lower edge of the conduction band N_c, multiplied by a shifted Boltzmann distribution [Eq. (27.11)]

$$n = N_c f_B(\xi) \tag{27.15}$$

with $\xi = -(E_c - E_F)/(kT)$. Using Eqs. (27.8) and (27.11), which is justified for $E_c - E_F > 3kT$, we obtain

$$n = \frac{1}{2\pi^2} \left(\frac{2m_n}{\hbar^2} \right)^{3/2} \int_{E_c}^{\infty} \sqrt{E - E_c} \exp\left(-\frac{E - E_F}{kT} \right) dE, \tag{27.16}$$

which can be transformed into

$$n = \frac{1}{\pi^2} \left(\frac{2m_n kT}{\hbar^2} \right)^{3/2} \exp\left(-\frac{E_c - E_F}{kT} \right) \int_0^{\infty} \xi^{1/2} \exp(\xi) d\xi. \tag{27.17}$$

The integral is tabulated and given as $\Gamma(3/2) = \sqrt{\pi}/2$. This yields for the pre-exponential factor ($= N_c$) in Eq. (27.17) for a simple parabolical band with spherical cross section near $k = 0$

$$\boxed{N_c = 2 \left(\frac{m_n kT}{2\pi \hbar^2} \right)^{3/2}} \tag{27.18}$$

or

$$N_c = 2.5 \cdot 10^{19} \left(\frac{m_n}{m_0} \right)^{3/2} \left(\frac{T(\text{K})}{300} \right)^{3/2} \quad (\text{cm}^{-3}). \tag{27.19}$$

The hole density can be obtained using the same arguments as for electrons, with

$$p = \int_{-\infty}^{E_v} g_p(E)(1 - f_{\text{FD}}(E)) dE = N_v f_B(\tilde{\xi}), \tag{27.20}$$

where $\tilde{\xi} = -(E_F - E_v)/(kT)$ and $1 - f_{\mathrm{FD}}$ is the probability of finding a hole there:

$$f_{\mathrm{FD}}^{(p)}(E) = 1 - f_{\mathrm{FD}}^{(n)}(E)$$

$$= 1 - \frac{1}{1 + \exp\left(\dfrac{E - E_F}{kT}\right)} = \frac{1}{1 + \exp\left(\dfrac{E_F - E}{kT}\right)}. \quad (27.21)$$

Here, N_v is the effective level density for holes at the upper edge of a single parabolic valence band of spherical symmetry:

$$N_v = 2\left(\frac{m_p kT}{2\pi\hbar^2}\right)^{3/2}. \quad (27.22)$$

For real valence band dispersion, see Section 27.1.1D. The total electron density in the conduction band and the total hole density in the valence band are consequently given by

$$n = N_c \exp\left(-\frac{E_c - E_F}{kT}\right) \quad \text{and} \quad p = N_v \exp\left(-\frac{E_F - E_v}{kT}\right). \quad (27.23)$$

27.1.1D Density-of-State Effective Mass   In the conduction band of *anisotropic semiconductors*, the mass has tensor properties. In general, the equi-energy surface can be described by an ellipsoid with three different axes, hence with three different curvatures, and therefore three different effective masses in the direction of the main axes, resulting in a density-of-state mass

$$m_{\mathrm{nds}} = \sqrt[3]{m_1 m_2 m_3}. \quad (27.24)$$

As discussed in Section 9.1.5, the *satellite valleys* of cubic semiconductors are described by a rotational ellipsoid with $m_3 = m_{nl}$ (the longitudinal effective mass along the $\langle 100 \rangle$ direction in Si) and $m_1 = m_2 = m_{nt}$ (the transverse effective mass, perpendicular to the main axis), with spherical symmetry—see Fig. 9.11. With a coordinate transformation in momentum space, we can reduce these ellipsoids to spheres and apply the commonly used calculation for the density of states. If there is more than one ellipsoid with identical energy (six in the example of conduction bands for Si), a *degeneracy* *factor ν_D* must be employed. This results in a *density-of-state mass for electrons* of

$$m_{\text{nds}} = \nu_D^{2/3}(m_{nl}m_{nt}^2)^{1/3}.$$

(27.25)

Sometimes an *anisotropy factor $K_a = m_{nl}/m_{nt}$* is used, rendering the density-of-state effective mass for electrons $m_{\text{nds}} = m_{nl}K_a^{-2/3}\nu_D^{2/3}$. For warped conduction bands, we have

$$m_{\text{ds}\pm} = \frac{m_0}{A \pm B'}(1 + 0.0333\Gamma + 0.0106\Gamma^2 + \dots);$$

(27.26)

for A, B', and Γ —see Section 30.3.2A.

The distribution of holes between the l (light) and h (heavy) hole bands is proportional to the density of states:

$$N_{\text{vl}} = 2\left(\frac{m_{\text{pl}}kT}{2\pi\hbar^2}\right)^{3/2} \quad \text{and} \quad N_{\text{vh}} = 2\left(\frac{m_{\text{ph}}kT}{2\pi\hbar^2}\right)^{3/2}.$$

(27.27)

There are more states per energy interval in a heavy than a light band near the band edge because of the lower curvature of $E(k)$. Hence, when filling the band to a certain energy, more holes are in the heavy band.

Since both densities are additive, these equations may be used to introduce an effective *density-of-state mass for holes*

$$m_{\text{pds}}^{3/2} = m_{\text{pl}}^{3/2} + m_{\text{ph}}^{3/2},$$

(27.28)

neglecting the contribution from the spin-orbit split-off band and any deviations from spherical $E(k)$ behavior in the valence band. For the influence of such warping, see Section 30.3.2A.

With increasing temperature, higher energy states are filled with a decreased curvature of $E(\mathbf{k})$, and therefore with an increased effective mass of the carriers. The corresponding density-of-state masses are shown in Fig. 27.5.

The density of states at the edge of the valence or conduction band [Eqs. (27.18) and (27.22)] is then obtained by replacing m_p or m_n with m_{pds} or m_{nds}:

$$N_v = 2\left(\frac{m_{\text{pds}}kT}{2\pi\hbar^2}\right)^{3/2} \quad \text{and} \quad N_c = 2\left(\frac{m_{\text{nds}}kT}{2\pi\hbar^2}\right)^{3/2}.$$

(27.29)

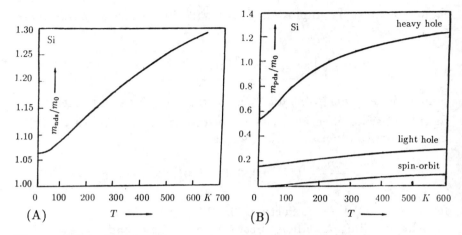

Figure 27.5: Density-of-state mass for Si as a function of the temperature, (A) for electrons after Barber (1967) and (B) for holes (after Lang et al., 1980).

Use of the Fermi Integral. When the Boltzmann approximation cannot be used in evaluating Eq. (27.14) because of the close proximity of E_F to the band edge, one must use the *Fermi integral*

$$F_{\tilde{n}}(\xi) = \int_0^\infty \frac{x^{\tilde{n}}\, dx}{\exp(x - \xi) + 1};$$ (27.30)

with $\tilde{n}$ as the order of the Fermi integral. One now obtains

$$n = \frac{1}{2\pi^2} \left(\frac{2m_n kT}{\hbar^2} \right)^{3/2} F_{\tilde{n}} \left(\frac{E_c - E_F}{kT} \right).$$ (27.31)

The Fermi integral is tabulated by McDougall and Stoner (1938) or Blakemore (1962). For $\tilde{n} = 1/2$ [see Eq. (27.17)], it can be approximated (Ehrenberg, 1950) by:

$$F_{1/2}(\xi) = \frac{2\sqrt{\pi} \exp \xi}{4 + \exp \xi}.$$ (27.32)

A comparison between the approximation (27.17) and (27.32), and the exact form (27.30) is shown in Fig. 27.6. They agree for $\xi < -2$, but deviate substantially from each other when the Fermi level moves closer than $2kT$ toward or beyond the band edge. For a review of approximations, see Blakemore (1982a).

27.1.1E The Fermi Level In order to obtain the position of the Fermi level, additional input is needed. In an undoped ideal

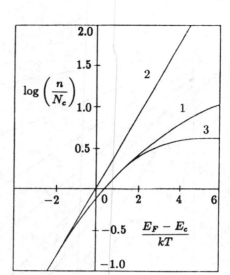

Figure 27.6: Relative carrier density as a function of the normalized energy difference $\xi = -(E_c - E_F)/(kT)$ for the exact solution, the classical approximation, and the Ehrenberg approximation (curves 1–3, respectively). ©Pergamon Press plc.

homogeneous semiconductor, electrons and holes are created in pairs. Therefore, the density of electrons in the conduction band must be equal to the density of holes in the valence band:

$$n = p. \tag{27.33}$$

Using this *neutrality condition* [Eq. (27.33)], we obtain with Eq. (27.23) the relation

$$E_F = \frac{E_c + E_v}{2} + kT \ln \left(\frac{m_{\text{nds}}}{m_{\text{pds}}} \right)^{3/4}, \tag{27.34}$$

which puts E_F essentially in the middle of the band gap for a sufficiently wide band gap, except for a shift due to the ratio of, $m_{\text{nds}}/m_{\text{pds}}$.

27.1.2 Intrinsic Carrier Generation

In intrinsic semiconductors, both electrons and holes contribute to the electronic properties.

The generation of these carriers by thermal excitation (Section 42.2) across the band gap is termed *intrinsic carrier generation—* intrinsic ionization. The carrier density is obtained by introducing

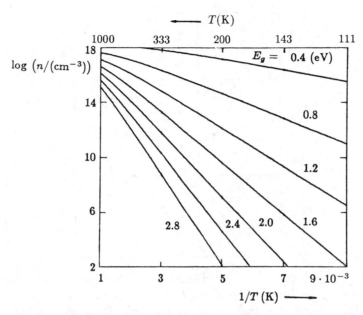

Figure 27.7: Intrinsic electron density in a semiconductor as a function of the reciprocal temperature. The band gap is the family parameter; $m_n = m_0$.

E_F from Eq. (27.34) into Eq. (27.17). For sufficiently wide band gap material, with E_g in excess of $\sim 6kT$ at room temperature, where the "1" in the denominator of the Fermi-Dirac distribution can be neglected, one obtains

$$n = p = n_i = N_c \left(\frac{m_p}{m_n}\right)^{3/4} \exp\left(-\frac{E_c - E_v}{2kT}\right) \qquad (27.35)$$

or, in a more symmetrical form, for the *intrinsic carrier density*

$$n_i = \sqrt{N_c N_v} \exp\left(-\frac{E_c - E_v}{2kT}\right). \qquad (27.36)$$

The intrinsic carrier density increases exponentially with a slope of one half of the band gap. This density is shown in Fig. 27.7 as a function of temperature with the band gap as family parameter. For a review, see Blakemore (1962), or Shklovskii and Efros (1984).

27.2 The Extrinsic Semiconductor

Extrinsic semiconductors contain lattice defects that determine the density of electrons or holes in the conduction or valence band since the intrinsic densities at room temperature are very small: $n_i = 2.4 \cdot 10^{13}$, $1.45 \cdot 10^{10}$, and $1.79 \cdot 10^6$ cm^{-3} for Ge, Si, and GaAs, respectively. The usual densities of electrons or holes of about $10^{15} \ldots 10^{18}$ cm^{-3} are generated almost exclusively by donors or acceptors. We will discuss here only such defects with levels within the band gap, which either donate an electron to, or accept an electron from, the host lattice, depending on energy and original occupancy of this state. The first defect is therefore called a *donor*; the second, an *acceptor*. The introduction of such donors or acceptors into the host lattice is mostly done by adding a small density of the appropriate impurities to the semiconductor. This process is referred to as *doping*.

27.2.1 n- or p-Type Semiconductors

Doped semiconductors have one predominant carrier. With donor doping, $n \gg p$, the semiconductor is termed *n-type*. In homogeneous materials, essentially all transport properties are determined by these *majority carriers*, here electrons in the conduction band. The influence of *minority carriers*, here holes in the valence band, is negligible with homogeneous doping. This is no longer true in inhomogeneous materials, where the influence of minority carriers can play an important role. For example, the current in most solar cells is almost entirely provided by minority carriers.

Semiconductors with acceptor doping ($p \gg n$) are called *p-type*. In these, the influence of electrons is negligible.

A rather simple way to determine experimentally whether a semiconductor is n- or p-type is based on minority carrier *injection* from a point contact. Here a high field causes such injection; whereas, at a large area contact, the field is insufficient for injection (Henisch, 1957). Therefore, for a bias that injects minority carriers (*forward bias*) a large current will flow, while with opposite polarity (*reverse bias*) only the small number of already-present minority carriers can be collected, resulting in a much smaller current (Fig. 27.8).

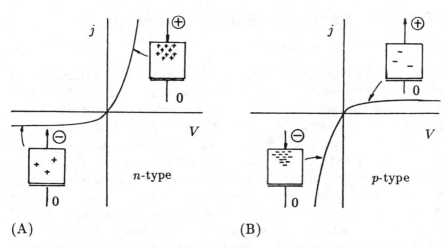

Figure 27.8: Schematics of semiconductors with point (top) and large area (bottom) contacts, showing rectifying characteristics of (A) n-type and (B) p-type semiconductors. The current is carried by minority carriers which are shown in the figure.

Other means to identify n- or p-type conductivity employ the Hall effect (Section 31.2.2A) or thermo-emf* (Section 31.2.1) measurements.

27.2.2 Carrier Densities in Doped Semiconductors

The density of electrons and holes in a semiconductor, with given densities of donors and acceptors of a known energy and charge character, can easily be determined by again using statistical arguments.
As shown in Section 27.1.1C, the carrier density is given by

$$n = N_c \exp\left(-\frac{E_c - E_F}{kT}\right) \quad \text{and} \quad p = N_v \exp\left(-\frac{E_F - E_v}{kT}\right).$$

$$(27.37)$$

In contrast to the intrinsic behavior discussed in Section 27.1.1C, both of these densities are no longer equal to each other. Due to an excess of either donors or acceptors, the Fermi level moves substantially away from the middle of the gap toward the conduction or valence band, respectively.

---

* When a hot wire touches the semiconductor, the wire becomes charged oppositely to the carrier type. For example, when contacted to an n-type semiconductor, the wire becomes positively charged with respect to the semiconductor since electrons are emitted into the semiconductor.

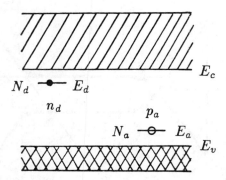

Figure 27.9: Band model with donors and acceptors.

27.2.2A The Position of the Fermi Level The position of the Fermi level can be determined from the neutrality condition:

$$n + n_a = p + p_d, \tag{27.38}$$

where n_a is the density of negatively charged acceptors, i.e., acceptors which have accepted an electron, and p_d is the density of positively charged donors, i.e., donors that have donated their electron—see Fig. 27.9:

$$n_a = N_a - p_a \quad \text{and} \quad p_d = N_d - n_d, \tag{27.39}$$

with N_a and N_d as the total density of acceptors or donors, independent of their occupation.

The degree of filling of donors (n_d/N_d) and acceptors (p_a/N_a) is given by the Fermi-Dirac distribution functions

$$n_d = \frac{N_d}{1 + \dfrac{1}{\nu_{Dd}} \exp\left(\dfrac{E_d - E_F}{kT}\right)} \quad \text{and} \quad p_a = \frac{N_a}{1 + \dfrac{1}{\nu_{Da}} \exp\left(\dfrac{E_F - E_a}{kT}\right)}$$

$$\tag{27.40}$$

with ν_{Dd} and ν_{Da} as the degeneracy factors of the donor and acceptor. The value of the degeneracy factors depends on the electronic structure of the defect—see Landsberg (1982). In the simplest case, ν_{Dd} or $\nu_{Da} = 2$ for defects which have an initially empty or paired state and a final state with an unpaired electron. Alternatively, ν_{Dd} or $\nu_{Da} = \frac{1}{2}$ if the initial state was occupied with the unpaired electron, while the final state is empty or is occupied with paired electrons—see also van Vechten, 1980.

Equations (27.37) through (27.40) present a set of algebraic equations for determining E_F. It is helpful to use a graphical method for obtaining E_F as a function of doping and temperature (Shockley,

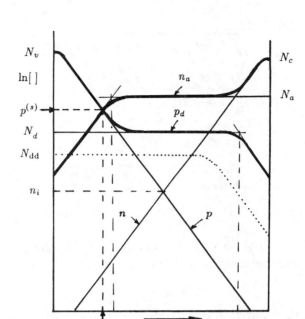

Figure 27.10: Carrier densities in bands (n and p) and in donors or acceptors (p_d and n_a) as a function of the Fermi energy. A graphical solution (s) of Eq. (27.38) is indicated by the arrow ($E_F^{(s)}$) with the corresponding hole density in the valence band $p^{(s)}$. N_{dd} is the density of a deep donor not affecting the solution.

1950). In Fig. 27.10, n, n_a, p, and p_d are plotted in a semilogarithmic representation as a function of E_F for given temperature and doping densities—here, predominant acceptor doping is assumed. The intersection of $n + n_a$ and $p + p_d$ (heavy curves) yields the solution $E_F^{(s)}$ (arrows at abscissa and ordinate in Fig. 27.10).

From the example shown in Fig. 27.10, the Fermi level is determined by p and n_a only, and lies between the acceptor level and the valence band. The material in this example is p-type.

The determining equation here can be simplified to $p = n_a$ and, neglecting the "1" in the denominators of Eq. (27.40), we obtain

$$N_v \exp\left(\frac{E_v - E_F}{kT}\right) = \nu_{Da} N_a \exp\left(-\frac{E_a - E_F}{kT}\right), \qquad (27.41)$$

yielding

$$E_F = \frac{E_v + E_a}{2} - \frac{kT}{2} \ln\left(\frac{N_v}{\nu_{Dd} N_a}\right). \tag{27.42}$$

When the density of donors is increased for partial compensation, keeping all other parameters constant, essentially no changes in E_F are observed until N_d approaches N_a. Then E_F moves away from the valence band until, when N_d reaches N_a, the Fermi level jumps to the middle of the gap; here compensation is complete. When N_d even slightly exceeds N_a, E_F further moves to a position near E_d. The semiconductor has turned from p-type to n-type, and the Fermi level is determined by

$$E_F = \frac{E_d + E_c}{2} + \frac{kT}{2} \ln\left(\frac{N_c}{\nu_{Dd} N_d}\right). \tag{27.43}$$

Figure 27.10 also shows that, in general, additional doping, indicated for a deeper donor level N_{dd} by the dotted curve, has little influence on E_F as long as the density is substantially below that of the most prevalent level of the same type. Observe the addition of concentrations on the logarithmic density scale in Fig. 27.10.

27.2.2B Temperature Dependence of the Fermi Level The temperature dependence can be obtained from the graphical representation, shown in Fig. 27.11 for three temperatures. At low temperatures, curve set 1, the carrier densities between acceptor and donor interact. The governing quasineutrality equation can be simplified to $n_a = p_d$, or, as seen from Fig. 27.11,

$$N_d = \frac{\nu_{Da}}{\nu_{Dd}} N_a \exp\left(-\frac{E_a - E_F}{kT}\right), \tag{27.44}$$

resulting in a Fermi level that in this example lies close to the acceptor level; this is a partially compensated case:

$$E_F = E_a - kT \ln\left(\frac{\nu_{Da}}{\nu_{Dd}} \frac{N_c}{N_d}\right). \tag{27.45}$$

With increasing temperatures, curve set 2, the carrier densities in the acceptors and in the valence band interact, yielding the set of equations discussed in the previous section. The effect of compensation is reduced. The Fermi level lies about halfway between the defect level and the corresponding band.

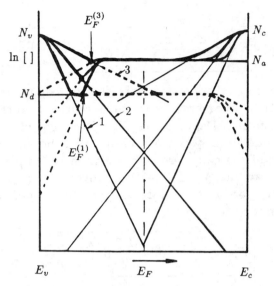

Figure 27.11: Carrier densities as in Fig. 27.10, but for three different temperatures, $T_1 < T_2 < T_3$, for curve sets 1, 2, and 3, respectively.

At still higher temperatures, curve set 3, the acceptor becomes depleted and the quasineutrality equation becomes

$$p = N_v \exp\left(\frac{E_v - E_F}{kT}\right) = \nu_{Da} N_a, \qquad (27.46)$$

with the Fermi level shifting closer to the center of the band gap:

$$E_F = E_v - kT \ln\left(\frac{N_v}{\nu_{Da} N_a}\right). \qquad (27.47)$$

Finally, at still higher temperatures, not shown in Fig. 27.11, electron and hole densities from the two bands are the dominating partners; this is the intrinsic case, described in Section 27.1.1C and given by Eq. (27.34). The Fermi level has now reached the center of the gap.

A typical temperature dependence of the Fermi level in a doped semiconductor is summarized in Fig. 27.12 for n- and p-type Si. This figure also indicates the dependence of the band gap on the temperature—see Section 9.2.1B.

For donor doping, one of the upper curves is selected, for acceptor doping, one of the lower curves is selected, according to the donor or acceptor density.

27.2.2C Defect Compensation To summarize: In the presence of both donors and acceptors, surplus electrons from the donors re-

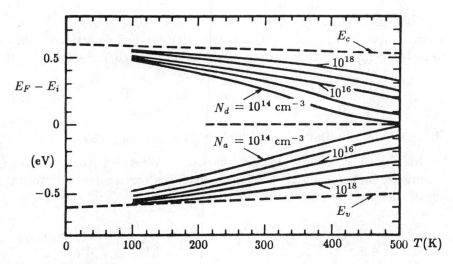

Figure 27.12: Fermi level in Si as a function of the temperature for five different doping levels, showing the decrease of the band gap at higher temperatures (modified from Grove, 1967).

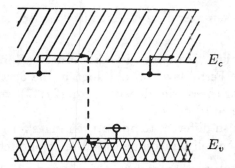

Figure 27.13: Recombination of surplus electrons and holes from donors and acceptors in a partially compensated semiconductor.

combine with holes from the acceptors, with the overall effect of reducing the surplus density of electrons and holes, i.e., bringing the concentration closer to the intrinsic, undoped case. Such a semiconductor is called a *partially compensated semiconductor* (Fig. 27.13); the effective density of defects is given by the difference of the concentration of donors and acceptors:

$$N_{\text{eff}} = N_d - N_a. \qquad (27.48)$$

This effective defect density will be donor-like if $N_d > N_a$, or acceptor-like if $N_a > N_d$. The carrier density for a partially compensated semiconductor is then given by an implicit equation:

$$\frac{n(n + N_a)}{N_d - N_a - n} = \nu_{Dd} N_c \exp\left(-\frac{E_c - E_d}{kT}\right), \qquad (27.49)$$

For example, see Blakemore (1962) and Landsberg (1982).

Inclusion of Excited Donor States.   When excited states of donors are included in the carrier balance, the carrier density can be obtained from the implicit equation,

$$\frac{n(n + N_a)}{N_d - N_a - n} = \frac{\nu_{Dd} N_c}{(1 + F^*)} \exp\left(-\frac{E_c - E_d}{kT}\right) \qquad (27.50)$$

with

$$F^* = \sum_i \frac{\nu_{Ddi}}{\nu_{Dd}} \exp\left(\frac{E_d - E_{di}}{kT}\right) \qquad (27.51)$$

with the sum taken over discrete, localized excited states E_{di} (Landsberg, 1956).

27.2.2D Carrier Density in Extrinsic Semiconductors
Introducing the Fermi level, as obtained in the Section 27.2.2B, into the equations for the carrier densities [Eq. (27.37)] yields the extrinsic branches in the $n(T)$ or $p(T)$ curves.

We distinguish different temperature ranges for partially compensated or for essentially uncompensated semiconductors.

1) Low temperatures: From Eqs. (27.37) and (27.47) we obtain for partially compensated semiconductors

$$n = \frac{\nu_{Dd}}{\nu_{Da}} \frac{N_c(N_d - N_a)}{N_a} \exp\left(-\frac{E_c - E_d}{kT}\right) \qquad (27.52)$$

and for uncompensated semiconductors

$$n = \sqrt{\nu_{Dd} N_c N_d} \exp\left(-\frac{E_c - E_d}{2kT}\right). \qquad (27.53)$$

The factor $1/2$ in the exponent is characteristic of uncompensated semiconductors.

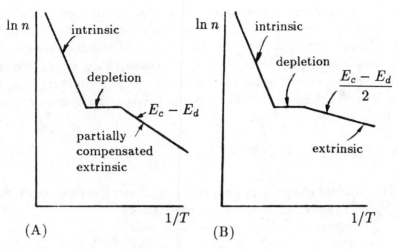

Figure 27.14: Carrier density as a function of the temperature in (A) compensated and (B) uncompensated semiconductors.

2) Medium temperatures: From Eqs. (27.37) and (27.43), we obtain for compensated or uncompensated semiconductors

$$n = \nu_{Dd}N_d - \nu_{Da}N_a, \tag{27.54}$$

i.e., carrier depletion from donors.

3) High temperatures: one reaches the intrinsic branch with

$$n = \sqrt{N_v N_c} \exp\left(-\frac{E_c - E_v}{2kT}\right). \tag{27.55}$$

This behavior is depicted in Fig. 27.14 for partially compensated and uncompensated materials.

A similar set of equations is obtained for p-type semiconductors by replacing N_c with N_v and $(\nu_{Dd}N_d,\ E_d)$ with $(\nu_{Da}N_a,\ E_a)$ in Eqs. (27.52)–(27.55).

Defects in Gapless Semiconductors. Donors or acceptors also influence the conductivity of gapless semiconductors, although the defect states lie inside one of the bands and therefore become resonant states—see Section 21.1.6. These defects cause a shift of the Fermi level away from the touching point in $E(k)$ (see, e.g., Fig. 9.5A, center diagram) making the material n- or p-type. The position of the Fermi level is determined by the quasineutrality condition [Eq. (27.38)]. See Tsidilkovski et al. (1985) for a review.

27.2.2E Intrinsic and Minority Carrier Densities The density of *minority carriers*, i.e., electrons in a p-type or holes in an n-type semiconductor, is obtained from the equilibrium equation once the position of the Fermi level is known. For instance, in the example given in the previous section for an n-type semiconductor, with E_F given by Eq. (27.43) and n given by Eq. (27.53), the minority carrier density p is given by

$$p = N_v \exp\left(-\frac{E_F - E_v}{kT}\right). \tag{27.56}$$

The product of minority and majority carrier densities in *thermal equilibrium* yields the *intrinsic carrier density*

$$np = n_i^2 = N_c N_v \exp\left(-\frac{E_c - E_v}{kT}\right), \tag{27.57}$$

In graded junctions or highly doped inhomogeneous materials (see Sections 9.2.1C and 28.3.1A), a similar relationship holds except that $E_g = E_g(x) = E_{go} + \Delta E_g(x)$; hence $n_i^2 = n_i^2(x)$ with

$$n_i^2(x) = N_c N_v \exp\left(-\frac{\Delta E_g(x)}{kT}\right) \exp\left(-\frac{E_{go}}{kT}\right), \tag{27.58}$$

and thus is independent of doping. With the knowledge of n_i, we can easily obtain the corresponding equilibrium density of minority carriers for any given majority carrier density. For example

$$p = n_i^2/n, \tag{27.59}$$

with n as given in Eq. (27.52). This relation, however, no longer holds when another excitation such as light, is employed, or when the minority carrier density is frozen-in—see Section 27.2.3A. Then $np > n_i^2$—see Chapter 45. For deviations in degenerate semiconductors, see Blakemore (1962).

27.2.3 Self-Activated Carrier Generation

Doping with acceptors and donors to provide extrinsic conductivity, as discussed in the previous section, is usually done at a constant level, independent of temperature. However, vacancies and interstitials can also act as donors or acceptors, and thereby contribute to the carrier generation. Since the generation of these defects from Schottky or Frenkel disorder is an intrinsic process, they

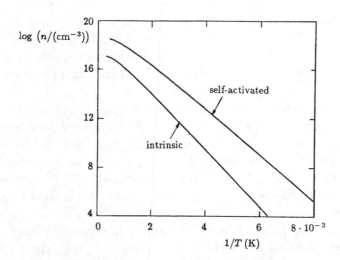

Figure 27.15: Self-activated carrier density as a function of the inverse temperature. Example is computed for a band gap of 2 eV, $E_S = 1.5$ eV, and $E_a - E_v = 0.1$ eV; $N_L = 10^{22}$ cm^{-3} and $N_v = 10^{19}$ cm^{-3}.

may be considered as influencing an intrinsic carrier generation (i.e., a generation which is not doping-dependent).

In a monatomic semiconductor with thermodynamic vacancy generation, the influence on the carrier generation is unambiguous. When vacancies act as acceptors, the density of the acceptors increases with temperature according to

$$N_a = N_L \exp\left(-\frac{E_S}{kT} \right). \tag{27.60}$$

where N_L is the density of lattice sites and E_S is the Schottky energy.

Introducing this relationship into the carrier density relation Eq. (27.56), we obtain

$$p = \sqrt{N_v N_L} \exp\left(-\frac{E_a - E_v + E_S}{2kT} \right). \tag{27.61}$$

The pre-exponential factor of Eq. (27.61) is larger than for an extrinsic carrier generation [Eq. (27.56)]. The behavior resembles electronic intrinsic generation [Eq. (27.55)], since $E_a - E_v + E_S$ can be on the order of the band gap and is not determined by doping except for compensation, which is neglected here.

Such conductivity was introduced as *self-activated semiconductivity* (Böer and Boyn, 1959). Furthermore, this conductivity can be distinguished from a purely electronic, intrinsic carrier generation by a larger carrier density at the intercept for $T \rightarrow \infty$, since N_L is typically on the order of 10^{22} cm^{-3}, while N_v and N_c are on the order of 10^{19} cm^{-3}. In addition, one has to consider a temperature dependence of the band gap, $E_g = E_{go} - |\alpha| T$, which causes a contribution of $\exp(|\alpha|/k)$ in the pre-exponential factor.

An experimental determination of self-activated semiconductivity by comparing the slope with the optically measured band gap E_g, is handicapped by ambiguities. These ambiguities are due to levels in the Urbach tail (see Chapter 24) and by the unknown magnitude of the Franck-Condon shift (see Section 20.3.3A), which permits thermal ionization with a lower energy than the optical band edge transition.

In addition, pairs of intrinsic defects in binary or higher compounds are created at densities which do not yield exact compensation—see Section 19.1.1G. This does not eliminate the influence of intrinsic defects, as an exact compensation would, but makes an analysis more complex.

27.2.3A Frozen-In Carrier Densities The minority carrier density for sufficiently doped, wider band-gap materials becomes unreasonably small. For instance, in n-type GaAs with a band gap of 1.424 eV, the intrinsic carrier density is $1.8 \cdot 10^6$ cm^{-3}. With a density of shallow donors of 10^{17} cm^{-3} yielding $n \simeq 10^{17}$ cm^{-3} at room temperature, the density of minority carriers in thermal equilibrium would be $p \simeq 3 \cdot 10^{-5}$ cm^{-3} [Eq. (27.57)]. This density is substantially below a value probably maintained by background cosmic radiation, and it would take an extremely long time to approach equilibrium without such radiation. Therefore, a different approach is necessary to describe the actual behavior for minority carriers in wider gap semiconductors.

The minority carrier response-time can be estimated from

$$\tau_p = \frac{1}{\nu_p} \exp \left(\frac{E_F - E_v}{kT} \right), \tag{27.62}$$

with ν_p as the attempt-to-escape frequency (Section 42.4.1A) of a *hole* from a shallow *donor*. With ν_p typically on the order of $10^{10} \ldots 10^{13}$ s^{-1} (see Sections 42.4.1A and 42.2.3, Table 42.1), one estimates that it will take more than 10^5 s to achieve equilibrium when the Fermi level is farther away than ~ 1 eV from the band edge. For most electronic experiments, the present state is considered *frozen-in* if a delay of more than one day ($\sim 10^5$ s) is necessary to achieve a *steady state*; that is, whenever the Fermi level is more than 1 eV away from the respective band edge. This means that minority carrier densities at room temperatures are not expected to drop below a frozen-in (f-i) density of

$$p_{\text{f-i}} \simeq N_v \exp\left(-\frac{1\,(\text{eV})}{\text{kT}}\right) \simeq 10^2\,\text{cm}^{-3}. \tag{27.63}$$

When, for consistency, one wants to maintain the Fermi level concept, then it requires the introduction of a *quasi-Fermi level* for minority carriers, here E_{Fp} for holes, which is formally introduced by

$$p_{\text{f-i}} = N_v \exp\left(-\frac{E_{Fp} - E_v}{kT}\right). \tag{27.64}$$

This concept can be justified as reasonable under certain conditions*—see Section 28.5.2. It is a helpful approximation for a variety of discussions dealing with an additional to thermal excitation.

When defining one quasi-Fermi level, it is customary to also convert the notation of the Fermi level relating to the majority carrier, although essentially unchanged from E_F,† to another quasi-Fermi level: here $E_{Fn} \simeq E_F$. Deviations from the thermal equilibrium then result simply in a split of E_F into E_{Fn} and E_{Fp}. Extensive use of this concept will be made in Sections 28.5.2 and 43.2.2.

It will suffice here to summarize that quasi-Fermi levels in normal observation of semiconductors cannot be farther separated from the respective bands than by at most $\sim 40kT$, i.e., about 1 eV for room temperature due to freezing-in:

$$\boxed{(E_c - E_{Fn}) < 40kT} \quad \text{and} \quad \boxed{(E_{Fp} - E_v) < 40kT.} \tag{27.65}$$

In wider band-gap materials ($E_g > 1$ eV), or at lower temperatures, freezing-in becomes an important consideration.

Freezing-in also applies to thermal excitation from deep defect levels in the band gap. However, since the frequency factor depends

---

* Justification in steady state, e.g., with optical excitation, is reasonable. The use, however, of quasi-Fermi levels close to a frozen-in situation becomes questionable when different types of defects are involved (see Section 42.2.3).

† Through mutual generation and recombination of electrons and holes, a change in minority carrier density also causes a change in majority carrier density; but, because of the much larger density of majority carriers, the change is truly negligible.

on the type of defect center (see Table 42.1), the freezing-in depth varies between 20 and $40kT$ for Coulomb-repulsive to -attractive centers, respectively, for a freezing-in time of 10^5 s.

In devices, usually much shorter response times are required. For freezing-in, this causes a much smaller distance of the quasi-Fermi level to the corresponding bands, and requires a more detailed discussion regarding which of the trap levels can follow with their population and which cannot.

Summary and Emphasis

Electrons and holes are the carriers of currents in semiconductors. The density of these carriers can be obtained from equilibrium statistics—the Fermi-Dirac statistics. The key parameter of this statistic is the Fermi level E_F, which can be obtained from quasi-neutrality. E_F lies near the middle of the band gap for intrinsic—and near the donor or acceptor level for doped—semiconductors.

The carrier density is determined by the product of the joint density of states, i.e., the product of defect level and effective band level densities, and the Boltzmann factor, with an activation energy of the difference between band edge and Fermi energy.

In doped semiconductors, both majority and minority carrier densities are of interest for semiconducting devices. The minority carrier density of wider gap semiconductors is often frozen-in at densities in excess of 10^2 cm^{-3}, represented by a quasi-Fermi level of typically not more than 1 eV above the minority carrier band edge at room temperature.

In materials with dominating intrinsic defects, i.e, Schottky or Frenkel disorder, self-activated semiconductivity must be considered at elevated temperatures. Here, the density of these defects, as well as the carrier densities, are temperature dependent.

The determination of the carrier density of semiconductors is one of the most fundamental issues for semiconductor devices. Although well-understood for majority carriers, it needs further exploration for frozen-in minority carriers, and for self-activated semiconductivity of materials with a high density of intrinsic defects.

Exercise Problems

1.(e) Give the level densities in the valence bands centered at $2\,kT$ from E_v for a width of kT at 300 K with $m_l = 0.1m_0$,

$m_h = 0.5m_0$, and $m_{so} = 0.2m_0$, and a spin-orbit splitting of 0.02 eV. Assume parabolic bands and no warping.

2.(e) Calculate the effective level density at the edges of the conduction and valence bands for Ge, Si, and GaAs by properly accounting for the effective masses and the degeneracies of the bands.

3.(e) In an uncompensated semiconductor with incomplete donor ionization, the position of the Fermi level as a function of temperature and donor density can be given explicitly. Using $n = N_d - n_d$, yielding

$$N_c \exp\left(-\frac{E_c - E_F}{kT}\right) = N_d\left\{1 - \frac{1}{1 + \frac{1}{2}\exp\left(\frac{E_d - E_F}{kT}\right)}\right\},$$

(27.66)

and introducing $\alpha = \exp[E_F/(kT)]$, $\beta_c = \exp[E_c/(kT)]$, and $\beta_d = \exp[E_d/(kT)]$, one can solve Eq. (27.66) for α, and thus explicitly for E_F. An elegant equation for $E_F(T, N_d)$ can be obtained by using the equality $\ln(x + \sqrt{x^2 + a^2}) = \ln a + \sinh^{-1}(x/a)$. Discuss $E_F(T, N_d)$.

4.(e) What is the intrinsic carrier density for GaAs at 100, 300, and 400 K?
(a) for constant E_g?
(b) for $E_g(T)$?

5.(*) In Section 27.1.1A we derived an expression for the level distribution near the edge of the conduction band in an isotropic crystal. Derive the level distribution for an orthorhombic crystal.

6. Calculate the electron density in Si doped with 10^{16} cm^{-3} P donors and $2 \cdot 10^{15}$ cm^{-3} B acceptors at 20 K.

7. How wide is the T range for depletion in Si doped with 10^{16} cm^{-3} P donors?

8.(e) Calculate the minority carrier density in Si and Ge doped with 10^{16} cm^{-3} P at 100 and 300 K. When does it become necessary to introduce a quasi-Fermi level? Why?

9.(*) There are several "simple" densities (cm^{-3}) commonly used for describing the carrier transport.
(a) What is the meaning of N_c and N_v?
(b) Why can we use a single quantity for these effective densities and of carrier densities rather than a density

distribution (in energy), and under what circumstances does this approximation break down?

(c) Which, if any, of the assumptions and parameters used to determine N_c would change when comparing a $k = 0$ minimum of $E(k)$ with a satellite minimum at $k \neq 0$ (e.g., in Ge or Si)? What consequences does this have for the numerical value of N_c at these two types of minima?

10.(e) Careful use of the proper statistics is essential in semiconductors.

(a) When do we have to use Fermi-Dirac statistics in a semiconductor? For the discussion of which phenomena relating to semiconductivity is it sufficient to use Boltzmann statistics?

(b) How do we normalize both so that the results are nearly identical?

11.(*) Quasineutrality is an important condition controlling many solid-state properties.

(a) Why is the quasi-neutrality condition such a strong condition to determine the electronic behavior of a semiconductor?

(b) If this neutrality condition in a semiconductor with 10^{18} cm^{-3} donors and of 1 cm^3 would be violated by 1%, what electrostatic energy would be accumulated in this cubic centimeter? Compare this with the energy contained in 1 cm^3 TNT.

(c) What electric field would develop between this 1 cm^3 charged by statistical fluctuations ($\sim \sqrt{n}$) if these electrons would be permitted to transfer to another semiconductor of 1 cm^3 kept at a distance of 1 cm (without back-transfer caused by the field)?

(d) What would be the fluctuating field if back-transfer is permitted?

(e) What would be the result if the semiconductor in the above *Gedanken* experiment is exchanged with a metal with $n = 10^{22}$ cm^{-3}?

(f) What prevents such giant fluctuation in *one* piece of metal of 2 cm^3 volume? How does this relate to electronic noise? Which type of noise? See Chapter 38.

PART VI

TRANSPORT

Chapter 28

Basic Carrier Transport Equations

Mobile charged particles gain energy under the influence of an external electric field, and show a preferential velocity component in field direction in addition to their thermal random motion. Scattering opposes such an energy gain, and acts as a friction component to yield a constant carrier drift velocity and heat.

Electrons and holes in the conduction and valence bands are quasi-free to move in space and energy, and to accept energy from an external field.

Carrier transport in nonideal* semiconductors is subject to scattering with a mean free path λ between scattering events. Such scattering reduces the effective volume to a value of λ^3, in which there is coherence of the electron wave. Only within such limited volume does any one electron experience lattice periodicity and is nonlocalized. Scattering introduces a loss, i.e., a damping mechanism counteracting the energy gain from an electric field. All scattering events change the carrier momentum, i.e., its direction of motion.

---

* A nonideal lattice is such that contains material lattice defects (e.g., impurities) and oscillatory defects (i.e., phonons).

However, only some of them, the *inelastic scattering* events, *significantly* change the energy of the carrier. Usually, several *elastic scattering* events are followed by one inelastic event after the carrier has gained sufficient energy from the field to permit inelastic scattering, usually by generating optical phonons. A large variety of scattering events can be distinguished, and is discussed in Chapter 32. The sum of all of these determines the carrier motion, its *mobility*, which will be defined later.

A quasiclassical picture will be used in this chapter to describe the basic elements of the carrier motion in a semiconductor. The quantum-mechanical part is incorporated by using an effective mass rather than the rest mass, i.e., by dealing with Bloch electrons or Bloch holes as quasi-particles.

Before we continue the discussion about carrier transport, we need to specify in a more refined model what we understand about carriers in semiconductors.

28.1 Carriers in Semiconductors

The excitation of electrons from the valence band (e.g., by absorption of photons), creates a certain concentration (n) of electrons in the conduction bands. These electrons interact with lattice defects, such as phonons, impurities, or other deviations from an ideal periodicity.

This interaction is termed a *scattering event*. The scattering tends to bring the electrons into *thermal equilibrium* with the lattice, and in doing so, to the lowest valley of the lowest conduction band, if not already there by either thermal or near-band gap optical excitation. This state of equilibrium will be discussed in this chapter.

28.1.1 Bloch Electrons or Holes

Near the bottom of the band, the electron is described as a *Bloch electron*,* i.e., as an electron with an effective mass given by the curvature of $E(k)$, as defined in Section 7.2.3, Eq. (7.20).

In an analogous description, the hole is described as a Bloch-type quasi-particle, residing near the top of the uppermost valence bands, with an effective mass given by their curvatures—see Section 9.1.1A.

---

* We are adopting here the picture of a localized electron. Such localization can be justified in each scattering event. In this model we use a gas-kinetic analogy with scattering cross sections, e.g., for electron-phonon interaction. This is equivalent to a description of the interaction of delocalized electrons and phonons when calculating scattering rates.

The band picture, however, which is the basis for this discussion, results from a series of approximations listed in Chapter 8. The most severe one is the adiabatic approximation, which limits the electron-phonon interaction. Within the band model, the Bloch electron interacts with a *static potential* of the nuclei, while only the electrons surrounding each nucleus are polarized dynamically. With sufficient coupling (i.e., a large enough coupling constant α_c— see Section 28.1.2B), this is no longer justified. For these cases, therefore, the Bloch electron picture needs to be augmented. The corresponding quasi-particle derived from a higher approximation is the *polaron*.

28.1.2 The Polaron

The previous discussion of free carriers with an effective mass was based on an ideal periodic lattice. There are numerous reasons why a real crystal lattice shows perturbation from this periodicity. The perturbations, which cause major changes in the carrier trajectories, can be described as local scattering centers, and will be discussed in Chapter 32.

Another interaction that *accompanies* the carrier throughout its entire motion,* however, is better incorporated in the effective mass picture. It also has an influence on the scattering effectiveness, and thus on the mobility, which will be discussed in Chapter 36. This interaction involves lattice polarization and causes deviations from the *Bloch electron* picture discussed above. Deviations are significant in lattices with a large coupling constant, such as in ionic crystals, typically alkali halides, or crystals with a strong ionic character and large band gap, e.g., transition metal oxides. These materials are not of main interest here, since they usually have inferior semiconducting properties, such as low mobilities. Many typical semiconductors, however, contain a fraction of ionic bonding, especially compound semiconductors with higher ionicity, e.g., II-VI compounds. Moreover, a stronger polarization of the lattice occurs in the neighborhood of certain impurities or in highly disordered semiconductors. The interaction of electrons with lattice polarization

---

* This interaction is also commonly described as an electron-phonon interaction, however, of a different kind than that responsible for scattering. As a lattice deformation relates to phonons, the interaction of electrons with the lattice causing a specific deformation can formally be described by continuously absorbing and emitting phonons (see the following sections).

then becomes important. Such interaction can be static, for an electron at rest, or dynamic, accompanying a moving electron.

The interaction involves the part of the Hamiltonian not considered in the band theory when describing the Bloch electron. This adiabatic approximation neglects the interaction of electrons with the induced motion of lattice atoms. In order to include the motion due to the lattice polarization by the electron, several approaches can be taken, all of which relate to the coupling of the electron with the lattice. Such coupling can be expressed as (Hayes and Stoneham, 1984):

- Fröhlich coupling, i.e, via interaction with longitudinal optical phonons,
- piezoelectric coupling, i.e., via the electric field produced by acoustic phonons in piezoelectric semiconductors, or
- deformation potential coupling, i.e., via the electric field produced by the strain field in acoustic oscillations.

The first coupling effect is dominant. It can be described as absorption or emission of virtual LO phonons by the electron, which thereby lowers its eigenstate.

Another approach to describe the interaction is a static one in which the electron induces a shift of the surrounding ions due to its own Coulomb field. The eigenstates of the electron in this Coulomb funnel are hydrogen-like, similar to that of a hydrogen-like donor (Section 9.1.1A); the ground state describes its *self-energy*. This lowers the energy of a Bloch electron accordingly.

For a sufficiently large coupling constant, the self-energy is large enough so that the electron is actually trapped in its own potential well; it becomes a *self-trapped electron*. This was suggested by Landau (1933) and Frenkel (1936).

Pekar (1954) pointed out, however, that when the electron moves to a neighboring lattice position, the same trap level appears. Thus, one can describe this as a continuous virtual level or band below the conduction band in which the electron can move. This picture applies for lattices with smaller coupling constant.

The state of the electron with its surrounding polarization cloud can then be described as a quasi-particle, the *polaron*. The distinguishing parameter between the self-trapped and the mobile polaron is the strength of the electron-lattice interaction, which also can be related to the polaron size. The *small polaron* is tightly bound and self-trapped; it moves via hopping between neighboring ions. The *large polaron* moves much like an electron described by the Boltzmann equation with scattering events, but with a larger effective mass caused by the polarization cloud carried along by the polaron (for a review, see Christov, 1982). One can also describe the polaron as an electron surrounded by a cloud of virtual phonons, which represents its surrounding polarization. One refers to these polarons as *phonon-dressed electrons*, since attached to them is the "fabric" of the surrounding lattice. They can be described by a Fröhlich Hamiltonian, which explicitly accounts for the electron-phonon (LO phonons at $k = 0$) interaction (Fröhlich et al., 1950; advanced by Lee et al., 1953). Alternatively, it can be described with the Feynman Hamiltonian (*Feynman path integral*, Feynman, 1955), which simulates the virtual phonons by a fictitious particle that interacts with the electron via a harmonic potential (Peeters and Devreese, 1984).

The size of the polaron is measured by the extent of the lattice distortion caused by the electron. Large polarons extend substantially beyond nearest-neighbor distances; small polarons do not.

28.1.2A Large Polarons   In the conventional band model, the polarization of Bloch electrons is included by using the *optical* dielectric constant, which describes the interaction with the electrons of each lattice atom. In the discussion of polarons, their polarization includes a shift in the position of the nuclei; that is, for the Coulomb interaction, one uses the *static* dielectric constant to account for the more intensive shielding:

$$\left(\frac{e^2}{8\pi\varepsilon_{\text{opt}}\varepsilon_0 r} \right)_{\text{el}} \rightarrow \left(\frac{e^2}{8\pi\varepsilon_{\text{st}}\varepsilon_0 r} \right)_{\text{pol}} = \frac{e^2}{8\pi r\varepsilon_0} \left\{ \frac{1}{\varepsilon_{\text{opt}}} - \left(\frac{1}{\varepsilon_{\text{opt}}} - \frac{1}{\varepsilon_{\text{st}}} \right) \right\}.$$
(28.1)

The net difference unaccounted for in the Bloch electron picture describes the Coulomb energy of the polaron:

$$E_{\text{pol.Coul}} = \frac{e^2}{8\pi r_{\text{pol}}\varepsilon^*\varepsilon_0} \quad \text{with} \quad \frac{1}{\varepsilon^*} = \frac{1}{\varepsilon_{\text{opt}}} - \frac{1}{\varepsilon_{\text{st}}}.$$
(28.2)

The resulting net Coulomb potential of the polaron is shown in Fig. 28.1.

Within this potential funnel, the polaron self-energy can be expressed as the $1s$ quasi-hydrogen ground-state energy

$$E_{\text{pol}} = E_{\text{qH}} \left(\frac{\varepsilon_{\text{st}}}{\varepsilon^*} \right)^2 \frac{m_{\text{pol}}}{m_n}.$$
(28.3)

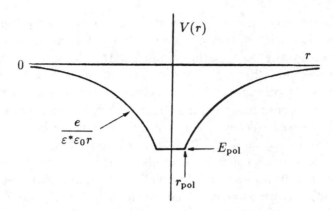

Figure 28.1: Net potential distribution assigned to a polaron, indicating the polaron radius, below which the uncertainty relation precludes further extrapolation of the quasi-Coulomb potential.

It is distinguished from the quasi-hydrogen ground-state energy by replacing ε_{st} with ε^*, and the electron effective mass with that of the polaron.

The radius of the polaron can be defined as the corresponding $1s$ quasi-hydrogen radius

$$r_{pol} = a_{qH}\frac{\varepsilon^*}{\varepsilon_{st}}\frac{m_n}{m_{pol}} = \frac{4\pi\varepsilon^*\varepsilon_0\hbar^2}{m_{pol}e^2}. \qquad (28.4)$$

When the polaron eigenfunctions are overlapping, the *polaron self-energy* [Eq. (28.3)] broadens from a sharp level into a polaron band within which polarons can move through the lattice. The width of this band can be estimated from the uncertainty by absorbing or emitting virtual LO phonons as $\pm\hbar\omega_{LO}$. The corresponding uncertainty radius of such a polaron is given by the uncertainty distance of finding a particle that interacted with an LO phonon: it has an energy $\hbar^2 k^2/(2m_{pol})$ with uncertainty $\pm\hbar\omega_{LO}$. Therefore, it has an uncertainty in wavenumber of $\pm\sqrt{2m_{pol}\omega_{LO}/\hbar}$, the reciprocal of which is its corresponding uncertainty in position.

$$r^*_{pol} = \frac{\hbar}{\sqrt{2m_{pol}\hbar\omega_{LO}}}, \qquad (28.5)$$

which is also used as the large polaron radius.

28.1.2B The Fröhlich Coupling Constant   For interaction with LO phonons, it is convenient to express the Fröhlich coupling in terms of the conventional *coupling constant* α_c. It is given as

Table 28.1: Coupling constants for various solids.

Solid	α_c	Solid	α_c	Solid	α_c
AlSb	0.02	PbSe	0.21	KCl	3.60
GaP	0.13	PbS	0.32	KBr	3.15
GaAs	0.03	PbTe	0.15	KI	2.50
GaSb	0.02	ZnTe	0.33	AgCl	1.91
InP	0.08	ZnSe	0.43	AgBr	1.60
InAs	0.05	ZnS	0.71	TlCl	2.61
InSb	0.02	ZnO	0.90	TlBr	2.05
CdS	0.65	Cu_2O	0.21	$SrTiO_3$	4.50
CdSe	0.46	CuCl	2.01	LiF	5.73
CdTe	0.39			RbBr	6.6

the ratio of the Coulomb energy of a polaron, which describes the electron-phonon interaction to the energy of the LO phonon, i.e., the predominantly interacting phonon:

$$\alpha_c = \sqrt{\frac{\frac{1}{2}\left(\frac{e^2}{4\pi\varepsilon^*\varepsilon_0 r_{\text{pol}}}\right)}{\hbar\omega_{\text{LO}}}};\qquad (28.6)$$

Entering the expression for r_{pol} into Eq. (28.6), one obtains for α_c

$$\alpha_c = \frac{\sqrt{2}e^2}{8\pi\varepsilon^*\varepsilon_0\hbar}\sqrt{\frac{m_{\text{pol}}}{\hbar\omega_{\text{LO}}}} = 22.88\sqrt{\frac{m_{\text{pol}}}{m_0}}\sqrt{\frac{300\text{ K}}{\Theta}}\frac{1}{\varepsilon^*},\qquad (28.7)$$

which is < 1 for good semiconductors. Herer α_c can be interpreted as twice the number of virtual phonons surrounding, i.e., interacting with, a slowly moving carrier in the respective band.

Some values of α_c for typical semiconductors are given in Table 28.1. They are much larger for ionic than for covalent semiconductors, and increase with increasing effective charge and decreasing strength of the lattice binding forces, i.e., decreasing Θ.

Polaron Energy and Effective Mass. With α_c one can express the polaron energy as a fraction of the LO phonon energy*:

$$E_{\text{pol}} = E_c - \frac{p^2}{2m_{\text{pol}}} = -(\alpha_c + 0.01592\alpha_c^2 + \ldots)\hbar\omega_{\text{LO}}. \qquad (28.8)$$

Assuming a small perturbation of the parabolic band, Lee et al. (1953) obtain for the energy dispersion within the band

$$E(k) = \frac{\hbar^2 k^2}{2m_n} - \alpha_c\left(\hbar\omega_{\text{LO}} + \frac{\hbar^2 k^2}{12m_n} + \ldots\right) = -\alpha_c\hbar\omega_{\text{LO}} + \frac{\hbar^2 k^2}{2m_{\text{pol}}}, \qquad (28.9)$$

which yields for the polaron mass:

$$m_{\text{pol},l} = \left(\frac{m_n}{1 - \dfrac{\alpha_c}{6}}\right). \qquad (28.10)$$

Equation (28.10) may be used for small α_c, i.e., a weak electron-phonon interaction. For larger values of α_c up to $\alpha_c \lesssim 5$, the following approximation is used for the polaron mass

$$m_{\text{pol},l} = \frac{1 - 0.08\alpha_c^2}{1 - \dfrac{\alpha_c}{6} + 0.0236\alpha_c^2}m_n. \qquad (28.11)$$

Materials with large polarons that show a significant increase in the effective mass are silver halides. These have an intermediate coupling constant.

The variational method of Lee et al. (1953) is used to compute intermediate coupling. Many II-VI and some III-V compounds have α_c-values that make these large polarons sufficiently distinct from electrons (see Evrard, 1984). Their effective mass can be determined

---

* This result is obtained from Fröhlich et al. (1950) and with a variational method from Lee et al. (1953). It can be used up to $\alpha_c \gtrsim 1$. Earlier results from Pekar, using an adiabatic approximation, yielded

$$E_{\text{pol}} = -\frac{\alpha_c^2}{3\pi}\hbar\omega_{\text{LO}}$$

which gives a lower self-energy than Eq. (28.8) in the range of validity: $\alpha_c < 1$.

Table 28.2: Polaron parameters.

Material	r_{pol} (Å)	E_{pol} (meV)	$\frac{m_{pol}}{m_0}$	$\frac{m_n}{m_{pol}}$	Material	r_{pol} (Å)	E_{pol} (meV)	$\frac{m_{pol}}{m_0}$	$\frac{m_n}{m_{pol}}$
AlSb	89.9	0.85	1.00	0.011	LiF	9.29	323	22.2	17.3
GaAs	39.5	2.6	1.01	0.068	KCl	18.2	92.2	2.26	0.97
GaSb	52.2	0.89	1.01	0.047	KBr	22.1	64.3	2.03	0.75
GaP	21.1	6.5	1.02	0.175	KI	25.6	44.8	1.71	0.56
InP	33.8	5.2	1.02	0.078	RbCl	19.8	86.2	2.76	1.18
InAs	73.9	1.51	1.01	0.023	RbI	27.7	42.3	2.13	0.79
InSb	105	0.49	1.00	0.014	CsI	27.9	42.7	2.59	1.08
ZnO	14.8	64.8	1.18	0.276	CuCl	17.8	54.7	1.50	0.66
ZnS	17.7	30.8	1.14	0.31	AgBr	31.6	27.2	1.35	0.30
ZnSe	27.0	13.1	1.08	0.183	AgCl	23.4	45.3	1.49	0.45
ZnTe	30.5	8.4	1.06	0.169	TlBr	38.5	29.3	1.52	0.27
CdS	25.3	20.2	1.10	0.126	TlCl	21.8	56.1	1.77	0.65
CdSe	33.1	12.2	1.08	0.14	CdF	13.0	16.0	2.14	1.01
CdTe	42.8	8.1	1.07	0.107	SiC	11.7	30.3	1.04	0.24
PbS	41.9	8.5	1.06	0.086	Cu_2O	9.1	15.8	1.04	0.63
PbSe	66.6	3.8	1.04	0.049					
PbTe	90.6	2.0	1.02	0.035					

by cyclotron resonance (Peeters and Devreese, 1984). Landau levels are shifted by $\Delta E \simeq \alpha_c \hbar \omega_{LO}$ for $\alpha_c \ll 1$; when the cyclotron resonance frequency approaches ω_{LO}, this level splits into two peaks indicating the strength of the electron-lattice interaction. For a more exact approximation using a path integral formulation, see Feynman (1955). For reviews, see Kartheuser et al. (1979), Bogoliubov and Bogoliubov, Jr. (1986), and Devreese (1984).

In Table 28.2, the properties of large polarons for a number of crystals are listed, as obtained from Eqs. (28.5), (28.8), and (28.10).

Large Polarons in a Magnetic Field. In the presence of a magnetic field, one has cyclotron behavior of polarons within the corresponding Landau levels. For a transition from Landau level n to $n + 1$, Bajaj (1968) obtained for polarons

$$\omega_{c,pol} = \omega_c \left(1 - \frac{\alpha_c}{6}\right) - \frac{3}{20}\alpha_c \frac{\omega_c^2}{\omega_{LO}} \left(\frac{\hbar k^2}{2m_n \omega_c} + n + 1\right) \quad (28.12)$$

where ω_c is the cyclotron resonance frequency of electrons.

28.1.2C Small Polarons Small polarons were introduced by Tjablikov (1952) and further analyzed by Holstein (1959). They are observed when the coupling constant is larger than 5. This strong interaction causes self-trapping: the wave function corresponds to a localized electron; the tight-binding approximation is appropriate for a mathematical description (Tjablikov, 1952). Small polarons are distinguished from electrons by a rather large polaron self-energy E_{pol}. The *optical energy* necessary to bring a small polaron into the band, i.e., to free the self-trapped electron, was estimated by Pekar to be

$$E_{pol,opt,s} \simeq 0.14\alpha_c^2 \hbar\omega_{LO}. \tag{28.13}$$

Their *thermal ionization energy* $E_{pol,th,s}$ is considerably smaller, typically $\sim (1/3)E_{pol,opt,s}$, because of the strong lattice coupling— see Section 20.1.2A. For further discussion of E_{pol}, see the review of Devreese (1984). The effective mass of small polarons is given by (see Appel, 1968):

$$m_{pol,s} \simeq \frac{m_n \alpha_c^2}{48}. \tag{28.14}$$

Examples for materials with small polarons are narrow band semiconductors with large α_c-values (~ 10), e.g., transition-metal oxides such as NiO or molecular crystals. A review for small polarons in these materials is given by Emin (1973).

A summary of polaron mobility due to various scattering mechanisms is given by Appel (1968) and Evrard (1984).

28.1.3 Bipolarons, Cooper Pairs

When the coupling to the lattice is strong enough so that the polarizing electron produces a significant Coulomb funnel, a second electron with opposite spin may be trapped within the same funnel (Chakraverty and Schlenker, 1976). This *bipolaron* again can move through the lattice at an energy below that of a free electron, and with an effective mass somewhat larger than that of two free electrons (Böttger and Bryksin, 1985).

A significant difference, compared with two free electrons or two independent polarons, is the fact that this new quasi-particle has zero spin and consequently acts as a boson. It is also referred to as a *Cooper pair*, and its formation is used to explain superconductivity— see Section 35.1.1.

28.1.4 Existence Criteria for Different Polarons

There are three different energies, the relative magnitudes of which determine the preferred existence of large polarons, small polarons, or bipolarons. These are*

- the relaxation energy $E_R = \alpha_c \hbar \omega_{LO}$,
- the transfer energy $J = \hbar^2 / (\hat{m} m^* a^2)$, and
- the Hubbard correlation energy $U = U_0 - \lambda^2 / \beta$ (Section 22.4.10)

where $\hat{m}$ is the coordination number, m^* is the effective mass, a is the nearest neighbor distance, λ is an electron-lattice coupling constant, and β is an elastic restoring term.

One distinguishes

(1) the competition between band formation and lattice relaxation

$$
\begin{array}{lll}
\text{a:} & E_R \ll J & \text{: large polaron} \\
\text{b:} & E_R \gg J & \text{: small polaron}
\end{array}
$$

(2) the competition between lattice relaxation and carrier correlation

$$
\begin{array}{lll}
\text{a:} & E_R \ll U & \text{: large polarons remain at separate sites} \\
\text{b:} & E_R \gg U & \text{: electrons prefer to share sites}
\end{array}
$$

(3) sign change in carrier correlation energy

$$
U < 0 \qquad \text{: bipolaron formation}
$$

For further discussion, see Toyozawa (1981).

28.1.5 Electrons or Polarons in Semiconductors

In a rigorous presentation, all electrons or holes near the band edges in equilibrium with the lattice should be replaced by polarons. Since for most semiconductors the difference between electrons and polarons is very small, it is justified to proceed with a conventional description, using instead Bloch electrons and holes. This is no longer sufficient in the neighborhood of lattice defects with strong electron coupling. Here, the Huang-Rhys factor S, rather than the Fröhlich coupling constant α_c, is used; and the polaron picture is applied in order to obtain reasonable results for the defect and lattice relaxation

---

* More precisely, these three energies are: E_R, the energy given to phonons during lattice relaxation; J, the bandwidth of a band created by free, uncoupled carriers of the given density; and U, the energy necessary to put two carriers with opposite spin on the same lattice site.

in agreement with the experiment—see Section 20.1.2A.2. Polaron effects are also significant in some amorphous semiconductors (Cohen et al., 1983).

28.2 Carriers and Their Motion

Carrier transport proceeds under external forces,* resulting in drift, and under internal quasiforces, resulting in diffusion. This may involve different charged particles which contribute additively to the current, or have an indirect effect when it involves neutral particles, e.g., excitons—see Section 15.1).

Carrier transport occurs in bands near the band edges, i.e., near E_c for electrons and near E_v for holes. For materials with a large enough defect density, carrier transport may proceed also via tunneling between trapping states. It may also involve carriers hopping from traps into the band. These carriers then travel a short distance in the band and later are recaptured, then reemitted, and so on—see Chapter 40.

All of these processes add up to produce the total current, and usually have vastly different magnitudes. Ordinarily, only one transport process predominates in homogeneous semiconductors. In non-homogeneous materials, however, at least two, and frequently four contributions are important in different regions of the devices. These are drift *and* diffusion currents of electrons *and* holes.

First, a rather simple picture of the carrier transport is presented, which serves as guidance for a more sophisticated approach in later chapters.

At finite temperatures, carriers are found above the edge of the respective band according to their statistical distribution function. In semiconductors, they usually follow the Boltzmann distribution function within the bands when they are not degenerate (see Section 24.3.1), i.e., when the carrier densities are below $0.1 N_c$ or $0.1 N_v$ [Eq. (27.18)].

---

* Strictly speaking, steady-state carrier transport is due to external forces *only*. The diffusion current originates from a deformed density profile due to external forces, and is a portion of the conventionally considered diffusion component. The major part of the diffusion is used to compensate the built-in field, and has no part in the actual carrier transport: both drift and diffusion cancel each other and are caused by an artificial model consideration—see Section 29.

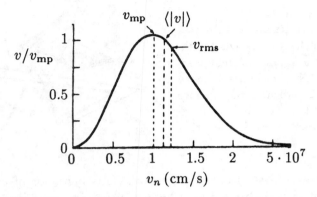

Figure 28.2: Classical velocity distribution, with root mean square, average, and most probable speed identified.

Their thermal velocity, the root mean square velocity,* is obtained from the equipartition principle, i.e., kinetic energy $= \frac{1}{2}kT$ per degree of freedom:

$$\frac{m_n}{2}\langle v^2 \rangle = \frac{3}{2}kT; \qquad (28.15)$$

hence

$$\boxed{\sqrt{\langle v^2 \rangle} = v_{\mathrm{rms}} = \sqrt{\frac{3kT}{m_n}}} \qquad (28.16)$$

or

$$v_{\mathrm{rms}} = 1.18 \cdot 10^7 \sqrt{\frac{m_0}{m_n}\frac{T(\mathrm{K})}{300}} \quad (\mathrm{cm/s}). \qquad (28.17)$$

In thermal equilibrium and in an isotropic lattice, the motion of the carriers is random.

The quantum-mechanical model of a periodic potential teaches that, in contrast to a classical model, an ideal lattice is transparent for electrons or holes within their respective bands. That is, the carriers belong to the entire semiconductor, and as waves are not localized: their position cannot be identified, except stating that n carriers (per cm^3) are within the given crystal. *There is no scattering*

---

* The *rms (root mean square) velocity*, which is more commonly used, should be distinguished from the slightly different *average velocity* and from the *most probable speed*. Their ratios are $v_{\mathrm{rms}} : \langle |v| \rangle : v_{\mathrm{mp}} = \sqrt{3} : \sqrt{8/\pi} : \sqrt{2} = 1.2247 : 1.1284 : 1$, as long as the carriers follow Boltzmann statistics. For a distinction between these different velocities, see Fig. 28.2.

Figure 28.3: Random walk of a carrier with (heavy lines) and without (thin lines) an external electric field. The triple-lined arrow from B to C indicates the relative displacement in field direction after the indicated nine scattering events.

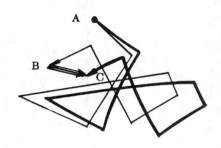

of carriers within such an ideal crystal. This behavior of carriers is unexpected in a classical model, which visualizes the filling of space with atomic spheres and expects only very limited possibilities for an electron traversing between these spheres without being scattered.

The introduction of crystal defects, including phonons, provides centers for scattering. In this way, a carrier motion results, which can be described as a *Brownian motion** with a *mean free path* commensurate with the average distance between scattering centers. This distance is several hundred Ångströms in typical crystalline semiconductors, i.e., the mean free path extends to distances much longer than the interatomic spacing.

When a carrier responds to an external field, it is accelerated in the direction of the electric field. Many important features of the carrier motion can be explained by assuming only inelastic scattering. Since it takes place at defects, which themselves are in thermal equilibrium with the lattice, the carrier tends to lose the excess energy gained between the scattering events. Figure 28.3 illustrates the typical motion of a carrier with and without an external electric field. The changes due to the field are exaggerated; under normal external fields, the changes from the random walk without field are very small perturbations.

28.2.1 Sign Conventions

In previous chapters, the elementary charge is used as $e = |e|$. When the *transport* of electrons and holes is discussed, it is instructive to discuss the proper signs: $-e$ for electrons and $+e$ for holes. This has an influence on derived parameters, e.g., the mobility, as will be discussed in Section 28.2.4.

---

* This motion resembles a random walk (Chandrasekhar, 1943).

The electric field* F is defined as the negative gradient of the vacuum level. The bias V is conventionally labeled $+$ for the anode and $-$ for the cathode, while the electrostatic potential ψ has the opposite signs: electrons have a larger potential energy at the cathode than at the anode. The relation to the field is therefore

$$F = \frac{dV}{dx} = -\frac{d\psi}{dx}.$$ (28.18)

With $E_c = |e|\psi + \text{const.}$, it yields a positive field when the band slopes downward from the cathode toward the anode, giving the visual impression that electrons "roll downhill" and holes "bubble up." We will use the proper signs in the following sections, however, reverting back to the commonly used $e = |e|$ later in order to avoid confusion in comparison to familiar descriptions.

When expressing forces, we need to distinguish the sign of the carrier; therefore, for an accelerating force we have

$$\mathcal{F} = (-eF)_n = (+eF)_p,$$ (28.19)

with subscripts n and p for electrons and holes.

28.2.2 Electronic Conductivity

In following the arguments introduced by Drude (1900; later refined by Lorentz, 1909, and Sommerfeld, 1928), electrons are accelerated in an electric field by the force $\mathcal{F} = -e\mathbf{F}$:

$$m_n \frac{d\mathbf{v}}{dt} = -e\mathbf{F}.$$ (28.20)

During a free path, the electron gains an incremental velocity, for an arbitrarily chosen field in x direction $\mathbf{F} = (F_x, 0, 0)$:

$$\Delta v_x = -\frac{e}{m_n} F_x \tau_{\text{sc}}.$$ (28.21)

After averaging the *incremental velocity* between collisions, and replacing τ_{sc} with the average time $\bar{\tau}$ between scattering events, we obtain the *drift velocity* v_D:

$$v_D = -\frac{e}{m_n} \bar{\tau} F_x.$$ (28.22)

---

* In chapters dealing with carrier transport, F is chosen for the field, since E is used for the energy.

With an electron density n and a charge $-e$, we obtain for the *current density* for electrons

$$\boxed{j_n = env_D,}$$ (28.23)

or, introducing the *electron conductivity* σ_n,

$$j_n = \sigma_n F \quad \text{with} \quad \sigma_n = \frac{e^2}{m_n}\bar\tau n.$$ (28.24)

In a homogeneous semiconductor, the external field is given by the bias V, divided by the electrode distance d, yielding *Ohm's law*

$$\boxed{j_n = \sigma_n \frac{V}{d} = \frac{V}{AR},}$$ (28.25)

with A as the area of the semiconductor normal to the current and the resistance

$$R = \frac{d}{A\sigma_n} = \frac{\rho_n\, d}{A},$$ (28.26)

with the *specific resistivity* $\rho_n = 1/\sigma_n$.

28.2.3 Joule's Heating

The additional energy from the external electric field is delivered during inelastic collisions, generating phonons. This *Joule's heating* can be calculated by accounting for the additional energy obtained from the field between scattering events $(m_n/2)(\Delta v_x)^2$, and the number of scattering events per second $n/\Delta t$, yielding the thermal energy gain Q of the lattice:

$$Q = \frac{n}{\Delta t} \cdot \frac{m_n}{2}(\Delta v_x)^2 = \frac{n}{\bar\tau_n} m_n \left(\frac{e}{m_n}\bar\tau_n F\right)^2.$$ (28.27)

with $\bar\tau = \Delta t$. This yields Joule's law after some reordering:

$$Q = \frac{\left(en\dfrac{e}{m_n}\bar\tau_n F\right)^2}{en\dfrac{e}{m_n}\bar\tau_n} = \frac{j_n^2}{\sigma_n}.$$ (28.28)

28.2.4 Electron Mobility

The quantity

$$\boxed{\frac{e}{m_n}\bar\tau_n = \mu_n}$$ (28.29)

is the *electron mobility*, since carriers are more *mobile* when they experience less scattering, i.e., the time between collisions is larger, and when their effective mass is smaller, i.e., they can be accelerated more easily. With $-e$ for electrons and $+e$ for holes, the mobility is negative for electrons and positive for holes, while the conductivity ($\propto e^2$) is always positive. Conventionally, however, $\mu_n = |\mu_n|$ is used and we will follow this convention here.

The *electron conductivity* and *hole conductivity* are given by

$$\boxed{\sigma_n = e\mu_n n} \quad \text{and} \quad \boxed{\sigma_p = e\mu_p p.} \tag{28.30}$$

28.2.5 Gas-Kinetic Model for Electron Scattering

Different types of lattice defects are effective to a differing degree in carrier scattering. In a simple gas-kinetic model, *scattering centers* have a well-defined *scattering cross section* s_n; a *scattering event*, i.e., a marked deflection* from an otherwise straight carrier path with exchange of momentum and/or energy takes place when the carrier approaches the scattering center within its cross section. A *mean free path* can then be derived by constructing a cylinder of cross section s_n around an arbitrary straight carrier path, and computing the average distance from the last scattering center to which this cylinder will extend until it incorporates the centerpoint of the next scattering center. At this length the cylinder volume $\lambda_n s_n$ equals the average volume that one of these centers occupies: $1/N_{\text{sc}}$ (cm^3); hence,

$$\lambda_n = \frac{1}{s_n N_{\text{sc}}}. \tag{28.31}$$

Here, N_{sc} is the density of scattering centers. Consequently, the time between scattering events is given by

$$\tau_{\text{sc}} = \frac{1}{v_{\text{rms}} s_n N_{\text{sc}}}. \tag{28.32}$$

This time is used to obtain an estimate for the carrier mobility in the Drude approximation:

$$\boxed{\mu_n = \frac{e}{m_n}\tau_{\text{sc}} = \frac{e}{m_n}\frac{\lambda}{v_{\text{rms}}},} \tag{28.33}$$

---

* Often a minimum scattering angle of 90° is used to distinguish scattering events with loss of memory from forward scattering events—see Section 30.3.3A).

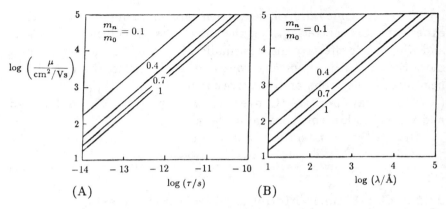

Figure 28.4: Carrier mobility as a function of (A) τ_{sc} and of (B) λ with m_n as family parameter.

or

$$\mu_n = 1.8 \cdot 10^{15} \frac{m_0}{m_n} \tau_{\text{sc}}(s) \qquad (\text{cm}^2/\text{Vs}), \qquad (28.34)$$

or, using the expression [Eq. (28.16)] for v_{rms}:

$$\mu_n = 1.5 \cdot \lambda \, (\text{Å}) \left(\frac{m_0}{m_n}\right)^{3/2} \left(\frac{300}{T(\text{K})}\right)^{1/2} \quad (\text{cm}^2/\text{Vs}). \qquad (28.35)$$

The application of this simple gas-kinetic model, however, has to be taken with caution because of its simplified assumptions. Generally, it yields too large densities of tolerable scattering centers.

28.3 The Drift Current

The drift current is the product of the elementary charge, the carrier mobility, the single carrier density (derived in Chapter 27), and the electric field. For electrons or holes, it is

$$\boxed{j_{n,\text{drift}} = en\mu_n F} \quad \text{or} \quad \boxed{j_{p,\text{drift}} = ep\mu_p F.} \qquad (28.36)$$

28.3.1 The Electric Field

In *homogeneous semiconductors*, disregarding space charge effects near the contacts, and for steady-state conditions—assumed with few exceptions throughout this book—the electric field is given by the applied voltage (*bias*) divided by the distance between the electrodes in a one-dimensional geometry—see Fig. 28.5:

$$F = \frac{V}{d}. \qquad (28.37)$$

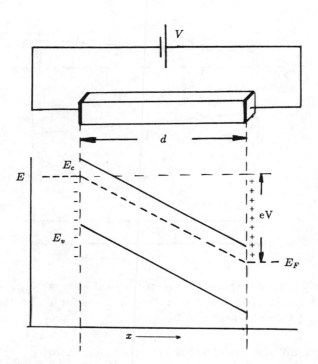

Figure 28.5: Preferred quasi-one-dimensional geometry with band diagram subject to an external bias V, resulting in band-tilting.

For the field concept to apply, the distance between electrodes d must also be large compared to the interatomic spacing. The field can then be expressed by the macroscopic sloping of the bands:*

$$F = \frac{1}{e}\frac{dE_c}{dx} = \frac{1}{e}\frac{dE_v}{dx}. \qquad (28.38)$$

It is also given, and more importantly so, by the slope of the Fermi potential (see Section 28.5.1) which, within the homogeneous material, is the same as the slope of the bands:

$$F = \frac{1}{e}\frac{dE_F}{dx}. \qquad (28.39)$$

The bias is expressed as the difference of the Fermi levels between both electrodes.

---

* As a reminder: here and in all following sections $|e|$ is used when not explicitly stated differently.

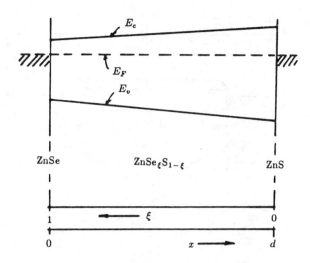

Figure 28.6: Band diagram for a mixed crystal with varying composition ξ along the x-axis. At $x = 0$, the material is ZnSe; at $x = d$, the material is ZnS with ξ varying linearly in between.

When using the the electrostatic potential ψ_n with

$$-\frac{d\psi_n}{dx} = F, \qquad (28.40)$$

the drift current can be expressed as a product of the electrical conductivity and the negative gradient of this potential

$$\boxed{j_{n,\mathrm{drift}} = -\sigma_n \frac{d\psi_n}{dx}} \quad \text{and} \quad \boxed{j_{p,\mathrm{drift}} = -\sigma_p \frac{d\psi_p}{dx}.} \qquad (28.41)$$

For reasons to become apparent in Section 28.3.1A, two electrostatic potentials are introduced: ψ_n and ψ_p for conduction and valence bands, respectively, with $e(\psi_n - \psi_p) = E_g$.

In a homogeneous semiconductor in steady-state, and with vanishing space charge, these drift currents are the total currents, and the slopes of both potentials are the same.

There are special cases, however, in which the band edges of the valence and conduction bands are no longer parallel to each other. One of these will be mentioned briefly in the following section.

28.3.1A Fields in Graded Band-Gap Semiconductors A semiconductor with a graded composition produces a position-dependent, varying band gap. If this composition varies smoothly without steps, e.g., without phase segregation, one or both bands are sloped without an applied bias, representing built-in fields—see Chapter 29. As an example, in $ZnSe_\xi S_{1-\xi}$, there is complete miscibility in the entire range $(0 \leq \xi \leq 1)$, with the S-Se sublattice being a statistical alloy—Section 24.2. The band gap changes linearly* from 2.45 eV for ZnSe at the left side of the crystal shown in Fig. 28.6 to 3.6 eV for ZnS at its right side—see also Fig. 9.22. ZnSe and ZnS are both n-type materials. Depending on doping, the Fermi level in ZnSe can be shifted easily between 0.8 and 0.2 eV below E_c, and in ZnS between 1.0 and 0.4 eV. Depending on the doping profile in the mixed-composition region, a wide variety of relative slopes, including nonmonotonic slopes, of valence and conduction bands can be designed for a vanishing bias, i.e., for a horizontal Fermi level. In Fig. 28.6, an example with opposite and linear sloping of $E_c(x)$ and $E_v(x)$ is shown, resulting effectively in a built-in field (see Chapter 29) of opposite sign for electrons and holes. In thermodynamic equilibrium, however, there is no net current in spite of the sloping bands. This is accomplished by exact compensation of finite drift currents with opposing diffusion currents, which self-consistently determine the slopes of the bands.

The change in the band gap can be expressed as

$$E_c(x) = E_v(x) + E_{g0} + \Delta E_g(x) \qquad (28.42)$$

or, using a conventional asymmetry factor A_E, which measures the fraction of the band gap change $\Delta E_g(x)$ occurring in the conduction band relative to the horizontal Fermi level, we obtain

$$E_c(x) = E_c(x = 0) + A_E \Delta E_g(x) = e\psi_n(x) \qquad (28.43)$$

and

$$E_v(x) = E_v(x = 0) - (1 - A_E)\Delta E_g(x) = e\psi_p(x). \qquad (28.44)$$

The corresponding built-in fields for electrons and holes are given by

$$F_n = -A_E \frac{\partial \Delta E_g(x)}{e \partial x} = -\frac{\partial \psi_n}{\partial x} \qquad (28.45)$$

---

* Major deviations from linearity of E_g with composition are observed when the conduction band minimum lies at a different point in the Brillouin zone for the two end members. One example is the alloy of Ge and Si. Other deviations (bowing—see Section 9.2.1A) are observed when the alloying atoms are of substantially different size.

and

$$F_p = -(A_E - 1)\frac{\partial \Delta E_g(x)}{e\partial x} = -\frac{\partial \psi_p}{\partial x}, \qquad (28.46)$$

justifying the introduction of separate electrostatic potentials for electrons and holes: with $\Delta E_g \neq 0$ and $A_E \neq 1/2$, we have $F_n \neq F_p$.

28.4 Diffusion Currents

Carrier diffusion by itself can be observed when the external field vanishes and a concentration gradient exists. An example in which these conditions are approximately fulfilled is the diffusion of minority carriers created by an inhomogeneous optical excitation. The *diffusion current* is proportional to the diffusion coefficient D and to the carrier density gradient; for electrons or holes it is

$$\boxed{j_{n,\text{diff}} = eD_n \frac{dn}{dx}} \quad \text{or} \quad \boxed{j_{p,\text{diff}} = -eD_p \frac{dp}{dx}.} \qquad (28.47)$$

The negative sign of the hole current is due to the fact that in both equations $\pm e = |e|$ is used. The diffusion current can be derived as the difference between two currents caused by a completely random motion of carriers originating in adjacent slabs with slightly different carrier densities (Fig. 28.7). The current, originating at $x_0 + dx/2$ and crossing the interface at x_0 from right to left, is caused by the Brownian motion of electrons of a density $n_0 + dn/2$. It is given by

$$\overleftarrow{j}_{n,\text{diff}} = e\left(n_0 + \frac{dn}{2}\right)\frac{v_{\text{rms}}^2}{3}\frac{\tau_n}{dx}; \qquad (28.48)$$

the current crossing the boundary from left to right is given by

$$\overrightarrow{j}_{n,\text{diff}} = e\left(n_0 - \frac{dn}{2}\right)\frac{v_{\text{rms}}^2}{3}\frac{\tau_n}{dx}. \qquad (28.49)$$

The current is proportional to the carrier velocity v_{rms} and the carrier mean free path λ_n. In turn, λ_n is given by $v_{\text{rms}}\tau_n$. The factor $\frac{1}{3}$ arises from gas-kinetic arguments when the root mean square velocity is obtained from an isotropic velocity distribution: $\overline{v^2} = v_{\text{rms}}^2 = v_x^2 + v_y^2 + v_z^2$. With $v_x^2 = v_y^2 = v_z^2$, we then obtain for the x-component used in Eq. (28.48) $v_x^2 = \frac{1}{3}v_{\text{rms}}^2$.

The difference of both currents [Eqs. (28.48) and (28.49)] is the net diffusion current

$$j_{n,\text{diff}} = \overrightarrow{j}_{n,\text{diff}} - \overleftarrow{j}_{n,\text{diff}} = e\frac{v_{\text{rms}}^2 \tau_n}{3}\frac{dn}{dx}, \qquad (28.50)$$

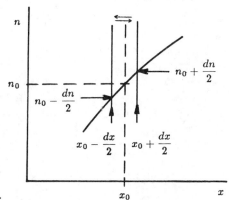

Figure 28.7: Illustration of the derivation of the diffusion current.

with the diffusion coefficient given by

$$D_n = \frac{v_{\mathrm{rms}}^2 \tau_n}{3}.$$
(28.51)

By using $v_{\mathrm{rms}}^2 = 3kT/m_n$ [Eq. (28.15)], we obtain the more commonly used equation for the diffusion current

$$\boxed{j_{\mathrm{n,diff}} = \mu_n kT \frac{dn}{dx}} \quad \text{and} \quad \boxed{j_{\mathrm{p,diff}} = -\mu_p kT \frac{dp}{dx}.}$$
(28.52)

Both diffusion currents for electrons and holes have the same negative sign for a positive gradient of $n(x)$ or $p(x)$ when recognizing that μ_n is negative and μ_p is positive. However, since the conventional notation with $\mu_n = |\mu_n|$ is used, the difference in signs appears.

28.4.1 Maximum Diffusion Currents

As the carrier density gradient increases, the diffusion current increases proportionally to it [Eq. (28.47)]. However, this proportionality is limited, when the density gradient becomes so steep that the reverse current [Eq. (28.49)] becomes negligible compared to the forward current [Eq. (28.48)].

When increasing the distance dx to the mean free path λ_n, we obtain from Eq. (28.50) with $\overrightarrow{j}_{\mathrm{n,diff}} \ll \overleftarrow{j}_{\mathrm{n,diff}}$ for the maximum possible diffusion current through a planar surface

$$j_{\mathrm{n,diff,max}} = en_0 \frac{v_{\mathrm{rms}}^2}{3} \frac{\tau_n}{\lambda_n}$$
(28.53)

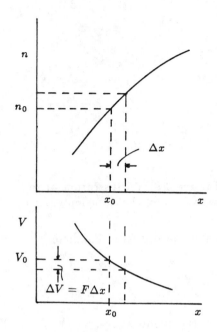

Figure 28.8: Electron density and electrostatic potential distribution in the Boltzmann region in thermal equilibrium (schematic).

or, for carriers following Boltzmann statistics and within a device with planar geometry,

$$j_{\mathrm{n,diff,max}} = \frac{en}{\sqrt{6\pi}} v_{\mathrm{rms}}.$$

(28.54)

This current is known as the *Richardson-Dushman current* (Dushman, 1930). It is equal to the *thermionic emission current* into the vacuum if the semiconductor is cut open at x_0 (Fig. 28.7), and if a vanishing workfunction is assumed; that is, if all electrons in the conduction band at x_0, with a velocity component toward the surface, could exit into the vacuum.

28.4.2 Einstein Relation

Comparing the diffusion equations, Eq. (28.47) with Eq. (28.52), we obtain a relation between the diffusion constant and the carrier mobility:

$$D_{n,p} = \frac{\mu_{n,p} kT}{e},$$

(28.55)

which is known as the *Einstein relation* and holds for systems that follow Boltzmann statistics. This can be seen from the following arguments. In thermal equilibrium, the total current, as well as each carrier current, vanish: $j \equiv j_n \equiv j_p \equiv 0$. The electron current is composed of drift and diffusion currents—see Eq. (28.59); hence,

$$\mu_n n F + D_n \frac{dn}{dx} = 0, \tag{28.56}$$

which can be integrated to yield

$$n(x_0 + \Delta x) = n(x_0) \exp\left(-\frac{\mu_n F \Delta x}{D_n}\right). \tag{28.57}$$

On the other hand, electrons obey the Boltzmann distribution in equilibrium in the conduction band of a semiconductor. Their surplus energy, obtained in an electric field at a distance Δx, is $\Delta E = e\Delta V = eF\Delta x$ (Fig. 28.8), yielding a density

$$n(x_0 + \Delta x) = n(x_0) \exp\left(-\frac{eF\Delta x}{kT}\right). \tag{28.58}$$

A comparison of the exponents in Eqs. (28.57) and (28.58) yields the Einstein relation [Eq. (28.55)]. In case the Boltzmann distribution is not fulfilled (degeneracy), a generalization of Eq. (28.55) can be derived involving Fermi integrals (Landsberg, 1952).

The assumptions used beyond the Boltzmann distribution are that of a one-carrier model near equilibrium and that the total current is small compared to drift and diffusion currents; hence, Eq. (28.56) holds. At high fields, one or more of these conditions are no longer fulfilled. Consequently, the Einstein relation needs to be modified—see Section 33.1.1. For nonparabolic bands, see Landsberg and Cheng (1985).

28.5 Total Currents

The *total current* is given as the sum of drift and diffusion currents; for electrons, one has:

$$j_n = j_{n,\text{drift}} + j_{n,\text{diff}} = e\mu_n n F + e D_n \frac{dn}{dx}, \tag{28.59}$$

and for holes,

$$j_p = j_{p,\text{drift}} + j_{p,\text{diff}} = e\mu_p p F - e D_p \frac{dp}{dx}. \tag{28.60}$$

The *total carrier current* is the sum of both:

$$\boxed{j = j_n + j_p.}$$

(28.61)

In homogeneous semiconductors, only one of the four components is usually predominant, while in a *pn*-junction with sufficient bias each one becomes predominant within a different region (Böer, 1985).

28.5.1 The Electrochemical Fields

For the total current, we need to consider the gradient of $\psi(x)$ and the gradient of $n(x)$. In thermal equilibrium, n is given by the Fermi distribution [Eq. (27.10)]. When the Fermi level is separated by several kT from the band edge, we can disregard the 1 in the denominator of Eq. (27.10) and approximate this equation with the Boltzmann distribution:

$$E_c(x) - E_F(x) = kT \ln \frac{N_c}{n(x)}.$$

(28.62)

Replacing $E_c(x)$ with $-e\psi_n(x) + c$, and differentiating both sides of Eq. (28.62) with respect to x, we obtain after division by e:

$$-\frac{d\psi_n}{dx} - \frac{1}{e}\frac{dE_F}{dx} = -\frac{kT}{e}\frac{1}{n}\frac{dn}{dx}.$$

(28.63)

After multiplying both sides with $\sigma_n = e\mu_n n$ and rearranging, we obtain

$$\sigma_n \frac{1}{e}\frac{dE_F}{dx} = -\sigma_n \frac{d\psi_n}{dx} + \mu_n kT \frac{dn}{dx}.$$

(28.64)

The right-hand side is the total electron current; thus, the left-hand side must also be equal to j_n:

$$j_n = -\sigma_n \frac{1}{e}\frac{dE_F}{dx}.$$

(28.65)

Since in thermal equilibrium, i.e., with vanishing external field, the Fermi level must be horizontal, we conclude that the electron, *and* hole, current must vanish separately in equilibrium:

$$\boxed{\text{for} \quad \frac{dE_F}{dx} \equiv 0 \quad \rightarrow \quad j_n \equiv j_p \equiv 0.}$$

(28.66)

28.5.2 Quasi-Fermi Levels

In *steady state*, e.g., with an external excitation—see Chapter 42, the electron and hole densities deviate from thermodynamic equilibrium

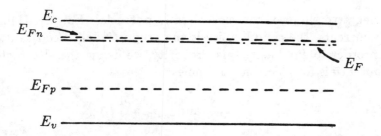

Figure 28.9: Band model with external excitation resulting in a split of the Fermi level into two quasi-Fermi levels; zero-field case.

values. Nevertheless, we may use the Fermi distribution to describe their density in the bands, using the *quasi-Fermi levels* E_{Fn} and E_{Fp} according to the definition equations

$$n \overset{\text{def}}{=} N_c \frac{1}{\exp \dfrac{E_c - E_{Fn}}{kT} + 1} \tag{28.67}$$

and

$$p \overset{\text{def}}{=} N_v \frac{1}{\exp \dfrac{E_{Fp} - E_v}{kT} + 1}, \tag{28.68}$$

with $E_{Fn} \neq E_{Fp}$. This rather useful approximation introduces errors which may or may not be acceptable depending on the cause for deviation from the thermal equilibrium. In general, the error is quite small for optical excitation and for low external fields. For high external fields, the distribution function is substantially deformed (see Section 42.4), and a more sophisticated approximation is required.

With optical or field-induced carrier generation (Sections 42.3 and 42.4), n and p are increased above their thermodynamic equilibrium value; hence, $E_{Fp} < E_F < E_{Fn}$, resulting in a decreased distance of both quasi-Fermi levels from their corresponding bands—see Fig. 28.9. In certain cases, the recombination may be increased above the equilibrium value, as for instance in a *pn*-junction in reverse bias; here, E_{Fn} can drop below E_{Fp}.

In using the same algebraic procedure as described in the previous section, we have for the total electron current in steady state:

$$j_n = -\sigma_n \frac{1}{e} \frac{dE_{Fn}}{dx}; \tag{28.69}$$

that is, the *total electron current* is proportional to the negative slope of the quasi-Fermi potential, as the drift current is proportional to the negative slope of the electrostatic potential [Eq. (28.41)]. For both currents, the conductivity is the proportionality constant.

In order to emphasize this similarity, we define the electrochemical potentials for electrons and holes:

$$\varphi_n = \frac{1}{e}E_{Fn} \quad \text{and} \quad \varphi_p = \frac{1}{e}E_{Fp}. \tag{28.70}$$

The total currents can now be expressed as

$$\boxed{j_n = -\sigma_n \frac{\partial \varphi_n}{\partial x}} \quad \text{and} \quad \boxed{j_p = -\sigma_p \frac{\partial \varphi_p}{\partial x}.} \tag{28.71}$$

For homogeneous semiconductors with homogeneous generation of carriers, these currents become the drift currents, and Eq. (28.69) becomes equal to Eq. (28.40).

In steady state, the total current is divergence-free, i.e., $j_n + j_p = $ const (see Sections 45.3.2 and 45.3.3). Therefore,

$$\sigma_n \frac{\partial \varphi_n}{\partial x} + \sigma_p \frac{\partial \varphi_p}{\partial x} \equiv \text{const} \tag{28.72}$$

or

$$\mu_n n(x) \frac{\partial \varphi_n}{\partial x} + \mu_p p(x) \frac{\partial \varphi_p}{\partial x} \equiv \text{const.} \tag{28.73}$$

Since a semiconductor is predominantly n- or p-type, except for the inner part of a junction, we usually can neglect one part of the sum. For example, for the n-type region,

$$n(x) \frac{\partial \varphi_n}{\partial x} \equiv \text{const;} \tag{28.74}$$

i.e., if there is a gradient in the carrier density, then the highest slope in $\varphi_n(x)$ is expected where the carrier density is lowest for an inhomogeneous $n(x)$ distribution.

From Eq. (28.73), we also conclude that for vanishing currents in steady state, the slopes of the quasi-Fermi potentials must be opposite to each other. The lower the corresponding carrier densities, the higher the slopes:

$$\mu_n n(x) \frac{\partial \varphi_n}{\partial x} = -\mu_p p(x) \frac{\partial \varphi_p}{\partial x}. \tag{28.75}$$

28.5.3 Summary: Potential Gradients and Currents

The various currents in a semiconductor can be expressed in a similar fashion. They are proportional to the negative gradient of electrostatic or electrochemical potentials with the conductivity as a proportionality factor:

$$
\begin{aligned}
j_n &= -\sigma_n \frac{\partial \varphi_n}{\partial x} \\
j_p &= -\sigma_p \frac{\partial \varphi_p}{\partial x} \\
j_{n,\text{Drift}} &= -\sigma_n \frac{\partial \psi_n}{\partial x} \\
j_{p,\text{Drift}} &= -\sigma_p \frac{\partial \psi_p}{\partial x} \\
j_{n,\text{Diff}} &= -\sigma_n \frac{\partial (\varphi_n - \psi_n)}{\partial x} \\
j_{p,\text{Diff}} &= -\sigma_p \frac{\partial (\varphi_p - \psi_p)}{\partial x}
\end{aligned}
\tag{28.76}
$$

Summary and Emphasis

Electrons and holes are carriers of the current in semiconductors. Although modified by polarizing the lattice, this influence, expressed as a change to polarons, is negligible for most semiconductors. It is contained in the effective mass, obtained by cyclotron resonance, and commonly is listed as an effective mass of the carriers, i.e., of electrons or holes.

The current through a semiconductor is composed of a drift and a diffusion current of electrons and holes. In homogeneous semiconductors, only one of these four components is dominant.

The drift current is determined by the electric field, which acts as a slight perturbation of an essentially random walk of carriers, except for very high fields. The additional energy obtained by carrier acceleration from the field is given to the lattice by inelastic scattering, causing Joules heating.

The diffusion current is proportional to the carrier gradient up to a maximum diffusion current, which is limited by the thermal velocity of carriers.

Proportionality factors of both drift and diffusion currents are the carrier mobility, which is proportional to a relaxation time and inversely proportional to an effective mass tensor, the mobility effective mass.

Drift and total currents are proportional to negative potential gradients: the first one being the electrostatic potential, and the second the electrochemical potential. The proportionality factor of both is the conductivity.

Simple classical models permit the description of the basic behavior of drift and diffusion currents. A more sophisticated analysis is necessary to explain the one key parameter of the currents—namely the carrier mobility.

Exercise Problems

1.   In the classical Drude theory, all collisions are assumed to be inelastic ones. Calculate Joule's heat under the assumption that, on the average, only every fifth scattering is inelastic, while the four events in between are elastic.

2.(*)   Calculate the mobility in a simple gas-kinetic model for a material that contains 10^{18} cm^{-3} scattering centers with a cross section of 10^{-13} cm^2, and 10^{19} cm^{-3} centers with a cross section of 10^{-15} cm^2. Assume $m_n \simeq 0.1\, m_0$. Discuss the results in relation to more realistic scattering events.

3.(e)   Calculate the carrier density gradient for which the limiting diffusion current is reached for $\mu = 10^3$ cm$^2/Vs$ and $T = 300$ K. How large is the electric field to produce a current of an equal amount?

4.(r)   Why are electron and hole contributions additive to the electric current of a semiconductor? Discuss the influence of the sign of $E(k)$, effective mass, mobility, charge carrier, direction of field, slope of electrostatic potential, and bands.

5.(e)   Discuss the difference between Coloumb radius and the uncertainty radius of a large polaron [Eq. (28.18) and (28.19)]. A similar relation can be obtained between the Coloumb energy of the polaron [Eq. (28.17)] and the energy related to α_c [Eq. (28.23)]. Explain.

Chapter 29

External and Built-In Fields

There are substantial differences between an external and a built-in field. The most significant being that an external field can heat a carrier gas, while a built-in field cannot.

The *external field* is created by an external bias resulting in a surface-charge on the two electrodes with no space-charge within the semiconductor (Fig. 28.5). Within a typical semiconductor, however, space-charge regions exist because of intentional or unintentional inhomogeneities in the distribution of charged donors or acceptors.[*] This charge density ϱ causes the development of an *internal field* according to the Poisson equation:

$$\boxed{\frac{dF_i}{dx} = \frac{\varrho}{\varepsilon \varepsilon_0}.}$$

(29.1)

The acting field is the sum of both internal, subscript i, and external, subscript e, fields:

$$F = F_i + F_e.$$

(29.2)

External and internal fields result in the same slope of the bands. Therefore, this distinction between internal and external fields is usually not made, and the subscripts at the fields are omitted.

We will indicate in this chapter some of the basic differences between external and internal fields as they relate to carrier transport. For some additional discussion, see Section 33.9.

---

[*] *pn*-junctions are the best studied intentional space-charge regions. Inhomogeneous doping distributions—especially near surfaces, contacts, or other crystal inhomogeneities—are often unintentional and hard to eliminate.

29.1 Penalties for a Simple Transport Model

There are, however, penalties one must pay for a general description of fields, which can best be seen from carrier heating in an electric field. *Carrier heating* is used to describe the field dependence of the mobility (see Section 33.2.1) in a microscopic model. Carriers are shifted up to higher energies within a band. Consequently, their effective mass changes, it usually increases, and the scattering probability changes—most importantly, due to the fact that it becomes easier to create phonons. For all of these reasons, the mobility becomes field-dependent; it usually decreases with increasing field.

The heating *is absent in thermal equilibrium*: the carrier gas and the lattice with its phonon spectrum is in equilibrium within each volume element; thus, carrier and lattice temperatures remain the same (Stratton, 1969). No energy can be extracted from an internal field, i.e., from a sloped band, due to a space charge *in equilibrium*.* This situation may be illustrated with an example replacing electrical with gravitational forces: a sloping band due to a space-charge region looks much like a mountain introduced on top of a sea-level plane, the Fermi level being equivalent to the sea level. As the introduction of the mountain does little to the distribution of molecules in air, the introduction of a sloping band does little to the distribution of electrons in the conduction band. Since there are fewer molecules above the mountain, the air pressure is reduced, just as there are fewer electrons in a band where it has a larger distance from the Fermi level (Fig. 29.1).

However, when one wants to conveniently integrate over all altitudes (energies) in order to arrive at a single number, the air pressure (or the electron density), one must consider additional model consequences to prevent winds from blowing from the valleys with high pressure to the mountain top with low pressure by following only the pressure gradient. Neither should one expect a current of electrons from the regions of a semiconductor with the conduction band close to the Fermi level, which results in a high electron density, to a region with low electron density in the absence of an *external field*.

---

* This argument no longer holds with a bias, which will modify the space-charge; partial heating occurs, proportional to the fraction of external field. This heating can be related to the tilting of the quasi-Fermi levels (Böer, 1985a).

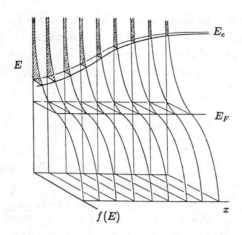

Figure 29.1: Fermi distribution for different positions in a semiconductor with a built-in field region (junction) and zero-applied bias.

To prevent such currents in the electron-density model, one uses the *internal fields*, i.e., the built-in fields, and balances the diffusion current with an exactly compensating drift current. The advantage of this approach is the use of a simple carrier density and a simple transport equation. The penalty is the need for some careful definitions of transport parameters, e.g., the mobility, when comparing external with built-in fields, and evaluating the ensuing drift and diffusion currents when the external fields are strong enough to cause carrier heating—see the following sections.

29.2 Built-In External Fields

The carrier distribution and mobility are different in built-in or external fields, as discussed in the following sections.

29.2.1 Distributions in Built-In or External Fields

The carrier distribution is determined relative to the Fermi level. For vanishing bias, the distribution is independent of the position; the Fermi level is horizontal. The distribution remains unchanged when a junction with its built-in field is introduced.* The sloping bands cut out varying amounts from the lower part of the distribution,

---

* With bias, the Fermi level in a junction is split into two quasi-Fermi levels which are tilted, however, with space-dependent slope. Regions of high slope within the junction region will become preferentially heated. The formation of such regions depends on the change of the carrier dis-

much like a mountain displaces its volume of air molecules at lower altitudes (Fig. 29.1). The carrier concentration n becomes space-dependent through the space dependence of the lower integration boundary, while the energy distribution of the carrier $n(E)$ remains independent in space:

$$n(x) = \int_{E_c(x)}^{\infty} n(E)dE. \qquad (29.3)$$

This is similar to the velocity distribution of air molecules, which is the same at any given altitude, whether over a mountain or an adjacent plane; whereas the integrated number, i.e., the air pressure near the surface of the sloping terrain, is not. This does not cause any macroscopic air motion, since at any stratum of constant altitude the molecular distribution is the same; hence, the molecular motion remains totally random.

In a similar fashion, electrons at the same distance above the Fermi level are surrounded by strata of constant electron density; within such strata their motion must remain random. During scattering in thermal equilibrium, the same amount of phonons are generated as are absorbed by electrons, except for statistical fluctuations: on the average, all events are randomized. Electron and hole currents both vanish in equilibrium *for every volume element*. Figure 29.2A gives an illustration of such a behavior.

In an *external field*, however, Fermi level and bands are tilted parallel to each other; that is, with applied bias, the carrier distribution becomes a function of the spatial coordinate (Fig. 29.2B). When electrons are accelerated in the field, they move from a region of higher density $n(E_1 - E_F)_{x1}$ to a region of lower density $n(E_1 - E_F)_{x2}$. These electrons can dissipate their net additional energy to the lattice by emitting phonons and causing lattice (Joule's) heating. In addition, while in net motion, electrons fill higher states of the energy distribution, thereby causing the carrier temperature to increase. The carrier motion in an *external field* is therefore no longer random; it has a finite component in field direction; the *drift velocity* $v_D = \mu F$ and the collisions with lattice defects are at least partially *in*elastic. A net current and lattice heating result.

---

tribution with bias and its contribution to the electrochemical potential (the quasi-Fermi level). Integration of transport-, Poisson-, and continuity equations yields a quantitative description of this behavior (Böer, 1985a).

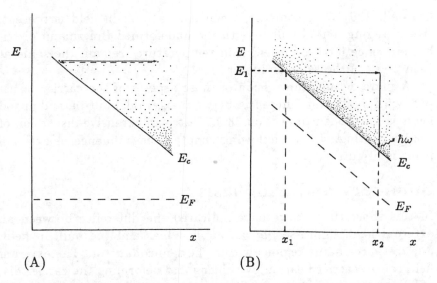

(A)   (B)

Figure 29.2: Sloping band due to (A) an internal (built-in) field with horizontal Fermi level; and (B) due to an external field with parallel sloping bands and Fermi level. The electron distribution is indicated by a dot distribution, and the action of field and scattering by arrows.

29.2.2 Mobilities in Built-In or External Fields

At higher fields the *carrier mobility* becomes field-dependent. The difference between the built-in and the external fields relates to the influence of carrier heating on the mobility, since the *averaging process* for determining the mobility uses the corresponding distribution functions. For instance, with an electric field in the x-direction, one obtains for the drift velocity of electrons

$$v_D = \mu_n F_x = \bar{v}_x = \frac{\iiint v_x f(v)g(v)d^3v}{\iiint f(v)g(v)d^3v}, \qquad (29.4)$$

where $g(v)$ is the density of states in the conduction band per unit volume of velocity space, and d^3v is the appropriate volume element in velocity space. If F_x is the built-in field F_i, then the distribution function is the Boltzmann function $f_B(v)$. If F_x is the external field F_e, the distribution function is modified due to carrier heating according to the field strength $f_{F_e}(v)$—see Section 30.3.

The averaging process involves the distribution function, which is modified by both scattering and effective mass contributions. This is addressed in numerous papers dealing with external fields (for a review, see Nag, 1980; see also Jacoboni et al., 1977; Seeger, 1973;

Conwell, 1967). In contrast, when only a built-in field is present, the averaging must be done with the undeformed Boltzmann distribution, since lattice and electron temperatures remain the same at each point of the semiconductor.

A more detailed discussion of those aspects of the carrier transport with significant differences between external, built-in, and mixed fields is postponed to Section 33.9.2, after an extensive discussion of carrier heating. In the following chapters the influence of *external* fields is discussed.

Summary and Emphasis

In this short chapter we have indicated the difference between an external field, impressed by an applied bias and the built-in field, due to space-charge regions within the semiconductor. The external field causes carrier heating by shifting and deforming the carrier distribution from a Boltzmann distribution to a distorted distribution with more carriers at higher energies within the band.

In contrast, the built-in field leaves the Boltzmann distribution of carriers unchanged; the carrier gas remains unheated at exactly the same temperature as the lattice at every volume element of the crystal, except for statistical fluctuations.

The important consequence of the difference between external and built-in fields is the difference in determining the field dependence of the mobility, which requires an averaging over carriers with different energies within the band. For a built-in field, the averaging follows a Boltzmann distribution; in an external field, there are more electrons at higher energies, and the distribution is distorted accordingly. This can have significant impact for the evaluation of device performances when high fields are considered.

Exercise Problems

1.(r) Discuss the microscopic behavior of an electron in the field produced by an internal space-charge and by an external surface-charge. Discuss the differences?

2.(*) Discuss scattering of electrons in a built-in and in an external field.

3.(*) Why is there no change of the distribution function in a built-in field?

4.(l) Obtain a recent paper analyzing the difference between built-in and external fields, and summarize its findings.

Chapter 30

The Boltzmann Equation

The Boltzmann equation provides the general tool to analyze the carrier transport in semiconductors.

There are several simplifying assumptions in the basic Drude-Sommerfeld approach that are not generally valid.

First, most of the scattering events are not inelastic. A carrier usually accumulates energy during several mean free paths. Each path is interrupted by a mostly elastic collision until it dissipates the increased energy in one inelastic collision; then it continues to accumulate energy, and so on. Different kinds of collisions must be distinguished. This requires replacing the average time between collisions with a relaxation time, which is typical for the decay of a perturbation introduced by an applied external force, e.g., an external electric field.

Second, the interaction with a scattering center often depends on the energy of the electron, e.g., the scattering cross section of ions is energy-dependent, and only more energetic electrons can be scattered inelastically. Therefore, the assumption of a constant, energy-independent time between scattering events needs to be refined.

In this chapter we will introduce the basic Boltzmann equation which permits an analysis of a more advanced description of the carrier scattering.

30.1 The Boltzmann Equation for Electrons

A formalism that permits a refinement of the carrier transport analysis must account for the change in the population of carriers in space and energy or momentum when exposed to external forces. This population is described by a distribution function: in equilibrium these are the Boltzmann or Fermi functions. Under the influence of a field, this distribution is modified. It is the purpose of an advanced theory to determine the modified distribution function. From it, other transport parameters can be derived.

The formalism first proposed by Liouville (1838—see Ferziger and Kaper, 1972), and known as the Liouville equation, is too cumbersome for the evaluation of the carrier transport. A more useful approach can be derived from the Liouville equation as a zeroth-order approximation;* this was suggested by Boltzmann, based on empirical arguments and will be described below.

Conventionally, one uses an accounting procedure for carriers in *phase space*, i.e., in a six-dimensional space and momentum representation (x, y, z, k_x, k_y, k_z). The population of electrons in phase space is given by the distribution function $f(\mathbf{r}, \mathbf{k}, t)$; it changes with time. A group of electrons within a volume element of phase space will move and reside in different volume elements as time progresses. Such motion is described by df/dt. To express the total differential by the local differential, one must consider the deformation due to the time dependence of $\mathbf{r}$ and $\mathbf{k}$ and obtains, using only the first term of a Taylor expansion:

$$\frac{df}{dt} = \frac{\partial f}{\partial t} + \dot{\mathbf{k}} \cdot \nabla_k f + \dot{\mathbf{r}} \cdot \nabla_r f = \left(\frac{\partial f}{\partial t}\right)_{\text{coll}}. \qquad (30.1)$$

The first term accounts for the local change of the distribution in time, the second term for the change in momentum space, and the third term for the change of the distribution in real space. The sum of these changes must be equal to the changes of the distribution caused by collisions, as shown in Eq. (30.1). This simplified relation is called the *Boltzmann equation*.

In *steady state* ($\partial f/\partial t \equiv 0$), the Boltzmann equation reads

$$\left(\frac{\partial f(\mathbf{r}, \mathbf{k}, t)}{\partial t}\right)_{\text{coll}} = \dot{\mathbf{k}} \cdot \nabla_k f + \dot{\mathbf{r}} \cdot \nabla_r f \qquad (30.2)$$

with the first term determined by the forces acting on free electrons, and $\dot{\mathbf{k}}$ given by

$$\dot{\mathbf{k}} = -\frac{e}{\hbar}\mathbf{F}. \qquad (30.3)$$

---

* The most severe approximation is the linear relation in time, which eliminates memory effects in the Boltzmann equation (Nag, 1980).

Here $\mathbf{F}$ is the electric field. The second term is proportional to the spatial gradient of the carrier distribution, and to the group velocity:

$$\dot{\mathbf{r}} = \frac{1}{\hbar} \nabla_k E(\mathbf{k}) = \mathbf{v}. \tag{30.4}$$

This basic Boltzmann equation contains all the dependencies necessary for analyzing carrier transport.* Some of these dependencies, such as the temperature dependency, are contained implicitly. The important part for the carrier transport is the innocent-looking, left-hand side of the Boltzmann equation (30.2), which contains the contribution of the more or less inelastic collisions that provide the "friction" for the carrier transport.

The collision term, also referred to as the *collision integral*, describes the transition of an electron from a state $E_k, \mathbf{k}$ to a state $E_{k'}, \mathbf{k}'$. This can be expressed as the difference between electrons scattered from the state $\mathbf{k}$, occupied according to the Fermi-Dirac distribution function $f_{\mathrm{FD}}(\mathbf{k})$, into the state $\mathbf{k}'$, unoccupied according to $1 - f_{\mathrm{FD}}(\mathbf{k}')$, minus the reverse process, and integrated over all possible states $\mathbf{k}'$, into and from which such scattering is possible:

$$\left(\frac{\partial f(\mathbf{k})}{\partial t} \right)_{\mathrm{coll}} = \frac{\mathcal{V}}{8\pi^3} \int \left\{ f_{\mathrm{FD}}(\mathbf{k}) [1 - f_{\mathrm{FD}}(\mathbf{k}')] S(\mathbf{k}, \mathbf{k}') \\ - f_{\mathrm{FD}}(\mathbf{k}') [1 - f_{\mathrm{FD}}(\mathbf{k})] S(\mathbf{k}', \mathbf{k}) \right\} d\mathbf{k}', \tag{30.5}$$

where $\mathcal{V}$ is the crystal volume, and S is the scattering probability

$$S(\mathbf{k}, \mathbf{k}') = \frac{2\pi}{\hbar} |M(\mathbf{k}, \mathbf{k}')|^2 \delta(E_k - E_{k'} \pm \Delta E). \tag{30.6}$$

$M(\mathbf{k}, \mathbf{k}')$ is the matrix element for the scattering event and ΔE is the fractional change in electron energy during the partially inelastic scattering. The matrix elements can be expressed as

$$\mathbf{M}(\mathbf{k}, \mathbf{k}') = \int_{\mathcal{V}} \psi_{q', k'}^* \Delta V \psi_{\varsigma, k} d\mathcal{V}, \tag{30.7}$$

where ΔV is the perturbation potential inducing the scattering event, $d\mathcal{V}$ is the volume element, and $\psi_{q,k}$, $\psi_{q',k'}$ are the wavefunctions

---

* Here discussed for electrons, although with a change of the appropriate parameters, it is directly applicable to holes, polarons, etc. The influence of other fields, such as thermal or magnetic fields, is neglected here; for such influences, see Chapter 32.

before and after scattering. The perturbation potential depends on the type of scattering event, and could be the deformation potential for scattering on acoustic or optical phonons—see Sections 32.2.1C and 32.2.1E:

$$\Delta V = \begin{cases} \Xi_c \nabla \cdot \mathbf{u} & \text{acoustic phonons} \\ D_0 u & \text{optical phonons} \end{cases} \tag{30.8}$$

with $\mathbf{u}$ as the displacement of the lattice atoms, and Ξ_c or D_0 as the appropriate deformation potentials. Other examples will be given in Chapter 32. The scattering potential and matrix elements for some of the most important scattering centers are tabulated in Table 32.3— see Section 32.4.

30.2 The Boltzmann Equation for Phonons

A similar Boltzmann-type equation can be set up for the phonon system, which interacts with the electron system. Since the only driving forces for the phonon system are that of diffusion due to thermal gradients (neglecting drag effects—see Section 30.3.5), one has for steady state

$$\left(\frac{\partial f(\mathbf{q})}{\partial t} \right)_{\text{coll}} = \dot{\mathbf{r}} \cdot \nabla_r f(\mathbf{q}) \quad \text{with} \quad \dot{\mathbf{r}} = \nabla_r \omega(\mathbf{q}); \tag{30.9}$$

here $\dot{\mathbf{r}}$ is the group velocity of phonons, i.e., the sound velocity in the low $\mathbf{q}$ acoustic branch. The gradient of the phonon distribution function $f(\mathbf{r}, \mathbf{q}, T(\mathbf{r}), t)$ contains the thermal gradient. If undisturbed, the phonon distribution is described by the Bose-Einstein function $f_{BE}(\mathbf{q})$—see Eq. (27.12). The collision term contains all phonon-phonon and phonon-electron interactions. We will regard the former as less important to the present discussion. Interaction of optical with acoustic phonons, however, can become quite important, e.g., for cooling of a heated electron ensemble—see Section 49.1. In a fashion similar to that given for the electron collision term [Eq. (30.5)], one obtains the transition rate by taking the product of the densities of the occupied and the empty states, and the matrix element for each transition, integrated over all possible transitions for absorption and a similar term for emission of phonons

$$
\left(\frac{\partial f(\mathbf{q})}{\partial t}\right)_{\text{coll}} = \frac{\mathcal{V}}{8\pi^3} \int_{\mathcal{V}} \Big\{ S\left(\mathbf{k}+\mathbf{q}, \mathbf{k}\right) \left[1 + f_{\text{BE}}(\mathbf{k})\right] f_{\text{BE}}(\mathbf{k}+\mathbf{q})
$$

$$
- S(\mathbf{k}, \mathbf{k}+\mathbf{q}) \left[1 + f_{\text{BE}}(\mathbf{k}+\mathbf{q})\right] f_{\text{BE}}(\mathbf{k}) \Big\} d\tau_q.
$$

(30.10)

In equilibrium, the right side vanishes as transitions from $\mathbf{k}$ to $\mathbf{k}+\mathbf{q}$ equal those from $\mathbf{k}+\mathbf{q}$ to $\mathbf{k}$. Only when a perturbation is introduced, either from the electron ensemble interacting with phonons or from a temperature gradient, will the right side remain finite. The collision term dealing with interaction of phonons and electrons can be evaluated after linearization.

In order to obtain numerical values, however, one needs to introduce specific assumptions about the microscopic collision process between phonons and electrons. The analysis of such collisions will fill the major part of Chapter 32 as well as Section 40.4.

30.3 The Deformed Boltzmann Distribution

In order to further discuss carrier transport, we have to solve the Boltzmann equation; that is, we have to obtain an expression for $f(\mathbf{r}, \mathbf{k}, t)$. Since the Boltzmann equation is a nonlinear integro-differential equation, it cannot be integrated analytically, and requires the use of approximations or of numerical methods. Both will be mentioned later—Section 30.3.1. However, in order to see some of the important relations, a simplified approach is introduced first: the *relaxation time approximation*.

Balance between gain due to all forces and loss due to collisions of a perturbation, induced by external forces, produces a steady-state, with a deformed electron distribution. When such forces are suddenly removed, the distribution rapidly returns to its unperturbed state according to

$$
\frac{\partial f}{\partial t} = \left(\frac{\partial f}{\partial t}\right)_{\text{coll}}.
$$

(30.11)

Assuming that this collision term is linear in the deviation from the unperturbed distribution f_0,

$$
\left(\frac{\partial f}{\partial t}\right)_{\text{coll}} = -\frac{f - f_0}{\tau_m}.
$$

(30.12)

Equation (30.11) can be integrated. It yields an exponential return from the steady-state, perturbed function $f = f_0 + \delta f$ to the undisturbed distribution in equilibrium with the *momentum relaxation time* τ_m as the characteristic time constant—see Sections 30.3 and 30.3.3A:

$$f(t) - f_o = \delta f \exp\left(-\frac{t}{\tau_m}\right). \tag{30.13}$$

In this linearized form, the deformed distribution function will be used first.

In the following sections, an example with zero-magnetic field and vanishing gradients in n and T is discussed. Here the second term of Eq. (30.2) vanishes.

In a homogeneous semiconductor with a force produced by a constant electric field $\mathcal{F} = -e\mathbf{F}$, we obtain from Eqs. (30.12) and (30.3) for a small perturbation of the distribution function

$$\frac{e\mathbf{F}}{\hbar} \cdot \nabla_{\mathbf{k}} f = \frac{\delta f}{\tau_m}, \tag{30.14}$$

or, with $\hbar\mathbf{k} = m_n\mathbf{v}$, hence $\nabla_{\mathbf{k}} = (\hbar/m_n)\nabla_{\mathbf{v}}$, we now have:

$$\frac{e}{m_n}\tau_m\, \mathbf{F} \cdot \nabla_{\mathbf{v}}\, f = \delta f. \tag{30.15}$$

This shows that the change in the distribution function is proportional to the drift velocity $\mathbf{v}_D = (e/m_n)\tau_m\mathbf{F}$. With f_0 given by the Boltzmann distribution

$$f_0 \propto \exp\left(-\frac{\dfrac{m_n v^2}{2}}{kT}\right), \tag{30.16}$$

we obtain for a small perturbation

$$\nabla_{\mathbf{v}} f \simeq \nabla_{\mathbf{v}} f_0 = -\frac{m_n\mathbf{v}}{kT} f_0, \tag{30.17}$$

and with Eq. (30.15), we have as the final result for the *deformed Boltzmann distribution* due to an external field, in the *relaxation time approximation*:

$$\boxed{f = f_0\left(1 - \frac{e}{kT}\tau_m\mathbf{F} \cdot \mathbf{v}\right).} \tag{30.18}$$

30.3.1 Carrier Scattering and Boltzmann Equation

For a homogeneous semiconductor $(\nabla_r f \equiv 0)$, we obtain from Eqs. (30.1) and (30.5):

$$\frac{\partial f(\mathbf{k})}{\partial t} = -\frac{e\mathbf{F}}{\hbar} \cdot \nabla_k f(\mathbf{k}) - \frac{\mathcal{V}}{8\pi^3} \int \Big\{ f(\mathbf{k})\,[1 - f(\mathbf{k}')]\,S(\mathbf{k}, \mathbf{k}')$$
$$- f(\mathbf{k}')\,[1 - f(\mathbf{k})]\,S(\mathbf{k}', \mathbf{k}) \Big\}\, d\mathbf{k}',$$

(30.19)

with the field $\mathbf{F}$ producing a deformation from the equilibrium distribution, and the collision integral, i.e., the second term in Eq. (30.19), counteracting this deformation. For an analysis, see Haug (1972).

In equilibrium, the solution of the Boltzmann equation is the Boltzmann or Fermi function $f_0(\mathbf{k})$. With an applied field, the distribution is shifted by the drift velocity and is slightly deformed. It is conveniently expressed by a series development using Legendre polynomials (P_n)

$$f(\mathbf{k}) = f_0(k) + \sum_{n=1}^{\infty} f_n(k) P_n(\cos\theta),$$

(30.20)

where θ is the angle between $\mathbf{F}$ and $\mathbf{k}$. After introducing $f(\mathbf{k})$ into Eq. (30.19), we obtain for the steady-state $(\partial f/\partial t \equiv 0)$ a set of n equations to determine $f(\mathbf{k})$.

For small fields, only the first two terms of the development of Eq. (30.20) are taken and yield

$$\frac{1}{3}\frac{e\mathbf{F}}{\hbar}\Big(\nabla_k f_1 + \frac{2f_1}{k}\Big) = -\frac{\mathcal{V}}{8\pi^3} \int \Big\{ f_0(\mathbf{k})\,[1 - f_0(\mathbf{k}']\,S(\mathbf{k}, \mathbf{k}')$$
$$- f_0(\mathbf{k}')\,[1 - f_0(\mathbf{k})]\,S(\mathbf{k}', \mathbf{k}) \Big\}\, d\mathbf{k}' + \dots$$

(30.21)

and

$$\frac{e\mathbf{F}}{\hbar} \cdot \nabla_k f_0 = \frac{\mathcal{V}}{8\pi^3}\cos\theta \int \Big[f_1(\mathbf{k})\big\{ [1 - f_0(\mathbf{k}')]S(\mathbf{k}, \mathbf{k}') + f_0(\mathbf{k}')S(\mathbf{k}', \mathbf{k}) \big\}$$
$$- \frac{\cos\theta'}{\cos\theta} f_1(\mathbf{k}')\big\{ f_0(\mathbf{k})S(\mathbf{k}, \mathbf{k}') + [1 - f_0(\mathbf{k})]\,S(\mathbf{k}', \mathbf{k}) \big\} \Big] d\mathbf{k}',$$

(30.22)

where θ' is the angle between $\mathbf{F}$ and $\mathbf{k}'$ after scattering. For low fields, the left side of Eq. (30.21) vanishes; then, the solution of Eq. (30.21) yields the equilibrium distribution function.

Equation (30.22) is the *linearized Boltzmann equation* to determine

$$f(\mathbf{k}) = f_0(\mathbf{k}) + \cos\theta\, f_1(\mathbf{k}). \qquad (30.23)$$

The perturbation term $f_1(\mathbf{k})$ of the distribution function is often expressed in terms of a function $\phi(\mathbf{k})$

$$f_1(\mathbf{k}) = \frac{e\hbar}{m_n}\mathbf{F}\cdot\mathbf{k}\frac{\partial f_0(E)}{\partial E}\phi(E), \qquad (30.24)$$

with $E = E(\mathbf{k})$. This permits a simplified expression for the collision integral

$$\left(\frac{\partial f(\mathbf{k})}{\partial t}\right)_{\mathrm{coll}} = -\frac{e\hbar}{m_n kT}\mathbf{F}\cdot\mathbf{k}\frac{\nu}{8\pi^3}\int f_0(E)\left[1 - f_0(E')\right]\times$$
$$\times\left[\phi(E) - \frac{k'\cos\theta_k}{k}\phi(E')\right]S(\mathbf{k},\mathbf{k}')\,d\mathbf{k}', \qquad (30.25)$$

where θ_k is the angle between $\mathbf{k}$ and $\mathbf{k}'$.

Any further simplification of the collision integral requires assumptions of the specific scattering event, which will be listed in Section 32.1 and dealt with sequentially in the following sections. However, some general remarks here will assist in categorizing the different scattering types.

30.3.1A Elastic Scattering   Elastic scattering keeps the electron energy during the scattering event unchanged: $E' = E$. This simplifies Eq. (30.25) to

$$\left(\frac{\partial f(\mathbf{k})}{\partial t}\right)_{\mathrm{coll}} = -\frac{e\hbar}{m_n kT}\mathbf{F}\cdot\mathbf{k}\, f_0(E)\left[1 - f_0(E)\right]\frac{\nu}{8\pi^3}\times$$
$$\times\int(1-\cos\theta_k)S(\mathbf{k},\mathbf{k}')\,d\mathbf{k}'. \qquad (30.26)$$

Elastic scattering events are:

(1) all acoustic phonon scattering events, such as
   (a) deformation potential and
   (b) piezoelectric scattering;
(2) all defect scattering events, such as scattering at
   (a) neutral impurities,
   (b) ionized impurities, and
   (c) larger defect scattering; and
(3) alloy scattering,

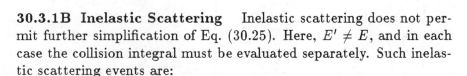

30.3.1B Inelastic Scattering Inelastic scattering does not permit further simplification of Eq. (30.25). Here, $E' \neq E$, and in each case the collision integral must be evaluated separately. Such inelastic scattering events are:

(1) optical phonon scattering, such as
   (a) nonpolar; and
   (b) polar optical scattering, and
(2) intervalley scattering.

The total scattering term is given as the sum over the different scattering types

$$\left(\frac{\partial f}{\partial t} \right)_{\text{coll}} = \sum_i \left(\frac{\partial f}{\partial t} \right)_{\text{coll},i}. \tag{30.27}$$

30.3.1C The Carrier Current The carrier current can be obtained from the deformed Boltzmann distribution by summation over all carriers and velocities

$$\mathbf{j} = \sum_{i=1}^{n} \sum_{\mathbf{v}} e\, \mathbf{v}\, \delta f = e^2 \sum_{i=1}^{n} \sum_{\mathbf{v}} \frac{\mathbf{v}\mathbf{F} \cdot \mathbf{v} f_0 \tau_m}{kT}, \tag{30.28}$$

using Eq. (30.18). Assuming a spherical equi-energy surface for $E(\mathbf{k})$, the summation over $\mathbf{v}\,\mathbf{F} \cdot \mathbf{v}$ can be taken, using for $\mathbf{v}\,\mathbf{v}$ the averages $\overline{v_x v_y} = \overline{v_y v_z} = \overline{v_z v_x} = 0$ and $\overline{v_x^2} = \overline{v_y^2} = \overline{v_z^2} = v^2/3$, yielding

$$\mathbf{j} = \frac{e^2}{3kT} \sum_{\mathbf{v}} v^2 \tau_m f_0 \mathbf{F}. \tag{30.29}$$

Considering that

$$n = \sum_{i=1}^{n} \sum_{\mathbf{v}} f_0, \tag{30.30}$$

one obtains

$$\mathbf{j} = en \frac{e}{3kT} \frac{\sum_{\mathbf{v}} v^2 \tau f_0}{\sum_{\mathbf{v}} f_0} \mathbf{F} = en \frac{e}{3kT} \langle v^2 \tau \rangle \mathbf{F}, \tag{30.31}$$

which gives the electron mobility as

$$\mu_n = \frac{e}{3kT} \langle v^2 \tau \rangle. \tag{30.32}$$

With $\langle E \rangle = (3/2)kT = m_n \langle v^2 \rangle /2$, one can replace $3kT$ by $m_n \langle v^2 \rangle$, yielding

$$\mu = \frac{e}{m_n} \frac{\langle v^2 \tau \rangle}{\langle v^2 \rangle} = \frac{e}{m_n} \frac{\langle E\tau \rangle}{\langle E \rangle}. \qquad (30.33)$$

This result replaces the average time between scattering events obtained from the Drude theory with the energy-weighted average of the relaxation time.

Dropping the requirement of spherical equi-energy surfaces, the end result [Eq. (30.33)] remains the same, except that m_n is replaced by the anisotropic mobility effective mass—for more detail, see Conwell (1982).

30.3.2 Mobility Effective Mass

Following external forces, the carriers are accelerated proportionally to their effective masses [Eq. (28.20)]. Their anisotropy is taken into consideration by introducing a *mobility effective mass*. For a three-axes ellipsoid, this effective mass is given by the inverse average of the effective masses along the main axes:

$$\frac{1}{m_{n\mu}} = \frac{1}{3} \left(\frac{1}{m_1} + \frac{1}{m_2} + \frac{1}{m_3} \right). \qquad (30.34)$$

In general, a mobility tensor is introduced for each of the ν_D satellite valley $E(k)$ ellipsoids, identified by the index i:

$$\underline{\mu}^{(i)} = \frac{e\bar{\tau}_m}{\hbar^2} \begin{pmatrix} \dfrac{\partial^2 E^{(i)}}{\partial k_x^2} & \dfrac{\partial^2 E^{(i)}}{\partial k_x \partial k_y} & \dfrac{\partial^2 E^{(i)}}{\partial k_x \partial k_z} \\[2mm] \dfrac{\partial^2 E^{(i)}}{\partial k_y \partial k_x} & \dfrac{\partial^2 E^{(i)}}{\partial k_y^2} & \dfrac{\partial^2 E^{(i)}}{\partial k_y \partial k_z} \\[2mm] \dfrac{\partial^2 E^{(i)}}{\partial k_z \partial k_x} & \dfrac{\partial^2 E^{(i)}}{\partial k_z \partial k_y} & \dfrac{\partial^2 E^{(i)}}{\partial k_z^2} \end{pmatrix}. \qquad (30.35)$$

In Si, one has three pairs of ellipsoids with different orientation. All of these ellipsoids have their $E(\mathbf{k})$ minima at equal energies and are therefore equally populated at vanishing external forces with

$N_0^{(i)} = \frac{1}{6}N_0$. Their effect on the total mobility is obtained by adding its components, which results in an isotropic mobility tensor

$$\sum \mu_n^{\{100\}}$$

$$= \frac{e\bar{\tau}_m}{\hbar^2} \begin{pmatrix} \frac{1}{3}\left(\frac{2}{m_{\mathrm{nt}}} + \frac{1}{m_{\mathrm{nl})}}\right) & 0 & 0 \\ 0 & \frac{1}{3}\left(\frac{2}{m_{\mathrm{nt}}} + \frac{1}{m_{\mathrm{nl}}}\right) & 0 \\ 0 & 0 & \frac{1}{3}\left(\frac{2}{m_{\mathrm{nt}}} + \frac{1}{m_{\mathrm{nl}}}\right) \end{pmatrix}.$$

(30.36)

Consequently, the *electron mobility effective mass* (for Si) is given by

$$\boxed{\frac{1}{m_{n\mu}} = \frac{1}{3}\left(\frac{2}{m_{\mathrm{nt}}} + \frac{1}{m_{\mathrm{nl}}}\right).}$$

(30.37)

The *hole mobility effective mass* can be derived in a similar fashion. Assuming spherical $E(\mathbf{k})$ surfaces around $k = 0$ (for warped bands, see Section 30.3.2A), and disregarding the deeper spin-orbit band, one obtains

$$\boxed{\frac{1}{m_{p\mu}} = \frac{1}{2}\left(\frac{1}{m_{\mathrm{pl}}} + \frac{1}{m_{\mathrm{ph}}}\right).}$$

(30.38)

Here, m_{pl} and m_{ph} represent the light and heavy hole masses in the corresponding bands.

30.3.2A Hole Mobility Mass in Warped Bands   The valence bands are significantly warped—see Section 9.1.5B. Here, the $E(k)$ surfaces can be represented by an empirical expression [Eq. (9.15)]. Using the abbreviations $B' = \sqrt{B^2 + \frac{1}{6}C^2}$ and $\Gamma_{\pm} = \mp C^2/\{2B' \times \times(A \pm B')\}$ in Eq. (9.15), one obtains, after taking the second derivative of $E(k)$ for the effective masses (see Section 9.1.5B), a useful approximation

$$m_{p\pm} = \frac{m_0}{A \pm B'}(1 + 0.0333\Gamma + 0.0106\Gamma^2 + \ldots) \qquad (30.39)$$

for the mobility effective mass. The variable $m_{\pm}$ represents the *light* or *heavy hole mass* if the upper or lower sign, respectively, is used in Eq. (30.39).

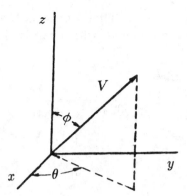

Figure 30.1: Spherical coordinate system.

The values of the constants A, B, and C can be obtained from the Luttinger parameters [Eqs. (9.16)], which are given in Table 9.6 for a number of typical semiconductors.

30.3.3 Momentum and Energy Relaxation

With each scattering event, momentum is exchanged; the carrier changes the direction of its path. In addition, more or less energy is exchanged, with the carrier losing or gaining energy from the scattering center.

30.3.3A The Average Momentum Relaxation Time The average momentum relaxation time is defined as [see Eq. (30.17)]

$$\langle \tau_m \rangle = -\frac{\int \tau_m(v) v_x \dfrac{\partial f}{\partial v_x} \, d^3 v}{\int f_0 \, d^3 v} = \frac{m_n}{kT} \frac{\int \tau_m v_x^2 f_0 \, d^3 v}{\int f_0 \, d^3 v}. \qquad (30.40)$$

It can be obtained from the net increment of the electron momentum, which is proportional to the average drift velocity,

$$\bar{v}_x = \frac{\int v_x f(v) g(v) \, d^3 v}{\int f(v) g(v) \, d^3 v}, \qquad (30.41)$$

where $g(v)$ is the density of states. Assuming only small changes from the thermal distribution, $g(v)$ can be expressed as the effective density of states at the edge of the band [Eq. (27.18)] and cancels out in Eq. (30.41).

In spherical coordinates (Fig. 30.1), with $d^3v = v^2 \sin\theta \, dv \, d\theta \, d\phi$, one obtains from Eq. (30.41)

$$\bar{v}_x = \frac{\int_0^{2\pi} d\phi \int_0^\infty \int_0^\pi f_0(v)\left(1 - \frac{e}{kT}\tau_m(v)F_x v \cos\theta\right) v^3 \cos\theta \sin\theta \, dv \, d\theta}{\int_0^{2\pi} d\phi \int_0^\infty \int_0^\pi f_0(v)\left(1 - \frac{e}{kT}\tau_m(v)F_x v \cos\theta\right) v^2 \sin\theta \, dv \, d\theta},$$

(30.42)

which can be integrated over ϕ and θ, and, after some reordering, yields

$$\bar{v}_x = -\frac{eF_x}{3kT} \frac{\int_0^\infty v^4 \tau_m f_0(v) \, dv}{\int_0^\infty v^2 f_0(v) \, dv}$$

(30.43)

or

$$\bar{v}_x = -\frac{eF_x}{3kT}\langle v^2 \tau_m(v)\rangle.$$

(30.44)

Using the equipartition law for a Boltzmann gas of electrons, $\frac{m}{2}\langle v^2 \rangle = \frac{3}{2}kT$, one obtains for the average drift velocity

$$\bar{v}_x = -\frac{eF_x}{m_n}\frac{\langle v^2 \tau_m(v)\rangle}{\langle v^2 \rangle} = -\frac{e}{m_n}\langle \tau_m \rangle F_x.$$

(30.45)

This result is closely related to the Drude equation, however, having replaced $\bar{\tau}$ in Eq. (28.22) with the *average momentum relaxation time*

$$\boxed{\langle \tau_m \rangle = \frac{\langle v^2 \tau_m(v)\rangle}{\langle v^2 \rangle}.}$$

(30.46)

For an evaluation of Eq. (30.46), one needs the distribution function f_0 and the actual scattering mechanism to determine $\tau_m(v)$—see Chapter 32 and Seeger (1973).

After a collision, the electron path changes by an angle θ, and the fractional change of angle per collision is on the average $\langle 1 - \cos\theta\rangle$. The momentum relaxation time is the time after which the electron path is totally randomized, i.e., its "memory" is lost; hence,

$$\boxed{\tau_m = \frac{\tau_{sc}}{\langle 1 - \cos\theta\rangle},}$$

(30.47)

where τ_{sc} is the average time between two collisions. Scattering with $\theta \geq 90°$ is memory-erasing—Section 32.2.4. Only the collisions in which all angles θ are equally probable result in $\langle \cos\theta\rangle = 0$ and, therefore, yield $\tau_m = \tau_{sc}$. For small-angle scattering events, one needs several scatterings before the momentum is relaxed: $\tau_m > \tau_{sc}$.

30.3.3B The Average Energy Relaxation Time

The energy loss or gain due to scattering of electrons with phonons is given by

$$\frac{dE}{dt} = \frac{\mathcal{V}}{8\pi^3} \int \left\{ [\hbar\omega_q S\,(\mathbf{k},\mathbf{k}')]_{\text{eq}} - [\hbar\omega_q S\,(\mathbf{k}',\mathbf{k})]_{\text{aq}} \right\} d\mathbf{k}'. \quad (30.48)$$

The subscript "eq" stands for emission, "aq" for absorption of a phonon.

When multiplying the Boltzmann equation with E and integrating over $\mathbf{k}$, one obtains (Seeger, 1973)

$$\frac{d\langle E \rangle}{dt} = e\mathbf{F}\langle\mathbf{v}\rangle - \frac{\langle E \rangle - E_L}{\tau_e} \quad (30.49)$$

with E_L the equilibrium energy at lattice temperature,

$$\langle E \rangle = \frac{\int E f d\mathbf{k}}{\int f d\mathbf{k}}, \quad \text{and} \quad \langle\mathbf{v}\rangle = \frac{\int \mathbf{v} f d\mathbf{k}}{\int f d\mathbf{k}}. \quad (30.50)$$

The *energy relaxation time* is then obtained from Eq. (30.49) after switching off the field, yielding:

$$\tau_e = \frac{\langle E \rangle - E_L}{\dfrac{d\langle E \rangle}{dt}}; \quad (30.51)$$

that is, τ_e is given by the ratio of the average surplus energy to the rate of *energy loss* due to scattering and is a function of E; the energy loss rate is not a simple exponential function. It shows a maximum when the electron energy equals the optical phonon energy.

The rate of momentum or energy loss depends on the actual scattering mechanism. From gas-kinetic arguments, one obtains for collisions between an electron and a *lattice defect* of mass M an energy exchange rate of

$$\left(\frac{\tau_e}{\tau_{\text{sc}}} \right)_{\text{ion}} = \frac{m_n}{M}, \quad (30.52)$$

The energy loss is negligible in one scattering event if the scattering center is a defect atom, since $M \gg m_n$.

The fraction of energy lost by an electron in a collision with *acoustic phonons* can also be obtained from an effective mass ratio. Using the *equivalent phonon mass*

$$m_{\text{ph}} = \frac{kT}{v_s^2}, \quad (30.53)$$

where v_s is the sound velocity; one obtains

$$\left(\frac{\tau_e}{\tau_{sc}}\right)_{\text{ac ph}} = \frac{m_n}{m_{\text{ph}}} = \frac{m_n v_s^2}{kT} = \frac{3v_s^2}{v_{\text{rms}}^2}, \qquad (30.54)$$

which is on the order of 10^{-3}. In other words, only 0.1% of the electron energy can be lost to an acoustic phonon during any one scattering event.

In contrast, the ratio of energy relaxation time to scattering time for *optical phonons* is

$$\left(\frac{\tau_e}{\tau_{sc}}\right)_{\text{op ph}} = \frac{\hbar\omega_0}{kT}, \qquad (30.55)$$

which is on the order of 1 at room temperature.

This means that many scattering events usually pass before the accumulated energy obtained from the field can be dissipated to the lattice by emitting one optical phonon, while the momentum is relaxed after one, or only a few collisions. This modifies the rather crude model given in Sections 28.2.2 and 28.2.3 by introducing the momentum relaxation time for evaluating the mobility and the energy relaxation time for Joule's heating.

For a more detailed discussion, however, one must await an analysis of the different scattering mechanisms, and a better estimate of the magnitude of energy obtained from the field. This discussion will be continued in Chapters 32 and 33.

30.3.4 The Mean Free Path of Carriers

Between collisions, the carrier traverses one free path. The mean free path is obtained by averaging:

$$\bar{\lambda} = \frac{\langle v^2 \lambda(v) \rangle}{\langle v^2 \rangle}. \qquad (30.56)$$

$\bar{\lambda}$ is related to the momentum relaxation time

$$\bar{\lambda} = \langle \tau_m \rangle \frac{\langle v^2 \rangle}{\langle v \rangle} = \sqrt{\frac{3\pi}{8}} \langle \tau_m \rangle v_{\text{rms}} = 1.085 \langle \tau_m \rangle v_{\text{rms}}. \qquad (30.57)$$

30.3.5 Phonon and Electron Drag

Interacting electrons and phonons exchange energy and momentum. A drift motion superimposed on the random motion of one ensemble transfers part of the net momentum to the other ensemble during scattering. This means that electrons drifting in an external field tend to push phonons in the same direction, which causes a slight temperature gradient in the field direction, superimposed on the homogeneous Joule's heating. This process is called *electron drag* (Hubner and Shockley, 1960).

Similarly, a temperature gradient tends to push electrons from the warm to the cold end of a semiconductor. This is known as *phonon drag*. The drag effect can be quite large, e.g., up to a factor of 6 compared to simple thermopower in p-Ge at 20 K, as shown by Herring (1954).

When phonons propagate as acoustic waves, ac-electric fields can be induced; or, vice versa, when sufficiently high electric fields are applied, coherent phonon waves can be generated when the drift velocity of electrons surpasses the (sound) velocity of the phonon waves (McFee, 1966). These *acousto-electric effects* have technical application for creating current oscillators (Bray, 1969).

Summary and Emphasis

The Boltzmann equation permits a more sophisticated analysis of the carrier transport, including the concurrent changes in the carrier distribution. Such change in the distribution substantially influences the averaging, which is necessary to arrive at well-defined values for a number of transport parameters—most importantly, the relaxation times.

Significant differences can be defined between the time between two scattering events τ_{sc}, the momentum relaxation time τ_m, and the energy relaxation time τ_e.

Although the Boltzmann equation cannot be integrated in closed form except for a few special cases, the deformed distribution function can be given explicitly in an approximation for small applied fields. It provides the basis for further investigation of various scattering processes.

Such scattering can be divided into essentially elastic processes with mainly momentum exchange, and, for carriers with sufficient accumulated energy, into inelastic scattering with energy relaxation. The latter becomes more prevalent at elevated temperatures and higher electric fields.

The Boltzmann equation is a very useful tool to solve problems with energy transmitted to or from the carrier gas while carrier transport is considered.

Exercise Problems

1. Derive the solution Eq. (30.3) of the Boltzmann equation (30.1) for a small perturbation by electrical and thermal fields in a relaxation-time approach

$$\left(\frac{\partial f}{\partial t}\right)_{\text{coll}} = (\mathbf{v}\cdot\nabla_r)f_0 + \frac{1}{\hbar}(e\mathbf{F}\cdot\nabla_k)f_0 = -\frac{\delta f}{\tau} \text{ with } f = f_0 + \delta f$$

in this linearized approximation (neglecting $\nabla_r \delta f$ and $\nabla_k \delta f$). Express

$$\nabla_r f_0 = \frac{\partial f_0}{\partial T}\nabla_r T \quad \text{and} \quad \nabla_k f_0 = \frac{\partial f_0}{\partial E}\nabla_k E$$

with $f_0 = \{1 + \exp[(E - E_F)/(kT)]\}^{-1}$, the Fermi-Dirac function, and $\nabla_k E = \hbar\mathbf{v}$ (see Haug, 1972).

2.(r) A careful discrimination between different characteristic times relating to carrier scattering is important for its analysis.

(a) Identify and discuss the differences among the time between scattering events, the mean scattering time, the dielectric relaxation time, the momentum relaxation time, and the energy relaxation time.

(b) Which of these times is used in carrier mobilities for cold, warm, and hot carriers, and which in relaxation of fast electrons?

3.(*) Determine how many elastic scattering events are necessary in a semiconductor with $\mu_n = 1000 \text{ cm}^2/\text{Vs}$ and a field of 10 V/cm before an inelastic scattering event with an optical phonon of $\hbar\omega_{\text{LO}} = 30$ meV takes place.

(a) Assume $T = 0$ K.

(b) Assume a carrier distribution at $T = 300$ K, and derive the proper averages.

4. Draw and discuss the deformed Boltzmann distribution in the relaxation-time approach for an isotropic semiconductor with $\mu_n = 1000 \text{ cm}^2/\text{Vs}$ and $F = 10^2$ V/cm. Assume $m_n = 0.1\,m_0$.

5.(e) Compute the mobility effective mass in Si

(a) taking only light and heavy hole band into consideration.

(b) Also consider the spin-orbit split-off band and a Boltzmann distribution for holes at $T = 400$ K.

Chapter 31

Carriers in Magnetic Field, Temperature Gradient

Important information about the carrier transport can be obtained when, in addition to the electric field, a magnetic field is applied or a temperature gradient is acting.

We will now introduce a magnetic field or a temperature gradient into the Boltzmann equation, and analyze their influences on the electron transport.

31.1 Boltzmann Equation

The steady-state Boltzmann equation, which describes the transport of electrons under the influence of external fields

$$\dot{\mathbf{k}} \cdot \nabla_k f + \dot{\mathbf{r}} \cdot \nabla_r f = \left(\frac{\partial f}{\partial t}\right)_{\text{coll}}, \tag{31.1}$$

(see Chapter 30) can easily be expanded to include the *magnetic field** when expressing the forces acting on the electrons [see Eq. (30.3)] by the sum of field and Lorentz forces:

$$\dot{\mathbf{k}} = \frac{e}{\hbar}(\mathbf{F} + \dot{\mathbf{r}} \times \mathbf{B}) \tag{31.2}$$

where $\dot{\mathbf{r}}$ is the group velocity of the electron wave packet and $\mathbf{B}$ is the magnetic induction—Eq. (30.4). The distribution function

---

* In the following sections the *magnetic induction* $\mathbf{B}$ is used, which is connected to the *magnetic field* $\mathbf{H}$ by $\mathbf{B} = \mu\mu_0\mathbf{H}$, with μ_0 the *permeability of free space* and μ the *relative permeability*. Occasionally, the *magnetization* $\mathbf{M}$ is used, which, similar to the polarization, is introduced via $\mathbf{B} = \mu_0\mathbf{H} + \mathbf{M}$ with $\mathbf{M} = \chi_m\mu_0\mathbf{H}$ and $\mu = 1 + \chi_m$, with χ_m the *magnetic susceptibility*.

$f = f(\mathbf{r}, \mathbf{k}, T, t)$ contains the temperature, which also can include temperature gradients.

Using a relaxation-time approximation and setting $f = f_0 + \delta f$, where δf is a small perturbation of the Fermi distribution f_0, we have

$$\left(\frac{\partial f}{\partial t}\right)_{\text{coll}} = -\frac{\delta f}{\tau(E)}; \qquad (31.3)$$

assuming that each collision probability is independent of the collision angle, the Boltzmann equation can be integrated. This can be done in a closed form when $B = 0$:

$$f = \frac{f_0(1 - f_0)\tau}{kT} \left(\mathbf{v} \cdot \left[e\mathbf{F} - \nabla_r E_F - \frac{E - E_F}{T} \nabla_r T \right] \right); \qquad (31.4)$$

the derivation of Eq. (31.4) is left as an exercise problem. Here and in the following equations, the electrochemical energy E_F (Fermi energy) is used, which includes the potential energy as well as changes in the carrier density, and permits a simplified expression. When applying the equation to a deviation from thermal equilibrium, E_F must be replaced by the quasi-Fermi energies E_{Fn} or E_{Fp} for electrons or holes, respectively—see Section 28.5.2.

When the magnetic induction is included, its influence can no longer be treated as a small perturbation.* In contrast to the electrical and thermal conductivities, which are observed at small fields, typical magnetical effects, such as the Hall effect and magneto-resistance, require rather large fields to become observable. Mathematically, this means that the term proportional to the gradient of δf must also be taken into consideration. This yields

$$\delta f = \frac{f_0(1-f_0)\tau}{kT} \left(\mathbf{v} \cdot \left[e\mathbf{F} - \nabla_r E_F - \frac{E - E_F}{T} \nabla_r T \right] \right) + \frac{e}{\hbar}\tau(\mathbf{v} \times \mathbf{B})\nabla_k \, \delta f, \qquad (31.5)$$

which can be solved by iteration. Evaluating the solution near the bottom of the conduction band, where, for spherical equi-energy

---

* If $\mathbf{v} \times \mathbf{B}$ is on the order of $\mathbf{F}$, its influence becomes negligible since, from Eq. (31.2), a term $\mathbf{v} \cdot (\mathbf{v} \times \mathbf{B})$ would appear in Eq. (31.4) which vanishes, since $\mathbf{v}$ is orthogonal to the vector $\mathbf{v} \times \mathbf{B}$.

surfaces, the electron velocity can be expressed as $\mathbf{v} = \hbar\mathbf{k}/m_n$ one obtains

$$\boxed{\delta f = \frac{f_0(1 - f_0)}{kT}\left[\frac{\tau}{1 + b^2}\left\{\mathbf{v}\cdot\mathbf{f} + \mathbf{v}\cdot(\mathbf{b}\times\mathbf{f}) + (\mathbf{v}\cdot\mathbf{b})(\mathbf{b}\cdot\mathbf{f})\right\}\right]}$$

(31.6)

with the abbreviations

$$\mathbf{b} = \frac{e}{m_n}\tau\mathbf{B} \quad \text{and} \quad \mathbf{f} = e\mathbf{F} + \nabla_r E_F + \frac{E - E_F}{T}\nabla_r T. \quad (31.7)$$

The distribution function $f = f(\mathbf{r}, \mathbf{k}, T(\mathbf{r}), B, t)$ now contains the influence of electric, thermal, and magnetic fields—for more detail, see Haug (1972) and Madelung (1981).

The deformed distribution function causes changes in the transport properties, i.e., changes in the electrical or thermal currents as a result of the interacting fields. Rather than following a stringent development of the transport from the Boltzmann equation, a task first solved for carrier conduction by Bloch, the following section will take an alternative, semiempirical approach by describing the different currents with proportionality constants. These constants are later interpreted by a microscopic model.

31.2 Transport Equations

The two governing transport equations, dealing a with carrier current $\mathbf{j}$ and an energy (heat) current $\mathbf{w}$, are given in their general form as

$$\mathbf{j} = -\frac{e}{\hbar}\int \nabla_k E(\mathbf{k})\, g(\mathbf{k}) f(\mathbf{r}, \mathbf{k}, T(\mathbf{r}), \mathbf{B}, t)\, d\mathbf{k} \quad (31.8)$$

$$\mathbf{w} = \frac{1}{\hbar}\int \nabla_k E(\mathbf{k})\, E(\mathbf{k})\, g(\mathbf{k}) f(\mathbf{r}, \mathbf{k}, T(\mathbf{r}), \mathbf{B}, t)\, d\mathbf{k}, \quad (31.9)$$

and contain the density of states and the distribution functions developed in the previous section.

In addition, one needs two conservation laws to describe the transport behavior in a homogeneous semiconductor: the *conservation of the number of carriers*

$$e\frac{\partial n}{\partial t} + \nabla \cdot \mathbf{j} = 0, \quad (31.10)$$

and the *conservation of energy*[*]

$$\rho \frac{\partial u}{\partial t} + \nabla \cdot \mathbf{w} = -\mathbf{F} \cdot \mathbf{j}, \tag{31.11}$$

where ρ is the density and u is the specific internal energy.

The solution of these transport equations in steady state can be expressed as a linear combination of transport parameters and driving forces. For example, when only electric fields act

$$\begin{aligned} \mathbf{j} &= \alpha_{11} \nabla \varphi \\ \mathbf{w} &= \alpha_{21} \nabla \varphi; \end{aligned} \tag{31.12}$$

when electric and thermal fields act

$$\begin{aligned} \mathbf{j} &= \alpha_{11} \nabla \varphi + \alpha_{12} \nabla T \\ \mathbf{w} &= \alpha_{21} \nabla \varphi + \alpha_{22} \nabla T; \end{aligned} \tag{31.13}$$

and when incorporating a magnetic field

$$\begin{aligned} \mathbf{j} ={}& \alpha_{11} \nabla \varphi + \alpha_{12} \nabla T + \beta_{11}(B \times \nabla \varphi) + \beta_{12}(B \times \nabla T) \\ &+ \gamma_{11} B \cdot (B \nabla \varphi) + \gamma_{12} B \cdot (B \nabla T) \\ \mathbf{w} ={}& \alpha_{21} \nabla \varphi + \alpha_{22} \nabla T + \beta_{21}(B \times \nabla \varphi) + \beta_{22}(B \times \nabla T) \\ &+ \gamma_{21} B \cdot (B \nabla \varphi) + \gamma_{22} B \cdot (B \nabla T). \end{aligned} \tag{31.14}$$

Here, φ is the electrochemical *potential*, distinguished from $E_F = e\varphi$, the electrochemical *energy*. For steady state, one has to replace φ with φ_n for electrons and φ_p for holes—see Section 28.5.2. The coefficients α_{ik}, β_{ik}, and γ_{ik} are the well-known transport coefficients, e.g., $\alpha_{11} = \sigma_c$, the electrical conductivity, and $\alpha_{22} = \kappa_c$, the thermal conductivity involving the respective carriers (with subscript $c = n$, or p). The other parameters will be explained below. The total thermal conductivity is related to all four coefficients $\kappa = (\alpha_{11}\alpha_{22} - \alpha_{12}\alpha_{21})/\alpha_{11}$ (Beer, 1963).

Important relations connect the different transport coefficients, such as the *Onsager relations*, obtained from the reciprocity of the effects

$$\alpha_{ik}(\mathbf{B}) = \overline{\alpha}_{ki}(-\mathbf{B}), \tag{31.15}$$

---

[*] Although the electric and magnetic fields act as external forces, and one has $-(\mathbf{F} + \mathbf{v} \times \mathbf{B})$ as total force, the scalar product of $(\mathbf{v} \times \mathbf{B}) \cdot \mathbf{j}$ is zero since the vectors $\mathbf{v} \times \mathbf{B}$ and $\mathbf{j}$ are perpendicular to each other; in first approximation, there is no energy input into the carrier gas from a magnetic field.

where $\bar{\alpha}_{ki}$ is the transposed tensor of α_{ik}. In anisotropic semiconductors, each of the transport parameters is a *tensor*, e.g., $\sigma_n = \sigma_{ik}^{(n)} = en\mu_{ik}$.

The transport coefficients are directly accessible through experimental observation. Their magnitudes depend on the relative orientation of the different fields and, for an anisotropic semiconductor, also on the relative crystallographic orientation. In samples that permit electrical currents in the x direction only, one also distinguishes *isothermal* and *adiabatic galvanomagnetic effects*, depending on whether $\nabla_y T = 0$ or $w_y = 0$ (see Madelung, 1981).

An overview of the different possibilities in an isotropic semiconductor is given in Tables 31.1 and 31.2. Some of the effects listed have gained technical interest or are used extensively for analytical purposes; the Peltier effect and the Hall effect are examples.

In the following sections, some of the more important effects will be discussed in some detail and the corresponding transport parameters will be analyzed in a microscopic model to yield information about the basic transport properties.

31.2.1 Thermoelectric Effects

There are four experimentally accessible constants that describe the relations between the electric and thermal fields and the electric and thermal currents given in Eq. (31.13). It is convenient to invert these equations, which directly yields the four conventional parameters (electrical resistivity $\varrho = 1/\sigma$, thermoelectric power α, Peltier coefficient π, and thermal conductivity κ):

$$\begin{aligned} \mathbf{F}^* &= \varrho\mathbf{j} + \alpha\nabla T \\ \mathbf{w}^* &= \pi\mathbf{j} - \kappa\nabla T, \end{aligned} \tag{31.16}$$

with $\mathbf{F}^* = \mathbf{F} - \nabla\varphi$ and $\mathbf{w}^* = \mathbf{w} - \mathbf{j}\varphi/e$. The coefficients can be obtained by solving the Boltzmann equation for a small perturbation. The results are listed in Table 31.1 (see Conwell, 1982).

One relation between the Peltier coefficient and the thermopower, called the *Kelvin relation*, is often useful:

$$\alpha = \frac{\pi}{T}. \tag{31.17}$$

Table 31.1: Electric and thermoelectric effects

	Given	It results	Coefficient	Name
Homogen. Effects — Electric	Electric field	Current	$\sigma = en\mu = \dfrac{e^2 n}{m^*}\dfrac{\langle E\tau\rangle}{\langle E\rangle}$	Conductivity
Homogen. Effects — Thermoelectric (1. Order effects)	Generation gradient	Photo-emf		Dember effect
	Temperature Gradient	Heat flow	$\kappa = \dfrac{n}{m^* T}\left(\dfrac{\langle (E-E_F)\,E\tau\rangle}{\langle E\rangle} - \dfrac{\langle (E-E_F)^2\,E\tau\rangle}{\langle E\rangle}\right)$ Heat conductivity	Seebeck effect (1. Benedick's effect)
		Thermo-emf	$\alpha = -\dfrac{1}{e}\dfrac{T\langle (E-E_F)\,E\tau\rangle}{\langle E\tau\rangle}$ Thermopower	
Homogen. Effects — Thermoelectric (2. Order effects)	Temperature gradient in semiconductor with neck	Temperature gradient near neck		2. Benedick's effect
	Temperature gradient parallel to electric current	Change of temperature gradient (heat evolution)	$\pi = \dfrac{1}{e}\dfrac{\langle (E-E_F)\,E\tau\rangle}{\langle E\tau\rangle}$ Peltier coefficient	Thomson effect
	Changing temperature gradient	Thermo-emf		Inverse Thomson effect
Inhomogen. Effects — Electric	Optical carrier generation between two materials	Photo-emf		Photovoltaic effect
Inhomogen. Effects — Thermoelectric	Temperature difference between two materials	Thermo-emf	$\Delta V = (\alpha_2 - \alpha_1)\Delta T$ $\alpha_2 - \alpha_1$ Seebeck coefficient	Seebeck effect
	Current through sequence of two materials	Temperature difference between contact points (heat evolution)	$\pi_2 - \pi_1$ Peltier coefficient	Peltier effect

Table 31.2: Galvanomagnetic and thermomagnetic effects

Applied	Transversal Effects — B⊥(j or ∇T)				Longitudinal Effects — B∥(j or ∇T)	
	Result	Name	Result	Name	Result	Name
Galvanomagnetic Effects — Electric field	emf (diagram)	Hall effect $R_H = \dfrac{F_y}{j_x B_z} = \dfrac{1}{ne}$	Change of conductivity (diagram)	Transversal magneto-resistance $\varrho_{xx} = \dfrac{F_x^*}{j_x}$	Change of conductivity (diagram)	Longitudinal magneto-resistance $\varrho_{zz} = \dfrac{F_z^*}{j_z}$
	Temperature difference (diagram)	Ettingshausen effect $P = \dfrac{\nabla_y T}{j_x B_z} = \dfrac{\pi_{yz}}{\kappa_{xx} B_z}$	Longitudinal temperature difference (diagram)	Nernst effect $\nabla_x T = j_x \dfrac{\pi_{zz}}{\kappa_{xx}}$	Longitudinal field temperature difference (diagram)	Longitudinal Nernst effect $\nabla T_z = j_z \dfrac{\pi_{zz}}{\kappa_{zz}}$
Thermomagnetic Effects — Temperature field ∇T	Temperature difference (diagram)	Righi-Leduc effect $S = \dfrac{\nabla_y T}{\nabla_z T \cdot B_z}$	Change of thermal conductivity (diagram)	Maggi-Righi Leduc effect $\kappa_{xx} = \dfrac{w_x^*}{\nabla_z T}$	Change of thermal conductivity (diagram)	Longitudinal Maggi-Righi-Leduc effect $\kappa_{zz} = \dfrac{F_x^*}{\nabla_z T}$
	emf (diagram)	1. Ettingshausen Nerst effect $Q = \dfrac{F_y^*}{\nabla_z T \cdot B_z}$	Longitudinal emf (diagram)	2. Ettingshausen Nernst effect $\alpha_{xx} = \dfrac{w_x^*}{\nabla_z T}$	Longitudinal emf (diagram)	Longitudinal 2. Ettinghausen Nernst effect $\alpha_{zz} = \dfrac{F_z^*}{\nabla_z T}$

Another relation is the *Wiedemann-Franz law*, which holds for metals, i.e, as long as the thermal conductivity is determined by the electron gas alone and the lattice conductivity is negligible:

$$\kappa = L\sigma T \quad \text{with} \quad L = \frac{1}{3}\left(\frac{\pi k}{e}\right)^2 = 2.45 \cdot 10^{-8}\left(\frac{\text{WV}}{\text{AK}^2}\right); \quad (31.18)$$

where L is the Lorentz number. For a comprehensive review, see Beer (1963).

When exposed to a temperature gradient, the electron gas at the hotter end obtains a higher kinetic energy. Therefore, some of these electrons in a "simple metal," i.e., an alkali metal, move preferentially to the cooler end, charging it negatively. The thermoelectric power can be obtained classically by setting equal to each other the currents caused by an electric field and by a thermal gradient, yielding (Drude) $\alpha = c_v^{(e)}/(3ne)$. When replacing the specific heat of the electron gas with $c_v^{(e)} = (\pi^2/2)(kT/E_F)nk$, one obtains

$$\alpha = -\frac{\pi^2}{3}\frac{k}{e}\frac{kT}{E_F}; \quad (31.19)$$

except for a factor of 2 due to insufficient consideration of scattering. With $k/e \simeq 86 \ \mu\text{V/K}$, one has α typically on the order of 1 μV per degree. The Seebeck coefficient α at $T = 300$ K is -8.3, -15.6, -4.4, and $+1.7$, $+11.5$, and $+0.2 \ \mu\text{V/K}$ for Na, K, Pt; Au or Cu, Li, and W, respectively.

For semiconductors, the thermoelectric power is usually much larger, and is approximated by

$$\alpha_n = -\frac{k}{e}\left[r - \ln\left(\frac{N_c}{n}\right)\right] \quad \text{or} \quad \alpha_p = \frac{k}{e}\left[r - \ln\left(\frac{N_v}{p}\right)\right] \quad (31.20)$$

for *n*-type or *p*-type semiconductors, respectively. Here r is a parameter depending on the scattering mechanism:

> $r = 1$ for amorphous semiconductors (Fritzsche, 1979)
> $r = 2$ for acoustic phonon scattering
> $r = 3$ for (polar) optical phonon scattering
> $r = 4$ for ionized impurity scattering, and
> $r = 2.5$ for neutral impurity scattering.

For an ambipolar semiconductor, one obtains

$$\alpha = \frac{\alpha_n\sigma_n + \alpha_p\sigma_p}{\sigma_n + \sigma_p}. \quad (31.21)$$

See, for example, Smith (1952) and Tauc (1954).

31.2.1A The Inhomogeneous Thermoelectric Effect　The thermo-emf is usually measured between two endpoints of a metal wire that is connected to a second metal wire. The two connecting points are kept at different temperatures. The resulting *thermo-emf* is then given by the difference of the thermoelectric power of the two metals, times the temperature difference between the two connecting points

$$\Delta\varphi = (\alpha_2 - \alpha_1)(T_2 - T_1). \qquad (31.22)$$

When measured against a metal with exceptionally small α, e.g., lead, one obtains the value for an *absolute thermoelectric power* which is tabulated. For a review, see Pollock (1985).

31.2.2 Magneto-Electric Effects

With a magnetic induction, the Lorentz force results in a curving of the electron path. When the magnetic induction is small enough, so that between scattering events only a small deviation from the straight path occurs, the superposition of electric field and magnetic induction results in a bending of the electron path independently. This means that in the relaxation-time approach of the Boltzmann equation, two components must be distinguished: from

$$e\left[\mathbf{F} + (\mathbf{v} \times \mathbf{B})\right]\nabla_p f = -\frac{\delta f}{\tau_m} \quad \text{with} \quad \delta f = \delta f_1(\mathbf{F}) + \delta f_2(\mathbf{F}, \mathbf{B}),$$
$$(31.23)$$

one obtains

$$\delta f_1(\mathbf{F}) \simeq -e\,\tau_m F \nabla_p f = \frac{e}{kT}\,\tau_m\,\mathbf{F}\cdot\mathbf{v} f_0 \qquad (31.24)$$

as previously discussed [Eq. (30.18)], and

$$\delta f_2(\mathbf{F}, \mathbf{B}) \simeq -e\,\tau_m(\mathbf{v} \times \mathbf{B})\cdot\nabla_p f = \frac{e^2\tau_m^2}{m_n kT}(\mathbf{v} \times \mathbf{B})\cdot\mathbf{F} f_0. \quad (31.25)$$

From Eq. (31.25) it follows that a magnetic induction parallel to the electric field has no effect ($\delta f_2 = 0$); whereas with a magnetic induction component perpendicular to $\mathbf{F}$, the contribution of δf_2 becomes finite. This contribution determines the Hall effect.

31.2.2A The Isothermal Hall Effect　For definiteness, one assumes $\mathbf{F} = (F_x, 0, 0)$ and $\mathbf{B} = (0, 0, B_z)$ for the relative orientations of the electric field and magnetic induction. The current density is given by

$$\mathbf{j} = \frac{e^2}{kT} \left[\sum_{\mathbf{v}} \tau f_0 \, \mathbf{F} \cdot \mathbf{v} \, \mathbf{v} - \frac{e}{m_n} \sum_{\mathbf{v}} \tau^2 f_0 (\mathbf{v} \times \mathbf{B} \cdot \mathbf{F}) \, \mathbf{v} \right], \qquad (31.26)$$

which has the components

$$j_x = \frac{e^2}{kT} \left[\sum_{\mathbf{v}} \tau f_0 \overline{v_x^2} F_x + \frac{e}{m_n} \sum_{\mathbf{v}} \tau^2 f_0 \overline{v_x^2} B_z F_y \right] = \sigma_{xx} F_x + \sigma_{xy} F_y$$

$$(31.27)$$

$$j_y = \frac{e^2}{kT} \left[\sum_{\mathbf{v}} \tau^2 f_0 \overline{v_y^2} B_z F_x + \sum_{\mathbf{v}} \tau f_0 \overline{v_y^2} F_y \right] = \sigma_{yx} F_x + \sigma_{yy} F_y.$$

$$(31.28)$$

The components of the *magneto-conductivity tensor* are (see Section 30.3.1C)

$$\sigma_{xx} = \sigma_{yy} = \frac{ne^2}{m_n} \frac{\langle E\tau \rangle}{\langle E \rangle} \qquad (31.29)$$

and

$$-\sigma_{yx} = \sigma_{xy} = \frac{ne^2}{m_n} \frac{eB_z}{m_n} \frac{\langle E\tau^2 \rangle}{\langle E \rangle}. \qquad (31.30)$$

A more general expression of the average is used here in terms of the energy distribution function, which is equivalent to the relation $\langle v^2 \tau \rangle / \langle v^2 \rangle$ for quasi-free electrons with $\langle E \rangle = m \langle v^2 \rangle / 2$. When more complex equi-energy surfaces are involved, the anisotropy of the effective carrier mass must be considered.

For a two-dimensional semiconductor, a platelet of the shape shown in Fig. 31.1, the initial **B**-induced current in the y direction causes a charging of the corresponding surfaces until the polarization field forces j_y to vanish. From Eq. (31.28) one obtains

$$\frac{F_y}{F_x} = -\frac{\sigma_{yx}}{\sigma_{xx}} = \frac{e}{m_n} B_z \frac{\langle E\tau^2 \rangle}{\langle E\tau \rangle}. \qquad (31.31)$$

The ratio of the resulting fields determines the *Hall angle*

$$\theta_H = \tan^{-1} \left(\frac{F_y}{F_x} \right) = B_x \frac{e}{m_n} \langle \tau_m \rangle = B_z \mu_H. \qquad (31.32)$$

This permits a direct measurement of the *Hall mobility* μ_H. The subscript H is used to distinguish the Hall mobility from the carrier

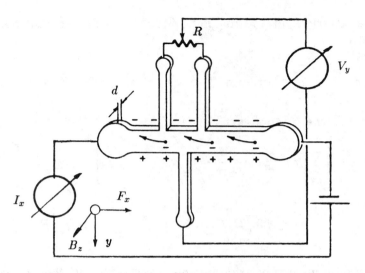

Figure 31.1: Experimental set-up for Hall effect measurement in a long two-dimensional sample. The Hall angle is determined by a setting of the rheostat R, which renders the current component $j_y = 0$. The Hall voltage is given by $V_y = R_H I_x B_z / d$, and is used to determine the Hall constant R_H.

mobility, which is usually slightly smaller than μ_H, namely $\mu_H/\mu = \langle E\tau^2 \rangle \langle E \rangle / \langle E\tau \rangle \simeq 3\pi/8$ for acoustic mode scattering, $\mu_H/\mu \simeq 1.7$ for ionized impurity scattering, and $\mu_H/\mu \simeq 1$ for higher defect densities and temperatures (Mansfield, 1956).

Often the *Hall constant* R_H is used instead, and is defined as

$$R_H = \frac{F_y}{j_x B_z} = -\frac{1}{B_z}\frac{\sigma_{yx}}{\sigma_{xx}\sigma_{yy}} = \frac{1}{en}\frac{\langle E\tau^2 \rangle \langle E \rangle}{\langle E\tau \rangle^2}. \tag{31.33}$$

The Hall constant is $\simeq 1/(en)$, except for a numerical factor that depends on the scattering mechanism and is on the order of 1—see the above remarks on μ_H/μ.

For ellipsoidal equi-energy surfaces, the Hall constant is given by (Herring, 1955)

$$R_H = \frac{1}{ne}\frac{\langle E\tau^2 \rangle \langle E \rangle}{\langle E\tau \rangle}\frac{3\left(\dfrac{1}{m_x m_y} + \dfrac{1}{m_y m_z} + \dfrac{1}{m_z m_x}\right)}{\left(\dfrac{1}{m_x} + \dfrac{1}{m_y} + \dfrac{1}{m_z}\right)^2}. \tag{31.34}$$

When electrons and holes are present in comparable densities (compensated semiconductors) or two types of carriers (electrons or holes in different bands, or polarons) are present, both types contribute to the Hall constant:

$$R_H = \frac{n_1 e_1 \mu_1 \mu_{H1} + n_2 e_2 \mu_2 \mu_{H2}}{(n_1 e_1 \mu_1 + n_2 e_2 \mu_2)^2}. \tag{31.35}$$

With $(e_1, e_2) = (-e, +e)$ for electrons and holes, respectively, the sign of the Hall constant indicates the *type* of majority carrier: it is negative for n-type and positive for p-type conduction. Here, the signs of e and μ are carried in accordance with the sign convention— see Section 28.2.1.

31.2.2B Transverse Magneto-Resistance For higher magnetic induction, one can no longer ignore second-order terms ($\propto B^2$). These terms cause a reduction in the conductivity with increased magnetic induction. This results from the fact that the Hall field compensates only for the deflection of electrons with average velocity, while slower or faster electrons of the distribution are more or less deflected, resulting in a less favorable path average for the carrier conductivity. Scattering itself, however, is not influenced by magnetic induction. This *magneto-resistance* effect, discovered by W. Thomson (1856), yields information about the anisotropy of the effective mass (Glicksman, 1958).

For a quantitative evaluation, an alternative method to the evaluation of the Boltzmann equation will be used (Seeger, 1973). It is based on the equation of motion for quasi-free electrons (Brooks, 1955):

$$m_n \frac{d\mathbf{v}}{dt} = e(\mathbf{F} + \mathbf{v} \times \mathbf{B}). \tag{31.36}$$

With $\mathbf{B} = (0, 0, B_z)$ in Eq. (31.36), one has two components

$$\left.\begin{aligned} \frac{dv_x}{dt} &= \frac{eF_x}{m_n} + \omega_c v_y \\ \frac{dv_y}{dt} &= \frac{eF_y}{m_n} - \omega_c v_x \end{aligned}\right\} \quad \text{with } \omega_c = \frac{eB_z}{m_n}, \tag{31.37}$$

where ω_c is the cyclotron frequency. These two components can be discussed in a complex plane:

$$v = v_x + iv_y \quad \text{and} \quad F = F_x + iF_y, \tag{31.38}$$

yielding from Eq. (31.37)

$$\frac{dv}{dt} = \frac{eF}{m_n} - i\omega_c v. \tag{31.39}$$

This equation can be integrated after both sides are multiplied by $\exp(i\omega_c t)$, yielding for the drift velocity (McKelvey, 1966)

$$v_D = v_0 \exp(-i\omega_c t) + \frac{eF}{i\omega_c m_n}\{1 - \exp(-i\omega_c t)\}, \qquad (31.40)$$

which shows oscillatory behavior. Scattering, however, interferes so that only a fraction of a cycle is completed for $\omega_c \tau_m < 1$. Considering a distribution of relaxation times, one obtains for the average drift velocity

$$\langle v_D \rangle = \frac{\displaystyle\int_0^\infty v_D(t) \exp\left(-\frac{t}{\tau_m}\right) dt}{\displaystyle\int_0^\infty \exp\left(-\frac{t}{\tau_m}\right) dt} = \frac{1}{1 + i\omega_c \tau_m}\left[v_0 + \frac{eF\tau_m}{m_n}\right]. \qquad (31.41)$$

The first term (v_0) of Eq. (31.41) drops out when averaging over all angles. Separating the real and imaginary parts of v and F, one obtains

$$v_x = \frac{e}{m_n}\left\{\left(\tau_m - \omega_c^2 \frac{\tau_m^3}{1 + \omega_c^2 \tau_m^2}\right) F_x + \omega_c \frac{\tau_m^2}{1 + \omega_c^2 \tau_m^2} F_y\right\}$$
$$v_y = \frac{e}{m_n}\left\{-\omega_c \frac{\tau_m^2}{1 + \omega_c^2 \tau_m^2} F_x + \frac{\tau_m}{1 + \omega_c^2 \tau_m^2} F_y\right\}. \qquad (31.42)$$

For the current densities, $j = en\bar{v}_D$ in the x and y directions, one must average these velocities, yielding

$$j_x = en\frac{e}{m_n}\left\{\left[\langle\tau_m\rangle - \omega_c^2 \left\langle\frac{\tau_m^3}{1 + \omega_c^2 \tau_m^2}\right\rangle\right] F_x + \omega_c \left\langle\frac{\tau_m^2}{1 + \omega_c^2 \tau_m^2}\right\rangle F_y\right\} \qquad (31.43)$$

and

$$j_y = en\frac{e}{m_n}\left\{-\omega_c \left\langle\frac{\tau_m^2}{1 + \omega_c^2 \tau_m^2}\right\rangle F_x + \left\langle\frac{\tau_m}{1 + \omega_c^2 \tau_m^2}\right\rangle F_y\right\}. \qquad (31.44)$$

For $\omega_c \tau_m \ll 1$, which is generally fulfilled, one can neglect the frequency dependence in the denominators. With $j_y = 0$, one then obtains from Eqs. (31.43) and (31.44) by eliminating F_y

$$j_x = en\frac{e}{m_n}\langle\tau_m\rangle F_x \left\{1 - \frac{e^2 B_z^2}{m_n^2} \frac{\langle\tau_m^3\rangle\langle\tau_m\rangle - \langle\tau_m^2\rangle^2}{\langle\tau_m\rangle^2}\right\}. \qquad (31.45)$$

This expression contains a second-order term that causes a decrease of the current j_x with increasing magnetic induction. With $\varrho = 1/\sigma$

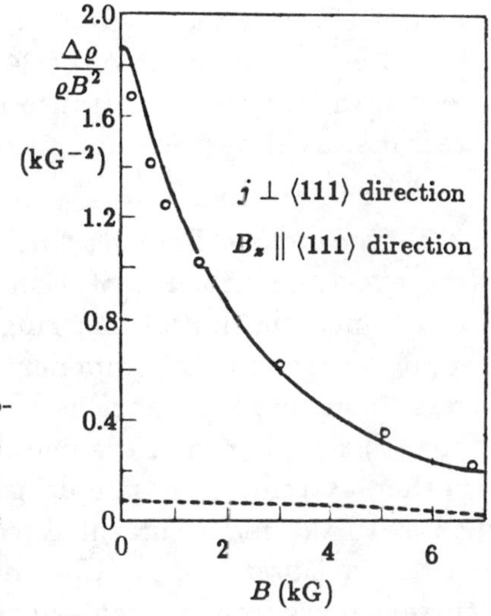

Figure 31.2: Transverse magneto-resistance of p-Ge at 205 K as observed (solid curve) and calculated for a single carrier heavy hole model (dashed curve) (after Harman et al., 1954).

and $e\langle\tau_m\rangle/m_n = \mu_n$, and thus $j_x = \sigma F_x\left\{1 - f(B_z^2)\right\}$, one obtains for the *magneto-resistance coefficient*

$$\frac{\Delta\varrho}{\varrho B_z^2} = \mu_n^2 \frac{\langle\tau_m^3\rangle\langle\tau_m\rangle - \langle\tau_m^2\rangle^2}{\langle\tau_m\rangle^4};\qquad (31.46)$$

that is, the coefficient is essentially equal to μ_n^2 except for the term containing the relaxation-time averages. This term represents a numerical factor that depends on the scattering mechanism, and lies between 0.38 and 2.15 (Seeger, 1973).

The case of magneto-resistance with two carriers is straightforward (McKelvey, 1966), and is additive for both carriers, even though they may be of opposite sign. The case of nonspherical equi-energy surfaces is rather involved, and is summarized by Conwell (1982); see also Beer (1963).

As an example, the magneto-resistance of a two-carrier semiconductor, p-type Ge, is given in Fig. 31.2. The two carrier types are light (p_l, μ_l) and heavy (p_h, μ_h) holes. Predominant carrier scattering is assumed to be due to acoustic phonons. The magneto-resistance coefficient is given by (Seeger, 1973)

$$\frac{\Delta\varrho}{\varrho B_z^2} = \mu_h^2 \cdot \frac{9\pi}{16}\left\{\frac{1 + \eta\beta^3}{1 + \eta\beta} - \frac{\pi}{4}\left[\frac{1 + \eta\beta^2}{1 + \eta\beta}\right]^2\right\},\qquad (31.47)$$

with $\eta = p_l/p_h$ and $\beta = \mu_l/\mu_h$. Although in p-Ge at 205 K only 4% of the holes are in the light hole band ($\eta = 0.04$), the large ratio of the effective mass $(\beta = 8)$ renders the numerical factor in Eq. (31.47) greater by a factor of 24 than for a single carrier model. As a result, the magneto-resistance is substantially enhanced by the light carriers, as shown in Fig. 31.2—compare the solid curve with the dashed curve.

31.2.2C Geometry Factors in Galvanomagnetic Effects
The Hall effect, described in Section 31.2.2A, depends on the sample geometry, since the initial charging of the sample surfaces in the y direction causes the y component of the current to vanish. The transverse magneto-resistance is usually determined in a long (x) and thin (y) sample rod or filament.

Another extreme is a sample geometry in which surfaces perpendicular to the main current direction do not exist. This can be achieved in a *Corbino disk* (Corbino, 1911), shown in Fig. 31.3A. The current flows from a circular hole in the center to its circumference. The magneto-resistance is maximized since no compensation of any curved electron path is possible. The change of resistance of the Corbino disk is given by

$$\frac{\Delta R}{R} = \left(\frac{\Delta \varrho}{\varrho}\right)_f + \mu_H^2 \frac{B_z^2}{1 + (\Delta \varrho/\varrho)_f}, \qquad (31.48)$$

where μ_H is the Hall mobility, and $(\Delta \varrho/\varrho)_f$ is the magneto-resistance change in a filament-type sample of the same material and at the same magnetic induction. Figure 31.3B presents the magneto-resistances of samples of different shapes; a thin filament sample shows the smallest effect.

31.2.3 Cyclotron Resonance
When the magnetic field is strong enough, and the mean free path is long enough for carriers to complete cyclic paths in the applied magnetic induction, strong resonances in an oscillating electromagnetic probing field are observed at the cyclotron frequency [Eq. (9.19)]:

$$\omega_c = \frac{eB}{m^*} = 17.84 \frac{m_0}{m^*} B \text{ (GHz/kG)}. \qquad (31.49)$$

For the derivation of resonance conditions, see McKelvey (1966). The cyclotron-resonance line width decreases rapidly the more cycles are completed before scattering occurs. Scattering, with its limiting relaxation time, acts as the damping parameter in the resonance equation [Eq. (31.41)], and with $\tau_m = 1/\gamma$ in Eq. (20.36) determines the resulting line shape.

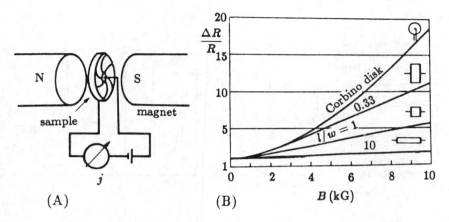

Figure 31.3: (A) Corbino disk with indicated current flow from the central electrode to the circumference electrode. (B) Relative change in resistance of n-InSb as a function of the magnetic induction for different sample geometries; length with (l/w) ratio indicated (after Welker and Weiss, 1954).

Cyclotron-resonance measurements are well suited for determining the effective mass in different crystallographic directions. This was discussed in Section 9.1.5D.

31.3 Carrier Quantum Effects in Magnetic Field

When the magnetic induction becomes large enough so that $\hbar\omega_c$ is no longer $\ll kT$, quantum-mechanical effects must be considered, i.e., splitting into Landau levels. The influence of a strong magnetic induction will be discussed in two steps: excluding scattering to obtain information on changes in the density of states, and, as in the following section, including scattering.

31.3.1 Quasi-Free Carriers in a Strong Magnetic Field

Assuming that the magnetic induction acts in the z direction ($\mathbf{B} = 0, 0, B_z$), the electron motion is described by the Schrödinger equation (Landau, 1930):

$$\left\{ -\frac{\hbar^2}{2m_n}\left(\frac{\partial^2}{\partial x^2} + \frac{\partial^2}{\partial y^2}\right) + \frac{m_n\omega_c^2}{8}(x^2 + y^2) - i\hbar\omega_c\left[x\frac{\partial}{\partial y} + y\frac{\partial}{\partial x}\right]\right\}\psi$$
$$= E\psi,$$

$$(31.50)$$

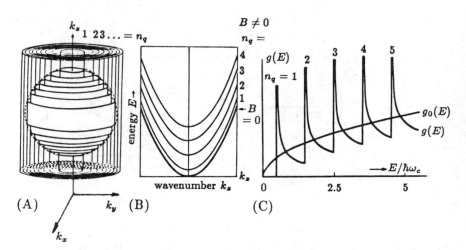

Figure 31.4: (A) Constant energy surfaces in k space for a given magnetic induction, resulting in concentric cylinders for each of the Landau levels. (B) $E(k_z)$ dispersion relation for zero-magnetic induction (lowest curve) and for a constant magnetic induction showing the split into a sequence of Landau bands. (C) Density of state without magnetic induction $[g_0(E)]$ and with magnetic induction $[g(E)]$ including the first five Landau levels and neglecting lifetime broadening. The total areas under both curves up to any given E are the same.

which has two additional terms caused by the magnetic induction, and depends on the *cyclotron frequency* ω_c. These terms impose a constraint on the electron motion in the xy plane due to the Lorentz force. The electron motion is given semiclassically by

$$x = x_0 + \sqrt{\frac{2\hbar}{m_n\omega_c}\left(n_q + \frac{1}{2}\right)}\,\cos(\omega_c t) = x_0 + r_{nq}\cos(\omega_c t)$$

$$y = y_0 + \sqrt{\frac{2\hbar}{m_n\omega_c}\left(n_q + \frac{1}{2}\right)}\,\sin(\omega_c t) = y_0 + r_{nq}\sin(\omega_c t),$$

$$(31.51)$$

which are circles with a radius r_{nq}, determined by the magnetic induction and the quantum number n_q. In k space, one consequently obtains, using $k_x = (m_n/\hbar)(dx/dt)$ and $k_y = (m_n/\hbar)(dy/dt)$,

$$k_x^2 + k_y^2 = k_{nq}^2 = \frac{2m_n\omega_c^2}{\hbar}\left(n_q + \frac{1}{2}\right),\qquad (31.52)$$

which is a set of cylinder surfaces determined by n_q with a radius proportional to B_z, as shown in Fig. 31.4A.

The energy of the electrons on these surfaces, obtained as eigenvalues of the Schrödinger equation (31.50), is given by

$$E = \frac{\hbar^2}{2m_n} k_z^2 + \left(n_q + \frac{1}{2} \right) \hbar\omega_c. \qquad (31.53)$$

In anisotropic semiconductors, or those with anisotropic effective mass, the relative direction of the magnetic field and the crystal orientation must be considered, and are included in the cyclotron frequency, as given in Eq. (9.20).

This shows that the application of a strong magnetic induction substantially changes the behavior of Bloch electrons from being quasi-free to being confined in the xy plane. It results in a splitting into magnetic subbands* or *Landau levels* at a given magnetic field according to the quantum number; whereas in the k_z direction, although the $E \propto k_z^2$ relation known for free electrons holds, it is offset by steps of the height of the cyclotron energy.

31.3.2 Diamagnetic and Paramagnetic Electron Resonance

The interactions between free electrons and a magnetic field due to the Lorentz force, leading to cyclotron resonances, are *diamagnetic interactions*. With a sufficient density of **free electrons**, the semiconductor becomes diamagnetic, i.e., its magnetic moment becomes negative: an oblong probe of the semiconductor suspended from a filament to permit free rotation, turns perpendicular to the magnetic flux. The induced magnetic momentum opposes its inducing force.

When including the electron spin in this discussion, one must consider an additional *paramagnetic interaction*. This interaction produces a positive contribution to the magnetic moment.

The eigenvalues of the Schrödinger equation, including spin interaction, (last term) are

$$E = \frac{\hbar^2}{2m_n} k_z^2 + \left(n_q + \frac{1}{2} \right) \hbar\omega_c \pm \frac{1}{2} g \mu_B B \qquad (31.54)$$

with − or + dependent on parallel or antiparallel spin, respectively. $\mu_B = e\hbar/(2m^*)$ is the *Bohr magneton*, and g is the *Landé g-factor*. For free electrons *in vacuo*, the cyclotron frequency is $\omega_c = eB/m_0$, which can also be expressed with the Bohr magneton $\mu_B = e\hbar/(2m_0)$

---

* In the k_z direction there are subbands; in the k_x and k_y directions, there are discrete levels in $E(k)$.

Table 31.3: Landé g-factor for selected semiconductors* (after Roth and Lax, 1959).

Material	g_c	Material	g_c	g_v^{so}	Material	g_c
Si	1.9989	GaSb	-7.68	-6.2	InSb	-50.6
Ge	-3.0	InP	1.48	-1.9	ZnSe	1.12
GaAs	-0.44	InAs	-15.6		CdTe	1.59

Material	$g_{c\parallel}$	$g_{c\perp}$	$g_{v\parallel}^{A}$	$g_{v\perp}^{A}$	$g_{v\parallel}^{B}$	$g_{v\perp}^{B}$
ZnS	2.2	1.9				
ZnTe	$-0.4\,[100]$	$-0.35\,[110]$				
CdS	1.774	1.787	1.25	0	1.8	1.8
CdSe	0.6	0.51	1.41			

*Indices c and v for conduction and valence bands; indices A and B for the respective valence bands; $\parallel$ and $\perp$ with respect to the c axis.

as $\omega_c = 2\mu_B B/\hbar$. The corresponding frequency in an atom is the *Larmor frequency* $\omega_L = g\mu_B B/\hbar$, which is equal to ω_c for $g = 2$. For electrons orbiting within a semiconductor, g can deviate substantially from 2, depending on the effective mass and the spin-orbit splitting energy Δ_0 (Lax et al., 1959):

$$g \simeq 2\left\{1 + \frac{m_n - m_0}{m_n}\left(\frac{\Delta_0}{E_g + 2\Delta_0}\right)\right\}, \qquad (31.55)$$

and may even become negative ($g \simeq -50$ for InSb). Since the effective mass has tensor properties, g is also a tensor. Some values of g are listed in Table 31.3—see also Roth and Lax (1959). Figure 31.5 illustrates the additional splitting of the Landau levels due to the spin.

With a sufficient density of **impurities** with uncompensated spins, one can observe resonant absorption when flipping the spin by electromagnetic radiation of a frequency equal to the difference between states with parallel and antiparallel spin. This is an *electron paramagnetic resonance* (EPR), also called *electron spin resonance* (ESR) and occurs at

$$\hbar\omega_s = g\mu_B B. \qquad (31.56)$$

These resonances provide information about the density of uncompensated spins through the strength of the resonance, and about the g factor through its resonance frequency.

band states   Landau levels   spin splitting

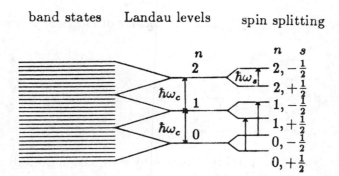

Figure 31.5: Splitting of band states into Landau levels, considering diamagnetic interaction by orbiting electrons and, in addition, paramagnetic interaction with the electron spin resulting in further splitting.

The resonance absorption can be measured directly by interacting with an electromagnetic field of appropriate frequency (ω_s) or optically by observing changes in the intensity or polarization of laser-excited luminescence—see review by Cavenett (1981).

Further information about **defect centers** can be obtained from the paramagnetic interaction with nuclear spins, which can be measured by inducing spin-flipping by absorption of electromagnetic radiation (*nuclear spin resonance*)—see Section 21.3.4B and Slichter (1963). There is a wide variety of interactions involving the nuclear spin of defects that can be used for analyzing certain defect properties—see Bagraev and Mashkov (1986) and Sections 21.3.4B–21.3.4D.

In addition to the paramagnetic interaction of electron spins, the *Pauli spin paramagnetism* at an impurity center, there is the diamagnetic part due to the orbital quantization: for bound electrons, this is the *Landau diamagnetism* that is 1/3 the magnitude of the Pauli contribution. For more detail, see Wilson (1954).

31.3.2A Density of States in Magnetic Fields The modified $E(\mathbf{k})$ relation described in the previous section is shown in Fig. 31.4A and B with an applied magnetic field. The density of states depends on the Landau quantum number, and is given for each subband by

$$g(k_z, n_q)\, dk_z = \frac{2}{(2\pi)^2} \frac{m_n \omega_c}{\hbar}\, dk_z. \qquad (31.57)$$

Using Eq. (31.53), one obtains the density of states as a function of the energy:

$$g(E, n_q) \, dE = \frac{1}{(2\pi)^2} \left(\frac{2m_n}{\hbar^2} \right)^{3/2} \frac{\hbar\omega_c}{\sqrt{E - \left(n_q + \frac{1}{2} \right) \hbar\omega_c}} \, dE. \quad (31.58)$$

The total density of states is obtained by summation over all possible quantum numbers n_q within the band: near the bottom of it, lifted to $E_c + \hbar\omega_c/2$, for $E_c + \hbar\omega_c/2 < E < E_c + 3\hbar\omega_c/2$ with a summation over only one subband, for $E_c + 3\hbar\omega_c/2 < E < E_c + 5\hbar\omega_c/2$ over two, and so on for the conduction band. This density-of-state distribution is compared in Fig. 31.4C with the undisturbed distribution for vanishing magnetic induction (g_0).

31.3.2B DeHaas-Type Effects In a metal or a degenerate semiconductor, the Fermi level lies within the conduction band. With increasing magnetic field, the spacing of the Landau levels increases and causes one after the other of these levels to cross E_F, thereby periodically changing the density of states (Fig. 31.4C) at E_F. This in turn causes the amplitude of certain properties which are determined near the Fermi surface to change periodically. These include the magnetic susceptibility, resulting in the *DeHaas-van Alphen effect*, shown in Fig. 31.6A, and the electrical conductivity, resulting in the *Shubnikov-DeHaas effect*, shown in Fig. 31.6B.

In actual materials, only rarely are the orbits at a spherical Fermi surface. In semiconductors, the orbits may be elliptical (in sidevalleys); in metals, they can have quite complicated shapes (see Sections 9.1.4A and 9.1.5C) that make the oscillations more complex (Fig. 31.6A), and in turn provide information about the shapes of the Fermi surfaces (first suggested by Onsager, 1952; see also Shoenberg, 1969; Ziman, 1972). As an example, the ratio of the wavelength of oscillation, shown in Fig. 31.6A, gives the ratio of the area of belly and neck (here 9)—see also Section 9.1.4A.

31.3.2C Magneto-Phonon Effects When the spacing between the Landau levels coincides with the energy of longitudinal optical phonons,

$$\hbar\omega_{\mathrm{LO}} = n_q \hbar\omega_c, \quad (31.59)$$

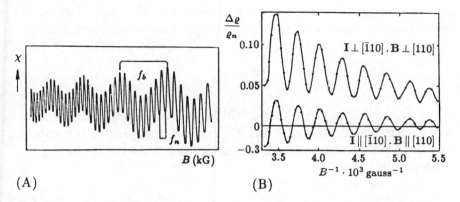

Figure 31.6: (A) DeHaas-van Alphen (1930) oscillation of magnetic susceptibility in silver with neck (high-frequency) and belly (low-frequency) oscillations when a Landau level passes through the Fermi surface (after Joseph and Thorsen, 1965). (B) Shubnikov-DeHaas (1930) oscillations of the relative resistivity as a function of the magnetic field in GaSb at 4.2 K. Hall coefficient $R_H = -4.8$ cm^3/As (after Becker and Fan, 1964).

electrons can be transferred more easily by scattering with these phonons between different Landau levels. This causes a more pronounced change in magneto-resistance, with a period length

$$\Delta \left(\frac{1}{B} \right) = \frac{e}{m_n \omega_{\mathrm{LO}}}, \tag{31.60}$$

which was first observed by Firsov et al. (1964) in InSb. The effect is small but observable at an intermediate range of the magnetic induction [Eq. (31.59)], temperature (to have sufficient optical phonons), and doping (the effect is sensitive to changes in scattering—Gurevich and Firsov, 1964). It can be used to obtain information about the effective carrier mass.

When higher electric fields are applied in addition to the magnetic field, carrier heating takes place, and distinct multiphonon transitions can be observed—see the review by Stradling (1984).

31.3.3 The Quantized Hall Effect

The quantized Hall effect requires a two-dimensional electron system at high magnetic fields and low temperatures. Such a two-dimensional sample may be formed by a quantum well structure. It requires the third dimension to be $\lesssim 100$ Å. To facilitate an understanding of this effect, we will first review the electron motion in orthogonal electric (F_x) and magnetic (B_z) fields. Without scattering, electrons move in circles when exposed to Lorentz forces $(e\,\mathbf{v} \times \mathbf{B})$, with the radius, frequency, and energy given by

$$r_c = \frac{m_n v}{eB_z}, \qquad \omega_c = \frac{eB_z}{m_n}, \qquad E = \frac{m_n}{2}\omega_c^2 r_c^2. \qquad (31.61)$$

With the addition of an electric field in the x direction, the electrons move perpendicularly to B_z and F_x in the y direction with constant velocity* of the center of each circle, forming *trochoids* (a flat spiral) as shown in Fig. 31.7:

$$v_y = \frac{F_x}{B_z}, \quad \text{or} \quad \sigma_{xy} = \frac{en}{B_z} \qquad (31.62)$$

while

$$v_x = 0 \quad \text{or} \quad \sigma_{xx} = 0. \qquad (31.63)$$

With substantial scattering, σ_{xy} decreases and σ_{xx} increases; the former is responsible for the Hall voltage, the latter for the magneto-resistance.

When the magnetic field is large enough to cause significant Landau-level splitting $(\hbar\omega_c \gtrsim kT)$, one has, instead of a continuum of states in the band, a set of discrete energy levels at

$$E_{nq} = E_c + \left(n_q + \frac{1}{2}\right)\hbar\omega_c \quad \text{with } n_q = 0, 1, 2, \ldots \qquad (31.64)$$

and with radii

$$r_{nq} = \sqrt{\frac{2\hbar}{eB_z}\left(n_q + \frac{1}{2}\right)}. \qquad (31.65)$$

Each of these Landau levels is degenerate, permitting occupation by

$$\tilde{n} = \frac{1}{2\pi r_{nq=0}^2} = \frac{eB_z}{h} \quad (\text{cm}^{-2}) \qquad (31.66)$$

electrons, which provides the highest packing without overlap for $n_q = 0$ of cyclotron orbits within the plane. Of n electrons in the conduction band, only $\tilde{n}$ can fill the first Landau level; then $\tilde{n}$ will fill the second level and so on, until all electrons are distributed, with the highest level at $T = 0$ partially filled, as long as n is not accidentally an integer multiple of $\tilde{n}$.

---

* In contrast to the case of vanishing magnetic induction where the motion proceeds in the x direction and, without scattering, is accelerated.

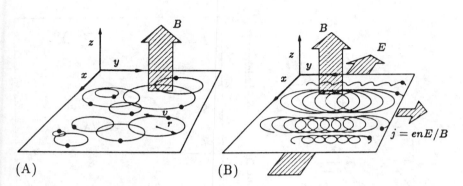

Figure 31.7: (A) Two-dimensional electron gas with magnetic field only. (B) Movement of these electrons in the y direction when additional electric field acts in the x direction.

Scattering can only occur for electrons in the highest Landau level, and only if this level is incompletely filled or $kT \ll \hbar\omega_c$. Consequently, the electron ensemble will follow unperturbed trochoids as shown in Fig. 31.7B.

The filling of the Landau levels can be done by electron injection* with increasing bias at constant magnetic induction, or by constant bias and increasing magnetic induction, which results in fewer Landau levels below E_F. This causes carrier rearrangement whenever a Landau level passes over the Fermi level, which tends to increase E_F and thereby changes the injection.

When the Fermi level coincides with a Landau level, the magneto-resistivity $\varrho_{xx} = 1/\sigma_{xx}$ vanishes, and the Hall resistance $\varrho_{xy} = 1/\sigma_{xy}$ shows a pronounced step (von Klitzing et al., 1980). This is measured by the voltage drop between the Hall probes of a two-dimensional sample—Fig. 31.8A. The results are shown in Fig. 31.8B.

From Eqs. (31.62) and (31.66), one can eliminate the incremental electron density per Landau step, and obtains a step distance of

$$\frac{1}{\Delta\sigma_{xy}} = \frac{B_z}{e\tilde{n}} = \frac{h}{e^2} = 25\,812.8\ \Omega. \qquad (31.67)$$

It is remarkable that the measured Hall resistance steps $\varrho_{xy} = V_H/I$ are not influenced by layer geometry, defects, or by the carrier

---

* Electron injection relates to electrode properties not discussed in this book. It provides an experimental means of increasing the carrier density by simply increasing the bias, thereby injecting more carriers from an appropriate electrode. For a review, see Rose (1978).

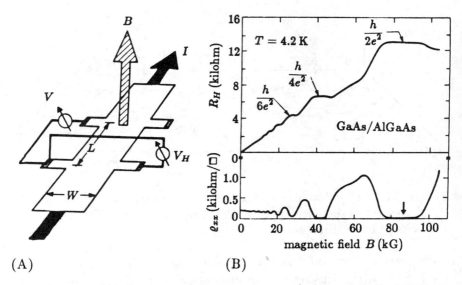

(A)   (B)

Figure 31.8: (A) Experimental set-up to measure the quantized Hall effect. (B) Measured Hall resistance and magneto-resistance in a GaAs/Al$_\xi$-Ga$_{1-\xi}$As quantum well at 4.2 K as a function of the magnetic induction, with Landau level steps at $n_q = 2$, 4, and 6 indicated (after von Klitzing et al., 1980).

effective mass. They are measured precisely to within 1 part in 10^8, and can be used to define an absolute standard of resistance, or to update the value of $\hbar/e^2$, or the fine structure constant* $\alpha = \hbar c/e^2$ (Tsui et al., 1982; von Klitzing, 1981, 1986).

The rounding of the step edges and the step width is due to localized defect states in the gap between the Landau levels, and the broadening of the Landau levels into narrow Landau bands. The persistence of the plateau for a substantial width is more difficult to understand (Störmer and Tsui, 1983).

31.3.3A Fractional Quantum Hall Effect

In high-quality GaAs/Al$_{1-\xi}$Ga$_\xi$As quantum well samples at high magnetic induction and low temperatures, one can reach the quantum limit, the first Landau level. When T is decreased below 4 K, one observes that in addition to the integer n_q steps, fractional quantum numbers appear. The first is $n_q = 1/3$, which is identified at 0.09 K to better than 3 parts in 10^5 (Chang et al., 1984). Recently, more and more

---

* The velocity of light contained in α is the best known of the three constants.

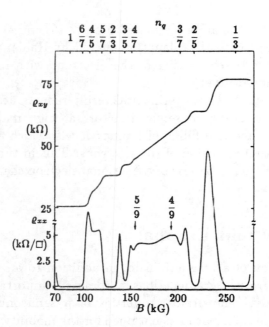

Figure 31.9: Fractional quantum Hall effect in GaAs/$Al_{1-\xi}Ga_\xi$As quantum well structure at 0.09 K: curve 1 Hall resistance, and curve 2 magnetoresistance in the F_x direction (after Chang et al., 1984).

fractional steps have been identified by plateaus in the Hall resistance and are listed below (Tsui and Störmer, 1986):

$$n_q = \frac{1}{3}, \frac{2}{3}, \frac{4}{3}, \frac{5}{3}, \left(\frac{7}{3}\right), \left(\frac{8}{3}\right)$$

$$n_q = \left(\frac{1}{5}\right), \frac{2}{5}, \frac{3}{5}, \left(\frac{4}{5}\right), \left(\frac{6}{5}\right), \frac{7}{5}$$

$$n_q = \left(\frac{2}{7}\right), \frac{3}{7}, \frac{4}{7}, \left(\frac{9}{7}\right), \left(\frac{10}{7}\right), \left(\frac{11}{7}\right)$$

$$n_q = \left(\frac{4}{9}\right), \left(\frac{5}{9}\right), \left(\frac{13}{9}\right).$$

The numbers in parentheses, however, are obtained from only a slight dip of the magneto-resistance. Some of these steps are shown in Fig. 31.9.

The reason for such fractional numbers is not completely understood. It is remarkable, however, that for $n_q = p/q$, q must be odd (3, 5, 7, 9) while p is a positive integer (1 ... 13). With one exception, however: $n_q = 5/2$ was recently observed at a GaAs/GaAlAs quantum well of exceptional perfection with $\mu_n > 10^6$ cm^2/Vs (Willett et al., 1987). A theoretical explanation offered by Haldane and Rezayi (1988) indicates boson pairing of the electrons with mixed spin in the second Landau level.

It is likely that this experimental result reflects new, yet unexplained electron states in semiconductors at high magnetic fields, possibly with fractional filling of Landau levels, which are stabilized by strong mutual repulsion of the electrons. This in turn can be described with pseudo-particles of a fractional electron charge (*Laughlin states*).

Summary and Emphasis

The application of a magnetic field, in addition to an electric field, or of a temperature gradient, has become an important tool with which to analyze numerous properties of a semiconductor. The most significant information is obtained for the mobility (Hall-effect), effective mass (cyclotron resonance), the origin of levels in a complex spectrum (spin-flip resonance), and the type of conductivity (Hall-effect or thermo-emf.). In addition, the quantum-Hall effect provides a means for high-precision measurement of elementary constants, and offers new insight into the quantum effects of a confined two-dimensional electron gas.

In contrast to the carrier transport in an electric field, the phenomena caused by magnetic fields almost exclusively require a high magnetic induction. At the low end of it, the quasi-free carrier transport can be investigated, and at still higher magnetic induction, carrier confinement into Landau levels or Landau bands provides additional means for analyses.

The addition of magnetic fields or thermal gradients provide opportunities for novel devices, in addition to the classical list, including Hall-effect magnetometers or thermocouples.

Exercise Problems

1.   In a semiconductor, the electron gas can be described similarly to a normal gas with its pressure and temperature related to each other by thermodynamics. Try to explain some of the thermo-electric effects listed in Table 31.1 with this model.

2.(e)   Calculate the Hall constant for Ge as a function of the electron density in the range between $n = 10^2$ and $n = 10^{18}$ cm^{-3}. Discuss the obtained curve.

3.   Plot schematically $E(k_x)$ and $E(k_y)$, which is shown for $E(k_z)$ in Fig. 31.4, and discuss the behavior of $E(\mathbf{k})$.

4.(r)   Explain the superstructure of the DeHaas-van Alphen oscillations shown in Fig. 31.6 with regard to the Fermi-surface in Ag.

5.(e)   In Fig. 31.7, circles for electrons with different radii are shown for the same magnetic induction. What can you conclude about the electron gas? Can you estimate its temperature?

6.(*)   Explain quantitatively, as best you can, the experimental results given in Fig. 31.8.

7.   Derive the density of states in the conduction band with a magnetic field, which is given by

$$g(E) = \frac{1}{4\pi^2} \left(\frac{2m_n}{\hbar^2} \right)^{3/2} \sum_{n_q=0}^{n_{qmax}} \frac{\hbar\omega_c}{\left(E - E_c - (n_q + \frac{1}{2})\hbar\omega_c \right)^{1/2}},$$

$$(31.68)$$

and show that the sum in Eq. (31.68) can be replaced with an integral between 0 and ∞, yielding the conventional density of states for vanishing magnetic induction.

8.(e)   Calculate the Hall constant for p-type Si in a two-carrier model by considering light and heavy hole bands, and assuming equilibrium distribution between both bands at 300 K. Here, neglect band warping.

Chapter 32

Carrier Scattering at Low Fields

Carrier scattering acts as a damping process for carrier motion. Both elastic and inelastic scattering influence the carrier transport and involve a large variety of scattering centers. Scattering determines the relaxation time, and with it, the carrier mobility.

Scattering at low fields will be discussed first. Here the deformation of the distribution function is very small and linearization, discussed in Section 30.3, is appropriate. The goal of this discussion is the estimation of the relaxation time, which is a measure of the carrier mobility.

When carriers are accelerated in an external electric field, their increased momenta and energies relax according to a multitude of scattering events. It is impossible to account for these events in a global fashion, and various approximations, which are different for different types of scattering centers, are required. We will first enumerate these centers, then will provide step by step some estimates of the various relaxation mechanisms, and will give the corresponding relation for the carrier mobilities.

32.1 Types of Scattering Centers

Any sufficiently large deviation from an ideal lattice periodicity can act as a scattering center. The more important ones are:

(1) Intrinsic lattice defects
    (a) phonons
        (i) acoustic, with deformation potential or piezoelectric interaction
        (ii) optical, with deformation potential (nonpolar) or polar interaction
    (b) intrinsic point defects (interstitials, vacancies, substitutionals, and antisite defects)

(c) alloys (statistical distribution of lattice atoms)
(2) Extrinsic point defects
    (a) neutral impurities
    (b) charged impurities
(3) Line defects (dislocations)
(4) Surface defects
    (a) grain boundaries
    (b) outer surfaces
    (c) hetero-interfaces
    (d) metal/semiconductor (electrode) boundaries
(5) Three-dimensional defects
    (a) small atomic clusters
    (b) microcrystalline, or colloidal, inclusions
(6) Secondary defects
    (a) electron-electron scattering
    (b) electron-hole scattering
    (c) electron-plasmon scattering.

Each of these defects will shorten the relaxation time. When estimating the effect of several types of scattering centers, each related to a specific τ_i, the total time between independent scattering events can be estimated from *Mathiessen's rule**

$$\frac{1}{\tau} = \sum_i \frac{1}{\tau_i},$$

(32.1)

since the collision term in the Boltzmann equation [Eq. (30.1)] is additive. Consequently, the inverse mobilities calculated for single, independent types of scattering are also added to result in the inverse total carrier mobility—see also Debye and Conwell (1954).

$$\frac{1}{\mu} = \sum_i \frac{1}{\mu_i}.$$

(32.2)

From Eq. (32.1), it is clear, however, that only the centers which influence the carriers the most need to be considered in an actual

---

* An error up to 20% can occur when applying Eq. (32.1) because of nonlinearities, interaction of different scattering events, as shown by Rode and Knight (1971).

crystal under a given condition. Depending on the type of crystal bonding, crystal preparation (growth and treatments), doping, temperature, and other external influences—such as light, strain, and electric and magnetic fields—the predominant scatterer may vary from sample to sample.

In addition, deviations from a simple lattice isotropy and other carrier-lattice interactions may create variations of the carrier scattering. For instance, one distinguishes:

(7) Semiconductors with
  (a) intravalley scattering
  (b) intervalley scattering
  (c) warped-energy surface effects

(8) Quasi-particles, such as
  (a) polarons directly contributing to the current instead of Bloch electrons and
  (b) excitons (exciton-polaritons) contributing to the transport of energy and to the creation of free carriers at places different from the originations.

In the following sections, we will discuss some of the more important scattering mechanisms. A review by Nag (1984) summarizes the different scattering mechanisms and gives tables for the expressions of the relaxation times. See also Seeger (1973) and Sections 32.4 and 33.6.

32.2 Intravalley Scattering

The scattering events described in the following sections leave the carrier within its valley. We will first assume that such scattering events take place with electrons near the Γ point.

32.2.1 Electron Scattering with Phonons

Most carriers have an energy near $(3/2)kT$. At low fields ($F \ll v_{\mathrm{rms}}/\mu$), they gain only a small fraction of additional energy compared to their thermal energy between scattering events. Carriers can interact with various types of phonons by absorbing or emitting a phonon. Before discussing this scattering in detail, a few general remarks will provide some overall guidance.

The phonon dispersion relation of most semiconductors shows that optical phonons have an energy larger than kT at room temperature, and therefore are scarce. This makes the creation of optical phonons unlikely since most of the carriers do not have sufficient energy. Therefore, carriers scatter predominantly with the lower energy acoustic phonons, which are plentiful at room temperature. During such a scattering event, the electron energy is changed by only a small fraction [Eq. (30.54)]; that is, the scattering is an essentially elastic event.

In *semiconductors with a direct band gap*, only phonons near the center of the Brillouin zone have a high probability of scattering. During such events, substantial changes in the direction of motion can occur: k can easily change its sign.

In *semiconductors with indirect band gap*, electrons are in a valley of a relatively large k value. Elastic scattering with low energy and momentum phonons tend to leave the electrons within their valley, with only small changes of their momentum, which changes their direction insignificantly. Intervalley scattering, which will be discussed in Section 32.3.1, requires a higher phonon momentum, i.e., higher energy acoustic or optical phonons except for Umklapp processes for electrons near the surface of the Brillouin zone where only small values of q are required to reverse the direction of the electron motion during scattering. This is most easily done in Ge, where the conduction band minimum lies at this surface, rather than in Si, where the minimum of $E(k)$ for conduction electrons lies at $0.8\pi/a$—for more detail, see Section 32.3.1.

32.2.1A Elastic and Inelastic Scattering When electrons scatter with phonons, energy and momentum conservation laws must be fulfilled. At room temperature, electrons have an average energy of kT ($\simeq 25$ meV). Near the Γ point, optical phonons are of the same order of magnitude, while acoustic phonons start from $\hbar\omega = 0$ at $q = 0$ and have energies $\hbar\omega \ll kT$ in its vicinity. Therefore, almost no energy is exchanged when scattering with such acoustic phonons; hence, such scattering is an *elastic scattering* event.

In contrast, only the faster electrons have enough energy to create an optical phonon, and thereby lose almost all of their excess energy. This type of scattering is an *inelastic scattering* event.

One needs, however, to be careful when scattering with phonons of higher momentum is considered. Since at higher $\mathbf{q}$ the energy of acoustic phonons approaches, in order of magnitude, that of optical phonons, both types of phonons cause *inelastic scattering*.

32.2.1B Phonon Generation and Annihilation When interacting with electrons, phonons can be generated, thereby cooling the electron ensemble, or annihilated, thereby heating it. This interaction is determined by the ion-electron interaction potential in the Hamiltonian [see Eq. (8.1)], which can be expressed as the sum over the individual contributions from each lattice atom:

$$H_{\text{ion,el}} \equiv V_{\text{ion,el}}(\mathbf{R}_1, \mathbf{R}_2, \ldots \mathbf{r}_1, \mathbf{r}_2, \ldots) = \sum_i V_{\text{ion,el}}(\mathbf{R}_i - \mathbf{r}_i). \quad (32.3)$$

This Hamiltonian is conventionally separated into one part that describes the electron-interaction with the periodic lattice $H_{\text{ion,el}}^{(o)}$ and another part that describes the interaction of the electron with the lattice oscillation $H_{\text{ph,el}}$

$$H_{\text{ion,el}} = H_{\text{ion,el}}^{(o)} + H_{\text{ph,el}}. \quad (32.4)$$

The second term involves the deviation from the periodic potential that causes the scattering. The simplest way of separating the periodic part of the potential from the oscillating perturbation is by an expansion of the potential, breaking it off after the linear term:

$$V_{\text{ion,el}}(\mathbf{R}_i - \mathbf{r}_i) = V_{\text{ion,el}}^{(o)}(\mathbf{R}_i^{(o)} - \mathbf{r}_i) - u_i \cdot \nabla V_{\text{ion,el}}(\mathbf{R}_i^{(o)} - \mathbf{r}_i), \quad (32.5)$$

where $u_i(t)$ is the displacement of the ith ion from its equilibrium position (see Section 5.1.1), which can be obtained from Eq. (5.3).

From the total Hamiltonian, one obtains the eigenvalues which, for the unperturbed system, are band states. With perturbation, transitions between the different eigenstates are initiated. The probability W for such transitions is proportional to the corresponding matrix elements [see Eq. (30.7), with ΔV given by the deviation from the periodic potential], and the population of initial and final states. For the absorption or emission of a phonon (see Section 30.2) one has:

$$W(\mathbf{k}, \mathbf{k} + \mathbf{q}) = \frac{2\pi}{\hbar} |M_{\mathbf{k},\mathbf{k}+\mathbf{q}}|^2 f(\mathbf{k}) [1 - f(\mathbf{k} + \mathbf{q})] \Big\{ f(\mathbf{q}) \delta [E(\mathbf{k} + \mathbf{q})$$
$$- E(\mathbf{k}) - \hbar\omega_{\mathbf{q}}] + [1 - f(-\mathbf{q})] \delta [E(\mathbf{k} + \mathbf{q}) - E(\mathbf{k}) + \hbar\omega_{\mathbf{q}}] \Big\}.$$
$$(32.6)$$

where f is the distribution function, and δ is the Dirac delta function. After integration over all possible transitions, one obtains the collision term of the Boltzmann equation [see Eq. (30.5)]:

$$\left(\frac{\partial f(\mathbf{k})}{\partial t} \right)_{\text{coll}} = \frac{\mathcal{V}}{8\pi^3} \int W(\mathbf{k}, \mathbf{k}') \, dk. \quad (32.7)$$

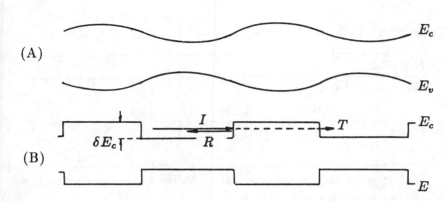

Figure 32.1: (A) Undulation of the band edges due to "pressure waves," e.g., in the long wave range of longitudinal acoustic phonons. (B) Step-like approximation of these undulations.

This integration, however, is difficult to perform for complex lattice oscillation. Therefore, the collision term is evaluated for one specific type of oscillation at a time, when approximations can easily be introduced. A simple example is the scattering of electrons on longitudinal acoustical phonons, which is discussed in the next section.

32.2.1C Longitudinal Acoustic Phonon Scattering The interaction of electrons with longitudinal acoustic phonons can be analyzed in a variety of models. For reviews, see Mitra (1969), Seeger (1973), Nag (1980), Madelung (1981), and Zawadzki (1982).

For pedagogical reasons, we use a classical approach with acoustic waves. In the acoustic branch at longer wavelengths, the lattice is alternatingly compressed and dilated. Consequently, the width of the band gap is modulated; it widens with compression—see Section 9.2.1B. Electrons are scattered at a wave crest of the modulated band edge—Fig. 32.1A.

In a further simplified model, the deformation wave is approximated by a potential step in both bands—Fig. 32.1B and McKelvey (1966). An electron wave impinging on such a step is partially transmitted and partially reflected. The reflection probability is estimated from the solution of the Schrödinger equation as the difference between the impinging wave and the transmitted wave. Their energies are given by

$$E_I = \frac{\hbar^2 k_I^2}{2m_n} \quad \text{and} \quad E_T = \frac{\hbar^2 k_T^2}{2m_n} = E_I - \delta E_c. \qquad (32.8)$$

Table 32.1: Deformation potentials* (eV).

Material				
Ge	Ξ_u^X	Ξ_d^X	Ξ_u^L	Ξ_d^L
	10.4	0.53	16.4	-6.4
GaAs	$\Xi_u^{\tau(111)}$	$\Xi_d^{\tau(111)}$		$\Xi_d^{\tau(100)}$
	16.5	-8		-11.2

*Subscript u stands for pure shear, and d for the diagonal component of the deformation potential tensor.

From continuity of the wavefunction and its derivatives at each step, we obtain for $k_I \simeq k_T$ for the reflection probability $R \simeq [(k_I - k_T)/(k_I + k_T)]^2$ (see Section 42.4.3), which yields

$$R \simeq \left(\frac{m_n \delta E_c}{2\hbar^2 k_I^2}\right)^2, \tag{32.9}$$

where δE_c is the step-height and k_I is the wave vector of the impinging electron. The step-height is related to the lattice compression by

$$\delta E_c = -\Xi \frac{\delta \mathcal{V}}{\mathcal{V}}, \tag{32.10}$$

with Ξ as the *deformation potential**, $\mathcal{V}$ as the volume, and $\delta \mathcal{V}$ as its change, which is related to the thermal energy by a simple thermodynamic analogy:

$$\frac{1}{2} \delta p \, \delta \mathcal{V} = ckT, \tag{32.11}$$

where c is a proportionality factor on the order of 1. Replacing the pressure increment δp from the compressibility (κ) relation

$$\kappa = \frac{1}{\mathcal{V}} \frac{\delta \mathcal{V}}{\delta p}, \tag{32.12}$$

we obtain for the probability of reflection, from Eqs. (32.9)–(32.12):

$$R \simeq \left(\frac{m_n}{2\hbar^2 k_I^2}\right)^2 \frac{c\kappa kT}{\mathcal{V}} \Xi^2. \tag{32.13}$$

---

* The deformation potential is defined as the change in band gap per unit strain, and is typically on the order of 10 eV. For a listing, see Table 32.1.

The probability of reflection can be connected with a mean free path λ by $\lambda = l/R$, with l as the length of the sample (of volume $\mathcal{V} = l^3$). We obtain for $k_I \simeq \pi/l$, i.e., for long wavelength acoustic phonons:

$$\lambda \simeq \frac{h^4}{4m_n^2 c \kappa k T \Xi^2}. \tag{32.14}$$

A somewhat more rigorous treatment (Bardeen and Shockley, 1950) yields a similar result:

$$\lambda = \frac{h^4 c_l}{m_n^2 k T \Xi^2}, \tag{32.15}$$

where c_l is the elastic constant for longitudinal deformation: $c_l = c_{11}$ for pressure in the $\langle 100 \rangle$ direction, $c_l = \frac{1}{2}(c_{11} + c_{12} + c_{44})$ in the $\langle 110 \rangle$ direction, $c_l = \frac{1}{3}(c_{11} + 2c_{12} + 4c_{44})$ in the $\langle 111 \rangle$ direction. Here, c_{ik} are components of the elastic tensor—see Section 4.1. Assuming an energy-independent $\lambda = \bar{\lambda}$ using $\bar{\tau} = \bar{\lambda}/v_{\mathrm{rms}}$, and replacing v_{rms} with Eq. (28.16), one obtains for the electron mobility due to acoustic phonon scattering

$$\mu_{\mathrm{n,ac}} = \frac{\sqrt{8\pi}}{3} \frac{e h^4 c_l}{m_n^{5/2} (kT)^{3/2} \Xi^2}, \tag{32.16}$$

$$\mu_{\mathrm{n,ac}} = 6.1 \cdot 10^3 \frac{c_l}{10^{12}(\mathrm{g\,cm/s^2})} \left(\frac{m_0}{m_n}\right)^{5/2} \left(\frac{300}{T}\right)^{3/2} \left(\frac{\mathrm{eV}}{\Xi}\right)^2 (\mathrm{cm^2/Vs}), \tag{32.17}$$

i.e., a $T^{-3/2}$ dependence at higher temperature, where this type of scattering is predominant. This is observed for direct, but not for indirect, band gap semiconductors.* Intervalley scattering has a significant influence in indirect gap semiconductors, and will be discussed in Section 32.3.1.

The effective mass used in Eqs. (32.13)–(32.17) requires the proper mix of density-of-state and mobility effective masses:

$$m_n^{5/2} = m_{\mathrm{nds}}^{3/2} m_{n\mu}. \tag{32.18}$$

The deformation potential used here has only slowly varying components in space. Another approach, suggested by Ginter and

---

* The experimentally observed exponent of T is -1.67 for Ge (Conwell, 1952) and not -1.5. The exponent of T for Si is still larger ($\simeq 2.5$). Inserting actual values for Si ($c_l = 1.56 \cdot 10^{12}$ dyn/cm², $m_n = 0.2 m_0$, and $\Xi = 9.5$ eV), one obtains $\mu_n = 5,900$ cm²/Vs, a value that is larger by a factor of ~ 4 than the measured $\mu_n = 1,500$ cm²/Vs at 300 K.

Mycielski (1970), contains a part of the potential varying with the lattice periodicity, which is more appropriate for shorter wavelength phonons. This approach is a more general one; still, it gives similar results in a number of examples.

32.2.1D Acoustic Phonon Scattering with Piezoelectric Interaction

In piezoelectric crystals, ion oscillations cause a dipole moment that interacts with carriers rather effectively. A dipole moment can be generated by alternating lattice compression and dilatation, which in turn are caused by longitudinal acoustic phonons. These create an electric field parallel to the propagation direction which has a similar interaction with carriers, although slightly stronger than the acoustic deformation potential discussed above. The resulting mobility shows a somewhat similar behavior (Meyer and Polder, 1953):

$$\mu_{n,pe} = \frac{16\sqrt{2\pi}}{3} \frac{\hbar^2 \varepsilon \varepsilon_0}{e m_n^{3/2} K^2 (kT)^{1/2}}, \tag{32.19}$$

where K is the electromechanical coupling constant,[*] which for most semiconductors is on the order of 10^{-3}. Numerically, one has

$$\mu_{n,pe} = 1.5 \cdot 10^4 \frac{\varepsilon}{10} \left(\frac{m_0}{m_n}\right)^{3/2} \left(\frac{10^{-3}}{K}\right)^2 \left(\frac{300\,\mathrm{K}}{T}\right)^{1/2} \quad (\mathrm{cm}^2/\mathrm{Vs}). \tag{32.20}$$

At low temperatures (observe the $T^{-1/2}$ relation compared with a $T^{-3/2}$ relation for acoustic deformation potential scattering—see Fig. 32.4), piezoelectric phonon scattering can be an important scattering mechanism for low density of ionized impurities which otherwise predominate. See Seeger (1973) and Zawadzki (1980); for an update, see Nag (1984).

32.2.1E Optical Phonon Scattering in Nonpolar Compounds

Low-energy electron scattering with optical phonons is predominantly elastic. This process can be understood as the annihilation of an optical phonon to create a high-energy electron, which in turn immediately creates an optical phonon in a highly

---

[*] K^2 can be expressed as the ratio of the mechanical to the total work in a piezoelectrical material: $K^2 = (e_{pz}^2/c_l)/[\varepsilon\varepsilon_0 + e_{pz}^2/c_l]$, with e_{pz} the piezoelectric constant (which is on the order of 10^{-5} As/cm^2), and c_l the longitudinal elastic constant (relating the tension T to the stress S and the electric field F as $T = c_l S - e_{pz} F$).

probable transition. Therefore, the electron energy is conserved in the turnaround, but not its momentum. When electrons have accumulated sufficient energy to *create* optical phonons, the scattering becomes very effective and is inelastic—this will be discussed in Section 33.3.2.

Optical phonon scattering in elemental nonpolar semiconductors couples both longitudinal and transverse optical modes with the scattering electron (Boguslawski, 1975). It can be estimated by using a deformation potential formalism for longitudinal optical phonons (see also Conwell, 1967). One obtains

$$\mu_{n,opt} = \frac{4\sqrt{2\pi}\ e\hbar^2 \rho\sqrt{k\Theta}}{3m_n^{5/2}D_o^2}\phi(T), \qquad (32.21)$$

where ρ is the density of the semiconductor, and D_o is the optical deformation potential (Meyer, 1958):

$$\delta E_c = D_o\delta r, \qquad (32.25)$$

with δ_r as the change in the interatomic distance, and $\phi(T)$ is a function that contains the temperature dependence of μ and the density of phonons (Seeger, 1973). At low temperatures, $\phi(T)$ is large (typically $10^4 \ldots 10^5$ at $T = \Theta/10$) and decreases rapidly to a value on the order of 1 near the Debye temperature Θ—see Fig. 32.2. The actual form of $\phi(T)$ depends on the approximation used, and is plotted for two approximations (ϕ_f and ϕ_g) in Fig. 32.2. They show a nearly exponential decrease with increasing temperature for $T < 0.3\Theta$.

Numerically, the mobility due to optical phonon scattering [Eq. (32.21)] can be expressed as

$$\mu_{n,opt} = 1.77 \cdot 10^3 \frac{\left(\dfrac{\rho}{\mathrm{g\ cm^{-3}}}\right)\left(\dfrac{\Theta}{300\ \mathrm{K}}\right)^{1/2}}{\left(\dfrac{m_n}{m_0}\right)^{5/2}\left(\dfrac{D_o}{10^8(\mathrm{eV/cm})}\right)^2}\phi(T) \qquad (32.26)$$

and shows a sufficiently low value near and above the Debye temperature, where $\phi(T) \leq 1$, to become the determining factor in high-purity semiconductors.

Nonpolar optical phonon scattering is the only electron-optical phonon interaction in nonpolar semiconductors, such as Si and Ge. It is important for Γ_8-bands, and, as was pointed out by Harrison

Table 32.2: Deformation potential* at Γ-points (in eV) (after Blacha et al., 1984).

	$a(\Gamma_1^c)$	$a(\Gamma_1^c)-a(\Gamma_{15}^v)$	b	d	$|d_o|$
C					90
Si	-15.3	-10.0	-2.2	-5.1	40
Ge	-19.6	-12.6	-2.3	-5.0	34
AlSb		-5.9	-1.4	-4.3	37
GaP	-19.9	-9.3	-1.8	-4.5	44
GaAs	-17.5	-9.8	-2.0	-5.4	48
GaSb		-8.3	-1.8	-4.6	32
InP	-18.0	-6.4	-2.0	-5.0	35
InAs		-6.0	-1.8	-3.6	42
InSb	-14.6	-7.7	-2.0	-5.0	39
ZnS	-14.5	-4.0	-0.62	-3.7	4*
ZnSe	-11.5	-5.4	-1.2	-4.3	12*
ZnTe	-9.5	-5.8	-1.8	-4.6	23
CdS		-3.1			
CdSe		-3.0			
CdTe	-9.5	-3.4	-1.2	-5.4	22
CuCl			-0.7	0.43	7
CuBr			-0.25	-0.65	3.8
CuIn			-0.64	-1.4	1.1

*These deformation potentials are defined by *hydrostatic strain*

$$a_i = \frac{dE_i}{d\ln \mathcal{V}} = B\frac{dE_i}{dp} \qquad (32.22)$$

with B as the bulk modulus, by *shear strain* along [111] or [100]

$$d = \frac{\delta E}{2\sqrt{3}\,\varepsilon[111]} \quad \text{or} \quad b = \frac{\delta E}{6\,\varepsilon[100]} \qquad (32.23)$$

where $\varepsilon[111]$ or $\varepsilon[100]$ is the strain in the [111] or [100] direction and by splitting of the Γ_{15}-state in the absence of spin-orbit interaction

$$d_0 = \frac{\delta E a_0}{u}, \qquad (32.24)$$

with δE as the observed energy shift, a_0 as the lattice constant, and u as the atomic displacement. The parameter b is related to the optical deformation potential D_o used in Section 32.2.1E by $D_o = -(3/2)b$.

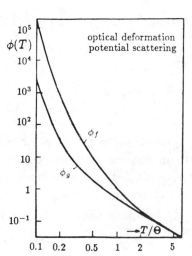

Figure 32.2: Auxiliary functions $\phi_f(T)$ and $\phi_g(T)$ for optical phonon scattering (after Seeger, 1973).

(1956), vanishes for the Γ_6-band. Therefore, it is unimportant for n-type InSb, but important for hole scattering in p-type InSb (Costato and Reggiani, 1972; Bir and Pikus, 1974).

32.2.1F Optical Phonon Scattering in Polar Semiconductors The scattering of carriers with longitudinal optical phonons (*Fröhlich interaction*) in a (partially) ionic lattice has a larger influence than the deformation potential interaction on the carrier mobility due to the larger dipole moment associated with such lattice vibration. This causes an induced field proportional to the polarization P:

$$F = -\frac{P}{\varepsilon \varepsilon_0}. \tag{32.27}$$

The polarization can be obtained from

$$P = \frac{e_c d_r}{V_o}, \tag{32.28}$$

where d_r is the change in the interatomic distance, V_o is the volume of the unit lattice cell, and e_c is the Callen effective charge [Eq. (11.39)].

The mobility was calculated by Ehrenreich (1961) using a variational method of Howarth and Sondheimer (1953), which accounts for the inelastic scattering. It yields for temperatures below the Debye temperature

$$\mu_{n,\text{ion,opt}} = \frac{e\hbar}{2m_n \alpha_c k\Theta} \exp\left(\frac{\Theta}{T}\right), \tag{32.29}$$

where α_c is the coupling constant given in Eq. (28.6). The Debye temperature is used here to account for the LO phonon energy at

$\mathbf{q} = 0$ with $\hbar\omega_{LO} = k\Theta_{LO} \simeq k\Theta$. The mobility is numerically given by

$$\mu_{n,\text{ion,opt}} = 355 \frac{\exp\left(\dfrac{\Theta}{T}\right)}{\alpha_c\left(\dfrac{m_n}{m_0}\right)\left(\dfrac{\Theta}{300\ \text{K}}\right)} \quad (\text{cm}^2/\text{Vs}). \qquad (32.30)$$

The mobility given by optical phonon scattering decreases linearly with increasing coupling constant. The mobility increases exponentially with decreasing temperature due to optical phonon freeze-out at lower temperatures.

At temperatures $T \gtrsim \Theta$, a different approximation (Seeger, 1973) yields:

$$\mu_{n,\text{ion,opt}} = 1.5 \frac{e\hbar}{2m_n\alpha_c k\Theta} \sqrt{\frac{T}{\Theta}} \exp\frac{\Theta}{T} \simeq 533 \frac{\sqrt{\dfrac{T}{\Theta}}\exp\dfrac{\Theta}{T}}{\alpha_c\left(\dfrac{m_n}{m_0}\right)\left(\dfrac{\Theta}{300\ \text{K}}\right)}.$$

$$(32.31)$$

Zawadzki and Szymańska (1971) used a Yukawa-type screened potential that results in a reduced effectiveness of the optical phonon scattering. At higher temperatures they obtained

$$\mu_{n,\text{ion,opt}} = \frac{\sqrt{2}}{8\pi} \frac{M_r a^3 (k\Theta)^2}{ee_c^2 m^{3/2} kT} \frac{\sqrt{E - E_c}}{F_{\text{op}}}, \qquad (32.32)$$

where F_{op} is a screening parameter* that depends on the Debye screening length, and is on the order of 1 (Zawadzki, 1980). This results in an increased mobility by a factor of 3.5 for InSb at room temperature, and for $n = 10^{19}$ cm^{-3}. When the Fermi level is shifted into the conduction band, the mobility becomes explicitly electron-density-dependent:

$$\mu_{n,\text{ion,opt}} = \frac{1}{4}\sqrt{\frac{3}{\pi} \frac{M_r a^3 (k\Theta)^2 \hbar}{ee_c^2 kT m_n (E_F)} \frac{\sqrt[3]{n}}{F_{\text{op}}}}. \qquad (32.33)$$

Figure 32.3 shows the electron mobility of InSb as a function of the *electron density* at 300 K, and identifies the most important

---

* Here, $F_{\text{op}} \simeq 1 + \frac{2}{\beta}\ln(\beta + 1) + 1/(\beta + 1)$, with $\beta = (2|\mathbf{k}|\lambda_D)^2$, and λ_D the Debye length given in Section 14.3.1, Eq. (14.11).

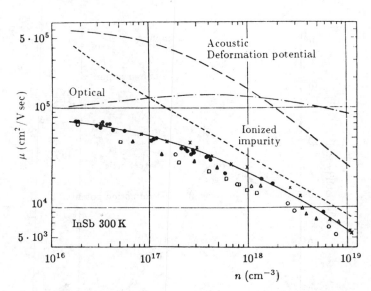

Figure 32.3: Electron mobility of n-InSb at 300 K as a function of the free-electron density. Dashed curves indicate the corresponding theoretical contributions (after Zawadzki, 1972).

branches of scattering by longitudinal optical phonons at low densities, by ionized impurities at high densities—see Section 32.2.4, and, of lesser importance, by acoustic phonons (after Zawadzki, 1972).

Another example, shown in Fig. 32.4, gives the electron mobility as a function of the *temperature* for n-GaAs. It indicates major scattering at ionized impurities (low T—see Section 32.2.4) and by longitudinal optical phonons (high T); whereas scattering by acoustic phonons, piezoelectric scattering, and scattering at neutral impurities are of lesser importance, in this order.

32.2.2 Scattering by Intrinsic Point Defects

Carrier scattering by intrinsic point defects, such as by interstitials, vacancies, or antisite defects (an A atom on a B site), is caused by deviations from the periodicity of the lattice potential in the equilibrium position of the disordered atoms. These changes can be seen as a local deformation of the bands. The interaction potential responsible for the scattering may be approximated as a Coulomb potential at charged localized centers, or by using the central core potential if the defect is not charged relative to the lattice. The intrinsic defects are similar to foreign atoms with respect to their

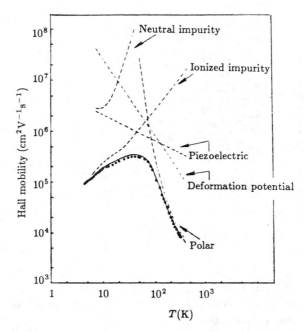

Figure 32.4: Electron mobility in n-GaAs as measured by Stillman et al. (1970). Dashed curve indicates the corresponding theoretical contributions (after Fletcher and Butcher, 1972).

behavior as scattering centers, which is discussed in the following sections.

All such scattering events at atomic point defects are considered elastic: the electron momentum is changed, but not its energy; the fraction of energy transferred is $\propto m_n/M \ll 1$. When its energy is changed, however, the electron becomes trapped. This is discussed in Section 43.2.1.

32.2.3 Scattering by Neutral Lattice Defects

Neutral lattice defects have a scattering cross section that is about the size of the defect atom, typically 10^{-15} cm^2. As described by Erginsoy (1950), they can become important scattering centers at low temperatures (at $T < 100$ K), when the density of ionic impurities has decreased by carrier trapping due to carrier freeze-out and the phonon scattering has decreased due to phonon freeze-out. In analogy to the scattering of electrons by hydrogen atoms, the scattering cross section s_n is estimated as $\propto \pi a_{qH}^2$:

$$s_n = \pi a_{qH}^2 \cdot \frac{\lambda_{DB}}{a_{qH}} = 2\pi^2 \frac{a_{qH}}{|\mathbf{k}|}, \qquad (32.34)$$

modified by a scattering correction factor λ_{DB}/a_{qH} (Seeger, 1973), where λ_{DB} is the De Broglie wavelength. Erginsoy estimated a similar relation $s_n \simeq 20 a_{qH}/|\mathbf{k}|$. Using the gas-kinetic estimate for the collision time

$$\tau_n^\times = \frac{1}{N^\times s_n v_{rms}} \qquad (32.35)$$

and $|\mathbf{k}| = 2\pi/\lambda_{DB} = mv/\hbar$, we obtain with $v \simeq v_{rms}$ for the mobility due to neutral impurity scattering

$$\boxed{\mu_n^\times = \frac{e}{m_n} \tau_n^\times = \frac{e}{2\pi^2 a_H \hbar} \frac{m_n/m_0}{\varepsilon_{st} N^\times}} \qquad (32.36)$$

or $\quad \mu_n^\times = 1.46 \cdot 10^3 \left[\frac{10^{16}}{N^\times (cm^{-3})} \right] \left(\frac{10}{\varepsilon_{st}} \right) \left(\frac{m_n}{m_0} \right). \qquad (32.37)$

This mobility is independent of the temperature. The Erginsoy approximation is valid for temperatures $T > 20$ K. For lower temperatures, the screening depends on the energy; Blagosklonskaya et al. (1970) obtained

$$\mu_n = \frac{e m_n^{3/2}}{\sqrt{2} \pi a_{qH}^2 N_n^\times \sqrt{E} \left(\dfrac{1}{E/E_i + 0.0275} + 10 \right)}, \qquad (32.38)$$

where E_i is the ionization energy of the impurity.

The scattering is different if the spin of the incident electron is parallel or antiparallel to the electrons in the scattering atom (triplet or singlet state, respectively). In considering also the multivalley structure of the conduction bands, Mattis and Sinha (1970) arrived at results similar to Blagosklonskaya et al. (1970) at low temperatures, with only a slight mobility reduction at temperatures above 10 K, as shown in Fig. 32.5—also see Norton and Levinstein (1972).

32.2.4 Scattering on Ionic Defects

When scattering occurs on *charged* defects, the carriers interact with the *long-range Coulomb forces*, resulting in a substantially larger scattering cross section that is typically on the order of 10^{-13} cm^2.

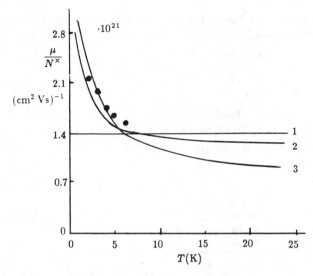

Figure 32.5: Electron mobility (normalized to constant neutral impurity density) as a function of the temperature: (1) for $\varepsilon = 10$ and $m_n = m_0$ (after Erginsoy, 1950); (2) (after Mattis and Sinha, 1970); and (3) (after Blagosklonskaya et al, 1970). Experimental data for Ge (after Baranskii et al, 1975).

The original scattering analysis on charged particles was done by Rutherford (1911) for α-particles, and is easily adapted to carriers scattered by ions of charge z in a solid; that is after introducing the screening of this potential in the solid by using the dielectric constant ε_{st}, and considering Bloch electrons by using the carrier effective mass. A differential cross section for the scattering of an electron of velocity v_{rms} under an angle θ into an element of solid angle $d\Omega = 2\pi \sin\theta\, d\theta$ (see Fig. 32.6) is then given (Leighton, 1959) by

$$s_n(\theta)\, d\Omega = \left(\frac{Z e^2}{8\pi\varepsilon\varepsilon_0 m_n v_{rms}^2} \right)^2 \sin^{-4}\frac{\theta}{2}\, d\Omega \qquad (32.39)$$

with

$$\frac{\theta}{2} = \tan^{-1}\frac{Z e^2}{4\pi\varepsilon\varepsilon_0 d m_n v_{rms}^2}, \qquad (32.40)$$

which shows a rather slow decrease of the scattering angle θ with increasing minimum distance d from the center—see Fig. 32.6. In order to totally randomize the angle after collision, however, only

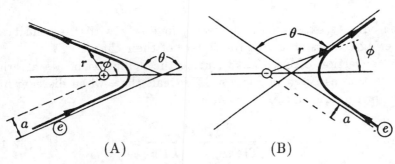

Figure 32.6: Electron trajectories for (A) a Coulomb-attractive, and (B) a Coulomb-repulsive, scattering center.

a fraction of $(1 - \cos \theta)$ of all scatterings describes the number of *memory-erasing collisions* [Eq. (30.47)]:

$$dn = N_I v_{\text{rms}} s_n(\theta)(1 - \cos \theta) \, d\Omega = d\left(\frac{1}{\tau_m}\right). \qquad (32.41)$$

This number is inversely proportional to the differential momentum-relaxation time.

Integration over all angles θ to obtain the total relaxation time requires a cutoff in order to avoid an infinite result, since $s_n(\theta = 0) = \infty$. Conwell and Weisskopf (1950) assumed that the closest distance d of the trajectory from the center to be considered must be smaller than a maximum distance, $d < d_{\max} = \frac{1}{2} N_I^{1/3}$, given by the average distance between ionized centers in the crystal. This yields for the time between collisions of carriers with velocity v_{rms}:

$$\tau = \frac{(4\pi \varepsilon_{\text{st}} \varepsilon_0)^2 m_n^2 v_{\text{rms}}^3}{2\pi Z^2 e^4 N_I} \frac{1}{\ln\left[1 + \left(\dfrac{\varepsilon_{\text{st}} \varepsilon_0 m_n v_{\text{rms}}^2}{2 Z e^2 N_I^{-1/3}}\right)^2\right]}. \qquad (32.42)$$

After averaging, one obtains the momentum-relaxation time $\overline{\tau_m}$, and with $v_{\text{rms}} = \sqrt{3kT/m_n}$, one has for the mobility, with $\mu_n = e\overline{\tau_m}/m_n$,

$$\mu_n = \frac{8\sqrt{2}(4\pi \varepsilon_{\text{st}} \varepsilon_0)^2 (kT)^{3/2}}{\pi^{3/2} Z^2 e^3 m_n^{1/2} N_I} \frac{1}{\ln\left[1 + \left(\dfrac{3 \varepsilon_{\text{st}} \varepsilon_0 kT}{2 Z e^2 N_I^{1/3}}\right)^2\right]}. \qquad (32.43)$$

This mobility increases with temperature $\propto T^{3/2}$, i.e, faster electrons are less effectively scattered. The logarithmic dependence is usually neglected. As expected, μ decreases inversely with the density of scattering centers and with the square of their charge eZ.

A conversion of Eq. (32.43), using the definition of a scattering cross section as given in Section 32, reveals that, when disregarding the logarithmic term, s_n is given by

$$s_n = \frac{\pi^{3/2} Z^2 e^4}{8\sqrt{6}(4\pi\varepsilon_{st}\varepsilon_0)^2 (kT)^2} = \pi r_i^2, \tag{32.44}$$

where r_i is the "scattering radius" of the ion. This scattering radius can be compared with r_C, the radius of a Coulomb well at a depth of kT:

$$\frac{Ze^2}{4\pi\varepsilon_{st}\varepsilon_0 r_C} = kT \quad \text{or} \quad r_C = \frac{Ze^2}{4\pi\varepsilon_{st}\varepsilon_0 kT}. \tag{32.45}$$

We now relate the scattering radius defined in Eq. (32.44) with the above defined Coulomb radius by

$$r_i = c_c r_C \tag{32.46}$$

where c_c is a correction factor. By comparison with Eq. (32.43), now including the logarithmic term, we obtain:

$$c_c = 2\left(\frac{2\pi}{3}\right)^{\frac{1}{4}} \frac{1}{\ln\left[1 + \left(\dfrac{3\varepsilon_{st} kT}{2Ze^2 N_I^{1/3}}\right)^2\right]}, \tag{32.47}$$

which is on the order of 1—see Fig. 32.7. One thereby sees that, except for this correction factor, the scattering cross section is equal to the square of the Coulomb radius at $E = E_c - kT$. Therefore, the cross section decreases $\propto 1/T^2$, or the increase in mobility is directly related to a decrease of r_C with T. This relationship is closely associated with the trapping of a charge at a Coulomb-attractive center when energy is dissipated—see Section 43.1.1 and Eq. (43.10). For such inelastic events to occur, the carrier must penetrate to $r < r_C$.

With the Coulomb radius introduced above, the Conwell-Weisskopf formula can be rewritten as the classical scattering relation

$$\mu_n = \frac{e}{m_n} \frac{c_c}{v_n s_n N_I}. \tag{32.48}$$

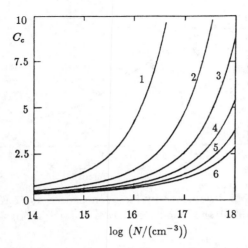

Figure 32.7: Correction factor for the Conwell-Weisskopf formula from the simple classical Coulomb well scattering of a cross section corresponding to a depth of kT/e, with the temperature as family parameter: $T = 100, 200 \ldots 600$ K for curves 1–6, respectively.

A somewhat refined approach was suggested by Brooks and Herring (Brooks, 1955). They assumed a cut-off in the integration over θ by replacing the Coulomb potential with a screened Yukawa potential, and used the Debye length λ_D as the screening length. In addition, they replaced the electron density in Eq. (14.11) with $n + (n + N_a)[1 - (n + N_a)/N_d]$, considering partial compensation. The ensuing result is similar to Eq. (32.43):

$$\mu_n = \frac{8\sqrt{2}(4\pi\varepsilon_{st}\varepsilon_0)^2(kT)^{3/2}}{\pi^{3/2}Z^2e^3m_n^{1/2}N_I} \frac{1}{\ln\left(1 + \beta - \dfrac{\beta}{1+\beta}\right)} \text{ with } \beta = (2|k|\lambda_D)^2.$$

$$(32.49)$$

Comparing the cut-off by Brooks and Herring with that of Conwell and Weisskopf, we observe a simple relation:

$$r_C\lambda_D^2 = n^{-1}, \qquad (32.50)$$

which introduces the carrier density dependence into the scattering. For a review, see Zawadzki (1980), and Chattopadhyay and Queisser (1981).

32.2.4A Coulomb Scattering in Anisotropic Semiconductors Ionized impurity scattering is sensitive to the anisotropy of the band structure, since low-angle scattering events dominate and inter-

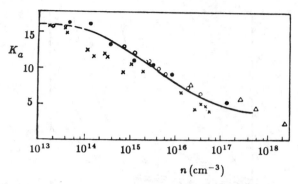

Figure 32.8: Mobility anisotropy factor measured at 77 K in n-Ge (after Baranskii et al., 1975) and calculated curve (after Dakhovskii and Mikhai, 1964).

fere with randomizing electron velocities (Herring and Vogt, 1956). One also requires that the screening length be substantially smaller than the mean free path in order to prevent successive collisions from occurring in the same defect-potential region.

The anisotropy of the effective mass and of the density of states, which are largest along the long axes of the valley ellipsoids, influences the mobility (Boiko, 1959). The mobility can be expressed as (Samoilovich et al., 1961)

$$\mu = e \left(\frac{2}{3} \frac{\langle \tau_\perp \rangle}{m_\perp^*} + \frac{1}{3} \frac{\langle \tau_\parallel \rangle}{m_\parallel^*} \right). \qquad (32.51)$$

with $m_{n\parallel}/m_{n\perp} = 19$ or 5.2 for Ge or Si, respectively.

With a screened Yukawa potential, the anisotropy factor $K_a = \mu_\perp/\mu_\parallel$ decreases with increasing carrier density as shown in Fig. 32.8. A smaller cross section of the scattering centers with higher carrier densities is caused by a decreasing $\lambda_D \propto 1/\sqrt{n}$; it also renders ion scattering more randomizing with increasing n.

32.2.4B Quantum Corrections for Ion Scattering   Several assumptions of the Brooks-Herring approximation are often not fulfilled (Moore, 1967). Corrections obtained by dropping these assumptions may be expressed in a linearized form:

$$\mu = \mu_0 \frac{1}{1 + \delta_B + \delta_m + \delta_d}, \qquad (32.52)$$

with three correction contributions:

1. The Born approximation, used to estimate the scattering probability, requires $|\overline{\mathbf{k}}|\lambda_D \gg 1$, with the average wave vector $|\overline{\mathbf{k}}| = m_n v_{\text{rms}}/\hbar$. When this condition is not fulfilled, a Born correction component δ_B is introduced with

$$\delta_B = \frac{2Q(\beta)}{\mathbf{k}^2 \lambda_D a_{\text{qH}}} \propto \begin{cases} \text{const.} & \text{for low } T \\ \sqrt{n}/T^{\frac{3}{2}} & \text{for high } T \end{cases} \qquad (32.53)$$

   with $Q(\beta)$ a slowly varying function: $0.2 < Q < 0.8$. One estimates $0.1 < \delta_B < 1$ for ion densities between 10^{16} and 10^{19} cm^{-3}. It is a more important correction for lower temperatures ($T < 100$ K; Moore, 1967).

2. When coherent scattering occurs from more than one ion, the mean free path becomes comparable to the screening length. The multiple-scattering correction factor δ_m has been estimated by Raymond et al. (1977), and is of minor importance at low temperatures.

3. A dressing effect can be expected to take care of the chemical individuality of the scattering center. The effect of the electron wavefunction of the scattering ion is rather small: δ_d is about 30–50% of δ_m. However, the defect individuality is observed at higher impurity densities for InSb doped with Se or Te (Demchuk and Tsidilkovskii, 1977), and for Si or Ge doped with As or Sb (Morimoto and Tani, 1962). It is suggested that the stress field surrounding the impurity of different sizes is the reason for the individuality of the scattering probability of such ions rather than central cell potential corrections (Morgan, 1972).

32.2.5 Carrier-Carrier Scattering

The scattering of carriers by other carriers within the same band does not change the total momentum of the carrier gas. Therefore, it does not influence the momentum relaxation time. Combined with other scattering mechanisms, however, it causes an accelerated relaxation—see Section 48.2.2C. This results in a *slight* decrease in mobility, which is usually on the order of only a few percent. Estimates made by Appel (1961) for covalent semiconductors, and by Bate et al. (1965) for ionic semiconductors, indicate a mostly negligible effect of low-field electron-electron scattering.

   Some larger effects in energy relaxation are discussed by Dienys and Kancleris (1975), and by Nash and Holm-Kennedy (1974). When a subgroup of electrons is excited to higher energies, electron-electron

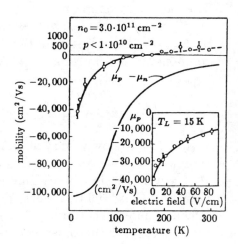

Figure 32.9: Minority (p) and (minus) majority (n) carrier mobilities as a function of the lattice temperature and of the electric field (at $T = 15$ K in GaAs/As$_{0.23}$ Ga$_{0.77}$As) multiquantum wells (after Höpfel et al., 1986).

scattering is important, as it tends to restore a thermal electron distribution.

When carriers of different bands are scattering with each other, a more pronounced influence can be observed, especially at higher fields (see Chapter 33), where the carriers are heated to differing degrees according to their effective mass. This can be a significant effect for holes in the light and heavy hole bands. Scattering reduces the difference between the two carrier temperatures.

32.2.5A Electron-Hole Drag Effect At low fields, the electron-hole scattering can be regarded as similar to the scattering of carriers at ionized impurities. It reduces the mobility of minority carriers (Prince, 1953; Ehrenreich, 1957), and can be described as a drag effect (McLean and Paige, 1960).

The exchange of momentum between the two sets of carriers also results in a *carrier drag* caused by the higher mobility majority carrier that, under extreme conditions, can lead to a drift reversal of the minority carriers. Minority holes are then dragged toward the anode, and minority electrons toward the cathode. Such a *negative absolute mobility* was observed in GaAs quantum wells by Höpfel et al. (1986)—Fig. 32.9.

32.2.6 Carrier Scattering on Plasmons

At higher carrier densities, one must consider the interaction of an electron with the collective electron plasma (see Section 12.1.1), characterized by its plasma frequency ω_p, as given in Eq. (12.4). A substantial energy loss occurs at resonance when the energy of the carrier equals the plasmon energy $\hbar\omega_p$. For thermal electrons, the corresponding carrier density is $< 10^{16}$ cm^{-3} [Eq. (12.5)]. With faster electrons, resonance occurs with a denser plasma, and losses increase.

In metals, plasmons have an energy of ~ 10 eV, and fast electrons penetrating a thin metal foil show distinct losses of a multiple of this plasmon energy.

32.2.7 Alloy Scattering

Mixing homologous elements forms semiconductor alloys in which these elements are statistically distributed. Examples have been mentioned previously (Sections 3.7.3, 9.2.1, and 22.4.7), and include $Si_\xi Ge_{1-\xi}$, $Al_\xi Ga_{1-\xi}S$, and $CdS_\xi Se_{1-\xi}$.

The random distribution of the alloying elements causes a fluctuation in the periodic potential of the lattice and an increased carrier scattering, known as alloy scattering (originally discussed by Wilson, 1965; see also Makowski and Glicksman, 1973).

The scattering probability, obtained by integrating $S(k, k')$ in Eq. (30.6) over all k':

$$S(k) = \frac{\mathcal{V}}{8\pi^3} \int S(k, k')\, dk',\qquad(32.54)$$

can be approximated for alloys by

$$S(\mathbf{k}) = \frac{\sqrt{2}}{\pi\hbar^4} N_0 \xi(1 - \xi)(E_{g1} - E_{g2})^2 m_n^{3/2} \sqrt{E(\mathbf{k}) - E_c},\qquad(32.55)$$

where E_{g1} and E_{g2} are the band gaps of the pure constituents of the alloy, and N_0 is the density of atoms in the alloy.

32.3 Multivalley Carrier Transport

Indirect band-gap semiconductors have several satellite minima at the conduction band edge in the different crystallographic directions. Scattering of carriers can occur within one of these valleys, *intravalley scattering*, or from one valley to another, *intervalley scattering*. The scattering described in the previous sections is a scattering that leaves each electron within its valley. The carrier distribution in all equivalent valleys remains equal.

32.3.1 Intervalley Scattering

Scattering from one valley to another requires a large exchange of momentum, which is provided by phonons of large $\mathbf{q}$ values, i.e., for optical phonons or for acoustic phonons of large energy—see Fig. 32.10. Such intervalley phonons ($\hbar \omega_i$) may transfer electrons in Si from a valley near $0.8 \frac{\pi}{a}$ to the equivalent valley at $-0.8 \frac{\pi}{a}$, requiring a momentum exchange* of $q = 0.4 \frac{\pi}{a}$—see Fig. 32.10. Another scattering into a nonequivalent valley at $0.8 \frac{\pi}{b}$ needs an even larger momentum exchange.

Polar and piezoelectric scattering does not provide for large momentum transfer and can be neglected for intervalley scattering.

Phonons with sufficient momentum have similar energies, usually within a factor of two, whether they are in the acoustic or the optical branch. Intervalley scattering can therefore be treated similarly to optical phonon scattering by replacing the longitudinal optical phonon energy $\hbar \omega_{\mathrm{LO}} = k\Theta$ in Eq. (32.21) with $\hbar \omega_i = k\Theta_i$, which requires typically a 20–40% less energetic phonon.

The intervalley scattering into a valley on the same axis is referred to as *g-scattering*. It results in a change of the sign of the electron momentum, i.e., "reflecting" the electron path by a sufficiently large angle. These scattering events are also known as *Umklapp processes*, indicating that the momentum has changed its sign without much change in value† (in German, *Umklapp* is a mirror operation). The other intervalley scattering processes, which transfer an electron into one of the four valleys on the other axes, are called *f-scatterings*. Selection rules determine which of the valleys can be reached depending on the symmetry of the scattering phonon (Bir and Pikus, 1974). Forbidden transitions, however, are also significant for carrier transport (Eaves et al., 1975).

All intervalley scattering events cause a reduction in the momentum relaxation time; the mobility is reduced. Since the Debye temperature enters exponentially into a polar optical scattering mode, a reduction from Θ to $\Theta_i < \Theta$ also reduces the electron mobility:

$$\mu_{\mathrm{iv}} = \frac{e\hbar}{2m\alpha_c k\Theta_i} \sqrt{\frac{T}{\Theta_i}} \exp \frac{\Theta_i}{T}. \qquad (32.56)$$

---

* See Fig. 32.10B, which shows two equivalent transitions: one requires $2 \cdot 0.8 \frac{\pi}{a} = 1.6 \frac{\pi}{a}$ in the extended $E(k)$ diagram, and one needs an actual momentum transfer of only $2 \cdot 0.2 \frac{\pi}{a} = 0.4 \frac{\pi}{a}$.

† Umklapp processes were introduced for scattering in metals, in which changes in the magnitude of the momentum after scattering from one to another side of the near-spherical Fermi surface are even smaller.

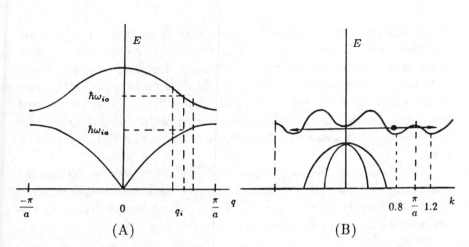

Figure 32.10: (A) Phonon $E(q)$ and (B) electron $E(k)$ diagrams indicating intervalley scattering.

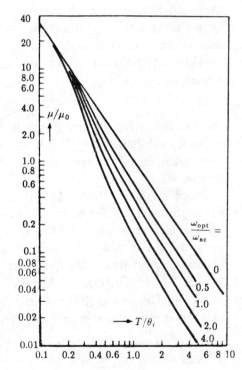

Figure 32.11: Intervalley scattering with acoustic and optical deformation potentials as a function of the temperature, and with the coupling-constant ratio as the family parameter (after Herring, 1955).

Therefore, intervalley scattering becomes the dominant optical phonon scattering mechanism: the increased phonon interaction with intervalley scattering reduces the relaxation time substantially at higher temperatures.

As shown before, intravalley scattering is insufficient to explain the observed lowering of μ_n with increasing temperatures: acoustic deformation potential scattering, although frequent enough, yields only a $T^{-3/2}$ decrease of $\mu_n(T)$; optical deformation potential scattering could give a steeper decrease of $\mu_n(T)$, but does not occur frequently enough to cause a marked reduction in the mobility: there are few optical phonons at temperatures below the Debye temperature.

The degree of the additional mobility reduction by intervalley scattering depends on the ratio of acoustic and optical deformation potential interaction rates, w_{ac} and w_{opt}, which are described by the ratio of the material-specific coupling constants:

$$\frac{w_{\text{opt}}}{w_{\text{ac}}} = \frac{1}{2}\left(\frac{D_i v_s}{\Xi_c \omega_i}\right)^2, \qquad (32.57)$$

where D_i is the intervalley optical deformation potential constant, Ξ_c is the acoustic deformation potential constant, v_s is the sound velocity, and ω_i is the phonon frequency for intervalley scattering. Considering the sum of acoustic and intervalley optical scattering relaxation

$$\frac{1}{\tau_m} = \frac{1}{\tau_{\text{ac}}} + \frac{1}{\tau_i}, \qquad (32.58)$$

one obtains a family of curves shown in Fig. 32.11. Adjusting the curves by selecting $\Theta_i = 720$ K and $w_{\text{opt}}/w_{\text{ac}} = 3$, one obtains for Si a much-improved agreement with the experiment (Herring, 1955)—see Fig. 32.12.

More recently, Rode (1972) computed the electron mobility for n-type Si and Ge for a combination of intervalley and intravalley acoustic deformation potential scattering. Excellent agreement with the experiment was obtained by using two adjustable parameters: the inter- and intravalley deformation potentials. The experimental slope is $T^{-2.42}$ (Putley and Mitchell, 1958) and $T^{-1.67}$ (Morin, 1954) for Si and Ge, respectively. Other intervalley scattering effects need a larger energy transfer and are discussed in Section 33.4.2.

32.3.1A Resonance Intervalley Scattering A very large additional scattering occurs when nonequivalent valleys coincide in energy. This can be achieved by applying hydrostatic pressure, e.g., 41 kbar for GaAs (see Section 21.3.1), with the result that the energy for the minima at Γ and X coincide. At this pressure, the mobility decreases markedly because of resonance intervalley scattering (Kosicki and Paul, 1966).

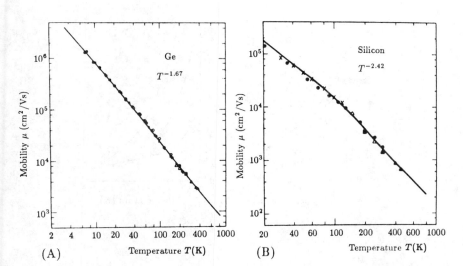

Figure 32.12: Temperature dependence of the electron mobility (A) in pure Si and (B) in pure Ge. Solid lines are calculated for deformation potential inter- and intravalley scattering (after Rode, 1972).

32.4 Summary of Low-Field Scattering Relaxation

A comparison of the relaxation times as a function of the temperature, and caused by the different scattering mechanisms, was made by Nag (1984) for a typical semiconductor with $m_n = 0.05 m_0$ and $\varepsilon = 15$. This is shown in Fig. 32.13 with the carrier density as family parameter.

A summary of the scattering potential, matrix elements, and the relaxation times with

$$\tau_i = \tau_{i,0} f(T, n) \tag{32.59}$$

is listed in Table 32.3. The dependence on the carrier density is contained in the screening length, which is part of the screening parameter S_c used in Table 32.3:

$$S_c = \frac{\hbar^2}{8 m_n \lambda_D^2 kT}. \tag{32.60}$$

Other parameters used in Table 32.3 are $x = E/kT$, $\Theta_0 = \hbar\omega_{LO}/k$, ρ = mass density, v_l = longitudinal acoustic velocity, Ξ_c and D_o = acoustic and optical deformation potentials, respectively, e_{pz} = piezoelectric constant, m^* = effective carrier mass, N_I = density of ionized impurities, n_q or n_i = phonon occupation number = f_{BE},

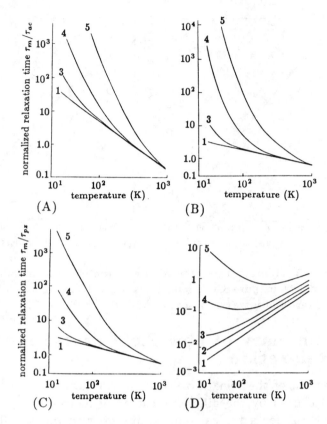

Figure 32.13: Momentum relaxation time as a function of the lattice temperature with the carrier density as family parameter for $n = 10^{13}$, 10^{15}, 10^{16}, 10^{17}, and 10^{18} cm^{-3} for curves 1–5, respectively. Scattering by: (A) acoustic phonons; (B) alloy disorder; (C) piezoelectric phonons; and (D) ionized impurities (after Nag, 1984).

$\mathcal{V}_0$ = volume per atom, ξ = fraction of alloying atoms, and H = Heaviside unit function. The temperatures used are in K.

Summary and Emphasis

Carriers are scattered by a multitude of crystal defects. Most of these events are elastic because of the large mass ratio between defect and carrier. Examples include the scattering at neutral or charged impurities or intrinsic defects, and intravalley scattering of acoustic phonons, as well as scattering by absorbing an optical phonon, followed by immediate reemission. During elastic scattering, only the momentum of the carrier is changed; its energy is not.

Table 32.3: Summary of matrix elements and relaxation times for different scattering mechanisms (after Nag, 1980). For parameters, see Sec. 32.4.

Scattering Mechanism	Scattering Potential	$\lvert M(k,k')\rvert^2$	$\tau_{r,0}$	$\tau_r = \tau_{r,0}\,f(\eta,T,F)$
Acoustic Phonon Deformation	$n\cdot\Delta\Xi$	$\Xi_c^2\hbar\omega\,\dfrac{n_q+\frac{1}{2}\pm\frac{1}{2}}{2V_0\rho v_l^2}$	$\dfrac{2\sqrt{2\pi}\,\rho\hbar^4 v_l}{3\Xi_c^2 m^{*3/2}(300\,k)^{3/2}}$	$\tau_{ac}=\dfrac{3\sqrt{\pi}}{4}\left(\dfrac{300}{T}\right)^{3/2}F_{ac}(z)\tau_{ac,0}$ $\quad$ $F_{ac}(z)=\dfrac{1}{\sqrt{z}}\dfrac{1}{1-4\frac{S_c}{z}+6\left(\frac{S_c}{z}\right)^2}\ln\left(1+\frac{z}{S_c}\right)-2\left(\frac{S_c}{z}\right)^2\left(1+\frac{S_c}{z}\right)$
Piezoelectric	$\dfrac{\varepsilon e_{pz}}{iq\varepsilon}\nabla\cdot u$	$e^2 e_{pz}^2\hbar\,\dfrac{n_q+\frac{1}{2}\pm\frac{1}{2}}{2V_0\rho\varepsilon^2\omega_q}$	$\dfrac{16\sqrt{2\pi}\,\rho\hbar^2 v_l^2\varepsilon^2}{3e^2 e_{pz}^2\sqrt{m^*}\sqrt{300\,k}}$	$\tau_{pz}=\dfrac{3\sqrt{\pi}}{8}\left(\dfrac{300}{T}\right)^{1/2}F_{pz}(z)\tau_{pz,0}$ $\quad$ $F_{pz}(z)=\sqrt{z}\,\dfrac{1}{1-2\frac{S_c}{z}+\ln\left(1+\frac{z}{S_c}\right)+\frac{S_c}{z}\left(1-\frac{S_c}{z}\right)}$
Optical phonon, non-polar	$D_0 u$	$D_0^2\hbar\,\dfrac{n_q+\frac{1}{2}\pm\frac{1}{2}}{2V_0\rho\omega_0}$	$\dfrac{2\sqrt{2\pi}\,\rho\hbar^3\omega_0}{3D_0^2 m^{*3/2}\sqrt{300\,k}}\left[\exp\left(\dfrac{\Theta_0}{300}\right)-1\right]$	$\tau_{op}=\dfrac{3\sqrt{\pi}}{2}\sqrt{\dfrac{300}{T}}\dfrac{\exp\left(\frac{\Theta_0}{T}\right)-1}{\exp\left(\frac{\Theta_0}{300}\right)-1}F_{op}(z)\tau_{op,0}$ $\quad$ $F_{op}(z)=\dfrac{1}{\sqrt{z+\frac{\Theta_0}{T}}+H\left(z-\frac{\Theta}{T}\right)\exp\left(\frac{\Theta}{T}\right)\sqrt{z-\frac{\Theta_0}{T}}}$
Optical phonon, polar	$\dfrac{e u_{LO}}{q}\sqrt{\dfrac{\rho}{\varepsilon}\left(\dfrac{1}{\varepsilon_{opt}}-\dfrac{1}{\varepsilon_{st}}\right)}\,u$	$e^2\hbar\omega_0\left(\dfrac{1}{\varepsilon_{st}}+\dfrac{1}{\varepsilon_{opt}}\right)\dfrac{n_q+\frac{1}{2}\pm\frac{1}{2}}{2V\varepsilon_0 q^2}$		
Intervalley phonon	$D_i u$	$D_i^2\hbar\,\dfrac{n_i+\frac{1}{2}\pm\frac{1}{2}}{2V_0\rho\omega_i}$		
Ionized impurity	$\dfrac{\lvert k-k'\rvert^2+\frac{1}{\lambda_D}}{e^4}\dfrac{1}{V_0^2\varepsilon^2}$		$\dfrac{128\sqrt{2\pi}\,\sqrt{m^*}\,\varepsilon^2(300\,k)^{3/2}}{N_i e^4}$	$\tau_{imp}=\dfrac{\sqrt{\pi}}{8}\left(\dfrac{T}{300}\right)^{3/2}F_{imp}(z)\tau_{imp,0}$ $\quad$ $F_{imp}(z)=\dfrac{z^{3/2}}{\ln\left(1-\frac{z}{S_c}\right)-\dfrac{1}{1+\frac{S_c}{z}}}$
Alloy	$2a^3(1-\xi)\dfrac{(E_{g2}-E_{g1})^2}{V_0}$		$\dfrac{\sqrt{2\pi}\hbar^4}{3a^3\xi(1-\xi)(E_{g2}-E_{g1})^2 m^{*3/2}\sqrt{300\,k}}$	$\tau_{all}=\dfrac{3\sqrt{\pi}}{4}\sqrt{\dfrac{300}{T}}F_{ac}(z)\tau_{all,0}$

Inelastic scattering causes a change in momentum and energy. Scattering by emitting optical phonons and intervalley scattering are the only inelastic scattering processes during which the carriers can lose a significant amount of their energy to the lattice. This occurs only to a marked degree after the carriers have accumulated a substantial amount of energy from external sources, i.e., electric field or electromagnetic radiation. Under low-field conditions, many elastic scattering events occur before an inelastic event can follow.

The carrier scattering determines its relaxation time for energy and momentum; both, inversely added, determine the carrier mobility.

Different types of scattering are dominant for various conditions of temperature, doping, and other material parameters. Usually, ionized impurity scattering is dominant at low temperatures, scattering at various phonons at high temperatures. At a given temperature, the ionized impurity scattering will become dominant at sufficient doping levels. Usually, the scattering at neutral impurities and with intravalley acoustic phonons are of lesser importance.

The carrier mobility is one of the key parameters of semiconducting materials. Its understanding is essential for selecting or designing of materials with attractive properties for efficient devices. By proper treatment, one is able to change the carrier mobility within certain limits with concurrent improvement in device performance.

Exercise Problems

1.(e) Calculate the probability of reflection [Eq. (32.9)] from a square potential barrier by using the continuity of the wavefunction $\psi = A \exp(ik_x x)$, its transmitted and reflected parts, and the continuity of its first derivatives.

2.(r) The influence of different types of phonons on carrier scattering varies widely.
(a) Which phonons are mostly involved in scattering of electrons at low fields and moderate or low temperatures?
(b) Why?
(c) Such scattering is referred to as elastic scattering. Why?

3.(*) In Joule's heating, is the relevant relaxation time the energy or momentum relaxation time? Careful! Explain your reasoning.

4.(*) Neutral impurity scattering can be reduced to a rather simple expression [Eq. (32.34)].

(a) What assumption is used for the wave vector of the electron to be scattered at a neutral impurity?

(b) Why is a quasi-hydrogen radius used for scattering rather than the atom radius of $\simeq 1$ Å?

5.(*) Discuss conditions under which carrier-to-carrier scattering has a more pronounced effect on carrier mobilities.

6.(r) Intervalley scattering shows interesting differences:

(a) How do you distinguish g- and f-intervalley scatterings? How would an experiment be designed to show the difference? What could be the results?

(b) Name semiconductors that are preferred for such experiments. Why?

7.(l) Intervalley scattering was used to explain better the actually observed $\mu(T)$ behavior of Ge and Si near room temperature (Herring, 1955). Review these results, update the information from the literature, and provide a critical evaluation.

8. Explain in your own words how the probability of reflection of an electron wave from an acoustic wave could be translated into a mean free path of the electron.

9. Scattering of neutral and charged defects is related to the quasi-hydrogen cross section. Compare the proportionality factors with each other and discuss your findings.

10. Give the physical reasons for the temperature dependence of scattering on charged defects. Be specific and detailed in your account.

Chapter 33

Carrier Scattering at High Fields

Carrier heating has a significant influence on scattering, usually increasing it and thereby limiting the drift velocity to values close to the thermal carrier velocity.

In the previous section, low electric fields were assumed; that is, the energy gain between scattering events was limited to a small fraction of kT. The incremental speed gained between collisions was much less than v_{rms}, and the linearization used in the Drude-Sommerfeld theory was justified; it yielded Ohm's law with field-independent mobility and carrier density. In this section, changes in the carrier mobility are analyzed for larger electric fields. These changes are caused by a substantial perturbation of the carrier path between scattering events (Fig. 33.1); they involve a marked increase in carrier energy. A discussion of changes in the carrier density induced by high fields will be presented in Section 42.4.

33.1 Transport Velocity Saturation

At sufficiently high fields, the carrier transport velocities, i.e., the drift and diffusion velocities, become subohmic and eventually saturate, as long as no *run-away currents* preceding the dielectric breakdown are initiated. The run-away current regime will be discussed in Sections 42.4 and 47.4.1. The reason for a decrease of the transport velocity and its saturation can be seen either from microscopic arguments, dealing with the specific scattering events or from a more general picture, as discussed below.

Microscopic arguments relate to the scattering events that become more plentiful or more efficient when carriers gain more energy from the field, since the additional energy must be dissipated to the lattice according to detailed balance considerations. The energy dissipation is rapidly increased when the additional energy gained from the field exceeds the LO phonon energy. These phonons interact strongly with

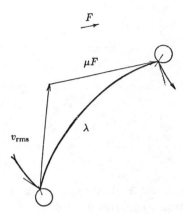

Figure 33.1: Carrier path between two scattering events at high fields, showing the comparable magnitudes of drift and thermal velocities.

electrons and effectively remove the surplus energy. Increased scattering results in a decreased mobility. Such a decrease of the mobility with sufficient fields is discussed in Sections 32.2.6–32.3. It leads into a range in which the carrier mobility decreases first proportionally to $F^{-1/2}$. At still higher fields, it decreases proportionally to $1/F$, and results in a *drift velocity saturation:* $\mu F = $ const.

General arguments identify a maximum transport velocity that can be deduced easily for a built-in field: for thermodynamic equilibrium, i.e., for zero bias and no light, the current vanishes at any position within a semiconductor. Therefore, throughout the semiconductor, the drift current here is equal and opposite in sign to the diffusion current. This diffusion current (as shown in Section 28.4.2) can never exceed the thermal velocity of the carriers, thus limiting the drift velocity to the thermal carrier velocity for these conditions. Since the carrier distribution function in thermal equilibrium is the undeformed Boltzmann distribution, this velocity is $v_{\rm rms}$, given by Eq. (28.16).

Although such an argument cannot be made for external fields, the mobility behavior is expected to result in a velocity saturation on the same order of magnitude.

33.1.1 Drift Velocity Saturation in External Fields

In an external field, the carriers are heated. The drift current is well understood at low fields (Section 28.3 and Chapter 32). At high fields, it requires the selection of the most important scattering mechanism related to material, doping, and temperature for the computation of the deformed distribution function appropriate for the field range under investigation—see Sections 32.2.6–32.3.

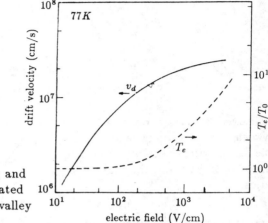

Figure 33.2: Drift velocity and electron temperature calculated for electrons in the central valley of GaAs at 77 K.

An example of the relation of the onset of field-induced changes of the mobility and of the onset of carrier heating, with consequent changes in the distribution function, can be found in Fig. 33.2, here obtained for GaAs with predominant polar-optical phonon scattering (Price, 1977).

An empirical formula describes the observed high-field behavior of the drift velocity (Jacoboni et al., 1977):

$$v_D = \mu F = \frac{\mu_0 F}{\left(1 + \left(\dfrac{\mu_0 F}{\hat{v}_m}\right)^{\beta}\right)^{1/\beta}}, \tag{33.1}$$

where $\hat{v}_m$ is the saturation velocity, β is a temperature-dependent parameter* on the order of 1, and μ_0 is the low-field mobility.

---

* The parameters $\hat{v}_m$ and β are obtained from curve-fitting, depending on materials, and for Si are (Canali et al., 1975)

$$\hat{v}_m = \begin{cases} \hat{v}_{mn} = 1.53 \cdot 10^9 \times T^{-0.87} \ (\text{cm/s}) & \text{for electrons} \\ \hat{v}_{mp} = 1.62 \cdot 10^8 \times T^{-0.52} \ (\text{cm/s}) & \text{for holes} \end{cases} \tag{33.2}$$

and

$$\beta = \begin{cases} \beta_n = 2.57 \cdot 10^{-2} \times T^{0.66} \ (\text{cm/s}) & \text{for electrons} \\ \beta_p = 0.46 \times T^{0.17} \ (\text{cm/s}) & \text{for holes,} \end{cases} \tag{33.3}$$

with the temperature in K.

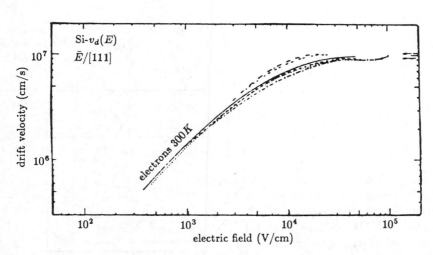

Figure 33.3: Drift-velocity limitation in Si at 300 K due to a decrease of the mobility $\propto 1/F$ at fields in the 10^4 V/cm range, applied in the [111] direction. The different curves represent different sets of measurements (collected by Jacoboni et al., 1977). ©Pergamon Press plc.

This saturation velocity is on the same order of magnitude as the rms velocity of carriers. For a specific scattering mechanism, it can be estimated; it is given in Section 33.3.2, Eq. (33.29) for optical deformation potential scattering. The drift current, e.g., for electrons, is thus limited to

$$j_n = en\mu_n(F)F \leq en\hat{v}_{mn}. \tag{33.4}$$

For a review, see Jacoboni and Reggiani (1979).

33.1.1A High-Energy Drift Measurements   The high-field carrier drift can be measured directly. Well-compensated semiconductors with good ohmic contacts are used at low temperatures to reduce the conductivity and thus Joule's heating. This, and a variety of other measurements, including microwave and time-of-flight techniques, are summarized by Jacoboni et al. (1977). Although each one has certain shortcomings (e.g., conductivity measurements rely on carrier densities obtained from low-field measurements, using Hall-effect data; time-of-flight techniques can hardly exclude shallow-level trapping), the resulting $\mu(F)$ dependencies are reassuringly similar (see Fig. 33.3) when care is taken to avoid, or to minimize, modifying effects.

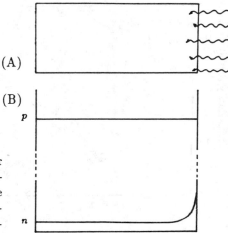

Figure 33.4: (A) Semiconductor with intrinsic optical excitation absorbed in a surface layer on the right. (B) Majority (p) and minority (n) carrier density distributions.

One method, which avoids contact effects, can be used when stationary high-field domains (Böer and Voss, 1968) can be obtained. Within such a domain, the semiconductor is space-charge free; the quasi-Fermi levels and bands are tilted parallel to each other, and the domain field is an external field. Within such a region, the Hall mobility can be measured and its field dependence can be directly obtained (Böer and Bogus, 1968), and shows current saturation near v_{rms}.

33.1.2 Carrier Diffusion Saturation

The equivalent of a drift current in an exclusively external field is the diffusion current without a space-charge region. It can be obtained in good approximation as a minority carrier current with the density gradient created in a homogeneous semiconductor by an inhomogeneous optical excitation, as indicated in Fig. 33.4. The resulting density gradient of the minority carriers causes a diffusion current of these carriers (n). This current is called the *Dember current* (Dember, 1931) in short-circuit conditions.

The minority carrier diffusion current is given by the conventional relation, using the Einstein relation

$$j_{\text{n,diff}} = \mu_n kT \frac{dn}{dx} \tag{33.5}$$

and is pictured in Fig. 33.5A. The figure also contains the two quasi-Fermi levels and shows both electrodes, at which the quasi-Fermi levels coincide. The collapse of both quasi-Fermi levels is due to

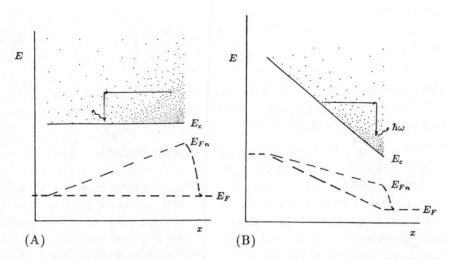

Figure 33.5: (A) Electron diffusion only, indicated by a sloping (quasi) Fermi level and a horizontal band. (B) Mixed case of built-in and external fields with carrier drift and diffusion, indicated by different slopes of the Fermi level and band. Carrier density distribution indicated by dot density.

perfect recombination at the metal surface. This diffusion current is also the total electron current as long as the field is small enough to neglect the drift component.

The diffusion current increases proportionally to the gradient of the carrier density. However, since it is derived as the difference of two random-walk currents [Eq. (28.50)] through an arbitrary surface, it is limited, as discussed in Section 28.4.1, by the Richardson-Dushman current

$$j_{n,\text{diff,max}} = e n_0 v_{\text{rms}}. \qquad (33.6)$$

The diffusion current therefore also saturates; the saturation velocity is the thermal electron velocity v_{rms}.

A comparison between Figures 33.5A and 33.5B indicates that, for a mixed field condition, carrier heating also occurs and causes a deformed carrier distribution. The root mean square velocity must then be calculated from this *deformed Boltzmann function* $f_{F_e}^*(v)$. Here, the electrochemical field gradient determines carrier heating with a similar result as for an *external field*. In order to identify this averaging with a modified distribution, the rms velocity v_{rms} is now replaced with a modified $\hat{v}_{\text{rms}}$ (Böer, 1985). Thus, one obtains

$$\boxed{j_{\text{n,diff}} = \mu_n kT \frac{dn}{dx} \le e n \hat{v}_{\text{rms}}.}$$ (33.7)

Both drift and diffusion currents are therefore limited at high-electric fields or high-density gradients, respectively, to similar velocities $\hat{v}_m$ and $\hat{v}_{\text{rms}}$ [Eqs. (33.4) and (33.7)].

33.1.2A High-Energy Diffusion Measurements The diffusion of carriers, which are excited to higher densities and thus extend within the band to higher energies, can be measured directly. These measurements involve transient effects, such as using field or light pulses, and electro-optical effects for detection. For instance, one uses an optical grating created by electron-induced bleaching, and the disappearance of the grating due to carrier diffusion after the initiating light is switched off—see also Section 36.1.

The density of carriers that determines the change in the optical absorption (see Section 12.2.1B) is given by

$$\frac{dn(x, y, t)}{dt} = g(x, y, t) - c_{\text{cv}} \left[n(x, y, t) \right]^2 + D\nabla^2 \left[n(x, y, t) \right], \quad (33.8)$$

with the optical generation rate $g(x, y, t)$ that produces the grating by bleaching of a grating design created by the interference pattern of two coherent light beams. The second term describes the recombination of these carriers [see Eq. (47.3) for $n = p$]. The third term gives the out-diffusion of carriers from the highly illuminated segments of the interference pattern, and thereby reduces its contrast.

When probing with a third light beam (Fig. 33.6), one observes a diffracted beam with a signal strength proportional to the contrast of the diffraction pattern, which provides a direct measure of the diffusion constant D—in Eq. (33.8). Such measurements were carried out by Smirl et al. (1982) for Ge, who obtained $D = 142$ cm^2/s at 135 K and 53 cm^2/s at 295 K for ambipolar diffusion under the given experimental conditions.

The results of these diffusion studies are complicated by several overlapping effects:

- the high density of carriers necessary for obtaining measurable optical absorption causes enhanced diffusion because of degeneracy;
- the many-body effects enhance diffusion due to carrier screening, and reduce it due to band gap narrowing, resulting in self-trapping;

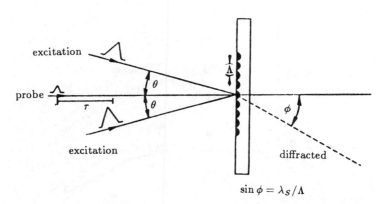

$$\sin\phi = \lambda_S/\Lambda$$

Figure 33.6: Schematics of a light-beam arrangement for creating a diffraction grating by induced electron absorption within a semiconductor, and consequent light diffraction by a low-intensity probe beam.

- the increased carrier energy, expressed by an increased carrier temperature, influences diffusion;
- the increased lattice temperature caused by optical excitation results in a decrease in diffusion; it increases scattering with phonons—see Smirl (1984) and van Driel (1985).

The diffusion coefficient of electrons as a function of an external field, heating the electrons, measured at different temperatures in Si, is shown in Fig. 33.7. The solid curves are obtained from Monte Carlo computation—see Section 33.8.1. The change of the diffusion coefficient at higher fields shows the expected $1/F$ behavior, as observed for the drift mobility in the same field range of up to $\sim 10^4$ V/cm. It is caused by a reduction in the carrier mobility.

33.2 Distribution Function at Higher Fields

We have indicated before that the carrier drift and diffusion currents saturate at a value close to the thermal rms velocity. The reasons for such saturation will be analyzed in a microscopic model in the following sections.

First, the concept of an electron temperature that is above the lattice temperature when an external field is applied will be introduced. For this purpose one starts with the Boltzmann equation. With an applied electric field, the electron distribution function $f(E)$ is deformed, as discussed in Section 30.3.

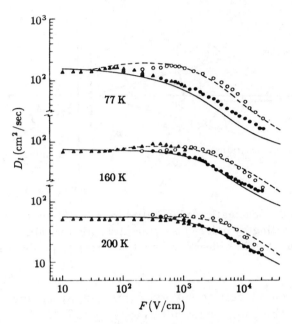

Figure 33.7: Longitudinal diffusion coefficient of electrons in Si as a function of an external field. • ⟨111⟩, ○ ⟨100⟩, time of flight measurements; △ electronic noise measurements. Curves: corresponding Monte Carlo computation (after Brunetti et al., 1981).

From Eq. (30.14), one obtains

$$f(v) = f_0(v) - \frac{e}{m_n}\tau(v)F_x\frac{\partial f(v)}{\partial v_x}. \tag{33.9}$$

For *small fields*, $f(v)$ in the derivative may be replaced by $f_0(v)$. This can be seen by forming the derivative of $f(v)$ from Eq. (33.9):

$$\frac{\partial f(v)}{\partial v_x} = \frac{\partial f_0(v)}{\partial v_x} - \frac{e}{m_n}\tau(v)F_x\frac{\partial^2 f(v)}{\partial v_x^2}; \tag{33.10}$$

when introduced into Eq. (33.9), it gives a third term that is proportional to F_x^2 and can be neglected for small fields.

When *higher fields* are applied, however, one must consider higher-order terms in Eq. (33.9); this will be discussed in the following section.

33.2.1 Warm and Hot Carriers

The expression describing the field-induced changes in the distribution function can be extended in two ways (see Seeger, 1973). The *first approach* uses an extension of the Fourier development given in Eq. (33.9):

$$f(v) = f_0(v) - \mu_n F_x \frac{\partial f_0}{\partial v_x} + (\mu_n F_x)^2 \frac{\partial^2 f_0}{\partial v_x^2} - \dots . \qquad (33.11)$$

This introduces higher terms in the definition equation for the drift velocity [Eq. (33.1)], which can now be expressed as

$$v_{nD} = \mu_{n0}(F_x + \tilde{\beta} F_x^3 + \tilde{\beta}_1 F_x^5 + \dots); \qquad (33.12)$$

the even terms vanish because f_0 is an even function.

Retaining the definition of the drift velocity as $v_D = \mu F_x$, one obtains a field-dependent mobility:

$$\mu_n = \mu_{n0}(1 + \tilde{\beta} F_x^2 + \tilde{\beta}_1 F_x^4 + \dots), \qquad (33.13)$$

with μ_{n0} as the field-independent mobility. Carriers are called *warm* when the second term in Eq. (33.13) is included, and *hot* when still higher terms are considered.

In the *second approach*, a deformed distribution function $f(E)$ is introduced, which is approximated by the Boltzmann distribution* at an elevated *carrier temperature*, $T_e > T$ (T = lattice temperature):

$$f(E) = f_0(E, T_e) \propto \exp\left(-\frac{E}{kT_e}\right). \qquad (33.14)$$

The increase in carrier temperature can be estimated from the incremental carrier energy in field direction—see also Section 42.4.2B, Eq. (42.37):

$$\frac{m_n}{2}\Delta v^2 = \mu_n e F^2 \tau_e = \frac{3}{2}k(T_e - T), \qquad (33.15)$$

where τ_e is the energy relaxation time. The field-dependent mobility can be expressed in terms of the increased electron temperature. For instance, for *warm electrons*

$$\mu_n = \mu_{n0}(1 + (T_e - T)\varphi(T) + \dots), \qquad (33.16)$$

where $\varphi(T)$ is a function of the lattice temperature alone and depends on the scattering mechanism. See Section 33.5 for the distinction between warm and hot *holes*.

---

* This approximation, however, is no longer acceptable for electrons exceeding the energy of LO phonons, as the tail of the distribution becomes substantially deformed—see Section 42.4.2B.

33.2.1A Displaced Maxwellian Distribution

In nondegenerate semiconductors, a so-called *displaced Maxwellian distribution* is often used (see Nag, 1980):

$$f(\mathbf{k}) \propto \exp\left\{ -\frac{\hbar^2(\mathbf{k} - \mathbf{k}_0)^2}{2m_n kT_e} \right\}, \qquad (33.17)$$

where $\mathbf{k}_0$ is the displacement vector of the momentum in field direction; $\hbar^2(k - k_0)^2/m_n^2$ is the drift velocity. A summary of relaxation times for the displaced Maxwellian distribution for the different scattering mechanisms is listed in Table 33.1.

For a more detailed description of the field dependence, the collision term in the Boltzmann equation must be analyzed. This will be done for a few examples in the following sections.

33.2.2 Elastic and Inelastic Scattering at High Fields

In Section 30.3.3B, it was indicated that the energy gained from an external field is transmitted to the lattice predominantly via scattering with longitudinal optical phonons. Intermediate scattering with acoustic phonons is substantially elastic. This means that the electron gains more energy from a sufficiently high electric field than it can dissipate by generating *acoustic* phonons.

At higher fields, the carrier temperature is increased markedly above the lattice temperature. However, when the carrier temperature approaches the Debye temperature, at which optical phonons can be generated in large quantities, a further rise in electron temperature is slowed down; that is, scattering increases substantially and the mobility decreases with increasing field.

A measure of the interaction between electrons and acoustic or optical phonons was obtained in Section 30.3.3B from the ratio between momentum and energy relaxation. This ratio is now changed according to the electron temperature, indicating the average energy gain of the electron ensemble from the field. For nondegeneracy, it can be estimated from (Seeger, 1973)

$$\frac{\langle \tau_e \rangle}{\langle \tau_m \rangle} \simeq \frac{3k(T_e - T)}{8m_n v_s^2}, \qquad (33.18)$$

where v_s is the sound velocity, and T is the lattice temperature. The ratio of elastic to inelastic scattering events is not significantly changed for the *average* electron, since T_e usually remains close to T, while the *fast* electrons in the high energy tail of the distribution are affected drastically (Brunetti and Jacoboni, 1984). See also Fig. 33.19.

33.3 Intravalley Scattering at High Fields

When electrons are heated only slightly, their main scattering will remain with the abundant low-energy acoustic phonons determining the momentum relaxation.

33.3.1 Scattering with Acoustic Phonons

The field dependence of the mobility can be estimated from the energy balance equation. In steady state, the collision term must equal the incremental electron energy between collisions:

$$\left\langle -\frac{\partial E}{\partial t} \right\rangle_{coll} = \frac{m}{2} v_D^2 \langle \tau_m^{-1} \rangle = e\mu F^2. \tag{33.19}$$

When this energy is dissipated with acoustic phonons, the collision term can be estimated (Seeger, 1973) as

$$\left\langle -\frac{\partial E}{\partial t} \right\rangle_{coll} = \frac{m}{2} v_{s,l}^2 \langle \tau_m^{-1} \rangle c_a \frac{T_e - T}{T}, \tag{33.20}$$

where $v_{s,l}$ is the velocity of longitudinal acoustic phonons, and c_a is a proportionality factor ($\simeq 32/(3\pi)$). The factor $(T_e - T)/T$ is introduced to account for the increased average energy of the phonons created by collisions with warm electrons at energy $(3/2)kT_e$.

Combining Eqs. (33.19) and (33.20), one obtains

$$\frac{T_e - T}{T} = \frac{1}{c_a} \left(\frac{\mu F}{v_{s,l}} \right)^2. \tag{33.21}$$

Introducing the approximation

$$\mu = \mu_0 \sqrt{\frac{T}{T_e}}, \tag{33.22}$$

one obtains for the field dependence of the mobility for *warm electrons* $(\mu_0 F \ll v_{s,l})$

$$\mu = \frac{\mu_0}{\sqrt{1 + \frac{1}{c_a} \left(\frac{\mu_0 F}{v_{s,l}} \right)^2}}. \tag{33.23}$$

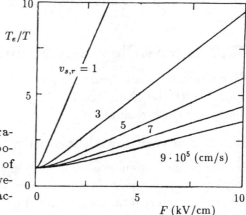

Figure 33.8: Electron temperature for acoustic deformation potential scattering as a function of the electric field with the sound velocity as the family parameter, according to Eq. (33.24).

At higher fields ($\mu_0 F \gg v_{s,l}$), one obtains for the electron temperature

$$\frac{T_e}{T} = \frac{1}{2}\left(1 + \sqrt{1 + \frac{4}{c_a}\left(\frac{\mu_0 F}{v_{s,l}}\right)^2}\right), \qquad (33.24)$$

which is shown in Fig. 33.8 as a function of the low-field drift velocity $\mu_0 F$, with the sound velocity as the family parameter. Eliminating T_e/T from Eqs. (33.22) and (33.24), one obtains the Shockley approximation (Shockley, 1951)

$$\mu = \mu_0 \left(c_a\right)^{1/4} \sqrt{\frac{v_{s,l}}{\mu_0 F}}; \qquad (33.25)$$

i.e., a square-root branch of the *field dependence of the drift velocity*

$$\boxed{v_D = \mu F = \left(c_a\right)^{1/4} \sqrt{v_{s,l}\mu_0 F}.} \qquad (33.26)$$

The field dependence of the mobility for scattering with *acoustic* phonons is given by Eq. (33.23) and shown in Fig. 33.9.

33.3.2 Scattering with Optical Phonons

At sufficiently high fields a large number of electrons have enough energy to dissipate this surplus energy through the creation of longitudinal optical phonons.

33.3.2A Scattering with Optical Deformation Potential

The scattering mechanism is similar to the one described before, except that the fraction of electrons involved in this type of scattering

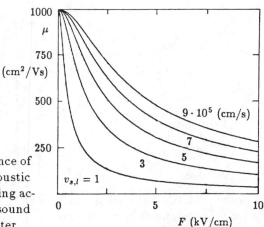

Figure 33.9: Field dependence of the electron mobility for acoustic deformation potential scattering according to Eq. (33.23). The sound velocity is the family parameter.

is larger. Therefore, a more substantial reduction of the average relaxation time $\bar{\tau}$ results, and $\mu = (e/m)\bar{\tau}$ decreases more rapidly with the electron temperature (Seeger, 1973)

$$\mu(T_e) = \frac{3(\pi)^{\frac{3}{2}}\hbar^2\rho\sqrt{k\Theta}}{2m_n^{\frac{3}{2}}D_o^2\Theta^{\frac{3}{2}}}\varphi_e(T_e),\qquad(33.27)$$

with $\varphi_e(T_e) =$

$$\frac{\left(\dfrac{T^{\frac{3}{2}}}{2\Theta}\right)\sinh\left(\dfrac{\Theta}{2T}\right)}{\cosh\left\{\left(\dfrac{T_e-T}{T_e}\right)\dfrac{\Theta}{2T}\right\}K_2\left(\dfrac{\Theta}{2T_e}\right)+\sinh\left\{\left(\dfrac{T_e-T}{T_e}\right)\dfrac{\Theta}{2T}\right\}K_1\left(\dfrac{\Theta}{2T_e}\right)}.$$

$$(33.28)$$

For K_1 and K_2, the Bessel functions, see Abramowitz and Stegun (1968). With increasing electron temperature, the electron mobility decreases. Introducing the field dependence of the electron temperature from Eq. (33.21), one obtains a drift velocity which first increases linearly with the field and then levels off at the saturation velocity

$$v_{Ds} = \sqrt{\frac{3k\Theta}{4m_n}}\tanh\left(\frac{\Theta}{2T}\right).\qquad(33.29)$$

The saturation drift velocity can be easily obtained by eliminating F from the energy balance $ev_D F = (3k\Theta/2)(\exp x - 1)$ and the momentum balance $eF = 2m^*v_D(\exp x +1)$ with $x = \Theta/T$. It is close to the thermal velocity at the Debye temperature: $(m_n/2)v_{Ds}^2 =$

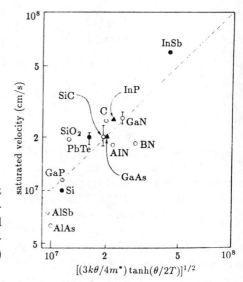

Figure 33.10: Saturation drift velocity determined from experiment (solid symbols) or detailed calculation (open circles), compared with the simple Eq. (33.29) (dashed line) (after Ferry, 1975).

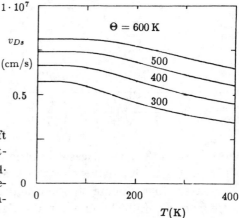

Figure 33.11: Saturation drift velocity as a function of the lattice temperature according to Eq. (33.29), for $m_n = m_0$, with the Debye temperature as family parameter.

$(3/2)k\Theta \cdot c_c$, where $c_c = 1/\{4\coth(\Theta/[2T])\}$ is a correction factor on the order of one; its temperature dependence is shown in Fig. 33.11.

The dependence of the drift velocity on the field is shown in Fig. 33.12 for various values of the lattice temperature T. It indicates that, above the Debye temperature, saturation is approached for $\mu_0 F \simeq 2v_{Ds}$. At lower temperatures, drift velocity saturation occurs more gradually.

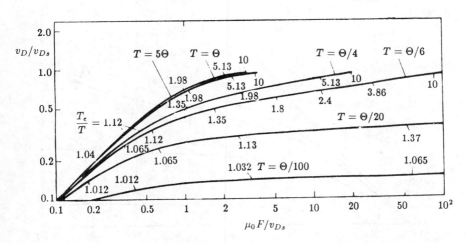

Figure 33.12: Drift velocity for optical deformation potential scattering as a function of the electric field, with T/Θ as family parameter and the relative electron temperature T_e/T given along the various curves (after Seeger, 1973).

The saturation velocities for different materials are given in Fig. 33.10, as compiled by Ferry (1975), and indicate a satisfactory agreement with Eq. (33.29).

33.3.2B Polar Optical Scattering at High Fields With polar optical scattering, the mobility as a function of the electron temperature is similar to the optical deformation potential scattering in the warm electron range. An equation similar to Eq. (33.27) is obtained, except that the order of both modified Bessel functions in the φ_e-dependence is reduced by one (Seeger, 1973).

The drift velocity does not saturate at high fields but shows an increase above the threshold field before the hot-electron range is reached. Here the onset of dielectric breakdown effects begin (Conwell, 1967—Fig. 33.13—see also Section 42.4.2).

33.3.2C High-Field Scattering with Acoustical Phonons and Ionized Impurities Since scattering by ionized impurities is elastic, its influence on the energy balance can be neglected. Its influence on the momentum relaxation can be related to the relaxation induced by acoustic scattering. The mobility due to combined scattering is given by (Seeger, 1973)

$$\mu_{\mathrm{ac,i}} = \mu_{\mathrm{ac}} \left(\frac{E}{kT} \right)^2 \left[\frac{6\mu_{\mathrm{ac}}}{\mu_{\mathrm{ion}}} + \left(\frac{E}{kT} \right)^2 \right], \qquad (33.30)$$

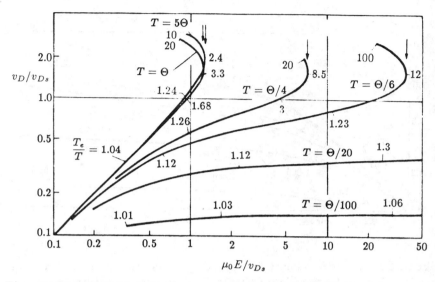

Figure 33.13: Drift velocity as a function of the field for polar optical scattering with the lattice temperature as the family parameter, indicating the onset of dielectric breakdown effects (arrows). Values of the relative electron temperature are indicated along each of the curves (after Seeger, 1973).

where E is the electron energy. At higher electron temperatures (warm electrons), one obtains after averaging

$$\mu_{\text{ac,i}} = \mu_{\text{ac}} \left[1 + \left(\frac{T}{T_e} \right)^2 \frac{6\mu_{\text{ac}}}{\mu_{\text{ion}}} \varphi_{\text{comb}} \right] \sqrt{\frac{T}{T_e}} \qquad (33.31)$$

with $\varphi_{\text{comb}} = \text{Ci}(\xi) \cos \xi + \text{Si}(\xi) \sin(\xi), \quad \xi = \frac{T}{T_e} \sqrt{\frac{6\mu_{\text{ac}}}{\mu_{\text{ion}}}},$

$$-\text{Ci}(\xi) = \int_\xi^\infty \frac{\cos t}{t} dt, \qquad \text{and} \qquad -\text{Si}(\xi) = \int_\xi^\infty \frac{\sin t}{t} dt.$$

The influence of the ionized impurity scattering is reduced with increasing electron temperature. It becomes negligible for hot electrons: faster electrons can penetrate closer to a Coulomb center, and thus experience a smaller scattering cross section.

33.4 Intervalley Scattering at High Fields

One distinguishes between intervalley scattering the equivalent and nonequivalent valleys. Equivalent valleys have their minima at the same energy, the conduction band edge. Nonequivalent valleys have a slightly higher energy, and hence need additional carrier energy to become populated.

33.4.1 Equivalent Intervalley Scattering

Intervalley scattering with optical phonons, using a deformation potential approximation, can be evaluated for warm electrons. The resulting mobility is given by (see Streitwolf, 1970)

$$
\mu_i = \frac{4}{3} \frac{\alpha_c k \Theta_i}{\hbar} \sqrt{\frac{T_e}{\Theta_i}} \frac{\sinh\left(\dfrac{\Theta_i}{2T}\right)}{\cosh\left[\left(1 - \dfrac{T}{T_e}\right)\dfrac{\Theta_i}{2T}\right] K_1\left(\dfrac{\Theta_i}{2T_e}\right)}, \tag{33.32}
$$

where K_1 is the modified first-order Bessel function.

At higher electric fields, the anisotropy of the valleys must be taken into consideration. The direction of the field relative to the valley axis identifies the relevant effective mass, which changes with different alignment. It is smallest in the direction of the short axis, and largest in the direction of the long axis of the rotational ellipsoid. Hence, electron heating, which is proportional to $\mu F = (e/m_n)\tau_m F$, is most effective when the field is aligned with the short axis of the ellipsoid. This ellipsoid is called the *hot ellipsoid*, and the corresponding valley, the *hot valley*; the other with the long axis in the field direction is called the *cool ellipsoid*, and the corresponding valley, the *cool valley*.

Scattering proceeds preferentially from the hot to the cool valley, since hot electrons have a higher average energy and consequently generate more intervalley phonons. This influences the average electron mobility—see Fig. 33.14.

The largest repopulation is observed when the field is applied in the direction of one of the main axes of the ellipsoids. No change in the population occurs when the field direction is 45° off the direction of the main axes.

33.4.1A The Sasaki-Shibuya Effect

The anisotropy of the intervalley scattering at high fields, which causes a repopulation from hot to cool valleys, also causes the effective conductivity to become anisotropic. The current prefers to flow in a direction closer to an alignment with the long axis of the ellipsoid; that is, the current may deviate from the direction of the applied field. As a result, the surfaces of the semiconductor, having a component in this preferred direction, become charged.

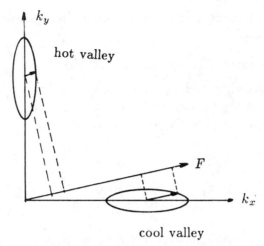

Figure 33.14: Intervalley scattering into nonequivalent valleys from hot to cool valleys. Hot valleys have their short axis aligned closer to the field direction.

Figure 33.15: Sasaki-Shibuya measurement of longitudinal and transverse currents in a semiconductor with elliptical valleys.

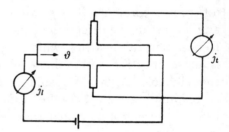

This produces a field component perpendicular to the applied field. It is similar to the Hall field, which is caused by surface charging in a magnetic field. Applied and induced field vectors define an angle ϑ, called the *Sasaki angle*:

$$\frac{F_t}{F_l} = \frac{j_t}{j_l} = \tan \vartheta \simeq \vartheta. \tag{33.33}$$

Subscripts t and l stand for *transverse* and *longitudinal*. The effect, called the *Sasaki-Shibuya effect*, can be used to obtain information about the anisotropy of the equi-energy surfaces in $E(\mathbf{k})$ (Shibuya, 1955; Sasaki et al., 1959).

33.4.2 Intervalley Scattering into Nonequivalent Valleys

A rather large effect is observed when electrons are scattered into higher valleys in which they have a substantially higher effective mass—as, e.g., in GaAs.

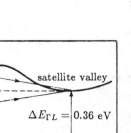

Figure 33.16: Electron scattering into a higher satellite valley with higher effective mass (GaAs).

In an external field of sufficient magnitude, electrons in a valley with a small effective mass are heated very efficiently. When scattered into higher satellite valleys (see Fig. 33.16) with a higher effective mass, heating is reduced and backscattering is lowered. Therefore, a substantial fraction of the conduction electrons can be pumped into the higher valley. As a result, the average electron mobility is reduced, and the current therefore increases less than ohmically with increased field (see Jacoboni et al., 1981 for Ge; and Alberigi-Quaranta et al., 1971 for compound semiconductors).

33.4.2A Negative Differential Conductivity Effects If the ratio of mobilities is high enough, one observes a *negative differential conductivity* as shown in curve 2 of Fig. 33.17A. Corresponding experimental results are shown in Fig. 33.17B for several III-V compounds. Here a field range exists in which the current decreases with increasing field. In this range, stationary solutions of the transport equation may not exist, and high-field domains may develop which can move through the semiconductor and cause current oscillations. These have been observed by Gunn (1963), and are termed *Gunn effect* oscillations.

If such a repopulation, with an increasing electric field, competes with thermal excitation at higher temperatures, only a change in the slope of the current-voltage characteristic is observed, without going through a maximum—curve 1 in Fig. 33.17. Increased doping in the high-doping range causes a decrease in the threshold field, and finally the disappearance of the negative differential conductivity (Seeger, 1973).

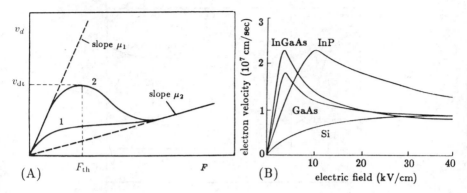

Figure 33.17: (A) Drift velocity as a function of the field when major repopulation of the higher satellite valley with substantially lower effective mass becomes effective at the threshold field F_{th} (schematic). (B) Measured field dependence of the drift velocity for several semiconductors (after Evans and Robson, 1974). ©Pergamon Press plc.

In GaAs, the L satellite minimum lies 0.36 eV above the Γ-point $E(\mathbf{k})$ minimum at $\mathbf{k} = 0$—see Fig. 33.16. The effective mass in the central minimum is $m_n = 0.07m_0$, and $m_n \simeq m_0$ in the satellite minima.* This causes a reduction of the mobility in n-type GaAs from ~ 8000 cm^2/Vs at low fields to ~ 200 cm^2/Vs at high fields, when most of the electrons are pumped into the satellite valley. Fields of 3 kV/cm are sufficient for achieving the necessary pumping.

33.5 Warm and Hot Holes

At higher fields, the mobility of holes decreases because of the re-population of the different valence bands. In typical semiconductors with light and heavy hole bands, the heavy bands become more pop-ulated at higher fields because of both the more effective heating of the holes in light bands and the more effective scattering. This causes a mobility reduction. At still higher fields, the more effective optical phonon scattering cuts off the distribution function at $\hbar\omega_{LO}$, and finally leads to a $1/F$ decrease of the mobility, i.e., to drift velocity saturation.

Drift velocity saturation is observed at lower temperatures, while at higher temperatures the onset of breakdown effects interferes (Seeger, 1973).

---

* The effective mass in one of the four satellite minima is $0.4m_0$; hence, one has $(0.4^{3/2} \times 4)^{2/3} m_n = 1.01m_0$ for the density of state mass.

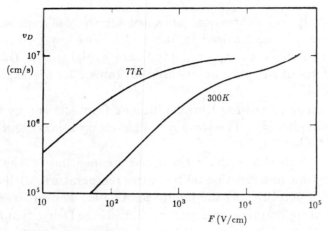

Figure 33.18: Saturation of the drift velocity of holes in p-type Ge (after Bray and Brown, 1960; Prior, 1960).

The Sasaki-Shibuya effect for p-type Ge confirms the existence of nonspherical (warped) $E(\mathbf{k})$ surfaces in the heavy hole valence band (Gibbs, 1962).

33.6 Summary of High-Field Relaxation

At sufficient external fields, the carrier distribution function is substantially deformed. It may be approximated by a Boltzmann distribution with a carrier temperature that significantly exceeds the lattice temperature $(T_e > T)$:

$$f_e = \frac{n}{I_n} \exp\left(-\frac{E}{kT_e}\right), \qquad (33.34)$$

where I_n is a normalization integral, given by

$$I_n = \frac{1}{4\pi^3} \int_{Bz} \left(-\frac{E(\mathbf{k})}{kT_e}\right) d\mathbf{k}, \qquad (33.35)$$

with the integration extending over Bz, the Brillouin zone.

At high electron densities, scattering between carriers is more frequent than with lattice defects. Here, a displaced Maxwellian distribution describes the electron ensemble:

$$f(\mathbf{k}) = c \exp\left(-\frac{\hbar^2 |\mathbf{k} - \mathbf{k}_0|^2}{2m^* kT_e}\right) = c \exp\left(-\frac{E - \hbar\mathbf{v}_D \cdot \mathbf{k}}{kT}\right). \qquad (33.36)$$

See also Eq. (33.17).

The analytical expressions obtained for the different scattering mechanisms are summarized in Table 33.1. The low-field relaxation times used here (e.g., τ_{ac}, τ_{pz}, etc.) are explained in Table 32.3. The meaning of the parameters used in Table 33.1 are explained in Section 32.4.

The *energy relaxation time* is obtained from the energy loss rate with optical phonons. The corresponding energy relaxation times are listed in Table 33.1.

Figure 33.19 shows the ratio of the momentum to the energy-relaxation time as a function of the lattice temperature, with different electron temperatures as the family parameter. It decreases rapidly with increasing temperature, with $\tau_m \simeq \tau_e$ at the Debye temperature.

33.7 Mobility Changes as Result of Optical Excitation

As a result of optical excitation, a number of defects will change their charge character. Charged centers may become neutral (e.g., a donor-type center after trapping an electron, or an acceptor-type center after trapping a hole), while other centers may become negatively or positively charged when trapping a photoexcited carrier or when ionized by light.

The redistribution of carriers, indicated by the split of the two quasi-Fermi levels with light, can be used to identify the degree of recharging: the centers between the two quasi-Fermi levels are preferentially recharged. Centers just above E_{Fp} tend to be depleted, whereas centers just below E_{Fn} tend to be filled by the optical excitation. The filling of centers between these quasi-Fermi levels is inverted from thermodynamic equilibrium—see Fig. 33.20; for further discussion, see Chapter 45.

The changes in the densities of neutral and charged centers consequently cause changes in the relative magnitude of scattering on these two types of defects, as described in Sections 32.2.3 and 32.2.4.

In addition, an optical generation initially creates carriers which are not in thermal equilibrium. Usually, they are excited to levels substantially beyond the band edge. Between continuous excitation and relaxation due to scattering, the resulting distribution function becomes severely deformed.

Both the defect center recharging and the change in the distribution function will cause the mobility to become a function of the optical generation rate g_o:

Table 33.1: Momentum and energy-relaxation times for heated carriers (after Nag, 1984).

Scattering Mechanism	Momentum Relaxation Time		Energy Relaxation Time
	Maxwell Distribution	Shifted Maxwell Distribution	
Acoustic Phonon Deformation Potential	$\tau_{acM} = \tau_{ac}\left(\dfrac{300}{T}\right)^{3/2}\sqrt{\dfrac{T}{T_c}}$	$\dfrac{9\pi}{32}\tau_{pM}$	$\dfrac{3\sqrt{\pi}}{8}\tau_{acE}\sqrt{\dfrac{300}{T_c}}\sqrt{\dfrac{T}{T_c}}$ $\tau_{acE} = \dfrac{\pi\varrho h^4}{2^{9/2}m^{*5/2}\Xi_c^2\sqrt{300}\,k}$
Piezoelectric	$\tau_{pzM} = \tau_{pz}\sqrt{\dfrac{300}{T}}\sqrt{\dfrac{T_c}{T}}$	$\dfrac{9\pi}{32}\tau_{acM}$	$\dfrac{3\pi}{4}\tau_{pzE}\sqrt{\dfrac{T}{300}}\sqrt{\dfrac{T}{T_c}}$ $\tau_{pzE} = \dfrac{\sqrt{2}\,\pi\varrho h^2\varepsilon_s^2\sqrt{300}\,k}{e^2\chi_{pz}^2m^{*3/2}}$
Optical Phonon, Nonpolar	$\tau_{opM} = \tau_{op}\dfrac{\exp\left(\dfrac{\Theta_0}{T_c}\right)-1}{\exp\left(\dfrac{\Theta_0}{300}\right)-1}$ $I_{op1} = 2\displaystyle\int_0^\infty \dfrac{z^{3/2}\exp(-z)\,dz}{\sqrt{z+\dfrac{\Theta_0}{T_c}}+\exp\left(\dfrac{\Theta_0}{T_c}\right)H\left(z-\dfrac{\Theta_0}{T_c}\right)\sqrt{z-\dfrac{\Theta_0}{T_c}}}$	$\dfrac{9\pi}{2}\tau_{opM}\dfrac{I_{op2}}{I_{op1}}$ $I_{op2}=\dfrac{1}{(T_c/\Theta_0)^2}\left[\exp\left(\dfrac{\Theta_0}{2T_c}\right)K_2^{+}+\exp\left(\dfrac{\Theta_0}{T}-\dfrac{\Theta_0}{2T_c}\right)K_2^{-}\right]$ $K_2^{\pm}=k_2\left(\dfrac{\Theta_0}{2T_c}\right)\pm k_1\left(\dfrac{\Theta_0}{2T_c}\right)$	$\dfrac{3\sqrt{\pi}}{2}\tau_{opE}\left[\left(\dfrac{T_c}{300}\right)^{3/2}\left(\dfrac{\Theta_0}{\Theta_0}\right)\left(\exp\left(\dfrac{\Theta_0}{T}-\dfrac{\Theta_0}{T_c}\right)-1\right)\dfrac{\exp\left(\dfrac{\Theta_0}{T}\right)-1}{\exp\left(\dfrac{\Theta_0}{300}\right)-1}\right]$ $\tau_{opE}=\dfrac{\sqrt{2}\,\pi\varrho h^2\sqrt{300}\,k}{D_0^2m^{*3/2}}$
Optical Phonon, Polar		$\dfrac{3}{2}\tau_{pop}\dfrac{I_{pop}}{\sqrt{\dfrac{T_c}{300}}}$ $I_{pop}=\dfrac{1}{\left[\exp\left(\dfrac{\Theta_0}{T_c}-\dfrac{\Theta_0}{2T_c}\right)K_1^{+}+\exp\left(\dfrac{\Theta_0}{2T_c}\right)\pm K_0\left(\dfrac{\Theta_0}{2T_c}\right)\right]}$ $K_1^{\pm}=k_1\left(\dfrac{\Theta_0}{2T_c}\right)\pm k_0\left(\dfrac{\Theta_0}{2T_c}\right)$	$\dfrac{3\sqrt{\pi}}{2}\tau_{popE}\left[\exp\left(\dfrac{\Theta_0}{T}-\dfrac{\Theta_0}{T_c}\right)-1\right]\exp\left(\dfrac{\Theta_0}{2T_c}\right)K_1\left(\dfrac{\Theta_0}{2T_c}\right)$ $\tau_{popE}=\dfrac{2^{3/2}\pi h^2\varepsilon_0\left\{\exp\left(\dfrac{\Theta_0}{300}\right)-1\right\}}{e^2\sqrt{m^*300}\,k}$
Ionized Impurity	$\tau_{impM}=\tau_{imp}\left(\dfrac{T}{300}\right)^{3/2}\left(\dfrac{T_c}{T}\right)^{3/2}I_{i1}$ $I_{i1}=\dfrac{1}{6}\displaystyle\int_0^\infty \dfrac{z^3\exp(-z)\,dz}{\ln\left(1+\dfrac{z}{S_{c1}}\right)-\dfrac{1}{1+\dfrac{S_{c1}}{z}}}$ $S_{c1}=S_c\dfrac{T}{T_c}$	$\dfrac{3\pi}{32}\tau_{impM}\dfrac{I_{i2}}{I_{i1}}$ $I_{i2}=\displaystyle\int_0^\infty\left\{\ln\left(1+\dfrac{z}{S_{c1}}\right)-\dfrac{1}{1+\dfrac{S_{c1}}{z}}\right\}\exp(-z)\,dz$	

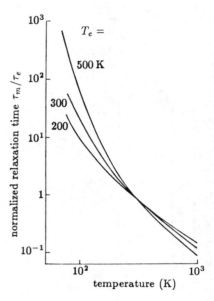

Figure 33.19: Ratio of momentum relaxation time to optical nonpolar phonon scattering as a function of the lattice temperature with electron temperature as family parameter (after Nag, 1984).

$$\mu = \mu(g_o). \tag{33.37}$$

33.7.1 Energy Relaxation of Optically Excited Carriers

The absorption into states higher in the conduction bands produces high-energy electrons which will relax into lower states according to their energy relaxation time. As a consequence, the electron gas is heated. Scattering with longitudinal optical phonons is the main energy-dissipation mechanism to the lattice.

On the other hand, if the optical excitation proceeds within less than kT of the band edge, the produced electron distribution is compressed and may be approximated by a Boltzmann distribution of a lower-than-lattice temperature. This means that the *electron gas is cooled*; this is referred to as *optical cooling*. The effect, however, is accompanied by a heating of the *lattice*, caused by the fraction of nonradiatively recombining electrons, which produce large amounts of phonons.

Further inside the band, scattering probabilities are different between the different branches, and vary widely within each branch. Again, scattering with LO phonons is the predominant energy-

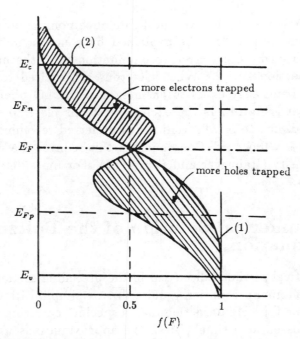

Figure 33.20: (1) Fermi distribution in thermal equilibrium, and (2) quasi-Fermi distribution in steady state with light. Ranges of redistributed carriers are hatched; the upper part is more negatively charged, the lower part more positively. The densities of recharged levels under both must be equal to maintain quasineutrality.

relaxation mechanism. Fröhlich and Paranjape (1956) estimate the energy-relaxation time for such scattering:

$$\tau_{e\mathrm{LO}} = \frac{1}{\alpha_c \omega_{\mathrm{LO}}} \frac{k}{\cosh^{-1}(ka)} = 10^{-13} \left(\frac{0.1}{\alpha_c} \frac{10^{14}}{\omega_{\mathrm{LO}}} \right) \frac{k}{\cosh^{-1}(ka)} \quad (\mathrm{s})$$

(33.38)

with $k = \sqrt{2m_n\omega_{\mathrm{LO}}/\hbar}$. For more recent estimations on energy relaxation of optically generated carriers, see Section 48.2.2C.

Using the Boltzmann equation, the surplus energy $E - E_c$, given by the optical excitation to an electron in the conduction band, can be equated to the surplus energy provided by an external electric field (Ferry, 1980). However, the influence of optical excitation on the carrier distribution is distinctly different; it can severely skew the distribution. Kinetic effects are helpful in distinguishing the different relaxation mechanisms—see Section 48.2.

An example of optical carrier heating is the oscillatory photoconductivity observed in InSb and GaSb, when monotonically increasing the photon energy of the exciting light. Such nonmonotonic changes in photoconductivity can be explained by a nonmonotonic carrier density caused by a preferred recombination from the band edge— see e.g., Section 20.1.2. When electrons are excited to an energy above the band edge equal to an integer of the LO phonon energy, the LO scattering brings the excited electrons close to the bottom of the band, i.e., to $k \simeq 0$, from which preferred recombination with holes near $k = 0$ occurs. Consequently, this reduces the steady-state carrier density (Habegger and Fan, 1964). For more detail, see Section 45.5.4.

33.8 Numerical Solution of the Boltzmann Equation

The analytical methods discussed in the preceding sections for solving the Boltzmann equation require drastic simplifications and permit only one type of interaction, i.e., a specific scattering mechanism to be discussed at a time. Often the approximations used are justified only in a very limited parameter range. This becomes critical at higher fields, when major deviations from the Boltzmann distribution are observed. Here, the truncation of an expansion of the distribution function, as given in Eq. (30.23), is no longer an acceptable approximation. Numerical solutions of the specific transport problems are now an advantage.

An *iterative technique* proposed by Budd (1966), extending the *variable path method* of Chambers (1952), yields solutions of the Boltzmann equation by stepwise processing the evolving carrier distribution function (Rode, 1970). A modification of this method, involving a fictitious *self-scattering* for the purpose of mathematical simplification, was proposed by Rees (1968—see also Rees, 1969, 1972).

Even more popular, however, is a simulation method of the actual carrier motion called the *Monte Carlo method*.

33.8.1 Monte Carlo Method

This method simulates the motion of a carrier under the influence of an electric field and the different scattering mechanisms by applying basic kinetic laws. Here, one stochastically selects the scattering process (phonons, impurities, carrier-to-carrier, etc.) and the final states with a probability distribution given by the density of states

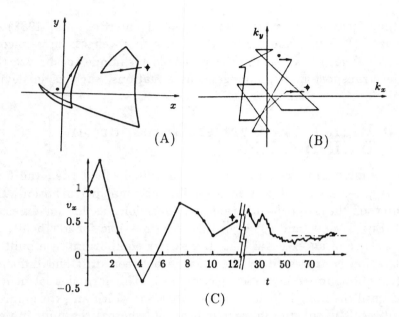

Figure 33.21: Schematics for Monte Carlo method. (A) Real space path of a carrier with large field in the x direction; (B) same path as in (A), but in momentum space; (C) carrier velocity averaged over all steps starting from step 1 up to the running step number as a function of simulation time (after Brunetti and Jacoboni, 1984).

times the squared matrix elements, determined from a microscopic theory of the different scattering centers (Kurosawa, 1966).

When successive scattering events of one electron are followed long enough, its behavior is equivalent to that of the average behavior of the entire electron ensemble (*ergodicity*).

Figure 33.21 illustrates the principle of the Monte Carlo technique. Subfigure A shows the actual electron path in two dimensions, under the influence of a large external field. Subfigure B shows the same eight events as heavy line segments in momentum space. These heavy lines are interfaced by light lines representing the momentum changes in each of the scattering events. Subfigure C gives the velocity of the carrier averaged at the nth point over all previous $(n-1)$ paths. This average velocity approaches the drift velocity (dash-dotted line) when enough paths are taken; 100 paths are shown in subfigure C. The drift velocity is a direct measure of the mobility, which then can be obtained by using the relation $v_D = \mu_D F$.

The Monte Carlo method permits the extraction of derived physical information from simulated experiments, and is a powerful tool for the discussion of stationary (Jacoboni and Reggiani, 1983) or transient (Lebwohl and Price, 1971) transport effects in semiconductors. It also has become the preferred technique to analyze the carrier transport in nonhomogeneous situations, such as in device simulation.

33.9 High-Field Carrier Transport in Built-In Fields

It is common practice to assume for a built-in field (i.e., the field in a space-charge region with negligible external perturbation) the validity of the same basic transport equations as for an external field [Eqs. (28.59) and (28.60)]; this is permissible for low fields. In the absence of an external bias, the carrier equilibrium in a built-in field region is expressed by the balance between drift and diffusion currents of opposite sign—see Section 29.1. This formal treatment is valid until the built-in field exceeds values for which an external field would result in substantial carrier heating, whereas the built-in field does not. In this field-range, the mobility becomes field-dependent and, for reasons indicated in Chapter 29, the difference between a built-in and an external field must be considered.

33.9.1 Saturation Currents with Built-In Fields

The limiting diffusion current is given by the Richardson-Dushman thermal emission current—Section 29.2.1,

$$j_{n,\text{diff},\max} = j_{n,\text{RD}} = env_{\text{rms}}^*, \tag{33.39}$$

with $v_{\text{rms}}^* = v_{\text{rms}}/\sqrt{6\pi}$ for planar geometry and Boltzmann statistics; for built-in fields, the carrier distribution function remains a Boltzmann distribution at lattice temperature.

Within a space-charge region, e.g., a *pn*-junction at zero bias, in thermodynamic equilibrium (no external excitation) the drift and diffusion currents must be equal with opposite sign to render the total electron and hole current separately zero in every volume element of the semiconductor. Therefore, even for high built-in fields, the maximum drift velocity cannot exceed the Richardson-Dushman velocity

$$v_{D,m}^* = v_{\text{rms}}^*. \tag{33.40}$$

Therefore, the drift current must also be limited to

$$j_{n,\text{drift},\max} = env_{\text{rms}}^*. \tag{33.41}$$

With applied bias, however, the situation becomes more complex and needs additional discussion.

33.9.2 High-Field Carrier Transport in Mixed Fields

Space-charge regions with external bias show a mixed, partially built-in, partially external field behavior—Fig. 33.5B. In different regions of the semiconductor, the field is predominantly external; whereas in others, it is predominantly a built-in field, as one can show in a numerical analysis of the solution curves of the conventional set of transport, continuity, and Poisson equations (Böer, 1985).

In contrast to a semiconductor with only an external field, the carrier transport controlled by the built-in part of the field is not accessible to direct measurements. Both drift and diffusion currents are highly inhomogeneous. In a typical *pn*-junction, these currents show a maximum on the order of 10 kA/cm^2, where the carrier gradient peaks, although the net current is usually a very small fraction thereof.

Since within the built-in field region the net current is small compared to the drift or diffusion current, one obtains from

$$j_{n,\text{drift}} \simeq -j_{n,\text{diff}},\qquad (33.42)$$

an equation that is *independent of the mobility*:

$$\frac{1}{n}\frac{dn}{dx} = \frac{eF}{kT} = -\frac{e}{kT}\frac{d\psi}{dx}.\qquad (33.43)$$

It yields the well-known *Boltzmann condition*, resulting in an exponential dependence of the carrier density on the electrostatic potential or band edge:

$$n = n_0 \exp\left(-\frac{e(\psi_n - \psi_{n0})}{kT}\right).\qquad (33.44)$$

This relation holds throughout most of the built-in field region in which Eq. (33.42) is sufficiently well fulfilled. This region is therefore referred to as the *Boltzmann region*. It represents the parts of the space-charge region with a sufficiently high majority carrier gradient, and, at zero bias, comprises the entire space charge region.

With bias, however, there are regions adjacent to the Boltzmann region in which Eq. (33.42) no longer holds. These are the regions of predominant drift or diffusion currents—see the quantitative discussion by Böer (1985a). These regions with predominant drift and diffusion currents are referred to as DRO- or DO-regions for "drift only" or "diffusion only," respectively. In semiconductors within a built-in field region, they represent well-distinguished regions of major drop in the minority carrier electrochemical potential, and lie adjacent to the part of a junction with dominant space charge.

Substantial carrier heating in space-charge regions (Böer, 1985) is restricted to minority carriers; that is, quasi-Fermi levels of minority carriers show the highest slopes when a bias is applied. Marked majority carrier heating in devices with space-charge regions will occur only when the external fields become comparable to the built-in field, and this occurs only close to breakdown.

Summary and Emphasis

High external electric fields cause a deformation of the carrier distribution with more carriers at higher energies. In a first approximation, such a deformed distortion can be approximated by a Boltzmann distribution, with an elevated carrier temperature, and with acceptable errors, when $T_e - T \ll T$ (warm carriers).

When carriers are accelerated further, the distribution is skewed substantially near and above $\hbar\omega_{LO}$, where substantial inelastic scattering occurs, and the concept of carrier heating with a further elevated T_e (hot carriers) becomes less satisfactory. A better description includes higher terms in the Fourier development of the distribution function, or mobility.

With increasing electric field, an increased scattering counteracts an increased accumulation of energy from the field. Such increased scattering is observed for interaction with phonons, causing a decrease in mobility, first $\propto 1/\sqrt{F}$ from mostly acoustical scattering, and at higher fields $\propto 1/F$ from LO-phonon scattering.

Scattering with ionized impurities, however, decreases with increasing field, as carriers can penetrate closer to the center before being scattered.

The dominating effect of LO-phonon scattering causes a saturation of the drift velocity close to the rms velocity of carriers. Similarly, a saturation of the diffusion current is observed at high carrier density gradients, with a maximum effective diffusion velocity also close to the rms velocity.

With additional thermal excitation, the saturation branch of the drift velocity may be hidden by dielectric breakdown effects.

When excitation of a carrier gas occurs because of photons, a substantially larger skewing of the carrier distribution is observed, with concurrent changes in carrier mobility.

High-field carrier transport needs consideration in many semiconducting devices which operate in the range of warm carriers. Moreover, most devices contain space-charge regions in which built-in fields in excess of 10 kV/cm are common. A more detailed analysis of the transport properties in the range of high built-in fields is required for a better understanding of device operation.

Exercise Problems

1.(r) Give a simple criterion for cases in which high-field effects must be taken into consideration for evaluating the carrier transport.

2. Evaluate the distinction between warm and hot *electrons*. What are the differences for warm and hot *holes*?

3. How does carrier heating influence the relation between elastic and inelastic scattering events?

4.(e) Describe explicitly, using the effective mass picture, the intervalley scattering between nonequivalent valleys.

5.(e) Justify the description of hot and cold valleys in intervalley scattering between equivalent valleys.

6.(*) When analyzing the classical path of an electron between scattering events, give a criterion at what fields the simple Sommerfeld-Drude description yielding Ohm's law breaks down. Assume properties for Si at T= 300 K, and a permissible error of 5%.

7.(a) Plot to scale the distribution function for GaAs at fields of 10^2, 10^3, $2 \cdot 10^3$, and $3 \cdot 10^3$ V/cm, assuming simple carrier heating, using Fig. 33.2.

8.(e) Calculate the bleached diffraction pattern created from two laser beams at 5000 Å incident under an offset angle of 5°, and give the angle of the first-order diffraction for the test beam—see Fig. 33.6.

9.(e) Discuss and draw in a double logarithmic scale the mobility as a function of the electric field for a hypothetical material with $\mu_0 = 100$ cm²/Vs, sound velocity $v_{s,l} = 10^5$ cm/s, and acoustic deformation potential scattering, using the approximation given by Eq. (33.22).

9.(*) Describe quantitatively the charging of the side-surfaces in the Sasaki-Shibuya effect for electrons in Ge.

10. Discuss in detail Fig. 33.20, which shows the redistribution of carriers from a Fermi-Dirac distribution to a deformed steady-state distribution with optical excitation.

11.(*) Discuss drift-field saturation at high electric fields
   (a) at room temperature for Si;
   (b) at room temperature for GaAs; and
   (c) at low temperatures, where in the low field range, predominant scattering occurs at ionized defects.

12.(*) Carrier diffusion saturates at the Richardson-Dushman limit. Discuss the influence of a changed carrier distribution function.
   (a) Why do you expect a change in electron distribution for larger electron gradients?
   (b) How does a changed distribution influence the diffusion current?

Chapter 34

Phase Transitions in Electron-Hole Systems

Phase transitions occur in semiconductors when the concentration of donors, carriers, or excitons exceed critical values. Then, processes can take place which turn the semiconductor into a metal—or, at sufficiently low temperatures, the condensation of these quasi-particles results in a new phase.

In Chapter 25 we saw that states in the gap lose their localization when they are close enough to each other in space and energy. These states then form a narrow band and, when partially filled, contribute to electronic conduction. When the Fermi-level remains pinned within such a band, the material continues to conduct for $T \to 0$ K, i.e., it behaves like a metal.

There are other ways to initiate metal-like behavior. For instance, rather than having the overlap of hydrogen-like *donors* creating such a band, an overlap of similarly bound *excitons* could create a critical condition. The excitons start to interfere with each other, causing their dissociation and producing free carriers. A sufficiently large density of excitons can be created by an adequately high optical excitation.

A similar critical process can be initiated by a sufficient optical generation of free electrons and holes. When their density is large enough, the interaction between electrons and holes becomes strong and they can no longer be regarded as a gas of independent particles; at low enough temperatures, condensation of the carrier ensemble into a liquid phase takes place, which causes substantial changes in the behavior of the semiconductor and renders it metal-like.

Finally, when a sufficiently high density of excitons , which are bosons, can be maintained at very low temperatures, Bose-Einstein

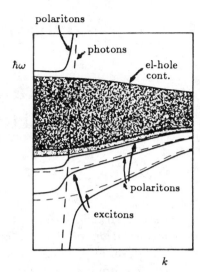

polaritons

photons

$\hbar\omega$

el-hole
cont.

polaritons

excitons

k

Figure 34.1: Dispersion relation of photons and excitons (dashed curve), exciton polaritons (solid curves), and electron-hole continuum (dark shaded). Higher excitons are in the lightly shaded region (after Egri, 1985).

condensation of these excitons is expected. This condensation process may be preceded by the formation of biexcitons.

All these processes require high carrier (exciton) densities and strong interaction. This occurs when their Coulomb energy becomes larger than their thermal energy:

$$\frac{e^2}{4\pi\varepsilon_{\mathrm{st}}\varepsilon_0\bar{r}} \simeq \frac{e^2}{4\pi\varepsilon_{\mathrm{st}}\varepsilon_0}n^{1/3} > kT, \qquad (34.1)$$

where $\bar{r}$ is the average distance between the carriers. Condition (34.1) is fulfilled for $n > 4\cdot10^{16}(\varepsilon_{\mathrm{st}}/10)^3(T/300\text{ K})^3$. However, this is usually not sufficient to initiate condensation, which will be discussed in the following sections. For a review, see Rice (1977) or Keldysh (1986).

For phase transitions involving traps and the adjacent band with changes in carrier densities induced by external parameters, e.g., the electric field, see Landsberg and Pimpale (1976) and Schöll (1987).

34.1 Electron-Hole Condensation

Condensations imply phase transitions. These are best understood for the electron-hole system, with ample experimental verification. Condensation within an electron-hole system is similar to the phase transition of normal gasses into liquids, with a well-known thermodynamic description. Therefore, we will discuss this case first.

At low temperatures and sufficient densities, achieved by intense optical excitation, electrons and holes can occupy a lower-energy state as interpenetrating fluids than as excitons or biexcitons. A schematic plot of the dispersion relation for the different excitons, photons, and exciton-polaritons is given in Fig. 34.1, together with the $E(k)$ range for the electron-hole continuum and the range in which higher excitons (biexcitons) are found.

At a high density, optically created electrons and holes thermalize with the lattice, typically within $\tau_{th} \simeq 1$ ns, before their carriers recombine, typically within $\tau_r \simeq 10$ μs. Therefore, these carriers can be regarded as having acquired lattice temperature. At low enough temperatures they tend to form excitons.

When the density of these excitons becomes large enough to interact with each other, i.e.,

$$n_{exc} = n_c \simeq a_{qH}^{-3}, \tag{34.2}$$

they could form exciton molecules or condense to a liquid, depending on the relative magnitude of the corresponding binding energies. Usually, condensation takes place when the thermal energy of the exciton gas has decreased below

$$kT_c \simeq 0.11 E_{qH} \tag{34.3}$$

(Landau and Lifshitz, 1976; Reinecke and Ying, 1979) with T_c as the critical temperature. Such condensation is similar to that of forming a liquid metal from its vapor phase: during the condensation, the orbits of valence electrons disappear, changing their behavior into that of free electrons. However, the heavy ion cores here are replaced by holes which have a mass on the same order of magnitude as the electrons, causing a breakdown of the adiabatic approximation. As a result, large zero-point vibrations with amplitudes of the exciton diameter prevent solidification of the electron-hole liquid at $T = 0$ K and yield a much lower binding energy of the liquid than for the core atoms in a corresponding metal.

An example is the condensation of mercury vapor with isolated atoms into the metallic liquid phase, which was discussed by Landau and Zel'dovich (1943). The gas-liquid and the insulator-metal transitions can be separated here.

The resulting electron-hole liquid has metallic properties where electrons and holes are free to move as interpenetrating Fermi liquids. The condensation into an electron-hole liquid was originally proposed

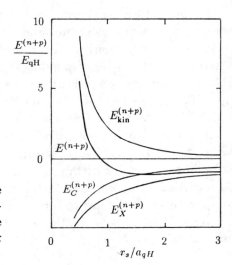

Figure 34.2: Dependence of the different energies for the electron-hole liquid as a function of the relative intercarrier distance (after Vashista et al., 1983).

by Keldysh (1968). We will describe the properties of this liquid in Section 34.1.3—see also Section 45.2.2B.

The condensation process is governed by the ground-state energy of the electron-hole system. It can be described with a liquid and gas-phase diagram, similar to the one for first-order phase transitions for vapor to liquid condensation.

34.1.1 Energy of the Electron-Hole Liquid

The energy of the electron-hole liquid consists of three components: the kinetic or Fermi energy, the Coulomb energy, and the exchange energy which arises from the Pauli principle.

The kinetic energy is determined by the sum of the Fermi energies of the electron and the hole gasses

$$E_{\text{kin}}^{(n+p)} = \frac{3}{5}(E_F^{(n)} + E_F^{(p)}) = \frac{3}{10}\hbar^2(3\pi^3 n)^{2/3}\left(\frac{1}{m_n} + \frac{1}{m_p}\right) = \frac{2.21}{r_s^2}E_{\text{qH}},$$
(34.4)

where r_s is a dimensionless parameter, measuring the average interparticle distance

$$\frac{1}{n} = \frac{4\pi}{3}r_s^3 a_{\text{qH}}^3.$$
(34.5)

The exchange energy is determined by the Coulomb term

$$E_C^{(n+p)} = -\frac{3}{2\pi}\frac{e^2}{4\pi\varepsilon_{\text{st}}\varepsilon_0}(3\pi^2 n)^{1/3} \simeq -\frac{1.83}{r_s}E_{\text{qH}}.$$
(34.6)

Table 34.1: Critical parameters for electron-hole droplets. E_ϕ is the binding energy per carrier pair (after Tikhodeev, 1985).

Material	E_ϕ (meV)	T_c (K)	n_c (cm^{-3})	$n(T = 0\,\text{K})\,\text{cm}^{-3}$
Si	23	28	$1.2 \cdot 10^{18}$	$3.5 \cdot 10^{18}$
Ge	6	6.7	$6 \cdot 10^{16}$	$2.3 \cdot 10^{17}$
GaP	6			$7 \cdot 10^{18}$
GaS	9			$4.5 \cdot 10^{20}$
AlAs	14.5			$12 \cdot 10^{19}$
GaAs	1	6.5	$4 \cdot 10^{15}$	$1 \cdot 10^{16}$
CdS	14	64	$7.8 \cdot 10^{17}$	$5.5 \cdot 10^{18}$
CdSe	5	30	$1.2 \cdot 10^{17}$	$8.3 \cdot 10^{17}$
CdTe	0.9	18	$4.4 \cdot 10^{16}$	$2.9 \cdot 10^{17}$
ZnO	22	70	$4 \cdot 10^{17}$	$2.8 \cdot 10^{19}$
ZnS	12	79	$1.4 \cdot 10^{17}$	$8.3 \cdot 10^{17}$
ZnSe	5			$3.2 \cdot 10^{18}$
ZnTe	3			$6.6 \cdot 10^{17}$
AgBr	30			$1 \cdot 10^{19}$

The correlation energy stems from the Pauli exclusion principle, which causes an increase in the distance between carriers of the same spin and consequently reduces the Coulomb term. It cannot be given explicitly (for more detail, see Vashista et al., 1983), but may be approximated for the sum of exchange and correlation energy by

$$E_{\text{xc}}^{(n+p)} = -\frac{4.8316 + 5.0879 r_s}{0.0152 + 3.0426 r_s + r_s^2} E_{\text{qH}}. \qquad (34.7)$$

The total binding energy per carrier pair of the electron-hole liquid is given by

$$E_\phi = E^{(n+p)} = E_{\text{kin}}^{(n+p)} + E_C^{(n+p)} + E_X^{(n+p)} = E_{\text{kin}}^{(n+p)} + E_{\text{xc}}^{(n+p)} \qquad (34.8)$$

which is an attractive energy because of the Coulomb interaction, but becomes repulsive when the interparticle distance decreases much below a_{qH}. For an example, see Fig. 34.2. The binding energy per electron-hole pair in the liquid phase is given in Table 34.1.

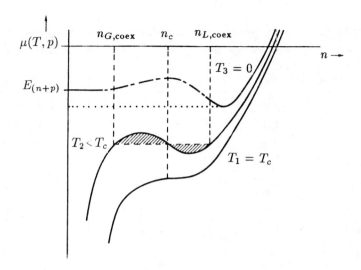

Figure 34.3: Schematics of the chemical potential as a function of the particle density with the temperature as family parameter. The lowest curve is shown for the critical temperature T_c, which has at n_c an inflection point with horizontal tangent (after Reinecke, 1982).

34.1.2 Phase Diagram for Electron-Hole Condensation

The properties of a system consisting of excitons, biexcitons, electrons, holes, and the electron-hole liquid at any given temperature can be described by the phase diagram.

When plotting the chemical potential $\mu = n(\partial F/\partial n)_T$, with F as the free energy, as a function of the particle density, one obtains a family of curves as shown in Fig. 34.3.

At temperatures below T_c, one observes condensation. For instance, when at T_2 the density increases above $n_{G,\text{coex}}$, a liquid forms of density $n_{L,\text{coex}}$, with a remaining vapor of density $n_{G,\text{coex}}$.

It is instructive to plot the phase diagram in a T vs. $\ln(n/n_c)$ representation (Fig. 34.4), which is rather universal when scaled to the individual N_c, T_c values. In any isothermal process below T_c, one observes at low densities a gas which, in equilibrium, contains excitons and excitonic molecules (biexcitons). At higher densities, it will appear as a mixture of electron-hole vapor plasma and liquid. Beyond the second-phase boundary, it will show only the liquid phase—Fig. 34.4.

The relevant critical values are listed in Table 34.1 for a number of semiconductors. The critical temperatures are on the order of $6\ldots60$ K for Ge and CdS, respectively. The critical densities are

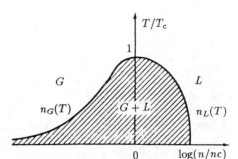

Figure 34.4: Phase diagram for an exciton gas G and an electron-hole liquid L (after Keldysh, 1986).

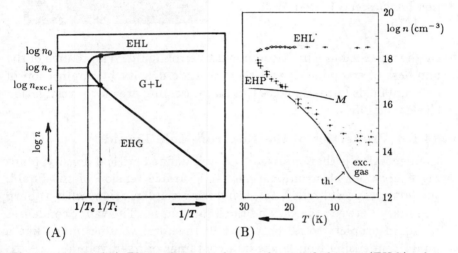

Figure 34.5: (A) Phase diagram for an electron-hole gas (EHG), electron-hole liquid (EHL), and the intermediate range of coexisting gas and liquid (G+L). (B) Experimental phase diagram for Si (after Diete et al., 1977). The diagram also contains the theoretical curve (*th*) and the curve for Mott condensation (*M*) (after Keldysh, 1986).

related to the quasi-hydrogen radius [see Eq. (34.10)], and the critical temperature to the binding energy as

$$n_c^{1/3} \simeq 0.2 a_{qH}^{-1} \quad \text{and} \quad kT_c \simeq 0.1 E_\phi. \qquad (34.9)$$

Typical values for n_c range from 10^{15} to 10^{20} cm^{-3} for GaAs and CuCl with II-VI compounds in between. Typical binding energies range from ≤ 1 meV for GaAs to ≥ 50 meV for AgBr and CuCl (Thomas and Timofeev, 1980). Binding energies, hence T_c, are increased in materials with highly anisotropic effective mass and in multivalley semiconductors. At low temperatures, electrons rarely exchange between different valleys; therefore, the electron-hole liquid

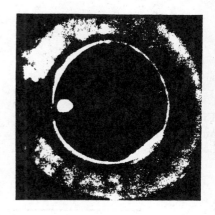

Figure 34.6: Electron-hole droplet in a circular Ge-platelet, shown as a small light spot at the left (observed by electron-hole recombination luminescence) (after Wolfe et al., 1975).

can be regarded as a multicomponent Fermi liquid. The change of the numbers of components and their relative density by application of external fields (stress) makes this an interesting area for investigation (Keldysh, 1986).

34.1.3 Properties of the Electron-Hole Liquid

Condensation is characterized by a well-defined critical density, binding energy per electron-hole pair and surface tension of the liquid; the latter causes droplets to form during condensation, with a sharp boundary between the gas and the liquid phase. The vapor condenses to liquid droplets to fill only part of the total volume, much like a liquid condensing from a gas in a container of fixed volume.

Such drops can be detected directly because of their different diffractive index, which gives rise to increased laser-light scattering within the band gap (Hensel et al., 1978), or by their specific luminescence,* shown in Fig. 34.6—see also Section 44.3.3C.

Electron-hole liquid drops are typically a few μm in diameter, although somewhat larger ones have been observed at low temperatures. Their limiting radius is ~ 10 μm for Ge and ~ 1 μm for Si (Keldysh, 1986). However, when strain-confined, larger drops up to 1 mm diameter are observed (Wolfe et al., 1975). The strain produces a band deformation which acts as a well for the electron-hole drops. With further decreasing temperature, the number of such droplets increases. Up to 10^8 drops/cm^3 have been observed.

---

* There is a typical luminescence signature of the electron-hole liquid, separated from the exciton luminescence peak (see Section 44.3.3D) that permits one to follow the development of condensation and evaporation of droplets.

In contrast to ordinary drops, which accumulate to a larger body of liquid, electron-hole droplets normally repulse each other, probably due to the phonon wind created by internal heat, which in turn is caused by electron-hole recombination. This limits the droplet size. Droplets are mobile and respond to the phonon wind interacting via the deformation potential (Damen and Worlock, 1976). A friction component in the droplet motion is the scattering with phonons.

Compared to any other known liquid, the electron-hole liquid has the lowest density $(m_n + m_p)n_c \simeq (10^{-9} \ldots 10^{-12})$ g cm^{-3}. It is distinguished from an ordinary liquid by the lack of a rest mass, although it has well defined effective masses; there is a lack of heavy particles (atomic nuclei). Both electrons and holes have a finite lifetime.

Carriers are confined within such drops as electrons are within a metal. They have to overcome a workfunction to exit from the drops. With increasing temperature, electron-hole pairs can form excitons by evaporating from the drops. With cessation of excitation, this goes on until all droplets have evaporated (see Section 48.2.2D): the recombination lifetime in indirect gap semiconductors is often longer than the time for evaporation.

The drops have a high mobility and can travel with velocities up to the velocity of sound. Since they contain electrons and holes in equal amounts, no electric current flows, but an energy current of $nE_g \simeq 1$ Jcm^{-3}. Such movement of drops can be induced by forces due to gradients of stress, electric and magnetic fields. For a review of strain-confined droplets, see Wolfe and Jeffries (1983); see also Bohnert et al. (1981), Kreissl et al. (1982), Klingshirn and Haug (1981), Tikhodeev (1985), and Keldysh (1986).

34.2 The Mott Transition

At temperatures and densities of shallow donors or acceptors near the critical values at which excitons would dissociate and form an electron-hole plasma with consequent electron-hole droplet formation, another process may take place. When the average distance between hydrogen-like impurities (Section 9.1.1A) approaches their diameter, they interfere with each other and dissociate. This produces more free carriers, which contribute to screening, consequently causing at low temperatures ionization of yet un-ionized hydrogen-like donors, in turn producing more free carriers. Such a transition does not require the external excitation necessary for electron-hole

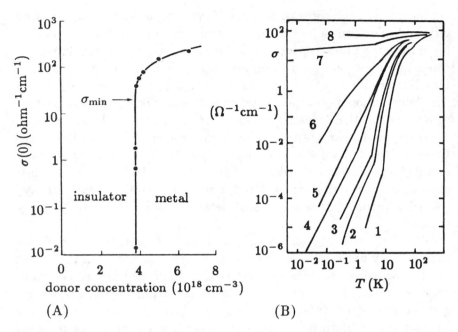

Figure 34.7: (A) Electrical conductivity of Si:P extrapolated to $T = 0$ K as a function of the donor (P) concentration (after Rosenbaum, 1980). (B) Resistivity of Si:P as a function of the temperature with donor density as family parameter: $N_d = 1.73$, 2.06, 2.21, 2.42, 2.50, 2.62, 3.34, and $3.58 \cdot 10^{18}$ cm^{-3} for curves 1–8, respectively (after Sasaki, 1980). A semiconductor-metal transition is accomplished when a finite conductivity is maintained at $T = 0$ K.

liquid condensation. As a consequence, a rather abrupt transition occurs, the *Mott transition* (Mott, 1974), which causes an increase in conductivity and makes the semiconductor metal-like. This transition can be achieved in thermal equilibrium: a semiconductor turns into a metal at high doping densities. After such a transition, the electrical conductivity remains high and nearly constant down to $T \to 0$ K, as shown in Fig. 34.7B for Si with a donor density above $\sim 3.2 \cdot 10^{18}$ cm^{-3}. The semiconductor now behaves like a metal.

The critical density for the Mott transition is closely related to the Anderson-Mott localization density [Eq. (25.17)] and is attained when the average distance between shallow donors or acceptors is approximately five times their quasi-hydrogen radius:

$$n_M \simeq 0.01 \frac{1}{a_{\text{qH}}^3}, \quad \text{or} \quad n_M^{-1/3} \simeq 5a_{\text{qH}} \quad (34.10)$$

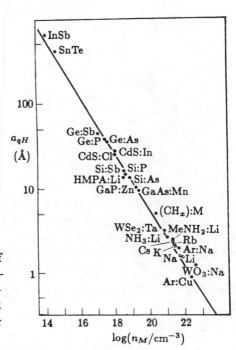

Figure 34.8: Mott density of donors for the insulator-metal transition as a function of the quasi-hydrogen donor radius. Straight line is given by Eq. (34.10) (after Edwards and Sienko, 1981). ©American Chemical Society.

(Mott and Davis, 1979). For more detail, see Section 40.2 and Mott (1987). Figure 34.8 indicates good agreement with the experiment for a variety of semiconductors. These critical densities are within the accuracy of the estimate on the same order of magnitude as the densities for electron-hole droplet formation—see Section 34.1.2.

34.3 Bose-Einstein Condensation of Excitons

Still another condensation process is the Bose-Einstein condensation of excitons,* which could take place at low temperatures and high densities of excitons (Blatt et al., 1962; Keldysh and Kopaev, 1965). In contrast to the electron-hole liquid, which is described by inter-penetrating Fermi-fluids, the excitons constitute a Boson gas. Excitons can condense when the De Broglie wavelength of the exciton at

---

* With strong photon coupling, however, one has polaritons rather than excitons. They cannot accumulate near $K = 0$ because of their photon nature. Therefore, Bose-Einstein condensation of these quasi-particles is impossible.

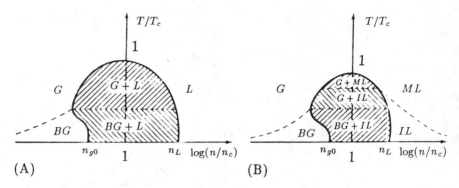

Figure 34.9: Phase diagram for exciton gas (EG) into electron-hole liquid (L) condensation; (A): including Bose-Einstein condensation to biexciton gas (BG), and (B): including also a transition from a metal-like (ML) to an insulating liquid (IL) (after Keldysh, 1986).

thermal velocity becomes equal to the interparticle distance (Blatt et al., 1962):

$$\lambda_{\mathrm{DB}}(T_{\mathrm{cr}}) = \frac{h}{\sqrt{2m_{\mathrm{exc}}kT_{\mathrm{cr}}}} \leq \frac{1}{\sqrt[3]{n_{\mathrm{exc}}}} \qquad (34.11)$$

where m_{exc} is the effective mass of the exciton, and T_{cr} is the critical temperature for Bose-Einstein condensation. This temperature is related to a sufficient exciton density:

$$T_c = 0.575 \frac{2\pi\hbar^2}{km_{\mathrm{exc}}} n_{\mathrm{exc}}^{2/3} = 3.17 \cdot 10^{-11} \frac{m_0}{m_{\mathrm{exc}}} n_{\mathrm{exc}}^{2/3}, \qquad (34.12)$$

and is on the order of 10 K at $n_{\mathrm{exc}} \simeq 10^{17}$ cm^{-3} for a typical exciton mass.

At low enough temperatures, excitons can associate to biexciton molecules which are also Bosons. When the energy E_ϕ to form an electron-hole liquid is higher than that to form an exciton (E_{qH}), and the dissociation energy of biexcitons E_B (typically $\simeq 0.1E_{\mathrm{qH}}$), i.e., when

$$E_\phi > E_{\mathrm{qH}} + E_B, \qquad (34.13)$$

then a Bose-Einstein condensation of biexcitons can take place, provided the temperature is low enough (Keldysh, 1968). There is some experimental evidence that Bose-Einstein condensation of excitons occurs in Cu$_2$O, and of biexcitons occurs in CuCl (Mysyrowicz et al., 1984).

Such quantum-condensation is preferred to a molecular-liquid condensation, since the lack of heavy particles suppresses molecular attraction necessary for liquification (Brinkman and Rice, 1973). It actually causes an intermolecular repulsion. Such repulsion also results in an increase of the energy per exciton molecule and, with increasing density, may reach a value to violate Eq. (34.13). Now, the electron-hole liquid becomes energetically the lowest state. Therefore, exciton dissociation and the condensation of the resulting electron hole plasma into an electron-hole liquid will proceed. The corresponding phase diagram is shown in Fig. 34.9A, indicating the phase boundary between the exciton gas (EG) and the Bose-Einstein condensed quantum-molecular gas (BG).

34.4 Condensation and Insulator-Metal Transition

We can now try to resolve some ambiguities between the condensation into an electron-hole liquid, and the Mott transition into a metallic phase.

At sufficiently low temperatures, a collective electron and hole interaction may be strong enough for a gap-formation around the Fermi-level. Consequently, the metal-like electron-hole liquid may become insulating (Keldysh and Kopaev, 1965; Des Cloizeaux, 1965; Halperin and Rice, 1968). Such a transition into an insulating liquid (IL) is indicated in Fig. 34.9B.

In addition, there are similarities between the Mott semiconductor-metal transition and the electron-hole condensation. How these two transitions are related to each other is not clear at this time. There is the possibility of two consecutive phase transitions; each one with its own critical temperature (7 K and 4.5 K in Ge) and a triple point in between (6 K in Ge) in which all three phases—insulating gas, insulating liquid, and metallic liquid—coexist (Schowalter et al., 1982).

Some evidence of an even more complex condensation phenomenon is obtained from a time-resolved analysis of the luminescence spectrum of an ultrapure Si crystal at very high excitation densities (Smith and Wolfe, 1986). It indicates the creation of a fixed-density electron-hole plasma ($n_{cp} = 2.3 \cdot 10^{17}$ cm^{-3}) before the electron-hole liquid is created with $n_{EHL} \simeq 2 \ldots 3 \cdot 10^{18}$ cm^{-3}.

The electron-hole droplet condensation is directly observed by light scattering and changes in the luminescence; the Mott condensation manifests itself by an abrupt increase in electrical conductivity (see Fig. 34.7A) at low temperatures. Such semiconductor-metal transitions, following Eq. (34.10), are observed for a variety of semiconductors (Edwards and Sienko, 1978). For a discussion of the conductivity relating to the statistical distribution of hydrogen-like defects and the resulting energy distribution, see Section 25.3.1 (Mott, 1984).

Summary and Emphasis

At low temperatures and high quasi-particle densities, a number of first- or higher-order phase transitions take place with substantial changes in the optical and electronic behavior.

One of these transitions involves excitons which can form exciton molecules (biexcitons). These biexcitons can in turn form a Bose-Einstein condensate with increasing density at very low temperatures. At higher density, they may condense to an insulating electron-hole liquid, which at still higher densities becomes metallic (conductive).

At slightly higher temperatures, the exciton gas may directly condense to an electron-hole liquid, which may be insulating, and at higher density becomes metallic.

At still higher temperatures—but below a critical temperature T_c—the condensation from the exciton gas may directly form conductive electron-hole droplets. These droplets can be optically observed; they are held together by surface tension, have a small maximum size due to phonon wind causing continuous evaporation, and can accelerate with up to the velocity of sound.

These drops are stable at somewhat higher temperatures in multi-valley and anisotropic semiconductors. With a uniaxial stress applied, such stabilization can take place in otherwise isotropic semiconductors, and permits the formation of much larger electron-hole drops.

The formation of electron-hole drops has a profound influence on electro-optical properties of semiconductors at low temperatures and high optical excitation. They provide a fertile ground for a large variety of exciting experiments relating to phase transitions of multi-component systems.

Exercise Problems

1.(e)   Plot the kinetic and the correlation energy for GaAs in a reduced scale E/E_{qH} vs. r/r_{qH}.

2.(*)   What changes occur in the plot when Si, rather than GaAs, is considered? Be specific as to proper scaling and considering the anisotropy of the effective masses.

3.(r)   Review the different critical processes with respect to energy, temperature, and density.

4.(r)   What effect does the mass ratio of electrons and holes have on the different condensates compared to a metal or an H_2-gas? Discuss the relevance of this question to the adiabatic approximation.

5.   What are the driving forces for electron-hole droplet motion? How can one stabilize a larger sized droplet? Explain quantitatively the stabilizing force.

6.(r)   How are electron-hole droplets observed? Identify their shape and follow their motion.

7.(*)   Discuss the Bose-Einstein condensation of excitons and biexcitons. Derive the critical temperature and identify limits when exciton dissociation would interfere.

Chapter 35

Superconductivity

High-temperature superconductivity is observed in layered semiconductors and is carried by hole-related quasiparticles.

Presently, there seem to be two classes of semiconductors that show superconductivity, low-temperature superconductors and high-temperature superconductors, which apparently follow different paths to become superconductive.

The low-temperature superconductors are highly doped semiconductors, or metals, with their superconductivity carried by electron-related quasi-particles. These materials become superconductive at very low temperatures. Most high-temperature superconductors are semiconductors normal to the layer plane of their highly anisotropic lattices at elevated temperatures. With sufficient carrier densities, they become superconductors at low temperatures (40...125 K), possibly after a semiconductor-metal transition has occurred—see Section 34.2. There is recent experimental evidence that the transition to the superconductive state can occur directly from a semiconductor state (Park et al., 1988). Their superconductivity is carried by hole-related quasi-particles.

We will first review some of the basic properties of the conventional low-temperature superconductors.

35.1 Low-Temperature Superconductors

According to the Bardeen-Cooper-Shrieffer (BCS) theory (Bardeen et al., 1957), the formation of electron pairs (*Cooper pairs*) can lead to superconductivity. For reviews, see Schrieffer (1964), deGennes (1966), or more recently, Allen and Mitrović (1982).

Such Cooper pairs follow Bose-Einstein statistics, and can undergo Bose-Einstein condensation when the denominator in the distribution function [Eq. (30.9)] vanishes. This condensation is believed to be responsible for superconductivity (Blatt, 1961; Onsager, 1961). The BCS theory predicts that the lowest energy state of such electron pairs is separated from higher energy states by an energy gap, which prevents any damping of the electron transport at low temperatures (for a general overview, see Rose-Innes and Rhoderick, 1978). For this damping, scattering events must occur with energy transfer. However, since all states below the band gap are occupied, no such scattering is possible for the Cooper pairs. They can move through the lattice as condensed bosons without energy loss, similarly to superfluid helium.

Although low-temperature superconductivity is observed in metals, and is therefore beyond the realm of this book, a short review of the basic concepts of the BCS theory is in order, as it may provide some guidance for designing new superconducting semiconductors. For a short summary, see Madelung (1981).

35.1.1 Cooper Pairs and Condensation

Despite the strong Coulomb repulsion, electrons can form pairs if they have opposite spin, and are assisted in pair formation by lattice polarization. One can understand this assistance by recognizing that an electron can create its own polarization well, a state known as the polaron—see Section 28.1.2. When this well is deep enough, resulting in a small polaron, it becomes possible at sufficiently low temperatures to trap another electron in the same well, forming an electron pair, which is sometimes referred to as a *doublon* or *bipolaron*—Section 28.1.3.

The electron interaction with the lattice, necessary for pair formation, can be analyzed mathematically by assuming emission and absorption of phonons from these electrons, which results in eigenstates below those of two independent electrons. Pairs formed between two electrons with opposite spin and wave vector directly at the Fermi surface of a metal or a highly degenerated semiconductor at $T = 0$ K are called *Cooper pairs*. With pair formation, they can lower their energy to occupy a state at Δ *below* the Fermi surface:

$$E_{\mathrm{Cp}} = (E_F - \Delta) \quad \text{with} \quad \Delta = \hbar\omega_D \exp\left\{ \frac{2}{g(E_F)V} \right\}, \quad (35.1)$$

where ω_D is the Debye frequency, $g(E_F)$ is the density of states at E_F, and $-|V|$ is the interaction energy, which is of the formi

$$V = |M_{kq}|^2 \frac{2\hbar\omega_q}{\{E(k+q) - E(k)\}^2 - (\hbar\omega_q)^2}, \qquad (35.2)$$

with M_{kq} as the electron-phonon matrix element. A reduction in energy is possible despite full occupation of every single energy state below E_F by electrons, since Cooper pairs are bosons and as such can condense at a lower ground-state energy.

One concludes that an energy Δ per electron is gained when Cooper pairs are formed. This renders the filled Fermi sphere unstable, and provides the driving force for Bose-Einstein condensation of such pairs.

When condensation of Cooper pairs takes place, the excited states for individual electrons are given by

$$E(\mathbf{k}) = \sqrt{(E - E_F)^2 + \Delta^2}, \qquad (35.3)$$

which has a minimum at $E_F + \Delta$, and creates a gap between the ground and excited states of

$$\Delta E_C = 2\Delta. \qquad (35.4)$$

This energy gap makes superconductivity plausible. It does not permit any scattering with an energy exchange of less than 2Δ. That is, an electron can only be scattered when the displacement of the Fermi sphere in an external field exceeds 2Δ:

$$\Delta E = \frac{\hbar^2}{2m}\left\{ (k_F + \delta k)^2 - (k_F - \delta k)^2 \right\} = \frac{2\hbar^2 k_F \delta k}{m} \geq 2\Delta. \quad (35.5)$$

Therefore, the current density, $\mathbf{j} = env = en\hbar\mathbf{k}/m$, must remain below a *critical current density*, above which superconductivity vanishes:

$$j \leq \frac{2en\Delta}{\hbar k_F} = j_{\text{crit}}. \qquad (35.6)$$

With typical values for $\Delta \simeq 1$ meV, $n \simeq 10^{22}$ cm^{-3}, and $k_F \simeq 10^8$ cm^{-1}, one has $j_{\text{crit}} \simeq 10^7$ A/cm^2 in an order-of-magnitude agreement with the experiment. In a classical, sense these high current densities signalize a very high drift velocity of superconducting electrons: from $j = env_D$, one concludes that v_D at j_{crit} must be on the order of 10^4 cm/s.

35.1.2 Critical Temperature, Two-Fluid Model

Cooper pair formation requires pairs of electrons with $\mathbf{k} \uparrow$ and $-\mathbf{k} \downarrow$ at the Fermi surface. The density of such pairs decreases with increasing temperature, following Fermi-Dirac statistics. Since the gap energy is a function of available electrons [$g(E_F)$ was used in Eq. (35.1) at $T = 0$ K], Δ shrinks with increasing temperatures and vanishes at a critical temperature, which can be estimated from a numerical integration of Eq. (35.1):

$$kT_c \simeq 0.57\Delta(T = 0). \tag{35.7}$$

Since $T_c \propto \Delta \propto \omega_D \propto 1/\sqrt{M}$, one expects an increase in the critical temperature with decreasing mass of the oscillating atoms. With different isotopes, only the mass changes; all other parameters are left unchanged. Such an *isotope effect* on T_c is indeed observed.

The superconductive state can be approximately described as consisting of a superfluid liquid of Cooper pairs, which is mixed with a liquid of normal electrons; the density ratio of both is determined by the temperature. Above T_c, no Cooper pair can exist and the gap vanishes. Therefore, all carriers above T_c are normal electrons. Below T_c, more and more electrons condense as Cooper pairs.

Evidence for such a two-fluid model can be obtained from optical absorption and ultrasound attenuation. Only the Cooper pairs have an energy gap, yielding a far-IR absorption edge. Only the free electrons provide ultrasound attenuation; Cooper pairs cannot be split by the low energy of ultrasonic phonons.

This model yields a temperature-dependent density ratio between electrons and Cooper pairs, and provides evidence for a temperature-dependent gap $\Delta(T)$, as shown in Fig. 35.1.

35.1.3 Meissner-Ochsenfeld Effect

In addition to a vanishing resistance, one observes that the magnetic field is expelled from within a superconductor when cooled below T_c, an effect referred to as the *Meissner-Ochsenfeld effect* (Meissner and Ochsenfeld, 1933). This perfect diamagnetic behavior cannot be explained solely by a vanishing conductivity. For this, one must modify Maxwell's equations of electrodynamics by the London phenomenological theory for superconductors, with the London equations

$$\nabla \times \mathbf{j}_s = -\frac{\mathbf{B}}{\mu_0 \lambda^2} \quad \text{and} \quad \frac{\partial \mathbf{j}_s}{\partial t} = \frac{\mathbf{E}}{\mu_0 \lambda^2}, \tag{35.8}$$

where λ is the London penetration depth of the magnetic field (London and London, 1935).

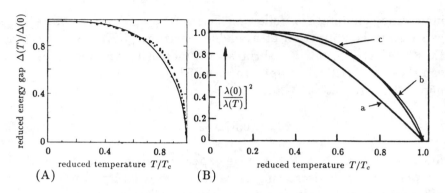

Figure 35.1: (A) Temperature dependence of the energy gap; solid curve: BCS theory; circles: experimental results from ultrasonic attenuation in tin (after Ryckayzen, 1965). (B) Penetration depth of the magnetic field after BCS theory and local or nonlocal approximation for curves a and b, respectively, and according to Eq. (35.9) for curve c (after Fetter and Walecka, 1971). ©John Wiley & Sons, Inc.

The penetration depth is on the order of 500 Å in typical metals. With increasing temperatures, the density of Cooper pairs decreases, and with it the penetration depth increases according to

$$\lambda(T) = \lambda(0) \frac{1}{\sqrt{1 - \left(\dfrac{T}{T_c}\right)^4}}, \qquad (35.9)$$

as shown in Fig. 35.1B. When the penetration depth exceeds the thickness of the wire, superconductivity has vanished.

35.1.3A Type-I or Type-II Superconductors Superconductors that exclude the magnetic flux as described in the previous section are called type-I superconductors. They are preferably diamagnetic. Above a critical magnetic field H_c, the entire superconductor becomes normal conductive. The phase boundary between the normal and superconductive state can be described by

$$H_c = H_0 \left[1 - \left(\frac{T}{T_c}\right)^2 \right] \qquad (35.10)$$

where H_0 is the magnetic field at $T = 0$ K, usually on the order of $0.1 \ldots 1$ kG—for Al and V, respectively, as examples. All superconductive metals are type-I, except for Nb.

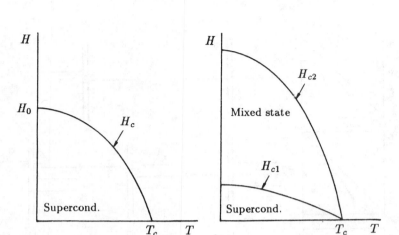

Figure 35.2: Phase boundaries between normal and superconductive states. (A) For type-I and (B) for type-II superconductors, with critical field boundaries indicated.

Type-II superconductors have two critical magnetic fields, H_{c1} and H_{c2}. Below H_{c1}, the superconductor displays type-i behavior. Above H_{c1}, the magnetic field penetrates into the superconductor and microscopic flux lines, also called *fluxoids* or *vortices*, are formed. Each flux line contains one magnetic flux quantum, and has a diameter of $\sim 10^{-5}$ cm, which fills the superconductor in a regular pattern—the flux lattice. Above H_{c2}, the type-II superconductor becomes normal-conductive. Typical fields H_{c2} are an order of magnitude higher than H_{c1} with the maximum known value at ~ 600 kG for $PbMo_6S_8$—*Chevrel phase* superconductors. A typical phase diagram for type-I and type-II superconductors is shown in Fig. 35.2.

The flux line pattern is influenced by crystal defects. Flux lines may move in the presence of a current that creates forces normal to these lines. Consequently, some minor power dissipation is observed in such type-II superconductors above H_{c1}. Some crystal defects are known to pin flux lines, and thereby create a more stable superconductor at high fields. This is important for NMR imaging, where extremely constant fields over long periods are required.

35.1.4 Josephson Tunneling

The most direct evidence for the energy gap in superconductors is obtained from tunneling between two metals through a very thin insulating layer.

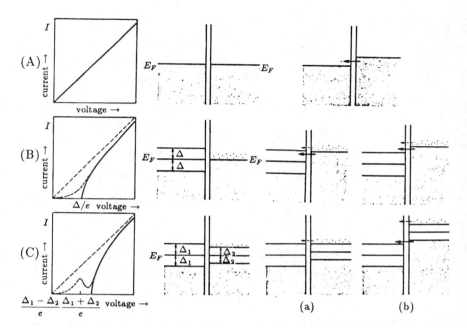

Figure 35.3: Tunneling through a thin insulating layer between two metals: (A) both metals are normal conductors; (B) one metal is a superconductor; and (C) both metals are different superconductors. Columns (a) and (b) for Giaever and Josephson tunneling, respectively. Solid line for the current-voltage characteristics at $T = 0$; dashed curve for Giaever tunneling at $T > 0$.

If both are normal metals, tunneling proceeds at $T \simeq 0$ K without threshold—Fig. 35.3A.

If one of the metals is a superconductor, no current is drawn near zero bias. Only when electrons are elevated above the gap does a small current flow (*Giaever tunneling*—Fig. 35.3Ba). With further increased bias, electrons from below the Fermi surface can be drawn across the barrier (*Josephson tunneling*—Fig. 35.3Bb), and the tunneling current increases steeply.

If both metals are superconductors, a maximum of the Giaever tunneling is observed when the bias is raised to $(\Delta_1 - \Delta_2)/e$ (Fig. 35.3Ca), while Josephson tunneling starts when the bias exceeds $(\Delta_1 + \Delta_2)/e$ (Fig. 35.3Cb).

Such a current-voltage characteristic thereby offers a simple way to measure the gap.

35.2 Low-Temperature Superconducting Semiconductors

Superconductivity in semiconductors was predicted by Gurevich et al. (1962) and by Cohen (1964), and in 1964 was observed in $Ge_{1-\xi}Te$, SnTe, and $SrTiO_3$ (Hein et al., 1964; Schooley et al., 1964). Transition temperatures are typically below 1 K, and depend on the stoichiometry, which influences the carrier density. The critical temperature increases with increasing carrier density—decreasing stoichiometry. When the electron density exceeds 10^{20} cm^{-3} in $SrTiO_3$, the transition temperature agains decreases (Appel, 1966). Many-valley semiconductors have a higher critical temperature as they have a larger density of states, since the density of state mass is increased—see Eq. (27.25).

35.3 High-T Superconductors

Transition temperatures, exceeding the highest observed in metal alloys, have been reported for several quaternary and quinternary compounds. These are layered compounds that are semiconductors at or above room temperatures.

An example is the Ba-La-Cu-O system, which at the appropriate composition can undergo a semiconductor-metal transition. At temperatures up to 40 K, some of these compounds show superconductivity (Bednorz and Müller, 1986).

Several other such quaternary oxides have been found which remain superconductive up to temperatures surpassing the boiling point of liquid nitrogen—Y-Ba-Cu-O, Y-La-Cu-O, La-Sr-Cu-O, for some of which the superconducting transition temperature approaches 100 K. Inclusion of Bi in a Bi-Ca-Sr-Cu-O system or Tl in a Tl-Ca-Sr-Cu-O system has recently pushed the transition temperature up to 125 K (Sheng and Hermann, 1988; Torardi et al., 1988; Subramanian, et al., 1988; Poole, 1988). All of these materials show p-type conductivity before becoming superconductive. However, layered n-type materials have also been found to become superconductive, although at somewhat lower temperatures ($\simeq$ 30 K).

The conductivity in these materials is highly anisotropic, with $\sigma_\parallel/\sigma_\perp = 10^2 \ldots 10^5$ near the transition temperature. The conductivity within the Cu-O layers is metallic, whereas the current perpendicular to these layers is substantially reduced, probably due to tunneling through barriers.

There is evidence from magnetic-flux quantization experiments that the quasi-particles responsible for high-temperature superconductivity are doubly charged like Cooper pairs.

Observed tunneling through Josephson junctions indicates that these are singlet pairs in an s-state. This is supported by a relatively small effect of chemical impurities on T_c; they should suppress T_c in p- or d-type state superconductors.

From critical magnetic field measurements, the size of these quasi-particles can be determined (Yamagishi et al., 1988). In $EuBa_2Cu_3O_7$, they display a pancake shape, with 35 Å diameter and 3.8 Å height lying in the Cu-O plane.

The bonding energy between the two carriers of the pair with a mediating partner, the phonon for the Cooper pair, can be obtained for weak coupling. Here one can use the BCS expression $kT_c = 1.14\hbar\omega_D \exp(-1/\lambda)$, where λ is the coupling constant, however, replacing the Debye energy with a larger, probably electronic energy $h\nu$, which is on the order of 0.3 eV.

Little (1988) argues on the basis of several experiments, relating the gap energy and the transition temperature, that for high-temperature superconductors the coupling can be described by the CBS weak coupling value

$$\frac{2\Delta(T = 0)}{kT_c} \simeq 3.52. \qquad (35.11)$$

However, more recently, direct evidence for a substantially larger gap was obtained from photoelectric emission spectroscopy. The photoelectrons emitted through the surface correspond to the energy distribution within the solid, and show a shift and a spike, expected from the CBS theory when a large enough gap appears beyond the experimental resolution. This is shown in Fig. 35.4 for a $Bi_2Sr_2CaCu_2O_8$ single crystal with $T_c = 82$ K. The width of the spike at 20 K indicates a gap of $\Delta = 24$ meV, which yields nearly twice the amount expected from the weak coupling theory—Eq. (35.11). For a recent review, see Margaritondo et al. (1989).

The mediating partner required to provide the coupling of the quasi-Cooper pair also needs to be a boson. As such, one may consider resonance valence bands (Anderson, 1987), magnons (Chan and Goddard, 1988), spin fluctuations (Schrieffer et al., 1988), plasmons, and excitons. There are strong arguments against a magnetic nature of the interaction (Little, 1988).

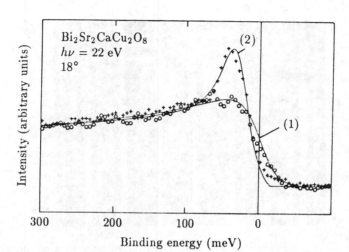

Figure 35.4: Energy distribution of photoemitted electrons from a $Bi_2Sr_2CaCu_2O_8$ single crystal at 90 and 20 K (open and solid circles, respectively), above and below $T_c = 82$ K. Curves (1) and (2) are best fits to the Fermi-Dirac and the BCS distributions (after Olson at al., 1989).

The possible coupling to excitons deserves some further explanation. Bose-Einstein condensation of excitons was first proposed by Blatt et al. (1962), and was more thoroughly investigated by Keldysh (1965), (1968) and Hanamura (1973). Recently, such condensation was suggested to trigger higher-temperature superconductivity (Little, 1964; Ginzburg, 1970; Allender et al., 1973; see also Varma et al., 1986). Excitons may couple to holes in these high-temperature superconductors, forming hole pairs with a stronger binding force than that achievable by phonon coupling for Cooper pairs.

Two-dimensionality of the lattice (Section 15.3) may help to increase the binding energy of excitons. Increasing ionicity* may aid the formation of still tighter bound excitons, which can survive higher temperatures.

A high-resolution transmission electron micrograph of superconductive $YBa_2Cu_3O_{7-x}$ shows the layer-like structure reminiscent of ultrathin superlattices. This superlattice-like (2D) structure could provide confinement with a consequent increase in exciton binding

---

* Here excitons, or polarons, become more tightly bound (Frenkel excitons). Earlier observation of heavy fermion superconductors (Steward, 1984) may point toward the assistance of more tightly bound polarons (with $m_n \simeq 200 m_0$) in forming superconducting compounds, e.g., $CeCu_2Si_2$.

Figure 35.5: High-resolution transmission micrograph of superconductive $YBa_2Cu_3O_{9-y}$ along the (100) direction. The black spots are barium and yttrium atoms, the light layers contain copper and oxygen. The unit cell (white rectangle) measures approximately 4 Å × 12 Å (after Shaw, 1987).

energy—see Section 15.3. However, a wider layer spacing may be required to provide a better defined offset of valence (and conduction) bands in order to obtain the necessary exciton confinement.

Recent investigations of numbers of quinternary systems ($Bi_2Ca_{1+x}Sr_{2-x}Cu_2O_{8+y}$) show an increased critical temperature with increased layer thickness (Ihara, et al., 1988). $Bi_2Sr_2Ca_{L-1}Cu_L O_{4+2L}$ or $Tl_2Sr_2Ca_{L-1}Cu_LO_{4+2L}$ compounds are identified with $L = 1 \ldots 4$, the number of CuO_2 layers in the unit cell, ignoring an important deviation from stoichiometry here. These compounds are referred to by the simplified notations 2201, 2212, 2223, and 2234, respectively. The numbers identify the ratio of the first four elements of the compound, e.g., of Bi:Sr:Ca:Cu. The increased layer thickness may provide better 2D confinement as it provides a better defined offset of the valence band for such thin layers.* The observed slight decrease of T_c, with further increased layer thickness, could signal a reduction in binding energy for lesser compressed excitons.

In the superconductive state, the critical current density is also anisotropic ($j_{c\parallel}/j_{c\perp} \simeq 10^3$). This indicates that superconductivity occurs within the plane of such layered crystals and may be interrupted by interfacing normal conducting layers.

---

* This may be one reason why p-type material is preferred for these high-temperature superconductors.

Interpolative and extrapolative estimates about possible transition temperatures have recently been made from basic principle calculation of the band structure, and assuming the validity of the Cooper pair estimate [Eq. (35.1)], although with an adjusted $\hbar\omega_D$ parameter to fit an experimental T_c value. This leaves open the question as to whether phonons, plasmons, or excitons are promoting such Cooper pairs. The estimates show an increase of T_c with increasing CuO_2 layer thickness (Herman et al., 1987a), and predict transition temperatures in excess of 150 K (Kasowski et al., 1988). See also Herman et al. (1987).

Recently, Anderson (1987) pointed out that a Mott transition (see Chapter 34) of a semiconductor with a resonant valence-bond state may be an alternative mechanism for high-temperature superconductivity. For the Mott transition, one needs a sufficient carrier density, hence sufficient doping. Electron-hole pairing occurs in the valence-bond state. Such a state can be seen by hybridization of the Cu^{2+} $s = \frac{1}{2}$ orbital with the p-levels of the surrounding oxygen atoms, and may be described as a tightly bound exciton when trapping an electron. Recent estimation on the bond energy, however, indicates a rather low upper limit of ~ 20 meV, which makes this mechanism a less likely one (Margaritondo et al., 1989).

Yet another mechanism was suggested by Phillips (1989), based on conduction in a defect-level band in the center of its narrow gap.

There is some evidence that high-temperature superconductivity occurs only in a fraction of the volume of several of the presently evaluated compounds, which may be responsible for low critical current densities due to thin current filaments of barely connecting superconductive regions (Cai et al., 1987; Ovshinsky et al., 1987). Careful growth of single crystal layers of $YBa_2Cu_3O_{7-x}$ has resulted in a substantial increase in critical current densities up to 10^5 A/cm^2 at 77 K and $2 \cdot 10^6$ A/cm^2 at 4 K (Chaudhari et al., 1987).

All high T_c compounds show type-II behavior, however, with low critical fields and substantial flux-line movements, causing some residual power dissipation. The flux-line lattice "melts" at temperatures well below T_c. This flux-line melting temperature is about 75 K for material with $T_c = 93$ K, and is somewhat lower for the $T_c = 125$ K material. Efforts are underway to pin the flux lines through changes in the defect structure, and thereby increase the critical current density. Some success has been reported recently after radiation damage induced by neutron bombardment on a $Ba_2Cu_3O_7$

crystal, increasing the critical current density by a factor of almost 100 (van Dover et al., 1989).

Summary and Emphasis

Until recently, superconductivity was believed to be restricted to metals and to temperatures below 25 K. It involves Cooper pairs of electrons, which can be formed due to phonon interaction. Bose-Einstein condensation of such pairs creates a band gap that eliminates damping of the transport which still occurs for the remaining normal electrons. An injection of electrons above the gap, via Giaever tunneling, shows conductivity of normal electrons in superconductors. The density of superconducting Cooper pairs increases with decreasing temperature at the expense of normal electrons, causing an increasing band gap, an increasing critical current (above which superconductivity vanishes), and a decreasing penetration depth for a magnetic field—the Meissner field-expulsion.

In certain layered p-type materials, superconductivity up to 125 K has recently been observed. The superconductivity transition is also related to pair-formation of carriers; however, the promotion of the pair-formation seems to follow a substantially different mechanism, for which a number of models are currently under discussion.

With the discovery of high-temperature superconductors, a dynamic field of new research was created with a large effort to find materials with even higher critical temperatures and critical current densities. The possible applications of such superconductors will open totally new opportunities, and may revolutionize a wide variety of present technologies—from new electronic devices to energy storage and mass transportation, e.g., superconductive train levitation.

Exercise Problems

1.(1)   Review the latest publications on high T_c superconductors and compare:

(a)   the experimental results (T_c and j_c) with the values listed in this book;

(b)   the development of new materials (layered?);

(c)   the development of models compared to the rather speculative descriptions given in this book; and

(d)   the development of p-type vs. n-type materials of higher T_c.

Chapter 36

Quasi-Carrier Mobilities

The transport of polarons proceeds similarly to that of common carriers and is drift-diffusive; whereas the transport of excitons is the transport of energy, which typically has a diffusion-like character.

Carrier transport in semiconductors can proceed via electrons and holes or via electronic quasi-particles, such as electrons, polarons, etc. Carrier transport via polarons is similar in several respects to normal carrier transport; it can be described by a polaron mobility. See Klinger (1979) and Peeters and Devreese (1984).

In contrast, excitons, for which field *gradients* are the driving force, have no contribution to the electric current. This is due to the fact that an exciton is neutral—see Section 12.2.1. Nevertheless, the diffusion of excitons from a region where they are produced, e.g., from an illuminated region, into another region, i.e., a dark region, where they can dissociate and produce free carriers, can influence the overall carrier transport. In this chapter we will discuss such quasi-carrier transport.

36.1 Exciton Diffusion

Excitons, like Bloch electrons, are properties of the entire crystal before scattering localizes the excited state. After elastic scattering, the excitonic state delocalizes again, but with a changed phase relation to the original excitation (Reynolds and Collins, 1981). It is therefore appropriate to deal with diffusion in a site representation that permits the definition of a probability $W_I(t)$ to find an exciton on site I at time t. With inelastic scattering, the exciton may lose part of its kinetic energy, or can dissociate into, or create other quasi-particles, such as carriers, phonons, or photons.

The transport of excitons through a lattice usually involves elastic scattering processes and proceeds as diffusion, first alluded to by

Frenkel (1936) and discussed by Förster (1948). Such diffusion was first measured in anthracene (Simpson, 1956; Avakian and Merrifield, 1964). It can, however, also involve an intermediate emission of exciton-luminescence, followed by reabsorption (Balkanski and Broser, 1957; Bleil and Broser, 1963). In both cases, the excitons penetrate farther into the semiconductor than the extent of the optically excited region (Simpson, 1956). Clouds of excitons can be guided into different parts of a semiconductor by application of strain fields (Markiewicz et al, 1977).

36.1.1 Time-Resolved Diffusion

The ambiguity of energy transfer between different locations by diffusion or by direct transfer requires a more detailed analysis. The direct transfer, also referred to as the Förster dipole-dipole interaction, is absent in triplet exciton diffusion. The problem becomes even more complicated when considering exciton trapping and release as additional mechanisms.

When assuming diffusion with a random walk, with a mean free path $\lambda \ll r_0$ as the distance for radiative energy transfer, one obtains for the change of the exciton density in time

$$\frac{dn_{\text{exc}}}{dt} = -\frac{n_{\text{exc}}}{\tau_{\text{exc}}} - n_{\text{exc}}^2 \left[4\pi D r_0 \left(1 + \frac{r_0}{\sqrt{\pi D t}} \right) \right], \qquad (36.1)$$

which reduces at sufficient elapsed time to

$$\frac{dn_{\text{exc}}}{dt} = -\frac{n_{\text{exc}}}{\tau_{\text{exc}}} - n_{\text{exc}}^2 \cdot 4\pi D r_0, \qquad (36.2)$$

where $4\pi D r_0$ is the rate of localization at a center of radius r_0 (Wolf, 1967). The solution of Eq. (36.1) can be written as

$$n_{\text{exc}}(t) = n_0 \exp(-\alpha t - b\sqrt{t}), \qquad (36.3)$$

which contains a $\sqrt{t}$-term caused by the sweeping-in of excitons when they approach the center to within r_0.

The competition between diffusion and direct transfer to a trap has been discussed by Tunitskii and Bagdasarian (1963) and Yokota and Tanimoto (1967). For a short, review see Knox (1986).

Exciton diffusion is usually small and requires sophisticated methods of measurement. One way of measuring the diffusion is by using changes of the refractive index near the exciton absorption wavelength (Aoyagi et al., 1982), thereby providing the possibility of making their distribution directly visible. Their steady-state density is given by a balance between generation and recombination. When the generation rate is spatially inhomogeneous, the exciton density is inhomogeneous; hence, the induced changes in the refractive index are inhomogeneous.

A closely spaced interference pattern can be produced by self-interference of one exciting beam with its split-off image. Its image in a semiconductor can cause interference of a third (test) beam. This method is similar to that described in Section 33.1.2A (Fig. 33.6), except that there the absorption of free carriers was used, whereas here the change in the index of refraction at high exciton densities is used. After the exciting beam is switched off, the diffusion of excitons and their decay cause the diffraction pattern to fade out, thereby reducing the diffracted test beam signal with a decay time t_t, given by

$$\frac{1}{t_t} = \frac{1}{\tau_{\text{exc}}} + \frac{4\pi^2}{\Lambda^2} D_{\text{exc}}, \qquad (36.4)$$

where τ_{exc} and D_{exc} are the lifetime and diffusion coefficients of the excitons, respectively, and $4\pi^2/\Lambda$ is the grating pitch (Aoyagi et al., 1982). The slope and crossing point of the $1/t_t$ vs. $4\pi^2/\Lambda^2$ plot yields $\tau_{\text{exc}} \simeq 280$ ps and $D_{\text{exc}} \simeq 330 \pm 200$ cm^2/s for the example of CuCl at 1.8 K, given in Fig. 36.1.

Much higher exciton diffusion constants have been observed recently in GaAs/AlGaAs superlattices because of the longer lifetime of these excitons, with $D = 10^6$ cm^2/s and a diffusion velocity

$$v_{\text{diff}} = \sqrt{\frac{D}{\tau}} \simeq 3.2 \cdot 10^8 \text{ (cm/s)} \qquad (36.5)$$

at 4 K (Junnarkar et al., 1987).

36.1.2 Diffusivity of Exciton Molecules

To create a bleached-out grating, as described in the previous section, one can also use excitonic molecules rather than excitons by applying a slightly different frequency of the bleaching light. The decay of a diffracted test beam yields the lifetime and diffusion coefficients of these molecules. In CuCl, e.g., one obtains at 1.8 K an exciton molecule lifetime of $\tau_{\text{mo}} \simeq 280$ ps and a diffusion constant of $D_{\text{mo}} \simeq 45 \pm 10$ cm^2/s—see the review of Aoyagi et al. (1982).

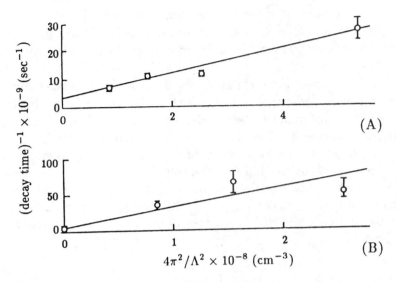

Figure 36.1: Inverse decay time of the diffracted signal vs. inverse grating pitch of a bleached-out transient grating for (A) an excitonic molecule and (B) for an exciton-polariton (after Aoyagi et al., 1982).

36.1.3 Superconductive Energy Flux, Condensed Excitons

At low temperatures and high densities of excitons, there is a possibility of Bose-Einstein condensation (Blatt et al., 1962; Keldysh and Kopaev, 1965) with a condensation temperature given by

$$T_c = 0.575 \frac{2\pi\hbar^2}{km_{\text{exc}}} n_{\text{exc}}^{2/3} = 3.17 \cdot 10^{11} \frac{m_0}{m_{\text{exc}}} n_{\text{exc}}^{2/3}. \qquad (36.6)$$

Such a condensate would have a fraction of

$$\frac{n_{\text{cond}}}{n_{\text{exc}}} = 1 - \left(\frac{T}{T_c}\right)^{3/2} \qquad (36.7)$$

in a state that could be compared with superfluid helium. Although no electric current would flow and no material transport is expected, as in [4] He, a resistance-free transport of energy is predicted.

The density necessary to cause condensation, $\sim 10^{17}$ cm^{-3} at 10 K, is low enough to occur before other condensation processes start—see Chapter 34.

36.2 Polaron Transport

The transport of large polarons is similar to that of electrons with similar scattering mechanisms. An equivalent relationship holds for hole-polarons compared to holes in the valence band.

The interaction of polarons with LO phonons has been extensively investigated, using the Boltzmann equation, for polarons with a temperature-dependent effective mass (Saitoh, 1970)

$$\frac{m_{\text{pol}}}{m_n} = \begin{cases} \left\{ \left(\frac{v}{m}\right)^2 + \left[1 - \left\{\left(\frac{w}{v}\right)^2 - 1\right\} \frac{6\Gamma}{v\Theta} + \ldots \right] & \text{for } T < \Theta \\ 1 + \frac{1}{60}\left(\frac{\Theta v}{T}\right)^2 \left[1 - \left(\frac{w}{v}\right)^2\right] + & \text{for } T > \Theta \end{cases}$$

(36.8)

where v and w are the *Feynman parameters* (Feynman, 1955). These parameters can be related to a frequency and mass ratio:

$$v = \frac{\tilde{\omega}}{\omega_{\text{LO}}} \quad \text{and} \quad \left(\frac{v}{w}\right)^2 = \frac{M}{m_n}.$$

(36.9)

Here, $\tilde{\omega}$ is the eigenfrequency of the polaron, assuming the electron is coupled to a fictitious quasi-particle, the phonon cloud, of mass m_f and spring constant β, yielding $\tilde{\omega} = \sqrt{\beta/m_r}$ and $m_r = m_n m_f/M$, where M is the *Feynman polaron mass*:

$$M = m_n + m_f$$

(36.10)

A reasonable approximation for the two branches was given by Saitoh (1970).

$$\frac{m_{\text{pol}}}{m_n} = \begin{cases} 1 + \frac{\alpha_c}{6}\left(1 - \frac{3T}{4\Theta} + \ldots\right) + \ldots & \text{for } T \ll \Theta \\ 1 + \frac{\alpha_c}{24}\sqrt{\frac{\pi\Theta}{T}}\left[1 - \frac{1}{96}\left(\frac{\Theta}{T}\right)^2 + \ldots\right] + \ldots & \text{for } T \gg \Theta \end{cases}$$

(36.11)

with α_c as the lattice coupling constant. The effective mass shows an increase of the polaron mass from the Feynman value [Eq. (36.10)] at $T = 0$, and after going through a maximum near $T = 0.4\,\Theta$, approaches the bare electron mass for $T \gg \Theta$.

The scattering with LO phonons at low temperatures is dominated by absorption, and for higher temperatures $(T \gtrsim \Theta)$ by emis-

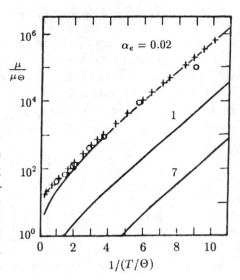

Figure 36.2: Normalized polaron mobility as a function of the reciprocal temperature ($\Theta =$ Debye temperature) with the coupling constant as family parameter. Curves calculated for relaxation-time approach, ○ from Monte Carlo calculation, and × after Devreese and Brosens (1981). Figure redrawn from Peeters and Devreese (1984).

sion of these phonons. Furthermore, the polaron mobility can be expressed in the relaxation-time approach by

$$\mu_p = \frac{e}{2\alpha_c n_{\mathrm{ph}}} \left(\frac{w}{v}\right)^3 \exp\left(\frac{v^2 - w^2}{vw^2}\right) \qquad (36.12)$$

which decreases exponentially with increasing temperature as the phonon density increases according to the Bose-Einstein distribution.

Figure 36.2 shows a typical temperature dependence as calculated for InSb with $\alpha_c = 0.02$. For larger coupling constants, the curves are essentially parallel-shifted to lower mobilities. For a review, see Peeters and Devreese (1984).

36.2.1 Small Polaron Transport

A small polaron is trapped within the ideal lattice, and needs thermal ionization from its self-trapped state to jump from neighbor to neighbor:

$$\mu_{\mathrm{pol,s}} \propto \exp\left(-\frac{E_{\mathrm{pol,th,s}}}{kT}\right). \qquad (36.13)$$

Its motion is a hopping motion; its "mean free path" is that of the nearest-neighbor distance (see Holstein, 1959).

At low temperatures ($T < \Theta/2$) the transport of small polarons proceeds in a narrow polaron band. The width of this band is determined by the overlap term, proportional to the resonance integral J between neighboring sites.

More recently, Holstein (1981) and Christov (1982) used a reaction-rate approach to describe the transport of small polarons in lattices with higher values of α_c, which allows for nonlinear electron-phonon interaction. In contrast to the large polaron mobility, which usually is on the same order of magnitude as the corresponding electron mobility, the mobility of small polarons is typically $\lesssim 10^{-2}$ cm^2/Vs. For a review, see Böttger and Bryksin (1985) and the extensive list of references therein.

Summary and Emphasis

Aside from electrons and holes, other quasi-particles are mobile and contribute to current or energy transport. Examples are phonons (discussed in Section 6.3), excitons, and polarons.

Excitons can diffuse, following a concentration gradient, or can be moved by a strain-field or an electric-field gradient. These processes represent an energy transfer in space. The diffusion can involve exciton scattering at lattice defects, energy transfer via dipole-dipole interaction, and intermediate trapping of free excitons. Time-resolved diffusion experiments permit distinction between the different processes.

Polarons behave similarly to normal carriers, except for an increased polaron mass. This mass first increases and then decreases with increasing temperature, approaching the free-carrier effective mass. The polaron mobility, when determined by LO-phonon scattering, decreases exponentially with increasing temperatures, and is shifted toward lower values with increasing electron-phonon coupling constant.

Usually, exciton diffusion and polaron transport in semiconductors are small effects, only contributing to a minor degree to the energy transport, or modifying the current transport described by normal carriers. However, an exception is seen in material with unusual semiconducting properties, e.g., high values of the coupling constant.

Exercise Problems

1.(e) Draw the polaron mobility as a function of the temperature for $\alpha_c = 0.1$, 0.5, and 2 in the low- and high-temperature branches. Refer to the review of Peeters and Devreese (1984).

2. Discuss the driving forces for excitons in a strain field.

Chapter 37

Carrier Mobility Influenced
by Larger Defects

*Carriers are scattered, or their path is severely per-
turbed, when they have to pass through regions of the
semiconductor that are disturbed by larger lattice defects.*

There are three types of large lattice defects that have a significant effect on carrier transport:

- dislocations, i.e., line defects,
- clusters, i.e., larger associates of point defects, and
- surfaces, such as grain boundaries between the same or different
  phases, or external surfaces.

The influence of these defects on carrier transport will be discussed briefly in this chapter.

37.1 Scattering at Dislocations

The edge dislocation presents a major disturbance in carrier transport within the lattice because of the surrounding stress field and
the charging of the reconstructed core states—see Section 18.3.4A.
Another disturbance can be expected from the charges induced in
piezoelectric crystals, and caused by the strain field around edge and
screw dislocations (Levinson, 1965; Pödör, 1977). These core states
attract electrons or holes, depending on the position of the Fermi
level, and are screened by free carriers surrounding the line as a cylin-
der of a radius equal to the Debye length [$L_D = \sqrt{\varepsilon_{\mathrm{st}}\varepsilon_0 kT/(e^2 p_0)}$),
where p_0 is the hole density in the bulk of the semiconductor, when
electrons are attracted to the dislocation core]. Core states in Ge and
Si are positively charged at low temperatures and negatively charged
at higher temperatures.

Bonch-Bruevich and Kogan (1959) discussed the carrier scattering at a charged cylinder. By treating the scattering cylinder similarly to a scattering center of spherical symmetry, Pödör (1966) obtained an expression for the mobility due to dislocation scattering:

$$\mu_{\text{disl}} = \frac{75\sqrt{p_0}\,a_\ell^2(\varepsilon_{\text{st}}\varepsilon_0)^{3/2}}{e^2 N m_n^{1/2}}\,kT, \tag{37.1}$$

where N (cm^{-2}) is the dislocation density, and a_ℓ is the distance between charged defect centers along the dislocation core line. Even without charge of the dislocation core at a "neutrality temperature" between negative and positive core charges, a lower mobility in deformed Ge is observed. This is caused by the scattering at the strain field.

A more refined theory takes into consideration the influence of the deformation potential, and the anisotropy, of the scattering for the dislocation fields with a preferred orientation (Düster and Labusch, 1973). The mobility, measured parallel to an array of aligned dislocations, is nearly equal to the mobility without dislocations; whereas when it is perpendicular to the array, the mobility is substantially reduced. This was experimentally confirmed by Schröter (1969) and Pödör (1974). For a short review, see Zawadzki (1980).

37.2 Scattering at Defect Clusters

Larger clusters of defects, such as associates of defects or small inclusions of a different phase, can interfere with the carrier transport beyond their own occupied volume via a space-charge cloud surrounding the defect. Depending on the charge, size, and distribution of these defects, the interaction can be modeled by a simple neutral center of a larger effective diameter, or by a charged center with a surrounding space charge extending up to a distance of several Debye lengths.

The charge of the defect associate may be due to carrier trapping. The charge of a different phase inclusion can be estimated from the difference in electron affinity between inclusion and host.

In addition, the strain field surrounding such a defect cluster influences the band edge via the deformation potential, and as such can act as an extended scattering center.

The influence on carrier transport may range from the scattering similar to that at point defects to the carrier repulsion from areas comparable to or larger than the mean free path.

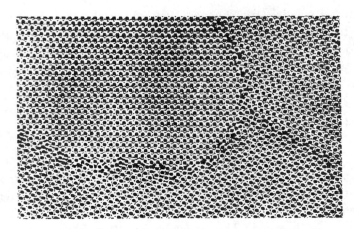

Figure 37.1: Soap-bubble model of a two-dimensional crystal with grain boundaries, indicating a high defect density at these grain boundaries.

37.3 Influence of Microcrystallite Boundaries

This influence is important for carrier transport in most microcrystalline semiconductors. A substantial reduction in mobility is observed when carriers must pass through the interface between crystallites on their way from one electrode to the other. These interfaces contain a high density of lattice defects, such as dislocations along small-angle grain boundaries and vacancies or clusters of defects with substantial lattice relaxation along other grain boundaries—see Fig. 37.1. Carrier transport through the interfaces is therefore subject to a high degree of scattering. For a first approximation, one may assume that a free path ends at such interfaces, with a consequent reduction of the effective mobility: the smaller the crystallite grains, the more the mobility is reduced compared with the mobility in large single crystals.

Usually, interfaces have a high density of traps, which can become occupied by carriers. As a consequence, interfaces are often charged. Screening charges are located on both sides of the interface, causing a space-charge triple layer (Fig. 37.2), which produces a potential barrier between each of the crystallites. The height of the barrier can be estimated from the integrated Poisson equation

$$\Delta V_b \simeq \frac{e}{\varepsilon \varepsilon_0} \Sigma_i L_D, \qquad (37.2)$$

where Σ_i is the surface charge density at the interface, and L_D is the Debye length. The actual surface charge density, however, is

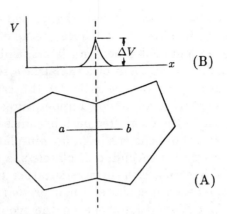

Figure 37.2: (A) Crystallite boundary and (B) potential barrier along line a–b, as indicated in (A).

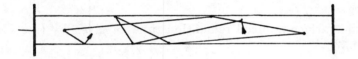

Figure 37.3: Specular and nonspecular scattering limited by surfaces in a thin platelet of high carrier mobility.

usually insufficiently known to make such an estimate meaningful. The barrier height $e\Delta V_b$ is deduced from experimental data. With such a barrier, an exponential dependence of the mobility vs. $1/T$ is observed:

$$\mu_b = \mu_0 \exp\left(-\frac{e\Delta V_b}{kT}\right), \tag{37.3}$$

when carriers must pass over these intergrain barriers, i.e., when the grain size is smaller than the distance between the electrodes.

The mobility μ_b is an effective mobility. Within each grain, the carrier mobility (μ_0) is larger than the effective mobility.

In the last three decades, carrier transport through grain boundaries has received much attention. However, as this topic deals with the inhomogeneous semiconductor, it is not a subject of this book. For a recent review, see Seager, 1985.

37.4 Influence of External Surfaces

External surfaces interact with carriers as perfect scattering surfaces, as surfaces for carrier recombination, and via their space charge. The space charge compensates the surface charge and extends into the bulk by a few Debye lengths. These space-charge effects are of technical importance, e.g., in field-effect transistors, and are discussed by Anderson (1970), Sze (1981), and others. Surface scattering reduces the mean free path to be on the order of the crystal dimensions (see below). It also plays a role at low temperatures in high-mobility semiconductors, where the mean free path becomes comparable to at least one of the crystal dimensions, e.g., in thin platelets—Fig. 37.3.

In *homogeneous* thin semiconductor platelets, a simple treatment of the influence of both platelet surfaces is rather transparent when one distinguishes between specular and nonspecular scattering at the surfaces.* The first has no influence on the mobility; the second causes a reduction in the average carrier relaxation time:

$$\frac{1}{\bar{\tau}} = \frac{1}{\tau_B} + \frac{1}{\tau_S}, \tag{37.4}$$

with the surface-induced relaxation time τ_S given by (see Many et al., 1965)

$$\tau_S \simeq \frac{\delta}{\lambda}\tau_B. \tag{37.5}$$

Here, τ_B is the bulk relaxation time, λ is the mean free path, and δ is the mean carrier distance from the surface. From Eqs. (37.4) and (37.5), one obtains for the ratio of actual-to-bulk mobility

$$\frac{\mu}{\mu_B} = \frac{\bar{\tau}}{\tau_B} = \frac{1}{1 + \dfrac{\lambda}{\delta}}. \tag{37.6}$$

Setting $2\delta = d$, where d is the platelet thickness, and using s as the fraction of the specular scattering events at the platelet surfaces, one obtains

$$\mu = \mu_s \frac{d}{d + 2(1 - s)\lambda}, \tag{37.7}$$

which is shown in Fig. 37.4 with s as the family parameter.

---

* More sophisticated estimates (e.g., Fuchs, 1938; Sondheimer, 1952) are applied for thin metal layers, where surface-induced effects are less complex. These yield results on the same order of magnitude as given here.

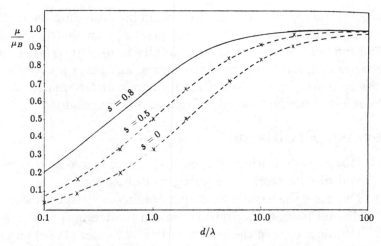

Figure 37.4: Carrier mobility as a function of the ratio of platelet thickness to mean free path with specular surface scattering fraction s as the family parameter. × is the result of the approximate theory (after Anderson, 1970). Curves calculated from Sondheimer (1952).

37.5 Influence of Metal-Semiconductor Boundaries

These boundaries are essential for semiconductor devices, for they serve as electrical contacts. Their electrical properties are determined by space-charge effects, and will be discussed extensively in *Transport Properties in Inhomogeneous Semiconductors* (in preparation). For extensive reviews on this subject, see Henisch (1957), Spenke (1958), Rhoderick (1978), and Sze (1981).

Summary and Emphasis

Carrier scattering at dislocations is usually anisotropic. It can make a substantial contribution to reducing the mobility for crystals with high dislocation densities or for carrier transport through heterointerfaces with a significant lattice mismatch.

Carrier scattering at defect centers can be influenced by trapping, by the space-charge surrounding a charged cluster, and by the deformation potential caused by the surrounding strain field.

Crystallite boundaries, when not passivated, have a large effect on perturbing carrier transport. This is due to a high density of disorder at the boundary, and to space-charge effects extending to several Deybe lengths beyond the boundaries. These space charges cause potential barriers, usually with a significant reduction in carrier transport, which in turn depends exponentially on the temperature.

The influence of larger crystal defects is usually related to the boundaries of crystallites, or of the device, and plays a major role in its performance. This becomes more important for smaller devices, and requires a careful analysis of spatial inhomogeneities.

Exercise Problems

1.(l) Obtain recent original literature, experimental or theoretical, and discuss carrier scattering on dislocations.

2.(r) Describe the strain field surrounding a defect cluster, and discuss its influence on the motion (scattering) of electrons.

3.(l) Obtain a copy of the review article by Seager (1985) and summarize the most important conclusions for carrier transport through grain boundaries in polycrystalline Si.

   (a) Discuss the influence of grain boundaries on the current, without optical excitation.

   (b) What is the influence of optical excitation on the carrier transport through such grain boundaries?

4.(*) Discuss the lateral carrier transport through a thin semiconductor layer.

Chapter 38

Noise

The fluctuations of carrier density, field and mobility are the cause for electronic noise and create a basic limitation of small signal detection.

In previous sections, carrier transport was regarded as stationary with a time-independent current. However, statistical fluctuations of carrier density, mobility, and field are cause for *current density fluctuation*, called *noise*, as in the acoustic impression one hears in a loudspeaker after sufficient amplification:

$$\delta j = e\mu n \delta F + e\mu F \delta n + enF\delta\mu. \qquad (38.1)$$

Field fluctuations are of importance for inhomogeneous semiconductors and will not be discussed here. We will first discuss the basic elements of current fluctuation, (see Van der Ziel, 1986) and then will give a brief review of the different noise mechanisms.

38.1 Elements of Fluctuation

The current in a semiconductor fluctuates about an average value. Such fluctuations are observable after amplification, which transforms the fluctuation $j(t)$ from the time (t) domain into the frequency (ν) domain $j(\nu)d\nu$, mathematically requiring a Fourier transformation.

Usually, one samples the random fluctuation of $j(t)$ shown in Fig. 38.1, or of $V(t) = RI(t)$ in a small frequency interval $\Delta\nu$, with its mean square variance given by

$$\overline{\left(j(t) - \bar{j}\right)^2} = \overline{\Delta j^2} = \int_{\nu_1}^{\nu_2} S_j(\nu)d\nu \simeq S_j(\nu)\Delta\nu. \qquad (38.2)$$

Here, $S_j(\nu)$ is the spectral intensity of the current fluctuation, which can be measured after amplification and is the subject of most theoretical investigation. Alternatively, sometimes the spectral intensity $S_V(\nu)$ of the fluctuation of an emf across the device is analysed.

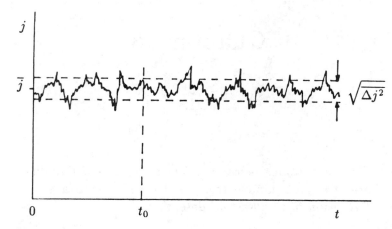

Figure 38.1: Fluctuating current with average value and mean square deviation identified. Probing occurs in specific time intervals of length t_0 after amplification through an amplifier of band width $\Delta\nu = 1/t_0$.

When expressing a fluctuating $j(t)$ by its Fourier components

$$j(t) = \sum_{n=-\infty}^{\infty} a_n \exp(i\omega_n t) \quad \text{with} \quad \omega_n = \frac{2\pi n}{t_0} \quad (38.3)$$

the amplitudes can be expressed as

$$a_n = \frac{1}{t_0} \int_0^{t_0} j(t) \exp(-i\omega_n t)\, dt \quad \text{with} \quad \frac{1}{t_0} = \Delta\nu \quad (38.4)$$

where $\Delta\nu$ is the sampling band width. The spectral density is given by

$$S_j(\nu) = \lim_{t_0 \to \infty} 2t_0 \overline{a_n a_n^*}. \quad (38.5)$$

In the following sections we will give a simplified picture of the general noise phenomenon. We then will present the results for the different types of noise for a homogeneous semiconductor, rather than deriving the spectral density in a vigorous manner. Such derivations can be found in several monographs (see, e.g., Ambrozy, 1982; Van der Ziel, 1986).

38.1.1 Electronic Noise

Electronic noise can be described as a train of pulses, which are generated by each movement of an electron between scattering events, or between generation and recombination. The averaged square of the deviation from the average current divided by $\bar{\imath}^2$ is inversely proportional to the number of current pulses $\hat{n}$. Hence, with $i = \hat{n}e/t$, and t as the length of each current pulse, one has, following the square root law of statistics:

$$\frac{\overline{\Delta i^2}}{\overline{i}^2} = \frac{1}{\hat{n}} = \frac{e}{\overline{i}\,t}, \quad \text{or} \quad \overline{\Delta i^2} = \frac{e\overline{i}}{t}. \tag{38.6}$$

After Fourier transformation of each current pulse, and integration over all amplitudes in a small frequency interval $\Delta\nu$ of amplification, one obtains the total current fluctuation within $\Delta\nu$.

Commonly one distinguishes

- thermal noise,
- shot noise,
- generation-recombination noise, and
- $1/f$ noise, here referred to as $1/\nu$ noise.

The first three classifications can be derived from basic principles and are well understood, while the fourth is still under discussion regarding its specific origin in semiconductors, although it is observed in many physical systems. For reviews, see van der Ziel (1970), Dutta and Horn (1981), Hooge et al. (1981), and Ambrozy (1982). One also distinguishes noise under equilibrium and nonequilibrium conditions.

38.2 Noise in Equilibrium Conditions

The equilibrium noise is given by the thermodynamic behavior of carriers. Little can be done to reduce its amplitude, except by changing temperature or band width of the amplifier.

38.2.1 Thermal Noise

Thermal noise, also referred to as *Johnson-Nyquist* noise, is caused by the random motion of carriers in a semiconductor. It was originally discussed by Nyquist (1928) from thermodynamic arguments and was later analyzed by Spenke (1939), and Bakker and Heller (1939) The mean square of the current is given by

$$\overline{\Delta j^2} = \frac{4\Delta\nu}{R} \frac{h\nu}{\exp\left(\dfrac{h\nu}{kT}\right) - 1} \simeq \frac{4kT}{R}\Delta\nu, \tag{38.7}$$

where R is the resistance of the semiconductor and $\Delta\nu$ is the band width of the amplifier. This noise is caused by the Brownian motion of the carriers. Equation (38.7) aptly describes the noise observed in metals and semiconductors in thermodynamic equilibrium, i.e., for vanishing current or optical excitation. However, the noise can be larger by many orders of magnitude due to density and field fluctuation in nonequilibrium conditions—see below.

38.2.1A Equivalent Noise Resistor, Noise Temperature

The actual noise in semiconducting devices is often described by an *equivalent noise resistor*, using Eq. (38.7), but describing the potential fluctuation

$$\boxed{\overline{V^2} = \overline{j^2}R^2 = 4kTR_n\Delta\nu} \qquad (38.8)$$

by an equivalent resistor R_n. When such a resistor becomes larger than given by the actual device resistivity, nonequilibrium noise is also included in this description.

Another way of assigning an elevated noise output to a device is by defining an *equivalent noise temperature* T_n, according to a similar formula

$$\overline{V^2} = 4kT_nR\Delta\nu. \qquad (38.9)$$

38.3 Nonequilibrium Noise

Nonequilibrium noise is generated by an injected current or by extrinsic excitation. This noise can be changed by doping, variation of interfaces, or contacts.

38.3.1 Shot Noise or Injection Noise

The *shot noise* is derived for vacuum diodes. It is caused by the discreteness of elementary charges crossing the diode, and is related to the statistical exit of electrons from the cathode. It may be applied to carrier injection from one (or two) of the electrodes (double injection) of a semiconductor—see Nicolet et al. (1975). The shot noise was first estimated by Schottky (1918) with a current fluctuation

$$\overline{I^2} = 2eI\Delta\nu. \qquad (38.10)$$

Except for a factor of 2, it is equal to the thermal noise* given in Section 38.2.1, when using thermal emission from the electrode $I = I_s\exp[-e(\psi_{MS} - V)/(kT)]$, where ψ_{MS} is the metal-semiconductor workfunction and I_s is the saturation current, linearizing, and setting

---

* In contrast to the vacuum diode, the current in a semiconductor is bidirectional, hence the factor 2.

$V/I = R$. For a review of the measurement techniques of the intrinsic noise, see Bittel (1959) and van der Ziel (1970).

38.3.2 Generation-Recombination Noise

Carriers can be generated by injection from the electrode, as discussed before, or within the semiconductor, e.g., by light, and annihilated by recombination or temporarily immobilized by trapping. This causes a fluctuation in carrier density. This contribution was first calculated by Gisolf (1949) and was later modified by van Vliet (1958, 1958a), yielding

$$\overline{\Delta I_n^2} = 4 \left(\frac{I}{N} \right)^2 \frac{\tau_n}{1 + 4\pi^2 \nu^2 \tau_n^2} \cdot \frac{N(N_A + N)(N_D - N_A - N)}{(N_A + N)(N_D - N_A - N) + N N_D} \Delta \nu$$

$$(38.11)$$

where τ_n is the electron lifetime in the conduction band, and N, N_D, and N_A represent the *number* (not concentration) of electrons, donors, and acceptors in the actual device (see also van Vliet and van der Ziel, 1958). An example of the spectral distribution for a device with two pronounced centers is shown in Fig. 38.2. The figure also contains two other noise components. The observed frequency distributions of the *generation-recombination noise* component more typically shows a dispersive behavior (Fig. 48.2), indicating a wide distribution of lifetimes (Böer and Junge, 1953; McWorther, 1955; Klaassen, 1961).

38.3.3 Fluctuation of the Mobility

The fluctuation of carrier mobilities due to a variety of scattering processes was evaluated by Kousik et al. (1985), and yields a $1/\nu$ noise for ionized impurity scattering or electron-phonon scattering. See also Pellegrini (1986).

38.3.4 1/f Noise

The $1/\nu$ noise,* in the early literature referred to as *flicker noise*, was first observed by Johnson (1925), and prevails in many semiconducting devices. The source of this noise, however, is still insufficiently known, despite several early attempts at understanding the nature of these fluctuations (see Van der Ziel, 1950). Some theo-

---

* In the noise literature, frequency is commonly denoted by f—hence, $1/f$ noise if the fluctuation spectrum decreases as $1/f$. For consistency, we use ν as frequency within this book.

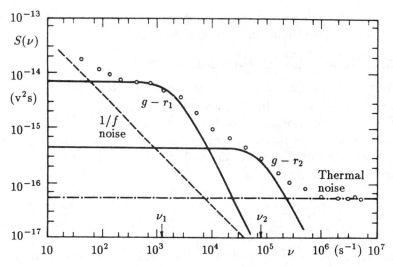

Figure 38.2: Spectral intensity of open circuit voltage fluctuation of a
p-type Si single injection device at a field of 83 V/cm and at 275 K. Solid
curves indicate two $g-r$ noise components with substantially different
characteristic frequencies. Also plotted are the limiting $1/f$ noise and
thermal noise contribution (after Bosman, 1981).

ries, which attempt explanation, include superposition of generation-
recombination noise, with a distribution of τ_n (McWorther, 1955),
and mobility fluctuation. Others include small-angle scattering with
phonons (Handel, 1980), and more general interference phenomena
of carriers with other quasi-particles (Sherif and Handel, 1982).

In collision-free devices (diodes and ballistic devices) the $1/\nu$ noise
can be explained by accelerated electrons emitting *Bremsstrahlung*,
which has a $1/\nu$-spectrum. This is caused by the feedback of the
emitted photon on the decelerating electron, yielding

$$\overline{I^2} = \frac{\alpha_H}{\nu} \frac{I^2}{N} \Delta\nu \qquad (38.12)$$

where N is the number of carriers in the system, and α_H is the Hooge
parameter (Hooge, 1969)

$$\alpha_H = \frac{4\alpha}{3\pi} \left(\frac{\Delta v}{c} \right)^2 . \qquad (38.13)$$

Here, α is the fine structure constant, Δv is the charge of carrier
velocity along its path, and c is the light velocity (see Handel,
1982; Van der Ziel, 1987). A similar formula can also be applied to

semiconductors in which the carrier transport is collision-determined, and α is modified by a charge-conglomerate factor (Van der Ziel, 1987, 1988; Van Vliet, 1989). This type of noise is of a basic nature, and is now referred to as *quantum 1/f noise*—see Handel, 1980 and Kousik et al., 1986.

For other reviews, see Van der Ziel (1979) and Dutta and Horn (1981).

38.3.5 Contact Noise

A major contribution to noise can be generated at the contact interface, or in the high-field region of a Schottky barrier. This *contact noise* also has $1/\nu$ behavior, and is part of the flicker noise. In devices with inhomogeneous contact areas, this contact noise can become the dominating noise.

Summary and Emphasis

Electronic noise in semiconductors has numerous sources. Noise is created by the random fluctuation of individual carrier motion, by the fluctuation of the carrier density and its locally determined mobility, and by field fluctuation, caused by local variation of the space charge in each volume element. When these sources are independent of each other, superposition holds, and each of the noise-frequency components is added to obtain the total noise: $\overline{j_\nu^2} = \Sigma_i \overline{j_{i\nu}^2}$.

Equilibrium noise, also described as thermal noise, is independent of the frequency (*white noise*) and directly proportional to the band width of the amplifier.

Nonequilibrium noise requires optical excitation and a current, and depends on doping or space-charge effects. It usually increases with decreasing frequency, with few features extending beyond the typical $1/\nu$ behavior. It is difficult to separate unambiguously the different components of nonequilibrium noise. However, avoidance of material and electrode defects has a major effect in reducing device noise, and in turn can be used as a sensitive tool to detect such defects.

Electronic noise presents a lower limit for signal detection in electronic devices, and therefore presents lower limits of reliable device operation for a wide variety of applications—such as sensors, detectors, and switching elements in micro-electronic circuitry, etc.

Exercise Problems

1.(l) Read a recent review article on $1/f$ noise and summarize the essential relationship of such noise as it relates to surface and bulk semiconductors.

2.(r) Review the different mechanisms responsible for electronic noise. Give the amplitude of electronic noise for a band width of 100 Hz, across a bulk semiconductor of cubic shape of side length 1 cm with a conductivity of 10^{-2} Ω^{-1} cm^{-1}, for thermal and for shot noise for $\bar{j} = 10$ mA/cm^2.

3.(*) The frequency factor is neglected in Eq. (38.7). What are the reasons for such neglection?

4. The description of a device with an equivalent noise resistor is only an approximation. For a better classification, what other parameters need to be known?

5.(e) In very small devices, the noise becomes more important as a limiting factor. Explain this for a hypothetical device with a volume of 1 μm^3 and a carrier density of 10^{15} cm^{-3}.

Chapter 39

Carrier Transport in Superlattices

*Partial confinement makes the carrier transport in su-
perlattices highly anisotropic. Parallel to the planes
of the superlattice, the mobility can exceed that of
the bulk semiconductor. Normal to the superlattice,
the carrier transport is non-ohmic, and shows features
conducive to device application.*

The carrier transport in superlattices is highly anisotropic. It is
rather low perpendicular to the sheet interfaces, where electrons have
to penetrate the barriers. In contrast, however, it is very high and
often much higher than in the bulk material within the plane of the
superlattice. Here, a high density of carriers can drift within layers of
high purity and lattice perfection, while dopants are confined to the
interspaced barrier layers. A recent review of this two-dimensional
(2D) carrier transport is given by Mendez (1986).

39.1 Carrier Mobility Normal to a Superlattice

The carrier transport from layer to layer of a superlattice occurs in
narrow minibands; carriers in these bands have a large effective mass.
The carriers also have to tunnel through the barriers. Therefore, the
effective carrier mobility is rather small.

In addition, the minibands are shifted with respect to each other
with increasing applied bias. Consequently, the current will vary
according to the density of states product in adjacent wells until the
minibands in adjacent wells no longer join. A sharp reduction in the
current (a *negative differential resistivity* regime) results, as shown in
Fig. 39.1C. When, with further increased bias, a match with the next
higher miniband is reached (Fig. 39.1A), the current increases again
until these bands are shifted away from each other and a second range
of negative differential resistance appears, and so on—Fig. 39.1B.

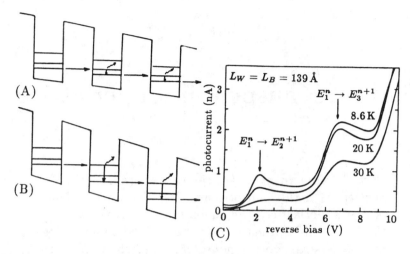

Figure 39.1: (A, B) Tunneling between quantum wells including phonon emission; and (C) corresponding current-voltage characteristics (after Capasso et al., 1986). ©IEEE.

The sharpness of the current maxima and their positions are a direct measure of miniband width and energy, and are in reasonable agreement with theoretical estimation using a Kronig-Penney potential and tunneling.

With broader minibands, occurring in superlattices with narrow barrier layers, one estimates the electronic conduction at very low fields similarly to the classical Drude theory. Because of the relatively small band width, however, the assumption of a constant effective mass is no longer justified.

39.1.1 Coherent and Sequential Tunneling

The tunneling from well to well can proceed within the same miniband at low bias over several barriers without losing its phase relation as coherent or *resonance tunneling*.

When interrupted by scattering, coherence of the electron wave is lost, and the tunneling to the next well is described as *sequential tunneling*. It yields a somewhat lower current—see also Section 39.1.4.

Sequential tunneling occurs when tunneling to an excited state. A carrier relaxation to the ground state is required before tunneling to the next well can proceed. Such "inelastic scattering" is accompanied by phonon emission (Fig. 39.1), and may act as the bottleneck for carrier transport through the superlattice when the gap between the minibands is larger than $\hbar\omega_{LO}$ (Capasso, 1987).

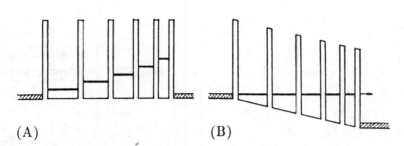

(A)   (B)

Figure 39.2: (A) Superlattice with decreasing well width; and (B) with excited states aligned for maximum tunneling current.

39.1.2 Tunneling Through Variable Width Wells

A special case of resonance tunneling through a superlattice can be achieved when the thickness of consecutive wells decreases in such a way that each of the resulting first excited levels lines up across the entire superlattice at a certain bias, as shown in Fig. 39.2. Here, the current through the superlattice is much increased, while it is very small at lower or higher bias values (see Brennan and Summers, 1987a).

The energy of electrons in thinner wells can exceed the impact ionization energy, rendering the behavior of such structures similar to that of a photomultiplier (Brennan and Summers, 1987b).

39.1.3 Domain-Type of Conductivity

In the negative differential conduction range, i.e., when the current decreases stronger than linearly with increasing bias, a homogeneous field distribution is not stable. At the same bias, with a slight fluctuation of the field, reducing the field for most of the superlattice while increasing it in one well (Fig. 39.3), another state can be achieved in which tunneling from a lower to the next higher miniband in the adjacent well is achieved. In doing so, the current through the minibands is increased: a lower field acts here, providing a higher field for the other wells; the higher current can be maintained through tunneling toward the upper band across the one well.

This process repeats itself at increased bias and causes a periodicity in the current-voltage characteristic equal to the separation of the first and second minibands (Fig. 39.3). This was observed in a GaAs/GaAlAs superlattice of 50 periods, with layer thicknesses of 45 and 40 Å, for wells and barriers, respectively (Esaki et al., 1972). The observed periodicity of 0.24 V agrees well with the calculated miniband gap in this superlattice.

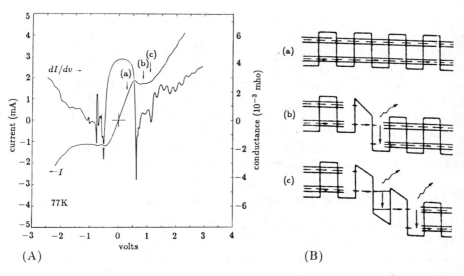

Figure 39.3: (A) Measured current-voltage characteristic in a 50-period superlattice (left scale) and derivative (conductance, right scale) to enhance the structure of the measured curve. The current oscillations are caused by successive localized tunnel conduction (b) and (c). (B) Miniband conduction with stepwise initiated tunneling (a), (b) and (c) corresponding to subfigure (A) (after Esaki et al., 1972).

For a review of the carrier transport normal to the superlattice, see Esaki (1986) and Capasso et al. (1986).

39.1.4 Hot-Electron Spectroscopy

One can use a quantum well adjacent to a bulk semiconductor to perform hot-electron spectroscopy. This can be achieved by shifting the narrow ground state of the quantum well across the lower edge of the conduction band, as shown in Fig. 39.4, and observing the tunneling of electrons from states within the band near its edge through the ground state of the well. One can use the luminescence (Capasso, 1987) or the resulting current to observe the tunneling. When the ground state is shifted beyond the filled levels in the conduction band, a steep decrease of the current is observed typically by a factor of 10 (Stone and Lee, 1986). From the increase of the current with bias, the carrier distribution near the band edge can be inferred.

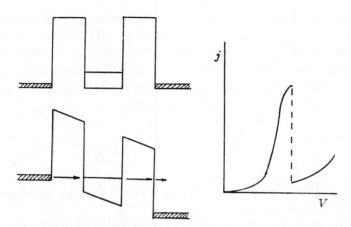

Figure 39.4: Hot-electron spectroscopy using a single quantum well between two highly doped n-type bulk semiconductors.

39.1.5 The Influence of Scattering

Superlattices result in minibands because of electron tunneling between the wells. The effective number of superlattice periods n_s that an electron experiences is reduced by scattering, it destroys the coherence of the wavefunction:

$$n = d/l \rightarrow n_s \simeq \lambda/l; \qquad (39.1)$$

for many superlattices, n_s is a rather small number, i.e., smaller than the actual number of layers. This limits the simple derivation of the miniband width, using the effective number rather than the actual number of layers. When $\lambda \simeq l$, one obtains rather sharp levels instead of minibands.

Broader minibands, however, can be explained despite the short mean free path by collision broadening caused by the carrier scattering—see Capasso et al. (1986). This collision broadening, the natural line width, can be estimated from Heisenberg's uncertainty relation—see also Stone and Lee (1985).

$$\Delta E_c \simeq \frac{\hbar}{\tau}. \qquad (39.2)$$

39.1.6 Carrier Localization and Hopping

When the mean free path normal to the superlattice is smaller than the period length of the superlattice, one observes localization of carriers within each of the wells.* Bloch waves of quasi-free carriers can no longer be defined. The localization within each superlattice layer is of the Anderson type (see Section 25.3.1; Anderson, 1958), and substantially reduces the carrier transport normal to the superlattice. It can be estimated by thermally activated tunneling—a special type of *hopping* (Calecki et al., 1984):

$$\mu \simeq \frac{el_B^2}{kT} \exp\left(-\sqrt{\frac{8m_n}{\hbar^2}\Delta E l_B}\right), \qquad (39.3)$$

where ΔE is the energy difference between levels in adjacent wells, and l_B is the barrier thickness.

Localization is a function of the effective carrier mass: for a given mean free path, the time between scattering events decreases with increasing thermal velocity ($\tau = \lambda/v_{\text{rms}}$), which in turn is a function of the effective mass. Therefore, collision broadening becomes larger for localized carriers with a smaller effective mass ($\tau \propto \sqrt{m_n}$; $\Delta E \propto 1/\sqrt{m_n}$); consequently, tunneling and, therefore, the mobility increase. This can be used to separate carriers of different effective masses, e.g., light and heavy holes, since heavy carriers remain more localized—*effective mass filtering* (Capasso et al. 1986).

39.2 Carrier Mobility Parallel to Superlattice Layers

Carriers within each well of a superlattice can be regarded as a 2D carrier gas. With selective doping of only the barrier layers, a high density of carriers in the wells can be achieved without doping of the wells (Mendez, 1986). Consequently, the mobility at low temperatures, where Coulomb scattering at impurities would dominate, can be substantially higher within the wells than in the bulk, even at high carrier densities. This yields very large conductivities in the layer direction (Dingle et al., 1978). In addition, the effective mass parallel to the well (x, y) is essentially equal to the very small mass in the bulk material of the well, rather than the large effective mass given by the minibands in the z direction—see Fig. 9.28.

---

* This is a localization only in the one direction perpendicular to the superlattice layers. Within such wells, i.e., in the layer direction, carriers are free to move—see Section 39.2.

The scattering mechanisms for a 2D system are modified* and yield a different temperature dependence than that of the bulk. The important scattering mechanisms for GaAs bulk material are shown in Fig. 39.5A. Polar optical scattering dominates at high temperatures, while ionized impurity scattering or piezoelectric scattering for ultrapure samples dominates at low temperatures. In Fig. 39.5B, the temperature dependences for the scattering mechanisms in a GaAs-well superlattice are given. Price (1981) calculates the scattering in a 2D gas with a slightly larger temperature slope (> 2) rather than $-3/2$ for the 3D case (see Section 32.2.1C) for polar optical scattering.

39.2.1 Ionized Impurity Scattering for 2D Case

Bulk mobilities decrease at low temperatures depending on the density of ionized impurities—see Fig. 39.5A and Section 32.2.4. Such a decrease of mobilities is *not* observed in superlattices (Morkoç, 1986). In the best presently made $GaAs/Al_\xi Ga_{1-\xi}As$ superlattices, the mobility continues to increase beyond the expected maximum at $\mu \simeq 10^5$ cm^2/Vs for the residual donor density in wells of 10^{14} cm^{-3}, and approaches $5 \cdot 10^6$ cm^2/Vs below 10 K—see Fig. 39.5B.

Screening by high-mobility carriers of these residual impurities can be responsible for the unusually high overall carrier mobility within such wells (Mori and Ando, 1980). Price (1982) has calculated the effect of carrier screening for such superlattices, and obtained an effective mobility of

$$\mu_{\text{ion}}^{(2D)} = \frac{8}{\pi} \frac{e}{\hbar} \frac{(k_F l)^3}{n_V} \qquad \text{with} \qquad k_F = \sqrt{2\pi n_V} \qquad (39.4)$$

or

$$\mu_{\text{ion}}^{(2D)} \simeq 6.1 \cdot 10^{16} \sqrt{n_V}\, l^3 \ (\text{cm}^2/\text{Vs}), \qquad (39.5)$$

where l is the well thickness, n_V is the carrier density, and k_F is the wave number at the Fermi surface parallel to the superlattice.

The hot-electron mobility at low temperatures decreases rapidly with increasing electric field, as shown in Fig. 39.6. In general, the peak drift velocities do not exceed the bulk velocities:

---

* The scattering from the walls of the thin well are essentially forward scattering events of specular surfaces with no influence on the mobility (see Section 37.4), since the potential barrier smoothes out small irregularities of the actual interfaces.

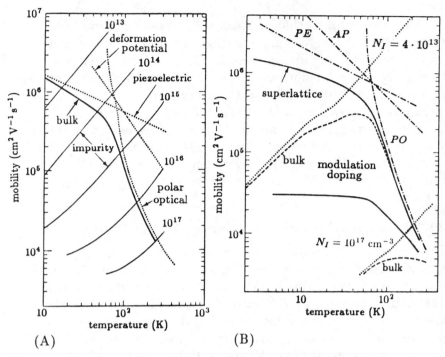

Figure 39.5: Electron mobility as a function of the temperature: (A) for bulk GaAs with major scattering mechanisms indicated and for different ionized impurity concentrations as family parameter. Heavy curve for $n_I < 10^{12}$ cm^{-3}; and (B) for GaAs/Al$_{0.3}$Ga$_{0.7}$As superlattice measured for different doping (heavy curves) and superlattice parameters (after Morkoç, 1986). ©IEEE.

$1 \ldots 3 \cdot 10^7$ cm/s are observed, the latter at 4 K (Inoue, 1985)—see Chapter 29 for the bulk limits.

An interesting effect was pointed out by Hess (1981), when with high electric fields, carrier acceleration is large enough that electrons can be transferred above the barrier and diffuse into the wide gap material, which contains most of the donors. Here, the electrons turn from a 2D into a 3D continuum, and the scattering increases substantially: a range of negative differential conductivity can be observed at high donor densities within the barrier layer.

39.2.2 Ultrathin Superlattices, Electron-Phonon

In superlattices with a layer width of less than six atomic layers, the phononic subsystem reacts to the 2D superlattice while the electronic subsystem does not—see Section 9.3.2. Moreover, the conduction band behaves more like that in a 3D lattice of periodic composition,

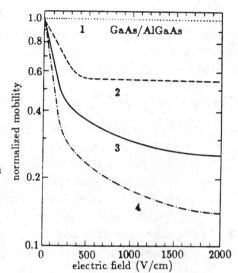

Figure 39.6: Normalized electron mobility in a GaAs/Al$_{0.24}$Ga$_{0.76}$ quantum well as a function of the applied electric field for $T = 300$, 170, 100, and 77 K for curves 1–4, respectively (after Keever et al., 1982).

whereas for the valence band, the critical number is somewhat lower, namely ~ 3 atomic layers for observing an offset between well and barrier.

A decoupling of the electron and phonon subsystems results in a somewhat reduced electron-phonon interaction (Ishibashi et al., 1986).

Summary and Emphasis

The mobility normal to a superlattice is much reduced from the corresponding bulk material. The reduction is caused by a substantially larger effective mass, a more tunneling-like transport, and possibly increased scattering.

In directions parallel to the superlattice, however, the mobility at low temperatures exceeds that of the corresponding bulk material, probably due to separation of dopants by doping only the barrier material. Specular forward scattering at well surfaces may be a contributing factor. Mobilities approaching 10^7 cm^2/Vs have been observed in GaAs wells at temperatures near 1 K.

Nonlinearities of the current, with increasing bias due to step-tunneling from a ground state in one well to the excited state in the adjacent well, are useful for various devices or analytical techniques.

The change in the electron-phonon coupling in ultrathin superlattices offers a new design parameter, which gives additional flexibility in tailoring new materials.

The anisotropy of semiconductive properties of superlattices makes them attractive candidates for new devices. The ultrahigh mobility, parallel to a superlattice at low temperature, may be useful to more exotic applications.

Exercise Problems

1.(r) Describe in your own words the carrier transport perpendicular to a superlattice.
   (a) Analyze the current as a function of the bias.
   (b) Describe domain-type conductivity in more detail.
   (c) What kind of device application can you imagine for such behavior?

2.(*) In directions parallel to the superlattice layers, the mobility in the GaAs wells is very high. Assume 10^6 cm^2/Vs at 10 K in a superlattice of 50 Å well thickness. Calculate from bulk properties the corresponding mean free path. What do you conclude about the scattering?

3.(e) Calculate the ratio of consecutive widths of five wells for the example given in Fig. 39.2, and give the field for which level matching is achieved in your example.

4.(*) Assume a current-voltage characteristic of the type given in Fig. 39.4, and derive explicitly from it the carrier density distribution in the conduction band.
   (a) Assume a single quantum well level with 1 meV width and a thermal distribution at 300 K in a band with $m_n = 0.1m_0$. Compute $j(V)$.

Chapter 40

Defect-Induced Carrier Transport

With a large density of impurities or other lattice defects, the carrier transport deviates substantially from the classical transport within the band. It is carried within energy ranges, which are determined by the defect structure.

In addition to causing scattering, lattice defects can contribute directly to the carrier transport in two ways. They permit direct quantum-mechanical exchange of carriers from defect to defect (i.e., tunneling from one trap to the next trap level), or by thermal ionization of a carrier from a trap level into the band, intermediate transport within the band, and then a retrapping as shown in Fig. 40.1. The first type of carrier transport is called *tunneling* or *impurity band conduction*; the second type is known as *phonon-activated*, or *hopping conduction*. These types of carrier transport are of major importance in highly disordered, highly doped, or amorphous semiconductors. For reviews, see Shklovskii and Efros (1984) and Mott (1987).

40.1 Impurity Band Width and Carrier Transport

In a rather simple model, the overlapping hydrogen-like donor states can be used to form a Hubbard band (Hubbard, 1963), which is centered about their ground-state energy. With partial compensation, there are $N_d - N_a$ free states in this Hubbard band, and conduction can occur (Adler, 1980). The band width ΔE_B is given by the overlap integral between equal centers at distance $1/\sqrt[3]{N}$. This band width is roughly equal to the interaction energy:

$$\Delta E_B \simeq \frac{e^2 \sqrt[3]{N}}{4\pi \varepsilon_{\mathrm{st}} \varepsilon_0}, \tag{40.1}$$

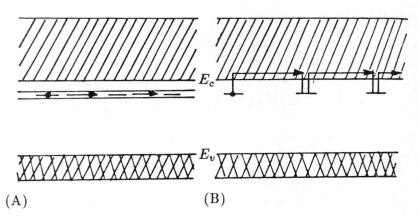

Figure 40.1: (A) Impurity-band conduction and (B) hopping conduction in highly doped semiconductors.

where N is the density of uncompensated donors $(= N_d - N_a)$. See also Section 22.4.10.

The effective mass within this narrow impurity band is much larger than in the adjacent carrier band; hence, the *impurity band mobility* is usually quite small $(<10^{-2}$ cm^2/Vs). This effective mass should not be confused with the effective mass of a Bloch electron within the conduction band, which is responsible for each quasi-hydrogen state of the donor.

The carrier transport within such narrow bands can no longer be described by the Boltzmann equation. The carrier transport must now be evaluated from the quantum-mechanical expectation value for the current, which is given by the *Kubo formula* (Kubo, 1956, 1957). A somewhat simplified version was developed by Greenwood (1958) in which the conductivity can be expressed as

$$\sigma = - \int \sigma_0(E) \frac{\partial f_0}{\partial E} \, dE \tag{40.2}$$

with $\quad \sigma_0(E) = \dfrac{\pi e^2 \hbar^2 \mathcal{V}}{m_0^2} g(E) \left| \int \psi^*(E') \nabla_i \psi(E) d\mathbf{r} \right|^2, \tag{40.3}$

being proportioned to $g(E)$, the density of states, and the matrix element describing the electron transitions from E to E'. Equation (40.2) is referred to as the *Kubo-Greenwood formula*, which, when evaluated for $E = E'$, gives the tunneling current between equivalent defect centers.

Since the distance between these impurities is not constant but fluctuates statistically, the impurity band is substantially undulated. It is broader where impurities are closer together, and narrower where they are more widely spaced—see Section 40.4.3.

Since there is no scattering during the tunneling between adjacent defects, the tunneling is essentially temperature-independent. Except for thermal expansion, which has a small influence on the average distance between defects, and except for the broadening of the defect levels with increased lattice oscillation, the *trap conductivity* is almost temperature-independent when the Fermi-level lies close to the extended states. Trap conductivity is important in highly doped semiconductors and in semiconducting glasses in which a high density of defects is present—see Section 40.4.

When carriers are provided by optical excitation, trap conductivity persists to low temperatures and has a quasi-metallic behavior (Mott and Davis, 1979). We will now describe in more detail the impurity band.

40.2 Impurity-Band Conduction

In semiconductors with high doping densities ($> 10^{18}$ cm^{-3}), shallow donors or acceptors can come close enough (< 100 Å) to each other so that their eigenfunctions overlap significantly and therefore permit the exchange of carriers directly, without the involvement of the adjacent bands. Consequently, the defect level is split and develops into a narrow *impurity band*—see Fig. 40.1A. Such impurity band formation is a basic effect that occurs whenever a defect level is present at sufficient density.

However, the term "band" should be used with caution, as it requires a more detailed density-of-states analysis and a distinction between localized and delocalized states—the latter are true band states. In principle, one could use the Anderson Model (Section 25.3.1) to obtain some information about the localization aspect of the states. We will first discuss this behavior in a rather general fashion.

40.2.1 The Lifshitz-Ching-Huber Model

In the Lifshitz model, a statistical distribution of N identical potential wells is analyzed to obtain a density-of-states distribution of these defect levels, and to identify a critical density at which the states within the center of the distribution become delocalized (Lifshitz, 1965). This model is a forerunner of the Mott version, which is used to distinguish localized and nonlocalized states in band tails— see Sections 25.3.1B and 40.4.2.

When two identical defect centers are brought together, they show a split of eigenstates of the form

$$\psi_s = \frac{1}{\sqrt{2}} (\phi_1 + \phi_2) \text{ and } \psi_a = \frac{1}{\sqrt{2}} (\phi_1 - \phi_2) \text{ with } E_s - E_a = 2I$$

$$(40.4)$$

where ϕ_1 and ϕ_2 are the wavefunctions of the two centers, E_s and E_a are the energies of the states ψ_s and ψ_a, and I is the transfer integral [Eq. (25.2)]. When a third center is approaching at an arbitrary distance, it will not, however, participate in the resonance splitting. This is due to by the fact that the doublet of the two centers closest to each other is far enough apart to be out of resonance with the third center.

Hence, the Lifshitz model yields a band of *localized* states for statistically distributed traps. Only when these defects are close enough to fulfill

$$N^{1/3} r_0 \simeq 0.3, \qquad (40.5)$$

with r_0 as the fall-off radius of the wavefunction of an isolated defect ($= a_{qH}$ for a hydrogen-like defect), can delocalization of the states in the center of the band occur (Ching and Huber, 1982). This result is close to the Mott-Anderson result for localization—see Eq. (25.17).

When interacting with each other, the splitting of the defect states also gives rise to a splitting of this defect band, causing the density of states to have a minimum near the center of the distribution.

40.2.2 The Coulomb Gap in an Impurity Band

The defect level minimum near the center of the density-of-states distribution may become complete in (partially) compensated semiconductors with a gap between filled and empty states. Such splitting is caused by the long-range Coulomb interaction of *localized* electrons (Pollak and Knotek, 1974; Efros and Shklovskii, 1975), and occurs at the position of the Fermi-level.

40.2.2A Delocalization and Coulomb Gap   It can be shown that such a Coulomb gap appears only for localized states. When the density of states becomes large enough, so that delocalization occurs (see below), then the Coulomb gap disappears—Aranov (1979) and Al'tshuler et al. (1980).

40.2.2B Mott-Transition The impurity band with localized states cannot contribute to the conductivity at $T = 0$ K since it has an energy gap between filled and empty states. When the density of impurities is increased to an extent that delocalization occurs at the Fermi-level, the gap disappears and quasi-metallic conductivity is observed. This transition within an impurity band is a Mott-transition, and is related to a critical conductivity that was termed by Mott as *minimum metallic conductivity.*

$$\sigma_{\min} \simeq 0.05 \frac{e^2}{\hbar} N_M^{1/3} \qquad (40.6)$$

where N_M is the critical doping density—see Section 40.2.2B.2 (Mott and Davis, 1979). The Mott-transition is observed to be smooth, rather than abrupt, probably because of density fluctuation of impurities.

For Si:P, the Mott-transition occurs at a critical donor density of $N_M = 3.7 \cdot 10^{18}$ cm^{-3}; for this example, one obtains $\sigma_{\min} \simeq 20$ Ω^{-1} cm^{-1} (Rosenbaum, 1980)—see also Section 34.2.

Scaling and Mott-Transition. When, for example, hydrogen-like donors are close enough to each other, the donors' electrons no longer belong to a certain donor, but are able to move freely between donors even at $T = 0$ K, like electrons in a metal, i.e., they belong to all of the donors. Critical to this transition are three units of length, the interdonor distance $1/\sqrt[3]{N}$, the quasi-hydrogen radius a_{qH}, and the mean free path λ, for a diffusion-type of carrier migration. Their relative magnitude determines the type of conductivity, and its discussion is a subject of the theory of scaling.

Abrahams et al. (1979) suggested to use a dimensionless conductance of a cube of dimension L rather than the conductivity

$$G = \sigma L \cdot \frac{2\hbar}{e^2}, \qquad (40.7)$$

measured in elementary units of $2\hbar/e^2$. They discussed the changes in G as a function of L; it should change when L approaches atomic dimensions. They argued that the scaling function

$$\beta(G) = \frac{\partial \ln G}{\partial \ln L} \qquad (40.8)$$

is a universal function (Thouless, 1977, 1980), which is $\simeq 1$ for large conductances, becomes $= 0$ at a critical conductance G_c, and turns negative for $G < G_c$—see Fig. 40.2. Within this theory, G_c is a

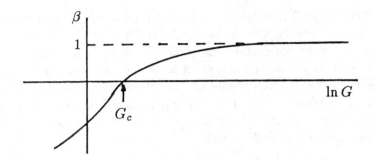

Figure 40.2: Scaling function vs. the dimensionless conductance [Eq (40.7)] for a 3D semiconductor.

universal constant,* and indicates the transition between metal-like and semiconductor-type conductivity. Here, Mott obtains for the critical conductivity

$$\sigma_c = 0.03 \frac{e^2}{\hbar a_{\mathrm{qH}}}, \qquad (40.9)$$

a value close to $\sigma_{\min}$ given by Eq. (40.6), here for $\sqrt[3]{N_M} \simeq 1/a_{\mathrm{qH}}$.

Ioffe-Regel Rule.   For further elucidation of the concept of minimum metallic conductivity, let us start from a metal and look for candidates of lower and lower mean free paths, i.e., reduced conductivities.   There are indications that with increased lattice disturbance lower conducting metals, such as liquid metals, have a lower mean free path, although, with a lower limit equal to the interatomic distance. One can argue that the electron wavefunction cannot lose phase memory faster than on the order of the interatomic distance a—*Ioffe-Regel rule* (Ioffe and Regel, 1960). This means that the conductivity of a metal cannot be smaller than $\sigma = e\mu n$, with $n = a^{-1/3}$ and $\mu = (e/m)\tau = (e/m)(a/v_F)$. The Fermi-momentum is given by $k_F = mv_F = m(3\pi^2 n)^{1/3}$; hence, one obtains as minimum metallic (Ioffe-Regel) conductivity

$$\sigma_{\mathrm{I\text{-}R}} = \frac{1}{\sqrt[3]{3\pi^2}} \frac{e^2}{\hbar a} = 0.32 \frac{e^2}{\hbar a} = 787 \cdot \frac{(1\text{Å})}{a} \ (\Omega^{-1}\mathrm{cm}^{-1}). \qquad (40.10)$$

In doped semiconductors two changes need to be introduced:

(1) instead of the interatomic distance, the quasi-hydrogen radius applies, and

---

* It should, however, depend on the microscopic atomic arrangement coordination (Mott and Kaveh, 1985).

(2) only a certain fraction of the impurity band states are extended states, which, after Mott and Kaveh (1985), is on the order of 8.5%, yielding Eq. (40.9) as critical conductivity in a semiconductor. For $a_{qH} \simeq 30$ Å, this critical conductivity is on the order of 20 $\Omega^{-1} cm^{-1}$.

40.2.3 Carrier Localization in Strong Electric Fields

When carriers are transported in narrow bands, independent of how such bands are produced, carrier localization can occur when the electric field is strong enough. Here, stationary electron states become localized in the direction of the electric field due to reflection at the boundaries of the Brillouin zone (Wannier, 1960). This causes a *Stark ladder*, with the possibility of phonon-induced jumps between the levels of this ladder (Hacker and Obermair, 1970). Resonance effects occur when the steps become equal to LO phonons (Mackawa, 1970), causing current oscillations.

Another possibility of carrier localization occurs for small polarons in strong electric fields, where the mobility decreases with increasing field in the tunneling regime (Böttger and Bryksin, 1979, 1980).

40.2.4 Phonon-Activated Conduction

For sufficiently high densities of impurities, the carrier transport within an impurity band occurs with a mean free path longer than the spacing of impurities. With lesser doping, the defect levels will become localized, and conduction can occur in one of two fashions:

- by tunneling from one defect to the nearest neighboring defect of the same type, or
- after thermal excitation into the adjacent band.

Competition between these two processes is exponentially dependent on the temperature. At sufficiently high temperatures the carrier transport via the conduction or valence band predominates.

If the mean free path of carriers is given by capture at impurities rather than by scattering, the conductivity can be described as a motion from one to another impurity center but with *electron transport through the conduction band**—see Fig. 40.1B (Fritzsche

---

* In highly disordered semiconductors, the motion may occur through excited states with greater overlap of their eigenfunctions. This is discussed in Section 40.4.5.

and Cuevas, 1960; Butcher, 1972). It can also be described as due to *inelastic* scattering at Coulomb-attractive centers, with phonon emission causing carrier capture. The corresponding carrier mobility is thermally activated

$$\mu = \mu_0 \exp\left(-\frac{\Delta E_t}{kT}\right),\qquad(40.11)$$

where μ_0 is an effective mobility given by equivalent scattering mechanisms of carriers within the band, and ΔE_t is the thermal activation energy.

40.2.4A Density Dependence of Hopping

When the density of impurities increases, tunneling from center to center becomes more probable. The tunneling transport is accomplished by *hopping* from one to the adjacent center, resulting in a conductivity

$$\sigma_{\text{hop}} = \sigma_{0,\text{hop}} \exp\left(-\frac{\Delta E_{\text{hop}}}{kT}\right)\qquad(40.12)$$

with ΔE_{hop} as the activation energy for hopping, which is described in the following section, and with the pre-exponential factor given by

$$\sigma_{0,\text{hop}} = \sigma_{00} \exp\left(\frac{2r_c}{r_0}\right)\qquad(40.13)$$

where r_0 is the fall-off radius of the impurity wavefunction ($= a_{\text{qH}}$ for quasi-hydrogen impurities). It is $\simeq 90$ Å for Ga in Ge, and $\simeq 47$ Å for Cu in Ge; Cu has a larger ionization energy of $\simeq 40$ meV. Also, r_c is the critical radius to establish a percolation path (see Section 40.4.2) from one electrode to the other, and can be estimated (McInnes and Butcher, 1979) as

$$r_c \simeq (0.865 \pm 0.015) N^{-1/3}\qquad(40.14)$$

where N is the density of the specific impurities between which hopping occurs.

In Fig. 40.3A, a computed relation of the resistivity vs. the mean separation of the impurities is shown. The corresponding relation of the measured resistivity vs. the density of corresponding donors in GaAs is given in Fig. 40.3B (Shklovskii and Efros, 1984). The solid curves give the theoretical estimate according to Eqs. (40.13) and (40.14).

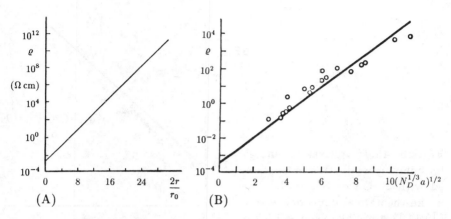

Figure 40.3: (A) Hopping conductivity as a function of the average separation between impurities. (B) Hopping resistivity as a function of the donor density in n-GaAs (after Shklovskii and Efros, 1984).

40.2.4B Activation Energy for Hopping

When impurities are spaced close enough to permit tunneling, the levels split to form a narrow band, as indicated in Section 40.2. Therefore, tunneling to arbitrary neighbors usually requires a slight thermal activation energy—see Section 40.2.1. The activation energy can be interpreted as the energy from the Fermi-level to the energy of the maximum of the density of empty state distribution. Typically, it is on the order of a few meV, and can be approximated for low compensation (Efros et al., 1972) by $\sim 60\%$ of the Coulomb energy at the average separation between the impurities:

$$\Delta E_{\text{hop}} \simeq 0.61 \frac{e^2}{4\pi\varepsilon_{\text{st}}\varepsilon_0} \left(\frac{4\pi}{3} N \right)^{1/3}. \tag{40.15}$$

The experimental values for ΔE_{hop} for Ge doped with P, Ga, or Sb are shown together with the theoretical curves [Eq. (40.15)] in Fig. 40.4.

With a distribution of defects in space and energy, the relation becomes more complex and is relevant for amorphous semiconductors— see Section 40.4.5. For an extensive review, see Mott and Davis, 1979, Shklovskii and Efros, 1984 and Mott, 1987.

A special type of hopping conduction relates to the hopping of small polarons, and is discussed by Holstein (1959) and Schnakenberg (1968)—see Section 36.2.1.

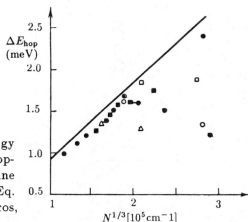

Figure 40.4: Activation energy for hopping conduction in Ge, doped with P, Ga or Sb. Solid line is the theoretical dependence [Eq. (40.15)] (after Shklovskii and Efros, 1984).

40.3 Highly Doped Semiconductors (HDS)

In highly doped semiconductors, the basic concepts discussed in the previous sections apply, however, in a modified fashion relating to the specific level distribution. This permits a number of more transparent theoretical approximations.

A semiconductor is heavily doped when the condition

$$N a_{\mathrm{qH}}^3 \geq 1 \qquad (40.16)$$

is fulfilled, which, dependent on the effective mass, is reached at vastly different doping densities in various semiconductors. For instance, $N a_{\mathrm{qH}}^3 = 1$ requires $N = 5 \cdot 10^{15}$ cm^{-3} in n-InSb and $N = 3 \cdot 10^{19}$ cm^{-3} in n-Ge. In several semiconductors, the highly doped regime cannot be obtained by diffusion doping, since clusters will form with limited solubility. Here, ion implantation or radiation damage can be used.

We will give a short review of the phenomena related to heavy doping in the following sections.

40.3.1 The Intermediate Doping Region

The distinction between light and heavy doping can be made in relation to the disappearance of the gap in the impurity band (Section 40.2.2A) and the transition from an activated semiconductivity to a quasimetallic conduction—see Section 40.2.2B. There is, however, a large intermediate a range between the Mott-transition at $N a_{\mathrm{qH}}^3 \simeq 0.02$ [Eq. (34.10)] and the HDS range which starts at

$Na_{qH}^3 \simeq 1$. In this intermediate range, some of the electrons are already delocalized.

One needs to recognize, however, that the transition is related to the statistical distribution of the defects, which are frozen-in or occur in the electron ensemble, even at low temperatures. We will give some insight into this relation in the next section. Other fluctuations are initiated at higher temperatures (*fluctuons*) and are reviewed by Krivoglaz (1974).

40.3.2 Density of States in HDS

In highly doped semiconductors there are two major contributions to the density of states: the states which are due to the extended eigenfunctions of the defects, and the states which are due to the perturbation in the surrounding host lattice. The latter may be described by analyzing the influence of heavy doping on free electrons. This influence can be expressed by band-edge perturbation, through the modulation of the band edges by the Coulomb potential of the defects (Kane, 1963; Halperin and Lax, 1966). In highly doped semiconductors, clusters of charged impurities often dominate. The charges of such clusters, however, are *not* Coulomb point charges.

In turn, the potential fluctuation near charged impurities results in an inhomogeneous distribution of electrons. When the potential fluctuation is smooth within the De Broglie wavelength of free electrons, the electron gas can be described classically. Its density varies according to the density of states, which is increased at positions near an attractive center where the conduction band is lowered. Near attractive centers there will be more carriers, while near repulsive centers there will be less of the corresponding type. With high doping densities, the potential fluctuations will have a higher amplitude. Complete state-filling of the valleys occurs at sufficiently low temperatures, whereas higher parts of the potential mountains extend above this "electron lake."

The density of states now becomes space-dependent

$$g(E, \mathbf{r}) \, dE = \frac{(2m_n)^{3/2}}{2\pi^2 \hbar^3} \sqrt{E + eV(\mathbf{r})}, \qquad (40.17)$$

with the fluctuating potential determined by a screened Coulomb potential

$$V(\mathbf{r}) = \frac{e}{4\pi \varepsilon_{st} \varepsilon_0 \mathbf{r}} \exp\left(-\frac{r}{r_0}\right). \qquad (40.18)$$

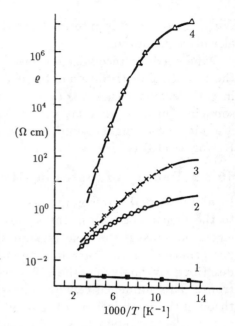

Figure 40.5: Resistivity of an uncompensated (1) to more and more compensated (2)...(4) samples of n-Ge:As $(N_D = 8 \cdot 10^{18} \text{ cm}^{-3})$ as function of the inverse temperature (after Gadzhiev et al., 1972).

When averaging over the space-dependent potential, one obtains from Eq. (40.17) the density of states tailing into the band gaps, as discussed in Section 24.1—Lifshitz tail.

40.3.2A The Degenerate HDS   The highly doped semiconductor with shallow impurities is usually degenerate, i.e., the Fermi-level is shifted to well within the band. Depending on its position, the "lake" of electrons rises within a hilly terrain to fill only the lowest valleys as little lakes, or with a rising level connects more and more lakes until navigation from one electrode to the other becomes possible. This behavior is similar to that of a percolation conductivity, as described in Section 40.4.2.

40.3.3 The Highly Compensated HDS

When compensating a highly doped semiconductor, the level of the carrier lake within the modulated band drops, which causes a substantial decrease of the conductivity.

With sufficient compensation, the semiconductor reverts from metallic conduction to one with thermal activation over saddle points in the hilly terrain, as shown in Figs. 40.5 and 40.6.

Here, carriers cannot contribute to percolation since they occupy only a small fraction of the volume, and tunneling is too expensive because of the high barriers between the remaining small puddles.

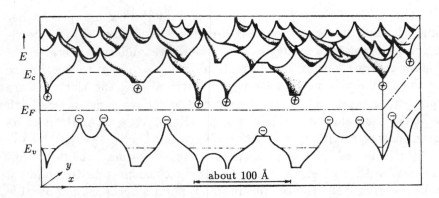

Figure 40.6: Two-dimensional representation of the band-edge fluctuation in highly doped semiconductors.

In completely compensated semiconductors, the potential fluctuation can increase further since the density of carriers is reduced below values, which are necessary for efficient screening. The maximum band fluctuation, however, is limited to $E_g/2$. Here, with penetration through the Fermi-level, the deep valleys again become filled and screening reappears, thereby limiting any further increase in fluctuation of the band edges.

There is a large body of experimental and theoretical research on highly doped semiconductors, including the influence of light (*persistent photoconductivity*—Ryvkin and Shlimak, 1973), of a magnetic field (*quantum screening*—Horring, 1969), and of low-temperature conductivity.

The carrier transport in such macroscopically fluctuating potentials is similar to that in semiconducting glasses (Ryvkin and Shlimak, 1973; Overhof and Beyer, 1981). In the following sections, we will analyze such transport in more detail.

40.4 Mobility in Semiconducting Glasses

The carrier transport in semiconducting glasses deserves a separate discussion because of the lack of long-range order and the high density of defects specific to the amorphous material. This does not permit simple translation of the effective mass picture, and requires a reevaluation of carrier transport and scattering concepts.

There are two aspects with direct influence on the carrier transport: the strong tailing of states into the band gap, and the absence of specific doping-induced defects in amorphous chalcogenides.

The *tailing of states* from the band into the band gap is rather smooth, and does not show a well-defined band edge. This necessitates a more careful analysis of the mobility at different energies. At higher energies within the conduction band, i.e., closer to the surface of the Brillouin zone (see below for justification), the electrons are quasi-free, except for scattering events, and have a mean free path λ larger than the interatomic spacing. Here, $\lambda k \gg 1$, and k may be used, however, with some caution (Mott and Davis, 1979).

With decreasing electron energy, the scattering probability increases, with scattering on potential fluctuations due to noncrystalline structures. Hence, the mean free path becomes comparable to the interatomic distance, and $\lambda k \simeq 1$. Here, k is no longer a good quantum number.* Substantial differences between the crystalline and the amorphous semiconductor become important. Therefore, the carrier transport must now be described in terms of a transport between *localized states*; the Mott-Anderson localization threshold is reached—see Section 25.3.1B.

In taking a slightly different point of view, one expects the band states near the *edge* to become *perturbed* with a concurrent widening or narrowing of the band gap, depending on the local degree of disorder. With charging of these defects, the Coulomb potential creates band undulations, as shown schematically in Fig. 40.6 (Böer, 1972). More recent experiments indicate that in amorphous semiconductors with a much lower density of charged centers, a similar mountainous profile of the near-edge band states results from the local stress and other defect-induced perturbations.

When the Fermi- or quasi-Fermi level is moved above the lowest valleys of this edge (to be better defined in Section 40.4.3), these valleys will fill up with carriers. Assuming that only near the surface of such "lakes" a carrier transport is possible, one recognizes that a continuous current can only flow when the Fermi-level rises enough to permit a *percolation path* from one to the other electrode

---

* In a crystalline structure, k, when closer to the center of the Brillouin zone, represents points in real space farther away from the unit cell; in this case long-range deviation from periodicity becomes important. In contrast, when $\lambda k \simeq 1$, the wave number is closer to the boundaries of the Brillouin zone; wheres in real space, the corresponding points are closer to the unit cell and the structure of the amorphous material resembles more that of a crystal.

(Fig. 40.7A to C); much below the "edge" carriers are trapped. We will now refine this roughly stated model.

40.4.1 Trap Depth and Overlap

Carriers are able to travel readily when the eigenfunctions of traps overlap. There are two arguments for a larger overlap of shallower traps: they usually have a larger fall-off radius of their eigenfunctions, and they are more plentiful, diminishing the intertrap distance.

With an exponentially decreasing trap distribution, and with an adequately large overlap of shallow traps, the carrier transport through such shallow centers is almost band-like.

40.4.2 The Mobility Edge

At slightly deeper trap levels the carrier transport proceeds from trap, to neighboring trap and has a diffusive character with a diffusion constant given by the exchange frequency ν_t:

$$D = \frac{\nu_t a_t^2}{6}. \tag{40.19}$$

Here, a_t is the distance between these traps.

Carriers in yet deeper traps will have to penetrate through increasingly thicker barriers via tunneling. Finally, such carrier transfer via tunneling becomes negligible, and requires thermal excitation into higher states.

In summary, the type of carrier transport depends on the depth of the traps between which such transport takes place. Carriers are significantly more mobile in shallower traps. There is a major step in the mobility of carriers between "localized" deeper and "extended" shallow trap states. This step is referred to as the *mobility edge*.

A material in which the Fermi-level at $T = 0$ K coincides with the mobility edge is called a *Fermi-glass*. This material displays metallic conductivity.

The distance between two defect centers at the mobility edge is approximately that of the nearest neighbors (Mott and Davis, 1979), yielding [see Section 19.3, Eq. (19.70)]:

$$D_\mu = \frac{\nu_\mu a^2}{6}, \tag{40.20}$$

with the tunneling frequency ν_μ approximated by the atomic electron frequency

$$\nu_\mu \simeq \frac{\hbar}{m_0 a^2}. \tag{40.21}$$

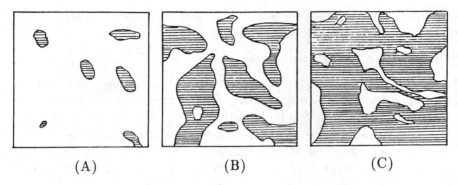

(A)   (B)   (C)

Figure 40.7: Percolation regions, which become larger and interconnect with increasing energy from (A)–(C), connect as puddles in a hilly terrain to form small and larger lakes when the water level rises, finally leaving only small islands near the highest points of the terrain.

We are *not* assuming hydrogen-like defects here. Therefore, tunneling is much reduced, and the distance of "defects" between which tunneling becomes significant can no longer be much larger than the normal interatomic distance a.

Using the Einstein relation, this yields, as an order of magnitude estimate for the mobility at the mobility edge,

$$\mu_{\text{edge}} = \frac{eD_\mu}{kT} \simeq \frac{e\hbar}{6m_0 kT} \simeq 7.5 \left(\frac{300\,\text{K}}{T}\right) \left(\frac{\text{cm}^2}{\text{Vs}}\right). \qquad (40.22)$$

Electrons that are excited much above the mobility edge contribute to the current, as they do in a crystal, by being scattered at defects, and usually have a mean free path that is much larger than the interatomic distance. Electrons closer to the mobility edge contribute via exchange interaction to neighboring traps, and electrons that have relaxed much below the mobility edge contribute through tunneling or after thermal activation.

Since the mobility decreases very steeply at the mobility edge, whereas the density of states does not, it is customary to identify the *band gap* in amorphous semiconductors as a *mobility gap*, i.e., the distance between the mobility edges for electrons and holes.

40.4.3 Diffusive Carrier Transport

We will now look a bit closer at the carrier transport around the mobility edge E_μ. With decreasing trap energy E_t, the trap density is reduced and the average distance between these defects is increased. At any given energy, the distance will fluctuate about an average number, making carrier transfer preferred in directions in which the distance is shortest. With further decreasing E_t, preferred paths become rarer. The carrier has to move in a diffusive path along preferred intertrap connections. This indicates that the carrier motion, which was randomly diffusive at higher energies, now becomes direction-selective toward the closest neighbor, thereby reducing the effective diffusion constant. Finally, the path connecting the two electrodes will be broken. From this point on, thermally activated conductivity becomes the sole possibility for carrier transport.

The selection of paths between neighboring sites at the mobility edge is significant in that it is a determinant of the *Hall mobility*. In amorphous semiconductors, the Hall effect cannot be calculated from Lorentz forces, but must be computed from quantum mechanical jump probabilities between localized states (Grünewald et al., 1981). Paths following the Lorentz force become slightly preferred. Because of this structure-determined path selection, the Hall-voltage becomes dependent on the average microscopic geometry of the atomic arrangement. For instance, preferred even- or odd-numbered rings (see Section 3.9.1) cause a sign reversal of the Hall effect for n- or p-type material (see Dresner, 1983).

40.4.4 Percolation

We will return once more to the carrier transport near the mobility edge. When filling traps by raising the Fermi-level, carrier diffusion is eased. This would appear homogeneously throughout the semiconductor if it were not for the mountainous profile of the potential, as mentioned in Section 40.4 and shown in Fig. 40.6. Here, in a mountain, the mobility edge is pushed above the Fermi-level; whereas in a valley, the Fermi-level lies above the mobility edge. In these lakes, the mobile electrons show diffusive motion along the surface of the lakes, but have to tunnel through the mountains. This type of transport, which can be understood from classical arguments (Broadbent and Hammersley, 1957), is commonly referred to as *percolation*. For a review, see Shante and Kirkpatrick (1971) or Böttger and Bryksin (1985).

The analysis of percolation was facilitated by the simple model of Miller and Abrahams (1960), using a network of random resistors and Kirchhoff's law to calculate the corresponding resistivity between the electrodes in a semiconductor with percolating conductivity.

Many aspects of carrier percolation can be discussed in the framework of *fractal networks*, i.e., a network of resistors in which a statistically increasing number of the interconnecting resistors are omitted.

40.4.5 Activated Mobility

Further below the mobility edge, the defect centers are sufficiently separated so that tunneling between them can be neglected compared to the thermal excitation into levels near the mobility edge. From here, electrons can be retrapped, excited again, etc. This process can be described as *thermally activated hopping*,* and requires a periodic interplay with phonons, i.e., carriers alternately absorb or emit phonons. Consequently, the hopping mobility depends exponentially on the temperature. For excitation from centers at an energy E_t, one obtains

$$\mu_{\mathrm{ht}} = \mu_0 \exp\left(-\frac{E_\mu - E_t}{kT} \right). \tag{40.23}$$

The thermally activated hopping mobility can be described in the form of a diffusion relation (Butcher, 1972):

$$\mu_{\mathrm{ht}} = \frac{e D_{\mathrm{hop}}^{(\mathrm{th})}}{kT} = \frac{e}{kT} \frac{\nu_{\mathrm{hop}}^{(\mathrm{th})} r_{\mathrm{hop}}^2}{6}. \tag{40.24}$$

The thermally activated effective hopping frequency is given by

$$\nu_{\mathrm{hop}}^{(\mathrm{th})} = \frac{\omega_{\mathrm{ph}}}{2\pi} \exp\left(-2\frac{r_{\mathrm{hop}}}{r_0} - \frac{W_{\mathrm{hop}}}{kT} \right), \tag{40.25}$$

where W_{hop} is the average energy difference between the two states for hopping, r_{hop} is the hopping distance, r_0 is the radius of the center, and ω_{ph} is an effective phonon frequency to match the energies of initial and scattered states. For hops of distance r_{hop}, the corresponding hopping energy is given by the band width of centers located at the Fermi-energy $\Delta E_B(E_F)$, which in turn is given by

$$W_{\mathrm{hop}} = \Delta E_B(E_F) = \frac{3}{4\pi r_{\mathrm{hop}}^3 N(E_F)} \tag{40.26}$$

---

* Hopping conduction can also involve small polarons which move by hopping from site to site, ions which hop from interstitial to interstitial site, electrons which hop between soliton bound states in one-dimensional conductors (acetylene) (Kivelson, 1982), or Frenkel excitons in molecular crystals (see references in Böttger and Bryksin, 1985).

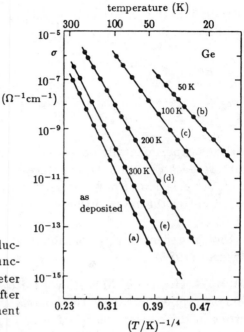

Figure 40.8: Electrical conductivity of amorphous Ge as a function of $\sqrt[4]{kT}$. Family parameter is the annealing temperature after damage with Ge$^+$ bombardment (b) (after Apsley et al., 1977).

with $N(E_F)$ (dimension $[\text{eV cm}^3]^{-1}$) is the density of defects at the Fermi-level from which such activation makes the largest contribution to the mobility (see Mott, 1969; Pollak, 1972).

40.4.5A Variable Range Hopping   In amorphous semiconductors with the Fermi-level below the mobility edge, thermal activation becomes essential to carrier transport. With reduced temperature, the width of the energy band decreases, thereby involving less centers, i.e., the distance between the active centers increases (Mott, 1968, 1969). The average hopping distance, which maximizes the hopping rate, is given by

$$\overline{r_{\text{hop}}} = \left(\frac{3r_0}{2\pi N(E_F)kT} \right)^{1/4}. \tag{40.27}$$

This results in a hopping frequency of

$$\nu_{\text{hop}} \propto \exp\left(-\frac{C}{(kT)^{1/4}} \right) \quad \text{with} \quad C = 2 \left(\frac{3}{2\pi} \right)^{1/4} \left(\frac{1}{r_0^3 N(E_F)} \right)^{1/4}. \tag{40.28}$$

Introducing this relation into Eq. (40.24), one obtains a hopping mobility

$$\mu_{\text{hop}} = \mu_0 \exp\left(-\frac{T_0}{T}\right)^{1/4}, \tag{40.29}$$

which is experimentally observed in some of the amorphous semiconductors—see Fig. 40.8. In thin layers, the $1/T^{1/4}$ relation changes to a $1/T^{1/3}$ relation (Knotek et al., 1973). Here, percolation paths are cut open by the layer surfaces normal to the current flow (Hauser, 1975).

For a recent review of hopping conduction, see Böttger and Bryksin (1985).

40.4.5B Hopping Mobility of Polarons The strong interaction of trapped carriers with phonons suggests the involvement of polarons in the carrier transport of amorphous semiconductors (Emin, 1975; Mott and Davis, 1979). In certain amorphous semiconductors (see Section 22.4.10A), the carrier transport may also be caused by hopping of bipolarons (Schlenker and Marezio, 1980; Elliott, 1977, 1978; Davis, 1984).

40.4.6 Dispersive Carrier Transport

One of the most convincing arguments about the carrier transport involving a quasi-exponential trap distribution stems from experiments with excess carriers, e.g., injected or photo-excited carriers. These are trapped, reemitted from shallow traps, retrapped, and so on; during the period between trapping, they are mobile and drift in an electrical field. The first carriers that traverse the device have not been trapped, followed by carriers that have been trapped once, twice, etc. Consecutive trapping causes further slow-down of carrier traversal. When being retrapped, energy is dissipated by emitting phonons, and successively deeper traps are filled; from here escape is much slower. This behavior results in a typical distribution of these excess carriers as a function of the time while in transit, in agreement with the experiment. This confirms the intimate involvement of a trap distribution in carrier transport. We will return to this important subject when we discuss carrier kinetics—Section 48.1.1D. For a review, see Tiedje (1984).

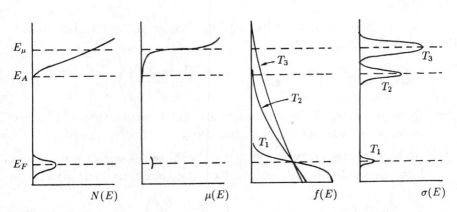

Figure 40.9: Typical distributions resulting in three different modes of conductivity at temperatures T_1, T_2, and T_3.

40.4.7 Conductivity in Amorphous Semiconductors

The relative magnitude of the different contributions to the conductivity is a function of the temperature. This is shown in Fig. 40.9, in which the Fermi-level remains pinned.

(a) At low temperatures (T_1), only carriers near E_F can contribute to the conductivity. Since pinning of the Fermi-level requires a high density of defect states at E_F, such conductivity is similar to the impurity conduction in crystalline semiconductors. One can distinguish two cases of this *impurity conductivity*:

   (α) The impurity density near E_F is large enough to permit sufficient tunneling within a band of width ΔE_1; then impurity conduction similar to a crystalline semiconductor dominates, with

$$\sigma = \sigma_{a1} \exp\left(-\frac{\Delta E_1}{2kT}\right). \tag{40.30}$$

   (β) The impurity density is smaller and its bandwidth is larger than kT; then *variable range hopping* occurs with the characteristic $1/\sqrt[4]{kT}$ dependence:

$$\sigma = \sigma_{a2} \exp\left(-\frac{C}{\sqrt[4]{kT}}\right), \tag{40.31}$$

   with C as given by Eq. (40.28). Such dependency is shown in Fig. 40.8.

(b) At medium temperatures (T_2), sufficient carriers are excited into tailing states near E_a, which show sufficient overlap for tunneling, so that hopping is activated. With ΔE_2 as the activation energy for hopping, one obtains

$$\sigma = \sigma_b \exp\left(-\frac{E_a - E_F + \Delta E_2}{kT}\right). \tag{40.32}$$

However, since $E_a - E_F$ is usually much larger than ΔE_2, one observes a constant slope in the $\ln \sigma$ vs. $1/T$ diagram.

(c) At higher temperatures (T_3), when sufficient carriers are excited into nonlocalized, i.e., band states, the conductivity is given by

$$\sigma = \sigma_c \exp\left(-\frac{\tilde{E}_c - E_F}{kT}\right). \tag{40.33}$$

(d) With further increasing temperatures, the mobility may increase sufficiently above the saddle point between the undulating band edges to provide yet one more significant contribution to the conductivity:

$$\sigma = \sigma_d \exp\left(-\frac{E'_g}{2kT}\right). \tag{40.34}$$

The pre-exponential factors are

$$\sigma_{a2} = e^2 \omega N(E_F)\overline{r^2} \tag{40.35}$$

for variable range hopping [see Eq. (40.27)],

$$\sigma_b = 0.03\frac{e^2}{\hbar \lambda_i} \tag{40.36}$$

for hopping from tailing states, and

$$\left.\begin{array}{c}\sigma_{a1}\\\sigma_c\end{array}\right\} = \sigma_{\min} \simeq \frac{e^2}{2\pi^2 \hbar a_{qH}} \simeq \frac{610}{a_{qH}[\text{\AA}]} \; \Omega^{-1}\text{cm}^{-1} \tag{40.37}$$

for band conductivity. Finally,

$$\sigma_d = \frac{e^2 N(E'_g)kT\tau_e}{m_n} \tag{40.38}$$

for conduction above the saddle points of the band edges. Here, $\sigma_{\min}$ is the *minimum metallic conductivity* (see Mott and Davis, 1979), τ_e is the energy-relaxation time, and $N(E'_g)$ is the joint density of states at E'_g, the shifted effective band gap.

The type of predominant conductivity depends on material preparation (Connell and Street, 1980). This becomes rather sensitive in tetrahedrally bound amorphous semiconductors, such as α-Ge:H or α-Si:H, where a wide range of $\sigma(T)$ behavior is observed, depending on deposition parameters, doping, hydrogenation, and annealing treatments (LeComber et al., 1972; Bullot and Schmidt, 1987).

Summary and Emphasis

As discussed in previous chapters, most carrier transport is defect-*influenced*. However, in highly doped or disordered semiconductors, the carrier transport becomes *induced* by defects.

If doping produces a well-defined predominant defect level with increasing density, it will split into two (bonding and antibonding) bands separated by a gap. Below a density to permit sufficient tunneling, excitation from the filled, lower impurity band into the conduction band is required for carrier transport.

When the impurity band is wider than $2kT$, and the Fermi-level lies in the middle of this band, "variable range hopping" occurs: with decreasing T, the predominant excitation occurs from a narrower range of width kT of these centers, and causes a semilogarithmic slope $\propto 1/T^{1/4}$.

With increased defect density, a diffusive transport within the upper impurity band becomes possible; the conductivity in this band requires a small activation energy to bridge the gap.

With further increase of the defect density, the carriers become delocalized, the gap disappears, and the conductivity within the impurity band becomes metallic.

A similar tunneling-induced carrier transport can take place in amorphous semiconductors within the tail of states, extending from the conduction or valence band into the band gap when the states are close enough to each other to permit significant tunneling. Here carriers become delocalized. The edge at which delocalizing occurs is referred to as the mobility edge. At this edge, major carrier transport starts; below the mobility edge, carriers are trapped rather than mobile.

With a statistical distribution of defects within semiconductors, at a given threshold, only some volume elements become conductive. With increasing temperature these volume elements will widen, will start to interconnect, and finally will provide an uninterrupted path from electrode to electrode. Such percolation character is typical for most of the conduction phenomena in highly doped or disordered semiconductors, which have a density-related threshold of conduction.

At sufficiently high temperatures, carrier transport higher within the conduction or valence band may compete significantly with the conduction mechanisms described above. This band conduction may have a mean free path compatible to the one in crystalline semiconductors.

The mobility of carriers in highly doped or disordered semiconductors is typically on the order of 10 cm²/Vs or lower. At low temperatures it is determined by tunneling (hopping) from neighbor to neighbor, and is very sensitive to the density of defects and their distribution in space and energy. With a variety of new devices based on amorphous semiconductors, it is essential to gain a better understanding of the microscopic processes involved in the carrier transport in such materials.

Exercise Problems

1.(r) Describe in your own words the creation of an impurity band with increasing density of shallow donors. Discuss the splitting of this band.

    (a) Discuss the influence of carrier delocalization. Identify the Mott transition.

    (b) Discuss the onset of conduction via percolation within such bands.

    (c) Discuss the difference between normal, donor-induced conductivity, and variable distance hopping.

2.(r) Why is scaling important in the discussion of conduction thresholds in highly doped semiconductors?

3.(*) What is the minimum metallic conductivity? Compare this concept for low-conductive metals, highly doped semiconductors, and compensated semiconductors.

4.(*) Discuss the Greenwood-Kubo formula in the context of other transitions already described.

5.(e) Develop the expression for the percolation mobility [Eq. (40.22)] and justify the use of Eq. (40.21).

6(r). Describe hopping mobility, and analyze the different regimes.

Chapter 41

Carrier Transport in Organic Semiconductors

A variety of carrier-transport phenomena is observed in various organic semiconductors. A few of them can be doped n- or p-type, and have attractive semiconductive properties. Photoconductive organic insulators are of major importance in electrophotography.

Organic solids are predominantly van der Waals bonded; some of them have other bonding superimposed, such as ionic, hydrogen, and charge-transfer bonding. Only a few of them have been obtained as single crystals, and purified sufficiently in order to study their intrinsic semiconductor properties. Most organic solids are excellent insulators and become semiconductive only after doping (Pope and Swenberg, 1982). Some organic polymers, however, show promising semiconducting properties (Goodings, 1976; Gill, 1976).

One distinguishes single-component and two-component (charge-transfer) semiconductors. The first group contains the classical organic semiconductors, such as anthracene; the second group includes highly conductive compounds, some of which show semiconductor-metal transitions and even superconductivity.

In this chapter, a short review is provided concerning the different types of carrier transport in organic semiconductors.

41.1 Single-Component Semiconductors

Single-component organic crystals are usually good insulators. Some become photoconductive with sufficient optical excitation. The class of aromatic hydrocarbons shown in Fig. 41.1 has been more thoroughly investigated. They have band-gap energies between 2 and 5 eV.

The band gap of organic semiconductors can rarely be determined by optical absorption, since valence-to-conduction band transitions

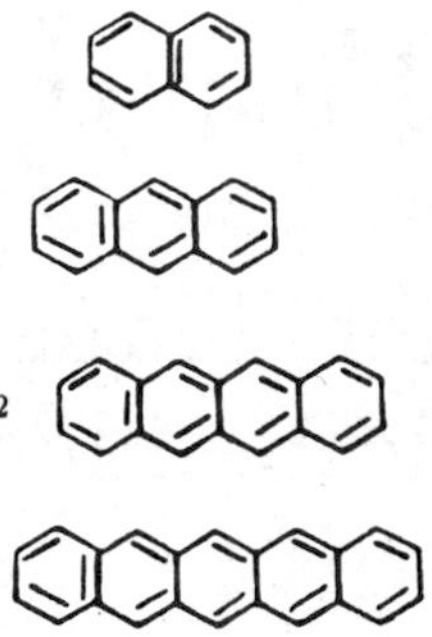

Naphthalene: $C_{10}H_8$ $E_g = 5.3$ eV

Anthracene: $C_{14}H_{10}$ $E_g = 4.0$ eV

Naphthacene: $C_{18}H_{12}$ $E_g = 2.5$ eV

Pentacene: $C_{22}H_{14}$ $E_g = 2.2$ eV

Figure 41.1: Structures and band gaps for lower polyacenes.

are masked by transitions to excited molecular states—excitons (Davidov, 1962), which are nonconductive; they lie within the band gap. The band gap can be estimated from

$$E_g = I_c - A_c, \tag{41.1}$$

where I_c is the workfunction, and A_c is the electron affinity of the crystal. E_g can be measured directly from the threshold of intrinsic photoconductivity (Marchetti and Kearns, 1970).

The mobility is usually low, for electrons and holes typically in the 10^{-2} to 10 cm^2/Vs range at 300 K, and falls with increasing temperatures.

Impurities play an important role in the electrical conductivity of organic semiconductors (LeBlanc, 1967), which, when purified, have resistivities usually exceeding 10^{16} Ωcm. Doping is especially important for *organic dyes*, which play a significant role for sensitization in photography (Meyer, 1974). Impurities with an ionization energy slightly less than the host act as donors, and those with an electron affinity slightly larger than the host act as acceptors (Karl, 1974). As most dyes dissociate before melting, they cannot be zone-refined, and in general are poor semiconductors (Gutman and Lyons, 1967).

41.2 Two-Component Semiconductors

Two-component semiconductors consist of pairs of complementary molecules with large differences in their redox properties. There are two types of representatives: the radical ion salts and the charge-transfer complexes (Soos, 1974).

41.2.1 Radical Ion Salts

Chemical stabilization in a two-component system, in which one is a donor and the other an acceptor, is the basis for forming an ordered structure. In radical ion salts,

(a) an organic radical *cation* (such as perylene$^+$) is combined with a counter anion (such as PF_6^-), or
(b) an organic radical *anion* (such as $TCNQ^-$) is combined with a counter cation.

41.2.2 Charge-Transfer Complexes

The combination of two organic molecules, one acting as an electron donor D and the other as an acceptor A, produces organic crystals that can show very low or vanishing activation energies and high conductivities. These conductivities are caused by an incomplete charge transfer between D and A, which results for the ground state in partially filled bands. Typical donor and acceptor molecules that form such charge-transfer crystals are given in Fig. 41.2.

Many such semiconductors have low-lying electronically excited states in which an electron is transferred from D to A. The charge-transfer transition in the excited state* may be written as

$$DA \rightarrow D^+A^- \quad \text{with} \quad E_{CT} = I_d - A_a - C, \qquad (41.2)$$

where I_d is the ionization energy of the donor, A_a is the electron affinity of the acceptor, and C is a Coulomb-binding energy of the excited state. E_{CT} is the "energy gap" between the ground state and the excited charge-transfer state (Mulliken, 1952).

The resulting structures are termed *neutral charge-transfer crystals*, typically with a one-dimensional array of $DADADA\ldots$ states.

The activation energy for semiconductivity is typically

$$E_s \simeq \frac{1}{2} E_{CT}, \qquad (41.3)$$

whereas the threshold energy for photoconductivity is usually (Gutman and Lyons, 1967):

$$E_{ph} \simeq E_{CT} + 0.2 \text{ eV}. \qquad (41.4)$$

This class of crystals contains generally poor semiconductors.

---

* The excited state has essentially ionic charge character.

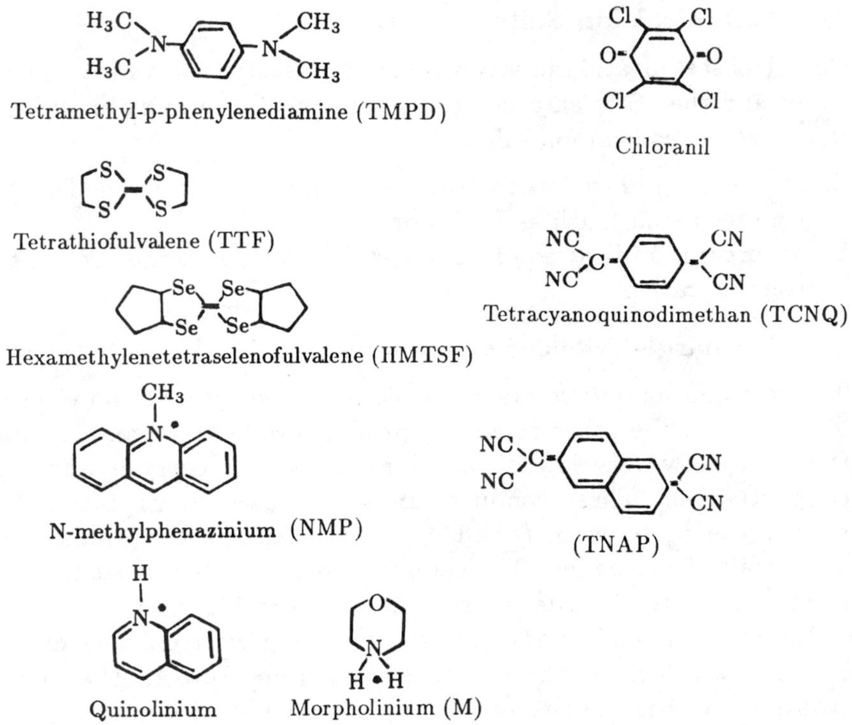

Figure 41.2: Structure of some typical organic donor and acceptor molecules in charge-transfer crystals (after Braun, 1980).

41.2.2A Partially Ionic Charge-Transfer Crystals   By far, the most interesting group of organic semiconductors is that in which the electron exchange from the donor to the acceptor molecule is not complete, and results in "partially filled bands" and consequently in a rather large semiconductivity—or even metallic conductivity.

The crystals are formed by a sandwich-like stacking of planar molecules—see Fig. 41.2. One distinguishes different types of such *charge-transfer crystals* (Soos, 1974), depending on whether:

(a) in a $D^{\delta+}A^{\delta-}D^{\delta+}A^{\delta-}$ structure $\delta = 1$ (*full charge transfer*) or $0 < \delta < 1$ (*fractional charge transfer*);

(b) the stacking contains $D^{\delta+}D^{\delta+}$ ... and $A^{\delta-}A^{\delta-}$ ... complexes, for *segregated stacking* or does not, for *mixed stacking*; and

(c) the distance of any neighbor within the stack is the same, in *regular stacking* or *alternating*, when the distance varies in a DA and AD sequence.

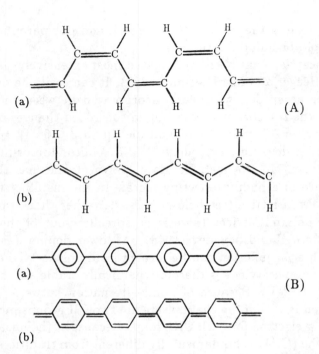

Figure 41.3: Structural formulae of linear organic polymers; (A) poly-acetylene as (a) *cis*- and (b) *trans*-isomer; and (B) polyparaphenylene in a (a) benzoid structure and (b) a quinoid structure.

The resistivity of these semiconductors lies between 10^2 and 10^6 Ωcm at room temperature with a transfer energy gap in the 0.1–0.4 eV range (Braun, 1980). The conductivity is usually highly anisotropic, with the electron transfer integral in the stacking direction typically a factor of 10 larger than in the direction perpendicular to the stacks (Keller, 1977). Trapping is of minor importance in these semiconductors with a high carrier density (Karl, 1984).

The charge-transfer crystals provide an opportunity for fine tuning of the semiconductive properties by replacing TTF-type donors and TCNQ-type acceptors with other similar molecules (Bloch et al., 1977), which could render such materials attractive for some technical applications.

41.3 One-Dimensional Organic Polymers

These polymers include *polyacetylene*, as the simplest member, shown in Fig. 41.3A. An example for an aromatic linear polymer is *polyparaphenylene*—Fig. 41.3B. These polymers can exist in two isomers:

polyacetylene in the *cis*- and *trans*- form, and polyparaphenylene in the benzoid or quinoid structure.

Polyacetylene has been investigated most extensively—see the review by Heeger and MacDairmid (1980). It can easily be doped with donors or acceptors, which are incorporated between the polymer chains of the *trans*-isomer, resulting in controlled changes of the conductivity over 13 orders of magnitude up to $3 \cdot 10^3$ $\Omega^{-1}\text{cm}^{-1}$. At high doping densities, i.e., above 1%, a semiconductor-metal transition occurs. In the metallic state, polyacetylene has the optical appearance of a highly reflecting metal. In the metallic state, however, a soliton lattice (see below) persists, rather than transforming it into a polaron lattice: there is no spreading-out of the localized state—Conwell and Jeyadev (1988). At lower doping densities, devices with a *pn*-junction can be formed from polyacetylene.

Trans-polyacetylene is thermodynamically stable and has a band gap of ~ 1.8 eV. Because of bond alternation between single and double bonds within the chain, there are two CH groups per unit cell, with one π electron for each CH group. Because of the unsaturated π system, the $(\text{CH})_x$ is fundamentally different from traditional organic semiconductors and is more akin to inorganic semiconductors, except for its high anisotropy—its chain-like character.

A bond alternation, shown in Fig. 41.4A, has attracted substantial interest as a manifestation of a *soliton*. Highly mobile, the soliton has a room temperature hopping rate in excess of 10^{13} s^{-1}, and can be induced by doping—Fig. 41.4B. Such a soliton can be described as a kink in the electron-lattice symmetry, rather than a spread-out transition,[*] and seems to be responsible for a wide variety of unusual electrical, optical, and magnetic properties of these polymers. A short review is given by Heeger (1981).

41.4 Mobility in Organic Semiconductors

The electron transfer in *single-component semiconductors* probably proceeds in rather narrow bands due to a small amount of wavefunction overlap of π electrons, and may be estimated from a tight-binding approximation (LeBlanc, 1967). The mobilities are on the order of $10^{-2} \ldots 10$ cm^2/Vs.

---

[*] Some spreading-out, however, over about 10 lattice constants is caused by minimum-lattice-energy consideration, and does not change during the kink motion.

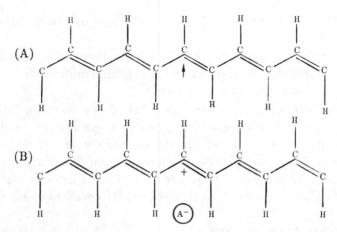

Figure 41.4: Bond alternation in polyacetylene: (A) as a neutral soliton, and (B) as a charged soliton with a compensating acceptor ion nearby.

Strong coupling to molecular oscillations suggests a polaron state contribution (Spear, 1974). A review of the theory of carrier mobility in organic semiconductors is given by Druger (1975).

Mobilities for a large number of organic semiconductors are tabulated by Schein (1977). They decrease with increasing temperature according to T^{-n} with $0 < n < 3$. Some better (charge-transfer) semiconductors show a mean free path that slightly exceeds the interlattice spacing. A critical review of mobility and other physical data for both one-component and mixed organic semiconductors is given by Karl (1984).

Summary and Emphasis

Organic semiconductors can be single-component materials, which can be doped, such as aromatic hydrocarbons, e.g., is anthracene. These materials are good insulators and show photoconductivity.

Another group of single-component semiconductors is comprised of certain linear polymers, such as polyacetylene. With doping, this group can change its conductivity up to 13 orders of magnitude and can become metallic in electrical behavior and optical appearance.

Two-component semiconductors contain molecules or molecular layers, which act as donors and others which act as acceptors. Variation of their donor/acceptor ratio can change the behavior from highly compensated to n- or p-type, with a wide range of conductivities, depending on the deviation from a donor to acceptor ratio of 1:1. There is a great variety of such crystals that exhibit a broad range of properties, including metallic conductivity and, at low temperatures, superconductivity.

The carrier mobility in organic semiconductors is substantially lower than in good semiconducting inorganic compounds, and typically is in the $10^{-2} \ldots 10$ cm^2/Vs range.

Organic semiconductors are attractive since they can be formed easily and offer a wide variety of possible candidates, only a few of which are presently being investigated. However, their relatively low mobility and tendency for carrier-trapping is a drawback to the development of devices. They are of technical interest because of their optical, exciton, and luminescence-related properties.

Exercise Problems

1.(r) Describe the different types of organic semiconductors.

2.(r) How is the band gap determined in organic semiconductors?
(a) Why does one observe major differences between the band gap determined by photoconductivity and that determined by semiconductivity?
(b) Why does one have to be careful in using an optical absorption spectrum to obtain the band gaps of organic semiconductors?
(c) What is the significance in obtaining the band gap from the workfunction and electron affinity?

3.(*) Search the literature for the highest mobility you can find for organic semiconductors at room temperature. Give proper citation.

PART VII

GENERATION-RECOMBINATION

Chapter 42

Carrier Generation

When the semiconductor is exposed to an external electromagnetic field or an electric or phonon field, free carriers can be generated, resulting in semiconductivity or photoconductivity.

The density of carriers in semiconductors can be determined from the difference between generation and recombination rates, and the net influx of carriers from surrounding regions:

$$\frac{dn}{dt} = g_n - r_n + \frac{1}{|e|} \operatorname{div}\mathbf{j}_n. \qquad (42.1)$$

In homogeneous semiconductors, this net influx vanishes and the electron density changes, just as a change in population occurs when there is a difference between birth and death rates.

Three types of **generation processes** can be distinguished. Each one of the generation processes needs energy, which can be supplied as

- thermal (phonon field),
- optical (electromagnetic field), or
- electrical (electric field) energy.

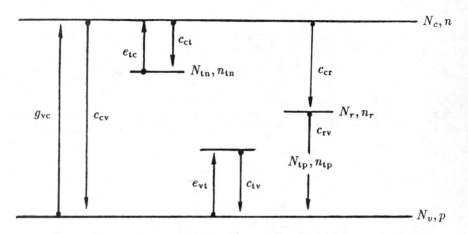

Figure 42.1: Electron transitions between localized (in band gap) and nonlocalized states (bands).

Two types of recombinations are distinguished. Each of the **recombination processes** releases energy in one of the following forms:

- thermal energy via nonradiative recombination or
- luminescence via radiative recombination.

The type of generation selected is usually the choice of the experimentalist, whereas the type of predominant recombination is mostly a function of the defect structure of the selected material. It is also influenced by the temperature, and sometimes by the electric field and other parameters. A number of examples will be discussed for each type of generation in the following sections, and for the recombination in the following chapter.

42.1 Typical Electron Transitions

In Fig. 42.1, typical transitions are shown among a variety of states. For consistency in the following descriptions, only electron transitions will be identified; hole transitions proceed in the opposite direction. The *transition coefficients* c_{ik} and e_{ik} can then be unambiguously defined by the first and second subscripts, indicating the initial and final states, respectively. The transition coefficients for 'e'xcitation are labeled e_{ik} to set them apart from the recombination or 'c'apture transitions c_{ik}.

A *transition rate* R_{ik} is defined as the product of the electron density in the original state, the hole density in the final state, and the transition coefficient:

$$R_{ik} = c_{ik} n_i p_k. \tag{42.2}$$

For example, the capture of an electron from the conduction band into an electron trap is given by

$$R_{ct} = c_{ct} n (N_{tn} - n_{tn}), \tag{42.3}$$

where N_{tn} and n_{tn} are the densities of electron traps and of captured electrons in these traps, respectively. In deviation from the above-given rule, the index c is left off from the electron density in the conduction band $(n_c \rightarrow n)$ to conform with common notation. This implies that an approximation is used here by describing the conduction and valence bands as a level at the band edges, and assuming that transition coefficients are independent of the actual carrier distribution within the bands.

The transition coefficients c_{ik} or e_{ik} have the units $\mathrm{cm^3 s^{-1}}$. The product of these coefficients with the hole density in the final state, $c_{ik} p_k$, is the *transition probability* $(\mathrm{s^{-1}})$. For further detail, see Section 42.2.2.

The following sections present a brief review of the different excitation mechanisms into higher energy states.

42.2 Thermal Ionization

Thermally induced *ionization* can occur from any type of lattice defect as *extrinsic ionization* or from lattice atoms as *intrinsic ionization*. Such ionization was discussed in Section 27.2.2, using an equilibrium approach without requiring an understanding of the microscopic mechanism involved in the actual process of ionization. The previous analysis required only the magnitude of the ionization energy and the density of levels in equilibrium. In this section, the discussion is extended to include information relating to the excitation process, which is necessary for a kinetic evaluation.

42.2.1 The Thermal Ionization Mechanism

Thermal ionization is a process requiring statistical consideration, since it usually needs the presence of several phonons simultaneously—or in a short time interval—to supply sufficient energy. Such a "simultaneous" supply of phonons is necessary for centers

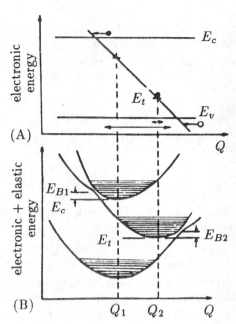

Figure 42.2: Excitation of a trapped electron and possible retrapping. (A) Band model with amplitude of normal and giant oscillations indicated by the double-headed arrows. (B) Corresponding configuration coordinate diagram; horizontal lines are spaced by the breathing-mode phonon energy.

that do not have eigenstates spaced closely enough, into which an intermediate thermal excitation can take place. The simultaneous phonon excitation can be interpreted as a process involving a transient giant oscillation of a lattice atom.

For *deep defect centers* that possess a phonon ladder, there exist two possibilities: a multiphonon absorption or a sequential absorption of phonons, a—cascade process. For the latter case, the absorption of the next phonon must occur within the lifetime of an excited state to accomplish the subsequent step of excitation, and so on, until the total ionization energy is supplied. In both cases, a transient giant oscillation of the defect center results, as indicated by the large double-headed arrow in Fig. 42.2A. This results in the total energy of the center exceeding that of the bottom of the conduction band, from which the electron can tunnel into the adjacent states of the conduction band.

Such a deep defect center can induce transitions from the valence to the conduction band. In this case, a giant oscillation of a lattice atom, i.e., a multiphonon process, can result in a transfer of an electron from the valence band into the defect center. If the lifetime of the electron within the center is long enough, a consequent giant oscillation of the center, as previously discussed, would bring this electron into the conduction band. The sum of these two processes is equivalent to a thermally induced band-to-band transition, which is more difficult to understand without this intermediate step.

For every excitation process, there exists an inverse deexcitation process. The lifetime of an electron in any intermediate state and the supply of phonons with the proper energy for accomplishing the next excitation step decide with what probability any higher excitation is accomplished. For further discussion of the actual microscopic phonon stepping, or "simultaneous" multiphonon excitation, see the review of Stoneham (1981). A semiclassical model of such ionization, presented below, will provide some insight.

42.2.2 Attempt-to-Escape Frequency

In a simple configuration coordinate diagram, one can represent the eigenstate of an electron as that of an oscillator—Fig. 42.2. With its surrounding lattice, it will oscillate between different electronic eigenstates. With each of the oscillations of variable amplitude, the electron has a varying probability of escaping from the center into the adjacent band.

Determining when enough phonons have been supplied for such escape is a matter of statistics. This probability is given by the Boltzmann factor: $\exp[-E_i/(kT)]$. Hence, the total ionization probability is

$$P_i = \nu_i \exp\left(-\frac{E_i}{kT}\right), \qquad (42.4)$$

where ν_i is the *attempt-to-escape frequency*:

$$\nu_i = \frac{\omega_i}{2\pi} \qquad (42.5)$$

which may be approximated by the breathing mode or a specific vector-mode phonon frequency ω_i of the defect center.

For Coulomb-attractive, shallow centers there is a different, rather crude way of estimating an electronic attempt-to-escape frequency from the ionization radius r_i of such a center, as well as the thermal velocity of the electron, assuming that it behaves like an effective mass particle:

$$\nu_i = \frac{v_{\mathrm{rms}}}{2\pi r_i}. \qquad (42.6)$$

This radius also determines the *capture cross section* $s_n = \pi r_i^2$; hence one obtains

$$\nu_i = \frac{v_{\mathrm{rms}}}{2\sqrt{\pi s_n}}. \qquad (42.7)$$

This approximation, however, is only useful for a Coulomb-attractive center, wherein the effective mass picture can be applied, and therefore the appropriate capture cross section can be determined. With an rms velocity of $\sim 10^7$ cm/s, and a capture cross section of $\sim 10^{-13}$ cm^2, one estimates the frequency factor of such centers to be $\sim 10^{13}$ s^{-1}.

This simple model, however, should *not* be applied to other types of centers, which would yield values that are substantially too high.

42.2.3 Thermal Excitation: Thermodynamic Arguments

Thermal excitation probabilities are easily obtained from thermodynamic arguments—see also Section 43.2.2B. This will be shown for a defect center that contains only the ground level, and interacts preferably with the conduction band: an electron trap.

In equilibrium, all transitions into the level must equal all transitions out of this level between each group of two states, since its population remains constant within each volume element. This fundamental **detailed balance principle** yields (see Fig. 42.1)

$$e_{tc} n_t p_c = c_{ct} n (N_t - n_t).$$ (42.8)

Equation (42.8) holds independently of the position of the Fermi-level. For ease of the following computations, let us assume that the Fermi-energy coincides with the energy of the trap level ($E_F = E_t$); then, in thermal equilibrium, the population of these traps is $N_t/2$, and one has $(N_t - n_t)/n_t = 1$. Therefore, with $p_c \simeq N_c$, one obtains

$$e_{tc} N_c = c_{ct} n \simeq c_{ct} N_c \exp\left(-\frac{E_c - E_t}{kT}\right),$$ (42.9)

yielding for the ratio of emission-to-capture coefficients,

$$\boxed{\frac{e_{tc}}{c_{ct}} = \exp\left(-\frac{E_c - E_t}{kT}\right).}$$ (42.10)

Although this condition was obtained for a specific case, namely in thermal equilibrium with $E_F = E_t$, the ratio holds true in general, since both coefficients are constant and do not change with trap population.

The *emission probability* $e_{tc} N_c$ can also be expressed in a microscopic model (see Section 42.4.1A) as the product of the attempt-to-escape frequency ν_i and the Boltzmann factor:

Table 42.1: Typical capture and ionization parameters

Center Type	Coulomb-attractive	Neutral	Coulomb-repulsive	Tight-binding	Dimensions
s_n	10^{-13}	10^{-16}	10^{-21}	10^{-18}	cm^2
c_{ct}	10^{-6}	10^{-9}	10^{-14}	10^{-11}	cm^3/s
ν_t	10^{13}				s^{-1}

$$e_{tc} N_c = \nu_i \exp\left(-\frac{E_c - E_t}{kT}\right). \qquad (42.11)$$

From gas-kinetic arguments, one can describe the capture coefficient as the product of the capture cross section s_n of the center and the rms velocity of the mobile carrier

$$c_{ct} = s_n v_{rms}. \qquad (42.12)$$

Along its path through the lattice, a carrier can be thought of as sweeping out the cylinder of a cross section of the capturing defect, as it recombines when it touches the defect at any point on its cross section. In combining Eqs. (42.8)–(42.12), one obtains for Coulomb-attractive centers a useful relation between the capture cross section s_n and an effective attempt-to-escape frequency

$$\nu_t = N_c v_{rms} s_n, \qquad (42.13)$$

or $\qquad \nu_t = 3 \cdot 10^{13} \left(\frac{s_n(cm^2)}{10^{-13}}\right)\left(\frac{m_n}{m_0}\right)\left(\frac{T}{300\,K}\right)^2 \quad (s^{-1}). \quad (42.14)$

The typical parameters characterizing an electron trap are given in Table 42.1. The use of Eq. (42.13) becomes problematic for centers with a capture cross section less than 10^{-14} cm^2; thus, the attempt-to-escape frequency for these centers is left open in Table 42.1.

In centers with a capture cross section less than the geometric cross section of the defect center, other arguments need to be considered, such as resonance transitions or tunneling.

The population of a defect center in thermal equilibrium is determined by its energy alone, while the attainment of this equilibrium, i.e., the time it takes to follow changes in excitation, is determined by the center's kinetic parameters c_{ct} and e_{tc}—i.e., by two parameters: the energy of the level, and either its capture cross section or its attempt-to-escape frequency.

42.3 Optical Carrier Generation

In Section 13.1, the optical excitation of electrons across the band gap was discussed as the product of the joint density of the initial and final states and the probability for each of these transitions. This probability is given by the product of the matrix element and the probability of finding the initial state occupied and the final state empty before the transition. The optical excitation is the part of the optical absorption that generates free carriers—electrons and holes.

One distinguishes bound-to-free and free-to-free excitation, depending on whether the excitation started from a center with an occupied state, i.e., below the Fermi-level, or from the valence band. In both cases the optical absorption has an edge-like character—see Sections 22.4.7 and 45.1. When the excitation proceeds in energy beyond the exciton spectrum, free carriers are produced—Section 45.1.

From the flux of impinging photons ϕ_o traveling in the direction z normal to the semiconductor surface, a certain fraction, depending on the wavelength, will be absorbed. The photon flux within the semiconductor is given by

$$\phi_\lambda(\lambda, z) = \phi_o(\lambda) \exp\left[-\alpha_o(\lambda)z\right]. \tag{42.15}$$

Here, $\phi_o(\lambda)$ is the *photon flux per unit wavelength* $(\Delta\lambda)$, which penetrates, after reflection is subtracted (see Section 45.1), through the top layer of the solid, measured in $cm^{-2}s^{-1}\Delta\lambda^{-1}$; and α_o is the *optical absorption coefficient*, measured in cm^{-1}. When polychromatic light is used, the total carrier-generating photon flux as a function of the penetration depth (z) is obtained by integration

$$\phi(z) = \int_{\lambda_1}^{\lambda_2} \phi_\lambda(\lambda, z)d\lambda, \tag{42.16}$$

where $\phi(z)$ is measured in $cm^{-2}s^{-1}$. Although $\phi_\lambda(\lambda, z)$ depends exponentially on the penetration depth, $\phi(z)$ usually does not, since, with polychromatic light of various absorption coefficients $\alpha_o(\lambda)$, the superposition of a wide variety of such exponential functions causes a dependence of ϕ on z of much lesser steepness than an exponential function.

42.3.1 Optical Generation Rate

The optical generation rate $g(x)$ is given by the absorbed photoelectrically active light in each slab of infinitesimal thickness; hence,

$$g(z) = -\frac{d\phi(z)}{dz}. \tag{42.17}$$

This rate *depends exponentially on z for monochromatic light*:

$$g(z, \lambda_0) = \alpha_o(\lambda_0)\phi_o(\lambda_0)\Delta\lambda \exp\left(-\alpha_o(\lambda_0)z\right), \tag{42.18}$$

where $\Delta\lambda$ is a small wavelength range in which $\alpha_o(\lambda)$ is constant. For polychromatic excitation, the decline of g with increasing depth is much more gradual; often a constant space-independent generation rate can be used as a better approximation (Böer, 1977):

$$g = g_o \qquad (\text{cm}^{-3}\text{s}^{-1}). \tag{42.19}$$

Optical excitation was discussed in Chapter 27; and consequent carrier-generation, causing photoconductivity will be described in Chapter 45, where more detail about the various types of optical excitation and related references can be found.

42.4 Field Ionization

The three field-ionization mechanisms are

- Frenkel-Poole ionization,
- impact ionization, and
- tunnel ionization.

All of these mechanisms produce free carriers predominantly by inducing bound-to-free transitions. At sufficiently high fields, band-to-band transitions can also be initiated.

42.4.1 Frenkel-Poole Ionization

The Frenkel-Poole effect requires the lowest field* for the ionization of Coulomb-attractive centers (Frenkel, 1938; Poole, 1921). This ionization is achieved by tilting the bands and thus lowering the thermal ionization energy of such a center. The ionization probability is thereby increased—see Fig. 42.3.

---

* Except for high-mobility semiconductors for excitation from shallow centers at low temperatures where impact ionization favorably competes— see Sections 42.4.2A and 42.4.2B.

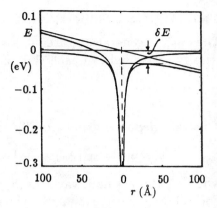

Figure 42.3: Lowering of the electron-binding energy by δE for a Coulomb-attractive center with an external electric field (Frenkel-Poole effect), assuming $\varepsilon = 10$ and $F = 50$ kV/cm.

The lowering of the potential barrier is obtained by superimposing an external field upon the Coulomb potential

$$V(x) = \frac{eZ}{4\pi\varepsilon_{st}\varepsilon_0 x} - Fx, \tag{42.20}$$

which shows a maximum where dV/dx vanishes:

$$x(V_{max}) = \sqrt{\frac{eZ}{4\pi\varepsilon_{st}\varepsilon_0 F}} = \frac{1.2 \cdot 10^{-4}}{\sqrt{F(V/cm)}} \cdot \sqrt{\frac{10}{\varepsilon_{st}}} Z \quad (cm); \tag{42.21}$$

at $x(V_{max})$ the barrier is lowered by $\delta E = eV_{max}$:

$$\delta E = e\sqrt{\frac{eFZ}{\pi\varepsilon_{st}\varepsilon_0}} = 2.4 \cdot 10^{-4}\sqrt{F(V/cm)}\sqrt{\frac{10}{\varepsilon_{st}}} Z \quad (eV). \tag{42.22}$$

The field-enhanced thermal ionization probability of such a center can now be approximated by

$$e_{tc} = \nu_{tc}^{(0)} \exp\left(-\frac{E_c - E_t - \delta E}{kT}\right), \tag{42.23}$$

where $\nu_{tc}^{(0)}$ is the frequency factor for thermal ionization—see Section 42.4.1A.

By setting $\delta E = kT$, a simple estimate yields a critical field for marked Frenkel-Poole ionization at

$$F_{FP} \simeq \frac{\pi\varepsilon_{st}\varepsilon_0}{eZ}\left(\frac{kT}{e}\right)^2 \simeq 1.08 \cdot 10^4 \left(\frac{\varepsilon_{st}}{10}\right)\left(\frac{T}{300}\right) \quad (V/cm). \tag{42.24}$$

In a more refined model (Dussel and Böer, 1970), one considers a three-dimensional Coulomb well that is deformed by the external

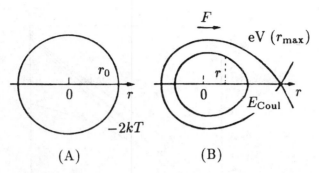

Figure 42.4: Critical equipotentials of a Coulomb-attractive center: (A) for $F = 0$, and (B) for $F > 0$.

field. The deformation causes a change in the capture cross section, the frequency factor, and the barrier height of the well. The net generation rate is expressed as the difference between the changed generation and recombination.

The change in the capture cross section can be obtained from a semiclassical approach. Electron capture occurs when its energy has decreased to $2kT$ below its barrier height. Without a field, this yields from

$$E_{\text{Coul}} = \frac{Ze^2}{4\pi\varepsilon_{\text{st}}\varepsilon_0 r_0} = 2kT \qquad (42.25)$$

as the critical radius r_0 of the excited state of the Coulomb-attractive center:

$$r_0 = \frac{Ze^2}{8\pi\varepsilon_{\text{st}}\varepsilon_0 kT} = 27.7\sqrt{\frac{10}{\varepsilon_{\text{st}}} \cdot \frac{300}{T}} Z \quad (\text{Å}). \qquad (42.26)$$

With an electric field, the critical radius for capture is reduced, following the implicit condition

$$eV\bigl(r(F)\bigr) = -(2kT + \delta E). \qquad (42.27)$$

With the field, the critical equipotential surface for electron capture is an ellipsoid, as indicated in Fig. 42.4. This ellipsoid shows a reduction in the capture cross section: the lowering of the barrier pushes the critical level down deeper into the Coulomb funnel—see also Fig. 42.2. The relative reduction of the capture rate with the field, normalized to $c_{\text{ct}}(r_0)$, is shown in Fig. 42.5 as curve 2. A reduction by a factor of ~ 10 is seen in the field range between 10^2 and 10^4 V/cm.

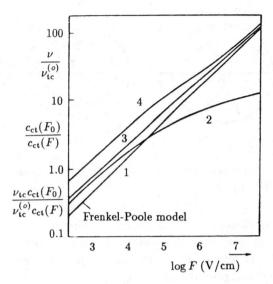

Figure 42.5: Curve 1: Frenkel-Poole ionization rate relative to the ionization rate at a field $F_0 = F(\delta E = 2kT)$ (Eq. (42.22)); curve 2: capture coefficient relative to this coefficient at r_0 (Eq. (42.26)); curve 3: ionization probability relative to this probability at r_0 and F_0; curve 4: the ratio of curve 3 to curve 2, which is the Frenkel-Poole ionization probability for the refined model computed for $\varepsilon = 10$, $Z = 1$, and $T = 200\,\mathrm{K}$ (after Dussel and Böer, 1970).

The dependence of the frequency factor on the field is estimated in a classical approximation from Kepler's law relating to $r(F)$ as

$$\nu_{tc} = \nu_{tc}^{(0)} \left(\frac{r(F)}{r_0} \right)^{3/2} \exp\left(\frac{\delta E}{kT} \right). \qquad (42.28)$$

Multiplying this factor with the Boltzmann term for the reduced barrier height yields curve 3 in Fig. 42.5. The decrease of the density of carriers trapped in this Coulomb-attractive center due to Frenkel-Poole ionization is obtained from reaction kinetic arguments [see Eq. (45.25)] and is given by

$$\frac{n_t}{N_t} = \frac{1}{1 + \dfrac{e_{tc} n_t}{c_{ct} n}} \simeq \frac{c_{ct}}{e_{tc} n_t} n; \qquad (42.29)$$

it depends on the ratio of ionization to the capture rate. The ratio increases by a factor of ~ 2.5 up to fields of 10^4 V/cm as compared with the classical Frenkel-Poole equation [Eq. (42.23)], shown as curve 4 in Fig. 42.5. The figure also shows that Frenkel-Poole ionization becomes marked (> 1) already near 1 kV/cm in a semiconductor with $\varepsilon = 10$ and at $T = 200$ K.

42.4.1A Inclusion of Local Fields The classical Frenkel-Poole model yields an ionization value which is often too large when compared to experimental observation. Furthermore, this model fails to provide a threshold field below which the field-enhanced ionization is not observed: the exponential dependence on $\sqrt{F}$ [Eqs. (42.22) and (42.23)] indicates a smooth onset of Frenkel-Poole excitation. However, a more pronounced threshold is obtained with competing optical excitation (see Dussel and Böer, 1970a).

At sufficient density, Coulomb-attractive centers will interact with each other due to their local field F_l, which needs to be added to the external field F (Dallacasa and Paracchini, 1986). The ionization probability consequently is given as

$$ e_{tc}g = e_{tc}^{(0)} \exp \left\{ \sqrt{\frac{e^3 Z}{\pi \varepsilon \varepsilon_0} \cdot \frac{\sqrt{(F + F_l)}}{kT}} \right\} ; \qquad (42.30) $$

the local field F_l is on the order of 10^4 V/cm for typical densities of charged defects of 10^{16} cm^{-3}. In Fig. 42.6, the relative ionization probability is given as a function of the normalized field for a number of different approximations.

42.4.2 Impact Ionization

An electron that gains enough energy between scattering events to exceed the ionization energy can set free an additional electron on impact. This ionization energy can be that of a defect center, e.g., a donor, or, at sufficient fields, the band gap energy. The two resulting electrons will in turn gain energy, each one setting free another electron. Sequentially, two, four, eight, etc., free electrons will result in an *avalanche formation*. In homogeneous semiconductors, it takes place over the length of the high-field region, i.e., throughout the entire distance between electrodes. With avalanche formation, the current increases dramatically and can lead to a *dielectric breakdown*.

In very thin semiconductors, however, there is not enough space from electrode to electrode to develop a breakdown avalanche. Consequently, breakdown fields at which the avalanche reaches critical values (see the following sections) are higher in such thin layers.

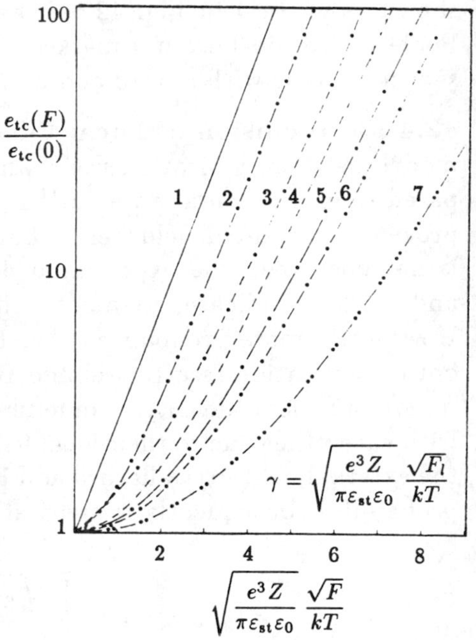

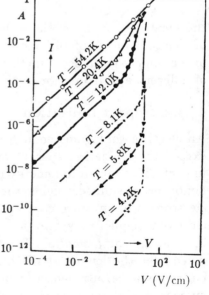

Figure 42.6: Relative ionization probability as a function of the normalized external field for: (1) the original Frenkel-Poole model (Poole, 1921, and Frenkel, 1938); (2) for the Hartke (1968) model; (3) the Hill (1971) model; (4) the Connell et al. (1972) model; (5) the Pai (1975) model; and (6, 7) the Dallacasa and Paracchini (1986) model, for $\gamma = 1$ and $\gamma = 10$, respectively (after Dallacasa and Paracchini, 1986).

In the upper figure, the vertical axis is labeled $\dfrac{e_{tc}(F)}{e_{tc}(0)}$ and the horizontal axis is labeled $\sqrt{\dfrac{e^3 Z}{\pi \varepsilon_{st} \varepsilon_0}} \dfrac{\sqrt{F}}{kT}$, with

$$\gamma = \sqrt{\frac{e^3 Z}{\pi \varepsilon_{st} \varepsilon_0}} \frac{\sqrt{F_l}}{kT}$$

Figure 42.7: Current-voltage characteristics of Ge with the temperature as the family parameter. Impact ionization is indicated near 10 V/cm, depleting electrons from shallow donors. Thermal depletion is completed at 54 K (after Lautz, 1961).

42.4.2A Impact Ionization of Shallow Donors

Shallow donors can easily be ionized by an impact with free carriers at relatively low fields, i.e., a few V/cm in the 4–10 K range in Ge, until all donors are depleted at ~ 50 V/cm. In Fig. 42.7, one sees two ohmic branches, with a steeply increasing branch in between, when impact ionization occurs. The magnitude of the shift between the low- and high-field branches is determined by the initial population of the shallow donors. This population is reduced at higher temperatures because of thermal ionization. For more detail, see Sclar and Burstein (1957) and Bratt (1977).

42.4.2B Impact Ionization Across the Band Gap Impact ionization can proceed by the impact of a conduction band electron with an electron from the valence band, thereby creating an additional electron and a hole. The critical energy for such ionization can be estimated from the conservation of momentum and energy. We will examine first a rather simple model that explains the main principles.

The momentum conservation between initial (k_{ni}) and final states $(2k_n + k_p)$, while conserving the group velocity after impact (Anderson and Crowell, 1972) $\mathbf{k}_n/m_n = \mathbf{k}_p/m_p$, yields

$$k_{ni} = k_n \left(2 + \frac{m_p}{m_n} \right), \tag{42.31}$$

and the energy conservation with $E_{ni} = 2E_n + E_p$ yields

$$E_{ni} = \frac{\hbar^2 k_n^2}{2m_n} \left(2 + \frac{m_p}{m_n} \right) + E_g. \tag{42.32}$$

Also, as long as a parabolic band approximation holds, one has

$$E_{ni} = \frac{\hbar^2 k_{ni}^2}{2m_n} = \frac{\hbar^2 k_n^2}{2m_n} \left(2 + \frac{m_p}{m_n} \right)^2; \tag{42.33}$$

for narrow band-gap material this may be a reasonable approximation. When combining Eqs. (42.31)–(42.33), one obtains for the threshold energy for impact ionization $E_{ii} = E_{ni}$:

$$E_{ii} = E_g \frac{2m_n + m_p}{m_n + m_p}, \tag{42.34}$$

which becomes the often-cited

$$E_{ii} \simeq \begin{cases} 1.5E_g & \text{for } m_n = m_p \\ E_g & \text{for } m_n \ll m_p \end{cases} \tag{42.35}$$

as the condition for the electron energy threshold. In typical semiconductors, the observed threshold energy is larger since the bands are nonparabolic and the effective mass increases with increasing energy; for example, for GaAs, one observes $E_{\text{ii}} \simeq 2.0 E_g$, and for Si, one has $E_{\text{ii}} \simeq 2.3 E_g$.

In indirect band-gap semiconductors, momentum conservation requires that intervalley phonons are involved in the ionization process: holes are located at the Γ point, while conduction electrons are accelerated from a side valley—see Anderson and Crowell (1972).

One must now evaluate how conduction electrons reach this threshold. This was originally done by Wolff (1954), who estimated the balance between energy gain from the field and losses due to scattering and ionization, in analogy to a gas discharge analyzed by Townsend. First however, we will briefly review the main concepts of Wolff's theory.

The characteristic parameter of the impact ionization is the ionization rate $\tilde{\alpha}_i$ per unit path length, measured in cm^{-1}. Electrons gain energy from the field and lose part of the incremental gain during scattering events. Since the scattering is a statistical process, there is a finite probability that an electron can accumulate enough energy for impact ionization over a sufficiently long path length (a "lucky" electron, as it is referred to later—Shockley, 1961).

The ionization rate increases rapidly with increasing fields. The incremental speed Δv of an electron gained from the electric field is given by

$$m_n \frac{dv_x}{dt} = m_n \frac{\Delta v_x}{\tau_m} = eF_x, \tag{42.36}$$

with τ_m as the *momentum* relaxation time. The average energy gain is therefore

$$\overline{\Delta E} = \frac{m_n}{2} \Delta v_x^2 = \frac{e^2 F_x^2 \tau_m^2}{2 m_n}. \tag{42.37}$$

The energy gain can be dissipated most efficiently by scattering with longitudinal optical phonons. Wolff estimates that the average gain during a free path λ is approximately equal to the optical phonon energy $\hbar\omega_{\text{LO}}$ near threshold fields:

$$\overline{\Delta E_{\text{LO}}} = \frac{e^2 F_x^2 \tau_e^2}{2 m_n} = \frac{e^2 F_x^2 \lambda^2}{2 m v_e^2} \simeq \hbar\omega_{\text{LO}}, \tag{42.38}$$

where v_e is the average electron velocity in the given electric field, i.e., a velocity *above* the rms velocity of electrons, not to be confused with the drift velocity. We have also used the energy-relaxation time here, rather than the momentum relaxation time at lower energies, in order to consider the strong interaction with LO phonons. With the kinetic energy of the accelerated electron $E = mv_e^2/2$, one obtains from Eq. (42.38)

$$\frac{\overline{\Delta E_{\text{LO}}}}{E} = \frac{e^2 F_x^2 \lambda^2}{4\hbar\omega_{\text{LO}}} \cdot \frac{1}{\frac{m}{2}v_e^2} \simeq 1 \qquad (42.39)$$

in the hot-electron field range. The field-modified electron distribution function can then be written as

$$f(E) \simeq A \exp\left(-\frac{E}{\Delta E_{\text{LO}}}\right) = A \exp\left(-\frac{BE}{F^2}\right) \qquad (42.40)$$

where A is a proportionality constant on the order of 1, and $B \simeq 4\hbar\omega_{\text{LO}}/(e^2\lambda^2)$. The calculation of the ionization rate results in a similar functional dependency on the field

$$\tilde{\alpha}_i = C \exp\left(-\frac{BE}{F^2}\right), \qquad (42.41)$$

Shockley (1961), however, argued that impact ionization is caused by a few "lucky" carriers that escape scattering altogether and are accelerated within one free path to the ionization energy E_{ii}. He obtains for the ionization rate

$$\tilde{\alpha}_i = C^* \exp\left(-\frac{B^* E}{F}\right). \qquad (42.42)$$

Baraff (1962) assumed a more general distribution function and obtained results that contain Shockley's $\tilde{\alpha}_i$ as the low-field limit and Wolff's $\tilde{\alpha}_i$ as the high-field limit.

All of these theories have a shortcoming in as much as they do not recognize the actual band structure, i.e., major deviation from the parabolic (effective mass) approximation of the conduction band. Except for very narrow-band semiconductors, the impact ionization energy lies high within the conduction band and requires a more accurate accounting of $E(k)$. See Curby and Ferry (1973) for a Monte Carlo analysis of impact ionization in InAs and InSb—see also Fig. 42.8.

Shichijo and Hess (1981) have used a Monte Carlo method to account for the competing field acceleration and polar optical or intervalley scattering within the Brillouin zone of GaAs. They observed characteristic anisotropies of scattering and ionization thresholds, and recognized that electrons can never gain sufficient energy

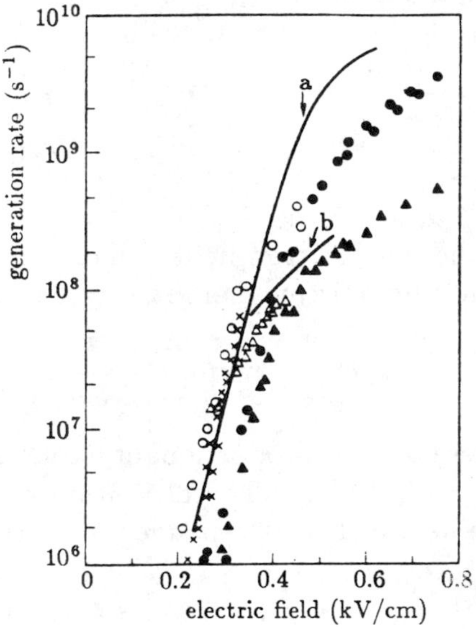

Figure 42.8: Computed electron generation rate due to impact ionization in InSb at 70 K for $n = 10^{14}$ and $2{\cdot}10^{16}$ cm^{-3} for curves (a) and (b), respectively. Measured values for various electron densities: $\circ\,6 \cdot 10^{13}$, $\times\,7.4 \cdot 10^{13}$, $\bullet\,2.7 \cdot 10^{15}$, $\triangle\,5 \cdot 10^{14}$, and $\triangle\,1.5 \cdot 10^{16}$ cm^{-3} (after Curby and Ferry, 1970).

within one ballistic path (Shockley's "lucky" electrons) while confined within the Brillouin zone. However, after an *elastic* scattering event, such electrons could start over again and attain sufficient energy for impact ionization—see Fig 42.9. At an energy sufficient for ionization, the electron in GaAs lies in the second conduction band.

The phonon-scattering rate is shown in Fig. 42.10 as a function of the electron energy. Above the impact ionization energy, the scattering rate increases dramatically, as given by Keldysh (1965)

$$e_{\text{ii}} = \frac{1}{\tau(E_{\text{ii}})} P \left(\frac{E - E_{\text{ii}}}{E_{\text{ii}}} \right)^2 \qquad (42.43)$$

where $e_{\text{ii}} = 1/\tau(E_{\text{ii}})$ represents the scattering (impact ionization) rate at the threshold energy, $P \simeq 50 \ldots 500$ is a dimensionless factor, and E is the electron energy.

Typical trajectories in the Brillouin zone near threshold fields are shown in Fig. 42.11, with corresponding variations of the electron energy computed by Shichijo and Hess during a sample of 1000 scattering events—Fig. 42.12A. This figure shows the typical behavior in which only a few events bring the electron to favorable starting positions from which they can be accelerated to reach the threshold energy. At lower fields, this spectrum is much more quiet and electrons stay well below the threshold energy—Fig 42.12B.

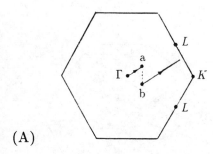

(A)

Figure 42.9: (A) Momentum vector trajectory in the T*KL*-plane of the Brillouin zone with an electric field in the ⟨111⟩ direction and one elastic scattering event (a → b). (B) Variation of the electron energy in time for the process shown in (A) (after Shichijo and Hess, 1981).

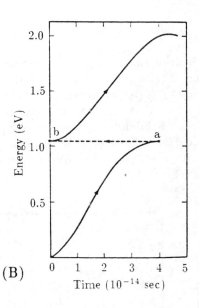

(B)

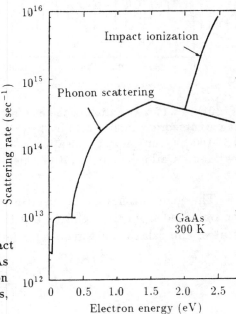

Figure 42.10: Phonon and impact ionization scattering rate in GaAs at 300 K as function of the electron energy (after Shichijo and Hess, 1981).

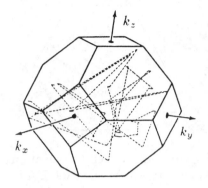

Figure 42.11: Typical electron trajectories within the Brillouin zone of GaAs at an electric field of 500 kV/cm in the ⟨100⟩ direction. Solid lines represent drift, and dashed lines represent the scattering event (after Shichijo and Hess, 1981).

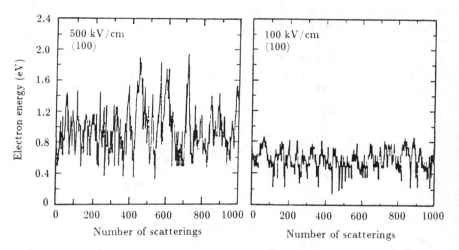

Figure 42.12: Variation of the electron energy in GaAs between scattering events for an applied field in the ⟨111⟩ direction of (A) 500 kV/cm, and (B) 100 kV/cm (impact ionization energy threshold is at 2 eV); obtained by Monte Carlo simulation (after Shichijo and Hess, 1981).

The generation rate of electrons due to impact ionization can be obtained from the deformed distribution function f_F, density of states, and relaxation time according to

$$g_i = \int \frac{f_F(E)g(E)}{\tau_e(E)} dE, \qquad (42.44)$$

and is shown for InSb in Fig. 42.8. The estimate includes the electron-electron interaction, which tends to shorten the relaxation time at higher electron energies (Curby and Ferry, 1970).

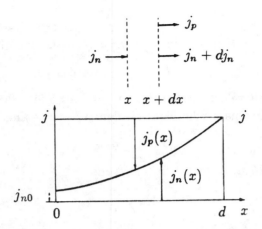

Figure 42.13: Current distribution as a function of the distance from the cathode (at left) with impact ionization. j_{n0} is the original current at the cathode; $\Delta j_n = j - j_{n0}$ is the current multiplied by impact ionization. Holes are minority carriers; thus $j_p(x = d) = 0$. Pair production by impact ionization makes the incremental currents for electrons and holes equal to each other: $\Delta j_n = \Delta j_p = j_p(x = 0) - j_p(x = d)$.

The ionization rate and the generation rate are related by

$$\tilde{\alpha}_i = \frac{g_i}{v_D}, \qquad (42.45)$$

with v_D as the drift velocity. For reviews, see Stillman and Wolfe (1977), Ridley (1983), and Dmitriev et al. (1987).

42.4.2C Avalanche Current and Multiplication Factor
With impact ionization, the electron current increases with increasing distance *from the cathode*. The increment of electrons in a distance dx is given by

$$dn = \tilde{\alpha}_n n dx, \qquad (42.46)$$

where $\tilde{\alpha}_n$ is the ionization rate of electrons. For band-to-band ionization, this increment is equal to the increment of holes ($dn = dp$) because of the mutual creation of electrons and holes during each ionizing impact.

In a similar fashion, the hole current increases with increasing distance *from the anode*, with an increment given by

$$dp = \tilde{\alpha}_p p \, dx. \qquad (42.47)$$

In an arbitrary slab of a semiconductor with planar electrodes at a distance x from the cathode (shown at the left in Fig. 42.13), the change in electron density is given by the change due to electrons coming from the left, $\tilde{\alpha}_n(n_0 + n_1)dx$, plus the change due to the holes coming from the right, $\tilde{\alpha}_p p_2\,dx = \tilde{\alpha}_n n_2\,dx$, yielding

$$dn_1 = \tilde{\alpha}_n(n_0 + n_1)dx + \tilde{\alpha}_p n_2\,dx, \qquad (42.48)$$

where n_0 is the electron density at the cathode. Shifting x to the anode, one obtains as the electron density at the anode

$$n_a = n_0 + n_1 + n_2, \qquad (42.49)$$

and a *multiplication factor* for electrons

$$M_n = \frac{n_a}{n_0}.$$

A similar multiplication factor for holes can be defined as $M_p = p_c/p_0$, where p_c is the hole density at the cathode.

After eliminating n_2 from Eqs. (42.48) and (42.49), one obtains

$$\frac{dn_1}{dx} = (\tilde{\alpha}_n - \tilde{\alpha}_p)(n_0 + n_1) + \tilde{\alpha}_p n_a, \qquad (42.50)$$

which is a linear differential equation. For the boundary conditions $n(x = 0) = n_0$ and $n(x = d) = n_0 + n_1 = n_a$, one obtains

$$M_n = \frac{1}{1 - \displaystyle\int_0^d \tilde{\alpha}_n[\exp\{-\int_0^x (\tilde{\alpha}_n - \tilde{\alpha}_p)dx'\}]dx}. \qquad (42.51)$$

The electron current increases from cathode to anode by

$$dj_n = j_n\tilde{\alpha}_n\,dx + j_p\tilde{\alpha}_p\,dx, \qquad (42.52)$$

and the hole current increases similarly from anode to cathode, so that the sum of both remains independent of x, as shown in Fig. 42.13. The total current increases with increasing M. When M becomes infinite, **dielectric breakdown** occurs. This occurs when the integral in Eq. (42.51) is equal to 1. In materials for which $\tilde{\alpha}_n \simeq \tilde{\alpha}_p$, e.g., in GaP, this breakdown condition

$$\int_0^d \tilde{\alpha}_n\left[\exp-\left\{\int_0^x (\tilde{\alpha}_n - \tilde{\alpha}_p)dx'\right\}\right]dx = 1 \qquad (42.53)$$

reduces to

$$\int_0^d \tilde{\alpha}\,dx = 1. \qquad (42.54)$$

This means that every electron or hole passing through the semiconductor creates another electron-hole pair, with a probability equal to 1. As this process goes on with more and more new carriers generated, breakdown occurs.

On the other hand, as long as the integral in Eq. (42.54) remains below 1, a stationary, increased carrier density is obtained and no breakdown occurs. For a recent review, see Dmitriev et al. (1987).

42.4.3 Electron Tunneling

Electrons can penetrate potential barriers if these barriers are thin and low enough, as pointed out by Oppenheimer (1928), Fowler and Nordheim (1928), and others. This quantum-mechanical phenomenon can easily be understood by recognizing that the wavefunction of an electron cannot immediately stop at a barrier, but rather decreases exponentially into the barrier with a slope determined by the barrier height. If the barrier is thin, there is a nonzero amplitude of the wavefunction remaining at the end of the barrier, i.e., a nonzero probability for the electron to penetrate, as indicated in Fig. 42.14. For a comprehensive review, see Duke (1969) or Wolf (1975). This phenomenon, called *electron tunneling*, will first be discussed for a simple one-dimensional rectangular barrier.

42.4.3A Tunneling Through Rectangular Barrier When an electron wave with an energy E impinges on a rectangular barrier with height eV_0 and width a, one distinguishes three wavefunctions in the three regions: before, in, and after the barrier—see Fig. 42.14:

$$\psi_1 = A_1 \exp\left(ik_0 x\right) + B_1 \exp\left(-ik_0 x\right) \qquad \text{for} \quad x < -\frac{a}{2} \quad (42.55)$$

$$\psi_2 = A_2 \exp\left(k_1 x\right) + B_2 \exp\left(-k_1 x\right) \quad \text{for} \quad -\frac{a}{2} < x < \frac{a}{2} \quad (42.56)$$

$$\psi_3 = A_3 \exp\left(ik_0 x\right) \qquad \text{for} \quad x > \frac{a}{2} \quad (42.57)$$

with amplitude coefficients A_i, for the incoming wave, and B_i, for the reflected wave at $-\frac{a}{2}$ and $+\frac{a}{2}$, respectively. The wave numbers k_0 and k_1 outside and inside the barrier are given by

$$k_0^2 = \frac{2m_n}{\hbar^2} E \quad \text{and} \quad k_1^2 = \frac{2m_n}{\hbar^2}\left(eV_0 - E\right). \quad (42.58)$$

Continuity of ψ and $d\psi/dx$ at $x = -\frac{a}{2}$ and at $x = +\frac{a}{2}$ provides four conditions for the coefficients A_i and B_i, from which one derives

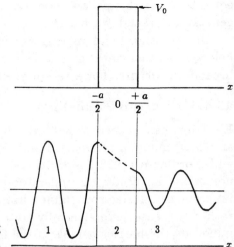

Figure 42.14: One-dimensional rectangular barrier with incoming and attenuated transmitted waves.

an expression for the transmission and reflection probabilities T_e and R:

$$T_e = \left(\frac{A_3}{A_1}\right)^2 \quad \text{and} \quad R = B_1^2 = 1 - \left(\frac{A_3}{A_1}\right)^2. \tag{42.59}$$

After solving the set of condition equations for A_i and B_i, one obtains

$$T_e = \frac{1}{1 + \left(\frac{k_0^2 + k_1^2}{4k_0 k_1}\right)^2 \sinh^2\left(\frac{k_1 a}{2}\right)} \simeq \left(\frac{4k_0 k_1}{k_0^2 + k_1^2}\right)^2 \exp\left(-k_1 a\right);$$

$$\tag{42.60}$$

this approximation is for $k_1 a \gg 1$. After introducing the expressions for k_0 and k_1, one obtains for the transmission probability:

$$T_e \simeq 16\left(\frac{E}{eV_0}\right)^2 \left(\frac{eV_0}{E} - 1\right) \exp\left(-a\sqrt{\frac{2m_n}{\hbar^2}(eV_0 - E)}\right). \tag{42.61}$$

For a barrier height eV_0 much larger than the kinetic energy of the tunneling electrons, Eq. (42.61) can be approximated as

$$T_e \simeq 16\frac{E}{eV_0} \exp\left(-0.512a\,(\text{Å})\sqrt{\frac{m_n}{m_0}(eV_0 - E)}\right), \tag{42.62}$$

For example, thermal electrons $[E = kT(300\ \text{K})]$ are attenuated by a factor of $2.2 \cdot 10^{-3}$ when impinging on a 10 Å thick barrier of 1 Volt height. The transmission probability also depends on the shape of the barrier.

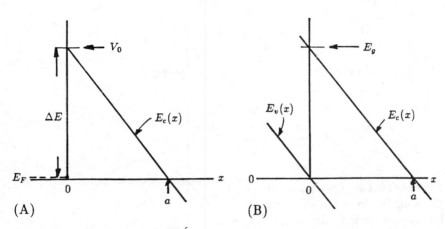

Figure 42.15: Triangular potential barrier at (A) a metal-semiconductor contact neglecting image forces, and (B) for a band-to-band transition.

42.4.3B Tunneling Through Triangular or Parabolic Barrier

In semiconducting devices, the potential barrier often can be approximated by a *triangular shape*, for instance, at a simplified metal-semiconductor (Fig. 42.15A) or heterojunction interface, or for a band-to-band transition (Fig. 42.15B). At sufficiently high fields, tunneling through such a barrier can become important.

The transmission probability T_e for a wide class of barriers is given in the WKB approximation by

$$T_e \simeq \exp\left\{-\int_0^a |k(x)|dx\right\}, \qquad (42.63)$$

neglecting the pre-exponential factor which is on the order of 1, and with the shape of the barrier contained in the wave vector:

$$k(x) = \sqrt{\frac{2m}{\hbar^2}\left(eV_0 - E(x)\right)}. \qquad (42.64)$$

For the triangular barrier, $k(x)$ is given by

$$k(x) = \sqrt{\frac{2m}{\hbar^2}\left(\Delta E - eFx\right)}. \qquad (42.65)$$

Integration of Eq. (42.63) with Eq. (42.65) yields

$$\int k(x)dx = \sqrt{\frac{2m}{\hbar^2}} \cdot \frac{2}{3}\frac{(\Delta E - eFx)^{3/2}}{eF/2}\bigg|_0^a, \qquad (42.66)$$

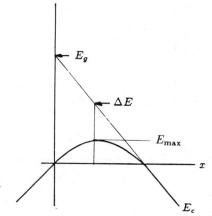

Figure 42.16: Parabolic barrier showing the lowering of a triangular barrier from E_g to E_{max}.

and with $\Delta E - eF \cdot 0 = \Delta E$, and $\Delta E - eF \cdot a = 0$, one obtains

$$\mathcal{T}_{e\Delta} = \exp\left\{ -\frac{4}{3}\sqrt{\frac{2m}{\hbar^2}}\frac{(\Delta E)^{3/2}}{eF} \right\} \qquad (42.67)$$

or $\qquad \mathcal{T}_{e\Delta} \simeq \exp\left\{ -6.828 \cdot 10^7 \frac{[\Delta E(\text{V})]^{3/2}}{F(\text{V/cm})} \right\}.$ $\qquad (42.68)$

The pre-exponential factor is similar in form to that given in Eq. (42.61); compared with the exponential, its ΔE- and F-dependencies are usually neglected. The barrier height/field relation is superlinear; thus, doubling the barrier height requires $2^{3/2} = 2.83$ times the field to result in the same tunneling probability.

For **band-to-band tunneling**, ΔE in Eq. (42.67) is replaced by the band gap E_g:

$$\mathcal{T}_{eg} \simeq \exp\left\{ -\frac{4}{3}\sqrt{\frac{2m}{\hbar^2}}\frac{E_g^{3/2}}{eF} \right\}. \qquad (42.69)$$

Although this transmission probability is multiplied by a large density of electrons acting as candidates for tunneling from band to band, one needs very high fields ($> 10^6$ V/cm) to produce significant tunneling currents, except for very narrow gap semiconductors.

A **parabolic barrier** is better suited as an approximation for barriers in which two fields overlap, e.g., the Coulomb-attractive field of a center or the image force of a metal-semiconductor barrier and the external field. This barrier type is shown in Fig. 42.16. The corresponding wave vector is given by

$$k(x) = \sqrt{\frac{2m}{\hbar^2}} \sqrt{\frac{(\Delta E)^2 - (eFx)^2}{\Delta E}}, \tag{42.70}$$

and yields after integration

$$\int_0^a k(x)\,dx = \sqrt{\frac{2m}{\hbar^2}} \frac{\pi}{8} \frac{(\Delta E)^{3/2}}{eF/2}. \tag{42.71}$$

Therefore, the transmission probability for a parabolic barrier is given by

$$\boxed{\mathcal{T}_{e\cap} = \exp\left\{ -\frac{\pi}{4} \sqrt{\frac{2m}{\hbar^2}} \frac{(\Delta E)^{3/2}}{eF} \right\}} \tag{42.72}$$

with an exponent that is reduced by a numerical factor of $\frac{3\pi}{16} = 0.589$ from the expression for the triangular barrier.

42.4.3C Tunneling in a Three-Dimensional Lattice
In the one-dimensional model, one is concerned only with the momentum in the direction of the barrier, which decreases exponentially during the tunneling transition.

In a three-dimensional lattice, however, there is a three-dimensional distribution of momenta. Assuming a planar barrier, only the component in the direction of tunneling is influenced by the tunneling process; the two components perpendicular to the tunneling are not: *these components are conserved.* The total transition probability is therefore reduced by a factor given by the fraction of electrons having a favorable momentum component to the ones that do not. This fraction η was calculated by Moll (1964):

$$\eta = \exp\left(-\frac{E_\perp}{\overline{E}} \right) \tag{42.73}$$

with $E_\perp$, the energy associated with the momentum perpendicular to the direction of tunneling, equal to $\hbar^2 k_\perp^2/(2m)$, and with $\overline{E}$ given by

$$\overline{E} = \sqrt{\frac{\hbar^2}{2m_n}} \frac{eF}{\pi\sqrt{\Delta E}} = 6.19 \sqrt{\frac{m_0}{m_n}} \frac{F\,(\text{V/cm})}{10^6} \frac{1}{\sqrt{\Delta E}} \quad (\text{meV}). \tag{42.74}$$

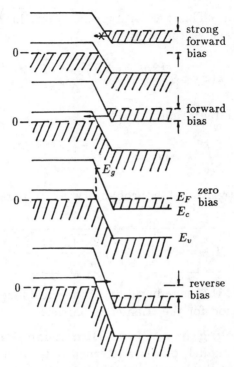

Figure 42.17: Band model with idealized tunnel junction for four different bias conditions. Current flow is indicated by a horizontal arrow.

Hence, for the tunneling probability through a flat plate barrier of parabolic shape, one obtains

$$\mathcal{T}_{e\cap,3} = \eta \mathcal{T}_{e\cap} = \exp\left(-\frac{E_g + 4E_\perp}{4\overline{E}}\right). \qquad (42.75)$$

42.4.3D Tunneling Currents   The current is proportional to the product of the tunneling transmission probability and the incident carrier flux. As an example, an idealized n^+p^+-junction is shown in Fig. 42.17 for various bias conditions.

Depending on the bias, the current can flow across this junction from filled states in the valence to empty states in the conduction band with reverse bias, or from filled states in the conduction band to the empty states in the valence band with forward bias. The net current is the difference of both (Moll, 1964): $j_n = \overrightarrow{j}_n - \overleftarrow{j}_n$. For any given bias, one has

$$\overrightarrow{j}_n = j_{\mathrm{n,vc}} = A \int_{E_{cr}}^{E_{vl}} \Big\{ N_v(E) f_n(E) \Big\}_l \Big\{ N_c(E) f_p(E) \Big\}_r \mathcal{T}_{e\Delta,3} dE \tag{42.76}$$

$$\overleftarrow{j}_n = j_{\mathrm{n,cv}} = A \int_{E_{cr}}^{E_{vl}} \Big\{ N_c(E) f_n(E) \Big\}_r \Big\{ N_v(E) f_p(E) \Big\}_l \mathcal{T}_{e\Delta,3} dE, \tag{42.77}$$

with $\mathcal{T}_{e\Delta,3} = \eta \mathcal{T}_{e\Delta}$ from Eqs. (42.67) and (42.73). The subscripts l and r stand for the left and right sides in Fig. 42.17; $N_v(E)$ and $N_c(E)$ are the respective density-of-state distributions in the valence and conduction bands; and f_n and $f_p = 1 - f_n$ are the Fermi distributions for electrons and holes. The first factor in parentheses under the integral identifies the density of available electrons; the second factor gives the density of holes into which tunneling can proceed. One therefore obtains

$$j_n = A \int_{E_{cr}}^{E_{vl}} N_v(E) N_c(E) \Big\{ f_{nr}(E) - f_{nl}(E) \Big\} \mathcal{T}_{e\Delta,3} dE. \tag{42.78}$$

The proportionality constant A can be obtained by accounting for the charge and the velocity of electrons in k space; with these, the current density is

$$j_n = \frac{e m_n}{2\pi^2 \hbar^3} \exp\!\left(\frac{\pi m_n^{1/2} E_g^{3/2}}{2\sqrt{2} e \hbar F} \right) \iint \Big\{ f_{nr}(E) - f_{nl}(E) \Big\} \exp\!\left(-\frac{2E_\perp}{\overline{E}} \right) dE\, dE_\perp. \tag{42.79}$$

assuming the same isotropic effective mass at the left and right sides of the junction. After integrating over $E_\perp$, which yields $\overline{E}/2$, and approximating the integral over E, one obtains

$$\boxed{j_n = \frac{e^3 m_n^{1/2} d}{2\sqrt{2}\pi^3 \hbar^2 E_g^{1/2}} F^2 \exp\left(-\frac{\pi m_n^{1/2} E_g^{3/2}}{2\sqrt{2} e \hbar F} \right),} \tag{42.80}$$

for $V_a \gg kT/e$ and $V_a \gg \overline{E}/e$; V_a is the applied voltage across the barrier. Equation (42.80) has the field dependence given by the Fowler-Nordheim formula (Fowler and Nordheim, 1928): $j \propto F^2 \exp(-F_0/F)$. Thus, Eq. (42.80) has the numerical value of

$$j_n = 1.011 \cdot 10^{-8} \left(\frac{F}{10^6}\right)^2 \left(\frac{m_n}{m_0}\right)^{1/2} \left(\frac{d}{100\text{Å}}\right) \frac{1}{\sqrt{E_g}} \exp\left(-\frac{F_0}{F}\right) \quad (\text{A}/\text{cm}^2)$$

(42.81)

with

$$F_0 = 4.04 \cdot 10^7 \left(\frac{m_n}{m_0}\right)^{1/2} E_g^{3/2} \quad (\text{V}/\text{cm}),$$

(42.82)

which indicates substantial tunneling ($> 10^{-3}$ A/cm^2) is expected for fields in excess of $1.5 \cdot 10^6$ V/cm for the given parameters. A reduced effective mass lowers the critical field by a factor of 3 for $m_n = 0.1\, m_0$.

42.4.3E Tunneling Spectroscopy The dependence of the tunneling current on the level density distribution [Eq. (42.78)] permits the use of this current for obtaining information about the distribution. There are several methods available; they involve a similar principle in shifting the Fermi-level or quasi-Fermi level on one side of a barrier with respect to the Fermi level on the other side (see Fig. 42.17), thereby permitting the tunneling of carriers through the barrier at variable energies. Thus, the level distribution is profiled near the top of the valence band or in the conduction band, depending on forward or reverse bias, respectively: the increment in tunneling current becomes larger when more levels become available.

The probing side of the barrier can be either a highly doped semiconductor, a metal electrode, or a superconductor.

A higher sensitivity of the probing can be achieved by the derivative technique (Thomas and Rowell, 1965; Adler and Jackson, 1966; Adler et al., 1971). In this method, a small ac-signal is applied on top of a varying dc-bias, $V(t) = V_0 + \Delta V \cos \omega t$, resulting in a current

$$I = I_0 + \frac{dI}{dV} \Delta V \cos(\omega t) + \frac{d^2 I}{dV^2} \left(\frac{\Delta V}{2}\right)^2 [1 + \cos(2\omega t)] + \dots, \quad (42.83)$$

which permits detection of first and second derivatives at ω and 2ω, respectively.

An example of probing the conduction band of Ge with a gold contact is shown in Fig. 42.18. Here, dI/dV, the incremental conduction, is plotted. It shows an increase in conduction as more states in the L_1-conduction band become available with increased reverse bias; it shows another edge at 154 meV above the conduction band edge when additional direct transitions into the Γ_2'-band become possible.

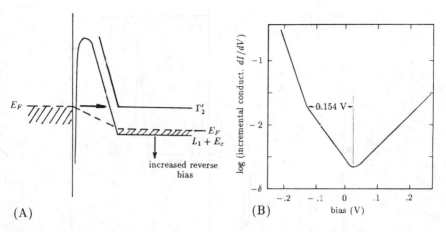

(A)   (B)

Figure 42.18: Incremental tunneling conduction of an Au/n-type Ge barrier at 4.2 K: (A) band model, and (B) experimental curve (after Conley and Tiemann, 1967).

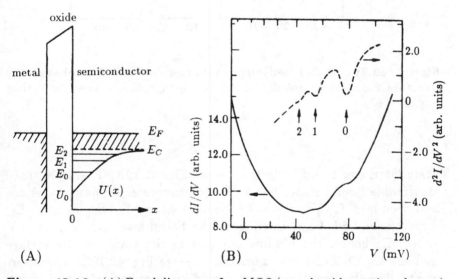

(A)   (B)

Figure 42.19: (A) Band diagram of an MOS (metal-oxide-semiconductor) junction with surface states in a potential well of depth U_0. (B) First and second derivatives of the tunneling current, showing transitions into the eigenstates of the surface well [indices correspond to those of the energy shown in subfigure (A)] for a Pb-oxide-n^+-PbTe junction (after Tsui et al., 1974).

Another example is the incremental current obtained by tunneling from a metal contact into the eigenstates of a quantum well, which is created at the surface of an n^+-PbTe semiconductor connected by an

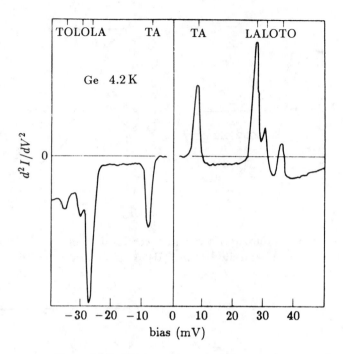

Figure 42.20: Second derivative of tunnel current to emphasize the structure for a Ge tunnel diode, indicating the phonon spectrum (after Payne, 1965).

oxide interlayer to a Pb-electrode—see Fig. 42.19A. There are three identifiable bound states below the continuum E_c. These states are best seen in d^2I/dV^2 at $V = 78$, 55, and 42 mV for E_0, E_1, and E_2, respectively, and are counted from the Fermi-level, which is 36 mV above E_c; hence, the binding energies of the states in the surface well are 42, 19, and 6 mV, respectively—see Fig. 42.19B. The insert of Fig. 42.19B shows this more clearly in an enhanced scale. Other examples of tunnel spectroscopy are listed in the following sections (see also Tsui, 1980; Hayes et al., 1986).

42.4.3F Tunneling with Phonon Assistance   Indirect bandgap materials show a much lower band-to-band tunneling probability, since an indirect transition requires an additional phonon to accomplish the change in momentum. An estimate of the tunneling probability by Keldysh (1958) shows a similar functional behavior as given by Eq. (42.69), but reduced by a factor on the order of 10^{-3}.

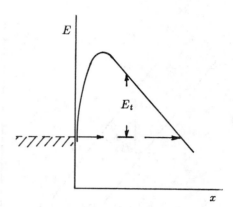

Figure 42.21: Tunneling from a contact into a semiconductor, which becomes marked when the trap level is lowered by the field to coincide with the Fermi-level of the metal.

Other changes in the tunneling probability are seen when the phonon energy is used in addition to the electron energy to bridge the barrier:

$$T_e = \exp\left\{ -\frac{4}{3}\sqrt{\frac{2m}{\hbar^2}}\frac{(E_g - \hbar\omega_0)^{3/2}}{eF} \right\}. \tag{42.84}$$

When the bias V reaches the phonon energy,

$$V \geq \frac{\hbar\omega_0}{e}, \tag{42.85}$$

the tunneling current increases measurably. This change can be used to identify the corresponding phonon energies as shown in Fig. 42.20 for Ge.

42.4.3G Tunneling with Trap Assistance Tunneling through a barrier into the conduction band can be assisted by tunneling first into a trap and then from the trap into the conduction band—a *two-step tunneling* process, as indicated in Fig. 42.21. When defect centers are spaced close enough to an interface, and are present at sufficient densities, the two-step tunneling can substantially increase the overall tunneling probability, which is calculated from the sum of the reciprocal individual probabilities:

$$T_{e,2\mathrm{st}} = \left(\frac{1}{T_{e1}} + \frac{1}{T_{e2}} \right)^{-1}. \tag{42.86}$$

42.4.3H Tunneling with Photon Assistance (Franz-Keldysh Effect) The Franz-Keldysh effect is an important tunneling phenomenon in which only a small fraction of the energy is supplied from the electric field. Most of the energy comes from an optical excitation from a state in the gap near the valence band edge to a

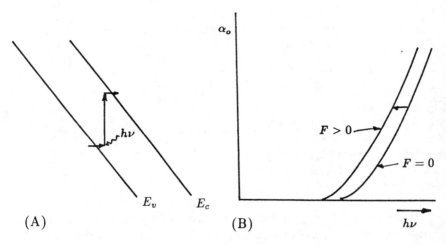

Figure 42.22: Franz-Keldysh effect: (A): photon-assisted tunneling from band to band; (B): resulting shift of the band edge.

symmetrical state close to the conduction band edge—Fig. 42.22A. This three-step process was suggested by Franz (1958) and Keldysh (1958), and was observed first by Böer et al. (1959); see also Böer and Kümmel (1960). It results in a shift of the absorption edge towards lower energies. The amount of the shift can be estimated from the photon-assisted tunneling—compare with Eq. (42.84):

$$\mathcal{T}_e = \exp\left\{ -\frac{4}{3}\sqrt{\frac{2m^*}{\hbar^2}}\frac{(E_g - h\nu)^{3/2}}{eF} \right\}. \qquad (42.87)$$

As a result, the optical absorption edge is shifted (Fig. 42.22B) by the same amount:

$$\alpha_o = \alpha_{o0}\exp\left\{ -\frac{4}{3}\sqrt{\frac{2m^*}{\hbar^2}}\frac{(E_g - h\nu)^{3/2}}{eF} \right\}. \qquad (42.88)$$

The shift $\Delta E_{g,\text{opt}}$ can be obtained from the condition that the exponent remains constant and $\simeq 1$, resulting in

$$\Delta E_{g,\text{opt}} = E_g - h\nu = \left(\frac{3}{4}eF\sqrt{\frac{\hbar^2}{2m^*}} \right)^{2/3} \qquad (42.89)$$

or

$$\Delta E_{g,\text{opt}} = 7.25 \cdot 10^{-6}[F(\text{V/cm})]^{2/3} \quad (\text{eV}); \qquad (42.90)$$

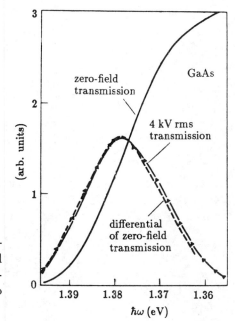

Figure 42.23: Optical transmission at the absorption edge and ac-component of the optical transmission with an ac-field applied to GaAs (after Moss, 1961).

that is, for a band edge shift of ~ 10 meV, equivalent to ~ 100 Å at a band gap of ~ 2 eV, one needs a field of ~ 50 kV/cm.

The photon-assisted tunneling is not restricted to the transition from the valence to conduction band near its principal edge; one can optically excite an electron to a state close to any higher band and complete the transition via tunneling. The Franz-Keldysh effect thereby provides a relatively simple method for measuring the energy of characteristic points in the $E(k)$ behavior of any band (Seraphin, 1964; Aspnes, 1967). At first view, this may suggest no advantage over a purely optical transition. When applying an ac-electric field, the maximum modulation signal of the optical response is observed where the absorption edge has the highest slope, i.e., at the inflection point of $\kappa(\lambda)$, as shown in Fig. 42.23. With overlapping higher bands, the optical absorption itself is not very structured. In contrast, however, the Franz-Keldysh modulation is highly structured and shows unusually sharp features with characteristic oscillations toward higher frequencies (Fig. 42.24), which have attracted numerous investigations. For reviews, see Frova and Handler (1965), Seraphin and Bottka (1965), and Aspnes (1980).

A quantum mechanical treatment reveals the characteristic features of the dielectric function. For a transition to an isotropic M_0

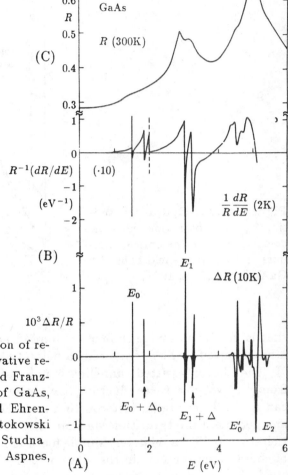

Figure 42.24: Comparison of reflectance (A), energy derivative reflectance (B), and low-field Franz-Keldysh reflectance (C) of GaAs, measured by Philipp and Ehrenreich (1963), Sell and Stokowski (1970), and Aspnes and Studna (1973), respectively (after Aspnes, 1980).

critical point, the imaginary part of the dielectric constant is given by

$$\varepsilon''(h\nu, F) = \frac{2\pi^2 e^2}{m^{*2}\nu^2}|\hat{\mathbf{e}} \cdot \mathbf{M}_{\mathrm{cv}}|^2 J_{\mathrm{cv}}(h\nu, F), \qquad (42.91)$$

where $\hat{\mathbf{e}}$ is the unit polarization vector of the light, $\mathbf{M}_{\mathrm{cv}}$ is the momentum matrix element, and J_{cv} is the joint density of states for transitions from the valence to the conduction band under the influence of a homogeneous electric field. Tharmalingan (1963) and

Aspnes (1966) obtain for the change in the complex dielectric function

$$\Delta\tilde{\varepsilon}(h\nu, F) = \frac{2e^2\hbar^2|\hat{\mathbf{e}} \cdot \mathbf{M}_{\text{cv}}|^2}{m^{*2}E^2}\left(\frac{2m_{\|}}{\hbar^2}\right)^{3/2}\sqrt{h\nu_F}\left[\tilde{G}(\eta) + i\tilde{F}(\eta)\right]$$

(42.92)

with the characteristic energy for the Franz-Keldysh effect (Cardona, 1969)

$$h\nu_F = \left(\frac{e^2F^2\hbar^2}{2m_{\|}}\right)^{1/3}$$

(42.93)

and with

$$\tilde{F}(\eta) = \pi\left[\text{Ai}'^2(\eta) - \eta\text{Ai}^2(\eta)\right] - \sqrt{-\eta}\,u(-\eta).$$

(42.94)

$$\tilde{G}(\eta) = \pi\left[\text{Ai}'(\eta)\text{Bi}'(\eta) - \eta\text{Ai}(\eta)\text{Bi}(\eta)\right] + \sqrt{\eta}\,u(\eta);$$

(42.95)

Here, Ai, Bi, and the primed functions are the Airy functions (Antosiewicz, 1972), $\eta = (E_g - h\nu)/(h\nu_F)$; u is the unit step function; and $m_{\|}$ is the projection of the effective mass in the field direction. $\tilde{F}$ and $\tilde{G}$ are proportional to the real and imaginary parts of the dielectric constant, i.e., proportional to reflection and absorption, respectively (see Bassani et al., 1975), and in Fig. 42.25A are plotted against a reverse η abscissa to indicate electric-field modulation. This figure also shows a sharp spike at the transition energy followed by typical *Franz-Keldysh oscillations* toward higher energies, which increase in amplitude and in period length (line width) with the electric field as

$$\text{Period} = \left(\frac{3\pi}{2}\right)^{2/3}h\nu_F = \left(\frac{3\pi}{2}\frac{e\hbar}{\sqrt{2m_{\|}}}F\right)^{2/3}.$$

(42.96)

Near the absorption edge (for $h\nu \ll E_g$), one obtains from Eq. (42.92)

$$\varepsilon''(\nu, F) = \frac{\sqrt{2}e^2}{2\pi\sqrt{m_{\|}}(h\nu^2)^2}|\hat{\mathbf{e}} \cdot \mathbf{M}_{\text{cv}}|^2\frac{\sqrt{h\nu_F}}{2(E_g - h\nu)}\exp\left[-\frac{4}{3}\left(\frac{E_g - h\nu}{h\nu_F}\right)\right],$$

(42.97)

which indicates the typical exponential extension of the optical absorption edge obtained by Franz (1958) and Keldysh (1958); this is shown in Fig. 42.26. The field dependence can in turn be used to probe for an electric field inside a semiconductor (Böer, 1960).

The electro-reflectance $\Delta R/R$, which is proportional to $\tilde{G}$ and is shown in Fig. 42.25B for Ge, has well-separated extrema and permits a comparison to the single-band transition computed in

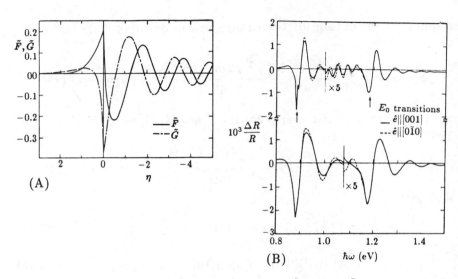

Figure 42.25: (A) Franz-Keldysh line shapes $\tilde{F}(\eta)$ and $\tilde{G}(\eta)$ [Eqs. (42.94) and (42.95)] plotted against $-\eta(\propto E_g - h\nu)$ for transition to an M_0-type critical point (after Aspnes, 1967). (B) Electro-reflectance of Ge at 10 K for $F = 46$ and 115 kV/cm for the upper and lower parts of the figure, respectively (after Aspnes, 1975).

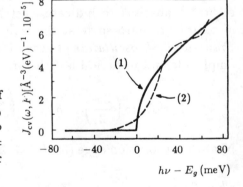

Figure 42.26: Joint density of states for a transition near an M_0 transition edge. Curve 1 for zero field; curve 2 calculated for $F = 5 \cdot 10^4$ V/cm and $m_\parallel = 0.1 m_0$ (after Aimerich and Bassani, 1968).

Fig. 42.25A (Aspnes, 1975). The increased amplitude and period length with increasing field is seen by comparison of the two parts of Fig. 42.25B. Some interferences above the characteristic points, identified by arrows, are due to the contributions from light- and heavy-hole bands, and are used to determine the complete set of interband effective masses (Hamakawa et al., 1968).

Further analysis of the Franz-Keldysh effect indicates the existence of three field ranges with substantially different behavior:

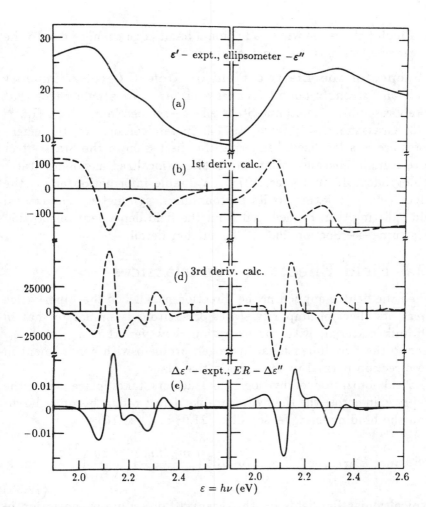

Figure 42.27: Comparison of the ε-spectrum and derivative line shape between experiment (top and bottom from electro-reflectance) and derivatives taken from independent ellipsometer data (top) (after Aspnes, 1972).

(1) the low-field range where the modulation of reflection $\Delta R/R$ is smaller than 10^{-3} and is employed for line-shape studies using a third-derivative analysis—Fig. 42.27;

(2) a medium-field range where the electron acceleration predominates* and determines line-shape broadening; and

---

* The electron is accelerated in the electric field and occupies a range of k states, thereby losing translational invariance, i.e., resulting in the loss

(3) a high-field range where additional band-edge changes due to the *Stark effect* must be considered.

Influence of the Electric Field on Defect Levels. *Impurity levels and strongly bound Frenkel excitons* show a field effect that closely resembles the atomic Stark effect—see Section 20.3.2B (Lüty, 1973; Grassano, 1977; Boyn, 1977). In semiconductors, the eigenstates are less localized, rendering the discussion of the Stark-effect more complex because of mixing between localized and band states (Vinogradov, 1973; Bauer, 1978). Valuable information about the defect center, relating to its surroundings, symmetry, and crystal field splitting, can be deduced from the field-dependent absorption spectrum—see Section 20.3.2B for further detail.

42.5 Field Effects in Superlattices

When the field is applied in the direction parallel to the superlattice layers, its effect on impurity states or excitons is similar to that in the bulk material; with the exception, however, that a_{qH} and R_H refer to the two-dimensional hydrogen problem with confinement in the direction normal to the superlattice plane.

Field ionization of hydrogen-like defects takes place when the energy gain across the the hydrogen-like defect radius becomes larger than the binding energy—see Eqs. (21.9) and (21.10):

$$eFa_{qH} \gtrsim E_{qH}, \quad \text{or} \quad F \gtrsim 2.76 \cdot 10^4 \left(\frac{m_n/m_0}{0.1} \right)^2 \left(\frac{10}{\varepsilon_{st}} \right)^3 \quad (\text{V/cm}).$$

$$(42.98)$$

Typical ionization fields for this Frenkel-Poole type of ionization in good semiconductors are on the order of 10 kV/cm (Miller et al., 1985)—see also Eq. (42.24).

In the direction normal to the superlattice plane, one has to consider confinement that permits much higher fields (typically > 10^5 V/cm) before ionization takes place (Bastard et al., 1983). Before ionization, two types of *quantum-confined Stark effects* (Miller et al., 1985) are observed: a *quadratic Stark shift*, in which the binding energy changes as

$$\Delta E_1 \propto m^* l^4 F^2 \qquad (42.99)$$

---

of **k** as a good quantum number. Different k-state mixing can describe the behavior and causes k-state broadening.

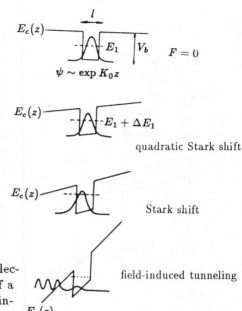

Figure 42.28: Influence of an electric field on the ground state of a quantum well with sequentially increasing field strength.

where l is the width of the quantum well; and a *Stark shift*, where the binding energy changes with a lesser slope

$$\Delta E_1 \propto F^\alpha \qquad \text{with} \qquad \alpha < 2. \qquad (42.100)$$

At still higher fields, tunneling out of the well becomes important (Bastard et al., 1983).

Figure 42.28 shows these cases schematically. The field-induced shift of the binding energy depends significantly on the position of the Coulomb-attractive defect within the well (see also Section 26.1.1) and is given in Fig. 42.29A. Changes in the exciton energy with field are more substantial and are given in Fig. 42.29B (see Brum and Bastard, 1985; Miller et al., 1985).

Summaries on the field effect in superlattices are given by Bastard and Brum (1986), and by Miller et al. (1986).

42.6 Ionization via Energetic Particles

Carriers can be generated by a wide variety of high-energy particles, such as x-ray photons, fast electrons, and various nuclear particles. Consequently, the use of certain solids, e.g., CdS, for the detection of such particles and for dosimetry was proposed early (Frerichs and

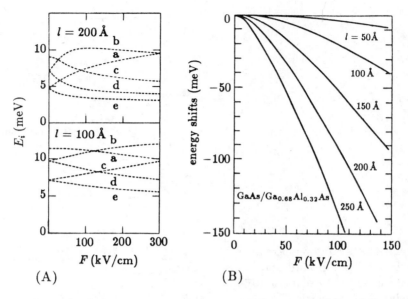

Figure 42.29: (A) Influence of an electric field on the impurity binding energy for wells of different widths and different positions of the defect within the well, identified by letters referring to the defect position at $-l/2$, $-l/4$, 0, $l/4$, and $l/2$ for a–e, respectively (see also Fig. 26.2A). (B) Influence of an electric field on the binding energy of excitons. The field direction is normal to the well plane; family parameter is the well width (after Brum and Bastard, 1985).

Warminski, 1946; Fassbender and Hachenberg, 1949; Broser et al., 1950).

In all of these excitation processes with high-energy particles, a large number of carriers are produced per incident particle. In addition, however, lattice damage is also produced depending on the energy and the mass of the particle—see Section 23.3. Such damage varies in severity from photochemical reactions with x-rays (Böer et al., 1966) to severe impact damage with protons and α-particles.

The excitation process depends on the type of particle. It may occur in the bulk via x-rays or γ-rays, or near the surface by fast electrons or α-particles; it may also be of a cascade character, i.e., the initially generated fast carriers may in turn create secondary carriers of sequentially lower energy. For a recent review, see Hayes and Stoneham (1984).

Summary and Emphasis

Free carriers are obtained by thermal, optical, or electrical field generation, or by a combination of these processes.

Thermal carrier generation from shallow centers requires the absorption of a phonon of sufficient energy, or at most a few phonons involving intermediate steps into excited states. Thermal generation from deep centers requires multiphonon-induced giant oscillations. Such oscillations cause a crossover of the defect level and levels of the conduction band, and provides sufficient probability for the defect-level electron to be ejected into band states. Similarly, a giant oscillation of an adjacent lattice atom could result in a replenishing of the emitted electron from the valence band. Thereby, a defect-assisted, band-to-band thermal excitation can take place.

Optical carrier generation proceeds from defect levels, or as band-to-band direct or indirect generation with photons of sufficient energy.

Electrical field generation of carriers can be caused by a field-enhanced thermal generation from Coulomb-attractive defect centers (Frenkel-Poole effect) at fields typically in the low 10 kV/cm range. Impact ionization from shallow impurity levels can be induced at very low fields, typically 10 V/cm at low temperatures. Impact ionization from deep centers is observed at much higher fields in materials of sufficient thickness and carrier density. Band-to-band impact ionization in typical semiconductors requires fields in the 10^5 V/cm range. At still higher fields, typically 10^6 V/cm, tunneling from deep defect centers or from the valence band takes place. These field-excitation mechanisms increase rapidly with increased bias and, when unchecked by a sufficiently large series resistance, cause dielectric breakdown.

Interaction between two or more of the generation processes provides a means of fine-tuning or sensitive detection. For instance, optically stimulated thermal or field generation of carriers yields information about defect levels or band spectra which are difficult to obtain otherwise. Field-enhanced thermal generation provides an ionization mechanism that acts selectively on Coulomb-attractive centers and can be controlled at relatively low-field strengths.

The generation of carriers, controlled by external parameters— such as temperature, optical excitation, or bias (electrical field)—is the primary process for many semiconducting devices. Changes in the carrier density caused by changes in the generation rate control the corresponding changes in currents, i.e., the output information of such devices. Electronic feedback can be used to amplify such signals.

Exercise Problems

1.(e)   Compare typical generation and recombination rates for various centers.

    (a)   What differences do you recognize by replacing attractive with neutral or repulsive centers?

    (b)   For a given generation rate, what are the results of replacing a certain density of attractive centers with the same density of neutral centers?

2.(*)   The attempt-to-escape frequency was estimated in two different ways—Eqs. (42.7) and (42.13).

    (a)   For what conditions do both agree with each other?

    (b)   Which is the more fundamental equation?

    (c)   Why are centers with small capture cross sections excluded from this discussion?

    (d)   Is there an appropriate way to define effective attempt-to-escape frequency for repulsive centers?

3.   Calculate the optical generation rate for an idealized solar radiation (black-body at $T = 6000$ K and total radiant flux of 100 mW/cm^2) for an Si slab of 1 mm thickness in the range $1.5 \leq h\nu \leq 4$ eV.

4.(*)   Discuss quantitatively the simple Frenkel-Poole effect for Coulomb-attractive donors in Si at 100 K. Compare the relevant quasi-hydrogen energies, phonon energies, and fields for obtaining marked changes in ionization rates.

5.   Explicitly develop the threshold energy for impact ionization [Eq. (42.34)] and carefully list all assumptions made during the derivation.

6.   Discuss in your own words the conditions for dielectric breakdown via avalanche formation.

7.(e)   Explicitly derive the transmission probability of tunneling through a rectangular barrier—Eqs. (42.60) and (42.61).

    (a)   Give the corresponding expression for the reflected wave. Discuss the relation.

    (b)   Analyze the corresponding expression for a triangular barrier.

(c) Compare the results of rectangular and triangular barriers.

8.(*) When tunneling from valence-to-conduction band, the effective masses of electrons and holes must be involved. Correct Eq. (42.67) to account for this fact. Assuming a rectangular barrier between the two bands, return to Eqs. (42.55) and (42.57) to derive the proper expression.

9. Describe an experiment which measures the field distribution within an inhomogeneously doped slab of a semiconductor using the Franz-Keldysh effect.

10. Describe quantitatively the Frenkel-Poole effect in a GaAs/GaAlAs superlattice of 30 Å well and barrier thickness when the electric field acts in the direction of the lattice plane.

11. Electron tunneling is an important phenomenon.

(a) Calculate the field that would result in a tunneling current through ultrapure Ge at 4.2 K in excess of its intrinsic current.

(b) What do you conclude about the chances of actually observing such tunneling? What other effects are competing?

(c) Cover the surface of this Si by oxidation with a 20 Å thick layer of SiO_2. Assume half its band gap as the barrier height and all of the voltage drop due to a bias of 1 V across such layer. What is the tunneling current, assuming Si is sufficiently doped n-type?

12. Is the Franz-Keldysh transition phonon-assisted? Explain.

13.(e) Plot the Fowler-Nordheim tunneling current in a semilogarithmic graph for $m_n = m_0$, with E_g as the family parameter for $E_g = 1$, 1.5, and 2 eV, and $d = 100$ Å.

14.(1) Tunnel spectroscopy is a very useful tool.

(a) Describe tunneling spectroscopy as a tool for investigating band structure.

(b) Can you distinguish a direct and an indirect band edge with this method?

(c) Which type of field is preferred for tunneling spectroscopy: (a) within a junction or (b) within a Schottky barrier? Explain.

Chapter 43

Carrier Recombination

Carriers can return immediately or after scattering to their original pre-excitation state (geminate recombination), or they can recombine with another state after diffusion away from their site of origin. On the average, for each act of generation there must be one inverse act of recombination.

Carrier recombination is the opposite process to carrier generation, which was discussed the previous chapter. For reasons of detailed balance, a process of recombination must occur for every process of generation in equilibrium. Although, there is a wide variety of such recombinations, depending on the type of energy released during recombination, one can distinguish two principle types

- radiative, and
- nonradiative recombination.

Radiative recombination releases a photon, is of great technical importance, and is discussed as *luminescence* in Chapter 44. Nonradiative recombination is discussed below.

43.1 Nonradiative Recombination

Nonradiative recombination is usually an undesired effect, since it converts high electronic energy into heat; that is, it increases the entropy of the system. It thereby decreases the performance of all but a few devices—such as bolometers, which are designed to measure the incident total energy; this is best accomplished by converting it into heat. Nonradiative recombination can occur by:

- single-phonon emission when recombining with a shallow level,
- a cascade emission of phonons,
- simultaneous multiphonon emission, and
- Auger generation of an accelerated carrier.

Recombination can occur from a band edge at

- $k = 0$, or at
- $k \neq 0$, or from an
- excited state of a defect level.

Recombination can proceed

- to another band,
- from a higher to a lower excited state, or
- to the ground state of a defect level.

Finally, recombination can proceed either
- directly back, or
- after some scattering back into the same state (geminate recombination), or
- after migration to another site.

43.1.1 Capture Cross g

A carrier, ignoring its kinetic energy, will be captured at a defect center with a phonon ladder if, while colliding with the center, it dissipates energy of at least kT. Its probability of reemission would be equal to its probability of further emitting a phonon, however, multiplied with the probability of finding a phonon for the reemission. This probability is < 1.

The capture cross section of such a center is

$$s_{\mathrm{nr}} = \pi r_0^2, \tag{43.1}$$

where r_0 is the radius of the electron eigenstate at $\gtrsim kT$ below the band edge. In the following section, we will present an estimate of r_0 for a Coulomb-attractive center.

Capture occurs with high probability when, during its Brownian path within a band, a carrier approaches the center

(a) within kT of the band edge *and*

(b) within a distance of less than r_0 of the recombination center.

The first condition is usually fulfilled (it occurs for carriers in thermal equilibrium), and only the second condition needs to be considered. Therefore, using a gas-kinetic model, carrier capture, like ordinary scattering, takes place after the electron has traveled a distance

$$\lambda_r = v_{\mathrm{rms}} \tau_r, \tag{43.2}$$

Table 43.1: Electron and hole capture cross sections of deep level impurities at room temperature in cm^2 and level depth in eV (collected from Landoldt Börnstein, 1982).

Dopant	Host Lattice	
	Si	CdS
Cu		$s_n(E_c - 0.09) = 4 \cdot 10^{-18}$
		$s_n(E_c - 0.2) = 3 \cdot 10^{-17}$
		$s_n(E_v + 1.1) = 3 \cdot 10^{-21}$
		$s_p(E_v + 0.34) \geq 10^{-15}$
Ag	$s_n(77\,\mathrm{K}) = 10^{-12}$	$s_n(E_c - 0.23) = 10^{-13}$
Au	$s_p(E_c - 0.55) = 10^{-15}$	$s_n(E_c - 0.065) = 10^{-19}$
	$s_n(E_v + 0.35) = 10^{-14}$	$s_n(E_c - 0.15) = 10^{-17}$
Zn	$s_p(E_v + 0.26) = 10^{-10}$	
S	$s_n(E_c - 0.59) = 2 \cdot 10^{-15}$	
W	$s_n(E_c - 0.22) = 10^{-16}$	
	$s_n(E_c - 0.3) = 6 \cdot 10^{-17}$	
	$s_n(E_c - 0.37) = 3 \cdot 10^{-18}$	
Co	$s_p(E_v + 0.29) = 2 \cdot 10^{-16}$	
	$s_p(E_v + 0.4) = 10^{-16}$	
Ni	$s_n(E_c - 0.35) = 10^{-15}$	$s_n = 10^{-17}$
	$s_n(E_v + 0.23) = 10^{-15}$	$s_p(E_v + 1.44) = 10^{-15}$
Rh	$s_n(E_c - 0.31) = 7 \cdot 10^{-16}$	
	$s_p(E_v + 0.52) = 6 \cdot 10^{-16}$	
Pd	$s_n(E_c - 0.22) = 3 \cdot 10^{-15}$	
	$s_p(E_c - 0.19) = 10^{-16}$	
Pt	$s_n(E_c - 0.23) = 5 \cdot 10^{-15}$	
	$s_p(E_v + 0.32) = 10^{-15}$	

where τ_r is the lifetime of a carrier between generation and capture and is given by

$$\tau_r = \frac{1}{v_{\mathrm{rms}}\, s_{\mathrm{nr}}(N_r - n_r)}. \qquad (43.3)$$

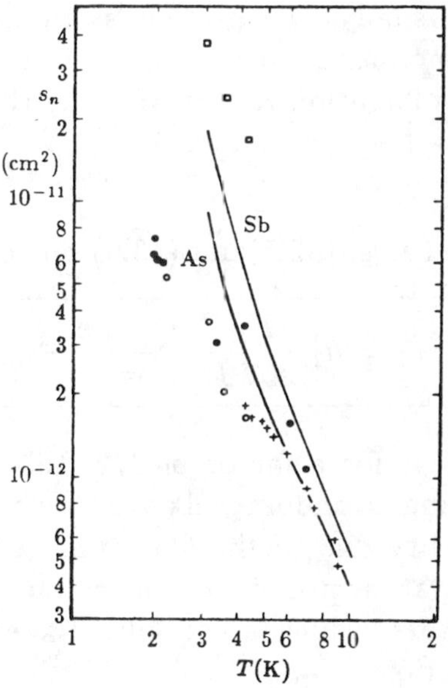

Figure 43.1: Giant capture cross section for electrons by positively charged donors in As- or Sb-doped Si as a function of temperature (after Ascarelli and Rodriguez, 1961).

with $N_r - n_r$ as the density of free recombination centers. A listing of capture cross sections for electrons or holes for some defect centers is given in Table 43.1. When nonequilibrium conditions are involved at higher fields or with optical excitation, an energy-dependent capture cross section must be considered that involves relaxation processes within the band—see Sections 33.2.1 and 43.1.1D.

43.1.1A Recombination at Coulomb-Attractive Centers

The energy spectrum of a Coulomb-attractive center is given by

$$E_{\mathrm{qH}} = \frac{m^*/m_0}{\varepsilon_{\mathrm{st}}^2 n_q^2} R_H \qquad (43.4)$$

where n_q is the quantum number. For capture to take place, one identifies a quantum number $\hat{n}_{kT}$ as the closest integer of n_q, for which $E_{n=\infty} - E_{\hat{n}} \gtrsim kT$. Then

$$\hat{n}_{kT}^2 \simeq \frac{m^*/m_0}{\varepsilon_{\mathrm{st}}^2 kT} R_H; \qquad (43.5)$$

one obtains as the radius of the corresponding eigenstate

$$\boxed{r_{kT} = a_H \frac{\varepsilon_{\mathrm{st}} m_0}{m^*} \hat{n}_{kT}^2.} \qquad (43.6)$$

Usually, the ground state is only slightly below kT at room temperature. Thus, $n_q = 1$ is often used for the quasi-hydrogen recombination radius, yielding the well-known approximation

$$r_{kT} = a_H \frac{\varepsilon_{st} m_0}{m^*}. \tag{43.7}$$

From Eqs. (43.5)and (43.6) one obtains

$$\boxed{r_{kT} = a_H \frac{R_H}{\varepsilon_{st} kT} = \frac{e^2}{8\pi \varepsilon_{st} \varepsilon_0 kT} = 27.8 \cdot \left(\frac{10}{\varepsilon_{st}} \frac{300}{T}\right) \text{ (Å).}} \tag{43.8}$$

Except for a factor of $3/2$, this is the same result as obtained by setting a random-walk velocity away from the center equal to a drift velocity due to the Coulomb potential (Bube, 1974, p. 488). The derivation presented here is not limited to lower-mobility semiconductors, which are implicitly required ($\lambda \ll r_{kT}$) in the velocity criterion.

The resulting recombination cross section is independent of m^*,

$$\boxed{s_{nr} = \frac{e^4}{64\pi(\varepsilon_{st}\varepsilon_0 kT)^2} = 2.42 \cdot \left(\frac{10}{\varepsilon_{st}} \frac{300}{T}\right)^2 \cdot 10^{-13} \quad (\text{cm}^2),} \tag{43.9}$$

and decreases with increasing temperatures $\propto 1/T^2$. It becomes very large at low temperatures, e.g., $\simeq 4 \cdot 10^{-12}$ cm^2 at 70 K, or $6 \cdot 10^{-10}$ cm^2 at $T = 5$ K. Measured capture coefficients are only slightly smaller (Fig. 43.1), possibly because of some overlap of the giant cross sections.

From Eq. (43.8), one obtains for the Coulomb potential:

$$2r_{kT} = \frac{e^2}{4\pi \varepsilon_{st} \varepsilon_0 kT}; \tag{43.10}$$

i.e., the radius of the Coulomb well for $E_{\hat{n}} \simeq kT$, is just twice the corresponding quasi-Bohr radius.

43.1.1B Geminate Recombination

In semiconductors with a relatively short mean free path, the excitation from a Coulomb-attractive center higher up into the band is followed by a number of scattering processes until the carrier is within kT of the band edge. If the carrier reaches the band edge within the same Coulomb funnel, it will recombine with the same center from which it was generated—see Fig. 43.2A. Such recombination is referred to as *geminate recombination*. It is observed in organic crystals and plays an important role in

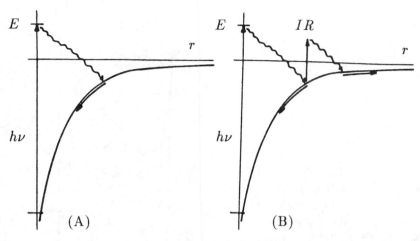

Figure 43.2: (A) Geminate recombination after excitation from a Coulomb-attractive center. (B) Out-diffusion after excitation with an additional IR excitation. The wavy arrow pointing down symbolizes scattering with phonon emission.

some of the amorphous semiconductors. A similar process also holds for geminate band-to-band recombination.

With an additional IR excitation, it is possible to increase the probability for out-diffusion and thereby reduce geminate recombination—Fig. 43.2.

43.1.1C Recombination in Amorphous Semiconductors

After optical excitation higher into the bands of amorphous semiconductors, inelastic scattering will relax these carriers into the tailing states, which extend from the band edges. Near the mobility edge (Section 40.4.2), the carriers continue to diffuse, thereby losing energy and consequently being trapped at deeper and deeper centers until thermal reemission and tunneling from center to center is no longer possible, as the deeper centers are more widely spaced—see Section 45.7.

43.1.1D Nonradiative Recombination at Deep Centers

Electron eigenfunctions in deep centers are strongly coupled with lattice oscillations—see Section 20.1.2. The energy of such a defect level depends on the relative position of the defect atom with respect to its surrounding lattice atoms. With vibrations of these atoms, the defect level moves up and down in the band gap about its equilibrium position.

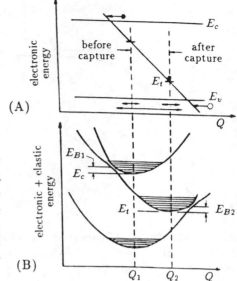

Figure 43.3: (A) Electron energy vs. configuration coordinate of a deep center before and after relaxation. (B) Electron plus elastic energy vs. coordination coordinate, shown for the deep level between valence and conduction bands (after Henry and Lang, 1977).

One mechanism of nonradiative capture is related to a sufficiently large lattice vibration, which moves the defect level into the conduction band. Here it can accept an electron from the conduction band. After the capture, the lattice is far from the equilibrium position, considering the recharging of the defect. This causes a violent vibration, which quickly relaxes with the emission of many phonons.

If the relaxation is very large, the defect level can move from the upper to the lower half of the band gap. Here, it can act now as a similar trap for holes, thereby completing the process of band-to-band recombination (Henry and Lang, 1977). This mechanism was alluded to in Sections 20.1.2 and 23.3, and is pictured schematically in Fig. 43.3A. The equilibrium position of the defect after electron capture is given by Q_2. The extent of average oscillations before and after capture (short arrows), as well as the extent of the giant oscillation directly after electron capture but before relaxation (large arrow), are indicated at the bottom of part (A) of this figure. The corresponding configuration coordinate diagram, which was given in Section 20.1.2, is redrawn here for convenience—Fig. 43.3B. The diagram identifies the activation energy E_{B1} for the electron trapping transition, and E_{B2} for hole trapping. The thermal binding energy of the electron in the center is $E_c - E_t$.

As an important consequence of this model, one obtains a temperature-dependent capture cross section:

$$s_n = s_\infty \exp\left(\frac{E_c - E_{B1}}{kT}\right).$$ (43.11)

The pre-exponential factor s_∞, obtained from detailed balance arguments, relates capture and emission, as shown in Eq. (42.8) for an unrelaxed trap. With lattice relaxation, care must be taken to account for the different activation energies; one then obtains (Sumi, 1983)

$$e_{tc} = \frac{\nu_c}{\nu_t} c_{ct} \exp\left(-\frac{S\hbar\omega_r}{kT}\right),$$ (43.12)

where ν_c is the number of equivalent valleys in the conduction band, ν_t is the degeneracy of the deep trap level, and S is the number of phonons emitted (Huang-Rhys factor) during the relaxation process after electron capture. Here, ω_r is the relevant defect eigenfrequency of a breathing mode.

With e_{tc} also given by (see Fig. 43.3B)

$$e_{tc} = \frac{\omega_r}{2\pi N_c} \exp\left\{-\frac{E_c - E_t}{kT}\right\},$$ (43.13)

one obtains with $c_{ct} = \nu_{rms} s_n$ and Eq. (42.13), setting $E_c - E_t = S\hbar\omega_r$, for the pre-exponential factor of the capture cross section (43.11):

$$s_\infty = \frac{\nu_t}{\nu_c} \frac{\omega_r}{2\pi} \frac{1}{N_c \nu_{rms}} \simeq 6.5 \cdot 10^{-15} \left(\frac{\hbar\omega_r}{40\,\text{meV}}\right)\left(\frac{300\,\text{K}}{T}\right)^2 \left(\frac{0.1}{m_n/m_0}\right) \text{cm}^2.$$ (43.14)

A more careful consideration of the approximations used to compute the capture requires the introduction of a factor η into Eq. (43.14), yielding a modified $s'_\infty = \eta s_\infty$. The correction factor η can be approximated as (Sumi, 1983)

$$\eta = \begin{cases} \dfrac{3\pi}{4}\gamma & \text{for } \gamma \ll 1 \\[2ex] 1 - \dfrac{5\pi}{9\sqrt{3}}\gamma^{-2/3} \simeq 1 & \text{for } \gamma \gg 1 \end{cases}$$ (43.15)

with the material parameter

$$\gamma = \frac{4\alpha_c kT\sqrt{\Delta E_c}}{3\hbar\omega_r}(E_c - E_t)^{3/2}$$ (43.16)

and ΔE_c as the width of the conduction band (see also Kayanuma and Fukuchi, 1984). Peuker et al. (1982) have given a short review of the different approaches to obtain transition probabilities between the states in the adiabatic approximation. Such probabilities in turn are proportional to the capture cross section discussed here.

The pre-exponential factor s_∞ for deep centers is on the order of 10^{-15} cm^2, and depends on the effective mass and degeneracies in agreement with measurements in typical III-V compounds (Henry and Lang, 1977). The thermal activation energies vary between a few meV and 0.6 eV for different defects; the larger energies represent a significant fraction of the level depth, and require substantial thermal activation for recombination. Such activation was observed in some early TSC measurements of deep centers in CdS (Böer and Borchardt, 1953).

43.1.1E Competition Between Radiative and Nonradiative Recombination

For transitions *within* a deep center, there is a simple rule as to whether radiative recombination from the upper minima to the lower curves in Fig. 43.4, or radiationless recombination from the upper curve via crossover to the lower curve is preferred. The Dexter-Klick-Russell rule (Dexter et al., 1956) states that radiative recombination occurs if the optical excitation E_n ends above the crossover E_B, i.e., for relatively weak coupling. Otherwise, the electron will cross over to the lower curve and reach the ground state in a nonradiative process via multiphonon emission. Depending on the strength of the coupling, such a crossover for nonradiative recombination may or may not require thermal activation; the nonthermal part is accomplished by tunneling to the lower curve.

This rule can be translated into the ratio of measurable energies, $E_n - E_0$ for optical excitation and $E_0' - E_0$ for thermal excitation:

$$\Lambda = \frac{E_n - E_0'}{E_n - E_0} \quad \begin{cases} 0 < \Lambda < 0.25 & \text{luminescence} \\ 0.25 < \Lambda < 0.5 & \text{weak luminescence} \\ 0.5 < \Lambda & \text{no luminescence} \end{cases} \quad (43.17)$$

with E_n, E_0, and E_0' as defined in Fig. 43.4, and as discussed by Bartram and Stoneham (1975). These authors have shown that, for F-centers in a variety of wide band-gap materials, this rule is fulfilled reasonably well. The energy E_B can be estimated as

$$E_B = \frac{(E_0' - E_n)^2}{4(E_n - E_0)}. \quad (43.18)$$

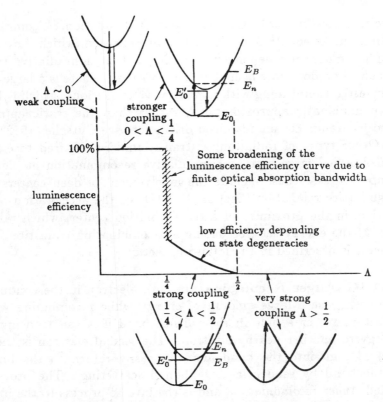

Figure 43.4: Luminescence efficiency as a function of the electron-lattice coupling expressed as parameter Λ [Eq. (43.17)]; with four typical configuration coordinate diagrams for ground and excited states of a deep level center shown in the insets for various degrees of coupling (after Hayes and Stoneham, 1984). ©John Wiley & Sons, Inc.

43.1.1F Nonradiative Multiphonon Recombination Nonradiative recombination into tightly bound centers, or from band to band with *simultaneous emission* of multiple phonons, i.e., typically ~ 30 phonons for a 1 eV band gap transition, are comparatively rare transitions. Haug (1972) estimates the transition probability for band-to-band recombination as

$$P_{\text{cvl}} \simeq A \exp\left(-\frac{(h\nu_l - S\hbar\omega_0)^2}{CkT}\right), \qquad (43.19)$$

where $h\nu_l$ is the electron energy to be dissipated, $\hbar\omega_0$ is the relevant phonon energy, and S, the Huang-Rhys factor, gives the average number of phonons emitted in the recombination process. From Eq. (43.19), one sees that P_{cvl} increases exponentially with temperature and decreases with increasing energy dissipation. A comparison with a similar equation for the radiative transition, which shows only minor temperature dependence, indicates that nonradiative transitions will predominate at higher temperatures. There is a large body of investigation dealing with nonradiative transitions, which is based on nonadiabatic approximations, rather than the static approach used by Haug. For an advanced discussion, see Gutsche (1982).

Other types of radiationless transitions are required to explain the observed large rate of nonradiative recombination at elevated temperatures. These include the involvement of deep centers with large lattice relaxation (Section 43.1.1D) or the acceleration of free carriers in the proximity of a recombination center, which take up part of the energy set free during a recombination transition. This process is described in the following section.

43.1.1G Auger Recombination An electron in the conduction band will lose a large amount of energy while recombining with a defect center or a hole in the valence band if it can transmit this energy to another nearby electron. The second electron is thereby excited high into the band and can easily return to the bottom of the band by sequential LO phonon scattering. The process is called *Auger recombination*, and is the inverse process to the impact ionization. It was originally proposed by Beattie and Landsberg (1958). For a recent review, see Landsberg (1987) and Haug (1988).

Energy and momentum need to fit the excited state higher in the band. When a phonon is provided to facilitate the momentum match, one speaks of *phonon-assisted Auger transitions* (Lochmann and Haug, 1980), which typically have a factor of 5 higher probabilities.

Several types of Auger recombination are possible, depending on whether the recombination occurs into an ionized defect center or into the valence band. The energy set free during the Auger recombination can be used to accelerate a second electron or a hole— Fig. 43.5. The smaller electron mass is the reason for a preference for the acceleration of a second electron (Landsberg and Willoughby, 1978; Landsberg, 1987).

A typical electron lifetime as a function of the electron density is given in Fig. 43.6; it shows three ranges. At low densities, τ is independent of n, then it decreases $\propto 1/n$, and finally $\propto 1/n^2$. We will concentrate first on the final range, which is determined by band-to-band Auger recombination.

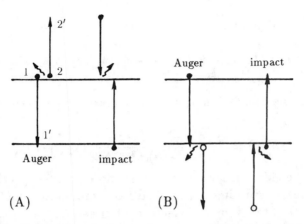

Figure 43.5: Comparison between Auger recombination and impact ionization involving (A) a second electron or (B) a second hole (open circles), with an additional phonon indicated for momentum matching.

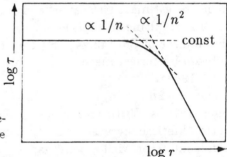

Figure 43.6: Schematic of the dependence of carrier lifetime on the carrier density.

The recombination rate and the electron lifetime, which are limited by Auger recombination, are given by:

$$C_{\rm cv}^{(A)} = Bn^2 p \quad \text{and} \quad \tau_A = \frac{1}{Bn^2};$$

(43.20)

B is typically on the order of $10^{-30} \ldots 10^{-22}$ cm^6 s^{-1}.

A quantum-mechanical derivation of the Auger recombination rate for band-to-band transition from thermally produced intrinsic carriers is given by

$$C_{\rm cv}^{(A)} = 2\frac{2\pi}{\hbar}\frac{\mathcal{V}^3}{(2\pi)^9}\iiint |\mathbf{M}|^2 f(E_1)f(E_2)|1 - f(E_1')||1 - f(E_2')|\times$$
$$\times \, \delta(E_1 + E_2 + E_1' - E_2')d^3k_1\,d^3k_2\,d^3k_1'$$

(43.21)

with the Auger matrix

$$M = \iint \phi_{k_1}^*(\mathbf{r}_1)\phi_{k_2}^*(\mathbf{r}_2)V(|\mathbf{r}_1 - \mathbf{r}_2|)\phi_{k_1'}(\mathbf{r}_1)\phi_{k_2'}(\mathbf{r}_2)d^3r_1\,d^3r_2;$$

(43.22)

where ϕ is the Bloch function, and $V(\mathbf{r})$ is the screened Coulomb potential

$$V(\mathbf{r}) = \int \frac{d^3q}{(2\pi)^3}\frac{4\pi e^2}{\varepsilon(q)(q^2 + \lambda^2)}\exp(\mathbf{q}\cdot\mathbf{r}).$$

(43.23)

Here, λ is the electron screening factor, and $\mathbf{q} = |\mathbf{k}' - \mathbf{k}|$ is the momentum transfer. Assuming parabolical isotropic bands and $m_n \ll m_p$, Haug (1972) obtains for the electron lifetime

$$\tau_A = \frac{\pi\hbar(4\pi\varepsilon_{\mathrm{opt}}\varepsilon_0\hbar)^2}{24e^4 m_n}\frac{\Delta E}{kT}\frac{\sqrt{2}}{0.01}\exp\left(\frac{\Delta E}{kT}\right).$$

(43.24)

Haug's formula also contains two overlap integrals, the values of which are estimated (Beattie and Landsberg, 1958) as $I_1 \simeq 1$ and $I_2 \simeq 0.1$, resulting in $(I_1 I_2)^2 = 0.01$, and as such are included in the denominator of Eq. (43.24). Further research is currently being done to obtain better estimates. For comments, see Haug (1988) and Laks et al. (1988).

$\Delta E = [(2m_n + m_p)/(m_n + m_p)]E_g$ is the energy dissipated in the Auger process. With an increasing band gap, the Auger-determined electron lifetime increases rapidly,* and reaches values not attained in a semiconductor with a band gap $E_g \geq 0.35$ eV; here, one eliminates $\tau_A \simeq 10^{-6}$ s.

Intrinsic Auger recombination at room temperature for thermally excited carriers is important only for narrow gap semiconductors. With high doping densities, however, sufficient carrier densities can be created to obtain Auger recombination rates in wider gap semiconductors. Such Auger recombination then dominates.

---

* It is interesting to see that Eq. (43.24) can be rewritten, using the quasi-hydrogen energy E_{qH}, as

$$\tau_A \simeq \frac{h}{E_{\mathrm{qH}}}\cdot\frac{E_g}{kT}\exp\left(\frac{E_g}{kT}\right).$$

(43.25)

The first part of Eq. (43.25) represents the Heisenberg uncertainty relation, indicating that τ_A cannot be smaller than the Heisenberg uncertainty time for an exciton, $\tau_A \stackrel{>}{\sim} 1.6\cdot 10^{-13}$ s. This presents a lower limit for $E_g \simeq kT$ for the approximation used.

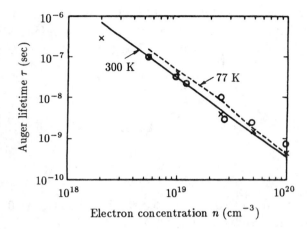

Figure 43.7: Auger lifetime in *n*-type Si as function of the electron density o and × experimental points at 77 and 300 K (after Dziewior and Schmid, 1977), and computation for the corresponding temperatures (by Laks et al., 1988).

For larger band-gap materials, however, the approximations used are too coarse. Recent computations, using the empirical pseudopotential method with self-consistent calculation plus Thomas-Fermi screening and $\varepsilon = \varepsilon(g)$, resulted in much improved results, as shown in Fig. 43.7 for *n*-type Si (Laks et al., 1988).

Similar results are obtained for Auger recombination with ionized defects, where again only shallow defects influence the observed lifetime.

With additional excitation, e.g., high-intensity optical carrier generation, sufficient carriers are available to render Auger recombination important. The Auger lifetime for recombination via recombination centers is given by (Haug, 1981)

$$\tau_A = \frac{1}{BN_r n} = \frac{v_{\mathrm{rms}}}{N_r n} \frac{m_n^2}{8\pi^2 e^4 \hbar^3} \frac{Q}{R} \sqrt{\frac{m_p}{m_n}} (E_c - E_r)^3, \qquad (43.26)$$

with $Q = 0.5(m_0/m_n)^{3/2}(1 + m_n/m_0)^2$ and where $R \simeq 2.6$ is an enhancement factor; E_r is the energy of the recombination center. This yields for typical values of $E_c - E_r \simeq 0.5$ eV and $m_n \simeq 0.1 m_0$ an Auger coefficient $B \simeq 10^{-26}$ cm^{6}s^{-1}. Somewhat lower values have also been suggested by Robbins and Landsberg (1980). For a density of recombination centers $N_r \simeq 10^{17}$ cm^{-3} and $n \simeq 10^{14}$ cm^3, one obtains an Auger lifetime of 10^{-5} s.

The Auger coefficient at recombination centers can be estimated from rough formulae with simple power law dependences in $E_c - E_r$ (Landsberg et al., 1976). In particular, for GaAs as (Haug, 1980)

$$B = \frac{2.5 \cdot 10^{-25}}{Q(E_c - E_r)^3} \, (\text{cm}^6 \text{s}^{-1}) \qquad (43.27)$$

The capture cross section for Auger recombination is given as

$$s_n = \frac{Bn}{v_{\text{rms}}} \qquad (43.28)$$

and is on the order of 10^{-11} cm^2 for shallow traps and 10^{-18} cm^2 for deep traps, when $n \simeq 10^{14}$ cm^3.

Evidence of intrinsic Auger recombination can be obtained from the dependence of the lifetime of minority carriers on the square of the density of majority carriers (Haynes and Hornbeck, 1955; Dziewior and Schmid, 1977).

43.1.1H Plasmon-Induced Recombination At very high excitation rates, the carrier density becomes high enough so that the plasmon energy $\hbar\omega_p$ equals the band gap energy; the recombination is much enhanced by such resonance transitions. Typical critical densities for a band gap on 1 eV are on the order of 10^{20}–10^{21} cm^{-3} [see Eq. (12.4)]. This *plasmon-induced recombination* can exceed the Auger recombination, which is also effective at high carrier densities (Malvezzi, 1987).

43.2 Statistics of Recombination

Except for stimulated lasing luminescent transitions (see Section 44.6.1), all recombination transitions are spontaneous and follow the rules of statistics.

In a statistical description of the recombination, one is not interested in how the carrier loses its energy, but rather with what state it recombines and what changes in carrier densities occur with changing rates of generation and temperature. The probability of recombination is described by a capture rate, which is linked to the capture cross section by Eq. (42.12).

43.2.1 Trapping or Recombination

When a carrier returns to an ionized state, one speaks of *recombination*. When a carrier is captured by a shallow level from which it can

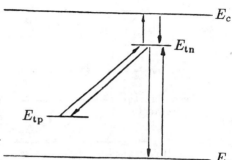

Figure 43.8: Various possible transitions to and from a localized state.

be thermally reemitted into the band before it finally recombines, the process is called *trapping*.

A more precise way to distinguish between trapping and recombination is given in the following section.

43.2.1A Electron and Hole Traps

There are several transitions possible between any center and other states. All such transitions are described by their corresponding rates. These rates are additive and describe the change in the population of the center. For instance, the change of the electron density in an electron trap is determined by excitation from the trap into the conduction band, by electron capture from the band, by recombination with holes from the valence band, and by electron transfer to other localized states of nearby defects to which such transitions are sufficiently probable.

For reasons of detailed balance, there are always pairs of transitions between two states (Fig. 43.8), which must be equal to each other in **thermal equilibrium**. The magnitude of transition rates *varies from pair to pair over a wide range.* For example, thermal excitation of an electron from the more distant valence band into an electron trap is much less probable than thermal excitation of a trapped electron into the closer conduction band.

In **steady state** (see Section 43.2.2), the total rate of transition from the center must equal the total rate into the center. Usually, one can *neglect all transitions compared to the one pair with the highest transition probability.* These pairs can now involve different states (see below). Thereby, one can identify different classes of centers according to the predominant types of transitions.

Centers close to the conduction band are identified as *electron traps*, and centers close to the valence band as *hole traps*, when these centers communicate predominantly with the adjacent bands.

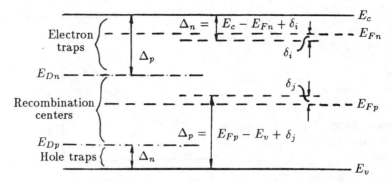

Figure 43.9: Band model with quasi-Fermi potentials and demarcation lines for one kind of electron trap, with corresponding capture cross sections for electrons and holes $(s_{\mathrm{ni}}, s_{\mathrm{pi}})$ and hole trap $(s_{\mathrm{nj}}, s_{\mathrm{pj}})$.

43.2.1B Recombination Centers   Centers close to the middle of the band gap readily communicate with both bands, since it is easier for a captured electron to recombine with a hole in the valence band than to be reemitted into the conduction band. These centers are called *recombination centers*.

Relaxation of the center after trapping a carrier is not included here. Such relaxation can assist significantly in bridging the band gap during recombination, and is discussed in Sections 20.3.3C and 43.1.1D—see also Henry and Lang (1977).

43.2.1C Demarcation Lines   A *demarcation line* between electron traps and recombination centers is defined (Rose, 1951) when the transition rates of electrons from these centers to the two bands become equal:

$$n_t e_{\mathrm{tc}} N_c = n_t c_{\mathrm{tv}} p. \qquad (43.29)$$

Using the expression for e_{tc} and p—see Section 42.2.3,

$$e_{\mathrm{tc}} = v_{\mathrm{rms}} s_{\mathrm{tv}} \exp\left(-\frac{E_c - E_t}{kT}\right) \quad \text{and} \quad p = N_v \exp\left(\frac{E_v - E_{F_p}}{kT}\right), \qquad (43.30)$$

one obtains an equation for this *electron demarcation line* from Eq. (43.29), defining a specific E_t [Eq. (43.30)] $= E_{Dn}$:

$$E_c - E_{D_n} = E_{F_p} - E_v + \delta_i, \quad \text{with} \quad \delta_i = kT \ln \frac{m_n s_{\mathrm{ni}}}{m_p s_{\mathrm{pi}}}. \qquad (43.31)$$

Although possible confusing at first glance, the reference to the *hole* quasi-Fermi level for determining the *electron* demarcation line is understandable, since the recombination path, competing with thermal ionization, depends on the availability of free holes. Figure 43.9 may help to clarify this dependency: the distance of the demarcation line for electrons from the conduction band, identified as Δp, is the same as the distance of the quasi-Fermi level for holes from the valence band plus a corrective δ_i or δ_j—see below. These correction terms are logarithmically related to the ratio of the capture cross sections for electrons and for holes at this center. The electron demarcation line defines the energy border between electron traps above, and recombination centers below, this center.

A similar relationship holds for the *hole demarcation line*:

$$E_{Dp} - E_v = E_c - E_{Fn} + \delta_j, \qquad \text{with} \qquad \delta_j = kT \ln \frac{m_p s_{pi}}{m_n s_{ni}}; \quad (43.32)$$

this line is also shown in Fig. 43.9. For n-type material with narrow $E_c - E_{Fn}$, there is a wide range of electron traps and a narrow range of hole traps, and vice versa.

The correction terms δ_i and δ_j depend on the ratio of capture cross sections for electrons and holes that change with the occupancy of the center. As an illustration, let us assume a simple example of a center that is neutral without an electron in it, having a cross section for an electron on the order of 10^{-16} cm^2. After it has captured the electron, it is negatively charged; its capture cross section for a hole has thus increased, say to $\sim 10^{-14}$ cm^2. For this example, $s_{ni}/s_{pi} \simeq 10^{-2}$ and $\delta_i \simeq -0.12$ eV. For a similar type of hole trap, the charge character changes from neutral to positive after hole capture, making $s_{nj}/s_{pj} \simeq 100$ and $\delta_j \simeq +0.12$ eV. The shifts δ_i and δ_j in Fig. 43.9 have been chosen accordingly.

Since the capture cross section varies from center to center, typically from $\sim 10^{-12}$ to $\sim 10^{-22}$ cm^2, δ_i varies for these different centers by as much as ~ 0.6 eV at room temperature. Hence, the demarcation lines of these centers are spread over a wide range within the band gap. Therefore, it is not customary to plot demarcation lines of all the possible centers, while it is still instructive to discuss those that provide the most important transitions in the given model.

Neglect of the other centers is often justified, because all transitions enter additively, spanning many orders of magnitude; therefore, usually only one kind of transition is important to each type of trap or recombination center for a given situation.

43.2.2 Thermal Equilibrium and Steady State

Thermodynamic (thermal) *equilibrium* is present when a semiconductor without any external excitation is kept long enough at a constant temperature to reach such equilibrium. Deviations from thermal equilibrium can occur because of nonthermal, additional excitation by light or an electrical field. When such deviations occur but have become stationary, a nonequilibrium *steady state* is reached.

43.2.2A Thermal Equilibrium In a semiconductor at constant temperature without optical or electrical excitation, thermal equilibrium becomes established. Electrons and holes are generated by thermal excitation alone. The same number of carriers which are generated in any volume element must recombine in the same volume element except for statistical fluctuations. There is no net transport of carriers. This also holds for space-charge regions, e.g., in a *pn*-junction, in which n and p are rapidly changing functions of the spatial coordinate, whereas j_n and j_p vanish independently in every volume element.

For thermal equilibrium the carrier distribution is uniquely described by the Fermi-level E_F. Consequently, when formally using quasi-Fermi levels, they must collapse to $E_{Fn} = E_{Fp} = E_F$. One sees from Eqs. (43.31) and (43.32) that the resulting demarcation lines then coincide: $E_{Dn} = E_{Dp} = E_d$, i.e., electron and hole traps join borders with *no* recombination center range in between. Here, thermal ionization (generation) and recombination on the average attain a balance within each volume element of the semiconductor.

43.2.2B Steady State When nonthermal carrier generation is introduced, the equilibrium density distribution is altered. As a consequence, the Fermi energy splits into two quasi-Fermi levels, and sets of two demarcation lines for each type of defect center appear. Thus, some levels, which previously acted as traps, will now act as recombination centers. Much of the content of the following sections deals with this steady-state condition.

43.2.2C The Hall-Shockley-Read Center Changing the external excitation will alter the demarcation lines between traps and recombination centers. Therefore, one needs to include all four transitions between the level and the two bands for such centers when variations of carrier distributions are considered (Hall, 1951; 1952; Shockley and Read, 1952).

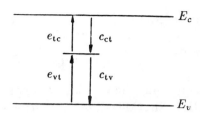

Figure 43.10: Hall-Shockley-Read
center with transitions to both bands.

For a center in the band gap that interacts only with the bands, the four transition rates are $e_{tc}n_t p_c$, $c_{ct}n(N_t - n_t)$, $e_{vt}n_v(N_t - n_t)$, and $c_{tv}n_t p$—see Fig. 43.10. In *equilibrium*, the sum of each pair of transitions to each band must vanish. In *steady state*, this is no longer necessary; there may be a net flow of carriers from one band through such a center to the other band, which is balanced with another transition, e.g., an optical band-to-band generation of carriers. The sum of all four transition rates, however, must vanish to maintain a time-independent electron population in the center:

$$c_{ct}n(N_t - n_t) - e_{tc}n_t p_c = c_{tv}n_t p - e_{vt}n_v(N_t - n_t). \qquad (43.33)$$

Equation (43.33) can be used to determine this population. After using Eq. (42.10) and an analogous condition for holes to convert e_{tc} and e_{vt} into the respective capture coefficients, and with the parameters $n_1 = N_c \exp[(E_t - E_c)/(kT)]$ and $p_1 = N_v \exp[(E_v - E_t)/(kT)]$, one obtains

$$n_t = \frac{N_t(c_{ct}n + c_{tv}p_1)}{c_{ct}(n + n_1) + c_{tv}(p + p_1)}. \qquad (43.34)$$

Introducing this steady-state density of the trapped electrons into the net rate equation, permitting a net flow U of electrons through such a center, yields

$$U = c_{ct}n(N_t - n_t) - e_{tc}n_t N_c = c_{tv}n_t p - e_{vt}n_v(N_t - n_t), \qquad (43.35)$$

and the equivalent rate equation for the same net rate of carrier transfer from the center to the valence band. By eliminating n_t from the net carrier flow rate, one obtains

$$U = \frac{c_{ct}c_{tv}N_t(np - n_1 p_1)}{c_{ct}(n + n_1) + c_{tv}(p + p_1)}. \qquad (43.36)$$

Using the condition

$$n_1 p_1 = n_i^2 \qquad (43.37)$$

and introducing the *intrinsic level* E_i

$$E_i = \frac{E_c - E_v}{2} + \frac{kT}{2} \ln \frac{N_v}{N_c}, \tag{43.38}$$

one obtains the well-known expression for the net carrier flow through a Hall-Shockley-Read center:

$$U = \frac{c_{ct} c_{tv} N_t (np - n_i^2)}{c_{ct} \left[n + n_i \exp \left(\dfrac{E_t - E_i}{kT} \right) \right] + c_{tv} \left[p + n_i \exp \left(\dfrac{E_i - E_t}{kT} \right) \right]}. \tag{43.39}$$

From Eq. (43.39), one sees immediately that U vanishes for thermal equilibrium, i.e., for $np = n_i^2$. In steady state, however, with optical generation and the Hall-Shockley-Read center acting as a dominant recombination center, U must equal the generation rate g_o in a homogeneous semiconductor.

Equation (43.39) becomes very valuable when deviations from the thermal equilibrium are analyzed. A separation into thermal generation rates

$$g_n = g_p = \frac{c_{ct} c_{tv} N_t n_i^2}{c_{ct}(n + n_i^+) + c_{tv}(p + n_i^-)}, \tag{43.40}$$

and recombination rates

$$r_n = r_p = \frac{c_{ct} c_{tv} N_t np}{c_{ct}(n + n_i^+) + c_{tv}(p + n_i^-)}, \tag{43.41}$$

is helpful. For brevity, the expression

$$n_i^{\pm} = n_i \exp \pm \frac{E_t - E_i}{kT} \tag{43.42}$$

is used above.

A deviation from thermal equilibrium may be caused by optical excitation. With band-to-band excitation, the optical generation term g_o is simply added to Eq. (43.40), yielding

$$g_{n,o} = g_{p,o} = \frac{c_{ct} c_{tv} N_t n_i^2}{c_{ct}(n + n_i^+) + c_{tv}(p + n_i^-)} + g_o; \tag{43.43}$$

the necessarily increased recombination in steady state is automatically included in Eq. (43.41) through the increase in n and p.

In *homogeneous semiconductors*, and for steady-state conditions, the generation rate will always be equal to the recombination rate; therefore, U will vanish.

Deviations from $U = 0$ will occur during kinetics (see Chapter 47) and when spatial inhomogeneities are considered. In space-charge regions, such as in Schottky barriers or *pn*-junctions, this formalism becomes most valuable. One aspect of such inhomogeneity deals with the current continuity in photoconductors, and is discussed in Section 45.3.2.

Summary and Emphasis

Nonradiative recombination of carriers releases energy in the form of phonons or by accelerating another electron, with Auger recombination, which consequently leads to LO phonon emission by the accelerated electron.

The release of phonons occurs in a single step, as a single phonon emission when trapping a carrier at a shallow defect center, or as a multiphonon emission when recombination at a tightly bound deep center occurs. A sequential release of a phonon cascade is possible during the lattice relaxation process of a deep center.

One distinguishes between carrier traps in which excitation into the adjacent band and trapping at the center dominate, and recombination centers through which carriers recombine from one band to the other. The former are located close to one band, the latter closer to the center of the gap. They are separated from each other by demarcation lines, which lie near the quasi-Fermi levels and appear when deviating from thermal equilibrium.

Most of the recombination occurs from free carriers after diffusion away from the place of generation. In low-mobility semiconductors, however, such separation from the place of generation can be impeded, causing geminate recombination at the same place.

As a characteristic parameter for carrier recombination, the capture cross section, which measures the "effectiveness" of a defect center in trapping or recombination, can be defined. The capture cross sections spread over more than 12 orders of magnitude, ranging from 10^{-10} cm^2 for Coulomb-attractive centers at low temperatures to less than 10^{-22} cm^2 for some Coulomb-repulsion centers or for deep centers with high lattice coupling.

Both Auger- and plasmon-induced recombinations become effective at high carrier densities, and transfer the relaxed recombination energy to single carriers or the collective of carriers, respectively.

Nonradiative recombination is one of the most important loss mechanisms in most semiconductor devices. It is almost always defect-center controlled, and can be reduced by proper doping with luminescence centers or with slow, sensitizing centers, avoidance of deep-level defect centers, donor-acceptor pairs, and other defect associates, each of which acts an efficient recombination center.

Exercise Problems

1.(*) Describe the ratio of radiative and nonradiative recombination through a deep recombination center with major lattice relaxation as a function of the temperature.

2.(e) Describe the position of the demarcation lines as a function of the generation rate for n-type photoconductors.

3.(e) The critical equipotential surface of a Coulomb-attractive center above which ionization occurs is a sphere in real space at $2kT$ below the conduction band. In two dimensions, it is a circle. Calculate this equipotential line in an external field.

4.(l) Auger recombination may involve electrons or holes (Fig. 43.5). Their relative involvement depends on the band structure. Explain.

5.(*) Determine the quasi-Fermi level and the demarcation line in GaAs with a deep donor of $N_d = 10^{17}$ cm^{-3} at 0.3 eV from the conduction band which relaxes by another 0.4 eV when occupied, at $T = 300$ K and $g_0 = 10^{20}$ cm$^{-3}s^{-1}$. Assume $s_n = 10^{-15}$ cm^2 and $s_p = 10^{-14}$ cm^2.

6.(e) Derive explicitly the expression for a carrier net flow through a Hall-Shockley-Read center [Eq. (43.39)].

7.(e) Plot the capture cross section of a Coulomb-attractive donor in Si as a function of the temperature in the range $1 < T < 300$ K.

8.(*) Give a quantitative description of the band-to-band Auger recombination of electrons in GaAs as a function of the electron density and temperature.

9.(*) Give the critical electron density in Ge for which plasmon-induced recombination can bridge the band gap.

Chapter 44

Radiative Recombination

The emission of light from semiconductors provides valuable information concerning the electronic structure of the material and its defects, and yields important insight into excitation and deexcitation mechanisms. Luminescence is of great technical interest.

In Chapter 43, carrier recombination was discussed in general terms. Recombination requires the dissipation of energy, which can proceed via the creation of phonons and can involve Auger collision of electrons, which results in *nonradiative recombination* or by the emission of photons as *radiative recombination.*

One distinguishes between several types of re-radiative processes, depending on their relationship to the optical excitation process and on intermittent processes:

- coherent re-radiation (reflection),
- thermal radiation (blackbody radiation),
- luminescence,
- stimulated emission (lasing), and
- phosphorescence (delayed luminescence).

These processes will be differentiated in this chapter, and the various types of luminescence will be discussed in more detail.

44.1 Reflection

Reflection of electromagnetic radiation involves the polarization of the semiconductor lattice, which is characterized by its index of refraction and extinction coefficient. The polarization appears in phase with the electric vector of the incoming radiation and produces *coherent re-radiation, essentially without delay*—see Section 49.3. Such re-radiation occurs from the undamped fraction of the excitation processes. It is observed from any medium with a refraction index larger than 1, as seen from the basic reflectance equation,

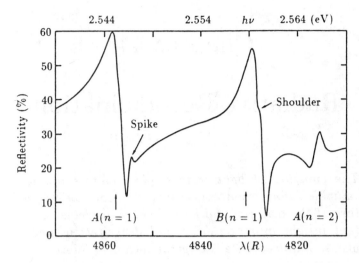

Figure 44.1: Reflectivity spectrum in the A and B polariton range of CdS for $\mathbf{E} \perp \mathbf{c}$ (after Evangelisti et al., 1974).

presented here for normal incidence and interface to vacuum—see Section 10.2,

$$R_o = \frac{(n_r - 1)^2 + \kappa^2}{(n_r + 1)^2 + \kappa^2}; \qquad (44.1)$$

the larger the index of refraction, the larger the reflectance.

Depending on the wavelength of the impinging radiation, electrons and heavier charged particles, e.g., ion cores, are involved in the polarization. Near resonances, the polarizability increases, and with it so does the material reflectance. The Kramers-Kronig relation describes the behavior relating $n_r(\nu)$ and $\kappa(\nu)$—Section 11.1.4. Such reflection is usually discussed in relation to the respective resonance phenomena. For examples, see Chapter 11 and Figs. 11.2 and 11.3; Section 12.1.1 and Fig. 12.1; or Section 12.3.3 and Fig. 12.7.

44.1.1 Spatial Resonance Dispersion

The reflectance spectrum near polariton resonances, however, shows characteristic deviations from a simple reflection spectrum, as given in Figs 11.2 and 11.3. These deviations are pointed out in Fig. 44.1 as an additional spike and a shoulder for the $A(n_q = 1)$ and $B(n_q = 1)$ polariton structure, and the reversal of the $A(n_q = 2)$ structure.

Difficulties in calculating this reflectance spectrum arise because the resonance becomes wave-vector-dependent:

$$hv = h\nu_t + \frac{\hbar^2 \mathbf{K}^2}{2M},$$ (44.2)

where ν_t is the transverse resonance frequency of the exciton, $\mathbf{K}$ is its wave vector, and $M = m_n + m_p$ is its effective mass. The dispersion is quadratic in $\mathbf{K}$, and therefore is a *spatial resonance dispersion*. The resulting dispersion relation is given by

$$\varepsilon(\nu, \mathbf{K}) = \varepsilon_{\text{opt}} + \frac{4\pi f_i \nu_t^2}{\nu_t^2 - \nu^2 + \dfrac{2\pi h\nu_t}{M}K^2 - 2\pi i\nu\gamma},$$ (44.3)

which contains a $\mathbf{K}$-dependent term (Skettrup, 1973); f_i is the oscillator strength, and γ is a phenomenological damping constant. The difficulties reside in the fact that the coupling between the external photon field and the bulk polariton must be well defined. This means that, in addition to Maxwell's boundary conditions, these $\mathbf{K}$-dependent polaritons require *additional boundary conditions* (ABC), which are insufficiently known. The $\mathbf{K}$-dependence means that a polariton with a certain frequency and polarization can propagate with different wave vectors.

Pekar (1958) suggested as a means to determine these ABCs the condition that the polarization vanishes outside the semiconductor. However, there are additional ambiguities, as excitons cannot exist near surfaces and the thickness of this *"dead layer"* without excitation varies (Broser et al., 1978). This causes substantial changes in the reflection spectrum, since the dead layer acts like an optical coating and can be influenced by surface fields or other surface treatments (Hopfield and Thomas, 1963). Various suggestions have been made to alter the ABCs, with results best described by a modified refractive index that can be inserted into the reflectance equation [Eq. (44.1)]. See Birman-Sein (1972), Ting (1975), and Kiselev (1974).

In early theories, the dead-layer problem is simply ignored or replaced by assuming empirical fitting parameters, which give reasonable results for Frenkel excitons but are unsatisfactory for Wannier excitons (Agranovich and Ginzburg, 1984).

Recently, a more direct approach has been taken by coupling the photon field with the electronic eigenfunctions rather than with the already derived polarization. Originally, this was done by using the adiabatic approximation (see Balslev, 1978). Later, the calculation from basic principles was successfully extended to include appropriate mass ratios, although with considerable computational effort. The results are impressive, e.g., now accounting for the spike in Fig. 44.1 within the experimental error (Gothard et al., 1984).

The additional boundary problem is thereby solved by a coherent wave approach, where the boundary condition at the surface is applied to the wavefunction. The dead-layer problem becomes mute, as it is included in the solution of the exciton eigenfunction near the surface (Balslev, 1988). For a review, see Stahl and Balslev (1987).

The absence of a time delay, except for extremely fast effects of polarization relaxation as discussed in Section 49.3, and the missing energy shift of the coherent re-radiation sets the reflection apart from any of the other re-radiation processes.

44.2 Thermal Radiation

The intrinsic band-to-band radiative recombination transition can be observed as thermal blackbody radiation. Electrons, thermally excited into the conduction band, can recombine with holes in the valence band with the emission of light. If the temperature is high enough, the emission becomes visible—the material will *glow*.

In the wavelength range of the intrinsic absorption, the semiconductor can be described as a blackbody. Here, the (thermally excited) emitted photon flux in the frequency range $\Delta\nu$ is given by Planck's formula

$$E_m = \frac{2\pi\nu^2}{\left(\frac{c}{n_r}\right)^2} \cdot \frac{\Delta\nu}{\exp\left(\frac{h\nu}{kT}\right) - 1} \qquad \text{for} \quad h\nu > E_g. \qquad (44.4)$$

This emission, originating within a thickness of $d \simeq 1/\alpha_o$ of the semiconductor, requires a volume generation rate of electron-hole pairs

$$g_o = \int_{E_g}^{\infty} \frac{E_m(\nu)}{d} \, d\nu, \qquad (44.5)$$

which must be equal to the radiative recombination rate of these electrons and holes

$$r_{\text{cv}} = c_{\text{cvr}} np. \qquad (44.6)$$

In equilibrium, one has

$$np = n_i^2 = N_c N_v \exp\left(-\frac{E_g}{kT}\right); \qquad (44.7)$$

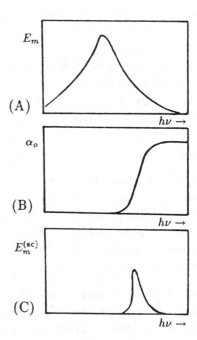

Figure 44.2: (A) Emission spectrum of a blackbody (schematic); (B) absorption spectrum of a semiconductor; (C) emission spectrum of the semiconductor obtained by multiplication of subfigures (A) and (B).

thus, one obtains from $g_o = r_{cv}$ for the radiative recombination coefficient

$$c_{cvr} \simeq \frac{\pi \hbar^3 E_g^2 \alpha_o}{c^2 (m_n m_p)^{3/2} (kT)^2}. \tag{44.8}$$

Equation (44.8) yields for the radiative band-to-band recombination cross section

$$s_{np} = \frac{c_{cvr}}{v_{rms}} = 3.38 \cdot 10^{-21} \left(\frac{E_g^2}{eV} \right) \frac{\alpha_o}{10^5} \left(\frac{m_0^2}{m_n m_p} \right)^{5/4} \left(\frac{1000}{T\,(K)} \right)^{5/2} (cm^2). \tag{44.9}$$

For typical values of $E_g = 1$ eV, $T = 1000$ K, and assuming $m_n = m_p = 0.1 m_0$, and $\alpha_o = 10^5$ cm^{-1}, one obtains $s_{np} \simeq 10^{-19}$ cm^2, which is a much smaller value than one would expect from a semi-classical model, i.e., an electron colliding with an ionized lattice atom with $s \simeq 10^{-16}$ cm^2. It should, however, be recognized that Eq. (44.8) is based on a rather crude estimate, and provides only an order-of-magnitude guidance for any specific transition.

44.2.1 Blackbody Radiation of Semiconductors

Compared to the Planck's emission distribution of a blackbody, the emission spectrum of the semiconductor is modified, since in the band-gap range there is essentially no absorption and thus little emission (*Kirchhoff's law*). Therefore, in a semiconductor, the long wave emission is suppressed—Fig. 44.2. If the band edge lies in the visible range, the thermal glow of the semiconductor is visibly different from the usual red or orange glow of a blackbody. For instance, ZnS with a band gap of $\sim$ 2.5 eV at 1000 K shows a green thermal glow: it is transparent for red and yellow light and thus shows no blackbody radiation there. This glow-light emission is closely related to the spontaneous band-to-band luminescence described in Section 44.3.1.

44.2.1A Thermal Radiation from Free Electrons In addition to the band-to-band emission, a band-gap emission can be observed when sufficient free electrons are created by doping, which in turn can recombine with the ionized donors and emit long wavelength radiation. An increase of thermal radiation has also been observed after the injection of excess electrons (Ulmer and Frankl, 1968).

44.2.1B Thermal Radiation from Lattice Vibration The thermal emission spectrum shows strong emission maxima in the *Reststrahl* range (Stierwald and Potter, 1967). Such emission is indicative of the direct coupling of the different types of phonons to the electromagnetic radiation, and can be used to obtain the corresponding spectrum of the optically active phonons.

44.3 Luminescence

Luminescence is defined as any radiative recombination from an excited electronic state, with the exception of reflection and blackbody radiation.* There are various possibilities for the excitation of luminescence:

- optical excitation,
- electron-beam excitation,
- excitation by high-energy nuclear radiation (α-, β-, γ-, or x-rays),
- excitation by carrier injection, and
- high-field excitation (e.g., by impact ionization).

---

* Other radiation due to nonlinear optical processes, such as photon scattering and higher harmonic generation, are also excluded in this discussion. These are discussed in Chapters 16 and 17.

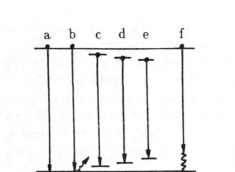 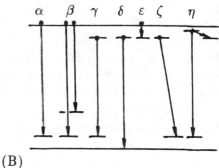

Figure 44.3: Typical transitions for (A) intrinsic and (B) extrinsic luminescence (see text for detail).

In all of these processes, excited states or free carriers are produced, a certain fraction of which in turn recombine with emission of a photon, yielding luminescence efficiency:

$$\eta_L = \frac{r_{\mathrm{r}}}{r_{\mathrm{r}} + r_{\mathrm{nr}}}, \qquad (44.10)$$

where r_{r} is the radiative, and r_{nr} is the nonradiative transition rate.

Radiative transitions can occur along a variety of paths, which are indicated in Fig. 44.3A for intrinsic and Fig. 44.3B for extrinsic luminescence. The intrinsic transitions are:

(a) direct band-to-band,
(b) indirect band-to-band,
(c) free exciton,
(d) free-exciton molecule,
(e) electron-hole liquid, and
(f) phonon-assisted band edge.

The extrinsic transitions are (with or without assistance of phonons):

(α) band-to-activator,
(β) band-to-multilevel activator,
(γ) excited state to ground state of a localized defect,
(δ) donor-to-valence band, or conduction band-to-acceptor,
(ε) conduction band-to-donor, or valence band-to-donor,
(ζ) donor-acceptor pair, and
(η) bound exciton.

During the luminescent transition, various types of photons or phonons can be released or absorbed (at higher temperatures), in addition to the emitted luminescence photon, which provides additional structure or broadening in the luminescence spectrum. A luminescent line spectrum can also be broadened due to a kinetic energy fraction of free carriers or excitons participating in the transition.

Other effects, too numerous to list here, influence the luminescence. They are described in many reviews of various fields relating to luminescence (see Voos et al., 1980). A few examples of some of the more important luminescence effects are given in the following sections.

44.3.1 Band-to-Band Luminescence

Direct band-to-band radiative recombination proceeds whenever an electron finds a hole with the same momentum vector within its recombination cross section. The highest probability, therefore, is realized when both the electron and hole are close to the band edge near $\mathbf{k} = 0$.

Band-to-band luminescence is called *spontaneous emission*. It is time-reversed to the optical excitation process, and thus the same matrix elements and selection rules apply—see Section 13.1.

The connection between optical absorption and luminescence emission can be seen by thermodynamic arguments. When a semiconductor is in equilibrium with a radiation field of brightness B (in photons/cm² per unit band width, unit time and solid angle 4π), this field loses entropy

$$\Delta S = -k \ln\left\{ 1 + \frac{8\pi\nu^2}{B} \left(\frac{n_r}{c}\right)^2 \right\}, \tag{44.11}$$

which is gained by the optical excitation of the semiconductor and is expressed in the spread of quasi-Fermi levels:

$$\Delta S = \frac{1}{T}\left\{ h\nu - (E_{\mathrm{Fn}} - E_{\mathrm{Fp}}) \right\}. \tag{44.12}$$

Equating (44.11) with (44.12) yields for the brightness of a radiation field in equilibrium with the semiconductor

$$B(\nu, \Delta E_{\mathrm{Fi}}, T) = 8\pi\nu^2 \left(\frac{n_r}{c}\right)^2 \frac{1}{\exp\left(\dfrac{h\nu - \Delta E_{\mathrm{Fi}}}{kT}\right) - 1} \tag{44.13}$$

with $\Delta E_{\mathrm{Fi}} = E_{\mathrm{Fn}} - E_{\mathrm{Fp}}$. For reasons of detailed balance, one can now express the radiative transition rate r_r for the spontaneous

Table 44.1: Recombination coefficients and cross sections for radiative band-to-band recombination (after Varshni, 1967).

Semiconductor	T	c_{cvr}	s_{nr}
	(K)	cm^3 s^{-1}	cm^2
Si	90	$1.3 \cdot 10^{-15}$	$2.4 \cdot 10^{-22}$
	290	$1.8 \cdot 10^{-15}$	$1.9 \cdot 10^{-22}$
Ge	77	$4.1 \cdot 10^{-13}$	$8.5 \cdot 10^{-20}$
	300	$5.3 \cdot 10^{-14}$	$5.5 \cdot 10^{-21}$
GaAs	90	$1.8 \cdot 10^{-8}$	$3.3 \cdot 10^{-15}$
	294	$7.2 \cdot 10^{-9}$	$7.6 \cdot 10^{-17}$
GaSb	80	$2.8 \cdot 10^{-8}$	$5.6 \cdot 10^{-15}$
	300	$2.4 \cdot 10^{-10}$	$2.5 \cdot 10^{-17}$

luminescence as the product of absorption and brightness of the impinging radiation field

$$r_r(\nu, \Delta E_{\mathrm{Fi}}, T) = \alpha_o(\nu, \Delta E_{\mathrm{Fi}}, T) B(\nu, \Delta_{\mathrm{Fi}}, T) \qquad (44.14)$$

which yields for weak optical excitation ($h\nu \ll \Delta E_{\mathrm{Fi}}$) the van Roosbroeck-Shockley (1954) relation:

$$r_r(\nu, \Delta E_{\mathrm{Fi}}, T) = \alpha_o(\nu, 0, T) 8\pi\nu^2 \left(\frac{n_r}{c}\right)^2 \exp\left(\frac{\Delta E_{\mathrm{Fi}} - h\nu}{kT}\right),$$
$$(44.15)$$

Equation (44.15) connects spontaneous luminescence emission with the optical absorption spectrum under the assumption that the electron distributions in conduction and valence bands are thermalized.

In a microscopic description, one obtains the intensity of the luminescence by multiplying the matrix element with the joint density of states and the product of the distributions of electrons and holes in the conduction and valence bands, respectively (see Yariv, 1975).

The recombination rate given above can be expressed as the product of electron and hole densities, with a recombination coefficient.

The *recombination coefficient* is given as the product of the capture cross section (of electrons finding the hole) and the thermal velocity of these carriers. In Table 44.1, the recombination coefficients are listed for a number of typical semiconductors at two temperatures, indicating recombination cross sections of 10^{-22}–10^{-15} cm^2 that decrease with increasing temperature. These cross sections are low for indirect band-gap semiconductors; they require a matching phonon for momentum conservation. In direct band-gap materials, e.g., in CdS (Reynolds, 1960) and GaAs (Shah and Leite, 1969), the recombination cross sections are more than three orders of magnitude higher, and are on the order of the atomic cross section.

A band-to-band luminescence spectrum is shown in Fig. 44.4 for Ge at 77 K. It shows two peaks: one due to indirect transition near 0.71 eV; and one, with six orders of magnitude *smaller* intensity, near 0.88 eV, due to direct transitions. The intensity relation can be easily understood from the smaller occupation for the higher valley: the Boltzmann factor indicates a ratio of $\sim 10^{-10}$; however, the fact that one compares the direct with the indirect transition makes the direct emission $\sim 10^4$ times more probable. Therefore, the observed ratio of 10^6 has a reasonable explanation.

The luminescence photoflux is given by

$$I_L = \vartheta r_r d, \qquad (44.16)$$

where ϑ is a geometry factor, taking into consideration that only a fraction of the light can exit the front surface where it can be observed. It is assumed here that reabsorption of the emitted light can be neglected. r_r is the radiative recombination rate, and d is the thickness of the device. The spectral distribution of the spontaneous luminescence intensity shows a steep rise at the band edge and an exponential decay toward higher energies, which reflects an exponential decrease of the distribution of thermalized electrons and holes with further distance from the band edges.

44.3.1A Spontaneous Luminescence from Hot Electrons

The high energy tail of the spontaneous luminescence, after sufficient relaxation (see Section 48.2.2C), is a direct measure of the electron temperature. When electrons are excited with light substantially above the band-gap energy, electron heating occurs and results in $T_e > T$.

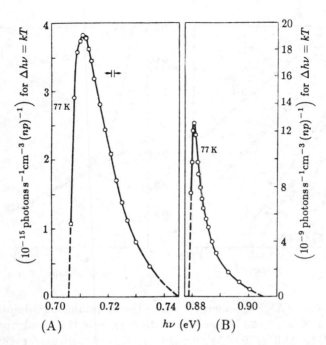

Figure 44.4: Band-to-band recombination luminescence spectrum for Ge at 77 K for (A) indirect and (B) direct recombination (after Haynes and Nilsson, 1964).

The intensity distribution of the spontaneous luminescence is given by—see Eq. (44.15):

$$I_{sL}(h\nu) \propto \alpha_o \exp\left(-\frac{h\nu - E_g}{kT_e}\right). \qquad (44.17)$$

The electron temperature can be obtained directly from the high energy slope of the luminescence peak, as shown in Fig. 44.5 for GaAs at a lattice temperature of 2 K. The hole distribution is little influenced by the optical excitation (see Section 48.2.2C) because of the larger hole mass.

The temperature of the electron gas increases logarithmically with the generation rate between 10^2 and 10^4 W/cm^2, and with a slope of $-k/\hbar\omega_{LO}$ due to a shift of the quasi-Fermi level in agreement with the experiment. At higher rates, the electron-hole interaction modifies the results (Shah, 1974); at lower rates, additional piezoelectric deformation-potential scattering dominates, which keeps the electron temperature low and explains the deviation from the straight line in Fig. 44.5B.

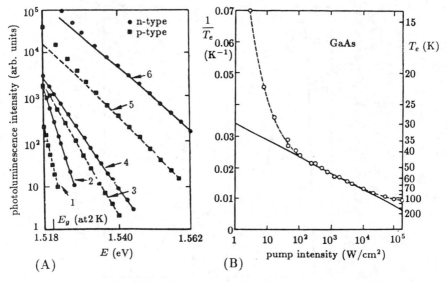

Figure 44.5: (A) High-energy tails of the spontaneous luminescence of GaAs at 2 K with the optical generation rate as the family parameter: $g = 0.0011, 0.0016, 0.0034, 0.0049, 0.64,$ and $0.76\,g_o$ for curves 1–6, respectively. Corresponding electron temperatures are $T_e = 14, 21, 36, 45, 64,$ and 76 K (after Shah and Leite, 1969). (B) Inverse electron temperatures as a function of the optical excitation rate in GaAs (after Shah, 1978). ©Pergamon Press plc.

44.3.2 Oscillatory Luminescence

When optical excitation is used to an energy level high within the band, rapid relaxation occurs with the preferred emission of LO phonons. Such relaxation is observed before the recombination of an electron with a hole occurs for spontaneous luminescence. Alternatively, a sufficiently slowed down electron can capture a hole to form an exciton with consequent exciton luminescence—see Section 44.3.3. The luminescence probability increases with decreasing kinetic energy of the carriers, hence causing oscillatory behavior with maxima when the LO phonon fits to bring the carrier close to the band edge. This is shown in Fig. 44.6. This condition yields luminescence maxima at energies given by

$$hv = E_g + \hat{n}\hbar\omega_{\text{LO}}\left(1 + \frac{m_n}{m_p}\right), \tag{44.18}$$

with $\hat{n}$ an integer number.

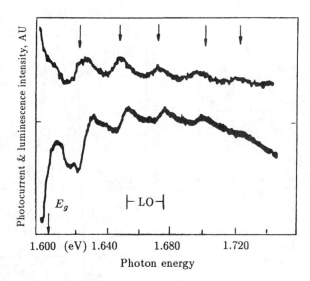

Figure 44.6: Photoexcitation spectrum of bound exciton luminescence (upper curve) and photoconductivity (lower curve) in CdTe at 1.8 K. Arrows according to Eq. (44.18). (After Nakamura and Weisbuch, 1978). ⓒPergamon Press plc.

44.3.2A Band-Edge Luminescence in Heavily Doped Semiconductors Heavy doping moves the Fermi level into the conduction band—see Section 24.3. Optical excitation then requires larger energies, labeled E_{g1} in Fig. 44.7A. Band-to-band luminescence, however, has a lower energy threshold, shown as E_{g2} in Fig. 44.7A. The width of the emission band is shown as ΔE_{g12}. In elemental indirect gap semiconductors, this luminescence band involves momentum-conserving TA or TO phonons with their corresponding replica, as shown for Si in Fig. 44.7B. In compound indirect semiconductors, LO phonons are employed in this process. In Ge, the LA replica are most prominent.

The *band filling*, obtained by heavy doping and observed by the Burstein-Moss effect, is obtained from the luminescent line shape. The band filling is shown in Fig. 44.8A. The corresponding line-broadening for Si is significant only for a carrier density in excess of 10^{18} cm^{-3}; it amounts to $\simeq 60$ mV at 10^{20} cm^{-3} and is slightly larger for holes than for electrons, since it relates to the density of states mass, with $m_{nds} \simeq 1.062 m_0$ and $m_{pds} = 0.55 m_0$ (see Barber, 1967).

The *decrease of the band gap* with band-tailing at higher doping levels (see also Section 24.3) is obtained by a square-root fitting of

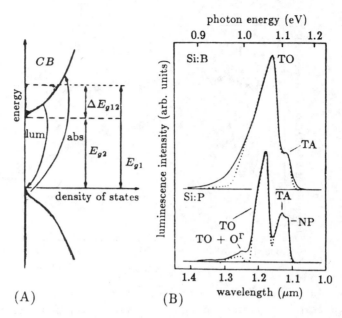

Figure 44.7: (A) Density of state distribution of a heavily doped n-type semiconductor with partially filled conduction band (schematic). (B) Band-to-band luminescence of n-type Si:P($n = 8 \cdot 10^{19}$ cm^{-3}—lower curve) and p-type Si:B ($p = 1.7 \cdot 10^{20}$ cm^{-3}—upper curve); dotted curve: theoretical fit (after Wagner, 1985). ©Pergamon Press plc.

the low-energy tail of the luminescence. It shows for Si an essentially constant band gap up to carrier densities of $\sim 10^{18}$ cm^{-3}, and a significant decrease of E_g starting near the critical Mott density, as shown in Fig. 44.8B. For further detail, see Wagner (1985).

At lower temperatures, free electrons and holes first recombine to form excitons. Consequent recombination radiation occurs from free excitons or, after trapping at lattice defects (impurities), from bound excitons. The intensity of this exciton recombination increases relative to the free electron-hole luminescence with decreasing temperature.

44.3.3 Exciton Recombination

One distinguishes free excitons in direct and indirect band gap materials. Their recombination luminescence is different.

44.3.3A Exciton Luminescence in Direct Gap Semiconductors Direct-gap semiconductors have excitons close to $\mathbf{K} = 0$, where mixing with photons is important. This creates *exciton*

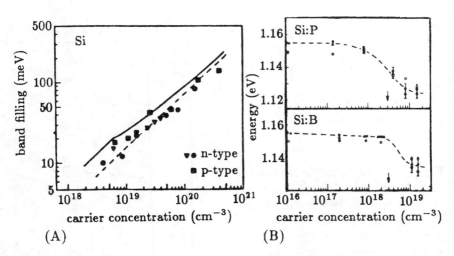

Figure 44.8: (A) Band filling of Si as a function of carrier density; calculated curves for *n*-type (dashed) and for *p*-type (solid curve) Si at $T = 20$ K. (B) Band gap reduction as a function of carrier density for Si obtained from optical absorption and from the photoluminescence cutoff at $T = 5$ K. Critical Mott density indicated by the arrows (after Wagner, 1985). ©Pergamon Press plc.

polaritons, which have been discussed in Section 11.3. The shape of the dispersion curves for $|\mathbf{K}| > 0$ near $\mathbf{K} = 0$ approaches a parabola (see Fig. 15.12B) for the different branches, since the exciton can acquire a *kinetic energy* that renders its total energy

$$E_{\mathrm{exc},(\mathrm{t},\mathrm{l})}(\mathbf{K}) \simeq E_g - E_{\mathrm{qH}(\mathrm{t},\mathrm{l})}^{(\mathrm{exc})} + \frac{\hbar^2 K^2}{2(m_n + m_p)}, \qquad (44.19)$$

where $E_{\mathrm{qH}}^{(\mathrm{exc})}$ is the ground state of the quasi-hydrogen exciton; the subscripts t and l are for transverse and longitudinal excitons—see below. The third term represents the kinetic energy of the exciton, which can be a substantial contribution. Therefore, the free-exciton line is relatively broad.

An example of a line spectrum for GaAs is given in Fig. 44.9, superimposed on the $E(k)$ diagram for identification of the upper and lower polariton branches; observe the stretched energy scale. The structure below the lower-branch exciton peak is due to impurities, which dominate the luminescence spectrum in bulk GaAs and will be discussed in Section 44.3.4.

The splitting of the exciton into longitudinal and transverse excitons is caused by the difference in polarizability; their energy differ-

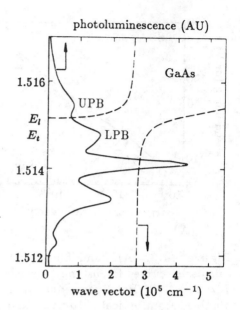

Figure 44.9: Exciton luminescence spectrum of the free-exciton polariton (upper two maxima); excitons bound at lattice defects (lower three maxima). UPB and LPB stand for upper and lower polariton branches, respectively. The dashed curve shows the $E(k)$ diagram (after Weisbuch, 1978).
©Pergamon Press plc.

ence is proportional to $\sqrt{\varepsilon_{\text{st}}/\varepsilon_{\text{opt}}}$. Only the transverse exciton couples with the photon (Knox, 1963). For a more extensive discussion, see Section 44.8.3. For phonon assistance or phonon scattering, see the review by Voos et al. (1980).

44.3.3B Excitons in Indirect-Gap Semiconductors In indirect band-gap semiconductors, the luminescence emission lines are usually phonon-assisted:

$$E_{\text{exc}} = E_g - E_{\text{qH}}^{(\text{exc})} - \hbar\omega. \qquad (44.20)$$

Selection rules determine which of the phonons is dominant, e.g., LA and TO phonons for Ge, as shown in Fig. 44.10. Peaks with TA and LO assistance also occur, but at a much lower intensity since they are dipole-forbidden (Lax and Hopfield, 1961).

44.3.3C Exciton Molecules Trions and biexcitons are formed at higher exciton density with binding energies E_{trexc} or E_{bexc}, which are on the order of a few meV. The resulting luminescence energy is, e.g., for biexcitons,

$$E_{\text{bexc}}(k) = E_g - E_{\text{qH}}^{(\text{exc})} - E_{\text{bexc}} + \frac{\hbar^2 K^2}{4(m_n + m_p)}. \qquad (44.21)$$

The existence of such excitonic molecules can be seen by observing the growth of this emission peak with increasing generation rate, and

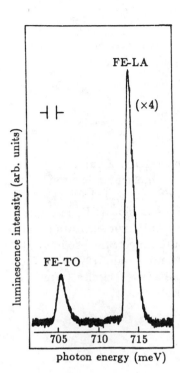

Figure 44.10: LA and TO phonon-assisted free-exciton (FE) luminescence of Ge at 4.2 K (after Etienne, 1975).

comparing its line shape with theoretical predictions—Fig. 44.11. Excitonic molecules have been observed in direct and indirect band-gap semiconductors, specifically in CuCl and AgBr, respectively.

An interesting alternative to exciton-molecule formation is the *Bose-Einstein condensation* of excitons at higher densities, originally suggested by Blatt et al. (1962)—see also Section 34.3. The condensation temperature depends on the exciton density N_{exc}, and is given by (Keldysh and Kozlov, 1968)

$$kT_c = 0.575 \frac{2\pi\hbar^2}{m_{exc}} n_{exc}^{2/3} = 32 \frac{m_0}{m_{exc}} \left(\frac{n_{exc}}{10^{18}} \right)^{2/3} \text{(eV)}. \qquad (44.22)$$

Condensation is expected to occur at 32 K for $n \simeq 10^{18}$ cm^{-3}, assuming $m_{exc} = m_0$. The low mass of excitons, compared to ^{4}He, facilitates Bose-Einstein condensation. More evidence has recently been compiled which suggests that such condensation indeed occurs and has profound effects on line shape narrowing (Hanamura and Haug, 1977); however, such narrowing may also be explained by exciton trapping.

44.3.3D Electron-Hole-Drop Luminescence
Electron-hole condensation can be observed from a band-gap reduction to E'_g and

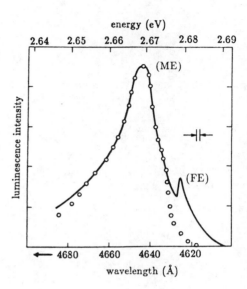

Figure 44.11: Luminescence spectrum of molecular excitons (ME) extending beyond the peak of free-exciton emission (FE) in AgBr at 4.2 K and high excitation rates. Measured spectrum—solid line (after Pelant et al., 1976); theoretical line shape—circles (after Cho, 1973).

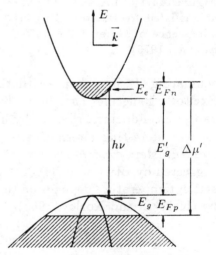

Figure 44.12: Electron-hole-drop luminescence transitions. $\Delta\mu'$ is the chemical potential of the electron-hole liquid.

from a broad line that appears at a well-defined temperature and at an intensity-related threshold—see Section 34.1. The electron-hole-drop related emission line occurs at

$$E_{eh} = E'_g + (E_{Fn} + E_{Fp}), \qquad (44.23)$$

where E_{Fn} and E_{Fp} represent the kinetic energies of free electrons and holes—see Fig. 44.12.

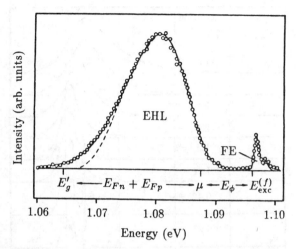

Figure 44.13: Luminescence spectrum of the electron-hole liquid at $n = 3.6 \cdot 10^{18}$ cm^{-3} and the free exciton in Si at 7 K (after Forchel et al., 1982).

The luminescence intensity is given by

$$I(h\nu) = A|\mathbf{M}|^2 \iint g_n(E_{Fn})g_p(E_{Fp})f_{\mathrm{FD}}(E_{Fn})[1 - f_{\mathrm{FD}}(E_{Fp})]\times$$
$$\times \, \delta(E_{Fn} + E_{Fp} + E_g' - h\nu)$$

$$(44.24)$$

with g as the density of states, f_{FD} as the Fermi-Dirac distribution, E_g' as the renormalized band-gap energy, and A as a proportionality factor. A typical electron-hole luminescence spectrum is shown in Fig. 44.13, which also contains a remaining free-exciton peak.

The characteristic energies can be obtained from this spectrum. Here, μ is the electrochemical energy of the electron-hole liquid, E_0 is the condensation energy, and $E_{\mathrm{exc}}^{(f)}$ is the energy of the free exciton below the band gap E_g. From amplitude and shape of the EHL-peak, one can obtain the pair density and carrier temperature. With decreasing temperature and pair density, the EHL-peak increases at the expense of the FE peak. For further discussion, see the reviews by Reinecke (1982), Rice (1977), and Hensel et al. (1977).

Luminescence and Mott Transition. At temperatures *above* the critical temperature T_c (Section 34.1) and at exciton densities approaching the *Mott transition density* n_M [Eq. (34.10)], the exciton gas becomes unstable and dissociates into a dense electron-hole plasma, indicating the semiconductor-metal transition (Mott, 1974). Consequently, a broadening of the recombination emission lines oc-

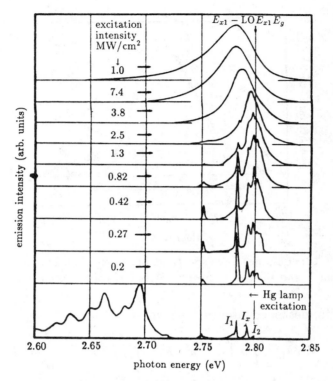

Figure 44.14: Luminescence spectrum of ZnSe at 4.2 K with excitation densities as family parameter. Lower spectrum obtained with Hg lamp excitation (after Era and Langer, 1969).

curs, as observed by Shah et al. (1977). An example of this broadening of emission lines with increased excitation density is shown in Fig. 44.14.

44.3.4 Bound Exciton Luminescence

Excitons can interact with point defects via an attractive dipole-monopole coupling when the defects are charged, or via a somewhat weaker dipole-dipole coupling to neutral defects. These excitons (X) are identified* as $(D^\bullet X)$, $(D^\times X)$, or $(A^\times X)$ when bound to a charged donor, a neutral donor, or an acceptor, respectively (Lampert, 1958).

---

* In the literature, these excitons are identified as D^+X, $D^\circ X$ and $A^\circ X$, with unambiguous reference to the charge of the donor or acceptor. For consistency, and to emphasize that this charge is relative to the neutral lattice, we have chosen the Kröger-Vink notation—see Section 18.2.

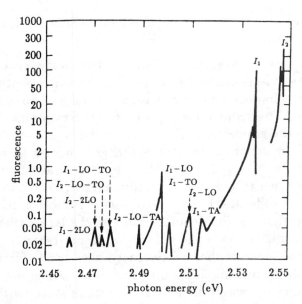

Figure 44.15: Bound exciton luminescence spectrum I_1 and I_2 of CdS at 1.6 K with phonon replica for $\hbar\omega_{LO} = 37.7$ meV, $\hbar\omega_{TO} = 34.4$ meV, and $\hbar\omega_{TA} = 20.6$ meV (after Thomas and Hopfield, 1962a).

In many materials, the line spectrum of bound excitons is abundant because of a variety of involved defects, various phonon replica, and the narrow line shape that permits the resolution of rather complex spectra at low temperatures. An example of such a spectrum is shown in Fig. 44.15.

Bound excitons show extremely sharp luminescence lines since the kinetic energy contribution, which causes the broadening for the free exciton, vanishes. The intensity of bound exciton lines is usually very large (lines I_1 and I_2 in Fig. 44.15), since their line width is narrow and they have very large oscillator strength (Thomas and Hopfield, 1962a). The oscillatory strength is proportional to the volume of the ground-state eigenfunction, typically to a_{qH}^3. Consequently, a giant oscillatory strength is found for materials of low m^* and high ε_{st} (Rashba and Gurgenishvilli, 1962).

Excitons, bound to neutral or ionized donors or acceptors, may be considered akin to an H atom, H^- ion, or H_2 molecule (Hopfield 1964). Not all of the possibilities give rise to luminescence.

Splitting under uniaxial stress (Benoit à la Guillaume and Lavallard, 1972), electric fields, and, especially, in a magnetic field, give im-

portant clues as to the nature and symmetry of the involved center—
see the following section.

44.3.4A Excitons Bound to Shallow Impurity States Shallow impurity states, i.e., hydrogen-like donors and acceptors, can be
detected with luminescence spectroscopy involving bound excitons.
There is a rich spectrum in high-quality single crystals involving transitions from the ground or excited state of the bound exciton leaving
the state of the donor or acceptor unchanged, or with a final state of
these impurities at $n = 2, 3 \ldots$. There is additional structure to the
luminescence due to the chemical individuality, i.e., the central-cell
potential that shifts the $1s$ state. This effect is more pronounced for
the acceptor than for the donor, because of the larger effective mass
of the hole, which renders the acceptor a deeper center.

In order to separate the closely spaced lines and to further reduce
their line width, one applies a magnetic field, which has the effect
of compressing the wavefunction. It also splits states with different
orbital angular momentum, and thereby assists in the identification
of specific features.

An example of the luminescent spectral distribution in GaAs with
C and Zn acceptors is shown in Fig. 44.16. The transitions involve
the annihilation of an exciton bound to these acceptors, with the
acceptor at the end of the transition in the excited $2s$, $3s$, or $4s$
states as indicated by $n_q = 2, 3$ and 4, respectively. The high degree
of resolution in the small range of the spectrum shown—here a total
of 6 meV—is emphasized (see Reynolds et al., 1985).

An even richer spectrum can be observed when excitons are bound
to distant pairs—Sections 22.1 and 44.5.1A.1. As an example, N-
pairs in GeAs can bind excitons yielding 47 identified lines as shown
by Skolnick (1986).

44.4 Optical Excitation Spectra for Luminescence

When luminescent transitions occur from lattice defects, it is instructive to measure the intensity of a specific luminescence line as a function of the wavelength of the *exciting light*, called *photoluminescence
excitation spectroscopy*. In this mode of excitation, the connection
between the excitation process and the luminescent center becomes
evident. One observes selective enhancement of the luminescence
when resonance transitions take place. This is important for the
analysis of bound excitons and the processes involved in the gener-

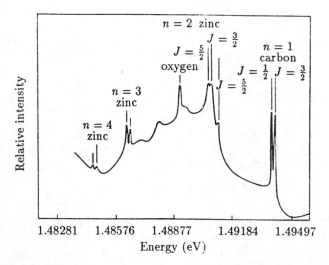

Figure 44.16: Luminescence of excitons bound to neutral C and Zn acceptors in GaAs at 2 K. Termination of the luminescent transition occurs in the $2s$, $3s$, or, $4s$ excited states of the acceptor. The fine structure is due to the initial state-splitting as indicated by $J = 1/2$, $3/2$, or $5/2$ as a result of combining two $J = 3/2$ holes and one $J = 1/2$ electron (after Reynolds et al., 1985). ©Pergamon Press plc.

ation of such excitations (see Dean, 1984). Examples of excitation spectra of luminescence are given in Figs. 44.21 and 44.26.

44.5 Defect Center Luminescence

A large variety of luminescent transitions occurs between free carriers and defect centers, or within a defect center from a higher to a lower excited state or to the ground state.

44.5.1 Hydrogen-Like Donors and Acceptors

Hydrogen-like donors or acceptors are known to have high luminescence efficiencies. In GaAs, for instance, these are Be_{Ga}, Mg_{Ga}, Zn_{Ga}, Cd_{Ga}, C_{As}, Si_{As}, Ge_{As}, and Sn_{As} as acceptors, and S_{As}, Se_{As}, Te_{As}, Si_{Ga}, Ge_{Ga}, and Sn_{Ga} as donors.

Luminescence transitions, involving these defects, occur from the band to an acceptor or a donor within such centers, or between *donor-acceptor pairs*.

With increasing temperature, more recombination transitions involving at least one of the bands, are favored since the lifetime of carriers in shallow donors is decreased. This is indicated by a

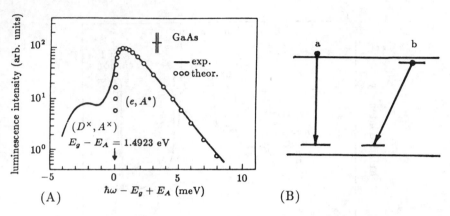

Figure 44.17: (A) Luminescence spectrum for transitions from the conduction band and from a donor into a hydrogen-like acceptor in GaAs at 1.9 K. Circles: theoretical line shape for $T_e = 14.4$ K (after Ulbrich, 1978). (B) Band model with corresponding transitions. ©Pergamon Press plc.

relationship of the type given in Eq. (44.29)—see Section 44.6.1C (William and Hall, 1978).

44.5.1A Free-to-Bound Transitions

Luminescence transitions from a band into a hydrogen-like donor or acceptor reflect the optical ionization energy. Because of a Franck-Condon shift (Markham, 1966), this optical ionization energy usually is different from the thermal ionization energy used for determining the semiconductivity. For hydrogen-like defects, however, such a shift is small because of a very small electron-lattice coupling, except for the transitions to ground state.

The line shape reflects the carrier distribution within the band. As an example, Fig. 44.17A shows the transition of an electron from the conduction band into an acceptor in GaAs at 1.9 K. This emission line shows a sharp cut-off at low energies corresponding to the energy from the band edge to the ground state of the acceptor— Fig. 44.17B,(a). Toward higher energies, a long exponential tail is observed, which reflects the electron energy distribution (f_n) within the band:

$$ I_L(h\nu) = A \, g_c(h\nu) N_a f_n \left[\frac{h\nu - E_g + E_a}{kT_e} \right], \qquad (44.25) $$

where $g_c(h\nu)$ is the density of states near the conduction band edge, N_a is the density of the acceptor, and A is a proportionality constant.

From the exponential slope of this tail, one estimates an electron temperature of 14.4 K. This increased electron temperature is due to the optical excitation, thermalized with phonons—see Section 45.5.4. Figure 44.17A also contains a lower energy luminescent feature, which is probably caused by the recombination of electrons from a hydrogen-like donor with a nearby acceptor—Fig. 44.17B(b).

Donor-Acceptor Pair Luminescence.   When donors and acceptors are both present in a semiconductor, there is some probability that a given donor has an acceptor *nearby*, with only certain separation possible, dictated by the relative positions of the lattice sites. The Coulomb interaction between donor and acceptor is thereby "quantized" according to the "shell" number, giving rise to a sequence of discrete lines in the recombination luminescence spectrum—see the following and Section 23.1.1.

Pairs of substitutionals of the same sublattice are referred to as *type I*; an example is the C_P-O_P pair in GaP. Pairs of substitutionals in *both* sublattices are referred to as *type II*. An example of this type is the Zn_{Ga}-O_P pair. In addition, there are pairs of the same *isoelectronic* defect; e.g., in GaP:N at a higher density of nitrogen, N_P-N_P pairs are present.

Such a pair could be a nearest neighbor pair or a more distant pair. The energy of the luminescent transition depends on the interdefect distance and is given by Eq. (23.3)—see Section 23.1.

The emission peaks are numbered sequentially according to their interdistance shell number $\hat{m}$—Eq. (23.4). In Fig. 44.18, the highly structured luminescence spectrum of GaP:S_P,Si_P is shown with donor-acceptor emission peaks identified up to $\hat{m} = 89$, indicating donor-acceptor interaction over at least seven lattice constants. At still higher separation, the peaks are more closely spaced, and can no longer be resolved; they produce a broad band near $h\nu \sim 2.22$ eV of rather intense luminescence.

The luminescence spectrum of GaP:N_P also shows numerous peaks due to pair formation of nitrogen substitutionals—see Fig. 44.19; the subscript again refers to the shell number identified in Eq. (23.3).

44.5.2 Isoelectronic Centers

Isoelectronic centers are of interest because they show a high luminescence efficiency at room temperature. This is caused by the fact that these centers, produced by neutral atoms of the same valency as the

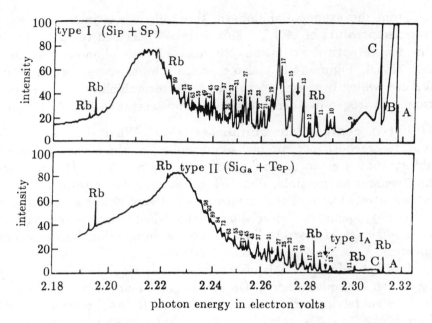

Figure 44.18: Luminescence spectrum of GaP type I, doped with S_P and Si_P $(E_A + E_D = 0.14\,\text{eV})$ in the upper diagram, measured at 1.6 K, and type II, doped with Si_{Ga} and Te_P $(E_A + E_D = 0.124\,\text{eV})$ in the lower diagram. The numbers in the figure refer to the shell number $\hat{m}$ in Eq. (23.3). Rb-lines are used for calibration (after Thomas et al., 1964.)

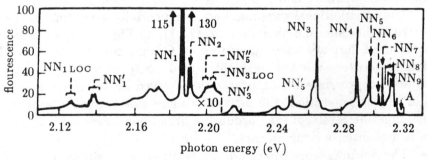

Figure 44.19: Luminescence spectrum of GaP doped with 2.10^{18} cm^{-3} nitrogen, forming N-N pairs, and measured at 4.2 K. The prime symbols identify phonon replicas, and LOC identifies local modes (after Thomas and Hopfield, 1966).

replaced lattice atoms, have a short-range potential, and therefore extend to rather large k values. This permits electron transitions from $k \neq 0$ without phonon assistance. This is important in indirect band-gap materials.

Examples of such isoelectronic centers are: N, As, or Bi in GaP (Thomas et al., 1965; Trumbore et al., 1966; Dean et al., 1969); O in ZnTe (Dietz et al., 1962; Hopfield et al., 1966; and Merz, 1968); or Te in CdS (Roessler, 1970). Several review papers have been written on this subject, e.g., Bergh and Dean (1976), and Crawford and Holonyak (1976).

At first view, the isoelectronic impurity, which replaces a lattice atom, seems to be a relatively simple defect; in its incorporation into the lattice, it resembles a shallow donor or acceptor. An estimate of the binding energy, however, is more difficult (Baldereschi and Hopfield, 1972). The high luminescence efficiency is suggested by an efficient capture of an exciton at such a center, and a high radiative recombination probability of this exciton. The exciton capture may be a two-step process by which the impurity first captures an electron (or hole) if it is a donor- (or acceptor-) like defect, and consequently, now charged, captures the oppositely charged carrier (Hopfield, 1967).

The radiative recombination from isoelectronic centers lies close to the band edge, and, at low temperatures, can be resolved as originating from two different exciton states, e.g., with $J = 1$ and $J = 2$* for emission lines A and B, respectively, in GaP:N$_P$ as shown in Fig. 44.20

44.5.2A Phonon Replica at Isoelectronic Traps In Fig. 44.20, pronounced replica of the primary luminescence (A,B) of excitons trapped at isoelectronic centers in GaP are visible; they are identified as A-TO, A-LO, and B-LO. These strong lines indicate that radiative transitions are likely to occur with phonon assistance, i.e., after LO or TO phonon emission.

In some materials, such phonon assistance at isoelectronic centers is especially pronounced, as, e.g., in CdS (see Fig. 44.15) and with even stronger electron-lattice coupling in ZnTe:O (Dietz et al., 1962) and in ZnSiP$_2$ with an unknown isoelectronic impurity (Shah and Buehler, 1971). In both examples, the first few phonon replica show a much higher intensity than the original zero-phonon transition— Fig. 44.21. The zero-phonon emission is observed in ZnTe:O but lies below the detection limit in ZnSiP$_2$.

---

* The two possible states that can be formed by coupling a spin $\frac{1}{2}$ electron with a spin $\frac{3}{2}$ hole.

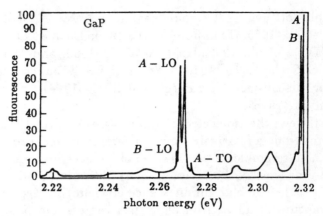

Figure 44.20: Luminescence spectrum from GaP:N with $5 \cdot 10^{16}$ cm^{-3} N$_P$ at 4.2 K, identified as an exciton transition bound at the isoelectronic center. The lines identified with LO and TO are phonon replicas (after Thomas and Hopfield, 1966).

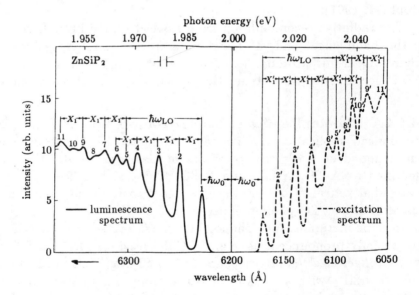

Figure 44.21: Luminescence (excited at 5850 Å) and luminescence excitation spectrum, related to an isoelectronic impurity in ZnSiP$_2$ at 1.8 K (after Shah, 1972).

The phonon interaction with such isoelectronic centers is so strong because of a severe lattice deformation that surrounds the impurity and provides an efficient coupling. The excitation spectrum (see Section 44.4) shows similar phonon replica and is almost mirror-symmetric to the luminescence spectrum except for slightly different phonon energies—see Section 20.3.3B and below for a classification according to the strength of the electron-lattice coupling.

The excitation spectrum is obtained by monitoring the amplitude of any of the first five luminescence lines, while scanning the exciting laser wavelength from 6050 to 6200 Å. There are three sets of phonons identifiable in emission: $\hbar\omega_0 \simeq 9.3$ mV, which is a zone-boundary optical phonon indicating that $ZnSiP_2$ is an indirect band-gap semiconductor, $\hbar\omega_{LO} \simeq 22.7$ mV, and $X_1 \simeq 6.5$ mV, both local modes, which appear in various combinations as shown in Fig. 44.21. In the excitation spectrum, the same combination of these phonons is observed, except that the X_1 phonon has a reduced energy of $X_1' \simeq 5.3$ mV, which is characteristic of local phonon modes and is caused by lattice relaxation between excited and ground states—see Section 20.1.2A.3.

44.5.3 Hot-Electron Luminescence into Shallow Acceptors

When the optical excitation is *substantially in excess of the band gap*, hot carriers result in a two-step process: first, hot electrons step down within the conduction band by emitting LO phonons. Then, when their energy has decreased below $\hbar\omega_{LO}$, further cooling of electrons, as well as that of hot holes, proceeds via scattering with acoustic phonons. This assumes that light holes and holes from the split-off spin-orbit band have relaxed to the heavy hole band via LO scattering. Figure 44.22A shows the high-energy tail of the hot-electron spontaneous band-to-band luminescence at 300 K. Spontaneous emission of photons from hot electrons, however, is rare since it requires finding an electron and a hole at the same position with the same k-vector—Section 44.3.1A.

The probability of photon emission increases dramatically if the hole is localized on an acceptor. With this localization, the hole wavefunction is spread out in k space to $1/a_{qH}$ for a hydrogen-like acceptor. The luminescence energy is given by

$$h\nu = \begin{cases} E_g + E_n(\mathbf{k}) - E_p(\mathbf{k}) & \text{for free-free transitions} \\ E_g + E_n(\mathbf{k}) - E_a & \text{for free-acceptor transitions.} \end{cases}$$

$$(44.26)$$

where $E_n(\mathbf{k})$ and $E_p(\mathbf{k})$ are the energies of the hot electron and hole within the respective bands, and E_a is the acceptor energy. This increase in luminescence is shown in Fig. 44.22B at 80 K, when most

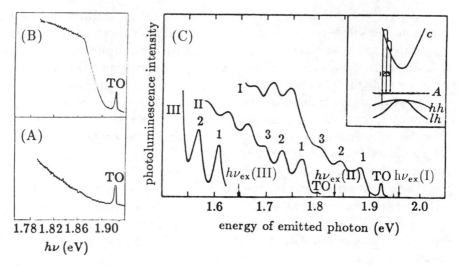

Figure 44.22: Hot free-electron luminescence spectra in GaAs: (A) electron recombination with h-holes at 300 K when all acceptors are ionized; (B) at 80 K when holes are trapped in acceptors A; the luminescence shows a steep edge 80 meV below $h\nu_{exc} = 1.96$ eV; acceptor density $N_a = 10^{18}$ cm^{-3} in (A) and (B); (C) at 30 K and $N_a = 10^{17}$ cm^{-3}, showing an oscillatory response; for three laser excitation energies, $h\nu_{exc}$ shown by arrows (I, II, and III). The Arabic numerals correspond to the transitions shown in the inset. The peak identified as TO is due to Raman scattering (after Mirlin et al., 1980).

holes are captured by the acceptor, to which the transition takes place.

Hopping-down from high-energy states in the conduction band in discrete LO-phonon steps is observed at 30 K and slightly reduced acceptor density—Fig. 44.22C. Here luminescence maxima occur between each of the scattering events; hence, the luminescence amplitude is modulated with maxima spaced by $\hbar\omega_{LO}$ below the monochromatic excitation. In Fig. 44.22C, a different excitation energy is used for each of the curves and is identified by arrows.

44.5.4 Far-Infrared Luminescence

Luminescence transitions from an excited to the ground state of a shallow donor lie in the far-infrared part of the spectrum. At low temperatures, the population of excited states can be achieved by impact ionization. Such luminescence has been observed in GaAs by Melngailis et al. (1969).

44.6 Luminophores

Materials that show highly efficient luminescence are called *lumino-phores*. Such materials include aluminates, oxides, phosphates, sulfates, selenides, and silicates of Ba, Ca, Cd, Mg, and Zn. They are of high technical interest for luminescent displays, such as TV screens, luminescent lamps, and paints. For reviews, see Curie (1963), Goldberg (1966), Landsberg (1967), and Cantow et al. (1981).

44.6.1 Luminescent Centers; Activators and Co-Activators

Doping of a semiconductor with certain elements can greatly enhance luminescence. Dopants with preferred radiative transitions are called *activators*. In contrast to the exciton and band-edge transitions, which were discussed in the previous sections, such activator-related luminescence is shifted from the band edge well into the band gap, where the optical absorption is reduced. These luminescent transitions have a very high transition probability, approaching 100% at room temperature.

An example of a good luminophore is ZnS, with a band gap at room temperature of 3.6 eV; with Cu as an activator, it yields a green luminescence peak at 2.3 eV. Other typical activators are Mn, Ag, Tl, Pb, Ce, Cr, Ti, Sb, Sn, and other transition metal ions except Fe, Ni, and Co—see Section 44.6.1C. The intensity of the luminescence can be obtained from the reaction kinetic considerations

$$\frac{dn}{dt} = g_o - n(c_{ca}p_a + c_{cr}p_r), \tag{44.27}$$

with $n = p_a + p_r$, and where $c_{ca}p_a$ is the luminescent and $c_{cr}p_r$ is the nonradiative transition. Here, g_o is the optical excitation rate; the other symbols are explained in Chapter 42. For steady-state conditions $(dn/dt \equiv 0)$, one obtains for the luminescent intensity I_L (proportional to the radiative recombination transition)

$$I_L \propto c_{ca}np_a = g_o - c_{cr}np_r \tag{44.28}$$

for reasons of detailed balance. Losses in luminescence efficiency relate to partial absorption, internal reflection, and to competing nonradiative recombination transitions. The luminescence intensity $(\propto c_{ca}np_a)$ increases also with increasing activator concentration.

Doping with a higher density of activators is usually restricted by their solubility, which is related to neutrality considerations—see Section 19.2.6. The neutrality is provided by intrinsic lattice defects, which are generated during the incorporation of the activator according to thermodynamic relations—see Section 19.2. Further *increased doping* can be achieved by the incorporation of atoms with an opposite charge character. In the example of ZnS doped with Cu, an oppositely charged defect is Cl, which results in ZnS:Cu,Cl. This additional dopant is called a *co-activator*.

44.6.1A Multiband Luminophores Heavily doped lumino-phores often show additional emission maxima. ZnS:Cu,Cl first shows a green emission. Increased densities of Cu and Cl result in a blue emission maximum. This is due to the fact that Cu can be incorporated at two different sites in the ZnS lattice, replacing one Zn atom as Cu_{Zn}, and as an interstitial Cu_i. The $Cu_{Zn}Cu_i$ associate is believed to be responsible for the blue emission (Riehl, 1958)—see Table 44.2.

Usually, one or the other of the transitions is preferred. A transition from one to the other can be achieved by changing the doping density, the intensity of the optical generation, or the temperature. A two-level activator model with successive filling with holes is able to account for the experimentally observed behavior—Fig. 44.23.

44.6.1B Sensitized Luminescence Here the light is absorbed by one type of center and the energy is then transferred to another center where luminescence occurs. Incorporation of the first *sensitizes* the luminescence at the second center, as it increases the absorption and avoids competing nonradiative recombination by carrier transport through the band to other recombination centers. An example is the $CaCO_3$:Mn phosphor used in fluorescent lamps, which can be sensitized by co-doping with Pb, Tl, or Ce, causing a substantial increase in light output and also changing the wavelength of the light at which the Mn can be excited. For the energy transfer to the Mn center, the sensitizer can be located beyond a next-neighbor site.

44.6.1C Quenching of Luminescence When holes are released into the valence band from electron-activated luminescent centers, the luminescence is reduced (*quenched*) since these holes can become trapped at other, competing centers, the *poisons* or *quenchers* with nonradiative transitions. Typical luminescent poisons are Fe, Ni, and Co.

The release of such holes can be caused by thermal ionization, resulting in thermal quenching of the blue luminescence of ZnS:Cu, which is thermally quenched already below room temperature, or by optical (IR) irradiation resulting in quenching also of the green

Table 44.2: Photoluminescence peaks, peak widths, and emission efficiencies of fluorescent lamp and ZnS phosphors (after Williams, 1963).

Phosphor	$h\nu_o$ (eV)	$h\Delta\nu$ (eV)	η	Comments
$(Ca,Zn)_3(PO_4)_2$:Tl	3.43	0.21	0.9	
$BaSi_2O_5$:Pb	3.56	0.22		
$Ca_3(PO_4)_2$:Ce	3.39	0.21	0.7	
$CaWO_4$	2.73	0.36	0.7	
$Ca_3(PO_4)_2$:Cu,Sn	2.56	0.22	0.7	
$Ca_3(PO_4)_2$:Cu	2.53	0.22		
$Ca_3(PO_4)_2 \cdot Ca(F,Cl)_2$:Sb,Mn	2.52	0.35	0.8-0.95	Evidence of additional Mn emission at small ν; relative intensities depend on Sb and Mn concentrations; ν_0 depends slightly on F/Cl composition
	2.08	0.14		
$MgWO_4$	2.48	0.36	0.9	
$3Sr_3(PO_4)_2 \cdot Ca(F,Ci)_2$:Sn,Mn	2.39	0.35		Relative intensities of Sb and Mn bands depend on Sb and Mn concentration
	2.18	0.13		
Zn_2SiO_4:Mn(willemite)	2.34	0.11	0.8	Slightly skewed to small ν $R = 0.42$ for overlapping Mn bands; R depends on Mn concentrations bands; R depends on Mn concentrations
$CaSiO_3$:Pb,Mn	3.46	0.41		
	2.15	0.13		
	1.48	0.13		
$(Sr,Mg)_2(PO_4)_2$:Sn	1.45	0.16		
$Cd_2B_2O_5$:Mn	1.45	0.14	0.7	

(excitation with $\lambda = 2537$Å)

Table 44.2: (con't.) Photoluminescence peaks, peak widths, and emission efficiencies of fluorescent lamp and ZnS phosphors (after Williams, 1963).

Phosphor	Band	$h\nu_o$ (eV)	$h\Delta\nu$ (eV)	Comments (all spectra at 300 K unless noted differently)
Hex,ZnS:Ag,Cl.	Silver blue	2.83	0.15	
Hex,ZnS:Cu,Cl.	Copper blue	2.74	0.19	Band skewed to small ν because of copper green
Hex,ZnS:I	Self-activated blue	2.71	0.23	
Hex,ZnS:Cu,I	Copper blue	2.71	0.20	Copper blue and self-activated blue resolvable at low temperature
Hex,ZnS	Self-activated blue	2.71	0.15	
Cub.ZnS:Ag,Cl.	Silver blue	2.73	0.18	
Cub.ZnS:Cu,Cl.	Copper blue	2.69	0.18	Band skewed to small ν because of copper green
Cub.ZnS	Self-activated blue	2.63	0.22	
Cub.ZnS:Al	Self-activated blue	2.58	0.23	
Hex,ZnS:Cu,Cl.	Copper green	2.37	0.15	
Cub.ZnS:Cu,Cl.	Copper green	2.32	0.15	
Hex.ZnS:Ag,In.	Silver red	1.93	0.28	Spectrum at 77 K
Cub.ZnS:Cu	Copper red	1.84	0.30	
Cub.ZnS:Cu	Copper red	1.72	0.25	Spectrum at 77 K
Hex.ZnS:Au,Ib.	Gold infrared	1.48	0.27	Spectrum at 77 K

luminescence. These transitions are similar to those that will be discussed in the section on photoconductivity—Chapter 45.

The intensity of luminescent transitions, competing with thermal quenching, can be described by—see Eq. (44.10)

$$I = I_0 \frac{I_0}{1 + \dfrac{\tau_{\mathrm{nr}}}{\tau_r}} = \frac{I_0}{1 + \gamma \exp\left(\dfrac{\Delta E_t}{kT}\right)}, \qquad (44.29)$$

where $\gamma = \tau_{\mathrm{nr0}} \exp[\Delta E/(kT)]/\tau_r$, and ΔE_t is the thermal activation energy for the quenching transition (William and Hall, 1978).

44.6.1D Concentration Quenching   With increasing concentration of activators, first an increase, and, at higher densities, a decrease of luminescence efficiency is observed. One reason for such

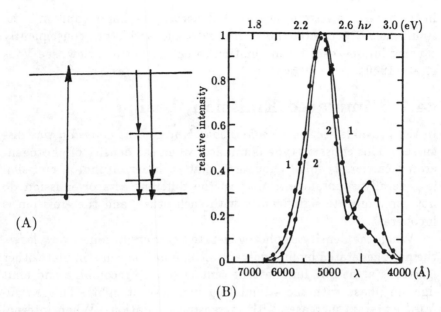

(A)

(B)

Figure 44.23: (A) Band model for ZnS:Cu luminophore with a two-level activator. (B) Blue and green luminescence in ZnS:Cu,Cl at 298 K and 77 K for curves 1 and 2, respectively (after Bowers and Melamed, 1955).

a concentration quenching is a carrier transfer from activator to activator, facilitated by their close proximity, until a poison center is encountered before luminescent recombination occurs.

44.6.1E Phosphorescence When deep traps are present, they may store optically excited or injected electrons for a substantial amount of time. These carriers are then slowly released by thermal excitation, and consequently recombine under radiative transitions. This phenomenon is called *phosphorescence*. Release of electrons from such traps can be enhanced by increasing the temperature.

The distinction between luminescence and phosphorescence is kinetic in nature; therefore, we will return to this subject in Section 47.3.1.

44.6.2 Photo-, Cathodo-, and Electroluminescence

The excitation of a luminophore can be achieved by light, electron bombardment, or sufficiently high applied fields; these excitations are referred to as *photo-, cathodo-, and electroluminescence*, respectively. Examples of these are luminescent light fixtures, the TV screen or CRT, and the light-emitting semiconductor diode, respectively.

In each case, electrons are excited across the band gap, and the simultaneously created holes move to the activator, consequently causing luminescent recombination to occur. For a review, see Voos et al. (1980).

44.7 Stimulated Emission, Lasing

In the previous sections, *spontaneous luminescent emission* was discussed. This emission type dominates when the density of photogenerated carriers is small enough so that each act of photon emission is generally spontaneous; that is, the different acts of emission do not communicate significantly with each other, and the emission is incoherent.

When the density of photogenerated carriers becomes very large, the light generated by luminescent recombination can stimulate other electrons at nearby luminescent centers to also recombine and emit light in phase with the stimulating luminescent light. This *stimulated emission* increases with increasing excitation. When internal reflection creates standing waves of the corresponding wavelength, and thereby identifies a well-defined phase relation, in-phase emission occurs, i.e., coherent emission. Lasing can finally be achieved with sufficient *pumping* to a critical intensity of the exciting light at this wavelength.

In a somewhat simplified fashion, such critical intensity can be deduced from the condition that the internal luminescent efficiency, given by Eq. (44.10): $\eta_L = 1/(1 + \tau_{sp}/\tau_{nr})$, must increase sufficiently to overcome the optical losses in the luminophore. The relaxation time for spontaneous luminescence τ_{sp} is usually much longer than that for nonradiative recombination, τ_{nr}; hence, η_L is usually $\ll 1$. When an *inverted population* of the excited and ground states of the luminescence center is produced, the luminescent lifetime decreases sharply from τ_{sp} to a stimulated emission lifetime τ_{st}; thus, η_L rapidly rises and lasing occurs. Such inverted population can be achieved when the excitation is strong enough so that there are more electrons in the excited than in the ground state.

With a luminescence generation per unit length of γ, one obtains a "generated" light flux, (usually $< I_0$) of

$$I_g = I_0 R_1 R_2 \exp(2\gamma d) \tag{44.30}$$

during one round trip within the semiconductor, including the reflection losses R_1 and R_2 at its two outer surfaces separated by a

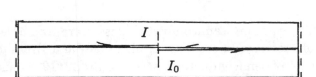

Figure 44.24: Semiconductor with semitransparent mirrors at opposing ends to create a cavity for light trapping.

distance d—Fig. 44.24) In addition, there is internal absorption proportional to $\exp(-\alpha_o x)$; therefore, during a round trip, only a fraction $\exp(-2\alpha_o d)$ of Eq. (44.30) arrives for a net flux of

$$I = I_0 R_1 R_2 \exp[2(\gamma - \alpha_o)d]. \qquad (44.31)$$

If this intensity becomes larger than I_0, light amplification in the semiconductor occurs, and lasing starts. As a threshold for lasing to occur, one obtains with $I = I_0$ from Eq. (44.31):

$$\gamma = \alpha_o + \frac{1}{2d} \ln \frac{1}{R_1 R_2}; \qquad (44.32)$$

i.e., the luminescent gain must be larger than the losses due to internal absorption and incomplete reflection at the outer surfaces.

Since γ and, to a lesser degree, α_o, R_1, and R_2 are functions of λ, this condition is fulfilled only for a narrow wavelength range. More stringent, within this wavelength range is the condition of standing waves, which identify a monochromatic emission. At the *lasing wavelength*, the intensity of this transition grows at the expense of all other competing luminescent transitions, resulting in a very intense emission in this wavelength range. For an *ideal* mirror surface, the phase condition for standing waves in the cavity forces a strictly monochromatic lasing.* For a more thorough treatment, see Loudon (1973) or Sargent et al. (1974).

The *pumping* of a laser, in order to achieve a sufficient population of excited states, can be performed by optical excitation. This was done first with a light flash exciting a ruby rod from its circumference (Maiman, 1960). However, carrier injection provides an efficient mode for pumping semiconductor lasers, originally carried out by Hall et al. (1962), Nathan et al. (1962), and Quist et al. (1962) in GaAs. Electron beam pumping was initially suggested by Hora

---

* A broadening occurs for lasing *pulses* due to the Heisenberg uncertainty relation, resulting in a natural line width of the lasing pulse.

(1965). Today, many semiconducting lasers are available; for short reviews and references, see Voos et al. (1980) and Bromberg (1988). For a review of semiconductor lasers, see Stern (1972) and Queisser (1973).

44.8 Luminescence in Superlattices

Luminescence from quantum wells or superlattices occurs from impurities, however, it also can significantly involve an intrinsic electron-hole recombination, whereas in bulk material one of the carriers is usually trapped at an impurity and recombines nonradiatively (Weisbuch et al., 1981).

The energy of the luminescent light is shifted beyond the band gap of the well material, and is determined by the eigenstates within the well, which in turn depend on the well width and well depth—see Eq. (9.50).

The luminescence is enhanced by the fact that the recombination from the states in the conduction to the valence band of the well are pseudo-direct transitions, even though the well material may have an indirect band gap (Gossard, 1983). In this sense, a narrow miniband can be regarded as a broadened energy state, i.e., localization in one dimension spreads the wave vectors about k_0, the satellite minimum, which can assist transitions to $k = 0$. The luminescence intensity is enhanced compared to the bulk material under comparable conditions due to confinement.

Such confinement also restricts the migration to nonradiative recombination centers, and concentrates electrons and holes in the wells rather than distributing them over the larger volume of a homogeneous bulk. In addition, the well material is more homogeneous than the bulk material due to better quality growth (Petroff et al., 1981), which improves the luminescence efficiency.

44.8.1 Influence of Magnetic Fields

The application of a microwave field, together with a magnetic field, heats free carriers in a mini-band when *cyclotron resonance conditions* are met (Romestain and Weisbuch, 1980). As a result, the line shape of the electron-hole luminescence, shown as the upper trace of Fig. 44.25A, is deformed in a characteristic manner: the luminescence near its undisturbed peak is reduced, and above it is increased, as shown in the differential trace of Fig. 44.25A,b. When the cyclotron orbits are larger than the well width, however,

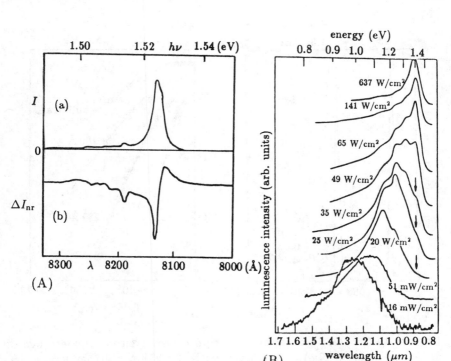

Figure 44.25: (A) (a) Luminescence (I) of GaAs/Al$_{0.33}$Ga$_{0.67}$As multi-quantum well at 1.8 K, and (b) change of it (ΔI_{nr}) with microwave-induced carrier heating showing a slight enhancement above $\hbar\omega_0$ and a steep reduction at the peak (after Cavenett and Pakulis, 1985). (B) Luminescence spectrum of an *n-i-p-i* GaAs doping superlattice with 700 Å lattice period ($l_n = l_p$) at room temperature, and doping with $N_a = N_d = 4 \cdot 10^{18}$ cm^{-3} in *p*- and *n*-type layers, respectively. Excitation intensity as family parameter (after Döhler et al., 1986). ©Pergamon Press plc.

such resonance heating is only observed when the magnetic field is nearly perpendicular to the superlattice plane, so that the orbit finds space within the well. This angle dependence directly confirms the confinement of the orbits within the plane of the superlattice (Cavenett and Pakulis, 1985).

44.8.2 Luminescence in Doping Superlattices

In *doping superlattices*, the wavelength of the luminescence peak can be *tuned* toward higher energy* with increased excitation intensity

---

* Here, the luminescent transition is assumed to proceed from the valley of the conduction band to an acceptor in the barrier range of the valence band.

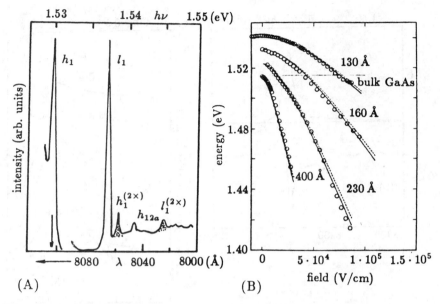

Figure 44.26: (A) Excitation spectrum of the exciton luminescence of an $Al_{0.35}Ga_{0.65}As/GaAs$ superlattice with 160/250 Å well/barrier widths, measured at 4.8 K with a field of 7.1 kV/cm. Detection wavelength indicated by arrow. (B) Stark shift of the h_1-peak as a function of the applied field for superlattices with various well widths. Barrier width is kept constant at 250 Å, the maximum Stark shift in bulk GaAs as indicated by the dotted line (after Viña et al., 1987.)

(see Section 45.6.4A), causing an increased carrier density and consequently a reduced barrier height. This increase in luminescence energy is limited when it approaches band-gap energy for a nearly flat-band conduction, i.e., vanishing barrier height. This is shown by Döhler et al. (1986) in Fig. 44.25B for a GaAs *n-i-p-i* doping superlattice.

44.8.3 Exciton Luminescence

As mentioned in Section 15.3, the confinement of excitons causes an increase in binding energy up to a factor of 4 for two-dimensional excitons (see Shinada and Sugano, 1966), and therefore renders excitons in superlattices observable at much higher temperatures than in bulk material. Even in high-quality MBE-grown undoped *bulk* GaAs (Miller and Kleinman, 1985), most of the luminescence stems from donor-acceptor pair and bound exciton emission and, in contrast, vanishes above ~ 20 K.

The exciton luminescence spectrum shows several distinct features related to heavy and light holes in the upper minibands of the valence band. Some of these features appear with applied uniaxial stress or an electric field, which permits otherwise forbidden transitions—Fig. 44.26A. The heavy and light hole excitons are labeled h_1 and l_1, respectively, with the first excited state by a superscript $(2\times)$ and the transition from the second heavy hole band by h_{12a}.

Positive identification can be made by polarization measurements. In identifying the possible absorption and emission transitions in GaAs, one recognizes that absorption with one sense of circular polarization σ^+, requiring $\Delta m_j = +1$, can proceed only from hole levels of $m_j = -3/2$ for the heavy hole and $m_j = -1/2$ for the light hole. However, emission that is not restricted in a polarization sense, i.e., with $\Delta m_j = \pm 1$, can proceed into all four hole states with m_j values of $\pm 3/2$ and $\pm 1/2$—see Fig. 44.27A. With circular polarized light at normal incidence to the superlattice layers, and a transition strength ratio of 1:3 between light and heavy holes, one obtains a strong in-phase polarization for the heavy hole excitation and a reversed polarization for the light hole exciton, as shown in Fig. 44.27B.

44.8.3A Stark Shift of Excitons With an applied electric field normal to the superlattice plane, one observes a Stark shift of the exciton which increases with increasing well width, i.e., with lower energy of the miniband, shifting it closer to the bulk band gap. The exciton becomes less tightly bound with increasing well width and hence becomes more polarized in the applied electric field.

In a superlattice the asymmetry of the eigenfunctions due to the electric field becomes more pronounced. This Stark shift is shown for the heavy hole exciton, in Fig. 44.26B, for a number of superlattices with different well widths. It is remarkable that a well width somewhat wider than the exciton diameter (300 Å) still provides sufficient confinement for a major Stark shift;[*] whereas such a shift is virtually nonexistent in bulk material (dotted line) where a Frenkel-Poole dissociation of the exciton takes place at fields of ~ 10 kV/cm before the Stark shift becomes marked—see Eq (42.24).

---

[*] Probably caused by a spatial distribution of *bound* excitons within the well, with the ones closer to the barrier layer still being confined.

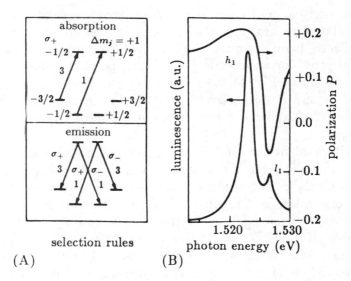

Figure 44.27: (A) Selection rules for absorption and luminescent emission from excitons formed from ground-state conduction electrons and heavy or light holes. (B) Luminescence spectrum and polarization for exciton luminescence excited with circular polarized light at 1.65 eV and 50 K in a GaAs/Al$_{0.3}$Ga$_{0.7}$As superlattice with 188 Å/19 Å well/barrier widths (after Miller and Kleinman, 1985).

The Stark shifts of the light and heavy hole excitons differ in magnitude. The light hole exciton shows a larger Stark shift: it has a larger diameter and therefore is more influenced by the electric field, and thus approaches the first excited heavy hole band—Fig. 44.28. Further increase in field strength permits a crossing of both peaks near 12 kV/cm. Here a strong mixing of these two valence bands is observed (see Section 9.3) with a significant increase in integrated intensity of the heavy hole peak at the intended crossing point, as shown in Fig. 44.28A. For further discussion on exciton binding and line shape, see Bastard (1985).

44.8.3B Stark Localization When thin minibands in a superlattice of appropriate widths of wells and barriers are tilted in an electric field to offset the bands sufficiently in adjacent wells, the electronic eigenfunction will no longer be delocalized. With increasing field, localization will extend over a few superlattice periods and will finally be reduced to a single period. This is illustrated in Fig. 44.30A for zero, medium, and high fields in subfigures (a)–(c), respectively.

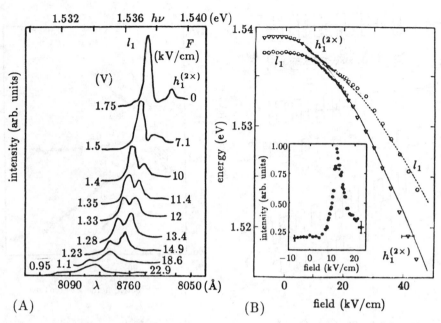

Figure 44.28: (A) Excitation spectrum of exciton luminescence showing anticrossing of l_1 and $h_1^{(2\times)}$ lines as a function of the applied field of an $Al_{0.35}Ga_{0.65}As/GaAs$ superlattice at 4.8 K (sample as in Fig. 44.26A). (B) Position of the l_1 and $h_1^{(2\times)}$ peaks as a function of the electric field. Inset shows the integrated intensity of the $h_1^{(2\times)}$ peak normalized to that of l_1 as a function of the field (after Viña et al., 1987).

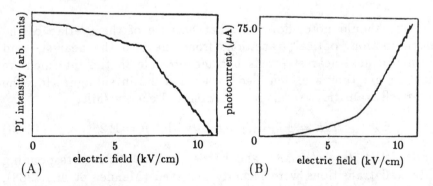

Figure 44.29: (A) Exciton luminescence in a $ZnSe/Zn_{0.55}Mn_{0.45}Se$ superlattice (36 periods 68 Å well, 115 Å barrier) at 10 K as a function of the applied field parallel to the lattice plane. (B) Corresponding photocurrent (after Das et al., 1987). ©IEEE.

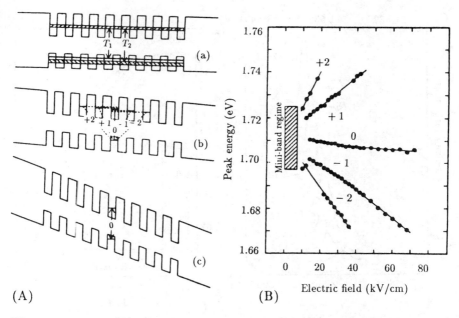

Figure 44.30: (A) Conduction and valence band profiles in GaAs/Ga$_{0.65}$Al$_{0.35}$As 30/35 Å superlattice with: (a) zero, (b) $2 \cdot 10^4$ V/cm, and (c) $1 \cdot 10^5$ V/cm applied. Transitions T_1 and T_2 from heavy and light hole bands, respectively. (B) Energies for heavy hole transitions as functions of the electric field measured from a photocurrent spectrum at 4.7 K. Numbers correspond to subfigure (A,b). The small shift of the original peak (0) is due to the exciton Stark shift discussed in Section 44.8.3A (after Mendez et al., 1988).

One should note, however, that because of the overlap of the eigenfunction, optical transitions from one into the nearest—and even the next-nearest—wells become possible in the intermediate field range; they are identified as +1 and +2 in subfigure b. The interwell transitions result in a pronounced energy shift

$$\Delta E = \pm \hat{n} e F (l_1 + l_2) \quad \text{with} \quad \hat{n} = 1, 2 \tag{44.33}$$

and with l_1, l_2 as the well and barrier widths. The corresponding interwell transitions were recently observed (Mendez et al., 1988), and are shown in Fig. 44.30B.

44.8.3C Field Quenching of Exciton Luminescence When the electric field is applied in the direction parallel to the plane of the superlattice, excitons can easily be dissociated by Frenkel-Poole ionization below 10 kV/cm. In this direction there is no confinement.

Consequently, the exciton luminescence decreases at the Frenkel-Poole threshold, while the photoconductivity increases at the same threshold, verifying the increasing density of set-free carriers as a result of the exciton dissociation—Fig. 44.29. The light hole exciton dissociates more easily because of its larger diameter.

44.8.3D Biexcitons and Localized Excitons   At high excitation rates, at which the density of excitons becomes large enough to interact and form biexcitons, a splitting of the h_1 exciton line is observed with a lower satellite about 1 meV below the h_1 exciton; this indicates a biexciton formation with a binding energy of ~ 1 meV. Above 10 K, the biexciton is no longer observed, as it is expected to dissociate for $k(T = 11 \text{ K}) = 1$ meV (Miller et al., 1982).

Heavy hole excitons bound to neutral donors have been discussed by Kleinman (1983). Excitons bound to acceptors have been observed in wells doped with 10^{17} cm^{-3} Be or Zn by Miller et al. (1982a). A doping density that is large enough to approach the exciton diameter (300 Å) with the dopant separation is needed to observe this exciton.

44.8.4 Exciton-Resonant Rayleigh Scattering

When the exciting laser energy approaches the exciton transition, a strong resonant Rayleigh scattering is observed in superlattices, with a scattering amplitude of more than 10^2 times the defect-scattering component (Hegarty et al., 1982). From the width and shape of the Rayleigh line, information on fluctuation of the well width and well composition can be obtained—see Section 15.3.

44.8.5 Superlattice Lasing

The increased quantum efficiency of luminescence in superlattices reduces the current threshold for lasing (Tsang, 1981). With changing well thickness, the laser wavelength can easily be tuned (Burnham et al., 1983). The high quality of the superlattice interfaces, especially for thin layers, is a contributing factor to the low current threshold for lasing. Another reason is the 2D confinement of the carriers, which reduces the density of states (see Section 9.3.3) within the miniband, as compared to a three-dimensional bulk lattice (Dingle and Henry, 1976).

Lasing at a current threshold as low as 250 A/cm^2 is observed in GaAs/Ga$_\xi$Al$_{1-\xi}$As superlattices (Tsang, 1981). Such lasing has also shown a reduced beam width normal to the superlattice plane, as

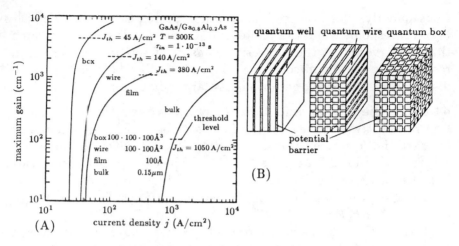

Figure 44.31: (A) Maximum gain as a function of the injection current density for the different quantum wells shown in (B) for GaAs/Ga.$_8$Al.$_2$As at 300 K, with lasing threshold indicated (after Asada et al., 1986).

well as a reduced temperature dependence. A review of quantum well lasers by Arakawa and Yariv (1986) evaluates the different contributions to the gain and other lasing properties.

44.8.5A Lasing in Quantum Wires, Quantum Boxes The density of states in one- and zero-dimensional quantum wells, often referred to as *quantum wires* or *quantum boxes* (Fig. 44.31B) become spike-like—see Fig. 9.34. Consequently, the threshold for lasing, which requires a population inversion in these states, becomes substantially lower: there are lesser states to be filled, as long as each quantum well is sufficiently separated from the other. The relationship of current density and gain is shown in Fig. 44.31A, with substantially reduced threshold current densities indicated for quantum wires and quantum boxes (Asada et al., 1986).

Summary and Emphasis

Radiative recombination can proceed in phase with the incoming radiation as reflection, or as an emission delayed by the lifetime of an excited state and changed in energy after relaxation of the excited state. Depending on the lifetime, one distinguishes luminescence and phosphorescence, the latter with a lifetime in excess of minutes.

When a phase relation to an intense excitation is established, coherent laser emission is stimulated as soon as the populations of excited and ground states become inverted.

Historically, thermal blackbody radiation has played a special role. Here, excitation is accomplished by phonons alone. Such radiation, however, has many similarities with the phonon-induced luminescence in the far IR emission range. At higher frequencies, the intrinsic coupling to a phonon distribution causes quantitative changes in the behavior, which makes thermal radiation distinctly different from luminescence.

The spectral distribution of the luminescence offers a wide variety of methods to analyze the electronic structure of the semiconductor and its defects.

Information from excitation or emission spectra is straightforward. The spectra of shallow-level defects in stress-free ultrapure crystals at low temperatures are extremely sharp and well understood. When lattice relaxation by phonon emission becomes involved, these spectra become broader and it is more difficult to unambiguously relate these to specific, deep-defect level centers.

Band-to-band excitation preceding spontaneous emission, the so-called *band-edge luminescence*, provides direct information concerning the carrier distribution near the edge. Through oscillatory luminescence of defect centers, information on preferred recombination from the band edge and an overlap of eigenfunctions in $E(k)$ can be obtained.

Convincing luminescence signatures are obtained from excitons and exciton-related particles, which assist in unambiguous identification. Electron-hole-drop and carrier plasma luminescence help to identify carrier phase transitions.

Most instructive is the analysis of the rich spectrum of defect center pairs and phonon replicas of many fundamental transitions, which identify the energies of the involved phonons.

Luminescence is of great technical importance for the production of a large variety of devices and appliances, ranging from luminescent lamps to television screens, and from luminescent diodes to luminescent paints. The field contains numerous challenging subjects, including the development of efficient luminophores of specific colors, fast and efficient luminescent diodes for fiber-optic communication, and efficient photoluminescent light concentrators of sufficient operating life for photovoltaic converters—to name just a few. New applications are constantly being discovered which abet the field, currently in a state of continuous aggressive research.

Exercise Problems

1. Verify Eq. (44.8) for thermal radiation. What assumptions were used?
   (a) Discuss in your own words s_{np} as a function of E_g and α_o.
   (b) What indicates the temperature dependence of s_{np}?

2.(*) How can the kinetic energy of excitons be measured? How can one create excitons with higher kinetic energy? Can excitons exist in indirect band-gap materials with holes near $k = 0$ and electrons near π/a? Discuss your reasoning.

3.(r) At low temperature (typically below 10 K), the near-band edge luminescence is highly structured and composed of several contributions.
   (a) Enumerate.
   (b) How do you distinguish between the different contributions?
   (c) Enumerate the different effects of line-broadening in luminescence.

4.(*) The most structured emission spectrum is that originating from recombination at donor/acceptor pairs. Usually, lines of up to $\hat{m} = 100$ can be distinguished.
   (a) Why not up to 200?
   (b) Why does one observe a broad emission *maximum* toward larger $\hat{m}$-numbers?

5.(r) Describe exciton luminescence and compare it with bound excitons, exciton molecules, and electron-hole liquid.

6.(r) Describe the difference of exciton luminescence in the bulk and in superlattices.

7.(*) For excitons and related quasi-particles:
   (a) Why are the spectral lines of free excitons, trions, and biexcitons broader than those of bound excitons?
   (b) How much broadening do you expect, and what is its temperature dependence?

8.(*) What special features do you obtain in the reflection spectrum due to spatial resonance dispersion of polaritons?

9.(*) Point out similarities and dissimilarities between thermal radiation and spontaneous emission (band-edge luminescence).

Focus your comparisons on the microscopic excitation and radiative recombination process.

10.(*) Can you investigate higher-band structures via luminescence? What are the limitations of this method?

11.(*) Describe the Stark effect and Stark localization in superlattices.

12.(r) What is the role of activators and co-activators in luminophores?

(a) What is the difference between a multiband and a sensitized luminophore?

(b) What is the role of quenchers in luminophores?

Chapter 45

Photoconductivity

Free carriers, generated by light and causing an increase in electrical conductivity, provide an important means of detecting light or measuring light-transmitted signals. This photoconductivity also permits a convenient analysis of carrier excitation and relaxation processes.

When the electric conductivity is *significantly* increased because of an increased *majority carrier density* as a result of an optical excitation, the resulting raised conductivity is called *photoconductivity*. With excitation from the valence to the conduction band, it is termed *intrinsic photoconductivity*; with excitation involving levels in the band gap, it is called *extrinsic photoconductivity*. Semiconductors that show strong photoconductivity (see below) are called *photoconductors*.

In this chapter, we will review first the different intrinsic and extrinsic carrier-generation processes. Then, we will give an overview of a reaction-kinetic analysis, which is the basis for a quantitative description of the change in carrier distribution due to optical excitation. Finally, we will provide a number of typical examples for the various processes of photon-induced changes in conductivity. For reviews see, Bube (1970, 1978), Mort and Pai (1976), and Rose (1978), as well as Volumes 5 and 12 of *Semiconductors and Semimetals* (R. K. Willardson and A. C. Beer, eds.; Academic Press, New York).

45.1 Carrier Generation

Band-to-band excitation was discussed in Section 13.1, in which most attention was given to an excitation near the fundamental edge. Photoconductivity with excitation beyond the band edge, i.e, extrinsic excitation, is rather homogeneous throughout the volume of the semi-

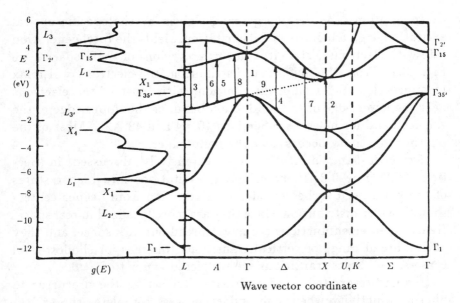

Figure 45.1: Band structure and density-of-state distribution of Si with competing direct electron transitions indicated at characteristic points (1, 2, and 3) along $E(k)$-ranges with parallel bands and comparatively high joint density of states (4 and 5), transitions into higher conduction bands (6 and 7), and from the light-hole valence band (8). The indirect band edge transition (9) is indicated by a dotted arrow.

conductor. At higher photon energies, however, the optical absorption becomes large and consequently the carrier generation occurs closer to, and often within the first $100\ldots 1000$ Å of the front surface. Most intrinsic transitions are direct (vertical) transitions which can proceed from any point in **k**-space with the appropriate photon energy—Fig. 45.1.

The generation rate is proportional to the impinging light flux ϕ_o (photon/cm² s) normal to a platelet surface (z), the square of the matrix element for each of the transitions at a given energy, and the joint density of states [neglecting reflection—it is included in Eq. (45.77)]. When expressing these in terms of the optical absorption coefficient, one has

$$g_o(\lambda)d\lambda = \phi_o(\lambda)\alpha_o(\lambda)\exp(-\alpha_o z)d\lambda \qquad (45.1)$$

$$\text{with}\qquad \alpha_o(\lambda) = \frac{e^2\lambda}{6\varepsilon_0 n_r c^2 m_0^2 \mathcal{V}}\,|M_{\mathrm{fi}}|^2 J_{c,v}. \qquad (45.2)$$

Integration over the energy range, given by the specific optical excitation, e.g., a band-to-band excitation, yields the total generation rate with electrons distributed in certain regions of the **k** space—see Fig. 13.7. From this deformed distribution, the electrons will relax to approach a Boltzmann distribution, with most of the electrons near the lower edge of the conduction band, or the upper edge of the valence band for holes—see Sections 45.5.4 and 48.2.2C. Most of the photoconductivity occurs near the band edges.

Defect-to-band excitation was extensively discussed in Sections 20.2 and 20.3.4B; here, again, excitations into higher states of the band need to be considered. Excitation from deeper centers shows a wider distribution, since they are more localized in real space. Hence, their eigenfunctions are more spread out in **k** space, and they offer a broad range of vertical transitions; whereas for shallow-level centers, the direct transitions are restricted closer to $k = 0$.

Ground-to-excited state excitation can act as the precursor to photoconductivity when a secondary process completes the excitation into the band. Such secondary processes can be

- tunneling,
- impact ionization,
- thermal ionization, or
- an excitation by additional photons.

Finally, an excitation of free carriers, from states near the band edge to higher states, influences the photoconductivity by changing the effective mass and relaxation time. Devices based on this effect are referred to as *free-electron bolometers*.

45.1.1 Photo-Ionization Cross Section

The excitation of electrons from deep-level centers into the conduction band can be described as transitions into resonant states. Such transitions are possible when the excited state contains Bloch functions of the same k value as the ground state—conservation of momentum. The cross section of localized deep centers extends over a large range in k-space; it therefore extends to a wide energy range of the band continuum. This cross section can be approximated (see Section 20.3.4A) by:

$$s_o(h\nu) \propto \frac{h\nu - |E_i|}{(h\nu)^3} g(h\nu - |E_i|), \tag{45.3}$$

Table 45.1: Oscillator strength (f_{ba}) for transitions of a hydrogen atom (after Bethe and Salpeter, 1957).

$1s \rightarrow 2p$	0.4162	$1s \rightarrow 5p$	0.0139
$1s \rightarrow 3p$	0.0791	$1s \rightarrow 6p$	0.0078
$1s \rightarrow 4p$	0.0290	$1s \rightarrow$ contin.	0.436

where E_i is the ionization energy, and $g(h\nu - |E_i|)$ is the density of states, which, for a simple parabolic band near its edge, can be expressed as

$$g(h\nu - |E_i|) \propto \sqrt{h\nu - |E_i|}. \qquad (45.4)$$

The cross section can be calculated when the deep center potential is known. When it is approximated by a square well function (Lucovsky, 1965), one obtains

$$s_o(h\nu) = \frac{1}{n_r} \left(\frac{F_{\text{eff}}}{F} \right)^2 \frac{32\pi^2 e^2 \hbar \sqrt{E_i}(h\nu - |E_i|)^{3/2}}{3m^* c (h\nu)^3}; \qquad (45.5)$$

where F_{eff}/F is the ratio of the effective local field [Eq. (20.24)] to the electromagnetic radiation field, which takes into consideration the local screening by charges near the defect center. This ratio is typically on the order of 1 for shallow levels; s_o is on the order of 10^{-16} cm^2 near the band edge. Further from the band edge, the expression becomes more complicated, causing a rapid reduction of the cross section (Jaros, 1977). Equation (45.5) describes rather well the IR absorption of GaAs:Au or GaAs:Ag (Queisser, 1971).

For excitation into an excited state, one obtains for the cross section

$$s_o(h\nu) = \frac{1}{n_r} \left(\frac{F_{\text{eff}}}{F} \right)^2 \frac{2\pi^2 e^2 \hbar}{m^* e} f_{ba} \simeq 1.1 \cdot 10^{-16} \frac{1}{n_r} \left(\frac{F_{\text{eff}}}{F} \right)^2 \frac{m_0}{m^*} f_{ba} \qquad (45.6)$$

with f_{ba} as the oscillator strength that is given for a hydrogen atom in Table 45.1 and applies to a quasi-hydrogen defect.

45.1.1A Photo-Ionization Cross Section Measurement
The relative cross section of different impurities can be easily obtained by a constant-photoconductivity method developed by Grimmeiss and Ledebo (1975).

One can show from detailed balance (Section 42.2.3) that the photo-ionization cross section is inversely proportional to the inten-

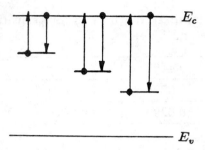

Figure 45.2: Photo-ionization from three independent centers into the conduction band.

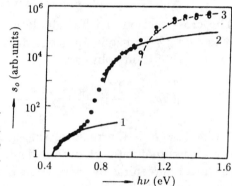

Figure 45.3: Relative photo-ionization cross section for excitation of three deep impurity centers into the conduction band of GaAs (after Grimmeiss and Ledebo, 1975).

sity of light $(I(h\nu)_\sigma)$ required to produce, with light of different photon energies, *the same* photoconductivity

$$s_o(h\nu) = \frac{\text{const}}{I(h\nu)_\sigma}. \tag{45.7}$$

This assumes that, with changing photon energy, photo-ionization occurs from different deep defect levels into the same band, and results in carriers with the same mobility. One must also assume that these levels do not communicate with other levels or bands; this is indicated in Fig. 45.2.

An example of results obtained from this method is given in Fig. 45.3, which shows the photo-excitation from three deep impurity levels into the conduction band of GaAs:O. The deepest level is at an energy of 1.03 eV below the conduction band edge, with a cross section almost 10^5 times larger than the shallowest level at 0.46 eV below E_c. The middle level, with a cross section close to the deepest level, lies at 0.79 eV below E_c.

Careful checking as to whether all of the assumptions used are fulfilled is imperative. The *photo-Hall effect** can be used to detect competing transitions involving the opposite carrier.

45.2 Reaction-Kinetics Evaluation

The density of carriers in the respective bands and in levels in the band gap is changed from thermal equilibrium with optical excitation. These densities can be obtained from reaction-kinetic arguments.

First, the carrier density in the respective bands is discussed for a number of rather simple reaction-kinetic models in order to identify the specific influence of certain defect levels.

45.2.1 Intrinsic Photoconductivity

The intrinsic photoconductivity involves only electrons and holes in the conduction and valence bands. They are generated at the same rate g_o, and recombine directly with each other (*intrinsic recombination*—Fig. 45.4A). The incremental carrier densities are equal. The change in carrier densities is given by the difference between generation (g) and recombination (r) rates:

$$\frac{dn}{dt} = g - r = g_o - c_{cv} n p = \frac{dp}{dt};\qquad(45.8)$$

for parameter identification in this chapter, see the introduction of Chapter 42. Equation (45.8) is written for high generation rates when the incremental carrier densities Δn and Δp are large compared to the thermally generated densities n_0 and p_0:

$$\Delta n = n - n_0 \simeq n \quad \text{and} \quad \Delta p = p - p_0 \simeq p.\qquad(45.9)$$

In steady state, with $dn/dt = dp/dt \equiv 0$, one obtains with $n = p$, from Eq. (45.8)

$$n = \sqrt{\frac{g_o}{c_{cv}}},\qquad(45.10)$$

which is usually referred to as the *bimolecular recombination* relation. Intrinsic recombination occurs at high generation rates when other

---

* That is, the Hall effect (Section 31.2.2A) measured for photogenerated majority carriers.

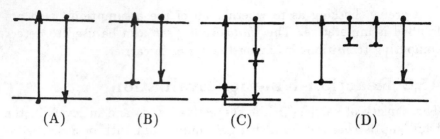

Figure 45.4: (A) Intrinsic photoconductivity; (B) extrinsic photoconductivity involving only one defect center, (C) involving a different recombination center, and (D) involving traps.

recombination paths are saturated. The intrinsic photoconductivity is *ambipolar* and is given by

$$\sigma_a = e(n\mu_n + p\mu_p) = en(\mu_n + \mu_p). \qquad (45.11)$$

45.2.2 Extrinsic Photoconductivity

Extrinsic photoconductivity involves levels in the band gap. It is either *n*- or *p*-type. A deep defect center from which carrier generation occurs is called an *activator*. One can distinguish three types of extrinsic photoconductivities involving such activators:

- excitation from an activator with direct recombination into the same type of level—Fig. 45.4B;
- excitation from an activator with carrier recombination through another level—Fig. 45.4C; and
- excitation into a band from which major trapping occurs before recombination—Fig. 45.4D.

When generation and direct recombination involve the same type of activator, and this activator is separated far enough from the valence band (Fig. 45.4) so that thermal ionization of the optically generated hole can be neglected, the carrier density follows the same bimolecular relationship as that for intrinsic carrier generation:

$$\frac{dn}{dt} = g_o - c_{ca}np_a = \frac{dp_a}{dt}, \qquad (45.12)$$

where the density of holes in activators is represented by p_a and the quasi-neutrality condition by

$$p_a = n. \qquad (45.13)$$

In the steady state, one obtains

$$n = \sqrt{\frac{g_o}{c_{ca}}};$$ (45.14)

reflecting *monomolecular recombination*. The photoconductivity is n-type with

$$\sigma_n = e n \mu_n.$$ (45.15)

At high intensities, depletion of these activators causes saturation of the photoconductivity induced by high-intensity lasers, which were pulsed in order to avoid thermal destruction of the material. This saturation has been observed, e.g., by Bube and Ho (1966) and Celler et al. (1975).

The following sections will analyze the two other types of extrinsic photoconductivity shown in more detail in Fig. 45.4C and D.

45.2.2A Influence of Traps on Photoconductivity   When photogenerated carriers are trapped intermittently before they recombine, the carrier balance is shifted and three *balance equations* must be considered:

$$\frac{dn}{dt} = g_o - c_{ct} n (N_t - n_t) + e_{tc} N_c n_t - c_{ca} n p_a,$$ (45.16)

$$\frac{dn_t}{dt} = c_{ct} n (N_t - n_t) - e_{tc} N_c n_t,$$ (45.17)

$$\frac{dp_a}{dt} = g_o - c_{ca} n p_a,$$ (45.18)

plus the *quasi-neutrality condition*:

$$n + n_t = p_a.$$ (45.19)

In steady state, all time derivatives vanish. As a consequence, the two terms related to trapping drop out of Eq. (45.16); hence, it becomes identical to the balance equation that was discussed in the previous section—Eq. (45.12):

$$g_o = c_{ca} n p_a.$$ (45.20)

The influence of the traps enters through the neutrality condition, yielding

$$g_o = c_{ca} n (n + n_t).$$ (45.21)

Deep Traps.   Deep traps tend to be completely filled, i.e.,

$$n_t \simeq N_t.$$ (45.22)

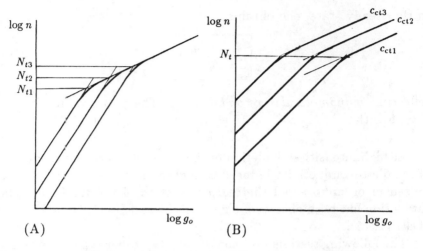

Figure 45.5: Electron density as a function of the optical generation rate with one deep trap level: (A) trap density as the family parameter; and (B) recombination coefficient as the family parameter (with $c_{ct1} > c_{ct2} > c_{ct3}$).

Introducing Eq. (45.22) into Eq. (45.21), one obtains for the electron density

$$n = \frac{1}{2}\left(-N_t + \sqrt{N_t^2 + 4\frac{g_o}{c_{ca}}} \right);$$ (45.23)

for small optical generation rates, i.e., for $g_o \ll N_t^2 c_{ca}/4$. Here, Eq. (45.23) reduces to

$$\boxed{n = \frac{g_o}{c_{ca}N_t},}$$ (45.24)

which is referred to as a *monomolecular recombination relation*, and indicates that each electron finds a constant density of available recombination sites with $p_a \simeq N_t$. For high optical generation rates, Eq. (45.23) converts back to Eq. (45.14); p_a increases and becomes $\gg N_t$, and therefore $n \simeq p_a$. Figure 45.5 shows the dependence of n on g_o, with a break between the linear and square-root branches at $n \simeq N_t$, permitting a determination of the trap density in a photoconductor with one dominating deep trap level.

Shallow Traps.   When the interacting electron trap is shallow, its degree of filling is incomplete; thus Eq. (45.22) must be replaced by the equation obtained from Eq. (45.17) in steady state:

$$n_t = \frac{c_{ct} n N_t}{c_{ct} n + e_{tc} N_c},$$ (45.25)

which yields Eq. (45.22) for high optical generation rates ($c_{ct}n \gg e_{tc}N_c$), and decreases proportionally to n for lower excitation rates:

$$n_t \simeq n \frac{c_{ct}}{e_{tc}} \frac{N_t}{N_c}. \tag{45.26}$$

The transition rates c_{ct} and e_{tc} can be replaced from detailed balance arguments [Eq. (42.10)], yielding

$$n_t = n \frac{N_t}{N_c} \exp\left(\frac{E_c - E_t}{kT}\right), \tag{45.27}$$

i.e., an exponential increase of electrons in traps with trap depth until saturation occurs. Then, Eq. (45.27) becomes identical to Eq. (45.22).

By introducing Eq. (45.27) into Eq. (45.21), one obtains, in the range of incomplete trap-filling, a quasi-bimolecular relationship

$$g_o = c_{ca}n^2(1 + \eta_t^*), \tag{45.28}$$

with $\eta_t^* = n_t/n$ as an effective *trap availability factor*

$$\eta_t^* = \frac{N_t}{N_c} \exp\left(\frac{E_c - E_t}{kT}\right); \tag{45.29}$$

thus, for higher generation rates, one obtains

$$n = \sqrt{\frac{g_o}{c_{ca}(1 + \eta_t^*)}}. \tag{45.30}$$

Since $\eta_t^* > 0$, one can see that the introduction of traps usually leads to a reduced photoconductivity:* part of the otherwise photoelectrically active electron population is stored in localized states, while the holes remain available for increased recombination.

Trap Distribution. In many photoconductors, more than one trap level actively influences the photoconductivity. The successive filling of these traps with increasing optical generation causes a successive spread of the two quasi-Fermi levels, and provides increasing storage capacity for carriers. This causes a successively increased re-

---

* Except when carrier excitation takes place from filled trap levels.

combination rate, since p_a increases with increasing n_t. The carrier density can be written similarly to Eq. (45.30) as

$$n = \sqrt{\frac{g_o}{c_{ca}\left[1 + \eta_t(g_o)\right]}},\tag{45.31}$$

where $\eta_t(g_o)$ is a monotonically increasing function indicating the increasing availability of traps:

$$\eta_t(g_o) = \frac{kT}{E_c - E_{Fn}} \int_{E_F}^{E_{Fn}} \frac{N_t(E)}{N_c} \exp\frac{E_c - E}{kT} dE.\tag{45.32}$$

The degree of filling can be described by the position of the quasi-Fermi level, which is given by the electron density according to

$$E_{Fn} = E_c - kT \ln\frac{N_c}{n(g_o)},\tag{45.33}$$

rendering Eq. (45.31) an implicit equation. The trap distribution can be reconstructed from the slope of $\eta_t(g_o)$ or $n(g_o)$. When a high-density trap level is being filled with increasing optical excitation, the quasi-Fermi level shifts by only a small amount; thus, $\eta_t(g_o)$ changes little, and the slope of n remains close to the square-root slope. When this level is filled, E_{Fn} moves rapidly up with increasing g_o, until the next level closer to the conduction band starts to become filled. This causes the slope of $n(g_o)$ first to increase, until it decreases again; E_{Fn} remains near the next-higher level when it is being filled, causing another square-root branch. Figure 45.6 illustrates qualitatively this behavior for a distribution with two trap levels.

In actual photoconductors, the trap-level distribution is less structured, and thus the $n(g_o)$ behavior has less pronounced steps. In addition, other interactions with different recombination centers interfere; this makes $n(g_o)$ an unattractive method for determining $N_t(E)$.

45.2.2B Recombination Centers in Photoconductors

Recombination of photogenerated carriers proceeds mostly via recombination centers. The activators from which the conduction electrons are optically generated can act as such centers. Their effectiveness as recombination centers depends on their capture cross section, which is affected by the charge character before and after photo-ionization. Such recombination can be very strong, e.g., into Coulomb-attractive centers, resulting in a low carrier lifetime; or weak, e.g., into Coulomb-repulsive centers, or centers with strong

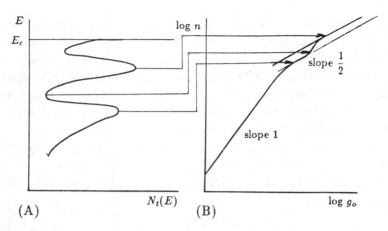

Figure 45.6: (A) Trap distribution with two trap levels of high density. (B) Carrier density as a function of the optical generation rate with linear and square-root branches and indication of the transition from the higher to the lower square-root branch when the quasi-Fermi level, with continuing trap depletion, moves through the minimum in $N_t(E)$.

lattice coupling, resulting in a long lifetime. A photoconductor with a long carrier lifetime is called a *sensitive photoconductor*.

An n-type photoconductor with doubly negatively charged activators, which become singly negatively charged when photo-ionized, is an example of a sensitive photoconductor in which a small intensity of light causes a large change in carrier densities. Such repulsive recombination centers have a very small cross section of 10^{-22} cm^2, as observed in sensitized CdS, resulting in $c_{cr} = s_{nr} v_{rms} \simeq 10^{-15}$ cm^{3}s^{-1}. With a density of recombination centers of $N_r = 10^{17}$ cm^{-3}, and for a generation rate* of 10^{18} cm^{-3}s^{-1}, one obtains from Eq. (45.24) an increase in carrier densities of

$$n = \sqrt{\frac{g_o}{c_{ca} N_a}} \simeq 10^{16} \text{ cm}^{-3}. \tag{45.34}$$

---

* For an extrinsic photoconductor with an absorption constant of 10^2 cm^{-1}, this is equivalent to a photon flux of 10^{16} cm^{-2}s^{-1}. For photons with an energy of ~ 1 eV, this is also equivalent to an illumination of 1 mW/cm^2, i.e., about 1% of solar radiation of 100 mW/cm^2; this example should be interpreted to obtain a rough estimate only, since monochromatic light is inappropriately compared here with polychromatic sunlight.

This increment is much larger than the carrier concentration of $< 10^{10}$ cm^{-3} in the dark for a wide gap material. Such a large ratio of carrier densities is typical of a *sensitive* photoconductor.

A photoconductor with neutral activators that becomes positively charged when ionized is an example of an *insensitive* photoconductor. A large recombination cross section prevents the buildup of a substantial density of photogenerated electrons.*

Insensitive photoconductors with a recombination cross section of $s_{nr} = 10^{-13}$ cm^2, and the same density of recombination centers of $N_r = 10^{17}$ cm^{-3} at a generation rate of 10^{18} cm^{-3}s^{-1}, yields an increment of

$$\Delta n \simeq 10^7 \text{ cm}^{-3}, \qquad (45.35)$$

which is usually only a small fraction of the already present carrier concentration in the dark.

Sensitive photoconductors typically have a carrier lifetime

$$\tau_n = \frac{1}{c_{cr} N_r} \qquad (45.36)$$

of $10^{-5} \ldots 10^{-2}$ s, whereas the carrier lifetime of insensitive photoconductors is on the order of $10^{-12} \ldots 10^{-8}$ s.

Thermal Ionization of Activators.   When the activator lies close enough to the valence band, the hole created by light in the activator can be thermally emitted into the valence band, from where it is redistributed with a shift of its population toward Coulomb-attractive hole traps—Fig. 45.7.

This redistribution causes a shift of the recombination of electrons from the activator to the center, which has the largest density times cross section product. This process has the tendency to decrease the electron recombination in sensitized photoconductors, and thus to increase the sensitivity, since it preferentially selects defects for hole storage that are more negatively charged than the activator. These centers then become the predominant recombination centers.

The carrier lifetime is determined by competition of recombination through the different centers—Fig. 45.7. Labeling the different recombination centers with an index f (fast) or s (slow) to indicate

---

* This recharging of recombination centers is given as a simple example of the sensitizing or desensitizing of a photoconductor. In actuality, deep centers with different degrees of lattice coupling must be considered.

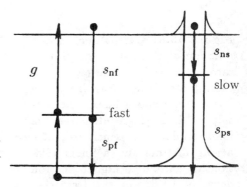

Figure 45.7: Competing recombination between slow and fast recombination centers.

their speed of electron capture, one has, neglecting electron traps in this example:

$$\frac{dn}{dt} = g_o - c_{crs} n p_{rs} - c_{crf} n p_{rf} = 0, \tag{45.37}$$

$$\frac{dp_{rs}}{dt} = -c_{crs} n p_{rs} + c_{rsv}(N_{rs} - p_{rs})p = 0, \tag{45.38}$$

$$\frac{dp_{rf}}{dt} = g_o - c_{crf} n p_{rf} + c_{rfv}(N_{rf} - p_{rf})p - e_{vrf} N_v p_{rf} = 0, \tag{45.39}$$

$$\frac{dp}{dt} = -c_{rfv} n_{rf} p + e_{vrf} N_v p_{rf} + e_{vrs} N_v p_{rs} = 0, \tag{45.40}$$

with the quasi-neutrality condition

$$n = p_{rf} + p_{rs} + p. \tag{45.41}$$

For a moderate generation rate, one obtains

$$n = g_o \tau_\infty \left(1 + \frac{N_{rf}}{p_v} \frac{c_{crs}}{c_{crs} + c_{rsv}}\right) \quad \text{and} \quad p = \frac{n}{1 + \dfrac{N_{rf}}{p_v}}, \tag{45.42}$$

with

$$p_v = N_v \exp\left(-\frac{E_{rf} - E_v}{kT}\right) \quad \text{and} \quad \tau_\infty = \frac{1}{N_{rs}}\left(\frac{1}{c_{crs}} + \frac{1}{c_{rsv}}\right). \tag{45.43}$$

For a higher generation rate, one can approximate

$$n = g_o^2 \frac{1}{p_v c_{rsv}^2 N_{rs}^2} \frac{c_{crs} + c_{rsv}}{c_{crs}} \quad \text{and} \quad p = \frac{g_o}{c_{rsv} N_{rs}}, \tag{45.44}$$

and for very high generation rates, one obtains

$$n = p = g_o \tau_\infty. \tag{45.45}$$

Figure 45.8: Carrier density as a function of the optical generation rate. At low intensities the photoconductor is not sensitized, so a low lifetime and a linear relationship hold. At high intensities the photoconductor is sensitized; here a long lifetime and a linear relationship shifted to higher electron densities hold. In the transition range the carrier density increases superlinearly with g_o.

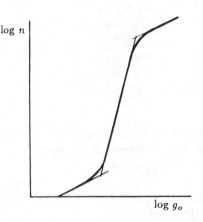

Equation (45.44) presents an intermediate range in which the carrier density increases *superlinearly* with the generation rate between two linear branches—Eqs. (45.42) and (45.45), as shown in Fig. 45.8; see also Section 45.2.2B. When traps are included, an additional sublinear branch at high intensities will appear.

The density of holes in slow centers controls the transition between the different ranges, with p_{rs} increasing for moderate generation rates related to p_v as

$$p_{rs} \simeq \frac{N_{rs}}{\dfrac{N_{rf}}{c_{rsv}p_v} + \dfrac{c_{cvs} + c_{rsv}}{c_{crs}c_{rsv}}}, \qquad (45.46)$$

and for higher generation rates related to p_v and n as

$$p_{rs} \simeq c_{rsv} N_{rs} \sqrt{\frac{p_v}{c_{crs}(c_{crs} + c_{rsv})n}}; \qquad (45.47)$$

for very high excitation levels, p_{rs} saturates at

$$p_{rs} \simeq \frac{c_{rsv} N_{rs}}{c_{rsv} + c_{crs}}. \qquad (45.48)$$

Sensitization of Photoconductors. Sensitization is accomplished by doping an n-type photoconductor with centers that have a large capture cross section for holes but, even after hole capture, only a very small cross section for electrons. Such centers, for example, could be doubly negatively charged defects or negatively charged tightly bound defects with large lattice coupling, requiring a large activation energy for electron capture—see Sections 43.1.1E and 34.4.

Doping with such centers causes an increase in the photogenerated carrier lifetime: it *sensitizes* the photoconductor.

An example of such sensitization is the doping of CdS with Cu. Copper on a lattice site, Cu_{Cd}, acts as a Coulomb-attractive center for holes and as a Coulomb-repulsive, slow recombination center for electrons. It produces CdS with an electron lifetime typically on the order of $10^{-4} \ldots 10^{-3}$ s.

Superlinearity of Photoconductivity. The sensitization obtained by doping with a higher density of sensitizing centers can also be obtained electronically if such centers are already present but their sensitizing action is yet inactive. This can best be seen by following the competing interaction of slow and fast recombination paths. With little optical excitation, only the fast path is open: every electron finds its hole at the fast recombination center.

With increasing excitation, the quasi-Fermi level for holes moves closer to the valence band; as the holes from the fast recombination centers are removed by thermal ionization, more and more holes are stored at slow centers. An equal density of electrons now has an increased lifetime, causing the average lifetime to increase with an increasing generation rate. This causes a transition range in which the current increases faster than linearly with an increasing generation rate, as is shown in Fig. 45.8. Such a superlinear range is observed for sufficiently sensitized photoconductors. It shifts to higher generation rates for lower temperatures, since thermal ionization becomes more difficult and must be counteracted by a higher degree of filling. Thus, the quasi-Fermi level for holes is shifted closer to the valence band. Such a shift is shown in Fig. 45.9 for sensitized CdS.

45.3 Photosensitivity and the Gain Factor

When photoconductors are used as sensitive devices to detect small amounts of light, the introduction of *photosensitivity* is sometimes instructive. It describes the changes of an electrical current in relation to the optical generation rate.

In the previous sections, such changes were discussed in terms of the majority carrier density. In n-type homogeneous photoconductors, one obtains for the incremental photocurrent

$$\Delta j_n = e g_o \mu_n \tau_n \frac{V}{d}, \qquad (45.49)$$

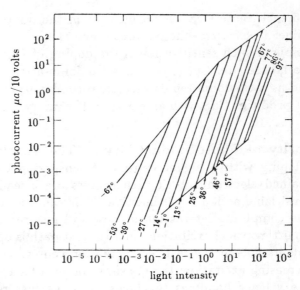

photocurrent $\mu a/10$ volts

light intensity

Figure 45.9: Shift of the superlinear range to higher optical generation rates with decreasing temperatures (in °C) as the family parameter (after Bube, 1957). ©Pergamon Press plc.

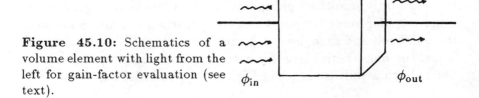

Figure 45.10: Schematics of a volume element with light from the left for gain-factor evaluation (see text).

ϕ_{in} ϕ_{out}

where V is the bias applied across the distance d between the electrodes, assuming ohmic contacts. Equation (45.49) shows the importance of a high $\mu\tau$-product to distinguish between sensitive and insensitive photoconductors: the increase in current Δj_n is directly proportional to the increase in carrier density

$$\Delta n = g_o \tau_n \qquad (45.50)$$

and to the mobility.

Another description is sometimes used to indicate similarities between a photoconductor and a vacuum photodiode or a photomultiplier, where each photon produces electrons according to the quantum efficiency and, in the latter, to a gain. In a photoconductor, the flux of electrons (per $cm^{-2}s^{-1}$) traversing the distance from cathode to anode is given by $nv_{Dr} = n\mu_n F$. The flux of photons (per $cm^{-2}s^{-1}$) absorbed in the volume of the photoconductor is given by $\Delta\phi = \phi_{in} - \phi_{out}$—see Fig. 45.10; hence, the *gain of a photoconductor* can be defined as

$$G_o = \frac{n\mu_n F}{\Delta\phi}. \tag{45.51}$$

Replacing n with $\Delta\phi\tau_n/d$, assuming homogeneous carrier generation [see Eq. (42.19)], and F with V/d, we obtain

$$G_o = \mu_n\tau_n\frac{V}{d^2}, \tag{45.52}$$

which again shows the importance of the $\mu\tau$-product for sensitive photoconductors. However, the influence of V/d^2, indicating a higher gain with increased V and decreased d, is intuitively unsatisfactory.

A better way to describe the gain involves the transit time of electrons from cathode to anode, $t_n = d/v_{Dr} = d^2/(\mu_n V)$; introducing t_n into Eq. (45.52), we obtain

$$\boxed{G_o = \frac{\tau_n}{t_n}.} \tag{45.53}$$

This relation can be readily interpreted as *chains of electrons* traversing the photoconductor, with each chain starting as an act of generation and terminating after τ_n with the act of recombination, i.e., assuming unhindered replenishment of electrons from an ohmic cathode. One can also regard τ_n as the lifetime of the immobile trapped hole, which is simultaneously created during the generation of an electron. Thus, after the initial electron has traveled to the anode, a new electron is pulled in from the cathode in order to maintain quasineutrality. This process continues until one of these electrons recombines with the immobile hole, thereby ending the chain of traversing electrons. This means that the gain factor compares the distance a photogenerated carrier can travel in a similar but longer photoconductor in the acting field before it recombines.

$$l_n = \mu_n F\tau_n, \tag{45.54}$$

with the distance d between the electrodes of the actual device:

$$\boxed{G_o = \frac{l_n}{d}.} \tag{45.55}$$

The length l_n is the *drift length*, and is an important parameter for various devices.

The use of the gain factor is limited, as is the similarity to the vacuum photodetector (diode or multiplier); the conduction mechanisms are different. A more useful quantity is the *photodetectivity*, which identifies the smallest light signal that can be detected with a photoconductor against the background of dark semiconductivity and its electronic noise. This will be discussed in the following sections.

45.3.1 Photodetector Figures of Merit

Photoconductors are used to detect light signals. Most of these require amplification. For this purpose the light signal is usually chopped, resulting in an ac-signal at chopping frequency ν_s, which in turn is amplified within a frequency band width $\Delta\nu$. The resulting power output is often referred to as root-mean-square amplitude $(=\sqrt{2^3}$ of the peak-to-peak amplitude of the fundamental component).

For comparison between different photodetectors, certain figures of merit are used, the most important of which are the responsivity, the detectivity, and the frequency response or time constant.

45.3.1A Responsivity   The responsivity $\mathcal{R}$ is defined as the ratio of the output signal, i.e., voltage V across the photoconductor in series to a load resistor, to the incident radiation power

$$\mathcal{R} = \frac{V}{\phi Ahc/\lambda} \ \left(\frac{V}{W}\right).$$ (45.56)

Here, A is the photoconductor area exposed to light, λ is the wavelength of incident light flux ϕ, and $h\nu/\lambda$ is the photon energy.

The voltage across a photoconductor in series with a load resistor and a battery (see Fig. 45.11) is proportional to $I\Delta n/n$, where I is the dc-current, and Δn and n are the photo-induced and dark carrier densities, respectively. Hence, the responsivity can be expressed as

$$\mathcal{R} = \frac{I\Delta n\lambda}{\phi Ahcn},$$ (45.57)

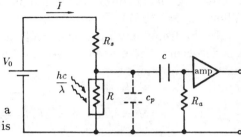

Figure 45.11: Schematic for a typical photodetector circuit. R is the photodetector.

where Δn is a function of the chopping frequency ν_s due to the sluggishness $f(\nu_s, \tau)$ of rise and decay; assuming a sinoidal photonflux variation, Δn is given by

$$\Delta n(t) = \Delta g(t) \cdot \tau f(\nu_s, \tau) = \Delta g_o \cdot \tau \, \frac{1 \pm \sin(2\pi\nu_s t - \psi)}{2\sqrt{1 + (2\pi\nu_s\tau)^2}}; \quad (45.58)$$

ψ is the phase shift between n and g. With $\Delta g_o \simeq \eta\phi/d$, and η as the fraction of light absorbed. d is the optical device thickness; this yields

$$\mathcal{R} = \frac{I\eta\lambda\tau}{Adhcn} \frac{1}{\sqrt{1 + (2\pi\nu_2\tau)^2}}. \quad (45.59)$$

The responsivity is a function of the wavelength of light, also contained in I, η, and τ, chopper frequency, bias (contained in I), device geometry, and temperature that is contained in n, I, and τ.

45.3.1B Detectivity For practical purposes, the output signal has to be compared with the total electronic noise. The *detectivity*, usually referred to a D^*, is the signal-to-noise ratio per Watt of radiation influx, normalized to unit area (1 cm^2) and band width ($\Delta\nu = 1$ Hz):

$$D^*(\nu_0) = \frac{\mathcal{R}(\nu_0)}{\mathcal{N}(\nu_o)} \sqrt{A\Delta\nu} \; \left(\sqrt{\frac{\text{cm}}{\text{s}}} / \text{W} \right), \quad (45.60)$$

where $\mathcal{N}$ is the root-power noise that is composed of generation/recombination, flicker ($1/f$), thermal (Johnson-Nyquist), and photonnoise. The first three noise components were discussed in Chapter 38. The photon noise is caused by the random arrival of photons generating the free electrons in the device. It has the same form as the generation/recombination noise. The photon noise will become important only for very long wavelength IR light detectors when competing background radiation needs to be considered (see Van Vliet, 1958a, 1967).

In the frequency range generally used for light detection with better detectors, the noise can be described by the generation/recombination component

$$\mathcal{N} = \sqrt{\overline{i^2}_{\text{g-r}}} = 2I\sqrt{\frac{\tau \Delta\nu}{ndA[1 + (2\pi\nu_s\tau)^2]}}. \qquad (45.61)$$

With Eq. (45.59), one obtains for the detectivity a rather simple expression:

$$D^* = \frac{\eta\lambda}{2hc}\sqrt{\frac{\tau}{nd}}. \qquad (45.62)$$

For more information, see Bratt (1977).

45.3.1C Frequency Response The frequency response $V(\nu_s)$ for ac-light signals depends on the photoconductor time constant τ—see Section 47.1.1. It is given by

$$V(\nu_s) = \frac{1}{\sqrt{1 + 2\pi\nu_s\tau}}. \qquad (45.63)$$

45.3.2 Current Continuity

The generation, recombination, extraction, and replenishment of carriers, described in Section 45.5, require a more detailed analysis of the current within the photoconductor. When an electric field is acting, electrons generated in one volume element are carried into another one by the current drift, where they recombine.

Consequently, the net change of the population in this volume element is given by the difference between birth (generation) and death (recombination) rates plus the net difference between the incoming and outgoing traffic (current density) from electrode to electrode in x direction—Fig. 45.12:

$$j_n(x + dx) = j_n(x) + eU\,dx, \qquad (45.64)$$

with U as the net flow rate per unit volume of carriers between valence and conduction bands—see Eq. (43.39). Therefore, one has for electron or holes

$$\boxed{\frac{dn}{dt} = g_n - r_n - \frac{1}{e}\frac{dj_n}{dx}} \quad \text{or} \quad \boxed{\frac{dp}{dt} = g_p - r_p - \frac{1}{e}\frac{dj_p}{dx}.} \qquad (45.65)$$

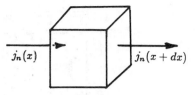

Figure 45.12: Current continuity with net generation/recombination within the volume element.

These are the basic equations* that permit an analysis of the reaction-kinetic behavior. For steady state, one obtains the *steady-state current continuity equations*:

$$\boxed{\frac{1}{e}\frac{dj_n}{dx} = g_n - r_n} \quad \text{and} \quad \boxed{\frac{1}{e}\frac{dj_p}{dx} = g_p - r_p.} \qquad (45.67)$$

45.3.3 Current Continuity in Ambipolar Photoconductors

In the previous sections, it was assumed that only one of the photogenerated carriers is mobile and that the other is trapped. The charge neutrality requires current continuity of the photogenerated carrier; therefore, for an n-type photoconductor,

$$\frac{dj}{dx} = \frac{dj_n}{dx} \equiv 0. \qquad (45.68)$$

Here, all generated carriers must recombine within the photoconductor, typically after passing several times through the external circuit.

In actuality, some of the generated holes are mobile, e.g., between thermal ionization from activators and trapping in recombination centers, although with substantially different lifetimes: for an n-type photoconductor with $\tau_p \ll \tau_n$. With any bias, some of these holes can be pulled into the electrode and recombine outside of the

---

* These relate to the basic Maxwell's equation with its condition for the conservation of electrons

$$\nabla \cdot \mathbf{j} = -\frac{\partial \varrho}{\partial t}; \qquad (45.66)$$

for equilibrium, with $\partial \varrho/\partial t \equiv 0$, it follows $\nabla \cdot \mathbf{j} \equiv 0$. In semiconductors, with the introduction of holes, one has two types of currents, j_n and j_p, and expects with $g_n = g_p$ and $r_n = r_p$ that $\nabla \cdot (\mathbf{j}_n + \mathbf{j}_p) \equiv 0$; the sign dilemma in comparing this equation with Eq. (45.67) can be resolved by replacing the conventional $e = |e|$ with $-e$ for electrons and $+e$ for holes. This is the condition for the conservation of charges. In actuality, however, only a fraction of the electrons and holes are mobile, others are trapped and do not contribute to the currents while participating in the total neutrality account.

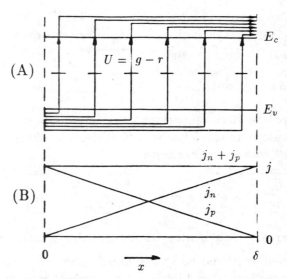

Figure 45.13: Current generation with sufficient bias to render $g \gg r$ for an ambipolar photoconductor with resulting current increments. (A) Band model with continuous current lines; (B) current distribution.

photoconductor in the external circuit with electrons supplied from the other electrode, thereby maintaining neutrality. This renders $dj_n/dx \neq 0$. In order to make the total current divergency-free, it requires a finite dj_p/dx of the same magnitude and opposite sign:

$$\frac{dj_n}{dx} + \frac{dj_p}{dx} \equiv 0. \tag{45.69}$$

For further explanation, we assume first an extreme case of symmetrical, ambipolar photoconductivity, $\tau_n = \tau_p$ and sufficient field resulting in a drift length much larger than the distance between the electrodes. Starting from the cathode, the electron current increases linearly with distance as increasingly more carriers are created on the way to the anode (see Fig. 45.13—with $r \ll g_o$):

$$\frac{dj_n}{dx} = eg_o. \tag{45.70}$$

Similarly, the hole current increases linearly with increasing distance from the anode with the same magnitude of the slope g_o) but with opposite sign. Although each carrier current is no longer constant, the sum must be constant in order to fulfill current continuity:

$$j = j_n(x) + j_p(x) = \text{const}, \tag{45.71}$$

as shown in Fig. 45.13, neglecting the dark current. At the electrodes, each current has also risen to the same value, and thus provides an equal flux of electrons and holes for recombination within the external circuit.

When the condition of a large drift length is not fulfilled, a similar increase of the current of each carrier within the photoconductor still holds, except that the slope is no longer constant but somewhat lower, as some of the electrons have recombined before leaving the photoconductor:

$$\frac{dj_n}{dx} = -e[g_o - r(n)] = -e\dot{U} \tag{45.72}$$

and consequently for holes

$$\frac{dj_p}{dx} = e[g_o - r(n)] = eU. \tag{45.73}$$

This current is called the *generation/recombination current.*

When recombination inside the photoconductor is no longer negligible, the slopes of the currents are no longer constant, since the recombination rate depends on n, or when there are inhomogeneities in doping.

When one carrier becomes the majority carrier, each current is composed of two components. One component is equal to the minority carrier current that exits the electrode; it is the generation/recombination current as given in Fig. 45.14. The other component is the divergency-free part of the majority carrier current, and contains the dark current caused by thermal ionization. Its magnitude is equal to the total current minus the maximum of the minority carrier current. This behavior is illustrated in Fig. 45.14.

In cases in which the minority carrier current is negligible, the generation/recombination current is also negligible, and Eq. (45.68) holds: all carriers generated also recombine within the semiconductor. With bias, carrier recombination in the external circuit of such photoconductors can be neglected.

45.3.4 Persistent Photoconductivity

An interesting consequence of current continuity in photoconductors with activators of extremely small capture cross section is a *persistent photoconductivity*. At low temperatures, photoconductivity has been observed which persists for days or even months after the optical excitation is terminated. The initially generated carriers are extracted

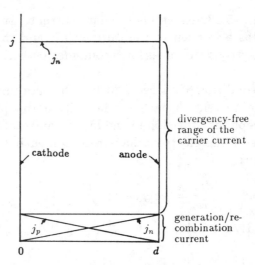

Figure 45.14: Current distribution in a photoconductor with dominant (*n*) majority carrier current indicating two current contributions: a divergency-free majority current, and a generation/recombination current.

from the photoconductor, and are repeatedly replenished. The current continues in order to maintain neutrality, since recombination is extremely slow. In CdTe:Cl at 77 K, the photocarrier lifetime can exceed 10^6 hours (Dmowski et al., 1977). Other examples include a donor at a thermal depth of 0.1 eV and an optical depth of 1.3 eV, thus with a giant lattice relaxation of 1.2 eV in $Al_\xi Ga_{1-\xi}As$, which shows a recombination cross section of $< 10^{-30}$ cm^2 at 77 K (Lang and Logan, 1977).

Such extremely small recombination cross sections are difficult to explain with Coulomb-repulsive centers. With large impurity-lattice coupling, however, a very large relaxation, e.g., for some *D-X* centers, is possible when the charge state of the impurity is changed—see Fig. 45.15. Such relaxation causes a large effective barrier for carrier capture, which cannot be overcome at low temperatures by thermal excitation. The tunneling probability through this barrier is sufficiently small to explain $s_n < 10^{-30}$ cm^2. For a recent review, see Bhattacharya and Dhar (1988).

45.4 Small Photoconductivity

In semiconductors in which the photo-induced increment in the majority carriers is smaller than, or of the same order as, the thermally generated carrier density, the approximations used before are

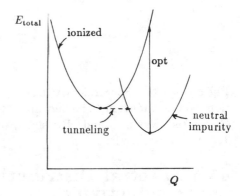

Figure 45.15: Configuration co-ordinate diagram for large impurity-lattice coupling to explain extremely small recombination cross sections.

no longer valid. The carrier-generation rate is a sum of the thermal, Shockley-Read-Hall type generation from a donor, and optical generation, as given in Section 43.2.2B—Eq. (43.43), i.e.,

$$g = e_{dc} N_c n_d + g_o = n_d \nu_d \exp\left(-\frac{E_c - E_d}{kT}\right) + g_o. \qquad (45.74)$$

In addition to the thermal and optical generation rate, impact ionization must be considered at low temperatures and for shallow levels that can compete at rather low field strength: ~ 10 V/cm in Ge at 4 K (Sclar and Burstein, 1957).

The recombination term includes recombination into the ionized donor and to the activator if the optical excitation is from such a deep center:

$$r = c_{cd} n (N_d - n_d) + c_{ca} n (N_a - n_a). \qquad (45.75)$$

The stationary carrier density is determined by $g = r$, and by the quasi-neutrality condition

$$n = p_d - n_a = N_d - n_d - (N_a - p_a). \qquad (45.76)$$

Since the density of photogenerated electrons is small compared to the already present holes for an optical transition with holes communicating with the valence band, only the *incremental* electron density is proportional to the generation rate.

With chopped light and ac amplification, the sensitivity of detection is substantially increased. This permits the detection of small light signals at far IR, which compete with thermal excitation from shallow donors.

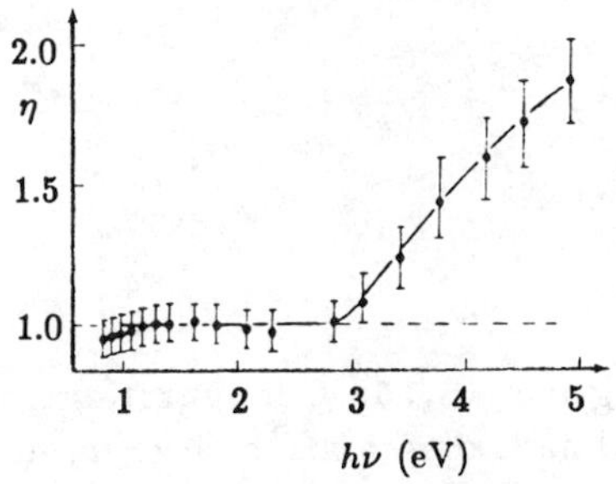

Figure 45.16: Spectral distribution of the quantum efficiency in Ge (after Vavilov and Britsin, 1958).

45.5 Spectral Distribution of the Photoconductivity

The wavelength dependence of the photoconductivity is determined by the wavelength dependence of the

- absorption coefficient and reflectivity,
- quantum efficiency,
- recombination rate,
- average effective mass and mobility caused by a deformed carrier distribution, and
- relaxation time.

The wavelength dependence can occur for a number of reasons:

- by a change between different levels from which excitation originates;
- by a change of the excited state between different discrete levels or to different heights within a band; and
- by a shift from primarily near-surface to bulk excitation.

The **quantum efficiency** η has been measured for various photoconductors (see Marfaing, 1980). It is close to 1 below and near the band edge, and increases linearly with the photon energy, starting at energies well in excess of twice the band gap (Fig. 45.16) for reasons of carrier multiplication due to impact ionization.

The **generation rate** depends on absorption $\alpha_o = \alpha_o(\lambda)$ and reflection $r = r(\lambda)$. For light, impinging on a photoconductor platelet, it is given by—see Sections 10.2.2 and 42.3.1

$$g_o = \alpha_o(1 - r)\frac{\exp(-\alpha_o x) + r\exp(\alpha_o x)\exp(-2\alpha_o d)}{1 - r^2\exp(-2\alpha_o d)}\phi_0\eta, \quad (45.77)$$

where ϕ_0 is the impinging photon flux, and d is the platelet thickness. The reflection r from front and back surfaces of a planar platelet is included.

The **wavelength dependence** of the absorption and reflection depends on the matrix element and the joint density of states between initial and final states, which were discussed in Sections 13.1 and 45.1. Only a short summary is given here for convenience.

Interband generation creates free electrons and holes with an absorption coefficient of $\sim 10^5 - 10^7$ cm^{-1} for direct transitions, and about a factor of 10^3 less for indirect transitions. Both carriers have usually *substantially* different lifetimes, making one of them the majority carrier.*

Extrinsic generation creates only one type of free carrier by excitation from a defect state into the band. Alternatively, excitation from a ground state can proceed into an excited state of a localized defect with consequent thermal or field ionization to produce photoconductivity. The absorption coefficient of extrinsic generation is typically $\sim 10^{-1} - 10^2$ cm^{-1}, and is suited for a homogeneous excitation of the photoconductor.

Intraband excitation, i.e., free electron or hole absorption, has a small effect on photoconductivity. It is measurable by causing a change in mobility while heating the carriers—electronic bolometer.

Exciton generation can have an influence due to the giant cross section for excitation of some of the excitons—see Section 44.3.4. A contribution to the photoconductivity occurs after dissociation of the exciton. Free excitons produce free electrons and holes. Bound excitons produce only one carrier; the other usually remains trapped.

45.5.1 Spectral Sensitivity of Photosensors

Depending on the type of photoconductivity employed, there are four types of photoconductive sensors:

(1) Intrinsic photosensors, depending on the band gap covering the range from UV to the near IR. Typical photosensor materials include CdS and GaAs for the visible spectrum to InSb, lead chalcogenides, and $Hg_x Cd_{1-x} Te$ for the IR range. For further examples, see Volume 5 of *Semiconductors and Semimetals* (R. K. Willardson and A.C. Beer, eds.; Academic Press, New York).

---

* This does not need to be the same as the majority carrier of the semiconductivity (in the dark), which is given by shallow donor or acceptor doping; whereas the photoconductivity of the same material is determined by generation, possibly from deep centers, and recombination at predominant recombination centers.

(2) Extrinsic photosensors, involving deep level centers, with high sensitivity in the visible and near-IR range. Sensitized CdS or GaAs are examples (Bube, 1960). Other examples include Ge:Au, Ge:Hg, and many other semiconductors with transition metal doping.

(3) Extrinsic photosensors involving shallow, hydrogen-like centers. Acceptor- or donor-doped Ge, and to a lesser extent Si and GaAs, are examples with sensitivities up to $\lambda \simeq 100$ μm for ground state to band excitation, and up to ~ 200 μm when excited states are involved (Bratt, 1977).

(4) Free-electron bolometers which employ free-electron absorption within the conduction band ($\propto \lambda^2$). Such electron heating can be observed as changes in the carrier mobility, and are sensitive into the submillimeter range, with InSb as a typical representative (Putley, 1977). A magnetic field substantially enhances the change in photoconductivity by photon absorption of free electrons because of a resonance-photoconductive effect near the cyclotron frequency (Kimmit and Niblett, 1963).

Photosensors in the visible and near-IR range can be used at room temperatures, while cooling to low temperatures is essential to reduce competition by thermal ionization for detection in the far-IR range. Careful shielding of background (300 K) illumination then becomes important. A restriction to low-bias levels, resulting in fields below 1 V/cm, is required to eliminate competing impact ionization.

Recent advances in deposition techniques, e.g., molecular beam epitaxy, have created a number of photoconductive materials with exciting photo-electric properties, e.g., various III-V compounds and their alloys (see Tsang, 1985).

45.5.2 Thermally Assisted Generation

The excitation processes listed in the previous section were caused exclusively by photons. A number of interesting effects can be seen when additional interaction with phonons is required.

Such thermally assisted generation can produce an excitation spectrum with many discrete lines. Such a spectrum occurs when photon-induced transitions of a defect center lead to excited states below the band edge from which an additional *thermal* ionization step creates the free carriers, or when a phonon-induced momentum adjustment is necessary to enter a band at $k \neq 0$. This method is known as *photothermal ionization spectroscopy*, and was observed

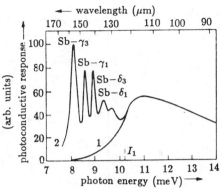

Figure 45.17: Photocurrent distribution of Ge:Sb at 4.2 K and at 10 K for curves 1 and 2, respectively (after Kogan and Sedunov, 1974).

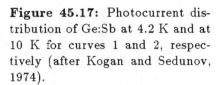

originally by Lifshitz and Nad' (1965). Figure 45.17 shows the broad optical transition from the donor ground state into the conduction band for an Sb-doped Ge crystal at 4.2 K. The additional line spectrum due to excitation from the ground state, first into excited states with consequent thermal ionization, becomes noticeable at temperatures sufficient for thermal ionization (10 K). These lines become narrower with higher purity of the sample to avoid overlap of defect eigenfunctions. Small amounts of neutral impurities do *not* destroy the peaks but instead broaden them. The amplitude of these lines is not markedly reduced, even down to extremely low donor densities* in the 10^{12} cm^{-3} range, which indicates giant cross sections for excitation. These giant cross sections are related to those for recombination, which were discussed in Section 43.1.1A (see also Lax, 1960).

The photothermal mechanism requires a photon to excite an electron from the ground state of the impurity to an excited state, and then the absorption of an LO phonon for the emission of the electron into the conduction band (Golka and Mostowski, 1975). Its probability increases exponentially with the temperature, i.e., proportional to the density of available phonons. A photoconductive response of these centers is expected for $T > 9$ K in agreement with the experiment, as shown in Fig. 45.17.

Since the excitation spectrum responds to the specific impurities with large cross sections, and appears in a wavelength range usually

---

* While the area under the line, which is proportional to the defect center density, is reduced, the line width is also reduced with lower donor densities until the natural line width is reached, keeping the amplitude essentially unchanged.

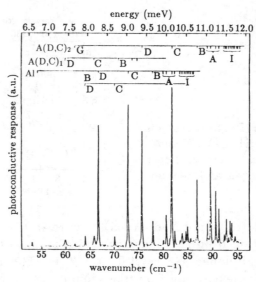

Figure 45.18: Photothermal ionization response spectrum of ultrapure Ge with residual acceptors B, Al, A (D,C), and A (D,C)$_2$ at 7.5 K (after Haller, 1986).

free from competing excitation, this method is very sensitive at detecting extremely small densities of impurities (Kogan and Lifshitz, 1977). Skolnick et al. (1974) detected dopants below 10^{10} cm^{-3} employing this analysis. The lines can be extremely sharp, offering a means for high-resolution spectroscopy—Fig. 45.18. With increased doping density, the width of the lines broaden by the random electric fields of other impurities (Larsen, 1973).

In general, the photothermal ionization spectrum reflects the optical absorption spectrum. It differentiates between the different types of absorption processes since the photoconductivity is more structured.

45.5.2A Phonon-Induced Conductivity

Conductivity, which is due to carrier generation *solely from single phonons*, is also referred to as *phonoconductivity*. It can be caused by phonon-induced ionization of shallow donors or acceptors, or by band-to-band ionization in narrow band-gap semiconductors. It is distinguished from normal semiconductivity, in which thermal ionization occurs by multiphonon processes.

An example spectrum of the phonon-induced conductivity of Te-doped silicon is given in Fig. 45.19. Several peaks are observed due to the ionization of shallow Te traps. An abundance of shallow levels,

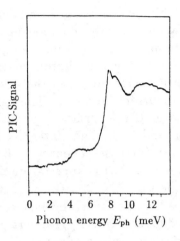

Figure 45.19: Phonon-induced conductivity (PIC) in Si:Te at 1 K with peaks at 8, 8.8, and 11.5 meV due to electrons trapped at Te-related centers (after Burger and Laßmann, 1986.

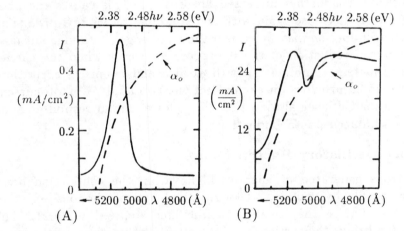

Figure 45.20: Optical absorption spectrum (α_o) and spectral distribution of the photocurrent (solid curve) of a CdS platelet with (A) high- and (B) low-surface recombination velocity (after Bragagnola and Böer, 1970).

which are due to the trapping of additional carriers in hydrogen-like defects, can be detected (Burger and Laßmann, 1986).

45.5.3 Edge- or Peak-Type Response

At higher temperatures, the photoconductivity spectrum near the band edge broadens and typically shows one of two types of responses: a shoulder, possibly with some structure, or a pronounced maximum at the band edge.

This behavior can be understood by an analysis of the balance between spatially inhomogeneous carrier generation and competing recombination in the bulk and at the surface.

Increasing the photon energy of the exciting light near the absorption edge increases the absorbed fraction of the light; thus, the photocurrent increases. The light is also absorbed closer to the front surface. If the surface is *passivated*, i.e., has low-surface recombination, and if the light intensity is sufficiently low so that the carrier density increases linearly with g_o, the photocurrent becomes independent of λ; the spectral distribution of the photocurrent shows a *shoulder*.

When, however, the surface recombination velocity is larger than the diffusion velocity, carriers generated closer to the surface have a higher probability of recombination, causing a reduction of the photocurrent with further increased absorption. This causes the photocurrent to decrease again with increased photon energy, *producing a photoconduction maximum* near the band edge. A similar but less-pronounced reduction of the photocurrent occurs when the carrier density increases sublinearly with g_o at a higher optical excitation. Figure 45.20 provides an example of the two kinds of photocurrent spectra for CdS; see also Section 45.5.5 for a similar response of the exciton-induced photoconductivity.

45.5.4 Oscillatory Photoresponse

Electrons generated at energies high within the conduction band relax rapidly by emission of longitudinal optical phonons—Section 42.4.2B; that is, they relax in well-defined steps—Fig. 45.21. The end step brings the carrier to an energy E_h within $\hbar\omega_{LO}$ of the lower band edge. Further relaxation via acoustic phonons is comparatively slow. Therefore, the energy distribution of the electrons shows periodicity: resulting in warm carriers, if this energy at the end of the last scattering is only slightly below $\hbar\omega_{LO}$, and cool carriers if this energy is near the band edge. This results in an *electron temperature modulation*.

Warm and cool electrons have different mobilities. Therefore, one obtains mobility oscillations depending on the initial monochromatic excitation of the photogenerated electrons, as observed by Godik (1968) in Ge and shown in Fig. 45.22.

In addition, the lifetime of these photogenerated electrons also varies with the final energy E_h after LO phonon relaxation; the lifetime is shorter for electrons near the band edge—see Section 44.3.2.

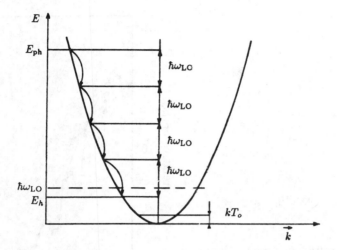

Figure 45.21: Relaxation of electrons by consecutive emissions of LO phonons to an energy E_h, which at low temperatures can be substantially above kT_L, and thereby produces hot electrons.

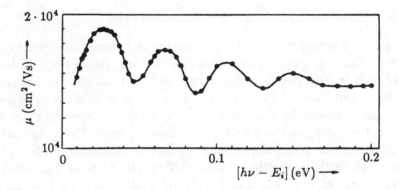

Figure 45.22: Mobility oscillations of electrons in Ge:Cu at 10 K with $F = 5$ V/cm. E_i is the threshold energy for impurity ionization (after Godik, 1968).

Therefore, oscillations in the capture rate and thus in the photocurrent are observed (Collins et al., 1968; Stocker et al., 1964), as shown for Ge:Ga in Fig. 45.23.

A slightly different relaxation mechanism, due to heavy and light hole bands, makes the phenomenon more complicated for holes (Shaw, 1971).

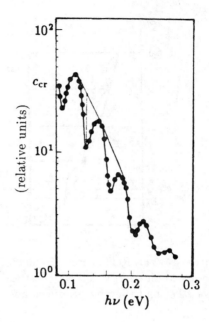

Figure 45.23: Spectral distribution of the recombination coefficient of Ge:Ga at 7 K (after Besfamilnaya et al., 1967).

45.5.5 Exciton-Induced Photoconductivity

The exciton absorption spectrum is highly structured at low temperatures—Section 15.1. Consequently, exciton absorption leads to structures in the photoconductivity spectrum, as shown originally by Gross and Novikov (1959), with its maxima or minima coinciding with the maxima in the absorption spectrum, depending on the crystal *type* (referred to as *class I or class II*, respectively, in CdS).

Excitons dissociate via thermal ionization (Novikov et al., 1976), interaction with impurities (Trlifaj, 1959), influence of high electric fields (Coret et al., 1971), and inelastic exciton-exciton interaction (Mengel et al., 1975).

The coincidence or anticoincidence of photocurrent and optical absorption maxima can be explained by inhomogeneous photosensitization (Gutsche et al., 1971), which is similar to the edge- or peak-type behavior described in Section 45.5.3: when the photoconductor is more sensitized near the surface, absorption and photoconductivity maxima will coincide; when there is large surface recombination, or the layer near the surface is less sensitive, maxima in absorption, which preferably excite this layer, will coincide with minima in photoconduction—Fig. 45.24. The sensitization is temperature-dependent; thus a change in temperature can reverse the relative position of the maxima, as shown in Fig. 45.24 between 77.3 K and

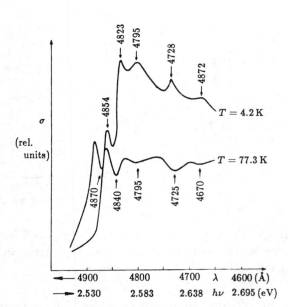

Figure 45.24: Photoconductivity spectra of a CdS platelet at 77.3 K and at 4.2 K in polarized light $\mathbf{e} \perp \mathbf{c}$ showing reversal between maxima and minima. Corresponding absorption maxima are indicated by arrows (after Gross and Novikov, 1962).

4.2 K. Surface sensitization depends on doping and surface treatment.

45.6 Photoconductivity in Superlattices

Compositional superlattices with large period length and barrier width have a number of discrete levels within the wells, as shown for a symmetrical well in Fig. 45.25A. With excitation, one populates these levels with electrons and holes, following the selection rule ($\Delta n_q = 0$). In an electric field, they contribute to a photocurrent (see Section 45.6.2) that exhibits maxima at the corresponding wavelengths of the exciting light—Fig. 45.25B (Nozik et al., 1985). The observed broadening of the expected narrow lines is probably due to fluctuations in the superlattice periodicity or to interface roughness.

45.6.1 Stark Effect in Superlattices

Excitation from light and heavy hole bands into the conduction minibands creates a spectrum of excitons which, after dissociation, contribute to the photoconductivity. The dissociation of these excitons in external fields was observed by Das et al. (1987). It is indicated

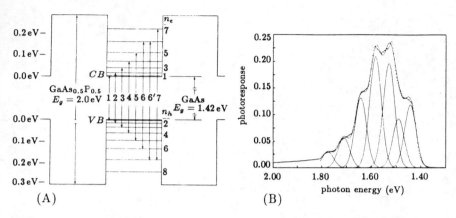

Figure 45.25: (A) Level spectrum in a symmetrical 250 Å/250 Å GaAs/-GaAs$_{0.5}$P$_{0.5}$ superlattice, and (B) photoresponse spectrum with deconvoluted Gaussian curves (after Nozik et al., 1985).
©American Chemical Society.

by a decrease in exciton luminescence while the photoconductivity concurrently increases—see Fig. 44.29. With an increasing electric field, the corresponding photoconductivity maxima shift to lower energies (Stark effect), and dipole-forbidden transitions become visible because of the field-induced deformation of the exciton eigenfunction; for example, transitions from $n_q = 1$ of the heavy hole band to $n_q = 2$ or $n_q = 3$ of the conduction band appear. See Collins et al. (1987); see also Section 44.8.3A for the corresponding observation of the luminescence.

45.6.2 Increased Carrier Lifetime with Field

With an applied external field, the wells are deformed as shown in Fig. 45.26. When the field is strong enough, the eigenfunctions for an electron and a hole in the ground state are displaced with respect to each other; consequently, their recombination probability decreases (Collins et al., 1987; Nurmikko, 1987).

45.6.3 Intersubband-Generated Photoconductivity

The excitation energy, here in the IR range, from the ground miniband state into an excited miniband state can be tuned by selecting the well width. As a direct transition, it results in a high optical absorption. Ground-state filling can be accomplished by doping the barrier material. In an applied electric field, the electrons from the excited state can tunnel through the top of the well as shown in Fig. 45.27B.

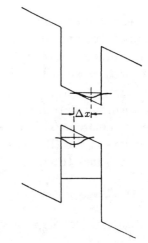

Figure 45.26: Relative displacement of the eigenfunction of an electron and a hole in a quantum well when a sufficiently large external field is applied.

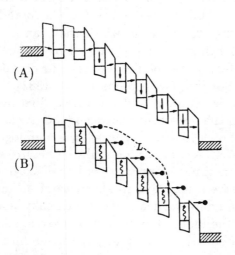

Figure 45.27: (A) Sequential resonant tunneling in the dark. (B) Photoconductivity due to IR absorption in the ground state, followed by tunneling out of the excited state of the well (after Levine et al., 1987).

With a GaAs(65 Å)/AlGaAs(100 Å) superlattice of 50 layers, an effective optical absorption of 10^4 cm^{-1} at 10.3 μm was reached for the excitation from the ground to the first excited state of the wells. With a tunneling efficiency in excess of 50% given by the ratio of carrier lifetime and tunneling time, Levine et al. (1987) obtained total quantum efficiencies in excess of 20%, with 600 mA photocurrent per Watt excitation and a 30 ps response time.

The photocurrent increases stepwise with applied voltage by dropping most of the bias first across one, then two, and finally all of the barriers, yielding maximum response by tunneling from all wells. The ratio of this photocurrent to the dark current—where sequential resonant tunneling occurs from the ground to the excited state of the next well as shown in Fig. 45.27A—can be increased by widening the barrier (hence reducing the dark current), while maintaining the photo-induced tunneling by keeping the excited state close to the top of the well. This offers the possibility of designing sensitive photosensors in the IR range, with a high light-to-dark current ratio and fast response time.

45.6.4 Persistent Photoconductivity in Superlattices

In compositional or doping superlattices, electrons and holes generated by band-gap excitation are separated from each other by the potential barriers or built-in fields—see Fig. 45.28 and Section 26.1. They relax into offset valleys from which they can no longer recombine without first being thermally excited over the barrier, or by tunneling into defect states of the adjacent layer from which such recombination is possible (see Collins et al., 1983). Consequently, the recombination is retarded and the carrier lifetime is increased. This description of the persistent photoconductivity is closely related to a description of dielectric after-effects in nonhomogeneous materials (see Section 47.4.1A) containing multiple heterojunctions. Such aftereffects are sometimes referred to as RC-relaxation. Charges at these interfaces have a long lifetime because of an insufficient supply of the corresponding carrier through nonleaky junctions. These charges maintain the induced current flow—Section 45.3.

Therefore, superlattices can reach a carrier lifetime which, at lower temperatures, can be many orders of magnitude larger than that in the host material. Hence, photoconductivity initiated with a short light pulse can persist for long periods of time until both carriers can recombine (Döhler, 1986).

45.6.4A Tunable Carrier Lifetime in Superlattices The barrier height in doping superlattices determines the carrier lifetime when thermal activation over, or tunneling into, the barrier is the critical step for carrier recombination. The barrier height can be lowered by decreasing the space charge in each of the superlattice layers. Since the original space charge without light is created by carrier depletion near each interface, it can be reduced by increasing the carrier densities in each of the layers by light or injection. This causes a split of the Fermi level into two quasi-Fermi levels, and

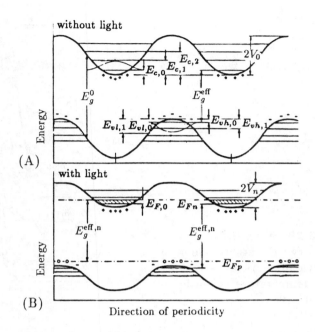

Figure 45.28: (A) Doping superlattice in thermal equilibrium. (B) The same superlattice, although with optical generation, which causes a spread of the quasi-Fermi levels and a reduced amplitude of $E_c(x)$ and $E_v(x)$ from $2V_0$ to $2V_n$ (after Döhler, 1986). ©IEEE.

thus decreases the amplitude of the band modulations—Fig. 45.28. Consequently, the carrier lifetime in such doping superlattices decreases with increasing optical excitation. This results in a sublinear photoresponse. A review of this field is given by Döhler (1986).

45.7 Photoconductivity in Amorphous Semiconductors

The photoconductivity of amorphous semiconductors decreases steeply below room temperature and remains constant below 50 K, as shown in Fig. 45.29. The photoconductivity is observed to be essentially independent of the photon-energy when excited above the mobility edge, and, when normalized to the same generation rate, falls into a narrow range of values, for most materials.

The photoconductivity is given by the optically generated density of carriers and their mobilities, which contribute to the conductivity. Both quantities are related to each other in amorphous semiconductors, and require some specific discussion.

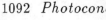

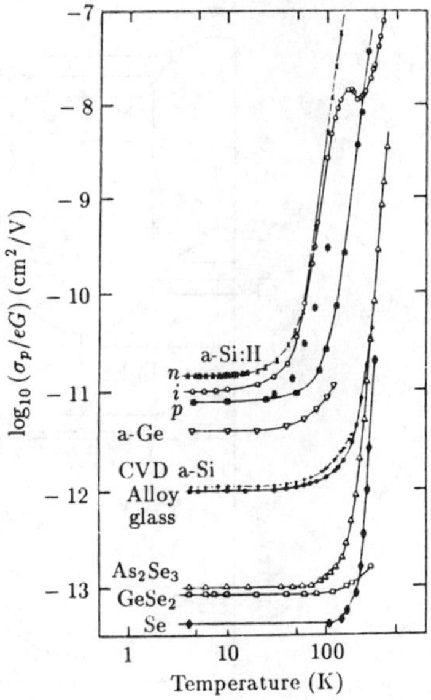

Figure 45.29: Normalized photoconductivity of various amorphous semiconductors as a function of the temperature (after Johanson et al., 1989).

The carrier density is determined by the balance between generation and recombination. The recombination process, however, is quite involved. With an exponentially tailing defect-level distribution into the band gap, photo-excited carriers relax rapidly into these levels. Here they move in a diffusive hopping pattern until they recombine. At low temperatures where phonons are frozen-out, the carriers can only lose energy when diffusing, thereby *emitting* phonons. This brings the carrier to lower and lower energy states, which are successively further apart from each other because of the exponentially tailing distribution, requiring progressively larger diffusion steps— Fig. 45.30. The number of defect centers available after each energy dissipating step is $N_{m+1} = N_m \xi_m$ with $0 < \xi_m < 1$ and with $\langle \xi_m \rangle = \frac{1}{2}$, independent of the actual level distribution (Shklovskii et al., 1989).

The probability to survive geminate recombination, while diffusing a distance R, is given by

$$\eta(R) \simeq \frac{R_c}{R} \qquad \text{with} \qquad R_c = \frac{a}{2} \ln(\nu_0 \tau_r) \qquad (45.78)$$

where R_c is the critical radius for geminate recombination, a is the localization radius for the defect states, ν_0 is the attempt-to-escape frequency from these defects, and τ_r is the recombination lifetime.

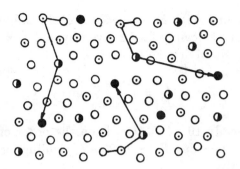

Figure 45.30: Diffusive (hopping) photoconduction in an amorphous semiconductor. Defect centers are identified by circles; the more filled in, the deeper the center is. The increasing length of consecutive diffusion steps is indicated, since energy is dissipated during each scattering event, making shallow levels no longer accessible.

For photoconductivity to occur, one requires nongeminate recombination,* which occurs when an electron meets a hole in real space, i.e., when

$$R \simeq \frac{1}{2}n^{-1/3} \qquad (45.79)$$

where n is the carrier density that can be obtained by equating the rate of generated carriers reaching the distance: $g_o\eta(R)$, with the rate of recombination $nc_{np} = (n/\tau_r)\exp(-2R/a)$, yielding:

$$n^{2/3}\exp\left(-\frac{n^{-1/3}}{a}\right) = g_o\tau_r a \ln(\nu_0\tau_r). \qquad (45.80)$$

The photocurrent is given by the product of nongeminate recombination $[g_o\eta(R)]$, with the average dipole moment of each drifted carrier:

$$j = g_o\eta(R)p_c \qquad \text{with} \qquad p_c = \frac{1}{3}\frac{e^2\overline{R}^2 F}{E_0} \qquad (45.81)$$

---

* Photoconductivity requires a finite drift length between carrier generation and recombination. Such drift length is canceled by the process of geminate recombination, i.e., recombination with the same center from which the electron-hole pair was generated.

in an external field F with E_0 as the slope of the tailing distribution—Eq. (24.12). With a scaling factor given by $n^{-1/3} = aL$, one obtains for the photoconductivity

$$\sigma = \frac{j}{F} = g_o \frac{e^2 a^2 L}{12 E_0} \ln(\nu_0 \tau_r) \qquad (45.82)$$

where L is typically $10 \ldots 15$, $\nu_0 \tau_r$ is on the order of 10^4, $a \simeq 10$ Å, and, for α-Si:H, the slope $E_0 \simeq 0.025$ eV, yielding for the normalized photoconductivity

$$\frac{\sigma}{e g_o} \simeq 10^{-11} \ \mathrm{cm}^2 V^{-1} \qquad (45.83)$$

which depends little on the actual semiconductor material (Shklovskii et al., 1989).

At temperatures above

$$T_c = \frac{3}{L} \frac{E_0}{k} \qquad (45.84)$$

upward hops in the trap spectrum become possible, absorbing rather than emitting phonons near an energy E_t—the *Monroe transport energy* (Monroe, 1985), at which the major carrier transport occurs. The width of this region, $\Delta E = \sqrt{6kTE_0}$, increases with increasing temperature, and thereby σ increases with T at higher temperatures—Fig. 45.29. In addition, changes in carrier density and mobility need to be taken into consideration (Fritzsche, 1989).

45.8 Negative Photoconductivity

Negative photoconductivity is defined as a *decrease* in conductivity with additional light. It can be observed when two light beams of different wavelengths are applied and the second beam causes a reduction in *carrier lifetime*. The reduction of photoconductivity is called *quenching*.

The reduction in majority carrier lifetime is induced by a redistribution of minority carriers over recombination centers of different capture cross sections. This redistribution can also be induced by thermal or field ionization. Consequently, one can distinguish among *optical*, *thermal*, and *field quenching*.

All quenching transitions can be regarded as a *desensitization*, i.e., a shift of minority carriers from slow to fast recombination centers. Therefore, quenching can only be observed in sensitized photoconductors.

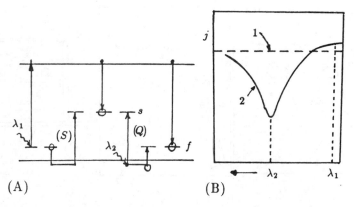

(A)   (B)

Figure 45.31: (A) Sensitizing, S, and quenching, Q, transitions in a sensitized photoconductor. (B) Corresponding spectral distribution of the photoconductor excited with a second light beam (2) of variable wavelength while the first beam (1) is kept constant (horizontal dashed line indicates the photocurrent of the first beam only).

45.8.1 Optical Quenching

The quenching transition is induced by the optical excitation of holes, for an n-type photoconductor, from the slow recombination center into the valence band. Although in a sensitized photoconductor these holes tend to fall back into the Coulomb-attractive slow recombination centers, their population can be significantly reduced when the quenching light is sufficiently intense. The fraction of holes that go to fast recombination centers causes an increase in the recombination of electrons through these centers, and thereby results in a decrease in photoconductivity.

On the left side of Fig. 45.31A is shown an optical excitation with consequent sensitizing (S). The recombination occurs predominantly through slow centers (s). On the right side of this figure the optical quenching transition (Q) is shown, which removes holes from slow centers and thereby reactivates some recombination traffic through fast centers (f). The spectral distribution of the optical quenching is given in Fig. 45.31B. The first beam of light at a constant wavelength λ_1 produces a photocurrent (1). The second beam with variable wavelength causes an increase of the photocurrent (2) near λ_1, and a decrease with a pronounced minimum below the photocurrent (1) at λ_2, corresponding to the transition into the slow recombination center.

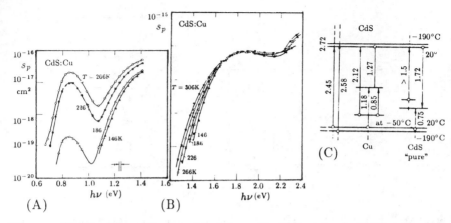

Figure 45.32: Thermal quenching from Cu impurities in CdS: (A) cross section for holes near the first quenching peak at ~ 0.85 eV; and (B) cross section for electrons near the second quenching peak at ~ 1.8 eV (after Grimmeiss et al., 1981). (C) Band model at 20°C and at −190°C (after Broser and Broser-Warminski, 1956).

In some photoconductors, e.g., ZnS and CdS, the sensitizing center has more than one level, as shown for the Cu center in CdS in Fig. 45.32A and B. Optical quenching can be observed from both of these levels at low temperatures; at higher temperatures only the higher-energy transition is observed since holes from the lower level are already thermally depleted. The corresponding band picture is shown in Fig. 45.32C.

45.8.2 Thermal Quenching

The quenching transition can be thermally initiated. Sensitizing (S) and desensitizing (quenching, Q) steps, are both thermal ionization steps and depend exponentially on the temperature. With little ionization from slow centers, holes are stored there until they are eliminated by recombination with electrons. When they are ionized faster than they can recombine with electrons, quenching is observed. A shift of the hole population occurs toward fast centers. This emphasizes the competitive nature of the slow and fast recombination processes, and indicates the light intensity dependency of thermal quenching. Thermal quenching is shown in Fig. 45.33 as a steep decrease in photocurrent with increasing temperature. Thermal quenching needs higher temperatures to become dominant for higher optical excitation.

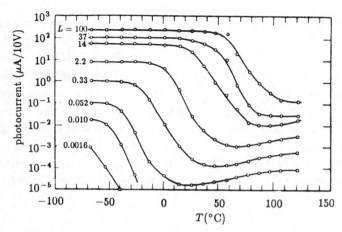

Figure 45.33: Thermal quenching shown as a steep reduction in photocurrent with increasing temperature as observed for CdSe with the light intensity in relative units as the family parameter (after Bube, 1960).

ⓒJohn Wiley & Sons, Inc.

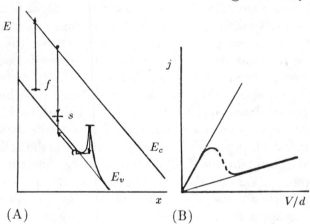

(A)   (B)

Figure 45.34: (A) Frenkel-Poole effect causing field-quenching by a depletion of slow centers. (B) Resulting negative differential photoconductivity induced in CdS by field-quenching (after Böer, 1965).

45.8.3 Field Quenching

Since sensitizing centers are Coulomb-attractive, they can be ionized by relatively low fields in the 10 kV/cm range due to the Frenkel-Poole effect—Fig. 45.34A.

As a result, one observes a reduction in photosensitivity when the recombination traffic is shifted from slow to fast recombination centers; thus, the photoconductivity decreases with increasing electric field. If this field quenching is strong enough at sufficient light intensity and fields (Dussel and Böer, 1970), one observes a decrease in the *photocurrent* with increased bias—Fig. 45.34B. In this range of *negative differential conductivity*, originally reported by Böer et al. (1959); see also Böer (1965), instabilities occur that cause current oscillations and moving high-field domains, similar to the Gunn effect (Gunn, 1963), but at lower frequencies due to the slower repopulation of centers. Under certain conditions, these high-field domains remain stationary and then can be used to determine reliably the field-dependent carrier mobilities or carrier densities (Böer, 1971).

Summary and Emphasis

Photoconductivity is initiated by the generation of free carriers directly by photon-absorption from free, band-to-band, or bound defect states, or indirectly with the assistance of phonons either for matching momenta or to supply additional energy. Here, the resultant photoconductivity can be used to measure the successful carrier generation, thereby obtaining direct information about the excitation process; that is, about the joint density-of-states distribution, the symmetry consideration distinguishing allowed and forbidden transitions, photon capture cross section, and lifetime of the excited state.

With excitation into higher states of the band, the relaxation of carriers and recombination from different energy distributions can be investigated. Various carrier distributions can be created and maintained in steady state; and the influence of such distributions on carrier transport or carrier mobility can be measured.

Information on surface vs. bulk-related generation and recombination can be obtained from the spectral distribution of the photoconductivity near the band edge, thereby varying the depth of the photoexcitation.

Photoconductivity is of high technical interest for the detection of electromagnetic radiation (light, IR). By doping with certain slow recombination centers, photoconductors can be dramatically sensitized, although, usually at the expense of time response. By proper doping, a wide range of response characteristics can be obtained. The maximum spectral sensitivity shifts from the vacuum UV to the far IR, with decreasing band gap of the photoconductor or with shallow defect level ionization.

Exercise Problems

1.(r) List and discuss the different mechanisms for carrier generation in the UV, visible, and IR ranges for a 2 eV band-gap
photoconductor with appropriate doping.

2.(r) Quantitatively describe the measurement of the photo-ionization cross section.

3.(e) Derive and plot the photocarrier (electron) density as a
function of the band-to-band generation rate for a photoconductor with a band-to-band recombination cross section
of 10^{-19} cm^2, an electron trap at $E_c - E_t = 0.5$ eV of
$N_t = 10^{16}$ cm^{-3}, and capture cross section of 10^{-13} cm^2
at $T = 300$ K for $10^{12} < g_o < 10^{20}$ cm^{-3} s^{-1}.

4.(*) Plot the electron density as a function of the band-to-band
generation rate as in **3.(e)**, however, with two traps, the
second at $E_c - E_t = 0.3$ eV of $N_{t2} = 10^{16}$ cm^{-3} and $s_n = 10^{-13}$ cm^2.

5.(*) Quantitatively describe the sensitization of the photoconductor given in **3.(e)**.

   (a) Discuss the parameter necessary for a dopant to accomplish an increase in electron lifetime by a factor of 100.

   (b) What form does $n(g_o)$ take after sensitization?

   (c) What do you need to assume for additional doping to
   make the photoconductor response superlinear?

6.(e) Discuss carrier generation and recombination within a circuit
containing a photoconductor with a gain factor of 10.

7.(*) Generation-recombination currents may cause a spatial slope
in electron and hole currents in a homogeneous semiconductor.

   (a) What are the conditions for such slopes to occur?

   (b) How would Fig. 45.13 have to be changed for grossly
   asymmetric electrodes (one blocking, one injecting)?

8.(e) Give and discuss at least two examples of persistent photoconductivity.

   (a) Think of an application for a device using persistent
   photoconductivity.

   (b) How can the lower dark-current be regenerated?

9.(e) Describe the relation between photoelectric gain factor and
persistent photoconductivity. Discuss its temperature dependency.

10.(*) Phonon-induced photoconductivity dies out at low temperatures. Figure 45.17 gives an illustrative example. Explain. How would a curve at 6 K and another at 8 K look?

11. The depth from the light-exposed surface of the optical excitation changes with changing photon energy.

(a) Explain.

(b) What effect does such a change have on the photoconductance of different photoconductors?

12.(*) Electron temperature modulation is achieved when energy relaxation is slowed down substantially below $\hbar\omega_{LO}$.

(a) Why is the amplitude of the mobility oscillations, reflecting the temperature modulation, decreasing with increasing energy of the optical excitation (Fig. 45.22)?

(b) What can you learn from the amplitude of the first oscillation? Give a quantitative answer.

13.(*) How can carrier recombination create oscillatory photocurrents?

(a) As a volume effect?

(b) As a surface-induced effect?

(c) Involving excitons? What is the difference to direct generation of free carriers?

14.(r) Review the three excitation mechanisms for photocurrent quenching.

(a) Which one is most specific in densensitization?

(b) In what types of experiments must we assure that quenching does not disturb an otherwise interesting behavior?

15.(*) Certain device parameters in doping superlattices are tunable. Which are these? Why? Explain.

16.(e) Explain the general behavior of photoconductivity in amorphous semiconductors with regard to carrier percolation near the mobility edge. Explain in your own words the dispersive carrier transport at low temperatures.

Chapter 46

Photochemical Reactions

Defect-chemical reactions, induced by light in pho-
toconductors, cause changes—usually degradation—in
their electronic properties. These changes are often
metastable and can be annealed out in the dark.

The photoconductivity depends to a large extent on the defect structure, mainly because of the carrier lifetime controlled by recombination, and to a lesser extent because of the carrier mobility: the carrier lifetime can change over many orders of magnitudes; it is on the order of 10^{-11} s for insensitive, to 10^{-2} s for sensitized photoconductors; the mobility may change over a few orders of magnitude at most in the better semiconductors.

In the previous sections, it was assumed that the defect structure is fixed and does not change during optical excitation. With light-induced recharging of defects and consequent lattice relaxation, however, local excitations occur. These consist of either purely electronic ionization, producing free electrons or holes, i.e., physical changes, or induced metastable changes in the defect structure, i.e, chemical changes. These light-induced structural changes are called *photo-chemical reactions*, and will be discussed in this chapter.

46.1 Defect-Chemical Reactions

Photochemical reactions involve the movement of atomic defects into a new, stable (rarely) or metastable position. Such motion requires sufficient temperature, or local heating, i.e., by a photo-induced giant oscillation, and often can be described as a reaction between two defects. It is optically induced by ionizing at least one of the partners, thus creating a favorable polarization for an atomic movement into the new position.

An example is the charging of a donor and an acceptor. They become Coulomb-attractive, and may move closer to each other and

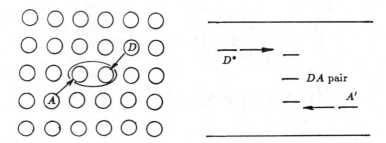

Figure 46.1: Ionized donors and acceptors moving together and creating a donor-acceptor pair, which acts as a recombination center (level drawn schematically).

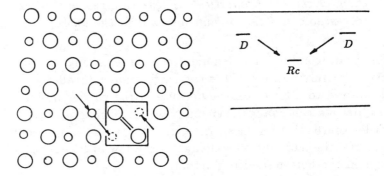

Figure 46.2: Two cation vacancies moving together, trapping two holes and creating an anion molecule that acts as a recombination center (Rc).

form a donor-acceptor pair (Fig. 46.1), which is sufficiently bound to resist separation at the given temperature.

Another example is the pair formation of cation vacancies, which is facilitated when these vacancies are ionized, i.e., a hole is trapped in each. When moving together, the two adjacent anions are neutralized and represent a strongly bound covalent molecule—Fig. 46.2. The cluster of the four lattice defects, two vacancies and the two-anion molecule, can be regarded as a new center with a substantially different scattering or recombination cross section, as compared to the original defects, i.e., two separate neutral anion vacancies.

Yet another example is the dissociation of a defect associate after it is ionized, with the ionization facilitating such dissociation.

When either one or both of the newly created defects act as a recombination center, the lifetime of photogenerated carriers will decrease; thus the photoconductivity is reduced when more and more of these centers are generated: usually, photochemical-reaction products have a larger recombination cross section (e.g., closer donor-acceptor pairs) than donors separated further from acceptors. Therefore, the photoconductivity decreases when photochemical reactions proceed.

In addition, the semiconductivity (*dark conductivity*) also changes since some of the donors or acceptors disappear. This results in a lowering of the dark conductivity.

In the dark, however, the photochemical-reaction products tend to dissociate again, by reversing the photochemical-reaction process. At sufficient temperatures, this reversal can be completed, returning the photoconductor to its original state—see Section 46.3.

46.1.1 Photochemistry by Giant Relaxation Oscillators

An example of how such defects can be created locally was suggested by Henry and Lang (1977), who have shown that the recharging of a deep center after trapping a carrier causes violent oscillations of the defect before relaxation by dissipation of phonons. Such oscillations can be regarded as strong local heating and can lead to a jump of the defect atom into a new metastable position with a different electronic character, i.e., the creation of a photochemical reaction product.

As an example, an exciton in an alkali halide has a very large electron-lattice coupling. This exciton can be trapped at a halogen lattice site, which in turn can relax to form a halogen molecule, contracting and moving it to the adjacent *single* halogen site where it forms an H-center. The electron from the original exciton becomes attached to the newly created adjacent vacancy and forms an F-center. This sequence of processes is shown in Fig. 46.3. Evidence of such simultaneous formation is provided by kinetic experiments (see Section 46.3.3), as reviewed by Itoh (1982). See also Stoneham (1979).

The given example is one of many possibilities for lattice-relaxation-induced giant oscillations, which are probably active in a wide variety of semiconductor defects in forming photochemical-reaction products. With such processes, recombination centers can be formed with an initiation energy much below a threshold at which an *impacting* electron can cause the *ballistic* creation of such a lattice defect. This creation can also take place at relatively low temperatures, depending on the actual shape of the configuration diagram. For example, the above-described F- and H-centers can be formed at temperatures far below room temperature, and require thermal ac-

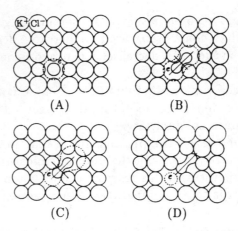

Figure 46.3: Creation of an F-center with adjacent H-center (a Frenkel pair) starting from (A) a free exciton, via (B) a relaxed, self-trapped exciton, via (C) an intermediate stage to the Frenkel pair, with (D) an anion vacancy adjacent to an anion interstitial, here in symmetrical configuration as a halogen molecule on a halogen lattice site (after Itoh, 1982).

tivation energies for relaxation between 0.02 and 0.2 eV, depending on the specific alkali halide (Itoh, 1982).

46.1.2 Photochromic Glasses

Photochemical reactions can be used to change the optical absorption of transparent glasses. If absorption in a wide wavelength range is desired as the result of exposure to light, e.g., for self-darkening sunglasses, scattering on larger colloids is employed. Such colloids can be formed, for instance, by incorporating silver-halide- ($AgCl$) enriched particles in oxide glasses, doped with Cu^+. During exposure to UV light, Cu^+ acts as a hole trap, forming Cu^{++} ions which in turn stimulate silver interstitial diffusion to form silver colloids with consequent increased visible light absorption. In the absence of UV light, thermal fading restores full transparency at room temperature. There is a wide variety of photochromatic processes that can be used commercially (see Armistaid and Stookey, 1964).

46.2 Photochemistry in Silver Halides

Photochemical processes in silver halides found an early application in the photographic process, which can be divided into three steps:

- the formation of the latent picture,

- the development, and
- the fixation of the picture.

Silver-bromide grains imbedded in a gelatine film are commonly used for photographic purposes.

The *latent picture* is formed in such a film when a grain is exposed to a minimum of ~ 10 photons, creating free electrons. When one of these electrons is trapped by a silver atom, e.g., of an Ag_2S center at the surface, it becomes negatively charged and attracts an interstitial silver ion that is highly mobile. The barrier for colinear interstitial jumps is 0.04 eV, causing an extremely high ionic conductivity. The newly created center has a deeper level and can in turn attract a second electron, charging this center again, and attracting another silver ion to form a dimer, Ag_2. Such a dimer is stable but not yet developable; it is called a *subimage*. The process with alternating electron capture and interstitial ion attraction repeats itself as long as free electrons are available, creating a cluster of silver atoms.

During the following *development*, the photographic film is exposed to a moderately reducing agent, with a reducing potential between 40 and 600 mV below that of a reference Ag/Ag^+ electrode. Agents with higher potential cause indiscriminate reduction of all silver grains, i.e., fogging of the film. Clusters of overcritical size (typically of four or more silver atoms) act as quasi-electrodes with catalytic reduction, when receiving electrons from the developer, causing growth of the silver cluster (Jaenicke, 1972). Now a metallic speck is formed which continues to grow, dependent on the length of the developing action until the entire light-exposed grain is converted into metallic silver. When stopping the developing process at an appropriate time, a wide range of gradation in grain darkening proportional to the initial light exposure can be achieved.

Individual grains may contain several specks of silver when more than one cluster of the latent image was formed per grain. Also, with progressing development, a negative-feedback mechanism slows down the reduction process, making it easier to time the appropriate length of the development.

The *fixation process* dissolves all remaining AgBr and thereby stabilizes the developed picture. For reviews, see Brown, (1976), James, ed. (1977), and Hamilton (1988).

46.2.1 The Print-Out Effect

When AgBr grains or single crystals are exposed to a larger dose of light, increasingly more silver nuclei are formed, and grow visibly as random spots. During this process Br_2 is released.

The motion of Ag to the Ag_n nuclei proceeds via interstitials which have a high mobility. Dislocations and grain boundaries assist in providing the space for Ag_n nuclei growth.

46.2.2 Film Sensitivity and Sensitization

The sensitivity of a photographic film is determined by a number of effects:

- grain size and shape;
- creation of a free electron from a photon in the spectral range of interest;
- diffusion of this electron to a nucleation site for latent image-cluster formation in competition with deep trapping and recombination with mobile holes (Mitchell, 1957); and
- critical size of the latent image cluster to permit growth during exposure to the developing agent.

The *grain size* influences in a most direct way the film sensitivity, since the same number of photons is sufficient to produce a latent image cluster, independent of grain size for typical grain sizes between $0.05 \ldots 2 \ \mu m$ diameter. One cluster is enough to turn the entire grain into metallic silver after sufficient development. In the range of AgBr absorption, the speed of a photographic film is therefore proportional to d^3, where d is the grain diameter.

The *diffusion of the free electron* to a nucleation site depends on its mobility, typically $50 \ cm^2/Vs$ at room temperature, intermediate trapping at shallow traps with $E_c - E_t \simeq 30 \ meV$, and recombination. For good films, the collection efficiency is typically $\sim 50\%$.

The *critical size of a latent image* cluster that can be developed is on the order of 4 to 7 atoms, and requires 8 to 14 photons for its creation. However, its size can be reduced by *sensitization*.

For such sensitization, a surface treatment—resulting in a deposit of gold ions, and the formation of Ag_2S—is commonly used; the latter is inadvertently done by incorporating the AgBr grains in sulphur-containing gelatine (for recent comments, see Keevert and Gokhale, 1987). As a result, the critical size of a developable cluster is reduced to ~ 3 to 4 gold/silver atoms. The clusters containing gold atoms are more stable than those composed of only silver atoms.

A slight, further sensitization can be achieved by a treatment in a reducing atmosphere (H_2) prior to exposure. Such treatment, however, reduces the safety margin to developing unexposed grains (fogging). Here, the threshold for creating developable clusters can be reduced to 2...3 photons.

The *spectral range* can be extended by absorbing certain organic dyes at the grain surface, which permit an easy electron transfer to the AgBr; best quantum efficiencies exceed 80%. See James (1977).

A shift of the absorption edge toward longer wavelength in AgBr is achieved by admixture of AgI, typically at the 1...10% range. For further information, see Hamilton (1988) and Mitchell (1989).

46.3 Time-Dependent Photochemical Reactions

A large group of photochemical reactions relates to a slow degradation of photoelectric properties during optical excitation, such as the Staebler-Wronski effect in α-Si:H or α-Si:F (Staebler and Wronski, 1980—for a recent review, see Hamakawa, 1988) or the slow decrease of photoconductivity in CdS (Böer et al., 1954; Borchardt, 1962).

Many other degradation effects, observed after exposure to sunlight, are identified as those caused by similar photochemical reactions, e.g., the chaffing of ZnO-based white paint (Hauffe, 1955).

46.3.1 Photochemical Degradation

As an instructive example, we will present here the kinetic analysis of photochemical reactions in CdS. It reveals the slow production of recombination centers with light, and the annealing of such centers in the dark (Fig. 46.4). The photocurrent first increases after light is switched on, with a rise time given by the filling of electron traps, as discussed in Section 47.1.1A. In addition, however, recombination centers are formed with a substantially slower reaction time. As a result of this photochemical reaction, the current goes through a maximum and decreases slowly until it reaches a steady-state condition. When the light is switched off, the current first decays with the electronic decay time. However, it undershoots below the original dark current, indicating that some donors were eliminated with the photochemical formation of recombination centers. With sufficient waiting in the dark, these recombination centers are annihilated, and the original density of donors is restored; consequently, the current returns to its original dark current.

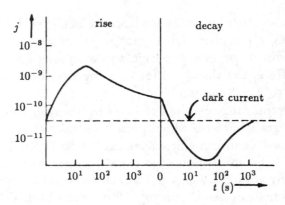

Figure 46.4: Rise and decay of the photocurrent for CdS at 100 °C, including photochemical reactions causing a nonmonotonic behavior (after Böer et al., 1954).

The two processes involved here are the redistribution of electrons over the bands and traps, and the photochemical formation of recombination centers; both have substantially different time constants and therefore are easily separated in the described kinetic experiment.

The general behavior can be described mathematically by assuming a creation of recombination centers with a density N_r proportional to the density of electrons, and an annihilation proportional to N_r (Böer et al., 1954):

$$\frac{dN_r}{dt} = \gamma_{pc} n - \delta_{pc} N_r. \tag{46.1}$$

The dissociation parameter δ_{pc} depends exponentially on the temperature:

$$\delta_{pc} = \delta_{pc0} \exp\left(-\frac{E_{pc}}{kT}\right); \tag{46.2}$$

the photochemical activation energy E_{pc} is observed to be typically on the order of 1 eV for CdS. E_{pc} depends on the doping elements (Borchardt, 1962).

The photoconductivity can be described with a reaction-kinetic model, including such recombination centers and optical as well as thermal generation rates—this includes the dark conductivity. The simplest of these models is given by

$$\frac{dn}{dt} = g_{th} + g_o - c_{cr} n N_r - c_{cr} n^2. \tag{46.3}$$

When augmented by traps and other recombination centers, the model becomes more realistic.

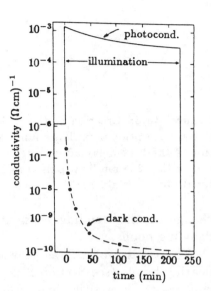

Figure 46.5: Photo- and dark conductivity of undoped α-Si:H at 300 K as a function of the time of illumination with 200 mW/cm^2 tungsten light (after Staebler and Wronski, 1977).

46.3.2 The Staebler-Wronski Effect

Similar photochemical reactions, as described in Sections 46.1.1 and 46.3.1 for CdS, are observed in α-Si:H (Staebler and Wronski, 1977) when it is exposed to prolonged illumination at sufficient intensity (sunlight) and temperature (300 K), as shown in Fig. 46.5. Parallel with the reduction in photocarrier lifetime, the dark conductivity, shown as the dashed curve in Fig. 46.5, and band-tail luminescence decrease and an increase in unpaired spin density and sub band-gap optical absorption is observed (Staebler and Wronski, 1980; Dersch et al., 1981; Moddel et al., 1980). The centers created by photochemical reaction also anneal out during heat treatment in the dark (Han and Fritzsche, 1983).

Despite a large body of experimental evidence, the photochemical reaction products in α-Si:H are not yet unambiguously identified. Three proposals are currently considered: that of *weak bond breaking* (Pankove and Berkeyheiser, 1980; Stutzmann, 1987); an *intimate charge transfer pair* with negative-U character (Adler, 1983); and the creation of pairs of dangling and floating bonds (Pantelides, 1978).

The first process involves the breaking of weak Si-Si bonds in the α-Si matrix with the possible motion of hydrogen to the dangling bond, thereby hindering bond healing.

The second process assumes the trapping of photogenerated electrons at sp^2-donors and of holes at p^3-acceptors, which causes a

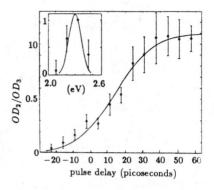

Figure 46.6: Creation of the F-center absorption in KCl as a function of the time-delay after excitation with a 266 nm light pulse (after Bradford et al., 1975).

change in the bonding configuration by creating a pair of sp^3-dangling bonds.

The third process was shown in Fig 25.3. With highly mobile floating bonds (see Section 25.2.3), they can react with Si:H bonds and create interstitial hydrogen: $FB + Si \rightleftharpoons H_i + Si_{Si}$. Interstitial hydrogen can also be used to anneal dangling bonds: $H_i + DB \rightleftharpoons SiH$. For recent reviews, see Pantelides (1988) and Hamakawa (1989).

46.3.3 Fast Photochemical Processes

Electronically induced photochemical processes in which a lattice ion only needs one step to a new metastable position can proceed very fast, typically in nano- or even picoseconds. As an example, we give the previously discussed formation of F- and H-centers in alkali halides—see Sections 22.4.1 and 46.1.1. As shown by Bradford et al. (1975), F-centers can be created in KCl by light pulses at 266 nm within ~ 10 ps (see Fig. 46.6). Higher excited states of the F-center are the precursors of the F-H-center pair formation (Williams, 1978; Stoneham, 1979; Itoh, 1982). Excitation with a nanosecond electron pulse forms F- and H-centers within 50 ns after the electron pulse (Kondo et al., 1972). This pair is annihilated within a few minutes after cessation of the excitation—see Fig. 46.7.

Summary and Emphasis

Photochemical reactions cause the motion of lattice or defect atoms into new metastable positions as a result of optical excitation. This motion is caused by giant oscillation of the involved atom after carrier capture and while relaxing.

In photoconductors, the photochemical reaction usually converts donors or acceptors into recombination centers, probably by associate formation. Some of these processes need thermal activation and

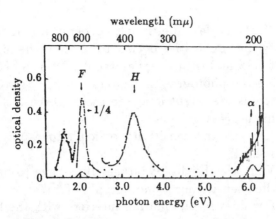

Figure 46.7: Transient optical absorption spectrum of KBr at 8 K measured 50 ns after excitation by an electron pulse, showing the created F- and H-absorption. A few minutes later, most of the induced absorption has disappeared except for a small F- and α-remnant at ~ 2 and ~ 6 eV (after Kondo et al., 1972).

cause a slow degradation of the photoconductive properties, e.g., in CdS or in α-Si. Other processes proceed very fast, as in the ns-range. For instance, the creation of an F-H center pair in alkali halides is initiated by a very fast relaxation of a self-trapped exciton.

There is a wide range of photochemical reactions, from the highly desired Ag-cluster formation in silver halides which forms the latent image, to the photochemical degradation (chaffing) of ZnO-based white paint, organic pigments, or certain polymers.

Many of the photochemical-reaction products can be annealed out with a heat treatment in the dark in CdS and α-Si:H. Other processes are irreversible and result in permanent material damage (chaffing of paint).

Photochemical degradation is a serious degradation process that limits the life of many products exposed to light. Such degradation is defect-structure-related, and can be minimized by proper doping.

Exercise Problems

1.(r) Can photochemical reaction change the donor or acceptor level density?

    (a) Do we expect a threshold to observe such a change?

    (b) Of what nature is such a threshold?

    (c) How can we observe it?

    (d) How are recombination centers involved in such a change?

2.(r) Quantitatively describe the photochemical reactions in CdS.

3.(*) Identify the similarities between the photochemical degradation of CdS and α-Si:H—the Staebler-Wronski effect.

4.(l) Describe the photographic process.

    (a) How is the latent image formed? Discuss the physics of this process.

    (b) How can the sensitivity of a photographic film be increased?

    (c) What are the disadvantages that have to be considered when sensitizing a photographic film?

    (d) How does recombination interfere with the latent-image formation?

    (e) Describe in detail the physics of the developing process.

    (f) What processes take place during fixation?

5.(*) Is there a barrier in the F-H associate formation from a self-trapped exciton in alkali-halides?

6.(e) Give a quantitative estimate for the reduction in donor density indicated in Fig. 46.4, obtained from a CdS platelet of $100\ \mu m \times 2$ mm cross section, with 5 mm electrode distance, and 10 V applied. Assume a donor depth of $E_c - E_d = 0.7$ eV and $T = 300$ K.

PART VIII

KINETICS

Chapter 47

Kinetics of Electron Distribution in Defects

Kinetic effects permit separate access to different solid-state processes that have substantially different time constants (relaxation times), and thus reveal predominant interactions.

At the beginning of this chapter, we present a few general remarks to introduce various kinetic effects in semiconductors. Six major systems with vastly different relaxation times can be distinguished:

- creation of atomic lattice defects,
- carrier redistribution in defect levels,
- phonon relaxation,
- orientation relaxation of anisotropic states,
- carrier-scattering relaxation (momentum and energy relaxation), and
- electron spin relaxation.

The corresponding relaxation times span time constants from geologic times down into the femtosecond range. The entire time range is now accessible to experimental observation.

We have touched on the creation of lattice point defects in Chapter 19, and briefly discussed some kinetic aspects in Section 19.1.2. The redistribution of carriers over defect levels in the band gap is the subject of this chapter, and the various relaxation effects will be discussed in the following chapters.

Kinetic effects are those in which any of the semiconductor variables change with time. These could be defect densities, carrier densities and mobilities, optical absorption, luminescence, or their polarization.

In order to induce such changes, external parameters are changed as a function of time, such as an external bias, magnetic field, optical generation rate, temperature, pressure, or other electromagnetic or particle irradiations.

These changes are induced as part of the operation of a semiconducting device, e.g., the bias or changing light intensity, or as a means of obtaining specific information about certain semiconducting parameters. Here and in the following chapters, a few examples of both will be presented. In the previous chapter, changes in the crystal-defect structure induced by photons were discussed (photochemical reactions). We will now analyze the change in carrier distribution over the energy levels in a semiconductor caused by changes in excitation.

When external excitation parameters are changed—such as light intensity, temperature, or electric fields—changes in internal parameters occur, e.g., in generation or recombination rates, and internal variables, such as the electron density distribution in traps and bands, will change. After a new steady-state distribution is reached and the external parameters are returned to the original value, the electron density distribution relaxes with time constants that are characteristic of each of the subsystems.

In the following sections, changes of the density distribution of electrons over defect levels in the band gap and of electrons and holes in conduction and valence bands are analyzed. The bands in this approximation are represented by levels at their band edges with an effective density of N_c and N_v [Eqs. (27.18) and (27.22), respectively], disregarding an energy distribution of the carriers within the band. This will be the subject of discussion in Section 48.2.2.

First, changes in the optical generation rate will be used to initiate the kinetic behavior.

47.1 Changes in Optical Excitation

Photoconductors are used in a wide variety of applications: street-light control, light meters in cameras, star trackers in satellites, and photoelectric counters, to name a few. In many applications, the photoconductor must respond to a rapidly changing light signal, which requires a sufficiently short rise and decay time.

47.1.1 Rise and Decay of Photoconductivity

The kinetics of a photoconductor is determined by the set of reaction-kinetic equations, one for each of the electron densities in a given level or band; we will neglect a redistribution of carriers within the band, and assume here that all transition coefficients are independent of such redistribution. In a homogeneous semiconductor, the kinetic expressions are of the form

$$\frac{dn_i}{dt} = g_i - r_i \quad i = 1, 2, \ldots, \tag{47.1}$$

where g_i and r_i are the generation and recombination rates for the ith level, respectively, defined in Chapter 42. The generation term contains all contributions that cause an increase in the population of this level; the recombination term combines all contributions that cause a decrease of the population. They can be written as

$$g_i = \sum_j n_j \, e_{ji} (N_i - n_i) \quad \text{and} \quad r_i = \sum_j n_i \, c_{ij} (N_j - n_j), \tag{47.2}$$

where e_{ji} and c_{ij} are the transition coefficients between level i and level j, N_i is the density of level i, and n_i is the density of electrons in the level, assuming single occupancy of each level; e transitions are upward (excitation) transitions within the band diagram, while c transitions are downward (capture) transitions with release of energy—Fig. 47.1.

Equation (47.1) represents a set of first-order, nonlinear differential equations, which usually cannot be solved in closed form except for some exceedingly simple cases, examples of which are given below.

47.1.1A Rise and Decay of Intrinsic Photoconductivity

The kinetics of an intrinsic photoconductor (Fig. 45.4A) is determined for electrons by

$$\frac{dn}{dt} = g_{\text{vc}} - c_{\text{cv}} np \tag{47.3}$$

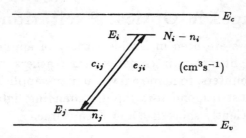

Figure 47.1: Electronic transition between two defect levels with the transition coefficient identified.

and by a similar equation for holes. With $g_{vc} = g_o$ (optical generation rate) and for sufficiently high generation with $n = p$, this equation has the solution

$$n(t) = \sqrt{\frac{g_o}{c_{cv}}} \tanh\left[\sqrt{g_o c_{cv}}(t - t_o)\right], \qquad (47.4)$$

with $n = g = 0$ for $t < t_o$ and $g = g_o$ for $t \geq t_o$. This shows a nearly linear rise of n for $t_0 < t < t + \tau_0$, with τ_0 as the photoconductive rise time,

$$\tau_0 = \frac{1}{\sqrt{g_o c_{cv}}}. \qquad (47.5)$$

According to the tanh function, the electron density approaches its steady state-value

$$n_0 = \sqrt{\frac{g_o}{c_{cv}}} \qquad (47.6)$$

exponentially. When, after reaching steady state, the optical excitation is switched off ($g = 0$ for $t > t_1$), the decay of the photoconductivity follows—see Eq. (47.3):

$$\frac{dn}{dt} = -c_{cv} n^2. \qquad (47.7)$$

This decay is hyperbolic

$$n(t) = \frac{1}{c_{cv}(t - t_1)} = n_0 \frac{\tau_1}{t - t_1} \qquad (47.8)$$

with a decay time constant of

$$\tau_1 = \frac{1}{n_0 c_{cv}}. \qquad (47.9)$$

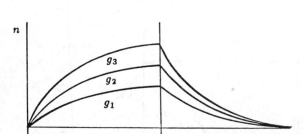

Figure 47.2: Rise and decay of intrinsic photoconductivity with the optical generation rate as the family parameter ($g_3 > g_2 > g_1$).

Introducing n_0 from Eq. (47.6), one sees that both the rise and decay time constants are equal:

$$\tau_1 = \tau_0. \tag{47.10}$$

Rise and decay proceed faster at higher light intensities and in materials with larger recombination cross sections $c_{cv} = s_{cv}v_{rms}$—see Section 42.2.3.

47.1.1B The Influence of Traps on Photoconductivity Kinetics

When traps are present, there are competing transitions. For instance, in the presence of electron traps, one obtains

$$\frac{dn}{dt} = g_o - \frac{dn_t}{dt} - c_{cv}n(n + n_t) \tag{47.11}$$

and

$$\frac{dn_t}{dt} = c_{ct}n(N_t - n_t) - e_{tc}N_c n_t. \tag{47.12}$$

Here, quasi-neutrality requires $p = n + n_t$. The transitions of electrons into traps compete with transitions into the valence band. As long as traps are mostly empty, the first transition predominates: with $n_t \ll N_t$, one has

$$\frac{dn_t}{dt} \simeq c_{ct}nN_t; \tag{47.13}$$

hence,

$$\frac{dn}{dt} \simeq g_o - c_{ct}nN_t \tag{47.14}$$

as long as $c_{ct}N_t \gg c_{cv}p(n + n_t)$. After separation of variables, Eq. (47.14) can be solved, yielding

$$n = n_0\left\{1 - \exp\left(-\frac{t - t_0}{\tau_t}\right)\right\} \quad \text{with} \quad \tau_t = \frac{1}{c_{ct}N_t}, \tag{47.15}$$

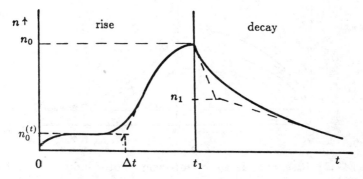

Figure 47.3: Rise of the photoconductivity at medium intensity optical excitation, which shows a plateau until deep traps are filled. In the decay of photoconductivity, two slopes are indicated.

and with τ_t as the time constant for trap-filling. The quasi-steady state between generation and trapping into nearly empty traps yields for the density of electrons in the conduction band while the traps are being filled:

$$n_0^{(t)} = \frac{g_o}{c_{ct} N_t}. \tag{47.16}$$

When the traps become filled, the trapping transition becomes clogged. Consequently, n rises again until steady state between generation and recombination is reached. With $N_t \gg n$, i.e., for low-generation rate, one obtains for the second rise

$$\frac{dn}{dt} = g_o - c_{cv} n N_t. \tag{47.17}$$

The kinetics for trap-filling is similar to the first rise of the electron density, except for a somewhat longer time constant, since usually $c_{ct} \gg c_{cv}$:

$$\tau_1 = \frac{1}{c_{cv} N_t}. \tag{47.18}$$

This behavior is presented in Fig. 47.3, which shows a first rise to a plateau while trap-filling occurs, and then another rise to reach a steady-state value at

$$n_0 = \frac{g_o}{c_{cv} N_t}. \tag{47.19}$$

The ratio of $n_0 / n_0^{(t)}$ can be used to determine the ratio of capture cross sections of electron traps and band-to-band transitions:

$$\frac{n_0^{(t)}}{n_0} = \frac{s_{ct}}{s_{cv}}. \tag{47.20}$$

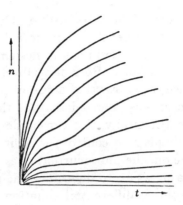

Figure 47.4: Rise of the photo-conductivity with trap-filling; light intensity as the family parameter. It shows an intermediate intensity range in which a plateau develops (after Böer and Vogel, 1955).

The length of the plateau Δt can be used to estimate the density of traps

$$g_o \Delta t \simeq N_t \qquad (47.21)$$

if recombination can be neglected during trap-filling (Böer and Vogel, 1955).

The decay after switching off the light proceeds inversely, first with a time constant given by the recombination transition, obtained from

$$\frac{dn}{dt} = -c_{cv} n N_t, \qquad (47.22)$$

which yields an exponential decay

$$n = n_0^{(t)} \exp\left(-\frac{t - t_1}{\tau_1}\right), \qquad (47.23)$$

where τ_1 is given by Eq. (47.18). Later, electrons are supplied by the emission from traps $(e_{tc} N_c n_t)$, which determines the slow tail of the decay of n. From

$$\frac{dn}{dt} = e_{tc} N_c N_t - c_{cv} n N_t \qquad (47.24)$$

for the first part of the decay, as long as n_t is still $\simeq N_t$, one obtains the onset of the slow decay when $n = n_1$ is reached, with a condition for

$$n_1 \simeq N_c \frac{e_{tc}}{c_{cv}} = N_c \frac{e_{tc}}{c_{ct}} = \frac{c_{ct}}{c_{cv}} N_c \exp\left(-\frac{E_c - E_t}{kT}\right), \qquad (47.25)$$

i.e., when the quasi-Fermi level just passes through the trap level. The decay is then determined by the slow depletion of the traps with a time constant given by the *net* emission from traps, i.e., emission minus retrapping.

47.1.1C The Method of Controlled Excitation The discussion of the reaction-kinetic behavior can be substantially simplified when the experiment is designed to eliminate one of the differential equations. Böer and Oberländer (1958) developed a method in which, after starting a rise of photoconductivity, the optical excitation rate is changed so that the electron density is kept constant. Experimentally, this can be done automatically by using a feedback mechanism to control the light intensity so that, from a preselected time on, the photocurrent does not change. Instead of $n(t)$, one now measures $g(t)$ to determine the parameters of interest. This converts the set of differential equations [Eqs. (47.11) and (47.12)] into the set

$$\frac{dn}{dt} = g(t) - \frac{dn_t}{dt} - c_{cv}n(n + n_t) \equiv 0 \tag{47.26}$$

$$\frac{dn_t}{dt} = c_{ct}n(N_t - n_t) - e_{tc}N_c n_t. \tag{47.27}$$

The order of this set is reduced by one, and it has the solution

$$g(t) = A + B \exp\left(-\frac{t - t_k}{\tau_k}\right), \tag{47.28}$$

with t_k as the start of the controlled optical excitation, and

$$\tau_k^{-1} = e_{tc}N_c + c_{ct}n_k, \tag{47.29}$$

$$A = c_{cv}n_o^2(1 + \tau_k c_{ct}N_t), \tag{47.30}$$

$$B = \left(\frac{1}{\tau_k} - c_{cv}n_0\right)\left(\tau_k n_k c_{ct}N_t - n_t(t_k)\right), \tag{47.31}$$

where n_k is the electron density for $t \geq t_k$. For $t > t_k$, the electron density is kept constant by changing the optical generation rate as necessary—Fig. 47.5B.

When the initiation time t_k is varied, the reciprocal time constant τ_k^{-1} of the changing generation rate varies linearly with the electron density n_k. From the slope of $n_k(\tau_k^{-1})$, one obtains the trapping coefficient c_{ct}, and from the intersect of the ordinate, the trap emission probability $e_{tc}N_c$—see Fig. 47.5C.

This method is useful when one set of traps predominates. Then the capture cross section of traps ($s_{ct} = c_{ct}/v_{rms}$) can be obtained and, when the experiment is done at different temperatures, the ionization energy and frequency factor of these traps can be determined:

$$e_{tc}N_c = v_t \exp\left(-\frac{E_c - E_t}{kT}\right). \tag{47.32}$$

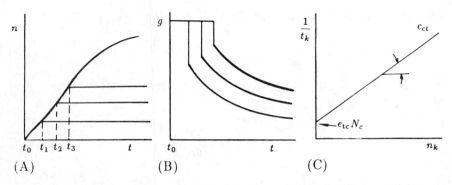

Figure 47.5: Method of controlled excitation. (A) Rise to a constant photoconductivity stopped at $t = t_1, t_2, ...t_k$ by changing the light intensity from this time on, according to (B) in three experiments. (B) Variation of the light intensity in order to achieve a constant photoconductivity. (C) Reciprocal time constant for changing light intensity as a function of constant electron density.

47.1.1D Photoconduction Time Constants The electron life-time in the conduction band is an important parameter in the discussion of photoconductivity:

$$\tau_n = \frac{n}{g_o}. \tag{47.33}$$

This lifetime depends on the electron distribution over levels in the band gap and thus depends on the light intensity $\tau_n(g_o)$. When only small changes from steady-state values are considered (see Section 47.1.1E), one may set this lifetime, here for extrinsic photoconductivity

$$\tau_n = (c_{ca}p_a)^{-1} \simeq (c_{ca}N_a)^{-1}, \tag{47.34}$$

as a constant, and obtains from Eq. (47.11)

$$\frac{dn}{dt} = g_o - \frac{n}{\tau_n} - \frac{dn_t}{dt}. \tag{47.35}$$

Assuming an exponential decay function for the photoconductivity, one has for the *photoelectron decay time* τ, which may be an implicit function of the electron density

$$\frac{dn}{dt} = -\frac{n}{\tau}, \tag{47.36}$$

and with it, one obtains from Eq. (47.35) with $g_o = 0$

$$\frac{dn}{dt} = -\frac{n}{\tau_n} - \frac{dn_t}{dt} = -\frac{n}{\tau}. \tag{47.37}$$

This presents an instructive relation between the decay time τ and the carrier lifetime τ_n:

$$\boxed{\tau = \tau_n\left(1 + \frac{dn_t}{dn}\right).} \tag{47.38}$$

The decay time is always longer than the carrier lifetime. It is increased by the change of trap population, which is caused by a changing electron density.

This is also an example for demonstrating the existence of *two relaxation times for the two subsystems:* conduction band/activator and conduction band/electron traps. When additional trap levels become involved, more relaxation times will appear—see Section 47.1.1F.

47.1.1E Small Sinusoidal Excitation

Another method designed to obtain reaction-kinetic parameters is based on the linearization of solutions, which can be applied if a small light signal is superimposed on top of a constant bias light:

$$g = g_o(1 + m\cos\omega t) \quad \text{with} \quad m \ll 1. \tag{47.39}$$

The constant light moves the quasi-Fermi level to a position around which the trap distribution is to be investigated, and the modulated light signal probes these trap levels. The method was originally used by Fassbender and Lehmann (1949) and further developed by Niekisch (1955).

As a result of this oscillating light signal, one obtains an electron density modulation

$$n = n_0 + n_1 \cos(\omega t - \phi). \tag{47.40}$$

Introducing the *Ansatz* (47.40) into the reaction-kinetic differential equation, one obtains the trap density from the phase shift:

$$N_t = \frac{g_o \tan\phi}{\omega}. \tag{47.41}$$

The coefficients are given by

$$n_0 = g_o\tau_n, \quad n_1 = \frac{g_o m \tau_n}{\sqrt{1 + \omega^2\tau^2}}, \quad \text{and} \quad \tan\phi = \omega\tau. \tag{47.42}$$

From the amplitude of the photocurrent modulation, one obtains the ratio of the time constants for $\omega\tau \gg 1$

$$n_1 = \frac{mg_o}{\omega}\frac{\tau_n}{\tau}. \qquad (47.43)$$

By changing the constant bias light, one can now move the quasi-Fermi level and probe a wide range of the trap distribution—see the following section. For the measured dc- and ac-photocurrent components, one consequently obtains

$$j_{\mathrm{dc}} = e\mu_n g_o \tau_n F \quad \text{and} \quad j_{\mathrm{ac}} = e\mu_n \frac{g_o \tau_n m}{\sqrt{2\omega\tau}} F. \qquad (47.44)$$

When the light intensity is high enough to fill all traps, dn_t/dn vanishes or the quasi-Fermi level falls in between two trap levels so that $dn_t/dn \ll 1$—see Eq. (47.38). Then τ_n becomes equal to τ, and Eq. (47.44) can be used to determine the carrier mobility:

$$\mu_n = j_{\mathrm{ac}}\frac{\sqrt{2\omega}}{eg_o m F}. \qquad (47.45)$$

47.1.1F Kinetics Influenced by Trap Distribution The competition between trapping and thermal reemission tends to fill (or deplete) a distribution of traps from lower to higher (or from higher to lower) traps for traps of similar capture coefficients. For sufficiently high temperatures and light intensities, a quasi-stationary approximation is justified; that is, one can use quasi-Fermi levels for the description of the gradual filling or depletion of the trap distribution

$$n(t) \simeq N_c \exp\left(-\frac{E_c - E_{Fn}(t)}{kT}\right) \qquad (47.46)$$

when changes in $n(t)$ proceed slowly enough so that the degree of trap-filling can follow.

The total trapped-electron density within such a trap distribution $N_t(E)$ [measured in $\mathrm{cm}^{-3}\,(\mathrm{eV})^{-1}$] is then given by

$$n_t = \int_{E_F}^{E_{Fn}} N_t(E)\,dE. \qquad (47.47)$$

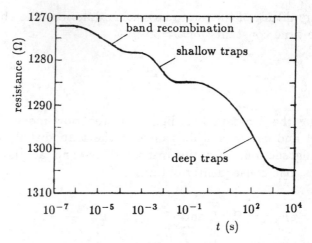

Figure 47.6: Decay of photoconductance in *p*-Si, indicating three distinct ranges of decay involving band-to-band recombination, and reemission from shallow and from deep traps (after Hornbeck and Haynes, 1955).

The change of trapped electrons within a trap distribution as a function of the electron density can be calculated from

$$
\frac{dn_t}{dn} = \int_{E_F}^{\infty} \frac{dn_t(E)}{dn} dE = \frac{1}{n} \int_{E_F}^{\infty} N_t(E) \frac{\exp -\dfrac{E - E_{Fn}}{kT} dE}{\left\{ 1 + \exp -\dfrac{E - E_{Fn}}{kT} \right\}^2}
$$

$$
= \frac{kT}{n} \left\{ N_t(E_{Fn}) + \frac{(\pi kT)^2}{6} \left. \frac{d^2 N_t}{dE^2} \right|_{E_{Fn}} + \dots \right\}.
$$

$$(47.48)$$

Therefore, the relation between decay time and electron lifetime can be written as

$$
\frac{\tau}{\tau_n} \simeq 1 + \frac{N_t(E_{Fn}) kT}{n}.
$$

$$(47.49)$$

This presents a relatively direct method for measuring the trap distribution, by obtaining τ_n from n/g_0 and τ from the decay immediately after switching off the optical excitation, provided the temperature and light intensity are high enough for the quasi-Fermi level to follow the changes in n in a quasi-steady-state condition—or in a more elegant way, from the methods discussed in the previous section.

A good example for the sequential depletion of shallow and deep traps is shown in Fig. 47.6, measured for *p*-Si. Here, the decay of the photoconductivity shows the typical ranges of substantial slowdown while the depletion of a specific level takes place.

For n distinct groups of levels, one would expect $n - 1$ relaxation times (Landsberg and Cole, 1966). Thus, one observes one relaxation time for the band-to-band transition, plus *several relaxation times* for the transitions between bands and different trap levels. A differentiation into distinct trap-related relaxations becomes possible when groups of trap levels are substantially separated from each other with only a few traps in between, as given in the example of Fig. 47.6.

47.1.2 Ultrafast Photodetectors

For many applications, such as in pattern recognition, fiber optics communications, and electro-optical computers, one desires sensitive and very fast photoconductive devices. Fast rise and decay times can be achieved by avoiding traps near the portion of the quasi-Fermi level under operating conditions, and by providing fast recombination centers. This reduces the lifetime, but also the photosensitivity, which can be partially compensated by selecting photoconductors with high carrier mobility.

An example of a reasonably sensitive ultrafast photoconductive detector is GaAs, when deposited at low temperatures (200 °C) by molecular beam epitaxy, which results in high dark resistivity and rise and decay times of ~ 1.5 ps in response to a short laser flash (Smith et al., 1988).

47.1.3 Competing Excitation Processes

The competing effects of optical excitation from different levels can be separated by a kinetic analysis whenever their relaxation times are sufficiently different from each other.

We will briefly review the various competing optical transitions first:

(a) Transitions with the same photon energy can proceed from an activator into the conduction band or from the valence band into an electron trap—Fig. 47.7A; or it can proceed from different deep levels into higher energies ($k \neq 0$) of the band when the deep center extends over a wider k range (Fig. 47.7B).

(b) Transitions with different photon energies can proceed from different sets of deep levels to the band edge—Fig. 47.7C, or from

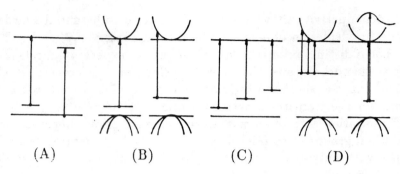

$$(A) \qquad (B) \qquad (C) \qquad (D)$$

Figure 47.7: Various extrinsic excitations as described in the text.

the same deep center into different parts of the same or a different band—Fig. 47.7D.

The resulting carrier distribution usually shows substantially different relaxation, depending on whether valence or conduction bands are involved, or carrier relaxation occurs from higher to lower band states. The latter will be discussed in Section 48.2.

A typical two-beam excitation, involving competing excitations into conduction and valence bands, is referred to as the *quenching transition* and will be discussed in the following section.

47.1.3A Optical Quenching   When one light beam induces sensitized photoconductivity and another causes desensitization, one speaks of *optical quenching*. Both processes proceed with a substantially different time constant: desensitization is usually the slower process.

A typical example is the quenching of the photoconductivity in CdS, as described in Section 45.8.1. Here, the primary photogeneration with an excitation at 2.3 eV from a fast center into the conduction band shows a fast rise of the photocurrent. When a steady state is reached, a second light pulse of 30 s duration and variable photon energy is added as shown in Fig. 47.8. A monotonic rise shows the additional excitation of conduction electrons when the photon energy is close to that of the primary beam. With decreasing photon energy of the secondary beam, however, an increasing contribution of the quenching transition is observed by a pronounced nonmonotonic behavior. For photon energies above ~ 1.75 eV, such a contribution is hidden in stationary measurements—Section 45.8.1: the resulting stationary photoconductivity is still larger when the second beam is applied in addition to the initial excitation at 2.3 eV—see Fig. 47.8.

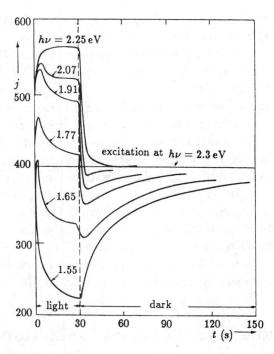

Figure 47.8: Rise and decay of the photocurrent in CdS with two light beams (primary beam at $h\nu = 2.25$ eV), indicating competition between electron generation and quenching when a nonmonotonic behavior becomes visible (after Taft and Hebb, 1952).

However, the nonmonotony observed in kinetic measurements indicates some quenching for energies below 2.1 eV.

The nonmonotonic behavior can also be observed in the decay of the additional photoconductive component that is induced by the second light beam at energies between 1.65 and 2.07 eV, resulting in an undershoot below the stationary value. The different relaxation times for the two subsystems, electrons/holes and holes/sensitizing centers, permit a distinction between the different processes.

In addition to the optical excitation described in the previous sections, there is a preferred excitation of anisotropic lattice defects with polarized light that relaxes by recombination. We will now briefly discuss this recombination relaxation.

47.2 Orientation Relaxation

The use of polarized light results in an excited state of anisotropic centers that is created in a subset of such centers with preferential orientation; the symmetry axis of the excited centers points in the direction of the electric vector of the exciting light. These excited states may involve, e.g., donor-acceptor pairs, other defect associates as discussed in Chapter 23, as and anisotropic molecular dopants in organic semiconductors.

The population of anisotropically excited states can be measured by observing bleaching, i.e., increased transmission, as soon as the population of the ground state becomes significantly reduced, while the population of the excited state approaches saturation. This bleaching is anisotropic and relaxes with a time constant given by the lifetime of the excited states. In addition, such anisotropic excitation can cause secondary effects which facilitate observation of its time evolution (Oudar et al., 1985), such as Faraday rotation— i.e., a polarization rotation. The relaxation of these anisotropic excited states occurs by recombination, except for rare cases in which resonance excitation of adjacent centers with similar energy states is possible.

47.3 Changes in Thermal Ionization

Changes in temperature cause an exponential change of the thermal ionization of carriers from traps into the respective bands. The free carriers obtained from thermal ionization contribute to an increased conductivity and luminescence. In addition, other transitions cause changes in sensitization or quenching. In a variety of experimental methods, the changes in temperature are usually made with a constant heating rate. An analysis of the resulting current or luminescence as a function of time permits a distinction of the different ionization effects.

47.3.1 Thermally Stimulated Luminescence

When a luminophore has deep traps, carriers are trapped at room temperatures and are frozen-in over long time periods. An estimate of the residence time in such traps can be made from the thermal emission probability

$$e_{tc} = \nu_t \exp\left(-\frac{E_c - E_t}{kT}\right). \tag{47.50}$$

A trap with an ionization energy of 1.5 eV, located close to the middle of the band gap in a wide gap luminophore (e.g., ZnS with $E_g = 3.6$ eV), has a residence time of millions of years at room

Table 47.1: Residence time of carriers in deep traps with an attempt-to-escape frequency $\nu_t = 10^{10}$ s^{-1}.

Trap depth	$T = 200$ K	$T = 300$ K	$T = 450$ K	$T = 600$ K
0.6 eV	1.3 d	1.2 s	$5 \cdot 10^{-4}$ s	$1 \cdot 10^{-5}$ s
0.8 eV	390 y	43 min	$8.5 \cdot 10^{-2}$ s	$5 \cdot 10^{-4}$ s
1 eV	$4.1 \cdot 10^7$ y	0.5 y	15 s	$2.5 \cdot 10^{-2}$ s
1.2 eV	$4.3 \cdot 10^{12}$ y	430 y	43 min	1.2 s
1.5 eV		$5 \cdot 10^7$ y	65 d	6.3 min

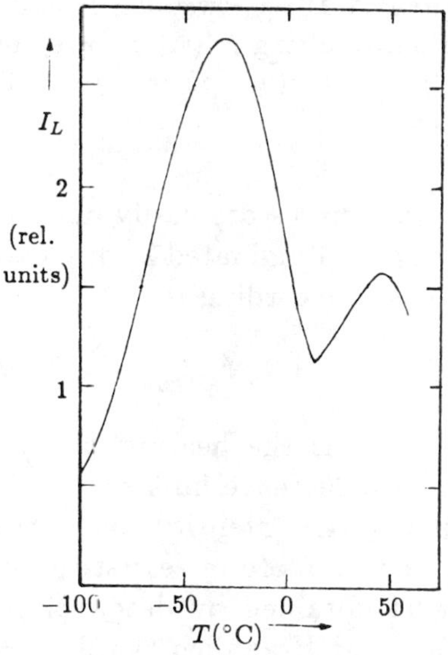

Figure 47.9: Glowcurve of a ZnS phosphor with two deep trap levels at 0.4 and 0.75 eV, resulting in peaks of the luminescence at -30 and $+45°$C, respectively (after Urbach, 1930).

temperature—see Table 47.1. Heating the luminophore to 600 K reduces this residence time to only a few minutes.

A luminophore with deep traps which store carriers for more than a second is called a *phosphorescent material* or a *phosphor*. Slowly released luminescence by electrons emitted from deep traps is called *phosphorescence*.

When the temperature is increased, the amount of phosphorescence will rapidly increase as more electrons are released from deep traps. It will then go through a maximum, and decrease again when the deep traps become depleted. Such an emission curve is called a *glowcurve* (Urbach, 1930)—Fig. 47.9.

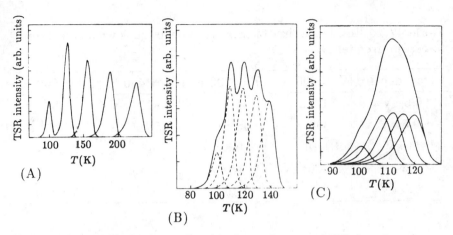

Figure 47.10: Glowcurves caused by several traps of different energy separation, resulting in (A) resolved or (C) unresolved glowcurve structures; (B) representation of an intermediate case (after Simmons and Taylor, 1971).

Glowcurves are easily obtained. Usually, the phosphor is cooled down and illuminated in order to fill traps. Then the temperature is increased according to

$$T = T_0 + a_T t, \qquad (47.51)$$

where a_T is the heating rate, and T_0 is the starting temperature. The luminescence intensity is then measured as a function of time. Plotting the intensity, one observes one or more maxima, which indicates a more or less structured trap distribution, which in turn can be obtained by deconvolution, as shown in Fig. 47.10. For a review, see Bräunlich (1979).

An instructive picture about the complexity of thermally stimulated luminescence is given in Fig. 47.11, which shows three peaks at 320, 360, and 530 K, plus a broad saddle between 360 and 480 K. The 360 and 530 peaks show three emission maxima near 3000, 3500, and 5000 Å.

An estimate about the trap depth in relation to the temperature of the glow maximum $T_{\max}$ was derived by Randall and Wilkins (1945) from the simple one-trap model by introducing $T(t)$ into the exponent of the trap-ionization coefficient:

$$e_{\text{tc}} = \nu_t \exp\left(-\frac{E_c - E_t}{kT(t)}\right). \qquad (47.52)$$

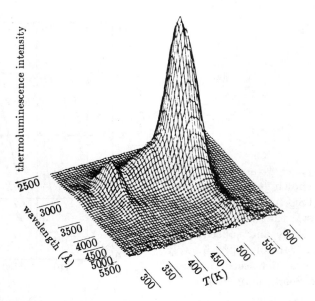

Figure 47.11: Thermoluminescence of KCl:Tl, showing its temperature and spectral distribution (after Mattern et al., 1970).

When we set the trap-ionization rate $e_{tc}n_t N_c$ equal to the luminescence transition rate $c_{ca}np_a$, we obtain for the time-dependent luminescence intensity

$$I_L = e_{tc}n_{t0}N_c \exp\left(-\int_0^t e_{tc}(t)dt\right),\qquad (47.53)$$

which can be solved for $E_c - E_t$:

$$E_c - E_t = kT_{\max}\{1 + f(\nu_t, a_T)\}\ln(\nu_t/D_\nu),\qquad (47.54)$$

where the function $f(\nu_t, a_T)$ for this simple model is small compared to 1, and D_ν is a dimension factor on the order of 1. With traps of a frequency factor of 10^{10} s^{-1}, Eq. (47.54) simplifies to the often-used estimate

$$\boxed{E_c - E_t \simeq 25kT_{\max}.}\qquad (47.55)$$

Phosphors with very deep traps have an important technical application for dosimetry; for instance, these phosphors can be excited by x-rays. The effect of such irradiation is a cumulative trap-filling. Trapped electrons can be stored over long time periods, and later checked by performing a glowcurve experiment. The area under the

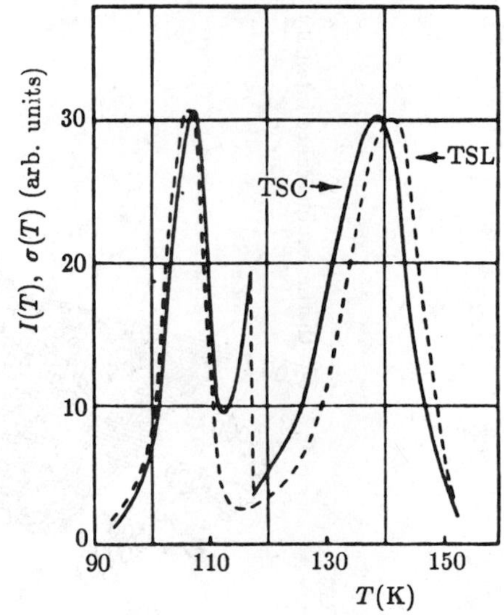

Figure 47.12: Example of a glow spectrum showing similar behavior between TSC and TSL for a LiF single crystal that was exposed to x-rays at 85 K and subsequently heated at a rate of approximately 0.05 degrees per second (after Böhm and Scharmann, 1969).

glowcurve is a measure of the dosage. See the review by DeWerd (1979).

One such dosimeter is the CaF:Mn phosphor, which shows linearity of the thermoluminescence peak after x-ray exposure over five orders of magnitude of the dosage from 0.1 to 10^4 R.*

47.3.2 Thermally Stimulated Currents

While the trapped carriers are released by an increase in temperature, one also observes an increase in electrical conductivity which shows a behavior similar to the glowcurves—Fig. 47.12. The corresponding current is called the *thermally stimulated current* (TSC).

When only one trap level is involved, both glow and TSC maxima occur at the same temperature, since both are proportional to the free carrier density. When other levels are involved, however, there are some differences in curve shape as well as the position of the maxima between TSC and thermally stimulated luminescence (TSL). There is competition between radiative and nonradiative transitions, and this competition is influenced by the temperature, most severely during

---

* R=Roentgen; 1 R is the amount of x-ray (or γ-ray) irradiation that produces 1 esu of charges ($\simeq 2.08 \cdot 10^9$ ions) per cm^3 of air. Natural background radiation is $\sim$ 120 mR per year. Diagnostic single x-ray exposures lie between 50 and 500 mR per exposure. Skin reddening occurs near 500 R.

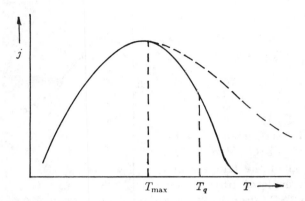

Figure 47.13: Competition of trap release centering at $T_{\max}$ and quenching at T_q.

thermal quenching. A cut-off of luminescence or current occurs at a temperature at which a quenching transition becomes significant. This is shown in Fig. 47.13, where a broad donor spectrum is assumed; the resulting TSC curve is shown as a dashed line. The steep decrease at T_q is caused by the release of holes from hole traps, causing a desensitization, i.e., a lowering of the carrier lifetime. Without additional information, such a decrease could be misinterpreted as a missing part in the electron-trap distribution.

A more detailed analysis shows that the logarithmic factor in Eq. (47.54) needs to be modified (Böer et al., 1958), which yields

$$ E_c - E_t = kT_{\max} \ln \left(\frac{\nu_t k T_{\max}}{a_T(E_c - E_t)} \frac{\eta}{(1-\eta)^2} \frac{c_{\text{ca}}}{c_{\text{ct}}} \right), \qquad (47.56) $$

with η as the relative trap-filling at $T_{\max}$. The result of the correction is twofold: it shows the influence of the competing recombination transitions, and the degree of trap-filling. When these traps are partially depleted,* the amplitude and temperature of the maximum change. When the carrier density at the TSC maximum, after sequentially increased fractional depletion, is plotted on a semilogarithmic scale as a function of $1/T_{\max}$, one obtains the trap depth from

---

* This can be done by a preceding, incomplete glowcurve run, which proceeds from run to run to progressively higher temperatures, followed by cooling to the starting temperature for the following run.

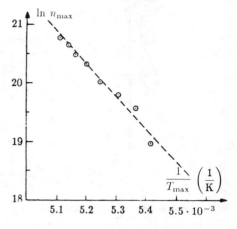

Figure 47.14: Current at the maximum of a TSC curve in CdS as a function of the temperature for partially depleted traps before starting the TSC (after Böer et al., 1958).

the slope (Fig. 47.14), providing only a single trap level is involved. This relationship can be written as

$$j_{max} = e\mu_n F N_c \frac{1-\eta}{\eta} \exp\left(-\frac{E_c - E_t}{kT}\right). \tag{47.57}$$

47.4 Changes in Field Ionization

In semiconductors with sufficiently low conductivity to avoid heating, trapped carriers can be freed by a high electric field. At relatively low fields, one observes impact ionization of shallow levels at low temperatures (see Section 42.4.2A), and then preferential depletion of Coulomb-attractive traps caused by Frenkel-Poole ionization; see Section 42.4.1 for a quantitative description.

When the field is increased, the carrier distribution is changed very similarly to the rise and decay of photoconductivity after changing the optical generation rate. Depending on the magnitude of the applied field, traps of various nature and depths are influenced, which causes the current to change with different time constants. Relatively slow rise or decay can be observed with a successive trap depletion by changing the applied voltage stepwise.

47.4.1 Field-Stimulated Current Curves

When increasing the bias as a function of time after previous trap-filling, one obtains nonstationary currents similar to TSC curves. Figure 47.15 shows a family of such curves of the electric conductivity with large maxima between 10 and 60 kV/cm, after the traps were filled by previous optical excitation. When shallow traps are partially

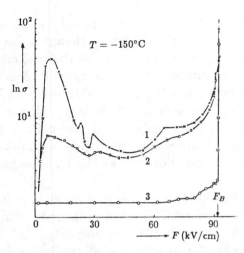

Figure 47.15: Field-stimulated conductivity curves (1 after filling traps, 2 after 2 min wait at 150 K, 3 after previous run of curves 1 or 2 without filling of traps), measured at CdS platelets (after Böer and Kümmel, 1954).

depleted by thermal ionization, the first, low-field maximum is much reduced. After the *field-stimulated conductivity curve* is traversed, the traps are depleted: a second curve does not show any significant increase in conductivity until $\simeq 80$ kV/cm is reached. The steep increase here in conductivity indicates the onset of the pre-breakdown range; F_B is the breakdown field.

47.4.1A Space-Charge Kinetics   An analysis of current kinetics must include the influence of space-charge effects due to influences of the electrodes and inhomogeneities within the crystal. All of these deal with inhomogeneities in the semiconductor, and are not the subject of this book. Consequently, they will only be mentioned briefly, with some reference to the extensive literature.

Changes in the current after varying the bias are observed due to polarization effects, which, in the absence of traps, are given by the *dielectric relaxation time*

$$\tau_\sigma = \frac{\varepsilon\varepsilon_o}{\sigma} = 8.8 \cdot 10^{-13} \frac{\varepsilon}{10} \frac{(\Omega\text{cm})}{\sigma} \quad \text{(s)}. \qquad (47.58)$$

With σ for typical semiconductors in the 10^{-4}–10^{+2} $\Omega^{-1}\text{cm}^{-1}$ range, τ_σ is on the order of 10^{-8}–10^{-14} s—i.e., simple polarization effects have a very short time constant. When space-charge layers (barriers and junctions) are present, the redistribution in these layers due to changing bias results in *dielectric aftereffects*, which cause changes in currents. When deep traps are present in the space-charge regions, such changes in current follow a change in bias slowly; this is due to the longer time constants for the thermal release of trapped carriers, time constants which usually exceed 1 s in large band-gap semiconductors (Böer and Kümmel, 1957; Shik and Vul', 1957).

With *injecting contacts* and sufficiently thin crystal platelets, *space-charge-limited currents* predominate and are trap-influenced. Release of trapped carriers by the field, or trapping at different locations within the platelet, causes kinetic changes of the current (Bube, 1960).

With *blocking contacts*, a high-field region is located near the cathode in *n*-type material. Its width changes with changing bias. Consequently, the space-charge boundary adjacent to the bulk must move, and thereby the distribution of trapped carriers in this region changes. Such changes result in transient currents with time constants relating to the carrier redistribution over traps. In crystals with an isolated trap, *one* relaxation time is observed, which depends exponentially on the temperature:

$$\tau_t = (\nu_t)^{-1} \exp\left(\frac{E_c - E_t}{kT}\right). \tag{47.59}$$

When measured as a function of the temperature, the trap depth, $E_c - E_t$, can be obtained from an Arrhenius plot (Matthews and Warter, 1966). However, a trap distribution is usually present and causes a time dispersion relation—see Section 48.1.1D.

Superlattices provide an effective means for charge separation between barriers and wells, which could lead to excessive time constants for recombination, observable as persistent photoconductivity—see Section 45.6.4.

47.4.1B Deep-Level Transient Spectroscopy

An important method, identified as *deep-level transient spectroscopy*, makes use of the fact that, after a small bias pulse is applied, the detecting current shows a steep rise and a longer decay. In a fixed-time window ($t_1 \ldots t_2$ in Fig. 47.16), the change of the current is largest in a narrow temperature range. From this temperature of maximum response, the depth of the level can be obtained (Lang, 1979).

In deep-level transient spectroscopy (DLTS), two major methods of analysis are distinguished: one is related to *current kinetics*, and the other is related to changes in the *capacitance*. The first method requires a knowledge of free-carrier densities *and* the space-charge

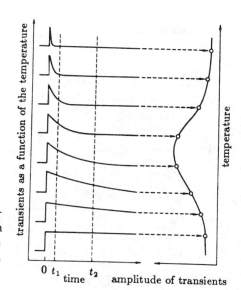

Figure 47.16: Amplitude of transients measured at window between t_1 and t_2, after a pulse at $t = 0$ as a function of the temperature (schematic).

distribution, which determines the field in the barrier, which in turn controls the current through the barrier.

The second method relates to the trapped carrier density alone, and is somewhat easier to interpret, although experimentally more involved. The related space-charge density directly determines the measured capacitance according to $C = \varepsilon_{st} A/W$, where A is the area of the capacitor, and W is the width of the space-charge region; the latter is given by $W = \sqrt{2\varepsilon\varepsilon_0 (V_D + V)/\varrho}$, with V_D as the (constant) diffusion voltage, and V as the (variable) bias voltage. In a simple depletion model of one trap level, the space charge ϱ is given by $\varrho \simeq e(N_t - n_t)$.

From the sign of the transient changes, one can distinguish whether electron or hole traps are involved—Fig. 47.17, inset. A typical DLTS spectrum, measured in GaAs, is given in Fig. 47.18. It indicates a high sensitivity for detecting low densities of traps in the 10^{13} cm^{-3} range, with a reasonable energy resolution in the 5 meV range. This method also permits spatial profiling of the doping distribution by variation of a dc-bias, placing the boundary of the barrier layer at various distances from the interface: the probing ac-signal modulates only this boundary.

Related to the above-mentioned kinetic analysis is the method of **field-induced thermally stimulated currents**, in which a well-defined heating program is employed while measuring the kinetics of polarization in the semiconductor. This analysis is more complex,

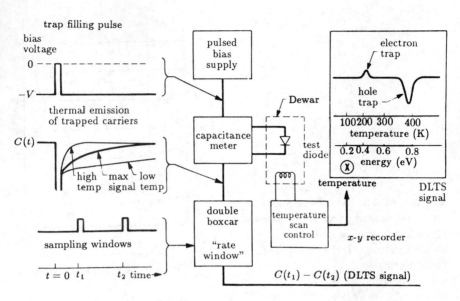

Figure 47.17: Block diagram showing the conventional features of the DLTS method.

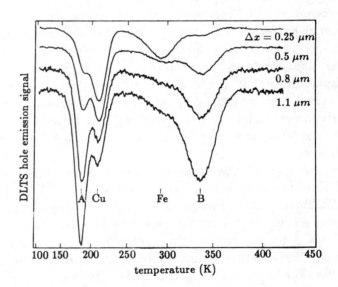

Figure 47.18: DLTS capacitance spectra of a GaAs barrier with four deep trap levels; two are identified as caused by Cu and Fe, which are present in the 10^{13} cm^{-3} range. The probing distance Δx from the barrier interface is determined by the dc-bias (after Lang and Logan, 1976).

as it is also designed to detect material inhomogeneities. For further detail, see Vanderschueren and Gassiot (1979).

Summary and Emphasis

A variation in the distribution of carriers over levels in the band gap results from a change in excitation. A differentiation of the occupation of various levels can be obtained by a kinetic analysis.

Such changes are induced most easily by light pulses of different photon energy and time profile. As a result, the induced photocurrent or luminescence changes in a form that is characteristic for the modified distribution of trapped carriers. A straightforward mathematical analysis of the shape of the current or luminescence signal permits the determination of the distribution of trapped carriers.

When this type of kinetic analysis is repeated at different light intensities and temperatures, one can stepwise obtain the entire trap distribution, according to the changes of the quasi-Fermi levels $E_{Fn}(g_o, T)$ and $E_{Fp}(g_o, T)$: the predominant changes occur in the neighborhood of the quasi-Fermi levels.

Various experimental methods have been developed, with different degrees of sophistication, which permit one to determine the level distribution. A rather simple method employs glowcurves, in which, after trap-filling, the temperature is increased linearly with time, and the changes in current are recorded. Characteristic maxima correspond to pronounced trap levels, with a simple relation between the temperature at which such maxima occur and the trap depth.

More sophisticated methods permit the distinction between different types of traps and recombination centers. These methods include: deep-level transient spectroscopy with current, and capacitance analysis, in which the carrier kinetics, initiated by a short light pulse, is investigated.

Instead of a light pulse, a field pulse can also be used by influencing the quasi-Fermi level in a space-charge layer, or by initiating field-ionization of traps.

Knowledge of the defect-level distribution in the band gap is essential for understanding the performance of electronic devices. The kinetic analysis yields easily obtainable information, and is widely used.

Exercise Problems

1.   What properties can best be evaluated by kinetic experiments?

2.(e) What are the criteria for hyperbolic or exponential photoconduction decay in a real crystal?

3.(e) What is the microscopic meaning of the fact that band-to-band recombination proceeds faster at higher generation rates?

  (a) How does this relate to capture cross sections?

  (b) What are the changes when a large (compared to ?) density of traps is introduced?

4.(r) How can a kinetic analysis help to distinguish between carrier generation and quenching?

5.   How can you extend this knowledge to use kinetic experiments for distinguishing between competing processes?

6.(*) Assume that for a certain phenomenon two groups of electron traps are involved. Design experiments to distinguish between them.

7.(r) What trap depths would result from a TSC peak at 300 K if the carriers are excited from attractive, neutral, or repulsive traps? What do you learn from the TSC curve shape?

8.(r) What do you have to observe as a precaution for evaluating field-stimulated current curves? What effect may interfere?

9.(*) Estimate the shape of a TSC curve with two electron traps; one with a trap depth of $E_c - E_t = 0.5$ eV, and the other trap at 0.7 eV. Assume an electron capture cross section for the first trap of 10^{-14} cm^2, and for the second trap at 10^{-16} cm^2. Obtain ν_t from reasonable arguments—see Section 42.2.3.

10.(e) Derive the solution given in Eq. (47.28) for the method of controlled excitation, and discuss the means to obtain the capture coefficient and the trap-emission probability from this method. How does the trap-emission probability relate to the thermal-ionization probability of the trap? Use detailed balance arguments with and without light.

11.(e) Discuss the change of the trapped-electron distribution with a change in optical generation rate [Eq. (47.48)] for a single trap and for a trap distribution. Give a trend analysis for various positions of E_{Fn} with respect to $N_t(E)$.

12.(l) Obtain recent literature on deep-level transient spectroscopy and describe the specific results obtained.

Chapter 48

Carrier Transit and Relaxation

Carrier transit limits some of the kinetic properties of electronic devices. Carrier relaxation within bands reveals their elastic and inelastic interaction with lattice defects and phonons. They give direct evidence of the rapid redistribution of carriers after cessation of a disturbance.

One can divide kinetic studies of carriers into transit through a device and carrier relaxation within a band. In addition, one can distinguish between global effects, which deal with all carriers, and the detailed analysis of the carrier distribution.

First, we will analyze the transit of carriers from electrode to electrode, which causes a specific transient behavior of injected current pulses and reveals transport and trapping properties.

In the second part of this chapter, we will discuss changes in carrier distribution within the band, which follow specific excitation pulses. The ensuing relaxation usually is analyzed by optical means, and reveals insight into the various relaxation mechanisms.

48.1 Carrier Transit Effects

When electrons are injected at a certain position of the semiconductor, and a bias voltage is applied, the electrons *drift* toward the anode. Their *drift velocity* can be determined by observing the increase in current as the drifting carrier cloud arrives at the anode—see Section 48.1.1C. This drift is slowed down by the intermittent trapping and release of these carriers, and can be used to obtain information about a number of typical relaxation mechanisms in the semiconductor.

48.1.1 Current Kinetics in a Semiconductor

When a bias pulse is applied and there is no trapping in a homogeneous semiconductor, the corresponding current pulse rises without delay, following the change in the electric field. However, if there is a space-charge region (a barrier at an electrode or a *pn*-junction), the changes in field distribution require a change in carrier distribution in the space-charge region; (without trapping) this change follows with a delay given by the dielectric relaxation time τ_σ—Eq. (47.58).

48.1.1A Characteristic Transport Times and Lengths
When electrons are injected and drift without being trapped until they reach the anode at a distance Δx, one observes a *transit time*

$$\tau_t = \frac{\Delta x}{\mu F},\qquad(48.1)$$

which is usually much longer than the dielectric relaxation time τ_σ.

The injected electrons have a *lifetime*

$$\tau_n = \frac{1}{c_{cr}(N_r - n_r)},\qquad(48.2)$$

where $N_r - n_r$ is the density of unoccupied recombination centers, and c_{cr} is the recombination coefficient. When this lifetime is smaller than the transit time, these carriers do not reach the electrode, but travel for a distance of the *drift length* (or *Schubweg*):

$$L_d = \mu_n F \tau_n.\qquad(48.3)$$

In the absence of an electric field, carriers diffuse in a random walk during their lifetime from the position of injection to a distance given by the *diffusion length*

$$L_n = \sqrt{\frac{\mu_n kT}{e}\,\tau_n} = 5.1\sqrt{\frac{\mu_n}{10^3\ (\text{cm}^2/\text{Vs})}\frac{T}{300\ (\text{K})}\frac{\tau_n}{10^{-8}\ (\text{s})}}\ (\mu\text{m}).$$

$$(48.4)$$

With a field applied, the distance traveled can be obtained by solving the transport and continuity equations for a homogeneous semiconductor with constant electric field, and is given by (Smith, 1978)

$$L_{n(u,d)} = L_n \frac{2L_n}{\sqrt{4L_n^2 + L_d^2} \pm L_d},\qquad(48.5)$$

where L_{nu} and L_{nd} are the *upstream* and *downstream diffusion lengths*, for which the sign in the denominator is positive or negative, respectively. This modified diffusion length is determined by a field opposed to, or in the direction of, the diffusion of the carriers. Consequently, the diffusion profile is compressed or stretched

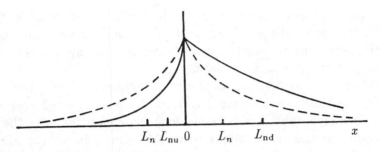

$L_n\ L_{\mathrm{nu}}\ 0\quad L_n\quad L_{\mathrm{nd}}\qquad x$

Figure 48.1: Diffusion profile of carriers injected at $x = 0$, for $F = 0$ (dashed curve), and for a sufficiently large field with upstream diffusion at the left and downstream diffusion at the right (schematic).

as shown in Fig. 48.1. For high fields ($L_d \gg L_n$), it yields for the *upstream diffusion length*

$$L_{\mathrm{nu}} = \frac{L_n^2}{L_d}, \tag{48.6}$$

and for the *downstream diffusion length*, it yields the drift length [Eq. (48.3)]:

$$L_{\mathrm{nd}} \simeq L_d. \tag{48.7}$$

These distances are often compared to the characteristic length of a space-charge region, the *Debye length* L_D—Eq. (14.11). When caused by a cloud of free carriers, L_D can be written in a fashion similar to the diffusion length, but with the dielectric relaxation time replacing the carrier lifetime:

$$L_D = \sqrt{\frac{\mu kT}{e}\,\tau_\sigma} = \sqrt{\frac{\varepsilon\varepsilon_0 kT}{e^2 n}} = 1205\sqrt{\frac{\varepsilon}{10}\frac{T}{300}\frac{10^{15}}{n}}\ (\text{Å}). \tag{48.8}$$

Often, however, the space charge is due to trapped carriers n_t. The carrier density n in Eq. (48.8) must then be replaced by n_t. The length that injected carriers can travel in a given time is given by the *drift velocity* $v_D = \mu_n F$, with

$$L_n = \frac{v_D}{\tau_n}; \tag{48.9}$$

the same drift velocity is used to describe the current:

$$j = env_D. \tag{48.10}$$

This is justified when one considers that only while in the conduction band do the electrons contribute to the current. The density of these electrons in steady state is n. In the bulk of a semiconductor, quasi-neutrality forces the establishment of a constant electron density, which is given by detailed balance between trapping and reemission of trapped carriers into the band.

When *intermittent trapping* and subsequent release from traps occur, the time required for a certain group of injected carriers to traverse a given distance becomes longer, since it includes the time of resting in the traps. This traversing time can be obtained by kinetic experiments— see Section 48.1.1B. Here, one accounts for this delayed arrival by defining a *modified drift velocity* according to

$$\tilde{v}_D = \mu_D F, \qquad (48.11)$$

where μ_D is the *drift mobility*, which includes trapping.

48.1.1B Current Kinetics with Traps Trapping and carrier release from traps modify the current kinetics of injected carriers. When one type of trap is present, trapping and release time are represented by the time constant τ_1—Eq. (47.10); an injected carrier pulse decays exponentially with τ_1.

When a trap distribution is present, a distribution of time constants determines the decay, which becomes *dispersive*. There are three types of current decays, which are shown in Fig. 48.2: one is box-like, with Gaussian spread shown as a dashed curve (see also Fig. 48.5) for carrier transport without trapping; one is a simple exponential for a single trap level, with a well-defined decay time τ_1; and one, which is dispersive, usually has two straight-line segments in a double logarithmic plot, with slopes below 1 and above 1 below and above the break, respectively—see Fig. 48.6 (Jonscher, 1983; Tiedje, 1984).

48.1.1C Shockley-Haynes-Type Experiments The modified drift velocity, which includes trapping, can be measured when the movement of an injected carrier cloud can be followed. Shockley and Haynes (1951) used the injection of a carrier cloud from a point contact into a long semiconducting sample (Fig. 48.3), and detected, with a field applied, the delayed arrival of this carrier cloud after a transit time $t_{tr} = t_2 - t_1 \simeq t_4 - t_3$. The drift mobility is then derived from

$$\mu_D = \frac{d}{t_{tr} F}. \qquad (48.12)$$

Other experiments use injected minority carriers from a short pulse of band-to-band exciting light. This light is absorbed in a thin

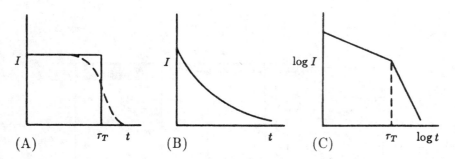

Figure 48.2: Shape of a current transient caused by a pulse of carriers injected near one electrode. (A) Ideal train of carriers without trapping; (B) with carrier trapping in a time shorter than the transit time; (C) response with a high degree of carrier dispersion.

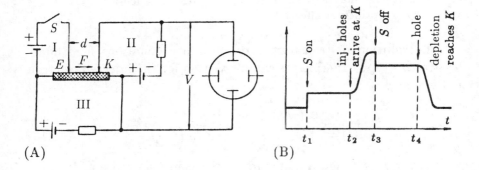

Figure 48.3: (A) Shockley-Haynes experiment to measure the drift time of minority carriers injected at E in a long semiconductor sample. (B) Schematic representation of the oscilloscope trace.

region near the front electrode, while a field is applied across the crystal. This causes a drift of the carriers through the bulk toward the opposite electrode at the rear side of the semiconductor.

Carrier drift can be made visible by the *print-out effect* in silver halides—see Section 46.2.1. Here, with sufficient exposure to light, enough metallic silver is segregated to become visible without developer. This occurs where electrons are trapped after traveling from the front surface into the crystal, following the external field— Fig. 48.4.

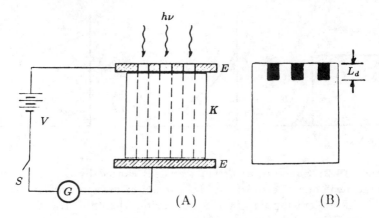

Figure 48.4: (A) Schematic of a carrier drift experiment in AgBr with perforated light mask and electrodes E. (B) Photograph of a AgBr single crystal platelet; the electron drift region with drift length L_d is made visible by the print-out effect.

The solution of transport and continuity equations for Shockley-Haynes-type experiments can be written for the injected electron pulse as

$$n(t) = \frac{\Delta n_0 \exp\left(-\dfrac{t}{\tau_n}\right)}{\sqrt{4\pi Dt}} \exp\left(-\frac{\Delta x^2}{4Dt}\right) \qquad (48.13)$$

with $\Delta x = \mu_D Ft$, and a halfwidth of the Gaussian pulse as

$$\Delta_{1/2} = 2\sqrt{\ln 2} \cdot \sqrt{4Dt} = 3.33\sqrt{Dt}, \qquad (48.14)$$

which permits one to determine the diffusion constant $D = \mu_D kT/e$, the drift mobility, and the minority carrier lifetime τ_n. The spreading of the pulse due to out-diffusion of carriers from the original confine within the narrow injected cloud is shown in Fig. 48.5:

48.1.1D Dispersive Carrier Transport in Amorphous Semiconductors   A wide distribution of a large density of traps in amorphous semiconductors causes a highly dispersive type of carrier transport. This can be measured by the current kinetics induced by a light flash absorbed in a thin near-surface layer of a photoconductive platelet. The carriers are then driven by an external field across the sample to the opposite electrode, in an arrangement similar to the one shown in Fig. 48.3. The measured current as a function of time is represented by two slopes on the order of 0.5 and 1.5 in the $\log j$ vs. $\log t$ diagram, as shown in Fig. 48.6 for As_2Se_3. These slopes

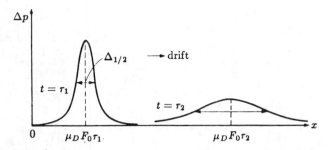

Figure 48.5: Diffusion profile for minority carriers injected at $x = 0$, after elapsed time intervals τ_1 and τ_2.

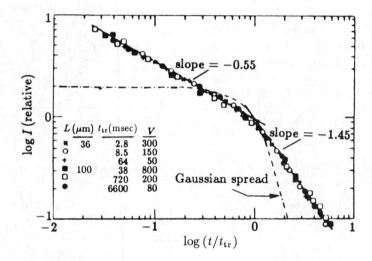

Figure 48.6: Dispersive transport of injected hole transitions in amorphous As_2Se_3 at 296 K. Plots are obtained by parallel shifting of the curves along the time and current axis to match the kinks at $(1,1)$. The kink identifies the transit time t_{tr}, which is listed together with the sample width L and bias V in the table insert of the figure (after Pfister and Scher, 1977).

remain unchanged, with changes in temperature (not shown), field, or sample thickness.

The dispersive carrier transport can be explained in an amorphous semiconductor with an exponentially tailing distribution of traps. Here, carriers that were generated near one electrode are rapidly trapped while diffusing toward the other electrode. With repeated retrapping, deeper traps will become progressively filled at the expense of the shallow traps.

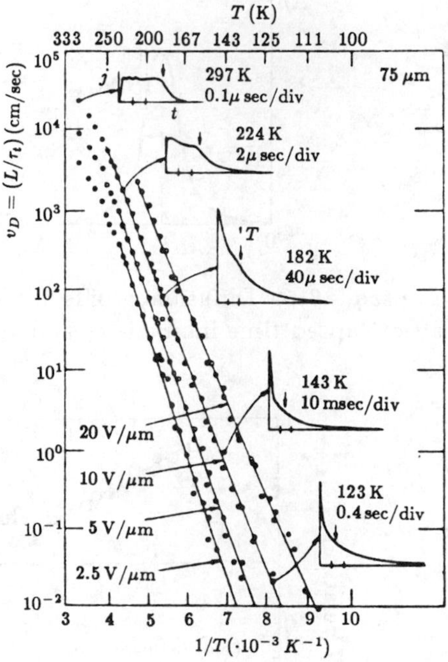

Figure 48.7: Temperature dependence of the drift velocity of injected minority carriers in α-Se with various electric fields as the family parameter. Insets show the current kinetics in linear scale that is dispersive for $T < 200$ K, nondispersive for $T > 250$ K (after Pfister, 1976).

It is observed that a drift current pulse decays after injection according to (Scher and Montroll, 1975):

$$
j(t) = \begin{cases} t^{\alpha-1} & \text{for} \quad t < t_{\mathrm{tr}} \\ t^{-\alpha-1} & \text{for} \quad t > t_{\mathrm{tr}} \end{cases} \tag{48.15}
$$

where α is the dispersion parameter, and t_{tr} is the transit time. This behavior can be explained by a progressive redistribution of carriers in an exponential trap distribution (Tiedje, 1984). The dispersion parameter is given by

$$
\alpha = \frac{T}{T_c} \quad \text{with} \quad N(E) = N_0 \exp\left(-\frac{E}{kT_c}\right), \tag{48.16}
$$

and provides a means to determine the slope of the trap distribution in the range where it is exponential.

In some amorphous semiconductors a transition between dispersive and nondispersive carrier transport becomes visible in a certain temperature range, where the trap distribution is not exponential and only few traps are active—Fig. 48.7; compare the insets of this figure with Fig. 48.2.

48.2 Relaxation of Carriers

Free carriers in thermal equilibrium follow the Fermi-Dirac distribution within their respective bands. They are coupled with phonons, which, as bosons, follow Bose-Einstein statistics. In equilibrium, both systems are described by the *same temperature*. The average momentum of the two systems is zero; the average energy is related to the common temperature, e.g., $3kT/2$ per quasi-particle for a Boltzmann ensemble.

When external forces such as a field, light, etc. are applied, the thermal equilibrium is disturbed, thermal conditions of the subsystems become variegated, and the distributions can no longer be described by exact thermodynamic distribution functions. Depending on the nature of the forces, either the electronic or phononic subsystem is disturbed primarily, while the other subsystem reacts by intersystem coupling—the electron-phonon scattering. This interaction has a dominating influence on the carrier or phonon transport determining the respective mobilities. This was discussed for steady-state conditions in Section 6.3.2 and Chapters 32 and 33. However, the addition of a multitude of effects often makes an unambiguous analysis in steady state difficult. A kinetic study permits differentiation of the involved subsystems when they have substantially different relaxation times.

In addition, steady-state experiments relating to carrier transport do not permit a direct analysis of the distribution of carriers as a function of energy. Instead, an average of the contribution of *all* electrons is obtained and expressed by the carrier mobility, from which only indirect conclusions about the distribution function can be deduced.

Relaxation studies deal with an analysis of the distribution functions of quasi-particles of the different subsystems after external forces are switched off. The momentum and energy relaxation of *electrons* or holes will be discussed first. As was done for steady state in Section 32, one needs to distinguish two ranges: one of a small perturbation, with the system not far removed from thermal equilibrium; and one of a large perturbation, somewhat similar to the discussion of warm and hot electrons in Section 33.

48.2.1 Momentum Relaxation of Electrons

Under the influence of an external electric field, the conduction electron gas is moved through the semiconductor with a momentum given by the product of its effective mass and the drift velocity $(\mu_n F)$. When the field is switched off, this motion of the entire electron gas will relax to zero with a characteristic time, the *momentum-relaxation time*. The momentum relaxation occurs through all types of scattering processes (see Chapter 32), i.e., the same types of scattering that are responsible for the steady-state distribution; however, not necessarily in the same sequence of importance. For instance, high-energy electrons are preferably scattered by LO phonons, or as optical intervalley scattering in Si. When equilibrium is approached, the dominant scattering processes involve preferably acoustic deformation potential and piezoelectric scattering. At the end of the relaxation process, the momenta of the electrons are distributed equally in all directions, and the electron gas has come to a standstill relative to the lattice.

A simple method of measuring this relaxation, wherein one observes the *decay of a current* after the field (bias) is switched off, runs into difficulties because of the extremely short relaxation times, which are on the order of $\tau_m \simeq \lambda/v_{\rm rms} \simeq 10^{-13}$ s ($= 100$ femtoseconds), with λ as the mean free path. Therefore, one works better in the frequency domain than in the time domain, and measures the complex conductivity at low IR frequencies. Here, the semiconductor responds to a harmonic excitation, i.e., to an electromagnetic sinusoidal wave. The time- and frequency-dependent behaviors are related to each other by the Fourier transformation:

$$\mathcal{G}(\omega) = \frac{1}{\sqrt{2\pi}} \int_{-\infty}^{\infty} G(t) \exp(-i\omega t)\, dt. \qquad (48.17)$$

The behavior can be analyzed by using the Boltzmann equation—Chapter 30,

$$\frac{\partial f}{\partial t} = -e\mathbf{F}\nabla_{\mathbf{k}} f - \frac{f - f_0}{\tau_m}, \qquad (48.18)$$

which, with

$$\mathbf{F} = \mathbf{F}_0 \exp(i\omega t) \quad \text{and the } Ansatz \quad f = f_0 + \delta f \exp(i\omega t), \quad (48.19)$$

yields

$$\delta f = -\frac{1}{i\omega}\left(e\mathbf{F}\nabla_{\mathbf{k}} f_0 \frac{-\delta f}{\tau_m}\right). \qquad (48.20)$$

With $\nabla_{\mathbf{k}} f_0 = v f_0/(kT)$ (see Section 30.3), one obtains the solution

$$\delta f = \frac{e(\mathbf{F}_0 \cdot \mathbf{v})\tau_m}{kT(1 + i\omega\tau_m)} f_0, \qquad (48.21)$$

which yields for the current (see Section 30.3.1C)

$$\mathbf{j} = \sum_{j=1}^{n} \sum_{\mathbf{v}} \frac{e^2 \mathbf{v} \, \mathbf{F}_0 \cdot \mathbf{v} \, \tau_m}{kT(1 + i\omega\tau_m)} \, f_0 \exp(i\omega t) = \sigma(\omega)\mathbf{F}_0 \exp(i\omega t) \quad (48.22)$$

with a frequency-dependent conductivity. Applying the arguments of Section 30.3.1C for averaging, one obtains for the complex conductivity

$$\sigma(\omega) = \frac{2}{3} \frac{ne^2}{m_n kT} \left[\left\langle \frac{E\tau_m}{1 + \omega^2\tau_m^2} \right\rangle - i\omega \left\langle \frac{E\tau_m^2}{1 + \omega^2\tau_m^2} \right\rangle \right], \quad (48.23)$$

showing that, for frequencies for which $\omega\tau_m \simeq 1$, the electrons are no longer in phase with the electric field. This results in a dispersion of the dielectric constant—see Section 10.1.1C,

$$\varepsilon(\omega) = \varepsilon_L - \frac{2}{3} \frac{ne^2}{m_n kT} \left\langle \frac{E\tau_m^2}{1 + \omega^2\tau_m^2} \right\rangle, \quad (48.24)$$

where ε_L is the dielectric constant of the lattice without free electrons. Here, $\varepsilon(\omega)$ shows a frequency dependency in the $\omega \gtrsim \tau_m^{-1}$ range, i.e., for far IR radiation with $\lambda \lesssim 0.1$ mm, which can be measured. For reviews, see Conwell (1982) and Jonscher (1983).

This analysis assumes an independence of τ_m on E. Usually, this is not fulfilled, and substantial errors are encountered when scattering is dominated by ionized impurities where a substantial $\tau_m(E)$ dependence exists. Errors are small, however, when acoustic or optical phonon scattering predominates (Nag, 1975).

Other means of measuring momentum-relaxation times relate to the width of a cyclotron-resonance peak, indicative of damping, magneto-resistance, or Faraday rotation measurements (Nag, 1984). The observed values for the momentum-relaxation time lie in the 1 ps and 0.1 ps ranges at liquid nitrogen and room temperatures, respectively, for typical semiconductors.

48.2.1A Ballistic Carrier Transport   When the dimensions of a semiconductor device become comparable to the mean free path between scattering events, or the time of observation becomes comparable to the relaxation time, there is a *drift overshoot* of the carrier velocity if the applied high electric field is high enough, as shown by Monte Carlo calculation (Malony and Frey, 1978) and given in Fig. 48.8.

For times and distances that are *shorter* than those between scattering events, a *ballistic transport* occurs, which can be compared to

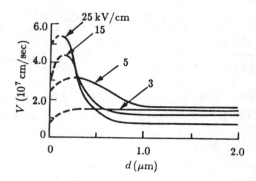

Figure 48.8: Drift-velocity overshoot as a function of the platelet thickness with the (homogeneous) field as the family parameter; obtained by Monte Carlo calculation in GaAs at 300 K (after Ruch, 1972).

the electron transport in a vacuum diode (Shur and Eastman, 1981): the electron is accelerated in the electric field without scattering until it hits the anode.

The current-voltage characteristic can be calculated similarly to that of a vacuum diode in a space-charge-limited case, using $j = env$ with $m_n v^2/2 = eV$, and V as the bias voltage. Integrating the Poisson equation [Eq. (29.1)] with $\varrho = e[n_0 - n(x)]$, one can approximate the current by

$$j_n \simeq \frac{4}{9}\sqrt{\frac{2e}{m_n}}\frac{\varepsilon\varepsilon_0}{l^2}V^{3/2}, \qquad (48.25)$$

where l is the length of the ballistic region.

The electrons exit from the cathode with a thermal energy distribution according to the lattice temperature of the metal. When the emission of electrons occurs through a thin insulating layer, however, as shown in Fig. 48.9, only very fast electrons can tunnel through this layer and start their ballistic path while already at high speed. In the semiconductor, they can be further accelerated if the bias is large enough.

Ballistic electron transport accounts for the improved high-frequency performance of small devices at high electric fields. In addition, it permits observation of quantum-mechanical interference phenomena, since the phase relation is not disturbed by scattering. Therefore, maxima and minima of the current flow are observed when the bias, and thereby the energy of the ballistic electrons, is varied. Ballistic transport was confirmed in GaAs by Heiblum et al. (1985). For a review, see Eastman (1982).

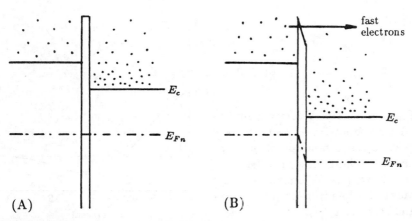

Figure 48.9: Heterojunction with thin insulating interlayer (A) in thermal equilibrium and (B) with applied bias injecting hot electrons into the narrow band gap material (schematics, neglecting space-charge effects).

Transient Carrier-Transport Analysis.   The transport dynamics in the ballistic, overshoot regime are no longer described by the classical Boltzmann equation, which assumes simultaneous response of the carriers to an applied force (Barker and Ferry, 1980). On a short time scale, there are memory effects from path to path. Retardation effects due to nonzero collision duration (Kreuzer, 1981) must also be taken into consideration.

One conceptual approach is to replace the mean additional velocity

$$\langle \Delta v \rangle = \frac{e}{m_n} F t \tag{48.26}$$

with a more appropriate time-dependent velocity function

$$v(t) = \frac{e}{m_n} F \int_0^t \phi_v(t', \theta) dt' \tag{48.27}$$

by introducing an autocorrelation function ϕ_v, which describes the variation of v in time (Zimmermann et al., 1981). This velocity is linear in time only if $\phi \simeq$ const.

One can then use the corresponding balance equations

$$m_n \frac{dv}{dt} = eF - m_n \int_0^t X_v(t') v_d(t - t') dt \tag{48.28}$$

and

$$\frac{d\overline{E}}{dt} = eF v(t) \left[1 - \phi_v(t, 0) \right] - \int_0^t \left[\overline{E}(t - t') - E_0 \right] X_E(t') dt', \tag{48.29}$$

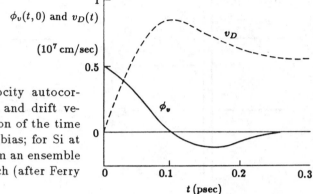

Figure 48.10: Velocity autocorrelation function ϕ_v and drift velocity v_D as a function of the time after switching on a bias; for Si at 300 K, computed from an ensemble Monte Carlo approach (after Ferry and Barker, 1981).

where X_E and X_v are the decay functions, which are intimately related to the energy and velocity autocorrelation functions ϕ_E and ϕ_v. After integration over a sufficiently long time, the time integrals over X_E and X_v yield the corresponding relaxation times τ_e and τ_m.

The correlation function is determined by a number of effects, including the influence of the field on the duration of each collision (Ferry, 1980; Barker, 1980) and on band-gap renormalization with injection of a large density of free carriers (Ferry, 1978). A band-gap narrowing is determined by the self-energy of electron-hole pairs (Inkson, 1976), and by a change in phonon frequencies due to free-carrier bond weakening (Brooks, 1955), as discussed in Section 6.2.4.

Typical behavior of the velocity autocorrelation function is shown in Fig. 48.10, as calculated by a Monte Carlo approach for Si (Ferry and Barker, 1981). The resulting transient drift velocity shows the velocity overshoot when the correlation function decreases in time; for more detail, see the review of Ferry et al. (1984).

48.2.2 Energy Relaxation of Electrons

In contrast to the momentum relaxation, there is no direct means of measuring the energy relaxation of hot electrons directly from carrier transport. The average energy of carriers cannot be measured directly, although it can be obtained indirectly from parameters that are a function of the carrier energy, such as the mobility.

However, more direct information can be obtained from optical transmission, reflection, luminescence, or Raman scattering experiments after carrier heating in a pulsed electric field (Bauer, 1978) or after an optical excitation pulse (Hearn, 1980). Modern optical tech-

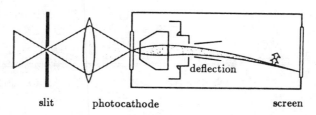

slit   photocathode   screen

Figure 48.11: Streak-image analyzer for short optical pulses using an image tube.

niques have extended relaxation studies into the femtosecond range (Shah and Leheny, 1984), and are reviewed in the following section.

48.2.2A Picosecond Spectroscopy   Traditional electronic methods are limited in their high-frequency resolution in the 100 ps range. This range can be extended to the 100 fs range (in this time a light pulse travels only 10^{-3} cm) by optical means, with light pulses of this duration produced by mode-locked lasers (Valdmanis et al., 1985).

These laser pulses have a large natural line width in the 10 meV range, requiring a laser cavity that permits an optical gain in a rather large band of frequencies. If the loss or gain of the cavity is modulated in this frequency range, an optical *wave packet* can be created, which produces a train of light pulses of extremely short duration (Laubereau and Kaiser, 1974).

Detection of light signals in the picosecond range requires fast shutters (*Kerr cells*) or a *streak camera*. In the latter, the light is focused onto an image converter (Fig. 48.11), and the electro-optical image is swept by electronic means across the screen with a speed near light velocity. The streak image, when it passes a narrow slit, is then read by a videcon coupled to a multichannel analyzer. The direct *electronic sequencing analysis*, however, is limited in time resolution by the electronic circuitry.

The detection limit can be extended into the 100 fs range by using *nonlinear optical means* to determine the optical pulse length and shape. Such means include the autocorrelation of a light pulse, with its delayed, split-off image, by detecting the second harmonics in a nonlinear crystal through which both beams have passed—Fig. 48.12A: only while both pulses are simultaneously present within the mixing crystal is the harmonic produced, which can then be detected by a properly tuned light sensor. The amplitude of this signal is proportional to the intensity product of both beams while

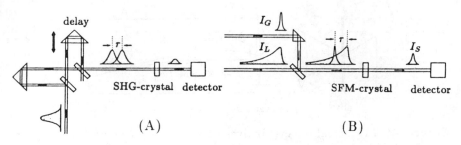

Figure 48.12: Nonlinear optics (mixing) analysis of (A) pulse length and (B) pulse shape. The detector is sensitive only to the mixed signal.

overlapping. The *pulse shape* of a longer pulse can be analyzed by using a shorter pulse as a gate, and varying its time delay by variation of the relative optical path length—Fig. 48.12B, as shown by Mahr and Hirsch (1975). Periodic pulse repetition permits detection above background noise. For a review, see von der Linde (1979).

48.2.2B Optical Studies of Carrier Distribution Information about the occupation of states within the bands can be obtained by using a high-intensity optical excitation from valence-band to conduction-band states, and consequent scanning of the optical absorption spectrum, which yields the energy distribution of carriers in these bands when significant filling is attained. Such filling causes a bleaching of the corresponding optical absorption. The kinetics of such a distribution is then observed by picosecond laser spectroscopy.

Optical absorption spectra yield information on the evolution of the carrier distribution near the band edge, from which the carrier relaxation can be derived. An example is given in Fig. 48.13 for GaAs. It shows three major features that are clearly visible in the difference spectra shown in subfigures B–E, compared to the initial absorption given in subfigure A:

(1) 1 ps after optical excitation with a 500 fs pulse, one observes an increase in absorption below the band edge ($E < 1.52$ eV) because of a renormalization of the gap, i.e., a decrease of the band gap due to exchange and correlation effects of the high-density free-carrier plasma (Section 9.2.1C), as summarized by Arya and Hanke (1981); see also Vahishta and Kalia (1982). These new states are rapidly filled by relaxing electrons; consequently, this enhanced absorption below the original band edge vanishes after 10 ps.

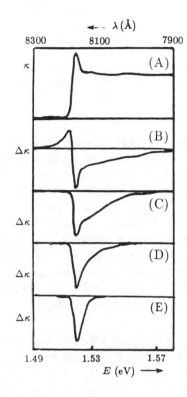

Figure 48.13: (A) Optical absorption spectrum of GaAs near the band edge at 10 K during excitation with a 500 fs broadband light pulse, and difference spectra after 1, 12, 62, and 242 ps delay in subfigures (B)–(E), respectively (after Leheny et al., 1979).
ⓒPergamon Press plc.

(2) A sharp decrease in absorption at the exciton peak ($E \simeq 1.52$ eV) is observed, due to the screening of the exciton states by free electrons, which persists for the length of the carrier lifetime (> 250 ps).

(3) The main feature indicating the dynamics of the energy distribution of the carriers is the tail of the bleaching within the band for $h\nu > 1.53$ eV, which decreases with a short relaxation time at higher energies (subfigures B and C) and a somewhat longer time at energies close to the band edge near 1.52 eV (subfigures D and E). This *relaxation of hot carriers* involves the effective scattering with longitudinal optical phonons (see Section 45.5.4), and will be discussed in the following section.

(4) Later, and not clearly discernible in Fig. 48.13, further relaxation takes place via scattering with acoustic phonons, a substantially slower process—see Section 48.2.2C; carrier recombination also becomes effective as seen by the area reduction under the difference-of-absorption curves.

Luminescence provides information about the *electron temperature* from the *slope of the high-energy tail* of the emission peak.

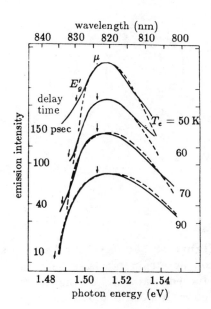

Figure 48.14: Spectra of spontaneous band-to-band luminescence of GaAs at 4.2 K with 4 mW/cm^2 excitation at 5320 Å, and the time delay after cessation of the excitation as the family parameter. Theoretical curves (dashed) are given for the listed electron temperatures, indicating a cooling from 90 K to $\sim$ 50 K within 150 ps delay (after Tanaka et al., 1980).

Use of a streak camera shows a successive sharpening of the emission peak, with proceeding energy relaxation of the excited electron gas (Graudßus and Göbel, 1981; Tanaka et al., 1980); this is shown in Fig. 48.14 in a somewhat longer time frame, and clearly indicates the successive cooling of the electron gas that was heated by the exciting light pulse.

48.2.2C Time Response of Energy-Relaxation Mechanisms
Picosecond spectroscopy provides direct information on the time response of the different scattering mechanisms of carriers. This response depends on the type of carrier, its density, and its initial energy distribution (Lyon, 1986). A large amount of data has been reviewed by Luzzi and Vasconcellos (1984), and is discussed in terms of a nonequilibrium thermodynamics approach, using as evolution equations the generalized balance equations. These are of the form

$$\frac{dE_c(t)}{dt} = J_{1L} - J_{LO}^1 - J_{1A} + J_{1R} \tag{48.30}$$

$$\frac{dn(t)}{dt} = J_{2L} - J_{2R} \tag{48.31}$$

$$\frac{dE_{LO}(t)}{dt} = J_{AN} - J_{LO}^1 \tag{48.32}$$

where J_{ik} represent the transition rates, which can be described as collision operators:

$J_{1L} =$ transition of energy from laser photons to the carriers
$J_{LO}^1 =$ energy loss of carriers due to LO phonon scattering
$J_{1A} =$ energy loss of carriers due to acoustic phonon scattering
$J_{AN} =$ energy loss of LO phonons due to anharmonic scattering with acoustic phonons
$J_{2L} =$ electron-hole pair generation rate
$J_{2R} =$ electron-hole recombination rate
$J_{1R} =$ change of carrier energy during pair recombination processes.

The different contributions depend on the material, experimental setup, and progressing relaxation, when one or the other transition rate becomes dominant, as will be discussed below. In addition, one must distinguish between hole and electron relaxation.

Hole Cooling Mechanisms. Holes usually have a larger effective mass than electrons, and follow different selection rules for the excitation of phonons: holes excite TO phonons, while electrons excite predominately LO phonons (for a review, see Wiley, 1975). In Fig. 48.15, the different relaxation processes are summarized: when electrons and holes are produced simultaneously by absorbing a photon (vertical transition), the electron obtains a much higher energy, typically 400 meV, before the hole is energetic enough (~ 35 meV) to excite an optical phonon—see also Fig. 13.7. Consequently, most of the *hole cooling* is caused by scattering with *acoustic phonons*, which is a relatively slow process—see below.

Electron Cooling Mechanisms. Electron cooling, starting from sufficiently high energy, can be divided into four regimes:

(a) Electron-electron interaction thermalizes the electron ensemble if the density of free carriers is high enough (Shah, 1978). The critical electron density at which electron-electron scattering exceeds interaction with phonons can be estimated from plasmon emission (Quinn, 1962), which yields for the energy loss

$$\left[\frac{dE}{dt}\right]_{e-e} \simeq \frac{2\pi n e^4}{\varepsilon_{\text{opt}}^2 m_n v_{\text{rms}}^2}. \tag{48.33}$$

At high electron densities, preferred energy loss occurs by hopping down in units of the plasmon energy (Lyon, 1986) when the plasmon energy exceeds the LO-phonon energy. Therefore,

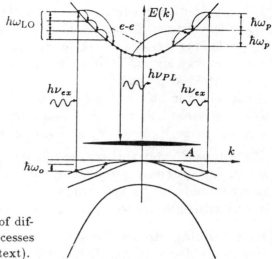

Figure 48.15: Schematics of different carrier-relaxation processes after optical excitation (see text).

from Eq. (12.4), one obtains as the critical electron density for preferred electron-electron interaction:

$$n_{e-e} = \frac{\omega_{LO}^2 \varepsilon_{opt} \varepsilon_0 m_n}{e^2} = 3.14 \cdot 10^{17} \left(\frac{\omega_{LO}}{10^{13}}\right)^2 \frac{\varepsilon_{opt}}{10} \frac{m_n}{m_0} \; (\text{cm}^{-3}). \tag{48.34}$$

This thermalization from a spike-like initial excitation takes place in the 100 fs range. Experimentally, one obtains as critical density a somewhat lower value of $n_{e-e} \simeq 8 \cdot 10^{16}$ cm^{-3} (Kash, 1989).

(b) Electron-LO-phonon interaction (Fröhlich interaction—see Fröhlich, 1937) is very strong.* This interaction produces step-like reductions of excess energy (see Section 44.5.2A—Mirlin, 1984), and proceeds in the Γ-valley with a time constant of 180 fs (Levi et al., 1986). An observed *reduced* rate of cooling via LO phonons at higher carrier densities (Leheny et al., 1979) can be explained by carrier screening of the Fröhlich interaction (Graudßus and Göbel, 1983; Yoffa, 1981). Another reason for such a reduced cooling rate may be the creation of "hot phonons," specifically, heating the LO mode, which consequently reduces the cooling efficiency of this mode until it has time to relax (Lyon, 1985).

---

* TO phonons can also interact with electrons and are coupled through their deformation potential. They are, however, forbidden to do so with carriers in *s*-like states (Wiley, 1975); such forbidden transitions have a factor of only 3 reduced probability, and are important for holes.

(c) Final cooling involves acoustic phonons and is a relatively slow process, extending into the nanosecond range, due to the small amount of energy transferred in any one transition (Ulbrich, 1978).

(d) Recombination of electrons with holes depends on the product of their densities for intrinsic recombination, and on the defect density for extrinsic recombination (see Dymnikov et al., 1978); it can extend from the nanosecond to the microsecond range or longer.

High-Energy Relaxation Mechanisms.   When optical excitation creates electrons with enough energy, **intervalley scattering** becomes important. Such scattering can be investigated with the help of $(e, A^{\times})$ luminescence as a probe (Fasol and Hughes, 1985); it is very fast.

An overview of the relaxation of higher energy electrons can be obtained from Fig. 48.16. When excitation occurs below 1.9 eV, only the main valley (Γ) is involved with relaxation via LO phonons or electron plasma interaction, as discussed in the previous section. The corresponding curve (1) shows the luminescence spectrum after excitation from heavy (HH) and light (LH) hole bands, with additional maxima given by LO-phonon relaxation.

At slightly higher energies ($1.9 \text{ eV} < h\nu < 2.3 \text{ eV}$), electrons can rapidly scatter from the Γ- into the L-valley with a time constant of ~ 540 fs. At energies above ~ 2.3 eV, an *intervalley scattering* from the Γ- into the X-valley now becomes dominant, with a time constant of ~ 180 ps. These time constants can be obtained from competition with the LO relaxation, and from the Γ-valley, with consequent $(e, A^{\times})$ luminescence, acting as an *internal clock* with 180 fs time constant. When electrons are scattered into side valleys, such luminescence does not occur: the extent of the $A^{\times}$ level in k does not reach to the minima of the side valleys. This is seen from curves (2) and (3) in Fig. 48.16, which are taken with excitation at energies at which intervalley scattering is important. Consequently, the direct HH and HL signatures are much reduced, giving a measure of the relative population of Γ and side valleys. In addition, a substantially shifted reentry signal (R), after back-scattering from the X- or L- into the Γ-valley, and its LO relaxation peaks, can be observed, again with much lower intensities (Ulbrich et al., 1989).

The time constant for the reentry (from L or X to Γ) is longer because of the larger effective mass in the satellite valleys. It is on

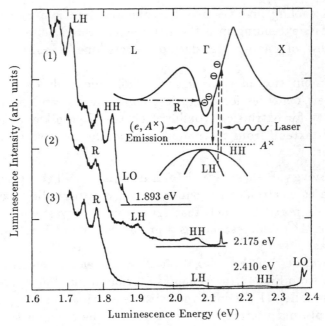

Figure 48.16: $(e, A^{\times})$ luminescence in GaAs: Mg at 25 K ($p = 1.2 \cdot 10^{17}$ cm^{-3}). Laser excitation at 1.893, 2.175, and 2.410 eV for curves (1)–(3), respectively. The peaks identified as LO are Raman scatterings. Inset: $E(k)$ with relevant transitions (after Ulbrich et al., 1989).

the order of 2.5 ps, and can be obtained from the *slow rise of the luminescence*—e.g., after excitation above 2 eV. Here, fast scattering into the L-valley provides a carrier storage, and consequently a slow supply path via reentry into the Γ-valley (Shah et al., 1987). When no reentry into the Γ-valley occurs, as, e.g., in InP, the rise of the luminescence is much faster (Shah et al., 1987).

At higher energy, excitation takes place further away from $k = 0$, and *warping of the valence bands* becomes marked. Consequently, there is a wider spread of initial electron energies from which LO-phonon relaxation starts, resulting in a broader luminescence peak: 1.5 meV at 1.57 eV, and 8 meV at 1.85 eV excitation (Ulbrich et al., 1989).

The scattering of holes between different valence bands becomes important for hole distribution relaxation. This distribution relaxes in the 10...100 ps range, as shown in Fig. 48.17 for CdSe at 4.2 K. With sufficient energy, both the *A* band and the split-off *B* valence band are populated. Changes in population occur through

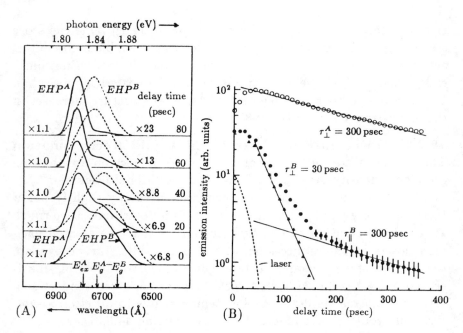

Figure 48.17: (A) Time-resolved spectral distribution of spontaneous luminescence of electron-hole pairs (EHP) in CdSe at 4.2 K excited by a 532 nm picosecond laser pulse of 50 MW/cm^2 and $\sim$ 20 ps duration (see dashed curve in B). Solid curves $E \perp c$, dashed curve $E \| c$ (superscript A and B in EHP refers to the A and B valence bands). E_{ex}^A gives the energy of the exciton associated with the A band; E_g^A and E_g^B indicates the two band gaps. (B) Time decay of the three luminescence peaks shown in (A). The straight line for $\tau_\perp^B$ is obtained after subtracting the $EHP_\perp^A$ signal (after Yoshida et al., 1981).

intervalence-band relaxation and via recombination of holes with electrons. Recombination with electrons occurs from both bands with a similar time constant of about 300 ps in the given example as long as both are in quasithermal equilibrium. Intervalence-band relaxation from $B_\perp$ to $A_\perp$ is much faster (30 ps). Since the split-off energy, 26.3 meV, is sufficient to accommodate an LO phonon with an energy of 26.1 meV, such scattering, in addition to LA phonon scattering, is probably responsible for the relaxation.

Lattice Heating Through Hot Electrons.   The electron-electron scattering does not remove energy from the electron ensemble, but causes thermalization within this ensemble: it smoothes out the distribution, which initially is shaped by the exciting laser pulse, and results in a Maxwell-type distribution with a well-defined electron temperature $T_e > T$ (lattice temperature T). One can store a substantial amount of energy in the electron ensemble before interaction with phonons cools the electron plasma. In fast-rising, high-energy laser pulses, the electron plasma can be heated to more than 10^3 K in the 100 fs range before the lattice temperature increases in the low ps range (Malvezzi, 1987).

48.2.2D Electron-Hole Plasma and Liquid Recombination

The energy relaxation and recombination of an electron-hole plasma are highly density-dependent. This needs additional discussion.

When the optical excitation occurs at an energy slightly *below band-gap* energy, initially excitons, or exciton-polaritons, are formed. When the generation rate is high enough so that the created excitons closely fill the semiconductor, and their orbits start to overlap, the exciton state becomes unstable and an electron-hole plasma is created—see Section 34.2. In typical indirect band-gap semiconductors, this plasma condenses at low temperatures to an *electron-hole liquid* of droplet shape (Rice, 1977)—see Section 44.3.3D.

Electron-Hole Liquid Kinetics. The electron-hole liquid and electron-hole plasma in indirect gap materials have a rather long recombination lifetime, typically in the microsecond range. The low probability of finding a phonon of the proper wave number at the same position with an electron and a hole causes the lower probability for recombination in indirect gap semiconductors. In direct band-gap semiconductors, however, no evidence of droplet formation is yet obtained (Saito and Göbel, 1984), while a high-density electron-hole plasma is formed (Haug, 1981), but with a substantially shorter recombination lifetime in the nanosecond range.

Electron-hole droplets evaporate, thereby creating excitons, which in turn recombine. After cessation of the optical excitation which created the droplets, droplet evaporation continues until all droplets have evaporated. This process can take much longer than the lifetime of excitons. Evidence for the slow evaporation of electron-hole drops, which replenishes the exciton gas, is shown in Fig. 48.18. When all the drops are evaporated, the density of the exciton gas decreases with its much faster decay time. When the exciting light pulse is stronger, more droplets are created, and it takes longer until all of them are evaporated (Manenkov et al., 1976).

During a short and intense light pulse, a sufficiently large density of electron-hole pairs is created near the surface, and a liquid layer can be formed. Consequent instabilities occur which complicate

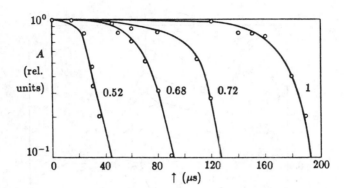

Figure 48.18: Decay of the exciton density after cessation of an optical excitation. While electron-hole drops slowly evaporate, the exciton gas is replenished and its density remains constant; thereafter, it decays with the exciton lifetime of $\sim 20~\mu s$. Family parameter is the intensity of initial light pulse, which controls the amount of initial droplets (after Manenkov et al., 1976).

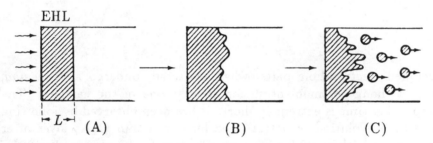

Figure 48.19: (A) Creation of a thick layer of an electron-hole liquid near the surface by strong optical excitation. (B) Growing capillary waves. (C) Breakup of these waves into electron-hole droplets, which are blown away by the phonon wind (after Keldysh, 1986).

the kinetics: the phonon wind generated within the layer creates capillary waves at the liquid surface, from which droplets of critical size (Section 34.1) break off and are then driven away, as shown in Fig. 48.19 (Keldysh, 1986). This can be experimentally observed since the index of refraction in droplets is different from that in a normal crystal, which makes the droplets visible (Damen and Worlock, 1976).

48.2.2E Electron-Hole Pair Dephasing and State-Filling

When the optical excitation occurs with an energy *exceeding* the band gap, *electron-hole pairs* are formed, as opposed to excitons, which exist only at energies below the band gap. These optically

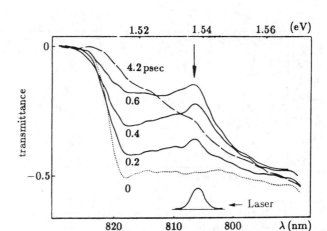

Figure 48.20: Time-resolved transmittance of a 0.75 μm thick GaAs layer at 15 K after excitation with a 100 MW/cm^2 100 fs laser pulse at 806 nm (lower insert). Arrow indicates hole burning (after Oudar et al., 1985).

coupled electron-hole pairs, when scattered, undergo a *dephasing*, i.e., a change of momentum for at least one of the carriers. The dephasing time is extremely short. It has been observed by following the time evolution of optical bleaching in a thin GaAs layer after excitation with a 100 fs high-power laser pulse (Oudar et al., 1985).

The intense laser pulse bleaches out the corresponding states in the valence and conduction bands because of *state-filling*. This is to be distinguished from *band-filling*, which is observed *after* thermalization, e.g., in steady state, and is known as the *Burstein-Moss effect*—see Section 24.3.1. State-filling is also referred to as *optical hole burning within the bands*, since a very narrow wavelength range of optical excitation is preferentially bleached. These states have an inverted population.

Figure 48.20 shows the transmission spectrum of GaAs and its changes over time at 15 K. Bleaching (arrow) occurs initially in the wavelength range of the laser pulse, shown in the lower insert. The bleached maximum at 1.54 eV then broadens and disappears, indicating a dephasing time of the electron-hole pairs of ~ 300 fs. With the progression of time, the bleaching spectrum becomes smooth, indicating thermalization after ~ 4 ps with $T_e \simeq 120$ K in the given example.

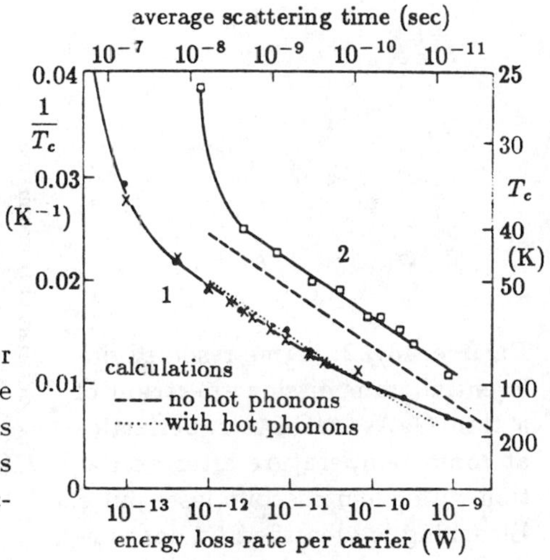

average scattering time (sec)

Figure 48.21: Inverse carrier temperature as a function of the energy loss rate in a GaAs/AlGaAs superlattice at 18 K for electrons and holes (curves 1 and 2, respectively) (after Shah et al., 1985).

48.2.3 Hot-Carrier Relaxation in Superlattices

When carriers are confined within quantum well superlattice structures, many electronic properties change, such as their density of states, carrier screening (it becomes weaker), and plasmon properties—they have zero energy and zero wave vector. However, the interaction of electrons with LO phonons in 2D structures has shown rather similar behavior as in 3D material, as discussed in several reviews—see Lyon (1986) and Shah (1986).

The average energy-loss rate from heated carriers to the lattice can be expressed much like that in a bulk semiconductor; it is given by

$$\left\langle \frac{dE}{dt} \right\rangle_{e,LO} = -\frac{\hbar\omega_{LO}}{\tau_{LO}} \exp\left(-\frac{\hbar\omega_{LO}}{kT}\right), \qquad (48.35)$$

where τ_{LO} is an effective electron/LO phonon relaxation time (Shah and Leheny, 1984).

A more advanced analysis shows only a negligible increase in this loss rate compared to that of bulk GaAs. Modification of the simple model with parabolic minibands, infinite potential steps, nondegenerate electrons, nonscreening to include degeneracy (insignificant up to 2D densities of 10^{12} cm^{-2}), slab modes (small effect; Shah et al., 1985), plasma effects, and screening (less than 40% influence; Das Sarma and Mason, 1985) has shown little effect on the energy transfer. Hot-phonon modes, however, may cause a more significant reduction in the energy-loss rate, as shown by Price (1985).

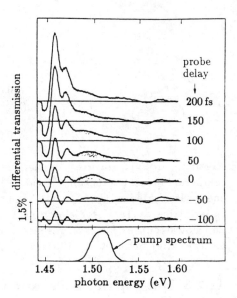

Figure 48.22: Time-resolved differential transmission spectrum of a thin GaAs/AlGaAs superlattice at room temperature after excitation with a femtosecond laser pulse. Bleaching is identified by dots under the new maximum of transmission (after Knox et al., 1986).

Optical probing by luminescence scanning while heating in an electric field shows larger energy-loss rates for holes than for electrons in such superlattices, as shown in Fig. 48.21. The reduction of energy losses from electrons could be explained by significant heating of LO phonons: a heated LO phonon gas is less effective in cooling than a cold phonon gas (Shah et al., 1985). Consequently, the electron temperature remains higher than the hole temperature.

The electron-hole scattering rate is $\sim 10^{-13}$ s^{-1} as shown by Höpfel et al. (1986). Typically, carriers thermalize within less than 100 fs; see review by Göbel et al. (1986). The most direct information is obtained by time-resolved bleaching experiments (Knox et al., 1986), which give direct evidence of *hole burning* in GaAs/AlGaAs superlattices and show electron relaxation with a relaxation time of ~ 50 fs at a carrier density of $\sim 10^{18}$ cm^{-3}—Fig. 48.22. A recent review of the field is given by Fouquet and Burnham (1986).

Summary and Emphasis

Carrier relaxation within the corresponding band may relate to carrier transport with kinetics responding to bias pulses. When such bias causes carrier injection, the response can relate to diffusion, drift, or a combination of both. The relaxation time of an injected cloud of carriers is given by the carrier lifetime within the band, and relates to a corresponding diffusion or drift length about which the cloud can be moved during its life.

The lifetime of an injected carrier cloud can be determined by trapping. When one type of trap is involved, a single exponential decay is observed; a trap distribution results in the more common dispersive decay, usually following a power law.

When the bias causes carrier heating, its energy relaxation can be measured by optical means as a damping contribution.

In thin layers, ballistic carrier transport becomes possible, avoiding scattering and permitting higher energy of the carriers and shorter transit times.

Substantial carrier heating can be initiated by light pulses. The additional energy, in excess of the thermal energy of electrons, is transmitted to the lattice mostly by LO phonons, and from holes primarily by TO phonons. At high optical excitation rates other coupling mechanisms become operative, and involve plasmons for a rapid first cooling step before LO phonons become dominant. In addition, the close interaction between electrons and holes in such a plasma has an influence of cooling the hotter electron gas because of $m_n < m_p$. Most of these fast processes are accessible either through ultrafast laser spectroscopy in the 100 fs range, using autocorrelation of a light pulse with its delayed image, or from the broadened spontaneous luminescence signal.

At low lattice temperatures, condensation into electron-hole droplets takes place, with droplet evaporation into an exciton gas and exciton annihilation determining the two relaxation mechanisms.

The relaxation of carriers within the respective bands has a determining influence on carrier transport and on recombination, as well as on the energy transport from the easily (laser-) heated carrier ensemble to the lattice, which is important for a wide range of laser-heating technologies. Fast relaxation in the 100 fs range has recently become accessible for direct observation, and awaits further aggressive exploration.

Exercise Problems

1.(e) Confirm the expression for upstream and downstream diffusion length, Eq. (48.5), by solving the transport equation $j_n = e\mu_n nF + \mu_n kT \, dn/dx$, and the continuity equation, which can be written as (Böer, 1977) $dj_n/dx = e(n_{10} - n)/L_n^2$ where $n_{10} = g_0 \tau_n$, g_0 is a constant generation rate, and $n = n(x)$.

2.(e)   Derive upstream and downstream diffusion lengths with superposition of a constant electric field. Discuss the transition between diffusion and drift velocity.

3.(r)   Describe the setup for transient optical grating experiments, and discuss the different relaxation effects which can be measured with this method.

4.   Discuss the relaxation of a group of monoenergetic electrons at 50 meV above the band edge, $\hbar\omega_{LO} = 35$ meV. Vary temperature and electron density.

5.(*)   How can you achieve fast laser beam heating of a thin Si crystal. Discuss wavelength dependence and relaxation.

6.(r)   Discuss the electron-hole liquid relaxation, including evaporation.

7.(r)   How can ballistic electrons be used in electronic devices? Give design features and discuss advantages and disadvantages.

8.(r)   Describe deep-level transient spectroscopy. Use additional literature. Discuss advantages and limitations.

9.(r)   Compare different experiments to measure diffusion and drift length. What experiments need to be performed to measure intermediate trapping?

10.   Describe in your own words optical mixing for ultrafast relaxation analysis.

11.(*)   What are the principal limitations to creating ultrafast light pulses?

12.(*)   Describe the different relaxation mechanisms for an electron gas that was excited by a monochromatic light pulse 0.1 eV above the band edge.

Chapter 49

Phonon, Exciton, Polariton Kinetics

The relaxation behavior of phonons, excitons, and polaritons reveals their interaction with other quasi-particles, and provides important clues for the understanding of thermalization after the cessation of a perturbation from equilibrium.

Quasi-particles can be excited by short light pulses, and following this excitation, reveal their relaxation behavior. These quasi-particles include excitons, phonons, and the corresponding polaritons. Recently, with faster detection techniques, the relaxation of simple lattice polarization has also become accessible to observation. In the following sections we will give examples of these relaxation processes.

49.1 Relaxation of Phonon Distributions

Nonthermal changes in phonon distributions can be achieved by selectively heating a specific phonon mode. This can be accomplished most directly by optical (IR) excitation of a specific atomic vibration, and by controlling the frequency and phase relation of the stimulated phonon branch. An indirect mode of phonon generation is that of inelastic scattering by hot electrons, which, when energetic enough, interact preferably with one phonon branch—the LO phonons.

The relaxation of a deformed phonon spectrum into its equilibrium distribution proceeds via inelastic three-phonon collisions: the optical phonon, following selection rules, decays into two acoustic phonons—one TA and one LA phonon of lower energy (Orbach, 1967).

In semiconductors with sufficient electron density, scattering with electrons provides an additional relaxation mechanism (von der Linde et al., 1980). When heated, electrons interact with each other; they thermalize with a relaxation time that can be shorter than the

electron-phonon relaxation at sufficiently high electron densities—
see Section 48.2.2C. Interaction with LO phonons accelerates their
relaxation.

49.1.1 TO-Phonon Relaxation After Optical Stimulation

Optical pumping of energy into a specific branch of the phonon spectrum requires intense illumination and can induce coherent oscillations due to nonlinear effects, or, below threshold, incoherent oscillations. Excitation can be accomplished by two synchronized laser
pulses of slightly different energy, so that the energy difference is
equal to the energy of the phonon branch which is to be stimulated—
see Chapter 16. The energy of such laser pulses is selected preferably
below the band gap, and away from any other resonances. Usually,
only one phonon mode, the one with the larger Raman cross section,
is stimulated (Gale and Laubereau, 1983).

When the intensity of the exciting light is high enough, and the
crystal geometry is appropriate for the stimulated phonon branch
so that a gain in excess of one is achieved (Laubereau, 1984),
amplification occurs from the input Stokes signal, and lasing occurs.
Transient excitation of a specific mode of lattice oscillations occurs
when the duration of the pumping pulse is comparable to the lifetime of these phonons, typically in the 10 ps range. For transient
pumping, a higher gain is needed to initiate stimulated scattering,
starting from quantum noise (Penzkofer et al., 1979).

The kinetics of the phonon population can be measured directly
with a delayed probing light beam, or indirectly from the line width
of the *Raman signal*, using the Heisenberg uncertainty relation.

A typical time evolution of the coherent anti-Stokes signal for
hot TO phonons in diamond is shown in Fig. 49.1. The signal decays
exponentially with a time constant of 2.9 ± 0.3 and 3.4 ± 0.3 ps
at room temperature and 77 K, respectively. The corresponding line
widths of *spontaneous Raman scattering* are 0.2 and 0.27 meV (Solin
and Ramdas, 1970), and agree reasonably well with the above-given
time constants for TO-phonon relaxation.

Similar measurements in GaP (Kuhl and von der Linde, 1982)
yield lifetimes for LO phonons at room temperature of 6.7 ± 0.3 ps.
The lifetime increases with decreasing temperatures, and at 5 K is
measured at 26 ± 2.5 ps. Theoretically, the lifetime should double
between 300 and 0 K if the decay into two acoustic phonons is
dominant. The experiment yields a factor of four for reasons yet
unknown (Laubereau, 1984).

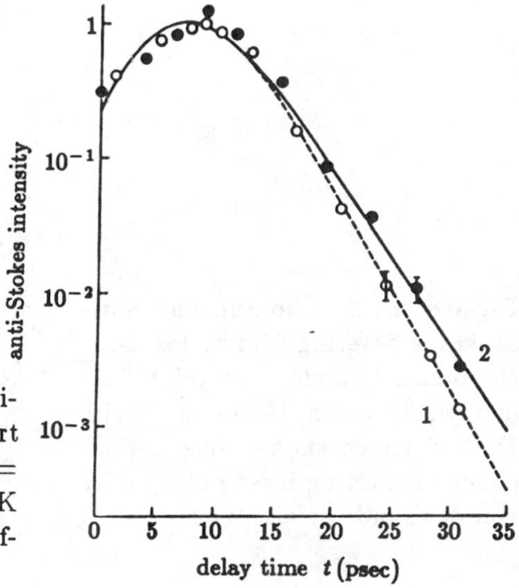

Figure 49.1: Coherent anti-Stokes signal, stimulated by a short laser pulse pair terminating at $t = 0$ in diamond at 295 K and 77 K for curves 1 and 2, respectively (after Laubereau et al., 1971).

49.1.1A Phonon-Polariton Relaxation Measurements of the decay of phonon-polaritons in GaP, which are connected with the TO mode, have also been performed by Kuhl and von der Linde (1982), with apparent lifetimes below 1.3 ps. The width of the polariton resonance is approximately 0.1 meV, and indicates a decay time of ~ 4 ps. The discrepancy is not yet resolved.

49.1.1B Phonon Dephasing Phonon scattering, as discussed in Section 49.1.1, is inelastic. In addition, elastic scattering events ("dephasing") change the momentum of the heated phonons but not their energy. The *dephasing time* is equal to the *momentum-relaxation time*. An example is the phonon scattering with impurities or phonon-phonon interaction within the same phonon branch, which results in a loss of phase information. This elastic scattering, following a preceding excitation, will produce a heated phonon ensemble distributed over a wide range within the Brillouin zone. The evolution of the signal from spontaneous anti-Stokes scattering provides a possibility of directly measuring the phonon distribution and its time evolution.

The time for an incoherent phonon ensemble to approach its thermal distribution is the *energy-relaxation time*. This should not be confused with the *lifetime of a mono-energetic phonon mode at a given wave vector*.

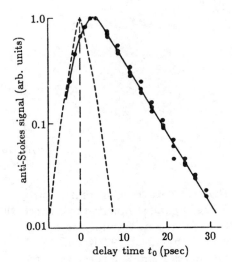

Figure 49.2: Spontaneous anti-Stokes scattering signal for nonthermal LO phonons stimulated by fast electrons in GaAs at 77 K. Dashed curve shows time dependence of exciting laser pulse (after von der Linde et al., 1980).

49.1.2 LO-Phonon Relaxation After Electron Stimulation

Optically excited electrons at sufficient energy within the conduction band stimulate LO phonons during scattering. Their heated distribution can be analyzed by the spontaneous Raman scattering induced by LO phonons with a time-delayed probing pulse.

Figure 49.2 shows the time evolution of the Raman signal in GaAs after cessation of the optical pulse (dashed curve) that causes the generation of hot electrons. The measured signal is proportional to the density of LO phonons, and rises steeply with the integrated energy input from the exciting light pulse (von der Linde et al., 1980).

The energy-relaxation time of the LO phonon is obtained from the exponential decay of the Raman signal, and yields 7 ± 1 ps in GaAs at 77 K—see also Kash et al. (1985). This relaxation time is within the experimental error, identical with the lifetime of LO phonons in GaAs obtained from the Raman line width of 0.1 meV.

49.1.3 Temperature Dependence of Optical Phonon Relaxation

The relaxation time of optical phonons is only weakly temperature-dependent, and increases by a factor of ~ 2 to 4 between 0 K and room temperature, as shown in Fig. 49.3. Experimentally, one observes a somewhat higher factor (4) than theoretically predicted (2)—see also Section 49.1.1 in which similar results are reported for GaAs. For further discussion, see Laubereau (1984).

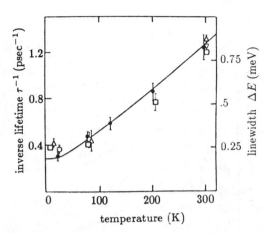

Figure 49.3: Temperature dependence of the inverse lifetime of optical phonons in quartz (after Laubereau, 1984).

49.2 Exciton Kinetics

Excitons or exciton polaritons and exciton molecules (biexcitons— Section 15.1.2E) have lifetimes that depend on the specific scattering or recombination mechanisms. At the end of their lifetimes, usually optical radiation is emitted (luminescence). Before recombination, excitons can change from one to the other exciton branch, or can associate with others to form exciton molecules.

In addition, excitons can be accelerated by a slightly more energetic optical excitation (spatial dispersion), and dissipate their excessive energy by inelastic scattering mechanisms.

Finally, excitons are oriented with polarized optical excitation, and elastic scattering causes dephasing—see Section 50.1.1.

49.2.1 Exciton Recombination in CuCl and CuBr

In this section we will discuss the results for CuCl and CuBr, which are well investigated because of the somewhat longer time constants of these materials. The electron and hole in the exciton state recombines with the emission of a photon. This radiative lifetime of excitons is approximately 150 ps in CuBr at 4.2 K (Unuma et al., 1982).

In addition, when exciting the material with a short laser pulse, the exciton molecule luminescence is observed, which is delayed by ∼ 100 ps (Fig. 49.4), indicating a reaction-kinetic process involving this formation—see following section.

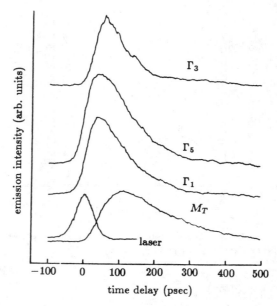

Figure 49.4: Time evolution of the exciton molecule (M_T) luminescence in CuBr at 4.2 K after excitation with a 100 ps laser pulse with band-gap energy E_g and for two-photon excitation (Γ_1, Γ_3, and Γ_5) (after Unuma et al., 1982).

Figure 49.5: Radiative lifetime of exciton polaritons in CuCl at 4 K as a function of the platelet thickness (after Matsumoto and Shionoya, 1982).

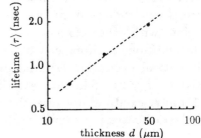

In CuCl, a somewhat longer exciton polariton lifetime is observed, which is usually limited by surface recombination (Toyozawa, 1959). It increases $\propto d^{0.75}$, as shown in Fig. 49.5, indicating the limiting effect of the proximity of the surface; d is the platelet thickness.

49.2.1A Exciton Molecule Radiative Lifetime

In Fig. 49.4, the time evolution of the exciton-molecule luminescence in CuBr is shown at higher exciton densities. There are three exciton molecules—Γ_1, Γ_3, and Γ_5—due to the hole-hole exchange interaction (Hanamura, 1975), which can be generated directly by two-phonon excitation. All three of these molecule states can dissociate to the exciton state (Phach and Levy, 1979). The rise of this luminescence

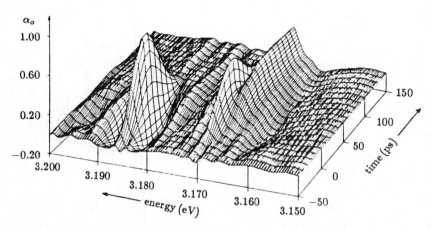

Figure 49.6: Time-resolved optical bleaching due to the creation and annihilation of cold and hot biexcitons in CuCl at 4.2 K. Laser pump pulse of 30 ps duration for two-photon absorption—see text (after Leonelli et al., 1986).

is proportional to the time-integrated exciting laser pulse with little delay. The delayed (by 100 ps) exciton molecule luminescence, when generated by a one-photon band-gap absorption, indicates an indirect generation of the exciton molecules from free excitons, which is governed by the following set of reaction-kinetic equations:

$$\frac{dn_{\mathrm{mo}}}{dt} = g_o - \frac{n_{\mathrm{mo}}}{\tau_{\mathrm{mo}}} + \beta n_{\mathrm{exc}}^2 \tag{49.1}$$

and

$$\frac{dn_{\mathrm{exc}}}{dt} = \frac{n_{\mathrm{mo}}}{\tau_{\mathrm{mo}}} - 2\beta n_{\mathrm{exc}}^2 - \frac{n_{\mathrm{exc}}}{\tau_{\mathrm{exc}}}, \tag{49.2}$$

where g_o is the optical excitation rate, β is the formation coefficient $\simeq 1/(n_{\mathrm{exc}}\tau_f)$ for exciton molecules of density n_{mo}; here, τ_{mo} and τ_{exc} are the respective radiative lifetimes. After switching-off the excitation and fitting the adjustable parameters to obtain agreement between Eqs. (49.1), (49.2) and the experiment, one obtains a radiative lifetime for exciton molecules of $\tau_{\mathrm{mo}} \simeq 60$ ps, and a lifetime for excitons before exciton molecule formation of $\tau_{\mathrm{exc}} \simeq 150$ ps in CuBr; for a recent review, see Matsumoto et al. (1984).

An instructive time-resolved absorption spectrum is given in Fig. 49.6, and shows the kinetics of the creation and annihilation of biexcitons. The generation of cold biexcitons from two-photon absorption near $k = 0$ is indicated by the peak at 3.186 eV. These biexcitons consequently decay into photon-like polaritons with a de-

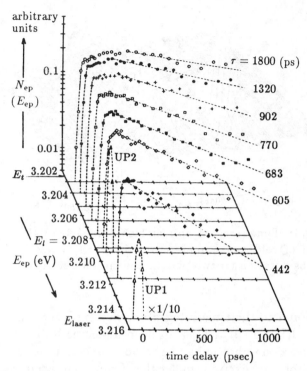

Figure 49.7: Time evolution of the distribution function of exciton polaritons in CuCl at 4.2 K after excitation with a 3.215 eV laser pulse; sampling light frequency as the family parameter (after Matsumoto and Shionoya, 1982).

lay of ~ 20 ps; these polaritons are observed by consequent formation of hot biexcitons from the test beam. The biexcitons can be seen by a delayed bleaching near 3.166 eV. These biexcitons then dissociate into hot exciton-like polaritons with a decay time of ~ 160 ps (Leonelli et al., 1986).

49.2.1B Polariton Redistribution Between Branches When laser excitation occurs into one of the upper polariton branches (UP1 and UP2 in Fig. 49.7), this branch rapidly decays after cessation of the exciting pulse with a decay time of 14 ps for CuCl at 4.2 K. The population in the lower branch has a substantially longer lifetime, as shown in Fig. 49.7, namely between 400 and 1800 ps.

49.2.1C Thermalization of Exciton-Polaritons The energy distribution in the lower exciton-polariton branch rapidly approaches a Boltzmann distribution at an elevated exciton temperature. These

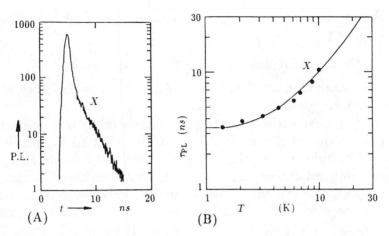

Figure 49.8: (A) Photoluminescence decay (tail) in ultrapure GaAs at 1.4 K after excitation at 8182 Å with 0.1 W/cm^2. (B) Radiative lifetime of free excitons as function of the sample temperature (after t'Hooft et al., 1988).

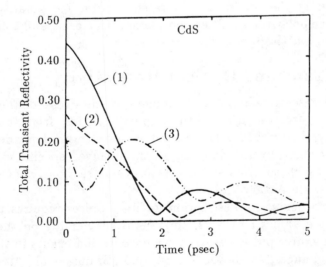

Figure 49.9: Total reflectivity in CdS as a function of time after cessation of illumination at the polariton energy $\hbar_t^{LO} = 2.55$eV. Curves: (1) Sommerfeld Brillouin local optics; (2) Birman ABC's; and, (3) Kinselev ABC's (after Branis et al., 1987).

polaritons relax, probably initially due to polariton-polariton interaction. They then relaxe slowly due to LA-phonon scattering (Matsumoto et al., 1984). From time-resolved absorption spectroscopy, one obtains a relaxation time of ~ 2.5 ns for the latter relaxation in CuCl (Levy et al., 1979).

49.2.2 Exciton Lifetime in GaAs

Another example in which a variety of processes are involved is the *recombination* of excitons in GaAs.

Normally, free excitons are trapped rapidly by defects from which they recombine with a short lifetime of 0.5 or 1 ns for donor- or acceptor-bound excitons, respectively (Henry and Nassau, 1970).

Only in specially prepared ultrapure GaAs can these recombination paths be saturated; the photoluminescence decay then yields a direct measure of the radiative lifetime of free excitons of 3.3 ns at 1.4 K, increasing to 10 ns at 10 K, as shown in Fig. 49.8. This corresponds to an oscillator strength of $\sim$ unity, i.e., it indicates an oscillator strength per exciton volume rather than per unit cell volume—t'Hooft et al. (1988).

Such lifetimes of free excitons can be explained by assuming a conversion of the originally generated exciton-like polaritons into photon-like polaritons via electron-phonon scattering and consequent recombination (Rappel et al., 1988).

49.3 Transient Optical Reflectivity

A special case of relaxation phenomena is the decay of the optical reflectivity after cessation of the impinging light. It relates to the specific type of polarization, i.e., it is wavelength-dependent, and it is determined by the damping of the involved specific excitation processes, such as discussed in the previous sections. For exciton-polaritons, see Askary and Yu (1985).

The mathematical description for light pulses requires the existence of *precursors* and oscillatory decay, depending on the degree of damping after pulse cessation when excitation occurs in the vicinity of resonances, as shown in Fig. 49.9 (Branis et al., 1987). For exciton-polariton resonance, the oscillatory decay of the total transient reflectivity proceeds typically with a time constant of 1 ps in CdS.

Summary and Emphasis

Optical phonons, excited directly by an IR light pulse (TO phonons) or indirectly by fast electrons (LO phonons) or holes (TO phonons), relax their momenta by elastic scattering with phonons in the same branch, or undergo an energy relaxation by scattering with a decay into two acoustical phonons. Energy relaxation of optical phonons can be measured by Raman spectroscopy, and is typically on the order of 10 ps.

Fast excitons can be slowed by inelastic scattering. The lifetime of free excitons is determined by trapping, biexciton formation, or emission of luminescence during annihilation. At sufficient exciton densities, biexciton formation dominates as a relaxation mechanism.

The radiative lifetime of excitons varies over a wide range. It is rather long in cuprous halides, on the order of 100 ps, and about two orders of magnitude smaller in GaAs. The radiative lifetime of exciton molecules is slightly smaller ($\sim$ 50 ps in CuCl).

The lifetime of polaritons is branch-dependent. Polaritons in the upper branches decay rapidly, while their lifetime in lower branches is much longer (e.g., $\sim$ 10 ps vs. $\sim$ 1000 ps in CuCl).

Finally, the polarization of the lattice, expressed by its complex dielectric constant, and, e.g., responsible for its optical reflection, is involved in various relaxation mechanisms, which are dependent on the energy of the impinging photons. This relaxation depends on the specific resonances, causing a kinetic behavior in the reflected light signal, which is pronounced close to such resonances and lies in the low ps range.

Major deviations from the thermal distribution of optical phonons can be created directly by laser excitation or indirectly via fast electron scattering. The ensuing relaxation has substantial influence on short time-scale lattice heating (annealing, damage, melting, or evaporation). Exciton kinetics is of interest for identifying material and defect-related decay mechanisms.

Exercise Problems

1.(*) Quantitatively describe the energy relaxation of optical phonons by creating two acoustical phonons.

2.(e) Estimate the density dependence of biexciton formation in CuCl from the data presented in Section 49.2.1A.

3.(*) What are the possible reasons for the lifetime of upper-branch polaritons to be much shorter than the polaritons in lower branches?

4.(r) What types of phonons can be stimulated by laser irradiation? Why?

Chapter 50

Orbital and Spin Relaxation

Excited states, created with polarized light, show a preferred orientation, which rapidly relaxes with scattering.

The orientation of orbitals or the spin of excited states can be aligned by using polarized light. These aligned states relax rapidly by scattering with phonons or other defects within the lattice. The recently developed femtosecond-spectroscopy opens this field to direct experimental observation.

50.1 Orbital Relaxation

When polarized light is used to excite electronic quasi-particles, such as excitons and electron-hole pairs, relaxation by collision with other quasi-particles tends to randomize the orientation of these states. The relaxation times of such states are rather short. Observation of the time evolution of such randomization may employ diffraction on a bleached-out grating, or four-wave mixing, in which nonlinear interaction between pump and test beams creates two additional photons (Lyon, 1986). These optical techniques provide an exceptionally powerful method to explore details of such scattering events. A few examples are given below.

50.1.1 Orientation Relaxation of Excitons

Excitons can be optically aligned by linear polarized optical excitation. The density of the so-generated excitons can be measured by producing an optical grating of polarized excitons, which is caused by the interaction of a first pulse with a slightly delayed second pulse of well-defined phase relation. One observes self-diffraction from the excitons of the second pulse with the coherent part of the original field of polarization—see also Section 33.1.2A, and Delfyett et al. (1987). The time evolution, measured by increasing the delay time τ_{12} of the self-diffracted signal, yields information about the relax-

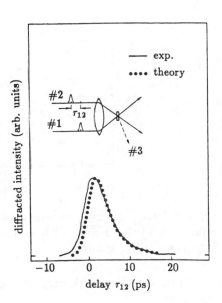

Figure 50.1: Signal intensity (beam #3) in a two-pulse, four-wave mixing experiment in a thin (1900 Å) GaAs layer at 2 K from oriented excitons, which lose their orientation by acoustic phonon scattering with a 7 ps time constant (after Schultheis et al., 1986).

ation of the *oriented* excitons. The method is illustrated in the insert of Fig. 50.1. The intensity of the diffracted beam #3 in this degenerate four-wave mixing experiment is a direct measure of the surviving phase coherence between beams #1 and #2.

Plotting the intensity of the diffracted signal (beam #3 shown in the insert of Fig. 50.1) as a function of the delay time, one obtains for GaAs at 2 K an orientation relaxation time of 7 ± 0.5 ps for relatively low exciton densities, $n_{exc} < 2 \cdot 10^{14}$ cm^{-3}. This relaxation time decreases linearly to ~ 2 ps with increasing temperature up to ~ 20 K; it is probably determined by elastic acoustic phonon and possibly residual impurity scattering (Schultheis et al., 1986). Above 15 K, thermal ionization of the exciton with a binding energy of 4 meV becomes important. Free-carrier scattering then substantially reduces the relaxation time into the femtosecond range—see also Schultheis et al. (1988).

When increasing the density of excitons by increased pumping intensity, or producing free carriers by excitation with another light pulse of higher than band-gap energy (1.520 eV), the orientation-relaxation time decreases as shown in Fig. 50.2. These results indicate the onset of exciton-exciton scattering as a dephasing mechanism at an exciton density of $\sim 10^{15}$ cm^{-3}. When sufficient free electrons are injected, the electron-exciton scattering becomes predominant already near 10^{13} cm^{-3}, and is approximately 10 times more efficient.

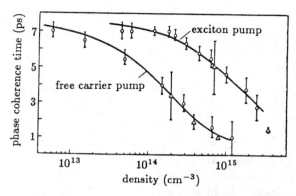

Figure 50.2: Orientation-relaxation time of excitons in GaAs at 2 K as a function of the density of free electrons for $n_{\text{exc}} = 2 \cdot 10^{14}$ cm^{-3} (after Schultheis et al., 1986).

From elastic scattering, one estimates for the exciton collision rate Γ

$$\tau_{\text{exc}} \simeq 2/\Gamma \quad \text{with} \quad \Gamma = s_{\text{exc,exc}} \, v_{\text{rms}} \, n_{\text{exc}}. \tag{50.1}$$

From $\tau = 3 \cdot 10^{-12}$ s at $n_{\text{exc}} \simeq 3 \cdot 10^{15}$ cm^{-3}, and with $v_{\text{rms}} = 1.6 \cdot 10^6$ cm/s at a mean kinetic energy of 0.3 meV, one obtains a very large exciton-exciton collision cross section of $1.5 \cdot 10^{-10}$ cm^2. With (Elkomoss and Munschy, 1981)

$$s_{\text{exc,exc}} = 12\pi a_{\text{qH}}^2, \tag{50.2}$$

there is reasonable agreement with the exciton radius of 130 Å for GaAs.

The increased collision rate with free carriers can be explained as being similar to the electron scattering at neutral defects (electron-hydrogen), using Erginsoy's estimate (Erginsoy, 1950):

$$\Gamma \simeq 20 \frac{a_{\text{qH}}}{k} n \tag{50.3}$$

—see Section 32.2.3. This results in a scattering cross section of approximately $5 \cdot 10^{-4}/v_{\text{rms}} \simeq 3 \cdot 10^{-9}$ cm^2 at 2 K, which is within a factor of 3 in agreement with the experiment.*

50.1.2 Orientation Relaxation of Carriers

At higher excitation intensities, excitons can no longer exist because of free-carrier shielding—see Section 34.2. As with excitons, however,

---

* Exciton-free-hole scattering can be neglected, compared to the scattering with free electrons, due to their substantially larger effective mass.

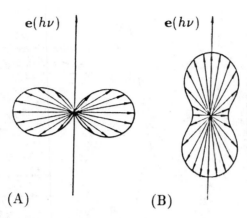

Figure 50.3: Momentum distribution functions of electrons excited with linearly polarized (e) light from (A) heavy hole and (B) light hole bands. The lengths of the vectors indicate population of states with the corresponding momentum directions (after Mirlin, 1984).

free electron-hole pairs are *preferably oriented* when excited by linear polarized light pulses (Oudar et al., 1985).

50.1.2A Momentum Polarization of Photon-Created Carriers Due to selection rules for direct band transitions, the momentum distribution of photogenerated carriers is anisotropic (Mirlin, 1984). The momentum-distribution function of electrons at the instant of excitation is given by (Zakharchenya et al., 1982)

$$f(\mathbf{p}) = f_o(p)\left[1 + \alpha_o^* P_2(\cos\vartheta)\right] \quad \text{with} \quad P_2(\cos\vartheta) = \frac{1}{2}(3\cos^2\vartheta - 1),$$
(50.4)

where $\alpha_o^* = -1$, 0, and $+1$ for excitation from heavy hole, spin-orbit, and light hole band, respectively, and ϑ is the angle between momentum and polarization of the light. The long axis of the momentum distribution of electrons, excited from the heavy hole band, are therefore perpendicular to the optical polarization vector; for electrons from the light hole band, they are predominantly parallel to it— see Fig. 50.3. The same relation holds true for the simultaneously created holes.

50.1.2B Momentum Relaxation of Electron-Hole Pairs
Due to the same selection rules, the spontaneous luminescence from these electrons with preferred orientation is also linearly polarized.

Relaxation of the *optical alignment* of electron momenta is extremely fast and needs excitation pulses in the sub-picosecond range

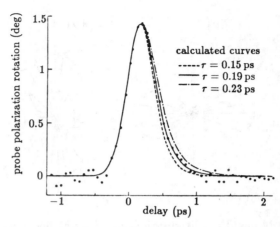

Figure 50.4: Angle of test pulse rotation as a function of time delay after an exciting pulse of 0.25 ps width, with a polarization angle of 30°, and after passing through a GaAs layer of 1.5 μm thickness (after Oudar et al., 1984).

for detection. *Time-resolved rotation* of a polarized test pulse produces a signal output as shown in Fig. 50.4 for GaAs, with an excitation of $6 \cdot 10^{17}$ electrons/cm^2. The signal shape is sensitive to the relaxation time. Comparison of the solution curves for the *optical Bloch equations* (see Yajima and Taira, 1979) with the experiment yields $\tau_{n,p} \simeq 190$ fs; the relaxation is probably caused by carrier-carrier scattering in this high electron density range—compare with Fig. 50.2.

50.1.2C Momentum Relaxation of Hot Electrons   The momentum relaxation of electrons, excited to higher energies within the conduction band before recombining from a lower energy near the bottom of the band, shows a surprising memory of the original momentum in spite of several inelastic scattering events. This can be seen from Fig. 50.5, where the degree of linear polarization of the $(e, A^{\times})$ luminescence in GaAs:Ge is shown with excitation at $h\nu = 1.96$ eV, generating electrons with an energy of $\Delta E \simeq 0.54$ eV above the band edge. These fast electrons are scattered with other carriers ($N_A = 6 \cdot 10^{17}$ cm^{-3}) and LO phonons and thereby rapidly lose their excess energy.

While exposed to such inelastic collisions, they also have a small probability to recombine with the neutral acceptor, generating the *hot photoluminescence* spectrum—i.e., the low-intensity, high-energy tail of the $(e, A^{\times})$ emission. This tail shows substantial *linear polar-*

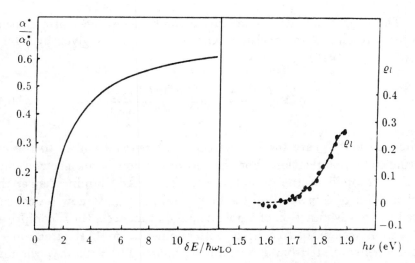

Figure 50.5: (A) Relative polarization factors [see Eqs. (50.4), (50.5), and (50.6)] as functions of the electron energy above the conduction band edge. (B) Degree of linear polarization [Eq. (50.7)] as a function of the hot $(e, A^\times)$ luminescence with excitation of GaAs:Ge at $h\nu = 1.96$ eV and $T = 80$ K. Sample plane (100), optical excitation e (110) (after Mirlin, 1984).

ization down to 0.3 eV below the excitation energy (Fig. 50.5B), i.e., after several inelastic LO scatterings.

It can be shown that the electron momentum of high-energy electrons survives several LO-phonon emissions, while electrons with an energy of only $\hbar\omega_{LO}$ above E_c are randomized by only one LO-phonon emission (Mirlin, 1985). This can be made plausible by recognizing that the $(e, A^\times)$ luminescence intensity in e' direction is given by

$$I_L(\mathbf{e}') \propto 1 + \alpha_o^* P_2(\cos \vartheta'), \qquad (50.5)$$

where ϑ' is the angle between polarization $\mathbf{p}$ and $\mathbf{e}'$, and P_2 is the second Legrendre polynomial. The anisotropy parameter α_o^* has the value of $+1$, -1, or 0 for the light, heavy, or split-off hole band, respectively. After the electrons relax with an energy loss of δE, the anisotropy parameter becomes energy-dependent, and the luminescence intensity is given by

$$I_L(\mathbf{e}', \delta E) \propto 1 + \frac{\alpha^*(\delta E)\alpha_o^*}{5} P_2(\cos \chi), \qquad (50.6)$$

where χ is the angle between the electric vector of the exciting light $\mathbf{e}$ and the luminescence $\mathbf{e}'$.

The degree of linear polarization for luminescence, propagating in the direction of the exciting light, is consequently given by (Dymnikov et al., 1976)

$$\varrho_l = \frac{I_\parallel - I_\perp}{I_\parallel + I_\perp} = \frac{3\alpha^*(\delta E)\alpha_0^*}{20 + \alpha^*(\delta E)\alpha_0^*} \tag{50.7}$$

where $I_\parallel$ and $I_\perp$ are the luminescence intersites $\parallel$ and $\perp$ to the exciting light polarization. For the recombination of an electron with a hole trapped at the acceptor, i.e., for $(e, A^\times)$ luminescence, the orientation of **e** in relation to the crystal axes must now be considered. The wavefunction of a hole trapped at an acceptor is comprised of wavefunctions of free holes with momenta in [111] direction—a consequence of HH-band warping with $m_p[111] = 0.8m_0$, whereas $m_p[100] = 0.4m_0$. In GaAs, the linear polarization of the $(e, A^\times)$ luminescence for light entering along [100] is then given by

$$\varrho_l = \frac{1}{4} \sin^2(2\varphi) \tag{50.8}$$

where φ is the angle between **e** and the [001] axis. Therefore, ϱ_l is largest when **e** is oriented in the [011] direction ($\varrho_l^{\max} = 0.25$) and zero when oriented along [001] (Mirlin and Reshina, 1977).

The dependence of the anisotropy factor α^* on the surplus energy δE while scattering with LO-phonons was estimated by Mirlin (1984), and is shown in Fig. 50.5A. Its relative value (α^*/α_0^*) decreases from $\simeq 0.6$ at $\delta E = 10\hbar\omega_{\mathrm{LO}}$ to 0 at $\delta E = \hbar\omega_{\mathrm{LO}}$.

In heavily doped p-type samples, the energy relaxation of optically generated hot electrons is dominated by scattering with free holes or charged acceptors with (D'yakonov et al., 1977):

$$\frac{\alpha^*}{\alpha_0^*} = \left(\frac{E_e}{\delta E}\right)^{\tau_e/\tau_{P2}} \quad \text{with} \quad \frac{d\alpha^*}{dt} = -\frac{\alpha^*}{\tau_{P2}}; \quad \frac{dE_e}{dt} = -\frac{E_e}{\tau_e} \tag{50.9}$$

and with E_e the initial (optically excited) surplus energy; for scattering with free holes, $\tau_e/\tau_{P2} \simeq 20$, where τ_{P2} is the relaxation of the momentum anisotropy given by P_2—see Fig. 50.5.

Relaxation of hot electrons with scattering at neutral acceptors yields $\tau/\tau_{P2} \simeq 3$ (Rebane, 1981). Other scattering mechanisms, including impact ionization of shallow impurities, will also influence this momentum relaxation. For a review, see Mirlin (1984).

50.1.3 Exciton Dephasing in Superlattices

In superlattices with a relatively large well width, rendering the properties almost 3D-like, the increased localization of 2D excitons causes a substantially increased orientation relaxation, which can be obtained from the line width of excitons. Dephasing is caused by a number of scattering mechanisms, such as scattering with phonons, other excitons, or free carriers, by recombination and field ionization. Typical dephasing times in GaAs/AlGaAs superlattices lie in the $10 \ldots 100$ ps range, and are dominated by scattering with acoustic phonons. Exciton recombination is observed with a somewhat larger time constant of ~ 400 ps. At higher excitation rates, exciton-exciton scattering, and with injection of carriers from the adjacent well, a dramatic increase of electron-exciton scattering is observed (Schultheis et al., 1986; Andrews and Harley, 1987).

50.2 Spin Relaxation

When a photon of sufficient energy creates an electron and a hole in a direct transition, the total spin and angular momentum of these two carriers equal the angular momentum of the absorbed photon. In right- or left-polarized light, the photons have an angular momentum of $+1$ or -1, respectively. This momentum is distributed between the electron and the hole, according to the selection rules within the band structure of the given semiconductor.

After a short excitation pulse, the spin orientation decreases due to a variety of relaxation processes. The fraction of carriers, which have not lost their original spin orientation, will cause a *circularly polarized luminescence* when recombining, and provide a measure of spin relaxation when carrier lifetime and spin-relaxation time are on the same order of magnitude. It can be analyzed further when the spin relaxation time changes as a function of external parameters (Parsons, 1969, 1971).

Other effects relating to spin orientation involve the nuclear spin system, which is strongly coupled to the electronic spin system. The latter can be influenced by optical excitation (see above) or by external magnetic fields. Changes can be detected by analyzing the degree of polarization of the luminescence (Ekimov and Safarov, 1972). For a review of the different relaxation effects involving the nuclear spin, see D'yakonov and Perel' (1984).

Spin relaxation can occur for several reasons:

- momentum relaxation through spin-orbit coupling (Elliot, 1954; Yafet, 1963);
- exchange interaction between electrons and holes (Bir et al., 1975);
- reabsorption of recombination luminescence (Kleinman and Miller, 1981); and
- dynamic narrowing of the magnetic resonance in spin-orbit split-off conduction bands (D'yakonov and Perel', 1971).

For a review, see Pikus and Titkov (1984).

50.2.1 The Hanle Effect to Measure Spin Relaxation

Measurements of the spin-relaxation time use the decrease of circular polarization of luminescent light induced by a transverse magnetic field (Parsons, 1969), known as the *Hanle effect* (Hanle, 1924)

The depolarization is caused by the precession of electrons in the magnetic field, which tends to move the direction of the spin away from the initial alignment that defines the degree of circular polarization of the luminescence.

When the magnetic field is perpendicular to the exciting light beam, i.e., $\mathbf{B} \perp S_0$, the initial spin, the component of the spin in beam direction (z), is given by

$$S_z(B) = S_0 \cos \omega_L t \quad \text{with} \quad \omega_L = \frac{\mu_B g B}{\hbar}, \qquad (50.10)$$

if all electrons have approximately the same descent time t. Here, the descent time may be replaced by τ_e, the energy-relaxation time; ω_L is the Larmor frequency,[*] μ_B is the Bohr magneton, and g is the Landé factor. The average of $S_z(B)$ determines the polarization of the luminescence. This average will be very small when, during the lifetime of the electron, it completes many Larmor revolutions.

---

[*] The Larmor frequency is equivalent to the cyclotron frequency for free electrons, with $\omega_c = (2/g)\omega_L$ and g as the Landé factor (g-factor), which is given for isolated atoms by

$$g = 1 + \frac{j(j+1) + s(s+1) - l(l+1)}{2j(j+1)}; \qquad (50.11)$$

for electrons in a semiconductor, the g-factor is influenced by the spin-orbit splitting of the valence band, and can have substantially different values—see Section 31.3.2, Eq. (31.55), and Table 31.3.

The average value of the S_z component depends on the spin lifetime, i.e., what fraction of a Larmor revolution can be completed before spin relaxation (τ_S) or electron recombination (τ_n) occurs: $t_S^{-1} = \tau_n^{-1} + \tau_S^{-1}$.

Therefore, one has to average Eq. (50.10) over the different realization of $t \to t_S$. Consequently, one obtains for the component of the electron spin

$$S_z(B) = \frac{S_z(B=0)}{1 + (\omega_L t_S)^2} \quad \text{and} \quad S_z(B=0) = \frac{S_0}{1 + \dfrac{\tau_n}{\tau_S}}. \tag{50.12}$$

Since the degree of circular polarization ϱ of the luminescence is equal to the component of the average spin in the direction of the exciting beam, one can obtain the *electron lifetime* from Eqs. (50.10) and (50.12) as

$$\tau_n = \frac{\varrho_0}{\varrho(B=0)} \frac{\hbar}{g\mu_B B_{1/2}}, \tag{50.13}$$

and the *spin relaxation time* as

$$\tau_S = \frac{\varrho_0}{\varrho_0 - \varrho(B=0)} \frac{\hbar}{g\mu_B B_{1/2}}, \tag{50.14}$$

where $B_{1/2}$ is the magnetic induction value at which the polarization is reduced to one half of its original value (at $B = 0$), and with

$$\varrho = \frac{I_+ - I_-}{I_+ + I_-} \tag{50.15}$$

where I_+, I_- are the intensities of the right and left polarized luminescence.

This is an interesting example of obtaining important relaxation times from steady-state experiments. The Hanle effect has been used to study the spin relaxation of electrons in a variety of samples (D'yakonov and Perel', 1984).

50.2.2 Spin-Relaxation Time in p-Type Semiconductors

When exciting a semiconductor with light, practically all minority carriers, i.e., *electrons* in p-type semiconductors, are created—here, by circular polarized light. By measuring the degree of circular polarization of the band-to-band (edge) luminescence, one measures the average spin of these electrons in the conduction band. The maximum value of the degree of circular polarization of this luminescence

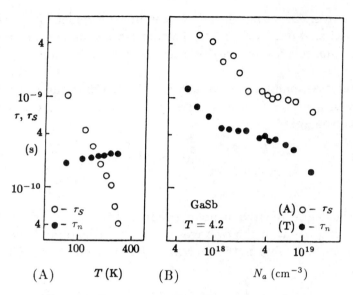

Figure 50.6: Electron lifetime and spin relaxation time (A) as a function of the temperature in AlGaAs for $N_a = 2 \cdot 10^{17}$ cm^{-3} (after Garbuzov et al., 1971) and (B) as a function of the doping density in GaSb at $T = 4.2$ K (after Aronov et al., 1979; Titkov et al., 1981).

in the absence of electron-spin relaxation for a Γ_8 (heavy and light hole bands) to Γ_7 (conduction band in GaAs) transition is equal to the initial average value of electron spin in such transitions (see D'yakonov and Perel', 1984)

$$\varrho = 0.25. \qquad (50.16)$$

With spin relaxation, the degree of circular polarization is reduced to

$$\varrho = \varrho_0 \frac{1}{1 + \dfrac{\tau_n}{\tau_S}}. \qquad (50.17)$$

The measured electron lifetime and spin-relaxation time in AlGaAs and GaSb are shown in Fig. 50.6 as a function of the temperature and acceptor density (at 4.2 K). These relaxation times vary widely, depending on the type of the most active relaxation mechanism, and lie typically between 10^{-8} and 10^{-11} s.

Spin relaxation can be used to show, that in highly doped degenerate p-type semiconductors, the interaction of electrons with thermalized, nonpolarized holes dominate: here, the spin relaxation

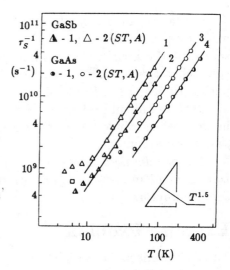

Figure 50.7: Spin-relaxation rate as a function of the temperature for degenerate p-type GaAs with $N_a = 1.8 \cdot 10^{18}$ cm^{-3}, $6.5 \cdot 10^{18}$ cm^{-3}, and in GaSb with $N_a = 5 \cdot 10^{18}$ cm^{-3} and $4 \cdot 10^{19}$ cm^{-3} in curves 1–4, respectively (after Safarov and Titkov, 1980; Aronov et al., 1983).

time is substantially reduced, and the experimentally observed spin-relaxation time (Fig. 50.7)

$$\tau_S \propto T^{-3/2} \tag{50.18}$$

has a temperature dependence consistent with the theory of electron-hole relaxation (Pikus and Titkov, 1984).

50.2.3 Spin Relaxation of Majority Carriers

When majority carriers are generated by circular polarized light, only a small fraction of these initially show a preferred spin orientation. However, accumulation of spin-oriented carriers via optical pumping of will occur, and gradually will replace electrons with unoriented spin, if the spin-relaxation time is much longer than the carrier lifetime. Such successive orientation of majority carriers can be observed even if their incremental density is small compared to the thermal equilibrium density. This accumulation depends on the field intensity, and can be used to determine the ratio of relaxation to carrier lifetime (Vekua et al., 1976).

Summary and Emphasis

Excitons, as well as electron-hole pairs, remain polarized after excitation in polarized light; and such polarization can be detected by the polarization of the ensuing recombination luminescence. The time decay of the degree of polarization offers direct information on dephasing when scattering occurs during their lifetime.

Similar to an orbital orientation, a spin alignment occurs with excitation in polarized light: the photon angular momentum is conserved by the excited electron hole pair. With consequent scattering, spin relaxation occurs, which can be measured by steady-state experiments using the Hanle effect.

Orbital and spin relaxation have short relaxation times, typically in the low ps range, since they are sensitive to any of the various scattering events.

A more detailed analysis of the dephasing of orientation polarization gives important information on various scattering events in semiconductors.

Exercise Problems

1.(r) Summarize the different time constants for kinetic processes in solids. List typical ranges and give relevant examples for major influence on some device performances.

2.(e) Describe orientation relaxation and experimental methods for an analysis.

4.(*) Discuss the use of four-wave mixing for an experimental analysis of oriented exciton dephasing.

5.(l) Describe in detail the Hanle effect and its use to measure spin relaxation in semiconductors.

  (a) Derive the expression for the z argument of the electron spin.

  (b) Discuss the Hanle curve, $B_{1/2} = \hbar/(g\mu_B \tau_s)$.

PART IX

APPENDIX, BIBLIOGRAPHY, INDEX

Appendix

A.1 Units

The constants used in this book are listed on the front cover and are given in SI units (Le Système International d' Unités), except that cm and g are used for units of length and mass, rather than m and kg. This system is used since these quantities are more familiar to semiconductor scientists. The temperature is mostly expressed in K (Kelvin) $= °C + 273.15$. In discussions relating to atomic dimensions, the Ångstrom unit is used: $1 \text{ Å} = 10^{-8}$ cm.

Other units used in this book are:

$1 \text{ T} = 10^{-4} \text{ Wb (Weber)}/\text{cm}^2 = 10 \text{ kG (Gauss)} = 10^4 \text{ Vs}/\text{cm}^2$

$1 \text{ N (Newton)} = 10^5 \text{ dyne}, \quad 1 \text{ dyne} = 1.0197 \cdot 10^{-3} \text{ g}^*$

$1 \text{ bar} = 10 \text{ N}/\text{cm}^2 = 1.01972 \text{ kg}^*/\text{cm}^2 = 750.06 \text{ torr} = 1.01972 \text{ at}$

$1 \text{ } = \text{ kg}^*/\text{cm}^2 = 1 \text{ (Technical atmosphere)}$

$1 \text{ atm} = 10.13 \text{ N}/\text{cm}^2 = 760 \text{ mm Hg}$

$1 \text{ Pa (Pascal)} = 1\text{N}/\text{m}^2$

$1 \text{ J (Joule)} = 1 \text{ Nm} = 10^7 \text{ ergs}$

$1 \text{ kg}^* = 9.807 \text{ N}$

$1 \text{ cal} = 4.19 \text{ Ws}$

$1 \text{ barn} = 10^{-24} \text{ cm}^2$

The units identified with an asterisk (g* and kg*) are weight rather than mass units. A word of caution is in order when using the mass

Table A.1: Table of coefficients in Eqs. (A.1)–(A.6) for the various unit systems.

Systems	χ	ν_c	ν_m	λ	ε_0	μ_0	γ
Part rational (Maxwell)	4π	1	4π	1			
Part rational (Gauss)	4π	4π	4π	1			
Nonrational (Schaeffer)	4π	4π	4π	4π			
Rational	1	1	1	1			
Electrostatic (3 parameters)					1	$1/c^2$	1
Electromagnetic (3 parameters)					$1/c^2$	1	1
Symmetric (3 parameters)					1	1	c
Four-parameter system					ε_0	μ_0	1

of a particle. Numerical results from equations containing electrical units can be most easily be obtained when using the equivalent

$$1 \text{ g} = 10^7 \text{ Ws}^2/\text{cm}^2.$$

There are other systems in use in the literature. The distinction between these systems can best be followed from the basic set of electrodynamic equations containing a number of parameters (χ, ν_e, ν_m, and λ) which are set equal to 1 or 4π, according to Table A.1, to distinguish the different rational and nonrational systems, and ε_0, μ_0, which have different values in the electrostatic and electromagnetic systems. The sets of equations are

$$\frac{\mathbf{D}}{\nu_e} = \varepsilon\varepsilon_0 \frac{\mathbf{E}}{\chi} = \varepsilon_0 \frac{\mathbf{E}}{\chi} + \frac{\mathbf{P}}{\lambda} \quad \text{and} \quad \frac{\mathbf{B}}{\nu_m} = \mu\mu_0 \frac{\mathbf{H}}{\chi} = \mu_0 \frac{\mathbf{H}}{\chi} + \frac{\mathbf{J}}{\lambda} \quad (\text{A}.1)$$

$$w_e = \frac{\mathbf{E} \cdot \mathbf{D}}{2\nu_e}, \quad w_m = \frac{\mathbf{H} \cdot \mathbf{B}}{2\nu_m}, \quad \text{and} \quad \mathbf{S} = \frac{\gamma}{\chi}(\mathbf{E} \times \mathbf{H}) \quad (\text{A}.2)$$

$$\vec{\mathcal{F}}_e = \frac{\chi e_1 e_2}{4\pi\varepsilon\varepsilon_0 r^2} \mathbf{r}^0 = e\mathbf{E} \quad \text{and} \quad \vec{\mathcal{F}}_m = \frac{\chi p_1 p_2}{4\pi\mu\mu_0 r^2} \mathbf{r}^0 = p\mathbf{H} \quad (\text{A}.3)$$

$$d\mathcal{F}_{em} = \frac{\chi}{\gamma} \mathbf{I} \left(\frac{d\mathbf{s} \times \mathbf{B}}{\nu_m} \right) = \frac{\chi p \mathbf{I}(d\mathbf{s} \times \mathbf{r}^0)}{4\pi\gamma r^2} \quad (\text{A}.4)$$

$$\psi = \frac{\chi e}{4\pi\varepsilon\varepsilon_0 r} \qquad \mathbf{A} = \frac{\nu_m}{\gamma} \frac{\mu\mu_0}{4\pi} \mathbf{I} \oint \frac{d\mathbf{s}}{r} \quad (\text{A}.5)$$

$$\nabla^2 \psi = -\chi \frac{\varrho}{\varepsilon\varepsilon_0} \qquad \nabla^2 \mathbf{A} = -\frac{\nu_m}{\gamma} \mu\mu_0 \mathbf{j} \quad (\text{A}.6)$$

where w_e and w_m are the electric and magnetic energy densities, **S** is the Pointing vector, ds is a line element, $\mathbf{r}^0$ is a unit vector in the field direction, **A** is the vector potential, **E** and **H** are the fields, **D** and **B** are the flux densities, **P** and **J** are the polarizations, **I** is the current, ψ is the electrostatic potential, and $\mathcal{F}$ is the force. The indices e and m stand for electric and magnetic, respectively. For further detail, see Stille (1955).

The SI system is a four-parameter rational system. The often-used Gaussian system is a three-parameter electrostatic system; observe the combination from the upper and lower half of Table A.1 to obtain all necessary parameters for any specific system.

A.2 Some Mathematical Tools

Space limitation forces us to select only a few examples, which are helpful when handy. For further detail, see, e.g., Madelung (1950), Spiegel (1966, 1968), Kamke (1944), Beyer (1981), or Abramowitz and Stegun (1972).

A.2.1 Series

Series are very helpful tools when approximations are of interest. Taylor Series:

$$f(x) = f(a) + f'(a)(x-a) + \frac{f''(a)(x-a)^2}{2!} + \dots \text{ with } f'(a) = \frac{\partial f(x)}{\partial x}\bigg|_{x=a}$$

Binomial Series:

$$(a+x)^n = a^n + \binom{n}{1}a^{n-1}x + \binom{n}{2}a^{n-2}x^2 + \dots \text{ with } \binom{n}{i} = \frac{n!}{(n-i)!i!}.$$

Special cases for $|x| < 1$, converging well for $|x| \ll 1$:

$$(1+x)^{-1} = 1 - x + x^2 - \dots;$$

$$(1+x)^{-2} = 1 - 2x + 3x^2 - \dots;$$

$$(1+x)^{1/2} = 1 + \frac{x}{2} - \frac{x^2}{8} + \dots;$$

$$(1+x)^{-1/2} = 1 - \frac{x}{2} + \frac{3}{8}x^2 - \dots;$$

$$\exp x = 1 + x + \frac{x^2}{2!} + \dots;$$

$$a^x = 1 + x \ln a + \frac{(x \ln a)^2}{2!} \dots;$$

$$\ln x = 2\left[\frac{x-1}{x+1} + \frac{1}{3}\left(\frac{x-1}{x+1}\right)^3 + \dots\right];$$

$$\ln(1+x) = x - \frac{x^2}{2} + \frac{x^3}{3} - \dots;$$

$$\sin x = x - \frac{x^3}{3!} + \frac{x^5}{5!} \dots;$$

$$\cos x = 1 - \frac{x^2}{2!} + \frac{x^4}{4!} - \ldots;$$

$$\tan x = x + \frac{x^3}{3} + \frac{2x^5}{15} + \ldots;$$

$$\cot x = \frac{1}{x} - \frac{x}{3} + \frac{x^3}{45} - \ldots;$$

$$\sinh x = x + \frac{x^3}{3!} + \frac{x^5}{5!} + \ldots;$$

$$\cosh x = 1 + \frac{x^2}{2!} + \frac{x^4}{4!} + \ldots$$

Gauss' error integral:

$$\int_0^x \exp(-t^2)\,dt = x - \frac{x^3}{3} + \frac{x^5}{10} - \frac{x^7}{42} + \ldots$$

A.2.2 Coordinate Systems

The choice of the coordinate system is governed by the symmetry of the problem. The **Cartesian coordinate system** is most widely used (x, y, z orthogonal to each other):

$$ds^2 = dx^2 + dy^2 + dz^2$$
$$\mathbf{a} = a_x\mathbf{i} + a_y\mathbf{j} + a_z\mathbf{k}$$

with $\mathbf{i}$, $\mathbf{j}$, and $\mathbf{k}$ the unit vectors in the x, y and z directions.

$$\mathbf{a} \cdot \mathbf{b} = a_x b_x + a_y b_y + a_z b_z$$
$$\mathbf{a} \times \mathbf{b} = (a_y b_z - a_z b_y)\mathbf{i} + (a_z b_x - a_x b_z)\mathbf{j} + (a_x b_y - a_y b_x)\mathbf{k}$$

$$\mathbf{a} \cdot \mathbf{b} = ab \cos \varphi \qquad \text{and} \qquad |\mathbf{a} \times \mathbf{b}| = ab \sin \varphi$$
$$\mathbf{a} \cdot \mathbf{b} = \mathbf{b} \cdot \mathbf{a} \qquad \text{and} \qquad \mathbf{a} \times \mathbf{b} = -\mathbf{b} \times \mathbf{a}$$

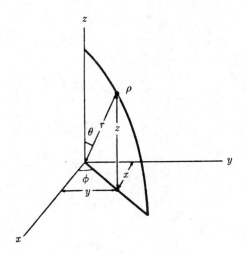

Figure A.1: Spherical coordinate system.

For **spherical coordinates** (Fig. A.1), one has with r, θ, and ϕ

$$ds^2 = dr^2 + r^2 d\theta^2 + r^2 \sin^2 \theta d\phi^2 \quad \text{(surface element)}$$

$$d\mathcal{V} = r^2 \sin \theta \, dr \, d\theta \, d\phi \qquad\qquad \text{(volume element)}$$

$$x = r \sin \theta \cos \phi \quad 0 \le r < \infty \, ; \quad r = \sqrt{x^2 + y^2 + z^2}$$

$$y = r \sin \theta \sin \phi ; \; 0 \le \theta < \pi \, ; \quad \theta = \arctan(\sqrt{x^2 + y^2}/z)$$

$$z = r \cos \theta \qquad 0 \le \phi < 2\pi \, ; \quad \phi = \arctan(y/x)$$

$$\nabla_r \psi = \frac{\partial \psi}{\partial r} \, ; \quad \nabla_\theta \psi = \frac{1}{r} \frac{\partial \psi}{\partial \theta} \, ; \quad \nabla_\phi \psi = \frac{1}{r \sin \theta} \frac{\partial \psi}{\partial \phi}$$

$$\nabla \cdot \mathbf{a} = \frac{1}{r^2} \frac{\partial}{\partial r} (r^2 a_r) + \frac{1}{r \sin \theta} \frac{\partial}{\partial \theta} (\sin \theta a_\theta) + \frac{1}{r \sin \theta} \frac{\partial a_\phi}{\partial \phi}$$

$$\Delta \psi = \frac{1}{r^2} \frac{\partial}{\partial r} \left(r^2 \frac{\partial \psi}{\partial r} \right) + \frac{1}{r^2 \sin \vartheta} \frac{\partial}{\partial \vartheta} \left(\sin \vartheta \frac{\partial \psi}{\partial \vartheta} \right) + \frac{1}{r^2 \sin^2 \vartheta} \frac{\partial^2 \psi}{\partial \varphi^2} .$$

A.3 Vector Analysis

With $\varphi = \varphi(\mathbf{r})$, one has

$$\nabla \varphi = \mathrm{grad}\varphi = \frac{\partial \varphi}{\partial x}\,\mathbf{i} + \frac{\partial \varphi}{\partial y}\,\mathbf{j} + \frac{\partial \varphi}{\partial z}\,\mathbf{k}$$

with $\mathbf{i}$, $\mathbf{j}$, and $\mathbf{k}$ as the unit vectors in the x, y, and z directions. The divergency, identified with a source or sink: $\nabla \cdot \mathbf{a} > 0$ or < 0, respectively, for $\mathbf{a} = \mathbf{a}(\mathbf{r})$:

$$\nabla \cdot \mathbf{a} = \mathrm{div}\,a = \frac{\partial a_x}{\partial x} + \frac{\partial a_y}{\partial y} + \frac{\partial a_z}{\partial z}$$

The curl

$$\nabla \times \mathbf{a} = \left(\frac{\partial a_z}{\partial y} - \frac{\partial a_y}{\partial z}\right)\mathbf{i} + \left(\frac{\partial a_x}{\partial z} - \frac{\partial a_z}{\partial x}\right)\mathbf{j} + \left(\frac{\partial a_y}{\partial x} - \frac{\partial a_x}{\partial y}\right)\mathbf{k}$$

(sometimes identified as $\nabla \times \mathbf{a} = \mathrm{rot}\,\mathbf{a} = \mathrm{curl}\,\mathbf{a} = \nabla \wedge \mathbf{a}$).

Every arbitrary continuous vector field that vanishes for $\mathbf{r} \to \infty$ can be composed of a sum of source-free and curl-free vector fields:

$$\mathbf{a} = \mathbf{a}_1 + \mathbf{a}_2 \quad \text{with} \quad \nabla \times \mathbf{a}_1 = 0 \quad \text{and} \quad \nabla \cdot \mathbf{a}_2 = 0$$

$$\Delta \varphi = \nabla \cdot \nabla \varphi = \nabla^2 \varphi = \mathrm{div} \cdot \mathrm{grad}\, \varphi = \frac{\partial^2 \varphi}{\partial x^2} + \frac{\partial^2 \varphi}{\partial y^2} + \frac{\partial^2 \varphi}{\partial z^2}$$

When the sources are electrical charges (electrons) then

$$\nabla^2 \varphi = 4\pi \varrho$$

which translates with proper introduction of the units used in this book into the well-known *Poisson equation*:

$$\nabla^2 \varphi = \frac{\varrho}{\varepsilon \varepsilon_0}. \tag{A.7}$$

Other vector relations are:

$$\nabla(\varphi\psi) = \varphi\nabla\psi + \psi\nabla\varphi$$
$$\nabla \cdot (\mathbf{a}\varphi) = \varphi\nabla \cdot \mathbf{a} + \mathbf{a} \cdot \nabla\varphi$$
$$\nabla \times (\mathbf{a}\varphi) = \varphi\nabla \times \mathbf{a} - \mathbf{a} \times \nabla\varphi$$
$$\nabla \cdot (\mathbf{a} \times \mathbf{b}) = \mathbf{b} \cdot (\nabla \times \mathbf{a}) - \mathbf{a} \cdot (\nabla \times \mathbf{b})$$
$$\nabla \times (\mathbf{a} \times \mathbf{b}) = (\mathbf{b} \cdot \nabla) \cdot \mathbf{a} - (\mathbf{a} \cdot \nabla) \cdot \mathbf{b} + (\mathbf{a} \cdot \nabla)\mathbf{b} - \mathbf{b}(\nabla \cdot \mathbf{a})$$
$$\nabla(\mathbf{a} \cdot \mathbf{b}) = (\mathbf{b} \cdot \nabla)\mathbf{a} + (\mathbf{a} \cdot \nabla)\mathbf{b} + \mathbf{a} \times (\nabla \times \mathbf{b}) + \mathbf{b} \times (\nabla \times \mathbf{a})$$
$$\nabla \cdot (\nabla \times \mathbf{a}) = 0; \ \nabla \times \nabla\varphi = 0 \quad \text{and} \quad \nabla \cdot \nabla\varphi = \Delta\varphi :$$

a field that can be expressed as the gradient of a scalar ($\nabla \varphi$) is curl-free.

A.3.1 Integral Transformation

Gauss' Theorem

$$\oint_s \mathbf{a} \cdot ds = \int_\mathcal{V} \nabla \cdot \mathbf{a} \, d\mathcal{V}$$

The first integral is over a closed surface surrounding volume $\mathcal{V}$; the second integral is over this volume; $a_n = \mathbf{n} \cdot \mathbf{a}$; the vector $\mathbf{n}$ is normal to the surface and is directed to the outside.

Stokes' Theorem

$$\oint \mathbf{a} \cdot d\mathbf{r} = \int_s (\nabla \times \mathbf{a}) \cdot ds$$

The first integral is over a closed curve; the second integral is over an arbitrary surface within this curve.

A.3.2 Tensor Analysis

The transformation of one vector into another vector (stretching and rotation) is accomplished by multiplication with a tensor T:

$$\mathbf{a} = T\mathbf{b} \quad \text{with} \quad T = \begin{pmatrix} \sigma_{xx} & \sigma_{xy} & \sigma_{xz} \\ \sigma_{yx} & \sigma_{yy} & \sigma_{yz} \\ \sigma_{zx} & \sigma_{zy} & \sigma_{zz} \end{pmatrix},$$

or explicitly in component form

$$\begin{aligned} a_x &= \sigma_{xx}b_x + \sigma_{xy}b_y + \sigma_{xz}b_z \\ a_y &= \sigma_{yx}b_x + \sigma_{yy}b_y + \sigma_{yz}b_z \\ a_z &= \sigma_{zx}b_x + \sigma_{zy}b_y + \sigma_{zz}b_z. \end{aligned}$$

Unit tensor $\mathcal{E}$

$$\mathbf{a} = T\mathbf{b}, \quad \mathbf{b} = T^{-1}\mathbf{a}, \quad TT^{-1} = \mathcal{E}$$

with

$$\mathcal{E} = \begin{pmatrix} 1 & 0 & 0 \\ 0 & 1 & 0 \\ 0 & 0 & 1 \end{pmatrix}$$

The sequential transformation of a vector can be expressed by a tensor product

$$\mathbf{a} = T_1\mathbf{b}, \quad \mathbf{b} = T_2\mathbf{c}, \quad \text{or} \quad \mathbf{a} = T_1 T_2 \mathbf{c}$$

but
$$\mathcal{T}_1 \mathcal{T}_2 \neq \mathcal{T}_2 \mathcal{T}_1.$$

Linearity requires
$$k\mathbf{a} = k\mathcal{T}\mathbf{b} = \mathcal{T}k\mathbf{b}; \quad \mathbf{a}_1 + \mathbf{a}_2 = \mathcal{T}\mathbf{b}_1 + \mathcal{T}\mathbf{b}_2 = \mathcal{T}(\mathbf{b}_1 + \mathbf{b}_2)$$
$$(\mathcal{T}_1 + \mathcal{T}_2)\mathbf{b} = \mathcal{T}_1\mathbf{b} + \mathcal{T}_2\mathbf{b}.$$

Symmetrical tensor:
$$\mathbf{a}\mathcal{T}_s\mathbf{b} = \mathbf{b}\mathcal{T}_s\mathbf{a}.$$

Antisymmetrical tensor:
$$\mathbf{a}\mathcal{T}_a\mathbf{b} = -\mathbf{b}\mathcal{T}_a\mathbf{a}.$$

It is always possible to express an arbitrary tensor as the sum of a symmetrical and antisymmetrical tensor:
$$\mathcal{T} = \mathcal{T}_s + \mathcal{T}_a.$$

An *adjunct tensor* can be expressed as the corresponding difference of these tensors:
$$\tilde{\mathcal{T}} = \mathcal{T}_s - \mathcal{T}_a,$$

and it always holds that
$$\mathbf{a}\mathcal{T}\mathbf{b} = \mathbf{b}\tilde{\mathcal{T}}\mathbf{a}.$$

An *orthogonal tensor* is defined by
$$\mathcal{O} = \tilde{\mathcal{O}}.$$

A.3.2A Matrices and Determinants

A system of $m \cdot n$ numbers in a rectangular scheme of m rows and n columns is called a **matrix:**

$$(a_{ik})_{m,n} = \begin{pmatrix} a_{11} & a_{12} \dots & a_{1n} \\ a_{21} & a_{22} \dots & a_{2n} \\ \dots & \dots & \dots \\ a_{m1} & a_{m2} & a_{mn} \end{pmatrix}.$$

Here, it is possible that $n \neq m$, and that some of the a_{ik} may be complex numbers.

Matrix definitions based on $\mathcal{A} = (a_{ik})$:

Transposed matrix:
$$\tilde{\mathcal{A}} = (a_{ki}).$$

Symmetrical matrix:
$$\tilde{\mathcal{A}} = \mathcal{A}.$$

Conjugate complex:
$$A^* = (a_{ik}^*).$$
Adjunct matrix:
$$A^\dagger = (a_{ki}^*).$$
Self-adjunct (hermitic) matrix:
$$A^\dagger = A.$$
Unitary matrix:
$$UU^\dagger = U^\dagger U = \mathcal{E}.$$
Orthogonal matrix:
$$O\tilde{O} = \tilde{O}O = \mathcal{E}.$$
Quadratic matrix:
$$A = (a_{ik})_{n,m} \quad \text{with} \quad n = m.$$

Similar matrices:
$$\left.\begin{array}{l} A = (a_{ik})_{n_1,m_1} \\ B = (b_{ik})_{n_2,m_2} \end{array}\right\} \quad \text{with } n_1 = n_2,\ m_1 = m_2.$$

For similar matrices, one has:
$$A + B = (a_{ik} + b_{ik}).$$

Products of rectangular matrices
$$\begin{aligned} AB &= (a_{ik})_{n_1 m_1}(b_{ik})_{n_2 m_2} \quad \text{only if } n_1 = m_2 \\ BA &= (b_{ik})_{n_2 m_2}(a_{ik})_{n_1 m_1} \quad \text{only if } n_2 = m_1. \end{aligned}$$

Products of quadratic matrices
$$AB = \left(\sum_{\nu=1}^{n} a_{i\nu}b_{\nu k}\right) = (c_{ik})_{n,n} = \begin{pmatrix} a_{11}b_{11} + a_{12}b_{21} & a_{11}b_{12} + a_{12}b_{22} \\ a_{21}b_{11} + a_{22}b_{21} & a_{21}b_{12} + a_{22}b_{22} \end{pmatrix}.$$

Direct product of dissimilar matrices
$$A \times B = (a_{ik})_{n_1 m_1} \times (b_{lm})_{n_2 m_2} = (c_{ik,lm}) = (a_{ik}b_{lm}),$$

for instance
$$(a_1 a_2) \times \begin{pmatrix} b_{11} & b_{12} \\ b_{21} & b_{22} \end{pmatrix} = \begin{pmatrix} a_1 b_{11} & a_1 b_{12} & a_2 b_{11} & a_2 b_{12} \\ a_1 b_{21} & a_1 b_{22} & a_2 b_{21} & a_2 b_{22} \end{pmatrix}.$$

Determinants of quadratic matrices are the sum of all products
$$\det(a_{ik})_{n,m} = \Sigma P(r_1,\dots r_n)a_{1r1}\dots a_{1rn} = \Sigma(-1)^J a_{1r1},\dots a_{1rn}$$

with $P(r_1, \ldots r_n) = \pm 1$ determined by even or odd permutation J of the numbers $1, \ldots n$. For instance:

$$\begin{vmatrix} a_{11} & a_{12} \\ a_{21} & a_{22} \end{vmatrix} = a_{11}a_{22} - a_{12}a_{21}$$

or

$$\begin{vmatrix} a_{11} & a_{12} & a_{13} \\ a_{21} & a_{22} & a_{23} \\ a_{31} & a_{32} & a_{33} \end{vmatrix} = \begin{aligned} & a_{11}a_{22}a_{33} - a_{11}a_{32}a_{23} + a_{21}a_{32}a_{13} \\ & - a_{21}a_{12}a_{33} + a_{31}a_{22}a_{13} - a_{31}a_{22}a_{31}. \end{aligned}$$

Determinants are used, e.g., in the process of solving sets of homogeneous linear equations. In matrix algebra this system

$$\begin{aligned}
a_1 &= b_{11}x_1 &+ \ldots& \quad b_{1n}x_n \\
a_2 &= b_{21}x_1 &+ \ldots& \quad b_{2n}x_n \\
\ldots & \quad \ldots & & \quad \ldots \\
a_m &= b_{m1}x_1 &+ \ldots& \quad b_{mn}x_n
\end{aligned}$$

can be written as

$$\begin{pmatrix} a_1 \\ a_2 \\ \vdots \\ a_m \end{pmatrix} = \begin{pmatrix} b_{11} & b_{12} & \ldots & b_{1n} \\ b_{21} & b_{22} & \ldots & b_{2n} \\ \ldots & \ldots & & \ldots \\ b_{m1} & b_{m2} & \ldots & b_{mn} \end{pmatrix} \begin{pmatrix} x_1 \\ x_2 \\ \vdots \\ x_m \end{pmatrix} \quad \text{or} \quad \mathbf{a} = \mathcal{B}\mathbf{x}$$

which has the solution, by multiplication with $\mathcal{B}^{-1}$ from the left

$$\mathcal{E}\mathbf{x} = \mathcal{B}^{-1}\mathbf{a}.$$

A determinant has the value zero when:

(a) all elements in one row or one column are zero; or
(b) all elements in one row or column are a multiple of the corresponding parallel row or column.

The value of a determinant remains unchanged:

(a) when the determinant is transposed (a_{ik} are exchanged with a_{ki});
(b) when a multiple of a parallel row is added to one row;
(c) except for a sign change, when two rows are exchanged;
(d) except for a factor of k, when all elements of a row are multiplied with k.

A.3.3 The Wave Equation

A harmonic oscillation in time is given by

$$s(t) = A \exp\left(2\pi i \frac{t}{\tau}\right) = A \exp(i\omega t); \qquad (A.8)$$

a harmonic oscillation in space is given by

$$s(\mathbf{r}) = A \exp\left(2\pi i \frac{\mathbf{n} \cdot \mathbf{r}}{\lambda}\right) = A \exp\left[i(\mathbf{k} \cdot \mathbf{r})\right]; \qquad (A.9)$$

a harmonic oscillation in time and space is given by

$$s(\mathbf{r}, t) = A \exp\left[i\left\{\omega t - (\mathbf{k} \cdot \mathbf{r})\right\}\right] = A \exp\left[ik\left\{vt - (\mathbf{n} \cdot \mathbf{r})\right\}\right],$$

where $A = |A| \exp(i\alpha)$ and where

$|A|$ the amplitude,
α the phase factor,
$\omega = 2\pi\nu$ the angular frequency,
$\mathbf{k} = \mathbf{n}\omega/v$ the wave vector,
$k = \omega/v = 2\pi/\lambda$ the wave number,
$\mathbf{n} = \mathbf{k}/k$ the normal unit vector,
$v = \omega/k$ the phase velocity (traveling wave velocity),
$\tau = 1/\nu = 2\pi/\omega$ the oscillation time,
$\lambda = 2\pi/k = 2\pi v/\omega$ the wavelength.

The oscillations given above are solutions of the wave equation

$$\nabla^2 s + k^2 s = \frac{1}{v^2} \frac{\partial^2 s}{\partial t^2};$$

Equation (A.8) follows for $k = 0$, Eq. (A.9) for $v = 0$.

A *wave packet* is obtained from a combination of waves of almost the same wave vector

$$s = \int A(\mathbf{k}) \exp\left[i\left\{\omega t - (\mathbf{k} \cdot \mathbf{r})\right\}\right] dV_k = A(\mathbf{r} - \mathbf{u}t) \exp\left[i\left\{\overline{\omega}t - (\overline{\mathbf{k}} \cdot \mathbf{r})\right\}\right]$$

with a group velocity (the velocity of this wave packet) of

$$\mathbf{v}_g = \nabla_k \omega = \frac{\partial \omega}{\partial k_x}\mathbf{i} + \frac{\partial \omega}{\partial k_y}\mathbf{j} + \frac{\partial \omega}{\partial k_z}\mathbf{k} = \mathbf{n}v + k\nabla_k v\Big|_{\overline{k}}$$

and where dV_k is the volume element in k space.

The amplitude $A(\mathbf{k})$ remains finite only near $\mathbf{k} \simeq \overline{\mathbf{k}}$ with frequencies near $\omega \simeq \overline{\omega}$ and within a relatively small volume ΔV.

A.4 Some Elements of Quantum Mechanics

We will present here a short overview of some of the essential elements of quantum mechanics, which are used in the main body of the text. For more detailed studies, we refer the reader to Dirac (1944), Blochinzew (1953), Landau and Lifshitz (1958), Sommerfeld (1953), Kittel (1963), Schiff (1968), Franz (1970), and Inkson (1984).

The basic facts that enter the quantum mechanical theory are:

(1) all energy transfer is composed of quantum steps, and
(2) all matter has a wave aspect as well as a particle aspect (and all waves have a particle aspect).

These facts require a modification of the classical physics, which holds well in the limits of large quantities (compared to the atomistic quanta). While in macroscopic dimensions matter and energy transfer is continuous, every phenomenon in the microcosmos is quantized.

This refers to particulate properties as well as to energy. Every elementary particle is indivisible; every quantum of energy is indivisible.

Experimental observation at large densities of such particles reveal their wave character as interference phenomena. When the intensity of the impinging beam is reduced, however, the particle character becomes imminent, fluctuation occurs, and single particle events can be distinguished. Only in a time average is a continuous flow observed. This dualism between particle and wave requires the introduction of an equivalence formalism as suggested by De Broglie.

A.4.1 Duality, the De Broglie Wavelength

In 1924 De Broglie postulated the extension of the then recognized dualism of light to all other atomic particles. For light, one has photons of a momentum

$$p = \frac{E}{c} = \frac{h\nu}{c} \quad \text{or relating to a wave} \quad p = \frac{h}{\lambda} \qquad (A.10)$$

The key proportionality factor in this equation is Planck's quantum constant h. In a similar fashion, De Broglie introduced for any other particle a dualism to a wave with a wavelength related to the particle momentum p

$$\lambda_{\mathrm{DB}} = \frac{h}{p} = \frac{h}{mv}; \qquad (A.11)$$

the only important difference here being that the velocity of a photon is the light velocity c, which is always larger than the particle or group velocity v.*

For electrons, this dualism was observed shortly thereafter by Davison and Germer, who found electron diffraction from crystals. From Eq. (A.11), one obtains a simple relation for charged particles, which can be accelerated in an electric field [with $(m/2)v^2 = eV$]:

$$\lambda_{\mathrm{DB}} = \frac{h}{\sqrt{2emV}} = \sqrt{\frac{150}{V}} \sqrt{\frac{m_0}{m}} \ (\mathring{\mathrm{A}}) \qquad (\mathrm{A.12})$$

For neutral particles, e.g. neutrons, the equivalent relation

$$\lambda_{\mathrm{DB}} = \frac{h}{\sqrt{2m_N E}} = 9.75 \sqrt{\frac{1}{T(\mathrm{K})}} \sqrt{\frac{m_0}{m_N}} \ (\mathring{\mathrm{A}}) \qquad (\mathrm{A.13})$$

is more handy, with m_N as the neutron mass.

When applied to solids, one sees that ~ 40 eV electrons, or neutrons at ~ 20 K (with corresponding De Broglie wavelength on the order of a typical lattice constant) will show major interference effects. This requires that, here, the wave nature of these particles needs to be considered, while at substantially lower energies, the particle nature, i.e., a classical picture, may be sufficient for the description. This, for instance, has a bearing on the description of electrons in solids. When electronic eigenstates are discussed, the total electron energy for conduction or valence bands is typically on the order of $10 \ldots 50$ eV; hence, the wave character needs to be considered. When Bloch electrons within the conduction band of a semiconductor are concerned, their energy is on the order of kT; hence, the classical particle model is often sufficient.

A.4.2 The Wavefunction

The wave nature of elementary particles requires a discussion of the corresponding wavefunction. This wavefunction is a solution of the basic wave equation, which deals with the wave nature of the

---

* However, the phase velocity $u = \lambda_{\mathrm{DB}} \nu$ of the particle wave is larger than c as can be seen from $\lambda_{\mathrm{DB}} \nu = E/(mv) = mc^2/mv = c^2/v$. This is not in contradiction to Einstein's theory of relativity, since the phase velocity cannot be used to transmit signals, whereas the group velocity can.

particles (Schrödinger, 1926). For single particles, one has as the wave equation the well-known time-dependent *Schrödinger equation*

$$\frac{\hbar}{i}\dot{\psi} - \frac{\hbar^2}{2m}\nabla^2\psi + V(\mathbf{r})\psi = 0.$$ (A.14)

The Schrödinger equation determines the wave mechanical properties of a particle in a potential field $V(\mathbf{r})$. When more particles are involved, one has

$$\frac{\hbar}{i}\dot{\psi} - \sum_k \frac{\hbar^2}{2m_k}\nabla_k{}^2\psi + V(\mathbf{r})\psi = 0$$ (A.15)

The solutions of the Schrödinger equation for specific boundary values are the eigenfunctions $\psi_n(\mathbf{r},t)$. The expectation values of the various measurable quantities are as follows:

Momentum:

$$\mathbf{p} = m\mathbf{v} = \int \psi^* \frac{\hbar}{i}\nabla\psi \, d\mathcal{V}.$$ (A.16)

Angular momentum:

$$\mathbf{L} = \int \psi^*(r \times \nabla\psi) \, d\mathcal{V}.$$ (A.17)

Force:

$$\mathcal{F} = \int \psi^*\psi\nabla V \, d\mathcal{V}.$$ (A.18)

Total energy:

$$\overline{E} = -\frac{\hbar}{i}\int \psi^* \frac{\partial\psi}{\partial t} \, d\mathcal{V}.$$ (A.19)

Potential energy:

$$\overline{V} = \int \psi^* V\psi \, d\mathcal{V}.$$ (A.20)

Kinetic energy:

$$\overline{E}_k = \frac{\hbar^2}{2m}\int |\nabla\psi|^2 \, d\mathcal{V}.$$ (A.21)

Probability density:

$$\varrho = \int \psi^*\psi \, d\mathcal{V}.$$ (A.22)

This density is normalized to $\varrho = 1$ when taken over the entire volume.

A.4.2A Time Dependence of the Schrödinger Equation

When the potential does not vary in time, the time-dependent Schrödinger equation can be solved by a product *Ansatz*

$$\psi(\mathbf{r},\mathbf{t}) = \varphi(\mathbf{r})\phi(t); \qquad (A.23)$$

one obtains from Eq. (A.15) for a time-independent potential $V(\mathbf{r})$

$$-\frac{\hbar^2}{2m}\phi\nabla^2\varphi + V(\mathbf{r})\varphi\phi = -\frac{\hbar}{i}\varphi\frac{d\phi}{dt} \qquad (A.24)$$

or, after division by ψ

$$-\frac{\hbar^2}{2m}\frac{\nabla^2\varphi}{\varphi} + V(\mathbf{r}) = -\frac{\hbar}{i}\frac{1}{\phi}\frac{d\phi}{dt} = E \qquad (A.25)$$

where E is the separation constant, which has the meaning of the total energy. This yields for the time-dependent part

$$\frac{d\phi}{dt} = -\frac{iE}{\hbar}\phi(t) \quad \text{with the solution} \quad \phi(t) = \exp\left(-\frac{iE}{\hbar}t\right) \quad (A.26)$$

and the *time-independent Schrödinger equation*

$$\boxed{\nabla^2\varphi + \frac{2m}{\hbar^2}\left[E - V(\mathbf{r})\right]\varphi(\mathbf{r}) = 0.} \qquad (A.27)$$

where $\varphi(\mathbf{r})$ is the stationary-state wavefunction. For a specific boundary condition, the solution of Eq. (A.27) are the eigenfunctions $\varphi_n(\mathbf{r})$ at certain values of the energy, the eigenvalue spectrum E_n, where n represents the permitted quantum states. The time-dependent wavefunction for this problem is simply

$$\psi(\mathbf{r},t) = \varphi(\mathbf{r})\exp\left(-\frac{iE}{\hbar}t\right). \qquad (A.28)$$

A.4.2B The Hamiltonian The *time-dependent Schrödinger equation* is sometimes written as

$$i\hbar\frac{\partial}{\partial t}\psi(\mathbf{r},t) = H\phi(\mathbf{r},t) \qquad (A.29)$$

with

$$H = H_0 + V(\mathbf{r},t) \quad \text{and} \quad H_0 = -\frac{\hbar^2}{2m_0}\nabla^2 + E(\mathbf{r}) \qquad (A.30)$$

where H is referred to as the *Hamiltonian* of the time-dependent Schrödinger equation.

Throughout much of this book we discuss the solutions of the time-independent Schrödinger equation. Conventionally, here one uses

$$H\psi(\mathbf{r}) = E\psi(\mathbf{r}) \tag{A.31}$$

with the corresponding Hamiltonian

$$H = -\frac{\hbar^2}{2m_0}\nabla^2 + V(\mathbf{r}). \tag{A.32}$$

The use of the Hamiltonian as an *operator* is convenient to shorten long expressions, and in separating a main part (H_0) from additional ones (often small perturbations). For example, $V(\mathbf{r},t)$ in Eq. (A.30) emphasizes the reference of a basic solution (here, the electron in *vacuo*) to the modified solution in a (small) perturbing electron potential.

It is especially helpful when the problem becomes more involved, requiring a matrix representation (see, e.g., Section A.4.10). Here, ψ becomes a vector (e.g., the two components of a spinor) and H becomes the corresponding matrix:

$$i\hbar\frac{\partial}{\partial t}\begin{pmatrix} \psi_a \\ \psi_b \end{pmatrix} = \begin{pmatrix} H_{11} & H_{12} \\ H_{21} & H_{22} \end{pmatrix}\begin{pmatrix} \psi_a \\ \psi_b \end{pmatrix}. \tag{A.33}$$

A.4.2C The Dirac Notation A further shortcut in writing long equations in quantum mechanics is the *braket notation* introduced by Dirac. Although we have not used the Dirac notation in this book, we will give a short summary here for the interested reader who finds it in the cited literature (Dirac, 1958).

In order to comprehend the notation, let us start from the time-dependent Schrödinger equation, which describes the evolution of the wavefunction $\Psi(\mathbf{r},t)$ in space and time. This wavefunction can be developed in a Fourier series, using a set of orthonormal* basis functions $u_i(\mathbf{r})$

$$\Psi(\mathbf{r},t) = \sum_i \psi_i(t)u_i(\mathbf{r}) \tag{A.35}$$

---

* Orthonormality requires

$$\int_{\mathcal{V}} u_i^*(\mathbf{r})u_j\,d\mathbf{r} = \delta_{ik} \tag{A.34}$$

where δ_i is the *Kronecker δ symbol*, defined as $\delta_{ij} = 0$ for $i \neq j$ and $\delta_{ij} = 1$ for $i = j$.

which can be interpreted as the description of an n-dimensional vector Ψ, with components ψ_i in the direction of each unit vector u_i—assuming that the sum is evaluated to its nth term.

The product of two such vectors Ψ and Φ is often of interest (e.g., if $\Phi = \Psi^*$, this product gives the probability density of finding a particle). Dirac identifies this scaler product by enclosing it in brackets of a special type

$$\langle \Psi | \Phi \rangle$$

which can be separated, identifying each quantitiy as a vector $\langle \Psi |$ and $| \Phi \rangle$; the first is referred to as *bra*, the second as *ket*.

With this notation, one writes Eq. (A.35) simply as

$$|\Psi\rangle = \sum_i \psi_i |i\rangle \qquad (A.36)$$

dropping the unessential information of the type of basis vector (i instead of u_i).

With Eq. (A.34), one obtains the inverse relation

$$\psi_i(t) = \int_{\mathcal{V}} u_i^*(\mathbf{r})\Psi(\mathbf{r},t)d\mathcal{V} \quad \text{or simply} \quad = \langle i | \Psi(t) \rangle. \qquad (A.37)$$

In a similar fashion, one has for the product

$$\langle \Psi | \Phi \rangle = \int \Psi^*(\mathbf{r})\Phi(\mathbf{r})d\mathcal{V}. \qquad (A.38)$$

For the Schrödinger equation, one writes

$$-\frac{\hbar}{i}\frac{\partial}{\partial t}|\Psi\rangle = \underline{H}|\Psi\rangle \qquad (A.39)$$

where $\Psi(\mathbf{r},t)$ is identified as a vector, and the Hamiltonian $\underline{H}$ consequently has matrix form.

The expectation value of a measurable quantity can be expressed as

$$\overline{\alpha} = \int_{\mathcal{V}} \Psi^* \alpha_{\text{op}} \Psi d\mathcal{V} \equiv \langle \Psi^* | \alpha | \Psi \rangle. \qquad (A.40)$$

When ψ_i is a complete set of components of the vector $|\Psi\rangle$, then a matrix element of the operator P is given as

$$P_{ik} = \int_{\mathcal{V}} \psi_i P_{\text{op}} \psi_k d\mathcal{V} \equiv \langle \psi_i | P | \psi_k \rangle. \qquad (A.41)$$

For instance, P_{ik} is a momentum matrix element when P_{op} is the momentum operator

$$P_{op} = \frac{\hbar}{i} \nabla. \qquad (A.42)$$

A.4.2D The Potential in the Schrödinger Equation   The key in describing phenomena with the Schrödinger equation is the potential $V(\mathbf{r}, t)$. The simplest case describes an electron *in vacuo* with $V(\mathbf{r}, t) \equiv 0$—see Section A.4.6. Another example deals with an electron in the potential of the Coulomb potential of the proton $V(\mathbf{r}, t) = e^2/(4\pi\varepsilon_0 r)$, describing the hydrogen atom—see Section A.4.8.

The electronic phenomena discussed in this book are much more complicated, since the electron is exposed to the potential created by many particles—the nuclei and electrons of all atoms in the semiconductor. In addition, a macroscopic potential due to an applied bias or a space charge may act. This hopelessly complex situation, however, can be simplified substantially by separation into parts. For instance, in a crystalline semiconductor, one can separate a part of the potential with lattice periodicity $V_L(\mathbf{r}, t)$, a part with random fluctuation, responsible for electron scattering $V_s(\mathbf{r}, t)$, and the externally created potential $V_{ext}(\mathbf{r}, t)$:

$$V(\mathbf{r}, t) = V_L(\mathbf{r}, t) + V_s(\mathbf{r}, t) + V_{ext}(\mathbf{r}, t). \qquad (A.43)$$

With appropriate mathematical manipulation, one can often reduce the problem. Such manipulation is relatively simple when one of the contributions to $V(\mathbf{r}, t)$ is small compared to another. Such a small perturbation can be dealt with by using a convenient expansion, and is described in the following section.

A.4.2E Perturbation Theory   First, we will discuss the *time-independent perturbation*. With $\mathbf{H}_0$ as the unperturbed Hamiltonian and $\delta \mathbf{V}(\mathbf{r})$ as the potential perturbation, one has for the perturbed system:

$$\mathbf{H}\psi(\mathbf{r}) = E\psi(\mathbf{r}) \quad \text{with} \quad \mathbf{H} = \mathbf{H}_0 + \delta\mathbf{V} \quad \text{and} \quad \mathbf{H}_0 = -\frac{\hbar^2}{2m}\nabla^2 + V(\mathbf{r}).$$
$$(A.44)$$

We assume that the solutions of $\mathbf{H}\psi_0 = E\psi$ are known and can be written as a set of eigenfunctions $\psi_n^{(0)}$, with a *discrete* and *non-degenerate* eigenvalue spectrum $E_n^{(0)}$.

The solution of the perturbed system [Eq. (A.44)] can be expressed as expansion in the eigenfunctions of the unperturbed system:

$$\psi(\mathbf{r}) = \sum_m c_m \psi_m^{(0)}(\mathbf{r}) \tag{A.45}$$

which permits the calculation of $\psi(\mathbf{r})$ in matrix form. When substituting this expansion into the Schrödinger equation (A.44), one obtains

$$\sum_m c_m \left(E_m^{(0)} + \delta V \right) \psi_m^{(0)} = \sum_m c_m E \psi_m^{(0)}. \tag{A.46}$$

After multiplication with $\psi_k^{(0)*}$, and integrating over the entire volume, one obtains

$$c_k \left(E - E_k^{(0)} \right) = \sum_m \delta V_{km} c_m \tag{A.47}$$

where δV_{km} is the matrix of the perturbation operator:

$$\delta V_{km} = \int \psi_k^{(0)*} \delta V \psi_m^{(0)} d\mathcal{V}. \tag{A.48}$$

When expressing the eigenvalue of the perturbed system as sum

$$E = E^{(0)} + E^{(1)} + E^{(2)} + \ldots \quad \text{and} \quad c_m = c_m^{(0)} + c_m^{(1)} + c_m^{(2)} + \ldots \tag{A.49}$$

with the superscript giving the order of the approximation, one can show that for $k = n$ one has*

$$E_n^{(1)} = \delta V_{nm} = \int \psi_n^{(0)*} \delta V \psi_n d\mathcal{V}, \tag{A.50}$$

i.e., the first order correction of the eigenvalues is equal to the mean value of the perturbation potential in the corresponding unperturbed state.

For the first-order correction of the perturbed eigenfunction, one obtains

$$\psi_n^{(1)} = \sum_k{}' \frac{\delta V_{kn}}{E_n^{(0)} - E_k^{(0)}} \psi_k^{(0)} \quad \text{with} \quad \psi_n = \psi_n^{(0)} + \psi_n^{(1)} + \ldots. \tag{A.51}$$

This perturbation approximation yields useful results as long as $\delta V_{kn} \ll E_n^{(0)} - E_k^{(0)}$, i.e., the perturbation potential is much smaller

---

* After the introduction of Eq. (A.49) into (A.47), neglecting higher than first-order terms, the off-diagonal elements (for $n \neq k$) vanish.

than the split between corresponding eigenvalues of the unperturbed system.

The second-order correction for the eigenvalue is given by

$$E_n^{(2)} = {\sum_k}' \frac{|\delta V_{kn}|^2}{E_n^{(0)} - E_k^{(0)}}. \tag{A.52}$$

When the unperturbed system has a *degenerate eigenvalue* spectrum, the perturbation removes part or all of the degeneracies. Such a split of the eigenvalues belonging to a certain state $E_n^{(0)}$ of the unperturbed system can be calculated from a modified Eq. (A.50):

$$E^{(1)} c_n^{(0)} = \sum_l \delta V_{nl} c_l \tag{A.53}$$

with $\psi_n^{(0)}$, $\psi_l^{(0)}$, ... as the eigenfunctions belonging to the same eigenvalue $E_n^{(0)}$ of the unperturbed system.

With vanishing nondiagonal terms of the matrix, one obtains the perturbed eigenvalues $E^{(1)}$ as solutions of the secular equation

$$\left| \delta V_{nl} - E^{(1)} \delta_{nl} \right| = 0 \tag{A.54}$$

where δ_{nl} is the Kronecker symbol. This equation is of νth degree, where ν is the degree of degeneracy of the level $E_n^{(0)}$.

Sometimes the split of the degenerate energy levels $E_n^{(0)}$ is so small that all $E^{(1)}$ contributions are zero. In such instances, one has to resort to the second-order approximation.

If the *perturbation is time-dependent* ($\delta V = \delta V(t)$), there are no stationary states of the perturbed system. Nevertheless, one can calculate approximate solutions of the perturbed system

$$i\hbar \frac{\partial \psi}{\partial t} = (H_0 + \delta V)\psi \tag{A.55}$$

from stationary solutions of the unperturbed system $\psi_n^{(0)}$, according to

$$\psi = \sum_n a_n(t) \psi_n^{(0)}, \tag{A.56}$$

although, now with time-dependent coefficients $a_n(t)$. Substitution of Eq. (A.56) into (A.55), and following similar operations as given before for the stationary case [Eqs. (A.46) and (A.47)], one obtains

$$i\hbar \frac{da_m(t)}{dt} = \sum_k \delta V_{mk}(t) a_k(t) \tag{A.57}$$

with

$$\delta V_{mk}(t) = \int \psi_m^{(0)*} \delta V(t) \psi_k^{(0)} d\mathcal{V} = \delta V_{mk} \exp(i\omega_{mk} t) \qquad \text{(A.58)}$$

and $\hbar\omega_{mk} = E_m^{(0)} - E_k^{(0)}$. With diagonal elements vanishing, this yields for the first-order expansion coefficient the condition

$$i\hbar \frac{da_k^{(1)}(t)}{dt} = V_{kn}(t) \qquad \text{(A.59)}$$

or, specifically, for the solution

$$\psi_n(t) = \sum_k a_{kn}(t) \psi_k^{(0)} \qquad \text{(A.60)}$$

with

$$a_{kn}^{(1)}(t) = -\frac{i}{\hbar} \int V_{kn}(t) dt = -\frac{i}{\hbar} \int V_{kn} \exp(i\omega_{kn} t) \, dt. \qquad \text{(A.61)}$$

In many cases the perturbation is periodic, e.g., produced by an electromagnetic radiation, of the form

$$\delta V(t) = \delta A \exp(i\omega t) + \delta B \exp(-i\omega t). \qquad \text{(A.62)}$$

From the Hermiticity of the operator δV, i.e., with $\delta V_{nm} = \delta V_{mn}^*$, one obtains a similar relationship between δA and δB:

$$\delta A_{nm} = \delta B_{mn}^* \qquad \text{(A.63)}$$

yielding

$$\delta V_{kn}(t) = V_{kn} \exp(i\omega t)$$
$$= \delta A_{kn} \exp\left[i\left(\omega_{kn} - \omega\right)t\right] + \delta A_{nk}^* \exp\left[i\left(\omega_{kn} + \omega\right)t\right]. \text{(A.64)}$$

With a similar manipulation as discussed previously, one obtains for the first-order expansion coefficients

$$a_{kn}^{(1)}(t) = -\frac{\delta A_{kn} \exp\left[i\left(\omega_{kn} - \omega\right)t\right]}{\hbar(\omega_{kn} - \omega)} - \frac{\delta A_{nk}^* \exp\left[i\left(\omega_{kn} + \omega\right)t\right]}{\hbar(\omega_{kn} + \omega)}$$
$$\text{(A.65)}$$

as long as the perturbing electromagnetic field avoids the resonance frequency $\hbar\omega = \pm(E_k^{(0)} - E_n^{(0)})$ of the unperturbed system, where the approximation for a small perturbation is no longer applicable. For further discussion of the perturbation theory, see Landau and Lifshitz (1958). A frequently used example of an application of the perturbation theory is the effective mass equation, discussed in Section A.4.3A.

A.4.3 Green's Functions

A less time-consuming approach to calculating the influence of perturbations is the use of Green's functions, which can be defined for the unperturbed problem in terms of the eigenfunctions ψ_n and eigenvalues E_n of the Hamiltonian operator H_0:

$$[H_0(\mathbf{r}) - E_n]\,\psi_n(\mathbf{r}) = 0 \qquad (A.66)$$

with $\psi_n(\mathbf{r})$ forming a complete orthonormal set of functions. The Green's function can then be given as

$$G(\mathbf{r}, \mathbf{r}', E) = \sum_{n,n'} G_{n,n'}\psi_n\psi_n^* \quad \text{with} \quad G_{nn'} = \frac{\delta_{nn'}}{E - E_n} \qquad (A.67)$$

yielding

$$G(\mathbf{r}, \mathbf{r}', E) = \sum_n \frac{\psi_n(\mathbf{r})\psi_n^*(\mathbf{r}')}{E - E_n}, \qquad (A.68)$$

which has poles at the eigenvalues of the system.

This Green's function is most useful for the calculation of the modified eigenfunction of the inhomogeneous equation

$$[H_0(\mathbf{r}) - E]\,\psi(\mathbf{r}) = f(\mathbf{r}) \qquad (A.69)$$

where $f(\mathbf{r})$ is the perturbation. Here, $\psi(\mathbf{r})$ is given by

$$\psi(\mathbf{r}) = -\int f(\mathbf{r}')G(\mathbf{r}, \mathbf{r}', E)d\mathbf{r}'. \qquad (A.70)$$

In a more common form, the perturbation is usually written as

$$[H_0(\mathbf{r}) + V(\mathbf{r}) - E]\,\psi(\mathbf{r}) = 0 \quad \text{with} \quad f(\mathbf{r}) = -V(\mathbf{r})\psi(\mathbf{r}), \quad (A.71)$$

yielding, for the perturbed wavefunction, the *integral equation*

$$\psi(\mathbf{r}) = \psi_0(\mathbf{r}) + \int G(\mathbf{r}, \mathbf{r}', E)V(\mathbf{r}')\psi(\mathbf{r}')d\mathbf{r}' \qquad (A.72)$$

where ψ_0 is the solution of the unperturbed Schrödinger equation.

We will now express $V(\mathbf{r})$ by its value at an infinite number of positions $\mathbf{r}_i$,

$$V(\mathbf{r}) = \{V(\mathbf{r}_1), V(\mathbf{r}_2), V(\mathbf{r}_3), \ldots\} = \tilde{\mathbf{V}}, \qquad (A.73)$$

i.e., by a vector in Hilbert space $\widetilde{\mathbf{V}}$, and will define the Green's functions in terms of the numbers $\mathbf{r}$ and $\mathbf{r}'$; i.e., $G(\mathbf{r},\mathbf{r}',E)$ can be written as a matrix

$$G(\mathbf{r},\mathbf{r}',E) = \widetilde{\mathbf{G}}. \qquad (A.74)$$

Then, the integral $\int G(\mathbf{r},\mathbf{r}',E)V(r')dr'$ reverts to a matrix product

$$\int G(\mathbf{r},\mathbf{r}',E)V(r')dr' = \widetilde{\mathbf{G}}\widetilde{\mathbf{V}} \qquad (A.75)$$

and $\psi(\mathbf{r})$ can be expressed as a perturbation series

$$\psi = \psi_0 + GV\left[\psi_0 + \widetilde{\mathbf{G}}\widetilde{\mathbf{V}}\psi_0 + \widetilde{\mathbf{G}}\widetilde{\mathbf{V}}\widetilde{\mathbf{G}}\widetilde{\mathbf{V}}\psi_0 + \dots\right], \qquad (A.76)$$

which is the *Dyson equation*

$$\psi = \psi_0 + \widetilde{\mathbf{G}}\widetilde{\mathbf{V}}\psi. \qquad (A.77)$$

In a similar form, one can express the Green's function of the perturbed problem by a Dyson equation, using the Green's function of the unperturbed system G_0:

$$\widetilde{\mathbf{G}} = \widetilde{\mathbf{G}}_0 + \widetilde{\mathbf{G}}_0 VG. \qquad (A.78)$$

For further detail, see Inkson (1984).

A.4.3A The Effective Mass Equation   The solution of the Schrödinger equation for an electron in the periodic lattice potential

$$\frac{\hbar}{i}\dot{\psi} + \frac{\hbar^2}{2m_0}\nabla^2\psi - V_L(\mathbf{r},t)\psi = o \qquad (A.79)$$

can be expressed in the form of Bloch functions [first two factors in Eq. (A.80)] with a time-dependent disturbance:

$$\psi_n(\mathbf{r},t) = u_n(\mathbf{k},\mathbf{r})\exp(i\mathbf{k}\cdot\mathbf{r})\exp\left[-\frac{i}{\hbar}E_n(\mathbf{k})t\right] \qquad (A.80)$$

where the amplitude function $u_n(\mathbf{k},\mathbf{r})$ has lattice periodicity (Bloch theorem). Here, n is the band index (see Section 8.1) and $\mathbf{k}$ is the wave vector.

When concerned with non-lattice-periodic contributions to the potential, one can, under certain circumstances,* simplify the problem by rewriting Eq. (A.79) as

$$\frac{\hbar}{i}\dot{\psi}_e + \frac{\hbar^2}{2m^*}\nabla_e^2\psi_e - \left[V_s(\mathbf{r},t) + V_{\text{ext}}^*(\mathbf{r},t)\right]\psi_e = 0. \qquad (A.81)$$

This equation is called the *effective mass equation* and is identical to Eq. (A.79), except that now the lattice potential is omitted. Its effect is included by replacing the electron rest mass m_0 with m^*, the effective mass, and V_{ext} with $V_{\text{ext}}^* = E_0 + V_{\text{ext}}$, the external potential (energy) added to the relevant band edge ($E_0 = E_c$ or $= E_v$).

The solution of Eq. (A.81) is referred to as the envelope function $\psi_e(\mathbf{r},t)$ (see Fig. 21.1 for an example), which gives a smoothed-out behavior. The solution of the exact Schrödinger equation near the edge of the specified band can be obtained in a reasonable approximation by multiplication of the envelope function with the lattice periodic amplitude $u(\mathbf{k},\mathbf{r})$ of the Bloch function

$$\psi(\mathbf{r},t) = u(\mathbf{k},\mathbf{r})\psi_e(\mathbf{r},t). \qquad (A.82)$$

When neglecting any scattering ($V_s = 0$), a specific solution of the effective mass equation can be written as

$$\psi_e(\mathbf{r},t) = c_k \exp(i\mathbf{k}\cdot\mathbf{r})\exp\left(-\frac{iE(\mathbf{k})t}{\hbar}\right) \qquad (A.83)$$

with the *dispersion relation*

$$E(\mathbf{k}) = E_c + \frac{\hbar^2}{2m^*}\mathbf{k}^2. \qquad (A.84)$$

The general solution is then given by superposition of all possible solutions Eq. (A.83):

$$\Psi_e(\mathbf{r},t) = \sum_k c_k \exp(i\mathbf{k}\cdot\mathbf{r})\exp\left(-\frac{iE(\mathbf{k})t}{\hbar}\right), \qquad (A.85)$$

which can form a wave packet for $\Psi_e(\mathbf{r})$, which is localized at a given time within a certain region. The wave packet travels with a group velocity

$$\mathbf{v}_g = \nabla_k E(\mathbf{k}). \qquad (A.86)$$

---

* That is, when electrons (or holes) are considered within one band only, close to the band edge.

The probability of finding an electron in a certain region $\mathcal{V}_0$ at a specific time is given by $\int_{\mathcal{V}_0} \psi^*(\mathbf{r}, t)\psi(\mathbf{r}, t)d\mathbf{r}$. Consequently, the density of electrons is given by

$$n(\mathbf{r}, t) = \sum \psi^*(\mathbf{r}, t)\psi(\mathbf{r}, t) = \sum |c_k|^2 \qquad (A.87)$$

with a summation over all electrons. The current density is given by

$$\mathbf{j}(\mathbf{r}, t) = -\frac{ie\hbar}{2m^*} \sum \left[(\nabla\psi)^*\psi - \psi^*(\nabla\psi)\right] = -\frac{ie}{m^*}\mathbf{k} \sum |c_k|^2. \qquad (A.88)$$

A.4.3B One-Dimensional Problems   When the perturbation potential within the effective mass equation varies only in one dimension (z), but is constant in the two other directions (x, y), then separation yields as a solution

$$\psi_e(\mathbf{r}, t) = c_k\phi(z)\exp\left[i(k_x x + k_y y)\right]\exp\left(-\frac{iE_0 t}{\hbar}\right) \qquad (A.89)$$

with ϕ being the solution of the one-dimensional Schrödinger equation

$$\frac{d^2\phi}{dz^2} + \frac{2m^*}{\hbar^2}\mathcal{E}(t)\phi = 0, \qquad (A.90)$$

and

$$\mathcal{E}(z) = E - E_c(z) - \frac{\hbar^2}{2m^*}(k_x^2 + k_y^2). \qquad (A.91)$$

This relationship becomes useful when discussing tunneling through a barrier layer, or the behavior of superlattices.

A.4.4 The Uncertainty Principle

The duality between particle and wave, which was addressed in the previous sections, permits us, e.g., to describe an electron as a wave, or as a superposition of waves, depending on the potential $V(\mathbf{r})$ and on the boundary conditions. When enclosed within a potential box with infinite walls, the amplitude of the De Broglie wave must be zero outside the box, and must have a node at the walls to avoid energy losses into the walls. The results are standing waves, with the wavelength λ being a multiple of the box dimension. An equivalent solution can be obtained by using cyclic boundary conditions at the walls.

When the box is large compared to λ, the position of the electron is defined only as being *within* the box, but not at any specific position. When the box becomes smaller, the uncertainty about the electron's position becomes reduced to the smaller size of the box.

When measuring a wave of finite length, one implicitly performs a Fourier analysis. The Fourier transform of such a wave train contains a range of frequencies, which have a wider band width the shorter the wave train.

Therefore, the location of a specific electron can be identified with increasing accuracy the wider the band width of the transmitted frequency spectrum, when dealing with the electron in the wave picture.

In other words, the electrons can be described as a wave packet, which is the better defined spatially the more frequency components are used for its composition; it is infinitely extended when represented by a single unrestricted wave.

One may therefore relate the uncertainty of the spatial localization Δx to the width of the wave number band:

$$\Delta x \simeq \frac{1}{\Delta(1/\lambda)}; \qquad (A.92)$$

this relation can be used with $\lambda = h/p$ to make the *Heisenberg uncertainty relation* plausible:

$$\boxed{\Delta x \Delta p \simeq h,} \qquad (A.93)$$

indicating that location and momentum of a particle cannot be determined at the same time to a better accuracy than given by Planck's constant.

An equivalent uncertainty relation can also be found between the canonic conjugates E and t: the time accuracy of identifying the passing of a wave packet of band width $\Delta \nu$ is given by:

$$\Delta t \simeq \frac{1}{\Delta \nu} \qquad (A.94)$$

and with $\Delta E = h \Delta \nu$, one obtains

$$\boxed{\Delta E \Delta t \simeq h.} \qquad (A.95)$$

A.4.5 Natural Line Widths

The relation (A.95) can be used to determine the natural line width of a transition with a lifetime τ:

$$\Delta E \tau \simeq h \quad \text{or} \quad \Delta \nu \tau \simeq 1. \qquad (A.96)$$

Here, the length of an optical transmission, i.e., the length of the wave train of the emitted signal is τ. Hence, it cannot be monochromatic but must contain Fourier components in a frequency band of band width $\Delta\nu$.

This has bearing on many optical effects: the spectral line width of an emission signal (luminescence) becomes broader when an atom becomes disturbed during emission, e.g., due to the collisions with other atoms, thereby interrupting the emission. In a similar fashion, laser pulses, which are intentionally shortened, also broaden in width.

A.4.6 The Electron in Vacuo

A free electron *in vacuo* follows the Schrödinger equation with zero potential:

$$\frac{\hbar}{i}\dot{\Psi} + \frac{\hbar^2}{2m_0}\nabla^2\Psi = 0, \tag{A.97}$$

which can be solved by the product *Ansatz*

$$\Psi(\mathbf{r}, t) = \psi(\mathbf{r})\phi(t), \tag{A.98}$$

yielding after insertion into Eq. (A.97)

$$\frac{\hbar}{i}\psi\dot{\phi} + \frac{\hbar^2}{2m_0}\phi\nabla^2\psi = 0 \tag{A.99}$$

or, after division by Eq. (A.98)

$$\frac{\hbar}{i}\frac{\dot{\phi}}{\phi} = -\frac{\hbar^2}{2m_0}\frac{\nabla^2\psi}{\psi}. \tag{A.100}$$

The left-hand side is a function of t only, the right-hand side is a function of $\mathbf{r}$ only. To hold for any $\phi(t)$ and $\psi(\mathbf{r})$, both sides must be independently equal to a constant, the *separation constant* , which has the dimension of an energy E. With

$$\frac{\hbar}{i}\dot{\phi} = -E\phi \tag{A.101}$$

one obtains as a solution

$$\phi(t) = \exp\left(-\frac{iEt}{\hbar}\right). \tag{A.102}$$

The space-dependent *wave function of the free electron* follows

$$\nabla^2\psi + \frac{2m_0}{\hbar^2}E\psi = 0, \tag{A.103}$$

which has solutions of the form

$$\psi(\mathbf{r}) = c_k\exp(i\mathbf{k}\cdot\mathbf{r}). \tag{A.104}$$

Insertion of Eq. (A.104) into (A.103) yields

$$E = \frac{\hbar^2}{2m_0} k^2, \qquad (A.105)$$

a simple, parabolic dispersion relation. A solution of Eq. (A.97) is consequently given by

$$\Psi = c_k \exp\left[i(\mathbf{k} \cdot \mathbf{r} - 2\pi\nu t)\right] \quad \text{with} \quad h\nu = E. \qquad (A.106)$$

To select specific solutions, boundary values, such as localizing the electron within a box of specific dimension, need to be introduced.

A.4.7 Free Electron in a Box

The confinement of an electron in a box limits the choice of $\mathbf{k}$ and c_k by forcing the amplitude to vanish at the box surfaces and outside of the box. This corresponds to standing waves with (for simplicity, here done in one dimension),

$$c_k \exp(ik_x a) = 0 \qquad (A.107)$$

where a is the width of the one-dimensional box—for a three-dimensional example, see Section 27.1.1A. Equation (A.107) can be fulfilled only when

$$ka = n_q\pi \quad \text{for} \quad n_q = 1, 2, \dots , \qquad (A.108)$$

yielding as eigenvalues [Eq. (A.105)]:

$$E_n = \frac{\hbar^2}{2m_0} \frac{n_q^2 \pi^2}{a^2}, \qquad (A.109)$$

i.e., permitting only discrete energy values.

The amplitude factor can be determined from the normalization condition, requiring that the electron must be in the box

$$\int_0^a \psi_n^* \psi_n \, dx = c_k^2 \int_0^a \exp\left(\frac{2n_q\pi x}{a}\right) dx = 1, \qquad (A.110)$$

yielding

$$c_k = \frac{1}{\sqrt{a}}. \qquad (A.111)$$

Consequently, the eigenfunctions of the electron in such a box can be written as

$$\psi_n = \frac{1}{2i\sqrt{a}} \left[\exp\left(i\frac{n_q\pi x}{a}\right) - \exp\left(-i\frac{n_q\pi x}{a}\right) \right], \qquad (A.112)$$

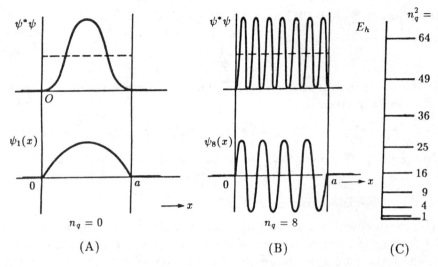

Figure A.2: Eigenfunctions and probability amplitude for $n_q = 1$ and $n_q = 8$ in subfigures (A) and (B), respectively. (C): Eigenvalue spectrum in units $\hbar^2 \pi^2 / (2m_0 a^2)$; all for an electron in a one-dimensional box of width $\sim a$ with infinite walls.

that is, as superposition of plane waves of equal amplitude propagating in opposite direction with perfect reflection at the box surface. Two solutions with eigenfunctions, probability amplitudes, and eigenvalues, are shown in Fig A.2 for $n_q = 1$ and $n_q = 8$.

A.4.8 The Hydrogen Atom

The hydrogen atom presents the simplest atomic system. Here, the electron moves in the Coulomb field of the positively charged nucleus, with

$$V(r) = \frac{e^2}{4\pi\varepsilon_0 r}. \tag{A.113}$$

The stationary-state Schrödinger equation for this system is therefore

$$\frac{\hbar^2}{2m_0}\nabla^2\psi + \frac{e^2}{4\pi\varepsilon_0 r}\psi = E\psi. \tag{A.114}$$

Since the problem is a spherical one, it is best to rewrite the Schrödinger equation in a spherical polar coordinate system with $\psi = \psi(r, \theta, \phi)$ and

$$\nabla^2\psi = \frac{1}{r^2}\frac{\partial}{\partial r}\left(r^2\frac{\partial\psi}{\partial r}\right) + \frac{1}{r^2\sin\theta}\frac{\partial}{\partial\theta}\left(\sin\theta\frac{\partial\psi}{\partial\theta}\right) + \frac{1}{r^2\sin^2\theta}\frac{\partial^2\psi}{\partial\phi^2}. \tag{A.115}$$

When using the product-*Ansatz*,

$$\psi(r, \theta, \phi) = R(r) \cdot \Theta(\theta) \cdot \Phi(\phi), \qquad (A.116)$$

the Schrödinger equation with separated variables can be written as

$$\frac{\sin^2 \theta}{R(r)} \frac{d}{dr}\left(r^2 \frac{dR(r)}{dr}\right) + \frac{\sin \theta}{\Theta(\theta)}\left(\sin \theta \frac{d\Theta(\theta)}{d\theta}\right) + \frac{2m_0}{\hbar^2}\sin^2\theta\left(E + \frac{e^2}{4\pi\varepsilon_0 r}\right)$$
$$= -\frac{1}{\Phi(\phi)}\frac{d^2\Phi(\phi)}{d\phi^2},$$

$$(A.117)$$

after recognizing that

$$\frac{\partial\psi}{\partial r} = \Theta(\theta)\Phi(\phi)\frac{dR}{dr}; \quad \frac{\partial\psi}{\partial\theta} = R(r)\Phi(\phi)\frac{d\Theta}{d\theta} \quad \text{and} \quad \frac{\partial\psi}{\partial\phi} = R(r)\Theta(\theta)\frac{d\Phi}{d\phi}.$$
$$(A.118)$$

The Schrödinger equation can now be separated into three parts, depending on r, θ, and ϕ only:

$$\frac{d^2\Phi(\phi)}{d\phi^2} = -m_l\Phi(\phi) \qquad (A.119)$$

$$\frac{1}{\sin\theta}\frac{d}{d\theta}\left(\sin\theta\frac{d\Theta(\theta)}{d\theta}\right) = \left(\frac{m_l^2}{\sin^2\theta} - \beta\right)\Theta(\theta) \qquad (A.120)$$

$$\frac{1}{r^2}\frac{d}{dr}\left(r^2\frac{dR(r)}{dr}\right) + \frac{2m_0}{\hbar^2}\left(E + \frac{e^2}{4\pi\varepsilon_0 r}\right)R(r) = \frac{\beta R(r)}{r^2}. \quad (A.121)$$

Here, m_l and β are separation constants. All three Schrödinger equations can be solved, yielding three independent sets of solutions, according to the three degrees of freedom of the hydrogen atom:

$$\Phi_m(\phi) = \exp(\pm im_l\phi) \quad \text{with} \quad m_l = 0, \pm 1, \pm 2, \ldots, l \ (A.122)$$

$$\Theta_{lm}(\theta) = \sqrt{\frac{(2l+1)(l-m_l)!}{2(l+m_l)!}}\, P_l^m(\theta) \text{ with } l=0, 1, \ldots n-1 (A.123)$$

$$R_{nl}(r) = -\left[\left(\frac{\rho}{r}\right)^3 \cdot \frac{(n-l-1)!}{(2n[n+l]!)^3}\right]^{1/2}\rho^l\exp(-\frac{\rho}{2})L_{n+1}^{2l+1}(\rho)$$
$$\text{with } n = 1, 2, \ldots \ (A.124)$$

and with $P_l^m(\theta)$ as the *associated Legendre functions* and L_{n+1}^{2l+1} as the *Laguerre polynomial*—for details see Pauling and Wilson (1935) or Leighton (1959). The normalized radius can be expressed as

$$\rho = \frac{r}{na_0} \quad \text{with} \quad a_0 = \frac{4\pi\varepsilon_0\hbar^2}{m_0\varepsilon} = 0.5292\,\text{Å} \quad (A.125)$$

and the energy for which the solutions (A.124) of Eq. (A.121) exist are the eigenvalues

$$E_n = \frac{m_0 e^4}{2(4\pi\varepsilon_0)^2\hbar^2} \cdot \frac{1}{n^2} = 13.607 \cdot \frac{1}{n^2}\ (\text{eV}). \quad (A.126)$$

The complete wavefunction, according to Eq. (A.116), is given by

$$\psi_{nlm}(r,\,\theta,\,\phi) = N_{nlm}R_{nl}(r)\cdot\Theta_{lm}(\theta)\cdot\Phi_m(\phi) \quad (A.127)$$

where N_{nlm} is a normalization factor:

$$N_{nlm} = -\left[\left(\frac{\rho}{r}\right)^3 \frac{(n-l-1)!(l-m_l)!(2l+1)}{4\pi\left[2n(n+l)!\right]^3(l+m_l)!}\right]^{1/2} \quad (A.128)$$

which renders

$$\int \psi_{nlm}^*\psi_{nlm}\cdot r^2\sin\theta\,dr\,d\theta\,d\phi = 1 \quad (A.129)$$

The wavefunctions (A.117) are orthogonal, i.e., $\int \psi_{nlm}^*\psi_{n'l'm'} = 0$, when at least one running number of the set $(n,\,l,\,m) \neq (n',\,l',\,m')$.

A.4.9 Quantum Numbers

In the previous section three running numbers were introduced to distinguish different possible solutions (*eigenfunctions*) of the Schrödinger equation for the hydrogen atom. These are the *quantum numbers*:

$n = 1,\,2,\,3,\,\ldots$ principle quantum number (n_q in the book)
$l = 0,\,1,\,2,\,\ldots n-1$ orbital angular momentum quantum number,[*]
$m_l = 0,\,\pm 1,\,\pm 2,\,\ldots \pm l$ magnetic quantum number, and
$m_s = \pm 1/2$ the spin quantum number.

---

[*] The states for different angular momentum quantum numbers are often referred to by letters: for $l = 0,\,1,\,2,\,3,\,4$ with s, p, d, f, and g, respectively; they are the first letters of an historic description of the corresponding

Table A.2: Wavefunctions of the hydrogen atom.

n	l	m_l	ψ_{nlm}	
1	0	0	$N_{100}\exp(-\frac{1}{2}\rho)$	1s
2	0	0	$N_{200}(2-\rho)\exp(-\frac{1}{2}\rho)$	2s
	1	0	$N_{210}\rho\exp(-\frac{1}{2}\rho)\cos\theta$	
		±1	$N_{211}\rho\exp(-\frac{1}{2}\rho)\sin\theta\exp(\pm i\phi)$	$2p^3$
3	0	0	$N_{300}(6-6\rho+\rho^2)\exp(-\frac{1}{2}\rho)$	3s
	1	0	$N_{310}\rho(4-\rho)\exp(-\frac{1}{2}\rho)\cos\theta$	$3p^3$
		±1	$N_{311}\rho(4-\rho)\exp(-\frac{1}{2}\rho)\sin\theta\exp(\pm i\phi)$	
	2	0	$N_{320}\rho^2\exp(-\frac{1}{2}\rho)(3\cos^2\theta-1)$	
		±1	$N_{321}\rho^2\exp(-\frac{1}{2}\rho)\sin\theta\cos\theta\exp(\pm i\phi)$	$3d^5$
		±2	$N_{322}\rho^2\exp(-\frac{1}{2}\rho)\sin^2\theta\exp(\pm 2\,i\phi)$	
4	0	0	$N_{400}(24-36\rho+12\rho^2-\rho^3)\exp(-\frac{1}{2}\rho)$	4s
	1	0	$N_{410}\rho\exp(-\frac{1}{2}\rho)(20-10\rho+\rho^2)\cos\theta$	
		±1	$N_{411}\rho\exp(-\frac{1}{2}\rho)(20-10\rho+\rho^2)\sin\theta\exp(\pm i\phi)$	$4p^3$
	2	0	$N_{420}\rho^2(6-\rho)\exp(-\frac{1}{2}\rho)(3\cos^2\theta-1)$	
		±1	$N_{421}\rho^2(6-\rho)\exp(-\frac{1}{2}\rho)\sin\theta\cos\theta\exp(\pm i\phi)$	$4d^5$
		±2	$N_{422}\rho^2(6-\rho)\exp(-\frac{1}{2}\rho)\sin^2\theta\exp(\pm 2\,i\phi)$	
	3	0	$N_{430}\rho^3\exp(-\frac{1}{2}\rho)(\frac{5}{3}\cos^2\theta-\cos\theta)$	
		±1	$N_{431}\rho^3\exp(-\frac{1}{2}\rho)(5\cos^2\theta-1)\sin\theta\exp(\pm i\phi)$	$4f^7$
		±2	$N_{432}\rho^3\exp(-\frac{1}{2}\rho)\sin^2\theta\cos\theta\exp(\pm 2\,i\phi)$	
		±3	$N_{433}\rho^3\exp(-\frac{1}{2}\rho)\sin^3\theta\exp(\pm 3\,i\phi)$	

An electron can be in any one of these states, identified by the triple n, l, m_l and described by the corresponding wavefunction. A few of these are listed in Table A.2. The corresponding electron density distributions are given in Fig. A.3, and show the wave-mechanical probability to find the electron in a certain mode of oscillation in the atom.

It should be noted, however, that in hydrogen, only modes with different *principal quantum numbers* have different energies (eigen-values), while all the different modes represented by different num-

---

spectroscopic lines: "sharp," "principal," and "diffuse," then following the alphabet.

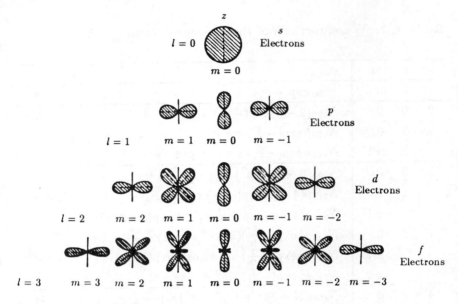

Figure A.3: Electron density distribution for various eigenstates of the hydrogen atom with the z axis in the plane of the figure, pointing up (after Leighton, 1959).

bers l and m_l have the same energy for the same n: these states are *degenerate*. When external electrical or magnetic forces are acting, this degeneracy is removed, and the corresponding energy levels will split.

The *angular momentum* of an orbiting electron is given by

$$\mathbf{L} = \mathbf{r} \times \mathbf{p} \tag{A.130}$$

with an operator representing the square of the total angular momentum

$$L^2 = L_x^2 + L_y^2 + L_z^2 = -\hbar^2 \left[\frac{1}{\sin\theta} \frac{\partial}{\partial\theta} \left(\sin\theta \frac{\partial}{\partial\theta} \right) + \frac{1}{\sin^2\theta} \frac{\partial^2}{\partial\phi^2} \right]. \tag{A.131}$$

When applied to the wavefunction ψ_{nlm}, one obtains from Eqs. (A.116), (A.120), (A.122), and (A.124)

$$L^2 \psi_{nlm} = \hbar^2 R_{nl}(r)\Phi_m(\phi) \cdot \beta\Theta_{lm}(\theta) = \hbar^2 l(l+1)\psi_{nlm}; \tag{A.132}$$

hence, the dynamic variable associated with the momentum operator is quantized; the angular momentum is conserved, and its expectation value is given by

$$\overline{L} = \sqrt{l(l+1)}\hbar. \tag{A.133}$$

The conservation of the z component of the angular momentum

$$L_z = \frac{\hbar}{i}\frac{\partial}{\partial \phi} \qquad (A.134)$$

can be obtained directly from

$$L_z \psi_{nlm} = \frac{\hbar}{i} R_{nl}(r)\Theta_{lm}(\theta) \cdot im_l \Phi_m(\phi) = m_l \hbar \psi_{nlm}, \qquad (A.135)$$

yielding for the *magnetic momentum*, the z component of the angular momentum:

$$L_z = m_l \hbar. \qquad (A.136)$$

This is a magnetic momentum because its degeneracy can be removed when a magnetic field is applied—Zeeman effect.

In addition, Dirac (1928) showed that as a consequence of relativity the electron possesses an intrinsic angular momentum, the *spin*, with its quantum number $s = \pm\frac{1}{2}$, yielding for the expectation value of the angular momentum—analogous to Eq. (A.133):

$$\overline{L_s} = \sqrt{s(s+1)}\hbar = \pm\frac{\sqrt{3}}{2}\hbar. \qquad (A.137)$$

Consequently, the magnetically important z component of L_s, which is analogous to Eq. (A.136), can have the value $+\frac{1}{2}$ or $-\frac{1}{2}$ with

$$\overline{L_{sz}} = m_s \hbar = \pm\frac{1}{2}\hbar, \qquad (A.138)$$

and with m_s as the *spin quantum number* of the electron in the hydrogen atom.

A.4.9A The Pauli Exclusion Principle   The *Pauli exclusion principle* states that in a multiparticle system no two particles of half-integer spin (electrons, protons, etc.) can occupy the same quantum state, i.e., have the same *quadruple* of quantum numbers n, l, m_l, and m_s within one atom; or, in systems of atoms, no more than N electrons can occupy a state which is N-fold degenerate.

This principle was first introduced by empirical judgment to explain the periodic system of elements, and can be justified by recognizing the antisymmetrical nature of the wavefunction dealing with half-integer spin particles.

The probability of finding particles 1 and 2 in position A and B, respectively, must be equal to finding them exchanged, since they are indistinguishable. This means $\psi^2[A(1)B(2)] = \psi^2[A(2)B(1)]$.

This can be achieved by particles with symmetric wavefunctions—
$\psi[A(1)B(2)] = \psi[A(2)B(1)]$ (*bosons*); or with antisymmetric ones:
$\psi[A(1)B(2)] = -\psi[A(2)B(1)]$ (*fermions*). The former is fulfilled for
particles with even spin (photons, phonons, excitons), the latter for
particles with half-integer spin (electrons, etc.). The antisymmetry is
characterized by writing the wavefunction of a multiparticle system
in the form of a determinant, the *Slater determinant*; for instance,

$$\psi[(1),\ (2)] = \frac{1}{2} \begin{vmatrix} \psi[A(1)] & \psi[A(2)] \\ \psi[B(1)] & \psi[B(2)] \end{vmatrix}, \qquad (A.139)$$

which gives a finite value only for antisymmetric wavefunctions
$= \psi^2[A(1)B(2)]$, and vanishes when two (or more) one-particle wavefunctions are equal to each other ($\psi^2[A(1)A(2)] = 0$)—the Pauli exclusion.

A.4.9B Multiparticle Systems
As multiparticle systems, one
distinguishes atoms which contain several electrons, and systems of
many atoms, such as molecules or condensed matter.

With the tools provided in the previous sections one can easily
explain the electronic structure within the periodic system of elements.

As shown for the hydrogen atom, there are a large number of
eigenstates, characterized by the quadruple of quantum numbers
$(n,\ l,\ m_l,\ m_s)$. Without excitation, only the lowest one is occupied;
for hydrogen it is $(1,\ 0,\ 0,\ \frac{1}{2})$.

When adding another electron (for helium), it must have another
eigenstate $(1,\ 0,\ 0,\ -\frac{1}{2})$. Continuation of the process yields for each
additional electron for Li $\rightarrow (2,\ 0,\ 0,\ \frac{1}{2})$, for Be $\rightarrow (2,\ 0,\ 0,\ -\frac{1}{2})$, for
B $\rightarrow (2,\ 1,\ 0,\ \frac{1}{2})$, and so on, following the rules specified for the four
quantum numbers in Section A.4.9.

In order to simplify the description of the arrangement for electrons in the ground state, one uses a sequence of numbers and
letters best explained with examples. $1s^2 2s$ describes the ground
state of Li with the "first shell" ($n = 1$)* filled with two electrons
(shown as superscript) and the next shell ($n = 2$) occupied by one
electron. All three electrons are in s-states (i.e., for $l = 0$). As
another example, $1s^2 2s^2 2p^5$ describes flourine, which has 7 electrons in the second shell ($n = 2$); however, since the s-state in

---

* Capital letters are conventionally assigned to these shells with K, L,
M, N, O, P, and Q for $n = 1, 2, \ldots 7$, respectively.

the second shell has only space for two electrons ($m_s = \pm\frac{1}{2}$), the next five electrons need to be placed in p-states ($l = 1$), according to $(2, 1, 0, \frac{1}{2})$, $(2, 1, 0, -\frac{1}{2})$, $(2, 1, 1, \frac{1}{2})$, $(2, 1, 1, -\frac{1}{2})$, and $(2, 1, -1, \frac{1}{2})$.

The filling of the electron shells follows the sequence $1s(2s\,2p)$ $(3s\,3p)$ $(4s\,3d\,4p)$ $(5s\,4d\,5p)$, etc. The insertion of $3d$, $4d$, etc., into the (s, p) sequence determines the formation of transition metals. The specific ordering depends on the electron interaction, and requires a more complex evaluation of the energy gain when adding the next electron (Hund's rule—see Ashcroft and Mermin, 1976). The actual ordering is shown in Fig. A.4.

A.4.9C Spectrum and Selection Rules   In the hydrogen atom, discussed in Sections A.4.8 and A.4.9, only one electron is involved in transitions between ground and excited states, resulting in a relatively simple spectrum as shown in Fig. A.5A.

Only changes in the principal quantum number determine changes in the electronic energy—Eq. (A.126). Depending on the initial state from which excitation occurs, one distinguishes different series according to

$$E = R_H \left(\frac{1}{n_1^2} - \frac{1}{n_2^2} \right) \qquad (A.140)$$

where R_H is the Rydberg energy ($= 13.607$ eV) given by Eq. (A.126) for $n = 1$. These series are identified in Fig. A.5B with $n_1 = 1, 2, 3, 4$, and 5 for the Lyman-, Balmer-, Paschen-, Bracket-, and Pfund-series, respectively, and $n_2 = n_1, n_1 + 1, n_1 + 2, \ldots$. All shown transitions are permitted.

In a multi-electron system, electron interaction results in a more complex behavior. This is accounted for by vector addition of the angular momenta and spin momenta of all outer electrons, i.e., all electrons in the outermost shell, or for transition metals in the partially filled shell.

In the one-electron problem, Fig. A.6 shows the relation between the orbital angular momentum $\mathbf{L}$, its expectation value $\overline{L} = \hbar\sqrt{l(l+1)}$, the length of $\mathbf{L}$, relating to the orbital quantum number l, and the projection of $\mathbf{L}$ in the z direction—L_z relating to the magnetic quantum number m_l.

When several electrons are involved, their angular momenta add as vectors, resulting in a total orbital angular momentum $\mathbf{L} = \sum_i \mathbf{L}_i$. In a similar way, one obtains the total spin vector as the sum over all individual spin vectors $\mathbf{S} = \sum_i \mathbf{s}_i$, which may be even

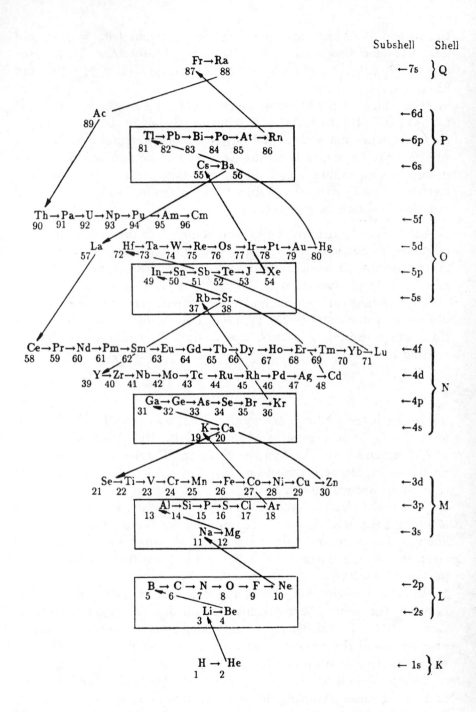

Figure A.4: Sequential filling of electron shells according to the *Aufbau* (assembly) principle of Pauli.

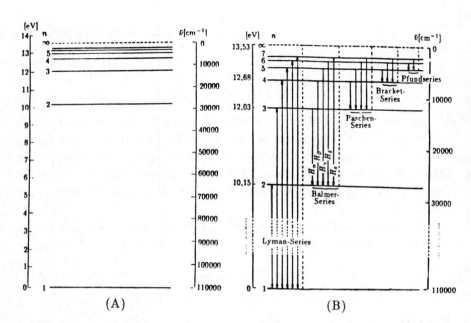

Figure A.5: (A) Level spectrum of the hydrogen atom. (B) Different spectral series (enlarged scale) of the hydrogen atom. The Lyman series lies in the vacuum UV, the Balmer series lies in the visible range of the spectrum, and the other series lie in the IR part of the spectrum.

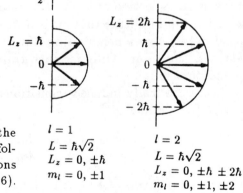

Figure A.6: Orientation of the angular momentum vector **L** following the quantum restrictions given in Eqs. (A.133) and (A.136).

$l = 1$
$L = \hbar\sqrt{2}$
$L_z = 0, \pm\hbar$
$m_l = 0, \pm1$

$l = 2$
$L = \hbar\sqrt{2}$
$L_z = 0, \pm\hbar \pm 2\hbar$
$m_l = 0, \pm1, \pm2$

or half-integer. From these vectors, one obtains the total angular momentum as

$$\mathbf{J} = \mathbf{L} + \mathbf{S}. \tag{A.141}$$

In a similar manner, one has in a one-electron system for the total angular momentum quantum number $\mathbf{j} = \mathbf{l} + \mathbf{s}$. The quantum states relating to multi-electron systems are then given by capital letters,

L, S, and J, and states with different orbital angular momentum are referred to according to

$$L = \quad 0 \quad 1 \quad 2 \quad 3 \quad 4 \quad 5 \quad 6 \quad 7 \quad 8 \quad 9 \quad 10 \quad \dots$$
$$ \quad S \quad P \quad D \quad F \quad G \quad H \quad I \quad K \quad L \quad M \quad N \quad \dots$$

When identifying a specific energy level, one places the *multiplicity* of the term as a superscript to the left of this letter, and one writes the total angular momentum J as a subscript to the right. For instance, the symbol $^3P_2$ identifies a triplet P term with total angular momentum 2; the symbol $^2S_{1/2}$ identifies a doublet S term with $J = 1/2$.

The multiplicity of a term, its degeneracy if no forces are acting to split the levels, is given as

$$r = 2(S, L)_{\max} + 1 \qquad (A.142)$$

which means that for $S > L$, it is given by $2S + 1$; and for $S < L$, it is given by $2L + 1$.

When an electron is excited to a higher level, it can return to lower states by emission of a photon only following certain conservation rules. These result in

$$\Delta l = \pm 1 \quad \text{and} \quad \Delta m_l = \pm 1 \quad \text{or} \quad = 0; \qquad (A.143)$$

there is no selection rule for the radial quantum number. These selection rules indicate that optical transitions (dipole radiation) can occur only between neighboring states, i.e., between s- and p-, p- and d-states, and so on. Only for such transitions is the electrical dipole momentum finite.

In magnetically induced transitions (Zeeman effect), one expects a triplet according to

$$\begin{aligned} \omega &= \omega_0 \pm \omega_c \quad \text{for} \quad \Delta m_l = \pm 1 \\ \text{plus} \quad \omega &= \omega_0 \quad \text{for} \quad \Delta m_l = 0 \end{aligned} \qquad (A.144)$$

with $\omega_0 + \omega_c$ ($\Delta m_l = +1$), it is right-polarized and $\omega_0 - \omega_c$ ($\Delta m_l = -1$), it is left-polarized.

A.4.10 Spinors

When a magnetic field is acting, degenerate electron eigenstates split. Such a magnetic field can be produced by external means, or by other orbiting electrons, or by the spin of nuclei, interacting with the specific electron under investigation. One way to incorporate such interaction is by writing spinor equations—see the following discussions.

A spinor is a two-component vector, which in spherical coordinates (see Fig. A.1) is described by its complex components in an up and down direction

$$\mathbf{s}(\theta, \phi) = \begin{pmatrix} \cos \dfrac{\theta}{2} \\ \exp(i\phi) \sin \dfrac{\theta}{2} \end{pmatrix}. \tag{A.145}$$

The wavefunction for electrons can be written as a two-component spinor

$$\psi(\mathbf{r}, t) \rightarrow \begin{pmatrix} \alpha(\mathbf{r}, t) \\ \beta(\mathbf{r}, t) \end{pmatrix} \tag{A.146}$$

with two coupled spinor equations

$$\begin{aligned} i\hbar\dot{\alpha} &= H_{11}\alpha + H_{12}\beta \\ i\hbar\dot{\beta} &= H_{21}\alpha + H_{22}\beta \end{aligned} \quad \text{or} \quad i\hbar\frac{\partial}{\partial t}\begin{pmatrix} \alpha \\ \beta \end{pmatrix} = \begin{pmatrix} H_{11} & H_{12} \\ H_{21} & H_{22} \end{pmatrix}\begin{pmatrix} \alpha \\ \beta \end{pmatrix}. \tag{A.147}$$

When $H_{12} = H_{21} = 0$, and $H_{11} = H_{22}$ is the conventional Hamiltonian,

$$H_{11} = H_{22} = H_0 = -\frac{\hbar^2}{2m_0}\nabla^2 + V, \tag{A.148}$$

the spinor equations decouple into two identical Schrödinger equations. One then accounts for the two-spin component by a simple twofold degeneracy. With a magnetic field **B**, the degeneracy is removed and another Hamiltonian H_B needs to be added

$$H_B = \mu_B \begin{pmatrix} B_z & B_x - iB_y \\ B_x + iB_y & B_z \end{pmatrix} \quad \text{with} \quad \mu_B = \frac{e\hbar}{2m_0}. \tag{A.149}$$

One now has to solve the two spinor equations.

The magnetic field Hamiltonian can also be written in component form

$$H_B = \mu_B(\sigma_x B_x + \sigma_y B_y + \sigma_z B_z) = \mu_B \sigma \cdot \mathbf{B} \tag{A.150}$$

with $(\sigma_x, \sigma_y, \sigma_z)$ the Pauli spin matrices

$$\sigma_x = \begin{pmatrix} 0 & 1 \\ 1 & 0 \end{pmatrix}, \quad \sigma_y = \begin{pmatrix} 0 & -i \\ i & 0 \end{pmatrix}, \quad \sigma_z = \begin{pmatrix} 1 & 0 \\ 0 & -1 \end{pmatrix}. \tag{A.151}$$

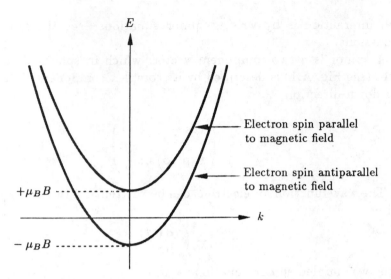

Figure A.7: Dispersion curves for a free electron in a homogeneous magnetic field.

A.4.10A Electrons in Homogeneous Magnetic Field   In the absence of an electric field $(V = 0)$, the Hamiltonian matrix reads

$$H = \begin{pmatrix} -\dfrac{\hbar^2}{2m_0}\nabla^2 + \mu_B B_z & \mu_B(B_x - iB_y) \\[2mm] \mu_B(B_x + iB_y) & -\dfrac{\hbar^2}{2m_0} - \mu_B B_z \end{pmatrix} ; \qquad \text{(A.152)}$$

with the trial solutions

$$\begin{aligned} \alpha(\mathbf{r}, t) &= \alpha_0 \exp(i\mathbf{k} \cdot \mathbf{r}) \exp\left(-i\frac{E(\mathbf{k})t}{\hbar}\right) \\[2mm] \beta(\mathbf{r}, t) &= \beta_0 \exp(i\mathbf{k} \cdot \mathbf{r}) \exp\left(-i\frac{E(\mathbf{k})t}{\hbar}\right) \end{aligned} \qquad \text{(A.153)}$$

substituted into the corresponding spinor equations, one obtains for the eigenvalues

$$E(\mathbf{k})\begin{pmatrix} \alpha_0 \\ \beta_0 \end{pmatrix} = \begin{pmatrix} \dfrac{\hbar^2\nabla^2}{2m_0} + \mu_B B_x & \mu_B(B_x - iB_y) \\[2mm] \mu_B(B_x + iB_y) & \dfrac{\hbar^2\nabla^2}{2m_0} - \mu_B B_z \end{pmatrix} \begin{pmatrix} \alpha_0 \\ \beta_0 \end{pmatrix}, \quad \text{(A.154)}$$

which yield $E(\mathbf{k})$ as solutions of

$$\begin{vmatrix} E - \dfrac{\hbar^2 k^2}{2m_0} - \mu_B B_z & -\mu_B(B_x - iB_y) \\[2ex] -\mu_B(B_x + iB_y) & E - \dfrac{\hbar^2 k^2}{2m_0} + \mu_B B_z \end{vmatrix} = 0 \qquad (A.155)$$

or

$$E_{1,2}(\mathbf{k}) = \frac{\hbar^2 k^2}{2m_0} \pm \mu_B \sqrt{B_x^2 + B_y^2 + B_z^2}, \qquad (A.156)$$

which can be recognized as the free-electron parabola, however, offset by $+\mu_B|B|$ when the electron spin is parallel to the magnetic field and by $-\mu_B|B|$ when it is antiparallel to the field—see Fig. A.7.

A.5 Tables

Some additonal tables, which would have interrupted the main text or relate to it in a subordinate fashion, are presented in this section.

Table A.3: Observed g-values in substitutional $3d$ impurities in Si and binary semiconductors (after Zunger, 1986).

State	System	g_{exp}
d^1, ^2E ($S=\frac{1}{2}$, $L=0$)	ZnS:^{45}Sc^{2+}, A^0	1.934
d^2, $^3A_2$ ($S=1$, $L=0$)	ZnS:^{47}Ti^{2+}, A^0	1.9280
	ZnS:^{51}V^{3+}, A^+	1.9433
	ZnSe:V^{3+}, A^+	1.9483
	ZnSe:Ti^{2+}, A^0	1.924
	CdS:V^{3+}, A^+	1.93
	CdS:Ti^{2+}, A^0	~1.92
	ZnTe:^{51}V^{3+}, A^+	1.917
d^3, $^4T_1$ ($S=\frac{3}{2}$, $L=1$)	ZnS:^{51}V^{2+}, A^0	$g_\parallel = 1.961$, $g_\perp = 1.913$
d^4, $^4T_2$	ZnS:Cr^{2+}, A^0	$g_\parallel = 1.94$, $g_\perp = 1.98$
	CdS:Cr^{2+}, A^0	$g_\parallel = 1.934$, $g_\perp = 1.970$
($S=2$)	ZnSe:Cr^{2+}, A^0	$g_\parallel = 1.961$, $g_\perp = 1.98$
	ZnTe:Cr^{2+}, A^0	$g_\parallel = 1.97$, $g_\perp = 1.99$
	CdTe:Cr^{2+}, A^0	$g_\parallel = 1.980$, $g_\perp = 1.980$
d^5, $^6A_1$ ($S=\frac{5}{2}$, $L=0$)	ZnS:^{53}Cr$^+$, A^-	1.9995
	ZnS:^{55}Mn^{2+}, A^0	2.0024
	ZnS:^{57}Fe^{3+}, A^+	2.019
	ZnSe:^{53}Cr$^+$, A^-	2.0016
	ZnSe:^{55}Mn^{2+}, A^0	2.0055-2.0069
	ZnSe:^{57}Fe^{3+}, A^+	2.0469

State	System	g_{exp}
d^2, $^3A_2$ ($S=1$, $L=0$)	GaAs:^{51}V^{3+}, A^0	1.957
	InP:^{51}V^{3+}, A^0	1.96
	GaP:Cr^{4+}, A^+	1.986
	GaAs:Cr^{4+}, A^+	1.994
	InP:Cr^{4+}, A^+	1.999
	Si:^{53}Cr^{4+}, A^0	1.9962
	Si:^{55}Mn^{5+}, A^+	2.0259
d^3, $^4T_1$	GaAs:Cr^{3+}, A^0	2.367, 5.154 1.636 $= (g_x, g_x, g_x)$
	GaP:Cr^{3+}, A^0	3.044, 4.736, 1.840 (g_x, g_y, g_z)
d^4, $^5T_2$	GaAs:Cr^{2+}, A^-	$g_\parallel = 1.974$; $g_\perp = 1.997$
	InP:Cr^{2+}, A^-	$g_\parallel = 1.978$; $g_\perp = 1.999$
	GaP:Cr^{2+}, A^-	$g_\parallel = 1.981$; $g_\perp = 2.010$
d^5, $^6A_1$ ($S=\frac{5}{2}$, $L=0$)	GaP:Cr$^+$, A^{2-}	1.999
	GaP:^{55}Mn^{2+}, A^-	2.002
	GaAs:^{55}Mn^{2+}, A^-	2.003
	InP:^{55}Mn^{2+}, A^-	1.997
	GaP:Fe^{3+}, A^0	2.026
	GaAs:^{57}Fe^{3+}, A^0	2.045, 2.053
	InP:Fe^{3+}, A^0	2.024

Table A.3: (Cont'd.) Observed g values in substitutional $3d$ impurities and binary semiconductors (after Zunger, 1986). ©John Wiley & Sons, Inc.

State	System	g_{exp}
	$ZnTe:^{53}Cr^+$, A^-	2.0026
	$ZnTe:^{55}Mn^{2+}$, A^0	2.0105
	$CdS:^{55}Mn^{2+}$, A^0	2.0029, 2.0018
	$CdS:Fe^{3+}$, A^+	2.018
	$CdSe:^{55}Mn^{2+}$, A^0	2.003
	$CdTe:^{53}Cr^+$, A^-	1.9997
	$CdTe:^{53}Mn^{2+}$, A^0	2.0069
	$CdTe:Fe^{3+}$, A^+	2.084
d^7, $^4A_2$ ($S=\frac{3}{2}$, $L=0$)	$ZnS:Fe^+$, A^-	2.2515
	$ZnS:Co^{2+}$, A^0	2.248
	$ZnS:Ni^{3+}$, A^+	2.1480
	$ZnSe:Co^{2+}$, A^0	2.270
	$ZnSe:Ni^{3+}$, A^-	2.1978
	$ZnTe:Fe^+$, A^-	2.280
	$ZnTe:Co^{2+}$, A^0	2.2972
	$CdS:Co^{2+}$, A^0	2.27
	$CdTe:Co^{2+}$, A^0	2.31
d^9, $^2T_2$ ($S=\frac{1}{2}$, $L=1$)	$ZnS:^{61}Ni^+$, A^-	1.3991
	$ZnSe:^{61}Ni^+$, A^-	1.4374
	$ZnS:Cu^{2+}$, A^0	$|g|=0.71$
	$ZnTe:Ni^+$, A^-	$|g|=1.367$
	$CdTe:Ni^+$, A^-	$|g|=1.425$
	$CdS:^{63}Cu^{2+}$, A^0	$g_\perp=2.24$, $g_\perp=1.75$

State	System	g_{exp}
	$InAs:Fe^{3+}$, A^0	2.035
	$Si:^{55}Mn^{2+}$, A^{2-}	2.0058
d^7, $^4A_2$ ($S=\frac{3}{2}$, $L=0$)	$GaP:Fe^{1+}$, A^{2-}	2.133
	$GaP:Co^{2+}$, A^-	2.164, 2.159
	$GaAs:Co^{2+}$, A^-	2.189, 2.182
	$InP:Co^{2+}$, A^-	2.193
	$GaP:Ni^{3+}$, A^0	2.089
	$GaAs:Ni^{3+}$, A^0	2.114, 2.106
	$InP:Ni^{3+}$, A^0	2.098
d^9, $^2T_2$ ($S=\frac{1}{2}$, $L=1$)	$GaP:Ni^+$, A^{2-}	0.94
	$GaAs:Ni^+$, A^{2-}	-1.16

Table A.4: Impurity diffusion coefficients in Si and Ge (after Casey and Pearson, 1975).

Impurity	Si		Ge	
	D_o (cm^2/s)	ΔE_i (eV)	D_o (cm^2/s)	ΔE_i (eV)
B	5.1	3.70	$1.1 \cdot 10^7$	4.54
Al	8.0	3.47	$1.6 \cdot 10^2$	3.24
Ga	3.6	3.51	40	3.15
In	16.5	3.91	33	3.03
P	10.5	3.69	2.5	2.49
As	60	4.20	10.3	2.51
Sb	12.9	3.98	3.2	2.42
Bi	$1.03 \cdot 10^3$	4.63		

Self- and impurity diffusion coefficients in III-IV compounds.

Impurity	GaAs		GaSb		InAs		InSb	
	D_o cm^2/s	ΔE_i (eV)	D_o (cm^2/s)	ΔE_i (eV)	D_o (cm^2/s)	ΔE_i (eV)	D_o (cm^2/s)	ΔE_i (eV)
As	0.7	3.2			$3 \cdot 10^7$	4·45		
Ga	$1 \cdot 10^7$	5.6	3200	3.15				
In	$5 \cdot 10^{-11}$	at 1000°C	$1.2 \cdot 10^{-7}$	0.53	$6 \cdot 10^5$	4.0	$1.76 \cdot 10^{13}$	4.3
Sb			$3.4 \cdot 10^4$	3.45			$3.1 \cdot 10^{13}$	4.3
Ag	25	2.27			$7.3 \cdot 10^{-4}$	0.26	$\sim 1 \cdot 10^{-7}$	~ 0.25
Au	29	2.64			$5.8 \cdot 10^{-4}$	0.65	$7 \cdot 10^{-6}$	0.32
Be	$7.3 \cdot 10^{-6}$	1.2						
Cd			$1.5 \cdot 10^{-6}$	0.72	$7.4 \cdot 10^{-4}$	1.15	$1.3 \cdot 10^{-4}$	1.2
Cr	4.300	3.4						
Cu	0.03	0.53			$3.6 \cdot 10^{-3}$	0.52	$9.0 \cdot 10^{-4}$	1.08
Li	0.53	1.0					$7.0 \cdot 10^{-4}$	0.28
Mg	0.026	2.7			$1.98 \cdot 10^{-6}$	1.17		
Mn	0.65	2.49						
O	0.002	1.1			6.78			
S	0.018	2.6			6.78	2.20	0.09	1.40
Se	0.008	4.16			12.6	2.20	1.6	1.87
Sn	0.038	2.7	$2.4 \cdot 10^{-5}$	0.80	$1.49 \cdot 10^{-8}$	1.17	$5.5 \cdot 10^{-8}$	0.75
Te	10^{-13}	at 1000°C	$3.8 \cdot 10^{-4}$	1.20	$3.43 \cdot 10^{-5}$	1.28	$1.7 \cdot 10^{-7}$	0.57

Table A.5: Refractive index of a few window materials as a function of the wavelength at 300 K (from *American Institute of Physics Handbook*, 1963).

	$\lambda(\mu m)$	n_r	$\lambda(\mu m)$	n_r	$\lambda(\mu m)$	n_r
SiO_2	0.34	1.47877	0.56	1.459561	1.50	1.444687
	0.35	1.47701	0.57	1.459168	1.60	1.443492
	0.36	1.47540	0.58	1.458794	1.70	1.442250
	0.37	1.47393	0.59	1.458437	1.80	1.440954
	0.38	1.47258	0.60	1.458096	1.90	1.439597
	0.39	1.47135	0.61	1.457769	2.00	1.438174
	0.40	1.470208	0.62	1.457456	2.10	1.436680
	0.41	1.469155	0.63	1.457156	2.20	1.435111
	0.42	1.468179	0.64	1.456868	2.30	1.433462
	0.43	1.467273	0.65	1.456591	2.40	1.431730
	0.44	1.466429	0.66	1.456324	2.50	1.429911
	0.45	1.465642	0.67	1.456066	2.60	1.428001
	0.46	1.464908	0.68	1.455818	2.70	1.425995
	0.47	1.464220	0.69	1.455579	2.80	1.423891
	0.48	1.463573	0.70	1.455347	2.90	1.421684
	0.49	1.462965	0.80	1.453371	3.00	1.41937
	0.50	1.462394	0.90	1.451808	3.10	1.41694
	0.51	1.461856	1.00	1.450473	3.20	1.41440
	0.52	1.461346	1.10	1.449261	3.30	1.41173
	0.53	1.460863	1.20	1.448110	3.40	1.40893
	0.54	1.460406	1.30	1.446980	3.50	1.40601
	0.55	1.459973	1.40	1.445845		

	$\lambda(\mu m)$	$n_{r,0}$	$n_{r,e}$
MgF_2	0.404656	1.38359	1.39566
	0.58937	1.37770	1.38950
	0.706525	1.37599	1.38771

Table A.5: (Con't.) Refractive index of a few window materials.

	$\lambda(\mu m)$	$n_{r,0}$	$\lambda(\mu m)$	$n_{r,0}$	$\lambda(\mu m)$	$n_{r,0}$
Sapphire	0.26520	1.8336	0.64385	1.7655	2.32542	1.7306
(ordinary	0.28035	1.8243	0.706519	1.7630	2.4374	1.7278
ray)	0.28936	1.8195	0.85212	1.7588	3.2432	1.7044
	0.29673	1.8159	0.89440	1.7579	3.2666	1.7036
	0.30215	1.8135	1.01398	1.7555	3.303	1.7023
	0.3130	1.8091	1.12866	1.7534	3.3293	1.7015
	0.33415	1.8018	1.36728	1.7494	3.4188	1.6982
	0.34662	1.7981	1.39506	1.7489	3.5078	1.6950
	0.361051	1.7945	1.52952	1.7466	3.70	1.6875
	0.365015	1.7936	1.6932	1.7437	4.258	1.6637
	0.39064	1.7883	1.70913	1.7434	4.954	1.6266
	0.404656	1.7858	1.81307	1.7414	5.1456	1.6151
	0.435834	1.7812	1.9701	1.7383	5.349	1.6020
	0.54607	1.7708	2.1526	1.7344	5.419	1.5973
	0.576960	1.7688	2.24929	1.7323	5.577	1.5864
	0.579066	1.7687				

	$\lambda(\mu m)$	n_r	$\lambda(\mu m)$	n_r	$\lambda(\mu m)$	n_r
LiF	0.1935	1.4450	0.366	1.40121	4.50	1.33875
	0.1990	1.4413	0.391	1.39937	5.00	1.32661
	0.2026	1.4390	0.4861	1.39480	5.50	1.31287
	0.2063	1.4367	0.50	1.39430	6.00	1.29745
	0.2100	1.4346	0.80	1.38896	6.91	1.260
	0.2144	1.4319	1.00	1.38711	7.53	1.239
	0.2194	1.4300	1.50	1.38320	8.05	1.215
	0.2265	1.4268	2.00	1.37875	8.60	1.190
	0.231	1.4244	2.50	1.37327	9.18	1.155
	0.254	1.41792	3.00	1.36660	9.79	1.109
	0.280	1.41188	3.50	1.35868		
	0.302	1.40818	4.00	1.34942		

Table A.6: Refractive index of polymers (after Brandrup and Immergut, 1975).

Groups	Polymers	$\bar{n}_r$
Thermoplastic polymers	Polyethylene High-density	1.545
	Low-density	1.51
	Polyvinyl chloride	1.54–1.55
	Polystyrene	1.59
	Polypropylene	1.47
	Polyamides (Nylon 66)	1.53
	Polytetrafluoroethylene (Teflon)	1.35–1.38
Elastomeric thermoplastics (unvulcanized)	Polyisoprene (Natural Rubber)	1.52
	Polybutadiene/Polystyrene Copolymer	1.53
	Polyisobutene/Polyisoprene Copolymer	1.51
	Polychloroprene	1.55–156
Thermosetting polymers	Phenolics (Phenol-formaldehyde)	1.47–1.50
	Epoxies	1.55–1.60
	Urethanes	1.5–1.6

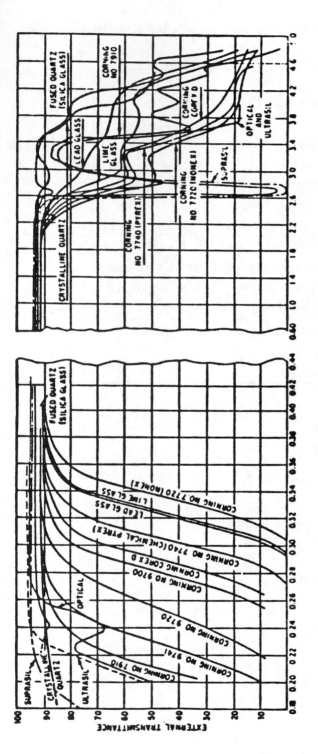

Figure A.8: Optical transmittance of fused SiO$_2$ and other glasses (after *Fused Quartz Catalog*, General Electric Co., 1957, and *Optical Fused Quartz*, Amersil Quartz Division, Engelland Industries, Inc.

Table A.7: Optical constants of evaporated metal layers: index of refraction n_r, extinction coefficient k, and calculated reflectance R_o (from *American Institute of Physics Handbook*).

	$\lambda(\mu m)$	n_r	k	R_0
Silver	0.3021	1.2	0.8	12.4
	0.3261	0.5	0.5	20.0
	0.3404	0.22	1.0	64.6
	0.40	0.075	1.93	93.9
	0.45	0.055	2.42	96.8
	0.50	0.050	2.87	97.9
	0.55	0.055	3.32	98.2
	0.60	0.060	3.75	98.4
	0.70	0.075	4.62	98.7
	0.80	0.090	5.45	98.8
	0.90	0.105	6.22	98.9
	1.0	0.129	6.83	98.9
	2.0	0.48	14.4	99.1
	4.0	1.89	28.7	99.1
	6.0	4.15	42.6	99.1
	8.0	7.14	56.1	99.1
	10.0	10.69	69.0	99.1
	12.0	14.50	81.4	99.2
Aluminum	0.2200	0.14	2.35	91.8
	0.240	0.16	2.60	92.1
	0.260	0.19	2.85	92.0
	0.300	0.25	3.33	92.1
	0.340	0.31	3.80	92.3
	0.400	0.40	4.45	92.6
	0.450	0.51	5.00	92.5
	0.492	0.64	5.50	92.2
	0.546	0.82	5.99	91.6
	0.650	1.30	7.11	90.7
	0.700	1.55	7.00	88.8
	0.750	1.80	7.12	87.7
	0.800	1.99	7.05	86.4
	0.900	1.96	7.70	88.5
	2.0	2.3	16.5	96.8
	4.0	6.1	30.4	97.5
	6.0	10.8	42.6	97.8
	8.0	17.9	55.3	97.9
	10.0	26.0	67.3	98.0
	12.0	33.1	78.0	98.2

Table A.7: (Con't.) Optical constants of evaporated metal layers: index of refraction n_r, extinction coefficient k, and calculated reflectance R_o (from *American Institute of Physics Handbook*).

	$\lambda(\mu m)$	n_r	k	R_0
Gold	0.1400	1.35	0.54	7.1
	0.1640	1.15	0.68	9.5
	0.2000	1.24	0.92	15.4
	0.450	1.40	1.88	39.7
	0.500	0.84	1.84	50.4
	0.550	0.331	2.324	81.5
	0.600	0.200	2.897	91.9
	0.700	0.131	3.842	96.7
	0.800	0.149	4.654	97.4
	0.900	0.166	5.335	97.8
	1.000	0.179	6.044	98.1
	1.75	1.1	9.59	95.4
	2.26	1.62	11.8	95.6
	3.43	3.49	18.6	96.2
	4.55	5.95	23.0	95.9
	6.65	12.9	35.5	96.4
	8.65	20.2	47.4	97.0
	11.1	29.0	62.5	97.6
	12.1	32.4	69.2	97.8

A.6 Collection of Important Formulae

Some of these formulae have only limited validity. The original equation numbers have been added here to facilitate tracing such formulae to the corresponding place in the text from which the limitations can be deduced.

A.6.1 General Formulae

Wave vector, electron energy (quasi-free electron):

$$\mathbf{k} = \frac{m_0 \mathbf{v}}{\hbar} = \frac{\mathbf{p}}{\hbar}; \qquad E = \frac{m_0}{2} v^2 = \frac{p^2}{2m_0} = \frac{\hbar^2 k^2}{2m_0}. \qquad (7.3)$$

Photon momentum:

$$p_{\text{phot}} = \frac{h}{\lambda} = \frac{h\nu}{c} = \hbar k. \qquad (11.47)$$

Group velocity v_g, phase velocity v_s:

$$v_g = \Delta_k \omega(\mathbf{k}) = \frac{1}{\hbar}\frac{dE}{dk} \qquad v_s = \frac{\omega}{k}. \qquad (7.17)$$

rms velocity of electrons:

$$v_{\text{rms}} = \sqrt{\frac{3kT}{m_n}} = 1.18 \cdot 10^7 \sqrt{\frac{m_0}{m_n}} \sqrt{\frac{T(\text{K})}{300}} \ (\text{cm/s}). \qquad (28.16),(28.17)$$

De Broglie wavelength:

$$\lambda_{\text{DB}} = \frac{h}{|p|} = \frac{h}{m_0}\frac{1}{v} = 7.27 \ \frac{1}{v(\text{cm/s})} \ (\text{cm}) \qquad (7.5)$$

$$\lambda_{\text{DB}} = \frac{h}{\sqrt{2m_0 E}} = 12.26 \ \frac{1}{\sqrt{E(\text{eV})}} \ (\text{Å}). \qquad (7.6)$$

A.6.2 Characteristic Lengths in Solids

Debye length:

$$\lambda_D = \sqrt{\frac{\varepsilon_{\text{st}}\varepsilon_0 kT}{e^2 n}} = 1205\sqrt{\frac{\varepsilon_{\text{st}}}{10}\frac{T}{300}\frac{10^{15}}{n}} \ (\text{Å}). \qquad (14.11)$$

Thomas-Fermi length:

$$\lambda_{\text{TF}} = \pi^{2/3}\frac{\hbar}{e}\sqrt{\frac{\varepsilon_{\text{st}}\varepsilon_0}{m_n(3n)^{1/3}}} = 1.16\sqrt{\frac{\varepsilon_{\text{st}}}{10}\frac{m_0}{m_n}\left(\frac{10^{18}}{n}\right)^{1/3}} \ (\text{Å}). \qquad (14.12)$$

Drift length or *Schubweg*:

$$l_c = \mu_c \tau_c F \qquad \text{with} \quad c = n \text{ or } p. \tag{48.3}$$

Diffusion length:

$$L_c = \sqrt{\frac{\mu_c kT}{e} \tau_c} = 5.1 \sqrt{\frac{\mu_c}{10^3 \,(\text{cm}^2/\text{Vs})} \frac{T}{300 \,(\text{K})} \frac{\tau_c}{10^{-8} \,(\text{s})}} \,(\mu\text{m}). \tag{48.4}$$

Upstream and downstream diffusion lengths:

$$L_{c(u,d)} = L_c \frac{2L_c}{\sqrt{4L_c^2 + l_c^2} \pm l_c}. \tag{48.5}$$

Mean free path:

$$\lambda_{\text{sc}} = v_{\text{rms}} \tau_{\text{sc}} = \frac{1}{s_n N_{\text{sc}}}. \tag{28.31}, (28.32)$$

A.6.3 Effective Masses

Effective carrier mass

$$m_n = \frac{\hbar^2}{\left.\frac{d^2 E}{dk^2}\right|_{E_c}}; \qquad m_p = -\frac{\hbar^2}{\left.\frac{d^2 E}{dk^2}\right|_{E_v}}. \tag{9.1}$$

Density of state masses:

$$m_{\text{nds}}^{3/2} = \nu_D (m_{n\|} m_{n\perp}^2)^{1/2}; \tag{27.25}$$

$$m_{\text{pds}}^{3/2} = m_{\text{pl}}^{3/2} + m_{\text{ph}}^{3/2}. \tag{27.28}$$

Mobility effective masses:

$$\frac{1}{m_{n\mu}} = \frac{1}{3}\left(\frac{2}{m_{n\perp}} + \frac{1}{m_{n\|}}\right); \tag{30.37}$$

$$\frac{1}{m_{p\mu}} = \frac{1}{2}\left(\frac{1}{m_{\text{pl}}} + \frac{1}{m_{\text{ph}}}\right). \tag{30.38}$$

Reduced effective mass:

$$\frac{1}{m_r} = \frac{1}{m_n} + \frac{1}{m_p}. \tag{13.14}$$

A.6.4 Phonons, IR Dispersion and Thermal Properties

Phonon dispersion equation:

$$\omega_{\pm}^2 = \frac{\beta_s(M_1 + M_2)}{M_1 M_2}\left[1 \pm \sqrt{1 - \frac{4M_1 M_2 \sin^2(qa)}{M_1 + M_2}}\right].\qquad(5.14)$$

Complex dielectric constant and conductivity:

$$\left.\begin{array}{rcl}\varepsilon' &=& n_r^2 - \kappa^2 \\ \varepsilon'' &=& \dfrac{\sigma}{\varepsilon_0 \omega} = 2n_r\kappa\end{array}\right\}\quad \varepsilon = \varepsilon' + i\varepsilon''\qquad(10.21)$$

$$\varepsilon'(\omega) = \varepsilon_{\text{opt}} + \omega_p^2 \frac{\omega_{TO}^2 - \omega^2}{\left(\omega_{TO}^2 - \omega^2\right)^2 + \gamma^2\omega^2}\qquad(11.31)$$

$$\varepsilon''(\omega) = \omega_p^2 \frac{\omega\gamma}{(\omega_{TO}^2 - \omega^2)^2 + \gamma^2\omega^2}.\qquad(11.32)$$

Kramers-Kronig relations:

$$\varepsilon'(\omega_a) = 1 + \frac{2}{\pi}P\int_0^\infty \frac{\omega\varepsilon''(\omega)}{\omega^2 - \omega_a^2}d\omega\ ,\ \varepsilon''(\omega_a) = -\frac{2\omega_a}{\pi}P\int_0^\infty \frac{\varepsilon'(\omega)}{\omega^2 - \omega_a^2}d\omega.$$
$$(11.19),(11.20)$$

Line width:

$$Q = \frac{\Delta\omega}{\omega_0} \simeq \frac{1}{\tan\delta} = \frac{\varepsilon'}{\varepsilon''} = \frac{\varepsilon\varepsilon_0\omega_0}{\sigma} = \omega_0\tau_\sigma.\qquad(11.33)$$

Complex conductivity:

$$\left.\begin{array}{rcl}\sigma' &=& \varepsilon''\varepsilon_0\omega = \sigma_0 \\ \sigma'' &=& \varepsilon'\varepsilon_0\omega = \varepsilon_0\omega(n_r^2 - \kappa^2)\end{array}\right\}\quad \sigma = \sigma' + i\sigma''.\qquad(10.27)$$

Lyddane-Sachs-Teller relationship:

$$\frac{\omega_{\text{LO}}}{\omega_{\text{TO}}} = \sqrt{\frac{\varepsilon_{\text{st}}}{\varepsilon_{\text{opt}}}}.\qquad(11.38)$$

Debye-specific heat:

$$C_v = \begin{cases} 233.78R\left(\dfrac{T}{\Theta}\right)^3 & \text{for } T \ll \Theta \\[4mm] 3R\left[1 - 0.05\left(\dfrac{\Theta}{T}\right)^2\right] & \text{for } T \gtrsim 0.5\Theta\ . \end{cases}\qquad(6.12)$$

Maximum thermal expansion:

$$\left(\frac{\Delta l}{l}\right)_{T_m} \simeq 0.07; \quad \alpha \simeq \frac{0.07}{T_m}. \tag{6.17}$$

Melting point:

$$T_m = \frac{k}{9\hbar^2} M\Theta^2 f_{\max}^2 a^2. \tag{6.20}$$

Thermal conductivity:

$$\kappa = \frac{1}{3} C_v v_s \lambda. \tag{6.30}$$

A.6.5 Electron-Lattice Interaction

Coupling constant:

$$\alpha_c = \sqrt{\frac{\dfrac{e^2}{8\pi\varepsilon^*\varepsilon_0 r_p}}{\hbar\omega_{LO}}} = \frac{e^2\sqrt{m_n}}{4\sqrt{2\pi}\hbar\sqrt{k\Theta}}\frac{1}{\varepsilon_0\varepsilon^*} = 22.88\sqrt{\frac{m_n}{m_0}}\sqrt{\frac{300}{\Theta\,(\text{K})}}\frac{1}{\varepsilon^*}. \tag{28.6), (28.7}$$

Huang-Rhys factor:

$$S = \frac{E_r - E_c}{\hbar\omega_b} = \frac{\frac{1}{2}M_r\omega_b^2(Q_2 - Q_1)^2}{\hbar\omega_b}. \tag{20.12}$$

Effective dielectric constant, reduced mass:

$$\frac{1}{\varepsilon^*} = \frac{1}{\varepsilon_{opt}} - \frac{1}{\varepsilon_{st}}; \quad \frac{1}{M_r} = \frac{1}{M_1} + \frac{1}{M_2}. \tag{14.9), (11.41}$$

Effective polaron mass:

$$m_{pol} = \frac{m_n}{\left(1 - \dfrac{\alpha_c}{6}\right)}. \tag{28.10}$$

Large polaron energy and polaron radius:

$$E_{pol} = \alpha_c\hbar\omega_{LO} = E_{qH}\left(\frac{\varepsilon_{st}}{\varepsilon^*}\right)^2\frac{m_{pol}}{m_n}; \tag{28.8), (28.3}$$

$$r_{pol} = \frac{4\pi\varepsilon_0\hbar^2}{m_0 e^2}\varepsilon^*\frac{m_0}{m_{pol}} = a_{qH}\frac{\varepsilon^*}{\varepsilon_{st}}\frac{m_n}{m_{pol}} = \frac{4\pi\varepsilon^*\varepsilon_0\hbar^2}{m_{pol}e^2}. \tag{28.4}$$

Yukawa potential:

$$V(r) = \left(\frac{e}{\varepsilon^* r}\right)\exp\left(-\frac{r}{\lambda_s}\right). \tag{14.10}$$

Callen effective charge:

$$e_c = \omega_{TO} \sqrt{\varepsilon_0 \frac{\varepsilon_{st} - \varepsilon_{opt}}{\varepsilon_{opt}^2} M_r a^3}. \tag{11.39}$$

Szigetti effective charge:

$$e_s = e_c \frac{3\varepsilon_{opt}}{\varepsilon_{st} + 2}. \tag{11.42}$$

A.6.6 Optical Properties

Optical absorption coefficient:

$$\alpha_0 = \frac{2\omega\kappa}{c} = \frac{4\pi}{\lambda}\kappa. \tag{10.23}$$

Normal incidence reflectance:

$$R_0 = \frac{(n_r - 1)^2 + \kappa^2}{(n_r + 1)^2 + \kappa^2}. \tag{10.53}$$

Average transmissivity and reflectivity:

$$\bar{T} \simeq (1 - R_0)^2 \exp(-\alpha d); \tag{10.61}$$

$$\bar{R} = R_0 \left(1 + \bar{T} \exp\left(-\alpha_0 d\right)\right). \tag{10.60}$$

Direct absorption edge:

$$\begin{aligned} \alpha_{o,vc} &= \frac{2\sqrt{2}\pi^2 e^2 m_r^{3/2}}{m_0^2 \varepsilon_0 h^2 c} \frac{\sqrt{h\nu - E_g}}{n_r} M_{vc} \\ &= 7.64 \cdot 10^5 \left(\frac{m_r}{m_0}\right)^{3/2} M_{vc} \frac{\sqrt{h\nu - E_g \text{ (eV)}}}{n_r} \quad (\text{cm}^{-1}). \end{aligned} \tag{13.16}$$

Indirect absorption edge:

$$\alpha_{o,ind} \propto f_{\hbar\omega}(h\nu + \hbar\omega - E_g)^2 + (1 - f_{\hbar\omega})(h\nu - \hbar\omega - E_g)^2. \tag{13.28}$$

Forbidden edge transitions:

$$\alpha_{o,dir,forb} \propto (h\nu - E_g)^{3/2}; \tag{13.30}$$

$$\alpha_{o,ind,forb} \propto (h\nu \pm \hbar\omega - E_g)^3. \tag{13.31}$$

Urbach tail:

$$\alpha_o = \alpha_{o0} \exp\left(\frac{h\nu - E_g}{E_0}\right). \tag{13.36}$$

Line shape: Lorentzian;

$$g(\nu - \nu_0) = \frac{\gamma}{2\pi} \frac{1}{(\nu - \nu_0)^2 + \left(\dfrac{\gamma}{2}\right)^2}. \qquad (20.36)$$

Gaussian;

$$g(\nu - \nu_0) = \frac{1}{\sqrt{2\pi}\gamma} \exp\left[-\left(\frac{\nu - \nu_0}{2\gamma}\right)^2\right]. \qquad (20.38)$$

A.6.7 Quasi-Hydrogen and Deep Well States

Quasi-hydrogen radius:

$$a_{qH} = \frac{4\pi\varepsilon_0\hbar^2}{m_0 e^2}\varepsilon_{st}\frac{m_0}{m_n}n_q^2 = 0.529\,\varepsilon_{st}\frac{m_0}{m_n}n_q^2 \quad (\text{Å}). \qquad (15.3)$$

Quasi-hydrogen energy:

$$E_{qH}^{(n_q)} = \frac{m_0 e^4}{2(4\pi\varepsilon_0\hbar)^2}\frac{m_n}{m_0}\frac{1}{\varepsilon_{st}^2}\frac{1}{n_q^2} = 13.6\left(\frac{m_n}{m_0}\frac{1}{\varepsilon_{st}^2}\right). \quad (\text{eV}) \qquad (15.2)$$

Wannier-Mott excitons:

$$E_{exc} = E_g + \frac{\hbar^2 k^2}{2m_r} - \frac{m_r}{\varepsilon_r^2}R_H\frac{1}{n_q^2}. \qquad (15.8)$$

Superlattice eigenstates:

$$E_n = \frac{\hbar^2 k^2}{2m_n} = \frac{\hbar^2\pi^2}{2m_n l_z^2}n_q^2, \quad \text{with } n_q = 1, 2, \dots . \qquad (9.50)$$

A.6.8 Equilibrium Densities, Energies, Velocities

Density of states (for parabolic bands):

$$g_n(E)\,dE = \frac{1}{2\pi^2}\left(\frac{2m_n}{\hbar^2}\right)^{3/2}\sqrt{E - E_c}\,dE. \qquad (27.8)$$

Effective level density:

$$N_c = 2\left(\frac{m_{nds}kT}{2\pi\hbar^2}\right)^{3/2} = 2.5\cdot 10^{19}\left(\frac{m_{nds}}{m_0}\frac{T\,(\text{K})}{300}\right)^{3/2} \qquad (27.18),\ (27.19)$$

$$N_v = 2\left(\frac{m_{pds}kT}{2\pi\hbar^2}\right)^{3/2} = 2.5\cdot 10^{19}\left(\frac{m_{pds}}{m_0}\frac{T\,(\text{K})}{300}\right)^{3/2}. \qquad (27.29)$$

2D Desity of states:

$$N_{2c} = \frac{m^* kT}{\pi \hbar^2}. \tag{9.53}$$

Intrinsic Fermi level, carrier density:

$$E_F = \frac{E_c + E_v}{2} - kT \ln \left(\frac{m_n}{m_p} \right)^{3/4}; \tag{27.34}$$

$$n_i = \sqrt{N_c N_v} \exp \left(-\frac{E_c - E_v}{2kT} \right). \tag{27.36}$$

Fermi gas:

$$\begin{aligned} E_F &= 50.1\,(\text{eV})\,(r_s/a_0)^{-2}, \\ k_F &= 3.63\,(\overset{\circ}{\text{A}}{}^{-1})\,(r_s/a_0)^{-1}, \\ \sigma_F &= 4.20 \cdot 10^8\,(\text{cm/s})\,(r_s/a_0)^{-1}, \\ T_F &= 5.82 \cdot 10^5\,(\text{K})\,(r_s/a_0)^{-2}, \\ &\text{with } \frac{r_s}{a_0} = 5.44 \left(\frac{n\,(\text{cm}^{-3})}{10^{22}} \right)^{-1/3}. \end{aligned} \tag{A.157}$$

Density of Frenkel or Schottky defects in equilibrium:

$$N_F = \sqrt{N_L N_i} \exp \left(-\frac{E_{Fr}}{2kT} \right); \qquad N_S = N_L \exp \left(-\frac{E_S}{2kT} \right). \tag{19.12),\ (19.11}$$

A.6.9 Transport Properties

Interstitial diffusion in cubic lattices:

$$D = D_0 \exp \left(-\frac{\Delta E_i}{kT} \right) \qquad \text{with} \qquad D_0 = a^2 \frac{\nu_0}{6}. \tag{19.72}$$

Boltzmann equation for electrons:

$$\frac{df}{dt} = \frac{\partial f}{\partial t} + \dot{\mathbf{k}} \nabla_{\mathbf{k}} f + \dot{\mathbf{r}} \nabla_{\mathbf{r}} f = \left(\frac{\partial f}{\partial t} \right)_{\text{coll}}. \tag{30.1}$$

Field and Lorentz force:

$$\vec{\mathcal{F}} = \hbar \frac{d\mathbf{k}}{dt} = e\,(\mathbf{F} + \mathbf{v} \times \mathbf{B}). \tag{9.17}$$

Deformed Boltzmann distribution (low fields):

$$f = f_0 \left(1 - \frac{e}{kT} \tau_m \mathbf{F} \cdot \mathbf{v} \right). \tag{30.18}$$

Randomizing scattering:

$$\tau_m = \frac{\tau_{\rm sc}}{\langle 1 - \cos\theta \rangle}. \tag{30.47}$$

Energy and momentum relaxation:

$$\left(\frac{\tau_e}{\tau_m}\right)_{\rm ion} = \frac{m_n}{M_i} \quad \text{or} \quad \left(\frac{\tau_e}{\tau_m}\right)_{\rm ac,ph} = \frac{3v_s^2}{v_n^2}. \tag{30.52}, (30.54)$$

Total current:

$$j_n = e\mu_n n F + e D_n \frac{dn}{dx}; \quad j_p = e\mu_p p F - e D_p \frac{dp}{dx}; \quad j = j_n + j_p.$$
$$\tag{28.59}, (28.60), (28.61)$$

Carrier currents:

$$
\begin{aligned}
j_n &= \sigma_n \frac{\partial \varphi_n}{\partial x} & j_p &= \sigma_p \frac{\partial \varphi_p}{\partial x} \\
j_{n,\rm drift} &= \sigma_n \frac{\partial \psi_n}{\partial x} & j_{p,\rm drift} &= \sigma_p \frac{\partial \psi_p}{\partial x} \\
j_{n,\rm diff} &= \sigma_n \frac{\partial(\varphi_n - \psi_n)}{\partial x} & j_{p,\rm diff} &= \sigma_p \frac{\partial(\varphi_p - \psi_p)}{\partial x}
\end{aligned}
\tag{A.158}
$$

Carrier mobility:

$$\mu = \frac{e}{m_n}\tau_{\rm sc} = \frac{e}{m_n}\frac{\lambda}{v_{\rm rms}} = \mu = 1.5 \cdot \lambda \,(\text{Å}) \left(\frac{m_0}{m_n}\right)^{3/2} \left(\frac{300}{T(\text{K})}\right)^{1/2} (\text{cm}^2/\text{Vs}).$$
$$\tag{28.33}, (28.35)$$

Nernst-Einstein relation:

$$\nu_i = \frac{e}{kT} D_i. \tag{28.55}$$

Carrier mobility:

$$\boxed{\mu_n = \frac{e}{m_n}\tau_{\rm sc} = \frac{e}{m_n}\frac{\lambda}{v_{\rm rms}},} \tag{A.159}$$

$$= \mu_n = 1.5 \cdot \lambda \,(\text{Å}) \left(\frac{m_0}{m_n}\right)^{3/2} \left(\frac{300}{T(\text{K})}\right)^{1/2} (\text{cm}^2/\text{Vs}). \tag{A.160}$$

Current continuity:

$$\frac{dn}{dt} = g_n - r_n - \frac{1}{e}\frac{dj_n}{dx}; \quad \frac{dp}{dt} = g_p - r_p - \frac{1}{e}\frac{dj_p}{dx}. \tag{45.73}$$

Saturation drift current and velocity:

$$j_{n,\text{diff,max}} = env_{Ds} \quad \text{with} \quad v_{Ds} = \sqrt{\frac{3k\Theta}{4m^*} \tanh \frac{\Theta}{2T_L}}. \qquad (33.29)$$

Poisson equation:

$$\frac{dF_i}{dx} = \frac{\varrho}{\varepsilon\varepsilon_0}. \qquad (29.1)$$

Anderson-Mott localization; semiconductor-metal: transition

$$N_{\text{AM}}^{1/3} a_{\text{qH}} \simeq 0.2. \qquad (25.17)$$

Minimum metallic conductivity:

$$\sigma_{\text{min}} = \frac{1}{2\pi^2} \frac{e^2}{\hbar} \frac{1}{a} \simeq \frac{12}{a(\text{Å})} \quad (\Omega^{-1}\text{cm}^{-1}). \qquad (40.6)$$

Electric fields:

$$F = F_i + F_e. \qquad (29.2)$$

A.6.10 Characteristic Times and Frequencies

Reststrahl frequency:

$$\omega_{\text{TO}} \simeq \sqrt{\frac{\beta}{M_r}}. \qquad (11.45)$$

Plasma frequency:

$$\omega_p = \sqrt{\frac{ne^2}{\varepsilon_{\text{opt}}\varepsilon_0 m_n}}; \qquad (12.4)$$

$$\hbar\omega_p = 38.9 \sqrt{\frac{n}{10^{16}} \frac{10}{\varepsilon_{\text{opt}}} \frac{m_0}{m_n}} \quad (\text{meV}). \qquad (12.5)$$

Cyclotron frequency:

$$\omega_c = \frac{eB_0}{m^*} = 17.84 \frac{m_0}{m^*} B_0 \ (\text{GHz/kG}). \qquad (9.19)$$

Dielectric relaxation time:

$$\tau_\sigma = \frac{\varepsilon\varepsilon_o}{\sigma} = 8.8 \cdot 10^{-13} \frac{\varepsilon}{10} \frac{(\Omega\text{cm})}{\sigma} \quad (\text{s}). \qquad (47.58)$$

Decay time (τ_l) and carrier lifetime:

$$\tau = \tau_n \left(1 + \frac{dn_t}{dn}\right); \qquad (47.38)$$

$$\frac{\tau_l}{\tau_n} = 1 + \frac{N_t(E)kT}{n}. \qquad (47.49)$$

Rise and decay times of photoconductivity:

$$\tau_t = \frac{1}{c_{ct} N_t}; \qquad \tau_1 = \frac{1}{c_{cv} n}. \qquad (47.18),\ (47.9)$$

Time between scattering events:

$$\tau_{sc} = \frac{1}{v_{rms}\, s_n\, N_{sc}}. \qquad (28.32)$$

Thermal noise, shot noise:

$$\overline{j_{th}^2} = \frac{4kT}{R}\,\Delta\nu; \qquad \overline{j_{sh}^2} = 2ej\,\Delta\nu. \qquad (38.7),\ (38.10)$$

A.6.11 Photoconductivity

Recombination cross section:

$$s_n = \frac{\omega_r}{2\pi}\frac{1}{N_c v_{rms}}\exp\left(-\frac{E_n}{kT}\right). \qquad (43.13),\ (43.14)$$

Emission and capture rates:

$$\frac{e_{tc}}{c_{ct}} = \exp\left(-\frac{E_c - E_t}{kT}\right) \qquad (42.10)$$

$$e_{tc} = \frac{\nu_t}{N_c}\exp\left(-\frac{E_c - E_t}{kT}\right) \qquad (42.11)$$

$$c_{ct} = s_n v_{rms}. \qquad (42.12)$$

Carrier density for photoconductivity:

$$n = \sqrt{\frac{g_o}{c_{ca}}} \qquad \text{or} \qquad n = \frac{g_o}{c_{ca} N_t}. \qquad (45.14),\ (45.24)$$

Photoelectric gain:

$$G_o = \mu_n \tau_n \frac{V}{d^2} = \frac{\tau_n}{t_n} = \frac{l_n}{d}. \qquad (45.52),\ (45.53),\ (45.54)$$

Carrier kinetics:

$$\frac{dn}{dt} = g_n - r_n - \frac{1}{e}\frac{dj_n}{dx}; \qquad \frac{dp}{dt} = g_p - r_p - \frac{1}{e}\frac{dj_p}{dx}. \qquad (45.65)$$

Net carrier flow through Shockley-Read-Hall center:

$$U = \frac{c_{ct} c_{tv} N_t (np - n_i^2)}{c_{ct}\left[n + n_i \exp\left(\dfrac{E_t - E_i}{kT}\right)\right] + c_{tv}\left[p + n_i \exp\left(\dfrac{E_i - E_t}{kT}\right)\right]}. $$
$$(43.39)$$

Rise of photocarrier density:

$$n(t) = n_0 \left\{1 - \exp\left(-\frac{t - t_0}{\tau}\right)\right\} \qquad \text{with } \tau = \begin{cases} \tau_t = \dfrac{1}{c_{ct} N_t} \\ \tau_1 = \dfrac{1}{c_{cv} n} \end{cases} \quad (47.15)$$

Decay of photocarrier density:

$$n(t) = n_0 \exp\left(-\frac{t - t_1}{\tau_1}\right) \quad \text{or} \quad n(t) = n_0 \frac{\tau_1}{t - t_1}. \quad (47.23), (47.8)$$

Trap depth from glow (TSC) maximum:

$$E_c - E_t \simeq 25 k T_{\max}. \qquad (A.161)$$

A.6.12 Field Effects

Impact ionization distribution function:

$$f(E) = A^* \exp - \left(\frac{4\hbar\omega_{LO}}{(e\lambda)^2} \frac{E}{F^2}\right). \qquad (42.40)$$

Fowler-Nordheim tunneling:

$$j_n = \frac{e^3 m_n^{1/2} d}{2\sqrt{2}\pi^3 \hbar^2 E_g^{1/2}} F^2 \exp\left(-\frac{\pi m_n^{1/2} E_g^{3/2}}{2\sqrt{2} e \hbar F}\right) \qquad (42.80)$$

$$j_n = 1.011 \cdot 10^{-8} \left(\frac{F}{10^6}\right)^2 \left(\frac{m_n}{m_0}\right)^{1/2} \left(\frac{d}{100\text{Å}}\right) \frac{1}{\sqrt{E_g}} \exp\left(-\frac{F_0}{F}\right) \ (\text{A/cm}^2)$$
$$(42.81)$$

$$F_0 = 4.04 \cdot 10^7 \left(\frac{m_n}{m_0}\right)^{1/2} E_g^{3/2} \quad (\text{V/cm}). \qquad (42.82)$$

Franz-Keldysh shift:

$$\Delta E_{g,\text{opt}} = E_g - h\nu = \left(\frac{3}{4} e F \sqrt{\frac{\hbar^2}{2m}}\right)^{2/3} = 7.25 \cdot 10^{-6} [F(\text{V/cm})]^{2/3} \ (\text{eV}).$$
$$(A.162)$$

A.7 Symbols Used

A	wave amplitude vector
A	atom A
A	amplitude factor, area, Auger coefficient, Madelung constant acceptor, warped valence band factor
A_a	electron affinity of acceptor
A_E	asymmetry factor for band gap
$A_{\delta F}$	factor quantifying gap narrowing
A_o, A_{oct}	surface area of spherical (octahedral) atomic cavity
a	lattice vector in x direction $(a = |\mathbf{a}|)$, acceleration
a	lattice constant for cubic lattice, or in x direction
a_H	Bohr radius of hydrogen atom
a_{qH}	quasi-hydrogen radius
a_ℓ	distance between defects on dislocation line
a_t	distance between traps
a_T	heating rate
B, B_0	magnetic induction, value of
B	atom B
B	amplitude factor, module of compression, exponential tunneling factor, Auger coefficient
$\tilde{B}$, B'	warped valence band factors
$B_{\delta e}$	factor quantifying gap narrowing
B_z	Brillouin zone
b	lattice vector in y direction $(b = |\mathbf{b}|)$, Burgers vector
b	gap bowing parameter
b/a	ionicity of bonding [Eq. (2.8)]
C	atom C
C	base-centered Bravais lattice, elasticity module, Coulomb-binding energy
C_v, C_p	heat capacity for constant volume or pressure
$\tilde{C}$	warped valence band factor
$C_{\delta e}$	factor quantifying gap narrowing
c	lattice vector in z direction $(c = |\mathbf{c}|)$
c	light velocity
c_l	elastic constant for longitudinal deformation
c_a, c_c	proportionality constant
$c_c(k)$, $c_n(k)$	coeff. of expansion of eigenfunction of band c or n
c_{cv}	recombination coefficient
c_a	numerical proportionality factor
c_{ik}	capture coefficient for carrier from i^{th} to k^{th} level,

	adiabatic elastic stiffness constant
c_{ijk}	third order elastic constants
c_{ijkl}	tensor moduli of elastic stiffness
$\mathbf{D}$	electric displacement
D	donor, diffusion constant
D_o	optical deformation potential
D_n, D_p	diffusion coefficient
D_{ik}	diffusion tensor
$D_{\delta p}$	factor quantifying gap narrowing
d	distance between electrodes or dislocations, platelet thickness
d_c	critical thickness for dislocation formation
$\mathbf{E}$	electric field vector in electromagnetic radiation
E	energy, elasticity (Young's) modulus
E_0	Urbach param., unit field in electromag. radiation
E_F	Fermi energy
E_{Fn}, E_{Fp}	quasi-Fermi energy
E_c, E_v	lower conduction and upper valence band edge
E_{qH}	quasi-hydrogen ionization energy
E_g	band gap
$E_{g,exc}$	gap energy for exciton generation
E_d, E_a, E_t	energy of donor, acceptor, or trap
$\tilde{E}_d$, $\tilde{E}_a$	distance of donor or acceptor from E_c or E_v
E_{da}	donor-acceptor pair level energy
E_{dd}	energy of deep donors
E_{Dn}, E_{Dp}	demarcation line
E_R, E_r	relaxation energy, energy of recombination centers
E_a	activation energy
E_S, E_{Fr}	Schottky energy, Frenkel energy
E_n	specific subband, energy of nth level
$E_{\delta p}$	factor quantifying gap narrowing
E_i	ionization energy
E_m	blackbody emission
E_μ, E_ν	energy levels for a transition from ν to μ
$\mathcal{E}$	electric field of an electromagnetic wave
e	elementary charge, electron
e^*	static effective charge
$\mathbf{e}$, $\hat{\mathbf{e}}$	unit electric vector of electromagnetic radiation
e_c, e_s, e_t	Callen, Szigetti, or transverse effective charge
e_i	reduced index strain components (Voigt notation)
e_{ik}	excitation coefficient from i^{th} to k^{th} level,

	strain components
$\mathbf{F}$	electric field vector
F	field, Helmholtz free energy
F^*	average field from ionized impurities
F, F'	F-center, F'-center
$F_{t,l}$	transversal and longitudinal field
F_l	local field
F_i, F_e	built-in (internal) or external field
F_n	Fermi integral
$\tilde{F}$	Franz-Keldysh function
F_{op}	screening parameter
$\mathcal{F}$	force
f	distribution function (preferably Fermi-Dirac), fractional displacement $= \bar{u}/a$
f_0	distribution function in equilibrium
f_n, f_p	Fermi function for electrons or holes
f_i	oscillatory strength
$f_{\mathrm{FD}}, f_{\mathrm{BE}}$	Fermi-Dirac or Bose-Einstein distribution function
f_B	Boltzmann distribution function
G	Gibbs free energy, shear modulus, number of states
$\tilde{G}$	Franz-Keldysh function
G_o	photoelectric gain factor
g	density of states, line shape function, Landé g-factor
g, g_n, g_p	generation rate
g_o	optical generation rate
$\mathbf{H}$	Hamiltonian, magnetic field vector
H_s	surface energy
H	Helmholtz free energy (or enthalpy), hexagonal Bravais lattice
H_0	Hamiltonian for unperturbed bands
h	Planck's constant; $\hbar = h/2\pi$, Miller index
$h^{\bullet}$	hole
I	current, transfer integral, body-centered Bravais lattice
I_d	ionization energy of donor
I, I_L	optical intensity, luminescence intensity
I_n	normalization integral
$J, J_{\mu v}$	joint density of states (J_{cv})
j	current density
j_i	diffusion current
j_n, j_p	electron and hole current

$\mathbf{K}$	reciprocal lattice vector, exciton momentum vector
K	mass reaction law constant, electromechanical coupling constant
K_a	anisotropy factor for mobility
$K_{1,2}$	modified Bessel functions
$\mathbf{k}$	momentum (wave) vector (for electrons)
$k_x,\ k_y,\ k_z$	components of momentum vector (= wave number)
k_F	wave number at Fermi level
k	Boltzmann constant, Miller index
$\mathbf{L}$	angular momentum operator
L	sample length, Kane parameter
L_D	Debye length
L_c, L_n, L_p	diffusion length
L_d	drift length
$\hat{\mathbf{L}}$	angular momentum matrix
l	length, angular quantum number, Miller index
l_B	superlattice barrier width
l_i	box dimensions ($i = x, y, z$)
$l_{1,2}$	superlattice layer widths
l_z	well width (superlattices or quantum wells)
l_{coh}	coherence length
l_n	*Schubweg*, drift length
l_ω	fracton localization length
M	mass of atom (ion), Kane parameter
M_i	critical points ($M_0,\ M_1,\ M_2,\ M_3$)
$\mathbf{M}$	momentum matrix element ($\mathbf{M}_{ik}$ for transitions from state i to state k)
M_r	reduced ion mass
M_r^*	effective atomic weight
m	mirror axis, coordination number, Born potential exponent
$\overline{m}$	average coordination number
$\hat{m}$	integral number, shell number, order number, coordination number
$\hat{m}_x,\ \hat{m}_y,\ \hat{m}_z$	effective mass-type parameters
$m_n,\ m_p$	effective mass of electron or hole
$m_l,\ m_t$	longitudinal or transverse effective mass (el)
m_0	restmass of electron
m^*	effective mass of either electron or hole
$m_{\mu n},\ m_{\mu p}$	mobility effective mass
$m_{\text{nds}},\ m_{\text{pds}}$	density of state effective mass

m_r	reduced effective mass (exciton)
m_x, m_y, m_z	effective mass in x, y, z directions
m_j	spin quantum number
N	density of atoms or ions or states, valency, Kane parameter
$\mathcal{N}$	number of particles
N_{AM}	critical Anderson-Mott density
N_{Av}	Avogadro's number
N_c, N_v	density of states at cond. and valence band edge
N_{2c}	density of state in two-dimensional structures
N_I	density of ionized lattice defects
N_L	density of lattice atoms
N_S, N_F	density of Schottky or Frenkel defects
N_{sc}	density of scattering centers
N_d, N_a	donor or acceptor density
N_{eff}	effective density of donors (acceptors) $= N_d - N_a$
N_t, N_{tn}, N_{tp}	trap density
N_V	density of vacancies
n_1, n_2, n_3	integer number
n_x, n_y, n_z	integer numbers
n	electron density, principal qu. number, band index
$\hat{n}, \hat{n}_i$	integral number, number of interstitial distances
n_q	running number, quantum number
$\hat{n}_q$	principal quantum number
n_d, n_a	electrons in donors or acceptors
n_i	intrinsic carrier density
n_r	index of refraction (n_{ra} of an external medium)
n_t, n_{tn}, n_{tp}	electron density in traps
n_M	Mott density
n'	electron
$n_\uparrow, n_\downarrow$	spin occupation number (0,1)
$\mathbf{P}$	polarization
$\mathbf{P}_{rc}$	momentum matrix element
P_l	angular projection operator
P	primitive Bravais lattice, normal momentum coordinate, optical transition probability, Ehrenreich parameter
P_i	probability for ionization
$\mathbf{p}$	momentum vector
p	momentum ($= mv = \hbar k$), pressure, hole density
p_A	momentum component in direction of electric vector

p_d, p_a	holes in donors or acceptors
p_{phot}	photon momentum
Q, ΔQ	heat of formation, heat flux, Joule heat
Q	configuration coordinate, normal coordinate
$\mathbf{q}$	momentum (wave) vector for phonons
q	wave number for phonons
$\mathbf{R}$	radius vector of atoms or ions
R	resistance, $R = Nk$ gas constant, microcrystallite radius, rhomboedric Bravais lattice, reflectance
R_A	Anderson distance
R_i	coordinates of atoms
R_{ik}	transition rate of carriers from i^{th} to k^{th} level
R_o, R_0	optical reflectivity, reflectance for normal incidence
R_H	Rydberg, Hall constant
$\mathbf{r}$	radius vector, vector in real space
$\mathbf{r}_e$, $\mathbf{r}_h$	radius vector to an electron or hole
r	interatomic distance
r_e	equilibrium distance between atoms
r_{da}	donor-acceptor pair distance
r_i	coordinates of electrons, scattering radius for ions
r_{ij}	distance from i^{th} to j^{th} particle (carrier)
r_o, r_0	optical reflectivity, reflectance for normal incidence
r_r	radiative recombination rate
r_A, r_B	atomic (ionic) radii of atoms (ions) A or B
r_c	Lorentz circle radius
r_{pol}, r_{pl}, r_{ph}	polaron radius (for electrons and holes)
r_n, r_p	recombination rate
r_{qH}	quasi-hydrogen radius
r_{kT}	quasi Bohr radius for $E = kT$
r_s	scattering radius of screened Coulomb center
$\mathbf{S}$	Poynting vector
S	entropy, scattering probability, Huang-Rhys factor
S_c	screening length
S_E	equi-energy surface in momentum space
S_k, S_ω	surface in k space or q space (in energy space)
S_s	structure factor
S_x, S_y, S_z	spin component in x, y, z-direction
S	stress
s	occupation number
s_i	sign matrix (± 1)
s_{ik}	elastic compliance constants

s_n, s_p	capture cross section
s_r	recombination cross section
s_o	optical excitation cross section
s_∞	cross section in gas kinetic model
T	temperature, lattice temperature, transmittance
$\mathcal{T}$	tension, force
$\mathcal{T}_{ik}$	stress tensor
$\mathcal{T}_e$	tunneling probability
T_L	lattice temperature
T_e	electron temperature
T, T_o, T_0	transmissivity, transmittance for normal incidence
T_c	critical temperature for superconductivity
T_m	melting point
$\mathbf{t}$	translational symmetry operation
t	time
t_n, t_p	transit time
U	energy, thermal lattice energy, Hubbard correlation energy, potential energy in exciton Hamiltonian
U_0	matrix element
$\mathbf{u}$	displacement vector
u	amplitude factor of Bloch function, specific internal energy, anion displacem. param., atomic displacem.
u_n	displacement of the nth atom
V	applied voltage (bias), potential, potential energy, vacancy
V_{cc}	central cell potential
V_0	barrier height potential
$\mathcal{V}$	volume
$\mathcal{V}_u$	volume of unit cell
$\mathcal{V}_m$	molar volume
$\mathbf{v}$	velocity vector
v	velocity, light velocity in dielectric medium
v_e, v_n	average electron velocity with field, without field
v_{diff}	diffusion velocity
v_l, v_t	longitudinal or transverse acoustic sound velocity
$v_{\text{rms}}, v_{\text{rms}}^*$	rms velocity of carriers ($v_{\text{rms}}^* = v_{\text{rms}}/\sqrt{6\pi}$)
v_g	group velocity ($= \nabla_q \omega(\mathbf{q}) = \nabla_k \omega(\mathbf{k})$)
v_D	drift velocity
v_{Ds}	saturation drift velocity
v_s	sound velocity ($= \omega/q$)
$v_m, \hat{v}_m, v_m^*$	saturation velocities

W	characteristic energy for impact ionization, energy, Debye-Waller factor, statistical probability
$W_{\text{opt,ac}}$	deformation-potential interaction rates
W	heat (energy) current
X	exciton
$x = x_1$	x direction
$y = x_2$	y direction
Z	ionic charge
$Z_{\text{imp}}(r)$	impurity charge function
$z = x_3$	z direction
z	number of ways to realize a defect, coordination number
α	angle between **b** and **c**, fine struct. const., thermal expansion coeff., thermoelectric power, bond binding force const., geometry factor, anisotropy factor
α_α	van der Waals attraction constant
α_β	Wannier function
$\tilde{\alpha}$	impact ionization rate
α_o	optical absorption coefficient (cm^{-1})
α_c	Fröhlich's coupling constant
α_A, α_{L-R}	sound attenuation
α_r	van der Waals repulsion constant
α_S	Seraphin coefficient
$\alpha_\tau = \alpha_0/\tau_m$	normalized optical absorption per τ_m
α_{ik}	transport coefficients
α_v	thermal volume expansion coefficient
β	angle between **a** and **c**, force (spring) const., nonlinearity const., hole mobility ratio (μ_l/μ_h), Born repulsion param. (also β'), bond-bending force const.
β, β_n, β_p	exponent for field dependence of μ, μ_n, μ_p
$\tilde{\beta}, \tilde{\beta}_1$	expansion coeff. for warm/hot electron mobility
β_c	compressibility
β_S	Seraphin coefficient
β_E	temperature coefficient of E_g
β_{ik}	transport coefficients
γ	angle between **a** and **b**, damping factor, Lindemann parameter, anisotropy factor, adiabatic parameters, penetration depth for Meissner-Ochsenfeld eff., Born repulsion parameter
$\gamma_{1,2,3}$	Luttinger parameters
γ_A, γ_{L-R}	sound attenuation

γ_a	anisotropy factor for $m^*\varepsilon$
γ_o, γ_c	optical auxiliary function
γ_t	tunneling rate
γ_{ik}	transport coefficients
γ_{pc}	photochemical reaction rate
Γ	central point in Brillouin zone, Grüneisen parameter
Γ_G	Gaussian shape factor
Γ'	warped valence band factor
δ	Dirac symbol, fractional charge, small change, dielectric loss: $\tan\delta$, lattice anharmonicity parameter
δ_{ik}	Kronecker Symbol
δ_o	phase shift
δ_E, δ_n, δ_μ	second-order temperature coefficient
δ_i, δ_j	shift of demarcation line
δ_{pc}	photochemical dissociation rate
Δ	displacement, direction along (100), energy gain for Cooper pairs
Δ_0	spin-orbit splitting energy
$\Delta_{1/2}$	halfwidth of line
Δ_1	crystal field splitting
ΔE_B	band width
ΔE_c	collision-broadening band width
ΔE_d	activation energy for variable range hopping
ΔE_i	impurity band width
ΔE_t	thermal activation energy
ΔQ	formation enthalpy
ΔV_b	intergrain barrier potential
ε, $\tilde{\varepsilon}$	dielectric constant, complex dielectric constant
ε_{ik}	deformation coefficients
ε_0	vacuum dielectric constant
ε_L	dielectric constant due to lattice oscillation
ε^*	effective dielectric constant
ε_{st}, ε_{opt}	static and optical dielectric constant
ε', ε''	real and imaginary parts of ε
ε_{ik}	dielectric constant tensor
ε_r	reduced dielectric constant
$\varepsilon_{\parallel}$, $\varepsilon_{\perp}$	dielectric const. $\parallel$ or $\perp$ to optical axis (c-axis)
ϵ	relaxation parameter
Θ	Debye temperature ($= \hbar\omega_0/k$)
Θ_i	Debye temperatures for intervalley phonons
Θ_U	characteristic temperature for Umklapp processes

θ	angle
θ_H	Hall angle
κ	extinction coefficient, isothermal compressibility, thermal conductivity
K_A	Anderson factor
K_c	thermal conductivity for electrons
λ	wavelength, mean free path, trans. matrix element
λ_{ph}	mean free path for phonons
λ_{scr}	screening length
λ_D	Debye-Hückel screening length (Debye length)
λ_{TF}	Thomas-Fermi screening length
λ_{DB}	De Broglie wavelength
Λ	direction along (111)
Λ_δ	factor to quantify band-gap narrowing
μ	transition matrix element
μ, μ_0	magnetic susceptibility, permeability of free space
μ_n, μ_p	mobility for electrons or holes
μ_B	Bohr magneton
ν	frequency (in optical spectrum), Poisson ratio
ν_o	optical resonance frequency
ν_p, ν_t, ν_i	attempt-to-escape frequency
ν_s	scattering efficiency
$\nu_D, \nu_{Dd}, \nu_{Da}$	degeneracy factor for the band, donor or acceptor
Ξ_c, Ξ_d, Ξ_s	deformation potential: index d and s for dilatation and shear stress
ξ	alloying composition, correlation coefficient for V fluctuations, reduced energy $(= -(E_c - E_F)/(kT))$
π	empirical line-shape parameter, Peltier coefficient
Π	Peltier coefficient
ρ	density of the material (g/cm^3)
ϱ	space-charge density, specific resistivity
σ	symmetry factor, electric conductivity
$\tilde{\sigma}$	complex electric conductivity
$\underline{\sigma}$	Pauli operator matrices, conductivity matrix
σ_n, σ_p	conductivity (for electrons or holes)
σ_{ik}	magnetoconductivity tensor
σ', σ''	real and imaginary part of σ
σ_{min}	minimum metallic conductivity
Σ_i	surface charge
τ	characteristic time
τ_e	energy-relaxation time

τ_m	momentum-relaxation time
$\tau_{\rm sc}$	time between collisions
τ_i	intervalley scattering relaxation time
$\tau_{\rm sp}, \tau_{\rm st}$	spontaneous and stimulated recombination times
$\tau_r, \tau_{\rm nr}$	radiative and nonradiative recombination times
τ_E	dielectric relaxation time
τ_t	transit time
τ_c, τ_n, τ_p	lifetime for carriers, electrons, or holes
τ_o, τ_1	rise and decay time constant
φ	electrochem. potential, wavefunction, gen. function
φ_n, φ_p	quasi-Fermi potentials
ϕ	angle of light beam in respect to reference surface
ϕ, ϕ_λ	photon flux, spectral photon flux
$\phi, \phi_{nq}, \phi_{n q \mathbf{k}}$	envelope function
Φ	wavefunction
χ	susceptibility, electronegativity, dielectric susceptibility
$\chi^{(1)}, \chi^{(2)}, \chi^{(3)}$	susceptibility tensor of i_1 rank
ψ	wavefunction, electrostatic potential $(= -V)$
ψ_o	optical auxiliary function
ψ_n, ψ_p	electrostatic potential for electrons or holes
$\psi_{\rm MS}$	relative workfunction (metal/semiconductor)
Ψ_i	impurity wave function
$\Psi_{\rm MS}$	metal-semiconductor workfunction
ω	angular frequency $(= 2\pi\nu)$ for phonons
ω_o	resonance frequency
ω_a	evaluating frequency for Kramers-Kronig relation
ω_c	cyclotron resonance frequency
ω_D	Debye frequency
ω_i	atomic oscillation or intervalley phonon frequency
ω_p	plasma frequency
ω_r	breathing-mode frequency
ω_R	Raman frequency
$\omega_{\rm TO}, \omega_{\rm LO}$	transverse and longitudinal optical phonon
$\omega_{\rm TA}, \omega_{\rm LA}$	transverse and longitudinal acoustic phonons
Ω	solid angle
η	efficiency, tetragonal distortion
η_L	luminescence efficiency
η_t	trap availability factor
ϑ	Sasaki angle
χ_{ijk}	polarization tensor

SUBSCRIPTS

c	carrier (electron or hole)
d	lattice defect
dd	deep level defect
dr, diff	drift, diffusion
exc	exciton
e	equilibrium
f	foreign atom
i, e	internal, external
fri	frozen-in
ph	phonon
r	reduced
t	traps
$t, l; T, L$	transverse, longitudinal
th	threshold
n	band index
O	original, basic
o, O	optical
n, p	for electrons or holes
h, l	heavy, light; high, low
l, s	large, small
$i, 1,$ or x	in x direction
$k, 2,$ or y	in y direction
$l, 3,$ or z	in z direction
$\parallel, \perp$	parallel, perpendicular
st, opt	static, optical
ds	density of states
dir, ind	direct, indirect
forb	forbidden
mp	most probable
nq	quantum number
v, c	valence or conduction band
i	interstitial, intrinsic
d, a, t	donor, acceptor, traps
eff	effective
f, s	fast, slow recombination centers
L	luminescence
U	Umklapp process
B, W	barrier, well

perc, hop	percolation, hopping
pol	polaron
0	undisturbed (in thermal equilibrium)
q_H	quasi-hydrogen
β	band index
$\pm$	solutions according to sign of square root

SUPERSCRIPTS

$'$	derived variable
$'$	negative
$\bullet$	positive
$\times$	neutral with respect to the lattice
$*$	effective
$\frown$	used for distinction
(R), (I)	real part, imaginary part

SPECIAL SYMBOLS

$\mathbf{i}$, $\mathbf{j}$, $\mathbf{k}$	unit vectors in x, y, z directions
∇	∇, differential operator $\frac{\partial}{\partial x}\mathbf{i} + \frac{\partial}{\partial y}\mathbf{j} + \frac{\partial}{\partial z}\mathbf{k}$
Δ	difference
Σ	sum
Σ'	sum, except for equal indices
Π	product
$[A]$	density of atoms A
$\langle x \rangle$	average value of x
$\bar{\tau}$	average of τ
$\underline{\mu}$	matrix of μ_{ik}
δE	small increment of E
Re, Im	real part, imaginary part
P	principal value
i	$\sqrt{-1}$
θ, ϕ	angle in spherical coordinates
Ω	solid angle
$\underline{1}$	unity matrix
$\vec{\mathcal{F}}$	vector of $\mathcal{F}$
$\mathbf{F}$	vector of F

Bibliography

A word of caution: This list of references is an incomplete assembly* of four classes of publications:

- *historic papers to honor early development;*
- *original literature to assist in further search, which is neither exhaustive nor necessarily most recent;*
- *review papers to facilitate a broader overview, which are often selected for pedagogical reasons rather than for comprehensiveness; and*
- *books which have assisted me and my students in gaining deeper knowledge.*

An independent search for additional references is strongly encouraged.

---

* With apologies to my colleagues who are insufficiently listed.

Abarenkov, I. V. and V. Heine (1965), Philos. Mag. **12**, 529.

Abragam, A. and B. Bleaney (1976), *Electron Paramagnetic Resonance of Transition Ions*, Clarendon Press, Oxford.

Abram, R. A. and S. F. Edwards (1972), J. Phys. **C5**, 1183.

Abram, R. A., G. J. Rees, and B. L. H. Wilson (1978), Adv. Phys. **27**, 799.

Abrahams, E., P. W. Anderson, D. C. Licciardello, and T. V. Ramakrishnan (1979), Phys. Rev. Lett. **42**, 673.

Abramowitz, M. and I. A. Stegun (eds.) (1968), *Handbook of Mathematical Functions*, Dover Publishing Company, New York.

Abrikosov, N. Kh., V. F. Bankina, L. V. Poretskaya, L. E. Shelimova, and E. V. Skudnova (1969), *Semiconducting II-VI, IV-VI, and V-VI Compounds*, Plenum Press, New York.

Abstreiter, G. (1986), Festkörperprobleme **26**, 41.

Abstreiter, G., H. Brugger, T. Wolf, H. Jorke, and H. J. Herzog (1985), Phys. Rev. Lett. **54**, 2441.

Abstreiter, G., R. Merlin and A. Pinczuk (1986), IEEE J. Quant. Electr. **QE22**, 1717.

Ackerman, R. and R. Maynard (1985), Phys. Rev. **B32**, 7850.

Adams, W.G. and R.E. Day (1876), Proc. Roy. Soc. (London) **25**, 113.

Adilov, K. A., et al. (1986), Fiz. Tverd. Tela **28**, 1918.

Adler, D. (1980), in *Handbook on Semiconductors*, Vol. **1**: *Band Theory and Transport Properties*, T. S. Moss and W. Paul (eds.), North Holland Publ. Co., Amsterdam, p. 805.

Adler, D. (1983), Solar Cells **9**, 133.

Adler, D. (1985), in *Physical Properties of Amorphous Materials*, D. Adler, B.B. Schwartz, and M.C. Steele (eds.), Plenum Press, New York and London.

Adler, D. and E. A. Joffa (1977), Canada. J. Chem. **55**, 1920.

Adler, J. G. and J. E. Jackson (1966), Rev. Sci. Instrum. **37**, 1049.

Adler, J. G., T. T. Chen, and J. Strauss (1971), Rev. Sci. Instrum. **42**, 362.

Agarwal, V. K. (1988), Physics Today **41**, 40.

Aggarwal, R. L. and A. K. Ramdas (1965), Phys. Rev. **A137**, 602 and **A140**, 1246.

Aggarwal, R. L., M. D. Zubek, and B. Lax (1969), Phys. Rev. **180**, 800.

Agranovich, V. M. and V. I. Ginzburg (1984), *Crystal Optics with Spatial Dispersion and the Theory of Excitons*, Wiley, New York.

Aimerich, F. and F. Bassani (1968), Nuovo Cimento **56B**, 295.

Akhieser, A. (1939), J. Phys. (USSR) 1, 277.

Akkerman, D. A. and R. Maynard (1985), Phys. Rev. B32, 7850.

Akimoto, O. and E. Hanamura (1973), J. Phys. Soc. Jpn. 33, 1357.

Alberigi-Quaranta, A., C. Jacoboni, and G. Ottaviani (1971), Nuovo Cimento 1, 445.

Allen, P. B. and B. Mitrović (1982), *Solid State Physics*, Vol. 37, H. Ehrenreich, F. Seitz, and D. Turnbull (eds.), Academic Press, New York, p. 1.

Allender, D., J. Bray, and J. Bardeen (1973), Phys. Rev. 7, 1020 and 8, 4433.

Altarelli, M. and G. Iadonisi (1971), Nuovo Cimento B 5, 21.

Altarelli, M., G. Bachelet, and R. Del Sole (1979), J. Vac. Sci. Technol. 16, 1370.

Al'tshuler, B. L. and A. G. Aranov (1979), Sovj. Phys. – JEPT 50, 968.

Al'tshuler, B. L., A. G. Aranov and P. A. Lee (1980), Phys. Rev. Lett. 44, 1288.

Ambrozy, A. (1982), *Electronic Noise*, McGraw-Hill Book Co., Inc., New York.

Amer, N. (1987), Proc. SPIE 794, 11.

Ammerlaan, C. A. and G. D. Watkins (1972), Phys. Rev. B5, 3988.

Anderson, A. C. and J. P. Wolfe (eds.) (1986) in *Phonon Scattering in Condensed Matter*, Vol. V, Springer-Verlag, Berlin.

Anderson, C. L. and C. R. Crowell (1972), Phys. Rev B, 2267.

Anderson, J. C. (1970), Adv. Phys. 19, 311.

Anderson, O. K. (1975), Phys. Rev. B12, 3060.

Anderson, P. W. (1958), Phys. Rev. 109, 1492.

Anderson, P. W. (1963), *Concept in Solids*, New York.

Anderson, P. W. (1972), Proc. Nat'l Acad. Sci. (USA) 69, 1097.

Anderson, P. W. (1975), Phys. Rev. Lett. 34, 953.

Anderson, P. W. (1987), Science 235, 1196.

Anderson, P. W., G. Baskaran, Z. Zou, and T. Hsu (1987), Phys. Rev. Lett. 58, 2790.

Anderson, P. W., B. I. Halperin, and C. M. Varma (1972), Philos. Mag. 25, 1.

Anderson, R. L. (1962), Solid State Electron. 5, 341.

Ando, T., A. B. Fowler, and F. Stern (1982), Rev. Mod. Phys. 54, 437.

Andrews, S. R. and R. T. Harley (1987), Proc. SPIE 792, 141.

Antončik, E. (1955), Czech. J. Phys. 5, 449.

Antosiewicz, H. A. (1972), in *Handbook of Mathematical Functions*, M. Abramowicz and I. A. Stegun (eds.), Dover Publ. Co., New York, p. 435.

Aoyagi, Y., Y. Segawa, and S. Namba (1982), Phys. Rev. **B25**, 1453.

Appapillai, M. and V. Heine (1972), *Tech. Rep. 5*, Solid State Group, Cavendish Laboratory, Cambridge.

Appel, J. (1961), Phys. Rev. **122**, 1760.

Appel, J. (1966), Phys. Rev. Lett. **17**, 1045.

Appel, J. (1968), Solid State Phys. **21**, 193.

Apsley, N., E. A. Davis, A. P. Troup, and A. D. Yoffee (1977), *Proc. 7th Int. Conf. on Amorph. and Liquid Semicond.*, CICL, Edinburgh, p. 447.

Arakawa, Y. and A. Yariv (1986), IEEE J. Quantum Electron. **QE 22**, 1887.

Armistaid, W. H. and S. D. Stookey (1964), Science **144**, 150.

Armstrong, J. A., N. Bloembergen, J. Ducuing, and P. S. Pershan (1962), Phys. Rev. **127**, 1918.

Aronov, A. G., D. N. Mirlin, L. P. Nitikin, I. I. Reshine, and V. F. Sapega (1979), Sov. Phys.-JETP Lett. **29**, 62.

Aronov, A. G., G. E. Pikus, and A. N. Titov (1983), Zh. Eksp. Teor. Fiz. **83**, 152.

Arushanov, E. K. (1986), Prog. Crystal Growth and Charact. **13**, 1.

Arya, K. and W. Hanke (1981), Phys. Rev. **B23**, 2988.

Asada, M., Y. Migamoto, and Y. Suematsu (1986), IEEE J. Quantum Electron. **QE 22**, 1915.

Ascarelli, G. and S. Rodriguez (1961), Phys. Rev. **124**, 1321.

Ashcroft, N. W. and N. D. Mermin (1976), *Solid State Physics*, Holt Reinhart and Winston, New York.

Askary, F. and P. Y. Yu (1985), Phys. Rev. **B31**, 6643.

Aspnes, D. E. (1966), Phys. Rev. **147**, 554.

Aspnes, D. E. (1967), Phys. Rev. **153**, 995.

Aspnes, D. E. (1972), Phys. Rev. Lett. **28**, 168.

Aspnes, D. E. (1975), Phys. Rev. **B12**, 2297.

Aspnes, D. E. (1980), in *Handbook on Semiconductors*, Vol. **2**: *Optical Properties of Solids*, T. S. Moss and M. Balkanski (eds.), North Holland Publ. Co., Amsterdam.

Aspnes, D. E. and A. A. Studna (1973), Phys. Rev. **B7**, 4605.

Aspnes, D. E. and A. A. Studna (1975), Appl. Opt. **14**, 220.

Aspnes, D. E. and A. A. Studna (1983), Phys. Rev. **B27**, 985.

Austin, B. J., V. Heine, and L. Sharn (1962), Phys. Rev. **127**, 276.

Avakian, P. and R. E. Merrifield (1964), Phys. Rev. Lett. **13**, 541.

Bacewicz, R. and T. F. Ciszek (1988), Appl. Phys. Lett. **52**, 1150.

Bachelet, G. B., D. R. Hamann, and M. Schlüter (1982), Phys. Rev. **B26**, 4199.

Bakker, C. J. and G. Heller (1939), Physica **6**, 262.

Bagraev, N. T. and V. A. Mashkov (1986), *Defects in Semiconductors*, H. J. von Bardeleben (ed.), Trans. Tech. Publ., Switzerland.

Bajaj, K. K. (1968), Phys. Rev. **170**, 694.

Bajaj, K. K. (1987), Proc. SPIE **794**, 2.

Baldereschi, A. (1970), Phys. Rev. **B1**, 4673.

Baldereschi, A. and F. Bassani (1970), *Proc. 10th Int. Conf. Semicond.*, Boston, U.S. Atomic Energy Comm., p. 191.

Baldereschi, A. and J. J. Hopfield (1972), Phys. Rev. Lett. **28**, 171.

Baldereschi, A. and N. O. Lipari (1973), Phys. Rev. **B8**, 2697.

Baldereschi, A. and N. O. Lipari (1976), *Proc. 13th Int. Conf. Semicond.*, Rome, F. G. Fumi (ed.), Tipografia Marves, Rome, p. 595.

Balkanski, M. (1972), *Optical Properties of Solids*, F. Abeles (ed.), North Holland Publ. Co., Amsterdam.

Balkanski, M. (1980), in *Handbook on Semiconductors*, Vol. **2**: *Optical Properties of Solids*, T. S. Moss and M. Balkanski (eds.), North Holland Publ. Co., Amsterdam.

Balkanski, M. and I. Broser (1957), Z. Electrochem. **61**, 715.

Balkanski, M. and P. Lallemand (1973), *Photonics*, Gauthiers-Villard, Paris.

Balslev, I. (1978), Phys. Stat. Sol. (b) **88**, 155.

Balslev, I. (1988), in *Excitons in Confined Systems*, R. DelSole, A. D'Andrea and A. Lapiccirella (eds.), Springer-Verlag, Berlin.

Baraff, G. A. (1962), Phys. Rev. **128**, 2507.

Baraff, G. A. and M. Schlüter (1979), Festkörperprobleme **19**, 303.

Baraff, G. A. and M. Schlüter (1980), Phys. Rev. **B19**, 4965.

Baraff, G. A., E. O. Kane, and M. Schlüter (1980), Phys. Rev. **B21**, 3563, and **B22**, 5662.

Baranowski, J. M. (1979), unpublished (see M. Jaros (1980), Advances in Physics **29**, 409.

Baranskii, P. I., W. P. Klochkov, and I. V. Potykievich (1975), *Semiconductor Electronics*, Naukova Dumka, Kiev.

Barber, H. D. (1967), J. Opt. Soc. **57**, 574.

Barber, H. D. (1967), Solid State Electr. **10**, 1039.

Bardeen, J. (1938), J. Chem. Phys. **6**, 367.

Bardeen, J. and W. Shockley (1950), Phys. Rev. **80**, 72.

Bardeen, J., F. J. Blatt, and L. H. Hall (1956), *Proc. Photoconductivity Conference*, John Wiley & Sons, New York, p. 36.

Bardeen, J., L. N. Cooper, and J. R. Schrieffer (1957), Phys. Rev. **108**, 1175.

Bardeleben, H. J. (ed.) (1986), *Defects in Semiconductors*, Material Science Forum **10-12**.

Barker, A. S., Jr. and A. J. Sievers (1975), Rev. Mod. Phys. **47**, Suppl. 2.

Barker, J. R. (1980), in *Physics of Non-Linear Transport in Solids*, D. K. Ferry, J. R. Barker, and C. Jacoboni (eds.), Plenum Press, New York.

Barker, J. R. and D. K. Ferry (1980), Solid State Electr. Dev. **23**, 519.

Barrett, C. and T. B. Massalski (1980), *Structure of Metals*, Pergamon Press, New York.

Barron, T. H. K. (1957), Ann. Phys. (N. Y.) **1**, 77.

Bartram, R. H. and A. M. Stoneham (1975), Solid State Commun. **17**, 1593.

Bassani, G. F. (1966), *Proc. Int'l School of Physics, "Enrico Fermi,"* Academic Press, New York, **34**, 35.

Bassani, G. F. (1966), *Semiconductors and Semimetals*, Vol. **1**, R. K. Willardson and A. C. Beer (eds.), Academic Press, New York.

Bassani, G. F. (1975), *Electronic States and Optical Transitions in Solids*, Pergamon Press, New York.

Bassani, G. F. and V. Celli (1961), J. Phys. Chem. Solids **20**, 64.

Bassani, G. F. and E. S. Giuliano (1972), Nuovo Cimento **8B**, 193.

Bassani, G. F. and G. Pastori Parravicini (1975), *Electronic States and Optical Transitions in Solids*, Pergamon Press, Oxford and New York.

Bassani, G. F., J. J. Forney, and A. Quattropani (1977), Phys. Rev. Lett. **39**, 1070.

Bassani, G. F., G. Iadonisi, and B. Preziosi (1969), Phys. Rev. **186**, 735.

Bassani, G. F., G. Iadonisi, and B. Preziosi (1974), Rep. Prog. Phys. **37**, 1099.

Bastard, G. (1985), J. Lumin. **30**, 488.

Bastard, G. and J. A. Brum (1986), IEEE J. Quantum Electron. **QE 22**, 1625.

Bastard, G., E. E. Mendez, L. L. Chang, and L. Esaki (1983), Phys. Rev. **B28**, 3241.

Bate, R. F., R. D. Baxter, F. J. Reid and A. C. Beer (1965), J. Phys. Chem. Solids **26**, 1205.

Bates, C. A. (1978), Physics Reports, Sect. C of Phys. Lett. **35**, 188.

Bates, C. A. and K. W. H. Stevens (1986), Rep. Prog. Phys. **49**, 783.

Batra, I. P., C. Ciraci, and J. S. Nelson (1987), J. Vac. Sci. Technol. **B5**, 1300.

Bauer, G. (1978), Solid State Electron. **21**, 17.

Bauer, R. S., P. Zucher, and H. W. Sang, Jr. (1983), Appl. Phys. Lett. **43**, 663.

Baumeister, P. W. (1961), Phys. Rev. **121**, 359.

Baur, J., K. Maschke, and A Baldereschi (1983), Phys. Rev. **B27**, 3720.

Beattie, A. R. and P. T. Landsberg (1958), Proc. Roy. Soc. **A249**, 16.

Beaumont, J. F., A. L. Harmer, and W. Hayes (1972), J. Phys. C **5**, 257.

Becker, W. M. and H. Y. Fan (1964), *Proc. 7th Int. Conf. Semicond.*, Dunod, Paris.

Bednorz, J. G. and K. A. Müller (1986), Z. Phys. B **64**, 189.

Beeler, F., O. K. Anderson, and M. Scheffler (1985), Phys. Rev. Lett. **55**, 1498.

Beeler, F., O. K. Anderson, and M. Scheffler (1986), *Proc. Int. Conf. Phys. Semicond.*, O. Engström (ed.), World Scientific Publ., Stockholm, p. 875.

Beer, A. C. (1963), *Galvanomagnetic Effects in Semiconductors*, Academic Press, New York.

Beïnikhes, I. L. and Sh. M. Kogan (1985), Zh. Eksp. Teor. Fiz. **89**, 722.

Beleznay, F. and S. Pataki (1966), Phys. Rev. **B6**, 1367.

Bell, D. G. (1953), Rev. Mod. Phys. **26**, 311.

Bell, R. J. and P. Dean (1972), Philos. Mag. **25**, 1381.

Bendt, P. and A. Zunger (1982), Phys. Rev. **B26**, 3114.

Benoit à la Guillaume, C. and P. Lavallard (1972), Phys. Rev. **B5**, 4900.

Benoit à la Guillaume, C., F. Salvan, and M. Voos (1970), J. Lumin. **1/2**, 315.

Bergh, A. A. and P. J. Dean (1976), *Light Emitting Diodes*, Clarendon Press, Oxford.

Berggren, K. F. and B. E. Sernelius (1981), Phys. Rev. **B24**, 1971.

Berggren, K. F. and B. E. Sernelius (1984), Phys. Rev. **B29**, 5575.

Bergman, K., G. Grossman, H. G. Grimmeiss, M. Stavola, C. Holm, and P. Wagner (1986), *Proc. Int. Conf. Phys. Semicond.*, O. Engström (ed.), World Scientific Publ., Stockholm, p. 883.

Bergman, K., M. Stavola, S. J. Pearton, and J. Lopata (1988), Phys. Rev. **B37**, 2770.

Bergmann, L. (1931), Phys. Z. **32**, 286.

Bergmann, L. (1934), Phys. Z. **35**, 450.

Berman, R. (1976), *Thermal Conductivity in Solids*, Clarendon Press, Oxford.

Bernard, J. E., L. G. Ferreira, S.-H. Wei, and A. Zunger (1988), Phys. Rev. **B38**, 6338.

Bernard, J. E. and A. Zunger (1987), Phys. Rev. **B36**, 3199.

Bernard, J. E. and A. Zunger (1988), Phys. Rev. **B37**, 6835.

Bernholc, J. and S. T. Pantelides (1977), Phys. Rev. **B15**, 4935.

Bernholc, J. and S. T. Pantelides (1978), Phys. Rev. **B19**, 1780.

Bernholc, J., S. T. Pantelides, and N. O. Lipari (1978), Phys. Rev. Lett. **41**, 895.

Bersuker, I. B. (1984), *The Jahn-Teller Effect and Vibronic Interaction in Modern Chemistry*, Plenum Press, New York.

Besfamilnaya, V. A., V. F. Elesin, I. A. Kurova, N. Normont, and V. V. Ostroborodava (1967), Phys. Stat. Sol. **24**, 757.

Besson, J. M., E. H. Mokhtari, J. Gonzales, and G. Weill (1987), Phys. Rev. Lett. **59**, 473.

Bethe, H. A. (1935), Proc. Roy. Soc. **A216**, 45.

Beyer, W. H. (1981), *CRC Standard Mathematical Tables*, CRC Press, Boca Raton.

Bhandari, C. M. and D. M. Rowe (1985), High Temp.-High Pressures (GB) **17**, 97.

Bhattacharya, P. (1988), Semicond. Sci. Technol. **3**, 1145.

Bhattacharya, P. K. and S. Dhar (1988), *Semiconductors and Semimetals*, R. K. Willardson and A. C. Beer (eds.) **26**, 143, Academic Press, New York.

Bienenstock, A. (1985) in *Physical Properties of Amorphous Materials*, D. Adler, B.B. Schwartz, and M.C. Steele (eds.), Plenum Press, New York.

Bilz, H., R. Geik, and K. F. Renk (1963), in *Proc. Int. Conf. on Lattice Dynamics*, Copenhagen, R. F. Wallis (ed.), Pergamon Press, Oxford, p. 355.

Bilz, H. and W. Kress (1979), *Phonon Dispersion Relations in Insulators*, Springer-Verlag, Berlin.

Bir, G. L., A. G. Aronov, and G. E. Pikus (1975), Sov. Phys.-JETP **42**, 705.

Bir, G. L. and G. E. Pikus (1972), *Symmetry and Deformation Effects in Semiconductors*, Nauka, Moscow.

Bir, G. L. and G. E. Pikus (1974), *Symmetry and Strain Induced Effects in Semiconductors*, Louvish (ed.), John Wiley & Sons, New York.

Birman, J. L. (1974), in *Encyclopedia of Physics*, Vol. **XXV**, **12b**, Springer-Verlag, Berlin.

Birman, J. L. (1982) in *Excitons*, E. I. Rasha and M. D. Sturge (eds.), North Holland Publ. Co., Amsterdam.

Birman, J. L. and J. J. Sein (1972), Phys. Rev. **B6**, 2482.

Birss, R. R. (1964), *Symmetry and Magnetism*, North Holland Publ. Co., Amsterdam.

Bittel, H. (1959), Erg. exakt. Naturwiss. **31**, 84.

Blacha, A., H. Presting, and M. Cardona (1984), Phys. Stat. Sol. **B124**, 11.

Blagosklonskaya, L. E., E. M. Gershenzon, Yu. P. Ladyshinskii, and A.P. Popova (1970), Sovj. Phys.-Solid State **11**, 2402.

Blakemore, J. S. (1962), *Semiconductor Statistics*, Pergamon Press, Oxford.

Blakemore, J. S. (1967), Phys. Rev. **163**, 809.

Blakemore, J. S. (1982), J. Appl. Phys. **53**, R123.

Blakemore, J. S. (1982a), Solid State Electron. **25**, 1067.

Blatt, J. M. (1961), Phys. Rev. Lett. **7**, 82.

Blatt, J. M., K. W. Böer and W. Brandt (1962), Phys. Rev. **126**, 1691.

Bleil, C.E. and I. Broser (1963), J. Phys. Chem. Solids, **25**, 11.

Bloch, A. N., T. F. Carruthers, T. O. Poehler, and D. O. Cowan (1977) in *Chemistry and Physics in One-Dimensional Metals*, H. J. Keller (ed.), Plenum Press, New York.

Bloch, F. (1928), Z. Phys. **52**, 555.

Blochinzew, D.J. (1953) *Grundlagender Quantenmechanik*, Deutscher Verlag der Wiss., Berlin.

Blodgett, K. B. (1935), J. Am. Chem. Soc. **57**, 1007.

Bloembergen, N. (1965), *Nonlinear Optics*, W. A. Benjamin, Inc., London.

Bloembergen, N. (1982), Rev. Mod. Phys. **54**, 685.

Blossey, D. F. (1970), Phys. Rev. **B2**, 507.

Blossey, D. F. (1970), Phys. Rev. **B2**, 3976.

Bodnar, I. V. and N. S. Orlova (1985), Phys. Status Solidi **91**, 503.

Böer, K. W. (1954), Ann. Phys. (Lpz.) **14**, 87.

Böer, K. M. (1960), *Proc. Internat'l Conf. Semicond. Phys.*, Prague, p. 828.

Böer, K. W. (1965), Phys. Rev. **139**, 1949.

Böer, K. W. (1969), Z. Naturforsch. **24a**, 1306.

Böer, K. W. (1971), *Proc. 3rd Int. Conf. on Photoconductivity*, E. M. Pell (ed.), Pergamon Press, New York.

Böer, K. W. (1972), J. Non-Cryst. Solids **8-10**, 586.

Böer, K. W. (1977), Phys. Stat. Sol. (a) **40**, 355.

Böer, K. W. (1983), J. Physics D **16**, 2379.

Böer, K. W. (1985), Ann. Phys. (Lpz.) **7**;**42**, 371.

Böer, K.W. (1985a), Phys. Stat. Sol. A **87**, 719.

Böer, K. W. and W. Borchardt (1953), Fortschr. Phys. **1**, 184.

Böer, K. W. and K. Bogus (1968), Phys. Rev. **176**, 899.

Böer, K. W. and R. Boyn (1959), Monatsber. D. Akad. Wiss. (Berlin) **1**, 396.

Böer, K. W. and K. Junge (1953), Z. Naturf. **8a**, 753.

Böer, K. W. and U. Kümmel (1954), Ann. Phys. (Lpz.) **14**, 341.

Böer, K. W. and U. Kümmel (1957), Ann. Phys. Lpz. **20**, 303, Natf. **12a**, 667.

Böer, K. W. and U. Kümmel (1960), Z. Phys. Chem. **214**, 127.

Böer, K. W. and S. Oberländer (1958), Z. Naturforsch. **13a**, 351.

Böer, K. W., S. Oberländer, and J. Voigt (1958), Ann. Phys. (Lpz.) (7) **2**, 130.

Böer, K. W. and H. Vogel (1955), Ann. Phys. (Lpz.) **17**, 10.

Böer, K. W. and P. Voss (1968), Phys. Rev. **171**, 899.

Böer, K. W., W. Borchardt, and E. Borchardt (1954), Z. Phys. Chem. (Lpz.) **203**, 145.

Böer, K. W., H. J. Hänsch, and U. Kümmel (1959), Z. Phys. **155**, 170.

Böer, K. W., J. C. O'Connell, and R. Schubert (1966), *Luminesc. Symposium*, K. Thiemig Verlag, Munich, p. 223.

Bogoliubov, N. N. and N. N. Bogoliubov, Jr. (1986), *Aspects of Polaron Theory*, World Sci., Singapore.

Boguslawski, P. (1975), Phys. Stat. Sol. B **70**, 54.

Bohm, D., Huang and D. Pines (1957), Phys. Rev. **107**, 71.

Böhm, M. and A. Scharmann (1969), Z. Phys. **225**, 44.

Bohnert, K., M. Amselment, G. Kobbe, C. Klingshirn, H. Haug, S. W. Koch, S. Schmitt-Rink, and F. F. Abraham (1981), Z. Phys. **B42**, 1.

Boiko, I. I. (1959), Sovj. Phys.-Solid State, **1**, 518.

Bonch-Bruevich, V. L. (1962), Sov. Phys. Solid State 4, 1953.

Bonch-Bruevich, V. L. and Sh. M. Kogan (1959), Sov. Phys.-Solid State **1**, 1118.

Boolchand, P. (1985), in *Physical Properties of Amorphous Materials*, D. Adler, B. B. Schwartz, and M. C. Steele (eds.), Plenum Press, New York.

Borchardt, W. (1962), Phys. Stat. Sol. **2**, 1575.

Born, M. (1919), Verh. Dtsch. Phys. Ges. **21**, 679.

Born, M. and Th. von Kármán (1912), Phys. Z. **13**, 297.

Born, M. and K. Huang (1954), *Dynamical Theory of Crystal Lattices*, Oxford University Press, London.

Born, M. and A. Landé (1918), Verh. Dtsch. Phys. Ges. **20**, 210.

Born, M. and I. R. Oppenheimer (1927), Ann. Phys. **84**, 457.

Bosman, G. (1981), Ph.D Thesis, University of Utrecht, The Netherlands.

Boswarva, I. M. (1970), Phys. Rev. **B1**, 1698.

Böttger, H. and V. V. Bryksin (1979), Phys. Status Solidi B **96**, 219.

Böttger, H. and V. V. Bryksin (1980), Philos. Mag. B **42**, 297.

Böttger, H. and V. V. Bryksin (1985), *Hopping Conduction in Solids*, VCH Verlagsgesellschaft/Akademic Verlag, Berlin.

Bouckaert, L. P., R. Smoluchowski, and E. Wigner (1936), Phys. Rev. **50**, 58.

Bourgoin, J. and J. W. Corbett (1972), Phys. Lett. **38A**, 135.

Bourgoin, J. and M. Lannoo (1983), *Point Defects in Semiconductors: Experimental Aspects*, Springer-Verlag, Berlin.

Bowers, R. and N. T. Melamed (1955), Phys. Rev. **99**, 1781.

Boyn, R. (1977), in *Proc. Autumn School Modul'n. Spectroscopy*, R. Enderlein (ed.), Zentralinstitut für Elektronenphysik, Berlin, p. 110.

Bradford, J. N., R. T. Williams, and W. L. Faust (1975), Phys. Rev. Lett. **35**, 300.

Bragagnolo, J. A. and K. W. Böer (1970), J. Luminesc. **1/2**, 572.

Bragg, W. L. (1940), Proc. Phys. Soc. (London), **52**, 54.

Brandt, N. B. and V. V. Moshchalkov (1984), Adv. Phys. **33**, 193.

Branis, S. V., K. Arya, and J. L. Birman (1987), Proc. SPIE **793**, 132.

Branis, S. V., K. Arya, and J. L. Birman (1989), Phys. Rev. **B39**, 8371.

Bratt, P. R. (1977), in *Semiconductors and Semimetals*, R. K. Willardson and A. C. Beer (eds.), Academic Press, New York.

Braun, C. L (1980), Organic Semiconductors in *Handbook on Semiconductors*, Vol. **3**: *Materials Properties and Preparation*, T. S. Moss and S. P. Keller (eds.), North Holland Publ. Co., Amsterdam.

Bräunlich, P. (1979), *Thermally Stimulated Relaxation in Solids*, Springer-Verlag, Berlin.

Bray, R. (1969), IBM J. Res. Dev. **13**, 487.

Bray, R. and D. M. Brown (1960), in *Proc. Int. Conf. Semic. Phys.*, Prague, Czech. Acad. Sci., Prague.

Brebner, J. L., E. Mooser, and M. Schlüter (1973), Nuovo Cimento B **18**, 164.

Bredikhin, S. I. and S. Z. Shmurak (1979), Sov. Phys.-JETP **49**, 520.

Brenig, W., R. Zeyher and J. L. Birman (1972), Phys. Rev. **B6**, 4617.

Brennan, K. F. and C. J. Summers (1987a), J. Appl. Phys. **61**, 614.

Brennan, K. F. and C. J. Summers (1987b), IEEE J. Quantum Electron. **QE23**, 320.

Brice, J.C. (1986) *Crystal Growth Processes*, Halstead Press, New York.

Brill, R., C. Hermann and C. Peters (1942), Ann. Phys. (Lpz) **41**, 37.

Brillouin, L. (1922), Ann. Phys. **17**, 88.

Brillson, N. J. (1978), Phys. Rev. Lett. **40**, 260.

Brinkman, W. F. and T. M. Rice (1973), Phys. Rev. **B7**, 1508.

Broadbent, S. R., and J. M. Hammersley (1957), *Proc. Cambridge Phil. Soc.* **53**, 629.

Brockhouse, B. N. (1959), Phys. Rev. Lett. **2**, 256.

Brockhouse, B. N., S. Houtecler and H. Stiller (1964), *Interaction of Radiation with Solids*, North Holland Publ. Co., Amsterdam.

Brockhouse, B. N. and P. K. Iyenger (1958), Phys. Rev. **111**, 747.

Brodsky, M. (1980), Solid State Commun. **36**, 55.

Bromberg, J. L. (1988) Physics Today **41:10**, 26.

Bromme, T. (1851), Atlas der Physik der Welt, Stuttgart.

Brooks, H. (1951), Phys. Rev. **83**, 879.

Brooks, H. (1955), Adv. Electr. Electron. Physics **7**, 85.

Brooks, H. (1963), Trans. Met. Soc. AIME **227**, 551.

Broser, I., R. Broser, E. Beckmann, and E. Birkicht (1981), Solid State Commun. **39**, 1209.

Broser, I. and R. Broser-Warminski (1956), J. Phys. Radium **17**, 791.

Broser, I., H. Oeser, and R. Warminski (1950), Z. Naturforsch. **5a**, 214.

Broser, I. and M. Rosenzweig (1979), Phys. Stat. Sol. B **95**, 141.

Broser, I. and M. Rosenzweig (1980), Solid State Commun. **36**, 1027.

Broser, I., M. Rosenzweig, R. Broser, M. Richard, and E. Birkicht (1978), Phys. Stat. Sol. B **90**, 77.

Brouwer, G. (1954), Philips Res. Repts. **9**, 366.

Brown, F. C. (1962), J. Phys. Chem. **66**, 2368.

Brown, F. C. (1967), *The Physics of Solids*, W. A. Benjamin, Inc., New York.

Brown, F. C. (1976), *Treatise in Solid State Chemistry*, Vol. 4, *Reactivity of Solids*, B. F. Hanum (ed.), Plenum Press, New York.

Brown, P. J. and J. B. Forsyth (1973), *The Crystal Structure of Solids*, Edward Arnold, London.

Brown, T. L. and P. A. Wolff (1972), Phys. Rev. Lett. **29**, 363.

Brueck, S. R. J. and A. Mooradian (1973), Opt. Commun. **8**, 263.

Brueckner, K. A. (1955), Phys. Rev. **100**, 36.

Brugger, K. (1964), Phys. Rev. A **133**, 1611.

Brum, J. A. and G. Bastard (1985), Phys. Rev. **B31**, 3893.

Brunetti, R., C. Jacoboni, L. Reggiani, G. Bosman, and R. J. J. Zijlstra (1981), J. Appl. Phys. **52**, 6713.

Brunetti, R., and C. Jacoboni (1984), *Semiconductors Probed by Ultrafast Laser Spectroscopy*, Vol. 1, Academic Press, New York.

Brus, L. (1986), IEEE J. Quantum Electron. **QE 22**, 1909.

Brust, D. (1972), Phys. Rev. **B5**, 435.

Brust, D., J. C. Phillips, and F. Bassani (1962), Phys. Rev. Lett. **9**, 94.

Bube, R. H. (1957), J. Phys. Chem. Solids **1**, 234.

Bube, R. H. (1960), *Photoconductivity of Solids*, John Wiley & Sons, New York.

Bube, R. H. (1970), *Phyical Chemistry*, H. Eyring, D. Henderson, and W. Jost (eds.), **10**: Solid State, p. 515.

Bube, R. H. (1974), *Electronic Properties of Crystalline Solids*, Academic Press, New York, London.

Bube, R. H. (1978), *Photoconductivity of Solids*, Krieger, Huntington, New York.

Bube, R. H. (1988), *Electrons in Solids*, Academic Press, New York.

Bube, R. H. and C. T. Ho (1966), J. Appl. Phys. **37**, 4132.

Buckley, H.E. (1951) *Crystal Growth*, Wiley, New York.

Budd, H. (1966), J. Phys. Soc. Jpn. Suppl. **21**, 420.

Buerger, M. J. (1956), *Elementary Crystallography*, John Wiley & Sons, New York.

Bührer, W. and Z. Iqbal (1984), in *Vibrational Spectroscopy of Phase Transitions*, Z. Igbal and F. J. Owens (eds.), Academic Press, New York.

Bullett, D. W. and M. J. Kelly (1975), Solid State Commun. **12**, 1379.

Bullot, J. and M. P. Schmidt (1987), Phys. Stat. Sol. B **143**, 345.

Burger, W. and K. Laßmann (1986), Phys. Rev. **B33**, 5868.

Burgers, J. M. (1939), *Proc. Kon. Ned. Akad. V. Wet. Amst.* **42**, 293, 315, and 378.

Burgers, J. M. (1940), Proc. Phys. Soc. (London) **52**, 23.

Burnham, R. D., C. Lindstrom, T. L. Padi, D. R. Scifres, W. Streifer, N. Holonyak, Jr. (1983), Appl. Phys. Lett. **42**, 937.

Burstein, E., C. Y. Chen, and S. Lundquist (1979), in *Light Scattering in Solids*, J. Birman, et al. (eds.), Plenum Press, New York, p. 479.

Butcher, P. N. (1972), J. Phys. C (London) **5**, 817.

Byer, R. L. and R. L. Herbst (1977), Topics in Applied Physics **16**, 81.

Bylander, D. M. and L. Kleinman (1987), Phys. Rev. **B36**, 3229.

Cahn, J. W., D. Shechtman, and D. Gratias (1986), J. Mater. Res. **1**, 13.

Cai, X., R. Joynt, and D. C. Larbalestier (1987), Phys. Rev. Lett. **58**, 2798.

Caldas, M., A. Fazzio and A. Zunger (1985), J. Electron. Material **14A**, 1035.

Calecki, D., J. F. Palmier, and A. Chomett (1984), J. Phys. C. (London) **17**, 5017.

Callaway, J. (1963), J. Phys. Chem. Solids **24**, 1163.

Callaway, J. (1976), *Quantum Theory of Solid State*, Academic Press, New York.

Callen, H. B. (1949), Phys. Rev. **76**, 1394.

Canali, C., G. Majni, R. Minder, and G. Ottaviani (1975), IEEE Trans. Electron Devices **ED 22**, 1045.

Cantow, H. J. et al. (ed.) (1981), Advances in Polym. Sci., Vol. **40**, *Luminescence*, Springer-Verlag, Berlin.

Capasso, F. (1987), Proc. SPIE **792**, 10.

Capasso, F., M. Khalid, and A. Y. Cho (1985), IEEE J. Quantum Electron. **QE 22**, 1853.

Car, R., P. J. Kelly, A. Oshiyama, and S. T. Pantelides (1984), Phys. Rev. Lett. **52**, 1814.

Car, R., P. J. Kelly, A. Oshiyama, and S. T. Pantelides (1985), Phys. Rev. Lett. **54**, 360.

Car, R. and A. Selloni (1978), *Proc. 14th Int. Conf. Phys. Semicond.*, Edinburgh, Wilson (ed.), The Inst. of Physics, Bristol.

Cardona, M. (1969), in *Solid State Physics*, Suppl. **11**, F. Seitz, D. Turnbull, and H. Ehrenreich (eds.), Academic Press, New York.

Cardona, M. (1972), *Proc. Enrico Fermi School of Physics*, Course 52, p. 514.

Cardona, M. and F. H. Pollak (1966), Phys. Rev. **142**, 530.

Carles, R. N., N. Saint-Cricq, M. A. Renucci and B. J. Bennucci (1977), *Lattice Dynamics*, M. Balkanski (ed.), Flammarion, Paris.

Carlson, A. E., A. Zunger, and D. M. Wood (1985), Phys. Rev. **B32**, 1386.

Carlson, D. E. and C. R. Wronski (1976), Appl. Phys. Lett. **28**, 671.

Carlson, R. O. (1957), Phys. Rev. **108**, 1390.

Carter, A. C., M. S. Skolnick, R. A. Stradling, J. P. Leotin, and S. Askenazi (1976), *Proc. 13th Intern'l Conf. Phys. Semicond*; Rome, F. G. Fumi (ed.), North Holland, Amsterdam, p. 619.

Casey, H. C., Jr. and G. L. Pearson (1975), in *Point Defects in Solids*, J. H. Crawford, and L. M. Stifkin (eds.), Plenum Press, New York.

Casey, H. C., Jr., D. D. Sell, and K. W. Wecht (1975), 'J. Appl. Phys. **46**, 250.

Casey, H. C., Jr. and F. Stern (1976), J. Appl. Phys. **47**, 631.

Casimir, H. B. G. (1938), Physica **5**, 495.

Castner, T. G., Jr., (1970), Phys. Rev. **B2**, 4911.

Castner, T. G., Jr., W. Känzig, and T. O. Woodruff (1958), Nuovo Cimento, Suppl. to VII, Series X, p. 612.

Cavenett, B. C. (1980), J. Phys. Soc. Jpn.n **49A**, 611.

Cavenett, B. C. (1981), Adv. Phys. **30**, 475.

Cavenett, B. C. (1984), Journ. Luminesc. **31/32**, 369.

Cavenett, B. C. and E. J. Pakulis (1985), Phys. Rev. B **32**, 8449.

Celler, G. K., S. Mishra, and R. Bray (1975), Appl. Phys. Lett. **27**, 297.

Chakraverty, B. K. and C. Schlenker (1976), J. Physique (Paris), Collog. **37**, 353.

Challis, L. J., V. W. Rampton, and A. F. G. Watt (1975), *Phonon Scattering in Solids*, Plenum Press, New York.

Chambers, R. G. (1952), Proc. Phys. Soc. (London) Ser. **A65**, 458.

Chandrasekha, R. (1943), Rev. Mod. Phys. **25**, 1.

Chang, A. M., P. Berglund, D. C. Tsui, H. L. Störmer, and J. C. M. Hwang (1984), Phys. Rev. Lett. **53**, 997.

Chang, I. F. and S. S. Mitra (1968), Phys. Rev. **172**, 924.

Chang, K. J., M. M. Dacorogna, M. L. Cohen, J. M. Mignot, G. Chouteau, and G. Martinez (1985), Phys. Rev. Lett. **54**, 2375.

Chang, L. L., L. Esaki, W. E. Howard, R. Ludeke, and G. Schul (1973), J. Vac. Sci. Technol. **10**, 11, and 655.

Chantre, A., A. Vincent, and D. Bois (1981), Phys. Rev. **B23**, 5335.

Charfi, F., M. Zuoaghi, C. Llinares, M. Balkanski, C. Hirlimann and A. Joullie (1977), *Lattice Dynamics*, M. Balkanski (ed.), Flammarion, Paris.

Chattopadhyay, D. and H. J. Queisser (1981), Rev. Mod. Phys. **53**, 745.

Chaudhari, P., R. H. Koch, R. B. Laibowitz, T. R. McGuire, and R. J. Gambino (1987), Phys. Rev, Lett. **58**, 2684.

Chelikowsky, J. R., D. J. Chadi, and M. L. Cohen (1973), Phys. Rev. **B8**, 2786.

Chelikowsky, J. R. and M. L. Cohen (1976), Phys. Rev. **B14**, 100, 139, and 556.

Chelikowsky, J. R. and J. C. Phillips (1978), Phys. Rev. **B17**, 2453.

Chemla, D. S. and J. Jerphagnon (1980), *Handbook on Semiconductors*, Vol. **2**: *Optical Properties of Solids*, T. S. Moss and M. Balkanski (eds.), North Holland Publ. Co., Amsterdam.

Chen, G. and W. A. Goddard (1988), Science **239**, 899.

Chen, J.-W. and A. G. Milnes (1980), Ann. Rev. Mat. Sci. **10**, 157.

Chernov, A. A. (1984), *Crystal Growth*, Springer-Verlag, Berlin.

Chiao, R. Y., C. H. Townes, and B. P. Stoicheff (1964), Phys. Rev. Lett. **12**, 592.

Chiarotti, G. and V. M. Grassano (1966), Nuovo Cim. **46B**, 78.

Childs, G. E., L. J. Erick, and R. L. Powell (1973), *Thermal Conductivity of Solids at Room Temperature and Below*, a review and compilation of literature, Nat. Bur. of Stand. monograph **131**, Washington, D. C.

Ching, W. Y. and D. L. Huber (1982), Phys. Rev. **B25**, 1096.

Chirgwin, B. H. (1957), Phys. Rev. **107**, 1013.

Cho, A. Y. (1971), Appl. Phys. Lett. **19**, 467.

Cho, K. (1973), Opt. Commun. **8**, 412.

Cho, K. (1979) in *Topics in Current Physics: Excitons*, K. Cho (ed.), Springer-Verlag, Berlin, p. 15.

Choy, M. M. and R. L. Byer (1976), Phys. Rev. **B14**, 1693.

Christov, S. G. (1982), Phys. Rev. **B26**, 6918.

Clerjaud, B. (1985), J. Phys. C (London) **18**, 3615.

Cochran, W. (1973), *The Dynamics of Atoms in Crystals*, Edward Arnold, London.

Cody, G. D., T. Tiedje, B. Abeles, B. Brooks, and Y. Goldstein (1981), Phys. Rev. Lett. **47**, 1480.

Cohen, E. and M. D. Sturge (1977), Phys. Rev. **B15**, 1039.

Cohen, M. H. (1963), Phys. Rev. **130**, 1301.

Cohen, M. H. (1970a), J. Non-Cryst. Solids **2**, 432.

Cohen, M. H. (1970b), J. Non-Cryst. Solids **4**, 391.

Cohen, M. H. and T. K. Bergstresser (1966), Phys. Rev. **141**, 789.

Cohen, M. H., E. N. Elonoman, and C. M. Soukoulis (1983), J. Non-Cryst. Solids **59&60**, 15.

Cohen, M. H., H. Fritzsche and S. R. Ovshinsky (1969), Phys. Rev. Lett. **22**, 1065.

Cohen, M. H. and V. Heine (1970), *Solid State Physics* **24**, H. Ehrenreich, F. Seitz and D. Turnbull (eds.), Academic Press, New York.

Cohen, M. L. (1962), Phys. Rev. **128**, 131.

Cohen, M. L. (1964), Phys. Rev. **134**, A511.

Cohen, M. L. (1984), Ann. Rev. Mater. Sci. **14**, 119.

Collins, A. T. (1981), J. Phys. C, Sol. St. Phys. **14**, 289.

Collins, A. T., E. C. Lightowlers and P. J. Dean (1968), Phys. Rev. **183**, 725.

Collins, D. M., D. E. Mars, B. Fisher and C. Kocol (1983), J. Appl. Phys. **54**, 857.

Collins, R. T., K. v.Klitzing and K. Ploog (1986), Phys. Rev. **B33**, 4378.

Collins, R. T., L. Viña, W. I. Wang, C. Malhiot, and D. L. Smith (1987), Proc. SPIE **792**, 2.

Colvard, C. (1987), Proc. SPIE **794**, 209.

Colvard, C., T. A. Gant, M. V. Klein, R. Merlin, R. Fisher, H. Morkoç, and L. A. Gossard (1985), Phys. Rev. **B 31**, 2080.

Combescot, M. (1988), Solid State Electron. **31**, 657.

Compaan, A. (1975), Solid State Commun. **16**, 293.

Compaan, A. and H. Z. Cummins (1973), Phys. Rev. Lett. **31**, 41.

Condon, E. U. and P. M. Morse (1929), *Quantum Mechanics*, McGraw-Hill Book Co., Inc., New York.

Condon, E. U. and G. H. Shortley (1959), *The Theory of Atomic Spectra*, University Press, Cambridge.

Conley, J. W. and J. J. Tiemann (1967), J. Appl. Phys. **38**, 2880.

Connell, G. A. N., D. L. Champhausen, and W. Paul (1972), Philos. Mag. **26**, 541.

Connell, G. A. N. and R. N. Street (1980), *Handbook on Semiconductors*, Vol. **3**: *Materials Properties and Preparation*, T. S. Moss and S. P. Keller (eds.), North Holland Publ. Co., Amsterdam.

Conwell, E. M. (1952), Proc. IRE **40**, 1327.

Conwell, E. M. (1958), Proc. IRE **46**, 1281.

Conwell, E. M. (1967), *High-Field Transport in Semiconductors*, Academic Press, New York.

Conwell, E. M. (1982), in *Handbook on Semiconductors, Vol. 1: Band Theory and Transport Properties*, T. S. Moss and W. Paul (eds.), North Holland Publ. Co., Amsterdam.

Conwell, E. M. and S. Jeyadev (1988), Phys. Rev. Lett. **61**, 361.

Conwell, E. M. and V. F. Weisskopf (1950), Phys. Rev. **77**, 388.

Corbino, O. M. (1911), Phys. Z. **12**, 561.

Coret, A., S. Nikitine, J. P. Zielinger, and M. Zonaghi (1971), in *Proc. 3rd Int. Conf. Photoconductivity*, M. Pell (ed.), Pergamon Press, New York.

Coret, A., J. Ringeissen, and S. Nikitine (1968), in *Localized Excitations in Solids*, R. F. Wallis (ed.), Plenum Press, New York.

Corson, E. M. (1951), *Perturbation Methods in Quantum Mechanics of n-Electron Systems*, Blackie and Son, Ltd., London.

Costato, M., F. Manchinelli, and L. Reggiani (1971), Solid State Commun. **9**, 1335.

Costato, M. and L. Reggiani (1972), J. Phys. C (London) **5**, 158.

Cottrell, A. H. (1958), *Dislocations and Plastic Flow in Crystals*, Oxford University Press, London.

Cottrell, A. H. (1964), *Theory of Crystal Dislocations*, Gordon and Breach, New York.

Coulson, C. A., L.B. Redei, and D. Stocker (1962), Proc. Roy. Soc. (London) **270**, 357.

Cracknell, A. P. and K. C. Wong (1973), *The Fermi Surface: Its Concepts, Determination, and Use in the Physics of Metals*, Oxford University Press.

Crank, J. (1953), *The Mathematics of Diffusion*, Oxford University Press.

Crawford, M. G. and N. Holonyak, Jr. (1976), *Optical Properties of Solids*, B. Seraphin (ed.), North Holland Publ. Co., Amsterdam, p. 187.

Curby, R. C. and D. K. Ferry (1970), Phys. Lett. A **35**, 64.

Csavinski, P. (1965), J. Phys. Soc. Jpn. **20**, 2027.

Curby, R. C. and D. K. Ferry (1973), Phys. Stat. Sol. A **15**, 319.

Curie, D. (1963), *Luminescence in Crystals*, John Wiley & Sons, New York.

Cuthbert, J. D. and D. G. Thomas (1967), Phys. Rev. **154**, 763 and J. Appl. Phys. **39**, 1573.

Czaja, W. and A. Baldereschi (1979), J. Phys. C (London) **12**, 405.

Dakhovskii, I. V. and E. F. Mikhai (1964), Fiz. Tverd. Tela **6**, 3474.

Dallacasa, V. and C. Paracchini (1986), Phys. Rev. **B34**, 8967.

Damen, T. C. and J. M. Worlock (1976), *Proc. 3rd Int. Conf. on Light Scattering in Solids*, Campinas, Brasil, Flammarion, Paris, p. 183.

Daniels, W. B. (1962), Phys. Rev. Lett. **8**, 3.

Dapkus, P. D. (1984), J. Cryst. Growth **68**, 345.

Das, B., D. R. Anderson, M. Yamanishi, T. C. Bonsett, S. Datta, R. L. Gunshor, and L. A. Kolodziejski (1987), Proc. SPIE **792**, 94.

Dash, W. C. and R. Newman (1955), Phys. Rev. **99**, 1151.

Das Sarma, S. and B. A. Mason (1985), Phys. Rev. **B31**, 5536.

Davidov, A. S. (1962), *Theory of Molecular Excitons*, McGraw-Hill, New York.

Davidov, A. S. (1971), *Theory of Molecular Excitons*, Plenum Press, New York.

Dawson, B. (1967) Proc. Roy. Soc. (G.B.) **A298**, 255.

Dawson, B. (1973) *Kinetics and Mechanisms of Reactions*, Methuen Education Ltd., London.

Dean, J. P. (1971), Phys. Rev. B **4**, 2596.

Dean, P. J. (1973), in *Luminescence of Crystals, Molecules and Solutions*, F. E. Williams (ed.), Plenum Press, New York, p. 538.

Dean, P. J. (1973), Progr. Solid State Chem. **8**, 48.

Dean, P. J. (1973a), J. Lumin. **7**, 51.

Dean, P. J. (1984), *Collective Excitation in Solids*, Plenum Press, New York, p. 247.

Dean, P. J. (1986), in *Deep Centers in Semiconductors*, S. T. Pantelides (ed.), Gordon & Breach, New York.

Dean, P. J. and D. G. Thomas (1966), Phys. Rev. **150**, 690.

Dean, P. J., J. D. Cuthbert, and R. T. Lynch (1969), Phys. Rev. **179**, 754.

Dean, P. J. and D. C. Herbert (1979), in *Excitons*, K. Cho (ed.) *Topics in Current Physics* **14**, Springer-Verlag, Berlin, p. 55.

Dean, P. J., M. S. Skolnik, Ch. Uihlein, and D. C. Herbert (1983), J. Phys. C **16**, 2017.

Debye, P. (1912), Ann. Phys. (Lpz.) **39**, 789.

Debye, P. (1914), *Vorträge über kinetische Gastheorie*, Teubner Verlag, Berlin.

Debye, P. and E. M. Conwell (1954), Phys. Rev. **93**, 693.

Debye, P., C. Y. Duh, and J. L. Moll (1967), IEEE Trans. Electron Devices **ED-14**, 46.

de Gennes, P. G. (1966), *Superconductivity in Metals and Alloys*, W. A. Benjamin, Inc., New York.

de Goër, A. M., J. Doulal, and B. Dreyfus (1965), J. Nucl. Mat. **17**, 159.

de Goër, A. M., M. Locatelli, and I. F. Nicolau (1982), J. Phys. Chem. Soc. **43**, 311.

De Haas, W. J. and P. M. van Alphen (1930), Leiden Commun. 208d, 212a, and (1933) 220d.

de Kock, A. J. R. (1982), in *Handbook on Semiconductors*, Vol. **3**, *Materials, Properties and Preparation*, S. M. Keller (ed.), North Holland, Amsterdam.

De Leo, G. C., W. B. Fowler, and G. D. Watkins (1984), Phys. Rev. **B29**, 1819.

Delfyett, P. J., R. Dorsinville, and R. R. Alfano (1987), Proc. SPIE **793**, 139.

Dember, H. (1931), Phys. Z. **32**, 554, 865.

Demchuk, K. M. and I. M. Tsidilkovskii (1977), Phys. Stat. Sol. B **82**, 59.

Denisov, V. N., B. N. Mavrin, and V. B. Podobedov (1987), Physics Reports **151**, 2.

Denteneer, P. J. H., C. G. Van de Walle, and S. T. Pantelides (1989), Phys. Rev. **B39**, 1809.

Denteneer, P. J. H., C. G. Van de Walle ans S. T. Pantelides (1989), Phys. Rev. Lett. **62**, 1884.

Dersch, H., J. Stuke, and J. Beichler (1981), Appl. Phys. Lett. **38**, 456.

Des Cloizeaux, J. (1965), J. Phys. Chem. Solids **26**, 259.

Deveaud, B., G. Picoli, B. Lambert, and G. Martinez (1984), Phys. Rev. **B29**, 5749.

Devreese, J. T. (1984), in *Polarons and Excitons in Polar Semiconductors and Ionic Crystals*, J. T. Devreese and F. M. Peeters (eds.), Plenum Press, New York.

Devreese, J. T. and F. Brosens (1981), Phys. Stat. Sol. **B108**, K29.

DeWerd, L. A. (1979), in *Thermally Stimulated Relaxation in Solids*, P. Bräunlich (ed.), Springer Verlag, Berlin, p. 275.

Dexter, D. L. (1958), in *Solid State Physics* Vol. **6**, F. Seitz and D. Turnbull (eds.), Academic Press, New York.

Dexter, D. L., C. C. Klick, and G. A. Russell (1956), Phys. Rev. **100**, 603.

Dexter, R. N., H. J. Zieger, and B. Lax (1956), Phys. Rev. **104**, 637.

DiBenedetto, A. T. (1967), *Structure and Properties of Materials*, McGraw-Hill Book Co., Inc., New York.

Dienys, V. and M. O. Kancleris (1975), Phys. Stat. Sol. B **67**, 317.

Dienys, V. and M. O. Kancleris (1975), Phys. Stat. Sol. B **67**, 317.

Diete, A. F., V. D. Kulakovskii, and V. B. Timofeev (1977), Zh. Eksp. Teor. Fiz. **72**, 1156.

Dietz, R. E., D. G. Thomas, and J. J. Hopfield (1962), Phys. Rev. Lett. **8**, 391.

Dimmock, J. O. (1971), Solid State Phys. **26**, 103.

Dimmock, J. O., I. Melngailis, and A. J. Strauss (1966), Phys. Rev. Lett. **16**, 1193.

Dingle, R. (1975), in *Festkörperprobleme/Advances in Solid State Physics*, Vol. 15, H. J. Queisser (ed.), Pergamon Press, Oxford, p. 21.

Dingle, R. and C. H. Henry (1976), U. S. Patent No. 3,982,207.

Dingle, R., W. Wiegmann, and C. H. Henry (1974), Phys. Rev. Lett. **33**, 827.

Dingle, R., H. L. Störmer, A. C. Gossard, and W. Wiegmann (1978), Appl. Phys. Lett. **33**, 665.

Dirac, P. A. M. (1944), *Die Prinzipien der Quantenmechanik*, Verlag Hirzel, Lpz..

Dirac, P. A. M. (1958), *The Principles of Quantum Mechanics*, Oxford, New York.

Dixon, J. R. (1960), *Proc. Int. Conf. Semicond. Physics*, Czech. Acad. Sci., Prague.

Dmitriev, A. P., M. P. Mikhailova, and I. N. Yassievich (1987), Phys. Stat. Sol. B **140**, 9.

Dmowski, L., M. Bay, A. Iller and S. Porowski (1977), *Proc. Int. Conf. on High Press. and Low Temp. Phys.*, Cleveland, Ohio 1977, Plenum Press, New York.

Döhler, H. G. (1972), Phys. Stat. Sol. B **52**, 79 and 533.

Döhler, H. G. (1986), IEEE J. Quantum Electron. **QE 22**, 1682.

Döhler, H. G. (1987), Crit. Rev. Solid State and Mat. Sci. **13**, 97.

Döhler, H. G. (1987), Proc. SPIE **792**, 45.

Döhler, H. G., O. Fasol, T. S. Low, J. N. Miller, and K. Ploog (1986), Solid State Commun. **57**, 563.

Dolling, G. (1974), *Dynamic Properties of Solids*, G. R. Norton and A. A. Maradudin (eds.), North Holland Publ. Co., Amsterdam.

Dollis, G. and J. L. T. Waugh (1965), in *Lattice Dynamics*, R. F. Williams (ed.), Pergamon Press, New York.

Dow, J. D. (1985), *Highlights of Condensed Matter Theory* **89**, 465.

Drabble, J. R. and H. J. Goldsmid (1961) *Thermal Conduction in Semiconductors*, Pergamon Press, New York.

Dresner, J. (1983), J. Non-Cryst. Solids (Netherlands) **58**, 353.

Dresselhaus, G. and M. S. Dresselhaus (1970), *Proc. 10th Int. Conf. Semicond.*, U.S. Atomic Energy Comm., p. 338.

Dresselhaus, G., A. F. Kip, and C. Kittel (1955), Phys. Rev. **98**, 368.

Drude, P. (1900), Ann. Phys. **1**, 566.

Druger, S. D. (1975), in *Organic Molecular Photophysics*, J. B. Birks (ed.), John Wiley & Sons, London, p. 313.

Duh, C. Y. and J. L. Moll (1968), Solid State Electron. **11**, 917.

Duke, C. B. (1969), *Tunneling in Solids*, Academic Press, New York.

Dulong, P. L. and A. T. Petit (1819), Ann. Chim. Phys. **10**, 399.

Dushman, S. (1930), Rev. Mod. Phys. **2**, 381.

Dussel, G. A. and K. W. Böer (1970), Phys. Stat. Sol. **39**, 375.

G. A. Dussel and K. W. Böer (1970a), Phys. Stat. Sol. **39**, 391.

Düster, F. and R. Labusch (1973), Phys. Stat. Sol. B **60**, 161.

Dutta, P. and P. M. Horn (1981), Rev. Mod. Phys. **53**, 497.

Dutton, D. (1958), Phys. Rev. **112**, 785.

D'yakonov, M. I. and V. I. Perel' (1971), Sov. Phys.-JETP **33**, 1053.

D'yakonov, M. I. and V. I. Perel' (1971), Zh. Eksp. Teor. Fiz. **60**, 1954.

D'yakonov, M. I. and V. I. Perel' (1984), in *Optical Orientation*, F. Meyer and B. P. Zakharchenya (eds.), Elsevier, New York.

D'yakonov, M. I., V. I. Perel' and I. L. Yassievich (1977), Sov. Phys. Semicond. **11**, 801.

Dymnikov, V. D., M. I. D'yakonov and V. I. Perel' (1976), Sov. Phys. Semicond. **11**, 801.

Dymnikov, V. D., D. N. Mirlin, V. I. Perel', and I. I. Reshine (1978), Sovj. Phys.-Solid State **20**, 1250.

Dziewior, J. and W. Schmid (1977), Appl. Phys. Lett. **31**, 346.

Eagles, D. M. (1960), J. Phys. Chem. Solids **16**, 76.

Eastman, D. E., M. H. Cohen, K. F. Freed, and E. S. Kirkpatrick (1974), in *Amorphous and Liquid Semiconductors*, J. Tauc (ed.), Plenum Press, New York, p. 101.

Eastman, L. F. (1982), in *Festkörperprobleme/Advances in Solid State Physics*, Vol. 22, P. Grosse (ed.), Friedr. Vieweg & Sohn, Braunschweig.

Eaves, L., R. A. Hoult, R. A. Stradlin, R. J. Tidley, J. C. Portal and S. Askenazy (1975), J. Phys. C (London) **8**, 1034.

Eberhart, M.E., K.H. Johnson, and D. Adler (1982), Phys. Rev. **B26**, 3138.

Ebert, J. J. (1789), *Unterweisung in den Anfangsgründen der Naturlehre*, Chr. Gottlieb Hertel, Lpz..

Eby, J. E., K. J. Teegarden, and D. B. Dutton (1959), Phys. Rev. **116**, 1099.

Edwards, D. F., T. E. Slykhouse, and G. A. Drickamer (1959), J. Phys. Chem. Solids **11**, 140.

Edwards, P. P. and M. J. Sienko (1978), Phys. Rev. **B17**, 2575.

Edwards, P. P. and M. J. Sienko (1981), J. Am. Chem. Soc. **103**, 2967.

Efros, A. L. and B. I. Shklovskii (1975), J. Phys. **C8**, L49.

Efros, A. L., B. I. Shklovskii, and I. Y. Yanchev (1972), Phys. Stat. Sol. **B50**, 45.

Egri, I. (1985), Physics Reports **119**, 364.

Ehrenberg, W. (1950), Proc. Phys. Soc. (London) Ser. **A63**, 75.

Ehrenreich, H. (1957), J. Phys. Chem. Soc. **2**, 131.

Ehrenreich, H. (1961), J. Appl. Phys. (Suppl.) **32**, 2118.

Ehrenreich, H. (1987), Science **235**, 1029.

Einstein, A. (1907), Ann. Phys. (Lpz.) **22**, 180 and 800.

Ekimov, A. I. and V. I. Safarov (1972), Zh. Eksp. Teor. Fiz. **15**, 453.

Elkomoss, S. G. and G. Munschy (1981), J. Phys. Chem. Solids **42**, 1.

Elliott, R. J. (1954), Phys. Rev. **94**, 266.

Elliott, R. J. (1961), Phys. Rev. **124**, 340.

Elliott, R. J. (1982), in *Polarons and Excitons in Polar Semiconductors and Ionic Crystals*, J. T. Devreese and J. M. Peeters (eds.), Plenum Press, New York.

Elliott, S. R. (1977), Philos. Mag. **36**, 1291.

Elliott, S. R. (1978), Philos. Mag. B **38**, 325 and 553.

Emin, D. (1973), Adv. Phys. **22**, 57.

Emin, D. (1975), Adv. Phys. **24**, 305.

Enderlein, R. (1977), in *Proc. Autumn School of Modul'n. Spectroscopy*, R. Enderlein (ed.), Zentralinstitut für Elektronenphysik, Berlin.

Era, K. and D. W. Langer (1969), J. Lumin. **1/2**, 514.

Erginsoy, C. (1950), Phys. Rev. **79**, 1013.

Erskine, D. J., A. J. Taylor, and C. L. Tang (1984), Appl. Phys. Lett. **45**, 54.

Esaki, L. (1958), Phys. Rev. **109**, 603.

Esaki, L. (1985), in *The Technology and Physics of Molecular Beam Epitaxy*, E. M. C. Parker (ed.), Plenum Press, New York, p. 143.

Esaki, L. (1986), IEEE J. Quantum Electron. **QE 22**, 1611.

Esaki, L., L. Chang, W. E. Howard, and V. L. Rideout (1972), *Proc. 11th Int. Conf. Phys. Semicond.*, Warsaw, p. 431.

Esaki, L. and R. Tsu (1970), IBM J. Res. Dev. 61.

Etherington, G., A.C. Wright, J.T. Wenzel, J.C. Dore, J.H. Clarke, and R.N. Sinclair (1982), J. Non-Cryst. Solids **48**, 265.

Etienne, B. (1975), Thesis, L'École Normale Supérieure.

Evangelisti, F., A. Frova and F. Patella (1974), Phys. Rev. **B10**, 4253.

Evans, A. G. R. and P. N. Robson (1974), Solid State Electron. **17**, 805.

Evrard, R. (1984), in *Excitons and Polarons*, B. DiBarbolo and J. Danko (eds.), Plenum Press, New York.

Fan, H. Y. (1951), Phys. Rev. **82**, 900.

Fan, H. Y. (1967), in *Semiconductors and Semimetals*, Vol. 3, R. K. Willardson and A. C. Beer (eds.), Academic Press, New York, p. 405.

Fan, H. Y., W. Spitzer, and R. J. Collins (1956), Phys. Rev. **110**, 566.

Fano, U. (1956), Phys. Rev. **103**, 1202.

Farge, Y. (1973), J. Phys. (Paris), **34**, C9–475.

Fasol, G. and H. P. Hughes (1986), Phys. Rev. **B33**, 2973.

Fasol, G., K. Ploog, and E. Bauser (1985), Solid State Commun. **54**, 383.

Fassbender, J. and O. Hachenberg (1949), Ann. Phys. (Lpz.) **6**, 229.

Fassbender, J. and H. Lehmann (1949), Ann. Phys. (Lpz.) (b), **6**, 215.

Faulkner, R. A. (1968), Phys. Rev. **175**, 991.

Faulkner, R. A. (1969), Phys. Rev. **184**, 713.

Faust, W. L. and C. H. Henry (1966), Phys. Rev. Lett. **17**, 1265.

Fehrer, G. (1959), Phys. Rev. **114**, 1219.

Feldman, A. and R. M. Waxler (1979), Proc. SPIE (USA) **204**, 68.

Fermi, E. (1934), Nuovo Cimento **II**, 157.

Ferry, D. K. (1975), Phys. Rev. **B12**, 2361.

Ferry, D. K. (1978), Phys. Rev. B **18**, 7033.

Ferry, D. K. (1980), in *Physics of Non-Linear Transport in Solids*, Plenum Press, New York.

Ferry, D. K. and J. R. Barker (1981), J. Appl. Phys. **52**, 818.

Ferry, D. K., H. L. Grubin, and G. J. Iafrate (1984), *Semiconductors Probed by Ultrafast Laser Spectroscopy*, Vol. 1, Academic Press, New York.

Ferziger, J. D. and H. G. Kaper (1972), *Mathematical Theory of Transport Processes in Gasses*, North Holland Publ. Co., Amsterdam.

Fetter, A. L. and J. D. Walecka (1971), *Quantum Theory of Many Particle Systems*, McGraw-Hill Book Co., Inc., New York.

Feynman, R. P. (1955), Phys. Rev. **97**, 660.

Feynman, R. P., R. W. Hellwarth, C. K. Iddings and P. M. Platzman (1962), Phys. Rev. **127**, 1004.

Feynman, R. P. and A. R. Hibbs (1965), *Quantum Mechanics and Path Integrals*, McGraw-Hill Book Co., Inc., New York.

Fiori, A. T., J. C. Bean, L. C. Feldman, and I. K. Robinson (1984), J. Appl. Phys. **56**, 1227.

Firsov, Yu. A., V. L. Gurevich, R. B. Parfeniev, and S. S. Shalyt (1964), Phys. Rev. Lett. **12**, 660.

Fischer, B. and J. Lagois (1979), in *Excitons: Topics in Current Physics*, K. Cho (ed.), Springer-Verlag, Berlin, p. 183.

Fitchen, D. B., R. H. Silsbee, T. A. Fulton, and E. L. Wolf (1963), Phys. Rev. Lett. **11**, 275.

Fletcher, G. C. (1971), *Electron Bond Theory of Solids*, North Holland Publ. Co., Amsterdam.

Fletcher, K. and P. N. Butcher (1972), J. Phys. C (London) **5**, 212.

Flohrer, J., E. Jaline, and M. Porsch (1979), Phys. Status Solidi **91**, 467.

Flubacher, P., A. J. Leadbetter, and J. A. Morrison (1959), Philos. Mag. **4**, 273.

Flynn, C. P. (1972), *Point Defects and Diffusion*, Clarendon Press, Oxford.

Fock, W. A. (1930), Z. Phys. **61**, 126.

Forchel, A., B. Laurich, J. Wagner, W. Schmid and T. L. Reinecke (1982), Phys. Rev. **B25**, 2730.

Fornari, B. and M. Pagannone (1978), Phys. Rev. **B17**, 3047.

Fouquet, J. E. and R. D. Burnham (1986), IEEE J. Quantum Electron. **QE 22**, 1799.

Fowler, R. H. and L. Nordheim (1928), Proc. R. Soc. (London), Ser. **A119**, 173.

Fowler, W. B. (1968), *The Physics of Color Centers*, Academic Press, New York.

Frank, F. C. (1949), Discuss. Faraday Soc. **5**, 48.

Frank, F. C. (1949), Proc. Phys. Soc. (London) **A62**, 202.

Frank, F. C. and W. T. Read, Jr. (1950), Phys. Rev. **79**, 722.

Frank, W. (1981), in *Festkörperprobleme/Advances in Solid State Physics*, Vol. 21, J. Treusch (ed.), Friedr. Vieweg & Sohn, Braunschweig, p. 221.

Franken, P. A. and J. F. Ward (1963), Rev. Mod. Phys. **35**, 23.

Franz, W. (1958), Z. Naturforsch. **13a**, 484.

Franz, W. (1970) *Quantentheorie*, Springer-Verlag, Berlin.

Frenkel, J. I. (1926), Z. Phys. **35**, 652.

Frenkel, J. I. (1931), Phys. Rev. **37**, 17, 1276.

Frenkel, J. I. (1934), Proc. Roy. Soc. Ser. **A146**, 930.

Frenkel, J. I. (1936), Sovj. Phys. **9**, 158.

Frenkel, J. I. (1938), Phys. Rev. **54**, 647.

Frerichs, R. and R. Warminski (1946), Naturwissenschaften **33**, 251.

Friedel, J. (1964), *Dislocations*, Addison Wesley, Reading, Massachusetts.

Friedel, J. and M. Lannoo (1973), J. de Phys. **34**, 155 and 483.

Friedman, B. (1956), *Principles and Techniques in Applied Mathematics*, John Wiley & Sons, New York.

Fritz, I. J., S. T. Picraux, L. R. Dawson, T. J. Drummond, W. K. Laidig, and N. G. Anderson (1985), Appl. Phys. Lett. **46**, 967.

Fritzsche, H. (1960), Phys. Rev. **119**, 1899.

Fritzsche, H. (1962), Phys. Rev. **125**, 1560.

Fritzsche, H. (1971), J. Non-Cryst. Sol. **6**, 49.

Fritzsche, H. (1976), *Electronic Phenomena in Non-crystalline Semiconductors*, B. T. Kolomiets (ed.), Nauka, Leningrad.

Fritzsche, H. (1979), J. Non-Cryst. Sol. **6**, 329.

Fritzsche, H. (1985), J. Non-Cryst. Sol. **77/78**, 273.

Fritzsche, H. (1989), *13th Int'l Conf. on Amorph. and Liquid Semicond.*, Nashville, N. C.

Fritzsche, H. and M. Cuevas (1960), Phys. Rev. **119**, 1238.

Fröhlich, D. H. (1970), *Festkörperprobleme/Advances in Solid State Physics*, Vol. **10**, O. Madelung (ed.), Friedr. Vieweg & Sohn, Braunschweig, p. 227.

Fröhlich, D. H. (1981), *Festkörperprobleme/Advances in Solid State Physics*, Friedr. Vieweg & Sohn, Braunschweig, Vol. **21**, 363.

Fröhlich, H. (1937), Proc. Roy. Soc. Ser. (London) **A160**, 280.

Fröhlich, H. (1953), Adv. Phys. **3**, 325.

Fröhlich, H. (1954), Adv. Phys. **3**, 325.

Fröhlich, H. and B. V. Paranjape (1956), Proc. Roy. Soc. (London) Ser. **A69**, 21.

Fröhlich, H., H. Pelzer, and S. Zienau (1950), Philos. Mag. **41**, 221.

Froyen, S. and M. L. Cohen (1983), Phys. Rev. **B28**, 3258.

Frova, A. and P. Handler (1965), Phys. Rev. **137**, A1857.

Froyen, S., D. M. Wood, and A. Zunger (1987), Phys. Rev. **B37**, 6893.

Fuchs, K. (1938), Proc. Cambridge Philos. Soc. **11**, 120.

Furdnya, J. K. (1982), J. Appl. Phys. **53**, 7637.

Furdnya, J. K. (1985), Solid State Commun. **53**, 1097.

Furdnya, J. K. (1986), J. Vac. Sci. Technol. A **4**, 2002.

Gadzhiev, A. R., S. M. Ryvkin, and I. S. Shlimak (1972), JETP **34**, 428.

Gal, M., B. C. Cavenett, and P. Smith (1979), Phys. Rev. Lett. **43**, 1611.

Gale, G. and A. Laubereau (1983), Opt. Commun. **44**, 273.

Galeener, F. L., A. J. Leadbetter, and M. W. Stringfellow (1983), Phys. Rev. **B27**, 1052.

Galeener, F. L., G. Lucovsky, and R. H. Geils (1979), Phys. Rev. **B19**, 4251.

Gammon, D., R. Merlin, W. T. Masselink and H. Morkoç (1986), Phys. Rev. **B33**, 2919.

Garber, J. A. and A. V. Granato (1975), Phys. Rev. **B11**, 3990 and 3998.

Garbuzov, D. Z., A. I. Ekimov, and V. I. Safarov (1971), Sovj. Phys.-JETP Lett. **13**, 24.

Geballe, T. H. and G. W. Hull (1964), Phys. Rev. **134**, 773.

Gebhard, K. F., P. D. Soper, J. Merski, T. J. Balle, and W. H. Flygare (1980), J. Chem. Phys. **72**, 272.

Gebhardt, W. and H. Kuhnert (1964), Phys. Rev. Lett. **11**, 15.

Gel'mont, D. L. and M. I. D'yakonov (1971), Sov. Phys. Semicond. **5**, 1905.

Gershenzon, M., D. G. Thomas, and R. E. Dietz (1962), *Proc. Int. Conf. Semicond.*, Exeter, Inst. Phys. Soc., London, p. 752.

Gibbs, W. E. K. (1962), J. Appl. Phys. **33**, 3369.

Gibson, A. F., M. F. Kimmitt, and A. C. Wagner (1970), Appl. Phys. Lett. **17**, 75.

Gilason, H. P., B. Monemar, P. J. Dean, D. C. Herbert, S. P. Depinna, B. C. Cavenett, and N. Killoran (1982), Phys. Rev. **B26**, 827.

Gilat, G. and L. J. Raubenheimer (1966), Phys. Rev. **144**, 390.

Gill, W. D. (1976) in *Photoconductivity and Related Phenomena*, J. Mort and D. M. Pai (eds.), Elsevier Scientific Publ. Co., Amsterdam, p. 303.

Ginter, J. and J. Mycielski (1970), J. Phys. C (London) **3**, 11.

Ginzburg, V. L. (1970), Sov. Phys. Usp. **13**, 335.

Gisolf, J. H. (1949), Physica **15**, 825.

Glassbrenner, C. J. and G. A. Slack (1964), Phys. Rev. **134**, A1058.

Glicksman, M (1958) in *Progress in Semiconductors*, Vol. 3, John Wiley & Sons, New York.

Glinskii, G. F., A. A. Kopylov, and A. N. Pikhtin (1979), Solid State Commun. **30**, 631.

Glyde, H. (1967), Rev. Mod. Phys. **39**, 373.

Gnutzmann, R. G. and K. Clauseker (1974) Appl. Phys. **3**, 9.

Göbel, E. O., J. Kuhl, and R. Hoger (1986), J. Lumin. **30**, 541.

Godik, E. E. (1968), Phys. Stat. Sol. **30**, K127.

Goede, O. and W. Heimbrodt (1988), Phys. Stat. Sol. B **146**, 11.

Goede, O., L. John, and D. Hennig (1978), Phys. Stat. Sol. B **89**, K183.

Goldberg, P. G. (ed.) (1966), *Luminescence of Inorganic Solids*, Academic Press, New York.

Goldschmidt, V. M. (1927), Skrifter det Norske Videnskaps

Golka, J. and J. Mostowski (1975), J. Phys. C. (London) **8**, L257.

Gomyo, A., T. Suzuki, K. Kobayashi, S. Kawabu, and I. Hindu (1987), Appl. Phys. Lett. **50**, 673.

Goodings, E. P. (1976), Chem. Soc. Rev. **5**, 95.

Goodman, C.H.L. (1978) *Crystal Growth: Theory and Techniques*, Plenum Press, New York.

Gorid'ko, N. Ya., P. P. Kuzmenko, and N. N. Novikov, (1961), Fiz. Tverd. Tela. **3**, 3650.

Gornick, E. and D. C. Tsui (1976), Phys. Rev. Lett. **37**, 1425.

Gösele, U. M. (1986), *Festkörperprobleme/Advances in Solid State Physics*, Vol. **26**, P. Grosse (ed.), Friedr. Vieweg & Sohn, Braunschweig, p. 89.

Gossard, A. C. (1983), Inst. Phys. Conf. Ser. **69**, Canterbury, p. 1.

Gossard, A. C. (1986), IEEE J. Quantum Electron. **QE 22**, 1649.

Gotthard, L., A. Stahl and G. Czajkowski (1954), J. Phys. **C17**, 4865.

Götze, W. (1981) in *Recent Developments in Condensed Matter Physics*, J. T. Devreese (ed.), Plenum Press, New York.

Gourley, P. L. and J. P. Wolfe (1978), Phys. Rev. Lett. **40**, 526.

Grassano, V. M. (1977), Nuovo Cimento B **39**, 368.

Graudßus, W. and E. O. Göbel (1981), J. Phys. Colloq., Orsay, France, **C7**, 437.

Graudßus, W. and E. O. Göbel (1983), Physica B **117**, 555.

Green, M. A. (1982), *Solar Cells*, Prentice Hall, Englewood Cliffs, New Jersey.

Greenaway, D. L. and G. Harbeke (1968), *Optical Properties and Band Structure of Semiconductors*, Pergamon Press, New York.

Greenwood, D. A. (1958), Proc. Phys. Soc. **A71**, 585.

Griffith, J. S. (1964), *The Theory of Transition Metal Ions*, University Press, Cambridge.

Grimmeiss, H. G. (1977), Ann. Rev. Mater. Sci. **7**, 341.

Grimmeiss, H. G. (1986), in *Proc. Int. Conf. Phys. Semicond.*, O. Engström (ed.), Stockholm, World Scientific, p. 589.

Grimmeiss, H. G. and L. Å. Ledebo (1975), J. Appl. Phys. **46**, 2155.

Grimmeiss, H. G., E. Janzén, and B. Skarstam (1980), J. Appl. Phys. **51**, 4212.

Grimmeiss, H. G., N. Kullendorf, and R. Broser (1981), J. Appl. Phys. **52**, 3405.

Grimmeiss, H. G. and E. Janzén (1986), in *Deep Centers in Semiconductors*, S. T. Pantelides (ed.), Gordon & Breach, New York.

Grondahl, L.O. (1926), Phys. Rev. **27**, 813.

Gross, E. F., and B. V. Novikov (1959), Sovj. Phys.-Solid State **1**, 321.

Gross, E. F., and B. V. Novikov (1962), in *Photoconductivity*, H. Levinstein (ed.), Pergamon Press, New York.

Grove, A. S. (1967), *Physics and Technology of Semiconductor Devices*, John Wiley & Son, New York.

Grüneisen, E. (1926), *Handbuch der Physik*, Vol. 10, Springer-Verlag, Berlin, p. 1.

Grunewald, M., P. Thomas and D. Wurz (1981), J. Phys. C **14**, 4083.

Guicher, G. M., C. Sébenne, F. Proix and M. Balkanski (1972), *Proc. 11th Int'l Conf. Phys. Semicond.*, PWN, Warsaw, p. 877.

Gunn, J.B. (1963), Solid State Commun. **1**, 88.

Gurevich, V. L., A. I. Larkin, and Yu, A. Firsov (1962), Sov. Phys.-Solid State **4**, 131.

Gurevich, V. L. and Yu. A. Firsov (1964), Sov. Phys.-JETP **20**, 489.

Gutman, F. (1948), Rev. Mod. Phys. **20**, 457.

Gutman, F. and L. E. Lyons (1967), *Organic Semiconductors*, John Wiley & Sons, New York.

Gutsche, E. (1982), Phys. Stat. Sol. (B) **109**, 583.

Gutsche, E., J. Voigt, and E. Ost (1971), in *Proc. 3rd Int. Conf. Photoconductivity*, M. Pell (ed.), Pergamon Press, New York.

Guyer, R. A. (1966), Phys. Rev. **148**, 789.

Habegger, M. A. and H. Y. Fan (1964), Phys. Rev. Lett **12**, 99.

Haber, F. (1919), Verh. Dtsch. Phys. Ges. **21**, 750.

Hacker, K. and G. Obermair (1970), Z. Physik **234**, 1.

Hagemark, K. I. (1976), J. Phys. Chem. Solids **37**, 461.

Haken, H. (1963) in *Polarons and Excitons*, Kuper and Whitfield (eds.), Oliver and Boyd, Edinburgh, p. 294.

Haken, H. (1976), *Quantum Field Theory of Solids*, North Holland Publ. Co., Amsterdam.

Haken, H. and S. Nikitine (eds.) (1975), *Excitons at High Densities*, Springer Tracts in Modern Physics, Springer Verlag, New York.

Haldane, F. D. M. and P. W. Anderson (1976), Phys. Rev. **B13**, 2553.

Hall, J. J. (1967), Phys. Rev. **161**, 756.

Hall, R. N. (1951), Phys. Rev. **83**, 228.

Hall, R. N. (1952), Phys. Rev. **87**, 387.

Hall, R. N., G. E. Fermer, J. D. Kingsley, T. J. Soltys, and R. O. Carlson (1962), Phys. Rev. Lett. **9**, 366.

Haller, E. E. (1986), in *Festkörperprobleme/Advances in Solid State Physics*, Vol. **26**, P. Grosse (ed.), Friedr. Vieweg & Sohn, Braunschweig, p. 203.

Haller, E. E. and W. L. Hansen (1974), Solid State Commun. **15**, 687.

Haller, E. E., W. L. Hansen, and F. S. Goulding (1981), Adv. Phys. **30**, 93.

Haller, E. E., H. Navarro and F. Keilmann (1986), *Proc. Int. Conf. Physics Semicond.*, Stockholm.

Halperin, B. I. and M. Lax (1966), Phys. Rev. **148**, 722.

Halperin, B. I. and M. Lax (1967), Phys. Rev. **153**, 802.

Halperin, B. I. and T. M. Rice (1968), Sol. St. Phys. **21**, 115.

Halsted, R. E. (1967) in *Physics and Chemistry of II-VI Compounds*, M. Aven and J. S. Prener (eds.), John Wiley & Sons, New York.

Halsted, R. E. and M. Aven (1965), Phys. Rev. Lett. **14**, C4.

Hamaguchi, C., S. Adachi and Y. Itoh (1978), Sol. State Electronics, **21**, 1585.

Hamakawa, Y. (1989), Adv. Solar Energy **5**, 1.

Hamakawa, Y., F. A. Germano and P. Handler (1968), Phys. Rev. **167**, 703.

Hamilton, B., A. R. Peaken and S. T. Pantelides (1988), Phys. Rev. Lett. **61**, 1627.

Hamilton, J. F. (1988), Adv. in Phys. **37**, 359.

Han, D. and H. Fritzsche (1983), J. Non-cryst. Solids, **59/60**, 397.

Hanamura, E. (1973), Solid State. Commun. **12**, 951.

Hanamura, E. (1975), J. Phys. Soc. Jpn. **39**, 1515.

Hanamura, E. (1976), in *Optical Properties of Solids*, B. Seraphin (ed.), North Holland Publ. Co, Amsterdam.

Hanamura, E. and H. Haug (1977), Phys. Rev. Rep. **33**, 210.

Handel, P. H. (1980), Phys. Rev. **A22**, 745.

Handel, P. H. (1982), Phys. Rev. A **26**, 586.

Hanke, W. R. and L. J. Sham (1974), Phys. Rev. Lett. **33**, 582.

Hanle, W. (1924), Z. Phys. **30**, 93.

Hanneman, R. E., M. D. Banus, and H. C. Gatos (1964), J. Phys. Chem. Sol. **25**, 293.

Hansen, G. L., J. L. Schmit, and T. L Casselman (1982), J. Appl. Phys. **53**, 7099.

Hanson, R. C., K. Helliwell, and C. Schwab (1974), Phys. Rev. **B9**, 2649.

Harman, T. C., R. K. Willardson, and A. C. Beer (1954), Phys. Rev. **95**, 699.

Harper, P. G., J. W. Hodby, and R. A. Stradling (1973), Rep. Prog. Phys. **36**, 1.

Harrison, W. A. (1956), Phys. Rev. **104**, 1281.

Harrison, W. A. (1966), *Pseudopotentials in the Theory of Metals*, W. A. Benjamin, Inc., New York.

Harrison, W. A. (1970), *Solid State Theory*, McGraw-Hill Book Co., Inc., New York.

Harrison, W. A. (1973), Phys. Rev. **B8**, 4487.

Harrison, W. A. (1977), J. Vac. Sci. Technol. **14**, 1016.

Harrison, W. A. (1980), *Electronic Structure and the Properties of Solids: The Physics of Chemical Bonds*, Freeman, San Francisco.

Hartke, J. L. (1968), J. Appl. Phys. **39**, 4871.

Hasegawa, A. and Y. Kodama (1981), Proc. IEEE **69**, 1145.

Hasegawa, H. (1969), *Physics of Solids in High Magnetic Fields*, D. Haidemenakis (ed.), Plenum Press, New York.

Hass, M. (1964), in *Semiconductors and Semimetals*, Vol. 3, R. K. Willardson and A. C. Beer (eds.), Academic Press, New York, p. 1.

Hauffe, K. (1955), *Reaktionen in und an Festen Stoffen*, Springer Verlag, Berlin.

Haug, A. (1972), *Theoretical Solid State Physics*, Pergamon Press, Oxford.

Haug, A. (1977), Solid State Commun. **22**, 537.

Haug, A. (1980), Phys. Stat. Sol. B **97**, 481.

Haug, A. (1981), *Proc. 15th Int. Conf. Semicond.*, Kyoto, Jpn. p. 503.

Haug, A. (1981), Phys. Stat. Sol. B **108**, 443.

Haug, A. (1988), J. Phys. Chem. Sol. **49**, 599.

Hauser, J. J. (1975), Solid State Commun. **17**, 1577.

Hayes, J. R., A. F. J. Levi, A. C. Gossard, and J. H. English (1986), Appl. Phys. Lett. **49**, 1481.

Hayes, T. M. and J. C. Boyce (1985), Solid State Phys. **37**, 173.

Hayes, W. and R. Loudon (1978), *Scattering of Light by Crystals*, John Wiley & Sons, New York.

Hayes, W. and A. M. Stoneham (1984), *Defects and Defect Processes in Nonmetallic Solids*, John Wiley & Sons, New York.

Haynes, J. R. (1960), Phys. Rev. Lett. **4**, 361.

Haynes, J. R. and J. A. Hornbeck (1955), Phys. Rev. **100**, 606.

Haynes, J. R. and N. G. Nilsson (1964), Comptes Rendus, *7th Congrès Int. Phys. des Semicond.*, Paris, Dunod, Paris.

Hearn, C. J. (1980), *Physics of Nonlinear Transport in Solids*, Plenum Press, New York.

Heeger, A. J. (1981), Comments Solid State Phys. **10**, 53.

Heeger, A. J. and A. G. MacDairmid (1980), *The Physics and Chemistry of Low Dimensional Solids*, L. Alcacer (ed.), Reidel, Dordrecht.

Hegarty, J., M. D. Sturge, C. Weisbuch, A. C. Gossard, and W. Weigmann (1982), Phys. Rev. Lett. **49**, 930.

Heggie, M. and R. Jones (1983), *Microscopy of Semiconduct. Materials*, Inst. of Phys. Conf. Series, **67**, 1.

Heiblum, M., M. I. Nathan, D. C. Thomas, and C. M. Knoedler (1985), Phys. Rev. Lett. **55**, 2200.

Hein, R. A., J. W. Gibson, R. Mazelsky, R. C. Miller, and J. H. Kulm (1964), Phys. Rev. Lett. **12**, 320.

Heine, V. (1980), Solid State Phys. **35**, 1.

Heine, V. and J. A. van Vechten (1976), Phys. Rev. B **13**, 1622.

Heine, V. and D. Weaire (1970), Solid State Phys. **24**, 250.

Heitler, W. and F. London (1927), J. Phys. **44**, 455.

Haldane, F. D. M. and E. H. Rezayi (1988), Phys. Rev. Lett. **61**, 1985.

Hellman, J. J. (1935), J. Chem. Phys. **3**, 61.

Hellwege, A. M. (1955), in *Landoldt-Börnstein, Atom and Molekularphysik*, Vol. 4, Kristalle, Springer-Verlag, Berlin.

Hellwege, K. H. (1970), *Einführung in die Festkörperphysik*, Heidelberger Taschenbücher, Springer-Verlag, Berlin.

Hemstreet, L. A. (1977), Phys. Rev. **B15**, 834.

Hemstreet, L. A. (1980), Phys. Rev. **97**, 969.

Henisch, H. K. (1949), *Metal Rectifiers*, Clarendon Press, Oxford.

Henisch, H. K. (1957), *Rectifying Semiconductor Contacts*, Clarendon Press, Oxford.

Henley, C. L. (1987), Comments Cond. Mat. Phys. **13**, 59.

Hennel, A. M. (1978), J. Phys. C (London) **11**, L389.

Henry, C. H. (1980), in *Relaxation of Elementary Excitation*, R. Kubo and E. Hanamura (eds.), Springer-Verlag, Berlin, p. 19.

Henry, C. H. and J. J. Hopfield (1965), Phys. Rev. Lett. **15**, 964.

Henry, C. H. and D. V. Lang (1977), Phys. Rev. **B15**, 989.

Henry, C. H. and K. Nassau (1970), Phys. Rev. **B1**, 1628.

Hensel, J. C., T. G. Phillips, and G. A. Thomas (1978), *Solid State Physics*, Vol. 32, H. Ehrenreich, F. Seitz, and D. Turnbull (eds.), Academic Press, New York, p. 88.

Herbert, D. C. (1977), J. Phys. C (London) **10**, 3327.

Herman, F. (1958), Rev. Mod. Phys. **30**, 102.

Herman, F. and S. Skillman (1963), *Atomic Structure Calculation*, Prentice Hall, Engelwood Cliffs, New Jersey.

Herman, F., R. V. Kasowski, and W. Y. Hsu (1987), Phys. Rev. **B36**, 6904.

Herman, F., R. V. Kasowski, and W. Y. Hsu (1987a), *Novel Superconductivity*, S. A. Wolf and V. Z. Kresin (eds.), Plenum Publ. Co., New York, p. 521.

Hermanson, J. and J. C. Phillips (1966), Phys. Rev. **155**, 652.

Herring, C. (1940), Phys. Rev. **57**, 1163.

Herring, C. (1954), Phys. Rev. **96**, 1163.

Herring, C. (1955), Bell Syst. Tech. J. **34**, 237.

Herring, C. and E. Vogt (1956), Phys. Rev. **101**, 944.

Herzberg, G (1937), *Atomic Spectra and Atomic Structure*, Dover Publications.

Hess, K. (1981), J. de Physique **42**, **C7**, 3.

Heydenreich, J., H. Blumtritt, R. Gleichmann and H. Johansen (1981), *Scanning Electron Microscopy*, Vol. 1, SEM, Inc., Illinois.

Hiki, Y. (1981), Ann. Rev. Mat. Sci. **11**, 51.

Hiki, Y. and K. Mukai (1973), J. Phys. Soc. Jpn. **34**, 454.

Hill, R. M. (1971), Philos. Mag. **23**, 58.

Hilsch, R. and R. W. Pohl (1930), Z. Phys. **59**, 812.

Hirsch, P. B. (1985), in *Dislocations and Properties of Real Materials*, M. H. Loretto (ed.), Inst. of Metals, London.

Hirth, J. P. and J. Lothe (1968), *Theory of Dislocations*, McGraw-Hill Book Co., New York.

Hjalmarson, H. P., P. Vogel, D. J. Wolford, and J. D. Dow (1980), Phys. Rev. Lett. **44**, 810.

Ho, K. M. and S. G. Louie (1979), Phys. Rev. **B19**, 1774.

Hobden, M. V. (1967), J. Appl. Phys. **38**, 4635.

Hohenberg, P. and W. Kohn (1964), Phys. Rev. **136**, 864.

Holden, A. (1965), *The Nature of Solids*, Columbia University Press, New York.

Holden, A. and P.S. Morrison (1982), *Crystals and Crystal Growing*, MIT Press Cambridge, Massachusetts.

Holden, A. and P. Singer (1960), *Crystals and Crystal Growing*, (1st Edition) Wesleyan University Press, Garden City, New York.

Holland, M. G. (1964), *Proc. 7th Int. Conf. Semicond.*, Dunod, Paris.

Holstein, T. (1959), Ann. Phys. (N.Y.) **8**, 325.

Holstein, T. (1981), Ann. Phys. (N.Y.) **132**, 212.

Holtzberg, F., S. von Molnar, and J. M. D. Coey (1980), *Handbook of Semiconductors, Vol. 3: Materials Properties and Preparation*, T. S. Moss and S. P. Keller (eds.), North Holland Publ. Co., Amsterdam.

Hönerlage, B., R. Lévy, J. B. Grun, C. Klingshirn, and K. Bohnert (1985), Physics Reports; Physics Letters **124**, 161.

Hooge, F. N. (1969), Phys. Lett. A **29**, 139.

Hooge, F. N., T. G. M. Kleinpfennig, and L. K. J. Damme (1981), Rep. Prog. Phys. **44**, 479.

Höpfel, R. A., J. Shah, and A. C. Gossard (1986), Phys. Rev. Lett. **56**, 765.

Höpfel, R. A., J. Shah, P. A. Wolf, and A. C. Gossard (1986), *Proc. Int. Conf. Phys. Semicond.*, World Scientific Publ., Stockholm, p. 1319.

Höpfel, R. A. and J. Shah (1988), Solid State Electron. **31**, 643.

Hopfield, J. J. (1958), Phys. Rev. **112**, 1555.

Hopfield, J. J. (1964), in *Proc. 7th Int. Conf. on the Physics of Semicond.*, Paris, Academic Press, New York.

Hopfield, J. J. (1967), in *Proc. Int. Conf. II-VII Semicond.*, W. A. Benjamin, Inc., New York, p. 786.

Hopfield, J. J. and D. G. Thomas (1963), Phys. Rev. **132**, 563.

Hopfield, J. J., D. G. Thomas, and R. T. Lynch (1966), Phys. Rev. Lett. **17**, 312.

Hora, H. (1965), Z. Naturforsch. **20a**, 543.

Hornbeck, J. A. and J. R. Haynes (1955), Phys. Rev. **97**, 311.

Horii, K. and Y. Nisida (1970), Proc. Phys. Soc. Jpn. **29**, 798.

Horring, N. J. (1969), Ann. Phys. **54**, 405.

Howarth, D. J. and E. H. Sondheimer (1953), Proc. Roy. Soc. (London) **A219**, 53.

Huang, K. and A. Rhys (1950), Proc. Roy. Soc. (London) Ser. **A204**, 406.

Hubbard, J. (1957), Proc. Roy. Soc. (London) Ser. **A240**, 539.

Hubbard, J. (1963), Proc. Roy. Soc. (London) Ser. **A276**, 238.

Huber, M. C. E. and R. J. Sandeman (1986), Rep. Prog. Phys. **49**, 397.

Huberman, M. and A. W. Overhauser (1982), Phys. Rev. **B25**, 2211.

Hubner, K. and W. Shockley (1960), Phys. Rev. Lett. **4**, 504.

Hull, D. (1975), *Introduction to Dislocations*, Pergamon Press, Oxford.

Hume-Rothery, W. (1936), *The Structure of Metals and Alloys*, Institute of Metals Monograph, London.

Hung, C. S. and J. R. Gleissman (1950), Phys. Res. **79**, 726.

Hutson, A. R. and D. L. White (1962), J. Appl. Phys. **33**, 40.

Hwang, C. J. and J. R. Brews (1971), J. Phys. Chem. Solids **32**, 837.

Iadonisi, G (1984), Rivista Nuovo Cimento **7**, 1.

Ibach, H. (1971), Phys. Rev. Lett. **27**, 253.

Igamberdiev, Kh. T., A. T. Mamadalimov, and P. K. Khabibullaer (1983), Izv. Akad. Nauk, Uzb. SSR. Fiz. Mat. **2**, 39.

Ihara, H., et al. (1988), Nature **334**, 510.

Ihm, J. and M. L. Cohen (1980), Phys. Rev. **B21**, 1527.

Ihm, J., A. Zunger, and M. L. Cohen (1979), J. Phys. C. **12**, 4409.

Inkson, J. C. (1976), J. Phys. C (London) **9**, 117.

Inkson, J. C. (1984), *Many-Body Theory of Solids*, Plenum Press, New York.

Inoue, M. (1985), Superlatt. Microstruct. **1**, 433.

Ioffe, A. F. and A. R. Regel (1960), Progr. Semicond. **4**, 237.

Ishibashi, A., Y. Mori, M. Itabashi, and N. Watanabe (1986), in *Proc. Int. Conf. Phys. Semicond.*, O. Engström (ed.), World Scientific Publ., Stockholm, p. 1365.

Isu, T., D.-S. Jiang, and K. Ploog (1987), Appl. Phys. A **43**, 75.

Itoh, N. (1982), Adv. Phys. **31**, 491.

Jäckle, J. (1986), Rep. Prog. Phys. **49**, 171.

Jackson, D. J. (1962), *Classical Electrodynamics*, John Wiley & Sons, New York.

Jacoboni, C., C. Canali, G. Ottaviani, and A. A. Cluaraula (1977), Solid State Electron. **20**, 71.

Jacoboni, C., F. Nava, C. Canali, and G. Ottaviani (1981), Phys. Rev. **B24**, 1014.

Jacoboni, C. and L. Reggiani (1979), Adv. Phys. **28**, 493.

Jacoboni, C. and L. Reggiani (1983), Rev. Mod. Phys. **55**, 645.

Jaenicke, W. (1972), J. Photogr. Sci. **20**, 2.

Jaffe, J. E. and A. Zunger (1984), Phys. Rev. B **30**, 741.

Jagannath, C., Z. W. Grabowski, and A. K. Ramdas (1981), Phys. Rev. **B23**, 2082.

Jagannath, C. and A. K. Ramdas (1981), Phys. Rev. **B23**, 4426.

Jahn, H. A. and E. Teller (1937), Proc. Roy. Soc. (London) Ser. **A161**, 220.

Jahne, E. (1977), Phys. Stat. Sol. **74**, 275 and **75**, 221.

James, R. W. (1954), *The Optical Principles of the Diffraction of X-rays*, Chs. 1, 6, and 8, G. Bell & Sons, Ltd, London.

James, T. H. (1977), *The Theory of the Photographic Process*, McMillian, New York.

Jandl, S., T. L. Brebner, and P. M. Powell (1976), Phys. Rev. **B13**, 686.

Jansen, R. W. and O. F. Sankey (1986), *Proc. Int. Conf. Phys. Semicond.*, O. Engström (ed.), World Scientific Publ., Stockholm, p. 814.

Jansen, R. W. and O. F. Sankey (1989), Phys. Rev. **B39**, 3192.

Janzén, E., G. Grossman, R. Stedman and H. G. Grimmeiss (1985), Phys. Rev. B **31**, 8000.

Jaros, M. (1977), Phys. Rev. **B16**, 3694.

Jaros, M. (1985), Rep. Prog. Phys. **48**, 1091.

Jaros, M. and S. F. Ross (1973), J. Phys. C (London) **6**, 1753.

Jaros, M. and S. Brand (1976), Phys. Rev. **B14**, 4494.

Jaros, M. and S. Brand (1979), J. Phys. C (London) **12**, 525.

Jaros, M., C. O. Rodriguez, and S. Brand (1979), Phys. Rev. **B19**, 3137.

Jaros, M. and P. J. Dean (1983), Phys. Rev. **B28**, 6104.

Jaswal, S. S. (1975), Phys. Rev. Lett. **35**, 1600.

Jaswal, S. S. (1977), in *Lattice Dynamics*, M. Balkanski (ed.), Flammarion, Paris.

Jaswal, S. S. (1978), J. Phys. C (London) **11**, 3559.

Jaswal, S. S. and T. P. Sharma (1973), J. Phys. Chem. Solids **34**, 509.

Jellison, G. E. and F. A. Modine (1983), Phys. Rev. **B27**, 7466.

Jen, H. R., M. J. Cherng, and G. B. Stringfellow (1986), Appl. Phys. Lett. **48**, 1603.

Johanson, R. E., H. Fritzsche, and A. Vomvas (1989), in *13th Int'l Conf. on Amorph. and Liquid Semicond.*, Nashville, N. C.

Johnson, F. A. (1959), Proc. Roy. Soc. (London), **B73**, 265.

Johnson, F. A. (1966), in *Semiconductors and Semimetals*, R. K. Willardson and A. C. Beer (eds.), Academic Press, New York.

Johnson, J. B. (1925), Phys. Rev. **26**, 71.

Johnson, K.H., H.J. Kolair, J.P. deNeufville, and D.L. Morel (1981), Phys. Rev. **B21**, 643.

Johnson, N. M. (1985), Phys. Rev. **B31**, 5525.

Jones, R. (1981), *Microscopy of Semiconduct. Materials*, Inst. of Phys. Conf. Series, **60**, 45.

Jones, R. L. and P. Fisher (1965), J. Phys. Chem. Solids **26**, 1125.

Jongler, J., C. Hetroit, P. L. Vuillermoz, and J. Triboulet (1980), J. Appl. Phys. **51**, 3171.

Jonscher, A. K. (1983), *Dielectric Relaxation in Solids*, Chelsea Dielectric Press, London.

Joos, G. (1945), Lehrbuch der Theoretischen Physik, Acad. Verl. Gesellschaft, Lpz..

Joseph, A. S. and A. C. Thorsen (1965), Phys. Rev. **138**, A1159.

Jouffrey, B. (1984), *L'ordre et le désordre dans les materiaux*, Les Edition de Physique, Paris, p. 308.

Joyce, B. A. (1985), Rep. Prog. Phys. **48**, 1637.

Junnarkar, M. J., K. Shum, S. Chao, R. R. Alfano, and H. Morkoç (1987), Proc. SPIE **793**, 6.

Jusserand, B., D. Paquet, A. Regreny, and J. Kervarec (1983), Solid State Commun. **44**, 499.

Jusserand, D., D. Paquet, J. Kervarec, and A. Regreny (1984), J. Phys. C (London) **45**, 145.

Kabler, M. N. (1964), Phys. Rev. A **136**, 1236.

Kahn, J. M., L. M. Falicov, and E. E. Haller (1986), Phys. Rev. Lett. **57**, 2077.

Kaiser, W., R. J. Collins, and H. Y. Fan (1953), Phys. Rev. **91**, 1380.

Kamimura, H. and T. Nakayama (1987), Comments Cond. Mat. Phys. **13**, 143.

Kaminska, M., J. M. Baranowski and M. Godlewski (1978), *Proc. 14th Int. Conf. Phys. Semicond.*, Edinburgh, Inst. of Physics, Bristol.

Kamke, E. (1944), *Differentialgleichungen*, Academische Verl. Ges., Leipzig.

Kane, E. O. (1956), J. Phys. Chem. Solids **1**, 245.

Kane, E. O. (1957), J. Phys. Chem. Solids **1**, 249.c

Kane, E. O. (1963), Phys. Rev. **131**, 79.

Kane, E. O. (1969), Phys. Rev. **180**, 852.

Kane, E. O. (1976), Phys. Rev. **B13**, 3478.

Kane, E. O. (1979), *Proc. Int. Summer School Phys. Narrow Gap Semicond.*, Nimes, Springer-Verlag, Berlin, p. 13.

Kaplan, R. (1970), in *Optical Properties of Solids*, D. Haidemenakis (ed.), Gordon & Breach, London.

Karl, N. (1974), in *Festkörperprobleme/Advances in Solid State Physics*, Vol. **14**, H. J. Queisser (ed.), Friedr. Vieweg & Sohn, Braunschweig, p. 261.

Karl, N. (1984), *Organic Semiconductors*, Landoldt-Börnstein, Springer-Verlag, Heidelberg.

Kartheuser, E., J. T. Devreese, and R. Evrard (1979), Phys. Rev. **B19**, 546.

Kash, J. A., (1989), Phys. Rev. **B40**, 3455.

Kash, J. A., J. C. Tsang, and J. M. Hvam (1985), Phys. Rev. Lett. **54**, 2151.

Kasowski, R. V., W. Y. Hsu, and F. Herman (1988), Phys. Rev. **B37**, 2309.

Kasowski, R. V., M. H. Tsai, T. N. Rhodin, and D. D. Chabliss (1986), Phys. Rev. **B34**, 2656.

Kasper, J. S. and S. M. Richards (1964), Acta. Crystallogr. **17**, 752.

Kastner, M., D. Adler, and H. Fritzsche (1976), Phys. Rev. Lett. **37**, 1504.

Katnani, A. D. and G. Margaritondo (1983), Phys. Rev. **B28**, 1944.

Kato, Y., C. I. Yu, and T. Goto (1970), J. Phys. Soc. Jpn. **28**, 104.

Katona, P. (1979), Hiradastech. Ip. Kut. Intez. Kozl. **19**(1), 57.

Kauffer, E., P. Pecheur, and M. Geri (1977), Phys. Rev. **B15**, 4107.

Kaufmann, U. and J Schneider (1983), Adv. Electr. Electron Phys. **58**, 81.

Kayanuma, Y. and S. Fukuchi (1984), J. Phys. Soc. Jpn. **53**, 1869.

Keating, P. N. (1966), Phys. Rev. **145**, 637.

Keever, M., W. Kopp, T. J. Drummond, H. Morkoç, and K. Hess (1982), Jpn. J. Appl. Phys. **21**, 55.

Keevert, J. and V. Gokhale (1987), J. Imaging Sci. **31**, 243.

Keldysh, L. V. (1956), Contemp. Phys. **27**, 395.

Keldysh, L. V. (1958), Sovj. Phys.-JETP **7**, 778.

Keldysh, L. V. (1958a), Sovj. Phys.-JETP **4**, 665.

Keldysh, L. V. (1958b), Sovj. Phys.-JETP **6**, 763.

Keldysh, L. V. (1958c), Zh. Eksp. Teor. Fiz. **34**, 1138.

Keldysh, L. V. (1965), Sov. Phys. JEPT **21**, 1135.

Keldysh, L. V. (1968), *Proc. 9th Int. Conf. Semicond.*, Moscow, Nauka Leningrad, 1303.

Keldysh, L. V. (1986), Contemp. Physics **27**, 395.

Keldysh, L. V. and Tu. V. Kopaev (1965), Phys. Tverd. Tela. **6**, 2791.

Keldysh, L. V. and G. P. Proshko (1964), Sov. Phys. Solid State **5**, 2481.

Keldysh, L. V. and A. N. Kozlov (1968), Sovj. Phys.-JEPT **27**, 521.

Keller, H. J. (ed.) (1977), *Chemistry and Physics of One-Dimensional Metals*, Plenum Press, New York.

Kelly, M. J. and R. J. Kelly (1985), Rep. Prog. Phys. **48**, 12.

Kerker, G. P., A. Zunger, M. L. Cohen, and M. Schlüter (1979), Solid State Commun. **32**, 309.

Khawaja, E. and S. Tomlin (1975), J. Phys. D (London) **8**, 58.

Kildal, H. and J. C. Mikkelsen (1973), Optics. Commun., **9**, 315.

Kildal, H. and G. W. Iseler (1976), Appl. Optics **15**, 3062.

Kimerling, L. C. and J. R. Patel (1979), Appl. Phys. Lett. **34**, 73.

Kimmit, M. F. and G. B. F. Niblett (1963), Proc. Phys. Soc. (London) **82**, 608.

Kirk, D. L. and P. L. Pratt (1967), Proc. Brit. Ceram. Soc. **9**, 215.

Kiselev, V. A. (1974), Sov. Phys.-Solid State **15**, 2338.

Kittel, C. (1949), Phys. Rev. **75**(6), 972.

Kittel, C. (1963), *Quantum Theory of Solids*, John Wiley & Sons, New York, p. 131.

Kittel, C. (1966), *Introduction to Solid State Physics*, John Wiley & Sons, New York.

Kittel, C. (1971), *Introduction to Solid State Physics*, John Wiley & Sons, New York.

Kittel, C. (1986), *Introduction to Solid State Physics*, John Wiley & Sons, New York.

Kivelson, S. (1982), Phys. Rev. B **25**, 3798.

Klaassen, F. M. (1961), Thesis, Vrije Universiteit to Amsterdam.

Klein, M. V. (1986), IEEE J. Quantum Electron. **QE 22**, 1760.

Kleinman, D. A. (1983), Phys. Rev. **B28**, 871.

Kleinman, D. A. and R. C. Miller (1981), Phys. Rev. Lett. **46**, 68.

Kléman, M. (1985), in *Dislocations and Properties of Real Metals*, M. H. Loretto (ed.), The Institute of Metals, London.

Klemens, P. G. (1955), Proc. Phys. Soc. **A68**, 1113.

Klemens, P. G. (1958), in *Solid State Physics*, Vol. **7**, F. Seitz and D. Turnbull (eds.), Academic Press, New York.

Klemens, P. G. (1986), Int. J. Thermophysics **7**, 197.

Klemens, P. G. and T. K. Chu (1976), *Thermal Conductivity*, Plenum Press, New York.

Klinger, M. I. (1979), *Problems of Linear Electron (Polaron) Theory in Semiconductors*, Pergamon Press, Oxford.

Klingshirn, C. and H. Haug (1981), Phys. Rep. **70**, 315.

Klug, H. P. and L. E. Alexander (1974), *X-Ray Diffraction Procedures*, John Wiley & Sons, New York.

Knaak, W., T. Hauß, M. Kummrow, and M. Meißner (1986), in *Phonon Scattering in Condensed Matter*, Vol. **V**, A. C. Anderson and J. P. Wolfe (eds.), Springer-Verlag, Berlin, p. 174.

Knights, J. C., D. K. Biegelsen, and I. Solomon (1977), Solid State Commun. **22**, 113.

Knotek, M. L., M. Pollak, T. M. Donovan, and H. Kurtzman (1973), Phys. Rev. Lett. **30**, 853.

Knox, R. S. (1963), in *Solid State Physics* Suppl. **5**, F. Seitz and D. Turnbull (eds.), Academic Press, New York.

Knox, R. S. (1984), in *Collective Excitations in Solids*, B. DiBartolo and J. Danko (eds.), Plenum Press, New York.

Knox, W. H., C. Hirliman, D. A. B. Miller, J. Shah, D. S. Chemla, and C. V. Shank (1986), Phys. Rev. Lett. **56**, 1191.

Kobiakov, I. B. (1980), Solid State Commun. **35**, 305.

Koeder, V. V., Yu. A. Osip'yan, W. Schröter, and G. Zoth (1982), Phys. Stat. Sol., **72**, 701.

Kogan, Sh. M. and T. M. Lifshitz (1977), Phys. Stat. Sol. A **39**, 11.

Kogan, Sh. M. and B. J. Sedunov (1974), Sov. Phys.-Solid State **8**, 1898.

Kohn, W. (1952), Phys. Rev. **87**, 472.

Kohn, W. (1957), *Solid State Physics*, Vol. 5, F. Seitz and D. Turnbull (eds.), Academic Press, New York, 257.

Kohn, W. (1957), Solid State Phys. **5**, 257.

Kohn, W. and J. M. Luttinger (1955), Phys. Rev. **98**, 915.

Kohn, W. and N. Rostoker (1954), Phys. Rev. **94**, 1411.

Kohn, W. and D. Schechter (1955), Phys. Rev. **99**, 1903.

Kohn, W. and L. J. Sham (1965), Phys. Rev. **140**, 1130.

Kondo, Y., M. Hirai, and M. Veta (1972), J. Phys. Soc. Jpn. **33**, 151.

Königsberger, T. and T. Weiss (1911), Ann. Phys. **35**, 1.

Kopylov, A. A. and A. N. Pikhtin (1975), Sov. Phys. Sol. State, **16**, 1200.

Korringa, J. (1947), Physica **13**, 392.

Kosicki, B. B. and W. Paul (1966), Phys. Rev. Lett. **17**, 246.

Kossel, W. (1952), *Zur Struktur und Materie der Festkörper*, Springer-Verlag, Berlin.

Koster, G. F. and J. C. Slater (1954), Phys. Rev. **95**, 1167 and **96**, 1208.

Koteles, E. S., C. Jagannath, J. Lee, and M. O. Vassell (1987), Proc. SPIE **792**, 168.

Kousik, G. S., C. M. van Vliet, G. Bosman, and P. H. Handel (1986), Adv. Phys. **34**, 663.

Krakauer, H., M. Posternak, A. J. Freeman, and D. D. Koelling (1981), Phys. Rev. **B23**, 3859.

Kramer, P. and R. Neri (1984), Acta. Crystallogr. A **40**, 580.

Kramers, H. A. and R. de L. Kronig (1929), Z. Phys. **30**, 521.

Krebes, E. S. (1977), Surf. Sci. **64**, 339.

Kreissl, A., K. Bohnert, V. G. Lyssenko, and C. Klingshirn (1982), Phys. Stat. Sol. B **114**, 537.

Kressel, H. (1981), Semicond. Semimet. **16**, 1.

Kreuzer, H. J. (1981), *Non-Equilibrium Thermodynamics and its Statistical Foundation*, Clarendon, Oxford.

Krieger, J. B. and G. J. Iafrate (1986), Phys. Rev. **B33**, 5494.

Krivoglaz, M. A. (1974), Sovj. Phys. Usp. **16**, 856.

Kroemer, H. (1975), Crit. Rev. Solid State Sci. (USA) **5**, 555.

Kroemer, H. (1983), Surf. Sci. **132**, 543.

Kröger, F. A. (1964), *The Chemistry of Imperfect Crystals*, North Holland Publ. Co., Amsterdam.

Kronig, R. de L. (1926), J. Opt. Soc. Am. **12**, 547.

Kronig, R. de L. and W. G. Penney (1931), Proc. Roy. Soc. (London) Ser. **A130**, 499.

Kuan, T. S., T. F. Kuech, W. I. Wang, and E. L. Wilkie (1985), Phys. Rev. Lett. **56**, 201.

Kuan, T. S., W. I. Wang, and E. L. Wilkie (1987), Phys. Rev. Lett. **51**, 51.

Kubo, R. (1952), Phys. Rev. **86**, 929.

Kubo, R. (1956), Can. J. Phys. **34**, 1274.

Kubo, R. (1957), J. Phys. Soc. Jpn. **12**, 570.

Kuhl, J. and D. von der Linde (1982), *Picosecond Phenomena* **III**, Springer-Verlag, Berlin.

Kulakovskii, V. D., V. G. Lysenko, and V. B. Timofeev (1985), Usp. Fiz. Nauk **147**, 3.

Kulakovskii, V. D. and V. B. Timofeev (1977), Pis'ma Zh. Eksp. Teor. Fiz. **25**, 487.

Kumar, G. S., J. W. Vandersande, T. Klistner, R. O. Pohl, and G. A. Slack (1985), Phys. Rev. **B31**, 2157.

Kunc, K., M. Balkanski, and M. A. Nusimovici (1975), Phys. Stat. Sol. B **71**, 341; **72**, 229 and 249.

Kuriyama, K. and F. Nakamura (1987), Phys. Rev. **B36**, 4449.

Kurosawa, T. (1966), J. Phys. Soc. Jpn. Suppl. **21**, 414.

Kurtz, S. K., J. Jerphagnon, and M. M. Choi (1978), Landoldt-Börnstein New Series III/II, K. H. Hellwege and A. M. Hellwege (eds.), Springer-Verlag, Berlin.

Labusch, R. and W. Schröter (1980), in *Dislocations in Solids*, Vol. 5, F. R. N. Nabarro (ed.), North Holland, Amsterdam.

Laks, D. B., G. F. Neumark, A. Hangleiter and S. T. Pantelides (1988), Phys. Rev. Lett. **61**, 1229.

Lampert, M. A. (1958), Phys. Rev. Lett. **1**, 450.

Landau, L. (1930), Z. Phys. **64**, 629.

Landau, L. D. (1933), Sovj. Phys. **3**, 664.

Landau, L. D. and G. Rumer (1937), Phys. Z. **11**, 18.

Landau, L. D. and Ya. B. Zel'dovich (1943), Acta. Phys. Chem. USSR **18**, 193.

Landau, L. D. and E. M. Lifshitz (1958), *Quantum Mechanics*, Pergamon Press, Oxford.

Landau, L. D. and E. M. Lifshitz (1958), *Statistical Physics*, English transl. by E. Peierls and R. F. Peierls, Addison-Wesley, Cambridge, Massachusetts.

Landau, L. D. and E. M. Lifshitz (1960), *Electrodynamics of Continuous Media*, Pergamon Press, Oxford.

Landau, L. D. and E. M. Lifshitz (1976), Teor. Fizika **5**, Nauka, Moscow.

Landoldt-Börnstein (1955), *Atom and Molekular Physik*, 4 *Kristalle*, Springer-Verlag, Berlin.

Landoldt-Börnstein (1982), New Series, III. 17 a and b, O. Madelung, M. Schulz, and H. Weiss (eds.), Springer-Verlag, Berlin, Heidelberg, New York.

Landodlt-Börnstein (1987), New Series, III, 22, O. Madelung, M. Schulz (eds.), Springer-Verlag, Berlin.

Landsberg, P. T. (1952), Proc. Roy. Soc. **A213**, 220.

Landsberg, P. T. (1956), Proc. Phys. Soc. (London) **B69**, 1056.

Landsberg, P. T. (1967), Solid State Electron. **10**, 513.

Landsberg, P. T. (1982), in *Handbook on Semiconductors*, T. S. Moss (ed.), Vol. **1**, *Band Theory and Transport Properties*, W. Paul (ed.), North Holland Pub. Co., Amsterdam.

Landsberg, P. T. (1987), in *Energy Transfer Dynamics*, T. H. Barrett and H. A. Pohl (eds.), Springer-Verlag, Berlin.

Landsberg. P. T. (1987), Solid State. Electron. **30**, 1107.

Landsberg, P. T. and S. G. Canagaratna (1984), Phys. Stat. Sol. (b) **126**, 141.

Landsberg, P. T. and H. C. Cheng (1985), Phys. Rev. **B32**, 8021.

Landsberg, P. T. and E. A. Cole (1966), Proc. Phys. Soc. **87**, 229.

Landsberg, P. T. and E. A. Pimpale (1976), J. Phys. **C9**, 1243.

Landsberg, P. T. and A. F. Willoughby (eds.) (1978), *Recombination in Semiconductors*, Solid State Electronics **21**, No. 11/12.

Landsberg, P. T., A. Neugroschel, F. A. Lindholm and C. T. Sah (1985), Phys. Stat. Sol. (b) **130**, 255.

Landsberg. P. T., C. Rhys-Roberts, and P. Lal (1964), Proc. Phys. Soc. **84**, 915.

Lang, D. V. (1979), in *Thermally Stimulated Relaxation in Solids*, P. Bräunlich (ed.), Springer Verlag, Berlin, p. 93.

Lang, D. V. (1986), in *Deep Centers in Semiconductors*, S. T. Pantelides (ed.), Gordon & Breach, New York.

Lang, D. V. and R. A. Logan (1976), J. Appl. Phys. **47**, 1533.

Lang, D. V. and R. A. Logan (1977), Phys. Rev. Lett. **39**, 635.

Lang, D. V., H. G. Grimmeiss, E. Meijer, and M. Jaros (1980), Phys. Rev. **B22**, 3917.

Langmuir, I. (1920), Trans. Faraday Soc. **15**, 62.

Lannoo, M. and J. Bourgoin (1981), *Point Defects in Semiconductors*, Springer-Verlag, Berlin.

Larach, S., R. E. Schrader, and C. F. Stocker (1957), Phys. Rev. **108**, 587.

Larsen, D. M. (1973), Phys. Rev. **B8**, 535.

Larsen, D. M. (1976), Phys. Rev. **B13**, 1681.

Lasher, G. and F. Stern (1964), Phys. Rev. **133**, A553.

Laubereau, A. (1984), *Semiconductors Probed by Ultrafast Laser Spectroscopy* Vol. **I**, Academic Press, New York.

Laubereau, A. and W. Kaiser (1974), Optoelectronics **6**, 1.

Laubereau, A., D. von der Linde, and W. Kaiser (1971), Phys. Rev. Lett. **27**, 802.

Laudise, R.A. (1970), *The Growth of Single Crystals*, Prentice Hall, Englewood Cliffs, New Jersey.

Laughlin, R. B. (1983), Phys. Rev. Lett. **50**, 1395.

Lautenschlager, P., P. B. Allen, and M. Cardona (1985), Phys. Rev. **B31**, 2163.

Lautenschlager, P., P. B. Allen, and M. Cardona (1986), Phys. Rev. **B33**, 5501.

Lautz, G. (1961), in *Halbleiterprobleme*, Vol. 6, F. Sauter (ed.), Friedr. Vieweg & Sohn, Braunschweig, p. 21.

Lax, B. (1963), *Enrico Fermi School of Physics Semiconductors*, Smith, (ed.), Academic Press, New York, p. 240.

Lax, B., L. M. Roth, and S. Zwerdling (1959), J. Phys. Chem. Soc. **8**, 311.

Lax, M. (1952), J. Chem. Phys. **20**, 1752.

Lax, M. (1960), Phys. Rev. **119**, 1502.

Lax, M. and J. J. Hopfield (1961), Phys. Rev. **124**, 115.

LeBlanc Jr., O. H. (1967) in *Phys. and Chem. of the Organic Solid State*, Vol. 3, D. D. Fox et al. (eds.), Interscience Publ., New York, p. 133.

Lebwohl, P. A. and P. J. Price (1971), Solid State Commun. **9**, 1221.

LeComber, P. G., A. Madan, and W. E. Spear (1972), J. Non-Cryst. Solids **11**, 219.

LeComber, P. G. and W. E. Spear (1976), *AIP Conf. Proc.* **31**, 284.

Lee, P. A., A. D. Stone, and H. Fukuyama (1987), Phys. Rev. **35**, 1039.

Lee, T. D., F. Low, and D. Pines (1953), Phys. Rev. **90**, 297.

Legros, R., and Y. Marfaing (1973), Phys. Stat. Sol. A **19**, 635.

Leheny, R. F., J. Shah, R. L. Fork, C. V. Shank, and A. Migus (1979), Solid State Commun. **31**, 809.

Leibfried, G. (1951), Z. Phys. **129**, 307.

Leibfried, G. and E. Schlömann (1963), Nachr. Akad. Wiss. Göttin gen, Kl. **2a**, 4; 71 (1951); English trans. AEC-tr-5892, 1 (1963).

Leighton, R. B. (1959), *Principles of Modern Physics*, McGraw-Hill Book Co., Inc., New York.

Leite, J. R., B. I. Bennett, and F. Herman (1975), Phys. Rev. **B12**, 1466.

Leite, R. C., J. F. Scott, and T. C. Damen (1969), Phys. Rev. Lett. **22**, 780.

Leonelli, R., J. C. Mathae, J. M. Hvam, and R. Levy (1986), *Proc. Int. Conf. Phys. Semicond.*, Stockholm, p. 1437.

Levi, A. F. J., J. R. Hayes and R. Bhat (1986), Appl. Phys. Lett. **48**, 1609.

Levine, B. F., K. K. Choi, C. G. Bethea, J. Walker, and R. J. Malik (1987), Appl. Phys. Lett. **50**, 273.

Levine, D. I. and P. J. Steinhardt (1984), Phys. Rev. Lett. **53**, 2477.

Levinger, B. W. and D. R. Frankl (1961), J. Phys. Chem. Solids **20**, 281.

Levinson, I. V. (1965), Sov. Phys.-Solid State **7**, 2336.

Levitt, A.P. (1970), *Whisker Technology*, Wiley-Interscience, New York.

Levy, R., B. Honerlage, and J. B. Grun (1979), Phys. Rev. **B19**, 2326.

Lietoile, A., A. Wakita, T. W. Sigmon, and J. F. Gibbons (1982), J. Appl. Phys. **53**, 4399.

Lifshitz, I. M. (1964), Adv. Phys. **13**, 483.

Lifshitz, I. M. (1965), Sovj. Phys. Usp. **7**, 549.

Lifshitz, T. M. and F. Ya. Nad' (1965), Sov. Phys. Dokl. **10**, 532.

Lightowlers, E. C. and G. Davis (1985), Solid State Commun. **53**, 1055.

Lindefelt, U. and A. Zunger (1982), Phys. Rev. **B26**, 846.

Lindefelt, U. and A. Zunger (1984), Phys. Rev. **B30**, 1102.

Lines, M. R. (1986), Ann. Rev. Mat. Sci. **16**, 113.

Liouville, J. (1838), Journ. de Math. **3**, 349.

Lipari, N. O. and A. Baldereschi (1970), Phys. Rev. Lett. **25**, 1660.

Lipari, N. O. and A. Baldereschi (1978), Solid State Commun. **25**, 665.

Lipari, N. O., J. Bernholc, and S. T. Patelides (1979), Phys. Rev. Lett. **43**, 1354.

Lischka, K. (1986), Phys. Stat. Sol. **133**, 17.

Little, W. A. (1964), Phys. Rev. **134**, A 1416.

Little, W. A. (1988), Science **242**, 1390.

Lochmann, W. and A. Haug (1980), Solid State Commun. **35**, 553.

Loferski, J. J. and P. Rappoport (1958), Phys. Rev. **111**, 432.

Logachev, Ya. A. and L. N. Vasilev (1973), Fiz. Tverd. Tela **15**, 1612.

London, F. and H. London (1935), Proc. Roy. Soc. (London) Ser. **A149**, 72.

Lorentz, H. A. (1909), *The Theory of Electrons*, Teubner Verlag, Lpz..

Loucks, T. L. (1965), Phys. Rev. **139**, A1333.

Loucks, T. L. (1967), *Augmented Plane Wave Method*, W. A. Benjamin, Inc., New York.

Loudon, R. (1963), Proc. Roy. Soc. (London) Ser. **A275**, 218.

Loudon, R. (1964), Adv. Phys. **13**, 423.

Loudon, R. (1973), *The Quantum Theory of Light*, Clarendon Press, Oxford.

Louie, S. G. (1986), *Proc. Int. Conf. Phys. Semicond.*, World Scientific Publ., Stockholm, p. 1095.

Louie, S. G., M. Schlüter, J. R. Chelikowsky and M. L. Cohen (1976), Phys. Rev. **B13**, 1654.

Louie, S. G., K. M. Ho, J. R. Chelikowsky, and M. L. Cohen (1977), Phys. Rev. B **15**, 5627.

Louie, S. G., K. M. Ho, and M. L. Cohen (1979), Phys. Rev. B**19**, 1774.

Löwdin, P. O. (1956), Adv. Phys. **5**, 1.

Lucovsky, G. (1965), Solid State Commun. **3**, 299.

Lucovsky, G. (1985), in *Physical Properties of Amorphous Materials*, D. Adler, B. B. Schwartz, and M. S. Steele (eds.), Plenum Press, New York.

Lucovsky, G., C. K. Wong, and W. B. Pollard (1983), J. Non-Cryst. Solids **59, 60**.

Ludwig, G. W. and H. H. Woodbury (1962), in *Solid State Physics*, Vol. 13, F. Seitz and D. Turnbull (eds.), Academic Press, New York.

Ludwig, W. (1974), *Proc. 2nd Int. Conf. on Solid Surf.*, Jpn. J. Appl. Phys. Suppl. **2**, 879.

Luffinger, J. M. and W. Kohn (1955), Phys. Rev. **97**, 969.

Lugli, P. (1988), Solid State Electron. **31**, 667.

Lundstrom, M. S., R. J. Schwartz, and J. L. Gray (1981), Solid State Electron. **24**, 195.

Luttinger, J. M. (1956), Phys. Rev. **102**, 1030.

Luttinger, J. M. and W. Kohn (1955), Phys. Rev. **97**, 869.

Lüty, F. (1960), Z. Phys. **160**, 1.

Lüty, F. (1973), Surf. Sci. **37**, 120.

Luzzi, R. and A. R. Vasconellos (1984), *Semiconductors Probed by Ultrafast Laser Spectroscopy*, Vol. 1, Academic Press, New York, p. 135.

Lyddane, R. H., R. G. Sachs, and E. Teller (1941), Phys. Rev. **59**, 673.

Lyon, S. A. (1986), J. Lumin. **35**, 121.

MacDonald, W. M., H. N. Rosenbluth, and W. Chuck (1957), Phys. Rev. **107**, 350.

MacFarlane, G. G., T. P. McLean, J. E. Quarrington, and V. Roberts (1957), Phys. Rev. **108**, 1377.

MacFarlane, G. G. and V. Roberts (1955), Phys. Rev. **97**, 1714, and **98**, 1865.

Mackawa, S. (1970), Phys. Rev. Lett. **24**, 1175.

Mackenzie, K. D., J. R. Eggert, D. J. Leopold, Y. M. Li, S. Lin, and W. Paul (1985), Phys. Rev. B**31**, 2198.

MacKinnon, A. (1981), in *Festkörperprobleme/Advances in Solid State Physics*, Vol. 21, J. Treusch (ed.), p. 149.

Madelung, E. (1909), Göttinger Nachr. 100.

Madelung, E. (1918), Physik. Z. **19**, 524.

Madelung, E. (1950) *Die Mathematischen Hilfsmittel des Physikers*, Springer-Verlag, Berlin.

Madelung, O. (1970), *Grundlagen der Halbleiterphysik*, Springer-Verlag, Heidelberg.

Madelung, O. (1973), *Festkörpertheorie*, Springer Verlag, Berlin, Heidelberg, New York.

Madelung, O. (1978), *Introduction to Solid State Theory*, Springer-Verlag, Berlin, Heidelberg, New York.

Mahan, G. D. (1980), J. Appl. Phys. **51**, 2634.

Mahan, G. D. and J. J. Hopfield (1964), Phys. Rev. Lett. **12**, 241.

Mahr, H. and M. D. Hirsch (1975), Opt. Commun. **13**, 96.

Maiman, T. (1960), Nature **187**, 493.

Makowski, L. and M. Glicksman (1973), J. Phys. Chem. Solids **34**, 487.

Maksym, P. A. and C. J. Hearn (1984), *Semiconductors Probed by Ultrafast Laser Spectroscopy*, Vol. 1, Academic Press, New York, p. 77.

Malony, T. J. and J. Frey (1978), J. Appl. Phys. **48**, 781.

Malvezzi, A. M. (1987), Proc. SPIE **793**, 49.

Mandelbrot, B. (1981), *The Fractal Geometry of Nature*, Freeman, San Francisco.

Manenkov, A. A., V. A. Miljaiv, G. N. Mihilova, V. A. Sanina, and A. S. Seferov (1976), Zh. Eksp. Teor. Fiz. **70**, 695.

Manley, J. and H. E. Rowe (1959), Proc. IRE **47**, 2115.

Mansfield, R. (1956), Proc. Phys. Soc. (London) Ser. **B69**, 862.

Many, A., Y. Goldstein, and N.B. Grover (1965), *Semiconductor Surfaces*, North Holland Publ. Co., Amsterdam.

Marchetti, A. P. and D. R. Kearns (1970), Mol. Cryst. Liq. Cryst. **6**, 299.

Marchetti, M.C., W. Cai and M. Lax (1988), Solid State Electron. **31**, 677.

Marfaing, Y. (1980), in *Handbook on Semiconductors, Vol. 2: Optical Properties of Solids*, T. S. Moss and M. Balkanski (eds.), North Holland Publ. Co., Amsterdam.

Margaritondo, G., D. L. Huber and C. G. Olson (1989), Science **246**, 770.

Markham, J. J. (1955), Phys. Rev. **103**, 508.

Markham, J. J. (1966), *F-centers in Alkali Halides*, Academic Press, New York.

Markham, J. J. (1966), in *Solid State Physics*, Suppl. 8, F. Seitz and D. Turnbull (eds.), Academic Press, New York, p. 370.

Markiewicz, R. S., J. P. Wolfe, C. D. Jeffries (1977), Phys. Rev. **B15**, 1988.

Marple, D. T. F. (1964), J. Appl. Phys. **35**, 1241.

Martienssen, W. (1951), Naturwissenschaften **38**, 482.

Martin, R. M. (1970), Phys. Rev. **B1**, 4005.

Martinez, G. (1980), in *Handbook on Semiconductors, Vol. 2: Optical Properties of Solids*, T. S. Moss and M. Balkanski (eds.), North Holland Publ. Co., Amsterdam.

Martinez, G., M. Schlüter, and M. L. Cohen (1975a), Solid State Commun. **17**, 5.

Martinez, G., M. Schlüter, and M. L. Cohen (1975b), Phys. Rev. **B11**, 660.

Martins, J. L. and A. Zunger (1984), Phys. Rev. **B30**, 6217.

Martins, J. L. and A. Zunger (1986), J. Mat. Res. **1**, 523.

Maruani, A., J. L. Oudar, E. Batifol, and D. S. Chemla (1978), Phys. Rev. Lett. **41**, 1372.

Mason, W. P. and T. B. Bateman (1964), Phys. Rev. A **134**, 1387.

Matsumoto, D. S. and A. C. Anderson (1981), J. Non-cryst. Solids **44**, 171.

Matsumoto, Y. and S. Shionoya (1982), J. Phys. Soc. Jpn. **51**, 181.

Matsumoto, Y., Y. Unuma, and S. Shionoya (1984), *Semiconductors Probed by Ultrafast Laser Spectroscopy*, Vol. 1, Academic Press, New York, p. 307.

Mattern, P. L., K. Lengweiler, P. W. Levy, and P. D. Esser (1970), Phys. Rev. Lett. **24**, 1287.

Matthews, N. F. J. and P. J. Warter, Jr. (1966), Phys. Rev. **144**, 610.

Mattis, D. and O. Sinha (1970), Ann. Phys. **61**, 214.

Mavroides, J. G. (1972), in *Optical Properties of Solids*, F. Abeles (ed.), North Holland Publ. Co., Amsterdam.

Mayer, N. I. and M. H. Jørgensen (1970), *Advances in Solid State Physics*, Pergamon Press, Oxford, and Friedr. Vieweg & Sohn, Braunschweig.

McLean, T. P. and M. Grosman (1963), *Polarons and Excitons*, Proc. Scottish Univ. Summer School, Edinburgh, C. G. Cooper and G. D. Whitfield (eds.), Oliver & Boyd, London.

McClure, D. S. (1961), *Solid State Physics*, Vol. 9, H. Ehrenreich, F. Seitz, and D. Turnbull (eds.), Academic Press, New York.

McCombe, B. D. and R. J. Wagner (1975), Adv. in Electrons and Electronic Phys. **37**, 1.

McDougall, J. and E. C. Stoner (1938), Phil. Trans. A **237**, 67.

McFee, J. H. (1966), *Physical Acoustics*, Vol. 4A, W. Mason (ed.), Academic Press, New York.

McInnes, J. A. and P. N. Butcher (1979), Phil. Mag. **B39**, 1.

McKay, K. G. (1954), Phys. Rev. **94**, 877.

McKelvey, J. P. (1966), *Solid State and Semiconductor Physics*, Harper & Row, New York.

McLean, T. P. and E. G. S. Paige (1960), J. Phys. Chem. Soc. **16**, 220.

McMurray, R. E., Jr. (1985), Solid State Commun. **53**, 1127.

McSkimmin, H. J. and P. Andreatch, Jr. (1964), J. Appl. Phys. **35**, 3312.

McSkimmin, H. J. and P. Andreatch, Jr. (1967), J. Appl. Phys. **38**, 2610.

McWeeny, R. (1957), Proc. Roy. Soc. (London) Ser. **A241**, 239.

McWorther, A. L. (1955), Lincoln Lab Report No. 80, Boston.

Mees, K. (1954) *Theory of the Photographic Process*, McMillian & Co., New York.

Meissner, W. and R. Ochsenfeld (1933), Naturwiss. **21**, 787.

Melngailis, I., G. E. Stilman, J. O. Dimmock and C. M. Wolfe (1969), Phys. Rev. Lett. **23**, 1111.

Mendez, E. E. (1986), IEEE J. Quantum Electron. **QE 22**, 1720.

Mendez, E. E., F. Agulló-Rueda, and J. M. Hong (1988), Phys. Rev. Lett. **60**, 2426.

Mendez, E. E., G. Bastard, L. L. Chang, L. Esaki, H. Morkoç, and R. Fischer (1982), Phys. Rev. **B26**, 7101.

Menelle, A. and R. Bellissent (1986), *Proc. Int. Conf. Phys. Semicond.*, O. Engström (ed.), World Scientific Publ., Stockholm, p. 1049.

Mengel, P., H. Liebing, and W. Ruppel (1975), Phys. Stat. Sol. B **72**, 525.

Mergel, D. and R. Labusch (1982), Phys. Stat. Sol. A **69**, 151.

Merz, J. L. (1968), Phys. Rev. **176**, 961.

Messmer, R. P. and G. J. Watkins (1973), Phys. Rev. **B7**, 2568.

Meyer, H. (1974), *Organic Semiconductors, Dark and Photoconductivity*, Verlag Chemie, Weinheim.

Meyer, H. J. G. (1958), Phys. Rev. **112**, 298.

Meyer, H. J. G. and D. Polder (1953), Physica **19**, 255.

Michaelson, H. B. (1972), *Handbook of Chemistry and Physics*, 52nd edition, Chemical Rubber Co., Cleveland, Ohio.

Midwinter, J. E. and J. Warner (1965), Br. J. Appl. Phys. **16**, 1135.

Miller, A. and B. Abrahams (1960), Phys. Rev. **120**, 745.

Miller, A., A. MacKinnon, and D. Weaire, (1981) in *Solid State Physics*, Vol. 36, H. Ehrenreich, F. Seitz, and D. Turnbull (eds.), Academic Press, New York.

Miller, D. A. B., D. S. Chemla, T. C. Damen, A. C. Gossard, W. Wiegmann, T. H. Wood, and A. C. Burrus (1985), Phys. Rev. **B32**, 1043.

Miller, D. A. B., J. S. Weiner, and D. S. Chemla (1986), IEEE J. Quantum Electron. **QE 22**, 1816.

Miller, R. C. (1964), Appl. Phys. Lett. **5**, 17.

Miller, R. C., D. A. Kleinman, W. A. Nordland, Jr., and A. C. Gossard (1980), Phys. Rev. **B22**, 863.

Miller, R. C., A. C. Gossard, W. T. Tsang, and O. Munteanu (1982), Solid State Commun. **43**, 519.

Miller, R. C., D. A. Kleinman, A. C. Gossard, and O. Munteanu (1982a), Phys. Rev. **B22**, 6545.

Miller, R. C. and D. A. Kleinman (1985), J. Lumin. **30**, 520.

Mills, R. G. J. and E. W. Montroll (1970), J. Math. Phys. **11**, 2525.

Milnes, A. G. (1983), Adv. Elec. Electron Phys. **61**, 63.

Minomura, S. and H. G. Drickamer (1962), J. Phys. Chem. Solids **139**, 451.

Minot, C., J. Chavignon, H. Le Person, and J. L. Oudar (1984), Solid State Commun. **49**, 141.

Mints, R. I., I. I. Milman, and V. I. Kryuk (1976), Sov. Phys. Usp. **19**, 697.

Mirlin, D. N. (1984), *Optical Orientation*, F. Meyer and B. P. Zakharchenya (eds.), Elsevier Science Publ. Co., New York.

Mirlin, D. N. and I. I. Reshina (1977), Sov. Phys. JEPT **46**, 457.

Mirlin, D. N., I. Ya. Karlik, L. P. Nikitin, I. I. Reshina, and V. F. Sapega (1980), Solid State Commun. **37**, 757.

Mitchell, J. W. (1958), Rep. Progr. Phys. **20**, 433.

Mitchell, J. W. (1989), J. Imaging Sci. **33**, 103.

Mitra, S. S. (1969), *Optical Properties of Solids*, S. Nudelman and S. S. Mitra (eds.), Plenum Press, New York.

Mitra, S. S. and M. E. Massa (1982), in *Handbook on Semiconductors, Vol. 1: Band Theory and Transport Properties*, T. S. Moss and W. Paul (eds.), North Holland Publ. Co., Amsterdam.

Mitra, T. K. (1969), J. Phys. C (London) **2**, 52.

Moddel, G., D. Anderson, and W. Paul (1980), Phys. Rev. **B22**, 1918.

Moldover, M. R., J. P. M. Trusler, and T. J. Edwards (1988), Phys. Rev. Lett. **60**, 249.

Moll, J. L. (1964), *Physics of Semiconductors*, McGraw-Hill Book Co., Inc., New York.

Mollwo, E. (1931), Nachr. Wiss. Ges. Göttingen, p. 97.

Mollwo, E. (1943), Reichsberichte **1**, 1.

Mooradian, A. (1968), Phys. Rev. Lett. **20**, 1102.

Mooradian, A. and G. B. Wright (1966), Solid State Commun. **4**, 431.

Moore, C. E. (1970), *Ionization Potentials and Ionization Limits Derived From the Analysis of Optical Spectra*, NSRDS-NBS 34 National Bureau of Standards, Washington D.C.

Moore, E. J. (1967), Phys. Rev. **160**, 607 and 618.

Moore, G. E. and M. V. Klein (1969), Phys. Rev. **179**, 722.

Moore, W. J. (1971), J. Phys. Chem. Solids **32**, 93.

Mooser, E. and W. B. Pearson (1956), J. Electronics, Vol **1**, 629.

Mooser, E. and W. B. Pearson (1962), Acta. Crystallogr. **12**, 1015.

Monroe, D. (1985), Phys. Rev. Lett. **54**, 146.

Morgan, T. N. (1972), *Proc. Int. Conf. Semicond.*, Polish Scient. Publ., Warsaw.

Mori, S. and T. Ando (1980), J. Phys. Soc. Jpn. **48**, 865.

Morimoto, T. and K. Tani (1962), J. Phys. Soc. Jpn. **17**, 1121.

Morin, F. J. (1954), Phys. Rev. **93**, 62.

Morita, A. and H. Nara (1966), J. Phys. Soc. Jpn., (Suppl.) **21**, 234.

Morkoç, H. (1986), IEEE J. Quantum Electron. **QE 22**, c185.

Morse, P. M. and H. Feshbach (1953), *Methods of Theoretical Physics*, McGraw-Hill Book Co., Inc., New York.

Mort, J. and D. M. Pai (eds.) (1976), *Photoconductivity and Related Phenomena*, Elsevier Scientific, New York, p. 502.

Moskalenko, S. A. (1958), Opt. Spektrosk. **5**, 147.

Mosley, L. E. and M. A. Paessler (1984), Appl. Phys. Lett. **45**, 86.

Moss, S.C. and J.F. Graczyk (1970), *Proc. 10th Int. Conf. Semicond.*, Washington, D.C., p. 658.

Moss, T. S. (1961), J. Appl. Phys. Suppl. **32**, 2136.

Moss, T. S. (1961a), *Optical Properties of Semiconductors*, Butterworth, London.

Moss, T. S., G. J. Burrell and B. Ellis (1973), *Semiconductor Optoelectronics*, John Wiley & Sons, New York.

Mott, N. F. (1938), Trans. Faraday Soc. **34**, 500.

Mott, N. F. (1968), J. Non-cryst. Solids **1**, 1.

Mott, N. F. (1969), Philos. Mag. **19**, 835.

Mott, N. F. (1969), in *Festkörperprobleme*, Vol. 9, O. Madelung (ed.), Friedr. Vieweg & Sohn, Braunschweig, p. 22.

Mott, N. F. (1974), *Metal-Insulator Transitions*, Barnes and Noble, New York.

Mott, N. F. (1977), Philos. Mag. **35**, 111.

Mott, N. F. (1984), Rep. Prog. Phys. **47**, 909.

Mott, N. F. (1987), *Conduction in Non-Crystalline Materials*, Clarendon Press, Oxford.

Mott, N. F. and H. S. W. Massey (1965), *The Theory of Atomic Collisions*, Clarendon Press, Oxford.

Mott, N. F. and M. Kaveh (1985), Adv. in Phys. **34**, 329.

Mott, N. F. and E. A. Davis (1979), *Electronic Processes in Non-Crystalline Materials*, Clarendon Press, Oxford.

Mueller, H. (1935), Phys. Rev. **47**, 947.

Mulliken, R. S. (1952), Proc. Nat. Acad. Sci. **38**, 160.

Musgrave, M. J. P. (1970), *Crystal Acoustics*, Holden-Day, San Francisco.

Nabarro, F. R. N. (1967), *Theory of Crystal Dislocations*, Clarendon Press, Oxford.

Nag, B. R. (1975), J. Appl. Phys. **46**, 4819.

Nag, B. R. (1980), *Electron Transport in Compound Semiconductors*, Springer-Verlag, Berlin.

Nag, B. R. (1984), *Semiconductors Probed by Ultrafast Laser Spectroscopy*, Vol. 1, Academic Press, New York, p. 3.

Nakamura, A. and C. Weisbuch (1978), Solid State Electron. **21**, 1331.

Nakayama, H. and H. Fujita (1985) in *Gallium Arsenide and Related Compounds*, M. Fujimoto (ed.), Inst. Phys. Conf. Ser. No. **79**, Hilger, Boston.

Namjoshi, K. V., S.S. Mitra, and J. F. Vetelino (1971), Phys. Rev. **B3**, 4951.

Nara, H. (1965), J. Phys. Soc. Jpn. **20**, 778.

Nara, H. and A. Morita (1966), J. Phys. Soc. Jpn. **21**, 1852.

Narita, S. (1985), Solid State Commun. **53**, 1115.

Nash, J. C. and J. W. Holm-Kennedy (1974), Appl. Phys. Lett. **24**, 139.

Nathan, M. I., W. P. Dumke, G. Burus, F. H. Dill, and G. Lasher (1962), Appl. Phys. Lett. **1**, 62.

Nazarewicz, W., P. Rolland, E. de Silva and M. Balkanski (1962), Appl. Opt. **1**, 369.

Nelin, G. and G. Nilsson (1972), Phys. Rev. **B5**, 3151.

Nelin, G. and G. Nilsson (1972a), Phys. Rev. **B6**, 3777.

Nelson, D. F. (1981), *Solitons in Condensed Matter Physics*, A. R. Bishop and T. Schneider (eds.), Springer-Verlag, Berlin.

Nelson, D. R. (1986), Sci. Am. **255**, 42.

Newnham, R. E. (1975), *Structure-Property Relations*, Springer-Verlag, New York.

Nguyen, V. T. and T. J. Bridges (1972), Phys. Rev. Lett. **29**, 359.

Nguyen, V. T. and E. G. Burkhardt (1976), Appl. Phys. Lett. **28**, 187.

Niekisch, E. A. (1955), Ann. Phys. (Lpz.) **15**, 279 and 288.

Niekisch, E. A. (1961), Z. Phys. **161**, 38.

Nichols, C. S., C. G. Van de Walle, and S. T. Pantelides (1989), Phys. Rev. Lett. **62**, 1049.

Nichols, J. F. (1951), Proc. Phys. Soc. (London) **64**, 953.

Nicolet, M.-A., H. R. Bilger, and R. J. J. Zijlstra (1975), Phys. Stat. Sol. **b70**, 9 and 415.

Nikitine, S., J. B. Grun, and M. Sieskind (1961), J. Phys. Chem. Solids **17**, 292.

Noras, J. M. (1980), J. Phys. C. **13**, 4779.

Norton, P. and H. Levinstein (1972), Phys. Rev. **B6**, 470.

Novika, S. I. (1966), in *Semiconductors and Semimetals*, R. K. Willardson and A. C. Beer (eds.), Academic Press, New York.

Novikov, B. V., N. S. Sokolov, and S. V. Gastev (1976), Phys. Stat. Sol. B **74**, 81.

Nozik, A. J., B. R. Thacker, J. A. Turner, and J. M. Olson (1985), J. Am. Chem. Soc. **107**, 7805.

Nurmikko, A. V. (1987), Proc. SPIE **793**, 19.

Nusimovici, M. A., M. Balkanski, and J. L. Birman (1970), Phys. Rev. **B1**, 595.

Nyquist, H. (1928), Phys. Rev. **32**, 110.

Olson, C. G., R. Liu, A.-B. Yang, D. W. Lynch, A. J. Arco, R. S. List, B. W. Veal, Y. C. Chang, P. Z. Jiang and A. P. Paulikas (1989), Science **245**, 731.

Onodera, Y. and M. Okazaki (1966), J. Phys. Soc. Jpn. **21**, 1273.

Onsager, L. (1952), Philos. Mag. **43**, 1006.

Onsager, L. (1961), Phys. Rev. Lett. **7**, 50.

Onton, A. (1971), Phys. Rev. **B4**, 4449.

Onton, A., P. Fisher, and A. K. Ramdas (1967), Phys. Rev. **163**, 686.

Onton, A. and R. C. Taylor (1970), Phys. Rev. **B1**, 2587.

Onton, A., Y. Yacobi and R. J. Chicota (1972), Phys. Rev. Lett. **28**, 966.

Onuki, M. and K. Shiga (1966), *Proc. Int. Conf. Semicond.*, Kyoto, p. 427.

Oppenheimer, J. R. (1928), Phys. Rev. **31**, 66.

Orbach, R. (1967), IEEE Trans. Sonics Ultrason. **14**, 140.

Orbach, R. (1984), J. Stat. Phys. **36**, 735.

O'Rourke, R. C. (1953), Phys. Rev. **91**, 265.

Orowan, E. (1934), Z. Phys. **89**, 634.

Osbourn, G. C. (1986), IEEE J. Quantum Electron. **QE 22**, 1677.

Osbourn, G. C., P. L. Gourley, I. J. Fritz, R. M. Biefeld, L. R. Dawson, and T. E. Zipperian (1986), in *Semiconductors and Semimetals*, R. K. Willardson and A. C. Beer (eds.), Academic Press, New York.

Osip'yan, Yu. A. (1983), Journ. de Physique, Colloque C4, Suppl. 9, **44**, 103.

Osip'yan, Yu. A. and I. B. Savchenko (1968), Sov. Phys.-JETP Lett. **7**, 100.

Osip'yan, Yu. A., V. F. Petrenko, and A. V. Zaretskiǔ (1986), Adv. Phys. **35**, 115.

Ottaviani, G. L. Reggiani, C. Canali, F. Nava, and A. Alberigi-Quaranta (1975), Phys. Rev. **B12**, 3318.

Otto, A. (1974) *Advances in Solid State Physics*, Vol. 4, p. 1.

Oudar, J. L., I. Abram, A. Migus, D. Hulin, and J. Etchepare (1985), J. Lumin. **30**, 340.

Oudar, J. L., D. Hulin, A. Migus, A. Antonetti, and F. Alexandre (1985), Phys. Rev. Lett. **55**, 2074.

Oudar, J. L., A. Migus, D. Hulin, G. Grillon, J. Etchepare, and A. Antonetti (1984), Phys. Rev. Lett. **53**, 384.

Ourmazd, A. and J. C. Bean (1985), Phys. Rev. Lett. **55**, 765.

Oussel, J. C., J. Leotin, S. Askenazy, M. S. Skolnik, and R. A. Stradling (1976), J. Phys. C (London) **9**, 2802.

Overhauser, A. W. (1978), Adv. Phys. **27**, 343.

Overhof, H. (1976), in *Fertkörperprobleme/Advances in Solid State Physics*, Vol. 16, 239.

Overhof, H. and W. Beyer (1981), Phil. Mag. **B43**, 433.

Ovshinsky, S. R. (1968), Phys. Rev. Lett. **21**, 1450.

Ovshinsky, S. R. (1976), *Int. Conf. Struct. Excit. Amorph. Solids*, *AIP Conf. Proc.* **31**, 31.

Ovshinsky, S. R. (1976b), Phys. Rev. Lett. **36**, 1469.

Ovshinsky, S. R. (1977), *Proc. 7th Int. Conf. Amorph. Liqu. Semicond.*, Edinburgh, p. 519.

Ovshinsky, S. R. (1980), J. Non-cryst. Solids **42**, 335.

Ovshinsky, S. R. and D. Adler (1978), Contemp. Phys. **19**, 109.

Ovshinsky, S. R., R. T. Young, D. D. Allred, G. DeMaggio, and G. A. Van der Leeden (1987), Phys. Rev. Lett. **58**, 2579.

Pai, D. M. (1975), J. Appl. Phys. **46**, 5122.

Pajot, B. and P. Clauws (1986), *Proc. Int. Conf. Phys. Semicond.*, O. Engström (ed.), World Scientific Publ., Stockholm, p. 911.

Palik, E. D. and J. Furdyna (1970), Rep. Prog. Phys. **33**, 1193.

Palik, E. D. (ed.) (1985), *Handbook of Optical Constants of Solids*, Academic Press, New York.

Palik, E. D., S. Teitler, B. W. Henvis, and R. G. Wallis (1962), *Proc. Int. Conf. Semicond.*, Exeter, The Institute of Physics, London, and the Physical Society, London.

Pamplin, B. (1980) *Crystal Growth* (2nd edition), Pergamon, New York.

Pandey, K. (1986), in *Defects in Semiconductors*, H. J. von Bardeleben (ed.), Material Science Forum Vol. **10–12**, Trans. Tech. Publ., Ltd., Switzerland.

Pankove, J. I. and J. E. Berkeyheiser (1980), Appl. Phys. Lett. **37**, 705.

Pankove, J. I., D. E. Carlson, J. E. Berkeyheiser and R. O. Wance (1985), Phys. Rev. Lett. **51**, 2224.

Pantelides, S. T. (1975), in *Festkörperprobleme/Advances in Solid State Physics*, Vol. 15, H. J. Queisser (ed.), F. Vieweg & Sohn, Braunschweig, pp. 45 and 149.

Pantelides, S. T. (1978), Rev. Mod. Phys. **50**, 797.

Pantelides, S. T. (1986), *Proc. Int. Conf. Phys. Semicond.*, O. Engström (ed.), World Scientific Publ., Stockholm, p. 988.

Pantelides, S. T. (1987), Phys. Rev. **B36** (rapid publ.), 3479.

Pantelides, S. T. (ed.) (1988), *Deep Centers in Semiconductors*, Gordon and Breach, New York.

Pantelides, S. T. (1989), in *Amorphous Silicon and Related Materials*, H. Fritzsche (ed.), World Scientific Publishing Co., New York.

Pantelides, S. T. and C. T. Sah (1974), Phys. Rev. **B10**, 638.

Pantelides, S. T. and W. A. Harrison (1976), Phys. Rev. **B13**, 2667.

Pantelides, S. T. and H. G. Grimmeiss (1980), Solid State Commun. **35**, 653.

Pantelides, S. T., A. Selloni, and R. Car (1985), Solid State Electron. **28**, 17.

Park, S. I., C. C. Tsuei, and K. N. Tu (1988), Phys. Rev. **37**, 2305.

Parrott, J. E. and A. D. Stuckes (1975), *Thermal Conductivity of Solids*, Pion, London.

Parsons, R. R. (1969), Phys. Rev. Lett. **23**, 1132.

Parsons, R. R. (1971), Can. J. Phys. **49**, 1850.

Parthé, E. (1964), *Crystal Chemistry of Tetrahedral Structures*, Gordon and Breach, New York.

Parthé, E. (1972), *Christallochimie des Structures Tétraédriques*, Gordon and Breach, Paris.

Patel, C., T. J. Parker, H. Jamshidi, and W. F. Sherman (1984), Phys. Stat. Sol. B **122**, 461.

Patel, C. K. N. and R. E. Slusher (1968), Phys. Rev. Lett. **22**, 282.

Patel, C. K. N. and A. C. Tam (1981), Rev. Mod. Phys. **53**, 517.

Patel, J. R. and A. R. Chaudhuri (1966), Phys. Rev. **143**, 601.

Patterson, B. D. (1988), Rev. Mod. Phys. **60**, 69.

Paul, W. (1966), in *Proc. Int. School Enrico Fermi*, Course 34, J. Tauc (ed.), Academic Press, New York, p. 257.

Pauling, L. (1960), *The Nature of the Chemical Bond*, Cornell University Press.

Pauling, L. and R. Hayward (1964), *Architecture of Molecules*, Freemann and Company, San Francisco.

Pauling, L. and E. B. Wilson (1935), *Introduction to Quantum Mechanics*, McGraw-Hill Book Co., New York.

Pauling, W. (1927), Proc. Roy. Soc. **114**, 181.

Pay-June Lin-Chung and S. Teitler (1972), Phys. Rev. **6B**, 1419.

Payne, R. T. (1965), Phys. Rev. **139**, A570.

Pearsall, T. P. (ed.) (1982), *GaInAsP-Alloy Semiconductors*, John Wiley & Sons, New York.

Pearsall, T. P., J. Bevk, L. C. Feldman, J. M. Bonar, J. P. Mannearts, and A. Ourmazd (1987), Phys. Rev. Lett. **58**, 729.

Pearsall, T. P., F. H. Pollak, J. C. Bean, and R. Hull (1986), Phys. Rev. **B33**, 6821.

Pearton, S. J., J. W. Corbett, and T. S. Shi (1987), Appl. Phys. A **43**, 153.

Peeters, F. M. and J. T. Devreese (1984), Solid State Physics, Vol. **38**, H. Ehrenreich and D. Turnbull (eds.), Academic Press, Orlando.

Peierls, R. E. (1929), Ann. Phys. (Lpz.) **3**, 1055.

Peierls, R. E. (1935), Ann. Inst. Henry Poincaré **5**, 177.

Peierls, R. E. (1955), *Quantum Theory of Solids*, University Press, Oxford.

Pekar, S. I. (1953), Uspekhi Fiz. Nauk. **50**, 193.

Pekar, S. I. (1954), *Untersuchungen über die Elektronentheorie der Kristalle*, Akademie Verlag, Berlin.

Pekar, S. I. (1958), Sov. Phys.-JETP **6**, 785, and **7**, 813.

Pelant, I., A. Mysyrowica and C. Benoit à la Guillaume (1976), Phys. Rev. Lett. **37**, 1708.

Pellegrini, B. (1986), Solid-State Electron. **29**, 1279.

Penn, D. R. (1962), Phys. Rev. **128**, 2093.

Penrose, R. (1974), Bull. Inst. Math. Its Appl. **10**, 266.

Penzkofer, A., A. Laubereau, and W. Kaiser (1979), Prog. Quantum Electron. **6**, 55.

People, R. (1986), IEEE J. Quantum Electron. **QE 22**, 1696.

Petrenko, V. F. and R. W. Whitworth (1980), Philos. Mag. A **41**, 681.

Petroff, Y. (1980), in *Handbook on Semiconductors, Vol. 2: Optical Properties of Solids*, T. S. Moss and M. Balkanski (eds.), North Holland Publ. Co., Amsterdam.

Petroff, P. M., A. C. Gossard, W. Weigmann, and A. Savage (1978), J. Cryst. Growth **44**, 5.

Petroff, P. M., A. C. Gossard, A. Savage, and W. Wiegmann (1979), J. Cryst. Growth **46**, 172.

Petroff, P. M., C. Weisbuch, R. Dingle, A. C. Gossard, and W. Wiegmann (1981), Appl. Phys. Lett. **41**, 635.

Peuker, K., R. Enderlein, A. Schenk, and E. Gutsche (1982), Phys. Stat. Sol. B **109**, 599.

Pfister, G. (1976), Phys. Rev. Lett. **36**, 271.

Pfister, G. and H. Scher (1977), Phys. Rev. **B15**, 2062.

Phach, V. D. and R. Levy (1979), Solid State Commun. **29**, 247.

Philipp, H. R. and H. Ehrenreich (1963), Phys. Rev. **129**, 1550.

Philipp, H. R. and E. A. Taft (1960), Phys. Rev. **120**, 37.

Phillips, J. C. (1956), Phys. Rev. **104**, 1263.

Phillips, J. C. (1964), Phys. Rev. **A136**, 1705.

Phillips, J. C. (1966), Solid State Phys. **18**, 56.

Phillips, J. C. (1970), Rev. Mod. Phys. **42**, 317.

Phillips, J. C. (1973), *Bonds and Bands in Semiconductors*, Academic Press, New York.

Phillips, J. C. (1980), Comments Solid State Phys. **9**, 191.

Phillips, J. C. (1989), *Physics of High*-T_c *Superconductors*, Academic Press, San Diego.

Phillips, J. C. and L. Kleinman (1959), Phys. Rev. **116**, 287.

Phillips, W. A. (1972), J. Low Temp. Phys. **7**, 351.

Pidgeon, C. R. (1962), Thesis, Univ. Reading.

Pidgeon, C. R. (1969) in *Electron Structure in Solids*, E. D. Haidemanakis (ed.), Plenum Press, New York, p. 47.

Pidgeon, C. R. (1980),in *Handbook on Semiconductors, Optical Properties of Solids*, M. Balkanski, (ed.), North Holland, Amsterdam, p. 1284.

Piesbergen, U. (1966), in *Semiconductors and Semimetals*, Vol. 2, R. K. Willardson and A. C. Beer (eds.), Academic Press, New York.

Pikus, G. E. and A. N. Titkov (1984), *Optical Orientation*, F. Meyer and B. P. Zacharchenya (eds.), Elsevier Science Publ. Co., New York, Ch. 3.

Pinczuk, A. and G. Abstreiter (1989), *Light Scattering in Solids* **V**, M. Cardona and J. Güntherodt (eds.), Springer-Verlag, Berlin.

Pinczuk, A. and J. M. Worlock (1982), Surf. Sci. **113**, 69.

Pine, A. S. (1972), Phys. Rev. **B5**, 3003.

Pines, D. (1956), Rev. Mod. Phys. **28**, 184.

Pines, D. (1963), *Elementary Excitations in Solids*, W. A. Benjamin, Inc., New York.

Pippard, A. B. (1965), *Dynamics of Conduction Electrons*, Gordon and Breach, London.

Planel, R., C. Benoit à la Guillaume, and A. Bonnot (1973), Phys. Stat. Sol. B **58**, 251.

Platzman, P. M. and P. A. Wolff (1973), in *Solid State Physics*, Suppl. **13**, H. Ehrenreich, F. Seitz, and D. Turnbull (eds.), Academic Press, N. Y.

Ploog, K. (1981), Ann. Rev. Mat. Sci. **11**, 171.

Ploog, K. (1987), J. Cryst. Growth **81**, 304.

Ploog, K. and G. H. Döhler (1983), Adv. Phys. **32**, 285.

Pödör, B. P. (1966), Phys. Stat. Sol. **16**, K167.

Pödör, B. P. (1974), *Proc. Summer School, Lattice Defects in Semicond.*, Erice.

Pödör, B. P. (1977), Phys. Stat. Sol. B **39**, K123.

Pohl, H. R. (1967), *Quantum Mechanics for Science and Engineering*, Prentice Hall, Englewood Cliffs, New Jersey.

Pohl, R. O., W. F. Love, and R. B. Stevens (1974), *Proc. 5th Int. Conf. on Amorph. and Liquid Semicond.*, J. Stuke and W. Brenig (eds.), Taylor and Francis, London, p. 1121.

Pohl, R. W. (1938), Phys. Z. **39**, 36.

Pokrovskii, Ya. E. (1972), Phys. Stat. Sol. A **11**, 385.

Pokrovskii, Ya. E. and K. I. Svishunova (1970), Sovj. Phys.-Semicond. **4**, 409.

Polk, D. E. (1971), J. Non-Cryst. Solids **5**, 365.

Pollak, F. H. (1965), Phys. Rev. A **138**, 618.

Pollak, M. (1972), J. Non-cryst. Solids **11**, 1.

Pollak, M. and M. L. Knotek (1974), Phys. Rev. **B9**, 644.

Pollmann, J. (1976), Solid State Commun. **19**, 361.

Pollock, D. D. (1985), *Thermoelectricity: Theory, Thermometry, Tool*, ASTM, Philadelphia, Pennsylvania.

Poole, C. P. (1967), *Electron Spin Resonance*, John Wiley & Sons, New York.

Poole, H. H. (1921), Philos. Mag. **42**, 488.

Poole, R. (1988), Science **240**, 146.

Pope, M. and C. E. Swenberg (1982), *Electronic Processes in Organic Crystals*, Oxford University Press, Oxford.

Pötz, W., M. A. Osman, and D. K. Ferry (1988), Solid State Electron. **31**, 673.

Poulet, H. (1955), Ann. Phys. (Paris) **10**, 908.

Poulet, H. and J. P. Mathieu (1970), *Spectres des Vibration et Symétrie des Cristeaux*, Gorden and Breach, London.

Pratt, G. W., Jr. (1957), Phys. Rev. **106**, 53.

Presser, E. (1925), Funkbastler 558.

Price, P. J. (1977), Solid State Electron. **21**, 9.

Price, P. J. (1981), Ann. Phys. (N.Y.) **133**, 217.

Price, P. J. (1982), Surf. Sci. **113**, 199.

Price, P. J. (1985), Physica B & C **134**, 164.

Priester, C., G. Allan and M. Lannoo (1984), Phys. Rev. B **29**, 3408.

Prikhot'ko, A. F. and A. F. Skorobogat'ko (1965), Ukr. Fiz. Zh. **10**, 350.

Prince, M. B. (1953), Phys. Rev. **91**, 271.

Prior, A. C. (1960), Proc. Phys. Soc. (London) Ser. **A76**, 465.

Prokofjew, W. (1929), Z. Phys. **58**, 255.

Putley, E. H. (1977) in *Semiconductors and Semimetals*, R. K. Willardson and A. C. Beer (eds.), Academic Press, New York.

Putley, E. H. and W. H. Mitchell (1958), Proc. Phys. Soc. (London) Ser. **A72**, 193.

Queisser, H. J. (1971), in *Festkörperprobleme/Advances in Solid State Physics*, Vol. 11, O. Madelung (ed.), F. Vieweg & Sohn, Braunschweig, p. 45.

Queisser, H. J. (1973) in *Electronic Materials*, N. B. Hannay and U. Colombo (eds.), Plenum Press, New York.

Queisser, H.J. (1985), *Kristalline Krisen*, R. Piper, GmbH & Co., München. English: *The Conquest of the Microchip*.

Quinn, J. J. (1962), Phys. Rev. **126**, 1453.

Quist, T. M., R. H. Rediker, R. J. Keyes, W. E. Krag, B. Lax, A. L. McWorther, and H. J. Zeiger (1962), Appl. Phys. Lett. **1**, 91.

Rabin, H. and C. L. Tang (1975), *Quantum Electronics, A Treatise* **1**, *Non-linear Optics*, Academic Press, New York.

Radhakrishnan, V., K. P. Roy, and P. C. Sharma (1982), *Proc. Nucl. Phys. Sol. St. Phys. Symp.* **25C**, 314.

Rajagopal, A. K. and R. Srinivasan (1960), Z. Phys. **158**, 471.

Randall, R. and M. Wilkins (1945), Proc. Roy. Soc. (London) **A184**, 366, 390.

Rappel, W. J., L. F. Feiner and M. F. H. Schuurmans (1988), in *Excitons in Confined Systems*, R. DelSole, A. D'Andrea and A. Lappiccirella (eds.), Springer-Verlag, Berlin.

Rashba, E. I. (1976), Izv. Akad. Nauk SSSR Ser. Fiz. **40**, 1.

Rashba, E. I. and G. E. Gurgenishvili (1962), Sovj. Phys.-Solid State **4**, 759.

Raymond, A., J. L. Robert, and B. Pistoulet (1977), *Proc. Int. Conf. on GaAs and Related Compounds*, C. Hilsum (ed.), Institute of Physics, London.

Read Jr., W. T. (1953), *Dislocations in Crystals*, McGraw-Hill Book Co., Inc., New York.

Rebane, K. K. (1984), J. Lumin. **31**, 744.

Rebane, Yu. T. (1981), Fiz. Tekhn, Poluprovodn. **15**, 2179.

Rees, H. D. (1968), Solid State Commun. **26A**, 416.

Rees, H. D. (1969), J. Phys. Chem. Solids **30**, 643.

Rees, H. D. (1972), J. Phys. **C5**, 64.

Reinecke, T. L. (1979), Phys. Rev. Lett. **43**, 1054.

Reinecke, T. L. (1982), in *Polarons and Excitons in Polar Semiconductors and Ionic Crystals*, J. T. Devreese and F. Peeters (eds.), Plenum Press, New York.

Reinecke, T. L. and S. C. Ying (1979), Phys. Rev. Lett. **43**, 1054.

Reiss, H., C. S. Fuller, and F. J. Morin (1956), Bell Syst. Tech. J. **35**, 535.

Reitz, J. R. (1955), Solid State Phys. **1**, 1.

Ren, S. Y. (1990), in *Lattice Dynamics in Semiconductor Physics*, Festschrift for Professor K. Huang, World Scientific Publ. Co., Singapore.

Ren, S. Y., J. D. Dow and D. J. Wolford (1982), Phys. Rev. **B25**, 766.

Ren, S. Y., J. D. Dow and J. Shen (1988), Phys. Rev. **B38**, 10677.

Reuszer, J. H. and P. Fisher (1964), Phys. Rev. **A135**, 1125.

Reynolds, D. C. (1960), Phys. Rev. **118**, 478.

Reynolds, D. C., K. K. Bajaj, and C. W. Litton (1985), Solid State Commun. **53**, 1061.

Reynolds, D. C. and T. C. Collins (1981), *Excitons: Their Properties and Uses*, Academic Press, New York.

Rhoderick, E. H. (1978), *Metal-Semiconductor Contacts*, Clarendon Press, Oxford.

Rice, T. M. (1977), in *Solid State Physics*, Vol. **32**, H. Ehrenreich, F. Seitz, and D. Turnbull (eds.), Academic Press, NY, p. 1.

Richardson, D. (1973), Ph.D. Thesis, University of Newcastle.

Richardson, O. W. (1912), Philos. Mag. **233**, 594 and **27**, 476.

Richter, W. (1986), in *Festkörperprobleme/Advances in Solid State Physics*, Vol. **26**, P. Grosse (ed.), Friedr. Vieweg & Sohn, Braunschweig, p. 335.

Ridley, B. K. (1983), J. Phys. C **16**, 4733.

Rieger, M., P. Cocevar, B. Bordone, P. Lugi, and L. Reggiani (1988), Solid State Electron. **31**, 687.

Riehl, N. (1958), *Halbleiter und Phosphore*, M. Schön and H. Welker (eds.), Friedr. Vieweg & Sohn, Braunschweig, p. 35.

Robbins, D. J. and P. T. Landsberg (1980), J. Phys. **C13**, 2425.

Roberts, G. G. (1985), Adv. Phys. **34**, 475.

Robertson, J. (1983), Adv. Phys. **32**, 361.

Rode, D. L. (1970), Phys. Rev. **B2**, 1012.

Rode, D. L. (1972), Phys. Stat. Sol. B **53**, 244.

Rode, D. L. and S. Knight (1971), Phys. Rev. **B3**, 2534.

Rodriguez, S., P. Fisher, and F. Barra (1972), Phys. Rev. **B2**, 2219.

Roessler, D. M. (1970), J. Appl. Phys. **41**, 4589.

Romestain, R., S. Geschwind, G. E. Devlin, and P. A. Wolff (1974), Phys. Rev. Lett. **33**, 10.

Romestain, R. and C. Weisbuch (1980), Phys. Rev. Lett. **45**, 2067.

Rose, A. (1951), RCA Rev. **12**, 326.

Rose, A. (1978), *Concepts in Photoconductivity and Allied Problems*, Robert E. Krieger Publ. Co., New York.

Rose, M. E. (1961), *Relativistic Electron Theory*, John Wiley & Sons, New York.

Rose-Innes, A. C. and E. H. Rhoderick (1978), *Introduction to Superconductivity*, 2nd ed., Pergamon Press, New York.

Rosenbaum, T. F., K. Andres, G. A. Thomas and R. N. Bhatt (1980), Phys. Rev. Lett. **45**, 1723.

Rössler, U. (1979), *Festkörperprobleme/Advances in Solid State Physics*, Vol. **19**, J. Treusch (ed.), Friedr. Vieweg & Sohn, Braunschweig, p. 77.

Rössler, U. (1984), Solid State Commun. **49**, 943.

Rotenberg, M. and J. Stein (1969), Phys. Rev. **182**, 1.

Roth, L. M. (1982), in *Handbook of Semiconductors*, T. S. Moss (ed.), Vol. **1**, W. Paul (ed.), North Holland Publishing Co., Amsterdam.

Roth, L. M. and B. Lax (1959), Phys. Rev. Lett. **3**, 217.

Roth, L. M., B. Lax, and S. Zwerdling (1959), Phys. Rev. **114**, 90.

Rowe, D. M. and C. M. Bhandari (1986), Prog. Cryst. Growth and Charact. **13**, 233.

Roy, R. and D. K. Agrawal (1985), Mat. Res. Soc. Symp. **40**, 83.

Ruch, G. (1972), IEEE Trans. Electron Devices **ED 19**, 652.

Ruden, P. P. (1987), Proc. SPIE **792**, 36.

Runnels, L. K. (1967), J. Math. Phys. **8**, 2081.

Ruppin, R. and R. Englman (1970), Rep. Prog. Phys. **33**, 149.

Rustagi, K. C. (1970), Phys. Rev. **B2**, 4053.

Rutherford, E. (1911), Philos. Mag. **21**, 669.

Ryckayzen, G. (1965), *Superconductivity*, Interscience Publisher, New York.

Rytov, S. M. (1956), Sov. Ph. Je. R. **2**, 466.

Ryvkin, S. M. and Shlimak (1973), Phys. Stat. Sol. **A16**, 515.

Safarov, V. I. and A. N. Titkov (1980), *Proc. 15th Int. Conf. Semicond.*, Kyoto, Komiyama Printing Co., p. 623.

Sah, C. T., R. N. Noyce, and W. Shockley (1957), Proc. IRE **45**, 1228.

Sah, C. T. and S. T. Pantelides (1972), Solid State Commun. **11**, 1713.

Sah, C. T., J. Y. C. Sun, and J. J. T. Tzou (1983), J. Appl. Phys. **54**, 5864.

Saito, H. and E. O. Göbel (1985), Phys. Rev. **31**, 2360.

Saitoh, M. (1970), Phys. Lett. A **33**, 44.

Sak, J. (1971), Phys. Rev. **B3**, 3356.

Samoilovich, A. G., I. Ya. Korenblit, I. V. Dakovskii, and V. D. Iskra (1961), Sovj. Phys.-Solid State **3**, 2148 and 2385.

Sargent, M., M. O. Scully, and W. E. Lamb (1974), *Laser Physics*, Addison-Wesley, New York.

Sari, S. O. (1972), Phys. Rev. **B6**, 2306.

Sarker, S. and E. Domany (1984), Phys. Rev. **B23**, 6018.

Sasaki, W. (1980), J. Phys. Soc. Jpn. **49**, 31.

Sasaki, W., M. Shibuya, K. Mizuguchi, and G. Hatoyama (1959), J. Phys. Chem. Solids **8**, 250.

Saxena, A. K. (1981), Phys. Stat. Sol. **B105**, 777.

Sa-yakanit, V. (1979), Phys. Rev. B **19**, 2266.

Sa-yakanit, V. and H. R. Glyde (1980), Phys. Rev. **22**, 6222.

Sa-yakanit, V. and H. R. Glyde (1987), Comments Cond. Mat. Phys. **13**, 35.

Sa-yakanit, V., W. Sritrakool, and H. R. Glyde (1982), Phys. Rev. **B25**, 2776.

Scarmozzino, R. (1971), Solid State Commun. **9**, 1159.

Scharman, B. L. (1978), *Metal-Semiconductor Schottky Barrier Junction and Their Applications*, Plenum Press, New York.

Schecter, D. (1962), J. Phys. Chem. Solids **23**, 237.

Schecter, D. (1975), Phys. Rev. B **11**, 5043.

Schein, L. B. (1977), Phys. Rev. **B15**, 1024.

Scher, H. and E. W. Montroll (1975), Phys. Rev. **B12**, 2455.

Schiff, L. I. (1968), *Quantum Mechanics*, McGraw-Hill Book Co., Inc., New York.

Schläfer, H. L. and G. Gliemann (1967), *Einführung in die Liegandenfeldtheorie*, Akademische Verlagsanstalt, Frankfurt.

Schlenker, C. and M. Marezio (1980), Phil. Mag. **B42**, 453.

Schmit, J. L. and E. L. Stelzer (1973), *Semiconductor Opto-Electronics*, Butterworths, Stoneham, p. 398.

Schmitt-Rink, S., K. Myake, and C. M. Varma (1986), Phys. Rev. Lett. **57**, 2575.

Schnakenberg, J. (1968), Phys. Stat. Sol. **28**, 623.

Schneider, J. (1967), *II-VI Semiconducting Compounds*, D. G. Thomas (ed.), W. A. Benjamin Inc., New York, p. 40.

Schöll, E. (1987), *Non-equilibrium Phase Transitions in Semiconductors*, Springer-Verlag, Berlin.

Schooley, J. F., W. R. Hosler, and M. L. Cohen (1964), Phys. Rev. Lett. **12**, 474.

Schottky, W. (1918), Ann. Phys. (Lpz.) **57**, 541.

Schottky, W. (1935), Z. Phys. Chem. **B29**, 335.

Schottky, W. and F. Stöckmann (1954), Halbleiterprobleme **1**, 80.

Schowalter, J. C., F. M. Steranka, M. B. Salomon, and J. P. Wolf (1982), Solid State Commun. **44**, 795.

Schrieffer, J. R. (1964), *Theory of Superconductivity*, W. A. Benjamin, Inc., New York.

Schrieffer, J. R., X. -G. Wenard and S. C. Zhang (1988), Phys. Rev. Lett. **60**, 944.

Schröder, U. (1973), in *Festkörperprobleme/Advances in Solid State Physics*, Vol. **13**, H. J. Queisser (ed.), Friedr. Vieweg & Sohn, Braunschweig, p. 171.

Schrödinger, E. (1926), Ann. Physik **79**, 361.

Schröter, W. (1969), Phys. Stat. Sol. B **31**, 177.

Schröter, W. (1979), J. de Phys., Colloque C6, Suppl. 6, **40**, 161.

Schubert, E. F., E. O. Göbel, Y. Horikoshi, K. Ploog, and H. J. Queisser (1984), Phys. Rev. **B30**, 813.

Schulman, J. H. and W. D. Compton (1962), *Color Centers in Solids*, Pergamon Press, New York.

Schulman, J. N. and T. C. McGill (1981), Phys. Rev. **B23**, 4149.

Schultheis, L., J. Kuhl, A. Arnold, and C. W. Tu (1986), Proc. ICPS.

Schultheis, L., J. Kuhl, A. Arnold, and C. W. Tu (1986b), Phys. Rev. Lett. **57**, 1635 and 1797.

Schultheis, L., J. Kuhl, A. Arnold and C. W. Tu (1988), in *Excitons in Confined Systems*, R. DelSole, A. D'Andrea and A. Lappiccirella (eds.), Springer-Verlag, Berlin.

Schultz, J. M. (1982) *Diffraction for Material Scientists*, Prentice Hall, Engelwood Cliffs, New Jersey.

Schuster, A. (1874), Philos. Mag. **48**, 251.

Schwarzschild, B. (1988), Physics Today **41**, 17.

Sclar, N. and E. Burstein (1957), J. Phys. Chem. Soild **2**, 268.

Scott, A. C., F. Y. F. Chu, and D. W. McLaughlin (1973), Proc. IEEE **61**, 1443.

Seager C. H. (1985), Ann. Rev. Mater. Sci. **15**, 271.

Seager, C. H., R. A. Anderson, and J. K. G. Panitz (1987), J. Mater. Res. **2**, 96.

Seccombe, S. D. and D. M. Korn (1972), Solid State Commun. **11**, 1539.

Seeger, K. (1973), *Semiconductor Physics*, Springer-Verlag Wien, New York.

Seitz, F. (1940), *The Modern Theory of Solids*, McGraw-Hill Book Co., Inc., New York.

Seitz, F. (1954), Rev. Mod. Phys. **26**, 7.

1335

Sell, D. D. and S. E. Stokowski (1970), *Proc. 10th Int. Conf. on Physics of Semicond.*, Cambridge, Massachusetts, USAEC. Div. Techn. Inf., Springfield, Illinois.

Seraphin, B. O. (1964), *Proc. 7th Int'l. Conf. Physics of Semicond.*, Dunod, Paris.

Seraphin, B. O. (1973), Surf. Sci. **37**, 1.

Seraphin, B. O. and N. Bottka (1965), Phys. Rev. **139**, A560.

Seymour, R. J., M. R. Junnarkar, and R. R. Alfano (1982), Solid State Commun. **41**, 657.

Shah, J. (1972), Phys. Rev. **B6**, 4592.

Shah, J. (1974), Phys. Rev. **B10**, 3697.

Shah, J. (1978), Solid State Electr. **21**, 43.

Shah, J. (1986), IEEE J. Quantum Electron. **QE 22**, 1728.

Shah, J. and R. C. C. Leite (1969), Phys. Rev. Lett. **22**, 1304.

Shah, J. and E. Buehler (1971), Phys. Rev. **B4**, 2827.

Shah, J., M. Combescot and A. H. Dayem (1977), Phys. Rev. Lett. **38**, 1497.

Shah, J. and R. F. Leheny (1984), *Semiconductors Probed by Ultrafast Laser Spectroscopy*, Vol. 1, Academic Press, New York.

Shah, J., A. Pinczuk, A. C. Gossard, and W. Wiegmann (1985), Phys. Rev. Lett. **54**, 2045.

Shah, J. and R. C. C. Leite (1986), Phys. Rev. Lett. **21**, 17.

Shah, J., B. Deveand, T. C. Damen, W. T. Tsang, A. C. Gossard and P. Lugli (1987), Phys. Rev. Lett. **59**, 2222.

Sham, L. J. and W. Cohn (1966), Phys. Rev. **144**, 390.

Shanker, J., G. G. Agrawal, and N. Duth (1986), Phys. Stat. Sol. B **138**, 9.

Shanker, J. and M. Kumar (1987), Phys. Stat. Sol. **142**, 325.

Shante, V. K. S. and S. Kirkpatrick (1971), Adv. Phys. **20**, 325.

Sharma, A. S. and S. Auluck (1981), Phys. Rev., **B24**, 4729.

Sharma, B. L. (1984), *Metal-Semiconductor Schottky Barrier Junctions and Their Application*, Plenum Press, New York.

Shaw, D. (1973), *Atomic Diffusion in Semiconductors*, Plenum Press, New York.

Shaw, R. W. (1971), Phys. Rev. **B3**, 3283.

Shaw, T. (1987), Phys. Today **40**, 4, 17.

Shay, L. and J. H. Wernick (1974), *Ternary Chalcopyrite Semiconductors*, Pergamon Press, Oxford.

Shechtman, D., I. Blech, D. Gratias, and J. W. Cahn (1984), Phys. Rev. Lett. **53**, 1951.

Sheng, Z. Z. and A. M. Hermann (1988), Nature **332**, 138.

Sherif, T. S. and P. H. Handel (1982), Phys. Rev.**A26**, 596.

Sherman, J. (1932), Chem. Revs. **11**, 93.

Shewmon, P. G. (1963), *Diffusion in Solids*, McGraw-Hill Book Co., Inc., New York.

Shibuya, M. (1955), Phys. Rev. **99**, 1189.

Shichijo, H. and K. Hess (1981), Phys. Rev. **B23**, 4197.

Shik, A. Ya. and A. Ya. Vul' (1957), Sovj. Phys. Semicond. **8**, 1085.

Shimuzu, I. (1965), Phys. Lett. **15**, 297.

Shinada, M. and S. Sugano (1966), J. Phys. Soc. Jpn. **21**, 1936.

Shklovskii, B. I., H. Fritzsche and S. D. Baranowski (1989), Phys. Rev. Lett. **62**, 2989.

Shklovskii, B. I. and A. L. Efros (1984), *Electronic Properties of Doped Semiconductors*, Springer-Verlag, Berlin.

Shockley, W. (1938), Phys. Rev. **52**, 866.

Shockley, W. (1950), *Electrons and Holes in Semiconductors*, D. van Nostrand Co., Inc., New York.

Shockley, W. (1951), Bell Syst. Tech. J. **30**, 990.

Shockley, W. (1953), Phys. Rev. **91**, 228.

Shockley, W. (1961), Solid State Electron. **2**, 35.

Shockley, W. and J. R. Haynes (1951), Phys. Rev. **81**, 835.

Shockley, W. and W. T. Read (1952), Phys. Rev. **87**, 835.

Shoenberg, D. (1969), *Physics of Metals, Vol. 1: Electrons*, J. M. Ziman (ed.), Cambridge University Press.

Shrivastava, G. P., J. L. Martins, and A. Zunger (1985), Phys. Rev. **B31**, 2561.

Shubnikov, L. and W. J. DeHaas (1930), Leiden Commun. 207a,c,d and 210a.

Shur, M. S. and L. F. Eastman (1981), Solid State Electron. **24**, 11.

Simmons, J. G. and G. W. Taylor (1971), Phys. Rev. **B5** 1619.

Simpson, O. (1956), Proc. Roy. Soc. **A238**, 402.

Singh, J. (1984), in *Solid State Physics*, Vol. **38**, H. Ehrenreich and D. Turnbull (eds.), Academic Press, New York, p. 295.

Singh, J. and K. K. Bajaj (1986), Appl. Phys. Lett. **48**, 1077.

Skettrup, T. (1973), Phys. Stat. Sol. B **60**, 695.

Skolnick, M. C. (1986), *Proc. Intern'l Conf. Semicond. Physics*, World Scientific Publ. Co., Stockholm.

Skolnick, M. S., A. C. Carter, Y. Couder, and R. A. Stradling (1974), J. Opt. Soc. Am. **67**, 947.

Skriver, H. L. (1983), *The LMTO Method*, Springer Series in Solid State Science, M. Cardona, P. Falk, and H.-J. Queisser (eds.), Springer-Verlag, Berlin.

Slack, G. A. (1979) in *Solid State Physics*, Vol. **34**, F. Seitz, D. Turnbull, and H. Ehrenreich (eds.), Academic Press, New York.

Slater, J. C. (1951), Phys. Rev. **81**, 385.

Slater, J. C. (1953a), Phys. Rev. **91**, 528.

Slater, J. C. (1953b), Rev. Mod. Phys. **25**, 199.

Slater, J. C. (1960), *Quantum Theory of Atomic Structure*, McGraw-Hill, New York.

Slater, J. C. (1980), Phys. Rev. **B21**, 643.

Slater, J. C. and K. H. Johnson (1972), Phys. Rev. **B5**, 844.

Smakula, A. (1930), Z. Phys. **59**, 603.

Smirl, A. L. (1984), *Semiconductors Probed by Ultrafast Laser Spectroscopy*, Vol. **1**, Academic Press, New York, p. 197.

Smirl, A. L., S. C. Moss, and J. R. Little (1982), Phys. Rev. **B25**, 2645.

Smith, A. H. (1949), Phys. Rev. **75**, 953.

Smith, D. Y. (1985), in *Handbook of Optical Constants of Solids*, E. D. Palik (ed.), Academic Press, New York.

Smith, F. W., H. Q. Le, V. Diadiuk, M. A. Hollis, A. R. Calawe, S. Gupta, M. Frankel, D. R. Dykaar, G. A. Monrou, and T. Y. Hsiang (1988), Appl. Phys. Lett **54**, 890.

Smith, L. M. and J. P. Wolfe (1986), *Proc. Int. Conf. Phys. Semicond.*, World Scientific Publ., Stockholm, p. 1409.

Smith, R. A. (1952), *The Physical Properties of Thermodynamics*, Chapman and Hall, London.

Smith, R. A. (1978), *Semiconductors*, Cambridge University Press, Cambridge.

Smith, S. D. (1967), Handbook der Physik **XXV**, **2a**, 234.

Smith, S. D., R. B. Dennis, and R.G. Harrison (1977), Prog. in Quantum Electron. **5**, 205.

Smith, T. F. and C. R. White (1975), J. Phys. C (London) **8**, 2031.

Sniadower, L. and R. Kavalczyk (1968), Phys. Status Solidi **25**, 233.

Sokoloff, J. B. (1985), Phys. Reports **126**, 189.

Solin, S. A. and A. K. Ramdas (1970), Phys. Rev. **B1**, 1687.

Solpangkat, H. P., P. Fisher, and S. Rodriguez (1972), Phys. Lett. **A39**, 379.

Soma, T., J. Sato, and H. Matsuo (1982), Solid State Commun. **42**, 889.

Sommer, A. H. (1968), *Photoemissive Materials*, John Wiley & Sons, New York.

Sommerfeld, A. (1928), Z. Phys. **47**, 1.

Sommerfeld, A. (1950) and (1953), *Atombau und Spektallinien*, Vol. I and II, Friedr. Vieweg & Sohn, Braunschweig.

Sommerfeld, A. and H. Bethe (1933), *Handbuch der Physik*, Vol. 24, Springer-Verlag, Berlin.

Sondheimer, E. H. (1952), Adv. Phys. 1, 1.

Soos, Z. G. (1974), Ann. Rev. Phys. Chem. 25, 121.

Soren, P. (1965), Phys. Rev. 132, A1706.

Spaeth, J. M. (1986), in *Defects in Semiconductors*, H. J. von Bardeleben (ed.), Trans. Tech. Publ., Switzerland.

Spear, W. E. (1974), Adv. Phys. 23, 523.

Spenke, E. (1939), Wiss. Veröff. Siemens, 18, 54.

Spenke, E. (1958), *Electronic Semiconductors*, McGraw-Hill Book Co., Inc., New York.

Spicer, W. E., I. Lindau, P. Skeath, and C. Y. Su (1980), J. Vac. Sci. Technol. 17, 1019.

Spiegel, M. R. (1964) *Complex Variables*, McGraw Hill, New York.

Spiegel, M. R. (1968) *Mathematical Handbook*, McGraw Hill, New York.

Spitzer, W. G. (1967), in *Semiconductors and Semimetals*, Vol. 3, R. K. Williardson and A. C. Beer (eds.), Academic Press, New York.

Spitzer, W. G. and H. Y. Fan (1957), Phys. Rev. 106, 882.

Sproul, R. L., M. Moss and H. Weinstock (1959), J. Appl. Phys. 30. 334.

Sramek, S. J. and M. L. Cohen (1972), Phys. Rev. B6, 3800.

Sritrakool, W., V. Sa-yakanit, and H. R. Glyde (1985), Phys. Rev. B32, 1090.

Sritrakool, W., V. Sa-yakanit, and H. R. Glyde (1986), Phys. Rev. B33, 1199.

Srivastava, G. P., J. L. Martins, and A. Zunger (1985), Phys. Rev. B31, 2561.

Staebler, D. L. and C. R. Wronski (1977), Appl. Phys. Lett. 31, 292.

Staebler, D. L. and C. R. Wronski (1980), J. Appl. Phys. 51, 3262.

Stahl, A. and I. Balslev (1987), *Electrondynamics of the Semiconductor Band Edge*, Springer-Verlag, Berlin.

Stark, J. (1914), Ann. Phys. 43, 965 and 983.

Stasiv, O. (1959), *Elektronen and Ionenprocesse in Ionenkristallen*, Springer-Verlag, Berlin.

Steigmeier, E. F. (1969), in *Thermal Conductivity*, R. P. Tye (ed.), Academic Press, London, New York.

Steinhardt, P. J. (1987), Science 238, 1242.

Steinhardt, P. J., R. Alben, and D. Weaire (1974), J. Non-Cryst. Solids **15**, 199.

Stephens, R. B. (1973), Phys. Rev. **B8**, 2896.

Stern, E. A. (1978), Contemp. Physics **19**, 289.

Stern, E. A. (1985), in *Physical Properties of Amorphous Materials*, D. Adler, B. B. Schwarz, and M. C. Steele (eds.), Plenum Press, New York.

Stern, F. (1963), in *Solid State Physics*, Vol. **15**, F. Seitz and D. Turnbull (eds.), Academic Press, New York, p. 300.

Stern, F. (1972), in *Laser Handbook, Vol. 1*, Arrechi and Schulz-Dubois (eds.), North Holland Publ. Co., Amsterdam.

Steward, G. R. (1984), Rev. Mod. Physics **56**, 755.

Stierwald, D. L. and R. F. Potter (1967), in *Semiconductors and Semimetals*, Vol. **3**, R. K. Willardson and A. C. Beer (eds.), Academic Press, p. 71.

Stille, U. (1955) *Messen und Rechnen in der Physik*, Friedr. Vieweg Verlag, Braunschweig.

Stillman, G. E., C. M. Wolfe, and J. O. Dimmock (1970), J. Phys. Chem. Solids **31**, 1199.

Stillman, G. E. and C. M. Wolfe (1977), in *Semiconductors and Semi metals*, Vol. **12**, R. K. Willardson and A. C. Beer (eds.), Academic Press, New York.

Stocker, H. J., C. R. Stannard, Jr., H. Kaplan, and H. Levinstein (1964), Phys. Rev. Lett. **12**, 163.

Stoicheff, B. P. (1968), *Proc. Int. Conf. Molec. Spectroscopy*, Hepple (ed.), Inst. of Petroleum, Brighton, U. K., p. 261.

Stolwijk, N. A., B. Schuster, J. Hölzl, M. Meher, and W. Frank (1983), Physica, **116B**, 335.

Stone, A. D. and P. A. Lee (1985), Phys. Rev. Lett. **54**, 1196.

Stone, A. D. and P. A. Lee (1986), Phys. Rev. Lett. **55**, 1622.

Stoneham, A. M. (1969), Rev. Mod. Phys. **41**, 82.

Stoneham, A. M. (1975), *Theory of Defects in Solids*, Clarendon Press, Oxford.

Stoneham, A. M. (1979), Adv. Phys. **28**, 457.

Stoneham, A. M. (1981), Rep. Progr. Phys. **44**, 1251.

Stoneham, A. M. (1986), in *Defects in Semiconductors*, H. J. Bardelebeu (ed.), Trans. Tech. Publication, Switzerland.

Störmer, H. L. and D. C. Tsui (1983), Science **220**, 1241.

Stradling, R. A. (1984), in *Polarons and Excitons in Polar Semiconductors and Ionic Crystals*, J. T. Devreese and F. M. Peeters (eds.), Plenum Press, New York.

Stradling, R. A. and R. A. Wood (1970), J. Phys. C (London) **3**, L94.

Stratton, R. (1969), J. Appl. Phys. **40**, 4582.

Street, R. A. (1978), Phys. Rev. **B17**, 3984.

Street, R. A. and N. F. Mott (1975), Phys. Rev. Lett. **35**, 1293.

Streitwolf, H. W. (1970), Phys. Stat. Sol. **37**, K47.

Stuke, J. (1976), *Electronic Phenomena in Non-Crystalline Solids*, B. T. Kolomietz (ed.), USSR, Acad. of Sci., Leningrad, p. 193.

Sturge, M. D. (1962), Phys. Rev. **127**, 768.

Sturge, M. D. (1967), in *Solid State Physics, Advances and Applications*, Vol. **20**, H. Ehrenreich, F. Seitz, and D. Turnbull (eds.), Academic Press, New York.

Stutzmann, M. (1987), Philos. Mag. **B56**, 63.

Subramanian, M. A., J. C. Calabrese, C. C. Torardi, J. Gopalakrishnan, T. R. Askew, R. B. Flippen, K. J. Morrissey, U. Chowdry, and A. W. Sleight (1988), Nature **332**, 420.

Suezawa, M. and K. Sumino (1983), Journ. de Physique, Colloque C4, Suppl. 9, **44**, 133.

Sumi, H. (1983), Phys. Rev. **B27**, 2374.

Sunagawa, I. (1975), *Crystal Growth and Characterization*, R. Ueda and J. B. Mullin (eds.), American Elsevier Publ. Co., Inc., NY.

Svenson, E. L., B. N. Brockhouse, and J. M. Rowe (1967), Phys. Rev.**155**, 619.

Sze, S. M. (1981), *Physics of Semiconductor Devices*, John Wiley & Sons, New York.

Sze, S. M. and J. C. Irwin (1968), Solid State Electron. **11**, 599.

Szigetti, B. (1949), Trans. Faraday Soc. **45**, 155.

Taft, E. A. and M. H. Hebb (1952), J. Opt. Soc. Am. **42**, 249.

Taguchi, R. and B. Ray (1983), Progr. Cryst. Growth and Charact. **6**, 103.

Takebe, T., J. Saraie, and H. Matsunami (1982), J. Appl. Phys. **53**, 457.

Talwar, D. N., M. Vandevyver, and M. Zigone (1980), J. Phys. **C13**, 3775.

Talwar, D. N., M. Vandevyver, K. K. Bajaj, and W. M. Theis (1986), Phys. Rev. **B33**, 8525.

Tam, A. C. (1986), Rev. Mod. Phys. **58**, 381.

Tan, S. I., B. S. Berry, and W. Frank (1973), in *Ion Implantation in Semiconductors and Other Materials*, B. L. Crowder (ed.), Plenum Press, New York, p. 19.

Tan, T. Y., U. Gösele, and F. F. Morehead (1983), J. Appl. Phys. **54**, 5864.

Tanaka, S., H. Kobayashi, H. Saito, and S. Shionaya (1980), J. Proc. Soc. Jpn. **49**, 1051.

Tanaka, S., H. Toshida, H. Saito and S. Shionaya (1984), *Semiconductor Probed by Ultrafast Laser Spectroscopy* Vol. 1, Academic Press, New York.

Tang, C. L. and D. J. Erskine (1983), Phys. Rev. Lett. **51**, 840.

Tauc, J. (1954), Phys. Rev. **95**, 1394.

Taylor, G. I. (1934), Proc. Roy. Soc. (London), **145**, 362.

Temkin, R. J. (1974), Solid State Commun. **15**, 1325.

Temkin, R. J., W. Paul, and G. A. N. Connell (1973), Adv. Phys. **22**, 581.

ter Haar, D. (1954), *Elements of Statistical Mechanics*, Holt, Reinhard and Winston, Inc., New York.

Tharmalingan, K. (1963), Phys. Rev. **130**, 549.

Theis, W. M., K. K. Bajaj, C. W. Litton, and W. G. Spitzer (1983), Physica **117** & **118B**, 116.

Thewalt, M. L. W. (1978), Solid State Commun. **25**, 991.

Thewalt, M. L. W., D. Labrie, and T. Timusk (1985), Solid State Commun. **53**, 1049.

Thomas, D. E. and J. M. Rowell (1965), Rev. Sci. Instrum. **36**, 1301.

Thomas, D. G. and J. J. Hopfield (1962a), Phys. Rev. **128**, 2135.

Thomas, D. G. and J. J. Hopfield (1962b), J. Appl. Phys. **33**, 3243.

Thomas, D. G. and J. J. Hopfield (1966), Phys. Rev. **150**, 680.

Thomas, D. G., M. Gershenzon, and F. A. Trumbore (1964), Phys. Rev. **133**, A269.

Thomas, D. G., J. J. Hopfield, and C. J. Frosch (1965), Phys. Rev. Lett. **15**, 857.

Thomas, D. G., J. J. Hopfield, and R. T. Lynch (1966), Phys. Rev. Lett. **17**, 312.

Thomas, G. A. and T. M. Rice (1977), Solid State Commun. **12**, 1183.

Thomas, G. A. and V. B. Timofeev (1980), in *Handbook on Semiconductors, Vol. 2: Optical Properties of Solids*, T. S. Moss and M. Balkanski (eds.), North Holland Publ. Co., Amsterdam.

t'Hooft, G. W., W. A. J. A. van der Poel, L. W. Molenkamp and C. T. Foxon (1988), in *Excitons in Confined Systems*, R. DelSole, A. D'Andrea and A. Lapiccirella (eds.), Springer-Verlag, Berlin.

Thouless, D. J. (1974), Phys. Rev. **C13**, 93.

Thouless, D. J. (1977), Phys. Rev. Lett. **39**, 1167.

Thouless, D. J. (1980), J. Non-Cryst. Solids **35**, 3.

Tiedje, T. (1984), in *Semiconductors and Semimetals*, R. K. Willardson and A. C. Beer (eds.), **21**C, J. I. Pankove (ed.), 207.

Tikhodeev, S. G. (1985), Usp. Fiz. Nauk. **145**, 3.

Ting, C. S. (1975), Solid State Commun. **17**, 1285.

Tinkham, M. (1964), *Group Theory and Quantum Mechanics*, McGraw-Hill Book Co., Inc., New York.

Titkov, A. N., E. I. Chaikina, E. M. Komova, and N. G. Ermakova (1981), Sov. Phys.-Semicond. **15**, 345.

Tjablikov, S. V. (1952), Zh. Eksp. Teor. Fiz. **23**, 381.

Toda, M. (1970), Progr. Theor. Phys. Suppl. **45**, 174.

Todd, J., R. Merlin, R. Clarke, K. M. Mohatny, and J. D. Axe (1986), Phys. Rev. Lett. **57**, 1157.

Toll, J. S. (1956), Phys. Rev. **104**, 1760.

Tomiki, T. (1969), J. Phys. Soc. Jpn. **26**, 738.

Torabi, A., K. F. Brennan, and C. J. Summers (1987), Proc. SPIE **792**, 153.

Torardi, C. C., M. A. Subramanian, J. C. Calabrese, J. Gopalakrishnan, K. J. Morrissey, T. R. Askew, R. B. Flippen, U. Chowdry, and A. W. Sleight (1988), Science **240**, 631.

Tosatti, E. and G. Pastori Parravicini (1971), J. Phys. Chem. Sol **32**, 632.

Touloukian, Y. S., P. E. Liley, and S. C. Saxena (1970), *Thermal Conductivity*, IFI/Plenum Press, New York.

Toyozawa, Y. (1959), Prog. Theor. Phys. Suppl. **12**, 111.

Toyozawa, Y. (1980), in *Excitons*, R. Kubo and I. Hanamura (eds.), Springer-Verlag, Berlin.

Toyozawa, Y. (1980), in *Relaxation of Elementary Excitations*, R. Kubo and I. Hanamura (eds.), Springer-Verlag, Berlin.

Toyozawa, Y. (1981), J. Phys. Soc. Jpn. **50**, 1861.

Toyozawa, Y., M. Inoue, T. Jnui, M. Okazaki, and E. Hanamura (1967), J. Phys. Soc. Jpn. **22**, 1337 and 1349.

Trlifaj, M. (1959), Czech. J. Phys. **9**, 446.

Troxell, J. R., A. P. Chatterjee, D. G. Watkins, and L. C. Kimerling (1979), Phys. Rev. **B19**, 5336.

Troxell, J. R., and G. D. Watkins (1980), Phys. Rev. **B22**, 921.

Truell, J. C. Elbaum, and B. B. Chick (1969), *Ultrasonic Methods in Solid State Physics*, Academic Press, New York.

Trumbore, F. A., M. Gershenzon, and D. G. Thomas (1966), Appl. Phys. Lett. **9**, 4.

Tsang, W. T. (1981), Appl. Phys. Lett. **39**, 786.

Tsang, W. T. (1985), in *Semiconductors and Semimetals*, R. K. Willardson and A. C. Beer (eds.), Academic Press, New York.

Tsidilkovski, I. M., G. I. Harus, and N. G. Shelushinina (1985), Adv. Phys. **34**, 43.

Tsui, D. C. (1980), in *Handbook on Semiconductors, Vol. 1: Band Theory and Transport Properties*, T. S. Moss and W. Paul (eds.), North Holland Publ. Co., Amsterdam.

Tsui, D. C., A. C. Gossard, B. F. Field, M. E. Cage, and R. F. Dziuba (1982), Phys. Rev. Lett. **48**, 3.

Tsui, D. C., G. Kaminski, and P. H. Schmidt (1974), Phys. Rev. **B9**, 3524.

Tsui, D. C. and H. L. Störmer (1986), IEEE J. Quantum Electron. **QE 22**, 1711.

Tunitskii, N. N. and Kh. S. Bagdasar'yan (1963), Optics and Spectrosc. (USA) **15**, 50.

Turbino, R., L. Piseri, and G. Zerbi (1972), J. Chem. Phys. **56**, 1022.

Turner, W. J. and W. E. Reese (1962), Phys. Rev. **127**, 126.

Tye, R. P. (1969), *Thermal Conductivity*, Academic Press, New York.

Uba, S. and J. M. Baranowski (1978), Phys. Rev. **B17**, 69.

Überall, H. and A. W. Säenz (eds.) (1985), *Coherent Radiation Sources*, Springer-Verlag, Berlin.

Ueta, M. and Y. Nishina (eds.) (1976), *Physics of Highly Excited States in Solids*, Lecture Notes in Physics, Vol. 57, Springer-Verlag, New York.

Ulbrich, R. G. (1978), Solid State Electron. **21**, 51.

Ulbrich, R. G., J. A. Kash and J. C. Tsang (1989), Phys. Rev. Lett. **62**, 949.

Ulmer, E. A. and D. R. Frankl (1968), *Proc. 9th Int. Conf. Semicond.*, Nauka, p. 170.

Unuma, Y., Y. Masumoto, and S. Shionoya (1982), J. Phys. Soc. Jpn. **51**, 1200.

Upit, G. P. and I. P. Manika (1978), Fiz. Tverd. Tela **20**, 3553.

Urbach, F. (1930), Sitzungsber. Akad. Wiss. Wien, Math. Naturw. Klasse Abt. IIa, **139**, 363.

Urbach, F. (1953), Phys. Rev. **72**, 1324.

Urbach, F., N. R. Nail, and D. Pearlman (1949), J. Opt. Soc. Am. **39**, 675.

Vagelatos, N., D. Wehne, and J. S. King (1974), J. Chem. Phys. **60**, 3613.

Vahishta, P. and R. K. Kalia (1982), Phys. Rev. **B25**, 6492.

Vakhabov, D. A., A. S. Zakirov, K. T. Igamberdiev, A. T. Mamadal-imov, S. O. Tursunov, and P. K. Khabibullaev (1985), Fiz. Tverd. Tela **27**, 3420.

Valdmanis, J. A., R. L. Fork, and J. P. Gordon (1985), Opt. Lett. **10**, 131.

Valentiner, S. and J. Wallot (1915), Ann. Physik **46**, 837.

van der Merve, J. H. (1978), CRC Crit. Rev. Sol. St. Mat. Sci. **7**, 209.

van der Pauw, L. J. (1958), Philips Techn. Rundsch. **20**, 230.

Vandersande, J. W. (1980), in *Phonon Scattering in Condensed Matter*, H. J. Marish, (ed.), Plenum Press, New York.

Vandersande, J. W. and R. O. Pohl (1982), Geophys. Res. Lett. **9**, 820.

Vandersande, J. W. and C. Wood (1986), Contemporary Physics **27**, 117.

Vanderschueren, J. and J. Gassiot (1979), in *Thermally Stimulated Relaxation in Solids*, P. Bräunlich (ed.), Springer-Verlag, Berlin.

van der Waals, J. D. (1873), *Die Konfigruität des gasförmigen und flüssigen Zustandes*, Leiden.

van der Ziel, A. (1950), Physica (Utrecht) **16**, 359.

van der Ziel, A. (1970), *Noise: Sources, Characterization, Measurement*, Prentice Hall, Inc., Engelwood Cliffs, New Jersey.

van der Ziel, A. (1979) Adv. Electron. Electron Phys. **49**, 225.

van der Ziel, A. (1986), *Noise in Solid State Devices and Circuits*, John Wiley & Sons, New York.

van der Ziel, A. (1987), J. Appl. Phys. **63**, 2456.

van der Ziel, A. (1988), Proc. IEEE **76**, 233.

Van de Walle, C. G., P. J. H. Denteneer, Y. Bar-Yam, and S. T. Pantelides (1989), Phys. Rev. **B39**, 1079.

van Dover, B., M. Gyorgy and L. Schneemyer (1989), Nature.

van Driel, H. M. (1985), *Semiconductors Probed by Ultrafast Laser Spectroscopy*, Vol. 2, Academic Press, New York, Ch. 16.

van Hove, L. (1953), Phys. Rev. **89**, 1189.

van Overstraeten, R. J. (1982), Proc. Phys. Semic. Dev., Wiley Eastern, Ltd., New Delhi, p. 69.

van Roosbroeck, W. and W. Shockley (1954), Phys. Rev. **94**, 1558.

van Vechten, J. A. (1980), in *Handbook of Semiconductors*, Vol. **3**: *Materials Properties and Preparation*, T. S. Moss and S. P. Keller (eds.), North Holland Publ. Co., Amsterdam.

van Vechten, J. A. and R. M. Martin (1972), Phys. Rev. Lett. **28**, 446.

van Vechten, J. A. and J. C. Phillips (1970), Phys. Rev. **B2**, 2160.

van Vleck, J. H. (1932), *The Theory of Electric and Magnetic Susceptibilities*, Oxford.

van Vliet, C. M. (1989), *Proc. 10th Internat'l Conf. Noise*, Budapest.

van Vliet, K. M. (1958), Phys. Rev. **110**, 50.

van Vliet, K. M. (1958a), Proc. IRE **46**, 1005.

van Vliet, K. M. (1967), Appl. Opt. **6**, 1145.

van Vliet, K. M. and A. Van der Ziel (1958), Physica **24**, 415.

Varma, C. M., S. Schmitt-Rink, and E. Abrahams (1986), Soild State Commun. **62**, 681.

Varshni, Y. P. (1967), Phys. Stat. Sol. **19**, 459 and **20**, 9.

Vashista, P., R. K. Kalia, and K. S. Singwi (1983), *Electron-Hole Droplets in Semiconductors*, C. D. Jefferies and L. V. Keldysh (eds.), North Holland, Amsterdam.

Vavilov, V. S. and K. I. Britsin (1958), Zh. Eksp. Teor. Fiz. **34**, 1354.

Vegard, L. (1921), Z. Phys. **5**, 17.

Vekua, V. L., R. I. Dzhioev, B. P. Zakharchenya, and V. G. Fleisher (1976), Fiz. Tekh. Poluprovodn. **10**, 354.

Velicki, B. and J. Sak (1966), Phys. Stat. Sol. **16**, 147.

Verie, C. (1967), *Proc. Int. Conf. II-VI Compounds*, Brown University, W. A. Benjamin, New York.

Verma, A. R. (1953), *Crystal Growth and Dislocations*, Butterworth's Scientific Publ., London.

Vigneron, J. P., M. Scheffler and S. T. Pantelides (1982), Physica **117B**, 137.

Villars, P. and L. D. Calvert (1985), *Pearson's Handbook of Crystallographic Data for Intermetallic Phases*, American Society for Metals, Metals Park, Ohio.

Viña, L., R. T. Collins, E. E. Mendez, W. I. Wang, L. L. Chang, and L. Esaki (1987), *Proc. Int. Conf. Excitons* (Rome).

Vinogradov, V. S. (1973), Sovj. Phys. Solid State **15**, 285.

Vogl, P. (1976), Phys. Rev. **B13**, 694.

Vogl, P. (1981), *Festkörperprobleme/Advances in Solid State Physics*, Vol. 21, J. Treusch (ed.), Friedr. Vieweg & Sohn, Braunschweig, p. 191.

Vogl, P., H. P. Hjalmarson and J. D. Dow (1983), J. Phys. Chem. Solids **44**, 365.

Voigt, J., F. Speigelberg, and M. Senouer (1979), Phys. Stat. Sol. B **91**, 189.

von der Linde, D. (1979), *Festkörperprobleme/Advances in Solid State Physics*, Vol 19, J. Treusch (ed.), Friedr. Vieweg & Sohn, Braunschweig, p. 387.

von der Linde, D., O. Bernecker, and W. Kaiser (1970), Opt. Commun. **2**, 149 and 215.

von der Linde, D., J. Kuhl, and H. Klingenberg (1980), Phys. Rev. Lett. **44**, 1505.

von Klitzing, K. (1981), Surf. Sci. **113**, 1.

von Klitzing, K. (1986), Rev. Mod. Phys. **58**, 519.

von Klitzing, K., G. Doreda, and M. Pepper (1980), Phys. Rev. Lett. **45**, 494.

von Laue, M. T. F. (1960), *Röntgenstrahlinterferenzen, 3. Ausg.*, Ch. 5, Akademische Verlagsgesellschaft, Frankfurt.

von Roosbroeck, W. and W. Shockley (1954), Phys. Rev. **94**, 1558.

Voos, M. and L. Esaki (1981), Springer Ser. Solid St. Sci. **24**, 292.

Voos, M., R. F. Leheny and J. Shah (1980), *Handbook on Semiconductors, Vol. 2: Optical Properties of Solids*, T. S. Moss and M. Balkanski (eds.), North Holland Publ. Co., Amsterdam.

Vrehen, Q. H. F. (1968), J. Phys. Chem. Solids **29**, 129.

Vul', A. Ya., G. L. Biraud, Ya. V. Shmartsev (1970), Sov. Phys. Semicond. **4**, 2005.

Wagner, J. (1984), Phys. Rev. **B29**, 2002.

Wagner, J. (1985), Solid State Electron. **28**, 25.

Wagner, P., C. Holm, E. Sirtl, R. Oeder and W. Zulehner (1984), *Festkörperprobleme/Advances in Solid State Physics* Vol. **24**, J. Treusch (ed.), Friedr. Vieweg & Sohn, Braunschweig.

Waire, D., M.F. Ashby, J. Logan, and M. Weins (1971), Acta. Metall. **19**, 779.

Wakita, K., M. Umeno, K. Tagaki, and S. Miki (1973), J. Phys. Soc. Jpn. **35**, 149.

Walter, J. P. and M. L. Cohen (1970), Phys. Rev. **B2**, 1821.

Walton, A. J. (1977), Adv. Phys. **26**, 887.

Walton, D. (1967), Phys. Rev. **157**, 720.

Wang, C. C. and N. W. Ressler (1969), Phys. Rev. **188**, 1291.

Wang, C. C. and N. W. Ressler (1970), Phys. Rev. **B2**, 1827.

Wang, Y., A. Suna, W. Mahler, and R. Kasowski (1987), J. Chem. Phys. **87**, 7315.

Wannier, G. H. (1937), Phys. Rev. **52**, 191.

Wannier, G. H. (1960), Phys. Rev. **117**, 423.

Warner, J. (1975), Quantum Electronics: A Treatise, Academic Press, New York, **1**, 103.

Warren, B. E. (1969), *X-Ray Diffraction*, Ch. 14, Addison-Wesley Publ. Co., Inc., Reading, Massachusetts.

Waseda, Y. and H. Ohta (1987), Sol. State Ionic Diff. and React. **22**, 263.

Washington, M. A., A. Z. Genack, H. Z. Cummins, R. H. Bruce, A. Compaan, and R. A. Forman (1977), Phys. Rev. **B15**, 2145.

Watkins, G. D. (1968), in *Radiation Effects in Semiconductors*, F. Vook (ed.), Plenum Press, New York.

Watkins, G. D. (1974), in *Lattice Defects in Semiconductors*, Inst. Phys. Conf. Ser. **23**, London, p. 1.

Watkins, G. D. (1976), *Defects and Their Structure in Nonmetallic Solids*, B. Henderson and A. E. Hughes (eds.), Plenum Press, New York, p. 203.

Watkins, G. D. (1977), *Proc. Int. Conf. Rad. Eff. in Semicond.*, Inst. Phys. London, p. 95.

Watkins, G. D. (1984), in *Festkörperprobleme/Advances in Solid State Physics*, P. Grosse (ed.), Friedr. Vieweg & Sohn, Braunschweig Vol. **24**, 163.

Watkins, G. D. (1986), in *Deep Centers in Semiconductors*, S. T. Pantelides (ed.), Gordon and Breach Sci. Publ., New York.

Watkins, G. D. and R.P. Messmer (1970), *Proc. 10th Int. Conf. Semicond. Physics*, Cambridge, Massachusetts.

Watkins, G. D. and R. P. Messmer (1973), in *Computational Methods for Large Molecules and Localized States in Solids*, F. Herman, A. D. McLean, and R. K. Nesbel (eds.), Plenum Press, New York.

Watkins, G. D. and J. R. Troxell (1980), Phys. Rev. Lett. **44**, 593.

Watson, G. N. (1944), *A Treatise of the Theory of Bessel Functions*, Cambridge University Press, London.

Watts, R. K. (1977), *Point Defects in Crystals*, John Wiley & Sons, New York.

Waugh, J. L. T. and G. Dolling (1963), Phys. Rev. **132**, 2410.

Weaire, D. and M. F. Thorpe (1971), Phys. Rev. **B4**, 2508 and 3518.

Weber, E. R. (1983), Appl. Phys. **A30**, 1.

Weber, E. R. and H. Alexander (1983), Journ. de Physique, Colloque C4, Suppl. 9, **44**, 319.

Weber, W. (1977), Phys. Rev. **B15**, 4789.

Wedell, R. (1980), Phys. Stat. Sol. B **99**, 11.

Weertman, J. and J. R. Weertman (1960), *Elementary Dislocation Theory*, MacMillan, New York.

Wei, S.-H. and H. Krakauer (1985), Phys. Rev. Lett. **55**, 1200.

Wei, S.-H. and A. Zunger (1986), Phys. Rev. **B35**, 2340.

Wei, S.-H. and A. Zunger (1987b), Phys. Rev. **B35**, 3952.
Weinreich, G. (1965), *Solids, Elementary Theory for Advanced Students*, John Wiley & Sons, New York.
Weinstein, B. A. (1987), Proc. SPIE **792**, 66.
Weisbuch, C. (1978), Solid State Electron. **21**, 179.
Weisbuch, C. and C. Hermann (1977), Phys. Rev. **B15**, 816.
Weisbuch, C., R. C. Miller, R. Dingle, A. C. Gossard, and W. Wiegmann (1981), Solid State Commun. **37**, 219.
Weisbuch, C. and R. G. Ulbrich (1978), *Proc. Int. Conf. Lattice Dynamics*, M. Balkanski (ed.), Flammarion, Paris.
Weißmantel, Ch. and C. Hamann (1979), *Grundlagen der Festkörperphysik*, Springer-Verlag, Berlin, Heidelberg, New York.
Welker, H. and H. Weiss (1954), Z. Phys. **138**, 322.
Wells, A. F. (1984), *Structural Inorganic Chemistry*, Clarendon Press, Oxford.
Wertheim, G. K. (1958), Phys. Rev. **110**, 1272.
Wessel, K. and H. Alexander (1977), Phil. Mag. **35**, 1523.
Wherrett, B. S. and W. J. Firth (1972), IEEE. J. Quantum Electron. **QE.8**, 865.
White, J. J. (1967), Canad. J. Phys. **45**, 2695.
White, T. J., J. H. Davis, and H. V. Waiter (1975), J. Appl. Phys. **46**, 11.
Whittingham, M. S. and A. J. Jacobson (eds.) (1982), *Intercalation Chemistry*, Academic Press, New York.
Wicksted, J., M. Matsushita, H. Z. Cummins, T. Shigenari, and X. Z. Lu (1984), Phys. Rev. **B29**, 3350.
Wigner, E. P. (1959), *Group Theory and its Application to Quantum Mechanics of Atomic Spectra*, Academic Press, New York.
Wigner, E. P. and H. Pelzer (1932), Z. Phys. Chem. **B15**, 445.
Wigner, E. P. and F. Seitz (1933), Phys. Rev. **43**, 804.
Wigner, E. P. and F. Seitz (1934), Phys. Rev. **46**, 509.
Wiley, J. D. (1975), in *Semiconductors and Semimetals*, Vol. 10, R. K. Willardson and A. C. Beer (eds.), Academic Press, New York.
Willett, R., J. Eisenstein, H. Störmer, D. Tsui, A. Gossard, and J. English (1987), Phys. Rev. Lett. **59**, 1776.
William, E. W. and R. Hall (1978), *Luminescence and the Light Emitting Diode*, Pergamon Press, New York.
Williams, F. E. (1963), in *American Institute of Physics Handbook*, McGraw-Hill Book Co., Inc., New York.
Williams, F. E. (1968), Phys. Stat. Sol. **25**, 493.

Williams, P. J., L. Eaves, P. E. Simmonds, M. O. Henry, E. C. Lightowlers, and C. Uihlein (1982), J. Phys. C (London) **15**, 1337.

Williams, R. T. (1978), Semicond. Insul. **3**, 251.

Williardson, R. K., T. Harman, and A. C. Beer (1954), Phys. Rev. **138**, 322.

Wilson, A. H. (1931), Proc. Roy. Soc. (London), Ser. **A133**, 458.

Wilson, A. H. (1954), *The Theory of Metals*, University Press, Cambridge.

Wilson, A. H. (1965), *Theory of Metals*, Cambridge University Press, London.

Wilson, D. K. and G. Feher (1961), Phys. Rev. **124**, 1068.

Wilson, J. A., F. J. DiSalvo, and S. Mahajan (1975), Adv. Phys., **24**, 117.

Winkler, H. G. F. (1950), *Struktur und Eigenschaften der Kristalle*, Springer-Verlag, Berlin.

Wolf, E. L. (1975), in *Solid State Physics*, Vol. 30, F. Seitz, D. Turnbull and H. Ehrenreich (eds.), Academic Press, New York.

Wolf, P. A. and G. A. Pearson (1966), Phys. Rev. Lett. **17**, 1015.

Wolfe, J. P. and C. D. Jeffries (1983), in *Electron-hole Droplets in Semiconductors*, C. D. Jeffries and L. V. Keldysh (eds.), North Holland, Amsterdam.

Wolfe, J. P., H. L. Hanson, E. E. Haller, R. S. Markiewicz, C. Kittel, and C. D. Jeffries (1975), Phys. Rev. Lett. **34**, 1291.

Wolfe, R. (1954), Proc. Phys. Soc. (London) **A67**, 74.

Wolff, H. (1955), Fortschr. Chem. Forschung **3**, 503.

Wolff, P. A. (1954), Phys. Rev. **95**, 1415.

Wolford, D. J. (1986), *Proc. Int. Conf. Phys. Semicond.*, O. Engström (ed.), World Scientific Publ., Stockholm, p. 1115.

Wolford, D. J., W. Y. Hsu, J. D. Dow, and B. G. Streetman (1979), J. Luminesc. **18/19**, 863.

Wolford, D. J., B. G. Streetman, W. Y. Hsu, R. J. Nelson, and N. Holonyak (1976), Phys. Rev. Lett. **36**, 1400.

Wood, D. M., S.-H. Wei, and A. Zunger (1988), Phys. Rev. **B37**, 1342.

Wood, D. M. and A. Zunger (1987), Phys. Rev. **B34**, 4105.

Wood, D. M. and A. Zunger (1988), Phys. Rev. Lett. **61**, 1501.

Wood, J. H. and G. W. Pratt (1957), Phys. Rev. **107**, 995.

Woodall, J. M. (1972), J. Cryst. Growth **12**, 32.

Woodruff, T. O. and H. Ehrenreich (1961), Phys. Rev. **123**, 1553.

Wright, G. B. and B. Lax (1961), J. Appl. Phys. **32**, 2113.

Wynne, J. J. (1971), Phys. Rev. Lett. **27**, 17.

Wynne, J. J. (1974), Comments on Solid State Physics, **6**, 31.

Yafet, Y. (1963), *Solid State Physics* **14**, 1, F. Seitz and D. Turnbull (eds.), Academic Press, New York.

Yajima, T. and T. Taira (1979), J. Phys. Soc. Jpn. **47**, 1620.

Yamagichi, A., H. Fuke, K. Sugiyama and M. Date (1988), Physica **C153-155**, 1459.

Yang, C. H., J. M. Carlsons, S. A. Lyons, and J. M. Worlock (1985), Phys. Rev. Lett. **55**, 2359.

Yariv, A. (1975), *Quantum Electronics*, 2nd edition, John Wiley & Sons, New York.

Yoffa, E. Y. (1981), Phys. Rev. **B23**, 1909.

Yokota, M. and O. Tanimoto (1967), J. Phys. Soc. Jpn.n **22**, 779.

Yoshida, H., H. Saito, and S. Shionoya (1981), J. Phys. Soc. Jpn. **50**, 881.

Yu, C. C. and J. J. Freeman (1986), in *Phonon Scattering in Condensed Matter*, Vol. **V**, A. C. Anderson and J. P. Wolfe (eds.), Springer-Verlag, Berlin, p. 20.

Yu, P. W., D. C. Reynolds, K. K. Bajaj, C. W. Litton, J. Klem, D. Huang, and H. Morkoç (1987), Solid State Commun. **62**, 41.

Yu, P. Y. (1979), Comments Sol. State Phys. **9**, 37.

Yu, P. Y. (1979a), in *Excitons*, K. Cho (ed.), Springer-Verlag, Berlin, p. 211.

Yu, P. Y., Y. R. Shen, Y. Petroff, and L. M. Falicov (1973), Phys. Rev. Lett. **30**, 283.

Zachariasen, W. H. (1932), J. Am. Chem. Soc. **54**, 3841.

Zachariasen, W. H. (1945), *Theory of X-Ray Diffraction in Crystals*, Dover Publ., New York.

Zachariasen, W. H. (1967), *Theory of X-Ray Diffraction in Crystals*, Dover Publ., Inc., New York.

Zakharchenya, B. P., D. N. Mirlin, V. L. Perel' and I. I. Reshina (1982), Sovj. Phys. Uspechi. **25**, 143.

Zawadzki, W. (1972), *Proc. 11th Int. Conf. Phys. Semicond.*, Polish Scient. Publ., Warsaw.

Zawadzki, W. (1974), Adv. Phys. **23**, 435.

Zawadzki, W. (1980), (1982), in *Handbook on Semiconductors*, Vol. **1**: *Band Theory and Transport Properties*, T. S. Moss and W. Paul (eds.), North Holland Publ. Co., Amsterdam.

Zawadzki, W. and B. Lax (1966), Phys. Rev. Lett. **16**, 1001.

Zawadzki, W. and Szymańska (1971), Phys. Stat. Sol. B **45**, 415.

Zdetsis, A. D. (1977), Phys. Lett. A **60**, 63.

Zdetsis, A. D. and C. S. Wang (1979), Phys. Rev. **19**, 2999.

Zeiger, H. J. (1964), J. Appl. Phys. **35**, 1657.

Zeeman, P. (1897), Phil. Mag. **44**, 55, 255.

Zener, C. (1947), Phys. Rev. **71**, 323.

Zernike, F. and J. E. Midwinter (1973), *Applied Non-linear Optics*, Wiley, New York.

Zeyher, R., J. L. Birman, and W. Brenig (1972), Phys. Rev. **B6**, 4613.

Ziman, J. M. (1964), *Principles of the Theory of Solids*, Cambridge University Press, London.

Ziman, J. M. (1969), *The Physics of Metals, Electrons*, Vol. **1**, Cambridge University Press, London.

Ziman, J. M. (1971), Solid State Phys. **26**, 1

Ziman, J. M. (1972), *Principles of the Theory of Solids*, Cambridge University Press, Cambridge.

Zimmermann, J., P. Lugi, and D. K. Ferry (1981), J. Phys. Colloq., Orsay, France, **42** (Suppl. 10), 95.

Zrenner, A. and F. Koch (1986), *Proc. Int. Conf. Phys. Semicond.*, O. Engström (ed.), World Scientific Publ., Stockholm, p. 1523.

Zrenner, A., H. Reisinger, F. Koch, and K. Ploog (1984), in *Proc. ICPS San Francisco*, J. P. Chadi and W. A. Harrison (eds.), Springer-Verlag, New York, p. 325.

Zunger, A. (1980), Phys. Rev. Lett. **44**, 582.

Zunger, A. (1981), in *Structure and Bonding of Crystals*, M. O'Keeffe and A. Navrotsky (eds.), Academic Press, New York, p. 155.

Zunger, A. (1983), Phys. Rev. Lett. **50**, 1215.

Zunger, A. (1985), Ann. Rev. Mat. Sci. **15**, 411.

Zunger, A. (1986), Int. J. Quant. Chem. **19**, 629.

Zunger, A. (1986b), *Solid State Physics*, Vol. **39**, H. Ehrenreich, F. Seitz, and D. Turnbull (eds.), Academic Press, New York, p. 275.

Zunger, A. and M. L. Cohen (1978), Phys. Rev. Lett. **41**, 53.

Zunger, A. and M. L. Cohen (1978b), Phys. Rev. **B18**, 5449, and **B20**, 4082.

Zunger, A. and U. Lindefelt (1983), Phys. Rev. **B27**, 1191 and 5989.

Zwerdling, S., B. Lax, and L. M. Roth (1957), Phys. Rev. **106**, 51, and **108**, 1402.

Word Index

degeneracy, 816.
  factor, 240, 676, 683.
  removal, 564.
degenerate
  eigenvalue perturb., 1215.
  HDS, 914.
  states, 1228.
degree of
  filling, 1062, 1067.
  ionicity, 552.
  linear polarization, 1187, 1188.
  warping, 219.
DeHaas-type effects, 770.
DeHaas-van Alphen effect, 770.
delamination, 453.
delayed exciton, 1177.
delocalization, 906, 907.
delocalized
  bonding, 20.
  states, 905.
demarcation lines, 996-998.
Dember current, 814.
density
  flucuations, 426.
  —, impurities, 907.
  functional, 185.
  —, method, 175.
  gradient,
  intrinsic defects, 462.
  mean free path, 719.
  steep, 719.
  modulation, 1122.
  of electrons, 1220.
  of-state, 103, 104, 149, 196, 198, 233, 247, 248, 332, 633, 634, 671, 770, 798, 913.
  —, band, 240.

—, distribution, 121, 197, 597.
—, effective, 744.
— —, mass, 676, 785.
—, joint, 326.
—, local (LDOS), 109.
—, magnetic field, 769.
—, mass, electron, 677.
— —, hole, 677.
—, phonon, 105.
—, staircase character, 247.
—, tailing, 914.
  profile, 16.
dephasing 1166.
  time, 1166, 1173.
derivative polarizability, 413.
descent time, 1190.
desensitization, 1094, 1126.
destruction thresholds, 395.
detailed balance, 995, 1033, 1055, 1061, 1144.
  principle, 940.
detectivity, 1071.
determinant, 1204
  value, 1205.
developable
  cluster, 1106.
  subimage, 1105.
devices, 1, 5.
  bipolar, 644.
Dexter-Klick-Russell rule, 988.
diamagnetic, 650, 862.
  behavior, superconductors, 861.
  interactions, 767.
  materials, 650.
  term, 569.
diamond, 593.
  lattice, 45, 61.

Table of Constants.*

	Symbol	Value (Units)	Item
GENERAL			
	c	$2.9979 \cdot 10^{10}$ (cms^{-1})	Light velocity
	e	$1.60217733 \cdot 10^{-19}$ (As)	Elementary charge
	$\hbar = h/(2\pi)$	$1.0545727 \cdot 10^{-34}$ (Ws^2)	Reduced Planck's constant
	m_0	$9.1093897 \cdot 10^{-35}$ (Ws^{3}cm^{-2})	Electron rest mass
		$9.1096 \cdot 10^{-28}$ (g)	
	k	$1.3806513 \cdot 10^{-23}$ (Ws/K)	Boltzmann constant
	V_0	22414 (cm^{3}mole^{-1})	Normal volume of ideal gas
	N_{Av}	$6.0221367 \cdot 10^{23}$ (mole^{-1})	Avogadro's number
	$N_{Lo} = N_{Av}/V_0$	$2.6873 \cdot 10^{19}$ (cm^{-3})	Loschmidt's number
	M_p/m_0	1836.152701	Proton/electron mass ratio
	M_p	$1.6725 \cdot 10^{-24}$ (g)	Proton rest mass
	M_n	$1.6748 \cdot 10^{-24}$ (g)	Neutron rest mass
	$T(ice)$	$273.150(K)$	
DERIVED			
	ε_0	$8.8543 \cdot 10^{-14}$ (AsV^{-1}cm^{-1})	Permittivity of free space
	$\mu_0 = 1/(\varepsilon_0 c^2)$	$1.2560 \cdot 10^{-8}$ (VsA^{-1}cm^{-1})	Permeability of free space
	$a_H = 4\pi\varepsilon_0\hbar^2/(m_0 e^2)$	0.5292 (Å)	Bohr radius
	$F = eN_{Av}$	96485.309 (As mole^{-1})	Faraday constant
	$R_H = m_0 e^4/[2(4\pi\varepsilon_0\hbar)^2]$	13.607 (eV)	Rydberg energy
	$\mu_B = e\hbar/2m_0$	$5.788 \cdot 10^{-6}$ (eV/kG)	Bohr magneton
	$\alpha^{-1} = c/v_n = \hbar c/e^2$	137.0359895	Reciprocal of fine structure constant
ENERGY RELATIONS	1 (VAs)	$2.38845 \cdot 10^{-4}$ (kcal)	1 Joule equivalent
	1 (eV)	$1.6022 \cdot 10^{-19}$ (Ws = J)	Energy associated with 1 eV
	1 (eV)	$1.6022 \cdot 10^{-12}$ (erg)	Energy associated with 1 eV
	1 (eV)	23.053 (kcal/mol)	Energy associated with 1 eV
	λ (1 eV)	1.23975 (μm)	Wavelength†
	$\bar{\nu}$ (1 eV)	8066.14 (cm^{-1})	Wavenumber†
	ν (1 eV)	$2.418 \cdot 10^{14}$ (s^{-1})	Frequency
	T (1 eV)	$11,605$ (K)	Temperature
	v (1 eV)	$5.931 \cdot 10^7$ (cms^{-1})	Velocity
	$k(T = 300\ °K)$	25.848833 (meV)	Thermal voltage
	$\bar{\nu} = 1$	0.1236 (meV)	Energy associated with 1 wavenumber
PREFIXES			

deca (da)	10	giga (G)	10^9	deci (d)	10^{-1}	nano (n)	10^{-9}
hecto (h)	10^2	tera (T)	10^{12}	centi (c)	10^{-2}	pico (p)	10^{-12}
kilo (k)	10^3	peta (P)	10^{15}	milli (m)	10^{-3}	femto (f)	10^{-15}
mega (M)	10^6	exa (E)	10^{18}	micro (μ)	10^{-6}	atto (a)	10^{-18}

* Updated: CODATA, 1986, Bulletin 63, Pergamon Press and Moldover, et al., (1988).

† in vacuo.